Peter M.A. Sloot David Abramson
Alexander V. Bogdanov Jack J. Dongarra
Albert Y. Zomaya Yuriy E. Gorbachev (Eds.)

Computational Science – ICCS 2003

International Conference
Melbourne, Australia and St. Petersburg, Russia
June 2-4, 2003
Proceedings, Part III

Volume Editors

Peter M.A. Sloot
University of Amsterdam, Informatics Institute, Section of Computational Science
Kruislaan 403, 1098 SJ Amsterdam, The Netherlands
E-mail: sloot@science.uva.nl

David Abramson
Monash University, School of Computer Science and Software Engineering
Wellington Road, Clayton, VIC 3800, Australia
E-mail: davida@csse.monash.edu.au

Alexander V. Bogdanov
Yuriy E. Gorbachev
Institute for High-Performance Computing and Information Systems
Fontanka emb. 6, St. Petersburg 191187, Russia
E-mail: bogdanov,gorbachev @hm.csa.ru

Jack J. Dongarra
University of Tennessee and Oak Ridge National Laboratory, Computer Science Dept.
1122 Volunteer Blvd., Knoxville, TN 37996-3450, USA
E-mail: dongarra@cs.utk.edu

Albert Y. Zomaya
The University of Sydney, School of Information Technologies, CISCO Systems
Madsen Building F09, Sydney, NSW 2006, Australia
E-mail: zomaya@it.usyd.edu.au

Cataloging-in-Publication Data applied for

A catalog record for this book is available from the Library of Congress

Bibliographic information published by Die Deutsche Bibliothek
Die Deutsche Bibliothek lists this publication in the Deutsche Nationalbibliographie;
detailed bibliographic data is available in the Internet at <http://dnb.ddb.de>.

CR Subject Classification (1998): D, F, G, H, I, J, C.2-3

ISSN 0302-9743
ISBN 3-540-40196-2 Springer-Verlag Berlin Heidelberg New York

Springer-Verlag Berlin Heidelberg New York
a member of BertelsmannSpringer Science+Business Media GmbH

http://www.springer.de

Printed in Germany

Typesetting: Camera-ready by author, data conversion by PTP-Berlin GmbH
Printed on acid-free paper SPIN: 10931172 06/3142 5 4 3 2 1 0

Preface

Some of the most challenging problems in science and engineering are being addressed by the integration of computation and science, a research field known as computational science.

Computational science plays a vital role in fundamental advances in biology, physics, chemistry, astronomy, and a host of other disciplines. This is through the coordination of computation, data management, access to instrumentation, knowledge synthesis, and the use of new devices. It has an impact on researchers and practitioners in the sciences and beyond. The sheer size of many challenges in computational science dictates the use of supercomputing, parallel and distributed processing, grid-based processing, advanced visualization and sophisticated algorithms.

At the dawn of the 21st century the series of International Conferences on Computational Science (ICCS) was initiated with a first meeting in May 2001 in San Francisco. The success of that meeting motivated the organization of the second meeting held in Amsterdam April 21–24, 2002, where over 500 participants pushed the research field further.

The International Conference on Computational Science 2003 (ICCS 2003) is the follow-up to these earlier conferences. ICCS 2003 is unique, in that it was a single event held at two different sites almost opposite each other on the globe – Melbourne, Australia and St. Petersburg, Russian Federation. The conference ran on the same dates at both locations and all the presented work was published in a single set of proceedings, which you hold in your hands right now.

ICCS 2003 brought together experts from a range of disciplines: mathematicians and computer scientists providing basic computing expertise, and researchers and scientists from various application areas who are pioneering advanced applications of computational methods in sciences such as physics, chemistry, life sciences, engineering, arts and humanities; along with software developers and vendors. The intent was to discuss problems and solutions in these areas, identify new issues, and shape future directions for research, as well as help industrial users apply advanced computational techniques.

Many of the advances in computational science are related to Grid Computing. The Grid has provided a way to link computation, data, networking, instruments and other resources together to solve today's complex and critical problems. As such, it is becoming a natural environment for the computational sciences. In these proceedings you will find original research in this new era of computational science and the challenges involved in building the information infrastructure needed to enable science and engineering discoveries of the future.

These four volumes, LNCS 2657, 2658, 2659 and 2660, contain the proceedings of the ICCS 2003 meeting. The volumes consist of over 460 peer-reviewed, contributed and invited papers presented at the conference in Melbourne, Australia and St. Petersburg, Russian Federation. The acceptance rate for oral presentations was 40% of the submitted papers. The papers presented reflect the aim of the scientific organization to bring together major players in the emerging field of computational science.

The conference included 27 workshops (10 in St. Petersburg and 17 in Australia), 6 presentations by Keynote speakers, and over 460 contributed papers selected for oral presentations and posters. Each paper/poster was refereed by at least two referees.

We are deeply indebted to all the authors who submitted high-quality papers to the conference, without this depth of support and commitment there would have been no conference at all. We acknowledge the members of the program committee and all those involved in the refereeing process, and the workshop organizers and all those in the community who helped us to convene a successful conference. Special thanks go to Dick van Albada, Martin Lack, Zhiming Zhao and Yan Xu for preparation of the proceedings; they did a marvelous job! Amitava Datta, Denis Shamonin, Mila Chevalier, Alexander Boukhanovsky and Elena Stankova are acknowledged for their assistance in the organization and all those 1001 things that need to be done to make a large (distributed!) conference like this a success!

Of course ICCS 2003 would not have been possible without the support of our sponsors, and we therefore gratefully acknowledge their help in the realization of this conference.

Amsterdam, June 2003

Peter M.A. Sloot,

on behalf of the co-editors:
David Abramson
Alexander Bogdanov
Jack J. Dongarra
Albert Zomaya
Yuriy Gorbachev

Organization

The conference was organized by the Section Computational Science, The University of Amsterdam, The Netherlands; the Innovative Computing Laboratory at The University of Tennessee, USA; the School of Computer Science and Software Engineering, Monash University, Victoria, Australia; the School of Information Technologies, The University of Sydney, New South Wales, Australia; and the Institute for High Performance Computing and Information Systems, St. Petersburg, Russian Federation.

Conference Chairs

Alexander Bogdanov, Chair of the St. Petersburg ICCS 2003 conference site
David Abramson, Chair of the Melbourne ICCS 2003 conference site
Jack J. Dongarra, Scientific and Overall Co-chair (The University of Tennessee, Knoxville, USA)
Peter M.A. Sloot, Scientific and Overall Chair (The University of Amsterdam, The Netherlands)

Workshops Organization and Program Chairs

Yuriy Gorbachev (IHPCIS, St. Petersburg, Russian Federation)
Albert Zomaya (The University of Sydney, Australia)

Local Organizing Committees

Martin Lack & Associates Pty. Ltd. (Australia)
Elena Stankova (IHPCIS, Russian Federation)
Alexander Boukhanovsky (IHPCIS, Russian Federation)
Mila Chevalier (NIP, Russian Federation)

Program Committee

Albert Y. Zomaya (The University of Sydney, Australia)
Alexander Bogdanov (IHPCIS, Russia)
Alexander Zhmakin (PhTI RAS, Russian Federation)
Alfons Hoekstra (The University of Amsterdam, The Netherlands)
Alistair Rendell (Australian National University, Australia)

Tarek El-Ghazawil (George Washington University, USA)
Vaidy Sunderam (Emory University, USA)
Valery Zolotarev (SPbSU, Russian Federation)
Vasil Alexandrov (The University of Reading, UK)
Vladimir P. Nechiporenko (Ministry of Industry, Science and Technologies, Russian Federation)
Xiaodong Zhang (National Science Foundation, USA)
Yong Xue (Chinese Academy of Sciences, China)
Yuriy Gorbachev (IHPCIS, Russian Federation)
Zdzislaw Meglicki (Indiana University, USA)

Workshop Organizers

Computer Algebra Systems and Their Applications
A. Iglesias (University of Cantabria, Spain)
A. Galvez (University of Cantabria, Spain)

Computer Graphics
A. Iglesias (University of Cantabria, Spain)

Computational Science of Lattice Boltzmann Modeling
B. Chopard (University of Geneva, Switzerland)
A.G. Hoekstra (The University of Amsterdam , The Netherlands)

Computational Finance and Economics
X. Deng (City University of Hongkong, Hongkong)
S. Wang (Chinese Academy of Sciences, China)

Numerical Methods for Structured Systems
N. Del Buono (University of Bari, Italy)
L. Lopez (University of Bari, Italy)
T. Politi (Politecnico di Bari, Italy)

High-Performance Environmental Computations
E. Stankova (Institute for High Performance Computing and Information Systems, Russian Federation)
A. Boukhanovsky (Institute for High Performance Computing and Information Systems, Russian Federation)

Grid Computing for Computational Science
M. Müller (University of Stuttgart, Germany)
C. Lee (Aerospace Corporation, USA)

Computational Chemistry and Molecular Dynamics
A. Lagana (Perugia University, Italy)

Recursive and Adaptive Signal/Image Processing (RASIP)
I.V. Semoushin (Ulyanovsk State University, Russian Federation)

Numerical Methods for Singular Differential and Differential-Algebraic Equations
V.K. Gorbunov (Ulyanovsk State University, Russian Federation)

Workshop on Parallel Linear Algebra (WoPLA03)
M. Hegland, (Australian National University, Australia)
P. Strazdins (Australian National University, Australia)

Java in Computational Science
A. Wendelborn (University of Adelaide, Australia)
P. Coddington (University of Adelaide, Australia)

Computational Earthquake Physics and Solid Earth System Simulation
P. Mora (Australian Computational Earth Systems Simulator)
H. Muhlhaus (Australian Computational Earth Systems Simulator)
S. Abe (Australian Computational Earth Systems Simulator)
D. Weatherley (QUAKES, Australia)

Performance Evaluation, Modeling and Analysis of Scientific Applications on Large-Scale Systems
A. Hoisie, (LANL, USA)
D.J. Kerbyson, (LANL, USA)
A. Snavely (SDSC, University of California, USA)
J. Vetter, (LLNL, USA)

Scientific Visualization and Human-Machine Interaction in a Problem Solving Environment
E. Zudilova (The University of Amsterdam, The Netherlands)
T. Adriaansen (Telecommunications & Industrial Physics, CSIRO)

Innovative Solutions for Grid Computing
J.J. Dongarra (The University of Tennessee, USA)
F. Desprez (LIP ENS, France)
T. Priol (INRIA/IRISA)

Terascale Performance Analysis
D.A. Reed (NCSA, USA)
R. Nandkumar (NCSA, USA)
R. Pennington (NCSA, USA)
J. Towns (NCSA, USA)
C.L. Mendes (University of Illinois, USA)

Computational Chemistry in the 21st Century: Applications and Methods
T.H. Dunning, Jr. (JICS, ORNL, USA)
R.J. Harrison (ORNL, USA)
L. Radom (Australian National University, Australia)
A. Rendell (Australian National University, Australia)

Tools for Program Development and Analysis in Computational Science
D. Kranzlmueller (Johannes Kepler University, Austria)
R. Wismüller (University of Vienna, Austria)
A. Bode (Technische Universität München, Germany)
J. Volkert (Johannes Kepler University, Austria)

Parallel Input/Output Management Techniques (PIOMT2003)
J.H. Abawajy (Carleton University, Canada)
Dynamic Data Driven Application Systems
F. Darema (NSF/CISE, USA)
Complex Problem-Solving Environments for Grid Computing (WCPSE02)
D. Walker (Cardiff University, UK)
Modeling and Simulation in Supercomputing and Telecommunications
Y. Mun (Soongsil University, Korea)
Modeling of Multimedia Sychronization in Mobile Information Systems
D.C. Lee (Howon University, Korea)
K.J. Kim (Kyonggi University, Korea)
OpenMP for Large Scale Applications
B. Chapman (University of Houston, USA)
M. Bull (EPCC, UK)
Modelling Morphogenesis and Pattern Formation in Biology
J.A. Kaandorp (The University of Amsterdam, The Netherlands)
Adaptive Algorithms for Parallel and Distributed Computing Environments
S. Moore (University of Tennessee, USA)
V. Eijkhout (University of Tennessee, USA)

Sponsoring Organizations

The University of Amsterdam, The Netherlands
Hewlett-Packard
Springer-Verlag, Germany
Netherlands Institute in St. Petersburg, (NIP)
Ministry of Industry, Science and Technologies of the Russian Federation
Committee of Science and High Education of the Government of St. Petersburg
St. Petersburg State Technical University
Institute for High Performance Computing and Information Systems, St. Petersburg
IBM Australia
Microsoft
Cray Inc.
Dolphin Interconnect
Microway
Etnus
ceanet
NAG
Pallas GmbH

Table of Contents, Part III

Australian Track

Track on Applications

Track on Clusters and Grids

Track on Models and Algorithms

Track on Parallel Methods and Systems

Track on Data Mining

Workshop on Parallel Linear Algebra (WoPLA03)

Workshop on Performance Evaluation, Modeling, and Analysis of Scientific Applications on Large-Scale Systems

Workshop on Scientific Visualization and Human-Machine Interaction in a Problem Solving Environment

Workshop on Innovative Solutions for Grid Computing

Table of Contents, Part I

Russian Track

Keynote Paper

Track on Scientific Computing in Mathematics, Mechanis, Physics, and Chemistry

Track on New Large-Scale Simulation Application Areas

Track on Parallel and Distributed Computing

Track on Grid Computing and Hybrid Computational Methods

Track on New Algorithmic Approaches to Existing Application Areas

Track on Advanced Numerical Algorithms

Track on Problem Solving Environments (Including: Visualisation Technologies, Web Technologies, and Software Component Technologies

---

Track on Computer Algebra Systems and Their Applications

Workshop on Computer Graphics

Workshop on Computational Science of Lattice Boltzmann Modeling

Table of Contents, Part II

Russian Track

Workshop on Computational Finance and Economics

Workshop on Numerical Methods for Structured Systems

Workshop on High-Performance Environmental Computations

Workshop on Grid Computing for Computational Science

Workshop on Computational Chemistry and Molecular Dynamics

Workshop on Recursive and Adaptive Signal/Image Processing (RASIP)

Workshop on Numerical Methods for Singular Differential and Differential-Algebraic Equations

Poster Papers

Table of Contents, Part IV

Australian Track

Workshop on Terascale Performance Analysis

Workshop on Computational Chemistry in the 21st Century: Applications and Methods

Workshop on Tools for Program Development and Analysis in Computational Science

Workshop on Parallel Input/Output Management Techniques (PIOMT2003)

Workshop on Dynamic Data Driven Application Systems

Workshop on Complex Problem-Solving Environments for Grid Computing (WCPSE02)

Workshop on Modeling and Simulation in Supercomputing and Telecommunications

Workshop on Modeling of Multimedia Synchronization in Mobile Information System

Workshop on OpenMP for Large Scale Applications

Workshop on Modeling Morphogenesis and Pattern Formation in Biology

Workshop on Adaptive Algorithms for Parallel and Distributed Computing Environments

Poster Papers

Track on Applications

A Bayes Algorithm for the Multitask Pattern Recognition Problem – Direct Approach

Edward Puchala

Wroclaw University of Technology, Chair of Systems and Computer Networks,
Wybrzeze Wyspianskiego 27, 50-370 Wroclaw, Poland
puchala@zssk.pwr.wroc.pl

Abstract. The paper presents algorithms of the multitask recognition for the direct approach. First one, with full probabilistic information and second one, algorithms with learning sequence. Algorithm with full probabilistic information was working on basis of Bayes decision theory. Full probabilistic information in a pattern recognition task, denotes a knowledge of the classes probabilities and the class-conditional probability density functions. Optimal algorithm for the selected loss function will be presented. Some tests for algorithm with learning were done.

1 Introduction

The classical pattern recognition problem is concerned with the assignment of a given pattern to one and only one class from a given set of classes. Multitask classification problem refers to a situation in which an object undergoes several classification tasks. Each task denotes recognition from a different point of view and with respect to different set of classes. For example, such a situation is typical for compound medical decision problems where the first classification denotes the answer to the question about the kind of disease, the next task states recognition of the stadium of disease, the third one determines the kind of therapy, etc. Let us consider the non-Hodgkin lymphoma as a common dilemma in hematology practice. For this medical problem we can utilise the multitask classification (this is caused by the structure of the decision process), which leads to the following scheme. In the first task of recognition, we arrive at a decision i_1 about the lymphoma type. After the type of lymphoma has been determined, it is essential for diagnosis and therapy to recognize its stage. The values of decision i_2 denote the first, the second, the third and the fourth stage of lymphoma development, respectively. Apart from that, each stage of lymphoma may assume two forms. Which of such forms occurs is determined by decision i_3. If $i_3=1$, then lymphoma assumes the form A (there are no additional symptoms). For $i_3=2$, lymphoma takes on form B (there are other symptoms, as well).

Decisions i_4 determines therapy, that is one of the known schemes of treatment (e.g. CHOP, BCVP, COMBA, MEVA, COP-BLAM-I). A therapy (scheme of treatment) cannot be used in its original form in every case. Because of the side

P.M.A. Sloot et al. (Eds.): ICCS 2003, LNCS 2659, pp. 3–10, 2003.

effects of cytostatic treatment it is necessary to modify such a scheme. Decision about modification is i_5. In the present paper I have focused my attention on the concept of multitask pattern recognition. In particular, so-called direct approach for problem solution will be taken into consideration.

2 Direct Approach to the Multitask Pattern Recognition Algorithm

Let us consider *N*-task pattern recognition problem. We shall assume that the vector of features $x_k \in X_k$ and the class number $j_k \in M_k$ for the *k*-th recognition task of the pattern being recognized are observed values of random variables $\mathbf{x_k}$ and $\mathbf{j_k}$, respectively [5]. When *a priori* probabilities of the whole random vector $\mathbf{j}=(\mathbf{j_1},\mathbf{j_2},...,\mathbf{j_N})$ denote as $P(\mathbf{j}=j)=p(j)=p(j_1,j_2,..j_N)$ and class-conditional probability density functions of $\mathbf{x}=(\mathbf{x_1},\mathbf{x_2},...,\mathbf{x_N})$ denote as $f(x_1,x_2,..x_N/j_1,j_2,..,j_N)$ are known then we can derive the optimal Bayes recognition algorithm minimizing the risk function [3], [4]:

$$R = \mathbf{E}\, L(\mathbf{i}, \mathbf{j}) \tag{1}$$

i.e. expected value of the loss incurred if a pattern from the classes $j=(j_1, j_2,..., j_N)$ is assigned to the classes $i=(i_1, i_2,..., i_N)$.

In the case of multitask classification we can define the action of recognizer, which leads to so-called direct approach. [1].

In that instance, classification is a single action. The object is classified to the classes $i=(i_1, i_2,..., i_N)$ on the basis of full features vector $x=(x_1, x_2,..., x_N)$ simultaneously. That we can see below (Fig.1).

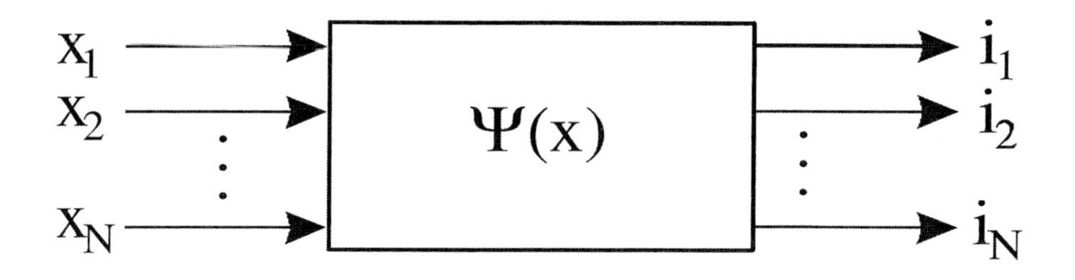

Fig. 1. Block scheme of the direct multitask pattern recognition algorithm.

Let $\Psi(x)$ denotes direct pattern recognition algorithm:

$$\Psi(x) = \Psi(x_1, x_2, ..., x_N) = (i_1, i_2, ..., i_N) \quad x_k \in X_k,\ i_k \in M_k \tag{2}$$

Minimization of the risk function *R:*

$$R[\Psi(x)] = E\{L((i_1, i_2, ..., i_N), (j_1, j_2, ..., j_N))\} \tag{3}$$

where *L* denotes the loss function, leads to the optimal algorithm Ψ^*.

$$R(\Psi^*) = \min_{\Psi} R(\Psi) \tag{4}$$

Average risk (3) expresses formula:

$$R(\Psi) = \int_X \{ \sum_{j_1 \in M_1} \sum_{j_2 \in M_2} \cdots \sum_{j_N \in M_N} L[(i_1, i_2, ... i_N), (j_1, j_2, ..., j_N)] * \\ * \ p(j_1, j_2, ... j_N / x)\} f(x) dx \tag{5}$$

where:

$$p(j_1, j_2, ... j_N / x)\} = \frac{p(j_1, j_2, ... j_N) f(x / j_1, j_2, ... j_N)}{f(x)} \tag{6}$$

denotes *a'posteriori* probability for the set of classes $j_1, j_2, ..., j_N$ As we can easily show the formula:

$$r(i_1, i_2 ... i_N, x) = E[L(i_1, i_2, ..., i_N), (j_1, j_2, ... j_N) / x] = \\ = \sum_{j_1 \in M_1} \sum_{j_2 \in M_2} \cdots \sum_{j_N \in M_N} L[(i_1, i_2, ..., i_N), (j_1, j_2, ..., j_N)] \times p(j_1, j_2, ..., j_N / x) \tag{7}$$

presents average conditional risk. Hence, the Bayes algorithm for multitask pattern recognition for direct approach may be derived. As we can see, it is result of optimization problem (4) solution. Thus, we have obtained optimal algorithm like below:

$$\Psi^*(x) = (i_1, i_2, ..., i_N) \text{ if}$$

$$r(i_1, i_2, ..., i_N, x) = \min_{i_1', i_2', ... i_N'} r(i_1', i_2', ..., i_N', x) \quad \textbf{(8)}$$

$$\Psi^*(x) = (i_1, i_2, ..., i_N) \text{ if}$$

$$\sum_{j_1 \in M_1} \sum_{j_2 \in M_2} \cdots \sum_{j_N \in M_N} L[(i_1, i_2, ..., i_N), (j_1, j_2, ..., j_N)] \times p(j_1, j_2, ..., j_N) \times$$

$$\times f(x / j_1, j_2, ..., j_N) = \quad \textbf{(9)}$$

$$= \min_{i_1', i_2', ..., i_N'} \sum_{j_1 \in M_1} \sum_{j_2 \in M_2} \cdots \sum_{j_N \in M_N} L[(i_1', i_2', .., i_N'), (j_1, j_2, ..., j_N) \times$$

$$\times p(j_1, j_2, ... j_N) \times f(x / j_1, j_2, ... j_N)$$

Let us consider characteristic form of loss function L. Value of this function depends on number of misclassification decisions:

$$L[(i_1, i_2, ..., i_N), (j_1, j_2, ..., j_N)] = n \quad \textbf{(10)}$$

Where n denotes number of pairs (algorithm's decision i_k and real class) for witch $i_k = j_k$. In this case, average conditional risk has the following form:

$$r(i_1, i_2, ..., i_N, x) = N - [p(i_1 / x) + p(i_2 / x) + ... + p(i_N / x)] \quad \textbf{(11)}$$

Because number of tasks N is constant for each practical problem and we are looking for minimum of average conditional risk, then optimal multitask pattern recognition algorithm for so called direct approach will be allowed to write like below:

$$\Psi^*(x) = (i_1, i_2, ..., i_N) \text{ if}$$

$$\sum_{k=1}^{N} p(i_k / x) = \max_{i_1', i_2', ..., i_N'} \sum_{k=1}^{N} p(i_k' / x) \quad \textbf{(12)}$$

The average risk function, for the loss function L (10), is the sum of the incorrect classification probabilities in individual tasks:

$$R[\Psi] = \sum_{n=1}^{N} P_e(n) = \sum_{n=1}^{N} [1 - P_c(n)]$$
$$P_c(n) = \sum_{j_1 \in M_1} \sum_{j_2 \in M_2} \cdots \sum_{j_N \in M_N} q(j_n / j_1, j_2, \ldots, j_N) \times p(j_1, j_2, \ldots, j_N) \qquad (13)$$

where $q(j_n / j_1, j_2, \ldots, j_N)$ is the probability of correct classification for object from classes $(j_1, j_2, \ldots, j_N)$ in n-th task:

$$q(j_n / j_1, j_2, \ldots, j_N) =$$
$$= \sum_{i_1 \in M_1} \cdots \sum_{i_{n-1} \in M_{n-1}} \sum_{i_{n+1} \in M_{n+1}} \cdots \sum_{i_N \in M_N} \int_{D_x^{(i_1,\ldots,i_N)}} f(x / j_1, \ldots, j_N) dx \qquad (14)$$

$D_x^{(i_1,\ldots,i_N)}$ - decision area for algorithm $\Psi(x)$.

3 Multitask Recognition with Learning

In the real world there is often a lack of exact knowledge of *a priori* probabilities and class-conditional probability density functions. For instance, there are situations in which only a learning sequence:

$$S_L = (x^1, j^1), (x^2, j^2), \ldots, (x^m, j^m) \qquad (15)$$

where:

$$x^k = (x_1^k, \ldots, x_N^k) \in X, \qquad j^k = (j_1^k, \ldots, j_N^k) \in M \qquad (16)$$

as a set of correctly classified samples, is known.

In this case we can use the algorithms known for conventional pattern recognition, but now algorithm must be formulated in the version corresponding to above concept. As an example let us consider α - nearest neighbour (α-NN) multitask recognition algorithm for direct approach.

Let us denote:

$$p_m(j) = \frac{I_j}{m} \tag{15}$$

estimator of the *a'priori* classes probability,
where:

I_j - number of objects from class *j* in learning sequence,

m – number of objects in learning sequence,

and

$$f_m(x/j) = \frac{\alpha_j(x)}{I_j \times V_j(x)} \tag{16}$$

estimator of the density function for α -NN algorithm,
where:

I_j - number of objects from class *j* in learning sequence,

α - number of object's *x* neighbours,

$\alpha_j(x)$ - number of objects from class *j* which are neighbours of *x* and belong to the area with $V_j(x)$ volume.

On the basis of (12) and (15), (16) final form of the multitask pattern recognition algorithm with learning is done:

$$\Psi^m(x_1, x_2, ..., x_N) = (i_1, i_2, ... i_N) \text{ if}$$

$$\sum_{k=1}^{N} \alpha_{i_k}(x) / V_{i_k}(x) = \max_{i_1', i_2', ..., i_N'} \sum_{k=1}^{N} \alpha_{i_k'}(x) / V_{i_k'}(x) \tag{17}$$

In the Fig.2 we can see values of probability of correct classification (for α - nearest neighbour (α -NN) multitask recognition algorithm for direct approach) depending on the length of learning sequence. Probability of correct classification rises, when numbers of elements in learning sequence rises too. When $m \cong 350$ or more, probability has values between 0,8 and 0,9. These results where obtained for computer simulated (generated) set of correctly classified samples.

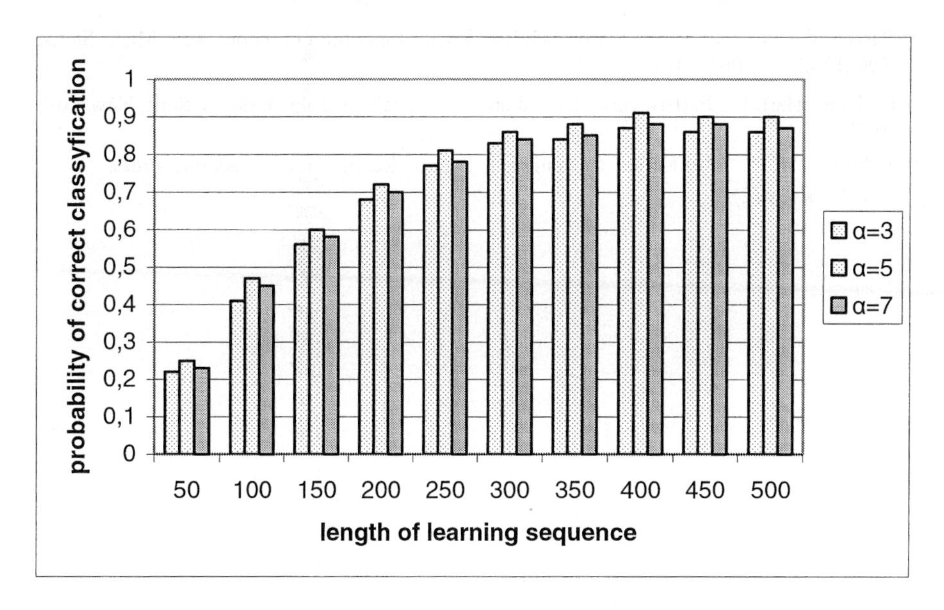

Fig. 2. Probability of correct classification as function of learning sequence's length, for various number of neighbors α (α - NN algorithm)

The superiority the multitask α -NN algorithm in direct version over the classical pattern recognition one demonstrates the effectiveness of this concept in such multitask classification problems for which the decomposition is necessary from the functional or computational point of view (e.g. in medical diagnosis). Direct approach to multitask recognition algorithms gives better results then decomposed approach because such algorithms take into consideration correlation between individual classification problems.

Acknowledgement.

The work presented in this paper is a part of the project *The Artificial Intelligence Methods for Decisions Support Systems. Analysis and Practical Applications* realized in the Higher State School of Professional Education in Legnica.

References

1. Kurzynski, M., Puchala, E., :Algorithms of the multiperspective recognition. Proc. of the 11th Int. Conf. on Pattern Recognition, Hague (1992)
2. Puchala, E., Kurzynski, M.,: A branch-and-bound algorithm for optimization of multiperspective classifier. Proceedings of the 12th IAPR, Jerusalem, Israel, (1994) 235-239

3. Parzen, E.,: On estimation of a probability density function and mode. Ann. Math. Statist., (1962) Vol.33, 1065-1076
4. Duda, R., Hart, P.,: Pattern classification and scene analysis. John Wiley & Sons, New York (1973)
5. Fukunaga, K., : Introduction to Statistical Pattern Recognition, Academic Press, New York (1972).

The Development of a Virtual Reality Environment to Model the Experience of Schizophrenia

Jennifer Tichon[1], Jasmine Banks[2], and Peter Yellowlees[1]

[1] Centre for Online Health, University of Queensland,
Brisbane 4072, Australia
jtichon@ccs.uq.edu.au
p.yellowlees@mailbox.uq.edu.au
http://www.coh.uq.edu.au/

[2] Advanced Computational Modelling Centre, University of Queensland,
Brisbane 4072, Australia
jbanks@maths.uq.edu.au
http://www.acmc.uq.edu.au/

Abstract. Virtual Reality (VR) techniques are increasingly being used for education about and in the treatment of certain types of mental illness. Research indicates that VR is delivering on its promised potential to provide enhanced training and treatment outcomes through incorporation of this high-end technology. Schizophrenia is a mental disorder affecting 1–2% of the population, and it is estimated 12–16% of hospital beds in Australia are occupied by patients with psychosis. Tragically, there is also an increased risk of suicide associated with this diagnosis. A significant research project being undertaken across the University of Queensland faculties of Health Sciences and EPSA (Engineering, Physical Sciences and Architecture) has constructed a number of virtual environments that reproduce the phenomena experienced by patients who have psychosis. Symptoms of psychosis include delusions, hallucinations and thought disorder. The VR environment will allow behavioral, exposure therapies to be conducted with exactly controlled exposure stimuli and an expected reduction in risk of harm. This paper reports on the current work of the project, previous stages of software development and the final goal to introduce VR to medical consulting rooms.

1 Introduction

Schizophrenia is a debilitating mental illness that affects 1–2% of the population at some point in their lives. It is estimated 12–16% of hospital beds in Australia are taken up with patients with schizophrenia. Tragically, suicide is more common among sufferers of this illness, which often strikes people in their prime. Psychotic symptoms associated with schizophrenia include delusions, hallucinations, and thought disorder [1]. Most people with psychotic symptoms "hear" voices, and a large proportion also "see" visual illusions. At present patients

P.M.A. Sloot et al. (Eds.): ICCS 2003, LNCS 2659, pp. 11–19, 2003.

have to describe their hallucinations, auditory and visual, to their therapists – there is no way that the therapists can either share the experiences or objectively evaluate them. As a consequence patients often feel that their therapists cannot really understand them, and the therapists themselves have difficulties learning about the exact nature of psychosis, as they have no personal experience of it.

The illusions experienced by patients with psychosis can be strange and often terrifying. Auditory hallucinations range in complexity from noises such as buzzes, bumps, screeches and muffled whispers, to ongoing multi- person discussions. The "voices" are often accusing or demeaning, and often giving paranoid suggestions to the patient. Auditory hallucinations include: audible thoughts that speak aloud what the patient is thinking; voices giving a running commentary on the patient's actions; two or more persons conversing, often about the patient who is referred to in the third person; and commands ordering the patient to do things. However they can also be frightening or dangerous, commanding acts of violence toward self or others.

Visual illusions range from simple flashes of light, subtle changes to objects, to elaborate visions. Visual hallucinations are usually seen nearby, clearly defined, in color, life-size, 3-dimensional, and moving. Subtle illusions include straight lines appearing curved, shimmering, slight movements, and objects appearing emphasized. People in pictures, on TV and in real life may appear to undergo subtle facial changes, or even to some extent morph into other entities, making their identity doubtful to the patient and increasing the patient's paranoia. Visual illusions will usually be accompanied by audio hallucinations, for example, pictures of people may morph into other entities and "talk" to the patient.

Virtual reality (VR) provides a real option for translating a person's psychological experience into a real experience others can share through the development of a virtual world that replicates the patient's world. VR techniques are increasingly being used in trial clinical programs and in the treatment of certain types of mental illness. The phenomenon of users of VR becoming immersed in virtual environments provides a potentially powerful tool for mental health professionals [2,3]. VR has been shown to help improve patient care in the form of advanced therapeutic options [4,5,6]. The research team at the University of Queensland have been collaborating for the past twelve months in the development of a VR tool to use in education, training and treatment of schizophrenia.

2 The VISAC Laboratory and Remote Visualisation

The Visualisation and Advanced Computing Laboratory (VISAC) at the University of Queensland consists of an immersive curved screen environment of 2.5 m radius and providing 150 degrees field of view. Three projectors separated by 50 degrees are used to project the images onto the curved screen. The curved screen environment is suitable for having small groups of people, eg patients and caregivers, to share the immersive experience.

The high quality infrastructure available at the University of Queensland provides a unique opportunity in Australia to undertake this collaborative project.

The university arguably has the most sophisticated advanced computing infrastructure in Australia. Within the last nine months it has established a two million dollar state-of-the-art virtual reality centre, which is linked to a 40 million dollar supercomputer and a 20 terabyte mass storage facility. These facilities are managed by the Advanced Computational Modelling Centre (ACMC). The Centre has around 20 academic staff with expertise in scientific modelling, advanced computing, visualisation and bioinformatics (see http://www.acmc.uq.edu.au).

Through the Queensland Parallel Supercomputing Foundation (QPSF), the University of Queensland is a participant in the recently funded GrangeNet. GrangeNet will provide an upgrade in bandwidth between Sydney, Melbourne, Canberra and Brisbane to approximately 2.5GB/sec. This will allow us to trial remote and collaborative visualisation between separate international locations.

Another aspect of this participation in GrangeNet is that the University of Queensland will build an area node within the next few months. This will enable us to perform multicasting between various international, national and regional sites with this facility and be able to display the visual images generated within VISAC around the world.

3 Aims of the Project

Using facilities of the VISAC laboratory, the project aims to artificially model the experience of psychosis in virtual environments, focusing on re-creating the audio hallucinations and visual illusions.

The ultimate goal of the project is to develop software that patients can use, in conjunction with their therapist, to re-create their experiences, in order that their therapists can better understand their symptoms, and allow for enhanced monitoring of their illness. It is envisaged that this environment could also be used as a structured and repeatable trigger for neurophysiological examinations of the central nervous system, and it may assist in understanding the intra-cerebral pathways of psychosis, and unraveling the mysteries of illnesses such as schizophrenia and psychosis in general.

To ensure that the re-created virtual world accurately models the patients' inner world, actual patients with schizophrenia have been interviewed and asked to describe their symptoms in detail. These descriptions have then been transcribed into models of the hallucinations that are personal to individual patients. What is being created are realistic environments where patients can be immersed into a virtual world created to specifically reflect the illusions for which they require treatment. Patients are interviewed for both feedback and evaluation on their individualised VR programme.

The next phase of the research project is to test its clinical potential. Case studies will be conducted where patients will be exposed to the specific hallucinations they have described and have been developed in the visualisation laboratory for them. The patient will be exposed to these hallucinations during therapy sessions. Measures of pre- and post-treatment on depression and symptoms of schizophrenia will be made to ascertain whether VR was an effective intervention tool.

In attempting to develop a realistic, virtual environment which can simulate a psychosis such as schizophrenia, the project also aims to provide remote display of the visual models over a high speed communications network. Therefore, web enabling of visual models for efficient and effective on-line access is also being examined.

4 Software Development

The project commenced in October 2001, and since then, has undergone a number of distinct development phases.

Phase 1. The initial work involved building a model of an everyday environment, in this case a living room, using a commercial 3D modelling package. This room contained standard furniture including a sofa, coffee table, dining table, chairs and bookshelf, as well as a number of objects including a TV, radio, pictures and painting, which were believed to trigger hallucinations. The hallucinations modelled included: a face in a portrait morphing from one person into another and also changing its facial expression; a picture on the wall distorting; the walls of the room contracting and distorting so that the straight edges of the walls appeared curved; the blades of a ceiling fan dipping down; and the TV switching on and off of its own accord. In addition, a soundtrack of auditory hallucinations, provided by the pharmaceutical company Janssen-Cilag, was played in the background as these visual hallucinations were taking place. This was in order to give a good approximation to the cacophony of voices that would be going on whilst a patient is trying to concentrate on everyday tasks.

Two screenshots of the living room model are shown in Fig. 1. Figure 2 shows the painting on the wall, before and after the distortion process.

This initial concept of modelling a set of psychotic experiences was shown to a number of patients. Feedback from patients and mental health care professionals

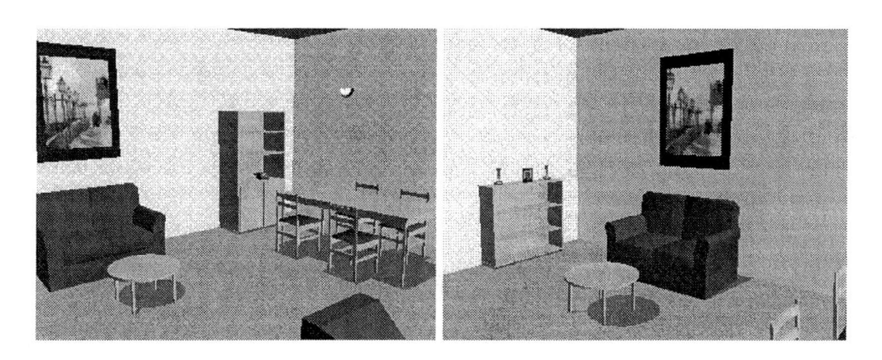

Fig. 1. Screen shots of the living room model

Fig. 2. Wall painting from the living room scene **a** original **b** distorted

was generally positive. However, due to the generic nature of the auditory and visual hallucinations portrayed, it was not possible to confirm how realistic the model was. A number of patients commented that they liked the idea, however, felt that the hallucinations modelled did not actually relate to them.

Phase 2. The second phase of the project involved modelling the experiences of one particular patient who agreed to be involved in the project. The models of the hallucinations were then implemented based on the patient's descriptions. In this way, a model of psychosis would be built from the patient's perspective.

The first challenge involved selecting a suitable environment in which the hallucinations could be modelled. A psychiatric ward was chosen, as it is an environment in which many patients would have experienced hallucinations, and could therefore be used as a basis for modelling the hallucinations of subsequent patients who become involved in the project. It also contains a variety of different rooms such as bedrooms, bathrooms, a common room, offices and hallways where hallucinations may occur.

Development of the virtual environment for Phase 2 comprised two main steps. The first step involved creating the model of the psychiatric ward, and models of static elements of the scene (eg. furniture) using a 3D modelling package. In order to build a model of a psychiatric ward and the objects within it, we visited the psychiatric unit at the Royal Brisbane Hospital, and took photographs of various textures and room layouts. This enabled the final model to be as realistic as possible. The static models of the psychiatric ward and objects were saved as VRML files for inclusion into the main program. Some views of the virtual psychiatric ward are shown in Fig. 3.

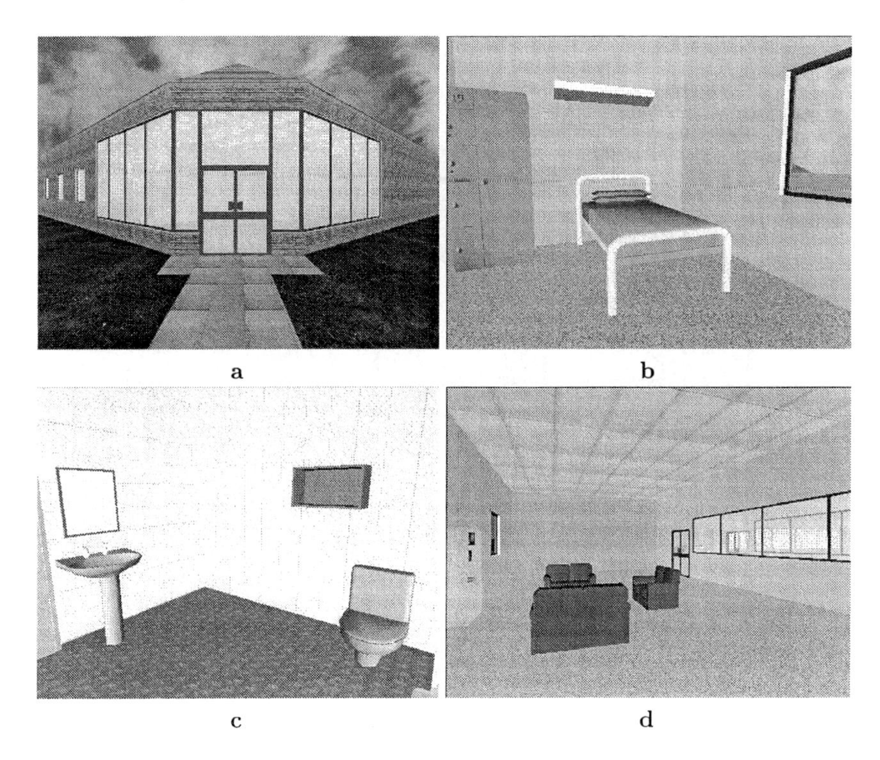

Fig. 3. Views of the psychiatric ward model **a** front **b** bedroom **c** bathroom **d** common room

The second stage of development involved writing the main program which loads, positions and displays the static elements, and also which implements the dynamic parts of the scene, such as sounds and movements of objects. The software was written in C/C++, in conjunction with an open source, cross platform scene graph technology. This method of implementation was chosen as it will allow us to eventually port the software from the current IRIX platform to a PC platform. This would enable the software to be used, for example, in a psychiatrist's office or in a hospital, making it more accessible to patients, caregivers and mental health workers.

The software was designed so that the user is able to navigate around the scene using the mouse and keyboard, and various hallucinations are triggered either by proximity to objects, by pressing hot keys, or by clicking an object with the mouse.

One particularly challenging hallucination modelled was a vision of the Virgin Mary, which would appear and "talk" to the patient. At first, it was thought that this would be implemented as a 3D model, however, this would have introduced a great deal more polygons and complexity in to the scene and slowed down performance. It was decided that a 2D film projected onto a billboard would give adequate performance and create an apparition- like effect. Therefore, the

Virgin Mary sequence was filmed using an actor, and the background set to transparent, so that only the apparition could be seen against the background of the psychiatric ward. A separate program was written to preprocess the movie, to remove some artifacts and to add a golden halo around the Virgin Mary figure. Other challenges with this hallucination involved making sure the movie and it's audio were in sync, and also achieving an adequate frame-rate.

Other hallucinations modelled include the word "Death" appearing to stand out of newspaper headlines, and random flashes of light. Some of these hallucinations are shown in Fig. 4. Audio hallucinations implemented included: a speech on the radio, which begins as a political speech, but changes to refer to the patient and concludes by commanding the patient to kill themselves; a chorus of music playing; voices discussing the patient in a derogatory manner; and other sounds and voices telling the patient they are evil; and laughter. The audio hallucinations can be triggered by proximity to a certain object (such as the stereo), or they can also play continuously in the background at random intervals as one "walks" through the virtual scene. In this way, it gives the effect of many different sounds and voices interjecting and happening simultaneously. It is also quite repetitive with the sounds and voices playing over and over again.

5 Feedback

The patient was positive about the simulation, stating "It's just such an extraordinary experience" and "It was a quite good approximation capturing the essence of the experience." The patient particularly liked how the visions and voices interlinked and were played over one another. This effect reportedly effectively re-created "the sensation of a lot going on in your head." Also, they liked the way that the hallucinations repeated themselves. "The fact that they went on and on and people would think, 'My god, why can't they turn off.' They go on and on – and how you just can't step out of the room and get away from them."

a **b**

Fig. 4. Hallucinations **a** Virgin Mary **b** "Death" headline

The patient commented that the virtual environment was effective in recreating the same emotions that they experienced on a day-to-day basis during their psychotic episodes. This is perhaps the most important goal of the simulation – more important than scene realism – if the software is to be used for education and for increasing empathy with sufferers of psychosis.

6 Further Work

At present, a second patient has been interviewed, and the hallucinations of this patient are being modelled. The model of the psychiatric ward is being again used as the basis for this. Many of these hallucinations are quite different from those of the first patient and will require development of new techniques in order for them to be re-created. These include working out and efficient way to add human figures to the scene, implementing reflective surfaces, implementing fog effects, and adding more realistic lighting and shadow effects to the scene.

The software currently works only in the VISAC lab and on machines running IRIX, and will need to be ported to a PC platform such as Linux or Windows. This will mean that some parts of the code, such as the interface to the audio and movie libraries, will need to be re-written.

7 Conclusion

To date no research exploring the clinical use of VR in psychosis can be located. As outlined, the main aims of the project are the development of the virtual reality software for use in 3D environments and to design it to be deliverable on consulting room PCs. This project has the potential to have a significance impact on the field of psychiatry in both the assessment and in the on-going monitoring of patients with schizophrenia.

It is expected that the virtual environment will also provide an effective 3-dimensional teaching and awareness environment, for mental health workers and students. It is known that a significant factor in the recovery of patients is the support they receive from family, caregivers and the community at large. It is hoped that this work will help family members of patients with schizophrenia to understand and empathise more with their loved one's experiences. Also, it is hoped that it will be used as an effective tool for improving the awareness and understanding of illnesses such as schizophrenia in the wider community, thus ultimately improving the quality of life and chances of recovery of patients.

Virtual Reality is changing the ways in which we learn about and treat medical conditions. Due to cost limitations the full potential of VR has not yet been realised. Projects such as this one are working to make state- of-the-art health technology more readily accessible and cost-effective to mental health practitioners.

References

1. H. Kaplan and B. Sadock. *Comprehensive Textbook of Psychiatry*, volume 1&2. Lippincott Williams Wilkin, Philadelphia, 7th edition edition, 2000.
2. L. Hodges, P. Anderson, G. Burdea, H. Hoffman, and B. Rothbaum. Treating psychological and physical disorders with VR. *IEEE Computer Graphics and Applications*, pages 25–33, Nov–Dec 2001.
3. M. Kahan. Integration of psychodynamic and cognitive-behavioral therapy in a virtual environment. *Cyberpsychology and Behavior*, 3:179–183, 2000.
4. G. Riva and L. Gamberini. Virtual reality in telemedicine. *Cyberpsychology and Behavior*, 6:327–340, 2000.
5. G. Riva. From telehealth to e-health: Internet and distributed virtual reality in health care. *Cyberpsychology an Behavior*, 3:989–998, 2000.
6. P. Anderson, B. Rothbaum, and L. Hodges. Virtual reality: Using the virtual world to improve quality of life in the real world. *Bulletin Menninger Clinic*, 65:78–91, 2001.

An Optimization-Based Approach to Patient Grouping for Acute Healthcare in Australia

A.M. Bagirov[1] and L. Churilov[2]

[1] Centre for Informatics and Applied Optimization, School of Information Technology and Mathematical Sciences, The University of Ballarat, Victoria 3353, Australia
a.bagirov@ballarat.edu.au

[2] School of Business Systems, Monash University, Victoria, 3800, Australia
leonid.churilov@infotech.monash.edu.au

Abstract. The problem of cluster analysis is formulated as a problem of nonsmooth, nonconvex optimization, and an algorithm for solving the cluster analysis problem based on the nonsmooth optimization techniques is developed. The issues of applying this algorithm to large data sets are discussed and a feature selection procedure is demonstrated. The algorithm is then applied to a hospital data set to generate new knowledge about different patterns of patients resource consumption.

1 Introduction

The subject of cluster analysis is the unsupervised classification of data and discovery of relationships within the data set without any guidance. The basic principle of identifying these hidden relationships are that if input patterns are similar, then they should be grouped together. Two inputs are regarded as similar if the distance between these two inputs (in multidimensional input space) is small. Due to its highly combinatorial nature, clustering is a technically difficult problem. Different approaches to the problem of clustering analysis that are mainly based on statistical, neural network and machine learning techniques have been suggested in [5,6,8,9,12]. An excellent up-to-date survey of existing approaches is provided in [4].

Mangasarian ([7]) describes an approach to clustering analysis based on the bilinear programming techniques. Bagirov et al ([2]) propose the *global optimization* approach to clustering and demonstrate how the supervised data classification problem can be solved via clustering. The objective function in this problem is both nonsmooth and nonconvex and this function has a large number of local minimizers. Problems of this type are quite challenging for general-purpose global optimization techniques. Due to a large number of variables and the complexity of the objective function, general-purpose global optimization techniques, as a rule, fail to solve such problems.

It is very important, therefore, to develop optimization algorithms that allow the decision maker to find "deep" local minimizers of the objective function. Such "deep" local minimizers provide a good enough description of the dataset

P.M.A. Sloot et al. (Eds.): ICCS 2003, LNCS 2659, pp. 20–29, 2003.

under consideration as far as clustering is concerned. The optimization algorithm discussed in this paper belongs to this type and is based on nonsmooth optimization techniques. The proposed approach has two distinct important and useful features:

- it allows the decision maker to successfully tackle the complexity of large datasets as it aims to reduce both the number of data instances (records) and the number of data attributes (so-called "feature selection") in the dataset under consideration without loss of valuable information
- it provides the capability of calculating clusters step-by-step, gradually increasing the number of data clusters until termination conditions are met.

The power of this approach is illustrated by conducting the clustering analysis of a hospital data set for the purposes of generating new knowledge about patient resource consumption. Knowledge about resource consumption and utilization is vital in modern healthcare environments. In order to manage both human and material resources efficiently, a typical approach is to group the patients based on common characteristics. The most widely used approach is driven by the Case Mix funding formula, namely to classify patients according to diagnostic related groups (DRGs). Although it is clinically meaningful, some experience suggests that DRG groupings do not necessarily present a sound basis for relevant knowledge generation ([10,11]).

The *objective* of this paper is, therefore, two-fold:

- to describe the new approach to clustering analysis based on nonsmooth nonconvex optimization techniques
- to demonstrate how optimization-based clustering techniques can be utilized to suggest an alternative grouping of the patients that generates homogeneous patient groups with similar resource utilization profiles.

Demographics, admission, and discharge characteristics are used to generate the clusters that reveal interesting differences in resource utilization patterns. Knowledge that is not available from DRG information alone can be generated using this clustering method since demographic and other data is used for the patient grouping. This knowledge can then be used for prediction of resource consumption by patients. A detailed case study is presented to demonstrate the quality of knowledge generated by this process. It is suggested that the proposed approach can, therefore, be seen as an evidence-based predictive tool with high-knowledge generation capabilities.

The rest of the paper is organized as follows: nonsmooth optimization approach to clustering is presented in Sect. 2; Sect. 3 describes an algorithm for solving clustering problem; the issues of the complexity reduction for clustering in large data set are discussed in Sect. 4; the description of the emergency dataset is given in Sect. 5, while Sect. 6 presents the discussion of the numerical experiments; Sect. 7 concludes the paper.

2 The Nonsmooth Optimization Approach to Clustering

In this section we present a formulation to the clustering problem in terms of nonsmooth, nonconvex optimization.

Consider set A that consists of r n-dimensional vectors $a^i = (a_1^i, \dots, a_n^i)$, $i = 1, \dots, r$. The aim of clustering is to represent this set as the union of q clusters. Since each cluster can be described by a point that can be considered as the center of this cluster, it is instrumental to locate a cluster's center in order to adequately describe the cluster itself. Thus, we would like to find q points that serve as centers of corresponding clusters.

Consider now an arbitrary set X, consisting of q points $x^1, \dots, x^q$. The distance $d(a^i, X)$ from a point $a^i \in A$ to this set is defined by

$$d(a^i, X) = \min_{s=1,\dots,q} \|x^s - a^i\|$$

where

$$\|x\|_p = \left(\sum_{l=1}^{n} |x_l|^p \right)^{1/p}, \quad 1 \le p < +\infty, \qquad \|x\|_\infty = \max_{l=1,\dots,n} |x_l|.$$

The deviation $d(A, X)$ from the set A to the set X can be calculated using the formula

$$d(A, X) = \sum_{i=1}^{r} d(a^i, X) = \sum_{i=1}^{r} \min_{s=1,\dots,q} \|x^s - a^i\|.$$

Thus, as far as optimization approach is concerned, the cluster analysis problem can be reduced to the following problem of mathematical programming

$$\text{minimize } f(x^1, \dots, x^q) \quad \text{subject to } (x^1, \dots, x^q) \in \mathbb{R}^{n \times q}, \tag{1}$$

where

$$f(x^1, \dots, x^q) = \sum_{i=1}^{r} \min_{s=1,\dots,q} \|x^s - a^i\|. \tag{2}$$

If $q > 1$, the objective function (2) in the problem (1) is nonconvex and nonsmooth. Note that the number of variables in the optimization problem (1) is $q \times n$. If the number q of clusters and the number n of attributes are large, the decision maker is facing a large-scale global optimization problem. Moreover, the form of the objective function in this problem is complex enough not to become amenable to the direct application of general purpose global optimization methods. Therefore, in order to ensure the practicality of the optimization approach to clustering, the proper identification and use of local optimization methods with the special choice of a starting point is very important. Clearly, such an approach does not guarantee the globally optimal solution the problem (1). On the other hand, this approach allows one to find a "deep" minimum of the objective function that, in turn, provides a good enough clustering description of the dataset under consideration.

Note also that the meaningful choice of the number of clusters is very important for clustering analysis. It is difficult to define *a priori* how many clusters represent the set A under consideration. In order to increase the knowledge generating capacity of the resulting clusters, the optimization based approach discussed in this paper adopts the following strategy: starting from a small enough number of clusters q, the decision maker has to gradually increase the number of clusters for the analysis until certain termination criteria motivated by the underlying decision making situation is satisfied. Further discussion on this issue with specific suggestions on the number of clusters is given in Sect. 4.

From optimization perspective this means that if the solution of the corresponding optimization problem (1) is not satisfactory, the decision maker needs to consider the problem (1) with $q+1$ clusters and so on. This implies that one needs to solve repeatedly arising global optimization problems (1) with different values of q - the task even more challenging than solving a single global optimization problem. In order to avoid this difficulty, a step-by-step calculation of clusters is implemented in the optimization algorithm discussed in the next section.

Finally, the form of the objective function in the problem (1) allows one to significantly reduce the number of records in a dataset. The way the proposed algorithm utilizes this feature is discussed in more detail in Sects. 4 and 6.

3 An Optimization Algorithm for Solving Clustering Problem

In this section we describe an algorithm for solving cluster analysis problem in a given dataset.

Algorithm 1. *An algorithm for solving cluster analysis problem.*

Step 1. (Initialization). Select a tolerance $\epsilon > 0$. *Select a starting point* $x^0 = (x^0_1, \dots, x^0_n) \in IR^n$ *and solve the minimization problem (5). Let* $x^{1*} \in IR^n$ *be a solution to this problem and* f^1 *be the corresponding objective function value. Set* $k = 1$...

Step 2. (Computation of the next cluster). Select a point $x^0 \in IR^n$, *construct a new starting point* $x^{02} = (x^{1*}, x^0) \in IR^{2n}$, *and solve the following minimization problem:*

$$\textit{minimize} \quad f^k(x) \quad \textit{subject to} \quad x \in IR^n \tag{3}$$

where

$$f^k(x) = \sum_{i=1}^{m} \min\{\|x^{1*} - a^i\|, \dots, \|x^{k*} - a^i\|, \|x - a^i\|\}.$$

Step 3. Let $x^{k+1,*}$ *be a solution to the problem (3). Take*

$$x^{k0} = (x^{1*}, \dots, x^{k*}, x^{k+1,*})$$

as a new starting point and solve the following minimization problem:

$$\text{minimize} \quad f^k(x) \quad \text{subject to} \quad x \in I\!R^{(k+1)\times n} \tag{4}$$

where

$$f^k(x) = \sum_{i=1}^{m} \min_{j=1,\dots,k+1} \|x - a^i\|.$$

Step 4. (Stopping criterion). Let $x^{k+1,*}$ *be a solution to the problem (4) and* $f^{k+1,*}$ *be the corresponding value of the objective function. If*

$$\frac{f^{k,*} - f^{k+1,*}}{f^{1*}} < \epsilon$$

then stop, otherwise set $k = k+1$ *and go to Step 2.*

Both problems (3) and (4) are nonsmooth optimization problems and we use the discrete gradient method developed in [1] to solve them.

4 Complexity Reduction for Large-Scale Data Sets

As was mentioned earlier in this paper, due to the highly combinatorial nature of clustering problems, two characteristics of a given data set can severely affect the performance of a clustering tool: a number of the data records (instances) and a number of data attributes (features). In simple terms, if each record in the data set is a separate row, the former is equal to the number of "rows", while the latter is equal to the number of "columns" of data in the data set. The natural priorities of a decision maker in this case are to reduce both the number of features and the number of instances without loss of knowledge generating ability.

In order to reduce the number of features, the feature selection procedure discussed below can be implemented.

4.1 Feature Selection

If in (1) $q = 1$, (3) becomes the following convex programming problem:

$$\text{minimize} \quad f(x) = \sum_{i=1}^{r} \|x - a^i\| \quad \text{subject to } x \in \mathbb{R}^n. \tag{5}$$

The solution x^* to the problem (5) is then the center of the set A. Definitely, one center cannot give good enough description of the set A from clustering perspective; however, it contains some information about the structure of this set.

On the other hand, the problem (5) is a classical convex programming problem and there exist effective and fast methods for its solution. Bagirov et al ([3]) discusses a feature selection algorithm for supervised data classification. This algorithm calculates the centers of each class in a dataset under consideration by solving problem (5) and removes closest coordinates in a step-by-step fashion as long as the structure of classes remains unchanged.

In the case of an unsupervised data classification, the feature selection is more difficult to achieve. One of the possible approaches to the feature selection for unsupervised classification is to use at least one of the features as an outcome and the rest of the features as inputs. In this case one can apply the feature selection algorithm from [3] to achieve the desired outcome.

4.2 Number of Instances Reduction

The form of the objective function in the problem (1) allows one to significantly reduce the number of vectors a^i, $i = 1, \dots, m$. Let D be a $m \times m$ matrix (i,j)-th element d_{ij} where $d_{ij} = \|a^i - a^j\|$. This is a symmetric matrix. For each i, $i = 1, \dots, m$, calculate

$$r_i = \min_{j \neq i} d_{ij}.$$

and

$$r_0 = \sum_{i=1}^{m} r_i.$$

Select $\epsilon = cr_0$ so that $c > 0$ is some number. Empirical results of numerical experiments show that the best values for c are $c \in [0, 2]$. Then, the following simple procedure to reduce the number of vectors a^i, $i = 1, \dots, m$ can be used: select first vector a^1, remove from the data set all the vectors for which $d_{1j} \leq \epsilon$, and assign to this vector a number of removed vectors. Then select the next remaining vector and repeat the above procedure for this vector, *etc.* Results of numerical experiments reported below suggest that such a procedure allows one to significantly reduce the number of instances in the data set.

5 Case Study

The data for this study was obtained from Frankston Hospital, a medium sized health unit located in the South-Eastern suburbs of Melbourne, Australia. It is part of a network of hospitals which serves nearly 290,000 people year-round, with a seasonal influx of visitors to the area of up to 100,000. The area is a prime seaside retirement location where there is a high proportion of elderly people. Demand for the services of Frankston hospital is exacerbated during holiday periods, when visitors impact heavily on emergency services.

The data used in this study consists of 15,194 records of admitted patients for years 1999/2000. It contains information about demographics of patients and their length of stay (LOS) in the hospital.

Age, gender and initial admissions information are known when the patients arrive at the hospital. Patients can be admitted to various wards including emergency, children's ward, cardiac care ward, short stay unit, post natal unit, hospital-in-the-home unit, intensive care unit, etc. Time spent in emergency data are obtained during the patient stay. Information about patients' DRGs, LOS and last discharge wards are determined once the patient leaves the hospital. There are about one thousand different DRG groups that are coded using the 3 digit ICD-9 (International Classification of Disease) codes. LOS of a patient is calculated as the difference between the time the patient is admitted to the hospital and the time the patient is discharged from the hospital.

As mentioned in earlier discussion, numerical experiments were carried out to determine the optimal number of clusters that could adequately distinguish the data. During the experiments LOS was not used as an input in order to determine whether each cluster would exhibit different LOS characteristics.

The analysis of the underlying decision making situation suggests the following criteria for selecting the "optimal" number of clusters:

- *The clusters themselves are distinct in terms of LOS* (if two groups of patients could independently come up with two different average LOS, then chances are these two groups are different from one another)
- *The variables that belong to each cluster make sense* (the variables that each cluster has should be distinct and carry some information of its own. When each cluster is analyzed, its profile should be unique and meaningful).
- *The sizes of clusters are comparable* (the size of each cluster needs to be monitored. If the cluster is too large then it is possible that more distinct groups could lie in the cluster. Likewise if it is too small, then there is high probability that the cluster is artificial).

Once the clusters are obtained, each cluster profile is examined in detail, and an examination of the DRGs with each cluster can then be performed.

6 Results and Discussion

As discussed earlier in the paper, in order to use the feature selection algorithm suggested in [3], one of the features has to be used as an output. LOS is the best candidate for such a job because it not only allows one to get a good enough partition in the datasets, but also was used in the similar capacity in [10].

The dataset is divided into separate classes using different values of LOS and then the feature selection algorithm suggested in [3] is used to reduce the length of the data records. Despite the fact that several different values of LOS have been used to divide the dataset, the subset of most informative features remains the same for all cases. This subset includes the following five features: time the patient spends in emergency department, patient's age, patient's gender, the "admitted to" ward, and the "discharged from" ward. Taking into consideration the fact that the initial statistics included 13 different related parameters, the feature selection procedure reduces the number of features more than twice, thus significantly reducing the complexity of the corresponding optimization problems.

During the next stage of investigation, the LOS and patient's diagnostic group (DRG) are added to this list and a new dataset is formed. Algorithm 1 is then applied to the latter dataset to identify clusters. We use Euclidean norm in our experiments. The tolerance is set to $\epsilon = 10^{-2}$ and $c = 1.5$ in numerical experiments. Four features including LOS, patient's time in emergency department, age, and gender are selected as inputs at this stage and then the distribution of patients with respect to DRG and hospital wards is analyzed for individual clusters. Since gender is a categorical feature and has only two values, the dataset can be divided into two subsets containing only males/females for which clustering is performed, and then the results for these two subsets are compared.

The "male" subset of the data set contains 7659 instances and the "female" contains 7509 instances. The number of instances is reduced using the procedure discussed in Sect. 4. Selecting $c = 1.5$ enabled this procedure to reduce the number of "male" and "female" instances to 2766 and 1363 instances respectively. Note that the total number of instances in the reduced data set is 4129, meaning that by applying the number of instances reduction procedure, the size of the data set is reduced by more than 3 times.

Then the clustering algorithm is applied to calculate clusters in both subsets. In both cases the algorithm identified 8 clusters. Further reduction in the tolerance parameter ϵ led to the appearance of two new very small and insignificant clusters. According to the principles for the number of clusters selection outlined in the previous section, the conclusion is that these data sets contain only 8 meaningful clusters. Different values of c can be used in numerical experiments. For $c \in [0, 1.5)$, the structure of clusters is very similar to the one generated for $c = 1.5$. On the other hand, for $c > 2$ quite a different structure is observed that does not satisfy the criteria for clusters specified in the previous section.

For the "male" subset the first cluster contains elderly patients (average age is 73) whose LOS is relatively small (average value is 13 hours) and who are admitted and discharged from the emergency department. The second cluster contains elderly patients (average age is 72) whose LOS is very large (average value is 697 hours). The third cluster contains young patients (average age is 26) with a relatively small time in the emergency department (average value is 0.8 hours). The fourth cluster contains young patients (average age is 28) whose time in the emergency department is average (average value is 3.9 hours). Note that in both cases the admitting and discharging wards are almost the same. The fifth cluster contains elderly patients (average age is 73) whose LOS (average value is 294 hours) and time spent in emergency department is large enough (average value is 5.5 hours) and who use a nonhomogeneous mix of admitting and discharging wards. The sixth cluster contains elderly patients (average age is 71) whose LOS is large enough (average value is 250 hours) whereas time in emergency department is very small (average value is 0.9 hours) and who also use a nonhomogeneous mix of admitting and discharging wards. The seventh cluster contains more middle-aged patients (average age is 59) whose LOS is not large (average value is 89 hours) while time in emergency department is large (average value is 7.0 hours), who are typically admitted to and discharged from the special care wards. Finally, the last cluster contains middle aged patients (average age is 61) whose LOS and time in emergency department are not large (average values are 121 and 2.1 hours respectively) and who use a nonhomogeneous mix of admitting and discharging wards.

A very similar situation is observed as the result of the analysis of "female" subset of the dataset.

It is very important to note that despite the fact that different clusters contain very different kinds of patients as far as their resource consumption is concerned, the distribution of DRGs for different clusters is very similar. This observation strongly suggests that DRG alone is not capable of adequately differentiating the patients based on their resource consumption and therefore should not be used as a single basis for hospitals funding.

7 Conclusions

In this paper a nonsmooth nonconvex optimization-based algorithm for solving cluster analysis problem has been proposed. As this algorithm calculates clusters step by step, it allows the decision maker to easily vary the number of clusters according to the criteria suggested by the nature of the decision making situation not incurring the obvious costs of the increased complexity of the solution procedure. The suggested approach utilizes the stopping criterion that prevents the appearance of small and artificial clusters. The form of the objective function allows one to significantly reduce the number of instances in a dataset - the feature that is extremely important for clustering in large scale data sets.

The power of this approach has been illustrated by conducting the clustering analysis of a hospital data set containing over 15,000 records for the purposes of generating new knowledge about patient resource consumption. Knowledge that is not available from DRG information alone has been generated using this clustering method. This knowledge can be used by hospital managers for prediction of resource consumption by different patients. The proposed approach can, therefore, be seen as an evidence-based predictive tool with high-knowledge generation capabilities.

References

[1] Bagirov, A.M.: Minimization methods for one class of nonsmooth functions and calculation of semi-equilibrium prices, In: Eberhard, A., Hill, R., Ralph, D. and Glover, B.M. (eds.): Progress in Optimization: Contribution from Australasia, Kluwer Academic Publishers (1999) 147–175

[2] Bagirov, A.M., Rubinov, A.M. and Yearwood, J.: Using global optimization to improve classification for medical diagnosis and prognosis. Topics in Health Information Management **22** (2001) 65–74

[3] Bagirov, A.M, Rubinov, A.M. and Yearwood, J.: A heuristic algorithm for feature selection based on optimization techniques. In: Sarker, R., Abbas, H. and Newton, C.S. (eds.): Heuristic and Optimization for Knowledge Discovery, Idea Publishing Group, Hershey (2002) 13–26
[4] Jain, A.K., Murty, M.N. and Flynn, P.J.: Data clustering: a review. ACM Computing Surveys **31(3)** (1999) 264–323
[5] Dubes, R. and Jain, A.K.: Clustering techniques: the user's dilemma. Pattern Recognition **8** (1976) 247–260
[6] Hawkins, D.M., Muller, M.W. and ten Krooden, J.A.: Cluster analysis, In: Hawkins, D.M.: Topics in Applied Multivariate Analysis, New York, Cambridge University press (1982)
[7] Mangasarian, O.L.: Mathematical programming in data mining. Data Mining and Knowledge Discovery **1** (1997) 183–201
[8] McLachlan, G.J.: Discriminant Analysis and Statistical Pattern Recognition. New York, John Wiley (1992)
[9] Michie, D., Spiegelhalter, D.J. and Taylor, C.C. (eds.): Machine Learning, Neural and Statistical Classification. Ellis Horwood Series in Artificial Intelligence, London (1994)
[10] Ridley, S. Jones, S., Shahani, A., Brampton, W., Nielsen, M. and Rowan, K.: Classification Trees. A possible method for iso-resource grouping in intensive care. Anaesthesia **53** (1998) 833–840
[11] Siew, E.-G., Smith, K. and Churilov, L.: A neural clustering approach to iso-resource grouping for acute healthcare in Australia. Proceedings of the 35th International Conference on Systems and Systemics, Hawaii (2002)
[12] Spath, H.: Cluster Analysis Algorithms. Ellis Horwood Limited, Chichester (1980)

Dynamic Parameterization to Simulate DIN Export Due to Gypsy Moth Defoliation

Ping Wang[1], Lewis C. Linker[2], and Keith N. Eshleman[3]

[1] University of Maryland Center for Environmental Science, Chesapeake Bay Program
410 Severn Avenue, Annapolis, MD 21403, USA
pwang@chesapeakebay.net

[2] US Environmental Protection Agency/CBPO, 410 Severn Ave., Suite 109.
Annapolis, MD 21403, USA
linker.lewis@epa.gov

[3] University of Maryland Center for Environmental Science, Appalachian Laboratory
301 Broddock Road, Frostburg, MD 21532, USA
eshleman@al.umces.edu

Abstract. A module of dynamic parameterization is added into the HSPF watershed software for simulation of dissolved inorganic nitrogen (DIN) export from forest associated with gypsy moth defoliation. It simulates a changing ecosystem following the breakout of defoliation, such as increasing mineralization and nitrification rates and soil temperature, and decreasing interception of precipitation, plant nitrogen uptake rate and evapotranspiration. These parameter values vary with the stages of a defoliation event, such as the progressive period, the peak period, and the recovery period, the simulated DIN export from a multi-occurrence defoliation area in Shenandoah National Park in Virginia, USA, is comparable with the observed data.

1 Introduction

Forest is one of the important landuses in the Chesapeake Bay watershed (Fig. 1), covering about 60 percent of the area. The load of nitrogen (mass per unit area) is usually much lower from forest than from other landuses due to low nitrogen input, and high uptake and storage potential. However, increased dissolved inorganic nitrogen (DIN) export from mid-Appalachian forest following insect defoliation by the gypsy moth caterpillar (*Lymantra dispar*) has been observed at numerous sites [1,2]. Efforts have been made to simulate such conditions in computer models in order to improve watershed calibration and loading computations, with varying degrees of success [3,4]. However, these efforts are not successfully associated within the widely applied watershed models, such as HSPF (Hydrologic Simulation Program – Fortran) [5]. The HSPF software is the primary tool to estimate nutrient load to the Chesapeake Bay, the largest estuary in the USA, from its watershed. The current Chesapeake Bay Watershed Model (CBWSM, which uses HSPF) does not consider nutrient processes involving gypsy moth defoliation [6]. It is desired for HSPF to have the capability in simulation of DIN export associated with gypsy moth defoliation, so that to provide more accurate information for nutrient management decisions.

P.M.A. Sloot et al. (Eds.): ICCS 2003, LNCS 2659, pp. 30–38, 2003.

The implementation of a watershed model with gypsy moth defoliation requires understanding of physical, chemical, and biological processes associated with the defoliation, and a model structure capable of simulating these processes. Wang et al. [7] discussed the consequences of ecosystem changes after defoliation and the adjustment of related parameters in HSPF for the ecosystem simulation. They obtained elevated DIN export, and demonstrated a prospective of using HSPF to simulate gypsy moth defoliation. However, the parameter adjustments were confined in the parameter blocks of the HSPF input-deck. The HSPF allows these parameters to vary among months but with the same values in every year for that months, and cannot specify a value in the middle of a month. Therefore, the simulated daily DIN export did not match well with the observed. The timing of outbreak, and the duration and intensity of gypsy moth defoliation are different in different occurrences. Therefore, it is important to have a dynamic parameterization for different stages in a defoliation event and for different occurrences of defoliation. Although the "Special Action" of the HSPF allows dynamic change in many parameters, its restrictions do not fully meet our requirements. We setup a module to change parameter values dynamically. This paper describes how we setup the module of dynamic parameterization in HSPF, and presents the simulated DIN export from a multi-occurrence defoliation area over a few years.

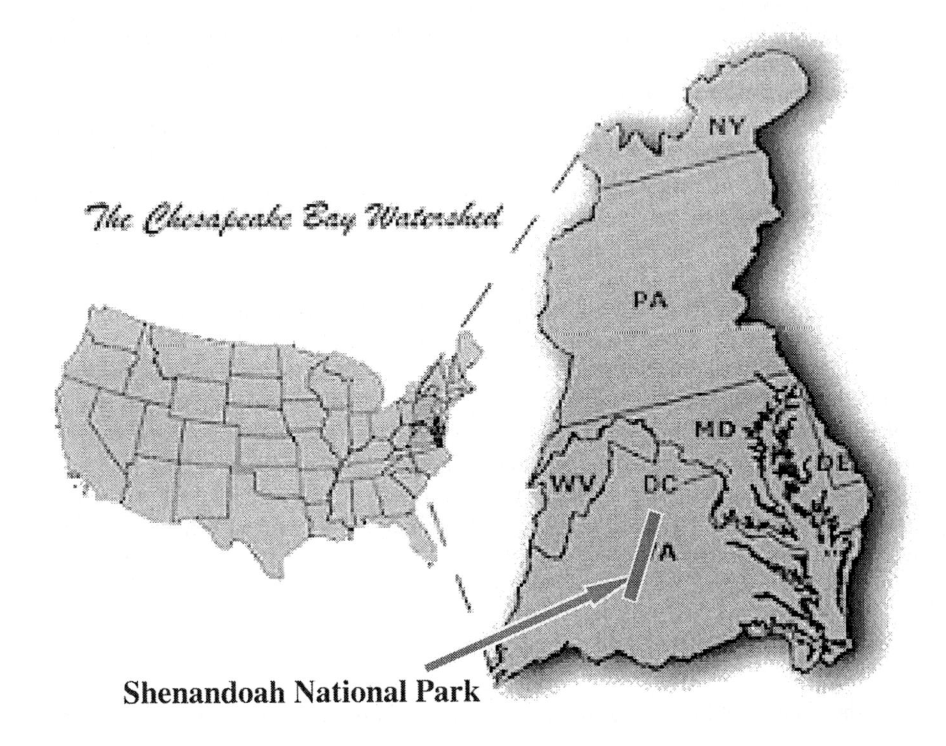

Fig. 1. Location of Shenadoah National Park and Chesapeake Bay watershed, USA

2 Method

HSPF version 11 is the base code for the model simulation. A dynamic parameterization module was added, so that parameter values can change with the changing ecosystem in different stages of a defoliation event. Note: the original HSPF version 11 code yielded lower DIN exports under higher flows when using the Michaelis-Menton saturation kinetics for plant nitrogen uptake [8]. Prior to adding the dynamic module, the associated code was corrected to yield a positive relationship between DIN export and flow in high flow conditions [9]. Most of the parameters used in this study are based on the CBWSM Phase 4.3 calibration for non-defoliation forest in the above-fall-line of the Piegmont area of the Rappahannack River basin, which is referred to as the "base condition" in the following text.

Defoliation could result in a variety of effects on the forest [10], such as 1) reduction in transpiration rate, 2) decrease in uptake of nutrient by plants, 3) higher soil temperature due to increased radiant energy flux to forest floor, and 4) alterations in microbiological activity. In the simulation of gypsy moth defoliation, the above changes in the ecosystem are considered. While other external conditions (i.e., precipitation, other meteorological conditions, and atmospheric nitrogen deposition) are set the same as in the base calibration. The dynamic parameterization module allows user to specify the following information related with a defoliation even: the beginning, peak and ending date of defoliation, percent coverage of defoliation area versus the forested area, intensity of defoliation (i.e., percentage of trees is defoliated in the defoliation area), parameter values such as plant uptake rate, mineralization and nitrification rates, evapo-transpiration rate, intercept of precipitation, and soil temperature factors in the peak and normal conditions. The lag time between the maximum change of some parameter values and the peak defoliation can be set by the user for ease of calibration, although it is not required.

The module converts the information to daily parameter values that vary with the stage of defoliation. During the recovery of the forest, the parameter values change back gradually to the normal condition. The model can also handle multi-occurrence of defoliation events. With the module, the HSPF updates these parameters in the simulation run. The module is, in fact, a bridge between the HSPF simulation and the info of dynamic changes in parameters associated with defoliation stages. The defoliation events in this study are referred to a forested watershed, Piney River watershed of Shenandoah National Park (Fig. 1), during the period 1987-1993. The area of mapped defoliation is 3.8%, 32.3%, 34.0%, 15.2%, 0.0%, 14.2% and 0.0% in 1987-1993, respectively. The intensity of defoliation is set at 50%, i.e., moderate heavy.

3 Result

With the above setting of parameters for gypsy moth defoliation, high DIN export was yielded (Fig. 2).

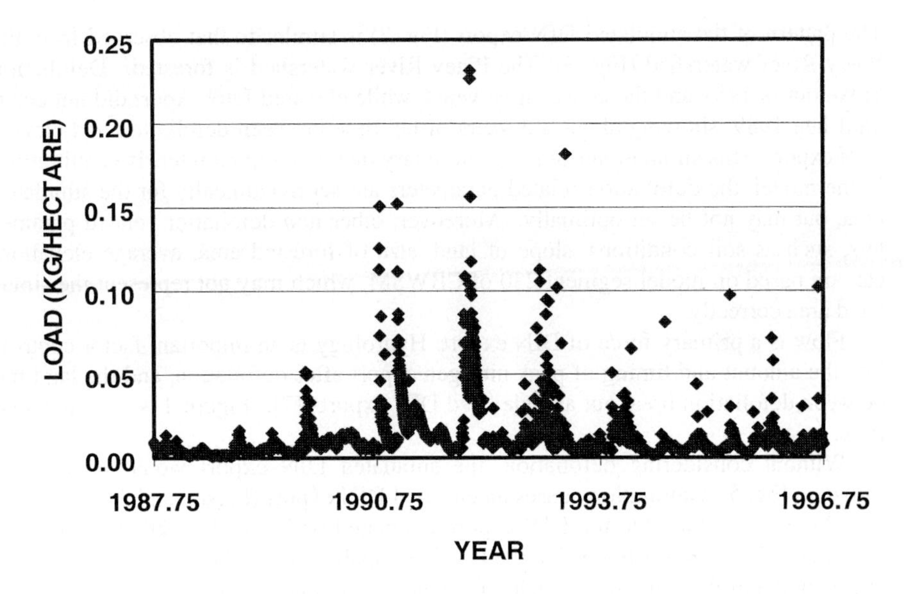

Fig. 2. Simulated DIN load from Piney River watershed in Shenandoah National Park using HSPF with a dynamic parameterization module

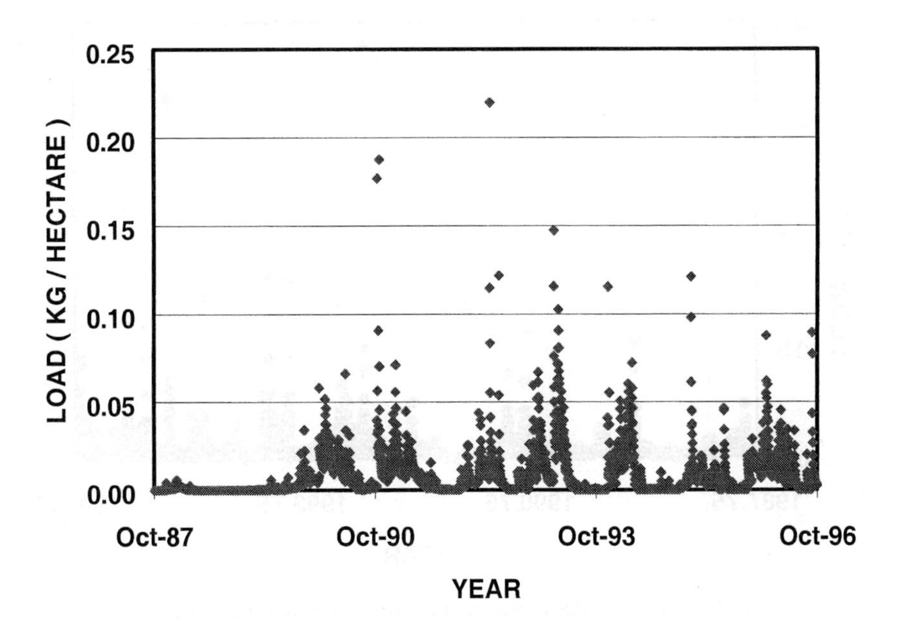

Fig. 3. Observed DIN loads from Piney River watershed in Shenandoah National Park, which undergone gypsy moth defoliation

The pattern of the simulated DIN export (Fig. 2) is similar to that observed from the Piney River watershed (Fig. 3). The Piney River watershed is forested. Defoliation broke out in 1987 and the consecutive years, while elevated DIN export did not occur until late 1989, showing about 2-3 years of lag time between defoliation and elevate DIN export. This simulation run is a preliminary one, without an intensive calibration. In the model, the defoliation-related parameters are set dynamically for the simulated area, but may not be set optimally. Moreover, other non-defoliation related parameters, such as soil conditions, slope of land, size of forested area, average elevation, etc, are based on model segment 230 of CBWSM, which may not represent the simulated area correctly.

Flow is a primary force of DIN export. Hydrology is an important factor controlling the amount and timing of peak nitrogen export after defoliation, and the lag time between defoliation breakout and elevated DIN exports [7]. Figure 4 is the simulated flow.

Without considering defoliation, the simulated DIN export would be low, as shown in Fig. 5. Defoliation causes an elevated DIN export (Figs. 2 and 3).

This work shows that the DIN export associated with a real defoliation case can be simulated. It demonstrates that the HSPF is capable of simulating the conditions of gypsy moth defoliation with the help of a dynamic parameterization module.

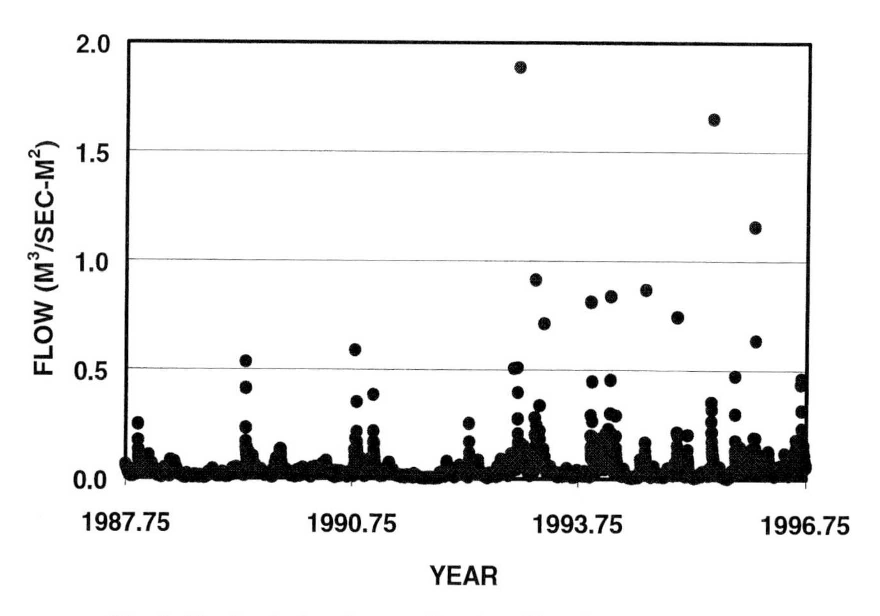

Fig. 4. Simulated edge of stream flow from Piney River watershed

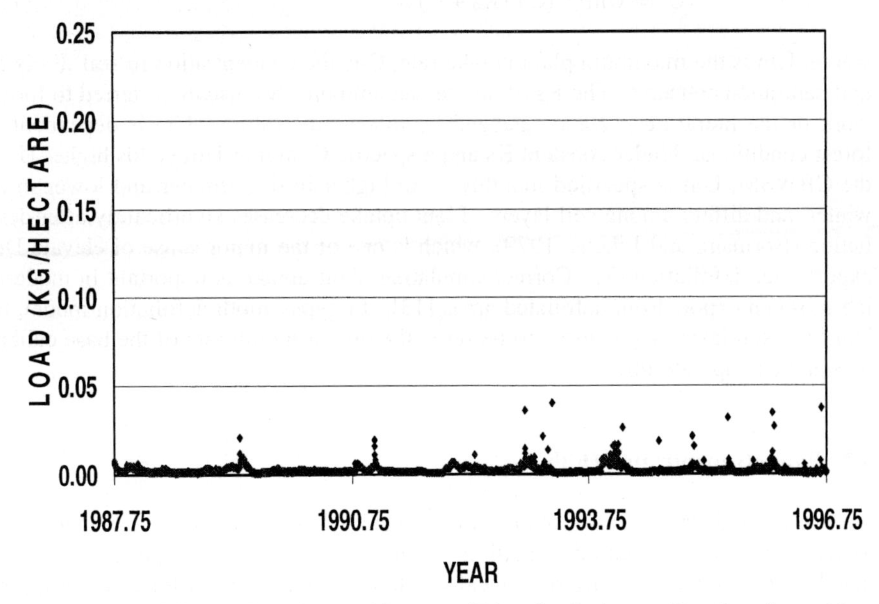

Fig. 5. Imulated DIN loads from forest of Segment 230 under the base calibration (i.e., without simulating defoliation)

4 Discussion

4.1 Transformation of Organic Nitrogen to Inorganic Nitrogen

The products of insect defoliation, such as insect biomass, frass, and litters, are important sources of organic nitrogen, and cause, through mineralization and nitrification, an increase in inorganic nitrogen availability [11][12]. The model simulates these sources as organic nitrogen and ammonia in the soil surface layer. These nitrogen constituents on the surface layer further transforms throughout the simulation period. Higher mineralization and nitrification rates after defoliation cause DIN more availability and higher export [7]. The mineralization rates of organic nitrogen and nitrification rates in the top soil layer are set as high as 10 times of the base calibration setting for a severe ecosystem change. The module allows a specification of lag time between the peak defoliation and the peak of tortured ecosystem. The rates and lag time can be adjusted through model calibration.

4.2 Plant Uptake

In the CBWSM, plant uptake rate (U)for forest is simulated with the Michaelis-Menten saturation kinetics:

$$U = Um * (C / (Ks + C) , \quad (1)$$

where, Um is the maximum plant uptake rate, C is the concentration in soil, Ks is the half saturation constant. The Ks of nitrate and ammonia are usually referred to former work or the literature [12]and finalized by model calibration. Um is dependent on forest conditions. Under constant Ks and a specific C, higher Um yields higher U. In the CBWSM, Um is specified monthly, with higher in the summer and lower in the winter, and differs among soil layers. Plant uptake decreases significantly after defoliation (Bormann and Likens, 1979), which is one of the major cause of elevate DIN export after defoliation [7]. Correct simulating plant uptake is important in moderating nitrogen export from defoliated areas [13]. In gypsy moth defoliation model, the Um for a soil layer is suggested to be set at the lowest month rate of the base calibration in the temperate forest.

4.3 Evapotranspiration (ET)

In the CBWSM, evapotranspiration is determined by various factors, such as meteorological, soil, and vegetation conditions. The potential evapotranpiration (i.e., demand) is determined by meteorological conditions. The sources for evapotranspiration are, according to moisture availability for the demand, from the storage or outflow of soil layers. The loss of leaves by gypsy moth defoliation could cause significant reduction in transpiration of moisture taken by roots from the lower zone. This can be simulated through adjusting the lower zone evapotranspiration parameter, LZET. Reduction in LZET after defoliation could cause higher discharge and increase DIN export. Our assessment in Shenandoah National Park suggested that setting LZET at about a half of the base calibration value be appropriate in peak defoliation.

4.4 Soil Temperature

In the CBWSM, surface soil temperature is estimated by regression from empirical air and soil temperatures:

$$Ts = Y + S * Ta , \quad (2)$$

where, Ts = surface layer temperature (degrees C), Y = y-intercept of regression, S = = slope of the regression, and Ta = air temperature (degrees C). The regression coefficients vary with months. The monthly y-intercept ranges from 2 to 16 degrees C from winter to summer in the base calibration. In the gypsy moth defoliation simulation, the y-intercepts range from 5 to 25 degrees C from winter to summer, while the slope is set the same as the base calibration. The elevated soil temperature in the simulation causes less organic nitrogen but slightly more DIN export [7]. This may be due to elevated soil temperature that promotes mineralization of organic nitrogen to DIN and subsequent nitrification. On the other hand, elevated soil temperature may also increase the evaporation rate, causing less flow and less

DIN export. The overall effect of elevated soil temperature appears to cause lower organic nitrogen (OrN) for a long period, higher DIN in the earlier period following defoliation, but less DIN in later period due to depletion of organic nitrogen in the soil storage [7].

4.5 Interception of Precipitation

In the CBWSM, interception is specified monthly, with highest values in the summer and lowest values in the winter. In modeling gypsy moth defoliation, the interception values were set as the January value in the base calibration to simulate the decrease of interception capacity of tree after defoliation. A decrease of interception after defoliation causes more water to infiltrate and slightly higher DIN exports based on a 10-year average [7]

Wang et al. [7] noticed that in some high flow days after the defoliation DIN export is slightly less than that by assuming a higher interception as the base calibration setting. This may be due to the fact that on high precipitation days the soil moisture storage meets ET demand in both defoliation and non-defoliation cases. In the defoliation case, less interception has more remaining ET demand for the lower layers and has lower ratio in groundwater/surface runoff, causing a relatively lower DIN/OrN ratio in export, because DIN export is related more with interflow or groundwater flow, and OrN export is related more with surface flow. Moreover, a high export of OrN in the early period of gypsy moth defoliation may allow less conversion of OrN to DIN in later days [7]. Nevertheless, in a 10-year average, both DIN and OrN exports are higher under the lower interception condition.

5 Conclusion

The above work shows that dynamic parameterization enables the HPSF watershed software to simulate gypsy moth defoliation. Under the defoliation conditions, insect biomass, frass, and leaf debris are important nitrogen sources for the forest soil. The changes in the forest ecosystem following defoliation play important roles in controlling rates of nitrogen export. The decrease of plant uptake is an important factor. The increase in biological activities causes higher availability of DIN in soil layer, leading to higher nitrogen export. Decrease in transpiration and interception increases flow and nitrogen export. The increase in soil temperature can increase evaporation rates and decrease flow and DIN export. However, the increase in soil temperature increases ammonification and nitrification rates, increasing DIN export. Hydrology pattern controls the amount and timing of nitrogen export. The availability of nitrogen in soil and hydrological forcing determine the lag time of peak DIN export after defoliation. The effect of gypsy moth defoliation can affect nitrogen export for several years.

References

1. Webb, J.R., Cosby, B.J., Deveney, F.A., Eshleman, K.N., Galloway, J.N.: Change in the acid-base status of an Appalachian mountain catchment following forest defoliation by the gypsy moth. Water Air Soil Pollut., (1995) 85, 535–540
2. Eshleman, K.N., Morgan R.P., Webb J.R, Deviney, F.A., Galloway, J.N.: Temporal patterns of nitrogen leakage from mid-Appalachian forested watersheds: Role of insect defoliation, Water Resour. Res. (1998) 34, 2005–2016
3. Castro, M.S., Eshleman, K.N., Morgan, R.P., Seagle, S.W., Gardner, R.H., Pitelka, L.F.: Nitrogen dynamics in forested watersheds of the Chesapeake Bay, Scientific & Technical Advisory Committee Publication 97–3, Chesapeake Bay Program, Annapolis, Maryland, USA. (1997)
4. Eshleman, K.N.: A linear model of the effects of disturbance on dissolved nitrogen leakage from forested watersheds, Water Resour. Res. (2000) 36, 3325–3335
5. Bicknell1, B.R., Imhoff1, J.C., Kittle, J.L, Donigian, A.D., Johanson, R.C.: Hydrological Simulation Program – FORTRAN, User's manual for version 11, U.S. Environmental Protection Agency, National Exposure Research Laboratory, Athens, Ga., EPA/600/R–97/080, (1997) p755
6. Linker, C.L, Shenk, W.G., Wang, P.,Storrick, J.M.: Chesapeake Bay Watershed Model Application & Calculation of Nutrient & Sediment Loading, Appendix B, EPA903–R–98–003, CBP/TRS 196/98 (1998)
7. Wang, P., Linker, L.C., Shenk, G.W., Eshleman, K.N.: Computer model of nitrogen export from disturbed forest by Gypsy moth defoliation. Proceeding of 5th International Conference on Diffuse/Nonpoint Polutionand Watershed Management, Milwaukee, WI, USA (2001) Sec.24, 7–15
8. Wang, P., Linker, L.C.: Effect of moisture on nitrogen export: comparison of two plant uptake saturation kinetics models. EOS Trans. AGU, 82(20), Spring Meet. Suppl., Abstract H41B–01, (2001) Boston, USA
9. Wang, P., Linker L.C.: Improvement of HSPF watershed model in plant uptake and DIN export from forest. EOS Trans. AGU, Fall Meet. Suppl., Abstract H12H–12, (2001) San Francisco, USA
10. Bormann, F.H., Likens, G.E.: Pattern and process in a forested ecosystem, Springer-Verlag, NY (1979) pp 253
11. Seastedt, T.R., Crossley, D.A.: The influence of arthropods on ecosystem. BioScience, (1984) 34, 157–161
12. Bowie, G.L, Mills, W.B., Porcella, D.B., Camplbell, C.L., Pagenkopf, J.R., Rupp, G.L., Johnson, K.M., Chan, P.W.H., Gherini, S.A., Chamberlin, C.E.: Rates, constants, and kinetics formulations in surface water quality modeling, NERL, US Environmental Protection Agency, Athens, GA, (1985) EPA/600/3–85/040, pp. 456
13. Swank, W.T.: Stream chemistry responses to disturbance, in: Forest Hydrology and Ecology at Coweeta, Swang W.T. and Croseley D.A. (ed), Springer-Verlag, New York, (1988) 237–271

Multi-model Simulations of Chicken Limb Morphogenesis

R. Chaturvedi[1], J.A. Izaguirre[1], C. Huang[1], T. Cickovski[1], P. Virtue[1], G. Thomas[2], G. Forgacs[3], M. Alber[4], G. Hentschel[5], S.A. Newman[6], and J.A. Glazier[7]

[1] Department of Computer Science and Engineering, University of Notre Dame, IN 46556
Corresponding author: `izaguirr@nd.edu`
Tel: (574) 631-7454 Fax: (574) 727-0873
[2] Department of Physics, University of Notre Dame, Notre Dame, IN 46556
[3] Departments of Physics and Biology, University of Missouri, Columbia, MO 65201
[4] Department of Mathematics, University of Notre Dame, Notre Dame, IN 46556
[5] Department of Physics, Emory University, Atlanta, GA 30332
[6] Department of Cell Biology and Anatomy, New York Medical College, Valhalla, NY10595
[7] Biocomplexity Institute and Department of Physics, Indiana University, Bloomington, IN 47405

Abstract. Early development of multicellular organisms (*morphogenesis*) is a complex phenomenon. We present COMPUCELL, a multi-model software framework for simulations of morphogenesis. As an example, we simulate the formation of the skeletal pattern in the avian limb bud, which requires simulations based on interactions of the genetic regulatory network with generic cellular mechanisms (cell adhesion, haptotaxis, and chemotaxis). A combination of a rule-based state automaton and sets of differential equations, both subcellular ordinary differential equations (ODEs) and domain-level reaction-diffusion partial differential equations (PDEs) models genetic regulation. This regulation controls the differentiation of cells, and also cell-cell and cell-extracellular matrix interactions that give rise to cell pattern formation and cell rearrangements such as mesenchymal condensation. The cellular Potts model (CPM) models cell dynamics (cell movement and rearrangement). These models couple; COMPUCELL provides an integrated framework for such computations. Binaries for Microsoft Windows and Solaris are available[1]. Source code is available on request, via email: compucell@cse.nd.edu.

1 Introduction

In the fields of bioinformatics and computational biology an aim is to link the wealth of data (*e.g.* genetic sequences and genetic regulatory networks) to the understanding of biological processes such as development of multicellular organisms (*morphogenesis*). Computational systems biology concerns itself with creating such integrated models. Relatively little integrated modeling of multicellular organisms exists; examples are: development models of the *Drosophila* embryo [1], and analysis of the gene

[1] `http://www.nd.edu/~lcls/compucell`

P.M.A. Sloot et al. (Eds.): ICCS 2003, LNCS 2659, pp. 39–49, 2003.

regulatory network of developing sea urchin [2]. These efforts are confined mainly to the modeling of gene regulation. For morphogenesis, we need to model cell behaviors like release and absorption of diffusible factors, adhesion, and motility [3], in addition to differential regulation of gene activity.

We have implemented the software framework COMPUCELL to model early morphogenesis. It allows interaction of the genetic regulatory network with cellular mechanisms; features include biosynthesis, cell adhesion, *haptotaxis* (the movement of cells along a gradient of a molecule deposited on a substrate) and *chemotaxis* (the movement of cells along a gradient of a chemical diffusing in the extracellular environment), and diffusion of *morphogens* (molecules released by cells that affect the behavior of other cells during development). The interplay of these factors results in arrangements of cells specific to a given organism, cf. [4]. We describe the model and apply it to simulate the skeletal pattern formation in avian (chicken) limb.

Biological Background: Skeletal pattern formation occurs within a mesenchymal tissue (mesenchyme are cells arranged loosely in a hydrated extracellular matrix (ECM); they make only minimal contact with one another) surrounded by a thin bounding layer, the ectoderm; Figure 1 shows a schematic of vertebrate limb. Our two-dimensional (2-d) simulation attempts to reproduce the pattern of development of bones in the forelimb of a chicken viewed from above when it lies palm down on a flat surface. The long axis from the body to the digits is the proximodistal direction (proximal means close to the body); from the thumb to the little finger is the anteroposterior direction. Thickness of the limb, from back to front (the dorsoventral direction) is one skeletal element throughout, justifying our 2-d simplification. Although asymmetry along the dorsoventral axis as seen in the typical paddle shape of limb is important to function, our initial simulations ignore it in favor of the more striking asymmetries along the two other axes. Apical ectodermal ridge (AER), the narrow strip of the ectoderm running along the apex of the growing limb bud, is necessary for elongation and patterning of limb. AER releases fibroblast growth factors (FGFs), which control mitosis (division) of mesenchymal cells. FGF concentrations increase in the distal direction. Rates of cell division are known experimentally for the avian limb [5]. Limb growth and pattern formation has been conceptualized by considering the space within the developing limb to comprise three zones – (1) apical zone where only growth takes place, (2) active zone where dynamics of pattern formation and cell rearrangement into precartilage condensations occur and (3) frozen zone where condensations have progressed to differentiated cartilage and no additional patterning occurs. In development in species with a bony skeleton, bone later isomorphically replaces the cartilaginous skeleton. Active zone may be the locus of a reaction-diffusion system in which one or more members of the TGF-β family of growth factors act as the activating morphogens [6]. Thus the active zone (2) is further classified into (2a) cell active zone where cells rearrange locally into precartilage condensations, (2b) reaction-diffusion active zone where reaction-diffusion mechanisms operate. Growth continues in both the active zones but not in the frozen zone. At sites of incipient condensation, cells bearing a second FGF receptor may release a laterally inhibitory morphogen, a necessary component of some pattern-forming reaction-diffusion schemes. Rather than use a reaction-diffusion scheme based on the actual biochemical interactions (they are still poorly characterized) we choose for heuristic purposes to use the well-studied Schnakenberg reaction-diffusion equations, Equation (5). We refer to 'u' in Schnakenberg equations as the reaction-diffusion activator, rather than identify it with a specific molecule TGF-β.

Cells sensing activator levels above a threshold secrete an adhesive material onto the substrate and become intrinsically stickier than surrounding cells. In the actual limb, TGF-β induces cells in active zones to produce the ECM glycoprotein fibronectin [7]. Fibronectin adheres to the cell surface, causing cells to accumulate at focal sites. Cells at these sites also become more adhesive to one another by producing a homophilic cell-surface adhesion protein N-cadherin [8]. The spatiotemporal pattern of activator thus results in a corresponding set of cell condensations. In the description of our model and simulation results for avian limb, we refer to the secreted substrate adhesive molecule that promotes haptotaxis as *SAM* (surface adhesion molecule) and the cell-cell adhesive molecule as *CAM* (cell adhesion molecule). The current model omits the mitogen (FGF) field due to AER, instead assuming that cell division is uniform throughout the limb bud [5].

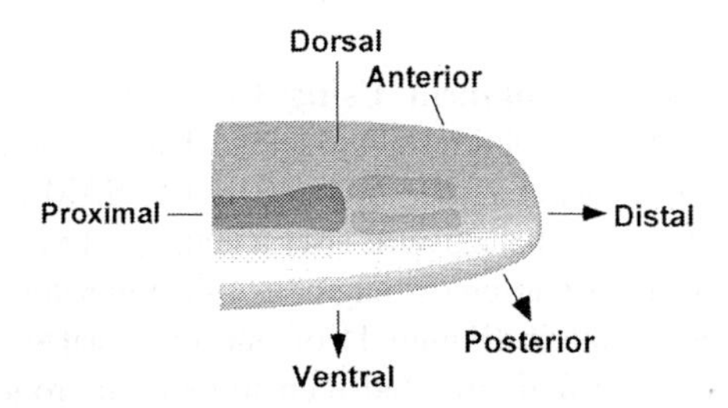

Fig. 1. Schematic representation of a developing vertebrate limb: We indicate the three major axes and the first two tiers of skeletal elements to form. In the chicken forelimb these elements are the humerus, shown as already differentiated (dark gray), followed by the radius and ulna, which are in the process of forming (light gray). Still to form are the wrist bones and digits. The apical ectodermal ridge runs along the distal tip of the limb approximately between the two points that the arrow indicating the anteroposterior axis intersects.

Brief Description of the Model: Steinberg's Differential Adhesion Hypothesis (*DAH*) uses cell-cell and cell-matrix adhesive interactions to explain cell rearrangement, including mesenchymal cell condensation [9]. This generic aspect of cell behavior is modeled using CPM [10], which we have extended to account for haptotaxis and chemotaxis. Cells may respond to morphogens they or their neighbors produce by altering their gene activity in continuous or discontinuous (switch-like) fashion; such nonlinear feedback loops may lead to *differentiation* into more specific cell types. We model the network of expressed genes and their products as a set of rules that trigger growth, cell division, cell death, secretion of morphogens and strength of adhesion between cells; in the limb they are mediated by level of activity and distribution of growth factors. [11] reviews these and other approaches to modeling gene regulatory networks. A reaction-diffusion mechanism may underlie limb skeletal pattern formation as well as other biological spatiotemporal patterning [12–14]. We propose a model that uses domain-level PDEs to simulate reaction-diffusion mechanisms. In our biological example, these PDEs generate morphogen concentration distributions that establish a prepattern for mesenchymal cell condensation. The rules also interact with the stochastic CPM which governs dynamics and geometry of cell growth and division. Domain growth follows from cell growth. Details are provided below.

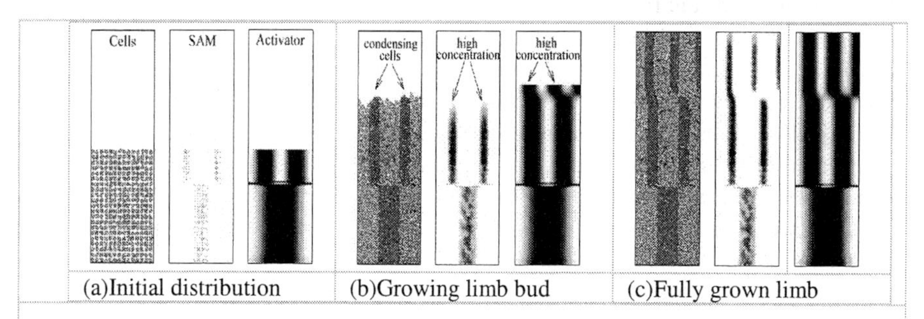

Fig. 2. Simulation of skeletal pattern formation in avian limb using COMPUCELL. For full limb, height to width ratio is 3 to 1. Figure not to scale.

Simulation of Avian Limb Development Using COMPUCELL: Figure 2 presents COMPUCELL simulation results of the biological model: (i) The pre-pattern formed by the morphogen (activator) chemical field (middle frame), (ii) The SAM concentration field produced by the cells in response to the activator after a delay (right frame), and (iii) The organization of mesenchymal cells into precartilaginous tissue in response to these fields (left frame). The typical bone pattern (Figure 1) of one proximal rod-shaped element followed by two elements of the mid-arm, and then three digits reflects the temporal progress of chicken limb morphogenesis. It is generally similar to formation of other vertebrate limbs. Our long-term goal is to explore developmental mechanisms such as avian limb pattern formation using realistic gene regulatory/reaction-diffusion networks and cell-cell adhesion measurements. In the absence of information on the interactions and quantities required for a biologically accurate simulation, the current simulations restrict themselves to describing the major biological issues involved in avian limb bud development. We have focused on simulating mesenchymal cell condensation and emergence of a skeletal pattern resembling that of chicken forelimb. COMPUCELL's modular architecture allows iteratively replacing the interactions and quantities with more detailed and realistic ones as they become available. Although our simulations use a rectangular, 2-d domain instead of a curvilinear, paddle-shaped domain, and differential equations that do not correspond exactly to the genetic processes in the living limb bud, they do adequately illustrate a general conceptual framework for vertebrate limb organogenesis. A similar statement can be made with regard to accurate measurements of tissue properties that serve as input to CPM. Figure 2 also shows growing zones and their interfaces moving in distal direction. Cells formed by division move downward (out of apical zone) and replenish active zones. Lowermost frozen zone grows as cells condense into a bone-like pattern.

2 Computational Model

Activator concentration in the internal space of the limb obeys a set of reaction-diffusion PDEs (we use the Schnakenberg equations). With an appropriate choice of parameters in the Schnakenberg reaction-diffusion equations Eq. (5), which includes the space in which the reaction-diffusion mechanism operates, the activator concentration produces a sequence of arithmetically increasing numbers of vertical stripes, roughly approximating the distribution of skeletal elements in a chicken limb, see Figs. 3a–c. Cells that sense a threshold level of activator concentration locally begin to produce SAM. This is modeled as a transition from a non-SAM producing mesenchymal cell state to a cell state capable of SAM production. Thus, we have three fields of chemical concentration in our computations – for the activator, the inhibitor and SAM. COMPUCELL allows definition of an arbitrary number of fields and state transition rules of interest. In our example, the first two fields evolve according to the Schnakenberg equations, the SAM field depends on local secretion by cells and the secretion rate itself obeys an ODE. Finally, the mesenchymal cells respond to elevated SAM concentration by upregulating their cell-cell adhesion parameters, and by responding to the 'gluing' effect of SAM through the haptotaxis term in the CPM. The secretion of SAM by cells reinforces condensation of mesenchymal cells in regions which already had high SAM concentrations. We now describe the submodels.

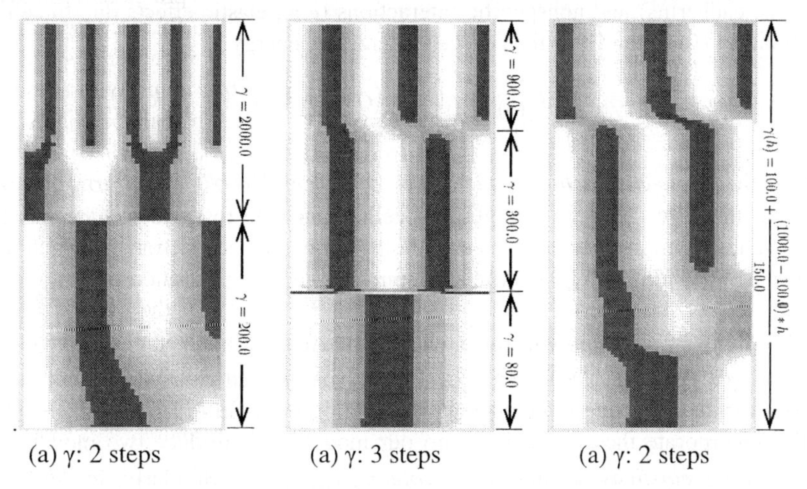

(a) γ: 2 steps (a) γ: 3 steps (a) γ: 2 steps

Fig. 3. Solutions of Schnakenberg equations (5) illustrating periodicity of stripes in stable solutions. Aspect ratio is 3:1. Figure is not to scale

2.1 The Cellular Potts Model (CPM)

CPM is an extension of the Ising model[2]. It can simulate condensation of cells based on differential adhesion [9,10]. We extend the model by (i) Adding an extra term to

[2] http://www.physics.cornell.edu/sethna/teaching/sss/ising/intro.htm

model haptotaxis of cells to SAM, (ii) Allowing time variation in the adhesivity of cells, and (iii) Allowing cell growth and division

Energy Minimization Formalism: CPM dynamics use an energy minimization formalism to simulate the spatially extended cells undergoing cytoskeletally driven fluctuations. Quantitative experiments show that the CPM successfully reproduces the behavior of simple cell aggregates [15,16]. Fundamental entities in the model are individual cells. An *effective energy*, *E,* and *"fields"* (*e.g.*, local concentrations of diffusants) describe cells' interactions, motion, differentiation and division. Effective energy mixes true energies, like cell-cell adhesion, and terms that mimic energies, *e.g.*, response of a cell to a chemotactic gradient. Cells move to minimize their total effective energy. CPM superimposes a lattice on the cells, with an *index* (also called a *spin*) associated with each lattice site (*pixel*). Value of the index at a lattice site is σ if the site lies in cell σ. *Domains* (*i.e.*, the collection of lattice sites with the same index) in the lattice represent cells; probability that such domains are connected is high.

Cell Adhesion: The net interaction between two cell membranes is phenomenologically described by a binding energy per unit area, $J_{\tau\tau'}$, which depends on the types of the interacting cells, τ and τ'. Binding energy incorporates both specific (*e.g.*, integrins, cadherins) and nonspecific interactions (*e.g.*, elastic effects due to cell deformations, [17]). In the CPM the cell-cell interaction energy is:

$$E_{\text{Contact}} = \sum_{(i,j,k)(i',j',k')} J_{\tau(\sigma),\tau'(\sigma')}\left(1-\delta\left(\sigma(i,j,k),\sigma'(i',j',k')\right)\right), \tag{1}$$

where the *Kronecker delta*, $\delta(\sigma,\sigma')=0$ if $\sigma\neq\sigma'$ and $\delta(\sigma,\sigma')=1$ if $\sigma=\sigma'$, ensures that only surface sites between neighboring cells contribute to the cell adhesion energy. Cell adhesion molecules may change both in quantity and identity, which we can model as variations in cell-specific adhesivity.

Cell Volume, Cell Division, and Cell Death: At any time, t, a cell of type τ has a volume $v(\sigma,\tau)$ and surface area $s(\sigma,\tau)$. Cell volume can fluctuate around a certain (*target*) value, *e.g.*, due to changes in osmotic pressure. In order to (phenomenologically) incorporate these behaviors into our model, we introduce two model parameters: *volume elasticity*, λ, and *target volume*, $v_{\text{target}}(\sigma,\tau)$. Similarly, for surface area fluctuations, we define a *membrane elasticity*, λ', and a *target surface area*, $s_{\text{target}}(\sigma,\tau)$. We incorporate these constraints in the energy minimization formalism of CPM by introducing energy penalties for variations in v and s of a cell from their target values, which we set to the average values for the particular cell types:

$$E_{volume} = \sum_{\text{all-cells}} \lambda_\sigma \left(v(\sigma,\tau) - v_{\text{target}}(\sigma,\tau)\right)^2 + \sum_{\text{all-cells}} \lambda'_\sigma \left(s(\sigma,\tau) - s_{\text{target}}(\sigma,\tau)\right)^2 \tag{2}$$

We model cell growth by allowing the values of $v_{\text{target}}(\sigma,\tau)$ and $s_{\text{target}}(\sigma,\tau)$ to increase with time. Cell division occurs when the cell reaches a fixed, type-dependent volume. For another form of cell division, see [17]. We model division by starting with a cell of average size, $v_{target}= v_{target,average}$; causing it to grow by gradually increasing v_{target} to $2v_{target,average}$; and splitting the dividing cell into two, each with a new target volume: $v_{target}/2$. One daughter cell assumes a new identity (unique value of σ). We can model cell death simply by setting the cell's target volume to zero.

Extracellular Matrix (ECM): We model ECM, liquid medium and solid substrates just like cells with a distinct index (spin). We must define the interaction energy between each cell type and the ECM.

Dynamics of Cell Evolution – Membrane Fluctuations: In mixtures of liquid droplets, thermal fluctuations of the droplet surfaces cause diffusion (*Brownian motion*) leading to energy minimization. The simplest assumption is that an *effective temperature, T*, drives cell membrane fluctuations. Fluctuations are described statistically using the Metropolis algorithm for Monte-Carlo Boltzmann dynamics; T defines the size of the typical fluctuation. If a proposed change in configuration (*i.e.*, a change in the spins associated with the pixels of the lattice) produces a change in effective energy, ΔE, we accept it with probability:

$$P(\Delta E)=1, \Delta E \leq 0; \qquad P(\Delta E)=e^{-\Delta E/kT}, \Delta E>0 \tag{3}$$

where k is a constant converting T into units of energy.

Chemotaxis and Haptotaxis: Chemotaxis requires additional fields to describe the local concentrations $C(\vec{x})$ of the molecules in extracellular space. Equation(s) for the field depend on the particular morphogen molecule. A reaction-diffusion model is used for the activator, which triggers production of SAM by cells. Cells respond to the SAM field by 'sticking' to it. An effective chemical potential, $\mu(\sigma)$ models chemotaxis or haptotaxis. Equation 4 incorporates the effective chemical energy into the CPM energy formalism. Thus, the cells execute a biased random walk averaging in the direction of the gradient. Experiments can measure $\mu(\sigma)$.

$$E_{Chemical}=\mu(\sigma)C(\vec{x}) \tag{4}$$

2.2 Reaction-Diffusion Equations

The Schnakenberg equations we use are the following [18]:

$$\begin{aligned}\frac{\partial u}{\partial t}&=\gamma(a-u+u^2v)+\nabla^2 u,\\ \frac{\partial v}{\partial t}&=\gamma(b-u^2v)+d\nabla^2 v.\end{aligned} \tag{5}$$

where u is the activator concentration at a location (x, y) at time t; v is the inhibitor concentration. γ is a parameter that affects the period (*wavelength*) of the (activator) pattern. Solutions in Fig. 3 demonstrate how controlling γ values can generate an activator pattern resembling chondrogenesis.

2.3 State Transition Model

The behavior of a cell depends on its state. State change rules depend on several chemical fields at the intra- and inter-cellular level. We assume all cells in the active zones are in a **mitosing** (*i.e.*, dividing) state. When a mitosing cell in the active zones senses a threshold local concentration of activator, it enters the **SAM producing** state. In this state, the cell also upregulates cell-cell adhesion (the parameter $J_{\tau, \tau'}$ ($J_{\text{cell,cell}}$) in the CPM decreases). Cells that have not experienced local threshold levels of activator are in the **mitosing and condensing** state; they undergo the dynamics of CPM. Such cells respond to local SAM concentration, but do not produce SAM on their own. This model of genetic regulation captures the formal, qualitative aspects of regulatory interactions and also allows fitting to quantitative experiments.

2.4 Modeling of Zones

For computational efficiency and biological realism, we apply the various dynamics (CPM, reaction-diffusion, state transitions) only in specific regions of growing limb bud. This zonal organization is typical of multicellular development. We thus describe zones, interfaces, and growth. The AER secretes signals that induce cell division in the proximal region; we simply assume that cell division takes place everywhere in the active zones. In the active zone, which allows for cell condensation and haptotaxis in response to SAM production, cells respond to threshold activator concentration by producing SAM and condensing into patterns governed by the activator pattern. In addition, we have an active zone for reaction-diffusion, which is slightly larger than the cells' active zone. In the reaction-diffusion active zone, activator concentration evolves to establish a pre-pattern for mesenchymal condensation. In the frozen zone, condensation into cartilaginous patterns has already occurred; no further evolution takes place here, saving on computation. In the absence of experimentally determined governing rules for these zones and their interfaces, we assume *ad hoc* rules for the motion of zones, based on the requirement that chemical concentration fields and cell clustering mechanisms have enough time to form distinctive patterns.

2.5 Integration of Submodels

We must integrate the submodels, in particular the stochastic CPM and continuum reaction-diffusion model to allow the various mechanisms to work in a coordinated fashion and simulate the full system. We must:

1. Match the spatial grid for continuum and stochastic models.
2. Define relative number of iterations for reaction-diffusion & CPM evolvers.

The section on "Software" below describes integration in more detail.

3 Software

COMPUCELL provides multiple user interfaces. A file-based input can be used to describe the parameters associated with *simulation*. Alternatively, the *model* to be simulated and the *simulation* can be controlled using BIOLOGO, a modeling language designed and implemented for COMPUCELL [3]. The front end includes: (1) A file based user interface for simulation parameters and (2) A graphical user interface for controlling the simulation and for presenting the results. Back end is split into two main engines that carry out the bulk of the number crunching: (1) A computational engine for the biological model and (2) A visualization engine for graphics.

Computational engine has three modules: (1) CPM engine (stochastic, discrete), (2) Reaction-diffusion engine (continuum, PDEs), and, (3) State transition model engine (rule based state automaton). Reaction-diffusion engine uses an explicit solver based on forward time marching. Results of these calculations are made available as objects of a 'Field' class. A Field object is also present for SAM concentration. CPM uses a field of pixels. For each of the fields an appropriate method for evolving it is used. The decision on which field to evolve is based on the criteria specified for interfacing the various grids and time scales. Grids are matched the using a simple interpolation; for time scales we specify the number of iterations for the evolution of field "1" before we update a field "2" that interacts with "1". In the CPM, the spins associated with pixels evolve according to the Metropolis algorithm. "State Transition Model" is used for the evolver governing cell differentiation for the genetically determined response of the cell.

Cell Division Algorithm: A cell capable of dividing splits after growing to twice its original size. To 'grow' the cell, its target volume is gradually increased so that the target volume doubles over a predetermined number of Metropolis steps (time). Assigning a new spin to half of the cell's constituent pixels then splits the cell. A modified breadth-first search is used to select pixels to be assigned a new spin; the split is approximately along the "diameter". Visualization ToolKit (VTK), available as freeware for various operating systems from the source URL[4], is used for visualization.

4 Discussion of Simulation Results for Chicken Limb Development

Figure 2 shows a simulation of the full model described above. Computational domain corresponds to the real anteroposterior width of 1.4 mm; patterning begins at stage 20 of chicken embryo development. The proximodistal length of 4 mm at stage 28 is about three times the width. A 100x300 grid covers the domain. Cells cluster subject to differential cell adhesion. The genetically governed response of cells to high activator concentration is to begin secreting SAM. Cells respond to SAM in two ways: (1) SAM causes cells to stick to the substrate; (2) SAM makes the cells more likely to condense by upregulating cell-cell adhesion. The parameter γ in Schnakenberg reaction-diffusion equations determines the periodicity of the pattern. Activator concentra-

[3] `http://www.nd.edu/~lcls/compucell`

[4] `http://public.kitware.com/VTK/get-software.php`

tion is shown at various times; formation of the pre-pattern directing later cell condensation into the chondrogenic pattern is clearly seen. Since cells exposed at some time to high activator concentration begin and continue to secrete SAM, and SAM in turn has the two effects described above, the pattern of SAM concentration resembles the activator pre-pattern. Finally, cells condense into the bone pattern of 1+2+3 (where 3 corresponds to the three digits) of the chicken limb. Growth of the limb bud depends on cell division rate and how fast cells move. New cells generated by cell division push the limb tip upward, making the growth look more natural.

Table 1. Runtimes for different grid sizes and number of cells

Grid	Number of Cells	Cell Density	Total iterations	Time, visualization	Time, no visualization
150X150	100	64%	700	31 minutes	2.5 minutes
300X300	900	64%	700	199 minutes	6 minutes
600X600	900	64%	700	329 minutes	18 minutes
150X150	325	52%	700	34 minutes	3.5 minutes

Simulations ran on a Sun-Blade-1000 with a 900MHz CPU and 512 Megabytes of memory. Table 1 presents data on runtimes for different grid sizes and numbers of cells. With the same grid size, the number of cells does not much affect the speed, demonstrating the scalability of our algorithms for quantities dependent on cell number. Visualizing the computed data is highly computation intensive and does not scale as well as the computation.

Acknowledgements. Support from NSF Grants IBN-0083653 and ACI-0135195 is acknowledged.

References

1. Mjolsness, E., Sharp, D. H. & Reinitz, J. (1991) A connectionist model of development. *J. Theor. Biol.,* **152,** 429–453.
2. Davidson, E., Rast, J. P., Oliveri, P., Ransick, A., Calestani, C., Yuh, C., Minokawa, T., Amore, G., Hinman, V., C.Arenas-Mena, Otim, O., Brown, C., Livi, C., Lee, P., Revilla, R., Rust, A., Pan, Z., Schilstra, M., Clarke, P., Arnone, M., Rowen, L., Cameron, R., McClay, D., Hood, L. & Bolouri, H. (2002) A genomic regulatory network for development. *Science,* **295**, 1669.
3. Marée, F. M. & Hogeweg, P. (2001) How amoeboids self-organize into a fruiting body: multicellular coordination in *Dictyostelium discoideum. Proc. Natl. Acad. Sci. USA,* **98**, 3879-3883.
4. Newman, S. A. & Comper, W. (1990) Generic physical mechanisms of morphogenesis and pattern formation. *Development,* **110**, 1–18.
5. Lewis, J. (1975) Fate maps and the pattern of cell division: a calculation for the chick wing-bud. *J. Embryol. exp. Morph.,* **33** , 419–434.
6. Newman, S. A. (1996) Sticky fingers: Hox genes and cell adhesion in vertebrate limb development. *BioEssays,* **18**, 171–174.

7. Downie, S. & Newman, S. (1995) Different roles for fibronectin in the generation of fore and hind limb precartilage condensations. *Dev. Biol.,* **172**, 519–530.
8. Oberlender, S. & Tuan, R. (1994) Expression and functional involvement of N-cadherin in embryonic limb chondrogenesis. *Development,* **120**, 177–187.
9. Steinberg, M. S. (1978). Specific cell ligands and the differential adhesion hypothesis: How do they fit together? In *Specificity of Embryological Interactions* (D. R. Garrod, Ed.), pp. 97–130. London: Chapman and Hall.
10. Graner, F. & Glazier, J. A. (1992) Simulation of biological cell sorting using a two-dimensional extended Potts model. *Phys. Rev. Lett.,* **69**, 2013–2016.
11. Jong, H. D. (2002) Modeling and simulation of genetic regulatory systems: a literature review. *J. Comp. Biol.,* **9**, 67–103.
12. Turing, A. (1952) The chemical basis of morphogenesis. *Phil. Trans. Roy. Soc. London,* **B 237**, 37–72.
13. Newman, S. A. & Frisch, H. L. (1979) Dynamics of skeletal pattern formation in developing chick limb. *Science,* **205**, 662–668.
14. Meinhardt, H., and Gierer, A. (2000) Pattern formation by local self-activation and lateral inhibition. *Bioessays* **22,** 753–760.
15. Mombach, J. & Glazier, J. (1996) Single cell motion in aggregates of embryonic cells. *Phys. Rev. Lett.,* **76**, 3032–3035.
16. Marée, S. (2000) *From Pattern Formation to Morphogenesis.* PhD thesis, Utrecht University Netherlands.
17. Drasdo, D. & Forgacs, G. (2000) Interplay of generic and genetic mechanisms in early development. *Developmental Dynamics,* **219**, 182–191.
18. Murray, J. D. (1993) *Mathematical Biology*, Second Corrected Edition (Biomathematics Vol. 19), Springer-Verlag, Berlin Heidelberg, pp. 156, 376, 406, 472, 739.

ProtoMol: A Molecular Dynamics Research Framework for Algorithmic Development

T. Matthey[1], A. Ko[2], and J.A. Izaguirre[2]

[1] Department of Informatics
University of Bergen
5020 Bergen, Norway
Thierry.Matthey@ii.uib.no
[2] Department of Computer Science and Engineering
University of Notre Dame
Notre Dame, IN, USA 46556-0309
izaguirr@cse.nd.edu

Abstract. This paper describes the design and evaluation of ProtoMol, a high performance object-oriented software framework for molecular dynamics (MD). The main objective of the framework is to provide an efficient implementation that is extensible and allows the prototyping of novel algorithms. This is achieved through a combination of generic and object-oriented programming techniques and a domain specific language. The program reuses design patterns without sacrificing performance. Parallelization using MPI is allowed in an incremental fashion. To show the flexibility of the design, several fast electrostatics (N-body) methods have been implemented and tested in ProtoMol. In particular, we show that an $\mathcal{O}(N)$ multi-grid method for N-body problems is faster than particle-mesh Ewald (PME) for $N > 8,000$. The method works in periodic and non-periodic boundary conditions. Good parallel efficiency of the multi-grid method is demonstrated on an IBM p690 Regatta Turbo with up to 20 processors for systems with $N = 10^2, 10^4$ and 10^6. Binaries and source code are available free of charge at http://www.nd.edu/~lcls/protomol.

1 Introduction

Molecular dynamics (MD) is an important tool in understanding properties and function of materials at the molecular level, including biological molecules such as proteins and DNA. The challenge of MD is related to the multiple length and time scales present in systems. For example, biological molecules have thousands of atoms and time scales that span 15 orders of magnitude. The MD research community continually develops multiscale integrators, fast N-body solvers, and parallel implementations that promise to bring the study of important systems within reach of computational scientists.

Although there are many programs for MD, most are very complex and several are legacy codes. This makes it harder for algorithm developers to incorporate their algorithms and disseminate them. The complexity usually arises from parallelization and other optimizations.

ProtoMol has been designed to facilitate the prototyping and testing of novel algorithms for MD. It provides a domain specific language that allows user prototyping of

P.M.A. Sloot et al. (Eds.): ICCS 2003, LNCS 2659, pp. 50–59, 2003.

MD simulation protocols on the fly, and it sacrifices neither performance nor parallelism. Its sequential performance is comparable to NAMD 2 [1], one of the fastest MD programs available, and it has good scalability for moderate numbers of processors.

PROTOMOL combines techniques from generic and object-oriented design in C++, but the lessons are applicable to other object oriented languages. It uses several design patterns, including some for simulations of dynamics of particles. These design pattern implementations have been sufficiently general to be reused in another framework called COMPUCELL, which models morphogenesis and other processes of developmental biology at the cellular and organism level [2].

PROTOMOL has been used in several applications, including simulation of ionic crystals, magnetic dipoles, and large biological molecules. It has also been used in courses on scientific computing simulations. As a proof of its flexibility, the design of PROTOMOL has allowed the prototyping of several new algorithms and effective implementation of sophisticated existing ones, cf. [3]. Substantial portions of this paper are in [4]. However, parallel results for multi-grid methods are reported here for the first time.

2 Physical and Mathematical Background

In classical MD simulations the dynamics are described by Newton's equation of motion

$$m_i \frac{d^2}{dt^2} \boldsymbol{x}_i(t) = \boldsymbol{F}_i(t), \tag{1}$$

where m_i is the mass of atom i, $\boldsymbol{x}_i(t)$ the atomic position at time t and $\boldsymbol{F}_i(t)$ the instant force on atom i. The force $\boldsymbol{F}_i$ is defined as a gradient of the potential energy

$$\boldsymbol{F}_i = -\nabla_i U(\boldsymbol{x}_1, \boldsymbol{x}_2, \ldots, \boldsymbol{x}_N) + \boldsymbol{F}_i^{\text{extended}}, \tag{2}$$

where U is the potential energy, $\boldsymbol{F}_i^{\text{extended}}$ an extended force (e.g., velocity-based friction) and N the total number of atoms in the system. Typically, the potential energy is given by

$$U = U^{\text{bonded}} + U^{\text{non-bonded}} \tag{3}$$

$$U^{\text{bonded}} = U^{\text{bond}} + U^{\text{angle}} + U^{\text{dihedral}} + U^{\text{improper}} \tag{4}$$

$$U^{\text{non-bonded}} = U^{\text{electrostatic}} + U^{\text{Lennard-Jones}}. \tag{5}$$

The bonded forces are a sum of $\mathcal{O}(N)$ terms. The non-bonded forces are a sum of $\mathcal{O}(N^2)$ terms due to the pair-wise definition. U^{bond}, U^{angle}, U^{dihedral} and U^{improper} define the covalent bond interactions to model flexible molecules. $U^{\text{electrostatic}}$ represents the well-known Coulomb potential and $U^{\text{Lennard-Jones}}$ models a van der Waals attraction and a hard-core repulsion.

2.1 Numerical Integrators

Newton's equation of motion is a second order ordinary differential equation. Its integration is often solved by the numerical leapfrog method, which is time reversible and symplectic. Despite its low order of accuracy, it has excellent energy conservation properties and is computationally cheap.

A complete MD simulation is described by Algorithm 1. It consists basically of the loop numerically solving the equation of motion, the evaluation of forces on each particle, and some additional pre- and post-processing.

MD Simulation:

1. Construct initial configuration of x^0, v^0 and F^0;
2. **loop** 1 **to** number of steps
   (a) Update velocities (by a half step)
   (b) Update positions (by a full step)
   (c) **Evaluate** forces on each particle
   (d) Update velocities (by a half step)
3. Post-processing

Algorithm 1. Pseudo-code of an MD simulation

2.2 Force Evaluation

Force evaluation in MD typically consists of bonded and non-bonded interactions. Bonded interactions (U^{bond}) are short-range and comparable cheap to compute. Non-bonded interactions are of a long-range nature and determined at run-time. They are the most computationally expensive. Thus, most MD program optimizations happen here.

One of the most common optimizations for non-bonded force computations is the use of cutoffs to limit the spatial domain of pairwise interactions. Closely related to this is the use of switching functions to bring the energy and forces smoothly to zero at the cutoff point. Furthermore, cutoff computation can be accelerated through the use of cell lists or pair lists to achieve $\mathcal{O}(N)$. PROTOMOL implements generic facilities supporting all these optimizations.

For systems with partial charges or higher multipoles, the electrostatic interactions play a dominant role, e.g., in protein folding, ligand binding, and ion crystals. Fast algorithms for electrostatic force evaluation implemented in PROTOMOL are described in Sects. 2.2–2.2. In particular, we describe the parallel implementation of a novel multi-grid summation method (MG) for $\mathcal{O}(N)$ fast electrostatics in periodic boundary conditions or vacuum.

Ewald Summation. A general potential energy function U of a system of N particles with an interaction scalar function $\phi(\boldsymbol{x}_{ij} + \boldsymbol{n})$ and periodic boundary conditions can be expressed as an infinite lattice sum over all periodic images. For the Coulomb potential energy this infinite lattice sum is only conditionally convergent.

The Ewald sum method separates the electrostatic interactions into two parts: a short-range term handled in the direct sum and a long-range smooth varying term handled approximately in the reciprocal sum using Fourier transforms. This splitting changes the potential energy from the slowly and conditionally convergent series into the sum of two rapidly converging series in direct and reciprocal space and a constant term.

This algorithm scales like $\mathcal{O}(N^2)$ unless some optimizations are performed. The splitting parameter β, which determines the relative rates of convergence between the direct and reciprocal sums, can be adjusted to reduce the computational time to $\mathcal{O}(N^{3/2})$, cf. [5].

Particle Mesh Ewald. Using the discrete Fast-Fourier transforms (FFT), the mesh-based Ewald methods approximate the reciprocal-space term of the standard Ewald summation by a discrete convolution on an interpolating grid. By choosing an appropriate splitting parameter β, the computational cost can be reduced from $\mathcal{O}(N^{\frac{3}{2}})$ to $\mathcal{O}(N \log N)$. The accuracy and speed are additionally governed by the mesh size and the interpolation scheme, which makes the choice of optimal parameters more difficult. This problem has been addressed by MDSIMAID, a recommender system that proposes possible optimal choices for a given system and a required accuracy [6].

At present, there exist several implementations based on the mesh-based Ewald method, but they differ in detail. Smooth particle-mesh Ewald (SPME) [7] is implemented in PROTOMOL. The mesh-based Ewald methods are affected by errors when performing interpolation, FFT, and differentiation [5]. Accuracy increases when the interpolation order or the number of grid points increase.

Multi-grid Summation. Multi-grid summation (MG) has been used used to solve the N-body problems by [3,8]. MG imposes a hierarchical separation of spatial scales and scales as $\mathcal{O}(N)$. The pair-wise interactions are *split* into a local and a smooth part. The local parts are short-range interactions, which are computed directly. The smooth part represents the slowly varying energy contributions, approximated with fewer terms – a technique known as *coarsening*. MG uses interpolation unto a grid for both the charges and the potential energies to represent its smooth – *coarse* – part. The splitting and coarsening are applied recursively and define a grid hierarchy (Fig. 1). For the electrostatic energy, the *kernel* is defined by $G(r) = r^{-1}$ and $r = ||\boldsymbol{y} - \boldsymbol{x}||$. $G(r)$ is obviously not bounded for small r. The interpolation imposes smoothness to bound its interpolation error. By separation, the *smoothed kernel* (smooth part) for grid level $k \in \{1, 2, \ldots, l\}$ is defined as

$$G^k_{\text{smooth}}(r) = \begin{cases} G_{s_k}(r) : r|| < s_k \\ G(r) \quad : \text{otherwise.} \end{cases} \tag{6}$$

Here, s_k is the softening distance at level k and $G_{s_k}(r)$ is the *smoothing function* with $G_{s_k}(s_k) = G(s_k)$. We define $s_k = a^{k-1}s$, typically $a = 2$. Corrections of the energies are required when the modified, smoothed kernel is used instead of the exact one.

3 Framework Design

MD programs may substantially differ in design, but they are essentially all based on Algorithm 1. Four main requirements were addressed during the design of the framework:

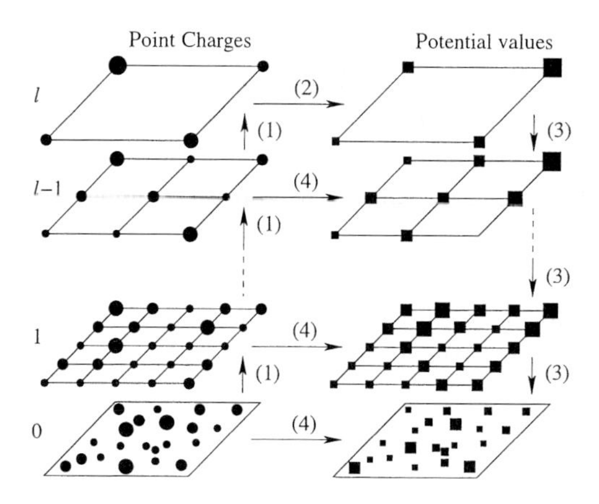

Fig. 1. The multilevel scheme of the MG algorithm. (1) Aggregate to coarser grids; (2) Compute potential energy induced by the coarsest grid; (3) Interpolate energy values from coarser grids; (4) Local corrections

1. Allow end-users to compose integrators and force evaluation methods dynamically. This allows users to *experiment* with different *integration schemes*. MTS methods require careful fine-tuning to get the full benefit of the technique.
2. Allow developers to easily integrate and evaluate *novel force algorithms* schemes. For example, the force design allows the incorporation of sophisticated multiscale algorithms, including mesh-based methods and MG.
3. Develop an encapsulated parallelization approach, where sequential and parallel components co-exist. This way, developers are not forced to consider the distributed nature of the software. Parallelism itself is based on range computation and a hierarchical master-slave concept [9].
4. Provide facilities to compare accuracy and run-time efficiency of MD algorithms and methods.

For the design of the component-based framework PROTOMOL, three different modules were identified. These are shown in Fig. 2 and are described next. The front-end provides *components* to compose and configure MD applications. The components are responsible for composing and creating the actual MD simulation set up with its integration scheme and particle configuration. This layer is strongly decoupled from the rest to the extent that the front-end can be replaced by a scripting language.

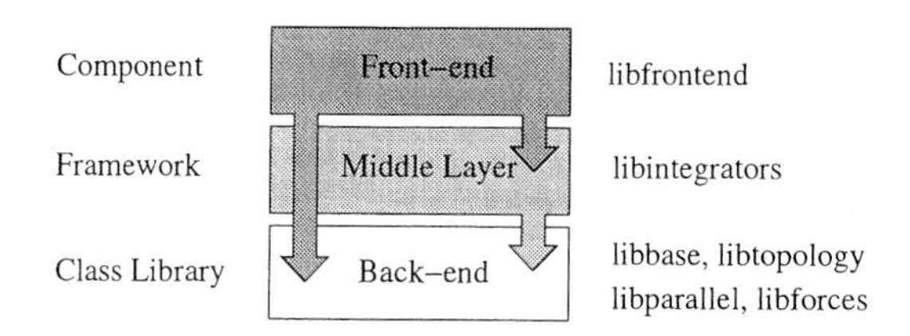

Fig. 2. The component-based framework PROTOMOL

The middle layer is a *white-box framework* for numerical integration reflecting a general MTS design. The back-end is a *class library* carrying out the force computation and providing basic functionalities (see Sect. 3.1). It has a strong emphasis on run-time efficiency.

The discussion of the framework has a strong emphasis on the design of force algorithms, since considerable time was spent to design and implement new force algorithms (e.g., standard Ewald summation, SPME, MG, etc.). The front-end is mainly the pre- and post-processing in Algorithm 1 and is not detailed in this paper, whereas the middle layer is briefly explained to give an overview of the collaboration of integrators and their associated forces. A complete domain analysis of the integrator library used in PROTOMOL is in [10], and some references to novel integrators developed and implemented using the program are [11,12].

3.1 Force Design

The forces are designed as separate components and part of the computational back-end. From an MD modeling point of view and from performance considerations, five different requirements (or customizable options) are proposed. These are discussed below:

R1 An algorithm to select an n-tuple of particles to calculate the interaction.
R2 Boundary conditions defining positions and measurement of distances in the system.
R3 A Cell Manager component to retrieve efficiently the spatial information of each particle. This has $\mathcal{O}(1)$ complexity.
R4 A function defining the force and energy contributions on an n-tuple.
R5 A switching function component to make the force and energy terms smoother.

Force Interface. In order to address these requirements, several design approaches can be chosen. To avoid an all inclusive interface with mammoth classes, we use multiple inheritance and generic programming. We combine templates and inheritance in the Policy or Strategy pattern [13, pp. 315-323]. This pattern promotes the idea to vary the behavior of a class independent of its context. It is well-suited to break up many behaviors with multiple conditions and it decreases the number of conditional statements.

The algorithm to select the n-tuples (R1) is customized with the rest of the four requirements (R2-R5). This allows the simultaneous evaluation of different types of forces with the same algorithm. Complex force objects stem from non-bonded forces. For example, to define an electrostatic force, we may choose a cutoff algorithm (R1) that considers only the closest neighboring atoms. To find the closest atoms in constant time, we use a cell manager based on a cell list algorithm (R3), defining some boundary conditions (R2), a function defining the energy and force contributions between two arbitrary atoms. We may even modify the energy and force by specifying a switching function. The forces are designed with a common interface, a deferred feature called `evaluate(...)` that does the evaluation of the force contributions based on its parameterization and policy choices.

Force Object Creation. Once the forces are designed and implemented, we need to create (or instantiate) the actual force objects needed by integrators. This can be solved by a sort of "just-in-time" (JIT) compiler that can transform a given input definition into a real force object.

The requirements of object creation are satisfied by the Abstract Factory [13, pp. 87-95] and the Prototype [13, pp. 117-126] patterns. The Abstract Factory pattern delegates the object creation, and the Prototype pattern allows dynamic configuration. At the end of the process, a fully featured force object with set parameters is created by the prototype. In order to make the dynamic configuration and the actual object creation independent, and the factory globally accessible, the force factory uses the Singleton pattern [13, pp. 127-134].

3.2 Performance Monitoring

Since one of the requirements of the framework is the ease to evaluate and compare different force algorithms, functionalities were added to the framework for this purpose. At present, pairs of forces can be compared to determine energy and force errors of new force methods. The comparison is performed on-the-fly, such that the reference force does not affect the current simulation. For example, one can compare a fast electrostatics method such as PME using two grid sizes, such that the more accurate one serves as an accuracy estimator. This is important to validate and verify a simulation.

For the benchmarking of forces, a timer function was implemented to measure the total and average time spent in dedicated force methods. Comparisons can be nested to evaluate accuracy and run-time performance simultaneously. The comparison of force pairs is based on the Count Proxy pattern [13, pp. 207-217] to link two forces together and calculate the actual errors.

4 Performance Evaluation

We describe the performance evaluation of the fast electrostatic methods implemented in PROTOMOL and described in this paper. Figure 3 shows parallel scalability of PROTOMOL applied on Coulomb Crystal systems [14,15], which are defined by a – computationally dominating – electrostatic part and an electric field with linear work complexity. The full electrostatics are solved by MG. The simulations were performed on an IBM p690 Regatta Turbo. Note that the sequential speedup for $N = 10^6$ is of order 10^2 or more compared to the direct method, and for lower accuracy a speedup of order 10^3 was observed.

Our initial parallel implementation of MG is based on global communication using MPI for the smooth part defined at particle level. In Fig. 1 the work of the anterpolation (1), the interpolation (3), the direct part (4) and the correction at grid level (4) can be estimated and distributed perfectly among the slaves at each level, given the grid dimensions, the softening distance and the number of particles. Accurate work estimates enables us to assign the work without any synchronization (i.e., a master node), and to reduce the idle cycles to a minimum when performing a global update of the local contributions at the grid levels. For the direct part at particle level (4) the work is

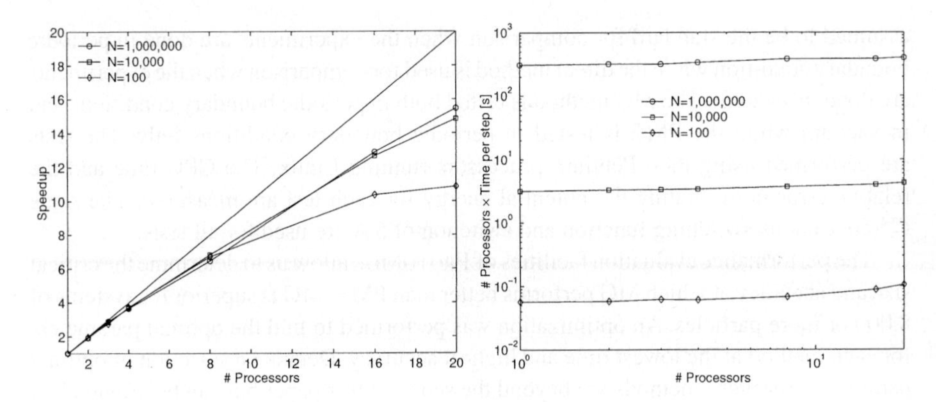

Fig. 3. Parallel scalability applied on Coulomb Crystal systems using MG electrostatic solver with relative error of order 10^{-5} or less; performed on an IBM p690 Regatta Turbo

distributed dynamically, since the work for a given spatial sub-domain can not be predicted statically, due to the fact that the distribution of particles is in general non-uniform. The master assigns ranges of work on demand to balance the work, which represents the force decomposition scheme with master-slave distribution. Additionally, the ranges of work are sent in pipeline to avoid slaves waiting for their next range. In general, a work range represents a collection of terms of a term of the sum U in Eq. (3). The ranges are based on the possible splittings provided by the force objects defined by the sum U. Furthermore, the work distribution can be performed using either a master-slave, as described above, or a static scheme. The static scheme distributes the work in a linear manner, which performs well for small number of nodes or uniform density systems.

Figure 4 compares the run-time of MG, which is extended to periodic boundary conditions for the first time here, and the smooth PME. The experiments are based on the TIP3 water model with atoms ranging from 1,000 to 100,000. The Ewald method is

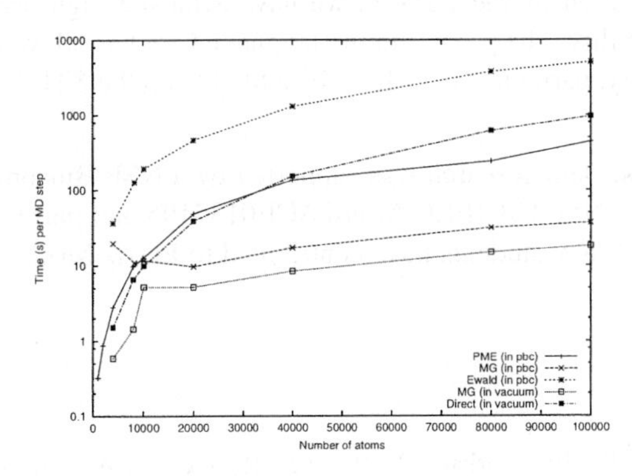

Fig. 4. Time per MD step for N-body electrostatic solvers implemented in PROTOMOL with relative error of order 10^{-4}

assumed to be the standard for comparison when the experiments are done in periodic boundary condition while the direct method is used for comparison when the experiments are done in vacuum. The MG method is tested both in periodic boundary conditions and in vacuum while the PME is tested in periodic boundary conditions only. The tests are performed using i686 Pentium processors running Linux. The CPU time and the relative error in evaluating the potential energy for each test are measured. The same C^2-continuous switching function and switchon of 5 Å are used for all tests.

The performance evaluation facilities of PROTOMOL allow us to determine the critical size and accuracy at which MG performs better than PME. MG is superior for systems of 8,000 or more particles. An optimization was performed to find the optimal parameters for each method at the lowest time and highest accuracy. Details on selection of optimal parameters for these methods are beyond the scope of this paper but can be provided on request. Some guidelines can be found in [8].

5 Discussion

The design of PROTOMOL has allowed the implementation of novel MTS integrators and fast N-body solvers. For example, the MG summation for fast electrostatic is 3-5 times faster than the particle-mesh Ewald for systems only 8,000 atoms. The parallel version of MG scales well for moderate numbers of processors: we have tested it with up to 20. Combination of these new methods have enabled simulations of million-particle systems with full electrostatics described above [15]. The facilities for performance monitoring have been very useful and general. Furthermore, the different algorithms and the comparison facilities give a unique opportunity to choose the best algorithm for a particular application and enable fair comparison of future novel algorithms.

The domain specific language makes our MD applications very flexible. By using the provide "JIT" compiler, users compose their own programs without having to touch the code. This was mainly achieved with help of the Abstract Factory pattern and the Prototype pattern, which also improves the extendibility on the developer level.

The object-oriented design of the framework along with the use of design patterns has eased the development of a fairly complex framework. By using object-oriented and generic implementation of PROTOMOL, we have achieved high performance without sacrificing extendibility. The programming language C++ has allowed us to achieve the goal of extendability, particularly we have benefited from the STL.

Acknowledgments. This research was supported by a NSF Biocomplexity Grant No. IBN-0083653 and a NSF CAREER Award ACI-0135195 and partly by the Norwegian Research Council. Many students have contributed to PROTOMOL. A list can be found in its webpage.

References

1. Kalé, L., Skeel, R., Bhandarkar, M., Brunner, R., Gursoy, A., Krawetz, N., Phillips, J., Shinozaki, A., Varadarajan, K., Schulten, K.: NAMD2: Greater scalability for parallel molecular dynamics. J. Comp. Phys. **151** (1999) 283–312

2. Chaturvedi, R., Izaguirre, J.A., Huang, C., Cickovski, T., Virtue, P., Thomas, G., Forgacs, G., Alber, M., Hentschell, G., Newman, S., Glazier, J.A.: Multi-model simulations of chicken limb morphogenesis. To appear in proceedings of the International Conference on Computational Science ICCS (2003)
3. Matthey, T.: Framework Design, Parallelization and Force Computation In Molecular Dynamics. PhD thesis, Department Of Informatics, University of Bergen (2002)
4. Matthey, T., Cickovski, T., Hampton, S., Ko, A., Ma, Q., Slabach, T., Izaguirre, J.A.: PROTOMOL: an object-oriented framework for prototyping novel algorithms for molecular dynamics. Submitted to ACM Trans. Math. Softw. (2002)
5. Petersen, H.G.: Accuracy and efficiency of the particle mesh Ewald method. J. Chem. Phys. **103** (1995)
6. Ko, A.: MDSimAid: An automatic recommender for optimization of fast electrostatic algorithms for molecular simulations. Master's thesis, University of Notre Dame, Notre Dame, IN (2002) Available from `http://www.nd.edu/~izaguirr/papers/KOthesis.pdf`.
7. Essmann, U., Perera, L., Berkowitz, M.L.: A smooth particle mesh Ewald method. J. Chem. Phys. **103** (1995) 8577–8593
8. Skeel, R.D., Tezcan, I., Hardy, D.J.: Multiple grid methods for classical molecular dynamics. J. Comp. Chem. **23** (2002) 673–684
9. Matthey, T., Izaguirre, J.A.: ProtoMol: A molecular dynamics framework with incremental parallelization. In: Proc. of the Tenth SIAM Conf. on Parallel Processing for Scientific Computing (PP01). Proceedings in Applied Mathematics, Philadelphia, Society for Industrial and Applied Mathematics (2001)
10. Izaguirre, J.A., Ma, Q., Matthey, T., Willcock, J., Slabach, T., Moore, B., Viamontes, G.: Overcoming instabilities in Verlet-I/r-RESPA with the mollified impulse method. In Schlick, T., Gan, H.H., eds.: Proceedings of 3rd International Workshop on Methods for Macromolecular Modeling. Volume 24 of Lecture Notes in Computational Science and Engineering. Springer-Verlag, Berlin, New York (2002) 146–174
11. Izaguirre, J.A., Catarello, D.P., Wozniak, J.M., Skeel, R.D.: Langevin stabilization of molecular dynamics. J. Chem. Phys. **114** (2001) 2090–2098
12. Skeel, R.D., Izaguirre, J.A.: An impulse integrator for Langevin dynamics. Mol. Phys. **100** (2002) 3885–3891
13. Gamma, E., Helm, R., Johnson, R., Vlissides, J.: Design Patterns. Elements of Reusable Object-Oriented Software. Addison-Wesley, Reading, Massachusetts (1995)
14. Hasse, R.H., Avilov, V.V.: Structure and Mandelung energy of spherical Coulomb crystals. Phys. Rev. A **44** (1991) 4506–4515
15. T. Matthey, J.P.H., Drewsen, M.: Bicrystal structures in rf traps of species with identical charge-to-mass ratios. Submitted to PRL (2003)

An Efficient Navigation Method for Virtual Endoscopy Using Volume Ray Casting

Byeong-Seok Shin and Suk Hyun Lim

Inha University, Department of Computer Science and Engineering
253 Yonghyeon-Dong, Nam-Gu, Inchon, 402-751, Korea
bsshin@inha.ac.kr
q2011498@inhavision.inha.ac.kr

Abstract. In virtual endoscopy, it is important to devise the navigation method that enables us to control the position and orientation of virtual camera easily and intuitively without collision to organ wall. We propose an efficient navigation algorithm that calculates depth information while rendering a scene for current frame with volume ray casting method, then computes new viewing specification of camera for the next frame using the depth information. It can generate a camera path in real-time, allows us to control the camera with ease, and avoids collision with objects efficiently. In addition, it doesn't require preprocessing stage and extra storage for maintaining spatial data structures. Experimental result shows that it is possible to navigate through human colon when we apply our method to virtual colonoscopy.

1 Introduction

Optical endoscopy is a less-invasive diagnosis method. We can directly examine the pathologies of internal organs by putting endoscopy camera into human body. It offers the highest quality images than any other medical imaging methods. However it has some disadvantages of causing patients discomfort, limited range of exploration and serious side effect such as perforation, infection and hemorrhage.

Virtual endoscopy is regarded as a good alternative of optical endoscopy. After acquiring cross-sectional images of human abdomen with CT or MRI, we can reconstruct three dimensional volume models and provide visualizations of inner surface of the human organ, which has pipe-like shape such as colon, bronchi, and blood vessels. Since the virtual endoscopy is non-invasive examination, there is no discomfort and side effects. It is adequate for mass screening of pathological cases and training tool for doctors and medical students.

In order to implement virtual endoscopy based on volume rendering technique, we should devise a method that generates high quality perspective images within short time. However, it is more important to detect collision between virtual camera and organ wall in real-time, and let the camera smoothly move along the center-line of human cavity. Since the organs to be examined such as colon and bronchus have

P.M.A. Sloot et al. (Eds.): ICCS 2003, LNCS 2659, pp. 60–69, 2003.

complex structures, it is difficult to move forward endoscopy camera and change its orientation even for experienced doctors. So, fast and accurate navigation is essential for efficient diagnosis using virtual endoscopy. Previous methods for navigation have some problems to demand a lot of computation cost in preprocessing step and it doesn't allow user to change the camera position and orientation during navigation, or to require extra storages for spatial data structures such as a distance map or a potential field.

In this paper, we propose an efficient algorithm that generates depth information while rendering a scene in current frame with volume ray casting, then determines camera orientation for the next frame using the information. Since the depth information can be estimated without additional cost during ray casting, it computes camera path in real-time. In addition, it doesn't require extra memory and preprocessing step.

Related works is summarized in the next section. In section 3, we present our algorithm in detail. Experimental results and remarks are shown in section 4. Lastly, we summarize and conclude our work.

2 Related Work

Volume ray casting is the most famous volume rendering method [1]. After firing a ray from each pixel on the view plane into volume space, it computes color and opacity on sample points along the ray using the corresponding voxel values and opacity transfer function. Then it determines final color for the pixel by blending them. Although it takes long time to make an image due to randomness in memory reference pattern, it produces high-quality images in comparison to the other methods. Also it can make perspective images such as endoscopic image. It is possible to improve its rendering speed using some acceleration technique such as template based rendering [2], shear-warp decomposition [3], and space-leaping [4].

In order to implement realistic virtual environment, several kinds of interactions between objects should be considered. Especially, collision detection is more important than any other components for visual realism [5]. We have to devise appropriate way to specify the surface of volumetric objects since they do not have explicitly defined surfaces unlike geometric objects. Usually we define object boundaries indirectly using the opacity transfer function, which assigns higher opacity values to voxels regarded as boundaries, and lower opacities to the others. This process is called *classification*. Several spatial data structure might be used to detect collision between an object and the indirectly specified surfaces.

Occupancy map has the same resolution as the target volume dataset and each cell of the map stores identifiers for objects that occupy the cell [6]. As the objects change its position, the values of each cell should be updated. If a cell has two or more identifiers, it is recognized as collision of those objects. This method requires large amount of storage as original volume dataset, and it should update its contents whenever an object change its position and size.

Three dimensional distance map can be used for collision detection and avoidance [7]. A distance map is a 3D spatial data structure that has the same resolution of its volume data and each point of the map has the distance to the nearest boundary voxel of an object. We can exploit the distance information generated in preprocessing time for collision detection without extra cost. The smaller the distance values the larger the collision probability and vice versa. Asymptotic complexity of collision detection is $O(1)$, since we have to refer the distance map only once. Due to the occupancy map, this method requires large amount of storage and considerably long preprocessing time.

Navigation method for virtual endoscopy can be classified into three categories; manual navigation [8], planned navigation [9] and guided navigation [10]. In manual navigation, we can directly control the virtual camera to observe anywhere we want to examine. However the user might feel discomfort since camera movement is entirely dependent on user's control, and collision with organ wall may occur if the user misleads the camera.

Planned navigation calculates entire navigation path in preprocessing time using several path generation algorithm, then moves through the camera along the pre-calculated path in navigation step. It can fly through the desired area without user intervention. However it requires a lot of computation cost in preprocessing step and it doesn't allow users to control the camera position and orientation intuitively.

Guided navigation is a physically-based method, which makes an spatial data structures such as potential field in preprocessing step and determines camera orientation by considering attractive force that directs from starting point to target point, repulsive force from organ surface, and user's input. It guarantees that a camera arrives to target point and move the camera to anywhere the user want to place intuitively without collision against organ wall. However it is very hard to implement and it demands a lot of cost to make and maintain the vector volume just like a potential field.

3 Navigation Method Using Volume Ray Casting

In order to implement efficient navigation, it have to satisfy the following requirements. (1) It should avoid collision between organ surface and virtual camera. (2) It should guarantee smooth movement of camera, and camera orientation should not change abruptly. (3) It should allow users to control the camera conveniently and intuitively. (4) It should not require preprocessing step, or minimize that if needed. (5) It should not require extra storages.

Guided navigation algorithms mentioned in the previous section is regarded as *object-space* method since they exploit overall information of entire volume dataset. Even though it guarantees accurate navigation, this is a reason why it takes a long preprocessing time. In virtual endoscopy, it requires information only for a small part of entire volume since field of view is restricted in cavities of a volume data set. Our navigation algorithm computes camera orientation using voxels visible in current viewing condition, thus we define it as *image-space* method.

In rendering point of view, navigation is a kind of animation sequence that moves a camera along a smooth path. Here, we explain how to determine camera orientation for the next frame f_{i+1}, using the information obtained in rendering time for current frame f_i. We determine the direction of rotation using the largest depth value, and angle (magnitude) of rotation based on the smallest value. Figure 1 shows the procedure of our method. Rendering process is the same as in the conventional animation. Our method has another pipeline to specify the viewing condition of camera. It generates depth information while rendering and computes direction of rotation and rotation angle by using the depth information.

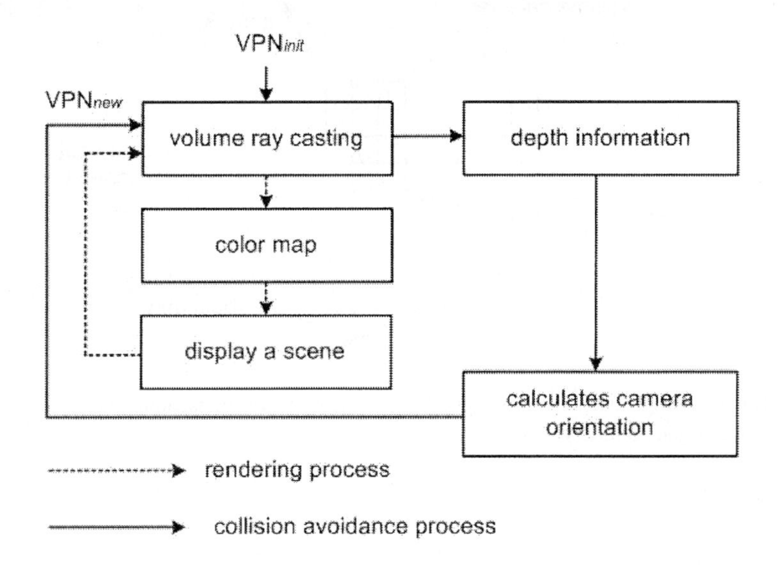

Fig. 1. A procedure for determining reliable navigation path

3.1 Calculating Depth Information

Volume ray casting fires a ray from each pixel on the view plane into volume space, then it computes color and opacity on sample points along the ray and determines final color for the pixel by blending them. Empty space (cavities) of volume data has low opacity. On the contrary organ wall has relatively high opacity value. While sampling voxel values along a ray, we can compute the distance from a pixel to the nearest non-transparent voxel using the number of samples and unit distance between two consecutive samples. Figure 2 depicts how to obtain distance value for each pixel. In our application, the distance can be regarded as the depth value from a camera to organ wall.

Actually, we need only the maximum and minimum depth value for the following steps. Let $d_{i,j}$ be a depth value of a pixel $p_{i,j}$, and $\mathbf{R}_{i,j}$ be the direction of ray for that pixel. Minimum and maximum depth value d_{min} and d_{max} can be defined as follows:

$$d_{\min} = \min(\bigcup_{i,j=0}^{M,N} d_{i,j}),\ d_{\max} = \max(\bigcup_{i,j=0}^{M,N} d_{i,j}). \tag{1}$$

where M and N are horizontal and vertical resolution of final image respectively. Also we store the direction of a ray fired from a pixel that has the maximum depth value as a form of unit vector, which is denoted as $\mathbf{R}_{max}$.

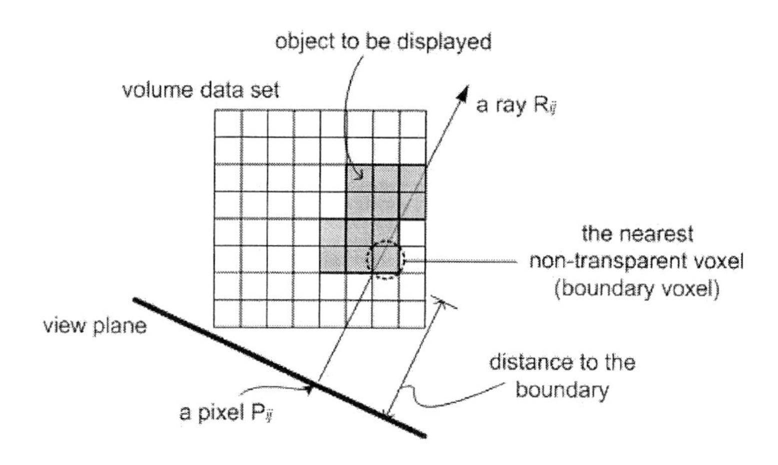

Fig. 2. Computing distance from a pixel to the nearest non-transparent voxel (is potentially regarded as organ boundary)

3.2 Determining the Direction of Rotation of Virtual Camera

Depth values are inverse proportional to the possibility of collision. Larger depth value means that obstacles are located far away from the current camera position. On the contrary, smaller the depth value, higher the collision possibility since the camera is near the organ wall. In this case, we have to rotate the camera rapidly to avoid collision. That is, we can minimize the collision possibility by altering the camera orientation to the direction of ray that has the maximum depth value.

We can define the direction of rotation $\mathbf{r}_i$ for the next frame f_{i+1} can be defined as follows:

$$\mathbf{r}_i = \mathbf{R}_{\max} - \mathbf{VPN}_i . \tag{2}$$

where $\mathbf{R}_{max}$ is a unit vector computed in the previous step, and $\mathbf{VPN}_i$ is also a unit vector that represents a view plane normal for current frame f_i. Figure 3 shows how to determine the direction of rotation using $\mathbf{R}_{max}$.

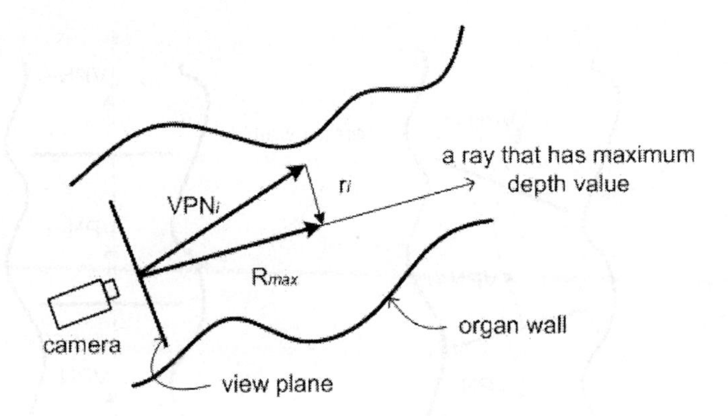

Fig. 3. Determination of the direction of rotation using the maximum distance value

3.3 Calculating the Rotation Angle

After determining the direction of rotation $\mathbf{r}_i$, we have to calculate rotation angle using minimum depth value d_{min}. As we mentioned before, smaller d_{min} implies that a camera is located near the organ wall, so we should increase the rotation angle along $\mathbf{r}_i$. On the contrary, larger d_{min} decreases the angle.

It is simple but has a problem of instability. Since our method changes the camera orientation in every frame, unnecessary changes might occur in open area as shown in Figure 4. This causes camera trembling, so it cannot satisfy the requirement (2). To solve the instability problem, we define threshold depth value d_{th}, then we do not alter camera orientation in sufficiently open area by considering the value. If d_{min} is larger than d_{th}, it is regarded as in open area and we have to keep the camera orientation unchanged.

We can determine the rotation angle θ based on d_{th} and d_{min} as follows:

$$\theta = \begin{cases} 0 & \text{if } d_{\min} > d_{th} \\ \arctan\left(\frac{d_{th} - d_{\min}}{d_{th}}\right) & \text{otherwise} \end{cases} . \tag{3}$$

Figure 5 shows how to determine the rotation angles according to the difference between d_{th} and d_{min}.

In consequence, we can obtain the final camera orientation $\mathbf{VPN}_{i+1}$ for the text frame as in the follow equation:

$$\mathbf{VPN}_{i+1} = \frac{\mathbf{r}_i \tan\theta + \mathbf{VPN}_i}{|\mathbf{r}_i \tan\theta + \mathbf{VPN}_i|} . \tag{4}$$

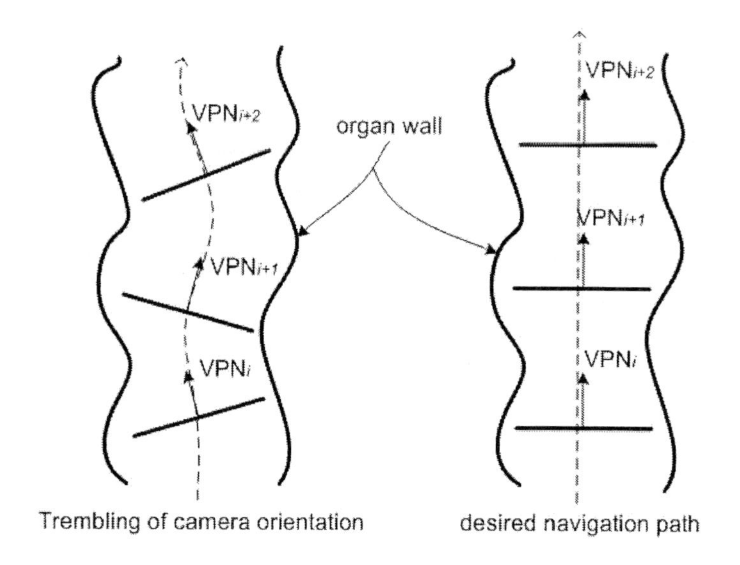

Fig. 4. An example of trembling of camera orientation in open area. It should be fixed along the desired navigation path.

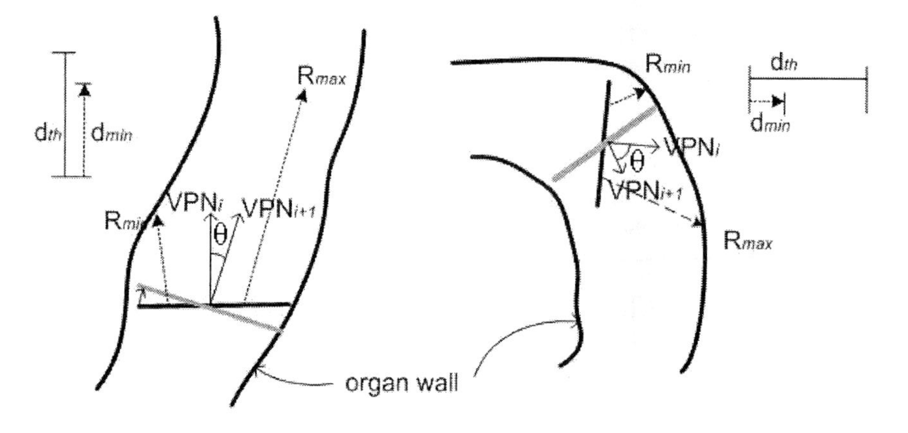

Fig. 5. Determination of the rotation angle using the minimum distance value

4 Experimental Results

In this section we will check whether our method offers reliable navigation through human colon with collision avoidance. Also we will compare the preprocessing and rendering time of a typical object-space navigation method based on distance transformation and our navigation algorithm. All of these methods are implemented on a

PC which has Pentium IV 1.7GB CPU, 1GB main memory, and ATI RADEON 9000 graphics accelerator. The volume data used for the experiment is a clinical volume obtained by scanning a human abdomen with a spiral CT (computed tomography) with resolution $512\times512\times541$.

Figure 6 depicts the result of the method that does not perform collision detection and avoidance. In this case, a camera would bump against colon wall unless a user manually changes its direction. Figure 7 shows the example of using collision avoidance feature of our method in virtual colonoscopy. As you can see, the camera could move forward along the colon cavity automatically.

Table 1 shows the comparison of preprocessing time and rendering time of a method that does not perform any kind of collision avoidance (NONE), an object-space guided navigation method using distance map (DIST) and our image space navigation method (OURS). While an object-space method spends long time for generating a distance map in preprocessing step, an image-space method doesn't need preprocessing time at all. In rendering time, the difference between NONE and OURS is very small (about 6.51 %), which implies that our method requires small amount of additional cost to calculate camera orientation during rendering. Also our navigation method is not so slow (only 4.91 %) even in case of comparing it with DIST.

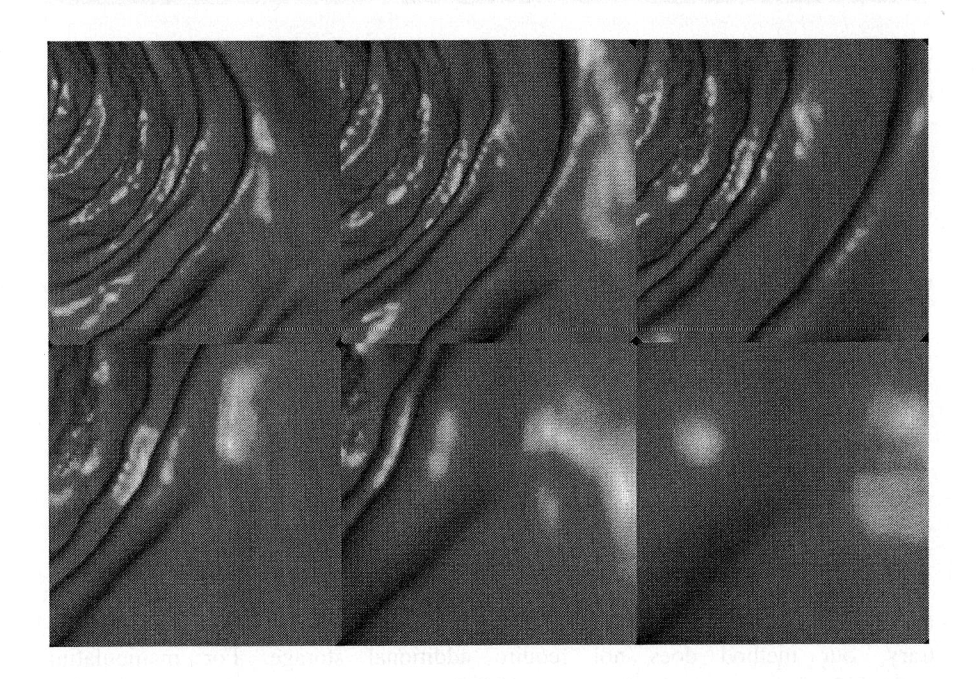

Fig. 6. This is an example of not using collision detection and avoidance feature. A camera would bump against colon wall unless a user manually changes its direction

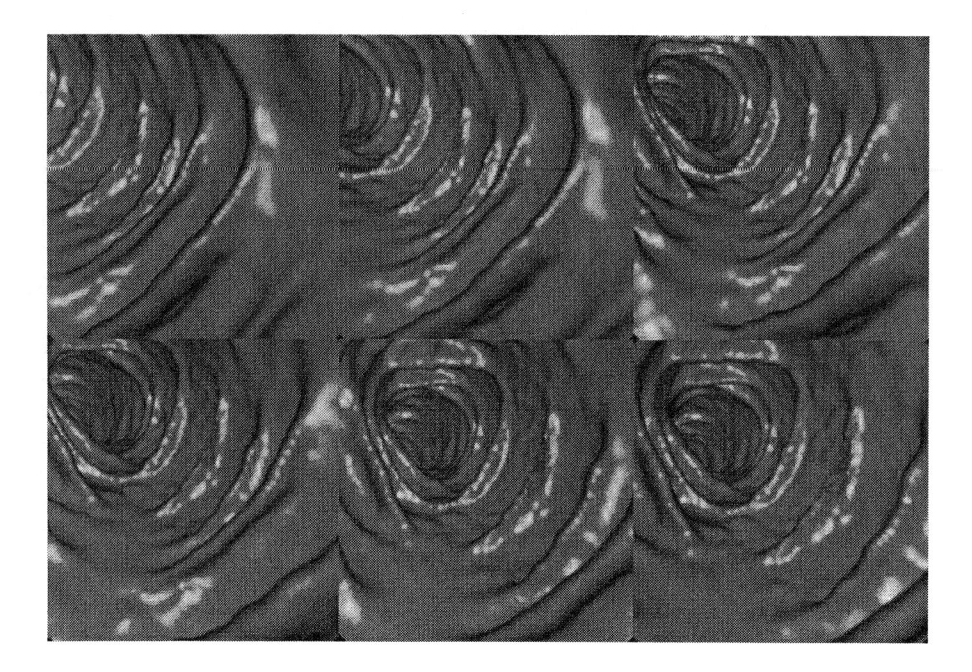

Fig. 7. This is a result of our navigation method. The camera could move forward along the colon cavity automatically

Table 1. comparison of preprocessing time and rendering time of a method that does not perform any kind of collision avoidance (NONE), an object-space guided navigation method using distance map (DIST) and our image space navigation method (OURS) for $512 \times 512 \times 541$ volume data set

	DIST (msec)	NONE (msec)	OURS (msec)	OURS / DIST (%)	OURS / NONE (%)
Preprocessing time	186880	–	–	n/a	n/a
Rendering time	265	261	278	**4.91**	**6.51**

Object space method doubles up the memory consumption since it should maintain the distance map that has the same size as the original volume data set. On the contrary, our method does not require additional storage. For manipulating $512 \times 512 \times 541$ volume dataset, about 142Mbytes memory is need for volume data and the sample size storage is required for the distance map.

5 Conclusion

We propose an efficient navigation method that reflects user's control and prohibits a camera from bump against organ wall, when the camera moves through a human cavity such as colon and bronchus. This method makes it possible to perform collision-free navigation and gives us smooth camera movement without trembling. Since it exploits depth information generated in the previous rendering stage, it doesn't require preprocessing step as well as extra storage at all. Using this method, we can go forward and backward anywhere we want to see without consideration about collision between camera and organ wall. Experimental results show us that our method satisfies most of requirement for guided navigation. Future work should be focused on improving this method to obtain more smooth and natural path for camera movement.

References

1. Levoy, M.: Display of Surfaces from Volume Data,• IEEE Computer Graphics and Applications, Vol. 8, No. 3 (1988) 29–37
2. Yagel, R. and Kaufman, A.: Template-based volume viewing, Computer Graphics Forum (Eurographics 92 Proceedings), Cambridge, UK (1992) 153–167
3. Lacroute, P. and Levoy, M.: Fast volume rendering using a shear-wqrp factorization of the viewing transformation, Computer Graphics (SIGGRAPH 94 Proceedings), Orlando, Florida (1994) 451–458
4. Yagel, R. and Shi, Z.: Accelerating volume animation by space-leaping, Proceedings of IEEE Visualization 1993 (1993) 62–69
5. He, T. and Kaufman, A.: Collision detection for volumetric objects, Proceedings of Visualization 1997 (1997)27–34
6. Gibson, S.: Beyond volume rendering: visualization, haptic exploration, and physical modeling of voxel-based objects, 6th Eurographics Workshop on Visualization in Scientific Computing, (1995)
7. De Assis Zampirolli, F. and De Alencar Lotufo, R.: Classification of the distance transformation algorithms under the mathematical morphology approach•, Proceedings XIII Brazilian Symposium on Computer Graphics and Image Processing (2000) 292–299
8. Gleicher, M. and Witkin, A.: Through-the lens camera control, ACM SIGGRAPH (1992) 331–340
9. Hong, L., Kaufman, A., Wei, Y., Viswambharan, A., Wax, M. and Liang, Z.: 3D virtual colonoscopy, IEEE Symposium on Biomedical Visualization (1995) 26–32
10. Hong, L., Muraki, S., Kaufman, A., Bartz, D. and He, T.: Virtual voyage: Interactive navigation in the human colon, ACM SIGGRAPH (1997) 27–34

Constructing a Near-Minimal-Volume Computational Box for Molecular Dynamics Simulations with Periodic Boundary Conditions

Henk Bekker[1], Jur P. van den Berg[1], and Tsjerk A. Wassenaar[2]

[1]Institute for Mathematics and Computing Science, University of Groningen, P.O.B. 800 9700 AV Groningen, The Netherlands. [2]Groningen Biomolecular Sciences and Biotechnology Institute (GBB), Department of Biophysical Chemistry, University of Groningen, Nijenborgh 4, 9747 AG, Groningen, The Netherlands.
bekker@cs.rug.nl, j.p.van.den.berg@wing.rug.nl, T.A.Wassenaar@chem.rug.nl

Abstract. In many M.D. simulations the simulated system consists of a single macromolecule in a solvent. Usually, one is not interested in the behaviour of the solvent, so, the CPU time may be minimized by minimizing the amount of solvent. For a given molecule and cut-off radius this may be done by constructing a computational box with near minimal volume. In this article a method is presented to construct such a box, and the method is tested on a significant number of macromolecules. The volume of the resulting boxes proves to be typically 40% of the volume of usual boxes, and as a result the simulation time decreases with typically 60%.

1 Introduction

Much CPU time is spent nowadays on the molecular dynamics (M.D) simulation of bio-macromolecules, notably proteins, with the goal to gain insight in the functioning of biophysical processes. Of these simulations a considerable part consists of the simulation of a single macromolecule m surrounded by a solvent, in most cases water. To prevent finite system effects, most M.D. simulations are done under periodic boundary conditions (PBC) which means that the computational box B is surrounded by an infinite number of replica boxes in a regular, space filling manner. In 3D there are five convex space filling shapes namely the triclinic box, the hexagonal prism, the dodecahedron, the elongated dodecahedron and the truncated octahedron, see figure 1.

Let s be the system formed by a computational box containing m and the water surrounding m, and let S be the infinite system formed by tessellating space with an infinite number of replica's of s, see figure 2. In M.D. simulations, interactions over a distance r_{co} are truncated. Moreover, in M.D. simulations replica's of m should not interact with each other. That means that in S no two replica's of m should be closer than r_{co}. This may be reformulated by introducing a shape M, see figure 2, defined as m dilated by a layer of width $\frac{1}{2}r_{co}$. Then, stating that in S no two replica's of m should be closer than r_{co} is equivalent with stating that in S no two replica's of M should overlap.

P.M.A. Sloot et al. (Eds.): ICCS 2003, LNCS 2659, pp. 70–79, 2003.

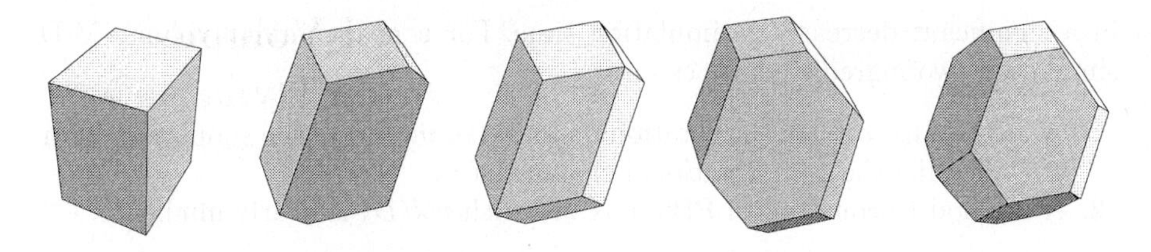

Fig. 1. The five box types used as computational box in current M.D. simulations with periodic boundary conditions. With each type, space may be tessellated in a space-filling way. From left to right: the triclinic box, the hexagonal prism, the dodecahedron, the elongated dodecahedron and the truncated octahedron

During an M.D. simulation m rotates in an erratic way, resulting in a number of full rotations during a typical simulation. Besides rotational motion m may also show conformational changes. Often these changes are of major interest and should not be restrained in any way. In current M.D. practice the computational box B is constructed by enclosing M by one of the five regular spacefillers such that m may rotate freely in B and may show some conformational change without leaving B. Often the additional space in B allowing for conformational changes of m is created by taking the width of the layer around m a bit larger than $\frac{1}{2}r_{co}$.

Let the space outside m and inside B be called C, so, $C = B - m$, and let the space outside M and inside B be called D, so, $D = B - M$, see figure 2c. After B has been constructed m is placed in B, and C is filled with water molecules. From the foregoing it will be clear that the water in D does not contribute to the simulation. Yet, per unit volume, simulating it takes approximately the same CPU effort as simulating the atoms in M, so, denoting the volume of D by $vol(D)$, we spend $\approx \frac{vol(D)}{vol(B)}$ of our CPU time on the simulation of irrelevant water.

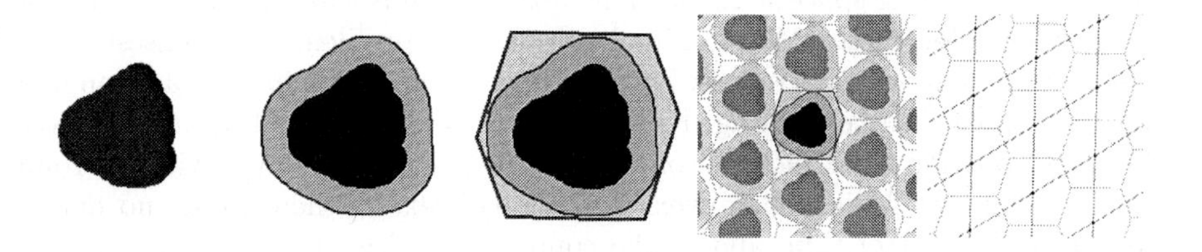

Fig. 2. a: A molecule m. b: m surrounded by a layer of width $\frac{1}{2}r_{co}$, giving M. c: A PBC box containing M. d: Part of an infinite M.D. system S, formed by tessellating space with a PBC box s. e: The lattice defined by the boxes in d.

In this article we present a technique to construct a near-minimal-volume PBC box around M, resulting in a significant decrease of $vol(D)$, so, resulting in a significant decrease of simulation time. For a near-minimal-volume M.D. simulation two ingredients are essential:

1. A method to restrain the rotational motion of m during the simulation without affecting the conformational changes of m.
2. A method to construct a PBC box of which $vol(D)$ is nearly minimal.

Fortunately, we do not have to take care of the first requirement. Some time ago such a method has been developed [1], amongst others with the goal to enable minimal volume simulations. The second requirement will be taken care of in this article.

The structure of this article is as follows. In section two we show that, instead of focussing on the shape of the computational box, we should focus on the lattice defined by box positions in S. In this way the problem of finding a box with minimal volume is reformulated as the problem of finding the densest lattice packing of M. We introduce a method to calculate this packing. In section three implementation issues are treated, and in section four our implementation is tested on some bio-macromolecules.

2 Boxes, Their Related Lattices and Calculating Lattice Packings

As mentioned earlier, there are five types of convex space-filling boxes. When s is such a box and S is formed by tessellating space with s, a lattice L is defined [2]. Here, a lattice L is the set of points $L = i * \mathbf{a} + j * \mathbf{b} + k * \mathbf{c} \quad i, j, k$ integer, and $\mathbf{a}$, $\mathbf{b}$ and $\mathbf{c}$ are the lattice vectors. So, in S at every point of L a box s is situated. Let the triclinic box spanned by the lattice vectors $\mathbf{a}$, $\mathbf{b}$ and $\mathbf{c}$ be called s_T. Now, instead of forming S by tessellating space with s, S may be formed just as well by tessellating space with s_T. So, a minimal volume simulation may be set up by devising a densest lattice packing of M. Let s_T^* be the triclinic box spanned by the lattice vectors of the densest lattice packing of M. Then S may be formed by tiling space with s_T^*. This observation is essential for our method.

Having reformulated the minimal-volume box problem as a densest lattice packing problem we have to look for a method which determines for a given body M the densest lattice packing. For polyhedral convex M such a method exists [3]. However, for most bio-macromolecules m and typical layer widths around m, the shape of M is non-convex. For non-convex M there exists no densest lattice packing algorithm, and in the computational geometry community this problem is considered as hard, so, in the foreseeable future, very probably, no such algorithm will be devised. For that reason we have to work with a heuristic method to find an approximation of the densest lattice packing of M, which we will call the near-densest lattice packing (NDLP) of M. The NDLP heuristic is based on the incorrect assumption that M is convex, and a check is added to filter out incorrect packings due to the non-convexity of M.

In principle, the NDLP method works for 3D bodies in general. However, to avoid discussions about degenerate problem instances, in this article we assume that the volume of M is non-zero and that M has no points at infinity. The first step in the NDLP algorithm is to construct the *contact body* of M, designated by N. It is constructed using the Minkowski sum [4]. The Minkowski sum $R = P \oplus Q$ of two bodies P and Q is another body R defined as $R \equiv \{\mathbf{a} + \mathbf{b} : \mathbf{a} \in P, \mathbf{b} \in Q\}$. Defining $-M$ as the body M inverted in the origin, N is given by

$$N \equiv M \oplus -M. \tag{1}$$

It can be shown easily that N is symmetric, and centered at the origin. Denoting by $M_{\mathbf{a}}$ the body M translated over the vector $\mathbf{a}$, N has the following property. *The boundary of N consists of all points* $\mathbf{a}$ *for which holds that M and* $M_{\mathbf{a}}$ *touch without overlapping.* See figure 3.

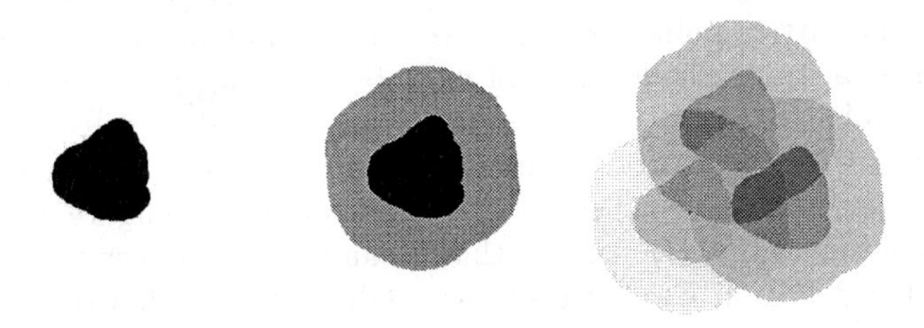

Fig. 3. 2D example of a body M (a), its contact body N (b), and N used to construct a situation where three copies of M touch each other (c). Note that N is point-symmetric.

Let us now explain how the NDLP heuristic works. We want to position M and three of its translates $M_{\mathbf{a}}$, $M_{\mathbf{b}}$ and $M_{\mathbf{c}}$ in such a way that *the volume of the triclinic cell spanned by* $\mathbf{a}$, $\mathbf{b}$, $\mathbf{c}$ *is minimal, without overlap between* M, $M_{\mathbf{a}}$, $M_{\mathbf{b}}$ *and* $M_{\mathbf{c}}$. In principle, for this we have to search through all combinations of $\mathbf{a}$, $\mathbf{b}$, $\mathbf{c}$, so, in principle we are dealing with a nine dimensional minimization problem. The key property of our NDLP method is that we reduce this nine dimensional problem to a three dimensional problem by making the following choice. *We only search through those combinations of* $\mathbf{a}$, $\mathbf{b}$, $\mathbf{c}$ *for which holds that every body of the set* $\{M, M_{\mathbf{a}}, M_{\mathbf{b}}, M_{\mathbf{c}}\}$ *is touched by the three other ones.* That this choice leads to a three dimensional minimization problem can be seen as follows. M is placed at an arbitrary location. For $M_{\mathbf{a}}$ $\mathbf{a}$ is chosen on the boundary of N, so, M and $M_{\mathbf{a}}$ touch. Because $\mathbf{a}$ is chosen on the boundary of N there are two degrees of freedom in choosing $\mathbf{a}$. Now we calculate the intersection of N and $N_{\mathbf{a}}$, which is a curve in 3D. On this curve we choose $\mathbf{b}$. So, $M_{\mathbf{b}}$ touches M and $M_{\mathbf{a}}$. Because $\mathbf{b}$ is chosen on a curve there is one degree of freedom in choosing $\mathbf{b}$. Finally we choose $\mathbf{c}$ on the intersection of N, $N_{\mathbf{a}}$ and $N_{\mathbf{b}}$, which is a small set of points, for typical M ranging from 2 to 10. So, the number of degrees of freedom in choosing $\mathbf{c}$ is zero. Herewith the number of degrees of freedom of the search problem proves to be $2 + 1 + 0 = 3$.

Now let us assume that we are searching through all combinations of $\mathbf{a}$, $\mathbf{b}$ and $\mathbf{c}$, according to our NDLP heuristic, with an appropriate search granularity to be discussed later. For every combination of $\mathbf{a}$, $\mathbf{b}$ and $\mathbf{c}$ we have to calculate $|det(\mathbf{a},\mathbf{b},\mathbf{c})|$ and store the $\mathbf{a}$, $\mathbf{b}$, $\mathbf{c}$ that give minimal $|det(\mathbf{a},\mathbf{b},\mathbf{c})|$. Obviously, $M_{\mathbf{a}}$, $M_{\mathbf{b}}$ and $M_{\mathbf{c}}$ do not overlap M. However, for non-convex M, possibly there exist one or more lattice points $\mathbf{d} = i*\mathbf{a}+j*\mathbf{b}+k*\mathbf{c}$ i,j,k integer, for which M and $M_{\mathbf{d}}$ overlap. That this may happen is not obvious to understand, there is no 2D analogy. To filter out these cases, for every $\mathbf{a}$, $\mathbf{b}$, $\mathbf{c}$ with minimal volume $|det(\mathbf{a},\mathbf{b},\mathbf{c})|$ we have to perform an additional test. That M and $M_{\mathbf{d}}$ overlap means that there are i,j,k not all 0, such that the lattice point $\mathbf{d} = i*\mathbf{a}+j*\mathbf{b}+k*\mathbf{c}$ i,j,k integer, is in the interior of N. So we have to test for all lattice points within a range $1/2\ diam(N)$ of the origin whether they fall in the interior of N, where $diam(N)$ is the diameter of N.

Having found a minimal volume box spanned by $\mathbf{a}$, $\mathbf{b}$, $\mathbf{c}$ that also passes the test that no lattice point lies in N, we have to put m in the triclinic box B_T spanned by $\mathbf{a}$, $\mathbf{b}$, $\mathbf{c}$. The location of m in B_T is completely free but the obvious choice is to locate m in the middle of B_T. Sometimes m will not fit entirely in B_T, it sticks out no matter where it is located in B_T. That does not matter, we simply locate m somewhere in the middle of B_T. See figure 4. Now for every atom of m protruding B_T it holds that it can be shifted over some lattice vector $\mathbf{d}$ such that it falls in B_T. For every protruding atom such a vector is calculated and the atom is translated over this vector. Now all atoms of m are in B_T but m is possibly fragmented. That is no problem because, when B_T containing a fragmented m is used to tessellate space, giving the infinite M.D. system S, in S complete molecules are formed from these fragments. Finally, all voids in B_T are filled with water molecules. Herewith we have constructed the near-minimum-volume triclinic system s_T^*. Summarizing, the complete NDLP algorithm outline is as follows.

```
from m and r_co construct M;
from M construct N;
forall a on boundary of N  do
  forall b on intersection of N, N_a  do
    forall  c on intersection of N, N_a, N_b do
      if |det(a,b,c)| < old_det_abc and not
      point_of_L_inside_N then store(a,b,c);
    end // c loop
  end // b  loop
end // a  loop
put m in box a,b,c
```

Let us briefly comment on our choice only to search through those combinations of $\mathbf{a}$, $\mathbf{b}$, $\mathbf{c}$ for which holds that every body of the set $\{M, M_{\mathbf{a}}, M_{\mathbf{b}}, M_{\mathbf{c}}\}$ is touched by the three other ones. It is shown that, for some convex bodies, there are other contact situations giving minimal volume [3]. Very probably this also holds for non-convex bodies, but little is known about that. However, searching through these situations would take much more CPU time than searching

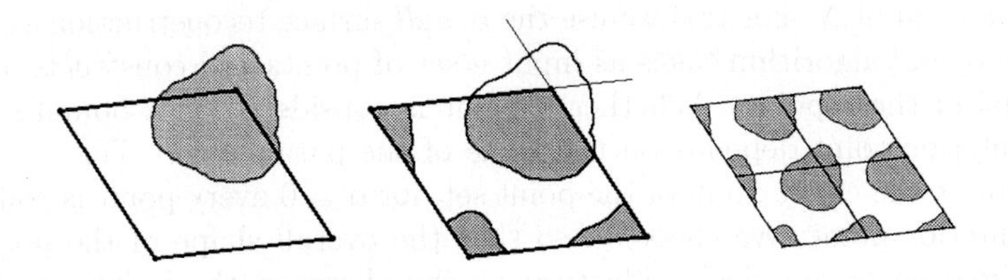

Fig. 4. a: 2D example of a minimal volume triclinic box B_T containing a protruding molecule m. b: The protruding parts of m have been reset in B_T by shifting them over lattice vectors, resulting in a fragmented molecule in B_T. c: Part of the infinite M.D. system formed by tessellating space with B_T with fragmented m. In the infinite system, whole molecules are formed by fragments from various boxes.

through the situations of our choice, probably without finding a considerably denser packing.

3 Implementation

In this section we explain how we transformed the NDLP algorithm into an efficient and robust program. In the previous section we did not specify how m, M, and N are represented. In the implementation m and M are point sets, and N is polyhedral. We start with a macro-molecule m from some library, for example from the Proteine Data Bank (PDB) [5]. For us, m is simply a point set, where every point represents an atom. In m we include all atoms of the macromolecule, so hydrogen atoms are included. From m we have to construct M. Recall that m dilated with a layer of width $\frac{1}{2}r_{co}$ gives M. We assume that r_{co} has been chosen big enough to allow for conformational changes. Now we construct a spherical point set *ball* by distributing $\approx$ 50 points more or less evenly on the boundary of a sphere with radius r_{co}. The point set M is obtained by taking the Minkowski sum of m and *ball*, so, $M \equiv m \oplus ball$. Obviously, the number of points in M is fifty times the number of points in m. Of M we only need boundary points, so we delete interior points. This strongly reduces the number of points in M.

From M we construct the contact body N by taking the Minkowski sum of M and $-M$, so, $N \equiv M \oplus -M$. The number of points in N is the square of the number of points in M. We use only part of these points by deleting all interior points of N and part of the boundary points. We want to have control over the number of remaining boundary points, which is done by using a grid-based selection method, that is, we construct a rectangular grid of $32 * 32 * 32$ cells, covering N, and determine of each point of N the cell it falls in. All points in interior cells are deleted, and of the points in a boundary cell only the one nearest to the boundary is kept. In this way the boundary is defined by typically 3000 points. Now we switch from the point set representation of N to a polyhedral representation of N. For this we use the *α-hull* surface reconstruction algorithm [6]. The α-hull algorithm takes as input a set of points and constructs an outer hull around these points. Whether a point is considered as a boundary point or an interior point depends on the value of the parameter α. For $\alpha = \infty$ the outer hull is the convex hull of the point set, for $\alpha = 0$ every point is considered as an interior point. We choose α so that the overall shape of the polyhedral approximation is practically identical to the shape of the point set. Besides constructing a polyhedral hull, the α-hull algorithm also generates a Delaunay triangulation of this hull, so, the result of the α-hull algorithm is a collection of triangles defining the hull of N.

In the NDLP method not only N but also translates of N are used. Translating N over some vector $\mathbf{a}$ is done by adding $\mathbf{a}$ to every coordinate of N, giving $N_\mathbf{a}$. To calculate the intersection curve of N and $N_\mathbf{a}$, represented by $N \cap N_\mathbf{a}$, we use the OBBTree algorithm [7]. This algorithm calculates of two sets of triangles in 3D which pairs of triangles intersect. For each pair of intersecting triangles we calculate the line segment that is in both triangles, so, the resulting set of intersection segments forms the intersection curve $N \cap N_\mathbf{a}$. Subsequently we calculate the intersection of $N \cap N_\mathbf{a}$ and $N_\mathbf{b}$.

As explained before, the NDLP method is in essence a search problem in three continuous parameters. To make the method practicable the parameter space has to be transformed into a finite set of discrete points, i.e. the granularity of the search process has to be determined. In our implementation the search granularity is dictated by the granularity of the triangulation of N. More precise, the vector $\mathbf{a}$ runs through all of the centers of the triangles of N. In the same way, $\mathbf{b}$ runs through all of the centers of the line segments of $N \cap N_\mathbf{a}$. As the triangulation of N depends on the number of points returned by the grid-based selection method, the search granularity may be controlled by the number of grid-cells.

We implemented the NDLP method in C++ using the computational geometry library CGAL[8], the α-hull algorithm and the OBBTree algorithm.

4 Results

We tested the NDLP method on seventeen macromolecules from [5]. The shape of these molecules ranges from almost spherical to complex, see figure 5. Every molecule was packed in two ways: with GROMACS [9] in the conventional way in a dodecahedron, and with the NDLP method. Subsequently, every molecule was simulated for 25000 timesteps of $2fs$ in two different triclinic boxes. To keep the comparison fair we did not do the simulation in the octahedron but in the triclinic box defined by the lattice of the octahedron, i.e. every molecule was simulated in two different triclinic boxes; one calculated via the truncated octahedron and one calculated by the NDLP method. We used the GROMACS M.D. simulation package, using rotational restraining for the simulation in the NDLP box. The simulations were done on one AMD Athlon 600 Mhz. In table 1 the molecules are given, the volume of their simulation boxes, the number of water molecules in the box, and the simulation time. From this table it is clear that on average *boxes calculated with the NDLP method have a volume of ≈40% of the corresponding GROMACS dodecahedron, and the speedup of the simulation is* ≈ 2.2.

Macro-molecules			Dodecahedron			NDLP triclinic box			
nr	PDB code	nr of atoms	volume [nm^3]	nr of water molecules	simulation time [hr:min]	volume [nm^3]	nr of water molecules	simulation time [hr:min]	speedup factor
1	1A32	1102	577.82	18610	06:51	118.93	3474	01:25	**4.83**
2	1A6S	805	142.43	4366	01:45	80.08	2254	00:58	**1.81**
3	1ADR	763	167.25	5137	01:59	80.73	2266	00:55	**2.16**
4	1AKI	1321	233.54	7027	02:48	93.99	2454	01:07	**2.50**
5	1BW6	595	130.27	3904	01:29	66.32	1874	00:45	**1.97**
6	1HNR	485	124.31	3865	01:29	59.30	1717	00:41	**2.17**
7	1HP8	686	177.10	5522	02:09	77.57	2203	00:53	**2.43**
8	1HQI	982	218.77	6764	02:38	103.71	2947	01:12	**2.19**
9	1NER	768	147.91	4498	01:45	85.35	2403	00:58	**1.81**
10	1OLG	1808	468.93	14537	05:37	203.44	5781	02:25	**2.32**
11	1PRH	11676	1337.80	38715	16:27	611.67	14554	07:47	**2.11**
12	1STU	668	190.32	5973	02:16	73.41	2074	00:50	**2.72**
13	1VCC	833	152.69	4612	01:48	69.77	1905	00:49	**2.20**
14	1VII	389	99.74	3093	01:08	46.96	1348	00:32	**2.12**
15	2BBY	767	159.26	4868	01:53	80.78	2271	00:56	**2.01**
16	1D0G*	1052	645.92	20805	07:42	112.78	3168	01:19	**5.84**
17	1D4V*	3192	1319.21	42202	15:51	451.23	13215	05:29	**2.89**

Table 1. Seventeen macro-molecules packed in simulation boxes, and simulated with GROMACS. Every molecule is packed in two ways: by the standard method of the M.D. simulation program GROMACS using the the dodecahedron, and using the NDLP method. In the NDLP method the width of the layer around m is $10\AA$. Subsequently, every molecule is simulated in two different triclinic boxes, namely the one defined by the lattice vectors of the space tessellation with the dodecahedron and the one calculated with the NDLP method. In both simulations $r_{co} = 14\AA$. Every box and molecule is simulated for 25000 timesteps of $2fs$ using the simulation package GROMACS running on a single 600 Mhz. AMD Athlon. For every box and molecule the box volume is given, the number of water molecules surrounding the macro-molecule and the simulation time. In the last column the speedup of the simulation time is given. The main result of this article is that on average the volume of the simulation box calculated with the NDLP method is ≈ 40% of the volume of the dodecahedron calculated with the current method of GROMACS, and that the average speedup of the simulation using the NDLP box is ≈ 2.2.

5 Discussion and Conclusion

- In the NDLP method, calculating N takes typically $\frac{1}{3}$ of the total time, and the actual search process takes typically $\frac{2}{3}$ of the total time. On our system

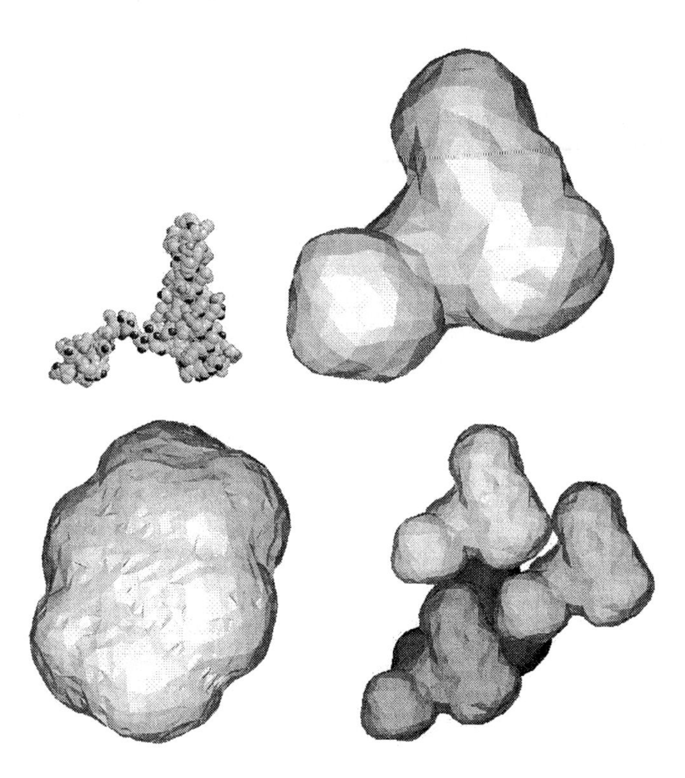

Fig. 5. The NDLP method applied to molecule $1a32$ from the PDB, consisting of 1102 atoms. The molecule is shown as a set of atoms, and M and N are shown as polyhedrals. The scaling of the figures varies. Top left: The molecule m. Top right: The body M, obtained by dilating m with a layer of width $10\mathring{A}$. Bottom left: The contact body N, defined as $N \equiv M \oplus -M$. Bottom right: The NDLP of M. The four copies of M in the NDLP configuration touch each other. Calculating N took 6 minutes, calculating the the NDLP took 9 minutes, done on a Pentium III 500 mhz. with 128 MB.

it takes $15 - 45$ minutes to calculate the NDLP of a molecule, depending on the complexity of its shape.
- The NDLP method is only useful when combined with rotational restraining.
- The overhead in CPU time introduced by the rotational restraining is negligible w.r.t. the speedup.
- The NDLP method works for a single macromolecule and for multiple macromolecules with more or less fixed relative positions. In table 1 the molecules 16 and 17 are of the latter type.
- We only compared the NDLP method with the GROMACS packing method, not with other M.D. simulation packages. However, because other packages use the same methods as GROMACS to calculate the computational box we expect that for other packages a similar gain in efficiency can be achieved.

- We will make the NDLP method available as an internet service on a page of the M.D. group in Groningen.
- The main conclusion of this article is that the speed of M.D. simulations using boxes constructed with the NDLP method is on average 2.2 times the speed of simulations using boxes constructed with conventional methods.

References

[1]A. Amadei, G. Chillemi, M. A. Ceruso, A. Grottesi, A. Di Nola, Molecular dynamics simulation with constrained roto-translational motions: Theoretical basis and statistical mechanical consistency. Journal of Chemical Physics, vol. 112, nr. 1, 1 jan. 2000.
[2]H. Bekker, Unification of Box Shapes in Molecular Simulations. Journal of Computational Chemistry, Vol. 18, No. 15, 1930-1942 (1997).
[3] U. Betke, M. Henk Densest lattice packings of 3-polytopes. Comput. Geom. 16, 3,pp. 157-186 (2000)
[4]M. de Berg, M. van Krefeld, M. Overmars, O. Schwarzkopf, Computational Geometry, Algorithms and Applications, Springer Verlag, Berlin (2000)
[5] H.M. Berman, J. Westbrook, Z. Feng, G. Gilliland, T.N. Bhat, H. Weissig, I.N. Shindyalov, P.E. Bourne: The Protein Data Bank. http://www3.oup.co.uk:80/nar/Volume_28/Issue_01/html/gkd090_gml.html
[6] H. Edelsbrunner, E. P. Muecke, Three dimensional alpha shapes. ACM Transactions on Graphics, Vol. 13, No. 1, pp. 43-72 (1994).
[7] S. Gottschalk, M. C. Lin, D. Manocha, OBBTree: A hierarchical structure for rapid interference detection. Computer Graphics, Vol. 30, Annual Conference Series, pp. 171-180 (1996)
[8] Computational Geometry Algorithms Library, http://www.cgal.org/Manual/doc_html/index.html
[9]GROMACS: The GROningen MAchine for Chemical Simulations, http://www.gromacs.org/

Recast of the Outputs of a Deterministic Model to Get a Better Estimate of Water Quality for Decision Makings

Ping Wang[1], Lewis Linker[2], Carl Cerco[3], Gary Shenk[2], and Richard Batiuk[2]

[1]University of Maryland Center for Environmental Science, Chesapeake Bay Program, 410 Severn Avenue, Annapolis, MD 21403, USA
pwang@chesapeakebay.net
[2]US Environmental Protection Agency/CBPO, 410 Severn Ave., Annapolis, MD 21403, USA
{linker.lewis,shenk.gary,batiuk.richard}@epa.gov
[3]US Army Corp of Engineers, 3909 Halls Ferry Road, Vicksburg, MS 39180, USA
cerco@erdc.usace.army.mil

Abstract. The outputs of a deterministic water quality model are retreated to recast the model simulation. A multi-variance regression method is used for daily model outputs versus observed data to assess the systematic errors in model simulation of individual model cell. The model outputs are re-adjusted to better represent the actual values, and to yield a better model calibration. Such a recast is important to model application for water quality analysis, especially for TMDL (Total Maximum Daily Load) which requires accurate simulation to exam criteria attainment. This paper addresses the recast method, its pre-requisites, and how the results are extended for the scenarios with various load reductions.

1 Introduction

Deterministic simulation models have been used to generate spatially distributed time-series data and applied in many fields, including the environmental field [1,2]. Imperfect calibration, i.e., deviations of the model prediction from what are actually observed, may be due to differences in the observed and simulated values and/or having a phase- (or time-) shift. Here, a phase-shift means that the model simulates a similar value as the observed, but it appears one or several days earlier or later. The deviations of model estimates affect model applications, particularly when model application is aimed at achieving water quality conditions above or below a particular water quality criterion. This is the case in the current Chesapeake Bay application when the model is applied against criteria values of dissolved oxygen (DO), chlorophyll and water clarity.

The Chesapeake Bay is one of the largest and most productive estuaries in the world (Fig. 1). Monitoring efforts, computer models, and researches indicated that the degradation of the Bay's water quality is primarily due to excessive nutrients and sediment inputs from its 166,000 km^2 watershed [3]. The US government and the states of the Chesapeake region committed to protect the Bay, to improve its water quality, and to remove the Bay from the list of impaired waters under the Clean Water Act by year 2010 [4]. The deterministic Chesapeake Bay Estuarine Model (CBEM)

P.M.A. Sloot et al. (Eds.): ICCS 2003, LNCS 2659, pp. 80–89, 2003.

has been used to simulate the responses of water quality to nutrient and sediment loads. In overall, its calibration to the observed data is good and acceptable according to the Chesapeake Bay Program Modeling Subcommittee (CBPMS) [5], though with certain under- or over-estimates and phase-shift. Accurate simulation of key water quality indices in certain value ranges is important to policy decisions. For example, the draft water quality criterion for DO in the designate-use-area of "Open Water" is daily minimum of 3.5 mg/L. If the observed DO on some days are close to the criteria, an under-estimate or over-estimate of the model in DO may result in non-attainment (i.e., below the criteria) or attainment (i.e., at or above the criteria) of the criteria, respectively.

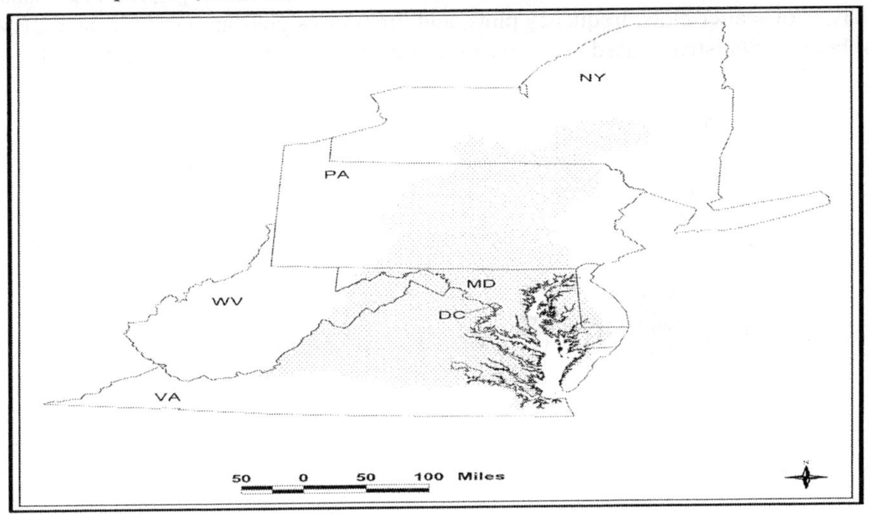

Fig. 1. Location of Chesapeake Bay and its watershed in the US mid-Atlantic states. Note: gray – the Chesapeake watershed; white within the gray – Chesapeake Bay

More problems may occur when running model scenarios, since such deviations will propagate. For instance, a DO in an "Open Water" is 2.8 mg/L, 0.7 mg/L lower than the criteria. A 30% nutrient reduction, which improves DO by 1.0 mg/L to be 3.8 mg/L, would comply with the criteria. If the model calibration is 0.5 mg/L lower, at 2.3 mg/L, the scenario of 30% nutrient reduction from the reference condition would probably yield a DO under 3.5 mg/L, resulting in a violation of the criteria. On the other hand, if the model calibration overestimates DO, it may give an impression that the water quality criteria have already been complied, and the model would not be able to provide meaningful information for management either. In short, although the CBEM simulates well the key water quality indices, the accuracy of the model around the critical values of these water quality indices may still not meet the requirements for decision making. It is desirable to have a better estimate in the interested water quality indices for the designate-use-area of the Bay. There are several data assimilation methods to get better model predictions [6,7], however, they cannot fulfil our requirements. This paper is to develop a method to retrieve the outputs of the deterministic CBEM to be better representative of the actual conditions.

2 Evaluation of CBEM Calibration

CBEM is a coupled three-dimensional finite-difference hydrodynamic model (based on CH3D [6]) and finite-volume water quality model (based on CE-QUAL-ICM [2]). The current model has 12960 cells with 2961 surface cells for the 11,000 km^2 surface area of the estuary. It simulates nutrient cycling and sedimentation, as well as interactions among key living resources and water quality parameters. Continuous daily values are output from model calibration and scenario runs. Discrete observations sampled on a specific cruise are used to represent the water conditions on that day.

There are many methods in model calibration and verification [9][10][11]. Visual analysis of scatter plots, frequency plots, and time-series plots are used for the CBEM calibration. Statistics is used to evaluate the confidence of the model by CBPMS [5].

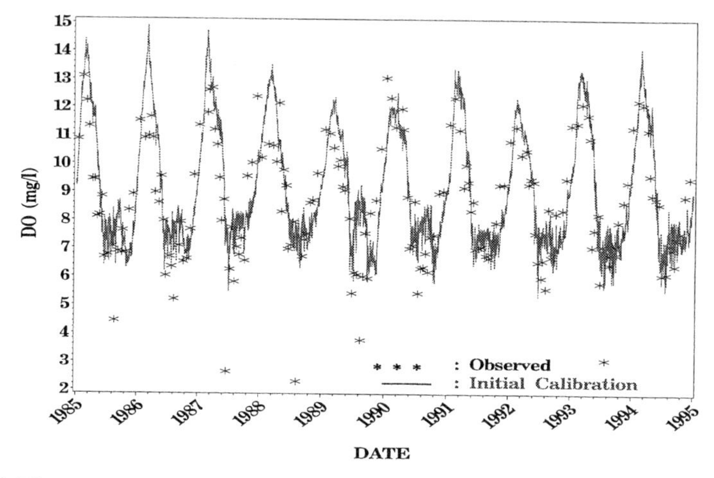

Fig. 2. Time series plot of initial calibration (-) and observed (*) DO in Cell 7563

The time-series plot of DO (Fig. 2), indicates that the model (the continuous line) is well calibrated for Cell 7563 (in "Open Water" of Segment 5), comparing to the observed (with stars). The days with low observed DO usually have low simulated DO, and the days with high observed DO usually have high simulated DO. The deviation of the model from the actual can be compared according to their values on the same date. The problem of phase-shift may also contribute the deviation of the model.

Figure 3 is a scatter plot of observed DO versus simulated DO for Cell 7563. The data are "paired", which means that only the simulations on the day having observations are used in the plot. A perfect calibration will yield the points to orient a line with a slope of one and intercept of zero, celled the ideal line. Since the plot is from paired observation and simulation, besides the deviation in absolute values, a phase-shift is another cause for the plot to deviate from the ideal line. Based on the paired values, we can apply statistics to assess the confidence of the model. The CBPMS set the following rules for a high confidence of DO simulation based on the statistics between the paired observed and modeled data: a) correlation coefficient > 0.5 (desirable), b) mean differences not greater than 1 mg/l (or roughly 10%), c) minimum concentration do not differ by more than 2 mg/l, and d) standard deviations do not differ by more than 0.5. Here the regression line is not used as a model to predict DO, but as the correlation of the paired data to evaluate the model. Since a phase shift and the complexity of the system, R^2 of 0.5 is considered to be an indication of good simulation. The statistics concludes the calibration of DO on Cell 7563 to be good.

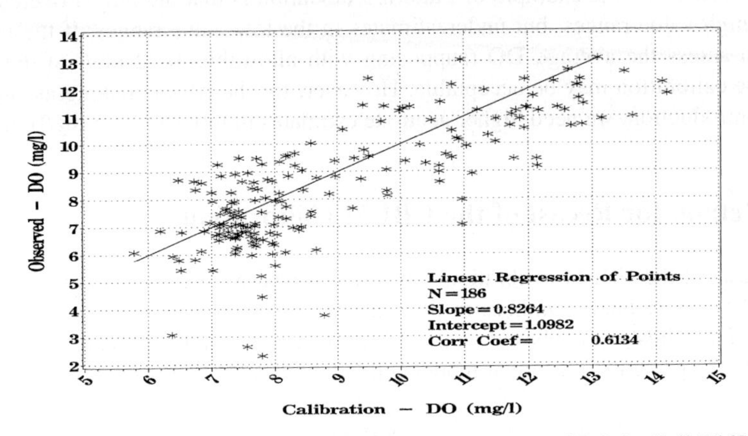

Fig. 3. Scatter plots for initial calibration and observed DO in Cell 7563

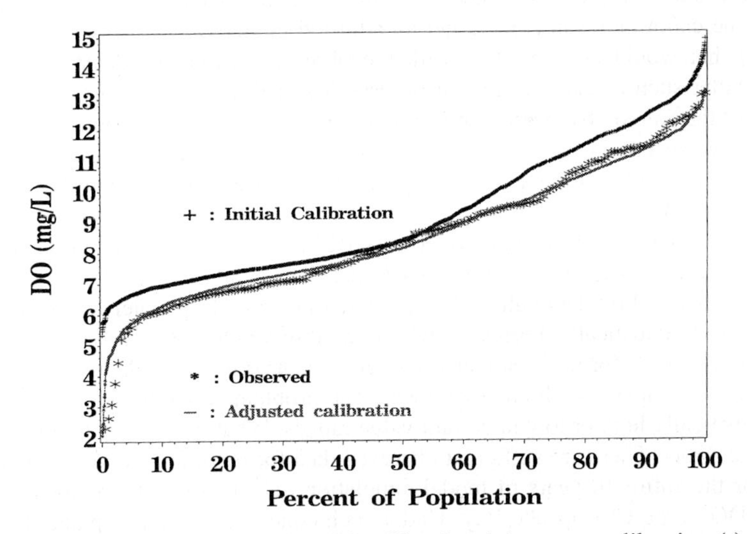

Fig. 4. Cumulative frequency of initial calibration (+), recast calibration (-), and observed (*) DO in Cell 7563

Figure 4 is a cumulative frequency plot of DO for the observed and initial calibration. A professional judgement based on the frequency plots by the CBPMS is also a key component to determine the confidence level of the model [5]. Note: the data for the cumulative frequency plot, both the modeled and the observed, are sorted, respectively, from low to high values. If the two curves are close to each other, then the calibration is regarded as good. It means that the model simulates similar high and low values as the observed, without considering whether they appear on the same date. A separation of the two curves reveals deviations of the model from the observations. The deviation has three basic patterns. The curve of the simulations can be either 1) above (overestimate), or 2) below (underestimate), or 3) across the curve of the observations. One example of Pattern 3 deviation is that the model overestimates in the high value ranges, but underestimates in the low value ranges. If the model is used to assess the average DO (supposing with allowable +/- 2 mg/l of deviation), then the calibration may be acceptable. However, for the policy-making as addressed in the introduction, we need higher accurate estimates in low DO (0–5 mg/l) values.

3 Method for Recast of the CBEM Calibration

The recast of the CBEM calibration is to use a stochastic method to adjust the output of the key water quality indices in the CBEM calibration run to be close to the observed. The adjustment is based on the relationship between the observed (e.g., y) and simulated (e.g., x) for individual model cell, which can be expressed as

$$y=f(x)\,, \tag{1}$$

and established by regression. Once such an equation is established, for a modeled value in the calibration run (x), we get an adjusted model value, x', (x' + residual = f(x) = y) that would be consistent with the observed within the desired accuracy. Theoretically, such a relationship can be established through a first-order linear regression of the paired observation and simulation that was used for Figure 3. Since the R^2 is too low, such a regressed equation is not suitable to recast the model calibration. The failure of such an approach is partly due to a phase-shift between the modeled and the observed.

Let's consider the following facts of the CBEM. The model overall simulates DO well. For example, it predicts a summer DO problem in areas where DO is observed to be problematic. Low DO values are simulated in some deep water in the summer, as is observed. Statistically, corresponding to an observed high (or low) value there is likely to have a high (or low) estimated model value near the day. The time-series and frequency plots support such an assessment. The problem is that the model estimates are systematically high or low in certain value ranges. What we want is to adjust such systematic errors. There are sufficient observed data for regression, about 1 or 2 each month for the entire 10 years of model simulation, which cover the periods of high and low DO in the Chesapeake Bay. Under such conditions, it is acceptable to order the data for the purpose of comparing the distribution of the two sets of data as we did for the cumulative frequency plots. The regression is basically on the two frequency curves, to make them to close to each other, with a focus on the low DO ranges. The days having low (or high) values before the adjustment would still have low (or high) values after the adjustment, and the recast values should be closer to the observed.
The regression can be performed on the ordered paired observations and modeled data that were used for a paired frequency plot or scatter plot. However, we suggest the

following additional work to improve the recast, since there usually exists a phase shift in the model. Corresponding to an observation, its most likely corresponding modeled data may not be on the same date. Therefore, it is better not to only use the observed and simulated which are paired on same dates, but to use all simulated daily values and the available observations. The modeled has about 20 times more values than the observed. We firstly order the modeled daily values, then split the population equally into groups according to the number of the observations. The median value of each group is picked out to form a set of sample, which has the same number as the set of observed data. We apply regression on these two sets of data. In this method, all the highest and lowest simulated values are considered in the regression, which provides an additional way to overcome the phase-shift problem.

4 Results of the Recast of CBEM Calibration

The time-series plot (Fig. 5) shows that the adjusted values are very close to the observed, especially in low DO levels. For a criterion of 3.5 mg/l, the initial calibration (Fig. 2) predicts no violation of the DO criteria, whereas, the recast (Fig. 5) shows certain violations, consistent with the observation. For the management of water quality (such as DO, chlorophyll, and water clarity) in the Chesapeake Bay, a few days of time shift may not be critical, whereas, absolution value is more important in evaluation of the attainment of criteria. Figure 4 also plots the cumulative frequency of the recast DO (with joined dots). It is very close to the observation curve. Five percent of tolerance is set for the DO criteria. Therefore, a 5% error in DO estimate is acceptable. The model recast meets the required accuracy of the model in DO simulation. The recast method was also applied to all other cells that have monitoring data of DO, chlorophyll, and water clarity. No matter whether the initial calibration is predominated with underestimate or overestimate, satisfactory recasts were resulted. This indicates that the recast method is acceptable.

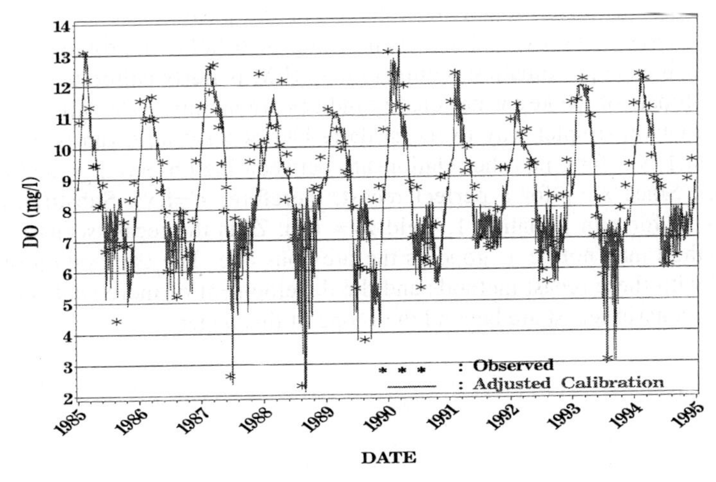

Fig 5. Time series plot of adjusted calibration (-) and observed (*) DO for Cell 7563

5 Recast for the Cells without Observed Data

The number of surface cells and monitoring stations varies among model segments, ranging from 4 to 100 and 1 to 10, respectively. The recast for the cells without observed data utilizes the recast results from the cells having observed data within the same model segment: use the relationships (i.e. the regressed coefficients) from three nearest calibration cells, and weigh them according to the inverse square distances to the cells (centroid).

We did such a check for a cell having a monitoring station. The recast by using its monitoring data directly is very close to that by using the weighed regression results from three nearby monitoring cells. We also did another type of check. For the cells having monitoring station, we aggregate all monitored and modeled DO in a certain depth range in a model segment to construct a regression, then applied the regression results uniformly to recast DO in each monitoring cell at the depth level. The result for each cell is close to the recast based on the individual cell monitoring info, although slightly worse. This indicates that the systematic errors in a segment are similar. We believe that the recast method for non-monitoring cells is acceptable.

6 Application to the Recast of Model Scenarios

A model scenario of nutrient and sediment reduction usually yields lower chlorophyll and higher dissolved oxygen and water clarity than the reference conditions (such as the calibration run). Since the model calibration deviates from the actual, the errors will propagate in model scenario runs. Such modeled information may not be used to determine the level of load reduction to improve water quality.

For DO of the initial scenario (z) and initial calibration (x), their relationship, $z = g(x)$, can be established through regression. The CBEM has good proportional response of water quality indices to load reduction. A nearly linear relationship between them was observed for most cells. For the ideal or recast scenario (z') and recast calibration (x') the above relationship also holds approximately, i.e., $z' = g(x')$. Such an approximation requires the CBEM is fairly calibrated, and the errors in the estimate of the key water quality indices are mainly systematic.
An alternative, simpler way to recast daily DO of a scenario run is by referring to Equation 1, $y = f(x)$, i.e., the relationship between the observed, y, and the calibration (x). Since $x'=y$ for a perfect recast, therefore, $x'=f(x)$. Substituting scenario outputs, z, for x in Equation 1, yield $z' = f(z)$. Z' is the recast scenario. However, this method may not be as good as the previous one. The exploration of the errors related with these recast methods and the development of more sophisticated methods for scenario recast are beyond the scope of this paper.

7 Discussion

7.1 Assumption and Validity of the Method

The requisite of the proposed recast method is that the deterministic model has good simulations of the physical, chemical, and/or biological processes for the modeled ecosystem, and is reasonably calibrated both temporally and spatially. The CBEM is one of the most sophisticated surface water quality models in the world, and meets these requirements. It simulates well the target water quality indices of DO, chlorophyll, and water clarity. High chlorophyll was simulated in spring and summer in high nutrient loading areas, consistent with the historical data of high chlorophyll. Low DO, especially in bottom cells in the summer, was simulated in the historically DO problematic areas. Similarly is for water clarity. The responses of DO, chlorophyll, and water clarity to the scenarios of nutrient reduction are well simulated: i.e., a lower loads yields lower chlorophyll level, more water clarity, and higher DO. This can also be proved by the surface analysis [12], which is based on regression of the responses of water quality to various loading scenarios. Therefore, we have high confidence on the model. The modeled high (or low) values of a constituent are likely to have high (or low) observed values around the simulate dates. What is needed to adjust for the model output is to reduce the systematical over- or under-estimates. The recast is to bring the simulated value to be close to the observed level, as seen from the frequency plot (Figure 4). We applied the same method for all cells with monitoring data, and similar recasts were yielded. This method does not intend to improve the correlation for the paired data as in the scatter plot, since this method does not regress to match the pair-values directly. Nevertheless, the recast usually improves their correlation, leading the points to be close to the ideal line.

7.2 Problem with Ordered Data

For any two sets of unrelated data, if they are ordered by values, they are likely to show a certain correlation (e.g., near linear). Therefore, it is usually no meaningful for a regression on two sets of ordered values of time-series data. However, here we are trying to transpose the distributions of the modeled versus observed for the frequency plots, not the individual points, and to obtain the relationships between the distributions of these two sets of data. The CBEM is well calibrated. Statistically, the modeled higher values correspond to the observed higher values, and the modeled lower values correspond to the observed lower values. After the recast, the corresponding values get closer to each other, as Fig. 5 versus Fig. 2. The fact of good recast for all monitoring cells supports this recast method. People may have noticed that the data preparation for the recast method described in Sect. 3 is similar to that for quantile-quantile (Q-Q) plot. We can expect that such recast results would look good in Q-Q plot, as shown in Figure 6, which has great improvements over the initial calibration. Q-Q plots have been recommended by many researchers for model evaluation and exploratory data analysis [13], therefore, we believe the recast method is plausible.

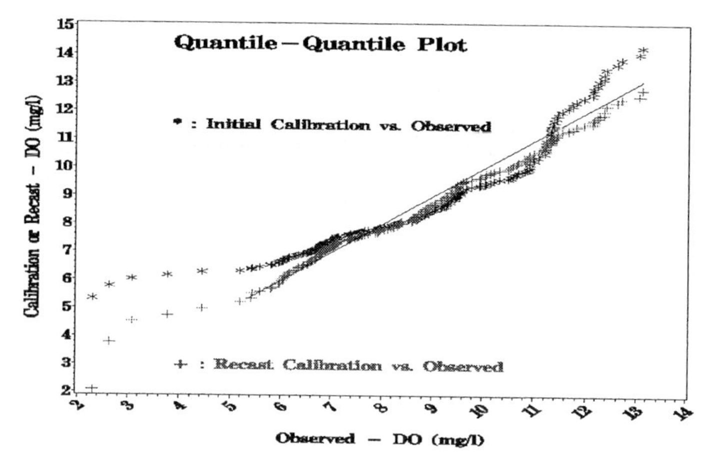

Fig. 6. Q-Q plot of initial calibration and recast calibration DO vs. observed DO

7.3 Regression on Paired or Non-paired Data

Our work reveals that a recast using paired observation and model data for regression provides a better estimate than the initial calibration. The regression using all daily model data (i.e., non-paired data) can remedy some problems associated with the phase-shift by a model. The latter has an even better recast than the method using paired values. The results of recast presented in this paper are by the latter approach.

7.4 Subdivision of Time Period for Regression

DO, chlorophyll, and water clarity during the period of May through September have contrast levels of water quality from those in the other period. In many cases, splitting data into these two periods improves regression. However, sometimes, for example, there are similar low DO values in the summers of two different years, but the model simulates somehow different levels of water quality in the two summers. In such a case, it is better to split data yearly. In fact, the model shows a strong correspondence of low-to-low or high-to-high values between simulated and observed within a year. Therefore, splitting data yearly for regression generally yields good results. Finer splitting data is not recommended, which may reduce sample population in the regression. It is useful to investigate the optimal subdivision of data. However, this is beyond the scope of this paper.

8 Conclusion

An accurate calibration of CBEM is required for the key water quality indices in the designate use area of the Chesapeake Bay. The method of this paper recasts the model calibration of the CBEM to yield more accurate daily estimates than the initial calibration. Care should be taken that the requirements of this method are satisfied -- the deterministic model should first be well calibrated with reasonable responses to seasonal change and load changes.

References

1. Thomann, R.V., Mueller, J.A.: Principles of surface water quality modeling and control, Harper Collins Pub., New York, (1987) 644pp
2. Cerco, C.F., Cole, T.M.: Three-dimensional eutrophication model of Chesapeake Bay. *J. Envir. Engrg.*, (1993) **119**(6): 1006–1025
3. Officer, C.B., Biggs, R.B., Taft, J.L, Cronin, L.B., Tyler, M.A., Boynton, W.R.: Chesapeake Bay anoxia: origin, development and significance. Science (1984) **233**: 22–27
4. Chesapeake Executive Council: *Chesapeake 2000*. Annapolis, MD, USA (2002)
5. CBPMS: A comparison of Chesapeake Bay Estuary Model calibration with 1985–1994 observed data and adjustment of model estimates of dissolved oxygen, clarity, and chlorophyll to better represent observed data. A report of the Modeling Subcommittee, Chesapeake Bay Program Office, Annapolis, MD, USA, October (2002) pp300
6. Malanotte-Rizzoli, P.: Modern approaches to data assimilation in ocean modeling. Elsevier, (1996) pp455
7. Emery, W.J., Thomson, R.E.: Data analysis methods in physical oceanography, Pergramon, (1998) pp634
8. Johnson, B., Kim, K, Heath, R., Hsieh, B., Butler, L.: Validation of a three-dimensional hydrohynamic model of Chesapeake Bay. J. Hydr. Engrg., ASCE (1993), **199**(1): 2–20
9. Thomann, R.V.: Verification of water quality modeles. J. Emvir. Eng. (1982) 108(5) 923–940
10. Rechhow, K.H., Clements, T.J., Dodd, R.C.: Statistical evaluation of mechanistic water-quality models. J. Envr. Eng. (1992) 116(2) 250–268
11. Wang, P., Storrick, J., Linker, L.: Confirmation of the Chesapeake Bay Watershed Model. Proceedings of the 16th International conference of the Coastal Society, Williamsburg, VA, USA (1998) 381–390
12. Wang, P., Linker, L.C., Batiuk, R.A.: Surface analysis of water quality response to load. in: Spaulding, M.L. (ed.) Proc. 7th Intern. Conf. in Estuarine and Coastal Modeling, Nov. 2001, St. Petersburg, FL, USA (2002) 566–584
13. Nosteller, F., Tukey, J.: Data analysis and regression, Addison-Wesley (1977)

A Fuzzy Approach for Overcurrent Relays Simulation

Hossein Askarian Abyaneh[1], Hossein Kazemi Karegar[1], and Majid Al-Dabbagh[2]

[1] Department of Electrical Engineering , Amir Kabir University of Technology, Hafez Ave, No 424
P.O. Box 15875-4413, Tehran, Iran
haskarian@yahoo.com
h_kazemi_ir@yahoo.com

[2] School of Electrical Engineering and Computer System of RMIT University, Victoria 3001
Melbourne, Australia
majid@rmit.edu.au

Abstract. Accurate models for Overcurrent relays (OC) with inverse time characteristics have an important role for efficient coordination of power system protection devices. This paper proposes a new method for modelling of OC relays based on fuzzy logic. The new model is more accurate than traditional models and has been validated by comparing it with the linear and nonlinear analytical models.

1 Introduction

Due to unforseen load expansion, many power systems are operated close to their design limits, therefore it is necessary to revise modelling of relays under these conditions [1]. OC relay models are defined in various ways. Time-Current (TC) curves of an OC relay is the most commonly used method [2]. There are two main methods for representing an OC relay i.e., by using digital computers: software models and direct data storage. Software models of OC relay characteristics play a major role in coordinating protection schemes in power systems [3].

A comprehensive review of computer representation of OC relays has been made in reference [4]. In this review, it is stated that Sachdev models are simple and useful polynomials for modelling OC relays for coordination purposes. It should be noted that any microprocessor relay abiding to the IEEE Std. C37.112 [5], should follow in this representation the equations provided by their standard. Furthermore in this context, the time dial setting provided by this standard is linear.

An alternative method for the representation of OC relays is based on direct data storage techniques. These techniques store data in memory of the computers for different TDS/TMS and then select operating points of a relay based on the stored data for different TDS/TMS. If the selected operating point does not match with one of the stored values an interpolation is necessary to determine the corresponding time. Therefore, this method requires storing large amount of data.

In this paper a new model based on fuzzy logic and is more accurate than analytical models and does not need look up tables is presented. The new methodology employed gives a more accurate model, compared with the model introduced in [6].

P.M.A. Sloot et al. (Eds.): ICCS 2003, LNCS 2659, pp. 90–99, 2003.

2 The New Method

The proposed fuzzy model is based on finding a simple mathematical equation with a fuzzy correction coefficient to calculate the operating time of OC relays.

In Fig. 1, (t_1, I_1), (t_2, I_2) and (t_3, I_3), sampled data are given is the operating time of OC relay for a given I^*. The simplest equation for fitting two points on a curve is the straight line equation. This mathematical equation does not need any complicated curve fitting technique. In reality, two adjacent points on OC relay characteristics are connected as a line. Therefore, the proposed fuzzy logic model finds a fuzzy correction coefficient to simulate the curve of the OC relay under consideration.

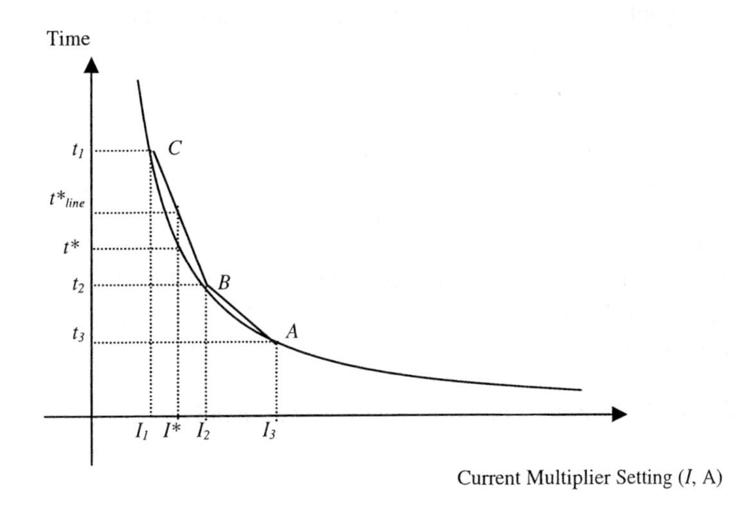

Fig. 1. A TC curve of an overcurrent relay

If the two points (t_1, I_1) and (t_2, I_2) of Fig. 1 are the considered sampled data, then the operation time associated with I* can be calculated by As follows:

$$t^*_{line} = (t_2 - t_1).(I^* - I_1)/(I_2 - I_1) + t_1 \quad (1)$$

The calculated operation time according to equation (1) is t^*_{Line} which differes from actual one, i.e. t*. Therefore, a correction factor must be added to equation (1). The variation of correction factor, r, is determined using fuzzy logic technique. The equation with correction factor is given as:

$$t^* = r.(t_2 - t_1).(I^* - I_1)/(I_2 - I_1) + t_1 \quad (2)$$

The fuzzy correction coefficient r varies between 0 and 1 when the location of I^* changes on the current multiplier setting axis of TC curves. The values of I^* and the slope (s) of the line between adjacent data, play an important role in calculating the variation of r. For example, when I^* gets large and goes near the tale of the TC curve, and the slope gets smaller. Subsequently, the curve between two neighbouring data is close to a direct line. Therefore, the value of r increases and approaches a value close to one.

Equations (3) and (4) describe the calculation of r and s for the sampled data.

$$r_i = [(I_{i+2} - I_i).(t_{i+1} - t_i)]/[(t_{i+2} - t_i).(I_{i+1} - I_i)] \quad (3)$$

$$s_i = (t_{i+1} - t_i)/(I_{i+1} - I_i) \quad (4)$$

For example, in considering Fig. 1, r_I can be obtained by Eq. (5).

$$r_1 = [(I_3 - I_1).(t_2 - t_1)]/[(I_3 - I_1).(t_2 - t_1)] \quad (5)$$

The membership functions of I, r and s of sampled data are necessary for calculating the value of r for other set of data. It is worth mentioning that the value of I can be obtained from the catalogues of OC relays. For example, I varies from 2 to 30 for RSA20, an electromechanical OC relay. However, the values of s and r must be calculated based on the stored data and then used for the calculation of the membership functions of r (μ_r) and s (μ_s). Because the values of I, r and s are positive, only Positive Small (PS), Positive Medium (PM) and Positive Big (PB) are effective.

The membership functions of I, r and s are shown in Fig. 2.

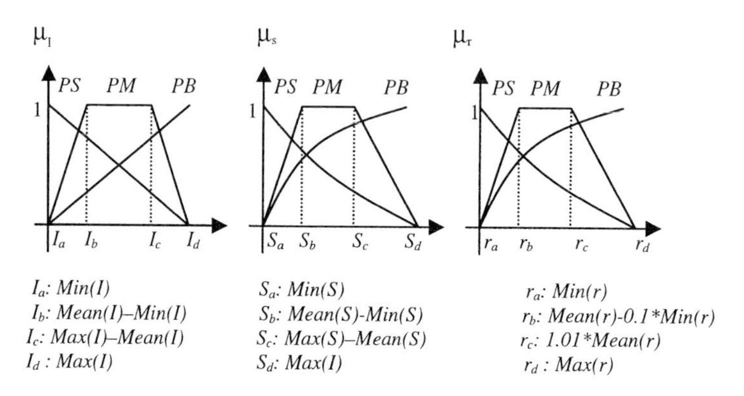

Fig. 2. Membership functions of I, r and s

It is found by trial and error that the trapezoid shape for PM produces better results than triangular shape, which is used in reference [6]. This feature is regarded as an advantage in the new fuzzy model, compared with the previous method,

Determining fuzzy rules are the next step in fuzzy modeling. As can be seen from Fig. 1, when I is small, near pick-up current, and s is large, because TC curve closes to its asymptotic, then the value of r must be selected large, i.e., 1. In other words, TC curves of OC relays near the pick-up currents are very straight like a direct line.

Based on above assumptions, the fuzzy rules for the proposed method are as below:

- If I is small and s is large then r is large.
- If I is small and s is medium then r is medium.
- If I is medium and s is small then r is medium.
- If I is medium and s is medium then r is small.
- If I is medium and s is large then r is medium.
- If I is large and s is medium then r is medium.
- If I is large and s is small then r is large.

The last step in fuzzy modelling is determining the value of r. In this paper, the centroid method is used to calculate r. Equations (6)–(8) show how to obtain r.

$$a_i = \mu_i \wedge \mu_s \tag{6}$$

$$\mu_r' = a_i \wedge \mu_r \tag{7}$$

$$r_i = centroid(\mu_r') \tag{8}$$

$$r = \sum_1^n a_i.r_i / \sum_1^n a_i \tag{9}$$

Equation (9) describes the normalized weight method to produce a single output corresponding to r [7].

2 Case Study

Two types of OC relays were used for evaluating the proposed model. The first one was RSA20, an electromechanical OC relay, where its TDS varies from 4 to 20. The second one was CRP9, a very inverse electromechanical OC relay. For both relays, the used sampled data are shown in Table 1 and Table 2 respectively, Additional sampled data which is shown in Tables 3 and 4 is used for testing and comparing the analytical and the proposed models.

Table 1. Operation time of RSA20 OC relay in ms when TDS=4, 8, 14 and 20

I	TDS=4	TDS=8	TDS=14	TDS=20
4	4238	5690	8840	11975
5.5	2235	3255	5439	7516
7	1648	2412	4012	5797
8.5	1373	2070	3522	5070
10	1240	1960	3283	4702
11.5	1155	1890	3146	4525
13	1112	1825	3076	4411
14.5	1075	1778	3025	4333
16	1061	1754	2993	4264
17.5	1045	1735	2954	4194
19	1028	1714	2911	4126
20.5	1017	1691	2874	4083
22	1010	1659	2846	4043
23.5	1008	1636	2799	4003
25	1006	1626	2771	3965
26.5	1004	1611	2747	3935
28	1002	1603	2734	3915
29.5	1000	1594	2697	3893

Table 2. Operation time of CRP9 OC relay in ms when TDS=4, 6, 7 and 10

I	TDS=4	TDS=6	TDS=7	TDS=10
3	2451	3521	4232	6495
4	1352	2107	2459	3935
5	953	1527	1853	2734
6	758	1243	1517	2175
7	653	1054	1302	1927
8	605	934	1153	1754
9	552	836	1051	1621
10	516	742	993	1523
12	459	658	894	1304
14	401	634	794	1221
16	375	618	744	1123
18	359	608	715	1059
20	358	585	669	1024

Table 3. Test operation time of RSA20 OC relay in ms when TDS=4, 8, 14 and 20

0l	TDS=4	TDS=8	TDS=14	TDS=20
4.5	2816	4220	6637	9097
6	1881	2751	4512	6496
7.5	1478	2196	3706	5319
9	1314	2005	3373	4863
10.5	1206	1930	3206	4617
12	1128	1860	3105	4461
13.5	1096	1802	3049	4370
15	1074	1765	3008	4303
16.5	1051	1743	2971	4228
18	1032	1720	2934	4158
19.5	1025	1702	2894	4106
21	1015	1685	2862	4066
22.5	1009	1652	2825	4023
24	1007	1634	2785	3982
25.5	1005	1616	2756	3951
27	1003	1607	2743	3929
28.5	1001	1597	2718	3901
30	998	1593	2689	3886

Table 4. Test operation time of CRP9 OC relay in ms when TDS=4, 6, 8 and 10

1l	TDS=4	TDS=6	TDS=8	TDS=10
3.5	1721	2832	3365	5241
4.5	1125	1746	2123	3289
5.5	849	1369	1648	2351
6.5	695	1124	1411	1986
7.5	634	1009	1213	1821
8.5	574	892	1098	1698
9.5	534	776	1035	1586
11	489	692	939	1401
13	425	642	835	1248
15	386	623	755	1167
17	367	612	732	1087
19	357	593	684	1048

The sampled data is obtained by performing experimental tests several times using an accurate computerised relay tester for RSA20 and CRP9 relays, to ensure the accuracy of the recorded expanded data.

3.1 Fuzzy Model Application

3.1.1 Fuzzy Model of RSA20 OC Relay

The recommended analytical model in reference [4] is selected for comparison between the new and the mathematical model. Due to the progresses in both calculating methods and software packages an equation with more polynomials coefficients is possible and provides more accurate results. MATLAB has provided a useful environment to calculate polynomials coefficients based on advanced non-linear curve fitting techniques [11,12]. Therefore, Eq. (10) with nine coefficients is chosen. It should be noted that when a polynomial equation with more than nine coefficients was used, ill-structure matrices and poor output results were reported by MATLAB. In addition, normalized operation time data were used to improve the accuracy of the obtained results by

$$t = a_0 + a_1/(I-1) + a_2/(I-1)^2 + \ldots + a_8/(I-1)^8 \tag{10}$$

The coefficients of Eq. (10), i.e. a_0 to a_8, are shown in Table 5 where data in columns TDS=4, 8 and 20 of Table 1 are selected as input. The first column of Table 5 shows the coefficients of Eq. (10) for the sampled data of TC curve when TDS=6.

Table 5. Polynomial coefficients of RSA20 OC relay when TDS=4, 8 and 20

Coefficient	TDS=6	TDS=8	TDS=20
a_0	0.2343	0.2386	0.2924
a_1	-0.1957	1.4592	1.0945
a_2	6.222	-9.0501	-4.7457
a_3	-0.0351	36.2787	26.2756
a_4	-2.9201	25.0367	14.3067
a_5	2.259	-39.6346	-33.2044
a_6	11.4195	-117.835	-87.3217
a_7	20.7218	-186.711	-133.39
a_8	28.6445	-239.456	-167.828

By calculating the average error percentage for each curve of RSA20, the two curves illustrating the mean error percentage of the fuzzy and the mathematical nine-coefficient model are shown in Fig. 4 and Fig. 5.

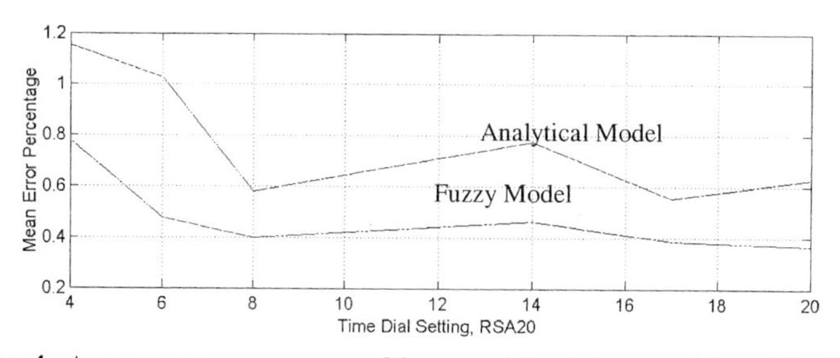

Fig. 4. Average error percentage of fuzzy and Analytical model for RSA20

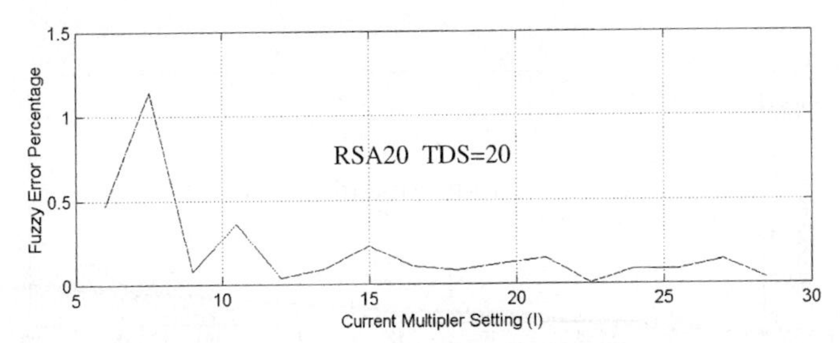

Fig. 5. Error percentage of fuzzy model for RSA20 when TDS=20

As can be seen from Fig. 4, the average of error percentages of the fuzzy model results are smaller than the polynomial form and in most cases is near 0.4 percent. In addition, Fig. 5 shows that the error percentage of the fuzzy model decreases when the fault current through the relay increases. This is an important feature, because high fault currents can cause more damages to the power systems components.

3.1.2 Fuzzy Model of CRP9 OC Relay

In this section, the fuzzy model is applied to CRP9 OC relay to find its characteristic. The relay is an electromechanical type and its data are given in Table 2. The data are obtained by experimental tests. The coefficients of the analytical model, i.e. a_0 to a_8 for the relay, are calculated and shown in Table 6.

Figures 6–7 show the obtained results after applying two different methods.

Table 6. Analytical polynomial coefficients for CRP9

Coefficient	TDS=4	TDS=6	TDS=10
a_0	0.0769	0.1401	0.0757
a_1	1.3063	0.1726	1.7057
a_2	-1.8269	5.262	-3.8979
a_3	4.7069	-3.3488	7.9226
a_4	6.6728	-5.9993	11.2491
a_5	1.3859	-0.9672	1.0027
a_6	-7.1234	6.6096	-15.5886
a_7	-15.6564	13.4513	-32.4335
a_8	-22.6173	18.796	-46.5238

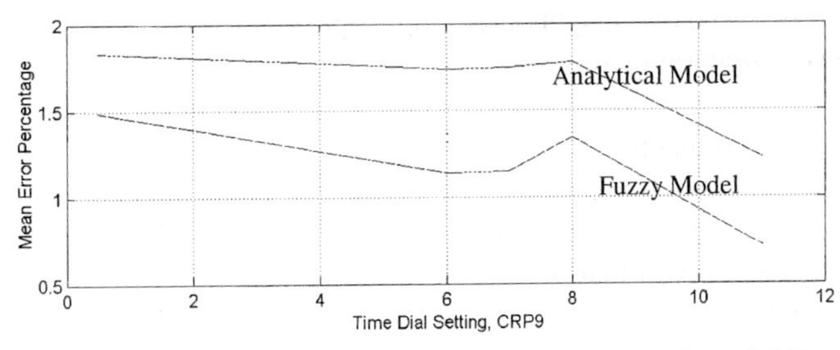

Fig. 6. Average error percentage of fuzzy and Analytical model for CRP9

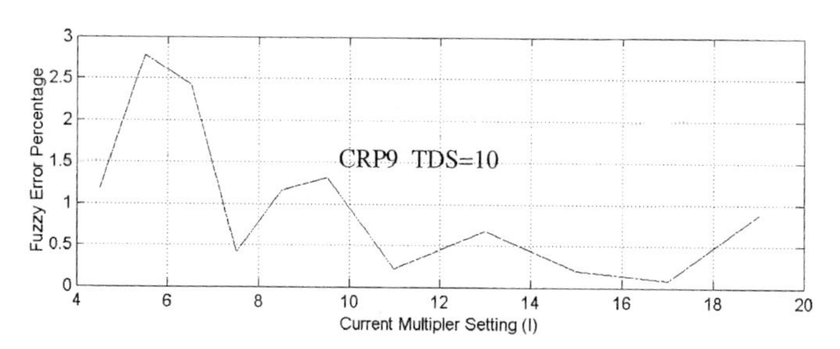

Fig. 7. Error percentage of fuzzy model for CRP9 when TDS=10

Figure 6 shows the fuzzy model of CRP9 is more precise than the analytical nine-coefficient model. Similar to RSA20, the average error percentage of proposed model is below 1.5 percent and usually decreases when current multiplier setting increases. On the other hand, the fuzzy method uses very simple mathematical equation and does not involve to complicated curve fitting techniques. Consequently, ill-structure matrices and poor results are not reported.

4 Conclusion

In this paper a new model for OC relays, based on fuzzy logic is presented. The new model was evaluated by experimental tests on two types of OC relays. The results revealed that the error percentages of fuzzy model for both relays were low. In comparing, the results of the new method with the analytical model, it was evident that the new model did not need any curve fitting technique. It has been shown that the method is flexible and can take into account different relay characteristics with linear and nonlinear features.

5 Glossary

t:	Operation time of OC relay
I:	Current multiplier setting
s:	Slope between two neighboured points on TC curve
r:	Fuzzy correction factor
μ_r:	Membership function of r
μ_s:	Membership function of s
μ_I	Membership function of I
f:	Asymetric activation function
y:	Internal activity of neuron
a,b:	Constants
x:	*TDS* or *TMS*

Refrences

1. T.S. Sidhu, M. hfuda and M.S. Sachdev, A technique for generating computer models of microprocessor-based relays, Proc. IEEE Communications, Power and Computing, WESCANEX'97, 1997, pp. 191–196.
2. P.G. McLaren, K. Mustaohi, G. Benmouyal, S. Chano, A. Giris, C. Henville, M. Kezunovic, L. Kojovic, R. Marttila, M. Meisinger, G. Michel, M.S. Sachdev, V. Skendzic, T.S. Sidue and D. Tziouvaras, Software models for relays, IEEE Trans. Power Delivery, 16(2001), 238–245.
3. M. S. Sachdev and T. S. Sidhu, Modelling relays for use in power system protection studies, Proc. IEE Development in Power System Protection, 2001, pp. 523–526.
4. IEEE Committee Report, Computer representation of overcurrent relay characteristics, IEEE Trans Power Deliver, 4(1989), 1659–1667.
5. IEEE Standard inverse-time characteristic equations for overcurrent relays, IEEE Std C37.112–1996.
6. H. Askarian, K. Faez and H. Kazemi, A new method for overcurrent relay (OC) using neural network and fuzzy logic, Proc IEEE Speech and Image Technologies for Telecommunications, TENCON '97, 1997, pp. 407–410.
7. H.T. Nguyen and E.A. Walker, A First Course in Fuzzy Logic, Chapman and Hall/CRC, New York, 2000, pp. 193~194.
8. L. Medsker and J. Liebowitz, Design and Development of Expert system and Neural networks, Macmillan, New York, 1994, pp.196~203.
9. L. Fausett, Fundamentals of Neural Networks, Architectures, Algorithms and Applications, Prentice-Hall, NJ, 1994, pp. 86~87.
10. S.S. Haykin, Neural Networks (A comprehensive foundation), Macmillan, New York, 1994, pp. 176–181.
11. T.F. Coleman and Y. Li , An interior, trust region approach for non-linear minimization subject to bounds, SIAM journal on Optimisation, 6(1996), 418–445.
12. T.F. Coleman and Y. Li , On the convergence of reflective Newton methods for large-scale nonlinear minimization subject to bounds, Mathematical Programming, 67(1994), 189–224.

A Knowledge-Based Technique for Constraints Satisfaction in Manpower Allocation

Khaireel A. Mohamed[1], Amitava Datta[2]*, and Ryszard Kozera[2]

[1] Institut für Informatik, Albert-Ludwigs-Universität Freiburg
Georges-Kohler-Allee, 79110 Freiburg, Germany
khaireel@informatik.uni-freiburg.de

[2] School of Computer Science & Software Engineering
University of Western Australia
Perth, WA 6009, Australia
{datta, ryszard}@csse.uwa.edu.au

Abstract. We show that knowledge-based techniques are as effective as mathematical techniques when satisfying constraints for solving manpower allocation problems. These techniques can be used to fulfill the corresponding local and global constraints based on the dynamic programming algorithm. It uses tools borrowed from genetic and simulated annealing algorithms, and fuzzy logic methodologies. The schedules produced by our algorithm match the best schedules produced by human experts.

1 Introduction

Manpower allocation (MA) problems are combinatorial optimisation problems with constraints that must be satisfied to arrive at feasible solutions (schedules). Traditionally, daily schedules in work places were prepared by hand [2, 6]. This class of problem arises in the management of manpower for organizations that provide round-the-clock services – assigning employees to the numerous available designated deployment posts. Consequently, scheduling manpower resources such as nurses, physicians, maintenance crew, and security personnel is important, as good schedules run entire organizations efficiently. Essentially, this task requires expertise and experience. It is necessary to ensure that the resultant schedules optimally match the skills of the available resources to the various conditions and requisites of the deployment posts.

Stochastically, there are many ways to achieve an algorithm that produces good schedules. Among the most promising are the methodologies that construe the use of linear and non-linear programming [7], genetic [4], simulated annealing [10], and artificial intelligence (AI) [1, 8] algorithms. These approaches attempt to find the lowest cost solutions that optimize given cost functions with respect to the associated constraints. Most researches on scheduling algorithms have concentrated mainly on regular performance measures in which an objective function is defined in terms of a set of resources (manpower) and a set of tasks (deployment posts). All the methods and techniques

* This author's research is partially supported by Western Australian Virtual Environments Centre (IVEC) and Australian Partnership in Advanced Computing (APAC).

P.M.A. Sloot et al. (Eds.): ICCS 2003, LNCS 2659, pp. 100–108, 2003.

portrayed in this paper contribute to coming up with realistic and feasible rosters for scheduling manpower to various available deployment posts. The rest of the paper is organized as follows.

In Section 2, we define the structuring of our constraints for interpreting MA problems using knowledge-based techniques. Optimal schedule determination on more global perspectives through Sauer's [12] *reactive* and *interactive* scheduling representations is discussed in Section 3. In Section 4, we discuss the rule-based methodologies for backtracking through fuzzy logic techniques. In Sections 5, we discuss our results.

2 Problem Formulation

Brachman's [1] description of AI suggests two ways in which knowledge-based techniques can be implemented to solve the constraints in MA problems. Firstly, each resource's abilities and the requisites of the various posts for allocation must form the knowledge representation that can be reasoned to reflect human intelligence. Secondly, it suggests that the only way such techniques can behave in this manner is if they contain formal mechanisms for representing knowledge and employ inference techniques that model computationally based structures.

We can define the MA problem as follows. There is a given set of manpower resources R, each of whom has zero or more skills, and is identified by their distinctive ranks of seniority. Let u denote a set of skills, and v a set of ranks. Let $R[u \cup v]$ represent the set of resources with skills u and rank v. The set R is made up of n categories representing n different types of employee management. For example, the resources employed by a particular company are either in the "full-time", "part-time", or "temporary" staff category.

There is also a given set of deployment posts P; each of which has one distinct requirement, and a range of possible ranks for accommodation. Let s denote a set of requirements, such that $P[s \cup v]$ represents the set of deployment posts with a requirement s, and an accommodation of ranks v. Some of these deployment posts can be further grouped by the descriptive nature of their job specifications. Let Q be a set of similarly grouped deployment posts, so that $Q \subseteq P$. It is noted here, that there also exists a set of optional deployment posts in P that do not require to be filled immediately.

The resources to be scheduled must match their skills against the requirements of the deployment posts, and their ranks must be within the range that can be accommodated by those posts. This *skill-rank* composition constraint is imposed so that we can make direct constraint matchings between the resources and the deployment posts. Also, further global constraint rules are necessary, and are described below.

- *Deployment Post Prioritisation Constraints.* Mandatory deployment posts must take precedence over optional ones.
- *Officer-In-Charge (OIC) Constraints.* There must be at least one OIC for any groups of similar posts in Q.
- *Categorical Constraints.* Each set of similar posts Q should have more than one category amongst the resources assigned

MA problems may be formulated as a grid network [6], where each node represents a slot in the roster. The deployment posts' references form the horizontal constraints, while the resources' references form the vertical constraints. This method relates closely to the standard Constraint Satisfaction Problem (CSP). Unfortunately, it has been proven that CSP is *NP*-complete, even though the underlying network is a grid.

The output is a feasible solution with all resources assigned to the various deployment posts for any given workday. An assignment is deemed *feasible* if it satisfies the *skill-rank* composition constraints without leaving out any resources. On top of that, the assignment is considered *practical* if it also satisfies the three global roster constraints defined in the previous section.

2.1 Knowledge Representations and Relations

We can develop sufficiently precise base-notations for representing knowledge through the use of *logical representation schemes*, which employ semantics to describe a world in terms of objects and their binary associations [13]. These association types relate objects (resources and posts) to their generic types (skills and ranks), and force a distinction between the objects and the generic types.

Inferences and deductions can be subsequently obtained through direct matching processes. Matching as a knowledge-based operation is used for classification, confirmation, decomposition, and correction. The utilization of a uniform representation of knowledge in this technique enables the construction of tools that can verify desired properties of the knowledge representation [9].

Each element in the sets $R[u \cup v]$ and $P[s \cup v]$ is represented as a 32-bit binary strain. In this paper, we assume the number of ranks and skills, related to the resources and the posts, to be fewer than 15. However, our method is not restricted to this assumption and can be applied to an arbitrary number of ranks and skills. The lowest 15 bits of this binary strain represent the various ranks a particular resource can have, while the highest 15 bits represent the skills. This exact configuration is used to represent the ranks that can be accommodated by the deployment posts, as well as the requirements needed to make up the job specifications. In addition to this pattern, one of the middle two bits is reserved to note a preferred flag for calculating a preferred probability value. This bit is set to follow the reference knowledge given by human experts, should the resource or the post referred to by the strain has an additional preference in the *categorical* requirements.

We can assume that u is directly related to s, since the skills of a resource can be matched up against the requirement of a deployment post. A relations-matrix W is referred to throughout this paper as a 2D matrix; with references to the deployment posts forming the rows, and references to the manpower resources forming the columns. The construction of W is detailed in the next section.

We start by creating feasible solutions with pre-assigned, randomly generated resources for a set of deployment. These are then shuffled and fed as input to a database to test the implementation of our prototype. These random data resources are controlled by variables that include the total number of resources in relation to each rank, the distribution of ranks, and the distribution of skills. The information for the deployment posts is also random. This ensures that the algorithm being tested is able to work on any set of data R, for any set of information P. We then set the final objective for the entire process, to minimize the number of un-assigned resources, while satisfying all local and global constraints.

2.2 Applying the Knowledge-Based MA Algorithm

Lau and Lua [6] gave a high level description of their rostering algorithm, on which they based their approach by using constraint set variables. Their algorithm has the following steps: **Step 1.** Perform feasibility checks; **Step 2.** Consider rostering constraints; **Step 3.** Generate a solution; **Step 4.** Apply local improvements.

Step 1 to 3 describes the Dynamic Programming (DP) algorithm, while *Step 4* is a backtracking rule-based mechanism that allows for additional improvements to be made to a feasible solution. We used the DP algorithm in the following context.

Firstly, we retrieve the information defined in R and P from the databases, and initialize their binary strains. Following which, we use the DP algorithm to perform the feasibility checks, solve the rostering constraints, and generate a feasible solution. This solution is checked to see if practical satisfaction is achieved *(see 2. Problem Formulation)*. Failing this, *Step 4* is executed to make local improvements by un-allocating resources that are perceived to have non-practical assignments. Lau and Lua termed this as the "relax & enrich" strategy of Schmidt in [6]. This is done through the use of additional constraints for an additional set of preferences to be met separately to avoid possible conflicts with the present ones. These constraints are part of the local improvement procedures. The process is then repeated until the number of unassigned resources reduces to zero.

3 Implementation and Methodologies

The DP algorithm above begins by building up the matrix W and shuffling the references to the resources and the posts for controlled fairness.

We used a knowledge-based technique to perform a local search for perfect matches between the row and column references of W. Each cell in W, referred to by a particular row and column, corresponds to a particular deployment post and resource respectively. A perfect match is found if the rank and skills possessed by a resource, matches that of the rank accommodations and the skill requirement of the deployment post as defined in Equation (1). The bit patterns defined in Figure 1 are used to contrast this matching process. A value of 1 is awarded to $W(i,j)$ should a perfect match be found, and 0 otherwise; here i and j are the row and column references to the relations matrix W.

$$W(i,j) = \begin{cases} 1, & \text{if } \{(u \cap s) \cup v\} > 0 \text{ where } s,u,v \in R,P \\ 0, & \text{otherwise} \end{cases} \tag{1}$$

The result of the perfect matches allows the building of a more global relation between the resources and the posts. The DP rule for this relation is defined by Equation (2), where i and j are the row and column references to the relations matrix W. The number in each cell of W describes the degree of relation between its rows and columns. The higher this number is, the better the relation.

$$W(i,j) = max\{max(\text{on row } i+1), max(\text{on column } j+1)\} + 1 \qquad (2)$$

Probability functions are used to overcome the situation when more than one solution becomes inevitable. Our probability function resolves the outcome of a decision making process through a set of pre-determined rules. It ensures optimality of the local results with respect to the global outcome. We split this decision making into two separate components contributing to the final probability: the *measure of suitability*, and the *preferred probability*. The measure of suitability refers physically as to whether or not a particular resource is suitable to handle a particular post. The preferred probability, on the other hand, tries to model the human perception of *preferred*-ness in allocating the best resource to a particular post.

To calculate the measure of suitability of a candidate for a deployment post, we can build a reference probability vector from the deployment post information through its nature of binary strains arrangement. A value of 0.45 is allotted at every occurrence of a positive value in the strain (*1* being positive, and *0* being negative). We then take the dot product of this reference vector with a candidate from the column element. The result of which will be the measure of suitability for the candidate to the deployment post.

In calculating the preferred probability, the value of the *preferred* bit in the strain is checked. If the *preferred* bit in the deployment post's strain information matches the category type of a particular candidate, a full 0.50 is awarded. Otherwise, a random number between 0.00 and 0.25 is given to the preferred probability value.

Starting at $W(1,1)$, the entire first row and column are searched for cells with the highest degree of relation. These identified cells are then used as guides in obtaining the best probability rate, before allocating a selected column j (the most suitable resource candidate from the identified range) to a selected row i, (deployment post) as a complete assignment. Following this, that entire row i and column j are crossed out from any further comparisons.

The next iteration starts on the immediate row and column neighbour, $W(i+1, j+1)$, and ends when either the row or column elements of W are exhausted. The whole process is then repeated; rebuilding a new relations matrix W, after removing the entries of all successfully allocated resources and posts.

All the simulations in the experiment ended with all candidate resources allocated a post. It was also noted that it took fewer than 20 iterations to generate a feasible solution for a deployment problem with 12 distribution of ranks and 340 manpower resources. However, the scope of this algorithm at this point is limited only to solving the local constraints of the MA problem. These feasible solutions have yet to go through *Step 4* of the MA algorithm for completeness.

4 Rule-Based Methodologies for Backtracking

A more global approach is imperative in coming up with a better overall solution than above. Sauer [12] classified the task of scheduling to be either *reactive* or *interactive*. In reactive scheduling, all events occurring in the scheduling environment have to be regarded, and then adapting the schedule to new situations with appropriate actions to handle each of the events. Interactive scheduling, on the other hand, means that several decisions about the constraints have to be taken by the expert human-scheduler. In the context of the MA problems, combining the two classes results in more effective and practical solutions. The knowledge-based techniques paired with the appropriate rule-based methodologies define the reactive and interactive representation to solving the local and global constraints respectively.

Gudes *et al.* [5] highlights the backtracking mechanism as a strategy to realize a new set of recommending rules. These rules suggest the de-allocation of one resource, and the allocation of another. The context here requires a different set of rule-lists to trigger the de-allocation procedure. In most backtracking scenarios, the completed roster is examined to identify assignments that are not "humanly" practical, that need to be taken off and reallocated elsewhere. For example, there can be too many officers-in-charge (OIC) for a group of similar deployment posts. The collective viewpoint to effect this de-assignment is to inspect the roster segregated into its associated groups, by the descriptive nature of the job specifications.

Zadeh's [14] conceptualization of fuzzy logic gives rule-based methodologies the ability to infer from the information resident in the knowledge-base. A membership function over a lattice of propositions can be formulated to encompass a wide range of logics of uncertainty. These logics make up the global rules for backtracking in our MA algorithm. These are in anticipation of the following *general rule of logic*:

If <*Number of resources*> AND <*Rank*> Then <*De-Allocate resources*>

Applying this general rule of logic requires the fuzzy logic concept of using *membership functions* with various associated degrees-of-freedom (DOF). A membership function is a graphical representation of the magnitude of participation of each input reference. It associates a weighting with each of the inputs that are to be processed, define functional overlaps between these inputs, and ultimately determine their influence on the fuzzy output sets of the final output conclusion. Once the functions are inferred, scaled, and combined, they are defuzzified into rationalized output values that determine the decision-making.

A triangular shaped membership function is the most common one used in many control systems, mainly because of its mathematical simplicity. Plugging in prescribed input parameters from the horizontal axis, and projecting vertically gives the rationalized value of its associated DOF. By computing the logical product of the membership weights for each active rule, a set of fuzzy output response magnitudes are produced. All that remains is to combine and defuzzify these output responses, as an interpretation to our implementation of the backtracking mechanism.

Going by human semantics, and preparing for the *general rule of logic*, we can tabulate the input linguistic rules to give the appropriate output action rule. These output rules are suggested by the experts, and are shown in Table 1. For example, if there are *several* resources with *high ranks*, then we need to *de-allocate several* of the already assigned resources.

Table 1. Rule-base relations between the input linguistic parameters and the output actions.

Resources	Low Rank	High Rank
Too Few	None	Few
Several	None	Several
Too Many	Few	A lot

4.1 Fuzzification and De-fuzzification

The Fuzzification procedure converts the input parameters to a related fuzzy value with the associated DOF. We then used Mamdani's Maximum-Minimum Centre of Area method to calculate the output equivalence with respect to the rule-base given in Table 1 [8]. This de-fuzzification results in the actual number of resources to be removed with an attached DOF.

Our global backtracking rules look through each individual group of the segregated deployment posts with their assignments. The backtracking procedure parses information to pick out the number of resources and their associated ranks. The rule-base is then inferred, through fuzzy logic techniques described in Section 4, to determine the number of resources to de-allocate, and to await future assignments. This algorithm will continue to allocate and de-allocate resources to and from their posts in a generalized swapping manner, due to its absolute random nature of selection. The iteration will stop only when all resources are completely assigned.

The combined knowledge- and rule-based methodologies were tested with numerous sample data that are typical in representing the real-world situation. Randomised data for resources and posts were also generated to test the firmness of the algorithm.

Figure 1 depicts a typical result at each iteration step of the MA algorithm. We observe the distinctive sections in the figure where the knowledge-base techniques successfully reduced the number of un-allocated resources. At stipulated intervals, the rule-base methodologies kicked-in to de-allocate the resources that were deemed impractical for the roster. In fewer than 30 iterations, the solution generated by the MA algorithm is converted from feasible to a practical one.

5 Implementation and Experimental Results

We tested the robustness of the combined MA algorithm with the new addition of the dynamic rules through various numbers of randomly generated resources and posts. Figure 2 illustrates a histogram of results obtained from simulating 209 resources and posts. The histogram depicts the resultant data obtained by running 1023 different samples.

The histogram also indicates that a large proportion of the samples took less than 60 iterations before arriving at feasible and practical solutions. Depending on the tightness-level of correlation between the random resources and the random posts, the behaviour of the algorithm displays variations in its number of iterations when generating a roster. MA-type problems are *NP*-complete in nature [6], and there exists multiple solutions for

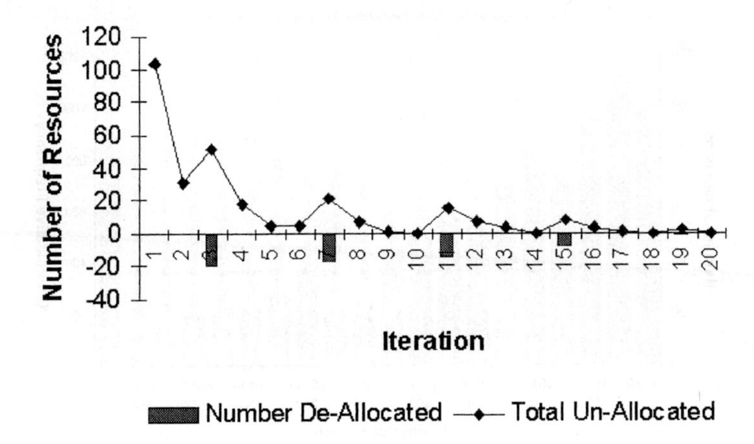

Fig. 1. Behaviour of the MA algorithm on the total un-allocated resources against the number of resources removed at each iteration. Data: Number of Posts = 151, Number of Resources = 103, Number of Ranks = 8.

any sample space that are optimal. Hence we can assume that the first optimal solution achieved is the accepted solution to be analyzed.

Mutalik *et al.* [10] and Ghoshray and Yen [3] used mathematical techniques in their approaches to solving constraints of optimisation problems. On the other hand, our knowledge- and rule-base improvement techniques use processor time for picking out identical strains, calculating probability functions (this is done only when more than one solution becomes inevitable), and resolving inference rules. All of the strains were preloaded into memory at initialization, thus making the comparison process simple but efficient.While the mathematical techniques in commercial scheduling softwares [13, 11] deal with non-integers to calculate the constraint function definitions, the error acquired at each function grows if the results obtained from one expression are reused again for another.Contrary to this, our knowledge-based techniques compare simple binary strains, and only deal with a one-off probability function for their decision-making. Clearly, the error rate here is not an issue.

References

1. R. J. Brachman. The basics of knowledge representation and reasoning. *AT&T Technical Journal*, 67(1):15, 1988.
2. Soumitra Dutta. Approximate spatial reasoning. In *Proceedings of the First International Conference on Industrial and Engineering Applications of Artificial Intelligence and Expert Systems*, pages 126–140. ACM Press, 1988.
3. S. Ghoshray and K. K. Yen. More efficient genetic algorithm for solving optimization problems. In *Systems, Man and Cybernetics. IEEE International Conference on Intelligent Systems for the 21st Century*, volume 5, pages 4515–4520. IEEE, Oct 1995.
4. Yu Gu. Genetic algorithm approach to aircraft gate reassignment problem. *Journal of Transportation Engineering*, 125(5):384–389, Sep 1999.

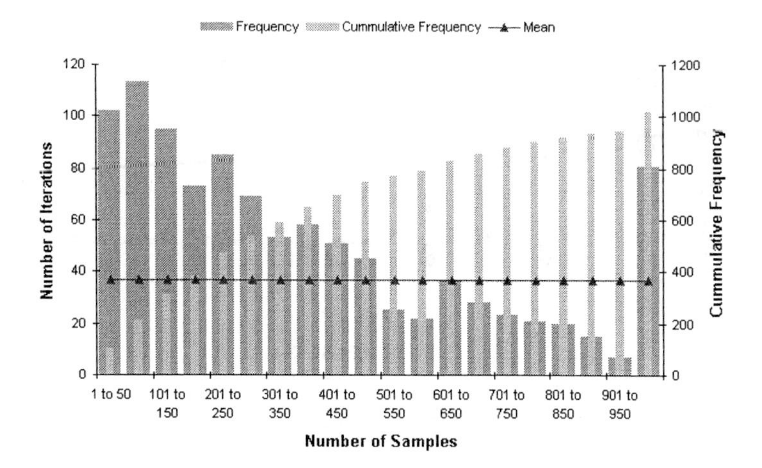

Fig. 2. Data obtained from simulating 1023 samples for 209 resources, 209 posts; 8 ranks, and 10 skills. The average number of iterations calculated is 367.152, with a standard deviation of 293.351. At $\alpha = 0.05$, the confidence interval is 17.977.

5. Ehud Gudes, Tsvi Kuflik, and Amnon Meisels. An expert systems based methodology for solving resource allocation problems. In *Proceedings of the Third International Conference on Industrial and Engineering Applications of Artificial Intelligence and Expert Systems*, pages 309–317. ACM Press, 1990.
6. H. C. Lau and S. C. Lua. Efficient multi-skill crew rostering via constrained sets. In *Proceedings of the Second ILOG Solver and Scheduler Users Conference*, July 1997.
7. David G. Luenberger. *Linear and Nonlinear Programming*. Addison Wesley Publishing Company, 2nd edition, 1989.
8. E. H. Mamdani. Application of fuzzy logic to approximate reasoning using linguistic synthesis. In *Proceedings of the Sixth International Symposium on Multiple-valued Logic*, pages 196–202, 1976.
9. L. J. Morell. Use of metaknowledge in the verification of knowledge-based systems. In *Proceedings of the First International Conference on Industrial and Engineering Applications of Artificial Intelligence and Expert Systems*, pages 847–857. ACM Press, 1988.
10. Pooja P. Mutalik, Leslie R. Knight, Joe L. Blanton, and Roger L. Wainwright. Solving combinatorial optimization problems using parallel simulated annealing and parallel genetic algorithms. In *Proceedings of the 1992 ACM/SIGAPP Symposium on Applied Computing*, pages 1031–1038. ACM Press, 1992.
11. Elmira Popova and David Morton. Adaptive stochastic manpower scheduling. In *Proceedings of the 30th Conference on Winter Simulation*, pages 661–668. IEEE Computer Society Press, 1998.
12. Jürgen Sauer. Knowledge-based scheduling techniques in industry. *Intelligent Techniques in Industry*, 1998.
13. Michel Turcotte, Gillian M. Mann, and Aaron L. Nsakanda. Impact of connection bank redesign on airport gate assignment. In *Proceedings of the 31st Conference on Winter Simulation*, pages 1378–1382. ACM Press, 1999.
14. Lotfi A. Zadeh. Coping with the imprecision of the real world. *Communications of the ACM*, 27(4):304–311, 1984.

A Symbolic Approach to Vagueness Management

Mazen El-Sayed and Daniel Pacholczyk

University of Angers, 2 Boulevard Lavoisier, 49045 Angers Cedex 01, France
{elsayed,pacho}@univ-angers.fr

Abstract. We study knowledge-based systems using symbolic many-valued logic and multiset theory. In previous papers we have proposed a symbolic representation of nuanced statements like "John is very tall". In this representation, we have interpreted some nuances of natural language as linguistic modifiers and we have defined them within a multiset context. In this paper, we continue the presentation of our symbolic model and we propose new deduction rules dealing with nuanced statements. We limit ourselves to present new generalizations of the Modus Ponens rule.

Keywords: knowledge representation and reasoning, imprecision, vagueness, many-valued logic, multiset theory.

1 Introduction

The development of knowledge-based systems is a rapidly expanding field in applied artificial intelligence. The knowledge base is comprised of a database and a rule base. We suppose that the database contains facts representing *nuanced statements*, like "Jo is very tall", to which one associates truth degrees. The *nuanced statements* can be represented more formally under the form "x is m_α A" where m_α and A are labels denoting respectively a nuance and a vague or imprecise term of natural language. The rule base contains rules of the form *"if x is m_α A then y is m_β B"* to which one associates truth degrees.

Our work presents a symbolic-based model which permits a qualitative management of vagueness in knowledge-based systems. In dealing with vagueness, there are two issues of importance: (1) how to represent vague data, and (2) how to draw inference using vague data. When imprecise information is evaluated in a *numerical way*, fuzzy logic which is introduced by Zadeh [11,12], is recognized as a good tool for dealing with aforementioned issues and performing reasoning upon common sense and vague knowledge-bases. In this logic, "x is m_α A" is considered as a fuzzy proposition where A is modeled by a fuzzy set which is defined by a membership function. This one is generally defined upon a numerical scale. The nuance m_α is defined such as a *fuzzy modifier* [3,10,12] which represents, from the fuzzy set A, a new fuzzy set "m_α A". So, "x is m_α A" is interpreted by Zadeh as "x is (m_α A)" and is regarded as many-valued statement. A second formalism, refers to a symbolic many-valued logic [4,10], is used when imprecise information is evaluated in a *symbolic way*. This logic is the logical counterpart

P.M.A. Sloot et al. (Eds.): ICCS 2003, LNCS 2659, pp. 109–119, 2003.

of a *multiset theory* introduced by De Glas [4]. In this theory, the term m_α linguistically expresses the degree to which the object x satisfies the term A. So, "x is m_α A" means "x (is m_α) A", and then is regarded as boolean statement. In other words, "m_α A" does not represent a new vague term obtained from A. In previous papers [7,8], we have proposed a symbolic-based model to represent nuanced statements. This model is based on the many-valued logic proposed by Pacholczyk [10]. Our basic idea has been to consider that some nuances of natural language can not be interpreted as satisfaction degrees and must be instead defined such as *linguistic modifiers.* Firstly, we have proposed a new method to symbolically represent vague terms of natural language. The basic idea has been to associate with each vague term a new *symbolic* concept called "*rule*". This symbolic concept is an equivalent to the membership function within a fuzzy context. By using the new concept, we have defined *linguistic modifiers* within a multiset context. In this paper, our basic contribution has been to propose deduction rules dealing with nuanced information. For that purpose, we propose deduction rules generalizing the *Modus Ponens* rule in a many-valued logic proposed by Pacholczyk [10]. We notice that the first version of this rule has been proposed in a fuzzy context by Zadeh [12] and has been studied later by various authors [1,3,9]. This paper is organized as follows. In Sect. 2, we present briefly the basic concepts of the M-valued predicate logic which forms the backbone of our work. Section 3 introduces briefly the symbolic representation model previously proposed. In Sect. 4, we study various types of inference rules and we propose new *Generalized Modus Ponens* rules in which we use only simple statements. In Sect. 5, we propose a generalized production system in which we define more Generalized Modus Ponens rules in more complex situations.

2 M-Valued Predicate Logic

Within a multiset context, to a vague term A and a nuance m_α are associated respectively a multiset $\mathbb{A}$ and a symbolic degree τ_α. So, the statement "x is m_α A" means that x belongs to multiset $\mathbb{A}$ with a degree τ_α. The M-valued predicate logic [10] is the logical counterpart of the multiset theory. In this logic, to each multiset $\mathbb{A}$ and a membership degree τ_α are associated a M-valued predicate **A** and a truth degree τ_α−true. In this context, the following equivalence holds: x is m_α A $\Leftrightarrow x \in_\alpha \mathbb{A} \Leftrightarrow$ "x is m_α **A**" is true $\Leftrightarrow$ "x is **A**" is τ_α−true. One supposes that the membership degrees are symbolic degrees which form an ordered set $\mathcal{L}_M = \{\tau_\alpha, \alpha \in [1, M]\}$. This set is provided with the relation of a total order: $\tau_\alpha \leq \tau_\beta \Leftrightarrow \alpha \leq \beta$, and whose smallest element is τ_1 and the largest element is τ_M. We can then define in $\mathcal{L}_M$ two operators $\wedge$ and $\vee$ and a decreasing involution $\sim$ as follows: $\tau_\alpha \vee \tau_\beta = \tau_{max(\alpha,\beta)}, \tau_\alpha \wedge \tau_\beta = \tau_{min(\alpha,\beta)}$ and $\sim \tau_\alpha = \tau_{M+1-\alpha}$. One obtains then a chain $\{\mathcal{L}_M, \vee, \wedge, \leq\}$ having the structure of De Morgan lattice [10]. On this set, an implication $\rightarrow$ and a T-norm T can be defined respectively as follows: $\tau_\alpha \rightarrow \tau_\beta = \tau_{min(\beta-\alpha+M,M)}$ and $T(\tau_\alpha, \tau_\beta) = \tau_{max(\beta+\alpha-M,1)}$.

Example 1. For example, by choosing M=9, we can introduce: $\mathcal{L}_9$={*not at all, little, enough, fairly, moderately, quite, almost, nearly, completely*}.

In the following of this paper we focus our intention on the management of statements which are nuanced by linguistic modifiers. So, we consider that m_α A represents a multiset derived from A, and "x is m_α A" is a many-valued statement.

3 Representation of Nuanced Statements

Let us suppose that our knowledge base is characterized by a finite number of concepts $\mathcal{C}_i$. A set of terms P_{ik}[1] is associated with each concept $\mathcal{C}_i$, whose respective domain is denoted as X_i. The terms P_{ik} are said to be the *basic terms* connected with the concept $\mathcal{C}_i$. As an example, basic terms such as *"small"*, *"moderate" and "tall"* are associated with the particular concept "*size of men*". A finite set of *linguistic modifiers* m_α allows us to define *nuanced terms*, denoted as "$m_\alpha P_{ik}$". In previous papers [7,8], we have proposed a symbolic-based model to represent nuanced statements of natural language. In the following, we present a short review of this model. We have proposed firstly a new method to symbolically represent vague terms. In this method, we suppose that a domain of a vague term, denoted by X, is not necessarily a numerical scale. This domain is simulated by a *"rule"* (cf. Fig. 1) representing an arbitrary set of objects. Our basic idea has been to associate with each multiset P_i a symbolic concept which represents an equivalent to the membership function in fuzzy set theory. For that, we have introduced a new concept, called "rule", which has a geometry similar to a membership L-R function and its role is to illustrate the membership graduality to the multisets. In order to define the geometry of this "rule", we use notions similar to those defined within a fuzzy context like the core, the support and the fuzzy part of a fuzzy set [12]. We define these notions within a multiset theory as follows: the core of a multiset P_i, denoted as $Core(P_i)$, represents the elements belonging to P_i with a τ_M degree, the support, denoted as $Sp(P_i)$, contains the elements belonging to P_i with at least τ_2 degree, and the fuzzy part, denoted as $F(P_i)$, contains the elements belonging to P_i with degrees varying from τ_2 to τ_{M-1}. We associate with each multiset a "rule" that contains the elements of its support (cf. Fig. 3). This "rule" is the union of three disjoint

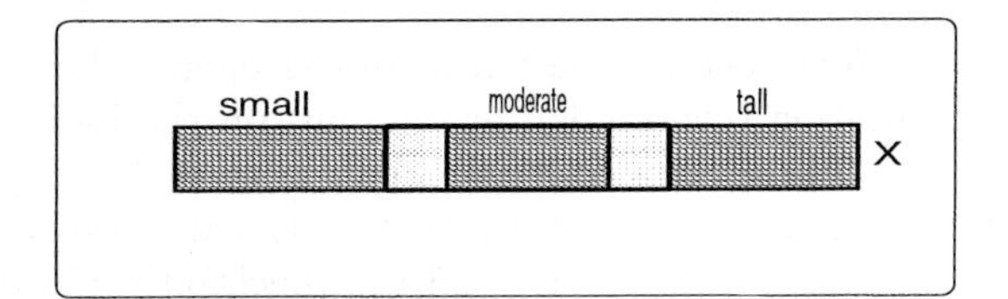

Fig. 1. Representation with *"rule"* of a domain X

[1] In the following, we use the same notation P_{ik} to represent either a vague term P_{ik}, the multiset $\mathbb{P}_{ik}$ and the predicate $\mathbf{P}_{ik}$ associated with it.

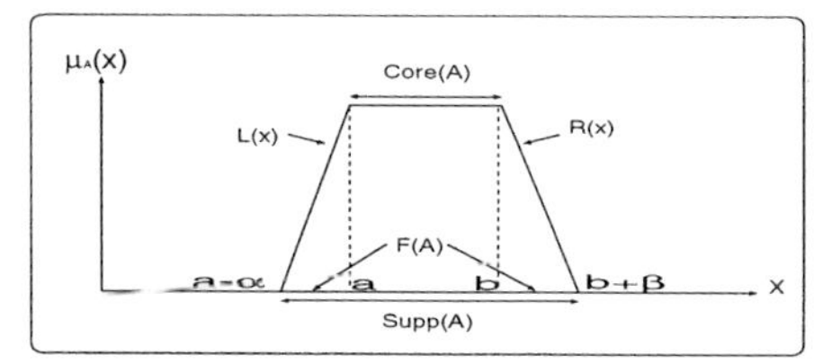

Fig. 2. A membership L-R function

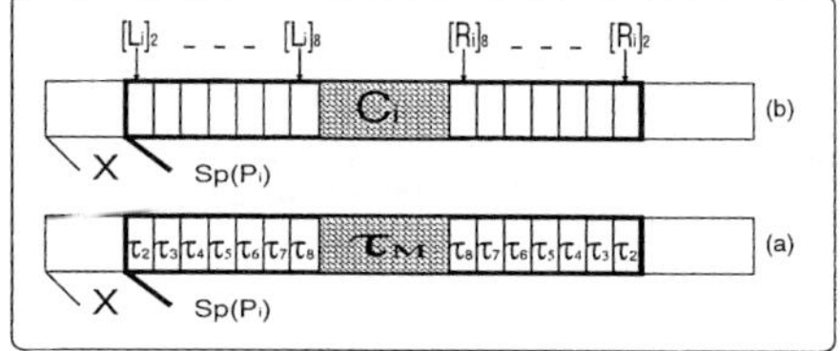

Fig. 3. Representation with *"rule"*

subsets: *the left fuzzy part*, *the right fuzzy part* and *the core*. For a multiset P_i, they are denoted respectively by L_i, R_i and C_i. We suppose that the left (resp. right) fuzzy part L_i (resp. R_i) is the union of M-2 subsets, denoted as $[L_i]_\alpha$ (resp. $[R_i]_\alpha$), which partition it. $[L_i]_\alpha$ (resp. $[R_i]_\alpha$) contains the elements of L_i (resp. R_i) belonging to P_i with a τ_α degree. In order to keep a similarity with the fuzzy sets of type L-R, we choose to place, in a "rule" associated with a multiset, the subsets $[L_i]_\alpha$ and $[R_i]_\alpha$ so that the larger α is, the closer the $[L_i]_\alpha$ subsets and $[R_i]_\alpha$ are to the core C_i (cf. Fig. 3). That can be interpreted as follows: the elements of the core of a term represent the typical elements of this term, and the more one object moves away from the core, the less it satisfies the term. Finally, we have denoted a multiset P_i with which we associate a "rule" as $P_i = (L_i, C_i, R_i)$, and we have introduced symbolic parameters which enable us to describe the form of the "rule" and its position in the universe X. These parameters have a role similar to the role of numerical parameters which are used to define a fuzzy set within a fuzzy context.

3.1 Linguistic Modifiers

By using the "rule" concept we have defined the linguistic modifiers. We have used two types of linguistic modifiers.

- *Precision Modifiers*: The precision modifiers increase or decrease the precision of the basic term. We distinguish two types of precision modifiers: contraction modifiers and dilation modifiers. We use $\mathbb{M}_6 = \{m_k | k \in [1..6]\}$ ={*exactly, really, ∅, more or less, approximately, vaguely*} which is totally ordered by $j \leq k \Leftrightarrow m_j \leq m_k$ (Fig. 4).
- *Translation Modifiers*: The translation modifiers operate both a translation and precision variation (contraction or dilation) on the basic term. We use $\mathbb{T}_9 = \{t_k | k \in [1..9]\}$ ={*extremely little, very very little, very little, rather little, ∅, rather, very, very very, extremely*} totally ordered by $k \leq l \Leftrightarrow t_k \leq t_l$ (Fig. 5). The translation amplitudes and the precision variation amplitudes are calculated in such a way that the multisets $t_k P_i$ cover the domain X.

In this paper, we continue to propose our model for managing nuanced statements. In the following, we focus our intention to study the problem of exploitation of nuanced statements.

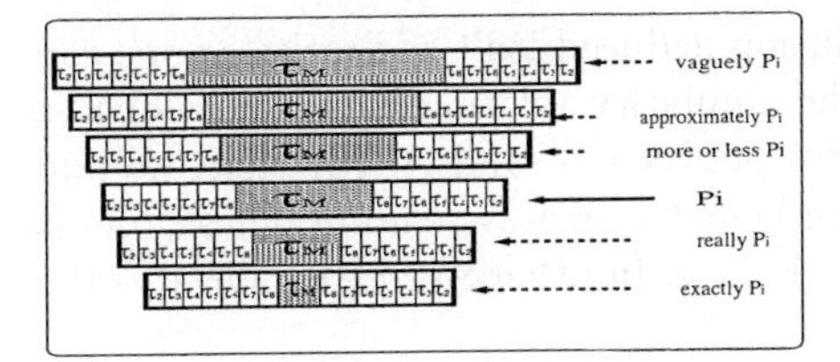

Fig. 4. Precision modifiers

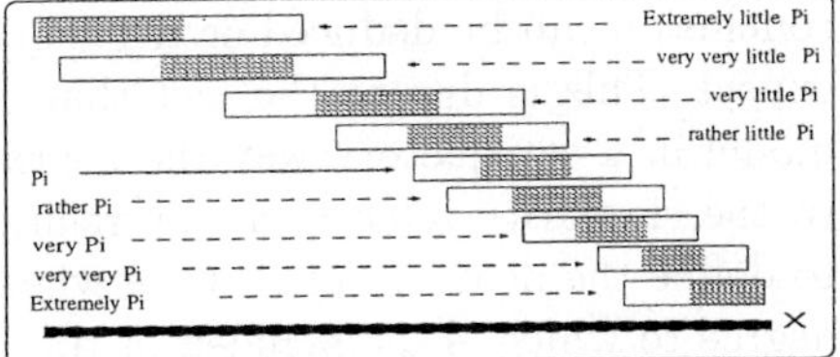

Fig. 5. Translation modifiers

4 Exploitation of Nuanced Statements

In this section, we treat the exploitation of nuanced information. In particular, we are interested to propose some generalizations of the Modus Ponens rule within a many-valued context [10]. We notice that the classical Modus Ponens rule has the following form: If we know that {*If "x is A" then "y is B" is true* and *"x is A" is true*} we conclude that "y is B" is true. Within a many-valued context, a generalization of Modus Ponens rule has one of the following forms:

F1- If we know that {*If "x is A" then "y is B" is* τ_β*-true* and *"x is* A'*" is* τ_ϵ*-true*} and that {A' is more or less near to A}, what can we conclude for "y is B", in other words, to what degree "y is B" is true?

F2- If we know that {*If "x is A" then "y is B" is* τ_β*-true* and *"x is* A'*" is* τ_ϵ*-true*} and that {A' is more or less near to A}, can we find a B' such as {B' is more or less near to B} and to what degree "y is B'" is true?

These forms of *Generalized Modus Ponens* (GMP) rule have been studied firstly by Pacholczyk in [10]. In this section, we propose new versions of GMP rule in which we use new relations of nearness.

4.1 First GMP Rule

In Pacholczyk's versions of GMP, the concept of nearness binding multisets A and A' is modeled by a similarity relation which is defined as follows:

Definition 1. *Let A and B be two multisets. A is said to be* τ_α*-similar to B, denoted as* $A \approx_\alpha B$*, if and only if:* $\forall x | x \in_\gamma A$ *and* $x \in_\beta B \Rightarrow min\{\tau_\gamma \to \tau_\beta, \tau_\beta \to \tau_\gamma\} \geq \tau_\alpha$.

This relation generalizes the equivalence relation in a many-valued context as the similarity relation of Zadeh [12] has been in a fuzzy context. It is (1) reflexive: $A \approx_M A$, (2) symmetrical: $A \approx_\alpha B \Leftrightarrow B \approx_\alpha A$, and (3) weakly transitive: $\{A \approx_\alpha B, B \approx_\beta C\} \Rightarrow A \approx_\gamma C$ with $\tau_\gamma \geq T(\tau_\alpha, \tau_\beta)$ where T is a T-norm.
By using the similarity relation to model the nearness binding between multisets, the inference rule can be interpreted as: {*more the rule and the fact are true*} and {*more* A' *and A are similar*}, *more the conclusion is true.* In particular, when A' is more precise than A ($A' \subset A$) but they are very weakly similar, any

conclusion can be deduced or the conclusion deduced isn't as precise as one can expect. This is due to the fact that the similarity relation isn't able alone to model in a satisfactory way the nearness between $A^{'}$ and A. For that, we add to the similarity relation a new relation called *nearness relation* whose role is to define the nearness of $A^{'}$ to A when $A^{'} \subset A$. In other words, it indicates the degree to which $A^{'}$ is included in A.

Definition 2. *Let A and B be two multisets such that $A \subset B$. A is said to be τ_α-near to B, denoted as $A \sqsubset_\alpha B$, if and only if $\{\forall x \in F(B),\ x \in_\beta A$ and $x \in_\gamma B \Rightarrow \tau_\alpha \rightarrow \tau_\beta \leq \tau_\gamma\}$.*

The nearness relation satisfies the following properties: (1) Reflexivity: $A \sqsubset_M A$, and (2) Weak transitivity: $A \sqsubset_\alpha B$ and $B \sqsubset_\beta C \Rightarrow A \sqsubset_\gamma C$ with $\tau_\gamma \leq min(\tau_\alpha, \tau_\beta)$. In the relation $A \sqsubset_\alpha B$, the less the value of α is, the more A is included in B. We can notice that the properties satisfied by the nearness relation are similar to those satisfied by the resemblance relation proposed by Bouchon-Meunier and Valverde [2] within a fuzzy context. Finally, by using similarity and nearness relations, we propose a first *Generalized Modus Ponens* rule.

Proposition 1. *Let A and $A^{'}$ be predicates associated with the concept C_i, B be predicate associated with the concept C_e. Given the following assumptions:*

1. *it is τ_β-true that if "x is A" then "y is B"*
2. *"x is $A^{'}$" is τ_ϵ-true with $A^{'} \approx_\alpha A$.*

Then, we conclude : "y is B" is τ_δ-true with $\tau_\delta = T(\tau_\beta, T(\tau_\alpha, \tau_\epsilon))$.
If the predicate $A^{'}$ is such that $A^{'} \sqsubset_{\alpha'} A$, we conclude: "y is B" is τ_δ-true with $\tau_\delta = T(\tau_\beta, \tau_{\alpha'} \longrightarrow \tau_\epsilon)$.

Example 2. Given that "really tall" $\approx_8$ "tall" and "really tall" $\sqsubset_8$ "tall", from the following rule and fact:
- if "x is tall" then "its weight is important" is true[2]
- "Pascal is really tall" is quite-true,
we can deduce: "Pascal's weight is really important" is almost-true.

4.2 GMP Rules Using Precision Modifiers

In the previous paragraph we calculate the degree to which the conclusion of the rule is true. In the following, we present two new versions of GMP rule in which the predicate of the conclusion obtained by the deduction process is not B but a new predicate $B^{'}$ which is more or less near to B. More precisely, the new predicate is derived from B by using precision modifiers[3] ($B^{'} = mB$). The first version assumes that the predicates A and $A^{'}$ are more or less similar. In other words, $A^{'}$ may be less precise or more precise than A. The second one assumes that $A^{'}$ is more precise than A.

[2] In our many-valued logic, "completely true" is equivalent to "true" in classical logic.
[3] The definitions of these are presented in appendix A.

Proposition 2. *Let A and $A^{'}$ be predicates associated with the concept C_i, B be predicate associated with the concept C_e. Let the following assumptions:*

1. *it is τ_β-true that if "x is A" then "y is B"*
2. *"x is $A^{'}$" is τ_ϵ-true with $A^{'} \approx_\alpha A$.*

Let $\tau_\theta = T(\tau_\beta, T(\tau_\alpha, \tau_\epsilon))$. If $\tau_\theta > \tau_1$ then there exists a $\tau_{n(\delta)}-$dilation modifier m, with $\tau_\delta \leq T(\tau_\alpha, \tau_\beta)$, such that: "$y$ is mB" is $\tau_{\epsilon'}$-true and $\tau_{\epsilon'} = \tau_\delta \longrightarrow \tau_\theta$. Moreover, we have: $B \subset mB$ and $mB \approx_\delta B$.

This proposition prove that if we know that $A^{'}$ is more or less similar to A, without any supplementary information concerning its precision compared to A, the predicate of the conclusion obtained by the deduction process (mB) is less precise than B (i.e. $B \subset mB$) and which is more or less similar to B. In the following proposition, we assume that $A^{'}$ is more precise than A.

Proposition 3. *Let A and $A^{'}$ be predicates associated with the concept C_i, B be predicate associated with the concept C_e. Let the following assumptions:*

1. *it is τ_β-true that if "x is A" then "y is B"*
2. *"x is $A^{'}$" is τ_ϵ-true with $A^{'} \sqsubset_\alpha A$.*

Let $\tau_\theta = T(\tau_\beta, \tau_\alpha \longrightarrow \tau_\epsilon)$. If $\tau_\theta > \tau_1$ then there exists a $\tau_{n(\delta)}-$contraction modifier m, with $\tau_\delta \geq \tau_\beta \longrightarrow \tau_\alpha$, such that: "$y$ is mB" is $\tau_{\epsilon'}$-true and $\tau_{\epsilon'} = T(\tau_\delta, \tau_\theta)$.
Moreover, we have: $mB \sqsubset_\delta B$.

This proposition prove that from a predicate $A^{'}$ which is more or less near to A we obtain a predicate mB which is more or less near to B. More precisely, if $A^{'}$ is more precise than A then mB is more precise than B. The previous propositions (2 and 3) present two general cases in which we consider arbitrary predicates A'. In the following, we present two corollaries representing special cases of propositions 2 and 3 in which we assume that the rule is completely true and that A' is obtained from A by using precision modifiers.

Corollary 1. *Let the following rule and fact:*

1. *it is true that if "x is A" then "y is B"*
2. *"x is m_kA" is τ_ϵ-true where m_k is a $\tau_{\gamma_k}-$dilation modifier.*

If $T(\sim \tau_{\gamma_k}, \tau_\epsilon) > \tau_1$ then we conclude:

$$\text{“}y \text{ is } m_kB\text{” is } \tau_{\epsilon'}\text{-true, with } \tau_{\epsilon'} = \sim \tau_{\gamma_k} \longrightarrow T(\sim \tau_{\gamma_k}, \tau_\epsilon).$$

Example 3. Given the following data:
- if "x is tall" then "its weight is important" is true,
- "Jo is more or less tall" is moderately-true.
Then, we can deduce: "Jo's weight is more or less important" is moderately-true.

Corollary 2. *Let the following rule and fact:*

1. *it is true that if "x is A" then "y is B"*
2. *"x is $m_k A$" is τ_ϵ-true where m_k is a τ_{γ_k}−contraction modifier.*

Then, we conclude that: "y is $m_k B$" is τ_ϵ-true.

Example 4. Given the following data:
- if "x is tall" then "its weight is important" is true,
- "Pascal is really tall" is moderately-true.

Then, we can deduce: "Pascal's weight is really important" is moderately-true.

These two corollaries present a particular form of graduality of inference. This form is known as graduality by means of linguistic modifiers [5]. It enables us to obtain, from a fact whose predicate A' is nuanced by linguistic modifiers, a conclusion whose predicate is also nuanced by linguistic modifiers.

5 Generalized Production System

In this section, we present some generalizations of Modus Ponens rule in more complex situations. More precisely, we study the problem of reasoning in 4 situations:

1. When the antecedent of the rule is a conjunction of statements.
2. When the antecedent is a disjunction of statements.
3. In presence of propagation of inferences. In other words, when the conclusion of the first rule is the antecedent of the second rule, and so on.
4. When a combination of imprecision is possible. In other words, when we have some rules which have the same statement in their conclusion parts.

So, we present the following 4 propositions representing inference rules in these situations.

Proposition 4 (Antecedent is a conjunction). *Given the following assumptions:*

1. *if "x_1 is A_1" and ... and "x_n is A_n" then "y is B" is τ_β-true,*
2. *for $i = 1..n$, "x_i is $A_i^{'}$" is τ_{ϵ_i}-true,*
3. *for $i = 1..n$, $A_i \approx_{\alpha_i} A_i^{'}$.*

Then, we can deduce: "y is B" is τ_δ-true with $\tau_\delta = T(\tau_\beta, T(\tau_{\alpha_1}, \tau_{\epsilon_1})) \wedge ... \wedge T(\tau_\beta, T(\tau_{\alpha_n}, \tau_{\epsilon_n}))$.
If, for $i = j$ .. k, the predicates $A_i^{'}$ are such that $A_i^{'} \sqsubset_{\alpha_i^{'}} A_i$, we can deduce: "y is B" is τ_δ-true with $\tau_\delta = \tau_{\delta_1} \wedge ... \wedge \tau_{\delta_n}$ and $\tau_{\delta_i} = T(\tau_{\alpha_i^{'}} \longrightarrow \tau_{\epsilon_i}, \tau_\beta)$ if $i \in [j, k]$ and $\tau_{\delta_i} = T(\tau_\beta, T(\tau_{\alpha_i}, \tau_{\epsilon_i}))$ if not.

Proposition 5 (Antecedent is a disjunction). *Given the following assumptions:*

1. *if "x_1 is A_1" or ... or "x_n is A_n" then "y is B" is τ_β-true,*
2. *for $i = 1..k$, "x_i is $A_i^{'}$" is τ_{ϵ_i}-true,*
3. *for $i = 1..k$, $A_i \approx_{\alpha_i} A_i^{'}$.*

Then, we can deduce: "y is B" is τ_δ-true with $\tau_\delta = T(\tau_\beta, T(\tau_{\alpha_1}, \tau_{\epsilon_1})) \vee ... \vee T(\tau_\beta, T(\tau_{\alpha_k}, \tau_{\epsilon_k}))$.
If, for $i = j$ .. L, the predicates $A_i^{'}$ are such that $A_i^{'} \sqsubset_{\alpha_i^{'}} A_i$, we can deduce: "$y$ is B" is τ_δ-true with $\tau_\delta = \tau_{\delta_1} \vee ... \vee \tau_{\delta_k}$ and $\tau_{\delta_i} = T(\tau_{\alpha_i^{'}} \longrightarrow \tau_{\epsilon_i}, \tau_\beta)$ if $i \in [j, L]$ and $\tau_{\delta_i} = T(\tau_\beta, T(\tau_{\alpha_i}, \tau_{\epsilon_i}))$ if not.

Proposition 6 (Propagation of inferences). *Given the following assumptions:*

1. *if "x is A" then "y is B" is τ_β-true,*
2. *if "y is B" then "z is C" is τ_γ-true,*
3. *there exists $\tau_\epsilon > \tau_1$ such that "x is $A^{'}$" is τ_ϵ-true,*
4. *there exists τ_α such that $A \approx_\alpha A^{'}$.*

Then, we can deduce: "z is C" is τ_δ-true, with $\tau_\delta = T(T(\tau_\beta, \tau_\gamma), T(\tau_\alpha, \tau_\epsilon))$.
If the predicate $A^{'}$ is such that $A^{'} \sqsubset_{\alpha'} A$, then we can deduce: "$z$ is C" is τ_δ-true, with $\tau_\delta = T(T(\tau_\beta, \tau_\gamma), \tau_{\alpha'} \longrightarrow \tau_\epsilon)$.

Proposition 7 (Combination of imprecisions). *Given the following assumptions:*

1. *for $i = 1..n$, if "x_i is A_i" then "y is B" is τ_{β_i}-true,*
2. *for $i = 1..n$, "x_i is $A_i^{'}$" is τ_{ϵ_i}-true,*
3. *for $i = 1..n$, $A_i \approx_{\alpha_i} A_i^{'}$,*

then we can deduce that: "y is B" is τ_δ-true with $\tau_\delta = T(\tau_{\beta_1}, T(\tau_{\alpha_1}, \tau_{\epsilon_1})) \vee ... \vee T(\tau_{\beta_n}, T(\tau_{\alpha_n}, \tau_{\epsilon_n}))$.
If, for $i = j$ .. k, the predicates $A_i^{'}$ are such that $A_i^{'} \sqsubset_{\alpha_i^{'}} A_i$, then we can deduce: "$y$ is B" is τ_δ-true with $\tau_\delta = \tau_{\delta_1} \vee ... \vee \tau_{\delta_n}$ and $\tau_{\delta_i} = T(\tau_{\alpha_i^{'}} \longrightarrow \tau_{\epsilon_i}, \tau_{\beta_i})$ if $i \in [j, k]$ and $\tau_{\delta_i} = T(\tau_{\beta_i}, T(\tau_{\alpha_i}, \tau_{\epsilon_i}))$ if not.

We present below an example in which we use the GMP rules presented in this section. In this example, we use index cards written by a doctor after his consultations. From index cards (IC_i) and some rules ($\mathcal{R}_j$), we wish deduce a diagnosis.

Example 5. Let assume that we have the following rules in our base of rules.

$\mathcal{R}_1-$ "If the temperature is high, the patient is ill" is almost true,

$\mathcal{R}_2-$ "If the tension is always high, the patient is ill" is nearly true,
$\mathcal{R}_3-$ "If the temperature is high and the eardrum color is very red, the disease is an otitis" is true,
$\mathcal{R}_4-$ "If fat eating is high, the cholesterol risk is high" is true,
$\mathcal{R}_5-$ "If the cholesterol risk is high, a diet with no fat is recommended" is true.

Let us assume now that we have an index card for a patient and we want to deduce a diagnosis.

$\mathcal{F}_1-$ "the temperature is rather high" is nearly true,
$\mathcal{F}_2-$ "the tension is always more or less high" is almost true,
$\mathcal{F}_3-$ "the eardrum color is really very red" is quite true,
$\mathcal{F}_4-$ "the fat eating is very very high" is moderately true.

Using the GMP rules previously presented, we deduce the following diagnosis:

$\mathcal{D}_1-$ "the patient is ill" is almost true,
$\mathcal{D}_2-$ "the disease is an otitis" is almost true,
$\mathcal{D}_3-$ "the cholesterol risk is high" is true,
$\mathcal{D}_4-$ "a diet with no fat is recommended" is true.

Let us assume that we have the following relations: "rather high" $\sqsubset_7$ "high", "more or less high" $\approx_8$ "high", "really very red" $\sqsubset_8$ "very red" and "very very high" $\sqsubset_2$ "high". Then, the diagnosis ($\mathcal{D}_1$ - $\mathcal{D}_4$) are obtained as follows.

- $\mathcal{D}_1$ is obtained by applying proposition 7 to $(\mathcal{F}_1, \mathcal{F}_2)$ and $(\mathcal{R}_1, \mathcal{R}_2)$,
- $\mathcal{D}_2$ is obtained by applying proposition 4 to $(\mathcal{F}_1, \mathcal{F}_3)$ and $\mathcal{R}_3$,
- $\mathcal{D}_3$ is obtained by applying proposition 1 to $\mathcal{F}_4$ and $\mathcal{R}_4$,
- $\mathcal{D}_4$ is obtained by applying proposition 6 to F_4 and $(\mathcal{R}_4, \mathcal{R}_5)$.

6 Conclusion

In this paper, we have proposed a symbolic-based model dealing with nuanced information. This model is inspired from the representation method on fuzzy logic. In previous papers, we have proposed a new representation method of nuanced statements. In this method, we have defined a vague term by symbolic parameters given by an expert in a qualitative way. By using this representation, we have defined some linguistic modifiers in a purely symbolic way. In this paper, we proposed some deduction rules dealing with nuanced statements and we presented new *Generalized Modus Ponens* rules. In these rules we can use either simple statements or complex statements.

References

1. J.F. Baldwin. A new approach to approximate reasoning using fuzzy logic. *Fuzzy sets and systems*, 2:309–325, 1979.
2. B. Bouchon-Meunier and L. Valverde. A ressemblance approach to analogical reasoning function. *Lecture notes in computer science*, 1188:266–272, 1997.
3. B. Bouchon-Meunier and J. Yao. Linguistic modifiers and imprecise categories. *Int. J. of intelligent systems*, 7:25–36, 1992.
4. M. De-glas. Knowledge representation in fuzzy setting. Technical Report 48, LAFORIA, 1989.
5. J. Delechamp and B. Bouchon-Meunier. Graduality by means of analogical reasoning. *Lecture notes in computer science*, 1244:210–222, 1997.
6. D. Dubois and H. Prade. Fuzzy sets in approximate reasoning: Inference with possibility distributions. *Fuzzy sets and systems*, 40:143–202, 1991.
7. M. El-Sayed and D. Pacholczyk. A qualitative reasoning with nuanced information. *8th European Conference on Logics in Artificial Intelligence (JELIA 02), 283–295, Italy*, 2002.
8. M. El-Sayed and D. Pacholczyk. A symbolic approach for handling nuanced information. *IASTED International Conference on Artificial Intelligence and Applications (AIA 02), 285–290, Spain*, 2002.
9. L.D. Lascio, A. Gisolfi, and U.C. Garcia. Linguistic hedges and the generalized modus ponens. *Int. J. of intelligent systems*, 14:981–993, 1999.
10. D. Pacholczyk. *Contribution au traitement logico-symbolique de la connaissance.* PhD thesis, University of Paris VI, 1992.
11. L.A. Zadeh. Fuzzy sets. *Information and control*, 8:338–353, 1965.
12. L.A. Zadeh. A theory of approximate reasoning. *Int. J. Hayes, D. Michie and L. I. Mikulich (eds); Machine Intelligence*, 9:149–194, 1979.

Development of Multiple Job Execution and Visualization System on ITBL System Infrastructure Software and Its Utilization for Parametric Studies in Environmental Modeling

Yoshio Suzuki[1], Nobuko Matsumoto[1], Nobuhiro Yamagishi[1], Kenji Higuchi[1], Takayuki Otani[1], Haruyasu Nagai[2], Hiroaki Terada[2], Akiko Furuno[2], Masamichi Chino[2], and Takuya Kobayashi[2]

[1] Center for Promotion of Computational Science and Engineering, Japan Atomic Energy Research Institute
6-9-3 Higashi-Ueno, Taito-ku, Tokyo 110-0015, Japan
{suzukiy,matsumoto,yama,higuchi,otani}@koma.jaeri.go.jp
http://www2.tokai.jaeri.go.jp/ccse/english/index.html

[2] Department of Environmental Sciences, Tokai Research Establishment, Japan Atomic Energy Research Institute
2-4 Shirakata-shirane, Tokai-mura, Naka-gun, Ibaraki 319-1195, Japan
{nagai,terada,furono,chino}@sakura.tokai.jaeri.go.jp
tkoba@popsvr.tokai.jaeri.go.jp
http://des.tokai.jaeri.go.jp/English/index.html

Abstract. Information-Technology Based Laboratory (ITBL) project has been propelled as one of e-Japan priority policy programs. The purposes of the project are to share intellectual resources such as remote computers, programs and data in universities and institutes and to support cooperative studies among researchers, building a virtual research environment, ITBL. Japan Atomic Energy Research Institute (JAERI) has been working on installation and management of hardware and development of infrastructure software and applications. As application software, researches on quantum bioinformatics and environmental sciences are carried out. This paper presents utilization of ITBL system infrastructure software for 'Numerical Environment System' which is developed for environmental studies. More effective job executions and visualization are expected by using Task Mapping Editor (TME) and AVS/ITBL, which are tools developed as infrastructure software.

1 Introduction

Information-Technology Based Laboratory (ITBL) project has been propelled as one of e-Japan priority policy programs (priority policy program for the infrastructure of network society of advanced information communication). By using recent technologies of information processing such as basic techniques of grid computing and infra-

P.M.A. Sloot et al. (Eds.): ICCS 2003, LNCS 2659, pp. 120–129, 2003.

structures of high-speed networks, Japan Atomic Energy Research Institute (JAERI) has been propelling ITBL project in cooperation with National Institute for Materials Science (NIMS), National Research Institute for Earth Science and Disaster Prevention (NIED), National Aerospace Laboratory of Japan (NAL), the Physical and Chemical Research Institute (RIKEN) and Japan Science and Technology Corporation (JST). This project aims to build virtual research environment for sharing intellectual resources such as remote computers, programs and data in universities and institutes and for supporting cooperative studies among researchers. JAERI has been working on installation and management of hardware and development of infrastructure software and applications. Office for ITBL promotion of Center for Promotion of Computational Science and Engineering (CCSE) is in charge of management of ITBL supercomputer. R&D group for parallel basic software and R&D group for parallel processing tools of CCSE have been developing the ITBL system infrastructure software. Studies of quantum biology and environmental sciences are appointed as its applications. Quantum bioinformatics group of CCSE and Department of Environmental Sciences of Tokai Research Establishment are building up the system respectively. The ITBL supercomputer, which is a PRIMEPWER made by Fujitsu Corporation, was installed in the ITBL center constructed at Kansai Research Establishment of JAERI in March 2003. This computer has 4 nodes consisted of 128 CPUs and its performance is 1 TFLOPS. It also has 1 TBYTE memory and 15 TBYTE disk.

The Department of Environmental Sciences, Tokai Research Establishment of JAERI is working on elucidation and prediction of material circulation in environment, started from studies on dynamic behavior of radionuclide of atmosphere, terrestrial, and oceanic environment [1], [2]. The study aims to clarify ways of preservation of environment and their effectiveness in the future as well as the effectiveness of nuclear power for preservation of environment by establishing comprehensive and numerical models which clarify and predict the circulation of materials including radioactive substances. Especially a numerical environment system has been developed to enable comprehensive numerical experiments by combination of models of regional atmosphere, land, and ocean and to enable synthetic prediction of the environmental pollution such as dispersion of radioactive substances caused by an accident of a nuclear power plant.

The numerical environment system consists of multiple numbers of codes such as atmospheric code, air pollution code, oceanic code, oceanic pollution code and so on. The atmospheric code predicts time-related changes of atmospheric variables such as wind velocity and temperature. The air pollution code calculates dispersion of pollutants based on the results of the atmospheric code. One example of use of these codes is to research which analysis is appropriate to identify real meteorological phenomena with multiple data sets for the same time and same area, which are obtained from some different analyses used to identify meteorological information. It is also possible to predict dispersion processes of pollutants in the atmosphere by coupled execution of atmospheric code and air pollution code. The oceanic code and the oceanic pollution code can be used similarly.

In this paper, an argument is advanced only within the case where these codes are used for such parametric studies. In such situations, each code is executed with specific input data repeatedly. In addition, each output data is manually visualized to compare results. Therefore, it is easily imagined to execute each code with prepared input data respectively and visualize each result automatically on multiple computers using grid-computing technique [3] as one of efficient ways to do job executions and visualization. To realize this, the ITBL system infrastructure software [4] is utilized to the numerical environment system. Especially it is aimed at:

- Enabling users to execute multiple simulation codes collectively, without executing them repeatedly.
- Enabling users to execute simulation codes efficiently using less busy computers in ITBL system automatically.
- Enabling users to manage data files easily without transferring data to do job executions and visualization.
- Enabling users to execute whole process from job executions to visualization collectively without visualizing results manually after simulations are completed.
- Enabling users to compare results easily by displaying multiple images for different parameters or different time stamps on a screen.

Here, Task Mapping Editor (TME) [5] and AVS/ITBL [6], which are tools included in infrastructure software, are used for the purpose. This paper presents the modifications and newly developed parts of the tools to apply for numerical environment system. In the next section, the background on ITBL, the ITBL system infrastructure software including TME and AVS/ITBL and some related studies are briefly given. The applied example and the contents of the development needed to its application are described in section 3. The functional specification is given in section 4. Finally in section 5, the advantages and problems of the application of infrastructure software are summarized.

2 ITBL System and Its Infrastructure Software

Figure 1 shows usage of the ITBL system infrastructure software. The functions of ITBL system infrastructure software enable users to execute jobs at a same time on multiple remote computers respectively and to execute a job on multiple machines at a time. It also allows users to set up a community among users and to share data and information, or to exchange their opinions in the community.

Figure 1 also shows the ITBL consists of sites. Each site has an ITBL server, which intermediates between user terminals and supercomputers. In addition, ITBL server works to get information needed for management of ITBL as much as possible and minimize load and influence on other computers at computational sites. The ITBL is available to be accessed through web browsers with Java applets on a PC or a workstation that has a certificate installed for the identification of users. Users can use computers on remote sites through ITBL even in environment where firewall is set. This is a specific characteristic of the ITBL.

Now there are some supercomputers available through the infrastructure software such as Fujitsu VPP5000 and Hitachi SR8000 at Tokai Research Establishment of

JAERI, SGI Origin3800 at Naka Fusion Research Establishment of JAERI, and Compaq ES40 at Kansai Research Establishment of JAERI. Fujitsu PRIMEPOWER at ITBL center and NEC SX6-i at Ueno Office of CCSE are under construction to connect to the ITBL system. Not only supercomputers of JAERI but also supercomputers of RIKEN and NAL are in the process of connection. Computers of all six organizations described before will have been connected by March 2004.

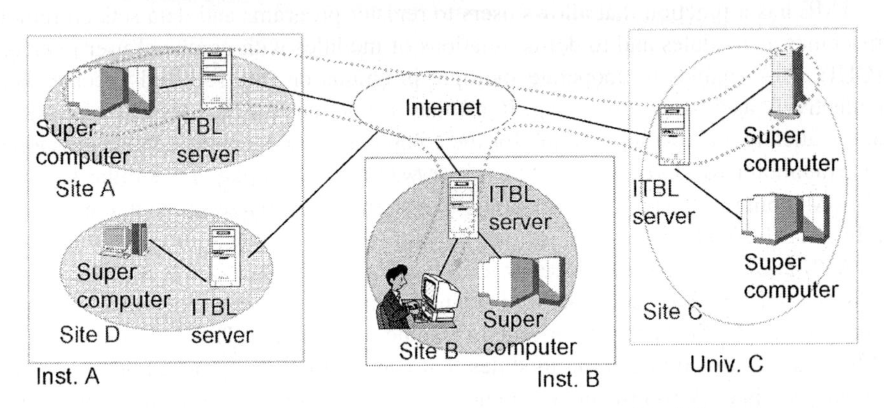

Fig. 1. Usage of the ITBL system infrastructure software. A user at Site B is using a supercomputer at Site A and a supercomputer at Site C

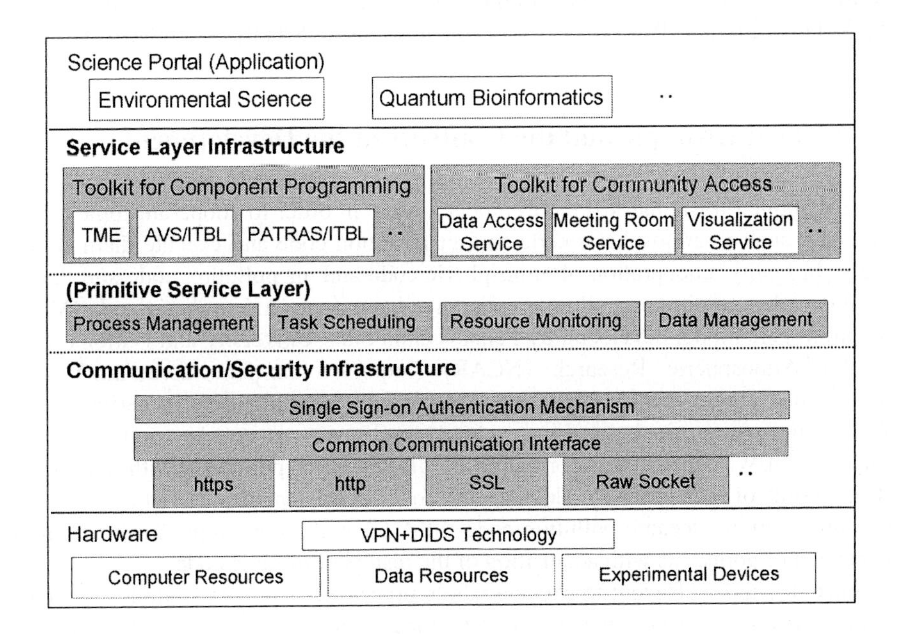

Fig. 2. The architecture of the ITBL

Figure 2 shows the architecture of the ITBL. Toolkits for component programming and community access as service layer infrastructure and single sign on authentication mechanism and common communication interface as communication/security infrastructure have been developed as infrastructure software. Specifically there are many functions developed for job executions. As it mentioned before, TME and AVS/ITBL have been developed for controlling job executions and for visualization of results, respectively.

TME has a function that allows users to register programs and data sets on remote machines as modules and to define relations of modules with graphical user interface (GUI). This enables to cooperate multiple programs or data sets and execute them collectively as a batch processing. It also has a function to find a less busy computer automatically and to allocate a job for the computer. Task Mapping Engine developed by Fraunhofer Resource Grid (`http://www/fhrg.fhg.de/`) is similar to TME. Task Mapping Engine aims for mapping of specific tasks to a resource that needs rapidly and effectively.

AVS/ITBL is a tool developed with a base of AVS/Express [7]. AVS/Express is widely used for post-visualization of scientific data but can read only data on a computer where AVS/Express is installed (graphic server). The ITBL system enhances AVS/Express by adding some modules that can read data sets on remote computers directly so that AVS/ITBL on a graphic server can read any data on any computer in the ITBL system. AVS/ITBL also has a function to visualize data on the web by designating a data file and a network file (v file). These developments have been achieved using STARPC library. STARPC, developed by CCSE/JAERI, is a RPC (Remote Procedure Call)-based communication library for a parallel computer cluster [8]. Reference 6 describes the details of functions on AVS/ITBL and related studies.

3 Applied Example and the Contents of the Development

TME is applied for numerical environment system in order to cooperate some different codes such as atmospheric code and air pollution code and execute them collectively. In reality, cooperation of atmospheric code and air pollution code and that of oceanic code and oceanic pollution code have been done. Atmospheric code 'MM5' is a mesoscale model developed by Pennsylvania State University and National Center for Atmospheric Research (NCAR) (http://www.mmm.ucar.edu/mm5/mm5-home.html). The execution of the atmospheric code enables users to derive atmospheric information, such as wind velocity and temperature. Air pollution code 'GEARN-NEW' can output information on dispersion of pollutants with acquisition of the results of atmospheric code. Oceanic code 'POM' enables users to derive oceanic information. Oceanic pollution code 'SEA-GEARN' can output information on dispersion of pollutants with acquisition of the results of oceanic code.

Since it is more efficient to execute whole process from job executions to visualization collectively, a new development is done to enable TME job flow to invoke AVS/ITBL with batch processing mode. AVS/ITBL is available for users to use it as an X Window System or web client, but it was not suitable to execute prepared procedures before its execution. Thus, AVS/ITBL has been enhanced by having a new function (batch visualization function) to execute AVS/ITBL in command mode, perform visualization process based on preserved visualization parameters and save image data to files. Execution of AVS/ITBL with command line options enables TME to register batch processing for visualization.

Numerical environment system adopts network Common Data Form (NetCDF) for output files, which is an interface for array-oriented data access and a library that provides an implementation of the interface, but was not accepted by AVS/ITBL. The new development on AVS/ITBL includes a development of a remote data reader for NetCDF.

In addition, an image viewer has been developed to display multiple images produced by parametric study and compare them to know how parameters and time-dependent change influence the results easily. The newly developed image viewer has three functions: a function to generate thumb-nail images from files saved on an graphic server, a function to display only designated thumb-nail images, and a function to display an image file whose thumb-nail is selected with an original size. ImageMagick used for generating thumb-nail images for X Window System are free software tools working on workstations (`http://www.imagemagick.org/`).

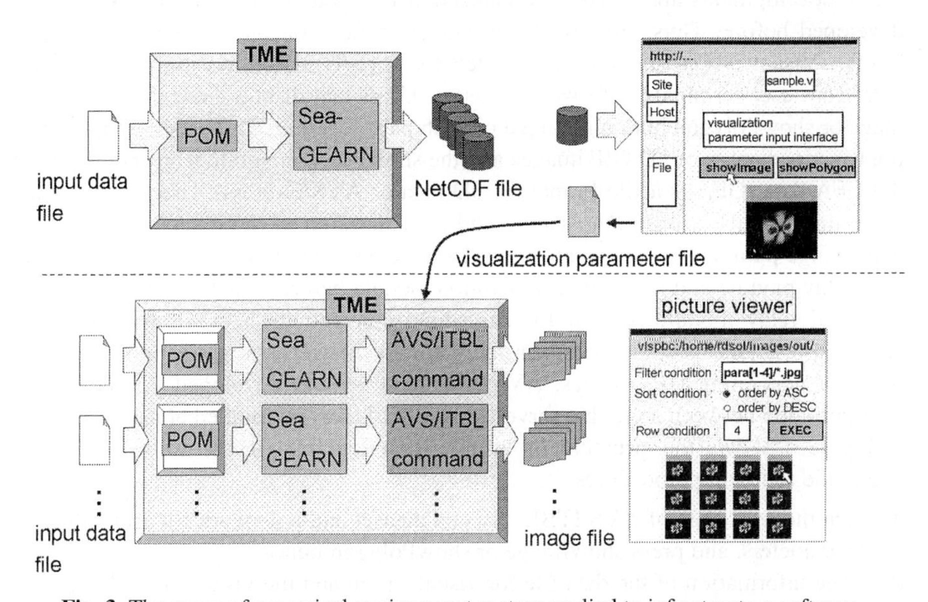

Fig. 3. The usage of numerical environment system applied to infrastructure software

Figure 3 shows the usage of numerical environment system applied to the ITBL system infrastructure software. As the preparation for performing parametric study efficiently, first of all, users need to make a TME job flow without AVS/ITBL batch processing, and then visualize the results with AVS/ITBL. Since batch processing for visualization does not allow users to operate it interactively, they need to generate a visualization parameter file in advance. It means users need to visualize data first and save methods, parameters and viewpoints to designate image files. After the procedure, users modify the TME job flow to apply it to multiple parameters and add a batch processing system for AVS/ITBL to execute visualization. The modified TME job flow can run simulations of parametric study and visualization with batch processing mode. The image viewer displays thumb-nail images made from output images so that comparative analysis can be done effectively.

4 Details of Functions

The developed functions to run AVS/ITBL in batch processing mode are as follows:

- A function to output visualization parameter file.
- A function to execute AVS/ITBL in command mode.
- A function to input visualization parameter file.
- A function to designate data file for visualization.
- A function to output visualized images with JPEG format.

These developments are based on visualization functions on the web for AVS/ITBL developed before. Thus, first of all, visualization function on the web is explained briefly. Visualization function on the web for AVS/ITBL allows users to visualize data from user terminals with web browsers. Users can display images with showImage or showPolygon button arranged on web page (see Fig. 3). The showImage button triggers display of 2D GIF images and the showPolygon button triggers display of 3D GFA data (GFA is a file format for 3D data of AVS/Express). Users can operate those images such as rotation, scaling and transportation of the objects interactively and change parameters for visualization. Users need to make network files including CGI relay module and image data generation module before visualization on the web. The CGI relay module is used for describing connections between parameters of AVS/Express and those appeared on the web. An image data generation module is used for generating GIF images. AVS relay process was developed to enable data transportation between a graphic server and an ITBL server installed at users' site.

Figure 4 shows the structure of visualization function on the web. The figure shows the following procedures:

1. On the web page of AVS/ITBL, select a dataset and a network file, set values to parameters, and press showImage or showPolygon button.
2. The information of the data file for visualization and the visualization parameter file is sent to AVS relay process (2nd) though a servlet on ITBL server (STARPC communication [8]).
3. AVS relay process (2nd) generates a visualization task from the information and the information of the data file for visualization and the visualization parameter file is sent to CGI relay module (socket communication).
4. The information affects AVS/Express.
5. When AVS/ITBL reads a dataset on a remote computer, it uses functions of remote modules. It means AVS relay process (1st) is used [6].

6. AVS/Express runs as the designated network file tells.
7. A generated image is sent through a servlet of ITBL server and displayed on the web browser.

Accordingly, AVS relay process (2nd) has been modified to add a function of batch processing to AVS/ITBL. While visualization on the web uses AVS relay process (2nd) to get information of a data file and a visualization parameter file with communication of STARPC through servlet, batch processing of AVS/ITBL runs AVS relay process (2nd) directly and passes the information used for visualization to the AVS relay process by revising a part of its functions. The enclosed area with doted line in Fig. 4 is equivalent to AVS/ITBL command in Fig. 3. The modification enables batch processing of AVS/ITBL to execute procedures that are similar to the ones from 3 to 6 in Fig. 4 and output image files. AVS/ITBL has a modification to be able to generate JPEG images for display of thumb-nail images.

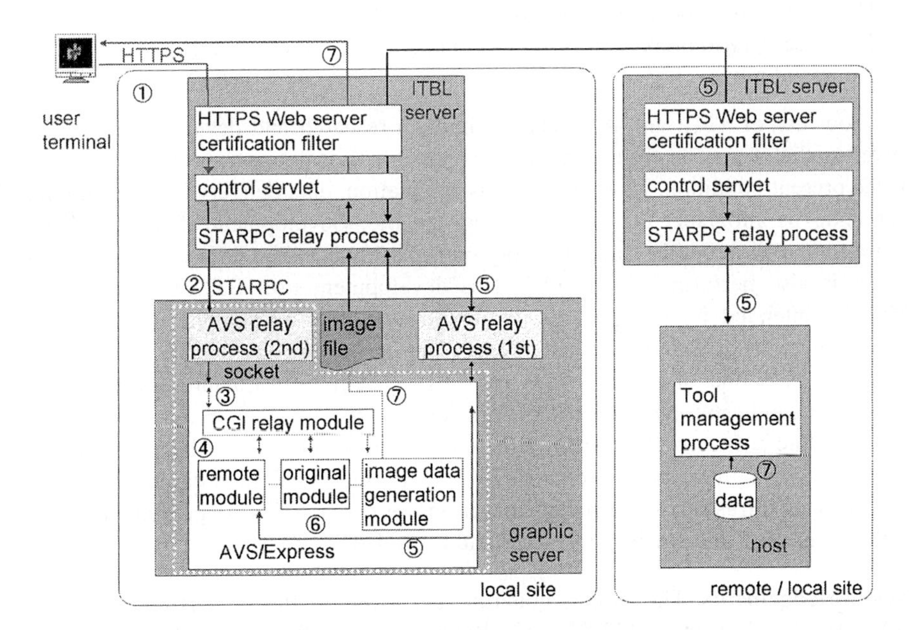

Fig. 4. The structure of visualization function on the web

5 Summary

It is expected that the integration of infrastructure software and application enables to do more effective job execution and analysis with visualization for parametric studies in numerical environment system. The following things are considered to be the real advantages and problems of the integrated system.

Advantages:
- Users need not to execute multiple simulation codes repeatedly but enable to execute them collectively.
- It is possible for users to execute simulation codes efficiently using less busy computers in ITBL system automatically. In addition, efficient usage of supercomputers is available.
- Users need not to transfer data to do job executions and visualization, which also makes it easy to manage data files.
- Users need not to visualize results manually after simulations are completed, but it is possible for users to execute whole process from job executions to visualization collectively.
- It is possible for users to compare results easily by displaying multiple images for different parameters or different time stamps on a screen.

Problems:
- Users need to know how to use TME and AVS/ITBL.
- It is necessary to prepare the visualization parameter file by running AVS/ITBL in command mode.

About use of TME and AVS/ITBL, although research and development are furthered so that it may be easy to use users as possible, it is not necessarily user-friendly as the present condition. Moreover, in this application, data delivery between simulation codes is altogether performed through the file. Development of the simulation codes for carrying out the direct memory transmission between different remote computers is also performed in parallel. This development uses STAMPI library [9], which is intended for carrying out communications among computers of different architectures in a heterogeneous environment developed by CCSE/JAERI. This library is also one of the functions of ITBL system infrastructure software. The current research and development of functions, which was not in the conventional infrastructure software, have been furthered in order to unite with the research style performed by numerical environment system. Furthermore, in order to utilize the functions of infrastructure software in various research fields, improvement united with the research style in each research field is needed.

Acknowledgements. The authors gratefully acknowledge the support of G. Yagawa (JAERI), T. Hirayama (JAERI), M. Yamagiwa (JAERI), M. Fukuda (JAERI), H. Maesako (JAERI), K. Sai (JAERI), K. Kimura (JAERI), Y. Hasegawa (JAERI), T. Takahashi (JAERI), Y. Tsujita (JAERI) and M. Fujisaki (Fujitsu Corporation). We also thank H. Kanazawa (Fujitsu Corporation), K. Suzuki (Fujitsu Corporation), N. Teshima (Fujitsu Corporation) and T. Arakawa (Fujitsu Corporation) for their insightful advice on this development.

References

1. Chino, M., Ishikawa, H., Yamazawa, H., Nagai, H. and Moriuchi, S.: WSPEEDI (World-wide Version of SPEEDI): A Computer Code System for the Prediction of Radiological Impacts on Japanese due to a Nuclear Accident in Foreing Countries, JAERI 1334, (1995)
2. Kobayashi, T., Lee, S., H. and Chino, M.: Development of Ocean Pollution Prediction System for Shimokita Region – Model Development and Verification –, Journal of Nuclear Science Technology. Vol. 39, No. 2, (2000) 171–179
3. Foster, I. and Kesselman, C. (ed.): The GRID, Blue Print for a New Computing Infrastructure. Morgan Kaufmann (1998)
4. Yagawa, G. and Hirayama, T.: Japan IT-Based Laboratory, to be published in Proceedings of the Global Grid Forum 5, Edinburgh (2002)
5. Takemiya, H., Imamura, T., Koide, H., Higuchi, K., Tsujita, Y., Yamagishi, N., Matsuda, K., Ueno, K., Hasegawa, Y., Kimura, T., Mochizuki, Y. and Hirayama, T.: Software Environment for Local Area Metacomputing, in Proceedings of 4th International Conference on Supercomputing in Nuclear Applications (SNA2000), Tokyo, (2000) CD-ROM
6. Suzuki, Y., Sai, K., Matsumoto, N. and Hazama, O.: Visualization system on Information Technology Based Laboratory, to be published in IEEE Computer Graphics and Applications. (2003)
7. Upson, C., Faulhaber, T., Jr., Kamins, D., Laidlaw, D., Schlegel, D., Vroom, J., Gurwitz, R. and van Dam, A.: The application visualization system: A computational environment for scientific visualization. IEEE Computer Graphics and Applications. Vol. 9, No. 4, (1989) 30–42
8. Takemiya, H. and Yamaghisi, N.: Starpc: a library for Communication among Tools on a Parallel Computer Cluster – User's and Developer's Guid to Starpc –, JAERI-Data/Code 2000-005 (2000)
9. Imamura, T., Tsujita, Y., Koide, H. and Takemiya, H.: An Architecture of Stampi: MPI Library on a Cluster of Parallel Computers. In Recent Advances in Parallel Virtual Machine and Message Passing Interface, LNCS 1908, Springer, (2000) 200–207

A Genetic Algorithm for Predicting RNA Pseudoknot Structures

Dongkyu Lee and Kyungsook Han*

School of Computer Science and Engineering, Inha University, Inchon 402-751, Korea

Abstract. An RNA pseudoknot is a tertiary structure element formed when bases of a single-stranded loop pair with complementary bases outside the loop. Computational determination of the RNA folding structure with pseudoknots from the linear sequence is a complex problem involving both spatial reasoning and the use of knowledge of chemistry and biology. We have developed a genetic algorithm for predicting the RNA folding structure with pseudoknots of any type. This paper analyzes the predictions by the genetic algorithm with different population generation methods and fitness functions, and compares the predictions to those by a dynamic programming method.

1 Introduction

An RNA pseudoknot is a tertiary structural element formed when bases of a single-stranded loop pair with complementary bases outside the loop. Pseudoknots are not only widely occurring structural motifs in all kinds of viral RNA molecules, but also responsible for several important functions of RNA. The RNA folding structure with pseudoknots is much more difficult to predict than the RNA secondary structure because prediction of pseudoknots should consider tertiary interactions as well as secondary interactions.

In the computational viewpoint, predicting RNA structure can be considered as an optimization problem. Many prediction algorithms obtain the RNA structure with the smallest free energy as an optimal structure [1]. Greedy algorithms and dynamic programming algorithms were also used to predict the optimal or suboptimal secondary structures [2, 3, 4]. In principle, dynamic programming algorithms can predict optimal RNA secondary structure with pseudoknots. But in practice, dynamic programming algorithms can be applied to small RNAs only, and cannot be applied to moderate or large RNAs due to computational complexity of the algorithm.

We have developed a genetic algorithm for predicting RNA secondary structure with pseudoknots. A genetic algorithm (GA) is a non-deterministic optimization procedure, derived from the concept of biological evolution. A genetic algorithm can solve an optimization problem, but it cannot guarantee to find optimal result (see references [5, 6, 7, 8, 9, 10] for previous works on predicting RNA structure using a

* To whom correspondence should be addressed. email: khan@inha.ac.kr

P.M.A. Sloot et al. (Eds.): ICCS 2003, LNCS 2659, pp. 130–139, 2003.

genetic algorithm). To assess the optimality of the predictions by our genetic algorithm, we used thermodynamic free energy as well as the number of base pairs of the predicted structures as fitness functions of the genetic algorithm.

As for the organization of this paper, section 2 describes a genetic algorithm we developed for predicting RNA structure with pseudoknots. Section 3 discusses the predicted RNA structures by the genetic algorithm and compares them with those by a dynamic programming algorithm. Analysis of the results and the general lessons learned from this study are summarized in the final, Conclusion section.

2 Genetic Algorithm

For RNA structure prediction, we used a steady-state genetic algorithm because its predictions are known better than others from our previous study [15]. After parsing an RNA sequence, all possible stems were identified as the initial populations of a stem pool. Thermodynamic free energy and the weighted value of the number of base pairs were used as fitness functions. We calculated the value of fitness function in each structure of initial populations and evolved using a genetic operator. RNA structure is represented in a binary string genome, as shown in Fig. 1. As a terminal condition we used both the number of generations and the convergence of each generation.

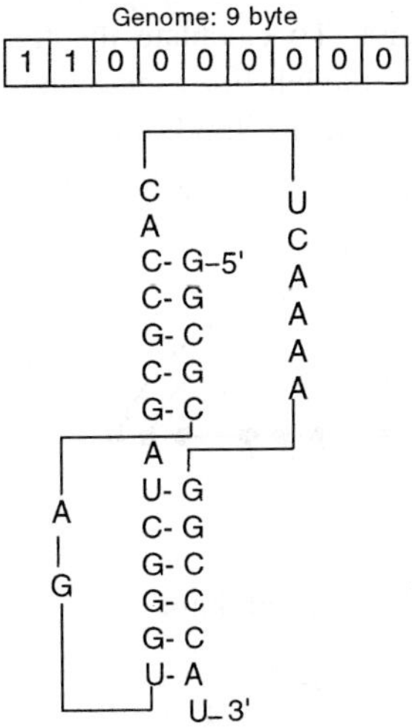

Genome: 9 byte

1	1	0	0	0	0	0	0	0

Stem pool

index	start	end	size	energy
0	7	32	6	-15.4
1	0	18	5	-12.5
2	5	21	5	-9.7
3	0	30	3	-6.7
4	0	12	3	-6.7
5	9	18	3	-6.7
6	16	29	3	-6.7
7	7	23	4	-5.7
8	6	17	3	-3.9

Fig. 1. Example of the representation of the pseudoknot of MMTV RNA in genome

2.1 Stem Pool Generation

We generate 3 stem pools after reading a RNA sequence. First we generate all possible stem lists using a covariation matrix. We consider a stem as having minimum 3 base pairs. We calculate the stacking energy of each stem in stem lists. We sort the stem lists in increasing order of their energy values. This is the first stem pool – what we call "fully zipped stem pool". We then remove consecutive wobble pairs at both ends of a stem since these wobble pairs are not stable enough. After removing all consecutive wobble pairs of each stem at both ends of a stem, we recalculate the stacking energy and removed irregular stems. This is the second stem pool – what we call "partially zipped stem pool". Finally we generate a pseudoknot stem pool by finding all possible pairs of stems that can form a pseudoknot. At this stage, we consider the number of connecting loops and the size of pseudoknot stems only.

These three stem pools can be selectively used in generating initial populations. The purpose of building the partially zipped stem pool is to consider all possible partially zipped stems. The pseudoknot stem pool is for predicting RNA structures with pseudoknots. Without the pseudoknot stem pool, the genetic algorithm predicts secondary structures only.

2.2 Initial Population

We generate structures that include every stem in the stem pools. As mentioned above, the three stem pools can be used selectively. After choosing a stem from stem pools, we insert all other stems that can coexist with the chosen stem topologically. If we use the pseudoknot stem pool, we first select 2 stems from the pseudoknot stem pool, and insert all possible stems in other stem pools. In topology test, we check overlapping relation of stems and the type of a pseudoknot. To calculate the free energy of a H-type pseudoknot we apply the known energy model. For pseudoknots of other types, we only check the type of the pseudoknots and use topology test in calculating the free energy of RNA structures. In the example of topology test shown in Fig. 2, cases of (A) and (C) are considered to fail the test.

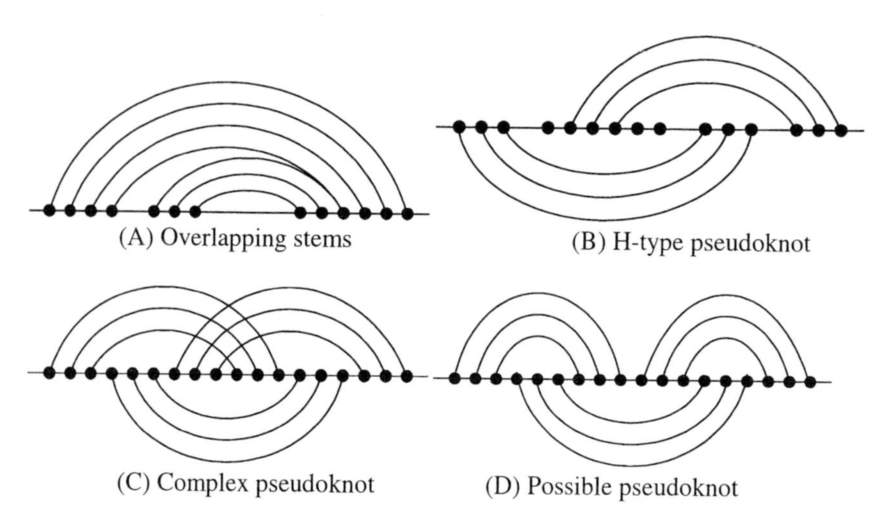

Fig. 2. Topology test of initial population

3 Method and Results

The genetic algorithm was implemented using C++ builder 5.0 of Inprise Company on 1.61 GHz Pentium 4 PC with 256 MB memory. To analyze the optimality of the genetic algorithm, we compared its predictions to the optimal structures predicted by a dynamic programming algorithm [4].

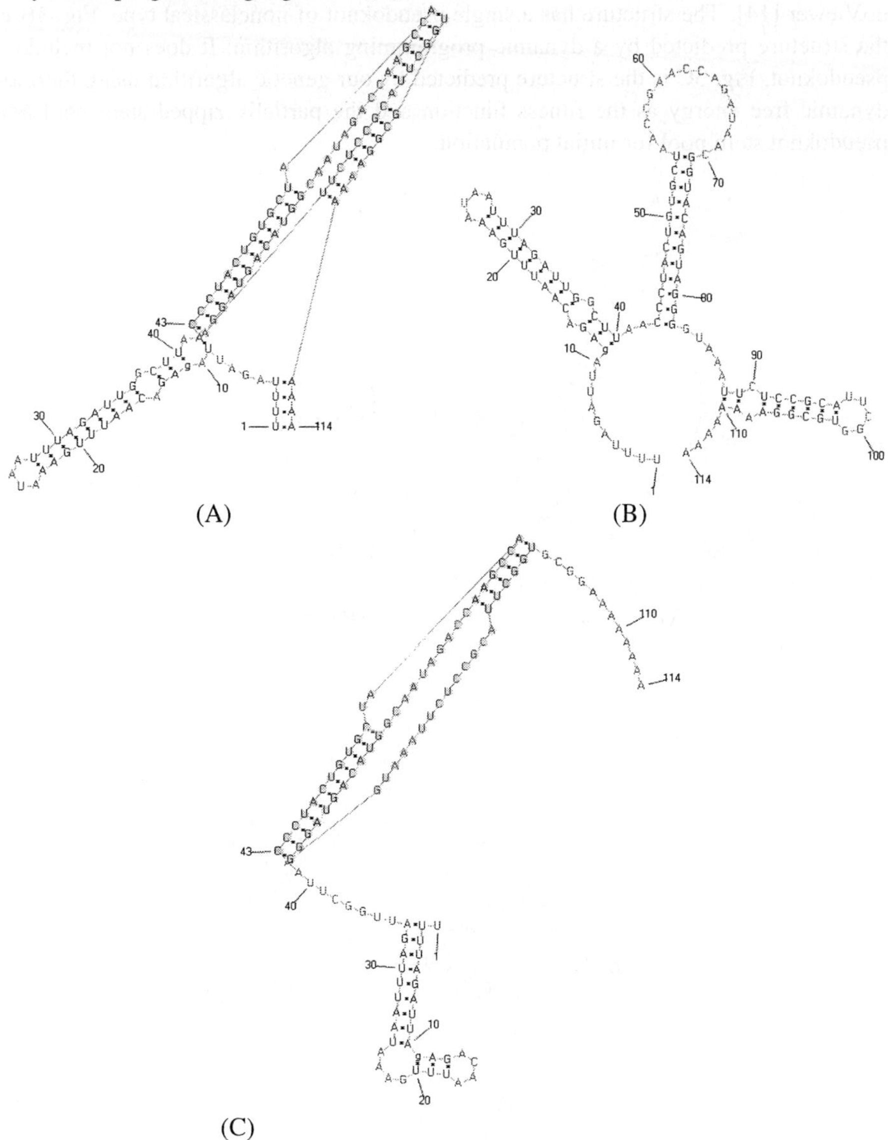

Fig. 3. (A) known structure of Coxsackie B3 virus. (B) structure predicted by the dynamic programming algorithm. (C) structure predicted by the genetic algorithm using thermodynamic free energy as the fitness function and the partially zipped stem pool and pseudoknot stem pool.

Since the genetic algorithm is nondeterministic by its nature, we ran the algorithm many times on a same test case. The mean, minimum, maximum and mode values of its predictions were used in comparison with the predictions by a dynamic programming algorithm. Different initial populations were tried using the partially zipped stem pool and fully zipped stem pool.

Fig. 3A shows the known structure Coxsackie B3 virus [11], visualized by PseudoViewer [14]. The structure has a single pseudoknot of nonclassical type. Fig. 3B is the structure predicted by a dynamic programming algorithm. It does not include a pseudoknot. Fig. 3C is the structure predicted by our genetic algorithm using thermodynamic free energy as the fitness function and the partially zipped stem pool and pseudoknot stem pool for initial population.

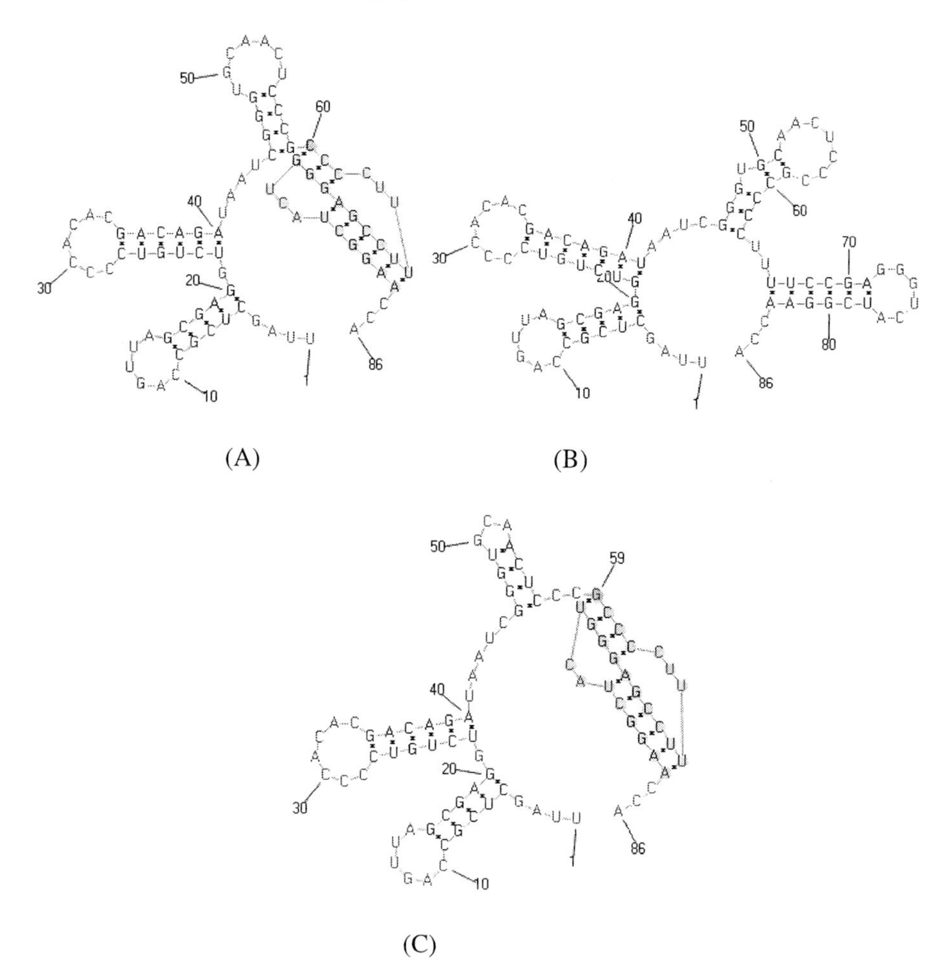

Fig. 4. (A) known structure of TYMV RNA. (B) structure predicted by the dynamic programming algorithm. (C) structure predicted by the genetic algorithm with thermodynamic free energy as the fitness function and the partially zipped and fully zipped stem pools.

Figure 4A shows the known structure of TYMV RNA [12], and Fig. 4B is the structure predicted by a dynamic programming algorithm. It does not include a pseudoknot. Figure 4C is the structure predicted by the genetic algorithm with thermodynamic free energy as the fitness function and the partially zipped stem pool and fully zipped stem pool for initial population. The structure predicted by the genetic algorithm is topologically similar to the known structure in Fig. 4A When the genetic algorithm runs with the fully zipped stem pool and the pseudoknot stem pool, its prediction becomes identical to the known structure.

Table 1 summarizes the analysis of the structures of TYMV and Coxsackie B3 virus predicted by two methods. The two methods were compared with respect to two measures: number of base pairs and free energy of the predicted structures. The exact free energy could not be computed for the known structure of Coxsackie since its pseudoknot structure is not a typical H-type. Although the dynamic programming algorithm predicted the structure with the smallest free energy value, its predictions did not include a pseudoknot at all. The genetic algorithm predicted suboptimal structures in terms of free energy, but the predicted structures are closer to known structures because they contain a pseudoknot.

Table 1. Comparison of predicted structures by a dynamic programming algorithm and a genetic algorithm. Free energy cannot be computed for the known structure of coxsackie

RNA	measure	known structure	prediction by dynamic programming	prediction by GA
TYMV	# base pairs	24.0	23.0	25.0
	free energy	-24.8	-29.6	-24.1
Coxsackie	# base pairs	41.0	32.0	29.0
	free energy	–	-34.4	-30.7

Table 2. Statistical values of the predictions by the genetic algorithm

stem pool	measure	Coxsackie RNA	TYMV RNA
partially zipped stem pool pseudoknot stem pool	# stems	132	62
	# predicted structures	1	11
	maximum energy	-30.7	-24.8
	minimum energy	-30.7	-13.0
	mean energy	-30.7	-20.4
	mode energy	-30.7	-24.8
partially zipped stem pool fully zipped stem pool pseudoknot stem pool	# stems	229	113
	# predicted structures	6	14
	maximum energy	-45.6	-26.2
	minimum energy	-25.7	-18.1
	mean energy	-29.9	-22.9
	mode energy	-30.7	-24.1

Table 2 summarizes the statistical values of the predictions by several runs of the genetic algorithm. In Table 2, we compare the predictions by the genetic algorithm with the partially zipped stem pool alone with those with both the partially zipped stem pool and the fully zipped stem pool. The pseudoknot stem pool was used commonly in both cases. By using the fully zipped stem pool additionally, we could predict many suboptimal structures, and also more stable structures. But as the size of stem pools increases, the genetic algorithm takes more time.

The execution time of the genetic algorithm depends on its termination condition: the number of generations or the convergence of a generation. The convergence of a generation was used to predict the structures shown earlier. Table 3 shows the effect of changing the number of generations when running the genetic algorithm.

Table 3. The effect of changing the number of generations when running the genetic algorithm

# generations	# predicted structures	maximum energy	minimum energy	mean energy	mode energy
50	8	-24.8	-13.9	-21.51	-24.8
150	9	-24.8	-13.9	-21.39	-24.8
250	6	-24.8	-13.9	-22.41	-24.8

We predicted the TYMV RNA using different numbers of generations. We found that there is little difference between the case using the convergence and that using the number of generations and that it converged to the most stable structure early.

We also predicted the RNA structures with long sequences. Longer sequences need more execution time. The genetic algorithm takes very long when the sequence length exceeds 300 bases. Fig. 5 and Fig. 6 show the structures of DIGIR1 RNA and TMV RNA, respectively [1, 13]. Fig. 5A is the known structure of DIGIR1 RNA and Fig. 5B is the predicted structure by the genetic algorithm. Fig. 6A is the known structure of TMV RNA with 4 H-type pseudoknots and 6 stems outside the pseudoknots. The genetic algorithm predicted a similar structure to the known structure with free energy as its fitness function. Both the partially zipped stem pool and the pseudoknot stem pool were used, but the fully zipped stem pool was not used due to the size of the RNA sequence. The algorithm terminated evolving process using the convergence of populations. The structures shown in Fig. 6B-D were obtained by the genetic algorithm with different fitness functions. The structure shown in Fig. 6C was predicted by the genetic algorithm with the weighted value of base pairs as the fitness function instead of energy value. Both structures of Fig. 6B and Fig. 6D were predicted by the genetic algorithm with free energy as the fitness function. Fig. 6B had a higher energy value (i.e., less stable) than the structure of Fig. 6D. The structure of Fig. 6D is the most similar to the known structure. The known structure of TMV has 14 stems, but 2 of them could not be generated at initial stage. The structure shown in Fig. 6C was obtained with the weighted value of base pairs, which is a rough approximation of free energy, but is similar to the known structure in Fig. 6A. For the structure with nonclassical pseudoknots, the genetic algorithm performs better with the weighted value of base pairs than with free energy since the exact energy model is not available for nonclassical pseudoknots.

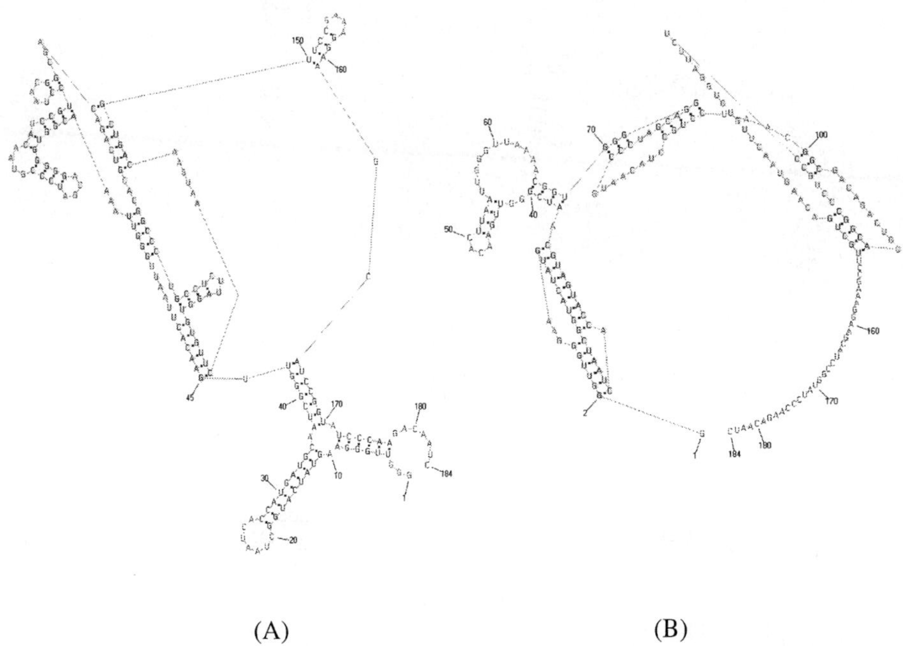

(A) (B)

Fig. 5. (A) known structure of DIGIR1 RNA. (B) predicted structure by the genetic algorithm.

4 Conclusion

We developed a genetic algorithm for predicting the RNA structures with pseudoknots, and compared the results from using different stem pools and fitness functions. We also compared the structures predicted by the genetic algorithm to the structures predicted by a dynamic programming algorithm.

The known RNA structures are often the suboptimal structures in terms of free energy, and therefore the optimal structures predicted by a dynamic programming algorithm may not correspond to the known structures. Genetic algorithms do not guarantee to find optimal structures but often predict better than dynamic programming algorithms. The difficulties of predicting RNA structures using a genetic algorithm come from several things. First, there exist stems of irregular types in natural RNAs, but all those irregular types cannot be considered during the population generation stage. This means that the quality of the prediction depends on the stem pools used. Second, no energy model is not available for nonclassical pseudoknots, and the energy model for H-type pseudoknot is not accurate, either. Therefore, simple and rough fitness criteria such as the weighted sum of base pairs can be a good fitness function when the exact energy model associated with nonclassical pseudoknots is not available.

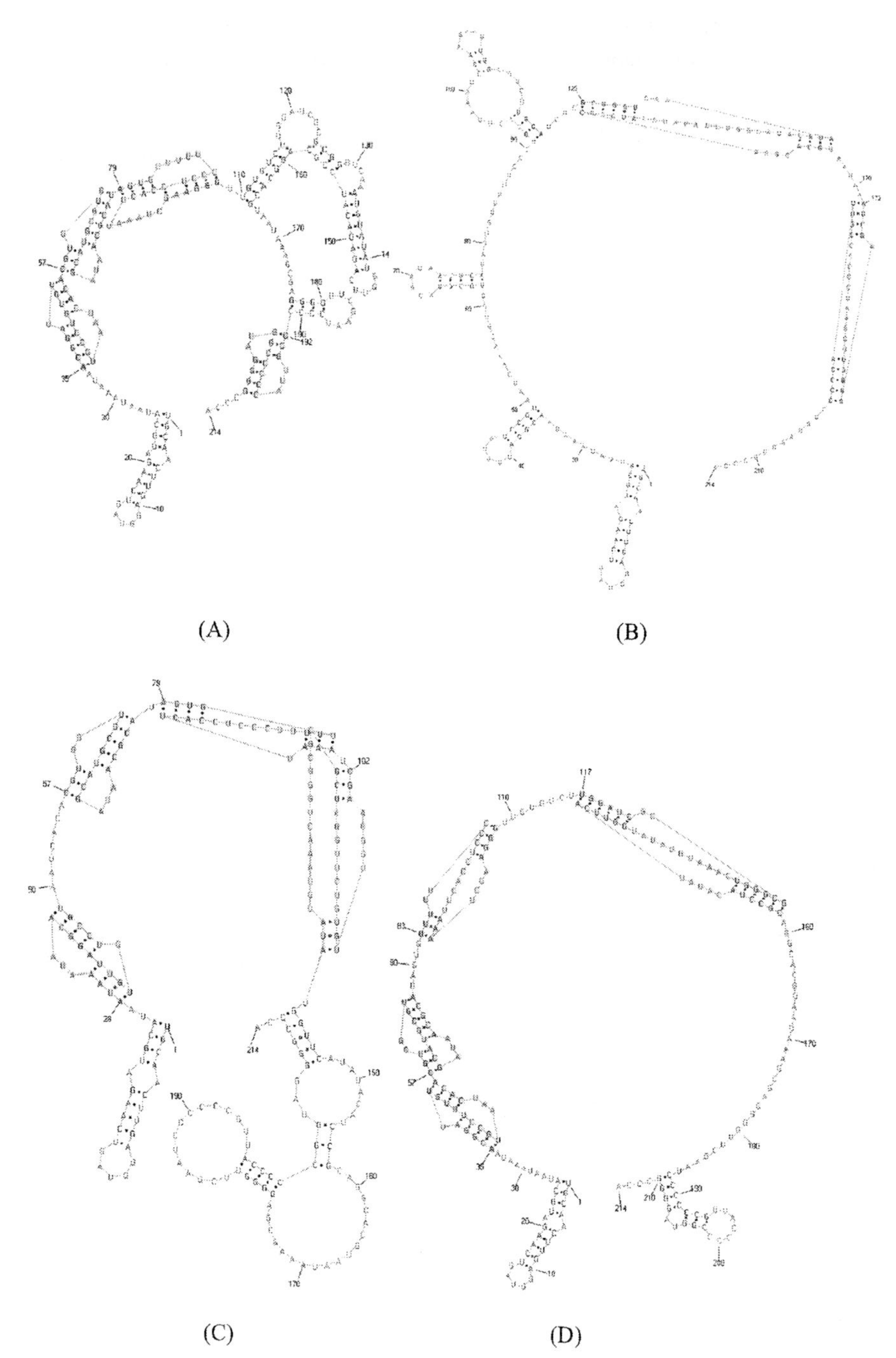

Fig. 6. (A) known structure of TMV RNA with 4 H-type pseudoknots. (B) structure predicted by the genetic algorithm with the weighted sum of base pairs as the fitness function. Structures of (B) and (D) were predicted by the genetic algorithm with free energy as the fitness function.

Acknowledgement. This work has been supported by the Korea Science and Engineering Foundation (KOSEF) under grant R05-2001-000-01037-0.

References

1. Chen, J.-H., Le, S.-Y., and Maizel, J. V.: A procedure for RNA pseudoknot prediction. Computer Applications in the Biosciences 8 (1992) 243-248
2. Abrahams, J. P., van den Berg, M., van Batenburg, E. and Pleij, C.: Prediction of RNA secondary structure, including pseudoknotting, by computer simulation. Nucleic Acids Res. 18 (1990) 3035-3044
3. Akutsu, T.: Dynamic programming algorithm for RNA secondary structure prediction with pseudoknots. Discrete Applied Mathematics 104 (2000) 45-62
4. Rivas, E. and Eddy S. R.: A dynamic programming algorithm for RNA structure prediction including pseudoknots. Journal of Molecular Biology 285 (1999) 2053-2068
5. Gultyaev, A. P., van Batenburg, F. H. D. and Pleij, C. W. A.: The computer simulation of RNA folding pathways using a genetic algorithm. Journal of Molecular Biology. 250 (1995) 37-51
6. Shapiro, B. A. and Wu, J. C.: Predicting RNA H-Type pseudoknots with the massively parallel genetic algorithm. Computer Applications in the Biosciences 13 (1997) 459-471
7. Shapiro, B. A. and Wu, J. C.: An annealing mutation operator in the genetic algorithms for RNA folding. Computer Applications in the Biosciences 12 (1996) 171-180
8. Shapiro, B. A., Wu, J. C., Bengali, D. and Potts, M. J.: The massively parallel genetic algorithm for RNA folding: MIMD implementation and population variation. Bioinformatics 17 (2001) 137-148
9. Benedetti, G. and Morosetti, S.: A genetic algorithm to search for optimal and suboptimal RNA secondary structures. Biophysical Chemistry 55 (1995) 253-259
10. Shapiro, B. A. and Navetta, J.: A massively parallel genetic algorithm for RNA secondary structure prediction. Journal of Supercomputing 8 (1994) 195-207
11. Deiman, B. A. and Pleij, C. W. A.: A vital feature in viral RNA. Seminars in Virology 8 (1997) 166-175
12. Hilbers, C. W., Michiels, P. J. A. and Heus, H. A.: New Developments in structure determination of pseudoknots. Biopolymers 48 (1998) 137-153
13. Einvik, C., Nielsen H., Nour, R. and Johansen, S.: Flanking sequences with an essential role in hydrolysis of a self-cleaving group l-like ribozyme. Nucleic Acids Res. 28 (2000) 2194-2200
14. Han, K., Lee, Y., Kim, W.: PseudoViewer: automatic visualization of RNA pseudoknots, Bioinformatics 18 (2002) S321-S328
15. Lee, D., Han, K.: Prediction of RNA pseudoknots–comparative study of genetic algorithms. Genome Informatics 13 (2002) 414-415

Computational Approach to Structural Analysis of Protein-RNA Complexes*

Namshik Han, Hyunwoo Kim, and Kyungsook Han**

School of Computer Science and Engineering, Inha University, Inchon 402-751, Korea
han_3567@hotmail.com, whytok@hanmail.net, khan@inha.ac.kr

Abstract. Analyzing protein-RNA binding structures depends on a significant quantity of manual work. Therefore, the protein-RNA binding structures are generally studied individually or on a small-scale. The task of analyzing the protein-RNA binding structures manually becomes increasingly difficult as the complexity and number of protein-RNA binding structures increase. We have developed a set of algorithms for automatically analyzing the structures of the protein-RNA complexes at an atomic level and for identifying the interaction patterns between the protein and RNA. The algorithms were implemented and tested on the actual structure data of 51 protein-RNA complexes. It is believed that this is the first structural analysis of a large set of protein-RNA complexes. The results of the analysis will provide insight into the interaction patterns between a protein and RNA, and will be useful in predicting the structure of the RNA binding protein and the structure of the protein binding RNA.

1 Introduction

Identifying how a protein molecule binds an RNA molecule with an affinity and specificity is a complex problem involving both spatial reasoning and an extensive knowledge of biochemistry. Individual or a small set of protein-RNA binding structures can be analyzed manually. However, as the number of protein-RNA complexes available in public databases is rapidly increasing a more systematic and automated method will be needed.

In contrast to the regular helical structure of DNA, RNA molecules form complex secondary and tertiary structures consisting of several structure elements, including stems, loops, and pseudoknots. Often, only specific proteins can recognize the structure elements arranged three-dimensional space. RNA structures display hydrogen bonding, electrostatic, and hydrophobic groups that can interact with small molecules to form specific contacts. However, it is unclear how the proteins interact with the RNA with specificity.

This paper presents a computational approach to analyze the interactions between the amino acids of a protein and the RNA nucleotides at the atomic level. The primary

* This work was supported by the Ministry of Information and Communication of Korea under grant number 01-PJ11-PG9-01BT00B-0012.

** To whom correspondence should be addressed.

P.M.A. Sloot et al. (Eds.): ICCS 2003, LNCS 2659, pp. 140–150, 2003.

focus of the work was to discover the conspicuous preferences in the pairing of amino acids with nucleotides. We have developed a set of algorithms for analyzing the protein-RNA binding structures and tested the algorithms on 51 protein-RNA complexes obtained from the PDB database [4]. The computational analysis attempted to address the following problems: (1) the hydrogen bonding propensity between the amino acids and the nucleotides, (2) the preferences between the main and side protein chains in the binding sites, and (3) the preferences between the base and RNA backbone in the binding sites.

2 Terminology and Notation

2.1 The Amino Acids: Main Chain, Side Chain

Proteins are sequences of up to 20 types of amino acids. Each amino acid contains the main chain of repeating units and one of 20 different "R" groups. It is the structure of the R group that determines the amino acid type. The "R" groups are referred to as the side chain and the other parts are called the main chain [1].

2.2 The Nucleic Acids: Base, Backbone, Base Pair, Base Step

Nucleic acids are sequences of nucleotides. A nucleotide consists of three parts: a five-carbon sugar, a nitrogen-containing ring structure called a base, and one, two, or three phosphate groups. Sugar and phosphorus are referred to as the backbone [1].

2.3 The Interactions: Single, Bidentate and Complex Interaction

As shown in Fig. 1, an interaction with one hydrogen bond between an amino acid and a nucleotide is called single interaction. An interaction with two or more hydrogen bonds with either a nucleotide or base paired nucleotides is known as a bidentate interaction. An interaction where an amino acid binds to more than one base step simultaneously is called a complex interaction. Our definition of the interactions is slightly different from Luscombe's definition [2]. Luscombe's definition considers the binding with the base part only, but we also consider the binding with the backbone part (that is, the sugar and phosphorus) [2]. Therefore, this study can show the difference in the binding propensities between the base, sugar and phosphorus.

2.4 The Hydrogen Bonds: Donor, Acceptor

A hydrogen bond is formed by three atoms: one hydrogen atom and two electronegative atoms (often N or O). The hydrogen atom is covalently bound to one of the electronegative atoms, and is called a hydrogen bond donor. The other electronegative atom is known as the hydrogen bond acceptor. The two electronegative atoms may take up some of the electron density from the hydrogen atom. As a result, each electronegative atom carries partial negative charge and the hydrogen atom carries a partial positive charge. Consequently, the hydrogen atom and the hydrogen bond acceptor can then have attractive interactions. The strength of the hydrogen bond depends on the donor and acceptor as well as their environment. The bond energy usually ranges from 1 to 5 kcal/mol. This energy is smaller than the covalent bond energy, but greater than the thermal energy (0.6 kcal/mol at room temperature). Therefore, a hydrogen bond can provide a significant stabilizing force in macromolecules such as proteins and nucleic acids [3].

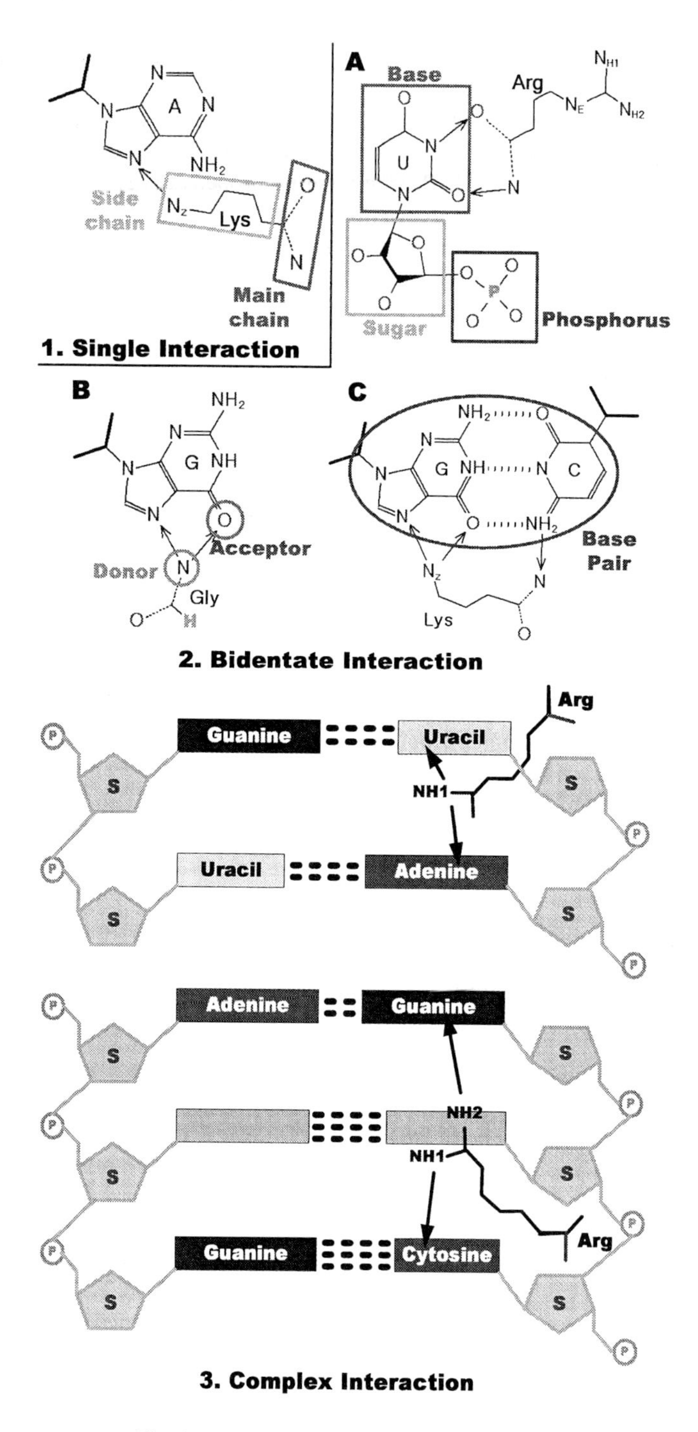

Fig. 1. Schematic diagram of three interactions

3 Framework for Analyzing Protein-RNA Binding Structures

3.1 Dataset

The protein-RNA complex structures were obtained from the PDB database [4]. The complexes, which were determined by X-ray crystallography with a resolution ≥3.0Å, were selected. As of September 2002, there were 188 protein-RNA complexes in PDB and the number of complexes with a resolution of 3.0 Å or better is 139. We used PSI-BLAST [5] for a similarity search on each of the protein and RNA sequences in these 139 protein-RNA complexes in order to eliminate the equivalent amino acids or nucleotides in homologous protein or RNA structures. 64 out of 139 protein-RNA complexes were left as the representative, non-homologous complexes after running the PSI-BLAST program with an E value of 0.001 and an identity value of 80% or below. We excluded 13 out of the 64 complexes that have no information of water molecules or are composed of artificial nucleotides. Therefore, the final data set was composed of 51 protein-RNA complexes. Table 1 shows the list of 51 protein-RNA complexes studied in our analysis.

Table 1. The 51 protein-RNA complexes in the data set

PDB code								
1B23	1B2M	1B7F	1C0A	1C9S	1CX0	1DFU	1DI2	1DK1
1E7X	1EC6	1EFW	1F7U	1F8V	1FEU	1FFY	1FXL	1G59
1GAX	1GTF	1GTN	1G2E	1H4Q	1H4S	1HC8	1HDW	1HE0
1HE6	1HQ1	1I6U	1IL2	1JBR	1JBS	1JID	1K8W	1KNZ
1KQ2	1L9A	1LNG	1MMS	1QF6	1QTQ	1SER	1URN	1ZDH
1ZDI	2BBV	2FMT	5MSF	6MSF	7MSF			

3.2 Hydrogen Bonds

The number of hydrogen bonds between the amino acids and nucleotides in the protein-RNA complexes was calculated using CLEAN, which is a program used for tidying Brookhaven files, and HBPLUS [6], which is a program to calculate the number of hydrogen bonds. The hydrogen bonds were identified by finding all proximal atom pairs that satisfy the given geometric criteria between the hydrogen bond donors (D) and acceptors (A). The positions of the hydrogen atoms (H) were theoretically inferred from the surrounding atoms, because hydrogen atoms are invisible in purely X-ray-derived structures. The criteria considered to form the hydrogen bonds for this study are: contacts with a maximum D-A distance of 3.9 Å, maximum H-A distance of 2.5 Å, and a minimum D-H-A angle and H-A-AA angle set to 90°, where AA is an acceptor antecedent.

All the hydrogen bonds were extracted from the HBPLUS output files. There were 1568 hydrogen bonds in the dataset. In order to compare the properties of a single interaction, a bidentate interaction and a complex interaction, separate experiments were conducted and the results were analyzed for the three classes of protein-RNA complexes: (1) single interaction, (2) bidentate interaction and (3) complex interaction. Figure 2 shows the sequence of classifying the three interaction types.

4 Algorithms

This section describes the seven algorithms, which were performed in sequence on the data shown in Fig. 2. The terminologies used in the algorithms are explained in Sect. 2.

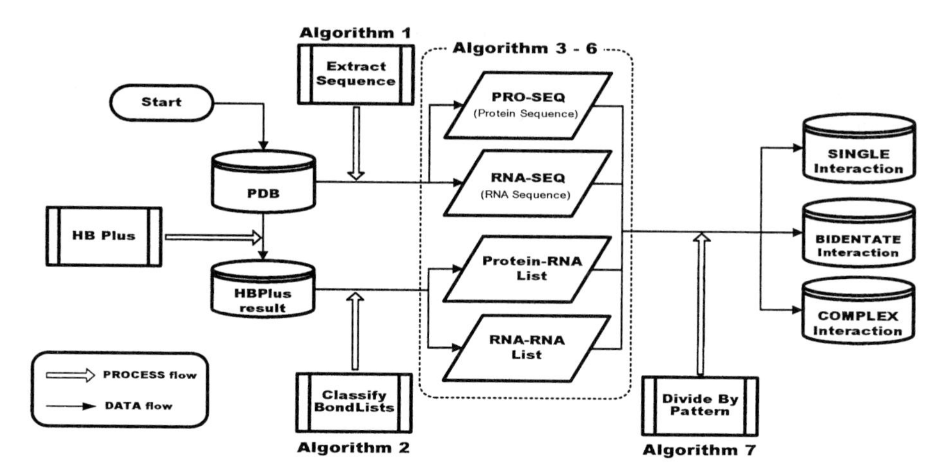

Fig. 2. Sequence of classifying three interaction types. The operation within a box labeled by an algorithm is explained in section 4. HBPLUS is explained in section 3.2.

Algorithm 1 constructs the PRO-SEQ and RNA-SEQ arrays to store the amino acid and RNA sequences, respectively. Algorithm 2 classifies hydrogen bonds into a P-R-List (list of hydrogen bonds between the protein and RNA) and R-R-List (list of hydrogen bonds between RNA and RNA). Algorithm 3 investigates the internal hydrogen bond relations of the RNA and marks the result of the investigation in a linked-list. Algorithm 4 investigates the hydrogen bonds between the protein and RNA and marks the result of the investigation in a linked-list. Algorithm 5 analyzes whether a nucleotide is paired with another nucleotide. It returns true if the nucleotide is paired. Algorithm 6 classifies the type of each amino acid into unary, binary and multi-bond based on the number of hydrogen bonds between the amino acid and the RNA. It calls Get_Max_Distance() to calculate the distance between a nucleotide and its binding amino acid. Algorithm 7 classifies the protein-RNA interaction types into three categories. The three categories are single interactions, bidentate interactions and complex interactions, as was explained in Sect. 2.

Algorithm 1

```
Extract RNA_Sequence from PDB file.
Extract PROTEIN_Sequence from PDB file.
RNA-SEQ = Make_RNA_Instance_Array(RNA_Sequence)
PRO-SEQ = Make_PROTEIN_Instance_Array(PROTEIN_Sequence)
```

Algorithm 2

```
PROCEDURE Classify_BondLists(hbfile)
 FOR EACH hbond in hbfile do
  IF (hbond.acceptor is amino_acid and hbond.donor is nucleo-
    tide) or
  (hbond.acceptor is nucleotide and hbond.acceptor is
    amino_acid)
  THEN P-R-List.append(hbond)
  ELSE IF hbond.acceptor is nucleotide and hbond.donor is nu-
    cleotide
  THEN R-R-List.append(hbond)
 endFOR
endPROCEDURE
```

Algorithm 3

```
PROCEDURE Mark_RNA_Internal_Bonds(RNA-SEQ, R-R-List)
  FOR EACH rr in R-R-List do
    RNA-SEQ[rr.acceptor.number].MarkContact(rr.acceptor.atom,
   rr.donor)
    RNA-SEQ[rr.donor.number].MarkContact(rr.donor.atom,
    rr.acceptor)
  endFOR
endPROCEDURE
```

Algorithm 4

```
PROCEDURE Mark_Protein_RNA_Bonds(RNA-SEQ, P-R-List)
  FOR EACH pr in P-R-List do
    PRO-SEQ[pr.aminoacid.number].MarkContact(pr.aminoacid.atom,
   pr.nucleotide)
  endFOR
endPROCEDURE
```

Algorithm 5

```
PROCEDURE Set_RNA_Struc(RNA-SEQ)
  FOR EACH nucleotide in RNA-SEQ do
    nucleotide.pair = false
    IF the base of the nucleotide binds with another THEN
      nucleotide.pair = true
    endIF
  endFOR
endPROCEDURE
```

Algorithm 6

```
PROCEDURE Classify_Protein_RNA_Bonds(PRO-SEQ)
  FOR EACH aminoacid in PRO-SEQ do
    IF SIZE(aminoacid.contact_list) == 0 THEN
      aminoacid.pattern = nothing
    ELSE IF SIZE(aminoacid.contact_list) == 1 THEN
```

```
      aminoacid.pattern = unary
    ELSE
      IF Get_Max_Distance(aminoacid) == 0 THEN
        aminoacid.pattern = binary
      ELSE
        aminoacid.pattern = multi
      endIF
    endIF
  endFOR
endPROCEDURE
PROCEDURE Get_Max_Distance(aminoacid)
  first = aminoacid.contact_list.pop()
  chain = first.chain
  min = max = first.number
  FOR EACH contact_nuc in aminoacid.contact_list do
    IF contact_nuc.chain != chain THEN
      return Great_Number
    ELSE IF contact_nuc.number > max THEN
      max = contact_nuc.number
    ELSE IF contact_nuc.number < min THEN
      min = contact_nuc.number
  endFOR
  return max - min
endPROCEDURE
```

Algorithm 7

```
PROCEDURE DivideByPattern(PRO-SEQ, P-R-List)
  FOR EACH aminoacid in PRO-SEQ do
    IF aminoacid.pattern == single THEN
      SINGLE-LIST.append(aminoacid)
    ELSE IF aminoacid.pattern == double THEN
      DOUBLE-LIST.append(aminoacid)
    ELSE IF aminoacid.pattern == multi THEN
      IF aminoacid bond with base pair THEN
         BIDENTATE.append(bin)
      ENSEIF aminoacid bond with base step THEN
         COMPLEX.append(bin)
      ELSE
        SINGLE.append(bin)
    endIF
  endFOR
endPROCEDURE
```

5 Results and Discussion

The protein-RNA interactions were computed for the 51 complexes. All interactions that occur repeatedly in the structurally related complexes were removed from the dataset in order to exclude any bias towards the proteins with multiple PDB entries. This filtering process resulted in a dataset of 1568 hydrogen bonds, which were the subject of this analysis.

5.1 Amino Acids

Among the three interaction types, single interactions are the most abundant, as shown in Table 2. Amino acids involved in the bidentate interactions predominantly interact with bases. Table 3 shows the ranking of bindings of the amino acids with nucleotides in each interaction type. The dominant bindings are shown in bold style. For example, PHE-G S in the bold style denotes that PHE binds mainly to the sugar part of guanine but not to the other parts. No hydrophobic residues were found in the complex interactions. In particular, ALA, MET, PRO and VAL were found in single interactions only.

Table 2. The number of hydrogen bonds in the nucleotides for each interaction type

	BASE	SUGAR	PHOSPHORUS	TOTAL
SINGLE	464 *(46%)*	267 *(26%)*	281 *(28%)*	1012 *(65%)*
BIDENTATE	286 *(60%)*	142 *(30%)*	48 *(10%)*	476 *(30%)*
COMPLEX	16 *(20%)*	29 *(36%)*	35 *(44%)*	80 *(5%)*
TOTAL	766 *(49%)*	438 *(28%)*	364 *(23%)*	1568 *(100%)*

Table 3. The top ten hydrogen bonds frequently participating in each interaction type. The bonds in the bold style represent they occur frequently

rank	TOTAL	rank	SINGLE	rank	BIDENTATE	rank	COMPLEX
1	GLU - G B 89	**1**	**THR - G B 37**	**1**	**GLU - G B 74**	1	ARG - C P 10
2	LYS - A B 69	**2**	**PHE - G S 29**	2	LYS - A B 41	2	ARG - U P 7
3	ASP - G B 60	2	ARG - U B 29	3	THR - A B 36	3	ARG - A P 5
4	THR - A B 58	4	LYS - A B 28	**4**	**ASP - G B 31**	3	ASP - G S 5
5	ARG - C P 42	4	LYS - C P 28	5	ARG - C B 23	3	ARG - C S 5
6	SER - G S 39	6	ASP - G B 27	6	ASN - U S 18	6	ARG - U S 4
7	THR - G B 38	7	SER - A B 25	7	SER - G S 13	7	ARG - G P 3
7	ARG - U B 38	7	SER - G S 25	**8**	**HIS - G S 10**	7	HIS - G S 3
9	ARG - C B 35	7	LYS - U P 25	8	SER - U S 10	7	LYS - G B 3
10	PHE - G S 31	10	LYS - A P 24	10	THR - A S 9	7	LYS - C P 3
10	LYS - C P 31	10	LYS - G B 24	10	ARG - U B 9		
		10	ARG - C P 24	10	ARG - U S 9		
		10	ASN - C- B24				

A: adenine, G: guanine, C: cytosine, U: uracil
B: base of nucleotide, S: sugar of nucleotide, P: phosphorus of nucleotide

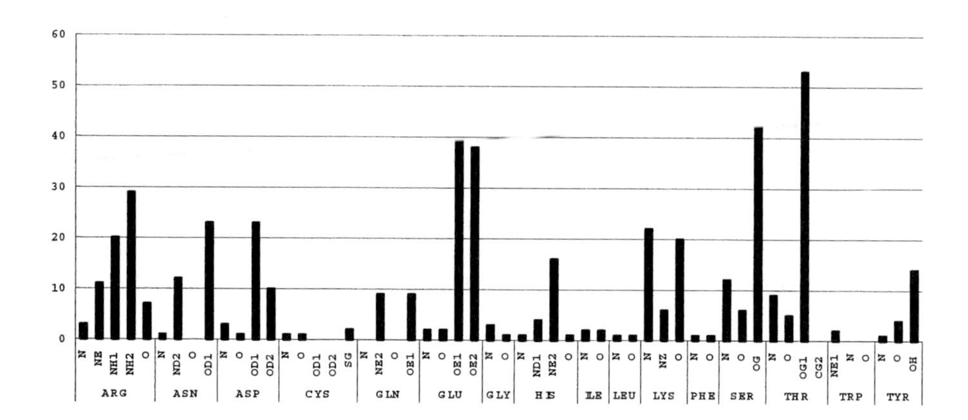

Fig. 3. The total number of hydrogen bonds of each amino acid, involved in the bidentate interactions with nucleotides. ALA, MET, PRO and VAL are not shown here because they were not involved in any bidentate interaction

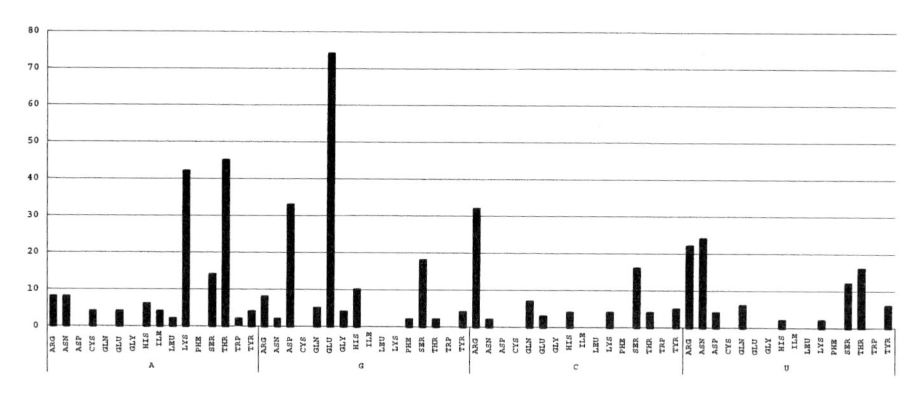

Fig. 4. The total number of hydrogen bonds of each nucleotide, involved in the bidentate interactions with amino acids

The number of hydrogen bonds in the amino acids is as follows. ARG (306), LYS (257), SER (164), THR (151), GLU (136), ASN (125), ASP (116), GLN (61), TYR (59) and GLY (40). But the ordering changes when the hydrogen bonds are classified based on whether they are found in the base, sugar or phosphorus of a nucleotide. LYS and THR frequently participate in the bonds with a base but not with sugar. In contrast, ARG and SER frequently participate in bonding with the backbone but not with the base. THR showed a significant dissimilarity from SER. THR is 2nd most common in binding to a base but the 10th in binding to a sugar. In contrast, SER is the 7th most common in binding to a base but 2nd in binding to a sugar. Fig. 3 shows the number of hydrogen bonds of each amino acid, involved in the bidentate interactions.

From an atomic point of view, the LYS NH binds to the phosphorus of the nucleotides 87 times and is the most frequent participator. Next are the ARG NH2-phosphorus and the THR O-base, which bind 58 times. The order of the others is as follows: ASP OD1-base (55 times), GLU OE2-base (51), SER OG-sugar (51), LYS NZ-base (49), SER OG-base (49), ARG NH2-base (45) and GLU OE1-base (45). This study made two interesting observations. The first observation was that 82 LYS NZ-phosphorus bonds out of a total 87 were observed in the single interactions. However, the THR OG1-base bond participates mainly in bidentate interactions. The second observation was that even though the THR O and OG1 consist of the same amino acids and bind to the same nucleotide part, this study found that the O is mainly involved in single interactions and the OG1 is involved in the bidentate interactions. This is partly, but not entirely, due to the difference in the side chain and the main chain. Both the THR OG1 and OG2 are located in the amino acid side chain. However, OG2 never binds while OG1 binds 46 times. Different amino acids showed different binding propensities depending on the interaction types. GLU and HIS participate mainly in bidentate interactions. In particular, GLU binds to a base frequently because GLU OE1 and OE2 have a strong binding propensity.

5.2 Nucleotide

On average, hydrogen bonds prefer the bases (49%) than the sugar (28%) and phosphorus (23%) of a nucleotide, but there is little difference in preference between the bases and the backbone (Table 2). The specific binding propensities change according to interaction types. In bidentate interactions, the hydrogen-bonding rate of the base increases to 60% and that of phosphorus decreases to 10%. In contrast, in complex interactions, the hydrogen-bonding rate of the base decreases to 20% and that of phosphorus increases to 44%. This difference can be explained by the structure. In complex interactions, an amino acid binds to at least two base pairs, which is a stem. Thus, an amino acid held within a stem has a great chance to bind to the backbone.

All the nucleotides bind frequently to ARG and LYS. However, guanine binds predominantly with GLU and ASP, which have an acidic side chain group, and adenine binds frequently to THR. This is because GLU OE1, OE2 and THR OG1 bind preferentially to the base in bidentate interactions. Figure 4 shows the number of hydrogen bonds of each nucleotide, which are involved in bidentate interactions.

6 Conclusion

Structural analysis of the protein-RNA complexes is labor-intensive yet it provides insight into the interaction patterns between a protein and RNA. The protein-RNA binding structures are studied either individually or on a small-scale. However, manual analysis of the protein-RNA binding structures is becoming increasingly challenging as the complexity and number of protein-RNA complexes increase. This study developed a set of algorithms for automatically analyzing the hydrogen bonds in the protein-RNA binding structures and for identifying the interaction patterns between a protein and RNA. The algorithms were used for analyzing 1568 hydrogen bonds in 51 protein-RNA complexes, which are the most representative set of protein-RNA complexes known today. This is the first computational approach for analyzing the structures of a large set of protein-RNA complexes. The interaction patterns discovered from this analysis will assist in the understanding of how proteins interact with RNA with specificity and to predict the structure of the RNA binding protein as well as the structure of the protein binding RNA.

References

1. Lesk, A.M.: Introduction to Protein Architecture (2nd Edition) Oxford University Press (2001)
2. Luscombe, N.M., Laskowski, R.A., Thornton, J.M.: Amino acid–base interactions: a three-dimensional analysis of protein–DNA interactions at an atomic level. Nucleic Acids Research 29 (2001) 2860–2874
3. Web-Book Home Page http://www.web-books.com/
4. Berman, H.M., Westbrook, J., Feng, Z., Gilliland, G., Bhat, T.N., Weissig, H., Shindyalov, I.N., Bourne, P.E.: The Protein Data Bank. Nucleic Acids Research 28 (2000) 235–242
5. Altschul, S.F., Madden, T.L., Schaffer, A.A., Zhang, J., Zhang, Z., Miller, W., Lipman, D.J.: Gapped BLAST and PSI-BLAST: a new generation of protein database search programs. Nucleic Acids Research 25 (1997) 3389–3402
6. McDonald, I.K. Thornton, J.M.: Satisfying Hydrogen Bonding Potential in Proteins. J. Mol. Biol. 238 (1994) 777–793

Improved Web Searching through Neural Network Based Index Generation

Xiaozhe Wang, Damminda Alahakoon, and Kate A. Smith

School of Business Systems, Faculty of Information Technology,
Monash University, Clayton, Victoria 3800, Australia
{catherine.wang,damminda.alahakoon,
kate.smith}@infotech.monash.edu.au

Abstract. In this paper we propose a method to improve web search results in search engines. The Self Organizing Map is used for clustering query logs in order to identify prominent groups of user query terms for further analysis. Such groups can provide meaningful information regarding web users' search interests. Identified clusters can further be used for developing an adaptive indexing database for improving conventional search engine efficiency. The proposed hybrid model which combines neural network and indexing for web search applications can provide better data filtering effectiveness and efficiently adapt to the changes based on the web searchers' interests or behaviour patterns.

1 Introduction

Understanding the needs, interests and knowledge of the web users has grown to be an important research area in the recent past [9]. Web mining is a popular method which is being used for this purpose. Web mining includes web usage mining, web content mining and web structure mining [2,12]. Web log files are the basic data source for extracting useful information in web usage mining, and web documents are used in the web content mining process [11]. Our previous work has described several useful techniques of web mining using soft computing techniques [14,15]. This paper describes our current work towards developing an integrated hybrid system which can make use of web mining results to provide improved functionality to web users.

With the hybrid approach, we propose to build an adaptive web searching system by combining neural network based clustering, cluster analysis and a dynamic indexing technique. The proposed system makes use of the discovered knowledge from data clusters of web query logs to develop an index for more efficient web searching. The novelty of this method is that the index would be updated routinely when the web query patterns change. In this paper we demonstrate the cluster generation and analysis techniques with experiments. In addition we propose an algorithm for identifying change (or shift) in web search patterns by recognizing *movement* and *shift* in data

P.M.A. Sloot et al. (Eds.): ICCS 2003, LNCS 2659, pp. 151–158, 2003.

clusters. Such change identification can then be used to update the index with current trends. The development of this complete adaptive system is presented as on-going work.

The rest of the paper is organised as follows. Section 2 describes the proposed hybrid system for adaptive query searching through web log mining. Experimental results on web query term clustering is provided and discussed in Sect. 3. Section 4 concludes the paper with details of ongoing and future work.

2 A Hybrid System for Improved Web Searching

The hybrid approach to developing the adaptive web searching system is as follows. The query log data are initially collected and pre-processed to convert to a suitable format for clustering and analysis. The pre-processing is described in Sect. 2.1. The pre-processed data are then clustered using the Self Organising Map (SOM) algorithm. The SOM is used due to its unsupervised nature and also its ability to handle high dimensional and large data sets [3]. The clustering is used to identify web searchers' query interest patterns from web query log data and further to create pattern files identifying different characteristics from clustered data results [8]. Results are used to develop a web documents indexing database. Since the web users search interests can change over time the index needs to adapt with the changes. We propose a technique for identifying change in data (Sect. 2.3) and such change can then be used to update the current user search terms index. The approach proposed in our research is represented in Fig. 1. As such the system could be dynamic instead of static with the new data and changes.

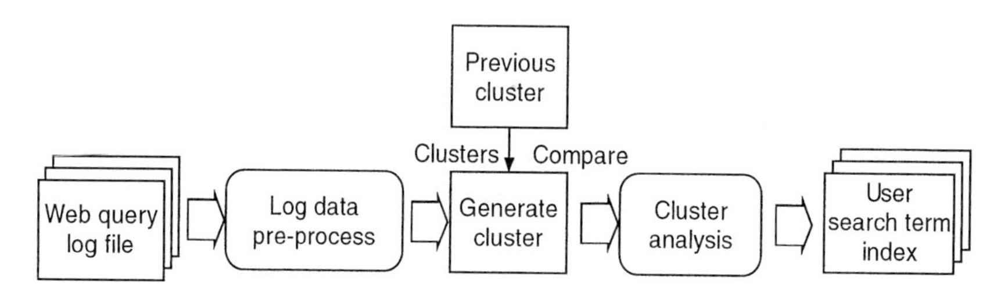

Fig. 1. Dynamic web searching system

2.1 Data Collection and Pre-processing

Query Log File Data Collecting. Experiments described in this paper were conducted with data acquired from Monash University's web query log files located at the main server in Melbourne [7]. The query logs record all past search query entries requested by web searchers on a weekly basis.

Query logs used cover the time range from 00:00:49 14 July 2002 to 00:00:04 25 August 2002 that represent 6 weekly log files in text format. The size of the weekly query log files varies from 2.6 megabytes to 4.1megabytes and consist of 35322 to 61745 query entries. Each entry in the initial log file records a single query, the *date and time stamp*, *number of documents retrieved* and *original query input*. The unprocessed logs contain data of the following form as shown in Table 1:

Table 1. Format of unprocessed query entries in web log files

Date	Time	Result	Query entry
2002/08/18	00:00:36	6	'http://www.adm.monash.edu.au/sss/handbook'
2002/08/18	00:00:39	75	'clayton map CSE2002'
2002/08/18	00:01:39	7854	'oversea fee'

Query Term Data Set Pre-processing. In order to construct a training set for the clustering algorithm, the data from the original log files is passed through the following three stages of pre-processing: *session identification, term identification* and *data set identification.*

Session Identification. Initially, non-null queries were extracted from the original log data and the data set session used for processing was identified. Time series data is a common feature for web log data. A session of the data set can be identified based on different time frames according to the difference in the analysis purpose. Since the university runs on a semester basis, and within each semester, the schedule is arranged on a weekly basis, in this research, a week is identified as the session time frame for data set determination. A frequency rule also was used to ensure that normal patterns could be obtained such as the distribution of data frequency on a weekly basis.

Term Identification. The text terms entered by web searchers represent their search topics. Each single term in the log data needs to be identified to process further analysis procedures. From the initial sectionalized log data, each text term and its frequency (number of appearances) is extracted, and the frequency is used as a weight to build up a vector for each term during the clustering process. For example, from a weekly log of 51474 query entries, 15735 text terms were extracted with occurrence values differing from 1 to 5084. From the huge amount of terms, there were only a limited number of terms representing common interests from the majority of web searchers. A high-frequency rule was applied in term identification by comparing it with a predefined threshold. If we denote the *total query number* as Q_t, the *null query number* (queries with no results) as Q_n and the *identifying parameter* as IP, the value of is experimentally set in the range $IP \in [0.0001, 0.01]$, so that 0.001 means that a term must appear in more than 0.1% of the query entries. In the experiments, by using a *frequency threshold* $T_f = (Q_t - Q_n) * IP$, for instance, if a log file has Q_t and Q_n numbers are 51474 and 48 respectively, the frequency threshold was 51 by using $IP = 0.1\%$. As a result, 271 terms survived after this selection.

After data pre-processing, each session had quite similar number of text terms with a stable trend after different steps as shown in Fig. 2, which reveals that there are a certain number of common query terms requested by most of web searchers all the time. It would be interesting to see whether same procedures could apply for other web environments as well.

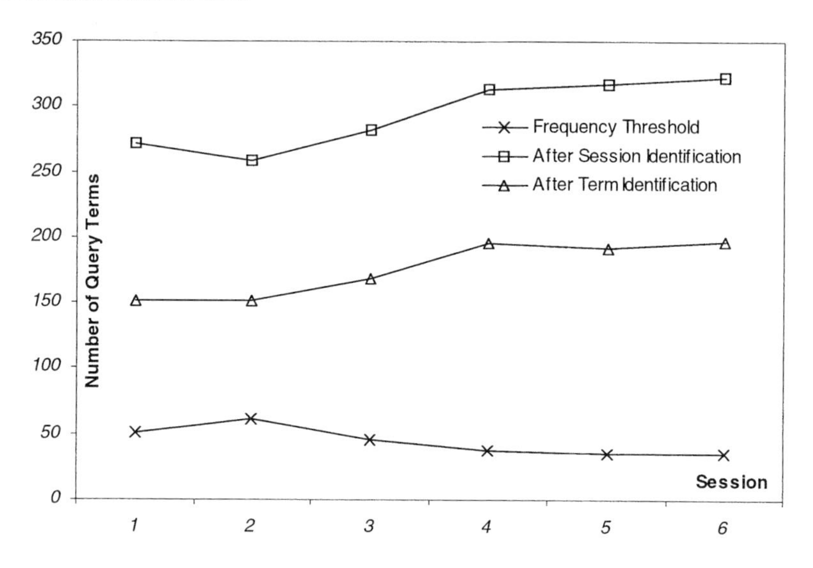

Fig. 2. Comparison of number of terms selected in different steps of data pre-processing

Data Set Identification. Since only the single text terms are extracted for analysis, noisy data including URLs and different versions of the same word are removed before forming the data set for further processing. From the experiments results, different sessions had a different number of query terms. After deciding the number of sessions used for forming a processing data set, all the terms were merged as a group in the identified data set. For instance, the number of query terms after the term identification step in 6 sessions of logs covered from 14 July 2002 to 18 August 2002 varies from 152 to 197. After the merging step, there were finally 218 terms selected in the identified data set. Then the query terms were represented as high-dimensional vectors of sessional *weights* w. The value of dimension was decided by the number of sessions used in data set. If n sessions of data were used in the data set, the *query term* (T_q)'s occurrence value in each session was w_i, and the query term vectors represented as $T_q = (w_1, w_2, ..., w_n)$. The value of the dimension increases when more sessions of data are used for the process.

2.2 Clustering of Web Query Terms

Clustering-based algorithms have gained a lot of attention lately and have been used in a variety of applications. SOM has been successfully used in a number of web mining projects [4][6] as one of the most appropriate techniques for the Web documents clustering process because of its strength in both flexibility of grouping and visualization for the clusters [3]. With web data, the related transaction entries are grouped into the same cluster and the relationship between different clusters is explicitly shown on the map.

2.3 Data Shift and Movement Identification Technique

Since the proposed system requires adaptation to user trend changes, the following method is suggested to identify such changes in data. It is proposed that any changes in data will be reflected in the subsequent clusters and as such an algorithm [1] for comparing clusters is made use of.

A measure called the cluster error (ERR_{cl}) between two clusters in two SOMs is defined as: $ERR_{cl}(Cl\_j(MAP1), Cl\_k(MAP2)) = \Sigma | A\_i(Cl\_j) - A\_i(Cl\_k) |$ where Cl_j and Cl_k are two clusters belonging to $MAP1$ and $MAP2$ respectively and $A\_i(Cl\_j)$, $A\_i(Cl\_k)$ are the attribute of clusters Cl_j and Cl_k.

We define two clusters similar when the condition $ERR_{cl}(Cl\_j, Cl\_k) \leq T_{ce}$ is satisfied, T_{ce} is the threshold of cluster similarity ($0 < T_{ce} < D/2$, where D is the dimension of the data) and has to be provided by the data analyst depending on the level of similarity required. If complete similarity is required, then $T_{ce} = 0$.

Since the similarity between two clusters depends on the T_{ce} value, we define a new indicator I_s called 'the measure of similarity' which indicates the amount of similarity when two clusters are considered to be similar. Then I_s is calculated as the fraction of the actual cluster error to the maximum tolerable error for two clusters to be considered similar $I_s = 1 - ERR_{cl}(Cl\_j, Cl\_k) / Max(T_{ce})$.

3 Experimental Results on Query Log Analysis

We used the Viscovery SOMine [13] to simulate the SOM. Data set contains 6 sessional frequencies were used as weights to form attributes for query term vectors from the pre-processing step was fed in to SOM algorithm [5] and clustering results were obtained after unsupervised learning.

3.1 Web Users' Query Interest Pattern Analysis

After feeding the pre-processed data set into the SOM, the most natural clustering results were discovered with a stable number of clusters. The 4 clusters were separated based on their own characteristics and the pattern of each cluster used to analyse for obtaining a better understanding of Web searchers' behavior (the fundamental knowledge to further build up online adaptive searching). Based on the degree of the query entry frequency and the time frame, we found interesting information representing each cluster, as illustrated in Fig. 3.

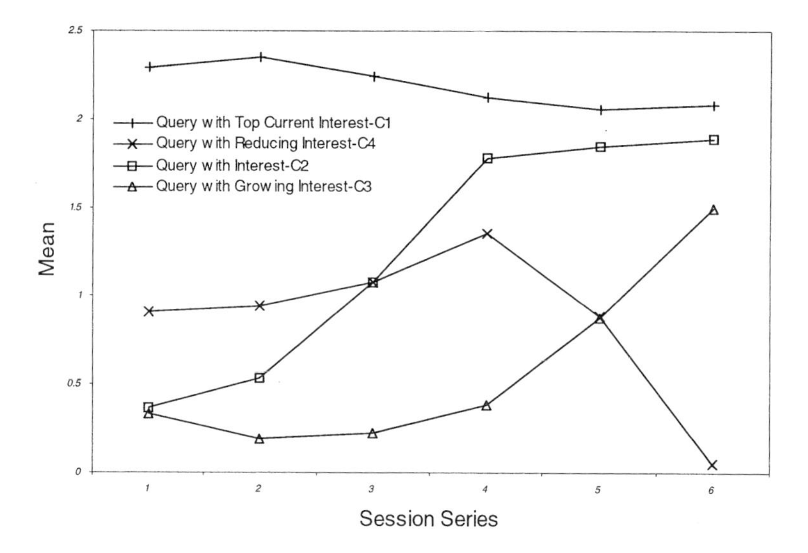

Fig. 3. Searching query term movement pattern of 4 categories from data set's clustering result

By computing the frequency degree and the time based trend change, the web searcher's interest patterns were categorized as 1) *Query with top in current interest*, 2) *Query with interest*, 3) *Query with growing interest* and 4) *Query with reducing interest.*

In the group 'Query with top in current interest', all terms have the same common feature that they remain the highest request rate most of the time within the whole period. These clustered terms directly reveal the search entries with highest frequency requested through the embedded search engine, and also represent the web users' current interests contents for the web site. Using this information, terms in this cluster with the highest priority can be used to generate the frequently requested areas index documents for searching.

With 'Query with interest', the frequency rate starts from a very low point and grows very fast to nearly close to the top level. By analysing detailed terms in the cluster, we found that this cluster could be also categorized as 'Query with growing interest'. Terms representing a growing interest trend from the web users for a longer period of time suggest a high possibility for them to move in to upgrade cluster in the coming period.

In the patterns for categories of 'Query with growing interest' and 'Query with reducing interest' are very similar. In the first half of time period, the frequency rate shows very minor change, but after the middle point, both the growing and reducing movements become significant until the end. For the growing interest trend, requests grow to almost equal to half of the top level, and the reducing trend moves to almost 0. The web documents' topics associated with those terms can be indexed in the temporary indexing database based on the movement of web user's interest change.

3.2 Adaptive Web Query Searching

To achieve dynamic and adaptive web query searching [10], 2 components are included in the hybrid system proposed in our research. Offline web query term pattern files construction and online web document indexing. From the clustered data, different features and characteristics are demonstrated as shown in Fig. 3. As such the experimental results demonstrate the offline part of the system and the associated possibilities.

Generating web document index is considered as the online part, since to be useful, the index has to dynamically adapt to the changes in the user search patterns. This online system uses the cluster comparison and change identification technique presented above and experiments are currently being carried out to ascertain the advantages of the method.

4 Current Work and Future Research Direction

In this paper we have proposed a system for analysing and improving web searching by using web query log data. The system described is in two parts as an offline data analysis part and an online dynamic index building and searching part. Experimental results were presented to demonstrate the potential of the system after detailed descriptions of the pre-processing of data has been presented. It is required to run similar experiments on several consecutive time (session) log files to confirm the apparent potential from this method. The terms in each cluster will need to be compared with terms in the clusters in the next session to identify the trends such as current interest terms losing interest over time.

References

1. Alakakoon, L. D.: Data Mining with Structure Adapting Neural Networks. PhD thesis, Monash University (2000)
2. Chang, G., Healey, M. J., McHugh, J. A. M., Wang, J. T. L.: Web Mining, Mining the World Wide Web. Kluwer Academic Publishers, Chapter 7 (2001) 93–104
3. Flakes, G. W., Lawrence, S., Giles, C. L., Coetzee, F. M.: Self-organization and Identification of Web Communities. IEEE Computer, vol. 35, no. 3 (2002) 66–71
4. Honkela, T., Kaski, S., Lagus, K., Kohonen, T.: WEBSOM – Self-organizing Maps of Document Collections. Proceedings of Workshop on Self-Organizing Maps (WSOM'97), Espoo, Finland (1997) 310–315
5. Kohonen T.: Self-Organizing Maps. 2nd edition, Springer, Heidelberg (1997)
6. Kohonen, T., Kaski, S., Lagus, K., Salojarvi, J., Honkela, J., Paatero, V., Saarela, A.: Self Organization of a Massive Documents Collection. IEEE Transactions on Neural Networks, Special Issue on Neural Networks for Data Mining and Knowledge Discovery, vol.11, no.3 (2000) 574–585
7. Monash University Web Site, <http://www.monash.edu.au>
8. Ng, A., Smith, K. A.: Web Page Clustering Using A Self-Organizing Map of User Navigation Patterns. Smart Engineering System Design: Neural Networks, Fuzzy, Logic, Evolutionary Programming, Data Mining, and Complex Systems, Missouri, USA (2000)
9. Paliouras, G., Papatheodorou, C., Karkaletsisi, V., Spyropoulous, C. D.: Clustering the Users of Large Web Sites into Communities. Proceedings of the 17th International Conference on Machine Learning (ICML'00), Stanford University (2000)
10. Perkowitz, M., Etzioni, O.: Towards Adaptive Web Sites: Conceptual Framework and Case Study. Computer Networks, Amsterdam, Netherlands, Artificial Intelligence (2000)
11. Pirolli, P., Pitkow, J., Rao, R.: Silk From a Sow's Ear: Extracting Usable Structures from the Web. Proceedings on Human Factors in Computing Systems (CHI'96), ACM Press (1996)
12. Srivastava, J., Cooley R., Deshpande, M. Tan, P. N.: Web Usage Mining: Discovery and Applications of Usage Patterns from Web Data. SIGKDD Explorations, vol. 1, no. 2 (2000) 12–23
13. Viscovery SOMine, <http://www.eudaptics.com/technology/somin4.html>
14. Wang, X., Abraham, A., Smith, K. A.: Soft Computing Paradigms for Web Access Pattern Analysis. Proceedings of the 1st International Conference on Fuzzy Systems and Knowledge Discovery, vol. 1 (2002) 631–635
15. Wang, X., Smith, K. A.: Clustering Web User Interests Using Self Organising Maps. Proceedings of the 2nd International Conference on Hybrid Intelligent Systems, Soft Computing Systems: Design, Management and Applications, IOS Press, the Netherlands (2002) 843–852

Neural Network for Modeling Non Linear Time Series: A New Approach

Chokri Slim[1] and Abdelwahed Trabelsi[2]

[1] University of Tunis, Bestmod Laboratory, 41rue de la liberté
2000 Le Bardo, Tunisia
chokri.slim@iscae.rnu.tn
[2] University of Tunis, Bestmod Laboratory, 41rue de la liberté
2000 Le Bardo, Tunisia
abdel.tabelsi@isg.rnu.tn

Abstract. Nonlinear modeling with neural networks offers a promising approach for studying the prediction of a chaotic time series. In this paper, we propose a neural net based on Extended Kalman Filter to examine the nonlinear dynamic proprieties of some financial time series in order to differentiate between low-dimensional chaos and stochastic behavior. Kalman filtering, because it can deal with varying unobservable states, provides an efficient framework to model these non-stationary exposures. A controlled simulation experiment is used to introduce the issues involved and to present the proposed approach. Measures of forecast accuracy are developed. The pertinence of this model is discussed from the Tunisian Stock Exchange database.

1 Introduction

Within the past decades, there has been a growing interest in applying nonlinear models to predict chaotic time series [1]. The major problem in these researches is the difficulty of distinguishing between deterministic chaos and purely random processes. An appreciable amount of recent literature deals with this subject [2] and [3].

By new neural network architecture we allow additional information flows between the different outputs and as a sequence we get a better representation of the underlying dynamic system. The net is based on multilayer feed forward architecture with random connections. The model can then convert to its equivalent state space representation. Using this state space form, a Bayesian inferential algorithm based on non linear Kalman Filter is derived to estimate the state. We analyze the net of a chaotic time series (logistic map) by using the dynamic invariant that characterizes the attractor, the largest Lyaponov exponent. This approach permits us to claim that the net capture "chaos" if it learns the dynamic invariant of a chaotic time series. A detailed step by step description of the methodology is presented to facilitate the use of this new method. Some issues related to the practical use of the proposed model in the context of the Tunisian Financial market are also considered.

P.M.A. Sloot et al. (Eds.): ICCS 2003, LNCS 2659, pp. 159–168, 2003.

2 Neural Net Architecture

A general form of nonlinear time series with functional coefficients can be given by:

$$Y_t = f\,(\mathbf{X}_t, \varphi(\mathbf{X}_t), \gamma(\mathbf{X}_t)) + \xi_t \tag{1}$$

Where f (.) : is unknown non linear function. Y_t : the time series.
$\mathbf{X}_t = (Y_{t-1}, Y_{t-2}, \ldots, Y_{t-p}, Z_{t-d-1}, Z_{t-d-2}, \ldots, Z_{t-d-m})$: is a vector of p-past values time series Y_t and m-past values input Z_t,. This vector reflect some changing environmental conditions that causes the system's parameters to vary; p, m, d : represent the orders and time delay of the mode; Z_t: the measured input; $\varphi(\mathbf{X}_t)$ and $\gamma(\mathbf{X}_t)$: are vectors of unknown parameters which are functions of $\mathbf{X}_t$; ξ_t: is a sequence of *iid* random variables with mean 0 and variance R_t.

The main difficulty in using the proposed model in (1) is specifying the functional coefficients. We consider in this section, a special class of neural networks called Tangent Hyperbolic neural Networks **THNN** to identify the model in (1). The most important feature of the **THNN** is the smooth output which is due to the shape of the tangent hyperbolic functions. It has been proved that this class of networks can approximate continuous functions at any arbitrary accuracy. A complete description of a Neural Networks theory and the application of neural networks to the problem of nonlinear system identification and prediction can be found in [5], [6]. The non linear coefficients φ_i $(\mathbf{X}_t)$ and γ_j $(\mathbf{X}_t)$ in (1) can be approximated by a sub-**THNN**, then:

$$\hat{\varphi}_i\,(\mathbf{X}_t) = \sum_{k=1}^{q} w_k^{\varphi_i}\, h_k\,(\mathbf{X}_t)\,. \tag{2}$$

$$\gamma_j\,(\mathbf{X}_t) = \sum_{k=1}^{q} w_k^{\gamma_j}\, h_k\,(\mathbf{X}_t)\,. \tag{3}$$

Representing each parameter of (2) and (3) as a neural network in figure1, the overall estimate model (1) can be represented in figure1, which is a two layered **THNN** structure. The first layer is composed of sub networks and the second layer is the output given by:

$$Y_t = h(\mathbf{X}_t\,, \hat{\varphi}(\mathbf{X}_t)\,, \hat{\gamma}(\mathbf{X}_t)\,) + \xi. \tag{4}$$

We call this new neural network a stochastic neural net (**SNN**). We propose, now, a Gradient Back propagation algorithm to train the free parameters which are the weights of the sub networks in the first layer of the **SNN**, while the "weights" of the second layer are viewed as fixed in every iteration of this algorithm. Denote:

$$\beta_t^{\varphi_i} = \left[w_1^{\varphi_i}, w_2^{\varphi_i}, \ldots, w_q^{\varphi_i} \right] \in IR^q,\ i = 1, \ldots, m\ . \tag{5}$$

$$\beta_t^{\gamma_j} = \left[w_1^{\gamma_j}, w_2^{\gamma_j}, \ldots, w_q^{\gamma_j} \right] \in IR^q,\ j = 1, \ldots, m\ . \tag{6}$$

$$\mathbf{B}_{t-1} = \left[\beta_t^{\varphi_1} \beta_t^{\varphi_2} \ldots \beta_t^{\varphi_p} \; \beta_t^{\gamma_1} \beta_t^{\gamma_2} \ldots \beta_t^{\gamma_m}\right]^{\mathrm{T}} \in IR^{qx(p+m)}. \tag{7}$$

$$\mathbf{h}_{t-1} = \left[h_1(\mathrm{X}_t), h_2(\mathrm{X}_t), \ldots, h_q(\mathrm{X}_t)\right]^{\mathrm{T}} \in IR^{\mathrm{q}}. \tag{8}$$

$$\mathbf{F}_{t-1} = \left[Y_{t-1}, Y_{t-2}, \ldots, Y_{t-p}, Z_{t-d-1}, \ldots, Z_{t-d-m}\right]^{\mathrm{T}} \in IR. \tag{9}$$

$$\mathbf{O}_{t-1} = \left[\varphi_1(\mathbf{X}_t), \ldots, \varphi_p(\mathbf{X}_t), \gamma_1(\mathbf{X}_t), \ldots, \gamma_m(\mathbf{X}_t)\right] \in IR^{\mathrm{p+m}}. \tag{10}$$

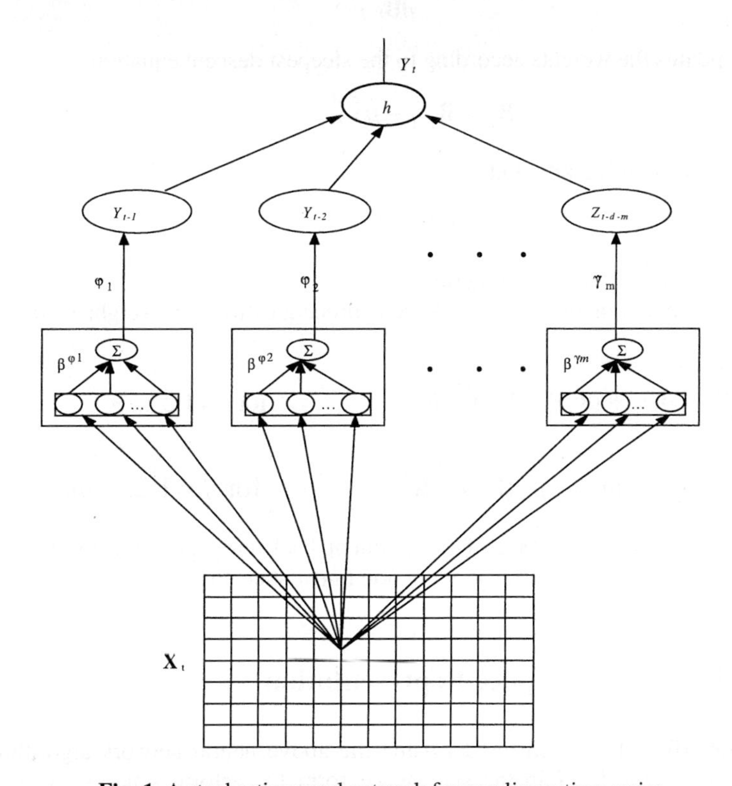

Fig. 1. A stochastic neural network for non linear time series

A time t, the input vector $\mathbf{F}_{t-1}$ is presented to the neural network shown in figure1. In the forward pass the networks calculates the output according to (2) and (3). The output of the first layer and final output of the **SNN** are given by:

$$\mathbf{O}_{t-1} = \mathbf{h}_{t-1}\,\mathbf{B}_{t-1}. \tag{11}$$

$$\hat{Y}_t = h(\mathbf{O}_{t-1}\mathbf{F}_{t-1}) = h(\mathbf{h}_{t-1}^{\mathrm{T}}\mathbf{B}_{t-1}\mathbf{F}_{t-1}). \tag{12}$$

The error function to be minimized is:

$$J = \frac{1}{2} \sum_{t-1}^{n} (Y_t - \hat{Y}_t)^2 \tag{13}$$

The algorithm computes the instantaneous gradient of the network's error function (J):

$$\hat{\nabla}_{t-1} = \frac{\partial J}{\partial \mathbf{B}_{t-1}} . \tag{14}$$

And the updates the weights according to the steepest descent equation:

$$\mathbf{B}_t = \mathbf{B}_{t-1} - \mu \hat{\nabla}_{t-1} \tag{15}$$

Where μ is the learning rate. Let:

$$\varepsilon_t = Y_t - \hat{Y}_t . \tag{16}$$

$h(.)$: is the tangent hyperbolic function.
We can now perform the steepest descent procedure to each weight in the second layer:

$$\beta_t^{\varphi_i} = \beta_{t-1}^{\varphi_i} + \mu \varepsilon_t Y_{t-i} (1 - \hat{Y}_t^2) \mathbf{h}_{t-1}^{\mathrm{T}} \qquad \text{for } i = 1,2,..., p \tag{17}$$

$$\beta_t^{\gamma_j} = \beta_{t-1}^{\gamma_j} + \mu \varepsilon_t Z_{t-d-j} (1 - \hat{Y}_t^2) \mathbf{h}_{t-1}^{\mathrm{T}} \qquad \text{for } j = 1,2,...,m. \tag{18}$$

This completes the description of the gradient back propagation algorithm and the next section discusses the corresponding non linear state space method.

3 Non Linear State Space Representation

Having identified the system model using the above neural network algorithm, it is useful to recast the **SNN** in the state-space form to perform state estimation. This requires the coefficients φ_i ($\mathbf{X}_t$) and γ_j ($\mathbf{X}_t$) as function of time. In this case the **SNN** can be in the following form:

$$Y_t = h\left(\sum_{i=1}^{p} \varphi_{i,t} Y_{t-i} + \sum_{j=1}^{m} \gamma_{j,t} Z_{t-d-j}\right). \tag{19}$$

The non linear state space of **SNN** is given by:

$$\Theta_{t+1} = \mathbf{H}(\Theta_t, \varphi_t, \overline{Z}_t) + \omega_t . \tag{20}$$

$$Y_t = h(\Theta_t, \overline{Z}_t) + \xi_t . \tag{21}$$

$$\Theta_t = \begin{pmatrix} \theta_{1,t} \\ \theta_{2,t} \\ \ldots \\ \theta_{p-1,t} \\ \theta_{p,t} \end{pmatrix}; \quad \mathbf{H}(\Theta_t, \varphi_t, \overline{Z}_t) = \begin{pmatrix} Y_{t-p+1} = \theta_{2,t} \\ Y_{t-p+2} = \theta_{3,t} \\ \ldots \\ Y_{t-1} = \theta_{p,t} \\ Y_t = h(\varphi_{1,t}\,\theta_{p,t} + \ldots + \varphi_{p,t}\,\theta_{1,t}) \end{pmatrix}$$

$$\text{And}: \overline{Z}_t = \sum_{j=1}^{m} \gamma_{j,t}\, Z_{t-d-j}$$

Equations (20) and (21) represent respectively the transition equation and the observation equation, with Θ_t is the state vector of the system at time t and Y_t is the observation vector at time t. The ω_t is of innovations, with zero mean and variance Q_t, ξ_t is an additive measurement noise with zero mean and variance R_t. We assume that the noise vectors are uncorrelated with covariance: $P^{\omega\xi} = \begin{pmatrix} Q_t & 0 \\ 0 & R_t \end{pmatrix}$.

4 Bayesian Inferential Algorithm Based on Extended Kalman Filter

The Kalman filer is a set of mathematical equations that provides an efficient computational (recursive) solution of the least-squares method. The filter is very powerful in several aspects: it supports estimations of past, present, and even future states, and it can do so even when the precise nature of modeled system is unknown. A complete discussion to the idea of the Kalman filter can be found in [7].

With the above non linear state space model of (20), the application of the Kalman filter is not straightforward because the model is not linear. We are established, a recursive algorithm for the measurement and the state update for the dynamic system represented by the **SNN**. This algorithm can be summarized by the update recursion equations.

$$\hat{\Theta}_{t+1/t} = H\left[(\hat{\Theta}_{t/t}, \overline{Z}_t)\right] \tag{22}$$

$$P^{\Theta}_{t+1/t} = \mathbf{A}_t P^{\Theta}_{t/t} \mathbf{A}_t^{T} + Q_t . \tag{23}$$

$$K_{t+1} = P^{\Theta}_{t+1/t} \mathbf{G}^{T}_{t+1} (\mathbf{G}_{t+1} P^{\Theta}_{t+1/t} \mathbf{G}^{T}_{t+1} + R_{t+1})^{-1}. \tag{24}$$

$$\hat{\Theta}_{t+1/t+1} = \hat{\Theta}_{t+1/t} + K_{t+1}(Y_{t+1} - h(\hat{\Theta}_{t+1/t}, \overline{Z}_{t+1})). \tag{25}$$

$$P^{\Theta}_{t+1/t+1} = P^{\Theta}_{t+1/t}(\mathbf{I} - K_{t+1}\mathbf{G}_{t+1}). \tag{26}$$

$$\hat{Y}_{t+1/t} = h(\hat{\Theta}_{t+1/t}, \overline{Z}_{t+1})). \tag{27}$$

Where $P^{\Theta}_{t+1/t}$: the prior estimate covariance of the state;

$$K_t \text{ : the gain filter ; } \mathbf{A}_t = \frac{\partial H\left[(\hat{\Theta}_{t/t}, \overline{Z}_t)\right]}{\partial(\hat{\Theta}_{t/t})} ; G_{t+1} = \frac{\partial h(\hat{\Theta}_{t+1/t}, \overline{Z}_{t+1})}{\partial(\hat{\Theta}_{t+1/t})} .$$

Equations (22) and (23) are the time update equations, these equations are responsible for projecting forward (in time) the current state and error covariance estimates to obtain the prior estimates for the next time step. Equations (24) to (27) are the measurement update equations, the first task during the measurement update is to compute the Kalman gain, K_t, the next step is to actually measure the process to obtain Y_t (27), and then to generate an a posterior state estimate by incorporating the measurement as in (25), the final step is to obtain an a posterior error covariance estimate via (26). After each time and measurement update pair, the process is repeated with the previous a posterior estimate used to project or predicts the new a priori estimates.

5 Simulation Studies

The largest Lyaponov exponent contains information on how far in the future predictions are possible. The Lyaponov spectrum has proven to be one of the most useful dynamic invariants that characterize chaotic dynamic systems. The Lyaponov exponents provide us with a measure of the averaged exponential rates of divergence or convergence of neighbor orbits in phase space [8]. If at least one positive Lyaponov exponent exists, the dynamic system is said to be chaotic, and the initial small differences between two trajectories will diverge exponentially.

We analyze the chaotic logistic map which has been hypothesized in a number of empirical studies of financial time series (see [9]). **SNN** is used with the coupled learning algorithm described in section2 and the **BNKF**. After training, the networks are used in an iterative mode in order to generate a series of estimate outputs, and the dynamic invariants measured for the original series. The data set is displayed in figure2. In this study the chosen model has parameters d=2 and p=2, q= 15 (number of hidden units) respectively, that is:

$$Y_t = h\left(\sum_{i=1}^{2} \varphi_i(\mathbf{X}_t) Y_{t-i}\right) + \xi_t . \qquad \textbf{(28)}$$

Where $\mathbf{X}_t$ is a function of $\{Y_{t-1}, Yt_{-2}\}$ which are the neural network's inputs. The data set from 1 to 500 is used to train the neural network and from 500 to 1000 to test the neural network performance (figure2).

The network, during the training, learns the correct value of the Largest Lyaponov exponent that is 0.52. Our analysis permits us to claim that the **SNN** capture "chaos" because it learns the dynamic invariant of a chaotic dynamic system. The simulated **SNN** and **BNKF** output are compared with the measured output in figure3. As can be seen, the agreements are very good.

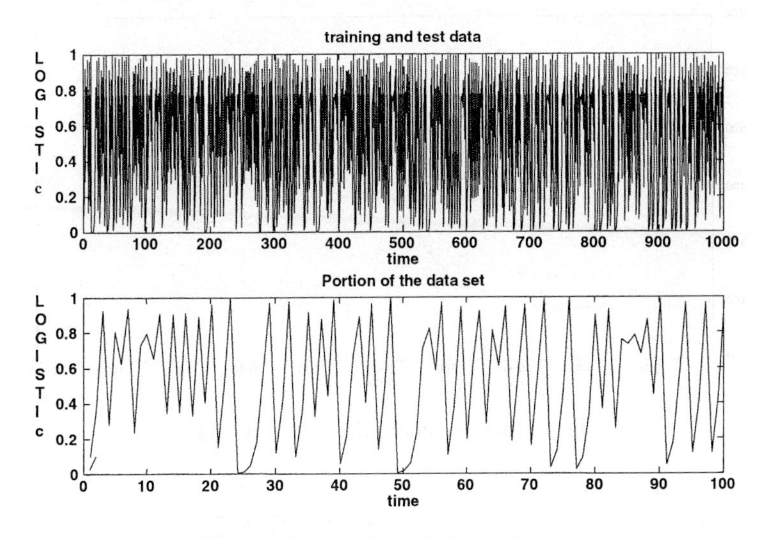

Fig. 2. . Data set from the Logistic map.

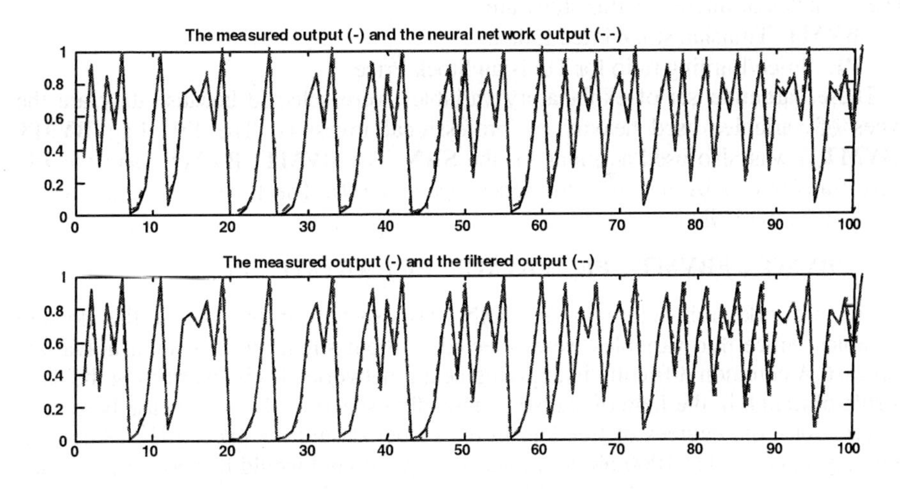

Fig. 3. Comparison of measured output and neural network model and filtered output.

6 Analysis of (BVMT)

To illustrate the application potential of the new **SNN** architecture on real data set, the case study, in this section, involved the modeling of the Tunisian Stock price index **BVMT**. The data set were daily closed values from Sept. 30, 1990 to Jul. 07, 1998. Figure4 shows a temporal plot of the Tunisian Stock price index

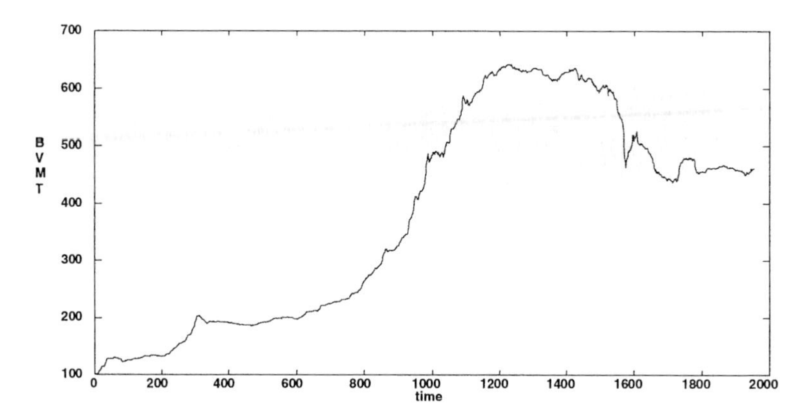

Fig. 4. Tunisian Stock Price Index (BVMT).

As we can see, the movements of the index not follow a random walk, and the system dynamics are not smooth, then the linear approximation of this dynamics is not appropriate. Identification of this non-linear structure by the **SNN** can clearly enable formulation of such system.

The variables of interest in this study are:

1. **BVMT**: Tunisian stock Price Index.
2. **PE** : Price/Earning ratio for Tunisian Stock Price.

These particular set of explanatory variables were selected because they are the types of variables used heavily by professional investors. The first Lag **BVMT** ($\mathbf{BVMT}_{t-1}$) was also used as input for the **SNN**. The $\mathbf{BVMT}_t$, $\mathbf{BVMT}_{t-1}$ and the $\mathbf{PE}_t$ were transformed by the first difference operator (D). The time series that can be modeled by the **SNN** is:

$$\mathbf{BVMT}_t = f(\mathbf{BVMT}_{t-1}, \mathbf{PE}_t, \varphi(\mathbf{BVMT}_{t-1}, \mathbf{PE}_t), \gamma(\mathbf{BVMT}_{t-1}, \mathbf{PE}_t)) + \xi_t$$

The network architecture of the **SNN** was determined in part by the domain variables. Since tree variables were selected as input stream to provide a univariate forecast. A common difficulty in applying neural networks lies in over fitting the data. A rule of thumb in the field of statistical modeling specifies that, for a case base of N observations, the degrees of freedom in the model should not exceed $N^{0.5}$. Since our training set is fixed at 1000 observations, an upper bound would be about 32 weights, each corresponding to a degree of freedom in the **SNN** to be trained. Initially, the configuration 3*8*1= (corresponding to 3*8+8=32 weights) was evaluated. However, the 3*7*1 architecture yielded slightly better performance in term of prediction error. In this paper, the value of 7 hidden units was chosen for the **SNN** first layer. The **SNN** was trained and validate by the gradient back propagation and the **BNKF** described in section 3 and 4 (for μ=0.1).

The performance among the predictive model is presented in figure5. The **SNN** output is closed to the desired output. Figure6 shows the evolution of the estimate Lyaponov exponent as function of the number of iterations via the **SNN**. It is important to remark that no overtraining is observed, the largest Lyaponov exponent is 1, 48. We conclude that the **BVMT** index is chaotic and stochastic.

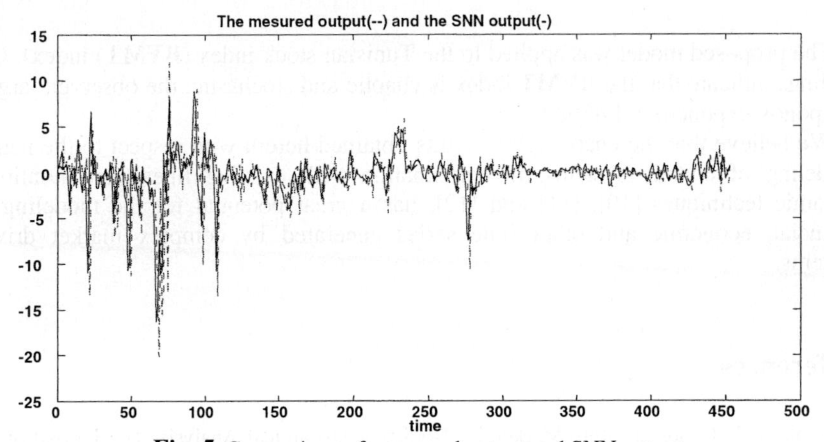

Fig. 5. Comparison of measured output and SNN output

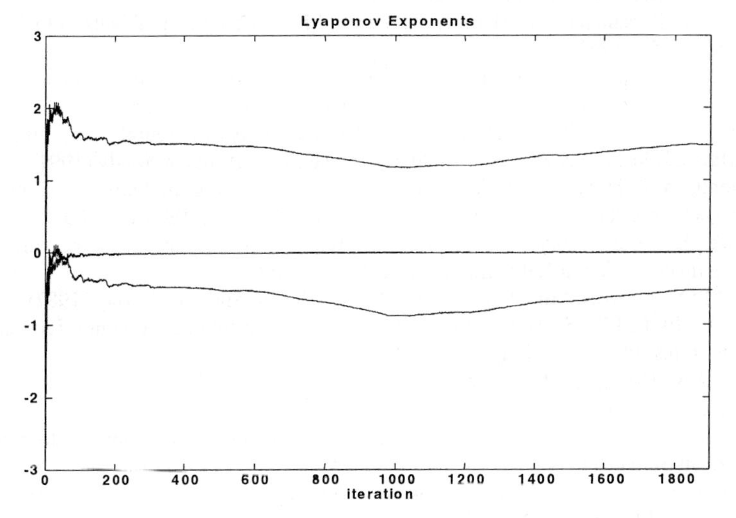

Fig.6. Lyaponov exponents as a function of the number of iteration for the SNN trained by the BNKF of the BVMT.

7 Conclusion

This paper proposed a new approach based on Extended Kalman filter to help diagnose the dynamic structure of non linear time series.

For a set of input and output, a **SNN** can be used to identify the nonlinear time series of (1). A non linear state space form for this net is presented and a Bayesian algorithm for this model is derived to estimate the state.

Our analysis permits us to claim that the **SNN** capture "chaos" because it learns the dynamic invariants of a chaotic dynamic system (Logistic map).

The proposed model was applied to the Tunisian stock index (**BVMT**) index). Our findings indicate that the **BVMT** index is chaotic and stochastic; the observed largest Lyaponov exponent is 1.48682.

We believe that the encouraging results obtained herein with respect to the neural modeling of chaotic systems, in combination with existing linear and nonlinear dynamic techniques [10], [11] and [12], has a great potential for the modeling of financial, economic and other time series generated by complex market driven systems.

References

1. Blank, S.C.: Chaos in future Markets? a Nonlinear Dynamical Analysis. The Journal of Future Markets. 11.6 (1991) 711-728
2. Albano, A., Passamante, A., Hediger, T., Farell, M. E.: Using Neural Nets to look for chaos. Physica D. 58 (1992) 1
3. Nychka, D., Ellner, S., Gallant, A.R., Maccaffrey, D.: Findings chaos in noisy systems. Journal of the Royal Statistical Society, Series B. 54 (1992) 399-426
4. Hertz, J., Krogh, A., Palmer, R.: Introduction to the Theory of Neural Computing, Santa Fe Institute Studies in Science of Complexity, Amsterdam: Addison-Wesly (1991)
5. Weigend, .A.S., Huberman, B.A., Rumelhart, D.E.: Predicting the Future: A Connectionnist Approach. International Journal of Neural Systems, Vol 1.3 (1991) 193-209
6. Hornik, K., Stinchcombe, M., White: Multilayer feed forward networks are universal approximators. Neural Networks, vol.2 (1989) 359-366.
7. Aoki, M.: State-Space Modeling of Times Series, Berlin: Springer-Verlag (1987)
8. Wolf, A., Swift, J.B., Swinney, H.L., Vastano, J.A.: Determining Lyaponov Exponents from a time series. Physica D. 16 (1985) 285-317
9. Hsieh, D.A.: Chaos and Nonlinear Dynamics:Application to Financial Markets. The Journal of Finance. 6.5 (1991) 1839-1877
10. Hillmer,S.C, Trabelsi, A.: Benchmarking of Economic Time Series. Journal of the American Statistical Association: Theory and Methods. 82 (1987) 1064-1071
11. Trabelsi, A., Hillmer, S.C.: A Benchmarking Approach to Forecast Combination. Journal of Business and Economic Statistics. (1989) 353-362
12. Trabelsi, A., Hillmer, S.C.: Benchmarking Time Series with Reliable Benchmarks. Journal of the Royal Statistical Society, Applied Statistics. 39:3 (1990) 367-379.

Export Behaviour Modeling Using EvoNF Approach

Ron Edwards[1], Ajith Abraham[2], and Sonja Petrovic-Lazarevic[3]

[1]Monash University, School of Business and Economics
McMahons Road, Frankston 3199, Australia
ron.edwards@buseco.monash.edu.au

[2]Department of Computer Science, Oklahoma State University
700 N Greenwood Avenue, Tulsa, OK 74106-0700, USA
ajith.abraham@ieee.org

[3]Monash University, Department of Management
McMahons Road, Frankston 3199, Australia
sonja.petrovic-lazarevic@buseco.monash.edu.au

Abstract. The academic literature suggests that the extent of exporting by multinational corporation subsidiaries (MCS) depends on their product manufactured, resources, tax protection, customers and markets, involvement strategy, financial independence and suppliers' relationship with a multinational corporation (MNC). The aim of this paper is to model the complex export pattern behaviour using a Takagi-Sugeno fuzzy inference system in order to determine the actual volume of MCS export output (sales exported). The proposed fuzzy inference system (FIS) is optimised by using neural network learning and evolutionary computation. Empirical results clearly show that the proposed approach could model the export behaviour reasonable well compared to a direct neural network approach.

1 Introduction

Malaysia has been pursuing an economic strategy of export-led industrialisation [3,6,7]. To facilitate this strategy, foreign investment is courted through the creation of attractive incentive packages. These primarily entail taxation allowances and more liberal ownership rights for investments [8,10,11]. The quest to attract foreign direct investment (FDI) has proved to be highly successful [9]. The bulk of investment has gone into export-oriented manufacturing industries.

Several specific subsidiary features identified in international business literature are particularly relevant when seeking to explain MNC subsidiary export behaviour. The location factors in attracting FDI to the country, the subsidiary's functional roles, size and age, and whether subsidiary products are targeted at niche or broader markets, have all been perceived to be determinants of export behaviour. This paper is concerned with the manner in which the structure and strategy of MNC that have invested in Malaysia affect the export intensity of their subsidiaries. Prior to going into the details of the study, it is important to explain that there are two related aspects of export behaviour. One aspect is the probability of a firm exporting at all. The other aspect is the relationship between the percentage of total sales exported and the size of the firm. According to the literature, larger firms are more likely to export. However,

P.M.A. Sloot et al. (Eds.): ICCS 2003, LNCS 2659, pp. 169–178, 2003.

there is no clear relationship between size of the firm and export intensity. For example Bonnaccorsi [4] found that although larger firms were more likely to export, there was no significant difference between the export intensity of small, medium, or large firms. Wolff et al [12] also found no significant difference in export intensity between small, medium and large firms. They argued that the type of resources available is a key factor, specifically, that with the appropriate type of resource, a small firm can use the same competitive patterns utilised by larger firms with the same effectiveness. Wagner (2001) notes that greater firm size is neither necessary nor sufficient for any industry or country.

In this paper we are concentrated on the MCS product manufactured, resources, tax protection, customers and markets, involvement strategy, financial independence and suppliers' relationship with a MNC. We use the EvoNF, an integrated computational framework to optimise FIS through neural network learning and evolutionary computation [1].

The paper is divided as follows: Section 2 explains the role of fuzzy inference systems in determining the export behaviour of MCS. Section 3 illustrates the experimentation results based on data provided by Malaysian MCS. The paper ends with concluding remarks.

2 The Role of FIS for Explaining the Export Behaviour of MCS

A FIS can utilize human expertise by storing its essential components in rule base and database, and perform fuzzy reasoning to infer the overall output value. The derivation of *if-then* rules and corresponding membership functions (MF) depends heavily on the researcher's *a priori* knowledge about the system under consideration. However, there is no systematic way of transforming experiences of knowledge of human experts to the knowledge base of a FIS. There is also a need for adaptability or some learning algorithms to produce outputs within the required error rate [2].

In this section, we define the architecture of EvoNF, as an integrated computational framework to optimise FIS by using neural network learning technique and evolutionary computation. The proposed framework could adapt to Mamdani, Takagi-Sugeno or other FIS. The architecture and the evolving mechanism can be considered as general framework for adaptive fuzzy systems. That is a FIS can change their MF (quantity and shape), rule base (architecture), fuzzy operators and learning parameters according to different environments without human intervention.

Solving multi-objective problems is, generally, a very difficult goal. In optimisation problems, the objectives often conflict across a high-dimension problem space and may also require extensive computational resources. The hierarchical evolutionary search framework could adapt MF (shape and quantity), rule base (architecture), fuzzy inference mechanism (T-norm and T-conorm operators) and the learning parameters of neural network learning algorithm. In addition to the evolutionary learning (global search) neural network learning could be considered as a local search technique to optimise the parameters of the rule antecedent/consequent parameters and the parameterised fuzzy operators.

Figure 1 illustrates the interaction of various evolutionary search procedures. For every type of FIS (for example Mamdani type), there exist a global search of learning algorithm parameter, inference mechanism, rule base and MF in an environment decided by the problem. Thus the evolution of FIS will evolve at the slowest time scale while the evolution of the quantity and type of MF will evolve at the fastest rate. The function of the other layers could be derived similarly.

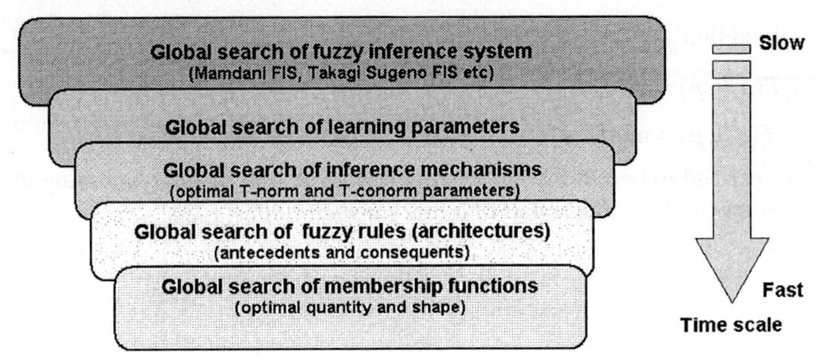

Fig. 1. General computational framework for EvoNF

Hierarchy of the different adaptation layers (procedures) will rely on the prior knowledge. For example, if there is more prior knowledge about the architecture than the inference mechanism then it is better to implement the architecture at a higher level. If we know that a particular FIS will suit best for the problem, we could also minimize the search space. For fine-tuning the FIS all the node functions are to be parameterised.

2.1 Parameterization of Membership Functions

FIS is completely characterized by its MF For example, a generalized bell MF is specified by three parameters (p, q, r) and is given by:

$$\mathrm{Bell}(x, p, q, r) = \frac{1}{1 + \left| \frac{x - r}{p} \right|^{2q}}$$

Figure 2 shows the effects of changing p, q and r in a bell MF. Similar parameterisation can be done with most of the other MF.

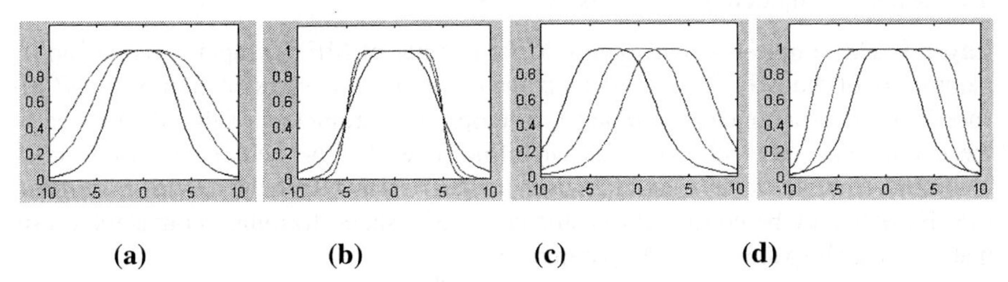

(a) (b) (c) (d)

Fig. 2. a Changing parameter p; **b** changing parameter q; **c** changing parameter r; **d** changing p and q

2.2 Parameterization of T-Norm Operators

T-norm is a fuzzy intersection operator, which aggregates the intersection of two fuzzy sets A and B. The Schweizer and Sklar's T-norm operator can be expressed as:

$$T(a,b,p) = \left[\max\{0,(a^{-p} + b^{-p} - 1)\}\right]^{-\frac{1}{p}}$$

It is observed that

$$\lim_{p\to 0} T(a,b,p) = ab$$

$$\lim_{p\to\infty} T(a.b,p) = \min\{a,b\}$$

which correspond to two of the most frequently used T-norms in combining the membership values on the premise part of a fuzzy *if-then* rule.

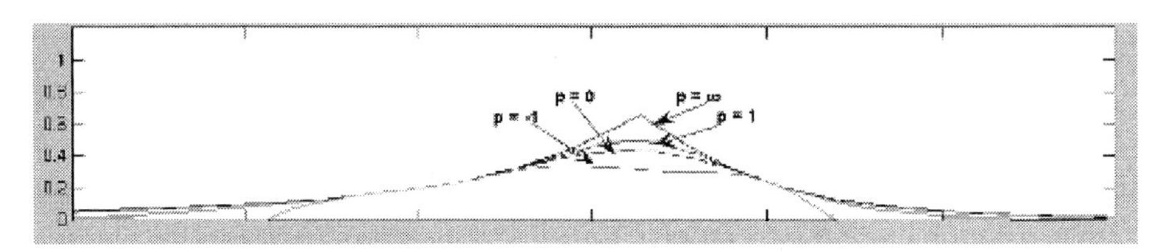

Fig. 3. Effects of changing p of T-norm operator for two Bell MF

To give a general idea of how the parameter p affects the T-norm operator, Figure 3 illustrates T-norm operator $T_{(a,b,p)}$ for different values of p.

2.3 Chromosome Modeling and Representation

The antecedent of a fuzzy rule defines a local region, while the consequent the behaviour within the region via various constituents. Basically the antecedent part remains the same regardless of the inference system used. Different consequent describes constituents result in different FIS. For applying evolutionary algorithms, problem representation (chromosome) is very important as it directly affects the proposed algorithm. Referring to Fig. 1 each layer (from fastest to slowest) of the hierarchical evolutionary search process has to be represented in a chromosome for successful modeling of EvoNF. A typical chromosome of the EvoNF would appear as shown in Fig. 5 and the detailed modeling process is as follows.

Layer 1. The simplest way is to encode the number of MF per input variable and the parameters of the MF. Figure 5 depicts the chromosome representation of *n bell* MF specified by its parameters p, q and r. The optimal parameters of the MF located by the evolutionary algorithm will be later fine tuned by the neural network-learning algorithm. Similar strategy could be used for the output MF in the case of a Mamdani FIS. Experts may be consulted to estimate the MF shape forming parameters to estimate the search space of the MF parameters.

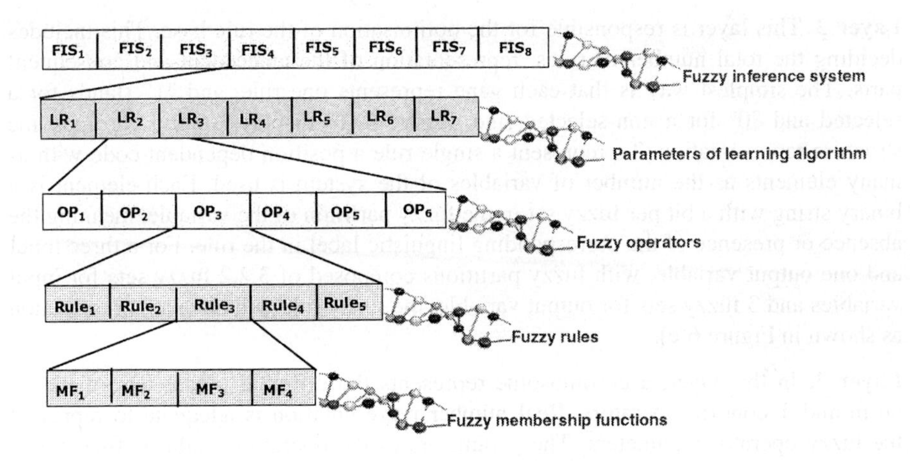

Fig. 4. Chromosome structure of the EvoNF model

MF1				MFn		
p_1	q_1	r_1	....	p_n	q_n	r_n

Fig. 5. Chromosome representing *n* MF for every input/output variable coding the parameters of a bell shape MF

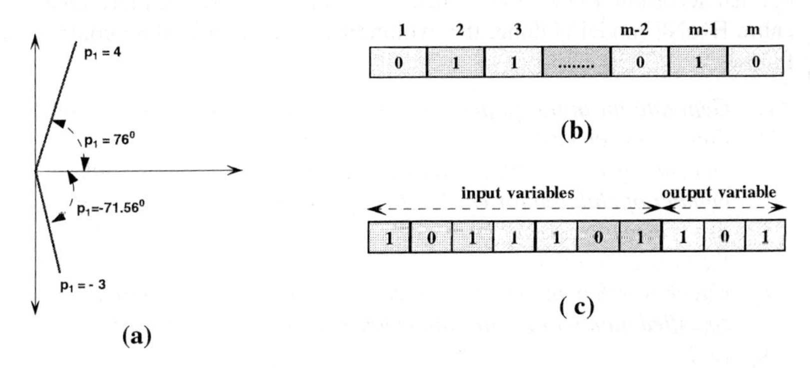

Fig. 6. (a) Angular coding technique (b) representation of *m* fuzzy rules (c) representation of I/O variables

We used the angular coding method proposed by [5] for representing the rule consequent parameters of the Takagi-Sugeno inference system. Rather than directly coding the consequent parameters, the "transformed" parameters represent the direction of the tangent $\alpha_i = arctan\ p_i$. The range for the parameters α_i is the interval (-90^0, $+90^0$), such that the parameters p_i can assume any real value. A single input Takagi-Sugeno system $Y = p_1 X + p_0$ defines a straight line. The real value p_1 is simply the gradient between this line and the *X*-axis. Parameter p_0 determines the offset of the straight line (intercept) along the *Y*-axis. The procedure is illustrated in Fig. 6.

Layer 2. This layer is responsible for the optimisation of the rule base. This includes deciding the total number of rules, representation of the antecedent and consequent parts. The simplest way is that each gene represents one rule, and "1" stands for a selected and "0" for a non-selected rule. Figure 6 (b) displays such a chromosome structure representation. To represent a single rule a position dependent code with as many elements as the number of variables of the system is used. Each element is a binary string with a bit per fuzzy set in the fuzzy partition of the variable, meaning the absence or presence of the corresponding linguistic label in the rule. For a three input and one output variable, with fuzzy partitions composed of 3,2,2 fuzzy sets for input variables and 3 fuzzy sets for output variable, the fuzzy rule will have a representation as shown in Figure 6(c).

Layer 3. In this layer, a chromosome represents the different parameters of the T-norm and T-conorm operators. Real number representation is adequate to represent the fuzzy operator parameters. The parameters of the operators could be fine- tuned using gradient descent techniques.

Layer 4. This layer is responsible for the selection of optimal learning parameters. Performance of the gradient descent algorithm directly depends on the learning rate according to the error surface. The optimal learning parameters decided by the evolutionary algorithm will be used to tune MF and the inference mechanism.

Layer 5. This layer basically interacts with the environment and decides which FIS (Mamdani type and its variants, Takagi-Sugeno type, Tsukamoto type etc.) will be the optimal according to the environment. Once the chromosome representation, *C,* of the entire EvoNF model is done, the evolutionary search procedure could be initiated as follows:

1. *Generate an initial population of N numbers of C chromosomes. Evaluate the fitness of each chromosome depending on the problem.*
2. *Depending on the fitness and using suitable selection methods reproduce a number of children for each individual in the current generation.*
3. *Apply genetic operators to each child individual generated above and obtain the next generation.*
4. *Check whether the current model has achieved the required error rate or the specified number of generations has been reached. Go to Step 2.*
5. *End*

3 Model Evaluation and Experimentation Results

For simulations we have used data provided from a survey of 69 Malaysian MCS. Each corporation subsidiary data set were represented by the following input variables:

- Product manufactured (1–5 scale representing fully independent from the parent and fully dependent)
- Resources (1–5 scale representing fully independent from the parent and fully dependent)

- Tax protection (1–5 scale representing tax protection and no tax protection) Customers and market (1 - 4 scale representing the geographical distribution of the customers)
- Involvement strategy (1–4 scale representing subsidiary, subsidiary and parent, parent alone and equal share)
- Financial independence (1–5 scale representing fully independent from the parent and fully dependent)
- Suppliers relationship (1–5 scale representing fully independent from the parent and fully dependent)

3.1 EvoNF Training

We used the popular grid partitioning method (clustering) to generate the initial rule base. This partition strategy requires only a small number of MF for each input. We used the 90% of the data for training and remaining 10% for testing and validation purposes. The initial populations were randomly created based on the parameters shown in Table 1. We used a special mutation operator, which decreases the mutation rate as the algorithm greedily proceeds in the search space 0. If the allelic value x_i of the i-th gene ranges over the domain a_i and b_i the mutated gene $x_i^{'}$ is drawn randomly uniformly from the interval $[a_i , b_i]$.

$$x_i^{'} = \begin{cases} x_i + \Delta(t, b_i - x_i), if\ \omega = 0 \\ x_i + \Delta(t, x_i - a_i), if\ \omega = 1 \end{cases}$$

where a represents an unbiased coin flip $p(a =0) = p(a =1) = 0.5$, and

$$\Delta(t,x) = x\left(1 - \gamma^{\left(1 - \frac{t}{t_{\max}}\right)^{b}}\right)$$

defines the mutation step, where γ is the random number from the interval $[0,1]$ and t is the current generation and t_{max} is the maximum number of generations. The function Δ computes a value in the range $[0,x]$ such that the probability of returning a number close to zero increases as the algorithm proceeds with the search. The parameter b determines the impact of time on the probability distribution Δ over $[0,x]$. Large values of b decrease the likelihood of large mutations in a small number of generations. The parameters mentioned in Table 1 were decided after a few trial and error approaches. Experiments were repeated 3 times and the average performance measures are reported. Figures 10 illustrates the meta-learning approach for training and test data combining evolutionary learning and gradient descent technique during the 35 generations.

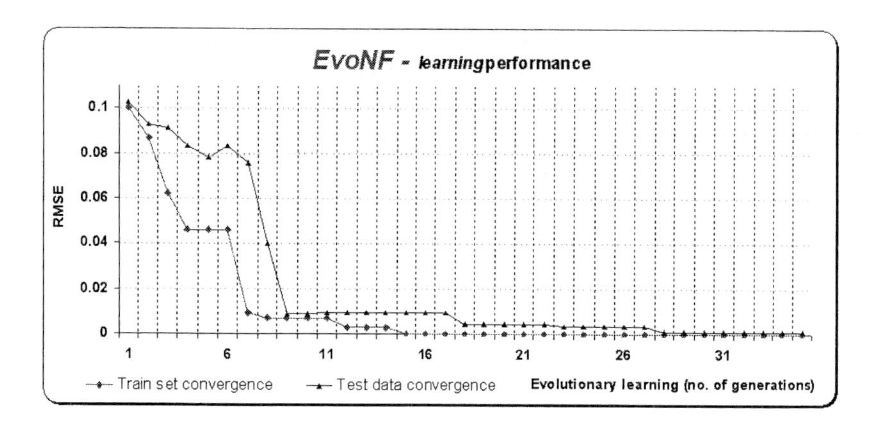

Fig. 7. Meta-learning performance (training and test) of EvoNF framework

The 35 generations of meta-learning approach created 76 *if-then* Takagi-Sugeno type fuzzy *if-then* rules compared to 128 rules using the grid-partitioning method. We also used a feed forward neural network with 12 hidden neurons (single hidden layer) to model the export output for the given input variables. The learning rate and momentum were set at 0.05 and 0.2 respectively and the network was trained for 10,000 epochs using BP [2]. The network parameters were decided after a trial and error approach. The obtained training and test results are depicted in Table 2 (CC=correlation coefficient).

Table 1. Parameter settings of EvoNF framework

Population size	40
Maximum no of generations	35
FIS	Takagi Sugeno
Rule antecedent MF	2 MF (parameterised Gaussian)/ Input
Rule consequent parameters	Linear parameters
Gradient descent learning	10 epochs
Ranked based selection	0.50
Elitism	5 %
Starting mutation rate	0.50

Table 2. Training and test performance of the different intelligent paradigms

Export output	**Intelligent paradigms**					
	EvoNF			**Neural network**		
	RMSE		***CC**	**RMSE**		***CC**
	Train	**Test**		**Train**	**Test**	
	0.0013	0.012	0.989	0.0107	0.1261	0.946

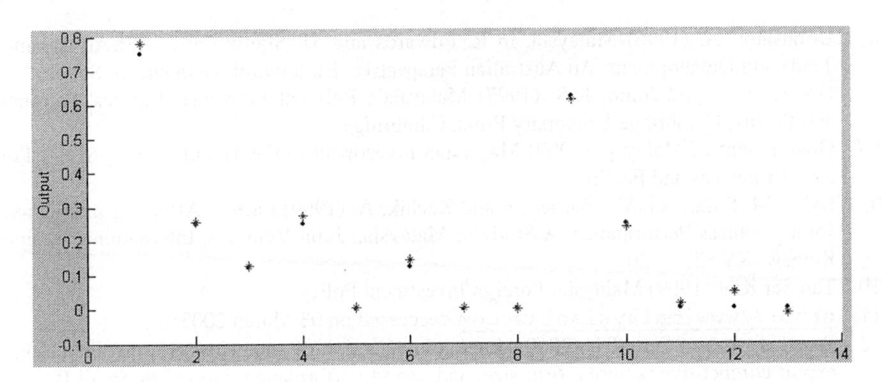

Fig. 8. Test results showing the export output (scaled values) for 13 MNC's with respect to the desired values (*)

4 Conclusions

Our analysis on the export behavior of Malaysia's MCS reveals that the developed EvoNF model could learn the chaotic patterns and model the behavior using an optimized Takagi Sugeno FIS. As illustrated in Figure 8 and Table 2, EvoNF could easily approximate the export behavior within the tolerance limits. When compared to a neural network approach, EvoNF performed better (in terms of lowest RMSE) and better correlation coefficient. Our experiment results also reveal the importance of all the key input variables to model the behavior within the required accuracy limits. These techniques might be useful not only to MNC's but also to administrators and Government for long-term strategic management of the economy.

As a future research, we also plan to incorporate more intelligent paradigms to improve the modeling aspects of the export behavior.

References

1. Abraham, A. (2002), EvoNF: A Framework for Optimization of Fuzzy Inference Systems Using Neural Network Learning and Evolutionary Computation, In Proceedings of 17th IEEE International Symposium on Intelligent Control, ISIC'02, IEEE Press, pp 327–332, 2002
2. Abraham, A. (2001), Neuro-Fuzzy Systems: State-of-the-art Modelling Techniques, Lecture Notes in Computer Science. Volume. 2084, Springer-Verlag Germany, Jose Mira and Alberto Prieto (Eds.), ISBN 3540422358, Granada, Spain, pp. 269–276.
3. Ariff, M. and Hill, H *(1985)* Export-Oriented Industrialisation: The ASEAN Experience, Allen and Unwin, Sydney.
4. Bonnaccorsi, A. (1992) On the Relationship between Firm Size and Export Intensity, Journal of International Business Studies, XXIII(4)" 605–635.
5. Cordón O., Herrera F., Hoffmann F. and Magdalena L. (2001), Genetic Fuzzy Systems: Evolutionary Tuning and Learning of Fuzzy Knowledge Bases, World Scientific Publishing Company, Singapore.

6. Doraisami, A. (1996) Malaysia, in R. Edwards and M. Skully (eds.) ASEAN Business Trade and Development: An Australian Perspective, Butterworth Heinemann, Sydney.
7. Gomez, E.T. and Jomo, K.S. (1997) Malaysia's Political Economy: Politics, Patronage and Profits, Cambridge University Press, Cambridge.
8. Government of Malaysia (1999) Malaysian Investment in the Manufacturing Sector Policies, Incentives and Facilities.
9. Lyles, M. Sulaiman, M., Barden, J. and Kechik, A. (1999) Factors Affecting International Joint Ventures Performance: A Study of Malaysian Joint Ventures, International Business Review, XV (2): 1–20.
10. Tan Ser Kiat (1999) Malaysia: Foreign Investment Policy
11. http://www.malaysianlaw.com <accessed on 03 March 2003>
12. Wolff, J. A. and Pett, T.L (2000), Internationalization of small firms: An examination of export competitive patterns, firm size, and export performance. Journal of Small Business Management, 38 (2), pp. 34–47.

Simulation Studies of a Multi-priority Dual Queue (MPDQ) with Preemptive and Non-preemptive Scheduling

Anthony Bedford and Panlop Zeephongsekul

Department of Mathematics and Statistics, RMIT University
Plenty Rd., Bundoora East, Victoria, 3083, Australia
{anthony.bedford,panlopz}@ems.rmit.edu.au

Abstract. Traffic management in communication environments like the Internet is vital in meeting the demand of users. The recent introduction of various dynamic scheduling algorithms has been to target improvements in QoS (Quality of Service). These new algorithms, such as dual queueing, aim to give better QoS to most traffic at the expense of a few rather than fairly give poor QoS to all customers. This is achieved through scheduling decisions that can change depending on the traffic conditions. The MPDQ introduces different classes into this type of scheme with the aim of enhancing higher-class packets QoS without great expense to lower class packets. In this paper we analyse the differences in loss and waiting time between preemptive and non-preemptive service disciplines for the MPDQ and then identify the best queueing regime for the pre-emptive MPDQ. We show that non-preemptive service dispensing is superior to packets of two classes than the preemptive scheme, and that highest class first (HCF) is the best queueing regime for either of these algorithms.

1 Introduction

The Multi-Priority Dual Queue (MPDQ) was designed to reduce congestion levels in communications models with finite buffers. By splitting a single finite buffer space into two queues, known as dual queues, and administering a priority scheme to arriving packets, we have shown distinct advantages to communications providers in Quality of Service (QoS) requirements [1]. The MPDQ was derived from the Dual Queue proposed in [3] and combined to include the prioritised packet approach proposed to relieve congestion in [5]. Their analysis in [3] involved the use of MPEG traces used also in [7], which consisted of 20 files. They analysed the delay characteristics of the Dual Queue against the pre-existing scheduling regimes FIFO (First in First Out) and a modified Deficit Round Robin technique (DRR) [8], and demonstrated distinct advantages using the Dual Queueing scheme. This work was extended to a wireless local area network where minor modifications were made to the Dual Queue (DQ) and it was shown to outperform standard Round Robin scheduling in this application for untethered environments [6]. These analyses were undertaken using a fixed amount of

P.M.A. Sloot et al. (Eds.): ICCS 2003, LNCS 2659, pp. 179–189, 2003.

traffic with no statistical assumptions made on the underlying distributions of the arrival or service processes. A simulation approach to Wireless Packet Discard Policies was undertaken in [9], and a simulation approach was undertaken to analyse the scheme in a network scenario. They looked at transition probabilities for the error model and a specific type of loss called Message Loss Ratio. Our contribution explores a multi-priority queueing system that is especially relevant in the Internet Engineering Task Force (IETF) Integrated Services Processes or Differentiated Services architectures. Our modified version of the DQ scheduling discipline provides a simple and effective mechanism for scheduling in these architectures. By introducing different classes into this scheme, we add additional complexity to the DQ with the aim of class-based QoS. Class-based packets within this system of greater importance (or higher class) have precedence over any lower classed packet with the aim of reduced waiting times. This MPDQ differs to the original DQ in that [3] used large finite buffers. As [3] deals with the handling of traffic streams, we change the scheduling to cope with individual prioritised packets, and instigate an abatement of both queues. The setting of the abatement threshold to be equal the capacity of the queues significantly changes the operation of the queues. This is as [3] deals with traffic streams and this paper deals with individual prioritised packets. The MPDQ is especially relevant for the Internet's development, and also has merits for use in certain manufacturing systems where a second holding area is available. In our model, loss is determined by available queue size, and we have not used a time-discard policy, in which packets are ejected from the system after a period of time is reached as in [3]. Through simulation we have shown the benefits of the non-preemptive MPDQ using a Highest Class First (HCF) queueing regime over other queueing regimes [1]. Furthermore, we have developed a detailed algorithmic procedure to explicitly solve the steady state distribution of the preemptive MPDQ for two classes of packets [2]. This technique required the matrix-geometric approach to define the state rate transitions and uniquely implemented them into an algorithm to obtain a solution. That work also highlighted the enormous task of analytically solving even the smallest MPDQ system.

Our aims in this paper are to firstly establish the accuracy of simulation by comparing them with known analytical results. We will then use simulation to explore the differences between the non-preemptive and preemptive service disciplines though extensive simulations using Arena, and explore in great detail the merits of different queueing regimes for the preemptive MPDQ. The results here provide a framework for communications providers in determining how service disciplines contribute to loss and delay, and how modelling can assist in determining how to relieve a network of this type from congestion. In Sect. 2, we define the MPDQ and the simulation design in detail, stating the scheduling algorithms used in the simulations. To investigate the accuracy and reliability of the simulations, in Sect. 3 we compare our analytical results for four models with the simulation results and describe the process of determining a suitable stopping time. In Sect. 4, the comparison of preemptive and non-preemptive MPDQ is undertaken by looking at the results of loss by queue size, waiting time in the system, and waiting time by queue size. This provides valuable insight into the behaviours of the two types of MPDQ's that can be used by service providers. In a similar analysis to our prior simulation studies we complete our investigations in Sect. 5 by looking at the preemptive MPDQ under various queueing regimes. Performance statistics and CDF of delay for four models are fitted, and the waiting time distribution is fitted for Class 1 packets under the HCF MPDQ.

2 MPDQ Design and Simulation

The MPDQ is designed with two finite buffer spaces and a server at the head of the first, or primary, queue. There is a single server (node) at the head of this queue dispensing an exponentially distributed service time with rate μ_i to traffic of class i, $i=1,..,k$ for k classes. If an arriving packet finds the primary queue full, they wait in the secondary queue (provided it isn't full) which has no service facility and only serves as a holding queue for the full primary queue. If both queues are full, an arriving packet is lost. In both queues, packets arrange themselves in order of priority, dependent upon the queueing regime. Here we assume class 1 packets (the high class) have priority ahead of class 2 packets (the low class). The queueing regime used here is, unless otherwise stated, Highest Class First (HCF). The arrivals are independent exponentially distributed arrival processes with mean arrival rate λ_k. Once a space is vacant in the primary queue, the head of the line in the secondary queue joins the primary queue. A schematic diagram describing the MPDQ is given in [1]. In Sect. 5, the queueing regimes investigated are HCF, LCF (Lowest Class First), FIFO, and LIFO (Last in First Out). The regimes stated hold for both queues in the system, so we cannot have a primary queue following a different regime to the secondary queue. For the priority schemes, packets of the same class are sorted on a FIFO basis. The non-preemptive case assumes no interruption to the packet in service, whereas the preemptive service discipline will interrupt service of a packet of lower class for one of a higher class. The lower class packet is returned to the head of the queue and awaits for the server to again become free. They resume requiring full service time.

2.1 Scheduling Algorithm

The simulations were conducted using the Arena Simulation Software package [4], a flexible program suitable for a vast area of statistical analysis with dedicated queueing subroutines. Arena has provided effective results of queueing disciplines in our prior work and was again chosen here as our tool of analysis. Arena allows us to define effective class based simulation due to its ability to generate arriving data and serve data from any statistical distribution. Further we can preempt packets from service if needed, and Arena gathers vast performance characteristics. In constructing the Arena model, we first needed to define the MPDQ in two independent algorithms (for the preemptive and non-preemptive cases). The preemption process required a separate procedure to check for the quantity of first class packets in the system and the preempt subroutine to expel a lower classed packet from service if a higher classed packet is in the primary queue. Both algorithms are simple to extend to multiple classes of packets. The simulation model developed for the non-preemptive simulation was substantially simpler than the preemptive model. Analytically the reverse holds, with the non-preemptive solution being more difficult to obtain than the preemptive solution using the matrix-analytic computational algorithm solution process.

3 Preemptive MPDQ

In [4], we have shown, using complex algorithms, that a steady-state solution for the dual queue with preemptive scheduling is possible. However, the process established is difficult to numerically evaluate due to division of large numbers for even small queueing systems. As the waiting space increases, the states in the system increase rapidly. This can be shown as follows. The steady state distribution π^t exists and is obtained by solving $\pi^t A$ where A is the infinitesimal generator matrix, and 0 is a vector of zeroes. The dimension of A is given by

$$\left(\frac{1}{2}\right)(c_1+1)(c_2^2+3c_2+c_1+2)\times\left(\frac{1}{2}\right)(c_1+1)(c_2^2+3c_2+c_1+2) \quad (1)$$

where c_i is the capacity of queue i. So if queue 1 is of size 4 and queue 2 is of size 6, A will have a dimension of 150. A matrix of this size will have 22,500 cells, well beyond a simple matrix multiplication, so simulation provides us with a valuable estimation tool.

The analytical results are a precise solution, based on the long-term steady-state Markovian birth and death processes. Therefore the choice of length for the simulations is vital in attempting to replicate a long-term steady-state result. From the general steady state solution of a birth death process, letting $t\rightarrow\infty$, $\Delta t\rightarrow 0$ solves the time invariant steady state probabilities in differential quotient form. In Arena simulations, there are three possible ways to cease the model running. The first is to set a finite simulation time, the second to define a statistic-dependent conditional statement, and the third is user interruption. Otherwise the simulation will run indefinitely. We use estimated average loss as the statistic-dependent conditional statement to determine a suitable finite simulation time.

3.1 Simulation and Analytical Results

For each of the models, we began with pilot simulations where the proportion of traffic lost is recorded. As expected this approached a constant value as the simulation time increased and will be explored in detail in the next section. We will begin by comparing analytical and simulation results for the probability of a busy server, an empty second queue and of loss. Due to the difficulty in initially finding these results analytically, we only consider queue 1 and 2 sizes of 2. We vary the inter-arrival rates and fix the service rates at 1 sec for Class 1 and 2 for Class 2.

We fixed the service rates in all simulations to the same values, with the details of four models (I–IV) to be used in all simulations given in Table 1. So using the rates used in our previous work [2], we used:

Table 1. Arrival and service rates

Model	Arrivals		Service		Load
	λ_1	λ_2	μ_1	μ_2	ρ
I	1	2	1	2	2
II	.8	1.6	1	2	1.6
III	.3	.6	1	2	.6
IV	.1	.2	1	2	.2

Our first interest was to compare the approximate probabilities for a busy server, π_B, an empty secondary queue, $\pi_{\{Q2=0\}}$, and loss, π_{Loss} in the simulations with our analytical results obtained in [2]. Our initial simulations were undertaken with the aim of obtaining close results to the precise analytical solutions.

Table 2. Comparison of steady-state and simulation statistics

Model	π_B		$\pi_{\{Q_2=0\}}$		π_{Loss}	
	Steady S.	Simul.	Steady S.	Simul.	Steady S.	Simul.
I	.9539	.9790	.2505	.2454	.5071	.5089
II	.9207	.9556	.3312	.3544	.4421	.4085
III	.5844	.5714	.7659	.8971	.1373	.0437
IV	.2079	.2084	.9742	.9977	.0160	.0001

As seen in Table 2, the results are close, with most probabilities within 0.03 of the actual result. This simulation result provides us with confidence in the reliability and validity of simulation to provide us with an accurate estimator of systems that are beyond current computational power.

To decide on an appropriate simulation time for the later analysis, a cumulative estimate of the mean loss was used. The continuous average probability of loss was calculated for both classes of packet. It is the class 1 loss that was used for the stopping rule as we have defined $\lambda_1<\lambda_2$. The reason for this is that we consider class 1 packets to be rarer arriving traffic that is of higher importance than class 2 packets. Further as the arrival rate is less, it is expected that the class 2 value should converge quicker to a steady-state. As the simulations were set up to accumulate extensive statistics, the aim was to run an efficient simulation that did not over-gather vast files of needless extra data. Since the traffic loss policy (Q 2 drop tail) is impartial to each class of traffic, it is expected that the probability of loss for both traffic classes should be the same. The loss rates will be proportional to the arrival rates of each traffic class. The pilot simulations for one replication were re-executed using a simulation time of 15,000 units. The generated average loss file for class 1 packet was 3Mb. It was projected that an average loss file over a simulation time of 50,000 units for 10 replications would be approximately 100Mb. The real-time execution period is sensitive to the system load, with the higher the load value, the faster the execution period. The stopping rule was based on the following, where the subscript t is the recorded simulation time as created by Arena, $\hat{\pi}_{Loss1}$ is the approximate probability of loss for a class 1 packet:

$$\left|(\hat{\pi}_{Loss1})_t - (\hat{\pi}_{Loss1})_{t-1}\right| \le 0.00005 \qquad (2)$$

Arena records statistics only if a change in value occurs. This avoids enormous files consisting of non-varying data being recorded. This assists the determination of a suitable simulation time, as the value t is only recorded to file if any of the statistics required have changed. It can be seen that the loss function settles down to a constant value by around 10,000 time units. Using the criteria (2) for stopping, the simulation times for each model were 9753.7, 10199.6, 10041.4 and 3917.5 for Models I, II, III, and IV. From these results a simulation length of 15,000 time units was used for all subsequent simulations. With the simulation design established, we begin our analysis.

4 Comparison of Scheduling Disciplines

The difference between preemption and non-preemption is simple enough to implement for people-type queueing problems but substantial in terms of algorithm design and analytical calculation. Let us define more formally the case where a packet of class i is always taken over a packet of class j. Consider a packet of class j being in service when a packet of class i arrives. Preemption is when the service of the jth class packet is interrupted and the server starts serving the ith class packet, based on a FCFS approach within classes. In non-preemption, when the service of the jth class packet is completed the server starts serving the ith class packet, based on a FCFS approach within classes. By this definition, we would expect Class 1 packets to be advantaged (in terms of expected waiting time) under the preemptive regime over Class 1 in non-preemptive and Class 2 packets under the non-preemptive regime over preemptive regime.

4.1 Loss by Queue Size

In this section, we analyse the effects on loss when we increase the sizes of the queues for each service regime. To undertake this, we use the models stated in Table 1. Since the traffic loss policy is impartial to each class of traffic, it was expected that the probability of loss for both traffic classes should be the same, and that was the case here. The loss rates will be proportional to the arrival rates of each traffic class, and we found that for most queue sizes, class 2 packet loss levels were marginally higher than class 1 loss levels of the same regime. The load gives us a clue as to why we found Model III and IV loss levels approaching zero. From Table 1, we have for Models III and IV $\rho<1$. Model IV has the smallest load and no loss. These two models shall see arriving packets find a system always with at least one vacant waiting space of the waiting room is extended sufficiently.

What is noticeable from all the models is the marginal improvement in loss levels as queue size increases. In packet based networks, if there are high volumes of traffic, a small decrease in loss levels would lead to a higher volume of packets receiving some service. For most elementary queueing models where $\rho>1$, the queue length will grow to infinity. In a finite buffer space, loss shall be experienced, and this will only decrease as queue size $\rightarrow\infty$. For packet traffic that follows the distributions stated here, loss levels slowly reduce when addition to the size of the queues is made. As an example, for Model I, in Table 3 below, loss for both classes reduces less than 1% for class 1 packets and less than 1.5% for class 2 packets for a ten-fold increase in queue size.

Table 3. Model I loss by queue qize

Total Queue Size	{Loss1}	{Loss2}
10	.4938	.4986
40	.4836	.4970
100	.4899	.4858

4.2 CDF of Delay

We use the CDF of delay time, or waiting time in the system W_s, to gain a picture of the overall QoS to packets for the two types of service regimes. Each model is investigated, with the expectation of shorter waiting times for class 2 in non-preemptive than preemptive and shorter waiting times for class 1 in preemptive than non-preemptive. In the four Figs. 1a-d, the CDF's follow the expected exponential curve.

In the single queue preemptive and non-preemptive infinite priority queueing models it can be derived from fundamental results that the expected waiting time in the queue for a class 1 packet is less in the preemptive priority case, and it is greater for class 2 packets in the preemptive case. (This is under the condition that $\rho<1$). From the CDF's for all models, the non-preemptive Class 1 packets outperform all other packets, however the advantage is marginal over the preemptive packets. Clearly, Class 2 packets wait longer as expected than Class 1 packets. For Models I and III, we found an increased chance of shorter waiting times towards the Class 2 over Class 1 packets (for Models I and III, $P(W_s^2 > 2) > P(W_s^1 > 2)$ for both priority schemes). For the Class 2 packets, the preemptive regime is the worst in all models but Model II. For communications providers, the MPDQ for two classes of packets, where $\rho\leq1$ we can expect Class 1 packets to receive superior waiting time over Class 2 packets. The regime to be implemented using the CDF's is the non-preemptive service regime. It is marginally better for each of the classes than its interrupt counterpart.

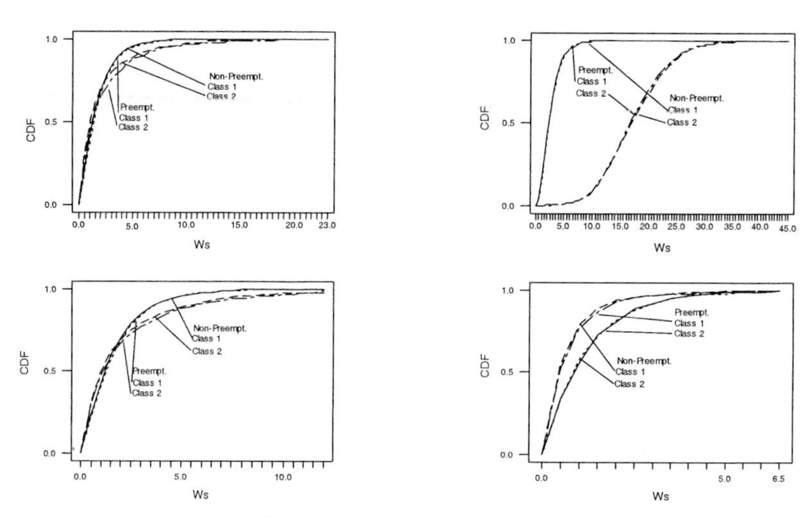

Fig. 1a–d. CDF of Ws for models I–IV

4.3 Waiting Time by Queue Size

What is interesting in our analysis is the link between buffer size and waiting time for class 2 packets. From the simulations undertaken here, in models where $\pi\{\text{Loss}\}\rightarrow 0$ both waiting times in the system reach constant values. Furthermore in models where loss is non-zero, waiting time in the system for class 2 traffic increases linearly as queue size increases, and loss for class 1 packets approaches a constant. As nearly all arriving packets will enter the system, a constant waiting time can be achieved for class 1 packets due to their class-based queue jumping. The preemptive class 2 packets is almost identical to the non-preemptive with slight deviations from a straight line the only difference. When we fitted a regression line for class 2 non-preemptive packets we achieved perfect positive correlation, indicating waiting time for class 2 packets is related linearly to queue size.

5 Regimes for Preemptive MPDQ

We now focus on the preemptive MPDQ with the expectation a priority regime is superior to the non-priority queueing regimes. In [1], we found the HCF regime in a non-preemptive model to provide superior QoS to packets over other priority/non-priority regimes. We will describe as in our prior work the waiting time in the system.

Table 4. Mean waiting time in the system by class and queue regime (c_1=c_2=2)

Regime	W_s^{class}	I	II	III	IV
HCF	1	2.28	2.20	1.57	1.17
	2	3.04	2.77	1.39	0.74
FIFO	1	3.10	2.96	1.77	1.23
	2	2.59	2.44	1.29	0.72
LCF	1	4.59	4.13	1.94	1.26
	2	1.87	1.77	1.09	0.69
LIFO	1	2.98	2.84	1.73	1.21
	2	2.60	2.38	1.22	0.69

From Table 4, for all of the models the waiting time in the system for Class 1 packets is lowest under the HCF regime. For Class 2 packets, the LCF regime is the best of the regimes. This further justifies the implementation of a priority scheme over a non-priority scheme. The margin between the times decreases as the model number decreases. It would appear that HCF again provides sufficient advantages over the non-priority schemes to warrant its use. Table 4 is the most important for packets as long waiting time is the cause of latency and jitter in many real-time web-based applications. An overall decrease in the difference in the statistics is noticeable as the Model number increases. This should come as no surprise as becomes smaller. So the smaller differences therefore should not be discounted as the traffic volume is high. In Model I, the Class 1 and 2 arrival rates are both one-tenth that of Model IV. Hence the traffic intensity in Model IV is forty times that of Model I. So a small difference in waiting may still lead to a large quantity of packets lost.

5.1 CDF of Class 1 Waiting Times

The CDF of waiting time (system) for Class 1 packets is now investigated for preemptive packets. In the models we analysed in the non-preemptive case [1], all of CDF's were superior under the HCF regime. Here we also have the same result. In Figures 7-10, the probability of waiting time in the system by waiting times is given. As ρ decreases, the difference between the curves decreases. The next best waits occur for LIFO, then FIFO and the poorest is LCF.

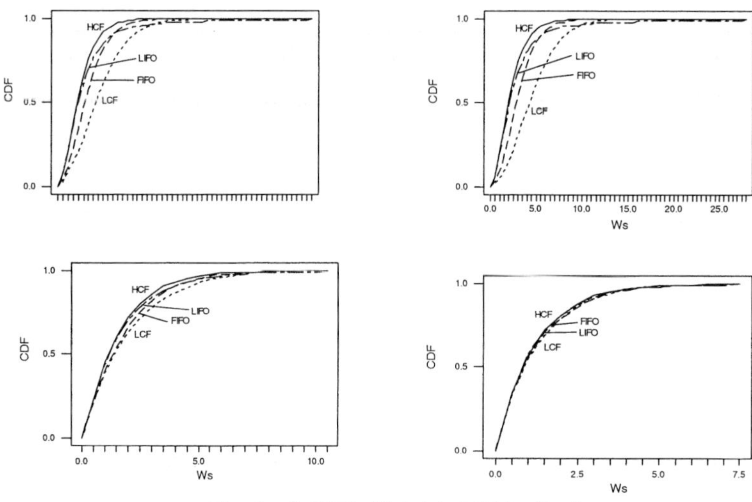

Fig. 2a-d. CDF of Models I-IV by Regime

It may be that service providers determine the difference is marginal and the priority system too complicated to introduce such a scheme over traditional FIFO. It has been shown in our prior work that the dual queueing system, irrespective of queueing regime, provides superior service over a single queueing scheme for non-preemptive.

6 Conclusion

Through analysis of loss and delay CDF's, we found the best service regime to be the non-preemptive regime. We looked further at the preemptive case and found the HCF the best regime, however as the traffic volume increased, the advantages over the non-priority schemes reduce. For many communications applications, a preemptive service regime is harder to implement as it may require more time to interrupt a packet in service, return it to the head-of-the-line, and recommence service for the higher classed packets than to just let it be served. Certainly, we found the differences small in many cases between preemption and non-preemption. We have provided a detailed framework with which analysts can apply a simulation to analyse a MPDQ system, and results that give tremendous insight into both service regimes for the MPDQ.

References

1. Bedford, A., Zeephongsekul, P.: Simulation Studies on the Performance Characteristics of Multi Priority Dual Queue (MPDQ) with Finite Waiting Room and Non-preemptive Scheduling. In: Kluev, V., Mastorakis, N. (eds.): Topics in Applied and Theoretical Mathematics and Computer Science. WSEAS Press, Greece (2001) 220–231
2. Bedford, A., Zeephongsekul, P.: Finite buffer queueing system with preemptive priority scheduling. Research Report No. 9, Dept. Mathematics and Statistics, RMIT University (2002)
3. Hayes, D., Rumsewicz, M., Andrew,L.: Quality of Service Driven Packet Scheduling Disciplines for Real-time Applications: Looking Beyond Fairness. Proc. IEEE Infocom (1999) 405-412
4. Kelton, W., Sadowski, R., Sadowski, D.: Simulation with Arena. 2nd edn. McGraw-Hill, (2002)
5. Odlyzko, A.: The Current State and Likely Evolution of the Internet. Proc. IEEE Globecom (1999) 1869–1875
6. Ranasinghe, R., Andrew, L., Hayes, D., Everitt, D.: Scheduling Disciplines for Multimedia WLANs: Embedded Round Robin and Wireless Dual Queue. Proc. IEEE Int. Conf. Commun. (2001) 1243–1248
7. Rose, O.: Statistical Properties of MPEG Video Traffic and their Impact on Traffic Modeling in ATM Systems. Proc. 20th Ann. Conf. on Local Computer Networks (1995) 397-406
8. Shreedhar, M., Varghese, G.: Efficient Fair Queuing using Deficit Round Robin. IEEE/ACM Trans. Networking, Vol. 4, No. 3 (1996) 375–385
9. Siripongwutikorn, P., Labrador, M., Znati, T.: A Wireless-aware Packet Dropping Policy for ATM Networks. Proc. Comm. Networks and Distributed Systems Modeling and Simulation Conf. (2000) 155–160

Visualization of Protein-Protein Interaction Networks Using Force-Directed Layout*

Yanga Byun and Kyungsook Han**

School of Computer Science and Engineering
Inha University, Inchon 402-751, Korea
quaah@hanmail.net, khan@inha.ac.kr
http://wilab.inha.ac.kr/protein/

Abstract. Protein interactions, when visualized as an undirected graph, often yield a nonplanar, disconnected graph with nodes of wide range of degrees. Many graph-drawing programs are of limited use in visualizing protein interactions, either because they are too slow, or because they produce a cluttered drawing with many edge crossings or a static drawing that is not easy to modify to reflect changes in data. We have developed a new force-directed layout algorithm for drawing protein interactions in three-dimensional space. Our algorithm divides nodes into three groups based on their interacting properties: biconnected subgraph in the center, terminal nodes at the outermost region, and the rest in between them. Experimental results show that our algorithm efficiently generates a clear and aesthetically pleasing drawing of large-scale protein interaction networks and that it is much faster than other force-directed layouts.

1 Introduction

Recent improvements in high-throughput proteomics technology such as yeast two-hybrid [4, 11] have produced a rapidly expanding volume of protein interaction data of an unprecedented scale. The interaction data is available either in text files or in databases. However, due to the volume of data (e.g., thousands of interacting proteins), a graphical representation of protein-protein interactions has proven to be much easier to understand than a long list of interacting proteins, prompting visualization studies of protein-protein interaction networks.

A Java applet program [6] has been developed for drawing protein interactions based on a relaxation algorithm and tested on the yeast two-hybrid (Y2H) data [11]. This program requires all protein-protein interaction data to be provided as parameters of the applet program in html sources. There is no way to save a visualized graph except by capturing the window. An image captured from the window is a static image and is of a generally low quality. It cannot be refined or changed later to reflect an update in data. A user can move a node

* This work was supported by the Ministry of Information and Communication of Korea under grant number IMT2000-C3-4.

** To whom correspondence should be addressed. email: khan@inha.ac.kr

P.M.A. Sloot et al. (Eds.): ICCS 2003, LNCS 2659, pp. 190–199, 2003.

but cannot select or save a connected component containing a specific protein for later use. Recent work on yeast proteome also utilized a relaxation algorithm to visualize a protein complex network [3].

Other visualization works on protein interactions do not have their own algorithms or programs developed for visualization, but use general-purpose drawing tools. PSIMAP [7], for example, displays interactions between protein families by comparing the Y2H data with the DIP data [12]. It was drawn by Tom Sawyer software (http://www.tomsawyer.com/) and then refined by significant amount of manual work to remove the edge crossings. From the perspective of graph drawing, PSIMAP is a static image and leaves several things to be improved. A research group of University of Washington [9, 10] has visualized the Y2H data using another general-purpose drawing tool called AGD (http://www.mpi-sb.mpg.de/AGD/). Although AGD is powerful, it is a general-purpose drawing tool and does not provide a function that we hold are necessary for studying protein-protein interactions.

Existing visualization studies in protein interactions suggest that the nature of protein interaction data require a new graph layout method for protein interaction networks. Protein interaction data can be characterized as follows:

1. The data yields a nonplanar graph with a large number of edge crossings that cannot be removed in a two-dimensional drawing.
2. Proteins have a very wide range of interacting proteins within the same set of data, so a graph visualizing the data contains nodes of very high degree as well as those of low degree.
3. When visualized as a graph, the data yields a disconnected graph with many connected components. The MIPS genetic interaction data (http://mips.gsf.de/proj/yeast/tables/interaction/), for example, contains 113 connected components.
4. The data often contains protein interactions corresponding to self-loops.

Considering these characteristics of protein interaction data, we have developed a new layout algorithm that divides nodes into three groups based on their interaction properties and layouts each group in three-dimensional space. The rest of this paper describes our algorithm and experimental results.

2 A Partitioned Approach to Graph Drawing

A common problem with many force-directed algorithms is that they become very slow when dealing with large graphs. We propose a new force-directed algorithm which divides nodes into three groups based on their interaction characteristics. Our layout is an extension of the algorithm by Kamada & Kawai [5] for drawing two-dimensional graphs. Their original algorithm has been modified not only for three-dimensional graph drawing but also for improvements in the efficiency and resulting drawings of the algorithm.

2.1 Finding Groups

Protein-protein interaction data can be visualized as an *undirected* graph $G = (V, E)$, where nodes V represent proteins and edges E represent protein-protein interactions. The *degree* of a node v_i is the number of its edges denoted by $deg(v_i)$. An edge $e = (v_i, v_j)$ with $v_i = v_j$ is a *self-loop.* A *cutvertex* in a graph G is a node whose removal disconnects G. A *path* in a graph G is a sequence $(v_1, v_2, \ldots, v_n)$ of distinct nodes of G, such that $(v_i, v_{i+1}) \in E$ for $1 \leq i \leq n-1$. A graph $G' = (V', E')$, such that $V' \subseteq V$ and $E' \subseteq E \cap (V' \times V')$, is a *subgraph* of graph $G = (V, E)$. We divide nodes V into three exclusive and exhaustive groups: V_1, V_2, V_3. The three groups are defined as follows:

- Group V_1 is a set of terminal nodes, i.e., nodes with degree 1.
- Group V_2 is a set of nodes in V-V1, which are in the subgraphs separated by cutvertices of degree ≥ 3, except the nodes in the largest subgraph separated by the cutvertices.
- Group V_3 consists of nodes which are members of neither V_1 nor V_2.

Example 1 Consider a graph $G = (V, E)$ displayed in Fig. 1. The nodes in G are separated into three groups. Six nodes belong to V_1 and these are separated into three sub-groups, $V_1 = \{\{v_1\}, \{v_5, v_9, v_{10}\}, \{v_{31}, v_{32}\}\}$. Each of the three sub-groups shares a neighbor. □

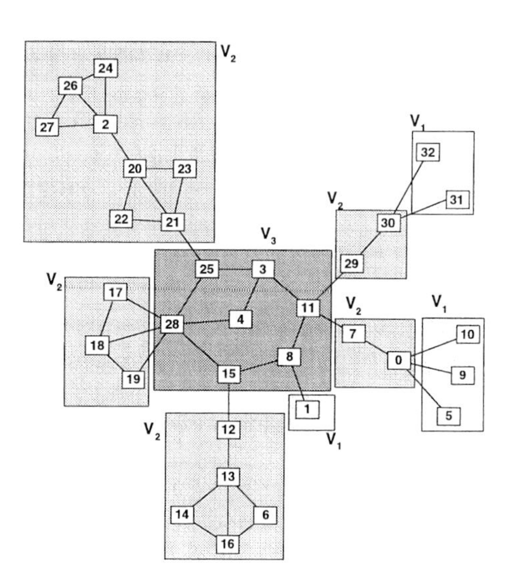

Fig. 1. Example of a partitioned graph. The nodes of V_1 are enclosed in yellow boxes, V_2 in green boxes, and V_3 in an orange box.

Example 2 In Fig. 1, two sub-groups $S_1 = \{v_0, v_7\}$ and $S_2 = \{v_{29}, v_{30}\}$ share a cutvertex v_{11}, so they are merged into one sub-group of V_2. Sub-groups $S_3 =$

$\{v_{24}, v_{26}, v_{27}\}$ and $S_4 = \{v_2, v_{20}, v_{21}, v_{22}, v_{23}, v_{24}, v_{26}, v_{27}\}$ do not share a cutvertex since the cutvertex of S_3 is v_2 and that of S_4 is v_{25}. However, S_3 is not counted as a sub-group of V_2 since $S_3 \subset S_4$. □

Algorithm 1 FindCutvertex

1: **for all** v_i such that degree($v_i > 2$) **do**
2: inputSet $= V - V_1 - v_i$
3: **while** P = IsCutvertex(inputSet) **do**
4: inputSet = inputSet -P
5: **end while**
6: **end for**

Algorithm 2 IsCutvertex(givenSet)

1: Randomly select a starting node $v_s \in givenSet$.
2: Insert v_s to a stack S.
3: **while** there is a top node v_t in S **do**
4: Popup v_t and assign v_t to a current node v_c.
5: **if** v_c is not marked **then**
6: Mark v_c and insert v_c to P_i $\{P_i$: a set of nodes in the path between v_s and $v_i\}$
7: Insert neighbors v_h of v_c to S, s.t. v_h are not marked.
8: **end if**
9: **end while**
10: Insert nodes not in P_i to P_i'.
11: **if** $(|P_i| > 0)$ and $(|P_i'| > 0)$ **then**
12: **if** $|P_i| \leq |P_i'|$ **then**
13: Insert $v \in P_i$ to V_2 and Return P_i.
14: **else**
15: Insert $v \in P_i'$ to V_2 and Return P_i'.
16: **end if**
17: **end if**
18: Return false

Nodes of each group are found in the order of V_1, V_2, and V_3. First, nodes with one neighbor are classified into V_1. Nodes of V_1 are further divided into sub-groups according to their shared neighbors. From $V - V_1$, nodes in V_2 are then found, and all remaining nodes constitute V_3.

After finding V_1, nodes of V_2 are determined by our heuristic algorithm *FindCutvertex* outlined in Algorithm 1. The initial input to the algorithm is nodes in $V - V_1$. For each input node v_i, the algorithm tests whether the node is a cutvertex (line 3 of Algorithm 1). Let P be the set of nodes in the path between v_i and the starting node and P' be the set of nodes not in the path. If neither of P and P' is empty, the node v_i is a cutvertex and the loop is repeated for the remaining nodes. The nodes in the smaller set between P and P' are included in V_2 (lines 11-17 of Algorithm 2). The nodes of V_2 are further separated into sub-groups based on their cutvertex. These sub-groups are merged into one if they have the same cutvertex. All remaining nodes after determining both V_1 and V_2 constitute V_3. V_3 corresponds to a *biconnected* subgraph (a connected graph with no cutvertices) in protein interaction data.

2.2 Force-Directed Layout for Three-Dimensional Graph Drawing

The algorithm by Kamada & Kawai [5] searches for a drawing in which the energy is locally minimal. The aim of our algorithm is to find a drawing in which the actual distance between two nodes is approximately proportional to the desirable distance between them. The global energy of a spring system with n nodes is defined by the equation:

$$\begin{aligned} E &= \sum_{i=1}^{n-1} \sum_{j=i+1}^{n} \frac{1}{2} k_{ij} (|p_i - p_j| - l_{ij})^2 \\ &= \sum_{i=1}^{n-1} \sum_{j=i+1}^{n} \frac{1}{2} k_{ij} \Big\{ (x_i - x_j)^2 + (y_i - y_j)^2 + (z_i - z_j)^2 \\ &\quad + {l_{ij}}^2 - 2 l_{ij} \sqrt{(x_i - x_j)^2 + (y_i - y_j)^2 + (z_i - z_j)^2} \Big\} \end{aligned} \tag{1}$$

where k_{ij} is the *stiffness* parameter of a spring, p_i is the position of a node v_i, and l_{ij} is the length at rest of the spring connecting v_i and v_j.

The algorithm seeks a position $p_m = (x_m, y_m, z_m)$ for each vertex v_m to minimize the potential energy in the spring system. Minima occur when the partial derivatives of E with respect to each variables x_m, y_m, and z_m are zero. This gives a set of $3|V| = 3n$ equations

$$\frac{\partial E}{\partial x_m} = \frac{\partial E}{\partial y_m} = \frac{\partial E}{\partial z_m} = 0, \ v_m \in V \tag{2}$$

In Kamada & Kawai's algorithm [5], a node is moved to a position that minimizes energy while all others remain fixed. The node to be moved is chosen as the one with the largest force acting on it, i.e., the one for which

$$\sqrt{\left(\frac{\partial E}{\partial x_m}\right)^2 + \left(\frac{\partial E}{\partial y_m}\right)^2 + \left(\frac{\partial E}{\partial z_m}\right)^2} \tag{3}$$

is maximized over all $v_m \in V$. However, this approach often yields unpleasant graphs and takes too much time for large-scale protein interactions. Thus, our algorithm moves all nodes to some level in each iteration until the difference between the current position and the previous position falls below a certain threshold value. For an initial layout we place nodes on the surface of a sphere instead of placing them randomly. Our algorithm yields more pleasant drawings than Kamada & Kawai's algorithm and is much faster for graphs with balanced groups.

Algorithm 3 ShortestPath

1: Compute adjacency matrix $A[a,b]$ in G, for $1 \le a < b \le n$.
2: Initialize shortest-path matrix $S[a,b]$ in G, for $1 \le a < b \le n$.
3: **for all** V_i, $i = 3, 2, 1$ **do**
4: **if** V_3 **then**
5: Call *FindShortestPath*(V_3,*null*) {in Algorithm 4}
6: **else if** V_2 **then**
7: **for all** each sub-group G_s **do**
8: Call *FindShortestPath*($G_s \cup v_c$, null) {v_c: shared cutvertex of G_s}
9: Call *FindShortestPath*($G_s \cup V_3, v_c$) {compute $S[p,q]$ not defined for $p \ne q \in G_s \cup V_3$}
10: **end for**
11: **else if** V_1 **then**
12: **for all** each sub-group G_s **do**
13: Call *FindShortestPath*($G_s \cup v_h$, null) {v_h: shared neighbor of G_s}
14: Call *FindShortestPath*($G_s \cup V_3 \cup V_2, v_h$) {compute $S[p,q]$ not defined for $p \ne q \in G_s \cup V_3 \cup V_2$}
15: **end for**
16: **end if**
17: **end for**

Algorithm 4 FindShortestPath (N, givenK)

1: **for all** k such that $k \in N$ or *givenK* **do**
2: **for all** l such that $l \in N$ **do**
3: $S[k,l] = A[k,l]$ ($S[k,l] = 2$ if $V_i = V_1$) {$S[k,l]$: shortest-path matrix}
4: **for all** m such that $m \in N$ **do**
5: **if** $S[l,m] > S[l,k] + S[k,m]$, for $(m \ne k)\&\&(m \ne l)$ **then**
6: $S[l,m] = S[l,k] + S[k,m]$
7: **end if**
8: **end for**
9: **end for**
10: **end for**

2.3 Finding Shortest Paths in Groups

The shortest path between every pair of nodes is computed for each group $V_i, i = 1,\ldots,3$. The algorithm for computing shortest paths is summarized in Algorithms 3 and 4. We first compute shortest paths between nodes in V_3. For V_2 and V_1, shortest paths are determined in each of their sub-groups. After computing shortest paths between nodes in each sub-group, shortest paths between nodes of V_2 and nodes of V_3 are computed using a shared cutvertex of each sub-group of V_2 (line 9 of Algorithm 3). Likewise, shortest paths between nodes of V_1 and nodes of V_2 and V_3 are computed using a shared neighbor of each sub-group of V_1 (line 14). For a sub-group of V_1, the initial shortest path between every pair of nodes is set to 2, since the distance between a node and its shared neighbor is 1 (line 3 of Algorithm 4).

3 Analytical and Empirical Evaluation

Here we briefly analyze the computational cost of our algorithm. Assuming that three groups are balanced, the total time for our algorithm is $(\frac{n}{3})^3 + (\frac{n}{3})^3 + (\frac{n}{3})^3 = \frac{n^3}{9}$ because we apply spring-embedder algorithm on each group. The asymptotic time complexity of our algorithm is the same as the time complexity $O(n^3)$ of Kamada & Kawai's algorithm [5]. But our algorithm is much faster Kamada & Kawai's algorithm in practice. Since the nodes in V_1 and V_2 are further divided into sub-groups, actual running times are reduced further for graphs with balanced groups. For a graph with unbalanced groups (for example, a graph in which the portion of V_3 is high due to few cutvertices or terminal nodes), the effect of dividing into three groups can be marginal, which is rare in protein interaction data. This has been supported by the experimental study to be discussed later.

Our algorithm was implemented in Microsoft C#. The program runs on any PC with Windows 2000/XP/Me/98/NT 4.0 as its operating system. We tested the program on five cases, the PSIMAP data [7], Helicobacter pylori data [8], Y2H data [11], and the data of protein interactions in yeast from the BIND (http://www.binddb.org/) and DIP (http://dip.doe-mbi.ucla.edu/) databases. In each of these data of protein interactions, the largest connected component was used for comparison. Table 1 shows the running times of our algorithm at each stage of partitioning nodes into three groups, finding shortest paths in all three groups, and layout and drawing.

Table 1. Running times of our algorithm at each stage. Hpylori: helicobacter pylori, P: partitioning nodes into V_1, V_2 and V_3, SP: finding shortest paths in all groups, LD: layout and drawing.

Data	Edges	Nodes			Running times			
		V_1	V_2	V_3	P (min: sec)	SP (min: sec)	LD (min: sec)	total $(= P + SP + LD)$
PSIMAP	661	187	83	171	00:00.06	00:00.04	00:02.02	00:02.12
Hpylori	1396	434	9	267	00:00.31	00:00.20	00:07.36	00:07.87
Y2H	12909	3066	28	642	00:26.78	00:34.38	09:12.25	10:13.41
BIND	8243	3039	45	893	00:15.67	00:26.02	08:21.82	09:03.51
DIP	14415	3278	126	1195	00:30.00	00:43.45	10:41.64	11:55.09

Fig. 2A shows an initial layout by our algorithm for the Y2H data with 12909 interactions between 3736 proteins. While we find groups in the order of V_1, V_2, and V_3, we layout them in reverse order; V_3 is first positioned in the center of the sphere, V_2 in the outer region of V_3, and V_1 in the outer region of V_2 and V_3. Groups for which node positions are fixed are shown in a rectangle. Nodes in remaining groups are relocated with modified polar coordinates in order to place them in the outer region of the groups that have been fixed. In Figs 2B

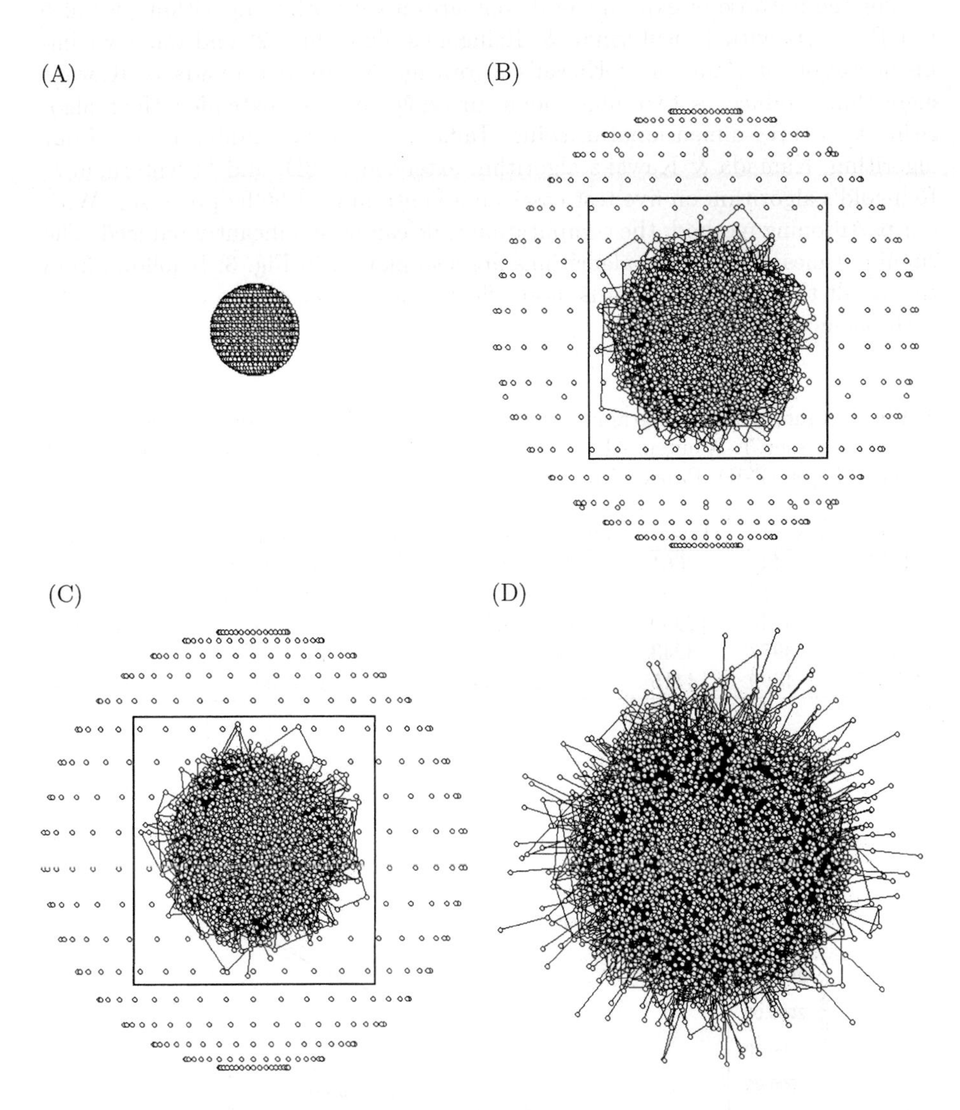

Fig. 2. Drawings of the Y2H data with 12909 interactions between 3736 proteins. (A) Initial layout, (B) After drawing the nodes in V_3 shown in the rectangle, (C) After drawing the nodes in V_3 and V_2 shown in the rectangle, (D) Final drawing.

and 2C, edges between nodes in the outer area are not shown for the clarity of the drawing.

For the purpose of experimental comparison with other algorithms, we also ran Pajek [1] with Fruchterman & Reingold's algorithm [2] and our own implementation of Kamada & Kawai's algorithm [5]. Since Kamada & Kawai's algorithm produces a two-dimensional drawing only, we extended their algorithm to a three-dimensional drawing. Table 2 shows the running times of our algorithm, Kamada & Kawai's algorithm extended to 3D, and Fruchterman & Reingold's algorithm on five test cases on a Pentium IV 1.5Ghz processor. With our partitioning method, the computation time can be significantly reduced. The running times of the three algorithms are also plotted in Fig. 3. It follows from this result that our algorithm is more effective for bigger graphs and for graphs with balanced groups.

Table 2. Running times of graph drawing algorithms on 5 test cases on a Pentium IV 1.5Ghz processor. K-K extended to 3D: Kamada & Kawai's algorithm extended for 3D drawing. Pajek (F-R): Pajek with Fruchterman & Reingold's layout.

Data	Nodes	Edges	Our algorithm	K-K extended to 3D	Pajek (F-R)
PSIMAP	441	661	00m 02.12s	00m 07.69s	00m 33s
Hpylori	710	1396	00m 07.87s	00m 22.14s	01m 14s
Y2H	3736	12909	10m 13.41s	13m 23.47s	32m 10s
BIND	3977	8243	09m 03.51s	14m 24.39s	41m 37s
DIP	4599	14415	11m 55.09s	19m 09.81s	55m 55s

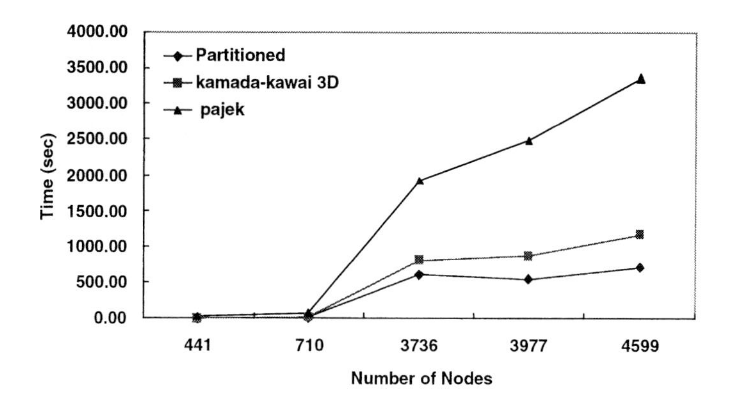

Fig. 3. Running times of three graph drawing algorithms.

4 Discussion and Conclusions

Many force-directed graph drawing algorithms are too slow to be used for visualizing large-scale protein interactions and they often yield unclear drawings with many edge crossings. This paper presented a new algorithm for drawing large-scale protein interaction networks in three-dimensional space. The algorithm divides nodes into three groups: a biconnected subgraph in the center of a graph (V_3), terminal nodes at the outermost region (V_1), and the rest in between them (V_2). These groups are identified from the outer group (V_1) to inner one (V_3) and are placed in reverse order. Experimental results demonstrate that the algorithm generates clear and aesthetically pleasing drawings of both large and small scale graphs and that for graphs with balanced groups it is significantly faster than other force-directed algorithms.

References

[1] V. Batagelj and A. Mrvar. Pajek - analysis and visualization of large networks. *Lecture Notes in Computer Science*, 2265:477–478, 2001.

[2] T.M.J. Fruchterman and E.M. Reingold. Graph drawing by force-directed placement. *Software-Practice and Experience*, 21(11):1129–1164, 1991.

[3] A.-C Gavin, M. Bosche, R. Krause, and et al. Functional organization of the yeast proteome by systematic analysis of protein complexes. *Nature*, 415(6868):141–147, 2002.

[4] T. Ito, K. Tashiro, S. Muta, R. Ozawa, T. Chiba, M. Nishizawa, K. Yamamoto, S. Kuhara, and Y. Sakaki. Toward a protein-protein interaction network of the budding yeast: A comprehensive system to examine two-hybrid interactions in all possible combinations between the yeast proteins. *Proc. Natl. Acad. Sci. USA*, 97(3):1143–1147, 2000.

[5] T. Kamada and S. Kawai. An Algorithm for Drawing General Undirected Graphs. *Information Processing Letters*, 31:7–15, 1989.

[6] R. Mrowka. A java applet for visualizing protein-protein interaction. *Bioinformatics*, 17(7):669–670, 2001.

[7] J. Park, M. Lappe, and S.A. Teichmann. Mapping protein family interactions: intramolecular and intermolecular protein family interaction repertoires in the PDB and Yeast. *J. Mol. Biol.*, 307:929–938, 2001.

[8] J.-C. Rain, L. Selig, H.D. Reuse, V. Battaglia, C. Reverdy, S. Simon, G. Lenzen, F. Petel, J. Wojcik, V. Schachter, Y. Chemama, A. Labigne, and P. Legrain. The protein-protein interaction map of helicobacter pylori. *Nature*, 409:211–215, 2001.

[9] B. Schwikowski, P. Uetz, and S. Fields. A network of protein-protein interactions in yeast. *Nature Biotechnology*, 18(12):1257–1261, 2000.

[10] C.L. Tucker, J.F. Gera, and P. Uetz. Towards an understanding of complex protein networks. *Trends in Cell Biology*, 11(3):102–106, 2001.

[11] P. Uetz, L. Giot, G. Cagney, and et al. A comprehensive analysis of protein-protein interactions in Saccharomyces cerevisiae. *Nature*, 403:623–627, 2000.

[12] I. Xenarios, E. Fernandez, L. Salwinski, X.J. Duan, M.J. Thompson, E.M. Marcotte, and D. Eisenberg. DIP: the database of interacting proteins: 2001 update. *Nucleic Acids Res.*, 29(1):239–241, 2001.

Track on Clusters and Grids

Counting Polyominoes: A Parallel Implementation for Cluster Computing

Iwan Jensen

Department of Mathematics and Statistics
The University of Melbourne, Victoria 3010, Australia
I.Jensen@ms.unimelb.edu.au
http://www.ms.unimelb.edu.au/~iwan/

Abstract. The exact enumeration of most interesting combinatorial problems has exponential computational complexity. The finite-lattice method reduces this complexity for most two-dimensional problems. The basic idea is to enumerate the problem on small finite lattices using a transfer-matrix formalism. Systematically grow the size of the lattices and combine the results in order to obtain the desired series for the infinite lattice limit. We have developed a parallel algorithm for the enumeration of polyominoes, which are connected sets of lattice cells joined at an edge. The algorithm implements the finite-lattice method and associated transfer-matrix calculations in a very efficient parallel setup. Test runs of the algorithm on a HP server cluster indicates that in this environment the algorithm scales perfectly from 2 to 64 processors.

1 Introduction

The enumeration of polyominoes is a classical combinatorial problem [1]. A polyomino is a connected set of lattice cells joined at their edges. The fundamental problem is the calculation (up to translation) of the number of *fixed* polyominoes, a_n, with n cells. If we also take into account rotation and reflection symmetries we arrive at *free* polyominoes. Thus □□ and □ count as different fixed polyominoes, but they are the same free polyomino, while □ isn't a polyomino because the two cells don't share an edge. The enumeration of polyominoes has traditionally served as a benchmark for computer performance and algorithm design [2,3,4,5,6,7,8,9].

Polyominoes are closely related to lattice animals. A *site* animal can be viewed as a finite set of lattice sites connected by a network of nearest neighbor bonds. Polyominoes are thus identical to site animals on the dual lattice. In the physics literature lattice animals are often called *clusters* due to their close relationship to percolation problems [10]. Series expansions for various percolation properties, such as the percolation probability or the average cluster size, can be obtained as weighted sums over the number of lattice animals, $g_{n,m}$, enumerated according to the number of sites n and perimeter m [11].

P.M.A. Sloot et al. (Eds.): ICCS 2003, LNCS 2659, pp. 203–212, 2003.

It has been proven [12] that the number of polyominoes grows exponentially

$$\lim_{n\to\infty} n^{-1} \ln a_n = \sup_{n>0} n^{-1} \ln a_n = \lambda \tag{1}$$

where λ is the *growth constant*. The best numerical estimate, based on an analysis of the series for the generating function up to order 46, is $\lambda = 4.062570(8)$ [13], while rigorous lower and upper bounds shows that $3.903184\ldots \le \lambda \le 4.649551\ldots$, where the lower bound was derived in [13] from the aforementioned series using a method developed by Rands and Welsh [14] and the upper bound is due to Klarner and Riverst [15]. It is generally believed, though not rigorously proven, that there is a power-law correction to the exponential growth

$$a_n \simeq A\lambda^n n^{-\theta}. \tag{2}$$

where θ is called the *entropic exponent*. It is generally believed [16] from theoretical arguments that $\theta = 1$, a prediction, which is overwhelmingly supported by the available numerical evidence [13].

As with most interesting unsolved combinatorial problems the enumeration of polyominoes is of exponential computational complexity. That is the time $T(n)$ required to calculate the first n terms grows like $T(n) \propto \kappa^n$. The finite-lattice method (FLM) reduces the value of κ for two-dimensional problems. The basic idea is to enumerate the problem on small finite lattices using a transfer-matrix (TM) formalism. Systematically grow the size of the lattices and combine the results in order to obtain the desired series for the infinite lattice limit.

Sykes and Glen [11] calculated $g_{n,m}$ up to $n = 19$ on the square lattice, and thus obtained the number of polyominoes, $a_n = \sum_m g_{n,m}$, to the same order. Redelmeier [4] presented an improved algorithm for the enumeration of polyominoes and extended the results to $n = 24$. This algorithm was later used by Mertens [6] to devise an improved algorithm for the calculation of $g_{n,m}$ and a parallel version of the algorithm appeared a few years later [7]. All of the above algorithms were variations on direct counting and their computational complexity thus grows as $T(n) = \lambda^n$. A major advance was obtained by Conway [8] who used the FLM to calculate a_n and numerous other series up to $n = 25$ [17]. For this algorithm the computational complexity was $T(n) = 3^{n/2}$. In unpublished work Oliveira e Silva [18] used the parallel version of the Redelmeier algorithm to extend the enumeration to $n = 28$. In [9] we used an improved version of Conway's algorithm to extend the enumeration to $n = 46$. This improved algorithm has complexity $T(n) = \kappa^{n/2}$ with $\kappa \simeq 2$. Knuth [19] improved the algorithm somewhat further and managed to obtain one further term.

In this paper we describe a parallel implementation of the TM algorithm for the enumeration of polyominoes. The algorithm implements the finite-lattice method and associated transfer-matrix calculations in a very efficient parallel setup. Test runs of the algorithm on a HP server cluster indicates that in this environment the algorithm scales perfectly from 2 to 64 processors.

2 Enumeration of Polyominoes

The method we use to enumerate polyominoes on the square lattice is based on the method used by Conway [8] for the calculation of series expansions for percolation problems, and is similar to methods devised by Enting for enumeration of self-avoiding polygons [20]. In the following we give a brief description of the algorithm used to count polyominoes. We then give some details of the parallel version of the algorithm.

2.1 Transfer Matrix Algorithm

The number of fixed polyominoes that span rectangles of width W and length L are counted using a transfer matrix algorithm. By combining the results for all $W \times L$ rectangles with $W \leq W_{\mathrm{max}}$ and $W + L \leq 2W_{\mathrm{max}} + 1$ we can count all polyominoes up to $N_{\mathrm{max}} = 2W_{\mathrm{max}}$. Due to symmetry we only consider rectangles with $L \geq W$ while counting the contributions for rectangles with $L > W$ twice. The maximal size N_{max} up to which one can count the number of polyominoes is limited by the available computational resources.

The transfer matrix technique involves drawing an intersection through the rectangle cutting through a set of W cells. Polyominoes in rectangles of a given width are counted by moving the intersection so as to add one cell at a time, as shown in Fig. 1. For each configuration of occupied or empty cells along the intersection we maintain a truncated generating function for partially completed polyominoes. Each configuration can be represented by a set of states $S = \{\sigma_i\}$, where the value of the state σ_i at position i must indicate first of all if the cell is occupied or empty. An empty cell is simply indicated by $\sigma_i = 0$. Since we have to ensure that we count only connected graphs more information is required if a cell is occupied. We need a way of describing which occupied cells along the intersection are connected to one another via a set of occupied cells to the left of the intersection. The most compact encoding of this connectivity is [8]

$$\sigma_i = \begin{cases} 0 & \text{empty cell,} \\ 1 & \text{occupied cell not connected to others,} \\ 2 & \text{first among a set of connected cells,} \\ 3 & \text{intermediate among a set of connected cells,} \\ 4 & \text{last among a set of connected cells.} \end{cases} \tag{3}$$

Configurations are read from the bottom to the top. As an example the configuration along the intersection of the partially completed polyomino in Fig. 1 is $S = \{102023404\}$.

Pruning. In the original approach [8] polyominoes were required to span the rectangle only length-wise and polyominoes of width $\leq W$ and length L were counted several times. It is however easy to obtain polyominoes of width exactly W and length L from such data [20]. In our algorithm we directly enumerate

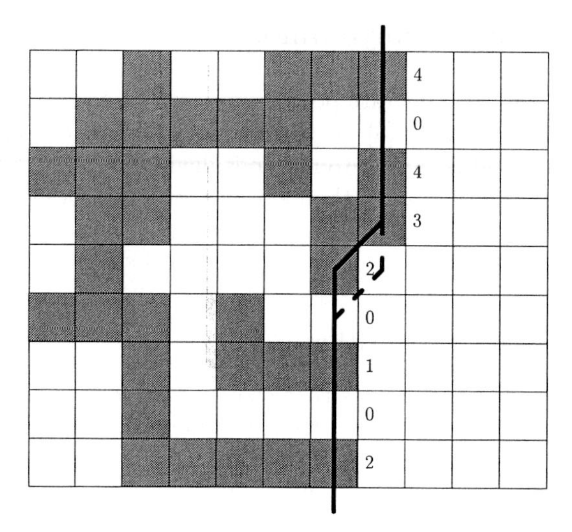

Fig. 1. A snapshot of the intersection (solid line) during the transfer matrix calculation on the square lattice. Polyominoes are enumerated by successive moves of the kink in the intersection, as exemplified by the position given by the dashed line, so that one cell at a time is added to the rectangle. To the left of the intersection we have drawn using shaded squares an example of a partially completed polyomino. Numbers along the intersection indicate the encoding of this particular configuration

polyominoes of width exactly W and length L. At first glance this would appear to be inefficient since for many intersection configurations we now have to keep 4 distinct generating functions depending on which borders have been touched. However, as demonstrated in practice [9] it actually leads to an algorithm which is both exponentially faster and whose memory requirement is exponentially smaller. Realizing the full savings in time and memory usage require enhancements to the original algorithm. The most important is what we call *pruning*. This procedure allows us to discard most of the possible configurations for large W because they only contribute to polyominoes of size greater than $N_{\max}$. Briefly this works as follows. Firstly, for each configuration we keep track of the current minimum number of cells $N_{\rm cur}$ already inserted to the left of the intersection. Secondly, we calculate the minimum number of additional cells $N_{\rm add}$ required to produce a valid polyomino. There are three contributions, namely the number of cells required to connect all sections of the partially completed polyomino, the number of cells needed (if any) to ensure that the polyomino touches both the lower and upper border, and finally the number of steps needed (if any) to extend at least W cells in the length-wise direction (remember we only need rectangles with $L \geq W$). If the sum $N_{\rm cur} + N_{\rm add} > N_{\max}$ we can discard the partial generating function for that configuration, and of course the configuration itself, because it won't make a contribution to the polyomino count up to the size we are trying to obtain.

The Updating Rules. In Table 1 we have listed the possible local 'input' states and the 'output' states which arise as the kink in the intersection is propagated by one step. The most important cell on the intersection is the 'lower' one situated at the bottom of the kink (the cell marked with the second '2' (counting from the bottom) in Fig. 1). This is the position in which the lattice is being extended. Obviously the new cell can be either empty or occupied. The state of the upper cell (the cell marked '3' in Fig. 1) is likely to be changed as a result of the move. In addition the state of a cell further afield may have to be changed if a branch of a partially completed polyomino terminates at the new cell or if two independent sections of a partially completed polyomino join at the new cell.

Details of how these updating rules are derived can be found in [9]. Here a few comments will have to suffice.

10: The lower cell is an isolated occupied cell and the new cell can be empty only if there are no other occupied cells on the intersection (otherwise we generate graphs with separate components) and if both the lower and upper borders have been touched. The result are valid polyominoes and the partial generating function is added to the total polyomino generating function.

14: This situation never occurs. The upper cell is last among a set of occupied cells, so the cell immediately to its left is also occupied, this in turn is connected to the lower cell, which therefore cannot be an isolated cell.

20: The lower cell is first among a set of occupied cells, so if the new cell is empty, another cell in the set changes its state. Either the *first* intermediate cell becomes the new first cell, and its state is changed from 3 to 2, or, if there are no intermediate cells, the last cell becomes an isolated cell, and its state is changed from 4 to 1. This relabeling of a matching cell is indicated in Table 1 by over-lining.

22: When the new cell is occupied two separate pieces of the polyomino are joined. The new cell remains the first cell in the joined piece while the upper cell becomes an intermediate cell. The last cell in the innermost set of connected cells also becomes an intermediate cell in the joined piece. We indicate this type of transformation by putting a hat over the string.

Table 1. The various 'input' states and the 'output' states which arise as the intersection is moved in order to include one more cell of the lattice. Each panel contains two 'output' states where the left (right) most is the configuration in which the new cell is empty (occupied)

Lower \ Upper	0	1	2	3	4
0	00 10	01 24	02 23	03 33	04 34
1	add 10	– 24	– 23	– 33	
2	$\overline{00}$ 20	$\overline{01}$ 23	$\overline{02}$ $\widehat{23}$	02 23	01 24
3	00 30	01 33	02 $\widehat{33}$	03 33	04 34
4	$\overline{00}$ 40	$\overline{01}$ 34	$\overline{02}$ 33	$\overline{03}$ $\widehat{33}$	

Computational Complexity. The algorithm has exponential complexity, that is the time required to obtain the polyominoes up to size n grows exponentially with n. Time and memory requirements are basically proportional to the maximum number of distinct configuration generated during a calculation. In [9] we s showed that the maximal number of configurations, N_{Conf}, grows with $W_{\max}$ as $N_{\text{Conf}} \propto \kappa^{W_{\max}}$, and from the numerical data we estimated that κ is a little larger than 2. Note that this is much better than a direct enumeration in which time requirements are proportional to the number of polyominoes and therefore has the growth constant, $\lambda \simeq 4.06\ldots$, of polyominoes. The price we have to pay is that the memory requirement also grows exponentially like N_{Conf}, whereas in direct enumerations the memory requirement grows like a polynomial in n.

Further Details. The integer coefficients occurring in the calculation become very large so we used modular arithmetic [21]. This involves performing the calculation modulo various integers p_i and then reconstructing the full integer coefficients at the end. The p_i are called moduli and must be chosen so they are mutually prime. The Chinese remainder theorem ensures that any integer has a unique representation in terms of residues. Since we are using a heavily loaded shared facility CPU time was more of a immediate limitation than memory and secondly more memory was used for the data required to specify the configuration and manage the storage than for storing the actual terms of the generating functions. So we used the moduli $p_0 = 2^{62}$ and $p_1 = 2^{62} - 1$, which allowed us to represent a_n correctly using just these two moduli.

2.2 Parallelization

In the past decade or so parallel computing has become the paradigm for high performance computing. The early machines were largely dedicated MPP machines which more recently have been superceded by clusters. The transfer-matrix algorithms used in the calculations of the finite lattice contributions are eminently suited for parallel computations.

The most basic concerns in any efficient parallel algorithm is to minimise the communication between processors and ensure that each processor does the same amount of work and use the same amount of memory. In practice one naturally has to strike some compromise and accept a certain degree of variation across the processors.

One of the main ways of achieving a good parallel algorithm using data decomposition is to try to find an invariant under the updating rules. That is we seek to find some property about the configurations along the intersection which does not alter in a single iteration. The algorithm for the enumeration of polyominoes is quite complicated since not all possible configurations occur due to pruning and an update at a given set of cells might change the state of a cell far removed as explained above. However, there still is an invariant since any cell not in the kink itself cannot change from being empty to being occupied and vice versa. Only the kink cell can change its occupation status. This invariant allows us to parallelise the algorithm in such a way that we can do the calculation completely independently on each processor with just two redistributions of the data set each time an extra column is added to the lattice. It should be noted that each redistribution results in a global synchronization of the processors.

The main points of the algorithm are summarized below:

1. With the intersection in an upright position distribute the data across processors so that configurations with the same occupation pattern along the *lower* half of the intersection are placed on the same processor.
2. Do the TM update inserting the top-half of a new column. This can be done *independently* by each processor because the occupation pattern in the lower half remains unchanged.
3. Upon reaching the half-way mark redistribute the data so that configurations with the the same occupation pattern along the *upper* half of intersection are placed on the same processor.
4. Do the TM update inserting the bottom-half of a new column.
5. Go back to 1.

The redistribution among processors is done as follows:

1. On each processor run through the configurations to establish the configuration pattern c of each configuration and calculate, $n(c)$, the number of configurations with a given pattern.
2. Calculate the global sum of $n(c)$ on say processor 0.
3. Sort $n(c)$ on processor 0.
4. On processor 0 assign each pattern to a processor $p(c)$ such that:
   a) Set $p_{id} = 0$.
   b) Assign the *most* frequent unassigned pattern c to processor p_{id}.
   c) If the number of configurations assigned to p_{id} is less than the number of configurations assigned to p_0 then assign the *least* frequent unassigned patterns to p_{id} until the desired inequality is achieved.
   d) set $p_{id} = (p_{id} + 1) \mod N_p$, where N_p is the number of processors.
   e) Repeat from (b) until all patterns have been assigned.
5. Send $p(c)$ to all processors.
6. On each processor run through the configurations sending each configuration to its assigned processor.

The calculations were performed on the facilities of the Australian Partnership for Advanced Computing (APAC) which is an HP Server Cluster with 125 ES45's each with 4 1 Ghz Alpha chips for a total of 500 processors in the compute partition. The cluster has a total peak speed of 1Tflop. Each server node has at least 4 Gb of memory. Nodes are interconnected by a fat-tree low latency (MPI < 5 usecs), high bandwidth (250 Mb/sec bidirectional) Quadrics network.

In Table 2 we list the time and memory use of the algorithm for $N_{\max} = 48$ at $W = 20$ using from 1 to 64 processors. The memory use of the single processor job was about 2Gb. As can be seen the algorithm scales perfectly from 1 to 64 processors since the total CPU time (column 2) stays almost constant

Table 2. Total CPU time, elapsed time, time cost of redistribution, and memory use for the parallel algorithm for enumerating polyominoes of maximal size 48 at width 20

Proc	CPU	Elapsed	Max Cost	Min Cost	Max Conf	Min Conf	Max Term	Min Term
1	15:06	15:08:04			47586716		121194174	
2	14:44	7:23:15	23:47	20:15	24773601	24248682	62848978	62119094
4	15:06	3:49:09	14:03	12:39	12438275	12247276	32034803	31180773
8	14:30	1:49:44	10:47	7:30	6291877	6070370	16270110	15592551
16	14:17	54:12	7:30	4:52	3298793	3024985	8472602	7541084
32	14:14	27:03	4:59	3:07	1753622	1570854	4513708	3834315
64	14:06	13:55	4:09	1:59	899144	851739	2464817	1923254

while the elapsed time is halved when the number of processors is doubled. We expect that the drop in CPU time at 32 or 64 processors is caused by better single processor optimization by the compiler. One would for example expect that the average time taken to fetch elements from memory drops as the problem size on individual processors drops from 2Gb for the computation using a single processor to just under 40Mb for the 64 processor computation. Another main issue in parallel computing is that of load balancing, that is, we wish to ensure that the workload is shared almost equally among the processors. This algorithm is quite well balanced. Even with 64 processors, where each processor uses only about 40Mb of memory, the difference between the processor handling the maximal and minimal number of configurations is less than 10%. For the total number of terms retained in the generating functions the spread is less than 20%. A simple timing of different sub-routines of the parallel algorithm shows that the typical time to do a redistribution is about the same as the average time taken in order to move the kink once. Further on this subject we have listed, in columns 4 and 5, the maximal and minimal 'redistribution cost' (total time spent in the redistribution sub-routine). Firstly we note that the typical overall cost of parallel execution is smaller than 10%, when the per processor problem size is large. As the number of processors is increased we are not surprised to see that the relative cost of parallel execution increases and as the problem becomes less well-balanced we also see an increase in the difference between the maximal and minimal cost of redistribution.

Another way of examining the efficiency of the parallel algorithm is to look at a fixed rectangle (in this case a 22×22 square) and grow the overall problem size by increasing $N_{\max}$. This means that more and more configurations and terms are retained. By increasing the number of processors we ensured that the problem size handled by individual processors remained relatively stable (we tried to make the number of configurations almost constant). In Table 3 we list the time and memory use of the algorithm as $N_{\max}$ is increased from 50 to 56 while using from 4 to 32 processors. Clearly we achieved the goal of keeping the number of configurations fairly constant. The elapsed time also stays fairly constant with changes largely reflecting the changes in workload as the number of configurations and terms increase or decrease. One thing we do note is that the difference between the maximum and minimum number of configurations

Table 3. Elapsed time, redistribution cost, and memory use for the parallel algorithm for enumerating polyominoes of maximal size $N_{\max}$ on a 22×22 square

$N_{\max}$	Proc	Elapsed	Max Cost	Min Cost	Max Conf	Min Conf	Max Term	Min Term
50	4	6:30:25	0:21:58	0:21:09	19148131	18767145	31440647	30611130
51	8	5:51:45	0:26:23	0:19:19	17250302	17004054	33497276	31924532
52	12	6:14:21	0:34:23	0:22:17	18141158	17262224	40396483	39119225
53	16	6:43:25	0:44:00	0:28:02	18808580	17856432	48789794	46255379
54	20	7:09:07	0:49:34	0:27:49	19899189	18821333	59065745	56473260
55	28	6:37:14	0:58:35	0:30:58	18402937	16789846	59699548	56015367
56	32	7:21:54	1:11:58	0:32:39	19525496	17883903	73484899	65934722

(and terms) increase significantly with problem size. In particular the difference in the maximal and minimal redistribution costs is markedly increased from the 4 to the 32 processor problem. Obviously this would indicate that there is some scope for further optimization of the parallel algorithm.

We have extended the series for square lattice polyominoes up to size 56. The calculations requiring most resources were at widths 22–24. These cases were done using 40 processors and took about 8-10 hours each. The total CPU time required was about 1500 hours per prime. Calculations for each width and prime are totally independent and several can be done simultaneously.

3 Results and Conclusion

We have presented a parallel algorithm for the enumeration of polyominoes on the square lattice. The computational complexity of the algorithm is exponential with time (and memory) growing as $\kappa^{n/2}$, where κ appears to be a little larger than 2. Implementation on the APAC server cluster has allowed us to count the number of polyominoes up size 56. In Table 4 we have listed the new terms obtained in this work for the number of polyominoes with perimeter 47–56. The number of polyominoes of length ≤ 46 can be found in [9]. Repeating the analysis of [13] we obtain an improved numerical estimate for the growth constant $\lambda = 4.0625696(5)$ and an improved lower bound $\lambda \geq 3.927378\ldots$.

Table 4. The number, a_n, of fixed n-cell polyominoes on the square lattice

n	a_n	n	a_n
47	2726808444249438406145386634	52	2731266600165191432933200262566
48	10850352851820877056853237388	53	10889336855593503008200959900300
49	43193315093445654875552706600	54	43429974696239331559427538990000
50	172014608812878717989424207366	55	173269870217379043849354343514900
51	685304131748455616181606049288	56	691507145625328969365744254802188

Acknowledgements. Financial support from the Australian Research Council is gratefully acknowledged. The calculations presented in this paper would not have been possible without a generous grant of computer time on the server cluster of the Australian Partnership for Advanced Computing (APAC).

References

1. Golomb, S.: Polyominoes: Puzzles, Patterns, Problems and Packings. Second edn. Princeton U P, Princetion, N. J. (1994)
2. Lunnon, W.F.: Counting polyominoes. In Atkin, A.O.L., Birch, B.J., eds.: Computers in Number Theory. Academic Press, London (1971) 347–372
3. Martin, J.L.: Computer enumerations. In Domb, C., Green, M.S., eds.: Phase Transitions and Critical Phenomna. Volume 3. Academic Press, London (1974)
4. Redelmeier, D.H.: Counting polyominoes: Yet another attack. Discrete Math. **36** (1981) 191–203
5. Martin, J.L.: The impact of large-scale computing on lattice statistics. J. Stat. Phys. **58** (1990) 749–774
6. Mertens, S.: Lattice animals: A fast enumeration algorithm and new perimeter polynomials. J. Stat. Phys. **58** (1990) 1095–1108
7. Mertens, S., Lautenbacher, M.E.: Counting lattice animals – a parallel attack. J. Stat. Phys. **66** (1992) 669–678
8. Conway, A.R.: Enumerating 2D percolation series by the finite lattice method. J. Phys. A: Math. Gen. **28** (1995) 335–349
9. Jensen, I.: Enumerations of lattice animals and trees. J. Stat. Phys. **102** (2001) 865–881
10. Stauffer, D., Aharony, A.: Introduction to Percolation Theory. 2 edn. Taylor & Francis, London (1992)
11. Sykes, M.F., Glen, M.: Percolation processes in two dimensions. I. Low-density series expansion. J. Phys. A: Math. Gen. **9** (1976) 87–95
12. Klarner, D.A.: Cell growth problems. Canad. J. Math. **19** (1967) 851–863
13. Jensen, I., Guttmann, A.J.: Statistics of lattice animals (polyominoes) and polygons. J. Phys. A: Math. Gen. **33** (2000) L257–L263
14. Rands, B.M.I., Welsh, D.J.A.: Animals, trees and renewal sequences. IAM J. Appl. Math. **27** (1981) 1–17
15. Klarner, D.A., Riverst, R.: A procedure for improving the upper bound for the number of n-ominoes. Canad. J. Math. **25** (1973) 565–602
16. Lubensky, T., Isaacson, J.: Statistics of lattice animals and dilute branched polymers. Phys. Rev. A **20** (1979) 2130–2146
17. Conway, A.R., Guttmann, A.J.: On two-dimensional percolation. J. Phys. A: Math. Gen. **28** (1995) 891–904
18. Oliveira e Silva, T.: See the web at http://www.ieeta.pt/˜tos/animals.html
19. Knuth, D.E.: Polynum. Program available from Knuth's home-page at http://Sunburn.Stanford.EDU/˜knuth/programs.html#polyominoes (2001)
20. Enting, I.G.: Generating functions for enumerating self-avoiding rings on the square lattice. J. Phys. A: Math. Gen. **13** (1980) 3713–3722
21. Knuth, D.E.: Seminumerical Algorithms. The Art of Computer Programming, Vol 2. Addison Wesley, Reading, Mass (1969)

Hyper-BLAST: A Parallelized BLAST on Cluster System

Hong-Soog Kim, Hae-Jin Kim, and Dong-Soo Han

School of Engineering
Information and Communications University
P.O. Box 77, Yusong, Daejeon 305-600, Korea
{kimkk,hjkim,dshan}@icu.ac.kr

Abstract. BLAST is an important tool in bioinformatics. It has been used to find biologically similar sequences to the given query sequence from the database of the annotated sequences. For high throughput processing of huge number of query sequences, there have been many studies on parallel batch processing of sequence similarity search using BLAST. As the number of sequences in the database increases at exponential rate, the search speed of BLAST itself becomes important. Although NCBI has developed a parallel BLAST using the thread on SMP machines for the speedup of BLAST, the speedup is still limited because the SMP machine has restricted the number of processors due to its architecture. In this paper, we present our parallelized BLAST on cluster systems for further speedup. The main strategy used is the exploitation of the inter-node parallelism, which can be extracted by logical partitioning of the database. For the inter-node parallelism, we have designed and implemented a logical database partitioning method, initiation and coordination of the BLAST on remote node and communication protocol for collecting remote node's result. According to our performance test with 2-way 8 node cluster system, roughly 12 times speedup has been achieved in terms of response time of similarity search for individual query sequence.

1 Introduction

For biologist, finding and annotating characteristics of novel genes and protein sequences is an extremely important mission. The most reliable way to determine a biological molecule's structure and function is direct experimentation. It is, however, much easier to obtain the DNA sequence of the gene corresponding to a messenger RNA or protein than to experimentally determine gene's function of structure. This fact provides a strong motivation for developing computational methods that can infer biological information from sequence alone [1].

BLAST (Basic Local Alignment Search Tool) [2,3,4] is one of the most widely used similarity search tools available to computational biologist. It rapidly identifies statistically significant matches by comparing newly sequenced segments of genetic material or proteins with already annotated nucleotide or amino acid

P.M.A. Sloot et al. (Eds.): ICCS 2003, LNCS 2659, pp. 213–222, 2003.

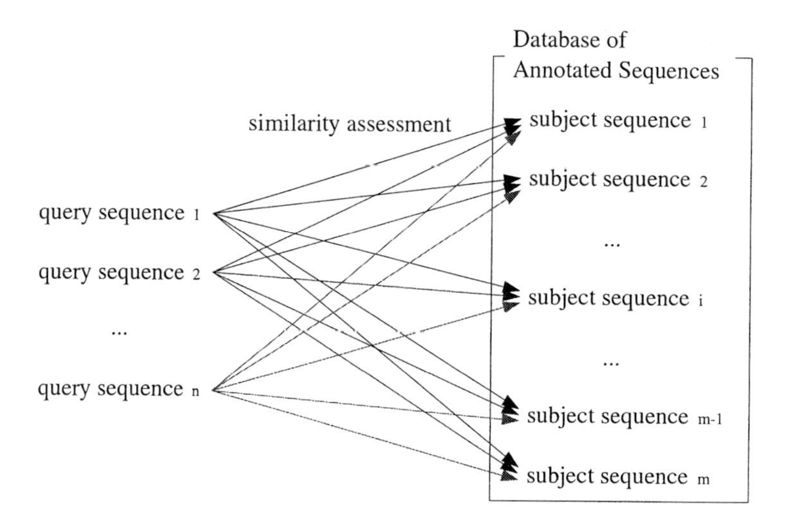

Fig. 1. Similarity search in BLAST

sequences in the database as shown in Fig. 1. This kind of search allows biologists to make inferences on the structure and function of the unknown gene or to screen new sequences for further investigation using more sensitive and computationally expensive methods. The information that BLAST provides in a few hours would otherwise takes months of laboratory work [5].

Similarity search with BLAST would be more useful, if there is more annotated sequences in the database. Because the time complexity of the similarity search algorithm used in BLAST is proportional to the size of the database [2], the response time of the BLAST becomes slow as the number of sequences in the database grows. GenBank, the primary repository for DNA sequence data, continues to grow at an exponential rate. It contains roughly 15,849,921,438 nucleotides in 14,976,310 sequences as of year 2001 [6]. Historically, GenBank has been doubling its size every 18 month, but that rate has accelerated to doubling every 15 months due primarily to the enormous growth in data from expressed sequence tags (ESTs). Keeping pace with the analysis of this data is difficult task for biologists.

As a result, sequence similarity analysis using BLAST is becoming a bottleneck [7]. A huge sequence database can also bring more serious problem. If the size of database exceeds the size of physical memory of the system, it may incur frequent paging while BLAST program scans the database and it could result in serious performance degradation. Therefore, as the size of the database grows, the speedup of BLAST becomes quite important.

Although a parallel version of BLAST, which exploits intra-search level parallelism explained in 3.1, has been developed by NCBI (National Center for Biotechnology Information) at the NIH (National Institute of Health) and Washington University, it is targeted only on SMP (Symmetric Multi-Processor) machines. Because SMP machine has certain limitation on the number of processors due to its architecture, the speedup of BLAST that can be achieved solely on one SMP machine has some limitation. Hence, the speedup improvement of BLAST on the SMP machine is not sufficient to cope with the current situation where enormous sequences are newly added in the database at exponential rate. In order to use more processors for more speedup of BLAST, we consider PC cluster system as an alternative to SMP machine.

In this paper, we present *Hyper-BLAST*, a parallelized BLAST on cluster system, which can provides scalable speedup in terms of response time. Hyper-BLAST adds inter-node parallel execution techniques to the intra-search level parallelism that is used by NCBI BLAST. Logical partitioning of database is used to prepare and enable the inter-node parallel execution of intra-search level parallelism. In our parallelized BLAST on cluster system, the master node drives remote nodes to search similar sequence from logically partitioned database and collects minimal but complete data for reporting the search results.

2 Related Work

There have been several researches on parallel processing of sequence similarity analysis on multiprocessor systems. Braun et al. studied parallel processing of sequence analysis using BLAST on workstation of cluster [8]. They provided a good classification scheme for parallel processing of BLAST and reported implementation of coarse grained level of parallelism (parallel batch processing of similarity search). They pointed out that the parallel batch processing approach has overhead for maintaining consistency of partitioned or distributed database.

The coarse grained level of parallelism is extracted from the fact that respective similarity searches for different query sequences can be done independently. We call this kind of parallelism *inter-search level parallelism* in order to distinguish the intra-search level parallelism used in NCBI BLAST.

Efforts to parallelize FASTA and BLAST using the coarse grained level of parallelism have been published since the early 1990s [9,10,11]. Recently, Disperse system and BeoBLAST system have been reported. Disperse system parallelized FASTA and SSEARCH (Smith-Waterman) program using Perl and UNIX utilities such as rcp (or scp) and rsh (or ssh) [12]. While Disperse system used UNIX utilities for initiating parallel execution of program on workers and tallying the results from workers, BeoBLAST used queuing system (GNU Queue) for executing BLAST and PSI-BLAST on a Beowulf cluster [13].

There also have been a few commercial products, such as High-Throughput BLAST (HT-BLAST) of SGI [14] and Turbo BLAST of TurboGenomics [15], based on coarse grained level of parallelism.

Albeit parallel batch processing of similarity analysis based on inter-search level parallelism can achieve higher throughput over a set of query sequences, it cannot improve the response time of the similarity search for individual query sequence because the response time of a similarity search itself is dependent on the computational power of each node or processor.

3 Parallelization of BLAST on Cluster System

In this section, we explain intra-search level parallelism that is used in NCBI BLAST, our extension of intra-search level parallelism for cluster system and the other implementation details.

3.1 Intra-search Parallelism in NCBI BLAST

The NCBI BLAST is parallelized using the thread facility for SMP machine. The parallelization of NCBI BLAST is based on task pool. The task is a similarity assessment of the subject sequence in database for given query sequence. In NCBI BLAST, each thread gets its task chunk (i.e., range of subject sequences in database), aligns the query sequence with subject sequences in its chunk, merges its search result into global search result and receives next chunk until all the sequences in database are searched. All threads repeat these steps. As SMP machine has one shared memory, access to subject sequences, assigning the task chunk to thread and collection of slave thread's search result are straightforward using mutex for exclusive access to global data structure.

The parallelism in NCBI BLAST originates from the fact that similarity assessments for every subject sequence in sequence database are mutually independent. We call it *intra-search level parallelism* in order to distinguish it from the inter-search level parallelism that was used in parallel batching processing approach. In terms of the granularity of parallelism, intra-search level parallelism is much finer than inter-search level parallelism.

Even though NCBI BLAST successfully extracted parallelism and achieved some speedup in the similarity search of individual query sequence using intra-search level parallelism, its speedup ratio is still limited by its underling parallel processor system.

3.2 Extension of Intra-search Parallelism in Hyper-BLAST

In Hyper-BLAST, we extend application of intra-search level parallelism from one SMP node to multiple nodes in cluster system by logically partitioning of the sequence database, in which every node calculates its database partition from the node configuration and confines its search space within its own database partition. The intra-search level parallelism is realized in two levels: inter-node and intra-node level.

In intra-node level, master thread of Hyper-BLAST on every node has its logical database partition that is assigned by inter-node level realization of intra-search parallelism. The master thread of Hyper-BLAST on each node allocates a chunk of subject sequence to its slave threads on demand. Because the search result from each node is only partial result, one designated node (master node) collects the search result from the rest of nodes and then makes similarity search report for the given query sequence from the collected results.

For inter-node level realization of intra-search parallelism, we implement mechanisms for specifying computation node set and logical database partition for each computation node, executing Hyper-BLAST on each computation node and controlling Hyper-BLAST program instances on the nodes. Since a cluster system is distributed memory multiprocessor system, we devise message communication protocol that can replace the read and write operation on shared memory of the SMP machine.

For logical database partitioning, we assume that all computation nodes access sequence database, query sequence and other internal data files for BLAST execution through NFS (Network File System).

Specification of Computation Node Set. For the specification of computation node set and logical database partition, we use a node configuration file that is specified in Hyper-BLAST command line option.

The node configuration file contains information on computation node set, role of each computation node (master/slave node), degree of parallelism (DOP) and the load weight for individual computation node. From the DOP and the load weight for individual node, the logical database partition for a node is calculated. In run-time, the master thread of Hyper-BLAST on computation node confines its search space to the calculated database partition instead of original database. Because NCBI BLAST uses memory mapped file I/O for database for fast access, logical database partitioning can reduce the possibility of paging if the whole sequence database is too large to fit into physical memory of a computation node. Paging is known to cause serious overhead in running application that accesses a large set of data.

Initiation of Hyper-BLAST on Slave Nodes. As multiple Hyper-BLAST process instances should be executed in different nodes of cluster system, it is necessary to explicitly execute Hyper-BLAST program on remote nodes. For the purpose of initiating Hyper-BLAST on remote nodes, we use rsh (remote shell) based approach. Except for the master node, initiation of the Hyper-BLAST program is performed using the rsh command in the below:

```
rsh node01 'hyperblastall -i sample.seq -o sample.html -p blastx
-d nr -e 10.0 -m 0 -b 15 -v 15 -I T -T T -c cluster_nodes.cfg' &
```

In the above rsh command for executing Hyper-BLAST on a remote node, all command line options are the same as those of NCBI BLAST except -c option added for specifying the node configuration file.

Since it could be a chore to make this kind of rsh command and node configuration file, we also have developed the GUI-based Hyper-BLAST execution environment (HBEE) program. In HBEE, users can check the node status and resources such as number of processors, processor types and physical memory size. While examining a node, users can add it to computation node set of Hyper-BLAST, designate it as master node or slave node, and specify the number of parallel processors used for Hyper-BLAST and its load weight. From the user's definition of computation node set, the HBEE generates node configuration file and series of rsh command for executing Hyper-BLAST on remote nodes that are designated as slave nodes in the node configuration file.

Control of Hyper-BLAST Program on Slave Nodes. Once Hyper-BLAST programs start on a master node and remote slave nodes, the master thread of the Hyper-BLAST program at every computation node first identifies its role as master or slave node and computes its database partition using the information in the node configuration file. Then, every Hyper-BLAST program carries out similarity search for the first query sequence in query sequence file that is specified by the command line option -i.

For nodes that have the DOP value of two or more, the master thread of the Hyper-BLAST program creates or reactivates as many slave threads as the DOP value. The master thread assigns the chunk of sequences to its slave threads on demand while the slave threads initiate the similarity search for the query sequence from its assigned chunk of sequences.

The final similarity search result at each node is kept within in-memory data structure. The in-memory data structure is a linked list of the sequence alignment information between the query sequence and the similar sequence found in the sequence database. Within the linked list of the sequence alignment information, the sequence alignment information is sorted by its rank order.

When the Hyper-BLAST programs at the slave nodes finish their respective similarity search and make a linked list of sequence alignment information, the respective master threads of Hyper-BLAST at the slave nodes make a message that corresponds to the sequence alignment information in the linked list, send the message to the master node and wait for the acknowledgment of the message. The master thread of Hyper-BLAST at the master node makes sequence alignment information from the message that is received from the slave node, insert it into the linked list of sequence alignment information and send acknowledgment to the slave node. If the sequence alignment information, which is made from the message from the slave node, is less than the least sequence alignment information in the linked list of the master node in terms of rank order, the sequence alignment information is discarded. The acknowledgment for discarded sequence alignment information makes the slave stop to send next data message.

When the master node receives the search results from all the slave nodes, the linked list of sequence alignment information in the master node becomes a complete similarity search result for the first query sequence in the query sequence file. Then, the master thread of Hyper-BLAST makes the report for the similarity search result and prints it. After printing the report, the master thread of Hyper-BLAST program at the master node sends the control message that orders slave nodes to start similarity search for the next query sequence in the query sequence file.

Figure 2 depicts the run time behavior of Hyper-BLAST program instances on the master node and the slave node. The master thread of Hyper-BLAST determines each node's logical database partition, controls the parallel search activity and communication for search results, and makes similarity search re-

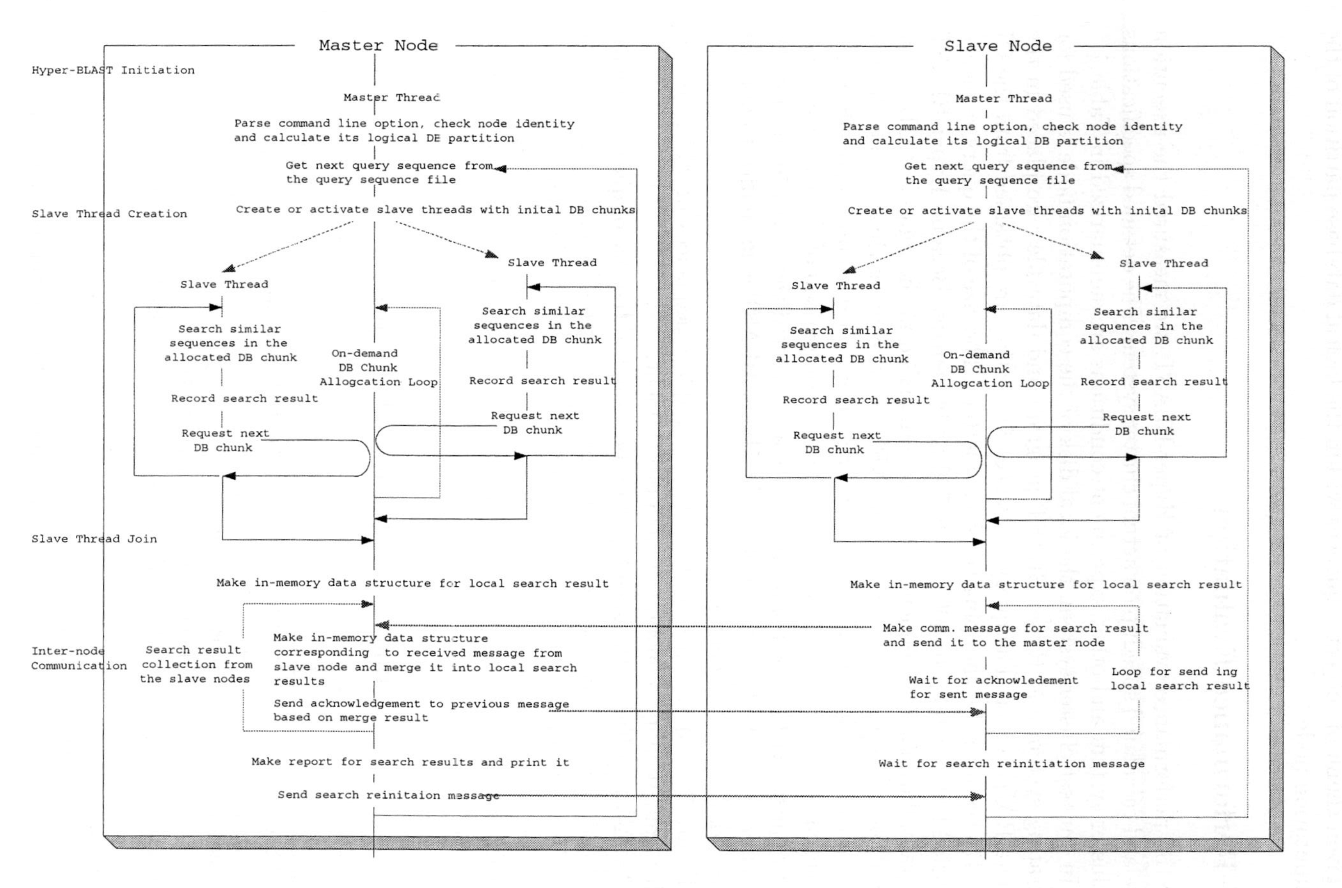

Fig. 2. Control flow of Hyper-BLAST program instances at computation nodes

port. The slave threads of Hyper-BLAST, which actually do the similarity search for given chunk of subject sequences within the logical database partition of the computation node.

4 Performance Evaluation

For the performance evaluation of Hyper-BLAST, we measured the execution time and calculated speedup on actual cluster system. The detailed specifications of cluster system and other experiment conditions are summarized in Table 1.

In our test cluster system, the local disk of the computation node is used for operating system and Hyper-BLAST program and data files are placed in the remote RAID server. The query sequences for testing are obtained from the GI 2054475 sequence by chopping the first n letters. The size of the query sequence is varied from 1000 bp to 5000 bp by 500 bp increment. Because Hyper-BLAST program, data file and sequence database are placed in the remote RAID server and all nodes access them through the NFS, the execution time can be affected by NFS cache, disk cache and memory cache. However disabling these effects for measurement is not simple. Hence, for the fair comparison of the response time, we have measured the response time twice and selected the second result for the measurements. The first execution of the Hyper-BLAST is for indirect enabling of the possible cache benefits.

Figure 3 compares the speedup ratio for query sequences of which length varies from 1000bp to 5000bp by 500bp increment.

As shown in Fig. 3, the speedup is proportional to the size of query sequence but it continuously increases as more processors are used. The maximum speedup using 16 processors in the cluster system ranges between 10.96 and 12.42. From the experimental results, we can conclude that our implementation of parallelized BLAST on cluster system gives scalable speedup as more processors are used for the similarity search.

Table 1. Specification of cluster system

	8 Node Cluster
Processors	Dual Intel Pentium III 1GHz per Node
Memory	1024KB per Node
Network	100Mbps Switch
O.S	Linux (Kernel version 2.4.18)
BLAST	Hyper-BLAST based on NCBI BLAST Version 2.2.1 blastx
BLAST DB	nr (929,420 sequences; 291,584,220 total letters)
Query sequence	GI 20544475 Homo sapiens adenylate cyclase 2 (brain) (ADCY2) (total length: 6551 letters)

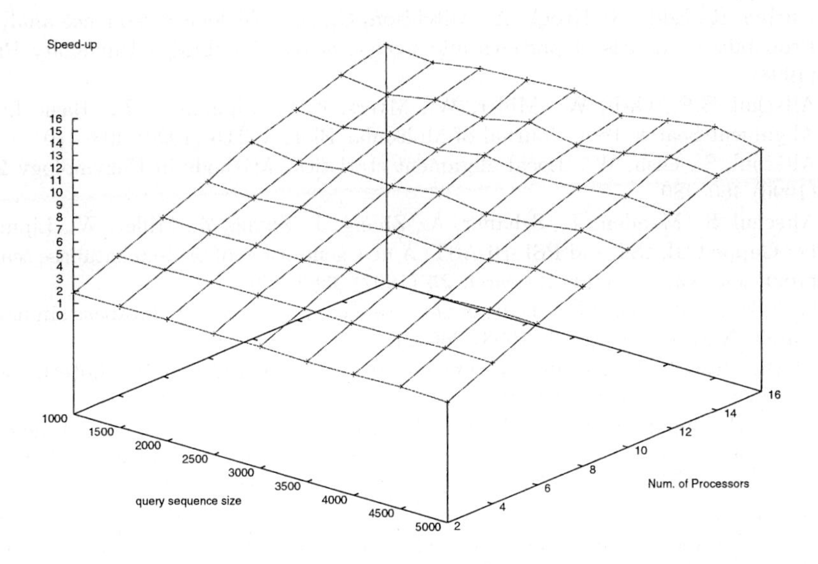

Fig. 3. Speedup comparison on 2-way 8 node cluster system

5 Conclusion and Future Work

BLAST is an important tool in bioinformatics. The exponential growth rate of the sequence database requires faster BLAST. Hence, the speedup provided by parallel NCBI BLAST on SMP machine becomes insufficient. To get more speedup of NCBI BLAST, we have designed and implemented Hyper-BLAST on cluster systems. Using logical partitioning of sequence database, Hyper-BLAST extends the intra-search level parallelism from one SMP machine to multiple computation nodes that are connected by network. A communication protocol and a message format is devised to control of Hyper-BLAST programs on slave nodes and collect final search results.

This extension of intra-search parallelism enables us to use more processors for similarity search using BLAST and gives more speedup of individual query sequence search. The performance evaluation result shows that Hyper-BLAST gives scalable speedup in terms of response time as more processors are used.

Currently, we are studying the performance of Hyper-BLAST on large-scale cluster systems. We anticipate that speedup might be saturated at some number of multiple processors. Hence, future work includes identification of speedup saturation point, modeling of speedup function and devising of combined approach that uses parallel batch processing technique on the top of Hyper-BLAST for more speedup and throughput.

References

1. Durbin, R., Eddy, S., Krogh, A., Mitchison, G., eds.: Biological sequence analysis: Probabilistic models of proteins and nucleic acids. Cambridge University Press (1998)
2. Altschul, S.F., Gish, W., Miller, W., Myers, E.W., Lipman, D.J.: Basic Local Alignment Search Tool. Journal of Molecular Biology **215** (1990) 403–410
3. Altschul, S., Gish, W.: Local alignment statistics. Methods in Enzymology **266** (1996) 460–480
4. Altschul, S., Madden, T., Schaffer, A., Zhang, J., Zhang, Z., Miller, W., Lipman, D.: Gapped BLAST and PSI–BLAST: A new generation of protein katabase search programs. Nucleic Acids Research **25** (1997) 3389–3402
5. Gish, W., States, D.: Identification of protein coding regions by database similarity search. Nature Genetics **3** (1993) 266–272
6. NCBI: Growth of GenBank. Technical report, National Center for Biotechnology Information (March 12, 2002)
7. Chi, E.H., Shoop, E., Carlis, J., Retzel, E., Ried, J.: Efficiency of shared-memory multiporceossors for a genetic sequence similiarity search algorithm. Technical report, Computer Science Dept., University of Minnesota (1997)
8. Braun, R.C., Pedretti, K.T., Casavant, T.L., Scheetz, T.E., Birkett, C.L., Roberts, C.A.: Parallelization of local BLAST service on workstation clusters. Future Generation Computer Systems **17** (2001) 745–754
9. Miller, P.L., Nadkarni, P.M., Carriero, N.M.: Parallel computation and FASTA: Confronting the problem of parallel database search for a fast sequence comparison algorithm. Bioinformatics (formerly CABIOS) **7** (1991) 71–78
10. Barton, G.J.: Scanning protein sequence databanks using a distributed processing workstation network. Bioinformatics (formerly CABIOS) **7** (1991) 85–88
11. Julich, A.: Implementations of BLAST for parallel computers. Bioinformatics (formerly CABIOS) **11** (1995) 3–6
12. Clifford, R., Mackey, A.J.: Disperse: A simple and efficient approach to parallel database searching. Bioinformatics **16** (2000) 564–565
13. Grant, J.D., Dunbrack, R.L., Manion, F.J., Ochs, M.F.: BeoBLAST: Distributed BLAST and PSI-BLAST on a Beowulf cluster. Bioinformatics **18** (2002) 765–766
14. Camp, N., Cofer, H., Gomperts, R.: High-Throughput BLAST. Technical report, Silicon Graphics, Inc. (1998)
15. Bjorson, R.D., Sherman, A.H., Weston, S.B., Willard, N., Wing, J.: TurboBLAST: A parallel implementation of BLAST based on the TurboHub process integration architecture. Technical report, TruboGenomics, Inc. (2002)

Parallel Superposition for Bulk Synchronous Parallel ML

Frédéric Loulergue

Laboratory of Algorithms, Complexity and Logic
University Paris XII, Val-de-Marne
61, avenue du général de Gaulle – 94010 Créteil cedex – France
Tel: +33 1 45 17 16 50 – Fax: +33 1 45 17 66 01
loulergue@univ-paris12.fr

Abstract. The BSMLlib is a library for Bulk Synchronous Parallel programming with the functional language Objective Caml. It is based on an extension of the λ-calculus by parallel operations on a parallel data structure named parallel vector, which is given by intention.
Those operations are *flat* and allow BSP programming in direct mode but it is impossible to express directly divide-and-conquer algorithms. This paper presents a new construction for the BSMLlib library which can express divide-and-conquer algorithms. It is called parallel superposition because it can be seen as the parallel composition of two BSP threads which can each use all the processors. An associated cost model derived from the BSP cost model is also given.

Keywords: Bulk Synchronous Parallelism, Functional Programming, Divide-and-Conquer, Cost model

1 Introduction

Some problems require performance carried out by only massively parallel computers of which programming is still difficult. Works on functional programming and parallelism can be divided in two categories: explicit parallel extensions of functional languages — where languages are either non-deterministic [24] or non-functional [11] — and parallel implementations with functional semantics [1] — where resulting languages do not express parallel algorithms directly and do not allow the prediction of execution times. Algorithmic skeleton languages [7, 9, 25, 6, 26, 8], in which only a finite set of operations (the skeletons) are parallel, constitute an intermediate approach. Their functional semantics is explicit but their parallel operational semantics is implicit. The set of algorithmic skeletons has to be as complete as possible but it is often dependent on the domain of application.

Among researchers interested in declarative parallel programming, there is a growing interest in execution cost models taking into account global hardware parameters like the number of processors and bandwidth. The Bulk Synchronous Parallel (BSP) [31, 22, 28] execution and cost model offers such possibilities and with similar motivations we have designed BSP extensions of the λ-calculus [20]

P.M.A. Sloot et al. (Eds.): ICCS 2003, LNCS 2659, pp. 223–232, 2003.

and a library for the Objective Caml language, called `BSMLlib` [19], implementing those extensions.

A BSP algorithm is said to be in *direct mode* [12] when its physical process structure is made explicit. Such algorithms offer predictable and scalable performance and BSML expresses them with a small set of primitives taken from the *confluent* BSλ calculus [20]: a constructor of parallel vectors, asynchronous parallel function application, synchronous global communications and a synchronous global conditional.

Those operations are *flat*: it is impossible to express directly parallel divide-and-conquer algorithms. Nevertheless many algorithms are expressed as parallel divide-and-conquer algorithms [29] and it is difficult to transform them into flat algorithms. In a previous work, we proposed an operation called parallel composition [18], but it was limited to the composition of two terms whose evaluations require the same number of BSP super-steps.

In this paper we present a new operation called parallel superposition and its associated cost model. It can be used to write divide-and-conquer algorithms. The presentation of those novelties needs the presentation of the flat `BSMLlib` library and its associated cost model.

We first present the BSP model (section 2). Section 3 is about the "flat" `BSMLlib` library and its cost model. Section 4 introduces a new operation called parallel superposition to the `BSMLlib` library and the cost model associated with this extended library. We then discuss related work (section 5) and conclude (section 6).

2 Bulk Synchronous Parallelism

Bulk-Synchronous Parallelism (BSP) is a parallel programming model introduced by Valiant [31, 28, 22] to offer a high degree of abstraction like PRAM models and yet allow portable and predictable performance on a wide variety of architectures. A BSP computer contains a set of processor-memory pairs, a communication network allowing inter-processor delivery of messages and a global synchronization unit which executes collective requests for a synchronization barrier. Its performance is characterized by 3 parameters expressed as multiples of the local processing speed: the number of processor-memory pairs p, the time l required for a global synchronization and the time g for collectively delivering a 1-relation (communication phase where every processor receives/sends at most one word). The network can deliver an h-relation in time gh for any arity h.

A BSP program is executed as a sequence of *super-steps*, each one divided into (at most) three successive and logically disjoint phases. In the first phase each processor uses its local data (only) to perform sequential computations and to request data transfers to/from other nodes. In the second phase the network delivers the requested data transfers and in the third phase a global synchronization barrier occurs, making the transferred data available for the next super-step. The execution time of a super-step s is thus the sum of the maximal local processing time, of the data delivery time and of the global synchronization time:

$\text{Time}(s) = \max_{i:processor} w_i^{(s)} + \max_{i:processor} h_i^{(s)} * g + l$ where $w_i^{(s)}$ = local processing time on processor i during super-step s and $h_i^{(s)} = \max\{h_{i+}^{(s)}, h_{i-}^{(s)}\}$ where $h_{i+}^{(s)}$ (resp. $h_{i-}^{(s)}$) is the number of words transmitted (resp. received) by processor i during super-step s. The execution time $\sum_s \text{Time}(s)$ of a BSP program composed of S super-steps is therefore a sum of 3 terms:$W + H * g + S * l$ where $W = \sum_s \max_i w_i^{(s)}$ and $H = \sum_s \max_i h_i^{(s)}$. In general W, H and S are functions of p and of the size of data n, or of more complex parameters like data skew and histogram sizes. To minimize execution time the BSP algorithm design must jointly minimize the number S of super-steps and the total volume h (resp. W) and imbalance $h^{(s)}$ (resp. $W^{(s)}$) of communication (resp. local computation).

3 "Flat" Functional Bulk Synchronous Parallel Programming

3.1 The BSMLlib Library

There is currently no implementation of a full Bulk Synchronous Parallel ML language but rather a partial implementation as a library for Objective Caml. The so-called BSMLlib library is based on the following elements.

It gives access to the BSP parameters of the underling architecture. In particular, it offers the function `bsp_p:unit->int` such that the value of `bsp_p()` is p, the static number of processes of the parallel machine. This value does not change during execution.

There is also an abstract polymorphic type `'a par` which represents the type of p-wide parallel vectors of objects of type `'a`, one per process. The nesting of `par` types is prohibited. A type system enforces this restriction.

The BSML parallel constructs operate on parallel vectors. Those parallel vectors are created by:

```
mkpar: (int -> 'a) -> 'a par
```

so that `(mkpar f)` stores `(f i)` on process i for i between 0 and $(p-1)$. We usually write `f` as `fun pid->e` to show that the expression `e` may be different on each processor. This expression `e` is said to be *local*. The expression `(mkpar f)` is a parallel object and it is said to be *global*.

A BSP algorithm is expressed as a combination of asynchronous local computations (first phase of a super-step) and phases of global communication (second phase of a super-step) with global synchronization (third phase of a super-step).

Asynchronous phases are programmed with `mkpar` and with:

```
apply: ('a -> 'b) par -> 'a par -> 'b par
```

`apply (mkpar f) (mkpar e)` stores `(f i) (e i)` on process i. Neither the implementation of BSMLlib, nor its semantics prescribe a synchronization barrier between two successive uses of `apply`.

We ignore the distinction between a communication request and its realization at the barrier. The communication and synchronization phases are expressed by:

```
put:(int->'a option) par -> (int->'a option) par
```

where `'a option` is defined by:

```
type 'a option = None | Some of 'a.
```

Consider the expression: `put(mkpar(fun i->fs`$_i$`))` (*)

To send a value `v` from process `j` to process `i`, the function `fs`$_j$ at process `j` must be such that (`fs`$_j$ `i`) evaluates to `Some v`. To send no value from process `j` to process `i`, (`fs`$_j$ `i`) must evaluate to `None`.

Expression (*) is evaluated to a parallel vector containing functions `fd`$_i$ of delivered messages. At process `i`, (`fd`$_i$ `j`) evaluates to `None` if process `j` sent no message to process `i` or evaluates to `Some v` if process `j` sent the value `v` to the process `i`.

The full language would also contain a synchronous conditional operation:

```
ifat: (bool par) * int * 'a * 'a -> 'a
```

such that `ifat (v,i,v1,v2)` will evaluate to `v1` or `v2` depending on the value of `v` at process `i`. But Objective Caml is an eager language and this synchronous conditional operation can not be defined as a function. That is why the core `BSMLlib` contains the function: `at:bool par->int-> bool` to be used only in the construction: `if (at vec pid) then... else...` where `(vec:bool par)` and `(pid:int)`. `if at` expresses communication and synchronization phases.

With the above operations, the global control cannot take into account data computed locally. Global conditional is necessary to express algorithms like:

Repeat
  Parallel Iteration
Until Max of local errors < epsilon

This framework is a good tradeoff for parallel programming because: we defined a *confluent calculus* so we designed a purely functional parallel language from it. Without side-effects, programs are easier to prove, and to re-use. An eager language allows good performances ; this calculus is based on BSP operations, so programs are easy to port, their costs can be predicted and are also portable because they are parametrized by the BSP parameters of the target architecture.

3.2 Cost Model

A formal parallel cost model can be associated to reductions in the BSλ-calculus [17]. "Cost expressions" are defined, and each rule of the semantics is associated with a cost rule on cost expressions. Given a particular strategy of reduction, a expression is always reduced in the same way. In this case costs can be associated with expressions rather that reductions.

For the BSλ-calculus it is possible to define two different reduction strategies for the two levels of the calculus. The `BSMLlib` library uses the strategy of the Objective Caml language, the same strategy for local and global reduction: weak call-by-value strategy.

We will not describe the cost of the evaluation of a local expression, it is the same as the cost of the evaluation of the expression by a strict functional language. We give the costs of the evaluation of global expressions.

The cost model associated with our expressions follows the BSP cost model. The evaluation of a parallel vector occurs in the first phase of the BSP super-step. The evaluation of `mkpar` f leads to the evaluation of $(f\ i)$ at each processor i, $0 \leq i < p$. If the sequential evaluation time of each $(f\ i)$ is w_i, the parallel evaluation time of the parallel vector is $max_{0 \leq i < p} w_i$.

Provided the arguments of the parallel application are values, the parallel evaluation time of `apply` $\langle\ f_0\ , \ldots,\ f_{p-1}\ \rangle\ \langle\ v_0\ , \ldots,\ v_{p-1}\ \rangle$ is $max_{0 \leq i < p} w_i$ where w_i is the evaluation time of $f_i\ v_i$ at processor i.

The evaluation of a put operation requires a full super-step. To evaluate `put` $\langle\ f_0\ , \ldots,\ f_{p-1}\ \rangle$, first each processor evaluates the p local expressions $f_i\ j$, $0 \leq j < p$ leading to p^2 values (p per processor) v_{ij} (first phase of the super-step). If the value v_{ij} of processor i is different from `None` it is sent to processor j (communication phase of the super-step). Once all values have been exchanged a synchronization barrier occurs (third and last phase of the super-step) making the values available for the next super-step. At the beginning of this second super-step, each processor i constructs the function (result of the put operation) from the v_{ji} values it has received. Thus provided the argument of the put operation is a value (in this case a parallel vector of values), the parallel evaluation time of `put` $\langle\ f_0\ , \ldots,\ f_{p-1}\ \rangle$ is

$$\max_{0 \leq i < p} w_i^1 \ + \ \max_{0 \leq i < p} h_i \times g \ + \ l \ + \max_{0 \leq i < p} w_i^2 \qquad \text{where}$$

- w_i^1 is the evaluation time at processor i of the p expressions $f_i\ j$, $0 \leq j < p$
- $h_i = \max\{h_{i+}, h_{i-}\}$ where h_{i+} (resp. h_{i-}) is the number of words transmitted (resp. received) by processor i. h_{i+} is the sum of the size of the v_{ij} values sent to other processors, and h_{i-} is the sum of the sizes of the values v_{ji} received by processor i from other processors
- w_i^2 is the evaluation time at processor i to construct the result function from the v_{ji} values.

The evaluation of `if` $\langle\ b_0\ , \ldots,\ b_{p-1}\ \rangle$ `at` n `then` E_1 `else` E_2 where n and b_i are respectively integer and boolean values is: first the processor n sends the value b_n to all other processors. A synchronization barrier occurs. If the value b_n is **true** (resp. **false**) then the evaluation of E_1 (resp. E_2) begins. The computation fails if the condition $0 \leq n < p$ does not hold. The parallel evaluation time is $(p-1) \times s_{\text{boolean}} \times g\ + l\ +\ T$ where s_{boolean} is the size in words of a boolean value and T the parallel evaluation time of E_1 (resp. E_2).

The costs (parallel evaluation times) above are context independent. The time required to evaluate a global expression E will be the same in (`fun` $x \to E'$) E, `put` E, `apply` $E\ E'$, *etc.* This is why our cost model is *compositional.*

4 Parallel Superposition

Objective Caml is an eager language. To express parallel superposition as a function we have to "freeze" the evaluation of its parallel arguments. So parallel superposition must have the following type:

```
super: (unit -> 'a par) -> (unit -> 'b par) -> 'a par * 'b par
```

The equational semantics of super is given by super E_1 E_2 $=$ $(E_1(), E_2())$.

To evaluate a parallel superposition, super E_1 E_2 the parallel machine will use a thread to evaluate E_1 and a thread to evaluate E_2. Nevertheless those two threads will not be independent. The communications and synchronization barrier will be shared.

Let consider the beginning of the evaluation of super E_1 E_2. First the asynchronous computation phases of E_1 and E_2 will be evaluated. This can be done using a thread mechanism but it also be done sequentially to avoid thread swapping overhead. Then the communications phases of E_1 and E_2 will be done together. The messages will simply be the concatenations of the messages of E_1 and E_2. Finally a single synchronization barrier will occur to end the super-step of E_1 and the super-step of E_2.

To determine the cost of the evaluation of super E_1 E_2 it is not sufficient to consider the total costs of E_1 and E_2 (in the form $W + H \times g + L$) but the list of the costs of each super-step of E_1 and E_2. Moreover the cost of a super-step have to be described by three vectors :

- the cost of the local computations times for each process: $\langle\ w_0\ , \ldots,\ w_{p-1}\ \rangle$,
- the size of the sent messages : $\langle\ h_0^+\ , \ldots,\ h_{p-1}^+\ \rangle$
- the size of the received messages : $\langle\ h_0^-\ , \ldots,\ h_{p-1}^-\ \rangle$

We will note $(\bar{w}, \bar{h}^+, \bar{h}^-)$ the cost of a super-step. The cost of an expression is thus a list of such triples of vectors. If the costs of E_1 and E_1 are respectively:

$$\begin{cases} (\bar{w}^0, \bar{h}^{0+}, \bar{h}^{0-}), \ldots, (\bar{w}^{k_1}, \bar{h}^{k_1+}, \bar{h}^{k_1-}) \\ (\bar{w}'^0, \bar{h}'^{0+}, \bar{h}'^{0-}), \ldots, (\bar{w}'^{k_2}, \bar{h}'^{k_2+}, \bar{h}'^{k_2-}) \end{cases}$$

then the cost of super E_1 E_2 is

$$\begin{pmatrix} \langle\ w_0^0 + {w'}_0^0\ , \ldots,\ w_{p-1}^0 + {w'}_{p-1}^0\ \rangle \\ \langle\ h_0^{0+} + {h'}_0^{0+}\ , \ldots,\ h_{p-1}^{0+} + {h'}_{p-1}^{0+}\ \rangle \\ \langle\ h_0^{0-} + {h'}_0^{0-}\ , \ldots,\ h_{p-1}^{0-} + {h'}_{p-1}^{0-}\ \rangle \end{pmatrix}, \ldots, \begin{pmatrix} \langle\ w_0^k + {w'}_0^k\ , \ldots,\ w_{p-1}^k + {w'}_{p-1}^k\ \rangle \\ \langle\ h_0^{k+} + {h'}_0^{k+}\ , \ldots,\ h_{p-1}^{k+} + {h'}_{p-1}^{k+}\ \rangle \\ \langle\ h_0^{k-} + {h'}_0^{k-}\ , \ldots,\ h_{p-1}^{k-} + {h'}_{p-1}^{k-}\ \rangle \end{pmatrix}$$

where $k = \max\{k_1, k_2\}$ and where w_i^n, h_i^{n+} and h_i^{n-} (resp. $w_i'^n$, $h_i'^{n+}$ and $h_i'^{n-}$) are considered equal to 0 if $n > k_1$ (resp. $n > k_2$).

The usual BSP cost of super E_1 E_2 is then

$$(\sum_{n=0}^{k} \max\{W^n\}) \ + \ (\sum_{n=0}^{k} h^n) \times g \ + \ k \times l$$

where $\begin{cases} W^n = \max_{0 \leq i < p}\{w_i^n + w_i'^n\} \\ h^n = \max_{0 \leq i < p}\{(h_i^{n+} + h_i'^{n+}) \ , \ (h_i^{n-} + h_i'^{n-})\} \end{cases}$

Using the above lists, the compositionality of our cost model is preserved.

The parallel superposition of E_1 and E_2 may be less costly than the evaluation of E_1 followed by the evaluation of E_2. Thus the parallel superposition is not only useful to express divide-and-conquer algorithms as shown in the next section, but it can also be used to efficiently program a parallel machine even without divide-and-conquer.

Using this operation, we can write a "scheduler" which composes several BSP programs (given for example by different users) in order to balance the sizes of the messages and decrease the number of synchronization barriers. If those programs are dynamically submitted to a queue, a formula describing their costs could be used by the "scheduler" to decide either to superpose several programs or to evaluate them sequentially. Superposing too many programs would lead to too high values of h. Moreover it is not a good idea to compose a program E_1 with a little amount of local computations and a high priority with a program E_2 with a lot of local computations because even if the superposition will be more efficient, the user who submitted the job E_1 will wait longer for its result. The same case occur if E_1 has few super-steps and E_2 many.

4.1 Examples

The example presented below is a divide-and-conquer version of the `scan` program. The network is divided into two parts and the scan is recursively applied to those two parts. The value held by the last processor of the first part is broadcast to all the processors of the second part, then this value and the value held locally are combined together by the operator op on each processor of the second part.

```
let scan op vec =
 let rec scan' fst lst op vec =
  if fst>=lst then vec else
  let mid = (fst+lst)/2 in
  let vec'= mix mid (super (fun()->scan' fst mid op vec)
                           (fun()->scan'(mid+1) lst op vec)) in
  let msg vec =
   apply (mkpar(fun i v->
      if i=mid
      then fun dst->if inbounds (mid+1) lst dst then Some v else None
      else fun dst-> None)) vec
  and parop = parfun2(fun x y->match x with None->y|Some v->op v y) in
  parop (apply(put(msg vec'))(mkpar(fun i->mid))) vec' in
 scan' 0 (bsp_p()-1) op vec
```

In this small program, we also use some functions which are parts of the current `BSMLlib` standard library:

```
let replicate f = mkpar(fun pid->f)
```

```
let parfun f v = apply (replicate f) v
let parfun2 f v1 v2 = apply (parfun f v1) v2
```

as well as the following functions which will be in the standard library of `BSMLlib` with parallel superposition:

```
let mix m (v1,v2) = let f pid v1 v2 = if pid<=m then v1 else v2 in
  apply (apply (mkpar f) v1) v2

let inbounds first last n = (n>=first)&&(n<=last)
```

This program was run on a sequential simulator. Nevertheless the BSλ-calculus with parallel superposition is confluent. Thus sequential and parallel evaluation give the same results.

5 Related Work

There are other libraries based on the BSλ framework. They are based either on the functional language Haskell [23] or on the object-oriented language Python [16]. They propose flat operations similar to ours but no parallel composition.

[30] presents a way to divide-and-conquer in the framework of an object-oriented language. There is no formal semantics and no implementation from now on. The proposed operation is close to our parallel *superposition* (several BSP threads use the whole network) but the programmer has less control over the use of those super-threads. The same author advocates in [21] a new extension of the BSP model in order to ease the programming of divide-and-conquer BSP algorithms. It adds another level to the BSP model with new parameters to describe the parallel machine.

[32] is an algorithmic skeletons language based on the BSP model and offers divide-and-conquer skeletons. Nevertheless, the cost model is not really the BSP model but the D-BSP model [10] which allows subset synchronization. We follow [13] to reject such a possibility. Another algorithmic skeletons language based on the BSP model [14] does not offer divide-and-conquer skeletons. [27] presents another model which allows subset synchronization.

In the BSPlib library [15] subset synchronization is not allowed as explained in [28]. The PUB library [4] is another implementation of the BSPlib standard proposal. It offers additional features with respect to the standard which follows the BSP* model [2] and the D-BSP model [10]. Minimum spanning trees nested BSP algorithms [5] have been implemented using these features.

We also previously worked on a parallel composition [18]. This operation cannot be added to the BSλ-calculus (the obtained systems is no longer confluent) because it is strategy dependent. Parallel superposition is thus the only one to propose a parallel composition which follows the simplest BSP model, which is compositional and which can be added to the BSλ_p-calculus.

6 Conclusions and Future Work

A parallel superposition has been added to the BSλ/`BSMLlib` framework. This new construction allows divide-and-conquer algorithms to be expressed easily, without breaking the BSP execution model.

Compared to a previous attempt [18], this new construction has not the drawbacks of its predecessor : the two sides of parallel superposition may not have the same number of synchronization barriers ; the cost model is a compositional one ; its semantics is not strategy dependent, it can be added to the BSλ-calculus.

The next released implementation of the `BSMLlib` library will include parallel superposition. Its ease of use will be experimented by implementing BSP algorithms described as divide-and-conquer algorithms in the literature.

Acknowledgments This work is supported by the ACI Grid program from the French Ministry of Research, under the project CARAML (www.caraml.org). The author wish to thank the anonymous referees for their comments.

References

1. G. Akerholt, K. Hammond, S. Peyton-Jones, and P. Trinder. Processing transactions on GRIP, a parallel graph reducer. In Bode et al. [3].
2. W. Bäumker, A. adn Dittrich and F. Meyer auf der Heide. Truly efficient parallel algorithms: c-optimal multisearch for an extension of the BSP model. In 3rd *European Symposium on Algorithms (ESA)*, pages 17–30, 1995.
3. A. Bode, M. Reeve, and G. Wolf, editors. *PARLE'93, Parallel Architectures and Languages Europe*, number 694 in Lecture Notes in Computer Science, Munich, June 1993. Springer.
4. O. Bonorden, B. Juurlink, I. von Otte, and I. Rieping. The Paderborn University BSP (PUB) Library - Design, Implementation and Performance. In *Proc. of 13th International Parallel Processing Symposium & 10th Symposium on Parallel and Distributed Processing (IPPS/SPDP)*, San-Juan, Puerto-Rico, April 1999.
5. O. Bonorden, F. Meyer auf der Heide, and R. Wanka. Composition of Efficient Nested BSP Algorithms: Minimum Spanning Tree Computation as an Instructive Example. In *International Conference on Parallel and Distributed Processing Techniques and Applications (PDPTA)*, 2002.
6. G.-H. Botorog and H. Kuchen. Efficient high-level parallel programming. *Theoretical Computer Science*, 196:71–107, 1998.
7. M. Cole. *Algorithmic Skeletons: Structured Management of Parallel Computation.* MIT Press, 1989.
8. M. Danelutto and D. Ratti. Skeletons in MPI. In S.G. Aki and T. Gonzalez, editors, *International Conference on Parallel and Distributed Computing and Systems*, Cambridge, USA, November 2002. ACTA Press.
9. J. Darlington, A. J. Field, P. G. Harrison, P. Kelly, D. Sharp, Q. Wu, and R. While. Parallel programming using skeleton functions. In Bode et al. [3].
10. P. de la Torre and C. P. Kruskal. Submachine locality in the bulk synchronous setting. In L. Bougé, P. Fraigniaud, A. Mignotte, and Y. Robert, editors, *Euro-Par'96. Parallel Processing*, number 1123–1124 in Lecture Notes in Computer Science, Lyon, August 1996. LIP-ENSL, Springer.

11. C. Foisy and E. Chailloux. Caml Flight: a portable SPMD extension of ML for distributed memory multiprocessors. In A. W. Böhm and J. T. Feo, editors, *Workshop on High Performance Functionnal Computing*, Denver, Colorado, April 1995. Lawrence Livermore National Laboratory, USA.
12. A. V. Gerbessiotis and L. G. Valiant. Direct Bulk-Synchronous Parallel Algorithms. *Journal of Parallel and Distributed Computing*, 22:251–267, 1994.
13. G. Hains. Subset synchronization in BSP computing. In H.R.Arabnia, editor, *PDPTA'98 International Conference on Parallel and Distributed Processing Techniques and Applications*, volume I, pages 242–246, Las Vegas, July 1998. CSREA Press.
14. Y. Hayashi. *Shaped-based Cost Analysis of Skeletal Parallel Programs*. PhD thesis, University of Edinburgh, 2001.
15. J.M.D. Hill, W.F. McColl, and al. BSPlib: The BSP Programming Library. *Parallel Computing*, 24:1947–1980, 1998.
16. K. Hinsen. Parallel Programming with BSP in Python. Technical report, Centre de Biophysique Moléculaire, 2000.
17. F. Loulergue. BSλ_p: Functional BSP Programs on Enumerated Vectors. In J. Kazuki, editor, *International Symposium on High Performance Computing*, number 1940 in Lecture Notes in Computer Science, pages 355–363. Springer, October 2000.
18. F. Loulergue. Parallel Composition and Bulk Synchronous Parallel Functional Programming. In S. Gilmore, editor, *Trends in Functional Programming, Volume 2*, pages 77–88. Intellect Books, 2001.
19. F. Loulergue. Implementation of a Functional Bulk Synchronous Parallel Programming Library. In 14th *IASTED International Conference on Parallel and Distributed Computing Systems*, pages 452–457. ACTA Press, 2002.
20. F. Loulergue, G. Hains, and C. Foisy. A Calculus of Functional BSP Programs. *Science of Computer Programming*, 37(1-3):253–277, 2000.
21. J. Martin and A. Tiskin. BSP Algorithms Design for Hierarchical Supercomputers. *submitted for publication*, 2002.
22. W. F. McColl. Scalability, portability and predictability: The BSP approach to parallel programming. *Future Generation Computer Systems*, 12:265–272, 1996.
23. Q. Miller. BSP in a Lazy Functional Context. In *Trends in Functional Programming*, volume 3. Intellect Books, may 2002.
24. P. Panangaden and J. Reppy. The essence of concurrent ML. In F. Nielson, editor, *ML with Concurrency*, Monographs in Computer Science. Springer, 1996.
25. S. Pelagatti. *Structured Development of Parallel Programs*. Taylor & Francis, 1998.
26. J. Serot and D. Ginhac. Skeletons for parallel image processing: an overview of the skipper project. *Parallel Computing*, 28(12):1685–1708, 2002.
27. D. B. Skillicorn. Multiprogramming BSP programs. Department of Computing and Information Science, Queen's University, Kingston, Canada, October 1996.
28. D. B. Skillicorn, J. M. D. Hill, and W. F. McColl. Questions and Answers about BSP. *Scientific Programming*, 6(3), 1997.
29. A. Tiskin. *The Design and Analysis of Bulk-Synchronous Parallel Algorithms*. PhD thesis, Oxford University Computing Laboratory, 1998.
30. A. Tiskin. A New Way to Divide and Conquer. *Parallel Processing Letters*, (4), 2001.
31. Leslie G Valiant. A bridging model for parallel computation. *Communications of the ACM*, 33(8):103, August 1990.
32. A. Zavanella. *Skeletons and BSP : Performance Portability for Parallel Programming*. PhD thesis, Universita degli studi di Pisa, 1999.

visPerf: Monitoring Tool for Grid Computing

DongWoo Lee[1], Jack J. Dongarra[2], and R.S. Ramakrishna[1]

[1] Department of Information and Communication
Kwangju Institute of Science and Technology, Republic of Korea
{leepro,rsr}@kjist.ac.kr
[2] Innovative Computing Laboratory
Computer Science Department, University of Tennessee, Knoxville, USA
dongarra@cs.utk.edu

Abstract. It is difficult to see the status of a working production grid system without a customized monitoring system. Most grid middleware provide simple system monitoring tools, or simple tools for checking system status. *visPerf* is a general purpose grid monitoring tool for visualizing, investigating, and controlling the system in a distributed manner. *visPerf* is a system based on a distributed monitoring sensor, *visSensor*, in which the sensor uses methods to monitor the status of grid middleware with little or no modifications to the underlying system.

1 Introduction

Primarily due to the emergence of high speed backbone network services such as vBNS and the Internet2, Grid Computing [6] has become one of the most exciting new trends in high-performance computing. Ease of use and total computing resource unification involving many kinds of computing tools, including new network facilities and new software, have contributed to the growing interest. The computing resources span large geographical areas ranging from inter-campus to international dimensions. As the number of resources increases, so does proneness to faults. Even though fault-tolerant mechanisms [9,12,10] exist for grid middleware, human intervention is still required for recognizing certain problems of the system. Because huge amounts of data are produced by many components of the system in the form of logs and trace events, it is often very difficult, especially in a large scale system, to find the proverbial "needle in a haystack" without human intervention. To maintain such large scale grid systems, we need the capability to monitor the system. Most grid middleware [2,4,1,7] have this capability to some extent. A monitoring system that is simple and that supports a heterogeneous environment is the need of the hour. *visPerf* is a kind of monitoring system for grid computing in which multiple computing entities are involved in solving a computational problem with parallel and distributed computing tools.

We describe the problems and the requirements of the monitor software for grid computing environment in Sect. 2. In Sect. 3, the system architecture of *visPerf* is presented with design concepts in detail, including its sensor architecture, monitor viewer, and monitor peer-to-peer network. In Sect. 4, we show

P.M.A. Sloot et al. (Eds.): ICCS 2003, LNCS 2659, pp. 233–243, 2003.

an example monitor, *visPerf* for NetSolve. Related work is presented in Sect. 5. Finally, we conclude the work in Sect. 6.

2 Support for Grid Computing

For grid computing, a monitoring system has to consider several requirements. It is also a design goal that the monitoring system be for grid computing. Most grid middleware consists of three parts: *client* (e.g. user application), *management* (e.g. resource scheduler), and *resource* (e.g. server, storage and etc). For the client part, users develop their application with grid middleware's programming interface. In the working phase, the user's grid application contacts the resource scheduler to get a resource for its computation. Then, the application can use the assigned resource. Because resources on a grid are located over a wide area (i.e. loosely coupled), it is difficult to be aware of errors and/or problems that result from a user's application. Grid middleware supports remote machines' standard output handles[1] to show the user the output of the application's execution. When the number of processes of a user application becomes large, it is often very difficult to track the errors/problems. If the interactions of the system can be visualized, then the user can better understand and maintain the system. Besides capturing the interactions of a grid application, it is useful to gather information such as the performance of a local machine's processor workload, disk usage and so on, which are related to remote machines in the grid. Users demand a tool for monitoring their system with little effort. To accommodate demands of a monitoring system for grid computing, there are several problems and requirements that must be met.

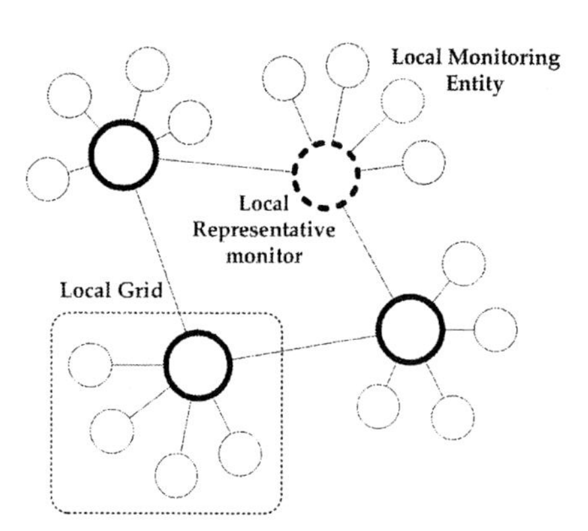

Fig. 1. Centralized+decentralized hybrid peer-to-peer network topology

[1] STDOUT/STDERR

2.1 Problems

While building a monitoring system for grid computing, we face several problems such as the following:

Heterogeneity. Hardware and software are configured to be used by grid middleware. In the case of a local system that is in the form of a cluster-like server system, the middleware does not need to consider heterogeneity support because it usually uses the same types of machines and software. But in the grid scale computing environment, the components of the system are located across a wide area and have various types of components, including grid middleware. We need a way to collect the information for the purpose of managing or investigating its working status with heterogeneity support.

Access Network. Due to the various network access policies including firewall, NAT, and so forth, sometimes it is difficult to monitor remote machines in a simple manner. In the case of grid middleware, it has its own mechanism for remote communication with a security mechanism (e.g. Kerberos). To get consistent access to a remote monitoring object, we have to use a tolerant protocol or interface to a network site having restricted access protocols.

Size of Information. The speed of transferring monitored information to an appropriate location is restricted by the capacity of the remote system and the communication channel. We have to devise a way to reduce the size of the monitoring data.

Interoperability. Monitored information can be shared with other applications such as grid resource scheduler and other monitoring systems. To accommodate such applications, an interoperable interface is required.

Scalability. In the grid network, we have to consider the scalability due to the multiple entities to be monitored. The NetSolve production grid [5], for example, sometimes exceeds 100 hosts. This is the case with just a local grid. But, if this grid extends into other external grids, the scalability issues of the monitoring system become more complicated.

2.2 Requirements

Support Various Systems. To monitor various grid systems, we have to have a way to get information from a target running system. We considered two ways for accomplishing this: *indirect* and *direct* interfaces. If grid middleware has no monitoring facility offering its internal status to an external program, we have to use indirect information, that is, *log* information as a form of an accessible data object. This information can be used by a monitoring system when the system is not integrated into the middleware, and hence the need for a program to process the log information generated by the grid in real time. If the grid middleware has a monitoring facility, we can use its interface directly. It also requires minimal effort to integrate it into the monitoring system.

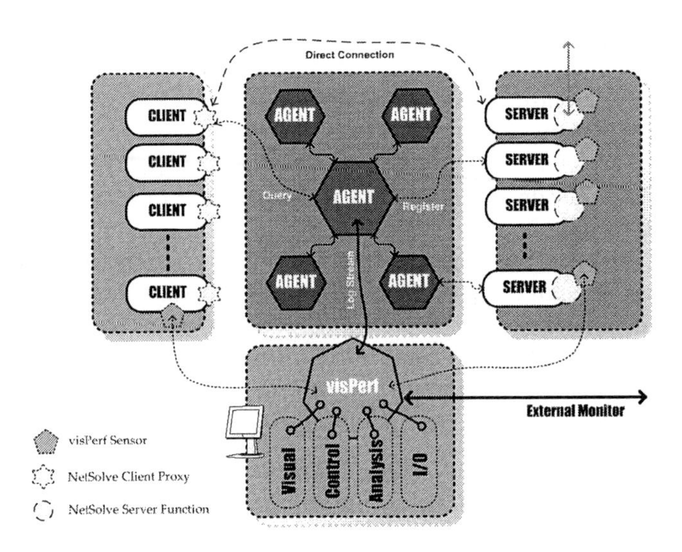

Fig. 2. Overview of *visPerf* monitoring system

Network Topology of Monitor. To deal with the problems listed above, a monitoring system should build a network for itself because it is necessary for a monitoring system to be efficient and simple. Because the centralized network of typical monitoring systems does not fit well with the nature of grid computing, we considered a peer-to-peer network topology [8]. Deploying this peer-to-peer network can cope with large amounts of data to be transferred to one point and can address the scalability problem due to the bottleneck effect when accessing the monitoring system. The monitoring system can maintain local information and then it can be shared with other outside monitoring systems. The appropriate P2P topology is the "centralized+decentralized hybrid P2P network topology" (shown in Fig. 1) because with this we can handle the local grid system in a centralized manner and the global grid in a decentralized manner. Each local representative sensor captures and maintains monitored information of a local grid [2].

Network Access Interface (Protocol). To solve the interface access problem mentioned before, we have to use some form of a flexible mechanism. We considered two network access interfaces (protocol): general TCP data communication (using monitor's own communication protocol) and XML-RPC [13] through a web server. Depending on the condition of a site to be accessed, we can use these two methods adaptively.

3 Monitoring System's Design Approach

With these problems and requirements in mind, we have designed a monitoring system architecture. Figure 2 presents the overview of the *visPerf* monitoring

[2] a.k.a local cluster or one unit of local machines using a same grid middleware

system[3]. Most grid middleware [4,2,1,7] consists of three parts as illustrated in the figure: client, management, and resource (server). This will be explained below, with the NetSolve system in mind. In the client portion, there is a client application equipped with a grid middleware-aware software library. This client application (grid middleware library) uses a resource scheduler of its own in order to get a resource to be used. For this grid configuration, we positioned a monitoring system in each part. The *sensor* is a service daemon for collecting a local host's information and propagating that information to a subscriber monitor viewer. The *main controller*, *visPerf*, is used to control the remote sensor as well as to analyze and visualize the collected information interactively by the user. As a remote sensor, *visSensor* works for monitoring a local machine.

3.1 Local Monitor: *visSensor*

visSensor is a monitor residing on a local machine. It is responsible for sensing a local machine's status by gathering that machine's performance information[4] and special purpose information that is tailored for a specific system like grid middleware (i.e. system interactions). A local machine's general performance information is gathered by performance measuring tools such as `vmstat`, `iostat`, `top` of UNIX-based system performance tools.

For specific system tailored monitoring, we support two methods: log-based monitoring and profiling API-based monitoring. As mentioned previously, there are various run-time systems of grid middleware. The simplest way to monitor a run-time system without any modification to the system is to use its log file because most run-time systems log their status into the log file for the purpose of debugging and monitoring. Although this method is simple, the log file of a grid middleware application has to have a formal form to depict its status in a consistent manner, that is, with a rule (or grammar) of the log file. To capture the running sequence in the case of NetSolve, logs have to have a meaningful format such as:

```
..<omitted>...
NS_PROT_PROBLEM_SUBMIT: Time 1017336647 (Thu Mar 28 12:30:47 2002),
Nickname inttest, Input size 12, Output Size 12, Problem Size 1,
ID leepro@anaka.cs.utk.edu

Server List for problem inttest:
neo15.sinrg.cs.utk.edu
neo9.sinrg.cs.utk.edu
..<omitted>...
NS_PROT_JOB_COMPLETED_FROM_SERVER: neo15.cs.utk.edu inttest 0
..<omitted>...
Server cypher12.sinrg.cs.utk.edu: latency: 1012   bandwidth: 870885
Server neo13.sinrg.cs.utk.edu: workload = 100
..<omitted>...
```

The log presented above, for example, is a part of the NetSolve log file that is produced by its central agent (the resource scheduler of NetSolve). We can

[3] The figure illustrates the NetSolve grid middleware as a representative grid system with *visPerf*

[4] CPU workload, memory status, disk I/O, and so forth.

figure out the sequence of the call from a client ("`ID leepro@anaka.cs.utk.edu`") to a server "`neo15.sinrg.cs.utk.edu`". The client sends a request ("`NS_PROT_PROBLEM_SUBMIT`") to the resource scheduler. Then, the scheduler presents the available resource list. The server ("`neo15.sinrg.cs.utk.edu`") is selected for the client. In this case, the first on the resource list is the selected resource because NetSolve's agent calculates collected performance information for each resource, applies a scheduling algorithm internally, and then sorts the lists in the order of the "most idle" (least loaded) machine. In addition to log-based monitoring, some systems have a profiling API-based interface for internal or external monitoring. For example, NetSolve has "Transactional Logging Facility" that is used by components of NetSolve to notify its activities to an information database server of NetSolve. In the case of NetSolve, all the components transfer their logs onto the agent. So, the rate of logging into a log file is very high when there are many computing resources. Consequently, the *visSensor*'s consumption of the log file in real time requires that large amounts of data be transferred to a subscriber. To alleviate this effect, *visPerf* has a preprocessing filtering function. This has two advantages: lightweight data size to be transferred to a subscriber and a standard format of a log to support various types of grid middleware. Figure 3 (right) shows components of *visSensor* in which there are several layers. The `Info/Log Collector` collects a local machine's general performance information (e.g, I/O and Kernel by using `/proc`) and filters a specific system's log (e.g. `NetSolve` log filter in the figure).

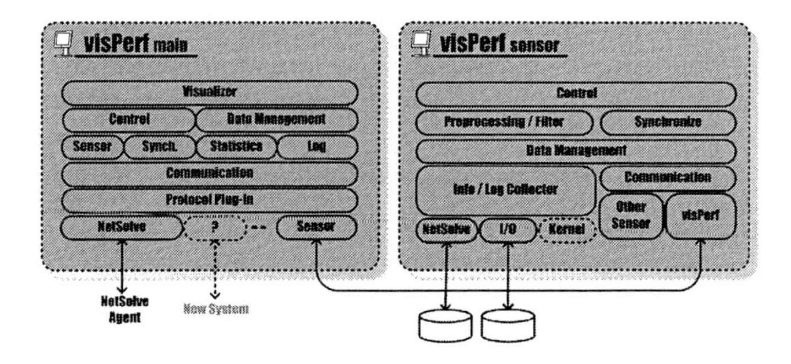

Fig. 3. Components of *visPerf*: sensor and main controller (visualizer)

3.2 Managing and Presenting Monitored Information: *visPerf*

visPerf is used for browsing and controlling the status of a local machine via a local sensor (*visSensor*). Figure 3 shows the relationship between the *visPerf* main controller and the local *visSensor* and components of each side. *visPerf* employs several protocols as a communication subsystem. For example, the `NetSolve` module at the bottom of the figure is used for communicating with the NetSolve agent to query its resource availability. The `Sensor` protocol module is the core communication module that provides two kinds of communication protocols: raw monitor protocol and XML-RPC protocol. The access protocol can be changed in accordance with the user environment. Also, a user can make its own visualizer through an XML-RPC interface of any programming language. Using the `Control` module, *visPerf* can control local sensors registered in the main controller or directly control a local sensor by specifying its network address. This module periodically sends a synchronization message to its local sensors. If a sensor is out of order or having a problem, a user can discover it. The monitored information is presented via multiple graphical presentations: the *performance fluctuation graph* and the *interaction map.* The interactions of the middleware are displayed in an animated resource map. The sequence of an interaction animation is based on the logs defined in the appropriate sensor filter in which the log runs from the beginning of a call (from a client application) to the end of the call.

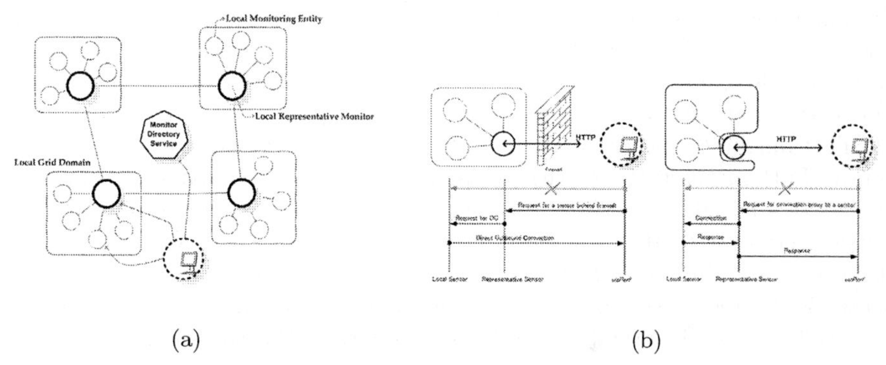

(a) (b)

Fig. 4. (a) Monitor directory service: *visPerf* can connect to a local sensor through MDS, (b) monitor proxy: behind firewall (left), private IP network with an access point (right)

3.3 Peer-To-Peer Monitor Network

As mentioned in Sect. 2.2, a network topology for a monitoring system is needed to manage the system effectively. Sensors are located across multiple grid domains to monitor local systems. Through the peer-to-peer network of a monitoring system, it is possible to see the representative sensor's summarized information as well as the specific local machine's status with a direct connection. Locally, the representative sensor maintains the status of its domain machines in which sensors also exist for each target machine in slave mode. To find out the appropriate sensor to be contacted, *visPerf* sends a query to the MDS (Monitor Directory Service)[5] (Fig. 4a) or to a local sensor with a unique domain name (e.g. UT-ICL[6]). When a local representative sensor is about to start service, the monitor registers itself with the specified MDS. When the user's *visPerf* uses the local monitor, the local sensor connected by the user forwards the query to the

[5] It's not the MDS of Globus.

[6] It is not the Internet domain name, but one managed by the monitor system.

MDS. The MDS returns the host name, its network port number, and the type of the target remote sensor. Because the sensor can be a representative sensor of a local domain, the type of the sensor is needed to determine its functionality. With the response from the query, *visPerf* can directly contact the local sensor to get its status.

3.4 Multiple Access Points: Monitor Proxy

For security, most organizations introduce a network firewall system as their front-end. In this case, it is impossible to connect to a local sensor directly from outside the network except for a secure inbound network port[7]. In addition, most local cluster systems use an internal network address scheme (e.g. private IP address or NAT) due to the lack of IP addresses for their working nodes except an exposed node connected to the Internet. This also makes it difficult for a monitoring system to connect to a local sensor from outside directly. Due to these reasons, we introduced a "*Monitor Proxy*" to support a secure connection from outside a network. The monitor proxy has two functions: (i) Delegate a connection request to a local sensor through a representative sensor that uses HTTP tunnelling, and (ii) Provide a connection proxy through a representative sensor. Both methods (i) and (ii) are applied to connect from outside the firewall. In case of a private IP network (or NAT), method (i) is applied to connect from outside the network. If a local sensor is not able to connect to an outside network, it has to use a kind of "*gateway host*" of its cluster or grid. If the representative sensor is located in the exposed node, it can create a connection on behalf of the inside sensor. The right side of Fig. 4b illustrates this situation. For security against a malicious connection, we use the md5 authentication method.

4 Case Study: visPerf for NetSolve

As an example of *visPerf*, we present the customized monitor for the NetSolve grid middleware[8] It provides remote access to computational resources including hardware and software and it supports different architectures and operating systems. NetSolve provides blocking and non-blocking calls to a problem on an available server. It also offers simultaneous multiple calls such as *farming* calls[9]. We can view the parallelism of NetSolve using our monitoring system. To view the interactions of a system, we have to prepare a log filter that is used for refining a specific system's log into a form of *visPerf* as mentioned in Sect. 3.1. To track the sequence of a NetSolve call, we added new logs to the NetSolve system, which are small modifications to the source code of NetSolve. This is just for completing visualization of *visPerf*. The parallelism of non-blocking calls can not

[7] e.g. SSH, HTTP port and so forth.

[8] The NetSolve project [2] is being developed at the Innovative Computing Laboratory of the University of Tennessee Computer Science Department.

[9] A user's data is divided into independent parallel NetSolve calls to use computing resources simultaneously

be shown because there is no log type to indicate the end of a NetSolve call. In addition to this log indicating the sequence of a NetSolve call, this log filter parses performance notification logs. Figure 5 (left) shows a snapshot of a user's problem being submitted between client application and server through one central NetSolve agent of a production grid, which occured during the middle of a NetSolve testing program (`Test` of current release of NetSolve 1.4). In addition to this interaction map, users can investigate a specific host using *visPerf* as in Fig. 5 (right). Figure 6 illustrates the performance fluctuation of a machine's CPU and disk I/O.

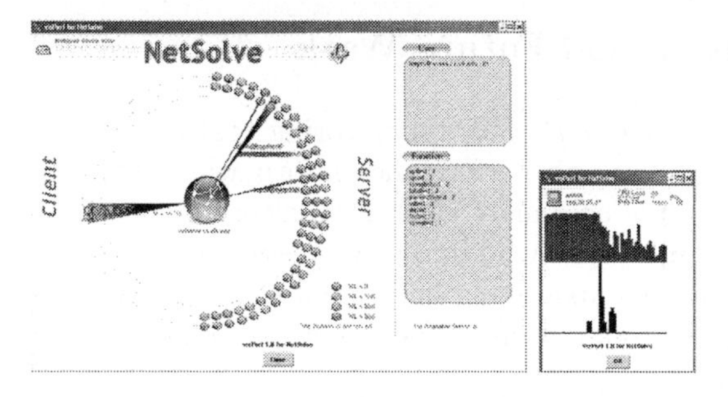

Fig. 5. Visual presentation: interaction map of NetSolve (left), performance graph (right): CPU workload (red bar) and I/O workload(blue bar)

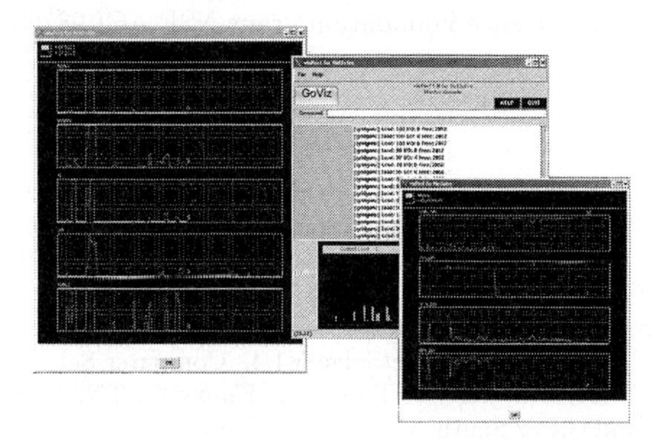

Fig. 6. Sensing local machine's CPU and I/O performance

5 Related Work

Globus HBM [10] is a monitor facility to detect faults of a computing resource involved in Globus. It checks the status of a target machine and reports it to a higher-level collector machine. GridMonitor Java Applet [3] is a kind of monitor for Globus system. It works by displaying the grid information and server status for all sites including Globus MDS. JAMM [11] is an agent-based monitoring system for grid environments that can automate the execution of monitoring sensors and the collection of event data. It supports not only the system's general performance including network and CPU workload, but also application sensors that are embedded inside applications to notify an overload by threshold variables. It is a type of a system that collects performance information of the grid environments. The difference from our work is that *visPerf* is designed for monitoring activities using grid middleware dependent information via direct and indirect interfaces. Our system works by visualizing interactions of the working system and showing useful information on the system including performance information in a simple, practical manner.

6 Conclusion and Future Work

This paper presented *visPerf* as a monitoring tool for grid middleware to show the running activities and performance information. This monitoring tool will be improved to serve as a more general monitoring system to support different types of grid middleware by adding new sensor functions and log filters. At the same time, using this underlying networked sensing system, we will attempt to make this monitor tool a useful information provider.

Acknowledgments. This work has been supported by the BK21 program in K-JIST, South Korea and the Innovative Computing Laboratory (ICL) and in part by the National Science Foundation grant NSF ACI-9876895.

References

1. James C. French Alfred C. Weaver Paul F. Reynolds Jr. Andrew S. Grimshaw, William A. Wulf. Legion: The next logical step toward a nationwide virtual computer. Technical Report CS-94-20, Department of Computer Science, University of Virginia, Charlottesville, Virginia, USA, June 1994.
2. D. Arnold, S. Agrawal, S. Blackford, J. Dongarra, M. Miller, K. Sagi, Z. Shi, and S. Vadhiyar. Users' Guide to NetSolve V1.4. Computer Science Dept. Technical Report CS-01-467, University of Tennessee, Knoxville, TN, July 2001.
3. Mark Baker and Garry Smith. Gridrm: A resource monitoring architecture for the grid. In *Springer-Verlag, LNCS (2536)*, page 268 ff, 2002.
4. I. Foster and C. Kesselman. Globus: A Metacomputing Infrastructure Toolkit. *The International Journal of Supercomputer Applications and High Performance Computing*, 11(2):115–128, Summer 1997.
5. SInRG (Scalable Intracampus Research Grid). `http://icl.cs.utk.edu/sinrg/`.
6. Carl Carl Kesselman (ed) Ian Foster (ed). *The Grid : Blueprint for a New Computing Infrastructure.* Morgan Kaufmann, 1998.
7. M. J. Litzkow, M. Livny, and M. W. Mutka. Condor : A hunter of idle workstations. In *8th International Conference on Distributed Computing Systems (IEEE)*, pages 104–111, Washington D.C, June 1988.
8. Nelson Minar. Distributed systems topologies. `http://www.openp2p.com`.
9. Anh Nguyen-Tuong. Integrating fault-tolerant techniques in grid application. Computer Science Dept. Dissertation, University of Virginia, Virginia, Auguest 2000.

10. P. Stelling, I. Foster, C. Kesselman, C.Lee, and Gregor von Laszewski. A Fault Detection Service for Wide Area Distributed Computations. In *Proceedings of the 7th IEEE International Symposium on High Performance Distributed Computing*, pages 268–278, Chicago, IL, 28-31 July 1998.
11. B. Tierney, B. Crowley, D. Gunter, J. Lee, and M. Andrew Thompson. A monitoring sensor management system for grid environments. *Cluster Computing Journal*, 4, 2001.
12. Job B. Weissman. Fault tolerant wide-area parallel computing. 2000. Proceedings of the International Parallel and Distributed Computing Symposium.
13. XMLRPC. `http://www.xmlrpc.com`.

Design and Implementation of Intelligent Scheduler for Gaussian Portal on Quantum Chemistry Grid

Takeshi Nishikawa, Umpei Nagashima, and Satoshi Sekiguchi

National Institute of Advanced Industrial Science and Technology
Grid Technology Research Center, Tsukuba Central 2
Tsukuba, Ibaraki 305-8568, Japan
{t.nishikawa,u.nagashima,s.sekiguchi}@aist.go.jp
http://unit.aist.go.jp/grid/

Abstract. We have developed a Quantum Chemistry Grid (QC Grid) for simple and effective use of computer resources for ab initio molecular orbital calculations. Ab initio is a powerful methodology in computational chemistry. Gaussian is one of the codes widely used in quantum chemistry research, and it is used not only by quantum chemistry experts, but also by non-experts. Since the CPU cycles of Gaussian jobs vary significantly with the input parameters, it is difficult for users to choose the most adequate computational resources in a local computing environment. By using grid technology on top of a high-speed network environment, QC Grid enables users to easily share the knowledge of experts and efficiently utilize costly computational resources without having to know the specific system environment. Gaussian Portal is the first implementation of the QC Grid concept. It consists of a Web interface, a meta-scheduler, computing resources, and archival resources on Grid infra-wares.

1 Introduction

Ab initio molecular orbital (MO) calculations are used to design novel materials and molecules in fields such as computer-aided chemistry, pharmacy, and biochemistry. The computation cost of the ab initio MO calculation strongly depends on the molecular size and calculation method. In the case of the Hartree-Fock method, which is the simplest of these methods, the computation cost is proportional to $O(n^4)$, where n is the number of atoms in the molecule, and thus it rapidly increases with system size. Therefore, it is very difficult for users to choose appropriate computer resources for an ab initio MO calculation because of the large variety of molecular sizes and calculation methods. The estimation of computational resource requirements is also difficult without extensive experience with ab initio MO calculations. Usually users choose a large high-performance computing (HPC) resource for execution of their jobs, even if the job is small enough to be executed on a PC. For example, more than 70% of the CPU time of the Tsukuba Advanced Computer Center [1] (TACC), National

P.M.A. Sloot et al. (Eds.): ICCS 2003, LNCS 2659, pp. 244–253, 2003.

Institute of Advanced Industrial Science and Technology (AIST), is occupied by Gaussian [2] jobs. Gaussian is one of the codes widely used in quantum chemistry research. Furthermore, a large variety of computer resources is required for the Gaussian jobs. The supercomputers at TACC are always busy with Gaussian jobs, even though more than 60% of these jobs are executable on standard desktop PCs.

Therefore, to improve the efficiency of utilization and to facilitate seamless use of different computer resources (supercomputers, servers and PC's), as well as to shorten the turn around time of Gaussian jobs, we have been developing the Quantum Chemistry Grid (QC Grid). The QC Grid adopts Grid technology for security and job scheduling in which the computational and quantum chemical knowledge of the ab initio MO calculation experts is used for computer resource allocation. Namely, the user doesn't need to be conscious of the specific computing resources on the QC Grid. A look-up function is incorporated in a database of the accumulated results so that already existing ab initio MO calculation results can be quickly accessed. The QC Grid also ensures security when accessing distributed HPC resources.

There are several systems [3,4,5] to enable the use of ab initio molecular orbital calculation programs via a Web-based interface or to enable access to a results database. However, in these systems, users select the computer resource by themselves and the Grid environment is insecure.

The QC Grid has an intelligent meta-scheduler and a computing resource allocation feature that is based on expert knowledge about the calculations and the characteristics and/or the state of its resources. We developed Gaussian Portal as the first implementation of the QC Grid.

The remaining sections of this paper describe the QC Grid's design concept and architecture. Our experience in using Gaussian Portal at TACC is also described.

2 Design Concept of the QC Grid

The design concept of this system is as follows. (1) Apply Grid technology to make the computing resource virtual and the computing service accessible at anytime from anywhere while maintaining security. (2) Build a simple and easy-to-use user interface by using Web technology. (3) Appropriately allocate computing resources according to the content of the calculation with the intelligent meta-scheduler. (4) Facilitate sharing of expert knowledge of quantum chemistry and the user's experience. (5) Archive all input and output (results) files. (6) Throughput must be a priority. The system should be inexpensive and have excellent cost performance.

Because the computing resources are physically distributed among several sites, we don't have the resources at each site that can endure the maximum predicted load. Instead, the maximum load of the entire system is the limitation of resource availability. Even if a situation occurs where a part of the computing resource becomes unavailable, scheduled operations may continue if they can be allocated to the remaining resources, thus maintaining availability and reliability.

Through the Web-based user interface, the user of the QC Grid can use the physically distributed computer resources as if they were a single system, i.e., without having to consider differences in authentication mechanism, operating system, job management system, hardware, computational chemistry application program, and so on.

The user does not have to worry about which computer to use. The meta-scheduler handles that. If a result for a new request exists in the archives and the retrieval time is shorter than for computing it anew, the user can obtain the result from the database instead of having computing resources assigned to do the calculation. Intelligent resource allocation with the meta-scheduler becomes possible because of expert knowledge sharing and because of the preserving of all input and result files. The meta-scheduler helps to optimize the entire computing resource allocation. The system is scalable to the number of jobs.

3 QC Grid Architecture

The QC Grid logical architecture is depicted in Fig. 3. The QC Grid consists of the Web-based user interface, meta-scheduler, knowledge database, results database, Input/Output(I/O) archives, and computing nodes with Grid infrastructure software (Grid infra-ware). Figure 3 shows the I/O flow and QC Grid components relationships controlled by meta-scheduler. The Grid infra-ware ensures security, management of resource allocation, access to remote data, and monitoring of remote resources. The following subsections describe the components.

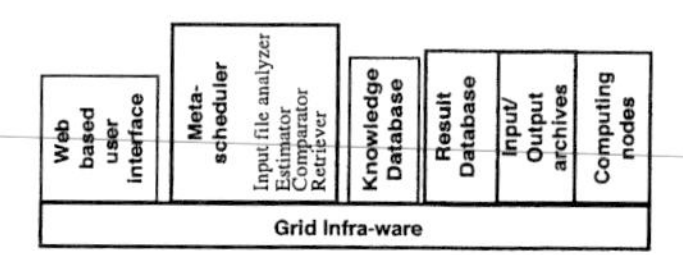

Fig. 1. QC Grid logical architecture

3.1 Web-Based User Interfaces

The QC Grid user has a Web-based interface that does all operations such as uploading thes input file, controlling the job, displaying job status, and getting results. The Web-based interface was created by using a toolkit that can quickly build the portal interfaces. It is served by an HTTP server daemon program. The HTTP server supports Secure Socket Layer (SSL). The Web interface is used to request job resources, control job status, visualize molecular structures, and display results.

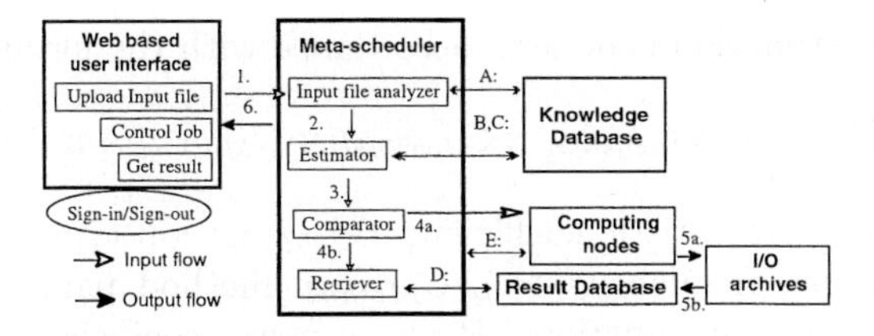

Fig. 2. Flow and relationship of the QC Grid components controlled by the meta-scheduler. 1. Upload input file. 2. Parse input file. 3. Estimate values. 4a. Submit new job. 4b. Retrieve computed data from database and archives. 5. Return computed result from computing node and store result in archives(a) and database(b). 6. Get results via Web interface. Queries are "A: Request system parameters using estimation." "B: Is computing time less than quick searching time ?", "C: Is computing time less than detailed searching time ?", "D: Retrieveing data from Results database.", and "E: Job control".

3.2 Meta-scheduler

The meta-scheduler consists of an input file analyzer, estimator, comparator, and retriever.

Input File Analyzer. The analyzer loads the input file, converts the end- of-line character, and assesses the resource requirements, the calculation method, basis sets, and molecular structure. Resource requirements, the calculation method, basis sets, and molecular structure are described in the input file. An example of the Gaussian input file is shown in Fig. 3.2. The calculation

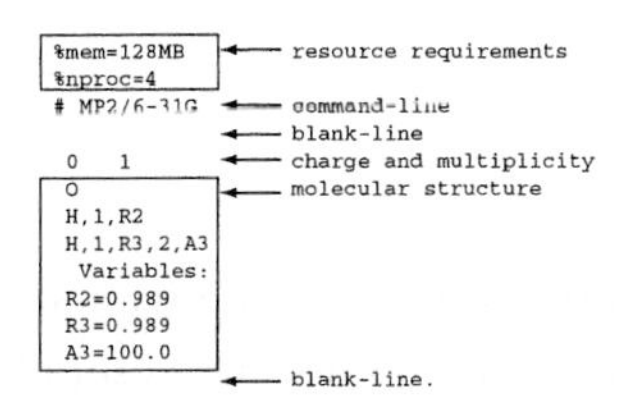

Fig. 3. Gaussian input file

method (HF, MP2, etc.) and basis sets (STO-3G, 3-21G, etc.) are specified on the command-line. The analyzer displays a Web page of its resource choices to the user.

Estimator. The Estimator predicts CPU time, time for retrieving archived results, amount of memory, and disk usage based on the values sent from the Input file analyzer and based on the information in the Knowledge database.

The CPU time estimation is done in accordance with the formula:

$$CPUtime = C_{system} \times C_{JobType} \times C_{Method} \times n^{P_{Method}} \quad (1)$$

C_{system}: system characteristic coefficient, $C_{JobType}$: job type parameter (single point, optimization, frequency, etc.), C_{Method}: method parameter (HF, MP2, etc.), n: Number of basis functions, P_{Method}: power coefficient of method.

Comparator. The Comparator compares the computing time and the retrieval time, and judges whether to retrieve or to compute. It compares the structures of existing results with the structure of the input data.

Retriever. The Retriever has a quick search function and a detailed search function. The former does an exact matching of the computing command-line and the molecule structure. The latter can narrow down results on more than one condition by checking the references and key words.

3.3 Knowledge Database

The Knowledge database contains parameters for computing the resource estimation, performance indices of the computing resources, a retrieval time database, an operation policy, and a list of the organizations facilities.

3.4 Results Database and I/O Archives

The Results database stores all the input files and the location of all output files including checkpoint files. I/O Archives exist on the local file server of each computing resource.

3.5 Computing Nodes

Computing Nodes are installed in the computational chemistry application and the job scheduler. These represent the normal computing environment to which the meta-scheduler can submit the job.

4 Gaussian Portal

Gaussian Portal is the first implementation of the QC Grid. Its system configuration is shown in Fig. 4.

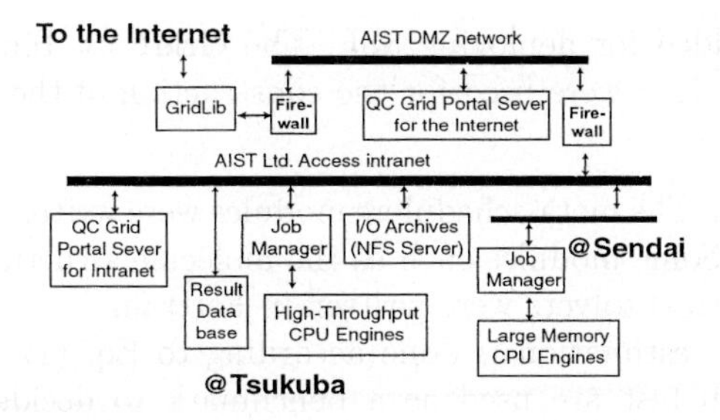

Fig. 4. The QC Grid/Gaussian Portal system configuration.

4.1 System Configuration

We have implemented the QC Grid/Gaussian Portal on geographically distributed servers and clusters in Tsukuba and in Sendai, Japan. These nodes are connected through a Fast-Ethernet switch LAN and uplink to the AIST backbone through Gigabit Ethernet. Two firewalls are installed: one between the Internet and the DMZ network of AIST and the other between the DMZ network of AIST and the AIST intranet. The GridLib server is installed on the Internet. The QC Grid Portal server has the Web server, meta-scheduler, knowledge database server, and Globus [6] Gatekeeper, all in one. There are two QC Grid Portal servers, which are operated under different policies. One is for the Internet, and is installed on the DMZ network of AIST. The other is for the AIST intranet, and is installed on the AIST intranet. The Results database and I/O Archives server (HP AlphaSever GS160), Globus Deploy servers, CPU Engines (IBM x330 cluster and IBM RS6000/SP), and the NFS server are also installed in the AIST intranet region. The GridLib [8] server, two QC Grid Portal servers, the High-Throughput CPU engines and Globus Deploy server for these are located in Tsukuba. Large-Memory CPU engines and Globus Deploy server are located in Sendai. The user can access the Gaussian Portal via the AIST intranet or via the Internet.

In the construction of this system, we gave priority to open source software. The exceptions are Gaussian, the PGI compiler [9] to build Gaussian, the software on the IBM RS6000/SP (AIX, XLF and LoadLeveler) and the software on the HP Alpha Server GS160 (Tru64 UNIX, Oracle). We use Linux for the OS of our machines with Intel architectures. Globus Toolkit, GridPort Toolkit [7], and GridLib prototype are used as the Grid infra-ware. Open-PBS [10] is used as the job manager for the High-Throughput CPU engines.

4.2 Implementations

Portal Interface. GridLib masks hardware and software differences with a common interface. The interface employs a self-signed Certificate Authority (CA) to grant access to the OpenSSL [11]. The CA certificate is installed in the servers and in the client's browsers. Apache [12] is used for the Web server, and mod_ssl is added for deploying SSL. The GridPort Toolkit and the codes written in Perl or Java were used for the construction of the Portal Interface.

Meta-scheduler. The meta-schedulers modules were written, for the most part, in Perl and Java. Some modules such as the molecule structure comparator and some of the numerical solvers were written in Fortran.

The CPU time estimation is done according to Eq. (1). The inputs of the standard Gaussian test are used as a benchmark to decide the value of the coefficients of Eq. (1). The memory and disk requirement are read directly from the input file. The amount of memory and/or disk space limit is specified in the input file. The meta-scheduler decides what resources to allocate to each job based on this information.

The meta-scheduler accesses the databases by using JDBC [13] technology. It provides cross-DBMS connectivity to a wide range of SQL databases. In this way, we dont have to be aware of the database differences. We use PostgreSQL as the DBMS of the knowledge database and Oracle as the DBMS of the results database.

Knowledge Database. The values extracted from the knowledge database store the meta-scheduler scripts or otherwise the response time would be too slow. For example, the values extracted from the knowledge database such as retrieval time for the database (4 seconds), Csystem for Large-Memory engines (half that of the High-Throughput CPU engines), and the memory requirement threshold (500MB) are embedded in the meta-scheduler scripts.

Results Database and I/O Archives. The Results database stores key words extracted from the input files and the location of all output files and checkpoint files. The I/O Archives exist on the NFS server.

Computing Nodes. Gaussian 98 Rev.A9 is installed in each computing node. The OS on the IBM x330 cluster is RedHat Linux 6.2J and the OS on the RS6000/SP is AIX 4.3.3. The job manager on the IBM x330 cluster is OpenPBS and the job manager on IBM RS6000/SP is LoadLeveler. The CPU Engines do not have an elapsed time limitation for a job. The single scratch file had to be under 2GB on the High-Throughput CPU engine. The scratch file space is limited to under 20GB on the High-Throughput CPU engine, and to under 18GB on the Large-Memory CPU engines. Jobs that require more than 500MB are executed on the Large-Memory CPU engines.

5 Experience

We report on practical operation experience weve gained from operation of the QC Grid/Gaussian Portal at the Tsukuba Advanced Computing Center

(TACC). The QC Grid/Gaussian Portal for the AIST intranet began operation in February 2002. Figure 5 compares the monthly normalized sum of CPU time on the IBM x330 cluster (Pentium III 1.2GHz 2way SMP/node, 108 nodes) and computing nodes of the QC Grid/Gaussian Portal from February to June, 2002 (black bars) with data for the same period of TACCs independently operated IBM RS6000/SP (POWER3 200MHz 2way SMP/node, for which 40 nodes are provided for serial applications such as Gaussian) (gray bars). The single CPU performance of the x330 is about two times greater than that of the RS6000/SP. Thus the sum of the CPU time of IBM x330 cluster was scaled by a factor of two in this @comparison. The monthly values were scaled at 100% of the RS6000/SP in February.

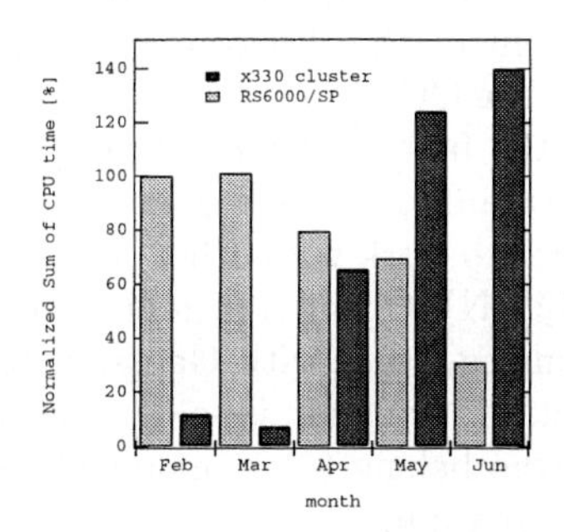

Fig. 5. The monthly normalized sum of CPU time. The ratio is based on the RS/6000SP CPU time in Feb. being equal to 100%. The normalized CPU time of the x330 cluster was scaled to twice that of the RS6000/SP because the single CPU performance of the x330 was two times that of the RS6000/SP

6 Discussion

The experience we gained during this period led us to switch from the IBM RS6000/SP to the IBM x330 cluster. As can be seen in Fig. 5, the series of normalized monthly sum of CPU time data on the IBM x330 cluster rapidly increased, whereas that on the RS6000/SP decreased. The usage ratio (RS6000/SP to IBM x330 cluster) became 1 : 5 in June. The users chose the Gaussian Portal to run their jobs. We couldnt evaluate the usefulness of the Results DB or the I/O Archives during this time, as these functions have just been installed and there still are not enough archived results to be useful to the general user.

7 Conclusion and Future Plans

We designed the QC Grid so that scientists from various fields can access quantum chemical computing services at anytime and from anywhere while maintaining security with Grid technology. We have developed and are operating the first implementation of the QC Grid as a Gaussian Portal. The meta-scheduler includes job scheduling functions and enables optimal allocation of the entire computing resource for a large number of jobs. It can be used with computers of various classes, although so far we have prepared only the Linux cluster and IBM RS6000/SP as the back-end computing resources for Gaussian. Our experience with Gaussian Portal at TACC shows that it can manage resources more efficiently than the previous system could. Introduction of the QC Grid/Gaussian Portal has made available surplus computing power in the computing resources occupied by Gaussian. It has increased the availability of the RS6000/SP, which has high-speed inter-connects for running parallel jobs.

Although the Gaussian Portal is fully implemented in the QC Grid, its implementation and concept still need tuning. We have to improve its estimation quality and implement automatic reconfiguration of parameters by using these estimations. A feature to control the disclosures of the results DB and I/O archives should be added to the QC Grid.

The Gaussian Portal is the first implementation of the QC Grid, and it is strongly dependent on Gaussian98. We will generalize the logic of the meta-scheduler to extend the portal to work with other computational chemistry programs such as GAMESS [14], NWChem [15], and ABINIT-MP [16]. Our first priority is to settle any licensing issues with Gaussian Inc., who may not have anticipated this implementation. Gaussian is sold under a single node license or site license that only users who belong to an organization with a single address are permitted to access the program.

Regarding the portal, we have to migrate from GridPort to GridLib to unify the user interface with other portals being developed by our colleagues, which include portals not only the computational chemistry calculations but also portals for calculation services of other fields, such as molecule dynamics, bio-infomatics, and weather forecasting.

The proper distribution of computing resources to various fields is necessary because the Linux cluster consisting of thousands of nodes becomes the main stream of the HPC resources. This can be readily seen by noting that a total of 63 PC clusters are in the TOP 500 supercomputer sites.

The general idea of the meta-scheduler developed with the QC Grid can be applied to meet the above resource demands. However, a cleanup of the software and packaging will be needed, before it can be introduced to the public.

Acknowledgements. We would like to thank the other members of the Grid technology research center, AIST, especially Dr. Osamu Tatebe, Dr. Yoshio Tanaka, Mr. Mototaka Hirano, Dr. Hidemoto Nakada, and Mr. Hiroshi Takemiya for their helpful discussions and encouragement. We also would like to thank Dr. Hikaru Samukawa, Japan IBM research center, for his helpful discussions and encouragement. The computing resources (IBM x330-108-node Linux cluster and IBM RS6000/SP cluster) are partly supported by the Tsukuba Advanced Computing Center (TACC). This research is partly (GridLib) supported with a grant for Research and Development Applying Advanced Computational Science and Technology, from the Japan Science and Technology Corporation (ACT-JST).

References

1. The mission of the TACC is to provide computing resources to the research units of the National Institute of Advanced Industrial Science and Technology(AIST). http://unit.aist.go.jp/tacc/
2. Gaussian 98 (Revision A.9), M.J. Frisch, G.W. Trucks, H.B. Schlegel, G.E. Scuseria, M.A. Robb, J.R. Cheeseman, V.G. Zakrzewski, J.A. Montgomery, Jr., R E. Stratmann, J.C. Burant, S. Dapprich, J.M. Millam, A.D. Daniels, K.N. Kudin, M.C. Strain, O. Farkas, J. Tomasi, V. Barone, M. Cossi, R. Cammi, B. Mennucci, C. Pomelli, C. Adamo, S. Clifford, J. Ochterski, G.A. Petersson, P.Y. Ayala, Q. Cui, K. Morokuma, D K. Malick, A D. Rabuck, K. Raghavachari, J.B. Foresman, J. Cioslowski, J V. Ortiz, A.G. Baboul, B.B. Stefanov, G. Liu, A. Liashenko, P. Piskorz, I. Komaromi, R. Gomperts, R.L. Martin, D.J. Fox, T. Keith, M.A. Al-Laham, C.Y. Peng, A. Nanayakkara, C. Gonzalez, M. Challacombe, P.M.W. Gill, B.G. Johnson, W. Chen, M.W. Wong, J.L. Andres, M. Head-Gordon, E.S. Replogle and J.A. Pople, Gaussian, Inc., Pittsburgh PA, 1998. Gaussian Inc.: http://www.gaussian.com/
3. CODATA SESSION PC2: Physics and Chemistry M. Aoyagi et al., The Development of a Distributed Computational Environment for Molecular Design on the World Wide Web, http://www.codata.org/newsletters/nl75.html, MO-SRV/Linux, http://cresta530.cc.kyushu-u.ac.jp/~planet/
4. Australian Computational Chemistry via the Internet Project, http://www.chem.swin.edu.au/
5. https://gridport.npaci.edu/GAMESS/, Kim Baldridge, Ph.D.: kimb@sdsc.edu, Jerry Greenberg, Ph.D.: jpg@sdsc.edu
6. Globus: A Metacomputing Infrastructure Toolkit. I. Foster, C. Kesselman. Intl J. Supercomputer Applications, 11(2):115–128, 1997. GLOBUS project: http://www.globus.org/
7. M. Thomas, S. Mock, J. Boisseau, M. Dahan, K. Mueller, D. Sutton. The GridPort Toolkit Architecture for Building Grid Portals. Proceedings of the 10th IEEE Intl. Symp. on High Perf. Dist. Comp. Aug 2001
8. Personal Communication. Yoshio Tanaka: yoshio.tanaka@aist.go.jp
9. The Portland Group, Inc.: http://www.pgroup.com/
10. Veridian Systems: http://www.openpbs.org/
11. The OpenSSL Project: http://www.openssl.org/
12. The Apache Software Foundation: http://www.apache.org/
13. http://java.sun.com/products/jdbc/
14. "General Atomic and Molecular Electronic Structure System" M.W. Schmidt, K.K. Baldridge, J.A. Boatz, S.T. Elbert, M.S. Gordon, J.H. Jensen, S. Koseki, N. Matsunaga, K.A. Nguyen, S. Su, T.L. Windus, M. Dupuis, J.A. Montgomery J. Comput. Chem., 14, 1347–63 (1993). http://www.msg.ameslab.gov/GAMESS/GAMESS.html
15. High Performance Computational Chemistry Group, NWChem, A Computational Chemistry Package for Parallel Computers, Version 4.1 (2002), Pacific Northwest National Laboratory, Richland, Washington 99352, USA. http://www.emsl.pnl.gov:2080/docs/nwchem/nwchem.html
16. T. Nakano, T. Kaminuma, T. Sato, K. Fukuzawa, Y. Akiyama, M. Uebayasi, K. Kitaura, Chem. Phys. Lett. 351 (2002) 475–480.

Extensions to Web Service Techniques for Integrating Jini into a Service-Oriented Architecture for the Grid

Yan Huang and David W. Walker

Department of Computer Science
Cardiff University
PO Box 916
Cardiff CF24 3XF
United Kingdom
{Yan.Huang,David}@cs.cardiff.ac.uk

Abstract. This paper discusses how to adapt Jini to create an OGSA-compliant system for Grid computing by introducing Web services techniques into the Jini system. Service Workflow Language (SWFL), an extension to WSFL for describing jobs composed of interacting Web services, is presented. SWFL provides a simple and succinct way of describing the conditional and loop constructs of Java, and supports more general data mappings than WSFL. In addition, a tool for automatically generating Java code to execute a composite job described in SWFL is described.

1 Introduction

This paper describes how Jini services can be integrated with Web services within a common Service-Oriented Architecture (SOA) for Grid computing. This common SOA permits the transparent interaction of Jini services and Web services, thereby extending the usefulness and applicability of both approaches. The main benefits of this work are: (1) Jini services will be accessible from outside of a Jini community; (2) Jini services will be invocable in the same way as other Web services; and, (3) Jini services will be able to be integrated with other Web services.

Web services have arisen as an essential component of the infrastructure of e-Business, and enable business-to-business transactions via the Internet. In general, these B2B transactions take place directly between computer programs, rather than between computer programs and users.

A Web service has five essential attributes [5]: It can be *described* using a standard service description language, usually Web Service Description Language (WSDL) [1]; it can be *published* to a registry of services, usually a UDDI (Universal Description, Discovery, and Integration) registry; it can be *discovered* by searching the service registry; it can be *invoked*, usually remotely, through a declared API; and, it can be *composed* with other Web services.

P.M.A. Sloot et al. (Eds.): ICCS 2003, LNCS 2659, pp. 254–263, 2003.

A WSDL document describes one or more Web services, each of which is made up of multiple messages, port types, and bindings. A message gives an abstract definition of the input and output data of a Web service. A port type is used to describe the functionality of a Web service in terms of a set of operations. A binding associates a concrete protocol and data format with a port type. The Simple Object Access Protocol (SOAP) is a widely used protocol for Web service messaging, and uses an XML data encoding and HTTP transfer protocol. The binding may also specify the security mechanism for the port's communications.

The Jini networking system aims to support the rapid configuration of devices and software within a distributed computing environment [2]. These devices and software are made available to remote clients as Jini services. Jini is one of the network technologies that are suitable for building the middleware for Grid computing. A central theme of Grid computing is the sharing of resources within a virtual organization through direct transactions between computer programs [4]. This has led to the emerging concept of a Grid service and to the Open Grid Services Architecture (OGSA) which is currently being developed on the basis of Web service concepts and technologies [3]. In this approach, Grid services are regarded as specialized extensions of Web services, and support new types of problem-solving applications that are composed of services. In this paper, a Jini-based Grid will be developed into a OGSA-compliant system. This is done by combining Web service and Jini service SOAs.

Jini's SOA is very similar to that used by Web services. By using Jini lookup services, Jini services can be published, discovered, and invoked and so possess three of the five essential attributes of a Web service enumerated at the start of this section. However, the other two of the five essential attributes, service description and composition, are not part of the basic Jini system. In addition, Web services use XML-based messaging which is not prescribed (nor prohibited) by the Jini communication model.

The simplest way to combine the Web service and Jini service SOAs is by registering services using both the service registry and lookup service mechanisms. The service registry works in the world external to the Jini community, and the lookup service works within the Jini community. Thus, correspondingly, there are two sorts of publish operations, two sorts of find or discovery operations, and two sorts of bind or invoke operations.

Henceforth, a request for a Jini service through the non-Jini mechanism will be called an *indirect service request*, and a request for a Jini service through a Jini lookup service will be called a *direct service request*. An indirect service request finds the service requested by sending a query based on the service description to the service registry and a task description based on a service flow language is formed, also based on the service description. An agent service called the *Workflow Engine* accepts the indirect service request, translates the request into a service request understandable by Jini, and makes sure the request is carried out and the results sent back to the requestor.

2 Service Workflow Language

In many cases it is desirable to create a job or application by composing multiple services. Such applications can be represented by a graph of interacting services that must be specified in a job description[1]. Thus, not only a service description language, but also a job description language, are needed to standardize the description of both services and composite jobs.

In May 2001, IBM's WSFL [6] and Microsoft's draft of XLANG [7] were released as languages for describing the composition of Web services. Although the intended uses of WSFL and XLANG are broadly the same, they have completely different structures: WSFL directly represents the control and data flow of an application in terms of its workflow graph; XLANG is closely based on the use of Java-like language constructs to describe a job. In August 2002, a combined version of WSFL and XLANG, called Business Process Execution Language for Web Services (BPEL4WS) [9], was published that largely inherited the programming-oriented approach of XLANG, rather than the graph-based approach of WSFL. BPEL4WS is effective for representing service-based composite applications for which the order of execution of the services is pre-defined. However, it does not have the same flexibility as WSFL, in which the only constraints on the order of execution of services are implicit in the workflow graph. Having a pre-defined service execution order, as in XLANG and BPEL4WS, is not suitable for representing a distributed application, where the ability to dynamically partition and schedule services at runtime is important in order to exploit potential parallelism and to make best use of the available distributed resources. WFSL, however, does allow this capability, and hence provides a flexible and effective basis for representing a Grid application.

Service Workflow Language (SWFL) extends WSFL in two important ways:

1. SWFL improves the representation of conditional and loop control constructs. Currently WSFL can handle *if-then-else*, *switch*, and *do-while* constructs and permits only one service within each conditional clause or loop body. SWFL also handles *while* and *for* loops, and permits sequences of services within conditional clauses and loop bodies.
2. SWFL permits more general data mappings than WSFL. SWFL can describe data mappings for arrays and compound objects.

SWFL can describe the conditional and loop control constructs of the Java programming language. The motivation for SWFL was to develop a tool for automatically converting the workflow description of a composite job into a Java program for running it. This tool, `SWFL2Java`, is also described in this paper.

As an example, consider loop control constructs. WSFL provides loop control flow through the optional `exitCondition` attribute of an activity. Exit conditions are represented with an XPath-based syntax. If on completion of the activity the exit condition evaluates to false, then the activity is run again. This continues until, after the activity has been run some number of times, the exit

[1] Here a job is a composition of interacting services.

condition evaluates to true, after which control flows to the next activity. For example:

```
<activity name="LoopActivity"
          exitCondition="XPath1 &gt; XPath2"/>
```

For clarity, attributes not relevant to the example have been omitted. The activity `LoopActivity` will be run repeatedly until the exit condition evaluates to true. This is equivalent to the following Java code:

```
do {
    LoopActivity ();
} while (X1 > X2)
```

Now consider the *for* loop. Since a *for* loop can be rewritten in terms of a *while* loop, and since the latter can be represented in WSFL, then it would appear that a *for* loop can also be represented in WSFL. However, there is a problem. Suppose we have a *for* loop that iterates on a certain activity. Then the body of the corresponding *while* loop contains that activity, followed by an new activity that simply updates the loop control variable. In WSFL an exit condition can be associated with only one particular activity, so there can be only one activity within a loop. Furthermore, if the activity in a *for* loop takes the loop control variable as one of its inputs, which is often the case, WSFL provides no way to represent the internal dataflow inside the body of the loop.

The difficulties in representing certain types of conditional and loop control constructs in WSFL have led us to develop SWFL. SWFL is a new job description language that makes three main extensions to WSFL, affecting the elements `wsfl:activityType`, `wsfl:controlLinkType`, and `wsfl:dataLinkType`, and the definition of a new element, `SWFL:jmapType`. In SWFL, conditional and loop processes are treated as special activities, which allows their data and control flow to be defined more clearly. They are added into the `jactivityType` element. The resulting `jactivityType` XML schema is as follows:

```
<xsd:complexType name="jactivityType">
    <xsd:choice>
        <xsd:element name="normal" type="wsfl:activityType"/>
        <xsd:element name="while" type="loopType"/>
        <xsd:element name="dowhile" type="loopType"/>
        <xsd:element name="for" type="loopType"/>
        <xsd:element name="if" type="controlType"/>
        <xsd:element name="switch" type="controlType">
            <xsd:key name="CasePortName">
                <xsd:selector xpath="case"/>
                <xsd:field xpath="@port"/>
            </xsd:key>
        </xsd:element>
    </xsd:choice>
    <xsd:attribute name="operation" type="NCName"/>
</xsd:complexType>
```

In the above `jactivityType` schema, the activity type is extended to six kinds of activities. These are: `normal`, `if`, `while`, `dowhile`, `for` and `switch`. The `normal` activity is the same as an activity defined in WSFL, and is of type `wsfl:activityType`. The others are newly-defined activity types corresponding to the conditional and loop constructs in the Java programming language.

The `for`, `while` and `dowhile` activities are all of type `loopType`. `loopType` is an extension of `wsdl:operationType` through the addition of a new element called `expression` and a new attribute called `setParallel`. A loop activity has `input` and `output` elements to specify its input and output data. In *while* and *do-while* loops the element `expression` is a Boolean expression based on the `input` message of the loop activity. However, in the case of a *for* loop, the `expresssion` element is represented by three statements separated by semicolons in the form: *initial statement ; Boolean expression ; increment statement.* This Java-based format currently allows only simple *for* loops of the type given is the following example. The loop control variable used in the statements is from the `input` message of the activity. Specifying the loop control variable as one of the parts of the `output` message of the loop activity allows the activity within the loop, as well as activities after the loop, to accept the loop control variable as input. As an example, consider the following `for` loop activity:

```
<wsdl:message name="forloopMessage">
    <wsdl:part name="out0" type="string"/>
    <wsdl:part name="out1" type="int"/>
    <wsdl:part name="index" type="xsd:integer"/>
</wsdl:message>

<activity name="forActivity">
    <for setParallel="no">
        <input message="forloopMessage"/>
        <output message="forloopMessage"/>
        <expression>
            <![CDATA[index=0;index<100;index++]]>
        </expression>
    </for>
</activity>
```

In this example there is a `part` element named `index` in the message input to the `for` activity which is used in the `expression` of the activity. The `setParallel` attribute of the activity is used to indicate whether the iterations of the `for` activity can be performed in parallel. If `setParallel` is set to "`yes`" a scheduler could arrange for the loop iterations to be done in parallel on different machines that provide the service in the loop body.

Normally an activity has one control input port and one control output port, but conditional and loop activities have multiple output control ports. A loop activity has two output control ports. As long as the loop condition is satisfied, control flows to the sequence of activities inside the loop body; otherwise, control flows to an activity after the loop body.

For if and switch activities there is a control output port for each conditional clause. The following is an example of a *switch* statement in Java:

```
switch(int0){
    case 10: ...; break;
    case 20: ...; break;
    case 30: ...; break;
    default: ...;
}
```

In this example, the switch has four cases so there are four output control ports to which control can flow on exiting the activity. The corresponding description in SWFL is as follows:

```
<SWFL:switch expression="int0">
    <SWFL:input name="switchInputMessage" message="switchMessage"/>
    <SWFL:output name="switchOutputMessage" message="switchMessage/">
    <SWFL:case port="0">10</SWFL:case>
    <SWFL:case port="1">20</SWFL:case>
    <SWFL:case port="2">30</SWFL:case>
    <SWFL:defaultCase port="default"/>
</SWFL:switch>
```

The attribute expression specifies the expression controlling the switch statement which must be of type byte, short, int, or char, and is from the input message of the switch activity. The value of each case must be a literal with the same type as the expression attribute. The port attribute of the case element is used to specify for that case the port of the switch activity that a control link flows out of.

The if conditional activity is very similar to the switch activity, as may be seen in the following example an if activity defined in SWFL:

```
<SWFL:if>
    <SWFL:input name="ifInputMessage" message="ifMessage"/>
    <SWFL:output name="ifOutputMessage" message="ifMessage/">
    <SWFL:case port="0"><![CDATA[int0 < 0]]></SWFL:case>
    <SWFL:case port="1">int0==0</SWFL:case>
    <SWFL:defaultCase port="default"/>
</SWFL:if>
```

An if activity does not have an expression attribute. Instead, the value of the case element is a Boolean expression composed from data in its input message and from literals.

The introduction of SWFL:controlType and SWFL:loopType activities having multiple control output ports, makes it necessary to specify which port of the source activity a control link flows out if the source activity is a conditional or loop activity. SWFL:jcontrolLinkType extends wsfl:controlLinkType by adding an optional attribute called controlPort to the controlLink element which is used only when the source activity of the control link is a conditional or loop activity. It specifies which control output port of the source activity is the source of the control link.

The WSFL syntax for data mapping is sufficient to define the mapping in a data link when the mapped data is a one-layer `complexType` (i.e., the elements in the `complexType` are `simpleType`), and is not an element of an array. This means that WSFL can handle only cases in which the input to an activity is a primitive datatype or an object containing primitive datatypes. Also WSFL allows only one data mapping in a data link. This means that an activity can accept data from only one source activity. SWFL provides a new definition of data mapping, `SWFL:jmapType`, that overcomes these limitations.

In the specification of `SWFL:jmapType` the attributes `sourceMessage` and `targetMessage` specify the source message and the target message of a data link to which the data map applies. A `SWFL:jmapType` can have multiple `part` elements, each corresponding to a different data mapping. A `part` has two sub-elements of type `SWFL:dataPartType`, named `sourcePart` and `targetPart`, which are used to specify a particular data element in a message. `SWFL:dataPartType` specifies a particular path leading to the final data element involved in the data mapping. It can contain any number of `item` elements. An `item` element has either a `field` element to specify a field of a complex type, or an `index` element to specify an element of an array.

In the example of `SWFL:mapType` below there is one data mapping: the data `u1.a.b[5]` is mapped to `u2.c` in Java notation. `u1` is a `part` of the `sourceMessage`. The mapped data is the 5th element of the array `b` which is a `field` of `a`. `a` itself is a `field` of part `u1`. `u2` is a `part` of `targetMessage` and `c` is a `field` of `u2`.

```
<SWFL:jmap sourceMessage="sourceMessage" targetMessage="targetMessage">
   <SWFL:part name="part0">
       <SWFL:sourcePart name="u1">
           <SWFL:item><SWFL:field name="a"/></SWFL:item>
           <SWFL:item><SWFL:field name="b"/></SWFL:item>
           <SWFL:item><SWFL:index dimension="0" index="5"/></SWFL:item>
       </SWFL:sourcePart>
       <SWFL:targetPart name="u2">
           <SWFL:item><SWFL:field name="c"/></SWFL:item>
       </SWFL:targetPart>
   </SWFL:part>
</SWFL:jmap>
```

3 Implementation Aspects

We have discussed how a job composed of interacting services can be represented by the SWFL job description language. The main motivations for developing SWFL were to describe Java-oriented conditional and loop constructs, to permit sequences of more than one service within conditional clauses and loop bodies, and to overcome limitations inherent in WSFL's data mapping approach. In addition, we have developed: (1) a tool called `SWFL2java` that converts the description of a job in SWFL into executable Java code; and, (2) a Workflow Engine that provides an execution environment to run the composite job.

The details of `SWFL2Java` are discussed elsewhere [8], however, we will give here an overview of the implementation. In `SWFL2Java` a SWFL document is not translated directly into a Java program but is stored in an intermediate form as a Java `FlowModel` object. This is made up of two Java `Graph` objects: `DataGraph` and `ControlGraph`. The former stores the data flow structure of the flow model, and the latter stores its control flow structure. One reason for storing the job description in this intermediate form is to be able to interact readily with Java-based tools for the visual composition of Web services. In such tools a composite job is represented as a graph in which activities/services correspond to nodes, and data links and control links correspond to different kinds of directed edges. The graph can be stored as a `FlowModel` object and converted to and from SWFL, as well as into a Java program. Another reason for using the `FlowModel` form is to reduce the overhead when the same job is used many times and scheduled on different resources. In such cases it is easier to generate the Java code from the intermediate form rather than starting from the original SWFL.

Given a graphical `FlowModel` of arbitrary structure to be transformed into a Java Jini-based distributed program, the following three issues have to be addressed: how to find all the services; how to decide the order of execution of services based on the flow model; and, how to automatically name all the variables, class names, and methods.

Normally the first thing a Jini-based distributed program does is to discover the services that are going to be used in the program. Thus, to automatically create a Java Jini-based program from a graphical flow model, service discovery is the first issue to be addressed. Service discovery is performed by an assistant class called *ServiceFinder*. The *ServiceFinder* thread takes a list of services and discovers each using a *ServiceDiscoveryManager*. Once a service is discovered, its *ServiceItem* is downloaded and stored.

After all the services needed in the program have been found and stored, the main program is generated. The problem here is determining the order in which tasks are processed by services. This order is deduced from the control graph using the Task Processing Sequential Order Generation algorithm described in [8].

In automatically generating a Java program several naming issues need to be decided. In the automatically produced program, the class name takes the form *FlowModel_XXX* where *XXX* is the name of the flow model. The class has a constructor and an *execute()* method that runs the job specified in the flow model. The parameters of the *execute()* method are determined by `flowSource` in the flow model. The parameter names are the same as the corresponding `part` names of the `flowSource` message. The return variable of an activity takes the form `XXX_return` in which `XXX` is the name of the activity.

Here we provide an example of a Java program automatically created by `SWFL2Java`.

```
import net.jini.core.lookup.ServiceItem;
```

```
import net.jini.core.lookup.ServiceTemplate;
import java.util.HashMap;
import SWFL2Java.XML2Graph.ServiceFinder;
public class FlowModel_Example1{
  private HashMap taskTemplateMap = null;
  private ServiceFinder serviceFinder = null;
  public FlowModel_Example1(){
      this.taskTemplateMap = new HashMap();
      this.setTaskTemplateMap();
      this.serviceFinder = new ServiceFinder(this.taskTemplateMap);
      Thread thread = new Thread(this.serviceFinder);
      thread.start();
      try{
          thread.join(2*60*1000);
      } catch (java.lang.InterruptedException e){
          System.out.println("Failure in finding services.");
          System.exit(-1);
      }
  }
  private void setTaskTemplateMap(){
      this.taskTemplateMap.put("task_1", new ServiceTemplate(null,
          new Class[]{services.Matrix.MatrixInterface.class}, null));
      this.taskTemplateMap.put("task_0", new ServiceTemplate(null,
          new Class[]{services.Math.MathInterface.class}, null));
  }
  public double[][] execute(int size, double[][] B)
                  throws java.rmi.RemoteException{
      for(int in0=0; in0<size; in0++){
          for(int in1=0; in1<size; in1++){
              services.Math.MathInterface task_0 =
                  (services.Math.MathInterface)
                  (this.serviceFinder.getServiceItem("task_0")[0].service);
              double task_0_return = task_0.function(in1 ,in0);
              B[in0][in1] = task_0_return;
          }
      }
      services.Matrix.MatrixInterface task_1 =
          (services.Matrix.MatrixInterface)
          (this.serviceFinder.getServiceItem("task_1")[0].service);
      double[][] task_1_B = task_1.inverse(B);
      return task_1_B;
  }
}
```

4 Conclusions

This paper has discussed several technical issues involved in updating a Jini system into a OGSA-compliant system and has focused on one of the most important issues – defining an XML-based description language for describing composite service-based Grid applications. SWFL, an extension to WSFL, is such a language that describes Java-oriented conditional and loop constructs and enhances WSFL's data mapping facility. Whereas WSFL can represent only *if-then-else*, *switch*, and *do-while* constructs, SWFL can also represent *while* and simple *for loop* constructs. SWFL also permits sequences of more than one activity within conditional clauses and loop bodies. SWFL enhances WSFL's data mapping capabilities by handling compound objects and arrays, and by permitting activities to accept input data from more than one source activity.

Given an SWFL job description, the `SWFL2Java` tool can generate a representation of the corresponding data and control link structure in the form of a Java `FlowModel` object. From this the corresponding Java program can be automatically generated.

SWFL can also be used to describe composite jobs made up of both Jini and Web services. The work described in this paper is part of a larger programme of research to introduce Web service technology into the Jini service architecture, thereby enabling Jini services and Web services to interoperate. Implementation details of this interoperation of services will be given in a subsequent paper.

References

1. Erik Christensen, Francisco Curbera, Greg Meredith, and Sanjiva Weerawarana, "Web Services Description Language (WSDL) 1.1," available as a W3C note at http://www.w3c.org/TR/wsdl/, March 2001.
2. W. K. Edwards and T. Rodden, "Jini Example By Example," published by Prentice Hall, 2001.
3. Ian Foster, Carl Kesselman, Jeffrey Nick, and Steven Tuecke, "The Physiology of the Grid: An Open Grid Services Architecture for Distributed Systems Integration," January, 2002. This is available online at http://www.globus.org/research/papers/ogsa.pdf.
4. Ian Foster, Carl Kesselman, and Steven Tuecke, "The Anatomy of the Grid: Enabling Scalable Virtual Organizations," The International Journal of High Performance Computing Applications, Vol. 15, No. 3, pages 200-222, Fall 2001.
5. K. Gottschalk, S. Graham, H. Kreger, and J. Snell, "Introduction to Web Services Architecture," IBM Systems Journal, Vol. 41, No. 2, 2002.
6. F. Leymann, "Web Services Flow Lauguage (WSFL 1.0)", available at http://www-4.ibm.com/software/solutions/webservices/pdf/WSFL.pdf, May 2001.
7. S. Thatte, "XLANG: Web Services for Business Process Design." Available at http://www.gotdotnet.com/team/xml_wsspecs/xlang-c/default.htm, 2001.
8. Yan Huang, "The Role of Jini in a Service-Oriented Architecture for Grid Computing," PhD thesis, Department of Computer Science, Cardiff University, December 2002.
9. F. Curbera, Y. Goland, J. Klein, F. Leymann, D. Roller, S. Thatte, S. Weerawarana "Business Process Execution Language for Web Services", available at http://www-106.ibm.com/developerworks/webservices/library/ws-bpel/, August 2002.

Multiple-Level Grid Algorithm for Getting 2D Road Map in 3D Virtual Scene

Jiangchun Wang[1], Shensheng Zhang[1], and Jianqiang Luo[1]

[1]CIT Lab, Department of Computer Science and Engineering
Shanghai Jiao Tong University, P.R. China
scott751111@sjtu.edu.cn

Abstract. It is a common requirement to build a 2D map of a 3D virtual scene for path planning, but it is difficult or tedious. So we propose a multiple-level grid algorithm to solve the problem. The original idea lies in the new method on distinguishing a grid unit's accessibility. We argue for an approach by employing two navigating lines, which are right-angle intersection and stand for the unit's accessibility in four directions. Moreover, another advantage of it is to dynamically divide the virtual scene into different sizes of grids at different precisions. With these multi-precision grids, we can approach objects at any granularity. Lastly, experiments are performed to test the effectiveness of the algorithm.

1 Introduction

Recent advances in graphic engines and software tools have facilitated the development of visual interfaces based on 3D virtual environments (VEs) [1]. These interfaces use interactive 3D graphics to represent visual and spatial information and allow natural interaction with direct object manipulation, such as VRML (Virtual Reality Modeling Language) [2]. People can get a lot of shared resource from the Internet to construct their own virtual world. Moreover, when a virtual scenic file is acquired, they usually also need assistant information, such as the scenic 2D map, for imbedding and navigating our own virtual objects. Unfortunately, it happens rarely. A solution probably uses the projection image, the screen snapshot, from top view in the virtual environment browser. But this simple way comes with so much misunderstanding that it is not reliable. For example, a room, which may be entered in, is displayed as impassable object in the top view projection image. To acquire a real 2D road map, this paper proposes a Multiple-level Grid Algorithm.

1.1 Related Work

Let us briefly talk about some previous work on this problem. To draw virtual scenic map is similar to drawing real world map. Many researches had been completed on how to get these practical maps [3,4,5]. The aerial survey mapping is efficient way for large area survey like the top view projection method in our virtual environment.

P.M.A. Sloot et al. (Eds.): ICCS 2003, LNCS 2659, pp. 264–274, 2003.

Many advantages had been presented in document [6]. But this method is unsuitable for our problem, for so much redundant information acquired probably. For example, an image of the top view projection would show some area as road area because of no objects here. But in fact, these areas maybe are embraced by other obstacles and can never be arrived at. We only want to know where can be reached in a virtual scene. The superabundance can only mislead us. To ransack a scene to find a road map belongs to the traversal problem. Many traversal algorithms had been proposed in the graph theory. Document [7] is a good reading material for this problem. Koucky proposed a universal traversal approach [8]. In some documents [9,10], this problem is called 'terrain acquisition'. Gonzalez had proposed a complementary regions algorithm for such a surface filling [11]. But their approaches are limited at precision for relying on the size of a gauge, which may be a mobile robot or a block. Our problem has partial similarity with the all of above, and its special attributes are described in next section.

Our method is based on the simulation of user's behavior. When a user is navigating in a virtual scene from a browser, such as a VRML browser, the interactive device is commonly a keyboard. The OS receives user input and only sends some keyboard messages to the browser. For example, Up-key message stands for user's idea of 'advancing', Left-key message means 'turning left', and so on. Therefore, we can substitute for OS to send same messages to the browser to simulate user behavior. Our program simulates a user movement in the virtual scene. With the help of embedded JAVA applet in the VRML file, the map recorder can receive the motion data of the user's viewpoint. Figure 1 illustrates this architecture. So the trail of user's viewpoint stands for the passage. How to compose a map with these points is our kernel problem. Because we have little interest in the difference of the motion trail on altitude, such difference is neglected and planar data are only kept, such as recorded x, z coordinates and neglected y (means height in the VRML).

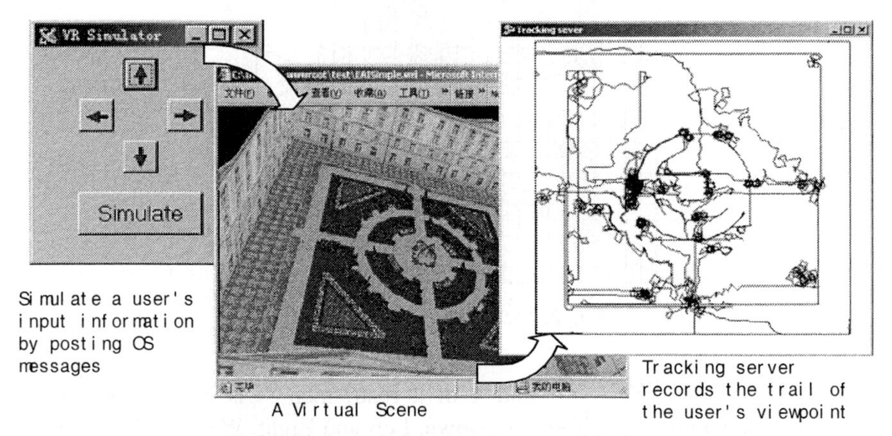

Fig. 1. The way for getting the accessible data

3 Virtual Terrain Acquisition Problem

In this section, we present the virtual terrain acquisition problem of the map in the virtual scene. In a 3D virtual scene, we are asked for to get out its 2D planar path map. Our problem is different from other path-plan problems [12,13], because the scenic environment is almost unknown and we need not only get out where are obstacles or roads, but must cover all position in it. The scenic information is very little except the scenic scope. While our avatar is walking, the data of viewpoint position can be acquired. The data don't stand for a set of small regions, but a set of sample points, which are infinitesimal. Due to the difficulty of walking on every point in the scenic map, it is impossible to use exhaustive approach to distinguish an obstacle or a road. So we propose a multiple-level grid algorithm to deal with it.

3.1 Problem Formulation

We assume that if a point x is of a road, its *neighborhood* $\Gamma(x)$, which is tightly close to it, is also of the road. Firstly, we divide the map into many small rectangles or blocks, which are of the same size. Figure 2 illustrates that.

Before we formulate the problem, let us define the following notation:

- n = Number of small rectangle in a dividing grid.
- a_i = The ith small rectangle in a dividing grid where $i = 1,2,...,n$.

If the n is big enough, in other words, there are enough many small rectangles divided from a scenic map, we can think the rectangle region as a road or an obstacle on the condition that whether or not a viewpoint can step into it from its adjacent rectangles. Here we define the neighborhood $\Gamma(x)$ of x, any point in the scenic map, as the small rectangle a_i:

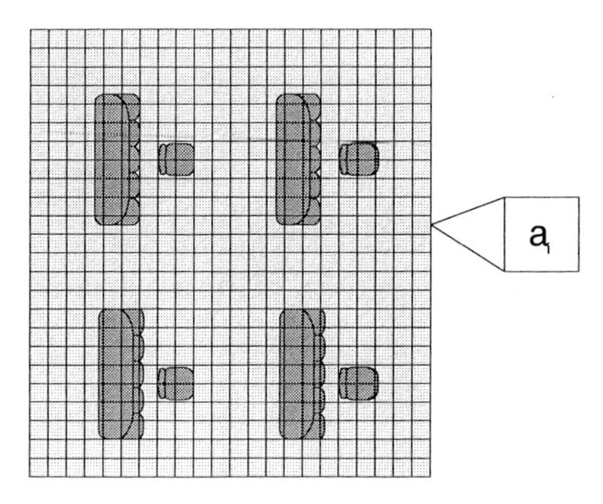

Fig. 2. The dividing grid for a map

If $x \in a_i, i = 1,..,n(n - \infty)$ and x of a road, then a_i is of a road, and vice versa. There are 4 edges in a_i — Up, Down, Left and Right. We statistically know if a user can steps into a small rectangle a_i in these four directions along with the mid-points of every edge, then, a_i is almost of a road. So we simplify the determinant with a definite precision. Figure 3 illustrates some samples to tell which rectangle region is a road or an obstacle.

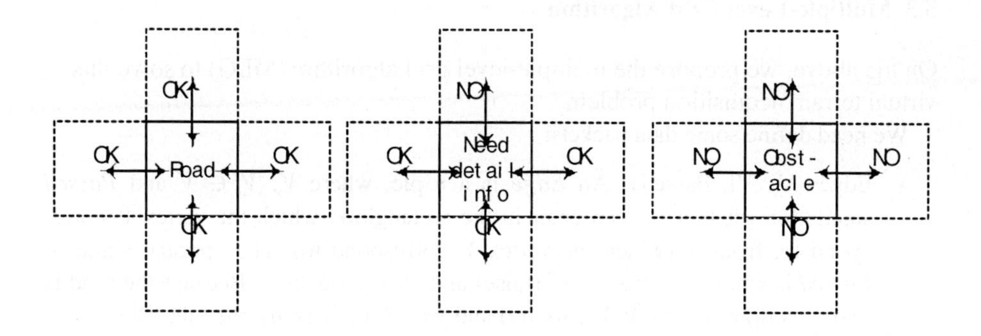

Fig. 3. Samples for describing the rectangle property of accessibility

Then, we use a graph notation to present a scenic map. Given a graph $G = (V, E)$, V represents the set of small rectangles in the dividing grid; E represents the set of edges such that an edge $e_{i,j} \in E$ represents that the small rectangle a_i is next to a_j. Because the number of small rectangles adjacent to a_i is 4 at most, we can construct such a graph like Fig. 4. The problem is converted to solve the accessibility of every edge in graph G.

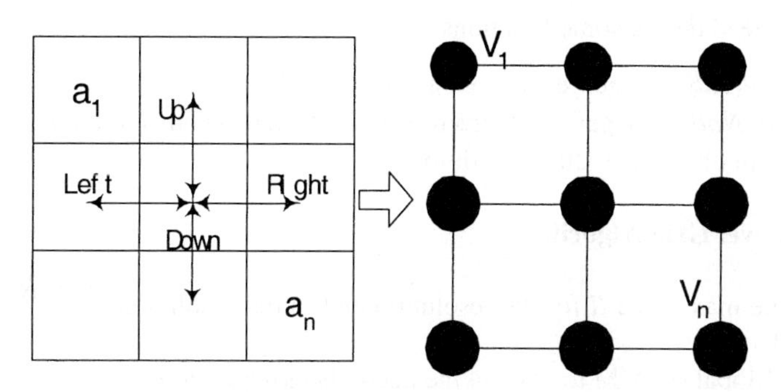

Fig. 4. Transferring from a grid to a graph

3.2 Random Walk Approach

One way to work out every small rectangles state is by random walk approach, that is, given n rectangles of the whole grid, which is a n-vertices graph, and then the view-point can start at any point and jump to any edge at every node. Let $|n|$ denote the total linked edges for detecting every node; thus, here is 4^n. If scenic scope is 100*100 and the required precision is 0.1, so $n = 10^6$, the random walk approach impossibly reach our anticipation.

3.3 Multiple-Level Grid Algorithm

On the above, we propose the multiple-level grid algorithm (MLG) to solve this virtual terrain acquisition problem.

We need define some data packets:

- Edge=(V_i, Vj, Passed) An *Edge* is a triple, where V_i, $Vj \in V$ and *Passed* stands for whether we can enter in the rectangle a_j, which the vertex V_j correspond to, from a_i (which the vertex V_i correspond to). The optional value of *Passed* is among '*T*'(true)□'*F*' (false) and '*U*'(untouched). Because the road is two directions, if $x=(V_i, V_j, Passed)$ and $y=(V_i, V_j, Passed)$, then the *Passed(X)* = *Passed(Y)*.
- Node=(V_i,Front_Edge,Back_Edge,Left_Edge, Right_Edge ,Status) An *Node* includes seven elements, where $V_i \in V$; *Front_Edg*, *Back_Edge, Left_Edge, Right_Edge* are the type of *Edge*. *Front_Edge* stands for passing-through nature between V_i and its adjacent vertex in the up direction, and *Down_Edge* is for the down direction, the third for left, the fourth for the right; the *Status* is for whether the rectangle is region of a road or an obstacle, which value is among '*R*'(road), '*O*'(obstacle), '*NFI*'(need detail information), or '*U*'(undone).

We also need define some functions:

- *Passed(Edge)* It get result from a *Edge* to tell if the *Edge* can be pass through.
- *Status(Node)* It get result from a *Node* to tell whether the rectangle is of a road, an obstacle or other conditions.

Multiple-Level Grid Algorithm:

1. begin

2. Define the max value *R* for the resolution and initial resolution $r_0 < R$, and iteration K = 1;

3. Create a global stack M to contain the need-checking region;

4. Let $r = 1, i = 0$ and push the whole scenic map as a graph, denoted by A_0, and a start point s_0 into stack M;

5. **while** ($r < R$){

6. $i = i - 1, r = r \cdot r_0, j = 0$;

7. **while** (Stack M is not empty){

8. Get out a graph A_j from the stack M;

9. Use the *grid-covering* algorithm to deal with A_j at the resolution r and start point s_i, then work out the status for every node of the graph.;

10. $j = j - 1$;}
11. $j = 0$;
12. **For** every node of all graphs {
  Uses the *adjacent-edge rule* to check them.;
  If a node status is NFI **then**
  { $j = j - 1$, mark it with A_j and push it and a point s_i ($s_i \in A_i$) into stack M;}
  } }
13. Integrate all nodes of every level to compose a whole road map *A* for the scene;
14. Scan all divided smallest rectangles in A to find a new start point s_K.
  { **If** found(s) **then**
    let $s_0 = s_K$ and K=K+1, goes to step 4
  Else
    final road map A is result;
  /* The smallest rectangle must be on border including the new start point. */ }
15. **end**

The *multiple-level grid algorithm* is iterative, the complexity in an iteration cycle is $O(n)$. The *algorithm* has two key components, namely, 1) the *grid-covering algorithm* (GC), and 2) the *adjacent edge rule* (AER). The GC algorithm uses a dived-and-conquer approach to filter out road blocks and obstacle blocks. The AER is based on the experiential assumption. In what follows, we describe each of these two components.

3.3.1 Grid Covering (GC) Algorithm

The main idea of the grid-covering algorithm is to divide the scene of a virtual environment into grid array $A(a_j \in A, j = 1,2,..., n)$ and then to construct the corresponding graph $G = (V, E)$. In the grid- covering algorithm, we first set the resolution r and a start position s_0, which are inputted from outside, for this level grid.

Grid Covering Algorithm:

1. **begin**
2. According to the scenic size and the resolution r, divide the scene with a set of small rectangles $a_i (i = 1,2,..., n)$;
3. According to a_i, construct a set of vertexes to compose an array, denoted by *V*, and create a new array of *Node*, denoted by M. The number of elements in V is n, and same to array M. V_i of $M[p]$ $(p = 1,2...,n)$ is belong to V;
4. Set initial value for array M;
  /*Set *Status* of $M[p](p = 1,2...,n)$ with '*U*';*/
5. From start vertex V_0 (It's a seed), which includes the start position s_0. Let $V_{curr} = V_0$;
6. From *Front_Edge* to *Right_Edge* of V_{curr}, check its *Edge* element *X*(*X*∈ *Front_Edge*, *Back_Edge*, *Left_Edge*, *Right_Edge*).

If *Passed*(*X*)=*U* then goes to step 6, or goes to step 8;

7. Check user can enter in V_{curr} 's neighbor V_{next} through the *Edge X*. If can not, set variable *Passed* of *Edge X* and V_{next} 's corresponding *Edge* element with value '*F*', and goes to step 5; set variable *Passed* of *Edge X* and V_{next} 's corresponding *Edge* element with value '*T*';
8. Let $V_{curr} = V_{next}$, goes to step 4;
9. If *Edge X* is *Right_Edge*, backwards to the previous *Node* V_{pre} . Let $V_{curr} = V_{pre}$ and goes to step 9, else goes to step 4;
10. If $V_{curr} = V_0$, goes to step 10, or goes to step 4;
11. **end.**

The purpose of the grid-covering algorithm is to get all real connections of every block in the grid. In the view of graph, we want to produce a connected graph so that we can perform the checking step by the adjacent-edge rule. This algorithm can check every node, which stands for the accessible area.

3.3.2 Adjacent-Edge Rule

The idea of the *adjacent-edge rule* is described as follows:

In the view of graph, if a node V_i can be reached from all the $e_{i,j} \in E$, we think that it stands for a road area; if just from some $e_{i,j} \in E$, we think it need-checking with detail information; if from none, it must be an obstacle area. We name the method as an adjacent-edge rule. Figure 5 illustrates these instances.

The function *Status(Node)* is based on above. After the implementation of GC algorithm, we set nodes state obeyed the *adjacent-edge rule.*

3.3.3 The Complexity of the Multiple-Level Grid Algorithm

The *multiple-level grid algorithm* is a recursive approach based on the GC and AER. GC algorithm makes a viewpoint move in the scenic graph along with a definite route. The most time cost exists in walking away from all edges. The dropping back step needs reset the viewpoint's position, which also shares cost, and then sometimes the viewpoint need backward many nodes to find a new acceptable edge. For discussing the complexity of our algorithm, we define these notations as following:

- $N_k = p_k \times q_k$ This means in the k th level grid, the whole region is divided into p_k rows and q_k columns of rectangles.
- a_k = The average cost of passing through an edge between two nodes in the k th level grid.
- φ_k = The average cost of the dropping back step in the k th level grid.

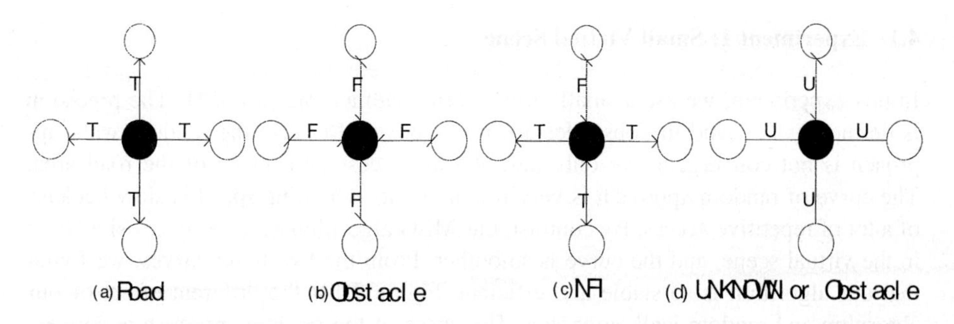

Fig. 5. (a) That all connections of a node are 'T' means it is a road area. (b) That no connections of a node are 'T' means it is an obstacle area. (c) That there are different statuses among connections of a node means it is NFI (need further information for checking). (d) That all connections of a node are 'U'(untouched) means that it is isolated as an obstacle in the smallest grid, or is unknown in other level grids

Lemma 1. *The complexity of Grid Covering Algorithm is* $O(N)$.

Proof. In the *i*th level grid, the all edges is $4N_k - 2p_k - 2q_k - 12$. When a collision happens, we need reset viewpoint to an acceptable edge. The number of reset times is same as the number of edges at most. The whole cost is based on the following formula:

$$C_k = (a_k - \varphi_k)\times(4N_k - 2p_k - 2q_k - 12) \tag{1}$$

Therefore, the complexity of Grid Covering Algorithm is $O(N_k)$.

GC algorithm is recursively used in the *multiple-level grid algorithm*. Therefore, the complexity of the *multiple-level grid algorithm* is $O(\sum N_k)$ at most. As we known,

$$\sum N_k = n \tag{2}$$

So the complexity of an iteration cycle in the *multiple-level grid algorithm* is $O(n)$; and in the worst case, the complexity is $O(n^2)$, which needs n iteration cycles. If we only check the untouched Edges in every iteration cycle with the help of status marks, the complexity will be decreased to $O(n)$. In fact, because of many similar areas, only a few rectangles need to be divided into small enough pieces. The convergence speed of our algorithm is higher than theoretic one.

4 Experiments

In this section, we select the VRML 97 and Cortona VRML Client 4.0 [14] on HP workstation x4000 as our test-bed. To evaluate the algorithm discussed in the previous section, we apply it to the different scales of virtual environment. For the small virtual scene, since the scenic space is manageable, we can give the comparison of the behavior of our algorithm with a random approach. In the large virtual scene, we compare the convergence speed of it with different number and size of obstacles.

4.1 Experiment 1: Small Virtual Scene

In this experiment, we use a small virtual scene with a dimension 1*1. The precision is defined as 0.1, and the obstacles are five boxes. Because the random walk approach is not convergent, we only describe the access percentage of the road area. The curve of random approach is very rough and its searching speed is slow because of a lot of repetitive access. By contrast, our MLG algorithm accesses every site faster in the virtual scene, and the curve is smoother. From the two trend curves, we found that our algorithm is so stable and efficient. Figure 7 list the different effect of our algorithm and random walk approach. The curve of the random approach is convex. With time elapsing, the curve is approximate to the end line (at 100%) infinitely. The approach works out the data of road area fast at the beginning, but later it is running more and more slowly. The curve of our MLG algorithm is concave and convergence. It arrives at the end line stably.

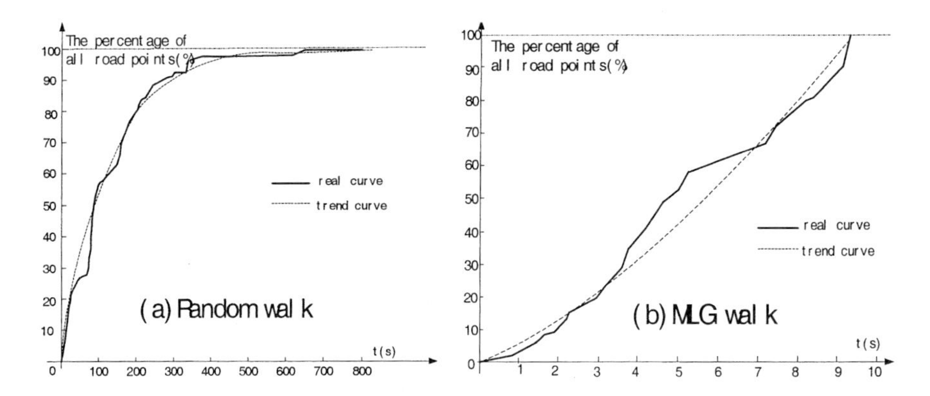

Fig. 6. The comparison of walk effects between two approaches in a small virtual scene

4.2 Experiment 2: Large Virtual Scene

Here, we use a large virtual scene with a dimension 100*100. The precision is also defined as 0.1. We vary the number and size of obstacles to compare the behavior of the multiple-level grid algorithm. In Fig. 7, we apply MLG as two levels of grid to a virtual scene. The horizontal axis stands for different ratio, which is 10^{-4}-10^{4}, between n_1 and n_2. (n_1 is the number of rectangles in the first level grid; n_2 is the number of rectangles in the first level grid.) The left vertical axis stands for its convergence time and the right vertical axis stands for accuracy ratio. The accuracy curve fluctuates in a limited range (96-100%). By contrast, the convergence-time curve changes much more (219-1250s). We recommend that a suitable r be in [10,100].

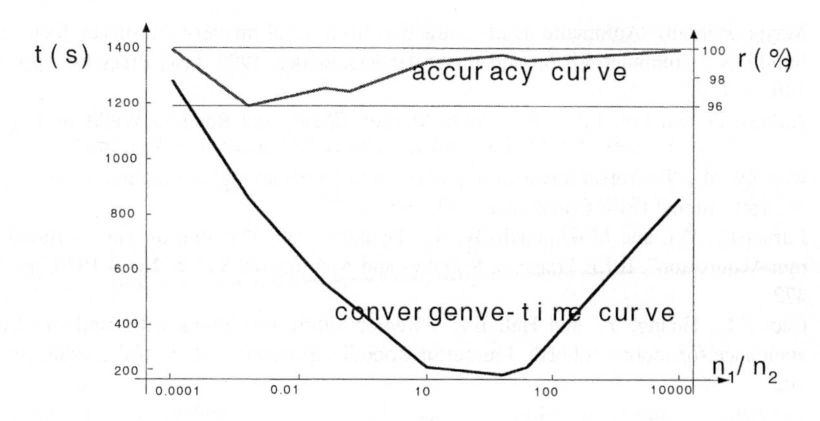

Fig. 7. The behavior of MLG algorithm in different large virtual scene

5 Conclusion and Future Work

This paper deals with the problem extracting the 2D planar map from 3D virtual scene to facilitate using the shared source of the virtual reality scene on Internet. An efficient algorithm is presented and experiments are carried out to testify its efficiency. We also adapt the algorithm to parallel domain. The algorithm is good at the convergence and reliable to versatile scenes with the acceptable recognition accuracy.

But our method only thinks about the same ground plane without caring for the little difference on the altitude, and we get the rough data of a map. Further researches on this project are how to composite a useful map with them and how to navigate the intelligent robot in a virtual scene with the map. The current research is good test-bed for robots navigate approach.

Acknowledgements. This work was supported by China National Science Foundation under grant No: 59789502, and by the National High Technology Plan 863/ CIMS under the grant No: 863-511-030-007-9.

References

1. Tiziana Catarci, Thomas, "Using 3D and Ancillary Media to Train Construction Workers", Multimedia at Work, April 2002, pp. 88-92
2. http://www.cs.nps.navy.mil/people/faculty/capps/4473/projects/VRML
3. Illert, A. "Automatic Digitization of Large Scale Maps", Technical Papers, ACSM ASPRS Annual Convention, 1991 (6)
4. Datta, A. and Parui, S.K., "A Robust Parallel thinning Algorithm for Binary Images". Pattern Recognition, 1994, Vol. 27, No. 9,
5. Krakiwsky, Edward J., Mueller, Ivan I., "Toward world surveying and mapping education". Report on the XIIIth North American surveying and mapping teachers conference, ACSM-ASPRS Annual Convention, Vol. 2., 1991, pp. 160–169

6. Mayer, Helmut, "Automatic object extraction from aerial imagery – a survey focusing on buildings", Computer Vision and Image Understanding, 1999 Acad Press Inc., pp. 138–149.
7. Aldous, D. and Fill, J.A., "Reversible Markov Chains and Random Walks on Graphs", http://www.stat.berkeley.edu/ users/aldous/book.html
8. Koucky, M., "Universal traversal sequences with backtracking", Computational Complexity, 16th Annual IEEE Conference, 2001, pp. 21–27.
9. Lurnelsky, V.J. and Mukhopadhyay, S., "Dynamic Path Planning on Sensor-Based Terrain Acquisition", IEEE Trans. on Robotics and Automation, Vol. 6, No 4,1990, pp. 462–472.
10. Cao, Z.L., Huang, Y., and Hall E.L., "Region filling operations with random obstacle avoidance for mobile robots", Journal of Robotics Systems, Vol. 5, No. 2,1988, pp. 87–102.
11. Gonzalez, E., Suarez, A., Moreno, C., and Artigue, F., "Complementary regions: a surface filling algorithm", IEEE International Conference, Vol. 1, 1996, pp. 909 –914
12. Alexopoulos, C, Griffin, P.M. "Path planning for a mobile robot". IEEE Trans. on System, Man, and Cybernetics, 1992, 22(2), pp. 318–322
13. Beom, H.B. "A sensor-based navigation for a mobile robot using fuzzy logic and reinforcement learning". IEEE Trans on SMC. 1995, 25(3), pp. 464–477.
14. http://www.parallelgraphics.com

Parallelisation of Nonequilibrium Molecular Dynamics Code for Polymer Melts Using OpenMP

Zhongwu Zhou[1], B.D. Todd[1], and Peter J. Daivis[2]

[1] Centre for Molecular Simulation, Swinburne University of Technology
PO Box 218, Hawthorn, Vic. 3122, Australia
[2] Dept. of Applied Physics, RMIT University
GPO Box 2476V, Melbourne, Vic. 3001, Australia

Abstract. The parallelisation of a sequential nonequilibrium molecular dynamics (NEMD) code for simulating polymer melts is presented. The issues impacting the efficiency of the parallel executable are probed. Various techniques, such as loop interchange, loop fusion and code restructure, have been applied to the incremental OpenMP parallelisation. Significant performance improvement and speed up are achieved for large sized systems when the parallelized code is compared to the existing sequential code. The parallelised code has successfully been applied to simulate the shear rheology of a polymer melt system.

1 Introduction

Nonequilibrium molecular dynamics (NEMD) has proven to be a useful tool in investigateing transport properties of materials. However, NEMD simulations of polymeric systems are often crippled by the excessively high computational effort required. The complexity arises mostly from the large number of atoms involved and the longer relaxation time of the materials. NEMD code for the simulation of polymer melts under shear and planar elongational flows at realistic flow conditions, recently developed in Fortran 90 by Matin, Daivis and Todd [1–3], was specifically optimised for high efficiency on vector architecture processors. However, it performs poorly on cluster supercomputers. The purpose of this work is to parallelise the NEMD code using OpenMP parallelism and explore the issues that impact the efficiencies of the parallelised code. Within the paper we also report the application of this parallelised code in successfully simulating the shear rheology of a polymer system.

OpenMP [4] is a portable programming model for shared memory architecture based on threads. It offers a small but efficient set of language constructs that support both fine- and coarse-grained parallelism paradigms. The fine-grained paradigm parallelises most of the loops in a code, which is simple and only requires a quick analysis of the loop in question. However, sometimes the number of loops is so large and the computation task in a loop is so small that the fine-grained scheme is probably not appropriate. The coarse-grained paradigm requires a parallelisation strategy similar to a MPI strategy and explicit synchronisation is required. The strength of the OpenMP approach lies in the possibility to proceed incrementally. It is easier to use OpenMP to convert an existing code. Although do-loop splitting with OpenMP is less efficient

P.M.A. Sloot et al. (Eds.): ICCS 2003, LNCS 2659, pp. 275–285, 2003.

and scalable than domain decomposition (DD) using MPI approaches, its fast implementation involves significantly less programming effort.

2 The NEMD Code

The NEMD code was developed for simulating linear chain polymers under shear and planar elongation [1–3, 5]. The polymer chain is treated as a freely joined chain (FJC) where a chain can be characterized by the number of beads (or sites) and the bond length between two adjacent sites. For performance gains, an efficient cell-code algorithm for constructing neighbour lists was implemented to calculate the forces [2]. The appropriate equations of motion for the positions and momenta are based on the SLLOD equations of motion for molecular fluids [6,7]. A fifth order Gear Predictor-Corrector Scheme was used for integrating the equations of motion [8]. Bond lengths between adjacent sites on the same molecule are held constant by bond-constraint forces, determined by using Gauss's principle of least constrain [9].

Normally, the effort exerted to generate the parallel code must be weighted against the speed-up reachable. Extensive performance analysis on the code and profiling tests reveal that various subroutines contribute cooperatively towards the total execution time, such as, the subroutines for advancing particles, calculating forces, and applying constraints. The amount of computational efforts involved in each subroutine can vary considerably depending on simulation parameters. For example, the force procedure becomes dominant in time consumption for systems of high density and short chains, while the time spent in computing bond constraints significantly increases as the number of sites per chain increases. The information from the code analysis and profiling tests suggests that most major subroutines have to be parallelised.

3 Parallelisation of the NEMD Code

When one writes parallel programs, one expects a linear speed up in performance. However, there are some hurdles to overcome in the efficient parallelisation of loops. For example, loops may have data dependencies among iterations caused by shared variables which result in some un-parallelisable code. There may be no sufficient work in a loop body and the performance suffers from the high parallel start-up costs. There may be too many references to shared variables and low cache affinity. Various techniques, such as loop interchange and loop fusion, were applied in the loop transformations to improve the parallel efficiency. In order to optimise data locality and efficient memory utilisation some procedures were restructured. To achieve this, it is necessary to maintain a global perspective of the program so that changes in one procedure have no side effects on others. The following sections describe the incremental parallelisation techniques used in detail and discuss the performance effects of the various techniques.

3.1 Data Dependencies

Data dependencies prevent relevant sections of code to be parallelised. The serial parts limit the performance of parallel code and pose an upper limit on the efficiency. In order for a loop to parallelise, access to the shared data must be mutually exclusive. Data dependency can be valid if all the iterations in a loop can be executed in any order and give the same result at the end of the execution. One example is the creation of neighbour lists, nlist[][]. The code fragment is shown as:

```
j = 1
do i = 1, n
  if (nmask(i))  then
    nlist(j, 1) = iindex(i)
    nlist(j, 2) = jindex(i)
    j = j + 1
  endif
enddo
```

nmask is a logical array that identifies the particle pair within a cut-off distance, iindex and jindex are arrays of indices of the paired particles. Obviously, the value of j in one iteration step depends on the results of previous iteration, therefore it can only be executed sequentially. Data dependency may also occur if a shared variable is written in one iteration and possibly read in another one. A typical example is the main force computing loop for accumulating forces as given below in pseudo code:

```
loop i <- 1 ... n
   compute k from i
   compute fijx[i]
   fx[k] <= fx[k] + fijx[i]
end loop i
```

fijx[i] is the force between i and j particles in the x direction and fx[k] is the accumulated force on the particle k in the x direction due to all other particles. From a parallelism perspective, when pair-wise interactions are treated using a neighbour list, complexities arise when atom k interacts with atom i, and simultaneously with atom j. In this case, the force exerted on k, fx[k], is updated from the contributions due to atoms i and j, which cannot be done concurrently on different threads. We had implemented synchronised code by using OpenMP 'atomic' updates or 'critical' [4] directives. However, testing revealed that the cost of the synchronisation is expensive due to significant synchronisation overhead. Thereafter, this data dependency problem was solved through moving the dependency part into a separate loop, and then only parallelising the main loop without the dependency. The distribution of loop iterations is often based on loop index, therefore, a clean relationship between the loop index variable and array indices is of fundamental importance. The data dependencies in the existing code were introduced from array linearisation in implementing the particular algorithm for the neighbour list. Array linearisation is a common efficient way to facilitate vectorisation of nested loops. However, it can obfuscate the relationship between array and loop indices, thereby foiling parallelisation efforts [10].

3.2 Loop Fusion and Interchange: Performance Improvement by Increasing Parallel Loop Granularity

The number of loops incrementally transformed in a code is sometimes so large that the parallelised code is too fine-grained. For example, the initial transformations of the force procedure produced about 30 parallel do loops. Rather than performance gains, the parallelised code ran slower on 4 processors. This is because many small loops have no sufficient work in their loop bodies, resulting in high parallel overhead when entering and exiting the loops. In later modifications, two techniques, loop fusion and loop interchange, were applied to increase the parallel loop granularity.

Loop fusion increases the work in a loop body by combining several loops. Fusion promotes software pipelining and reduces the frequency of branches, synchronisation and scheduling overhead. Loop fusion can be inhibited by statements between loops which may have dependencies with data accessed by the loops. To promote fusion, it is often necessary to reorder the code to get loops which are not separated by statements creating data dependencies. One example is the calculation of forces and potential energy. As the existing serial code needs to handle different flow types (e.g., shear flow, elongational flow, bulk compression), the force and potential calculations were scheduled into several loops. A segment of the pseudo code is given below:

```
loop i <- 1 ... n
   calculate rijx[], rijy[], rijz[]
end i loop
if not do_elongation
   loop j <- 1 ... n
      calculate PBC's rijx[], rijy[], rijz[]
      calculate rijsq[]
   end j loop
else
   loop j <- 1 ... nab
      transform rijx[]... to rijx_trans[] ...
      calculate PBC's rijx_trans[]...
      transform rijx_trans[] ... back to rijx[]...
      calculate rijsq[]
   end j loop
end if
loop i <- 1 ... nab
   calculate rijxc[],rijyc[],rijzc[],fmask[]
end i loop
loop j <- 1 ... nab
   if rijsq[j] ≤ rcutsq
      calculate uij[],fijx[],fijy[],fijz[]
   end if
end j loop
```

```
sum forces fx[],fy[],fz[]
sum potential uintra[],uinter[]
```

The first loop calculates distances over the whole neighbour list. Next, if the flow is not do_elongation the loop simply applies standard periodic boundary conditions (PBCs) and computes the square of distances, otherwise it first needs to transform distances, rijx, to the appropriate elongational flow PBC frame, rijx_trans [11-13]. The next loop computes centre of mass distances, rijxc, rijyc and rijzc. They are used later for computing the molecular pressure tensor. fmask is a logical array that identifies those pairs within the cut-off distance. After fmask is determined, the loop calculates forces and potential, uij, over all pair interactions within the cut-off distance. The accumulation of forces is conducted in three loops. To sum potential energy over all the particles, several additional loops are used first to separate intra and intermolecular interactions (uintra and uinter) and then to sum them separately.

The reorder and fusion made to the code combined most of the loops into a main loop. First, the *if* condition statement was moved before the computation of distances. This gives a higher level branch and allows distance computation, PBC and force calculations to be done in one main loop. The force accumulation may be combined into the main loop as well. However, as discussed in section 3.1, this can result in data dependency problems, therefore the three loops were modified into one nested loop. The several loops for computing potentials were simplified and then fused into the main loop, with only several lines of code. The loop for the computation of centre of mass distances was moved into other parallel sections. Now the force and potential calculations become two main loops for parallelization as shown in the following:

```
if not do_elongation
   loop i <- 1 ... nab
      calculate rijx[], rijy[], rijz[]
      calculate PBC's rijx[], rijy[], rijz[]
      calculate rijsq[], fmask[]
      if rijsq[j] ≤ rcutsq
         calculate uij[],fijx[]... uintra[],uinter[]
      end if
   end loop i
else
   loop j <- 1 ... nab
      calculate rijx[], rijy[], rijz[]
      transform rijx[]... to rijx_trans[] ...
      calculate PBC's rijx_trans[] ...
      transform rijx_trans[] ... back to rijx[]...
      calculate rijsq[], fmask[]
      if rijsq[j] ≤ rcutsq
         calculate uij[],fijx[]... uintra[],uinter[]
      end if
   end loop j
end if
sum forces fx[], fy[], fz[]
```

In the case of nested loops, once an array dimension and its corresponding loop have been selected for parallelisation, performance can be obtained by moving this loop to the outmost position. This loop interchange method reduces the frequency of entering and exiting a parallel loop and hence the parallel overhead. The loop interchange method was extensively used in parallelising the procedures involving the calculation of constraint forces, proportional feedback and linear equation solvers. Given below is an example of the code to calculate the symmetric dot product.

```
do i = 1, n
  do j = i, n
    a(i,j,:)=(x12(i,:)*x12(j,:)+y12(i,:)*y12(j,:) &
    &         + z12(i,:)*z12(j,:))*ka(i,j,:)
    a(j,i,:) = a(i,j,:)
  enddo
enddo
```

The innermost loop is an implicit one expressed as an array operation. The parallelisation of this nested loop includes the conversion of the implicit loop to an explicit one and loop interchange. The parallelised loop is shown below:

```
!$omp  parallel do default(shared) private(i, j, k)
    do k = 1, nm
      do i = 1, n
        do j = i, n
          a(i,j,k)=(x12(i,k)*x12(j,k)+y12(i,k)*y12(j,k) &
           &            + z12(i,k)*z12(j,k))*ka(i,j,k)
          a(j,i,k) = a(i,j,k)
        enddo
      enddo
    enddo
```

3.3 Parallel Regions: Performance Improvement by Reducing fork/joins

There are some cases in which loop fusion is not suitable. One case is when several loops have to be executed sequentially. Another case is when various loops have a different loop index or step. We can parallelise each individual do loop. However, this possibly produces parallel start-up overhead because of many thread forks and joins involved. OpenMP supports parallel regions. Several do loops can be put inside the parallel region. New threads are forked when entering the region and then joined when exiting the region. Parallel loops inside the parallel region are executed in a work-sharing fashion. The execution of associated statements is distributed among existing threads without new threads created when entering the next loop from the current one inside the parallel region. In the parallel NEMD code, performance improved when a series of parallel loops were enclosed within a parallel region. This replaced multiple fork/joins with a single fork/join. Another parallel region construction is parallel sections, which define a sequence of contiguous blocks. The beginning of each block is marked with a "section" directive and one block is assigned to one thread. This method was applied for computing the molecular pressure tensor.

3.4 Efficient Use of Memory

Effective parallelisation and efficient memory utilisation are tightly coupled. This requires maximising cache reuse and minimising cache misses. The way to efficiently use local memory (caches) is to use a memory stride of one. This means array elements are accessed in the same order they are stored in memory. Fortran uses "Column-major" order for storing array elements. When possible, nested loops in the code were interchanged to make the leftmost index of a multi-dimensional array in the inner loop to achieve the preferred order. This allows the inner loop to correspond to access in consecutive data elements and promotes the reuse of caches. The second example presented in Section 3.2 can also illustrate this point. In order to increase the data locality, most array variables used for holding intermediate data in the code were replaced with scalar variables. The parallel performance was improved by declaring these variables as private to threads. This also reduces the memory usage, particularly in the case of using the so-called 'brute force' method [5].

3.5 Parallelised Code Performance

In order to probe the efficiency of the parallelised NEMD code, a series of tests were performed. Most tests were conducted on systems at equilibrium and at a reduced density of 0.8, reduced molecular temperature of 1.0, and reduced time-step of 0.001. The data in Table 1 compare the performances of the existing sequential code and parallelised code running on 4 processors where the cell-code implementation of neighbour list construction is used [2]. It can be clearly seen that significant performance improvement has been achieved for big systems with large numbers of sites per chain. It should be noted that the restructure and optimization of the code during the parallelisation implementation also contribute to the high speed-up for the long chain polymers which is much higher than the processors used. The parallelised code also reduces the memory requirement for running the program. In the case of using the brute force method, nearly 80% memory reduction was achieved for large sized systems. More than 30% memory reduction can be achieved when the cell method is used.

Table 1. Execution time (in CPU hh:mm:ss) of the sequential and parallel codes for systems of fixed particle number (np = 40000) and various numbers of molecules (nm) and sites (ns). Note that nm = np/ns

Ns	1	4	10	20	50	100
Seq. code	00:17:19	00:19:12	00:35:11	00:53:15	05:25:46	22:07:52
Paral. code	00:06:02	00:06:16	00:06:42	00:09:09	00:20:45	01:09:41

For long chain polymers with large number of sites, ns, most of the execution time is consumed in the procedures to solve liner equations and calculate proportional feedback as a result of using Gaussian constraint forces to constrain bond lengths. These procedures contain few nested loops which are very intensive when the site number, ns, is large. The parallel strategies and optimisation techniques discussed in Section 3 are particularly efficient for these procedures, where linear and sometimes super-linear speed-up has been achieved in the case of large site and particle numbers. The super linear speed up may stem from an optimal usage of memory cache and hence the speed up obtained can be higher than the ratio of processors used.

However, the performance improvements for the systems of short chains or small number of particles is less significant. Various factors affect the parallel efficiency. One is the high parallel start-up cost due to the lack of sufficient work in a parallel loop, particularly when the system size is small. This may be attributed to the nature of the 'fork-join parallelism' of OpenMP and the structure of the code where some loops have less computational workload, so that the 'forking' and 'joining' processes are expensive. The serial parts in the procedures involving the formation of neighbour lists and implementation of the cell-code algorithm limit the performance of the parallel code as well. This limitation can become more significant if the neighbour lists need to be updated frequently. Another major factor is the less efficient use of memory caches. For example, the main force computation loop contains intensive computation tasks but with many references to various array variables, and memory access to the elements of some arrays is in an irregular manner. This results in low cache affinity and more cache missing. Despite the less significant performance improvement of the parallel code for small sized systems, the great improvement for large systems is of particular importance for our work, as future simulations will be concentrated on long chain polymers.

4 Simulation Results

In addition to the performance tests discussed in Section 3.5, we conducted simulations on a 50-site chain system of 400 molecules. Equilibrium and shear properties were examined. The simulations were performed at a site density of 0.84 and a molecular temperature of 1.0 with a time-step of 0.001. Due to the longer relaxation time of this system, each simulation typically requires over 1 million steps depending on the value of shear rate applied. This involves around one month of simulation time for a single task if the sequential code is used. The execution time for a simulation is now reduced to several days when the parallelized code runs on 4 processors.

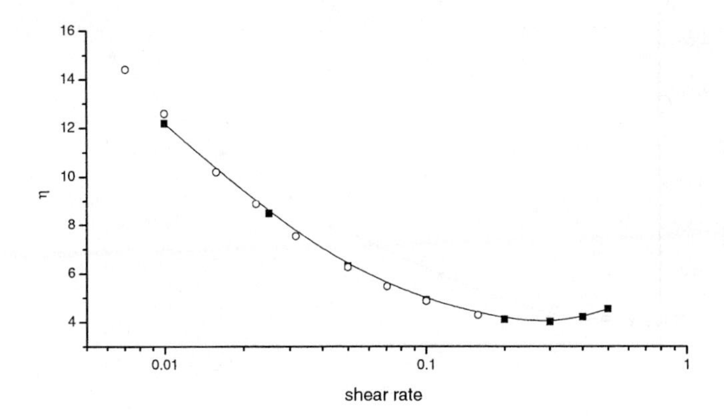

Fig. 1. Non-Newtonian shear viscosity as a function of the shear rate. The open circle data are taken from Matin [5], while our current parallelised simulation data are represented by solid squares

The shear viscosity versus the shear rate is plotted in Fig. 1 for constant volume simulations. A shear-thinning region is evident. The lower Newtonian regime was not reached because the examined shear rates were still not low enough. A slight shear-thickening can be observed from Fig. 1, due to performing the simulations at constant volume [14]. The data plotted in Fig. 1 also include the results from a system of 256 molecules, taken from Matin [5]. Very good agreement is obtained, which validates the accuracy and correctness of the parallel implementation of the code.

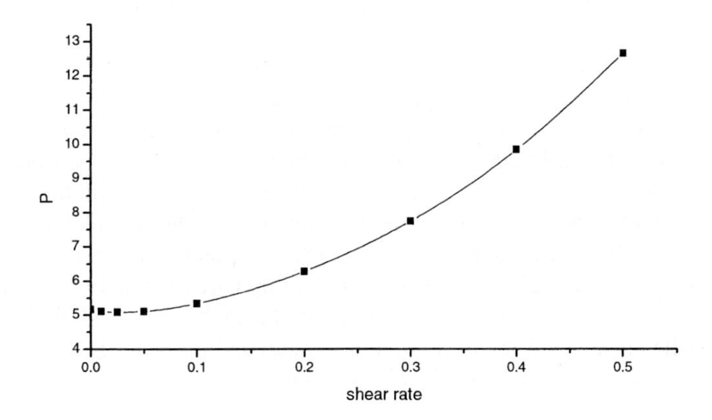

Fig. 2. Pressure versus shear rate

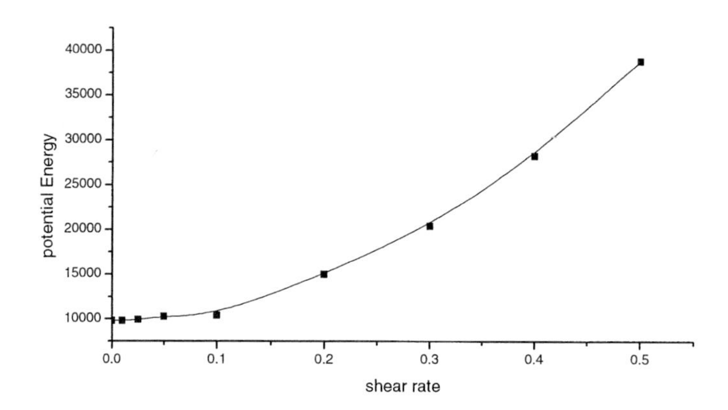

Fig. 3. System potential energy versus shear rate

The results for the molecular pressure and the potential energy are plotted against the shear rate respectively on Figs. 2 and 3. Both the molecular pressure and the potential energy change little when the shear rate is less than 0.1. After this value, they increase rapidly as the shear rate increases. For simple fluids, NEMD simulations have demonstrated that the pressure or energy as a function of shear rate under planar shear flow follows a power law [15, 16]. The results from this study seem unlikely to give such a simple relationship for long chain systems.

5 Summary

A sequential NEMD code has been successfully converted into an efficient parallel version using OpenMP directives. Major techniques, such as loop interchange, loop fusion, and code restructure, have been applied to the incremental OpenMP parallelisation. Performance studies of the parallel code are made. The speed up achieved in the present work is significant for large sized systems with large numbers of sites per chain. This is particularly important for the simulation of long-chain polymer melt systems. Linear or super-linear speed-up is achieved on some parallelised procedures in the case of large sized systems, while the speed up of other parallelised procedures is lower than the theoretical speed up value. The computational workload that can be parallelised varies, depending on the frequency of updating neighbour lists, and hence depends on parametric conditions. The parallelised code has successfully been applied to simulate the shear rheology of a polymer system and produces results well consistent with previously validated serial and vector code.

References

1. M.L. Matin, P.J. Daivis and B.D. Todd, *J. Chem. Phys.* **113** (2000), 9122. [Erratum: *J. Chem. Phys.*, **115** (2001), 5338]
2. M. L. Matin, P.J. Daivis and B.D. Todd, *Computer Physics Communications* **151** (2003), 35.
3. P.J. Daivis, M.L. Matin and B.D. Todd, *Nonlinear shear and elongational rheology of model polymer melts by non-equilibrium molecular dynamics*. Accepted, *J. Non-Newtonian Fluid Mechanics*.
4. `http://www.openmp.org`
5. M.L. Matin, *Molecular Simulation of Polymer Rheology*. Ph.D. thesis, RMIT University (Oct. 2001).
6. R. Edberg, D.J. Evans and G.P. Morriss, *J. Chem. Phys.* **84** (1986), 6933.
7. R. Edberg, G.P. Morriss and D.J. Evans, *J. Chem. Phys.* **86** (1987), 4555.
8. M.P. Allen and D.J. Tildesley, *Computer Simulation of Liquids*, Clarendon Press, Oxford, (1987).
9. D.J. Evans and G.P. Morriss, *Statistical Mechanics of Nonequilibrium Liquids*, Academic Press, New York (1990).
10. T.W.Clark, R. von Hanxleden, etc., '*Programming Issues for Molecular Dynamics*', Computational Biomedicine Symposium, Houston, Texas, Dec. 1997.
11. B.D. Todd and P.J. Daivis. *Phys. Rev. Lett.* **81** (5) (1998) 1118.
12. B.D. Todd and P.J. Daivis. *Computer Physics Communications* **117** (3) (1999) 191.
13. B.D. Todd and P.J. Daivis. *J. Chem. Phys.* **112** (1) (200) 40.
14. P.J. Daivis and D.J. Evans, *J. Chem. Phys.* **100** (1994), 541.
15. G. Marcelli, B.D. Todd and R.J. Sadus, *Phys. Rev. E* **63** (2001), 02 12041.
16. J. Ge, G. Marcelli, B.D. Todd and R.J. Sadus, *Phys. Rev. E* **64** (2001), 02 12011.

A Fault Tolerance Service for QoS in Grid Computing*

Hwa Min Lee, Kwang Sik Chung[2], Sung Ho Jin[1], Dae-Won Lee[1], Won Gyu Lee[1], Soon Young Jung[1], and Heon Chang Yu[1]

[1] Dept. of Computer Science Education, Korea University, Seoul, Korea
{zelkova,wingtop,ldw1996,jsy,yuhc}@comedu.korea.ac.kr
http://comedu.korea.ac.kr

Abstract. This paper proposes fault tolerance service to satisfy QoS requirement in grid computing. The probability of failure in the grid computing is higher than in a tradition parallel computing. Since the failure of resources affects job execution fatally, fault tolerance service is essential in grid computing. And grid services are often expected to meet some minimum levels of quality of service (QoS) for desirable operation. However Globus toolkit does not provide fault tolerance service that supports fault detection service and management service and satisfies QoS requirement. In order to provide fault tolerance service and satisfy QoS requirements, we expand the definition of failure, such as process failure, processor failure, and network failure. And we propose fault detection service and fault management service and show simulation results.

1 Introduction

Grid computing has emerged as an important new field, distinguished from conventional distributed computing by its focus on large-scale resource sharing, innovative applications, and high-performance orientation [1]. Computational grid [2] consists of large sets of diverse, geographically distributed resources that are grouped into virtual computers for executing specific applications. As the number of grid system components increases, the probability of failure in the grid computing is higher than in a traditional parallel computing [2].

The vast computing potentiality of computational grids is often hampered by their susceptibility to failures, which include process failures, machine crashes and network failures. In grid computing, the fault management is very important and difficult problem to grid application developers. Since the failure of resources affects job execution fatally, fault tolerance functionality is essential in grid computing. Grid services are often expected to meet some minimum levels of quality of service (QoS) for desirable operation. Appropriate mechanisms are needed for monitoring and regulation system resource usage to meet QoS requirements [5]. Therefore, grid applications require fault tolerance services that detect resource failures and resolve from detected failures.

---

* This work was granted by University Research Program supported by Ministry of Information & Communication in republic of Korea.

P.M.A. Sloot et al. (Eds.): ICCS 2003, LNCS 2659, pp. 286–296, 2003.

However Globus toolkit [10] that is an integrated set of basic grid services does not provide fault tolerance service to satisfy QoS requirements. Although the Heartbeat Monitor [4] was an attempt to provide some support for fault detection, it has been discontinued as a separate module. HBM (Heartbeat Monitor) only detects process failure and computer failure through periodic heartbeat messages to a centralized collector. In this paper, we provide fault tolerance service that detects resource failures, deviations from required QoS levels, and excessive resource usages and resolves detected failures. In order to provide fault tolerance service and satisfy QoS requirements, we expand the definition of failures, such as process failure, processor failure, and network failure. And we propose fault detection service using fault detector, and fault management service using fault manager.

This paper is organized as follows: In Section 2, we present related work about fault tolerance service. Section 3 presents a system model, failure definition and architecture of fault tolerance service. Section 4 discusses fault detection service and fault management service. Section 5 shows simulation results. Finally, the paper concludes in Section 6.

2 Related Work

In Globus, a noticeable flaw is the lack of support for fault tolerance [7]. To date, grid applications have either ignored failure issues or have implemented fault detection and response behavior completely within the application [6]. The support for fault tolerance has consisted mainly of fault detection services and monitoring system. The HBM [4] designs and implements local monitor and data collector for providing fault detection service of process and computer for applications developed with the Globus toolkit. The local monitor locates on each host, and monitors processes' state on each host through periodic "I-am-alive" message. It arranges result of monitoring and generates heartbeats message periodically. And the data collector identifies failed computers based on missing heartbeat messages. However, the HBM can detect the limited failures, i.e., a process failure and a computer failure and did not provide fault manage service. While the NWS (Network Weather Service) [9] monitors available network bandwidth, memory, fraction of CPU available, free memory size, and free disk space size, the NWS can not provide fault detection service and fault management service. The Big Brother [12] is the monitoring, fault notification, and web-based system and network administration tools for distributed systems. But the Big Brother can not support application-specific event notification.

To support a fault tolerance service and QoS, fault detection service and fault manage service considering QoS requirements are essential. However, the Globus doesn't provide any one at present. Thus we propose fault tolerance service that detects failures through monitoring process, processor, and network and manages the detected failures. Our fault tolerance service also satisfies the QoS requirements of grid application.

3 A Design of a Fault Tolerance Service

We discuss our system model, the definition of failures, and the architecture of fault tolerance service.

3.1 System Model

In this paper, we propose the fault tolerance service for applications developed with the Globus toolkit. The proposed fault tolerance service supports Linux, Unix, and Solaris operating systems. In common grid computing, resource components are process, processors within a computer, network interfaces, network connections, specific computers, entire site, and specific computers. If we would consider all these resource components, the complexity and overhead of fault tolerance service increase. For reasons of complexity and overhead, we define the resource failure components as process, processor, and network. When a process may fail, it loses its volatile state and stops execution according to the fail-stop model. To provide fault tolerance service, we propose a fault detector and a fault manager. The fault detector is responsible for monitoring resources state and detecting resource failure. The fault manager is responsible for managing detecting failures. Fig. 1 shows modified GRAM [8] architecture when fault detector and fault manager is employed in the existing GRAM.

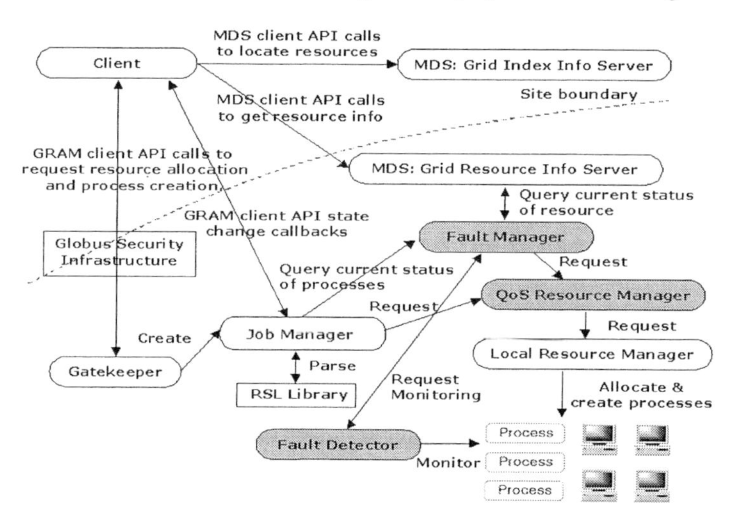

Fig. 1. A modified GRAM architecture when fault detector and fault manager is employed in the existing GRAM

3.2 Expansion of Failure Definition

In distributed computing, the resource sharing focuses on the process sharing. Thus the main target of fault tolerance is process in distributed computing and many fault-tolerance techniques for process have been studied in approach of checkpoint, message logging, and replication.

But computational grids shares processor, memory, storage, network and software as well as process. In computational grids, it is significant that fault tolerance service deals with various types of resources failure that dose not satisfy QoS requirements. Thus we expand definition of failure in order to provide fault tolerance service to satisfy QoS requirements.

[*Definition* 1] failure

It is a failure if and only if one of the following two conditions is satisfied.

1. resource stop due to resource crash
2. availability of resource does not meet the minimum levels of QoS

Table 1 shows the expanded failures that include process failure, processor failure, and network failure according to Definition 1.

Table. 1 Types of expanded failures

Process failure
1. a process stop (Process stop failure) 2. a starvation of process (Process QoS failure)
Processor failure
1. a processor crash (Processor stop failure) 2. a decrease of processor throughput due to burst job (Processor QoS failure)
Network failure
1. a network disconnection and partition (Network disconnection failure) 2. a decrease of network bandwidth due to communication traffic (Network QoS failure)

3.2 Architecture of Fault Tolerance Service

Figure 2 shows the architecture of proposed fault tolerance service in this paper.

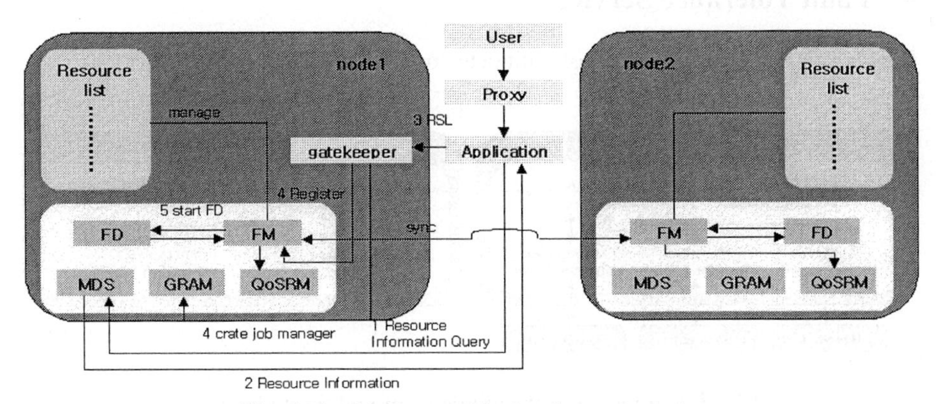

Fig. 2. Architecture of fault tolerance service

i) A fault detector (FD) is responsible for monitoring the states of process, processor, and network and deciding the occurrence of failures. The fault detector decides the occurrence of each resource failure from the detected state information and transmits the occurrence of resource failure to fault manager.

ii) A fault manager (FM) is responsible for displaying the states of resources and managing failures. The fault manager receives the state information of resources from fault detector and then displays the states of whole resources. And fault manager resolves the failures according to the recovery polices of each resource if failures occur.

iii) A QoS resource manager (QoS RM) provides advanced reservation and end-to-end management for QoS on various types of resources, such as processor, network, disks, and memory.

iv) GRAM (grid resource allocation management) [8] is responsible for allocation of computation resources and control of computation on those resources.

v) MDS (meta computing directory service) [9] provides a configurable information provider component called a Grid Resource Information Service (GRIS) and a configurable aggregate directory component called a Grid Index Information Service (GIIS).

vi) A resource list keeps the information of resources that the running jobs use.

In figure 2, the procedure of fault tolerance service is as follows. An application requests resource information to MDS and then MDS returns requested resource information. The application makes ground RSL using received resource information and requests resource allocation and process creation through gatekeeper. The gatekeeper authenticates mutually user and resource and starts a job manager that executes as that local user and actually handles the request. Simultaneously the gatekeeper registers resources information to fault manager. A fault manager requests monitoring of resource state to fault detector and then fault detector periodically reports resource state information to fault manager. The fault detector decides occurrence of failure through analyzing resource state information and transfers failure information to fault manager. If fault manager receives failure information, fault manager resolves failure according to the management policy of each resource.

4 Fault Tolerance Service

Figure 3 shows the components of fault detector and fault manager.

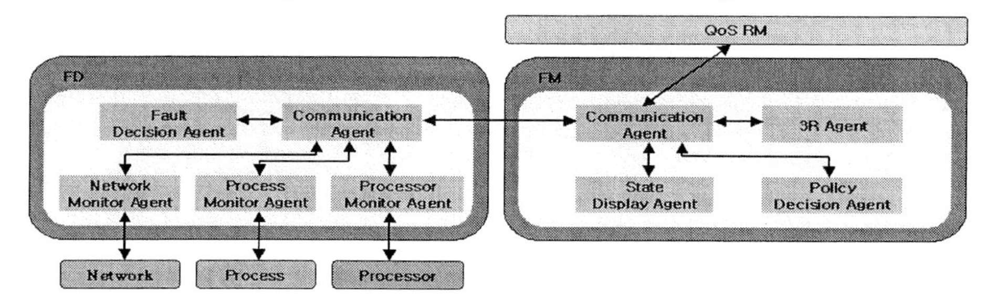

Fig. 3. Components of fault detector and fault manager

4.1 Fault Detection Service

Fault detector provides fault detection service that is as follows:

- Monitoring service, which monitors resource states of process, processor, and network
- Fault detection service, which decides the failure occurrence for each resource
- Communication service, which provides communication with each component

For fault detection service, the fault detector consists of process monitor agent, processor monitor agent, network monitor agent, fault decision agent, and communication agent.

i) A process monitor agent monitors process' state and starvation of process in job scheduler queue. The process monitor agent divides process' state into processing state, stop state, silent state, and unknown state.

ii) A processor monitor agent monitors process crash state and normal state. During processor's normal execution, processor monitor agent collects the used CPU utilization and the available CPU utilization.

iii) A network monitor agent monitors communication bandwidth, communication latency time, and network disconnection and partition between its own node and connected nodes.

iv) A fault decision agent decides the occurrence of failure through analyzing states information of each resource and then it identifies process failure, processor failure, or network failure. If failure occurs, fault decision agent reports failure information to fault manager. Fig.4 shows the failure decision algorithm of fault decision agent.

```
Fault Decision Agent
if receive resource_state_information message from communication agent
  then call DCT_OccurFailure();
Failure_state = OccurFailure; send Failure_state message to communication agent;

DCT_OccurFailure (CPU_state, Network_state, Process_state,
  Available_CPU_Utilization, Network_Bandwidth, Process_Waiting_Time)
{ if (CPU_state==failed) then return OccurFailure=yes;
  if (Network_state=disconnected) then return OccurFailure=yes;
  if (Process_state==suspend or Process_state=unknown)
    then return OccurFailure=yes;
  if (Available_CPU_Utilization < required_CPU_Utilization or 0 <= Available_
    CPU_Utilization <=20) then return OccurFailure=yes;
  if (Network_Bandwidth < requied_Network_Bandwidth)
    then return OccurFailure=yes;
  if (Process_Waiting_Time > Therhold_Waiting_Time)
    then return OccurFailure=yes;
  else return OccurFailure=no; }

Communication Agent
if receive Failure_state message from fault decision agent then if (Failure_ state=yes)
  then send resource_state_information message to policy decision agent;
```

Fig. 4. Fault decision algorithm

If states information of process is S (silent) or U (unknown), the fault decision agent detects the process stop failure. When fault decision agent receives the processor crash information, it detects the processor stop failure. If the CPU utilization is less than required quality of application or the CPU utilization is 0 ~ 20%, fault decision agent detects the processor QoS failure. And when fault decision agent receives the network disconnection or partition information, it detects the network disconnection failure. And if the amount of available network bandwidth is less than required quality of application or there is too much network traffic, fault decision agent detects the network QoS failure.

v) A communication agent manages communication between agents, FM and FD.

4.2 Fault Management Service

Fault manager provides the fault management service as like follows:

- Display service, which displays each resource state to user
- Policy service, which applies fault management method based on predefined policy according to kinds of resource failure
- Communication service, which provides communication with each component

```
Policy Decision Agent
if receive failure_state_information from communication agent then call CBR();
optimal_resource_allocation = best solution;
send optimal_resource_allocation message to communication agent;

CBR(failure_state_information)
{   retrieve best_matched_cases from Case_Library;
    while(evaluation_value >= threshold){ adapt best solution to current situation;
      evaluate evaluation_value; modify best solution; }
   apply best solution in real world; evaluate apply_results;
    add current case and best solution to Case_Library; }

Communication Agent
if receive optimal_resource_allocation from policy decision agent
  then send resource_reallocation_request message to 3R agent;
```

Fig. 5. Policy decision algorithm

For fault management service, the fault manager consists of state display agent, policy decision agent, resource reallocation request agent, and communication agent.

i) A state display agent displays each resource state to user using resource state information from fault detector and identified failure information from fault decision agent.

ii) A Policy decision agent examines predefined condition for failure management, when fault decision agent identifies QoS failure of each resource. Then, policy decision agent decides whether it requests job reallocation to QoS RM or continues doing the job at current node. Figure 5 shows the policy decision algorithm of policy decision agent.

iii) A 3R(resource reallocation request) agent requests job reallocation of QoS RM for failure recovery if process stop failure, processor crash, and network partition occur, and policy decision agent decides job reallocation for QoS failure recovery.

iv) A communication agent manages communication between agents, FM and FD, FM and QoS RM.

5 Simulation

In this paper, the goals of simulation experiment are as follows:

i) The fault detector monitors the resource state of process, processor, and network.

ii) The fault detector decides failure occurrence and identifies type of failure by monitoring resource state information.

iii) When failure occurs, fault manager decides which policy for failure recovery.

Figure 6 shows MDS structure for fault tolerance service simulation. Our testbed consists of the six hosts for simulation.

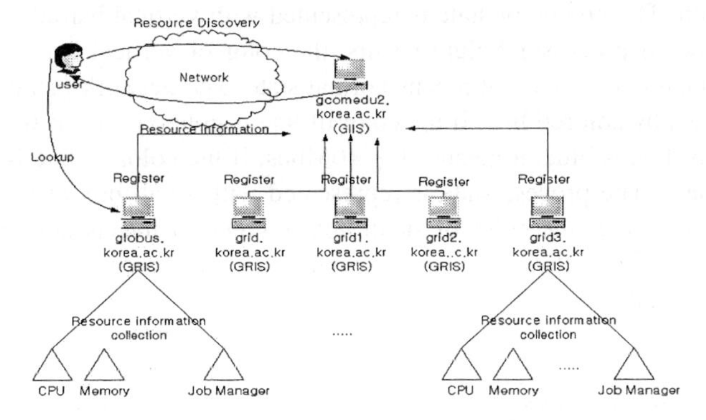

Fig. 6. MDS structure for simulation

For simulation of the fault tolerance service, we developed the simulation system with JAVA. The simulation system shows process state, processor state, and network state per a second. The features of our simulation system are as follows:

First, our simulation system displays the real-time information of the resource state information. The displayed information is running process' state, CPU utilization, network bandwidth between nodes. Second, our simulation system decides the QoS failure occurrence if resource state does not satisfy QoS requirement. And it notifies the failure occurrence to user.

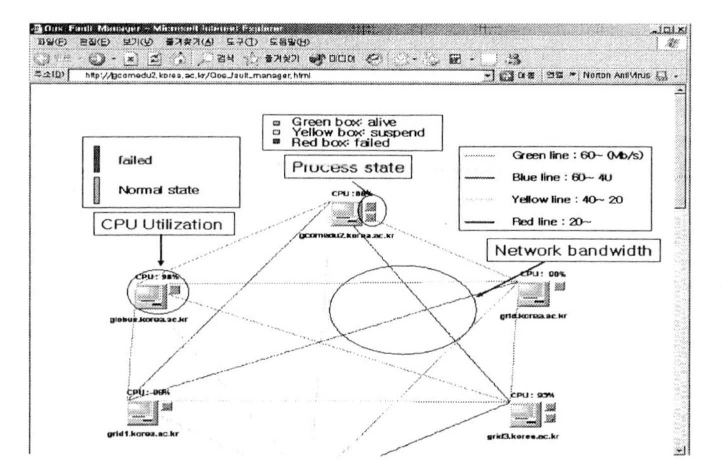

Fig. 7. The interface of our simulation system

Figure 7 shows the interface of our simulation system. Our simulation system supports GUI in order to check resource state easily. In Fig. 7, computer icon symbolizes the each component of MDS. CPU utilization is displayed by percentage (%) above computer icon. The processor state is represented with vertical bar at the right side of computer icon. If processor failure occurs, the color of vertical bar turns red. If the color of vertical bar is green, it means normal state. The network bandwidth between nodes is shown by colored line. If the color of line is green, it means 60Mbps or more. If the color of line is blue, it means 20 ~ 40Mbps. If the color of line is red, it means 20Mbps or less. The process state is represented with small box at the right side of processor state bar. If the color of small box is green, it means normal state. If the color of small box is yellow, it means suspended state. If the color of small box is red, it means failed state.

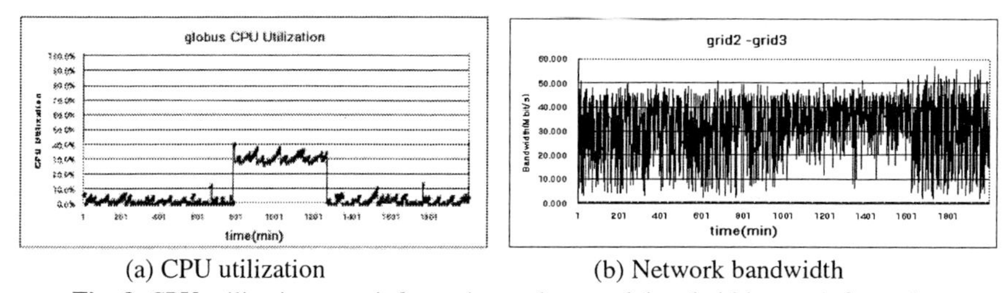

(a) CPU utilization (b) Network bandwidth

Fig. 8. CPU utilization state information and network bandwidth state information

In simulation system, we monitored resource states in components of our testbed for 2000 minutes. Figure 8a shows CPU utilization state information, one of the six hosts in our testbed. In Fig. 8a, the CPU utilization is increased rapidly from 800 min to 1270 min. At that time, we detect CPU QoS failure through simulation application and select predefined policy for resolving failures. Figure 8b shows network bandwidth state information between two nodes. Dynamism of network bandwidth is relatively reduced from 1000 min to 1600 in Fig. 8b.

We receive notification of network QoS failure through application, and execute predefined policy for resolving failures. But in case of process failure, we do not detect a native fault during monitoring. So we involve the failure injection through process kill on Unix System signal. Through simulation results, we verified that the proposed fault tolerance service monitors the resource state information of process, processor, and network, decides failure occurrence and identifies type of failure by monitoring resource state information. We also verified the need and usefulness of QoS fault tolerance service.

6 Conclusion

The vast computing potentiality of computational grids is often hampered by their susceptibility to failures, which include process failures, machine crashes and network failures. And grid services are often expected to meet some minimum levels of quality of service (QoS) for desirable operation. Therefore, grid applications require fault tolerance services that detect resource failures and resolve detected failures. Appropriate mechanisms are needed for monitoring and regulation system resource usage to meet QoS requirements. In this paper, we proposed fault tolerance service for process, processor, and network. In order to provide fault tolerance service and satisfy QoS requirements, we expanded the definition of failures and proposed fault detection service and fault management service. And we verified our fault tolerance service through the simulation.

In future work, we will implement our fault-tolerant service to satisfy QoS requirement in Globus toolkit. And we will implement the QoS resource manger for fault management service and quality of service on various types of resources.
If you wish to include color illustrations in the electronic version in place of or in addition to any black and white illustrations in the printed version, please provide the volume editors with the appropriate files.

References

1. I. Foster, C. Kesselman, S. Tuecke, The Anatomy of the Grid : Enabling Scalable Virtual Organizations, International J. Supercomputer Applications, 15(3), 2001.
2. Ian Foster, Carl Kesselman, The Grid : Blueprint for a New Computing Infrastructure, Morgan Kaufmann Publishers, 1998.
3. I. Foster, A. Roy, V. Sander, A Quality of Service Architecture that Combines Resource Reservation and Application Adaptation, 8th International Workshop on Quality of Service, 2000.
4. P. Stelling, I. Foster, C. Kesselman, C.Lee, G. von Laszewski, A Fault Detection Service for Wide Area Distributed Computations, Proc. 7th IEEE Symp. on High Performance Distributed Computing, pp. 268–278, 1998.

5. A. Waheed, W. Smith, J. George, J. Yan, An Infrastructure for Monitoring and Management in Computational Grids, In Proceedings of the 5th Workshop on Languages, Compilers, and Run-time Systems for Scalable Computers, March, 2000.
6. Anh Nguyen-Tuong, Integrating Fault-Tolerance Techniques in Grid Applications, Ph.D. Dissertation, August, 2000.
7. R.J. Allan, D.R.S. Boyd, T. Folkes, C. Greenough, D. Hanlon, R.P. Middleton, R.A. Sansum, Globus and Associated Grid Middleware, Consolidated Evaluation Report from UKHEC Sites, 2001.
8. K. Czajkowski, I. Foster, N. Karonis, C. Kesselman, S. Martin, W. Smith, S. Tuecke, A Resource Management Architecture for Metacomputing Systems, Proc. IPPS/SPDP '98 Workshop on Job Scheduling Strategies for Parallel Processing, pg. 62–82, 1998.
9. M. Swany and R. Wolski, Representing Dynamic Performance Information in Grid Environments with the Network Weather Service, 2nd IEEE International Symposium on Cluster Computing and the Grid (CCGrid2002), Berlin, Germany, May 2002.
10. I. Foster, C. Kesselman, Globus: A Metacomputing Infrastructure Toolkit, Intl J. Supercomputer Applications, 11(2): 115–128, 1997.
11. I. Foster, C. Kesselman, C. Lee, R. Lindell, K. Nahrstedt, A. Roy, A Distributed Resource Management Architecture that Supports Advance Reservations and Co-Allocation, International Workshop on Quality of Service, 1999.
12. Big Brother System and Network Monitor, available from `http://maclawran.ca/bb-dnld/bb-dnld.html`

Performing Grid Computation with Enhanced Web Service and Service Invocation Technologies

Gang Xue, Graeme E. Pound, and Simon J. Cox

Southampton Regional e-Science Centre
School of Engineering Sciences
University of Southampton
Highfield, Southampton, SO17 1BJ, UK
{gx,gep,sjc}@soton.ac.uk

Abstract. Exploitation of Web service technologies is being attempted in various areas of Grid computing. In our effort to perform Grid computation tasks based on the Web service enabled job submission system, a series of new technologies have been adopted and developed in order to achieve better functionalities, performance, and seamless integration with existing Grid computing environment. This paper presents work with detailed descriptions of new Web service and service invocation technologies, as well as demonstrations of how they are deployed in a job submission service. Experiences with these technologies are also provided, accompanied by results of tests on some of the enhanced service functionalities for evaluation and reference.

1 Introduction

Progress in the research and development of Grid computing has given rise to numerous innovative applications of computer technologies in various science and engineering domains [1]. A significant and elementary Grid functionality often exploited in these applications is the organisation and delivery of spare compute power from resource providers in heterogeneous computer environments. It is often used to solve computationally intensive problems, in a way similar to how electrical power is delivered and consumed. Such functionality is usually implemented as job submission services, through which computation tasks are dispatched to resource sites, performed, monitored, and managed. Well-known examples of this technology can be found in the GRAM system of Globus and the UNICORE system.

Early attempts of job submission services were often restricted to specific selections of programming tools and platforms, or particular usage scenarios. While the desired functionality was still achievable, there could only be limited successes as the basic requirement of interoperability for Grid computing was hardly met. New solutions were called for, which was answered by the emergence of open-standard XML Web service technologies in the Grid scenario.

The maturing and standardisation of the elementary Web service technologies – XML, SOAP, XSD and WSDL – have drawn great interest from the Grid community for their capability of standardising Grid technologies, which is essential to overcoming heterogeneity in the Grid environment. Extensive and intensive investigations were made on the applications of Web services for Grid computing, which in general

P.M.A. Sloot et al. (Eds.): ICCS 2003, LNCS 2659, pp. 297–306, 2003.

showed that such applications are viable [2,3]. The same conclusion can also be drawn from the announcement and acceptance of the Open Grid Service Architecture (OGSA) [4]. In earlier work [5], we demonstrated that XML Web service technologies could be applied to the construction of job submission service with high interoperability. Through further attempts on putting Web service based job submission into practical use, we identified a number of problems, which need to be solved with enhancement to the technologies before successful Grid computation can be performed.

One easily identifiable issue with a Web service based job submission service is the degraded performance. Compared to binary-based messages, the network overhead and the cost of message processing with the use of XML are significantly higher [3,6]. While performance penalties from submission requests and resource negotiations are arguably tolerable, especially for jobs with long execution times, huge costs brought by the transmissions of large job files and result data files in SOAP and base64 XML have caused great concerns. We believe that in order to reduce the unnecessary performance penalties, an additional standard message format is needed to facilitate the transmission of attachments along with SOAP messages. In our recent work, we have attempted to apply the latest progress in the development of such technologies to our job submission service, which is described in detail in the next section.

Like all Internet based computer applications, security management is one of the most common concerns for Grid computing and the Grid job submission service. However, no standard security mechanism has been defined by basic Web service technologies. In addition, in order to facilitate the use of the job submission service, the service needs to be compatible with GSI [7], the common security infrastructure in Grid computing. Our work demonstrates the adoption of the candidate Web service security standard for the job submission service, and shows how GSI or other security mechanisms can be integrated.

Experience with Grid computation through our job submission service shows that Web service technologies, especially the service invocation technologies, need to be improved in order to achieve the transparency and interoperability desired by Grid computing. Currently, Web services are normally consumed in a language API style: in most of today's Web service tools, the common practice is to get the WSDL file of target Web services and generate a service proxy based on it, which will then be called in the client program just like normal language APIs. While the development work is facilitated, the disadvantage is obvious: the service and the client are still tightly coupled by the interface definition, and it is impossible for client applications to consume other services with different interfaces, or adapt to changes on the service side without undergoing major modifications. In our client tool for the job submission service, we take advantage of the simple and standard SOAP protocol, which allows layered processing with independent intermediaries, to develop a new service invocation mechanism, which is implemented as a chain of SOAP filters. With this tool, client applications are provided with consistent and transparent access to Grid computation resources.

In this paper, work on performing Grid computation with Web services is described with two parts: firstly we consider the job submission service and secondly we discuss the implementation of the client tools. The descriptions were followed by evaluation and practical experience with the system. In the final part, we draw our conclusions and describe our future work briefly.

2 Enhancements to the Grid Job Submission Service

Our previous implementation of a job submission service with Web service technologies has provided users with basic functionalities for carrying out Grid computation tasks [5]. Users can submit computation tasks with specified resource requirements, monitor the job execution, perform basic job management operations, and retrieve job results through the exchanges of standard XML/SOAP messages. In order to integrate our job submission service with common Grid computing environment such as Globus, so that standard access to general Grid systems and resources can be provided, we have applied several recent new Web service technologies to enhance the original system.

2.1 Exploiting DIME for Data Transmission in Web Service Interactions

Direct Internet Message Encapsulation (DIME) [8] is a MIME-like new specification designed mainly for the transmission of SOAP messages together with additional attachments, such as binary files, XML fragments, and perhaps other SOAP messages, using standard transport protocols like HTTP and TCP. It defines a standard message structure in which data of various types that do not fit expediently or efficiently into the XML format can be contained and transmitted along with the SOAP messages. Compared to data transfer with SOAP and base64 XML, using DIME will bring significantly higher efficiency and flexibility. The DIME specification has been submitted to the Internet Engineering Task Force (IETF) [9].

Just like MIME, a DIME message is comprised of a number of DIME records, which are similar self-describing data sections with headers of binary information used for message parsing. The structure of a DIME record is shown in Fig. 1, and detailed descriptions of the record fields can be found in [10]. Compared with MIME, DIME defines a simpler message format, which does not allow the inclusion of extra metadata with custom message headers. It only contains information about the length and encoding of the message header fields and payload. While being less flexible, this ensures faster and more efficient processing of the messages. And when used together with SOAP, additional information could be delivered within the SOAP message.
When applying DIME for data transfers in the job submission service, the SOAP message signalling the data transfer operation is contained in the first DIME record. All attached data files are contained in the subsequent records and are identified by UUIDs (Universal Unique Identifier) in the ID fields of the corresponding DIME records. The attachments are cross-referenced in the SOAP message by the UUIDs, with the file names and sizes also specified. This enables a check outside the DIME protocol to make sure of the data integrity. In addition, in order to indicate the use of DIME, extensions to the WSDL file of the job submission service have been added with reference to the corresponding standard [11]. A sample DIME message used in the data transfer operation and the piece of WSDL with DIME extensions can be found in Figs. 2 and 3. In our service implementation, the layout attribute of the DIME message is set to *http://schemas.xmlsoap.org/ws/2002/04/dime/closed-layout*, which specifies that all parts of the DIME message should be referenced in the SOAP message in proper order.

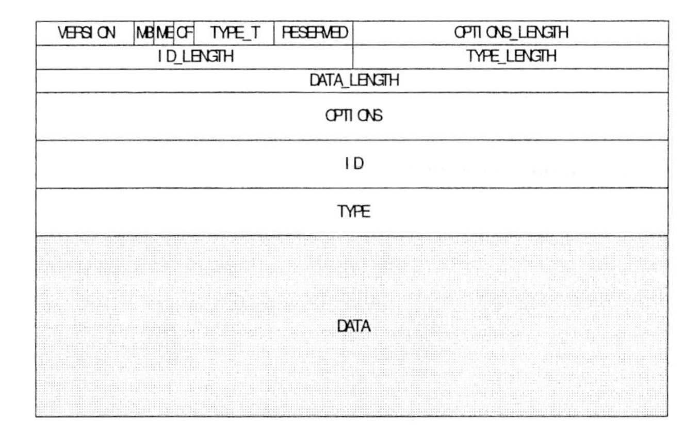

Fig. 1. The DIME record structure

```
00001100001000000000000000000000
00000000000000000000000000101000
00000000000000000000001001011011
http://schemas.xmlsoap.org/soap/envelope
<?xml version="1.0" encoding="utf-8"?>
<soap:Envelope
  xmlns:soap="http://schemas.xmlsoap.org/soap/envelope/"
  xmlns:ref="http://schemas.xmlsopa.org/ws/2002/04/reference"
>
  <soap:Body>
    <DIMEJobFileTransfer
      xmlns="http://draco.sesnet.soton.ac.uk/ComputationService2/">
      <JobID>8fa464fc-a300-424b-9827-45b3bb223127</JobID>
      <FileNames>
        <file ref:location="uuid:fd25bb9e-7fd2-4a0f-8cb4-
                            d6db005e443a">job.exe</file>
        <file ref:location="uuid:3c9d6fa9-50ab-4da3-b859-
                            db4934f3243f">library.dll</file>
      </FileNames>
      <FileSizes>
        <size>1024</size>
        <size>1048576</size>
      </FileSizes>
    </DIMEJobFileTransfer>
  </soap:Body>
</soap:Envelope>
00001000000100000000000000000000
00000000001010010000000000010111
00000000000000000000010000000000
uuid:fd25bb9e-7fd2-4a0f-8cb4-d6db005e443a
application/macbinhex40
<<1024 bytes of binary data for job.exe>>
00001010000010000000000000000000
00000000001010010000000000010111
00000000000100000000000000000000
uuid:3c9d6fa9-50ab-4da3-b859-db4934f3243f
application/macbinhex40
<<1 MB of binary data for library.dll>>
```

Fig. 2. Sample DIME message for job file transfer

```
<binding name="JobSubmissionSoapDIME" type="s0:JobSubmissionSoap">
  <soap:binding transport="http://schemas.xmlsoap.org/soap/http"
          style="document"/>
  <operation name="DIMEJobFileTransfer">
    <soap:operation
      soapAction="http://draco/ComputationService2/DIMEJobFileTransfer"
      style="document" />
    <input>
      <dime:message
          layout="http:/schemas.xmlsoap.org/ws/2002/04/dime/closed-layout"
          wsdl:required="true"/>
      <soap:body use="literal" />
    </input>
    <output>
      <soap:body use="literal" />
    </output>
  </operation>
</binding>
```

Fig. 3. WSDL extensions for DIME in the job submission service

In order to achieve a better understanding of the DIME performance, a number of tests were carried out, comparing data transfers with HTTP, HTTP+DIME, and SOAP with base64 XML. The results are shown in Fig. 4. The advantage of DIME over normal SOAP data delivery is clearly demonstrated. It is mainly because the data can be placed in DIME messages directly without additional encoding, which is unavoidable for SOAP because of the SOAP message encoding style.

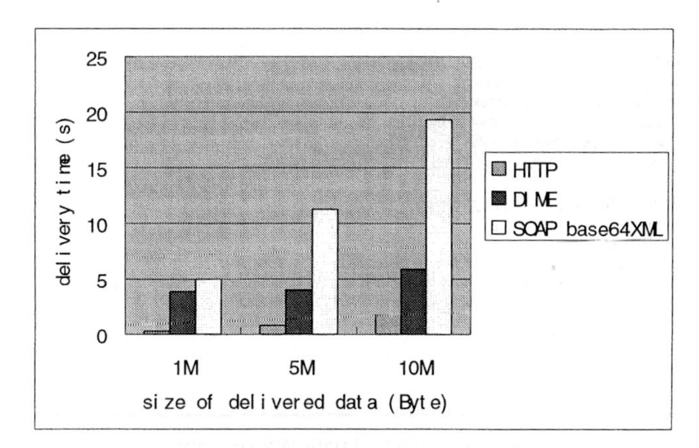

Fig. 4. Test of performance in data transmission

2.2 Using WS-Security with the Grid Security Infrastructure

Since no formal security framework was defined by the basic Web service technologies such as SOAP and WSDL, security mechanisms of the job submission service had to be built based on the underlying transport protocol – HTTP. While basic message authentication, integrity and privacy can be assured, such solution cannot provide satisfying features for Grid computing. One reason is that in some sophisticated operations, message exchanges may involve a route more complicated than the end-to-end HTTP connection, and may even involve different transport protocols. Another important reason is it does not integrate well with GSI. The Grid Security Infrastructure (GSI) [12] is a security mechanism based on public key cryptography, and is widely supported by the Grid community. In order to adopt GSI in the job submission service, a standard mechanism is needed for the delivery of GSI certificates and proxy certificates, which is the central part of GSI authentication. As a result, an implementation of the recently proposed Web services specification, WS-Security [13], has been provided in our service.

WS-Security defines a message level security mechanism, which is independent of the transport methods. Instead of defining a whole new solution, WS-Security focuses only on specifying how security information should be embedded in a SOAP message. It is therefore an open protocol, which allows existing security solutions, including Kerberos, X.509, and GSI to be leveraged.

When delivering GSI user credentials in WS-Security, the user certificates or proxy certificates are contained in a *BinarySecurityToken* element, and is placed in the *wsse:Security* SOAP header element. Since the GSI user certificates are encoded in the X.509 format, they can be directly treated as standard WS-Security binary security tokens[1]. As for GSI proxy certificates, an extension to WS-Security is defined for the job submission service, which is illustrated in Fig. 5.

```
<soap:Envelope
  xmlns:soap="http://schemas.xmlsoap.org/soap/envelope/"
  xmlns:ref="http://schemas.xmlsopa.org/ws/2002/04/reference"
>
  <soap:header>
  <wsse:Security
      xmlns:wsse=http://schemas.xmlsoap.org/ws/2002/07/secext
      soap:mustUnderstand="1">
    <gsi:BinarySecurityToken gsi:ValueType="GlobusProxy"
       Id="dab19bd1-a680-4a98-aa81-562e0cb48e70"
       xmlns:gsi="http://www.globus.org/gsi">
        MIIB3TCCAUYCAwnAiDANBgkqhkiG9w0BAQQFAD...
        ..........
      </gsi:BinarySecurityToken>
    </wsse:Security>
  </soap:header>
  <soap:Body>
    <JobSubmissionRequest
      xmlns="http://draco.sesnet.soton.ac.uk/ComputationService2/">
      ...........
    </JobSubmissionRequest>
  </soap:Body>
</soap:Envelope>
```

Fig. 5. Using GSI with WS-security

In addition to authentication, WS-Security also defines formal methods for encryptions of important messages parts. The service and users can therefore use information from the user accounts (the password) to encrypt/decrypt the security tokens, so that user privacy can be protected. The GSI user certificates can also be applied in asymmetric encryptions of important job information, such as the job handler, with the public key in the certificate for encryption and the private key held by the service client for decryption.

[1] The WS-Security specification only defines three value types for binary security token: *wsse:509v3*, *wsse:Kerberosv5TGT*, *wsse:Kerberosv5ST*.

2.3 Integration with General Grid Computing Environment

The purpose of the job submission service is to provide standard, transparent access to computation resources in various Grid computing environments. One of the currently most popular Grid environments is the Globus system. With the help of its middleware collection, which provides core Grid functionalities including security (GSI), resource allocation (GRAM), data transfer (GridFTP), and resource information service (MDS), the Globus Toolkit [14] enables the construction of computational grids through the aggregation of resources that are presented as Grid services. As compatibility with GSI has been achieved, it is feasible to integrate the job submission service with the Globus environment, and therefore make the resources managed by Globus accessible to client applications in a programming language and platform independent fashion. This integration also solves some of the firewall problems associated with Globus caused by proprietary network port settings in Globus system components such as GASS and GridFTP.

The integration with Globus was accomplished with the help of the Commodity Grid (CoG) kits [15], which provide core Globus functionalities as sets of client APIs in 'commodity technologies' including Java, Python, CORBA and Perl. In our work the Java CoG kit [16] is applied, as it is the most suitable one to work with our service.

3 The Client Tool for Service Oriented Grid Computing

As explained in the introduction, the language API-styled Web service invocation technologies are not suitable for the Grid environment, which is vast in scale and anarchic in nature. The true value of applying Web services for Grid computing is that it provides a way of interaction between independent components that share a common understanding of operational semantics, but are loosely-coupled at the interface level. Web service client tools used in Grid computing must therefore be compliant with this feature. As a result, our client tool for the job submission service has been implemented in two parts - the client utility and the message processor.

The client utility exposes job submission functionalities to application programs that need to perform computation tasks on the Grid. Different form normal service proxies, it only represents the minimum semantics of the job submission operations and procedures, and bears no information about the target service. The utility is therefore completely independent of the implementation of the computation resources, and can remain unchanged in spite of the highly changeable Grid environment.

When the client utility is called, the operation instructions are passed on to the message processor, which is responsible for the underlying interactions with targeted services, and feeding the results back to the utility. The message processor does not have a fixed composition. It is implemented as two dynamic chains of input and output message filters. Each filter is responsible for the process of a specific message part, or even the entire message. Important filters include the SOAP message handler, the WS-Security handler, and DIME handler, etc. The filter chains are formed during the runtime based on the information loaded from a configuration file, which is detached from the client applications and can be modified to add or remove message filters so as to adapt to any potential changes.

```
<?xml version="1.0" encoding="utf-8" ?>
<configuration>
   <soap>
     <outputfilters>
        <filter name="jobsubmissionservice.soapprocessor"
                  assembly="C:\windows\system32\SOAPFilters.dll" />
        <filter name="WSSecurity.GlobusCertificateOutputFilter"
                  assembly="C:\windows\system32\SOAPFilters.dll" />
     </outputfilters>
   </soap>
   <dime>
      <outputfilters>
        <filter name="dime.dimeoutputfilter"
                  assembly="C:\windows\system32\DIMEFilters.dll" />
      </outputfilters>
   </dime>
</configuration>
```

Fig. 6. Configuration file for the client tool message processor

Figure 6 shows a sample configuration file, which provides the full names of the filters and the locations of the libraries where the filter classes are contained. In the current implementation, the information is used by object reflection technology to dynamically generate filter instances and compose the filter chains.

A complete view of the client tool for the job submission service can be found in Fig. 7. In addition, it is necessary to point out that the filters for the message processor are not restricted to Web services messages. Custom filters can also be developed to enable access to other computation services using the same client tool.

4 Exemplar of Grid Computing with Enhanced Web Services

In order to examine how our enhancements to the job submission service and the integration with the Globus system work, the client tool, the job submission service and a Globus system were put together to create a sample Grid computing scenario, which is illustrated by the following figure.

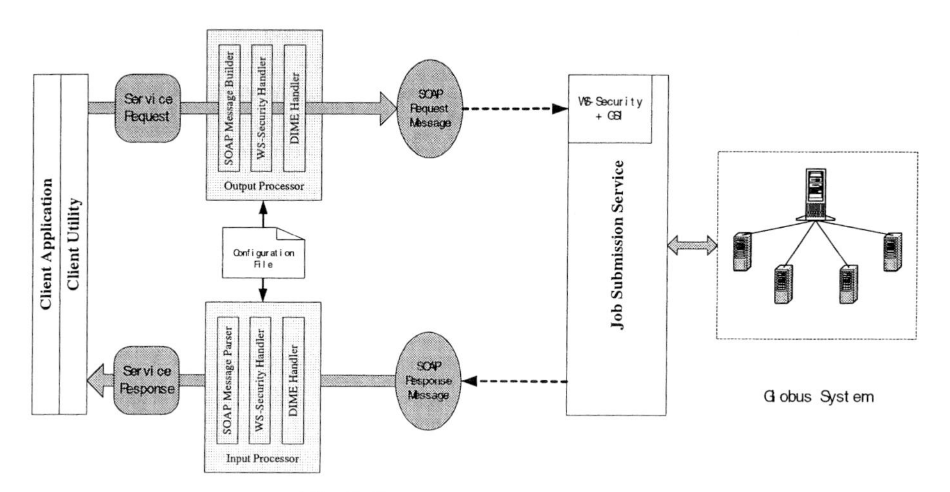

Fig. 7. The Grid computing scenario

A relatively simple computation task [2] was created and carried out under different conditions to evaluate the effects on job submission operations with various service technologies. The test results were shown in Fig. 8.

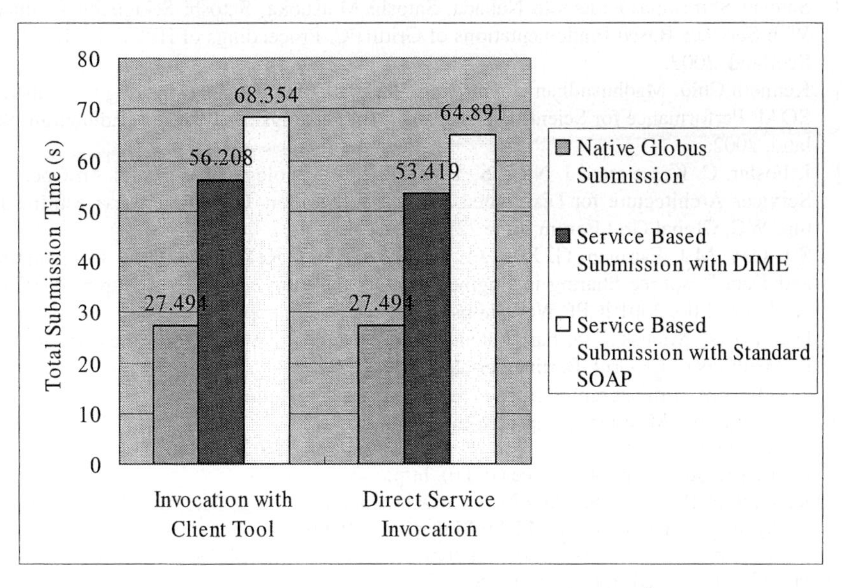

Fig. 8. Test results from sample job submissions [3]

Although the results might vary due to differences in network conditions, computer system statuses and service implementations, they in general show the cost of Web service based job submission, and the difference in performance of the new technologies. It also shows that the performance penalty of the filter-based service client tool is less than anticipated.

5 Conclusion and Future Work

Recent developments in Web service technologies have provided better solutions to issues unsolved by the basic standards. In this paper, we discussed and demonstrated how these new technologies can be applied to improve the job submission service for Grid computation. In the next stage of our work, we will try to extend the applications to other Web service enabled components and operations on the Grid, while keeping the services up-to-date with the latest developments in these technologies.

---

[2] The job has a typical composition of an executable, a runtime library and an input file with the total size of 8365961 bytes.

[3] The test is conducted on a 100M LAN, with the job submission service running on a server with a 900MHz CPU and 768M memory. The job submission procedure includes submission request, job file transfer and job-start notification.

References

[1] The GEODISE project: http://www.geodise.org

[2] Satoshi Shirasuna, Hidemoto Nakada, Satoshi Matsuoka, Satoshi Sekiguchi: Evaluating Web Services Based Implementations of GridRPC. Proceedings of HPDC-11, Edinburgh, Scotland, 2002.

[3] Kenneth Chiu, Madhusudhan Govindaraju, Randall Bramley: Investigating the Limeits of SOAP Performance for Scientific Computing. Proceedings of HPDC-11, Edinburgh, Scotland, 2002.

[4] I. Foster, C. Kesselman, J. Nick, S. Tuecke: The Physiology of the Grid: An Open Grid Services Architecture for Distributed Systems Integration. Open Grid Service Infrastructure WG, Global Grid Forum, June 22, 2002.

[5] S.J. Cox, M.J. Fairman, G. Xue, J.L.Wason, and A.J. Keane. The Grid: Computational and Data Resource Sharing in Engineering Optimisation and Design Search. IEEE Proceedings of the 2001 ICPP Workshops, Valencia, Spain, September 2001.

[6] Dan Davis, Manish Parashar: Latency Performance of SOAP Implementations. IEEE Proceedings of CCGrid, Berlin, Germany, May 2002.

[7] Grid Security Infrasturcture. http://www.globus.org/security/

[8] Direct Internet Message Encapsulation. http://www.ietf.org/internet-drafts/draft-nielsen-dime-02.txt

[9] Internet Engineering Task Force (IETF). http://www.ietf.org

[10] Jeannine Hall Gailey: Sending Files, Attachments, and SOAP Messages Via Direct Internet Message Encapsulation. MSDN Magazine, 12/2002. http://msdn.microsoft.com/msdnmag/issues/02/12/DIME/default.aspx

[11] Mike Deem: WSDL Extension for SOAP in DIME.

[12] http://www.gotdotnet.com/team/xml_wsspecs/dime/WSDL-Extension-for-DIME.htm

[13] Grid Security Infrastructure. http://www.globus.org/security/

[14] WS-Security. http://www-106.ibm.com/developerworks/library/ws-secure/

[15] The Globus Toolkit. http://www.globus.org/toolkit/

[16] Commodity Grid Kits. http://www-unix.globus.org/cog/

[17] Java CoG Kit version 0.9.13 . http://www-unix.globus.org/cog/java/index.php

Software Architecture and Performance Comparison of MPI/Pro and MPICH

Rossen Dimitrov and Anthony Skjellum

MPI Software Technology, Inc.
101 S. Lafayette St, Suite 33,
Starkville, MS 39759 USA
{rossen,tony}@mpi-softtech.com
http://www.mpi-softtech.com

Abstract. This paper presents a comparison of two implementations of the MPI standard [1] for message passing: MPI/Pro, a commercial implementation of the MPI standard from MPI Software Technology, Inc., and MPICH, an open source, high-performance, portable MPI implementation. This paper reviews key distinguishing architectural features of the two MPI implementations and presents comparative performance results from micro benchmarks and real applications. A discussion on the impact of MPI library architecture on performance is also offered.

1 Background

MPICH was developed by Argonne National Laboratory and Mississippi State University and was the first publicly available MPI implementation [2]. MPICH has been used as a basis for a large number of open source and vendor MPI libraries. MPICH has demonstrated that a portable MPI implementation can be used to achieve high performance and scalability on a variety of parallel platforms. MPICH has played an important role in popularizing the MPI standard, which presently is the predominant model for parallel programming of multi-computers and clusters of workstations.

MPI/Pro is a high-performance, scalable implementation of the MPI 1.2 standard for clusters with Linux, Windows, and MacOS operating systems [3,5]. MPI/Pro supports communication over a variety of high-speed networks, such as Myrinet, VI Architecture, and InfiniBand, as well as over traditional TCP/IP transports, such as Fast and Gigabit Ethernet. Efficient intra-box (SMP) communication is also supported. MPI/Pro has a number of architectural features that facilitate high-performance and scalability while imposing controlled processor overhead. MPI/Pro has been commercially offered on a number of platforms and operating systems since 1997. The code base of MPI/Pro has been developed from first principles and has no legacy limitations.

P.M.A. Sloot et al. (Eds.): ICCS 2003, LNCS 2659, pp. 307–315, 2003.

2 Architectural Characteristics

MPI libraries implement a standard programming interface but there is a large number of architectural choices that affect the library performance and its behavior as a whole. This section presents some of the important architectural decisions in MPICH and MPI/Pro and their impact on application performance.

Message Completion Notification

Message completion notification is the mechanism that the MPI library uses to identify the completion of a communication activity and to notify the user about the completion of the requested operation. MPICH uses polling notification, which relies on continuously querying the operating system or polling a memory flag (if the underlying communication infrastructure supports user-level messaging) to identify when a communication operation is completed. Polling notification requires the involvement of the system CPU, and burns cycles, which could otherwise be used for useful computation. As opposed to this, MPI/Pro uses blocking notification, which is based on interrupts and kernel objects for synchronization. The user thread that expects message completion is put to sleep until the system is notified that the requested communication operation is completed. This reduces the use of CPU resources for communication activities, which is a major goal of any parallel system.

Blocking notification increases the message-passing latency of short messages. This increase is a result of the interrupt-based software mechanisms and kernel objects used for synchronization. The type of notification has a negligible impact on the bandwidth of medium size and long messages. In fact, experiments have shown that under similar conditions, the blocking mode can sustain higher peak bandwidth than the polling mode [4,5]. In order to address this issue, MPI/Pro implements a polling notification mode for some of its devices. Using a run-time flag, users can select the library notification mode – blocking or polling. Although experiments on real applications and computational benchmarks, such as the NAS Parallel Benchmarks, have not demonstrated any advantage of the polling mode versus the blocking mode [4], many micro-benchmarks often emphasize low latency of short messages. Having user selectable polling and blocking message completion notification is a unique feature of MPI/Pro and allows users to adjust the behavior of the library to their needs.

Message Progress

The MPI standard has defined a rule for message progress that guarantees that if the user has started a communication operation, the library should complete this operation regardless of the subsequent (call-to-the-MPI-library) behavior of the user process. This rule requires that the MPI library uses a mechanism for independent message progress. MPI/Pro uses such a scheme. Independent message progress guarantees timely and predictable delivery of messages, regardless of their size. As opposed to that, polling progress relies on a progress engine that is invoked only when the user process calls the MPI library. Thus, the message transfers, especially the ones of long messages, can be significantly affected by the behavior and execution timeline of the user code. For example, if the code enters a long computation loop, this may directly affect the transmission of a pending long message, thus resulting in significant decrease of effective bandwidth. This effect cannot be observed and measured by microbenchmarks such as the ping-pong test and should be studied by other means.

Low CPU Overhead

Reducing the CPU involvement in communication activities is a major goal of message-passing middleware. The internal architecture of MPI/Pro is designed so that it employs system services and mechanisms that are asynchronous in nature and thus CPU conscious. As demonstrated in the performance results section below, MPI/Pro can achieve better performance than MPICH at lower CPU utilization. Application programmers are thus able to employ various techniques for communication overhead hiding, such as overlapping of communication and computation.

Overlapping of Communication and Computation

Asynchronous completion notification, independent message progress, and the low CPU overhead for message passing allow MPI/Pro to facilitate efficient overlapping of communication and computation. Overlapping is an important programming technique for improving overall parallel application performance and can be applied with a high degree of success to a variety of algorithms and platforms. As opposed to MPI/Pro, MPICH uses polling notification and polling progress, which cause high CPU overhead for communication and does not allow for concurrent communication and computation activities. This leads to a minimal, if any, benefit of overlapping even though the application algorithm can be written in a way to take advantage of overlapping.

Multi-device Architecture

MPI/Pro has efficient multi-device architecture. Multiple devices enable the MPI library simultaneously to utilize different communication media offered by hierarchical memory/network systems. Such hierarchical systems include clusters built from networked multiprocessor nodes. Modern operating systems provide efficient interprocess communication (IPC) mechanisms between processes on one node. MPI/Pro utilizes these mechanisms through its SMP devices for Windows and Linux. The multi-device architecture of MPI/Pro allows all active devices to operate in an independent manner, which removes the performance dependency of faster devices on slower devices. MPI libraries with polling progress, such as MPICH, do not provide such isolation of their devices, which leads to a negative impact on performance and scalability [6].

Thread Safety and Thread Awareness

MPI/Pro is one of only a few MPI implementations that support multithreaded MPI applications. Thread-safety allows for a programming model with multi-level concurrency. Parallel applications can be designed so that they perform multi-threaded processing within a cluster node in order to exploit local parallelism while message-passing level parallelism is achieved through the MPI library using network communication between nodes. Thread-safety also enables MPI/Pro to operate in a hybrid-programming environment using a combination of MPI and OpenMP. OpenMP provides intra-node compiler-level parallelism. MPI/Pro not only enables multi-threaded user programs to use MPI, but also facilitates multi-threading by providing the highest-degree of thread safety as defined by the MPI-2 standard. Multiple threads can communicate efficiently as MPI/Pro ensures concurrent progress of communication activities initiated by all threads. The quality to facilitate multi-threaded programs is referred to as thread-awareness, which is viewed as a more desirable quality than basic thread-safety.

Performance Optimizations

MPI/Pro has a number of optimizations that lead to improved performance. Among these optimizations are an efficient derived data type engine, collective operations algorithms, and the message de-multiplexing scheme. In addition, MPI/Pro offers a large set of tunable parameters that can be used by the users to adjust the library performance to specific run-time environments and applications.

3 Performance Results

Performance results in this section are presented in two groups. The first group is micro-benchmarks for point-to-point bandwidth and execution time of a collective operation. The second group contains results from LINPACK used for the Supercomputer Top500 list and two real applications. All MPI/Pro and MPICH results are obtained on the same equipment, using the TCP/IP devices of the two libraries for most accurate comparison. The test platform for the tests of the first group is a cluster of eight Dell Dimension 4100 workstations with a single Intel Pentium III processor at 800 MHz with 128 MB RAM and Linux RedHat 7.2; with the 2.4.7-10 kernel. The network interconnect is 100 MB/sec Ethernet with Cisco Catalyst 2950 switch. MPICH version 1.2.4 and MPI/Pro versions 1.6.3 and 1.6.4 were used.

Point-to-Point Bandwidth

The bandwidth test is based on a ping-pong message-passing pattern that is used for measuring the round-trip time. The bandwidth for each message size is calculated as the message size is divided by the half of the round-trip time. The experiment shows that MPI/Pro's peak bandwidth is about 10% higher for message sizes greater than 256 kilobytes.

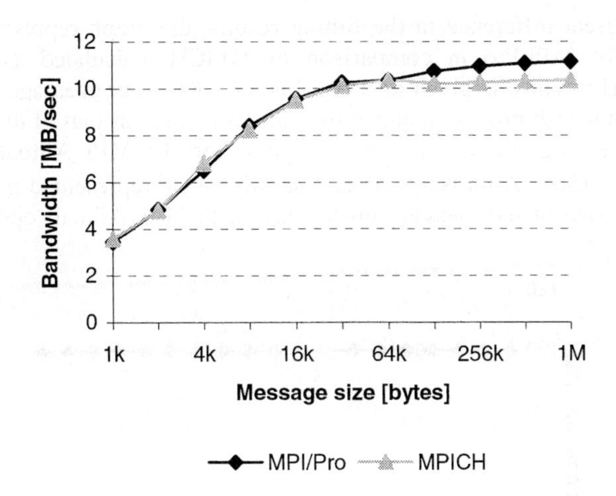

Fig. 1. Point-to-point bandwidth on a 100 Mbps FastEthernet network

The same ping-pong test was also performed on a pair of Dell PowerEdge 2650 servers with Intel Pentium 4 Xeon processors running at 2.4 GHz with 2GB RAM, interconnected with GigabitEthernet. This test was performed at Sandia National Laborary. It shows even greater advantage of MPI/Pro over MPICH, as it can be observed by the figure below – more than 40% difference in peak bandwidth in favor of MPI/Pro.

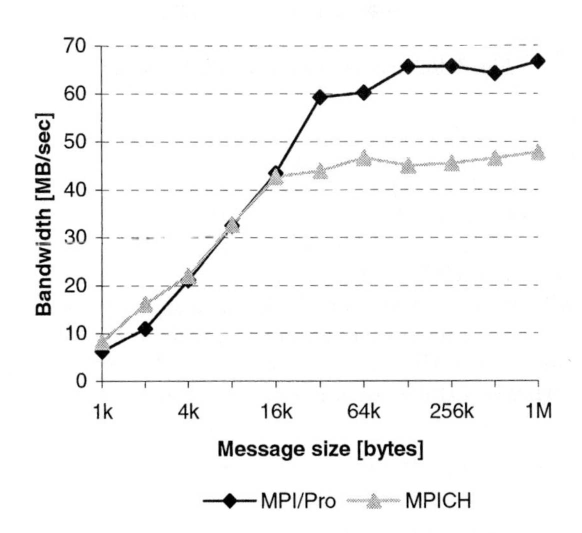

Fig. 2. Point-to-point bandwidth on a GogabitEthernet network

Collective Operations: Alltoall

This test measures the execution time of the MPI_Alltoall operation of the tested MPI library on eight processes, each executed on a single machine of the test cluster. Because of the great difference in the timing results, the graph represents the relative performance of MPI/Pro in comparison to MPICH, calculated as a percentage. MPI/Pro's performance is given as a baseline at 100%. The message size represents the message that each process sends to the other processes as part of the Alltoall transformation. The total size of the message passed to the MPI_Alltoall operations is CommSize**x**MsgSize. A message of size one Megabyte represented in the graph corresponds to a size of the message buffer passed to MPI_Alltoall operation of eight Megabytes.

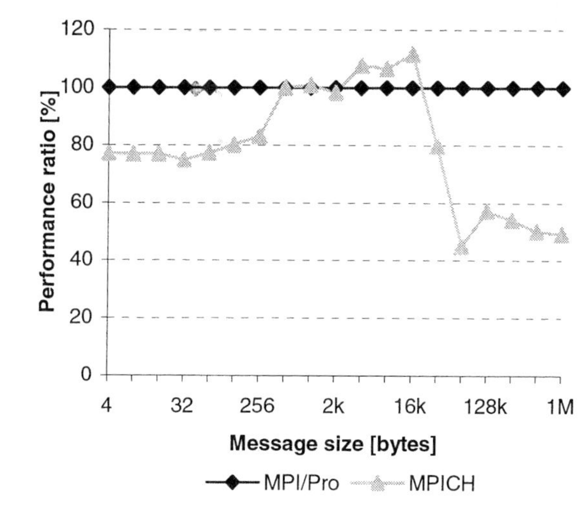

Fig. 3. Relative performance of MPI_Alltoall

During the Alltoall test, the CPU and memory utilization were observed. The results for message size of 1 Megabyte show that the average CPU utilization of MPICH was 24% while the CPU utilization of MPI/Pro was 9%. This demonstrates that MPI/Pro achieves a significantly better performance at a much lower CPU utilization, which leaves more CPU cycles for useful computation. The memory utilization shows that MPICH used more than 27 MB of memory, while MPI/Pro used less than 20 MB. The actual test memory requirements, without the MPI library, are about 17 MB. So, this shows that MPI/Pro uses only about 3 MB of system memory for internal purposes, whereas MPICH uses 10 MB, over 3 times more than MPI/Pro.

LINPACK

The results from the LINPACK benchmark were obtained in the NSF Engineering and Research Center at Mississippi State University. The test environment is as follows: 32 x IBM x330, 2 x Pentium III processors @ 1.0 GHz, 1.25 GB RAM (64 processors); Linux RedHat 7.2 with 2.4.9-13smp custom-built kernel; Cisco Catalyst 3548 FastEthernet switch. LINPACK was built with HPL 1.0 and BLAS was built with ATLAS 3.2.1, using gcc 2.95.3 compiler.

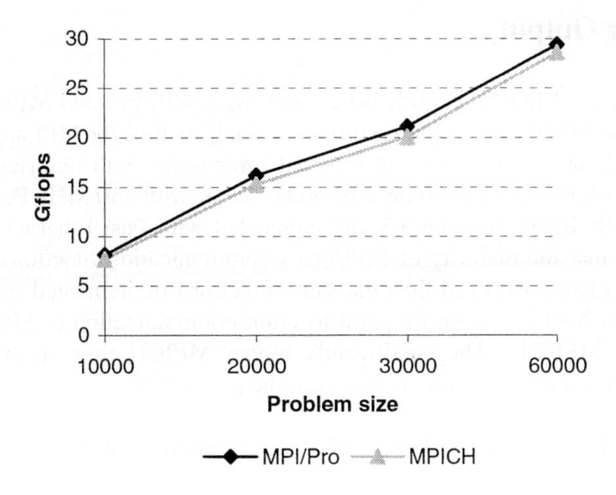

Fig. 4. LINPACK performance comparison

The LINPACK run with MPI/Pro shows between 3% and 6% higher performance than the run with MPICH.

Unstructured 3D Flow Simulation

The test results are obtained from an unstructured 3D flow solver of the incompressible Navier-Stokes equations using an implicit finite volume methodology. The code was executed at the NSF Engineering and Research Center at Mississippi State University for simulation of a Navy destroyer class combatant with and without propeller. The presented experimental results are the execution times for one iteration of the simulation. The simulation with propeller was run on 42 processors while the simulation without propeller was run on 32 processors. The test equipment used is the same as the one presented in the LINPACK benchmark above. The execution time of the simulation with propeller with MPICH was 222.76 seconds, while with MPI/Pro it was 168.93 seconds, which is an improvement of more than 31%. The simulation without propeller completed for 97.32 seconds when run with MPICH and 60.97 seconds with MPI/Pro, which is an improvement of almost 60%. So, MPI/Pro performed consistently better than MPICH on this code, showing a significant saving in time for the user.

NASA-Langley USM3D

This test was conducted at Lockheed Martin using a 64-way cluster with dual Intel Pentium III processors @ 850 MHz and 768 MB RAM, interconnected with a Foundry FastEthernet switch. The wall clock time obtained with MPICH was 3,213 seconds, while the wall clock time with MPI/Pro was 2,677 seconds, which is a performance improvement of more than 20%. Also, the user reported that occasionally random crashes were observed with MPICH while no such instabilities were seen with MPI/Pro.

4 Profiler Output

Executions of the LINPACK HPL benchmark with MPI/Pro and MPICH were compared using SeeWithin/Pro [7], a performance analysis tool for MPI applications. The table below shows a comparison of selected timings for both libraries for a test run with input problem size of 1000 on a 16-node cluster with 750 MHz Pentium III processors, 256 MB RAM, RedHat 7.3 and connected with Fast Ethernet. The profiling results reveal that the majority of MPI/Pro's communication functions incur less cumulative processing overhead than the same functions implemented in MPICH. The overall time of MPI/Pro spent on point-to-point communication is 25% less than the same time in MPICH. The significantly longer MPICH time spent in MPI_Wait could be attributed to the polling process engine of MPICH.

Table 1. Comparitive performance break-up revealed by SeeWithin/Pro analyzer

MPI Function	Total Time Spent (seconds)	
	MPI/Pro	MPICH
MPI_Irecv	65.39	10.20
MPI_Recv	1050.33	1516.77
MPI_Send	341.85	378.97
MPI_Wait	1027.09	1210.81
Total (Point To Point Communication)	*2484.66*	*3116.75*
MPI_Comm_free	0.00046	0.00066
MPI_Comm_split	1230.03	2069.03

5 Future Work

MPI Software Technology is currently developing a next generation MPI implementation under the product named ChaMPIon/Pro. ChaMPIon/Pro builds on the successes of MPI/Pro and specifically emphasizes ultra-scale parallel systems. Currently, ChaMPIon/Pro is being developed for the highly scalable DOE ASCI platforms. MPI-2 functionality is another major emphasis of ChaMPIon/Pro. At present, fully compliant support for parallel file I/O (MPI I/O) is available for NFS, GPFS, and PVSF. Also completed is one-sided communication. In development are dynamic process management and the remaining chapters of the MPI-2 standard.

6 Conclusions

MPI/Pro has a number of architectural solutions that distinguish it from MPICH and other open source public MPI implementations, among which are blocking completion notification, independent message progress, and efficient multi-device mode of operation and thread safety. These features facilitate high-performance programming mechanisms such as overlapping of communication and computation, early binding, multithreading, and exploitation of hybrid programming models such as MPI and OpenMP. MPI/Pro's architecture enables it to deliver high-performance and scalability to user applications and at the same time provide reliability and robustness. Also, MPI/Pro is well suited for environments where the timeliness and predictability of results are of critical importance.

References

1. Message Passing Interface Forum. 1994. MPI: A Message-Passing Interface Standard. *Int. J. of Supercomputer App.* 8 (3/4): 165–414.
2. Gropp, William, Ewing Lusk, Nathan Doss, and Anthony Skjellum. 1996. A High-performance, Portable Implementation of the MPI Message Passing Interface Standard. *Parallel Computing* 22 (6): 789–828.
3. Dimitrov, Rossen and Anthony Skjellum. 1999. An efficient MPI implementation for Virtual Interface Architecture – enabled cluster computing. In *Proceedings of the 3rd MPI developer's and user's conference*, Atlanta, Georgia, March 1999: 15–24.
4. Dimitrov, Rossen and Anthony Skjellum. 2000. Impact of latency on applications' performance. In *Proceedings of the 4th MPI developer's and user's conference,* Ithaca, New York, March 2000.
5. Dimitrov, Rossen. Overlapping of Communication and Computation and Early Binding: Fundamental Mechanisms for Improving Parallel Performance on Clusters of Workstations. *Ph.D. Dissertation*, Department of Computer Science, Mississippi State University, May 2001.
6. Protopopov, Boris and Anthony Skjellum. 2000. Shared-memory Communication Approaches for an MPI Message Passing Library. *Concurrency: Practice and Experience* 12(9): 799–820.
7. Krishna Kumar, C. R. et al. Automatic Parallel Performance Analysis and Tuning for Large Clusters. *High Performance Computing Conference,* Hyderabad, India, December 2001.

Virtual and Dynamic Hierarchical Architecture for E-science Grid

Lican Huang, Zhaohui Wu, and Yunhe Pan

Department of Computer Science and Engineering, Zhejiang University
Hangzhou 310027, PRC
{lchuang,wzh}@cs.zju.edu.cn, yhpan@sun.eju.edu.cn

Abstract. This paper presents an e-Science Grid architecture called as Virtual and Dynamic Hierarchical Architecture (VDHA). VDHA is a decentralized architecture with some P2P properties. VDHA has scalable, autonomous properties, and full and effective service discovery. The synchronization can be easily achieved by using VDHA. Further more, VDHA has the security architecture fulfilling the requirements of Grid. VDHA was adopted by Chinese University e-Science Grid as its architecture. In this paper, the advantages and several protocols of VDHA are also discussed.

1 Introduction

"e-Science is about global collaboration in key areas of science, and the next generation of infrastructure that will enable it" [1]. e-Science enables scientists to generate, analyze, share and discuss their insights, experiments and results in a more effective manner. The main characteristics of e-Science are coordinated resource sharing and problem solving in dynamic, multi-institutional virtual organizations called as VO [2], and dynamically involving a large number of nodes generally distributed globally in geography.

The computing architecture of e-Science is usually based on Grid [2]. The newly proposed Open Grid Services Architecture (OGSA) [3] of Grid integrates the so called computational/data Grid architecture [2] with Web services [4], and this architecture is a service-oriented architecture. From data exchange point of view, the architecture of e-Science can be classified from raw data to knowledge data as three-layered services architecture: computation/data service Grid, information service Grid and knowledge service Grid [5].

There are two major types of computing models. The prevalent client-server model is suitable for slim hosts as clients, especially mobile apparatus such as palm computers, but it may cause a performance bottleneck and an entire breakdown due to a single point of failure. Peer to Peer (P2P) model [6,7] can solve the scalable and fault tolerance problems, but it has some challenges such as security, network bandwidth, and architecture designs. We present Virtual and Dynamic Hierarchical Architecture (VDHA) (some ideas were formed in the paper [8]) to combine the advantages of the above two models and avoid their shortcomings.

P.M.A. Sloot et al. (Eds.): ICCS 2003, LNCS 2659, pp. 316–329, 2003.

VDHA is suitable for autonomous systems such as Internet, which are prerequisite for scalability; VDHA has easy authentication and authorization schema for the requirements of Grid; VDHA can fully and exactly discover services; VDHA makes synchronization easy. In this paper, we describe VDHA, its advantages, and several protocols related to this architecture.

The structure of this paper is as following: Sect. 2 describes VDHA, related protocols and several properties; Sect. 3 gives out an example about virtual cooperative research projects granted by China Educational Ministry, and finally we give out conclusions.

2 Overview of VDHA

We define the kind of Grids mainly for scientific research as e-Science Grid. Its nodes are usually located in the Universities or Institutes. The nodes are relative stable compared with other type Grids. The Universities are always formed into virtual organizers according to specific domains, and several virtual organizers share a more general common domain. Apart from general computers, there are many electronics equipment such as PDA, sensors and so on to access the e-Science Grid. According to these properties, we proposed the network architecture of e-Science Grid (see Fig. 1) and VDHA architecture to satisfy these requirements. In network architecture of e-Science Grid, there are a core circle formed by e-Science Grid nodes, and a surrounding circle, which is consisted of desktop computer, mobile computer, palm, PDA, sensors, other networks, etc. The core circle uses VDHA as its architecture. The users can use any apparatus such as PDA, palm, mobile phone or Grid node as host machines and can login into Grid system anywhere. More than one user can share a same computational apparatus. The other network such as telecommunication networks such as GSM and so on can access the system via an entrance grid node. The Grid node has sole IP address, which is used for its identification ID. The hierarchical virtual group is identified by its name. The virtual groups are generally hierarchically arranged according to the related domains.

2.1 Description of VDHA

VDHA is a virtual and dynamic hierarchical architecture (see Fig. 2) in which Grid nodes are grouped virtually. Nodes can join the group and leave the group dynamically. The groups are virtually hierarchical, with one root-layer, several middle-layers, and many leaf virtual groups (these groups are called VOs). Among these nodes of VOs, one (just one) node (called as gateway node) in each group is chosen to form upper-layer groups, from the nodes of these upper-layer groups to form upper-upper-layer groups in the same way, and this way is repeated until to form one root-layer group. In the same group all nodes are capable to be gateway node. Gateway node is the node which is not only in low-layer group, but also in up-layer group. Gateway nodes will forward the low-layer group's status information to all the nodes in the up-layer group, and distribute

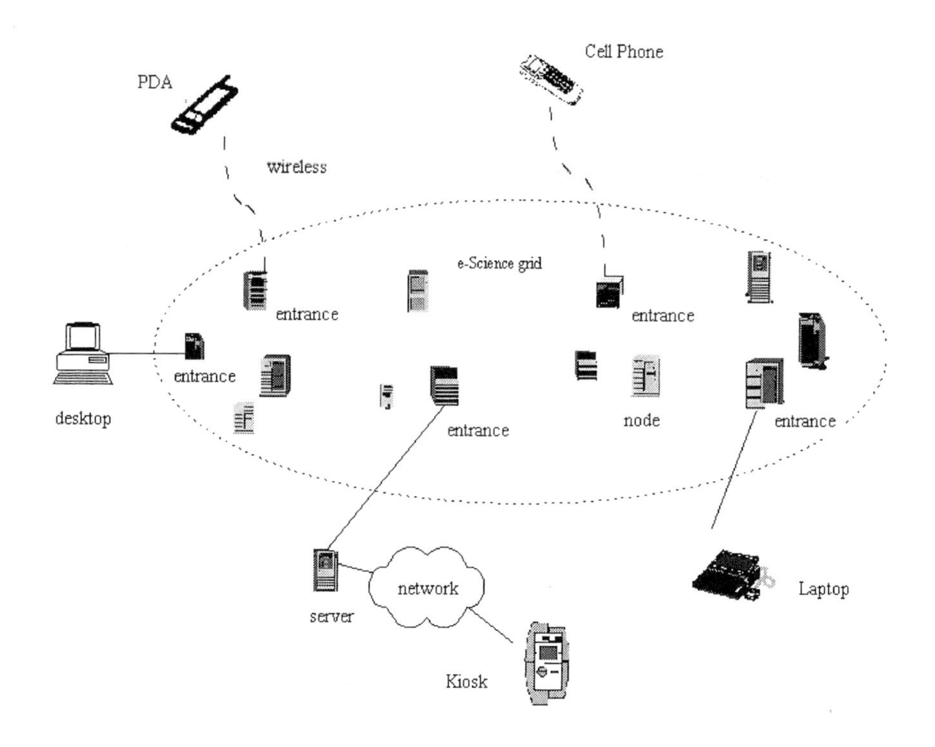

Fig. 1. Network architecture of e-Science grid

the upper-layer group's status information to all the nodes in the lower-layer group. The numbers of nodes in a VO can be dynamically changed by the way that the node can dynamically join and leave the VO. A VO may join and leave the Grid system as a whole, and this autonomous property makes the large scalable systems possible.

2.2 Formal Definition of VDHA

Definition 1. *Grid node (symbol as p) is the node in the Grid system. All p form a set PS, that is,* $PS = \{p_i | i \in N\}, N = \{1 \ldots n\}$*,here, n is the number of the Grid nodes, each* p_i *has ID (usually Internet IP address).*

Definition 2. *Entrance node (symbol as ent) is a Grid node, which is an entrance point for users to login into the Grid system.*

Definition 3. *Owned node (symbol as ow) is a Grid node, which manages the Users.*

Definition 4. *User (symbol as user) is the role which uses the Grid system. User is managed only by its owned node, not by the entire Grid. And it may be the same user which belongs to owned node before the owned node joins the Grid system.*

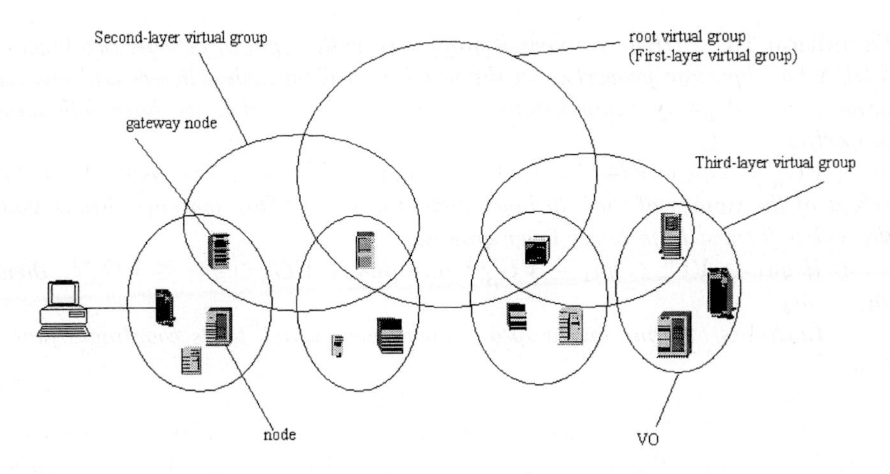

Fig. 2. Structure of VDHA. Note: There are 13 nodes in the grid system. These nodes are grouped as 4 VOs. The number of nodes in each VO is 4,3,3,3 respectively. From each VO we choose one node as gateway node to form two up-layer groups with each having 2 nodes. Then from these two groups, one node each was chosen to form a root group

Definition 5. *Client host (symbol as cli) is an apparatus (such as desktop computer, palm, PDA, mobile computer, etc), which are used by users to login into the Grid system and to do the business.*

Definition 6. *Gateway node (symbol as gn) is a Grid node which takes coordinate functions in several different layer virtual groups.*

Definition 7. *Virtual group (symbol as VG) is formed virtually by the Grid nodes. VG_{α}^{i} means the group is in the ith layer and the name of this virtual group is α. The virtual group is identified by its group name and layer number.*

Definition 8. *Coordinator of virtual group (symbol as cvg) is a gateway node taking coordinate functions in the virtual group. The symbol $cvg_{\alpha}^{i}(cvg_{\alpha}^{i} \in VG_{\alpha}^{i})$ means that it is a gateway node in the ith layer α – named virtual group which functions as coordinator.*

Definition 9. *Virtual group tree (symbol as VGT) is hierarchical tree formed by virtual groups. In VGT there is a root virtual group (symbol as RVG), many leaf VGs called as virtual organizer (symbol as VO). $VO_{\alpha_m}^{m}$ means that the virtual organizer is in the m-th layer and its name is α_m. The order of layers is counted from RVG, which is defined as the first-layer VG. VG except VO is formed purely by gateway nodes. VO is formed by Grid nodes with one (and just one) gateway node. RVG can not be a VO, and VO can be within all the layers except the first layer. N_{α}^{i} is the numbers of the nodes in VG_{α}^{i}. N_{g}^{i} is the number of virtual groups in the ith-layer of VGT.*

Definition 10. *VDHA is a virtual group tree with depth of at least two layers. VDHA has dynamic properties in the number of Grid nodes, layers and virtual groups, virtual group compositions, and so on. In VDHA, we have following properties:*

1. $VG^i_\alpha = \{gn \in VG^{i+1}_\beta | \beta \in A^i\}, i > 0, VG^i_\alpha$ *is not a VO, here,* A^i *is the subset of the names of the i-th layer virtual groups. (This sentence means that the VG is formed from lower-layer groups.)*

2. If $gn_1 \in VG^i_\alpha \cap gn_1 \in VG^{i+1}_\beta$ *and* $gn_2 \in VG^i_\alpha \cap gn_2 \in VG^{i+1}_\beta$*, then* $gn_1 = gn_2$.

3. Each VG has one and only one node (cvg) which takes coordinate functions.

4. Grid node p can join more than one VO

5. $PS = VO_1 \cup VO_2 \cup \ldots \cup VO_{n1}$*, Here, n1 is the number of virtual organizer.*

6. If p satisfies the following condition: $p \in VO^m_{\alpha_m} \cap p \in VO^{m-1}_{\alpha_{m-1}} \cdots p \in VO^{m-k}_{\alpha_{m-k}}$, $m \geq 2, k \geq 1$*, the p is gateway node. It is expressed with symbol* $gn(m, k, \alpha_{m-k} \ldots \alpha_{m-1}\alpha_m)$*. The meanings of parameter values are: m is the layer order of VO in VGT (*$gn \in VO$*); k is the number of layers in which the gateway node functions;* $\alpha_{m-k} \ldots \alpha_{m-1}\alpha_m$ *are the names of the virtual groups from* $V^m_{\alpha_m}$ *to* $VO^{m-k}_{\alpha_{m-k}}$*. Symbol* $gn^i_{\alpha_i} \in VG^i_{\alpha_i}$ *means that the gateway node is in the ith layer group with name* α_i*.*

2.3 Grid Group Management Protocol (GGMP)

GGMP is a protocol used to manage membership of virtual group and virtual group tree. Before giving out the protocol, we first define several functions and primitives as Table 1 shows and *gn.Reselect_GatewayNode_Coordinator*() – a function for selecting a new gateway node.

gn.Reselect_GatewayNode_Coordinator () {
if (gn$\neq \phi$){
if (Pnode(gn) == cvg$\in$ VGroup (gn))
$choose\_new\_cvg(cvg, VGroup(gn), (cvg \in VGroup(gn) \cap Pnode(cvg) \neq$
$Pnode(gn) \cap cvg_w$ =Maxium of $p_{iw}, p_i \in VGroup(gn)$ and online));
Remove(gn, VGroup (cvg));
If (Layer(cvg)$\neq$Layer(TOP_gn(gn))) add(cvg, VGroup(UP_gn(gn)));
Down_gn(gn).Reselect_GatewayNode_Coordinator (); } }

Now, we give out the algorithm of GGMP. (*Due to the length of paper, the synchronization is not described, and the algorithm is outline with a lot of details omitted and may not be an optimum one*)
/* p_w means that Grid node has weight value w. Weight values of nodes are assigned as several classes according to nodes' resources etc. Suppose gn at the layer gn^i_α, For all $p_i \in VGroup(gn^i_\alpha), p_i$ contains three tables. (If p_i is in the root group or leaf group (VO), then two tables). 1. State table which includes

Table 1. Primitives and functions

Description	Meanings
sender.send (message, receiver)	sender sends message to receiver
sender.send (message,receiver $\in Set$**)**	sender sends message to all the receiver belong to Set
remove(p,VO)	p is removed from VO
choose_new_cvg (p, VG, condition)	p is chosen from VG according to the condition
add (p, VG)	p is added into VG
Pnode(cvg)	returns node ID of cvg
group_name(VGroup(p))	returns this group's name
Layer $(var^m_{\alpha_m})$	returns the layer order of cvg or gn
UPcvg$(cvg^m_{\alpha_m})$	returns cvg in the up-layer
LOWcvg$(cvg^m_{\alpha_m})$	returns cvg in the low-layer
VGroup $(cvg^m_{\alpha_m})$	returns virtual group in which there is $cvg^m_{\alpha_m}$
Type(VG)	returns VG's type (VG or VO)
BOTTOM_gn(gn)	returns gn down to the bottom layer
TOP_gn(gn)	returns gn up to the top layer
Down_gn$(gn^i_{\alpha_i})$	downs gn into the low layer
UP_gn$(gn^i_{\alpha_i})$	ups gn into the up layer
gn.Down_Update ()	update status of all nodes in all the lower layer groups the gn involved
gn.Up_Update ()	update status of nodes in all the upper layer groups the gn involved

group member list,group_name, and cvg, etc . 2. Up state table which includes upper layer group member list,upper layer group_name, and upper layer cvg, etc. 3. Down state table which includes lower layer group member list, lower layer group_name, and lower layer cvg, etc . These tables are needed to be synchronized and consistent, that is, every node keeps a copy of the status*/

while(true) {
  switch(event) {
    case: a VO_α joins VDHA Grid system
      choose_new_cvg (gn, VO_α, $(gn \in VO_\alpha \cap gn_w =$ Maximum of p_{iw},
        $p_i \in VO_\alpha$ and online)); /* If p_{iw}, p_{jw}, etc are with the same value, random node is chosen. */
      set $cvg \in VO_\alpha$= gn;
      cvg uses QDP protocol (defined as Sect. 2.4) to find the
        interested parts of the structure of virtual group tree such as
          VG^k_β;
      cvg.send(JOIN_MESSAGE, cvg^k_β);
      if (cvg^k_β accepts the requisition) add (Pnode(cvg), VG^k_β);
      cvg.send(state_ table of VGroup(cvg)_message, UPcvg(cvg));
      UPcvg(cvg).send (state_ table of VGroup(cvg)_message,

```
        p ∈ VGroup(UPcvg(cvg)));
    UPcvg(cvg).send (state_ table of VGroup(UPcvg(cvg))
        _message, cvg);
    cvg.send(state_ table of VGroup(UPcvg(cvg))_message,
        p ∈ VGroup(cvg));
case: a VO_α leaves from VDHA Grid system
    gn = Pnode(TOP_gn(gn ∈ VO_α);
    gn. Reselect_GatewayNode_Coordinator();
    gn = Pnode(BOTTOM_gn(gn ∈ VO_α), gn.Up_Update ();
    gn = Pnode(TOP_gn(gn ∈ VO_α)), gn.Down_Update ();
    Delete VO_α;
case: gn leaves VDHA Grid system
    VG = VGroup(BOTTOM_gn(gn));
    gn. Reselect_GatewayNode_Coordinator ();
    set new gn = cvg in VG;
    gn = Pnode(BOTTOM_gn(gn∈ VG)), gn.Up_Update ();
    gn =Pnode(TOP_gn(gn ∈ VG), gn.Down_Update ();
case: cvg fails to receive messages from p ∈ VGroup(cvg),
        p ∈ VGroup(UPcvg(cvg)) and p ∈ VGroup(Lowcvg(cvg))
        exceeding a given times
    set gn = Pnode(cvg);
    VG = VGroup(BOTTOM_gn(gn));
    gn. Reselect_GatewayNode_Coordinator ();
    set new gn = cvg ∈ VG, add (Pnode(gn),VG);
    /*change previous gn to an ordinary node.*/
    gn = Pnode(BOTTOM_gn(gn ∈ VG)), gn.Up_Update ();
    gn = Pnode(TOP_gn(gn ∈ VG), gn.Down_Update ();
case: a node p joins a VO
    p.send(REQUISITION_MESSAGE, p_neighbor);
    /*p_neighbor is the neighboring node in the VO, which is known to p*/
    p_neighbor.send(REQUISITION_MESSAGE, cvg ∈ VO);
    if (cvg accepts the requisition) add(p, VO);
    cvg.send(p_joins_message, p_i ∈ VGroup(cvg));
    cvg.send(copy_all_table_message, p);
    cvg.send(p_joins_message, UPcvg(cvg));
    UPcvg(cvg).send (p_joins_message, p_i ∈ VGroup(UPcvg(cvg)));
case: a node p leaves from a VO
    p.send (REQUISITION_MESSAGE, cvg ∈ VO);
    if (cvg accepts the requisition) remove(p, VO);
    cvg.send(p_leaves_message, p_i ∈ VGroup(cvg));
    cvg.send(p_leaves_message, UPcvg(cvg));
    UPcvg(cvg).send (p_leaves_message,p_i ∈VGroup(UPcvg(cvg))); } }
```

2.4 Query and Discovery Protocols

In VDHA, query and discovery protocols are used for querying and discovering some entities such as resources and services, virtual group name, node status, etc. Every node has resources and services which are described by WSDL [9] or ontology languages, etc. Matching the request message is done by the agent of node which has the services. There are two kinds of QDP: Full Search Query and Discovery Protocol (FSQDP), which searches all nodes to find nodes that match the request message, and Domain-Specific Query and Discovery Protocol (DSQDP), which searches nodes in only specific domains. FSQDP first finds root virtual group, and then the coordinator of root virtual group forwards the query message to its all members. All of these members execute parallel forwards of the message down to the members of their low-layer groups until leaf virtual groups. FSQDP has time complex $O(\log N)$ (N is the numbers of nodes), space complex $O(_{Nvg})$ ($_{Nvg}$ is node numbers of each virtual group), and message-cost $O(N)$. FSQDP is effective, but may cause much traffic. Domain-Specific Query and Discovery Protocol (DSQDP) (see Fig. 3) is quite similar to FSQDP but it only searches the nodes whose catalogue matches the requested group keywords. To use this protocol, the object of virtual group must maintain the catalogue with classifying services from general to detail. It may be done by the nodes' joining the proper virtual group of Grid system. The protocol DSQDP has time complex $O(_{Nvg})$,space complex $O(_{Nvg})$, and message-cost $O(_{Nvg})$. This protocol is effective and message cost is low. The detains can be found in the paper [10].

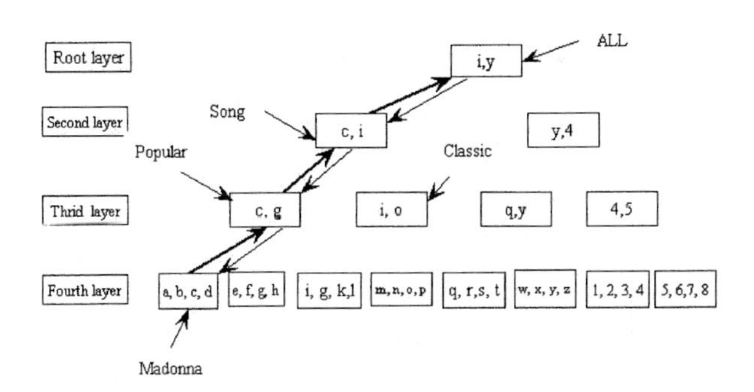

Fig. 3. DSQDP searching process

2.5 Authentication Protocol (AP)

Authentication Protocol in VDHA are based on public key infrastructure. In VDHA, the owned node takes as CA of the users and generates the user's public key and private key. The owned node keeps its owned users' public key, and also some information of the owned users such as password, etc., which are used to identify user in ordinary ways. So this authentication policy is compatible with the common authentication policy used before joining Grid system. Because the numbers of nodes are smaller than the numbers of users, and for security and easy implementation reason, all the nodes' public keys are authenticated by CA centers. This AP protocol is somewhat different with Globus GSI [11]. We use authentication ticket to solve the problems such as single-sign-on, etc. Meanwhile, because the client host's IP address is generally LAN IP address, not the Internet IP address, we use the entrance nodes as proxy stations to help the client to connect to the Grid system . There are four modes about user's login (remote/local ow via client/node cli). Details of Authentication Protocol can be found in paper [8].

2.6 Message-Based Implementation of VDHA

The implementation of VDHA is based on the broker of message/event, as Fig. 4 shows. One of the working scenarios is as following: The application layer requests a service A in the Grid by sending query message which indicates the service name, searching method, and so on. Then QDP locates the nodes which have the service A. Then Service Lifetime Management Service (SLMS) creates service A instance, and this service instance provides the services to the application layer. After finishing providing the service, SLMS will destroy service A instance. Monitor and Control Service (MCS) is an optional service which is used to monitor and control the status of the node and service instances.

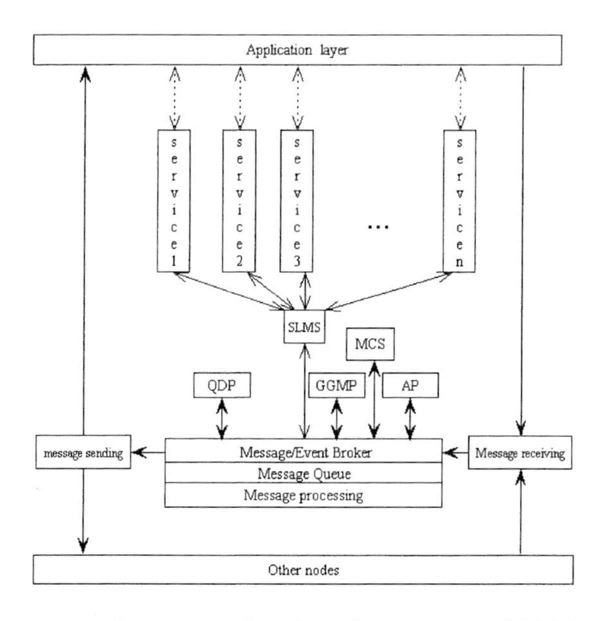

Fig. 4. The message-based implementation of VDHA

2.7 Grid Service Description Framework (GSDF)

As the above section, the service is the key in the VDHA Grid. The services can be dynamically appended into or deleted from the nodes. As infrastructures are different, MDS in Globus [12], and WSDL in Web Service are not enough to satisfy the needs. In VDHA, the services are used with three kinds of ways. (1) the service is simple, and the client end application can directly use it; (2) the client end application software uses the service's client-end API; (3) the client end program must be modified by programmer. So, the service description language must include service definitions which are understood by computer. This can be solved by the Ontology method. The language must include entities that are needed for implementing the service by SLMS. It also includes authorizations, accounting, protocol binding, and message format, etc. Therefore, GSDF must answer the following questions:
(1) How does QDP use GSDF to find the services?
(2) What protocol does the service bind and how does the protocol marshals?
(3) How does SLMS use GSDF to implement the services?
(4) How does the service account?
(5) How does the service grant access right (authorization)?
(6) What QoS does the service support?
And so on.
For satisfying those above requirements, the service description at least has the following items:

Service Definition:
    Natural language description: WHAT it is?
    Ontology definition: WHAT it is?
    Natural language description: WHAT it does?
    Ontology definition: WHAT it does?
    Natural language description: HOW does it do?
    Ontology definition: HOW does it do?
Protocol
    Natural language description: protocol document
    Ontology definition: protocol
    Port
        Transport protocol
        Port
    Security Message Format
    Message encoding
Method
    Method name
    Parameter,... Parameter
Authorization
Account
Qos
etc

We are engaging now on the draft of specification of GSDF and ontology-based Grid service description language (GSDL).

2.8 Synchronizing Schema

Because of the difficulty of implementing a distributed locking mechanism [13], synchronizing is still a problem hard to solve. But, VDHA has an easy schema to implement synchronization, because it has coordinator node in every virtual group. The coordinator of VG which is in the most top layer among the VGs with which a task involves will be taken as a center node to achieve mutual exclusion. The schema is similar to paper [14]. Whenever a node wants to enter a critical region it wants to enter and ask for permission. If no other node is currently in that critical region, the coordinator sends back a reply granting permission, as shown in Fig. 5. The schema may be explained more if coordinator is crashed down. When coordinator is crashed down, the GGMP will choose a new coordinator. Then coordinator by GGMP protocol will send message to notify all the nodes, and all the nodes which have requested or occupied the resource cancel their actions, and ask for permission of new coordinator again.

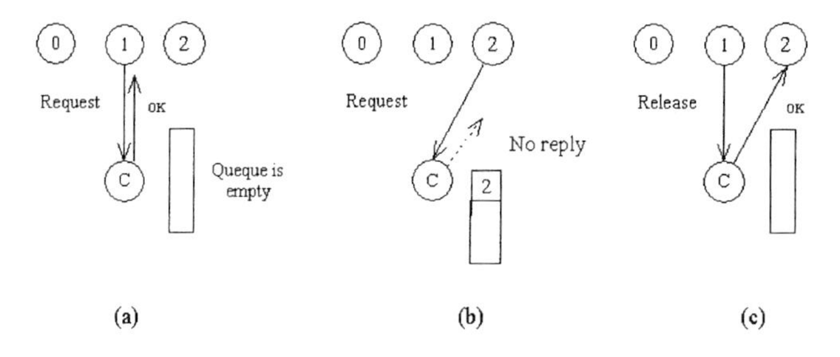

Fig. 5. synchronizing schema of VDHA. (a) Node 1 asks the coordinator for permission to enter a critical region. Permission is granted. (b) Node 2 then asks permission to enter the same critical region. The coordinate does not reply. (c) When node 1 exits the critical region. It tells the coordinator, which then replies to node 2

3 A Case Study

The virtual research projects granted by Chinese Educational Ministry are aimed to enhance the science and technology research by virtual cooperation via Internet. There are now 19 virtual organizers, each has a special domain. Each virtual organizer has average 6 nodes which are located in Universities or research institutes. In order to combine these 19 organizers into an e-Science Grid system, we use VDHA to model this e-Science Grid system prototype (called as Chinese University e-Science Grid CUEG). Nineteen nodes chosen from every 19 VOs each plus one Chinese Educational Ministry node form an up-layer

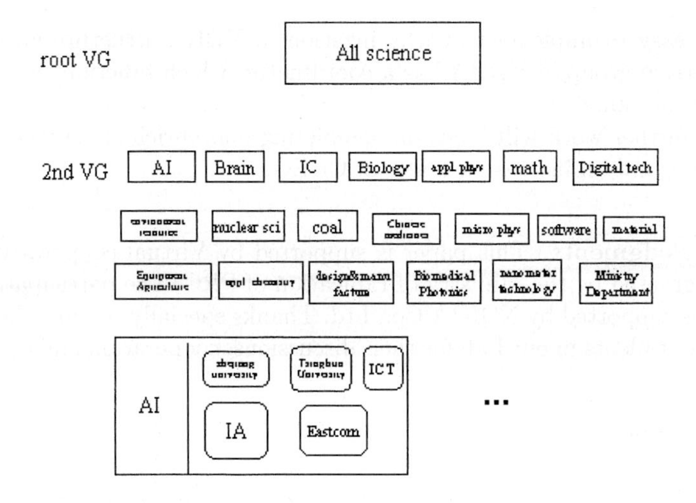

Fig. 6. VDHA architecture of CUEG (only nodes of AI are showed in the figure)

virtual group. Initially, the nodes located in the primary institutes of 19 VOs are chosen as gateway nodes (see Fig. 6). We have developed the prototypes of heterogenous information sharing service, literature resource service portal and knowledge service provider [8,15]

4 Conclusion

VDHA can solve the scale and autonomy problems. Some nodes can form a VO, and this VO can join the e-Science Grid without centralized administrator. In VDHA the messages are generally only concerned with the nodes of the three neighboring layer virtual groups, not with entire grid network. So, the e-Science Grid with VDHA has the possibility to become a huge net.

VDHA has high performance and exact discovery of resources and services. From the virtual group tree, we can know the detail information of every virtual group, so we can exactly and fully search the resources and services.

VDHA may easily manage privileges and roles of users. The users can be grouped, and the groups may be a member of a super group, and so on. The group can inherit privileges from super group. So, a user, who is the member of the group which inherits from a super group, can access the resource, if the privilege of accessing this resource is assigned to the super group by the authorization policy. This strategy has advantage for simplifying resource authorization policy if the Grid net is huge.

We have proposed the security architecture and authentication for VDHA. The security architecture fulfills the requirements of Grid such as single sign on, protection of credentials, compatible with local security solutions, and scalability.

It is easy to implement synchronization in VDHA structure Grid, because every virtual group of VDHA has a coordinator, which functions as a referee of locking/unlocking.

Our further work will focus on completing and enriching services of CUEG prototype, and on increasing nodes of CUEG.

Acknowledgments. This paper is supported by Virtual cooperative research project granted by the Ministry of Education of PRC. The participation of conference is supported by NOKIA Co., Ltd. Thanks specially to our colleagues and graduate students in our Lab for their discussions, cooperation and contribution.

References

1. John Taylor, e-Science definition, `http://www.e-science.clrc.ac.uk`
2. I. Foster, C. Kesselman and S. Tuecke, "The Anatomy of the Grid: Enabling Scalable Virtual Organizations", *International Journal of High Performance Computing Applications*, 15(3), 200–222, 2001,
   `http://www.globus.org/research/papers/anatomy.pdf`
3. I. Foster, Kesselman, J.M. Nick, S. Tuecke, "The Physiology of the Grid: An Open Grid Services Architecture for Distributed Systems Integration", 2002.2.17
   `http://www.globus.org/research/papers/ogsa.pdf`
4. Grid Web Services Workshop.2001,
   `http://gridport.npaci.edu/workshop/webserv01/agenda.html`
5. David De Roure, Nicholas Jennings and Nigel Shadbolt, "Research Agenda for the Semantic Grid: A Future e-Science Infrastructure", 2001, 9
   `http://www.semanticgrid.org/v1.9/semgrid.pdf`
6. D. Clark, "Face-to-Face with Peer-to-Peer Networking", *Computer*, Vol. 34, No. 1, January 2001, pp. 18–21
7. Krishna Kant, Ravi Iyer and Vijay Tewari, "A Framework for Classifying Peer-to-Peer Technologies", *Proceedings of the 2nd IEEE/ACM International Symposium on Cluster Computing and the Grid (CCGRID'02)*
8. Huang Lican, Wu Zhaohui and Pan Yunhe, "Virtual and Dynamic Hierarchical Architecture for Chinese University e-Science Grid", *In the proceedings of 2002 International workshop on Grid and Cooperative Computing (GCC2002)*, Publishing House of Electronics Industry, pp. 297–311
9. Christensen, E., Curbera, F., Meredith, G. and Weerawarana, "Web Services Description Language (WSDL) 1.1". W3C, Note 15, 2001,
   `http://www.w3.org/TR/wsdl.`
10. Huang Lican, Wu Zhaohui and Pan Yunhe, "Virtual and Dynamic Hierarchical Architecture: an Overlay Network Topology for Discovering Grid Services with High Performance", *submitted to Twelfth IEEE Symposium on High Performance Distributed Computing (HPDC12)*
11. Foster, I., Kesselman, C., Tsudik, G. and Tuecke, S., "A Security Architecture for Computational Grids", *In ACM Conference on Computers and Security*, 1998, 83–91
12. I. Foster and C. Kesselman, "Globus: A Metacomputing Infrastructure Toolkit", *International Journal of Supercomputer Applications*, 11(2): 115–128, 1997

13. B. Allcock, J. Bester, J. Bresnahan, A.L. Chervenak, I. Foster, C. Kesselman, S. Meder, V. Nefedova, D. Quesnal, S. Tuecke. "Data Management and Transfer in High Performance Computational Grid Environments". *Parallel Computing Journal*, Vol. 28 (5), May 2002, pp. 749–771.
14. Andrew S. Tanenbaum, "Distributed Operating Systems", *Prentice-Hall International, Inc.*, 1995, pp. 134–135
15. Huang Lican, Wu Zhaohui and Pan Yunhe, Zhou Xuezhong, "Knowledge Services Provider Model based on Virtual and Dynamic Hierarchical Architecture", *In the proceedings of 2002 International workshop on Grid and Cooperative Computing (GCC2002)*, Publishing House of Electronics Industry, pp. 245–253

Track on
Models and Algorithms

Algorithmic Entropy, Phase Transition, and Smart Systems

E.V. Krishnamurthy

Computer Sciences Laboratory,
Australian National University, Canberra, ACT 0200, Australia
abk@discus.anu.edu.au

Abstract. A smart system exhibits the three important properties: (i) interactive, collective, coordinated and parallel operation (ii) self-organization through emergent properties (iii) adaptive and flexible operation. A hierarchy based on metric entropy is suggested among the computational systems that transcend from the unsmart to the smart system through a phase transition like phenomenon. Understanding smart systems is useful to solve hard-optimization problem inspired by the self-organizing processes found in nature. Such systems will be valuable to create artificial systems made up of exotic matter to solve specific problems in particular domains of interest with a high efficiency.

1 Introduction

Smart systems have no formal definitions, although such a system seems to be a precursor to the living system, Brooks [4,5]. As such no suitable formal model is available for smart systems. Therefore, we define the smart systems as those systems having the following important properties:

1. Interactive, Collective, Coordinated, and Highly Efficient Parallel Operation
They can interact with the environment (hence called open). Also they collectively and cooperatively perform actions, coordinating their actions when there is competition, to obtain maximal efficiency.

2. Self Organization and Emergence
The total dynamic behaviour of the system cannot be inferred from the dynamic behaviour of its components and new properties emerge abruptly. These new properties of the system are not predictable, in advance from the properties of the individual interactions. In particular, under emergence, the many degrees of freedom arising due to its component parts collapse into a fewer new ones with a smaller number of globally relevant parameters; see Makishima [27].

3. Adaptive, Fault-Tolerant, and Flexible Operation
Smart systems are always flexible to change-they can modify their past behaviour by a learning process and adapt to environmental changes, available resources, as well as, tolerating failures or non-cooperation of some of their components.

Our purpose in this paper is to restrict our consideration to the following issues:

1. Is the behaviour of smartness analogous to the critical phenomenon (phase transition or percolation)? Can we obtain suitable parameters to describe this phenomenon?

P.M.A. Sloot et al. (Eds.): ICCS 2003, LNCS 2659, pp. 333–342, 2003.

2. Is there a hierarchy of degree of smartness among computational systems?
3. Can we simulate smart systems to solve intractable problems?

2 Conrad's Principles

Before we answer the above questions, we recall the principles proposed by Conrad [7]. These have a direct bearing on defining smart systems. According to Conrad, a general purpose computing system cannot have all of the three following properties:

1. Structural programmability (algorithmizability)
2. High computational efficiency
3. High evolutionary adaptability and flexibility

Properties 1 and 2 are mutually exclusive. That is, we cannot have high computational efficiency in a programmable system. Properties 1 and 3 are mutually exclusive in the region where maximum computational efficiency exists. That is, whenever maximal computational efficiency is required, the system should be highly adaptive rather than structurally programmable (algorithmic). Based on Conrad's principles, we can say that self-organization is not possible in an algorithmizable system.

A further support to Conrad's principles is provided by Wegner [35]. Wegner argues that the algorithmic machines are less powerful than machines that can interact with the environment and such interacting systems turn out to be much more efficient in coupling material resources to problem solving. As we know, efficient coupling to environment makes enormous difference for problems such as pattern recognition, survival etc. As a further support to Wegner's hypothesis, in this paper we will show that such interaction introduces positive entropy into the system to turn a rigid algorithmic system into a smart system. An example of a smart system is visual perception which is computationally most intensive. While a digital computer makes a large number of sequential decisions, the biological visual system uses a holistic approach, collective or Gestalt .The system makes use of its computational resources very efficiently, whether it is measured by speed of calculation or energy considerations. The computations are resistant to damage to hardware. Also emergent properties - such as self-awarencss seems to be present. Further, there is no structural hardware diagram or equivalently a structured program that does it, Haken [15], Hameroff [16], Hopfield [20].

3 Phase Transition Model for Smart System

As mentioned in the introduction, smart systems need to possess the three properties:

1. Interactive, collective and parallel operation
2. Self organization through emergence
3. Adaptive and flexible operation

Properties 1 and 2 hold near the critical point in a phase transition in a physical system [9] or in a percolation model, Stauffer and Aharony [33], near the percolation threshold. Note that the percolation model is more general than the phase transition model, since it deals with the more abstract geometric properties rather than the thermodynamic properties used in phase transition models. These two models lie at the border of order and disorder and are concerned with the collective and parallel interaction among the microscopic objects that result in scale invariance and universality. Scale invariance corresponds to a power law behaviour over a wide range of control parameter; the exponent involved in this power law is called the critical exponent.

By universality, we mean that the set of exponents found in diverse systems group themselves into distinct classes, with the property that all systems falling in the same class have the same exponents. This suggests that there are common features among the underlying microscopic mechanisms responsible for the observed scale invariant behaviour. The idea of universality is that apparently dissimilar systems show considerable similarities near their critical points. Accordingly, all percolation/ phase transition problems can be divided into a small number of different classes depending upon the dimensionality of the system and the symmetries of the order state. Within each class, all phase transitions have identical behaviour in the critical region, only the names of the variables are changed. Thus universality describes the relationship among different phase transitions.

Also percolation / phase transition models reflect the cooperative, as well as, competitive behaviour among the microscopic objects. The modelling uses a suitable geometric structure with a local computation that results in a global change. Further, the percolation / Phase transition systems have the property 3, namely, adaptive and flexible behaviour to tolerate failures, since they both deal with the formation of clusters or creating paths among distant neighbours, even if some of these neighbours are non-cooperative. This ultimately can lead to an emergent behaviour through self-organized criticality.

From the above arguments we see that to model a smart system, the phase transition or percolation model will be of value since the smart system seems to lie between order and disorder.This observation has been made earlier by Langton [24], Wolfram [31], who suggest that life exists at the edge of chaos, and by Zak et al [37] who suggest that instability is necessary to create intelligence.

Prigogine [30,31] emphasises why non-unitary evolution based on star- hermitean operator is needed to deal with open systems, as well as, systems which are far-from equilibrium and attain stability. Prigogine emphasises that the irreversible (non-unitary) process play a fundamental rolc in biological systems. According to Prigogine, future and past play the same role in time- reversible unitary systems; hence they cannot explain the emergence of new dynamical patterns involved in the intelligent or smart biological systems.

Hence, the three required properties of smart systems are difficult to realise within the unitary transformation or time reversible evolutionary systems. It looks as though hysteresis, long term memory and time arrow have a direct role in turning the systems smarter. In fact, it is due to the time arrow and memory, the smart system remembers and orders the temporal events as earlier and later, without the explicit sequential addressing mode used in conventional programming. This leads to a kind of self-awareness of past, present and future leading to a psychological time arrow. Breakdown of time symmetry seems essential for constructing special purpose systems made up of exotic phases of matter that can function similar to the Nature beyond the computable domain.

4 Role of Metric and Algorithmic Entropy

Metric entropy plays a similar role like the ordinary entropy does in thermodynamics or information theory to measure the disorder in a system. Metric entropy is defined as the average information per measurement obtained from an (denumerable) infinite sequence of finite precision – identical measurements made on a time evolving system from a time minus infinity to plus infinity. If the metric entropy is zero it implies that the measurement sequence, begun in the remote past, ceases to provide any more additional information after a finite time T. This means adequate information has been acquired to predict the properties of that system completely and unambiguously. This also means that the information gained by measurement equals the entropy decrease in the system so that no uncertainty remains. However, if the metric entropy is positive this means the entropy in the system has not decreased as much as the information obtained by measurement and still the uncertainty remains. It can also be due to the fact that the system is producing entropy faster than what we can measure.

Metric entropy is related to another entropy called "algorithmic entropy" or Kolmogorov-Chaitin entropy, Chaitin [6]. Unlike in ordinary entropy, in algorithmic entropy we use a rule or a program as the basis for the creation of order or disorder in objects that are created from a given input. We then express the order and disorder of the created objects in terms of the program length and the length of the output it generates. For this purpose, we measure the bit-length of the minimal program (or functional rule) that generates a required output sequence from a given input sequence.

As an example, consider the creation of a totally erratic infinite sequence. This sequence cannot be described as an output sequence of some input sequence by using a simple program whose length is smaller than the desired output length. Analogously, whenever the predictions of a physical theory can be obtained completely, such a theory is representable by a Turing machine [10]. The analogy between measurement procedure and algorithm is now clear. In both cases there is a well-defined input and we want a well-defined output after a finite time that can be described in finite time and space. This analogy permits us to relate metric entropy to the algorithmic entropy. Also it enables us to use the algorithmic complexity measure to describe the complexity of the measurement procedure. That is the bit-length of the minimal program that describes the functional rule to generate a required output from a given input, can be used to define the complexity of the measurement procedure, as well as, algorithms.

We can define the algorithmic entropy measure thus: Let K(n) define the bit length of a minimal program that can output any n bit finite subsequence using a functional rule. Also let $K = \text{Lim } n \rightarrow \infty [K(n)/n]$. We note that $0 \leq K \leq 1$. Thus when $K > 0$ the length of the program required to generate a desired output sequence goes on increasing or the rule is not finite and turns out to be more and more complex. However, for $K = 0$ the program length is much shorter than the required output thus defining a deterministic, orderly and predictable output according to an algorithmic rule. Thus K can be used to measure metric entropy.

In general, systems with zero entropy are tractable (predictable) and those with positive entropy are not predictable. Positive entropy systems cannot be described in any simpler way than they are and there no formal grammatical rules that can be used to understand them. Well- structured objects (e.g., Context free grammars, regular grammars and serial-parallel orders) provide for easy description through functional rules and hence have zero metric entropy. The systems with zero metric entropy are classified "Turing or algorithmically expressible". Such zero-entropy systems are amenable for accurate measurements in a finite time and their orbital sequences are algorithmically predictable leaving no uncertainty in measurement. Also, the information gained by measurement equals the entropy decrease in the system. In positive entropy systems, the rate of evolution can be faster than what we can measure to gain information.

Remark: In Gell-Mann [12], p.101, the algorithmic entropy is called **"depth"** and the metric entropy is called **"crypticity"**; the former measures the amount of (labour) time required to go from the compressed spatial information or minimal program to the full spatial description, while the latter measures the amount of labour (time) involved to compress the full spatial description into a minimal program.

4.1 Algorithmic Information and K Entropy

Consider $K = \lim_{t \to \infty} [K(t)/t)]$; where K(t) is the algorithmic information needed to record a piece of the trajectory in the interval of time t. Thus for positive K-entropy, in the long run the recording of information for evolution increases unbounded and the evolution cannot be followed deterministically, unless a disproportionately long or infinite time is devoted to this task beyond a critical time
t = T(c). As we approach T(c), the recording is critically slowed down -much like in phase transition phenomena. At this time the motion is no longer deterministic and the forward and backward evolution are not reversible resulting in a spontaneous breakdown of time - reversal symmetry [10]. In contrast, when K= 0 the evolution can be recorded deterministically, without breaking the time reversal symmetry.
Thus a phase -transition like situation arises between the Turing expressible and Turing non-expressible systems leading to a critical point behaviour. By the universality principle used in phase transition-like phenomenon it is possible that critical exponents and scaling factors exist for such systems so that we can say when the system can become smart. We will discus this aspect again later.

4.2 Hierarchy among Machines

Based on Metric-Algorithmic entropy we can establish a hierarchy among the computational systems (See Fig. 1). That is the metric entropy seems to play an important scale-free role as in percolation or phase transition in deciding whether the system is smart. Essentially, the zero entropy machines cannot cope up with the complexity of solving a problem leading to a critical point corresponding to intractability and non-computability. At this point a phase transition like phenomena can help towards emergence of new properties resulting in positive entropy and self -organization, Bak [1]. Thus we can classify the two major classes of machines, ordinary (O) and dissipative (P) based on metric entropy as below.

O. Ordinary or Zero Metric Entropy Machines:
Completely structured, Deterministic, Exact behaviour (or Algorithmic) Machines.
This class contains: Finite State machines, Push down-stack machines, Turing Machines (Deterministic) that halt and Exactly integrable Hamilton flow machines.

Such machines are information-lossless; their outputs contain all the required information as dictated by the programs and the information gained by running the program exactly equals the entropy decrease in the program, Gell-mann [12], p.224.

P. Positive Metric Entropy Machines:
These are Partially Structured, Non-deterministic, Probabilisitic, Average behaviour, ergodic systems that preserve shape and volume in phase space , mixing systems where shape in phase space is distorted, Kolmogorov -flow (K-flow) systems with the motion in phase space unpredictable, nonequilibrium systems exhibiting macro and emergent behaviour – such as Chemical and Biological machines and living systems (Fig. 1). The positive entropy machines do not obey the unitary transformation law. Their evolution results in increasing entropy. Their actions are irreversible. That is the time-reversal symmetry is spontaneously broken. Only such machines can become smart; but they will have to experience an arrow of time. The measurements on these systems will not reveal all the required information.

4.3 Combining Zero and Positive Entropy Machines to Create Smartness

Prigogine [30,31] suggests the use of non-unitary transformation, called star-Hermitean operators to extend the capabilities of computational systems to reflect average behaviour. That is we require the tools of both equilibrium and non-equilibrium quantum statistical mechanics to create smart systems that lie at the critical point between order and disorder. We also recall from Gal Or [11] that irreversibility is essential for creating smart matter. In his book Gal-Or asserts that all processes in nature can be understood by an optimal mixture of order and disorder. Thus to create smart systems we need to combine the above two different classes of machines, Zak et al [37], Gell-Mann[12].

5 Simulating a Mixture of Entropy in Systems

A way to simulate a mixture of the zero and positive entropy machines is by choosing the mode of application and the action set of a rule-based program to be either deterministic, nondeterministic, probabilistic or fuzzy. Rule application policy in a production system [23] can be modified by:

(i) Assigning probabilities/fuzziness for applying the rule
(ii) Assigning strength to each rule by using a measure of its past success
(iii) Introducing a support for each rule by using a measure of its likely relevance to the current situation.

The above three factors provide for competition and cooperation among the different rules. In particular, the probabilistic rule system can lead to emergence and self-organized criticality. Thus, we may attempt to enlarge the capabilities of class O machines by simulating special features of class P machines - using nondeterminism, randomness, approximation, probabilities, equilibrium-statistical mechanical (e.g. simulated annealing) and non-equilibrium statistical mechanical(e.g. genetic algorithms and Ant algorithm [8]).

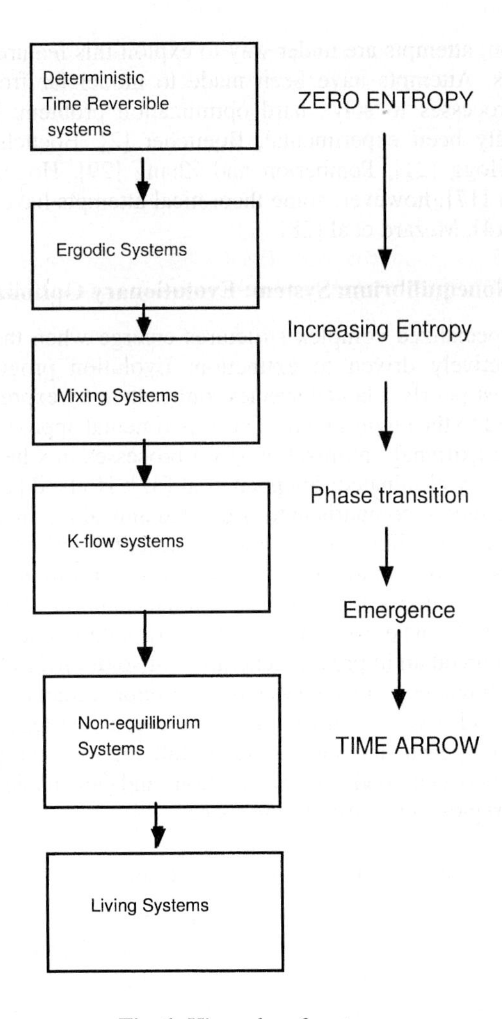

Fig. 1. Hierarchy of systems

6 Phase Transitions in AI Systems

Phase transition like situation have been observed experimentally in a wide variety of heuristics used for search problems in NP class or beyond. This phenomenon is called a knife-edge phenomenon (name given in AI); Hubermann and Hogg [21], Gent and Walsh [13] , Lau and Okagaki [25],Walsh [34], Zhang and Korf [38]. Here the over-constrained problems tend to become more constrained and under- constrained problem tends to become less constrained as the search becomes deeper. In between is the knife-edge, a region in which critically constrained problems (that is those at the solvability phase transition) independently of the depth at which one examines the search tree. This knife edge phenomenon is also known in Artificial life (AL) as the Edge of chaos, Langton et al [21]. Since self- organized criticality is associated with the phase transition, attempts are under way to exploit this feature to solve hard optimization problems. Attempts have been made to model far from equilibrium and nonequilibrium processes to solve hard optimization problem, Bak [1]. These attempts have largely been experimental, Boettcher [2], Boettcher and Percus [3]; Hubermann and Hogg [21], Pemberton and Zhang [29], Hogg and Williams[18], Hogg and Kephart [17]; however, some theoretical attempts have been made by Gomes and Selman [14], Mezard et al [28].

6.1 Simulating Nonequilibrium System: Evolutionary Optimization

In nature highly specialized complex structures emerge when their most inefficient elements are selectively driven to extinction. Evolution progresses by selecting against the few most poorly adapted species, rather than by expressly breeding those species best adapted to the environment. The experimental approach by Boettcher and Percus [3] uses the extremal optimization (EO) processes in which the least fit variables are progressively eliminated; Sneppen et al [32], Holland [19]. The EO process uses a different strategy in comparison to simulated annealing. In simulated annealing the system is forced to equilibrium dynamics by accepting or rejecting local changes. EO, however, takes the system to a far from equilibrium position and persistent selection against the worst fitness lead to near-optimal solution. Also EO differs from Genetic algorithm (GA); whereas GA keeps track of entire gene pools of states from which to select and breed an improved generation of solutions, EO operates only with local updates on a single copy of the system, with improvements achieved instead by elimination of the bad. EO also differs from the greedy strategy which aims at improving the solution at each step and as a result falls into a local optimum. EO, however, can fluctuate between good and bad solutions and can enable us to cross barriers and approach new regions in configuration space.

6.2 Boolean Satisfiability Problem and Phase Transition

Another important approach in this direction is due to Korkin [22]. Korkin considers the solution of the boolean satisfiability problem or SAT near the midpoint of the phase transition curve from satisfiable to the nonsatisfiable to illustrate the self-organization..Based on this we can create a self organized evolutionary process for solving K-SAT when the system is in nonequilibrium state near the phase transition.

6.3 Do Critical Exponents and Scaling Laws Exist for Smartness?

Although the different studies show the existence of a phase -transition like phenomena and power law relationships do hold, as yet, we cannot confidently predict the critical probability and power law exponents without experimentation for each individual complex problem. Although scaling and universality are widely accepted experimentally in many different areas (including Biology, Medicine, Physics and Social sciences and in random networks, including the World-Wide-Web), they are yet to be established mathematically. Also, practical computation for evaluating such parameters is difficult. Thus the only approach seems to be available at the present is simulation. This gives rise to various types of scaling laws depending upon the nature of the problem. Studies on nonequilibrium physics, Prigogine [31], Lebowitz [26] can help us to understand these aspects.

7 Matter and Life

Brooks [4,5] examines the more general questions than those raised in this paper, namely, the relationship between matter and life, what is that makes the matter alive and what do we lack in our understanding to create living machines. He argues that this inability to bridge matter and life may arise due to four factors:

1. Our models may not have adequately correct parameters
2. Our models may not be complex enough to bring out the distinction and may lack unimagined features of life.
3. Lack of computing power
4. Lack of "new stuff", namely mathematics and physics

Since smart system is a precursor to living system, we should expect a phase transition like situation arising to distinguish living systems from non-living systems.

8 Conclusion

We described some important properties that a smart system need to posses and the entropy conditions required for an ordinary system to become smart resembling a phase-transition. We also a presented a review of the current attempts to create smart heuristics by using the knife-edge phenomenon encountered in solving intractable problems. Although, percolation model is suitable for understanding a smart system, we are unable to obtain quantitative parameters, e.g., scaling laws and critical exponents that can help us distinguish a smart system from unsmart systems. Computational science along with Physics, Biology and Mathematics will play a major role in the development of a new science as envisaged by Wolfram [36].

References

1. Bak, P.: How Nature works: The Science of self-organized criticality, Springer Verlag, New York (1996)
2. Boettcher, S.: Extremal Optimization of graph partitioning at the percolation threshold, J.Phys A.: Math.Gen **32**, (1999) pp. 5201–5211.
3. Boettcher. S and A. Percus, A.: Nature's way of Optimizing, Artificial Intelligence, **119** (2000) 275–286.
4. Brooks, R.: The relationship between matter and life, Nature, **409** (2001) 409–411.
5. Brooks, R.: Steps towards living machines, pp. 72–94, in Lecture Notes in Computer Science, Vol. **2217**, Springer Verlag, New York (2001) 72–94,
6. Chaitin, G. "Algorithmic Information Theory, M.I.T Press, Cambridge, Mass (1987)
7. 7. Conrad, M.: Molecular Computing, Advances in Computers, **31**, Academic Press, New York (1990). 235–325
8. 8. Dorigo, M et al.: Antalgorithms, Lecture Notes in Computer Science, **2463** Springer Verlag, NewYork (2002)
9. Emch, G and Liu, C: The Logic of Thermostatistical Physics, Springer, New York, (2002)
10. Faisal, F.H.M: Quantum Chaos, Algorithmic Paradigm and Irreversibility, in Quantum Future, Ph. Blanchard and A. Jadczyk (ed), Springer Verlag, New York (1997), 47–57
11. Gal-Or, B: Cosmology, physics and philosophy, Springer Verlag, New York (1981)
12. Gell-Mann, M: The Quark and the Jaguar, W.H. Freeman, New York (1994).

13. Gent, I and Walsh, T: Phase transitions from real computational problems, Proc. 8th Symposium on AI, (1996) 356–364.
14. Gomes, C.P and Selman, B., Satisfied with Physics, Science, **297** (2002) 784–785.
15. 15 . Haken, H.: Synergetic Computers and Cognition, Springer Verlag, New York (1991)
16. Hameroff, S.R,: Ultimate Computing, North Holland, Amsterdam (1987)
17. Hogg,T and Kephart, J.O.: Phase transitions in higher dimensional pattern classification, Computer Systems Science and Engineering, **5 (1990)** 223–232.
18. Hogg, T and Williams, C.P : The hardest constraint problem: a double phase transition, Artificial Intelligence, **69** (1994) 359–377
19. Holland, J.H.: Emergence-chaos to order, Addison Wesley, Reading, Mass. (1998)
20. Hopfield, J: Physics, Biological Computation and Complementarity, in The lesson of quantum theory, North Holland, Amsterdam (1986) 295–314.
21. Hubermann, B.A. and Hogg T .: Phase transitions in Artificial Intelligence Systems, Artificial Intelligence, **33** (1987) 155–171.
22. Korkin, M.: Self-organized Evolutionary Process in sets of interdependent variables near the mid point of phase transitions in K-satisfiability, Lecture Notes in Computer Science, **2210**, Springer Verlag, New York (2001), 225–235
23. Krishnamurthy, E.V and Krishnamurthy,V.: Rule-based Programming paradigm: A formal basis for biological, chemical and Physical computation, Biosystems. **49** (1999) 205–228
24. Langton, C.E et al: Life at the Edge of chaos, in Artificial Life II, Addison Wesley, Reading,Mass., (1992) 41–91
25. Lau, M and Okagaki, T.: Applications of Phase transition theory in visual recognition and classification, Journal of visual communication and Image representation, **59**(1994) 88–94.
26. Lebowitz, J.L.: Microscopic Origins of irreversible Macroscopic Behaviour, Physica **A 263** (1999).516–527
27. Makishima, S. : Pattern Dynamics: A theory of Self-Organization, Kodansha Scientific, Tokyo (2001)
28. Mezard, M., Parisi, G., and Zecchina, R.:Analytic and algorithmic solution of random satisfiability problems, Science, **297** (2002) 812–815.
29. Pemberton, J and Zhang,W (1996); Epsilon Transformation: Exploiting phase transitions to solve combinatorial optimization problems, Artificial Intelligence, **81** (1996) 297–325
30. Prigogine, I: From being to becoming, W.H. Freeman and Co, San Fransisco, (1980)
31. Prigogine, I.: Laws of Nature, Probability and time symmetry breaking, *Physica* **A 263** (1999). 528–539.
32. Sneppen, K and Bak, P,: Evolution as a self-organized critical phenomena, Proc. Natl. Acad. Sci., **92** (1995) 5209–5213.
33. Stauffer, D and Aharony, A: Introduction to Percolation theory, Taylor & Francis, London (1993)
34. Walsh, T.: The constrainedness in knife edge, Proc. AAAI-98, (1996) 406–411.
35. Wegner, P.: Why Interaction is more powerful than algorithms, Comm. ACM **50**, (1997) 81–91.
36. Wolfram, S.: A new kind of science, Wolfram Media, Inc, Champaign, Ill (2002)
37. Zak, M, Zbilut, J.P and Meyers, R.E .: From Instability to Intelligence, Springer Verlag, New York (1997).
38. Zhang, W and Korf, R.: A study of complexity transitions on the asymmetric travelling salesman problem, Artificial Intelligence, **81** (1996) 223–239.

Computations of Coupled Electronic States in Quantum Dot/Wetting Layer Cylindrical Structures

Roderick V.N. Melnik[1] and Kate N. Zotsenko[2]

[1] University of Southern Denmark,
Mads Clausen Institute, DK-6400, Denmark, rmelnik@mci.sdu.dk
[2] Electronic Data Systems Australia,
60 Currington Street, Sydney, NSW 2000, Australia

Abstract. There exist several techniques allowing to confine the motion of an electron from all three dimensions. The resulting structures, known as quantum dots, promise new opportunities in optoelectronics, and a better understanding of such structures requires incorporating new effects into existing models which inevitably leads to the necessity of using efficient computational tools. In this paper we analyse computationally the electronic states in quantum dots of cylindrical shape taking into account the influence of the wetting layer. Coupling effects between quantum dot and wetting layer electronic states are demonstrated with numerical examples.

1 Introduction

Microelectronics has provided some of the most challenging problems in computational science [5]. Moving to nanosize devices opens new opportunities for computational scientists, and many important problems in nanoscience have been already attacked with computational tools. The major focus of the present paper is on modelling low-dimensional semiconductor nanostructures. At present, there exist several methodologies allowing to produce electronic states confined from all three dimensions. The resulting structures are known as quantum dots, and one of the most popular techniques of their growth is the Stranski-Krastanow methodology. This methodology is essentially a self-organised hetero-epitaxial growth during molecular beam epitaxy. The properties of the structure resulting from this growth process are dependent on thermal, elastic, piezoelectric, and other effects, and controlling the process is a difficult task [10] which can be assisted by computational experiments. The structures themselves can be grown in different shapes such as pyramidal, ellipsoidal, conical, cylindrical. The geometry of the structure influences substantially optoelectronic properties of the quantum device, as do different effective masses in the dot and the crystal matrix of such devices [4,9]. Recently, it was shown that the inclusion of the wetting layer, a "substrate" on which the dot is grown, is also essential since the wetting layer electronic states and the states of the quantum dot are essentially

P.M.A. Sloot et al. (Eds.): ICCS 2003, LNCS 2659, pp. 343–349, 2003.

coupled [6]. It is our purpose in this paper to examine further this influence, in particular for cylindrical quantum dots. Although cylindrically-shaped quantum dots have been analysed before (e.g., [8] and references therein), no prior systematic investigation of the influence of the wetting layer on electronic states of the whole structure have been carried out. A better understanding of this influence is important in evaluating optical properties of low-dimensional nanostructures (e.g., [1]).

2 Position-Dependent Effective Mass Model for Quantum Dot/Wetting Layer Nanostructures

In this paper we are considering InAs quantum dots embedded in GaAs matrices. We consider the full 3D model of the quantum dot by using the envelope function approximation. The discussion is focused on the one-band model as an example, in which case the Hamiltonian of the system can be represented in the following form (e.g., [7]):

$$H = -\frac{\hbar^2}{2}\nabla_{\mathbf{r}} \cdot \left(\frac{1}{m(\mathbf{r})}\nabla_{\mathbf{r}}\right) + V(\mathbf{r}), \tag{1}$$

where $m(\mathbf{r})$ is the position-dependent electron effective mass and $V(\mathbf{r})$ is the confinement (band-edge) potential. Although it is often assumed that the electron effective mass is a constant inside of the domain of interest, this assumption cannot be justified for any real considerations since $m(\mathbf{r})$ is a function of position and varies considerably between two major regions of interest, InAs and GaAs in this paper. Taking into account this difference is essential, but this is connected with the necessity to interpret the model

$$H\psi(\mathbf{r}) = E\psi(\mathbf{r}) \tag{2}$$

in a variational sense [3] with E being the electron potential energy which takes discrete levels. The quantum dot is grown as a few monolayer InAs structure of cylindrical shape. The structure rests on a thin InAs wetting layer, as it is depicted in Fig. 1. The whole dot/wetting layer InAs structure is embedded in a GaAs crystal matrix. For such geometries the resulting model is not amenable to analytical treatments which have been traditionally applied in the physical literature in this context under much simplified assumptions. Note also that taking into account additional effects, such as elastic, thermal, and piezoelectric to name just a few, would necessarily lead to an increasing attention to the development of efficient computational tools for the models discussed in this paper.

The growth direction, considered in this paper as an example, is (100), in which case the model (1), (2) allows a separation of variables, and can be reduced to

$$\bar{H}\bar{\psi}(r,z) = E\bar{\psi}(r,z), \quad \psi(\mathbf{r}) = \bar{\psi}(r,z)\bar{\Phi}(\phi), \tag{3}$$

where n is an integer making function $\bar{\Phi} = \exp(in\phi)$ a single-valued function such that $\bar{\Phi}(0) = \bar{\Phi}(2\pi)$, and the Hamiltonian of the problem can be represented in the form

$$\bar{H} = -\frac{\hbar^2}{2}\left[\frac{\partial}{\partial r}\left(\frac{1}{m(r,z)}\frac{\partial}{\partial r}\right) + \frac{\partial}{\partial z}\left(\frac{1}{m(r,z)}\frac{\partial}{\partial z}\right)\right] + \frac{\hbar^2 n^2}{2m(r,z)r^2} + V(r,z). \quad (4)$$

The boundary conditions can be derived based on the Ben-Daniel-Duke formulations (e.g., [2, 8]), relating the values of the wave function in the dot material and in the crystal matrix. In this paper we use the idea analogous to that proposed originally in [6].

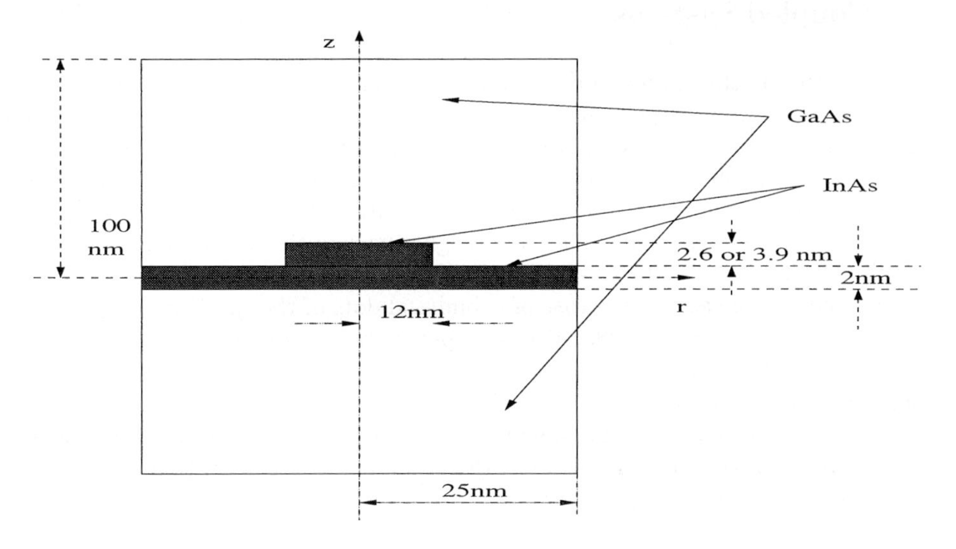

Fig. 1. Quantum dot geometries

3 Numerical Procedure

The model formulated in Section 2 is a partial differential equation eigenvalue problem which needs to be solved in order to determine energy levels and the corresponding wave functions of this coupled quantum dot/wetting layer system. The procedure applied in this paper is based on the spectral transformation Arnoldi iterations. This is a Krylov-subspace-based technique which allows us to deal effectively with large matrices resulting from the numerical discretisation of (3), (4). In particular, in solving the discretised problem

$$[A - ((E - V)I]\mathbf{x} = \mathbf{g} \quad (5)$$

with given matrix $A : n \times n$ and $\mathbf{g} : 1 \times n$ (resulting from the discretisation of (3) and (4)), and with n varying from 10^4 to 10^5, we construct iterations with respect to a sequence of Krylov subspaces

$$\mathbf{x}, A\mathbf{x}, A^2\mathbf{x}, ..., A^{m-1}\mathbf{x}, ... \quad (6)$$

Computations have been carried out on a sequence of grids for all numerical results reported in the next section.

Geometric characteristics of the structure under consideration are given in Fig.1. Other parameters used in computations include the conduction-band effective masses ($0.023m^*$ and $0.067m^*$ for InAs and GaAs, respectively, with m^* being the free electron mass) and band-edge energies (-0.697 and 0 for InAs and GaAs, respectively).

4 Computational Analysis of Electronic States in QD/WL Coupled Systems

We present in this paper four groups of experiments. Firstly, we consider a cylindrically shaped quantum dot with wetting layer. The size of the dot is 24 nm at the basis (the computational domain is half of this value due to the cylindrical symmetry). The height of the cylinder is 2.6 nm. In Fig. 2 we present the ground state and the first three excited states. Recall that for the conical quantum dots with the same base and height the resulting eigenvalues reported in [6] (-0.409, - 0.307, -0.253, -0.150) are higher compared to their counterparts in the cylindrical case. In the case of cylindrical dots of the specified dimensions they are -0.474, -0. 324, -0.289, -0.170, respectively (we rounded off all results to include the third digit after the dot only). In both cases, however, the ground state is clearly localisable.

In the second group of experiments we neglect the wetting layer and consider the "pure" quantum dot. In this case, the ground state (see Fig. 3 left) is -0.343 which is substantially larger compared to the result obtained for the quantum dot with wetting layer. The first (the only excited state in this case) is -0.186, and this state much better confined compared to the corresponding state in the quantum dot/wetting layer case. The wave functions of these two states are represented in Fig. 3.

Next, we consider the wetting layer only. In this case, the profile of the wave function has a zero-slope shape, typical for quantum wells (see Fig. 4 left), corresponding to the ground state eigenvalue which has the value of -0.310. In this case, the three excited states are -0.281, -0.212, and -0.110, respectively.

By increasing the height of the dot by factor of 1.5 and keeping its width the same, for the dot without wetting layer we obtain values of -0.437 and -0. 264 for the ground and the first three excited states, respectively, which are the only eigenvalues in this case. As expected, these values are smaller compared to the "pure" quantum dot of smaller height. The ground state wave function in this case is represented in Fig. 4 (right). Finally, we note that the ground state and the first three excited states for the new quantum dot/wetting layer structure correspond to the values of -0.348, -0.300, -0.182, -0.062, respectively. These values produce wave functions similar to those presented in Fig. 2, but, as it is expected, their values are larger compared to smaller height cylindrical quantum dots with wetting layers.

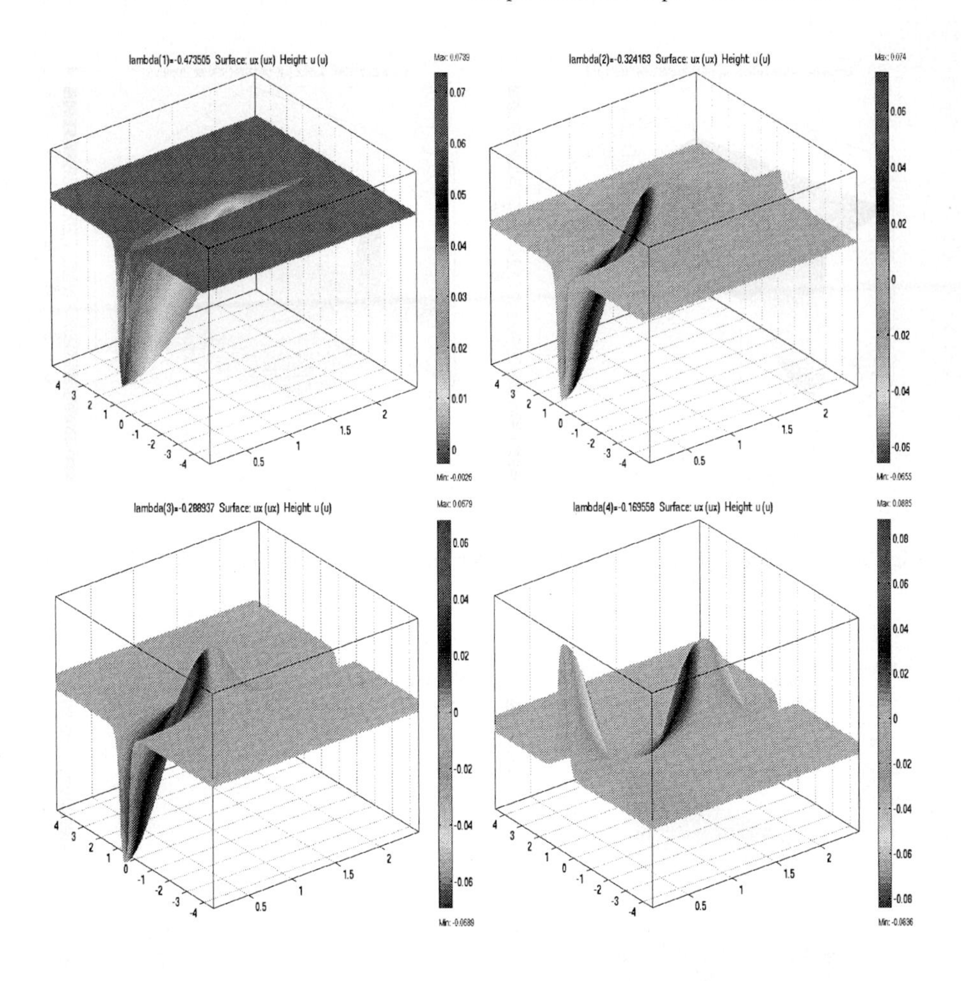

Fig. 2. Cylindrical QD with wetting layer: ground state and the first three excited states.

Acknowledgment.

The first author is grateful to Morten Willatzen for many fruitful discussions on the topics of this paper.

References

1. Adeler, F. et al: Optical Transitions and Carrier Relaxation in Self-Assembled InAs/GaAs Quantum Dots. J. Appl. Phys. **80** (1996) 4019—-4026
2. Bastard, G.: Wave Mechanics Applied to Semiconductor Heterostructures. Halsted Pressm (1988)

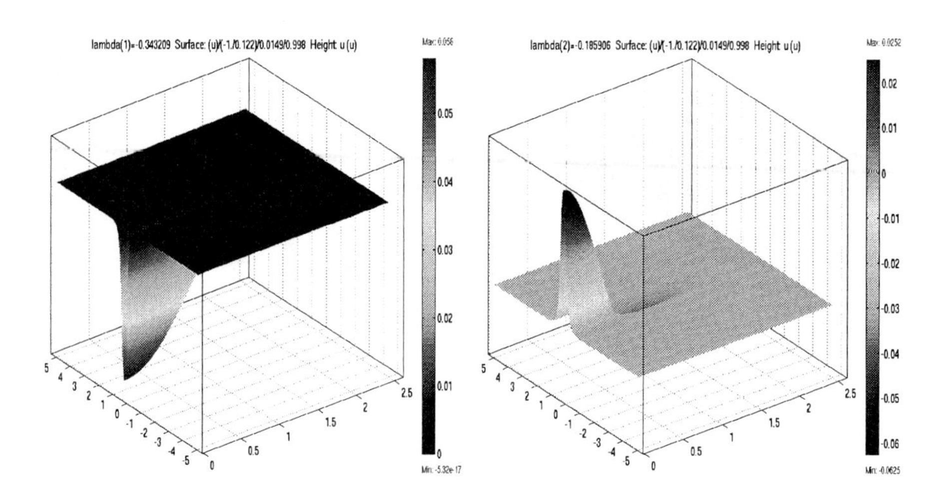

Fig. 3. Cylindrical QD without wetting layer: ground state and the first excited state.

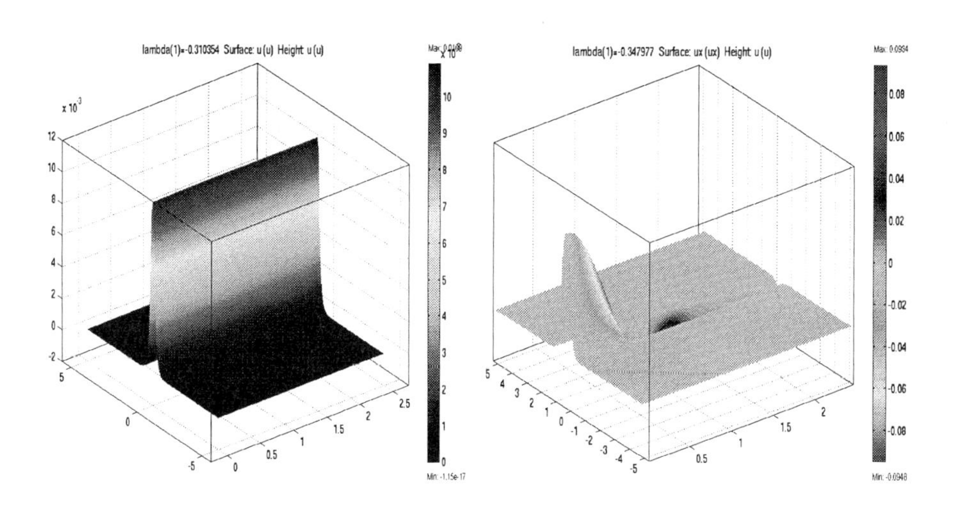

Fig. 4. Ground state of the wetting layer (left) and the ground state of the increased-height dot with wetting layer (right).

3. Gelbard, F. and Malloy, K.J.: Modeling Quantum Structures with BEM. J. Comp. Phys. **172** (2001) 19–39
4. Bimberg, D., Grundmann, M., and Ledentsov, N.N.: Quantum Dot Heterostructures. John Wiley and Sons, UK (1998)
5. Melnik, R.V.N., He, H.: Modelling nonlocal processes in semiconductor devices with exponential difference schemes. Journal of Engineering Mathematics. **38** (2000) 233–263
6. Melnik, R.V.N. and Willatzen, M.: Modelling Coupled Motion of Electrons in Quantum Dots with Wetting Layers. Proceedings of the Modeling and Simulation of Microsystems (MSM) Conference, April 21–25, 2002, USA. (2002) 506–509
7. Li, Y. et al: Electron Energy Level Calculations for Cylindrical Narrow Gap Semiconductor Quantum Dot. Computer Physics Communications. **140** (2001) 399–404
8. Li, Y. et al: Computer Simulation of Electron Energy Levels for Different Shape InAs/GaAs Semiconductor Quantum Dots. Computer Physics Communications. **141** (2001) 66–72
9. Stier, O., Grundmann, M., and Bimberg, D.: Electronic and Optical Properties of Strained QDs modeled by 8-band $k \cdot p$ Theory. Phys. Rev. B. **59** (1999) 5688–5701
10. Williams, R.L. et al: Controlling the Self-Assembly of InAs/InP Quantum Dots. J. of Crystal Growth. **223** (2001) 321–331

Deadlock Free Specification Based on Local Process Properties

D.P.Simpson and J.S.Reeve

The Department of Electronics and Computer Science,
University of Southampton, Southampton SO17 1BJ, UK
jsr@ecs.soton.ac.uk

Abstract. We present a design methodology for the construction of parallel programs that is deadlock free, Provided that the "components" of the program are constructed according to a set of locally applied rules. In our model, a parallel program is a set of processes and a set of events. Each event is shared by two processes only and each process progresses cyclically. Events are distinguished as input and output events with respect to their two participating processes. On each cycle a process must complete all output events that it offers to the environment, be prepared to accept any, and accept at least one, of its input events before completing any computations and starting a new cycle. We show that however the events are distributed among the processes, the program is deadlock free. Using this model we can construct libraries of constituent processes that do not require any global analysis to establish freedom from deadlock when they are used to construct complete parallel programs.

1 Introduction

In this paper we develop a programming design methodology for parallel programs that manifestly avoids deadlock. Deadlock is an issue in many different types of program and examples of explicit attempts at producing or establishing deadlock free formulations can be found in applications ranging from robotics[12] and control[6], to scheduling[1, 7] and graph reduction[19]. More general methodologies[2, 11] have also been published. The programming strategy that we establish in this paper is a general method of constructing algorithmically parallel[16] programs so that they cannot deadlock. Our methodology comes in two flavours, one is a synchronised model which is guaranteed to produce "live" programs but at the cost of some redundant communications. The other, a non-synchronised model, removes the event redundancy at the cost of "liveness". In practice however, as we show in §5 liveness is restored by a fair queueing scheme.
This paper shows how to structure and combine components, using only 'local' or component centred rules to ensure that the resulting program is free from deadlock, even with arbitrary cycles in the communication graph. No global requirements, other than each output must have a corresponding input, are necessary, and no overall synchronization is needed. The programming model is readily implementable with existing software, for example MPI [20].

P.M.A. Sloot et al. (Eds.): ICCS 2003, LNCS 2659, pp. 350–359, 2003.

A program is composed of components which cycle continuously. Each component is structured in pseudo code as

```
y = y0 INITIAL OUTPUT VARIABLES
x = x0 INITIAL INPUT VARIABLES
FOREVER
   DO IN PARALLEL
      OUTPUT y
      INPUT AT LEAST ONE x
   END DO
   COMPUTE y = f(x)
```

The details of the computation are application dependent, but we assume that it comprises actions internal to the local process only and has no bearing on the event sequence of the overall program.

Deadlock due to components waiting to send a message before they are willing to receive one is impossible because wanting to send a message implies being willing to read one, unless a message has been read from that source in this cycle already. Consequently closed cycles don't appear in the communication graph[13]. Furthermore, no fairness conditions, which specify "equal treatment" conditions, of any kind are required. Components can compute an appropriate subset of their output to exercise, which can be different on each cycle, without invalidating the result. We call this the asynchronous model. A related model ,the synchronous model, which also has the same freedom from deadlock, exercises all its output on every cycle.

In practice the asynchronous model is more efficient as fewer messages are being passed around the system but the synchronous model has long been used to ensure deadlock freedom in Petri-Net designs [5, 9] or other state transition based methods[3, 14, 15].

Figure 1 is a pictorial representation of a particular parallel program, the alphabet of which is $\{a, b, c, d, e, f\}$ and the constituent processes are $\{P_1, P_2, P_3, P_4, P_5\}$. This program can operate according to our asynchronous or synchronous rules, but not a mixture of both. The alphabet is partitioned into input events and output events with respect to each constituent process. For example the input event set (depicted by the inward arrows) for process P_1 is $I_1 = \{d, f\}$ and the output event set (depicted by the outward arrows) is $O_1 = \{a\}$. Similarly, $I_2 = \{g\}$ and $O_2 = \{e, f\}$, and so on. The precise definition of components is written in CSP. The proof establishes that it is impossible for any output to be permanently blocked. In §2 we give some background to Communicating Sequential Processes, CSP, the underlying methodology that we use, and in §3 and §4 we formally define the synchronous and asynchronous models respectively and establish the terms and conditions of their freedom from deadlock. Finally in §5 we show how these models are effectively used in practice.

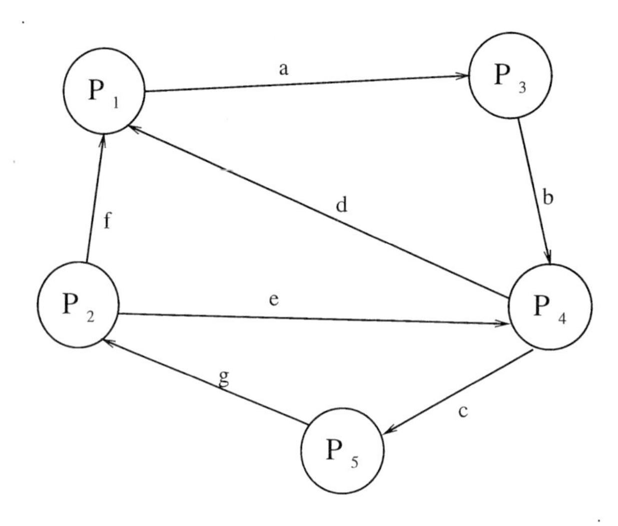

Fig. 1. A Diagrammatic Representation of a Program

2 Theoretical Background

In this paper we use the failures model of Communicating Sequential Processes (CSP) to formally define and verify our programming model . This is a process based model that focuses on the communications events that occur between processes. These events are assumed to be atomic and instantaneous and for an event a to happen, it must be engaged in simultaneously by all processes that contain a in their alphabet. The principle introductory references to CSP are the books by Hoare [8] and Roscoe [17].

A process P is a 3-tuple $(\alpha P, \mathit{failures}(P), \mathit{divergences}(P))$. The alphabet of P, αP, is the set of events that P can engage in . The list of events that a process has engaged in, is its *trace*, and the set of all possible traces of P is $\mathit{traces}(P)$. $\mathit{failures}(P)$ is a set of pairs, (s, R), such that P can engage in the action of $s \in \mathit{traces}(P)$ in order and then refuse to engage in all the actions in the *refusal* set R. After engaging in a trace $t \in \mathit{divergences}(P)$, P can engage in an unbounded sequence of events from its alphabet. This is *livelock*.

P can only affect events in its alphabet and in particular $P \parallel Q$ must synchronize on events in their common alphabet. After engaging in the trace t, P becomes P/t.

In the definitions in the following sections we make use of the special process, RUN, which is a process that never diverges nor refuses any event in its alphabet. This paper uses RUN to eliminate the effect of terminating processes which have done their job via the identity $RUN \parallel P = P$ for any process P provided that $\alpha RUN \subseteq \alpha P$. We also use the result $RUN_A \parallel RUN_B = RUN_{A \cup B}$, which is

readily proven by establishing the equivalence of the *alphabets*, *divergences* and *failures* of the processes ($RUN_A \parallel RUN_B$) and $RUN_{A \cup B}$.
Practical parallel programming processes communicate by passing messages or sharing data at some common location. CSP is not concerned with the 'direction' in which the data is flowing, only in the fact that all the processes with a given event in their alphabet are synchronised when that event happens. Later in this paper we will have to distinguish between *input* and *output* processes because events common to two processes are handled differently by those sharing processes.
The following standard[18] definition means that for our programming model we must establish that the *refusals* set is never the entire alphabet in order to prove our assertion of deadlock freedom.

Definition 1 (Deadlock Freedom) *A process P is deadlock free, if and only if* $\nexists\, t \in traces(P)$ *such that* $(t, \alpha P) \in \mathrm{failures}(P)$.

The definitions and proofs that establish deadlock freedom for our programming model in this paper are easily established, since the components of the model are constructed to do so. This means that programs constructed according to our prescription do not require a global analysis to establish deadlock freedom as do more general programming models[17].

3 The Synchronous Programming Model

We begin with a simplified model which we call the synchronous programming model. The characteristics of this model are that each component process (node in the graphical representation) executes cyclically and on each cycle engages in all its *input* and *output* events before computing new output variable values. Since every *output* and *input* event is allowed to happen in parallel there are no cycles of event dependency and so deadlock is manifestly not possible. Moreover, every event happens on every cycle. This is the standard way in which deadlock is avoided in state machine driven embedded systems [3].
In the following section we will relax the condition that every input event be satisfied on every cycle. This, while retaining the freedom from deadlock property, eliminates messages(in real systems) that are only there to establish that freedom from deadlock. This relaxation does however, introduce the possibility of some events being refused for an indeterminate number of cycles. In practice though, simple fairness implementations, like FIFO queueing [4] of messages at the constituent processes readily rectifys this.
A synchronous program *SP* is a finite collection of constituent processes $\{S_1, S_2, \ldots S_n\}$ that interact through a set of events IO. The event set IO has two partitions $\pi_I(IO) = \cup_{j=1,n} I_j$ and $\pi_O(IO) = \cup_{j=1,n} O_j$. I_j and O_j are termed the *input* and *output* events of S_j respectively. In addition each element $io \in IO$ is contained in the alphabet of two, and only two, processes. Hence $\forall\, io \in IO\, \exists$ unique I_i and O_j such that $O_j \cap I_i = \{io\}$

With these definitions in mind we define a synchronous program SP as

$$SP \equiv \|_j \ S_j$$
$$S_j \equiv OUT_j \parallel IN_j \parallel CYCLE_j \quad (1)$$

Each of the constituent processes S_j has a synchronising event s_j in its alphabet, which facilitates the cycling of the process once all *input* and *output* events have happened. Each S_j then has the structure:-

$$OUT \equiv \|_{o \in O} \ o \rightarrow RUN_s$$
$$IN \equiv \|_{i \in I} \ i \rightarrow RUN_s$$
$$CYCLE \equiv \Box_{io \in O \cup I} \ io \rightarrow CYCLE \ \Box \ s \rightarrow S \quad (2)$$

To accommodate sources and sinks we use the following special definitions when some or all of the constituent event sets are empty.

$$OUT \equiv RUN_s \text{ when } O = \{\}$$
$$IN \equiv RUN_s \text{ when } I = \{\}$$
$$CYCLE \equiv s \rightarrow S \text{ when } O \cup I = \{\} \quad (3)$$

The complete alphabet of S is $O \cup I \cup \{s\}$. The internal computation phase is assumed to happen "between" the s and S in the term $s \rightarrow S$ in the $CYCLE$ process. The *output* process OUT cannot engage in s until all *output* events have happened. Likewise, the *input* process IN cannot engage in s until all *input* events have happened.
We now establish that the OUT and IN processes must engage in all their events before engaging in the synchronising event s. In the following we use the notation $<[O]>$ to stand for a *trace* comprised of all the elements of O in some (unspecified) order.

Lemma 1 *All output events happen once before synchronisation. Provided that the environment is prepared to engage in the events of* O*, an output process,* OUT *as defined in equation 2 has a trace,* $t = <[O], s>$.

Proof *If* $O = \{\}$ *then* $OUT = RUN_s$ *and the lemma holds since* $t = <s>$. *When* O *is not empty,*

$$OUT = \|_{o \in O} \ o \rightarrow RUN_s$$

Because the constituent parallel processes have disjoint alphabets each can engage in its initial event before being ready to engage in s. OUT *can engage in the initial events* c *in any order so any trace of* OUT *is of the form* $t = <[O], s>$.

We note that the above result is a statement about the *trace* of OUT under the assumption that none of the events of O are in the refusal set of any other process. Also because the form of IN is the same as that of OUT, an equivalent result holds, namely:-

Lemma 2 *All input events happen once before synchronisation Provided that the environment is prepared to engage in the events of I, an input process, IN as defined in equation 2 has a trace, $t = <[I], s>$.*

Now we must show that a synchronous program SP provides the environment for lemma 1 and lemma 2 to hold, which will allow us to establish the cyclical nature of synchronous programs, prove that they are free from deadlock and moreover, show that they are "live", in the sense that every event happens on every cycle.

Theorem 1 *A synchronous program is deadlock free*
If SP is an asynchronous program as defined in equation 1, then
$\nexists\, t \in traces(SP)$ *s.t.* $(t, \alpha SP) \in failures(SP)$

Proof *The result required is that the $refusals(SP) \subset \alpha SP$ and we note that $refusals(P_1 \parallel P_2) = refusals(P_1) \cup refusals(P_2)$[8].*

$$\begin{aligned} refusals(SP) &= \cup_j refusals(S_j) \\ &= \cup_j (refusals(OUT_j) \cup refusals(IN_j) \cup refusals(CYCLE_j)) \end{aligned}$$

The refusals set $refusals(SP)$ is maximal when $refusals(S_j)$ are maximal $\forall\, j = 1, \ldots, n$. $refusals(OUT_j)$ and $refusals(IN_j)$ are maximal after all the O_j and I_j have happened, in which case:-

$$\begin{aligned} & refusals(S_j / <[I_j \cup O_j]>) = O_j \cup I_j \quad \text{and} \\ & refusals(CYCLE_j / <[I_j \cup O_j]>) = \{\} \end{aligned}$$

At this point only events in $SY = \{s_j \mid j = 1, \ldots n\}$ can happen. However

$$\begin{aligned} & OUT_j / <[O_j]> = RUN_{s_j} \\ & IN_j / <[I_j]> = RUN_{s_j} \quad \text{and} \\ & CYCLE_j / <[I_j \cup O_j]> = CYCLE_j \end{aligned}$$

Consequently $S_j / <[I_j \cup O_j], s_j> = S_j$ using lemma 1 and lemma 2 and so $SP / <[IO], [SY]> = SP$, and the maximal refusals set of $SP \subset \alpha SP$.

4 The Asynchronous Programming Model

We now seek to remove some of the restrictions of the synchronous model and in particular consider systems the component processes of which need not exercise all possible output events on every cycle. Our overall definition of an asynchronous program is the same as that of a synchronous program but the component processes behave differently.
Similar to a synchronous program, an asynchronous program AP is a finite collection of constituent processes $\{A_1, A_2, \ldots A_n\}$ that interact through a set of events IO, partitioned as in §3.

We define an asynchronous program AP as

$$\begin{aligned} AP &= \|_j\ A_j \\ A_j &= OUT_j \parallel IN_j \parallel CYCLE_j \end{aligned} \tag{4}$$

Each component component process has the form:

$$\begin{aligned} OUT &= KEY \ \|_{o' \in O'}\ o' \rightarrow RUN_s \\ KEY &= \|_{o^m \in O \setminus O'}\ (o^m \rightarrow RUN_s \ \Box\ RUN_s) \\ IN &= \|_{i \in I}\ (i \rightarrow RUN_s \ \Box\ RUN_s) \parallel LOCK \\ LOCK &= \Box_{i \in I}\ \left(i \rightarrow RUN_{I \cup \{s\}}\right) \\ CYCLE &= \Box_{io \in O \cup I}\ (io \rightarrow CYCLE) \ \Box\ s \rightarrow A' \end{aligned} \tag{5}$$

To accommodate sources and sinks we use the following special definitions when some or all of the constituent event sets are empty.

$$\begin{aligned} OUT &\equiv RUN_s \text{ when } O = \{\} \\ IN &\equiv RUN_s \text{ when } I = \{\} \\ CYCLE &\equiv s \rightarrow A' \text{ when } O \cup I = \{\} \end{aligned} \tag{6}$$

A is the overall component and s synchronises the various component processes. The complete alphabet of A is $O \cup I \cup \{s\}$. A computation phase is assumed to occur "between" the s_j and A'_j in the term $s_j \rightarrow A'_j$ but as the details of the computation phase of the cycle are irrelevant they are not modeled here. A process A_j, in general is allowed to select different active set of outputs $O'_j \subseteq O_j$ for each cycle on any basis. The term $s \rightarrow A'$ indicates that the "active" output set O' may be different on the next cycle. The *output* process OUT cannot terminate until all *active output* events have happened.
We now show that this formal definition has the attributes described in the introduction. The IN process ensures that at least one input events completes before IN is ready to engage in the synchronising event s. Now because each process can select the output events that must complete on every cycle, there remains the possibility that a process A_j could become isolated by virtue of not being offered any of its input events by any of the other processes. This scenario is circumvented by the presence of the KEY process in the OUT process.
The following lemma shows that for an output process OUT, all output events happen once and only once before the synchronising event s happens.

Lemma 3 *All output events happen once before synchronisation Provided that the environment is initially prepared to engage in O, an output process*

$$OUT = KEY \ \|_{o' \in O'}\ o' \rightarrow RUN_s$$

has a trace, $t = <[Q \cup O'], s>$, *where* $Q \subseteq O \setminus O'$.

Proof *If $O' = \{\}$ then $OUT = RUN_s$ and the lemma holds since $t =< s >$.*
When O' is not empty,

$$OUT = KEY \parallel_{o' \in O'} o' \rightarrow RUN_s$$

Because the constituent parallel processes have disjoint alphabets each can engage in its initial event before being ready to engage in s. OUT can engage in the initial events O' in any order and KEY can engage in any subset $Q \subseteq O \backslash O'$ so any trace of OUT is of the form $t =< [Q \cup O'], s >$.

The following lemma shows that for an input process IN, at least one input events happens before the synchronising event s happens. A particular input event can only happen once before s happens

Lemma 4 *At least one input event happens before synchronisation Provided that the environment is initially able to engage in I, an input process*

$$IN = \parallel_{i \in I} (i \rightarrow RUN_s \,\square\, RUN_s) \parallel LOCK$$
$$LOCK = \square_{i \in I} \left(i \rightarrow RUN_{I \cup \{s\}}\right)$$

has a trace, $t =< [Q], s >$, where $Q \subseteq I$.

Proof *If $I = \{\}$ then $IN = RUN_s$ and the lemma holds since $t =< s >$.*
When I is not empty,

$$IN = \parallel_{i \in I} (i \rightarrow RUN_s \,\square\, RUN_s) \parallel \square_{i \in I} \left(i \rightarrow RUN_{I \cup \{s\}}\right)$$

Both the process

$\square_{i \in I} \left(i \rightarrow RUN_{I \cup \{s\}}\right)$ *and the process* $\parallel_{i \in I} (i \rightarrow RUN_s \,\square\, RUN_s)$

can engage $Q \neq \{\} \subseteq I$, then

$$\square_{i \in I} \left(i \rightarrow RUN_{I \cup \{s\}}\right) / < [Q] >= RUN_{I \cup \{s\}}$$

and

$$\parallel_{i \in I} (i \rightarrow RUN_s \,\square\, RUN_s) / < [Q] >= RUN_s \parallel_{i = \in I \backslash Q} (i \rightarrow RUN_s \,\square\, RUN_s)$$

If s now occurs we have

$$RUN_{I \cup \{s\}} \parallel_{i \in I \backslash Q} (i \rightarrow RUN_s \,\square\, RUN_s) / < [Q], s >= RUN_{I \cup \{s\}}$$

then $t =< [Q], s >$ and the lemma is established.

Now we must show that a synchronous program AP provides the environment for lemma 3 and lemma 4 to hold, which will allow us to establish the cyclical nature of asynchronous programs, prove that they are free from deadlock. However asynchronous programs are not "live" in the sense that every event need not happen on every cycle.

Theorem 2 *An asynchronous program is deadlock free*
If AP is an asynchronous program as defined in equation 4, then
$\nexists\, t \in traces(AP)$ *s.t.* $(t, \alpha AP) \in failures(AP)$

Proof *The refusals set* $refusals(AP)$ *is maximal when* $refusals(S_j)$ *are maximal* $\forall j = 1, \ldots, n$. $refusals(OUT_j)$ *and* $refusals(IN_j)$ *are maximal after all the* O_j *and* I_j *have happened, in which case:-*

$$refusals(S_j / <[I_j \cup O_j]>) = O_j \cup I_j \qquad and$$
$$refusals(CYCLE_j / <[I_j \cup O_j]>) = \{\}$$

At this point only events in $SY = \{s_j \mid j = 1, \ldots n\}$ *can happen. However*

$$OUT_j / <[O_j]> = RUN_{s_j}$$
$$IN_j / <[I_j]> = RUN_{s_j} \qquad and$$
$$CYCLE_j / <[I_j \cup O_j]> = CYCLE_j$$

Consequently $A_j / <[I_j \cup O_j], s_j> = A_j$*, by lemma 3 and lemma 4 and so* $AP / <[IO], [SY]> = AP$*, and the maximal refusals set of* $AP \subset \alpha AP$.

5 Conclusion

We present a design methodology for the construction of parallel programs that is deadlock free, Provided that the "components" of the program are constructed according to a set of locally applied rules. In our model, a parallel program is a set of processes and a set of events. Each event is shared by two processes only and each process progresses cyclically. Events are distinguished as input and output events with respect to their two participating processes. On each cycle a process must complete all output events that it offers to the environment, be prepared to accept any, and accept at least one, of its input events before completing any computations and starting a new cycle. We show that however the events are distributed among the processes, the program is deadlock free. Using this model then, we can construct libraries of constituent processes that do not require any global analysis to establish freedom from deadlock when they are used to construct complete parallel programs[10].
In practice, parallel program construction according to our asynchronous model, is readily implemented and made more effective by incorporating local fairness. For example the MPI libraries provide means to input all waiting messages from unique sources, thereby assuring fair servicing of inputs while at the same time not allowing two inputs from the same source on the same program cycle.
We conclude that our programming model, although not suited to all possible parallel program communication patterns, particularly where the task graph is dynamic, does ensure deadlock freedom in algorithmic parallelism which is the principle paradigm used in the programming of embedded systems.

References

1. TF Abdelzaher and KG Shin. Combined task and message scheduling in distributed real-time systems. *IEEE Trans. Parallel Distrib. Syst.*, 10:1179–1191, 1999.
2. S Abramsky, SJ Gay, and R Nagarajan. A specification structure for deadlock-freedom of synchronous processes. *Theor. Comput. Sci.*, 222:1–53, 1999.
3. A. Arnold. *Finite Transition Systems.* Prentice Hall, Hemel Hempstead, 1992.
4. M. Ben-Ari. *Principles of Concurrent and Distributed Programming.* Prentice Hall, Hemel Hempstead, 1990.
5. DY Chao. Petri net synthesis and synchronization using knitting technique. *J. Inf. Sci. Eng.*, 15:543–568, 1999.
6. L Ferrarini, L Piroddi, and S Allegri. A comparative performance analysis of deadlock avoidance control algorithms for FMS. *J. Intell. Manuf.*, 10:569–585, 1999.
7. BP Gan and SJ Turner. An asynchronous protocol for virtual factory simulation on shared memory multiprocessor systems. *J. Oper. Res. Soc.*, 51:413–422, 2000.
8. C. A. R. Hoare. *Communicating Sequential Processes.* Prentice Hall, Hemel Hempstead, 1985.
9. K. Jensen and G. Rozenberg. *High-level Petri nets: Theory and Applications.* Springer-Verlag, Berlin, 1991.
10. J.S.Reeve and M.C.Rogers. MP: an application specific concurrent language. *Concurrency: Practice anmd Experience*, 8(4):313–333, May 1996.
11. D Kadamuddi and JJP Tsai. Clustering algorithm for parallelizing software systems in multiprocessors environment. *IEEE Trans. Softw. Eng.*, 26:340–361, 2000.
12. T Matsuura and Y Maeda. Deadlock avoidance of a multi-agent robot based on a network of chaotic elements. *Adv. Robot.*, 13:249–251, 1999.
13. T.G. Lewis nad H. EL-Rewini. *Introduction to Parallel Computing.* Prentice Hall, Hemel Hempstead, 1992.
14. TS Perraju and BE Prasad. An algorithm for maintaining working memory consistency in multiple rule firing systems. *Data Knowl. Eng.*, 32:181–198, 2000.
15. A Pnueli, N Shankar, and E Singerman. Fair synchronous transition systems and their liveness proofs. *IEEE Trans. Parallel Distrib. Syst.*, 1486:198–209, 1998.
16. D.J. Pritchard. Practical parallelism using transputer arrays. *Lecture Notes in Computer Science*, 258:278, 1987.
17. A. W. Roscoe. *The Theory and Practice of Concurrency.* Prentice Hall, Hemel Hempstead, 1998.
18. A. W. Roscoe and N. Daith. The pursuit of deadlock freedom. *Information and Computation*, 75(3):289–327, 1987.
19. W Sadiq and ME Orlowska. Analyzing process models using graph reduction techniques. *Inf. Syst.*, 25:117–134, 2000.
20. M. Snir, S.W. Otto, S. Huss-Lederman, D.W. Walker, and J. Dongarra. *MPI: The Complete Reference.* The MIT Press, Cambridge, Massachusetts, 1996.

On the Reconfiguration Algorithm for Fault-Tolerant VLSI Arrays

Jigang Wu and Srikanthan Thambipillai

Centre for High Performance Embedded Systems
Nanyang Technological University, Singapore, 639798
{asjgwu,astsrikan}@ntu.edu.sg

Abstract. In this paper, an improved algorithm is presented for the NP-complete problem of reconfiguring a two-dimensional degradable VLSI array under the row and column routing constraints. The proposed algorithm adopts the partial computing for the logical row exclusion so that the most efficient algorithm, cited in literature, is speeded up without loss of performance. In addition, a flaw in the earlier approach is also addressed. Experimental results show that our algorithm is approximately 50% faster than the above stated algorithm.

Keywords: degradable VLSI array, reconfiguration, fault-tolerance, greedy algorithm, NP-completeness

1 Introduction

The mesh-connected processor array has a regular and modular structure and allows fast implementation of most signal and image processing algorithms. With the advancement in VLSI (very large scale integration) and WSI (wafer scale integration) technologies, integrated systems for mesh-connected processor arrays can now be built on a single chip or wafer. As the density of VLSI and WSI arrays increases, the probability of the occurrence of defects in the arrays during fabrication also increases. These defects obviously affect the reliability of the whole system. Thus, fault-tolerant technologies must be employed to enhance the yield and reliability of VLSI/WSI arrays.

There are generally two methods for reconfiguration in fault-tolerant technologies, namely, the redundancy approach and the degradation approach. Various strategies to restructure a faulty physical system using the redundancy approach are described in many papers, e.g.,[1–12]. Degradation approach uses as many fault-free elements as possible to construct a target system. The final dimension is flexible and depends on the needs of the application. Usually, a maximum dimension is desirable. Literatures [13–15] have studied the problem of reconfiguring two-dimensional degradable arrays. They have shown that most problems that arise under the constraint *row and column rerouting* are NP-complete.

In this paper, we consider the reconfiguration problem of two-dimensional degradable VLSI/WSI arrays. It is defined as follows [13–15].

P.M.A. Sloot et al. (Eds.): ICCS 2003, LNCS 2659, pp. 360–366, 2003.

Given an $m \times n$ mesh-connected host array H with the fault density ρ ($0 \leq \rho < 1$), integers r and c, find a $m' \times n'$ fault-free subarray T under the row and column rerouting scheme such that $m' \geq r$ and $n' \geq c$.

The latest work for this problem is the algorithm in [15], which is denoted as *RCRoute* in this paper. This algorithm is dominated by two sub-procedures named *Logical_Row_Exclusion* (*LRE*) and *Greedy_Column_Rerouting* (*GCR*), respectively. The time complexity of each sub-procedure is $O((1-\rho) \cdot m \cdot n)$. The routing manner in *GCR* leads to reconfiguring two neighboring fault-free elements lying in same physical row into the same logical column. But the related architecture does not support this kind of the routing. In this paper we point out this flaw and repair it. In addition, we also present a partial computing approach for the logical row exclusion. The new approach reduces the time complexity of *LRE* from $O((1-\rho) \cdot m \cdot n)$ into $O((1-\rho) \cdot n)$. Thus, we improve *RCRoute* in running time, without loss of performance. Experimental results show that the improved algorithm is approximately 50% faster than *RCRoute*.

2 Preliminaries

This section gives the definitions and the notations used in this paper.

In this paper, the original VLSI/WSI array that has been manufactured is called a host array. This host array may contain faulty elements. A degradable subarray of the host array, which contains no faulty element, is called a target array or logical array. The rows (columns) in the host array and target array are called physical rows (columns) and logical rows (columns), respectively. $row(e)$ ($col(e)$) denotes the physical row (column) index of element e. H (S) denotes the host (logical) array. R_i denotes the ith logical row. Using the same assumptions as in [13][14][15], in this paper two neighboring elements in the host array are connected by a four-port switch. All switches and links in an array are assumed to be fault-free since they have very simple structure.

In a host array, if $e(i, j+1)$ is a faulty element, then $e(i, j)$ can communicate with $e(i, j+2)$ directly and data will bypass $e(i, j+1)$. This scheme is called row bypass scheme. If $e(i, j)$ can connect directly to $e(i', j+1)$ with external switches, where $|i' - i| \leq d$, this scheme is called row rerouting scheme, d is called row compensation distance. The *column bypass scheme* and the *column rerouting scheme* can be defined similarly. By limiting the compensation distance to 1, we essentially localize the movements of reconfiguration in order to avoid complex reconfiguration algorithm. In all figures of this paper, the shaded boxes stand for faulty elements and the white ones stand for the fault-free elements.

3 Algorithms

In this section we point out a flaw in *RCRoute*[15] and repair it. Then we present our algorithm denoted *New_RCRoute* in this paper.

3.1 Updating *RCRoute*

For the *row and column rerouting* scheme, the latest efficient work is the algorithm *RCRoute*[15]. This greedy algorithm consists of two procedures namely *Row_First* and *Column_First*. *Column_First* is invoked after running *Row_First*. *Column_First* is identical to *Row_First* except that the roles of rows and columns are interchanged. Therefore, *Row_First* plays a key role in the description of *RCRoute*. For the detail description of *Row_First*, see [15].

There are two key sub-procedures in *Row_First*, denoted as *LRE* and *GCR*. The sub-procedure *LRE* selects one row to be excluded from the set that was previously selected and uses it to compensate for faulty elements in its two neighboring rows. Let M_i denote the maximum number of logical columns that pass through two consecutive rows R_i and R_{i+1}, where $i \in \{0, 1, \ldots, k-1\}$. Let $M_\gamma = \min_{0 \leq i \leq k-1} M_i$. *LRE* first calculates M_γ and selects the row R_γ, and then decides whether R_γ or $R_{\gamma+1}$ will be excluded. Let X denote the maximum number of logical columns that pass through $R_{\gamma-1}$ and $R_{\gamma+1}$. Let Y denote the maximum number of logical columns that pass through R_γ and $R_{\gamma+2}$. If $X > Y$, then row R_γ is selected for exclusion, otherwise, $R_{\gamma+1}$ is excluded.

GCR is used for finding a target array that contains a set of selected logical rows. It reroutes the fault-free elements to form logical columns. The successor of the fault free element u in R_i is selected from $Adj(u)$ in a left-to-right manner, where $Adj(u) = \{v : v \in R_{i+1},\ v \text{ is fault-free},\ \ |col(u) - col(v)| \leq 1\}$. For the detailed description of the procedure, see [14, 15].

The sub-procedures *LRE* and *GCR* are executed iteratively until the row-based target array is found. *LRE* tests $O((1-\rho) \cdot m \cdot n)$ valid interconnections in the $m \times n$ host array in a row by row fashion, and *GCR* tests these valid interconnections column by column. Obviously, they have same time complexity $O((1-\rho) \cdot m \cdot n)$.

As can be seen from [15], the four-port switch model has a very simple architecture. But it is due to this simple construction that provides less functions that the switch model does not support reconfiguring two neighboring fault-free elements lying in same row into same logical column. Assume R_γ is to be excluded by subprocedure *LRE*. Then R_γ will be used to compensate for faulty elements in its two neighboring rows, $R_{\gamma-1}$ and $R_{\gamma+1}$, *i.e.*, each fault-free element $e(\gamma, j)$, $1 \leq j \leq n$, will be used to compensate $e(\gamma-1, j)$ or $e(\gamma+1, j)$ if they are faulty. After compensation, The subprocedure *GCR* will be executed on $\{R_1, R_2, \cdots, R_{\gamma-1}, R_{\gamma+1}, \cdots, R_m\}$ to find the current target array. However, it is possible that two neighboring fault-free elements in R_γ will be rerouted into same logical column by *GCR*. That is the flaw of *RCRoute*.

We correct *RCRoute* by adding the constraint $row(u) < row(v)$ into the definition of set $Adj(u)$, *i.e.*, let $Adj(u) = \{v : v \in R_{i+1}, v \text{ is fault-free}, |col(u) - col(v)| \leq 1 \text{ and } row(u) < row(v)\}$. The constraint limits the elements in the result logical column, in the strictly increasing order of their physical row indices. This prevents the above conflict when *GCR* runs.

3.2 Partial Computing for Logical Row Exclusion

Assume $R_1, R_2, \ldots, R_k$ (initially, $k = m$) are the previously selected logical rows. In order to select one row, say R_l, to be excluded in the current iteration, *LRE* takes $O((1-\rho) \cdot k \cdot n)$ time to calculate $M_1, M_2, \ldots, M_{k-1}$ and takes $O(k)$ to select the minimum M_γ resulting in $l = \gamma$ or $\gamma + 1$ according to the compensation approach. In *Row_First*[15], *LRE* does not reuse the previous calculation of $M_i, 1 \leq i < k$. In fact, except for M_{l-2}, M_l and M_{l+1}, these M_i calculated in the previous iteration are also available in the current iteration as the compensations by R_l only affect R_{l-1} and R_{l+1}. We only need to update M_{l-2}, M_l and M_{l+1}. Hence, we simplify the sub-procedure *LRE* into *New_LRE* (Fig. 1 (left)) in order to avoid the repeat calculations of M_i except for M_{l-2}, M_l and M_{l+1}. The initial values of each M_i can be calculated out of the body of while-loop in *Row_First*[15]. Obviously, this simplified approach saves the running time of *RCRoute* but does not affect its harvest since the strategy for the selection of the row to be excluded has not been changed. By simple analysis, the running time of *LRE* in qth iteration of *Row_First* is reduced from $O((1-\rho) \cdot q \cdot n)$ to $O((1-\rho) \cdot n)$ since only M_{l-2}, M_l and M_{l+1} need to be updated, where $1 \leq q \leq m$. In the other hand, *New_LRE* still uses the same compensation strategies as described in *RCRoute*[15] after the row R_l is excluded. The time complexity of the compensation is $O((1-\rho) \cdot n)$. Hence, the time complexity of *New_LRE* is $O((1-\rho) \cdot n)$, which is far lower than $O((1-\rho) \cdot m \cdot n)$, the time complexity of *LRE*. Figure 1 shows the formal description of the improved algorithm.

```
Procedure New_ Logical_Row_Exclusion ( H, S, l, m );
begin

    /* select the minimal M_γ */
    Min := ∞;
    for i := 1 to m-1 do
        If M_i < Min then begin Min := M_i; γ := i end;

    /* select R_l to delete and compensate */
    Greedy_Column_Rerouting (H, R_γ-1, R_γ+1, X);
    Greedy_Column_Rerouting (H, R_γ, R_γ+2, Y);
    if X < Y then l := γ+1  /* delete row R_γ+1 */
             else l := γ;   /* delete row R_γ */
    Row_Reroute(H, R_l-1, R_l, R_l+1)[6]; /* compensation with R_l*/

    /* update M_l-2, M_l-1, M_l and M_l+1 */
    Greedy_Column_Rerouting (H, R_l-2, R_l-1, M_l-2);
    M_l-1 := ∞;  /* M_l-1 will not be considered in the next iteration */
    Greedy_Column_Rerouting (H, R_l-1, R_l+1, M_l);
    Greedy_Column_Rerouting (H, R_l+1, R_l+2, M_l+1);

end.
```

```
Procedure New_Row_First( H, m, n, r, c, row ,col );
/* Rerouting based on rows */
begin
    row_first_fail:= false;
    S={ R_1, R_2, ..., R_m };
    for i=1 to m-1 do /* calculate each M_i */
        Greedy_Column_Rerouting (H, R_i, R_i+1, M_i );
    Greedy_Column_Rerouting (H, S, n' ); /* for initial solution*/
    /* n'= maximum number of logical columns through the rows in S */
    m':=m;
    while (m' ≥ r) and (n' < c) do
    begin
        New_Logical_Row_Exclusion(H, S, γ, m' );
        Delete row R_γ from S;
        Greedy_Column_Rerouting ( H, S, n' );
        m':=m'-1;
    end;
    if (m' ≥ r) and (n' ≥ c)
    then
        begin row:=m'; col:=n'; end
    else row_first_fail:=true;
end;
```

Fig. 1. The formal description of the procedure *New_LRE* and *New_Row_First*

3.3 Main Algorithm and Complexity

For the description of the main algorithm, we first describe a procedure called *New_Row_First*, which is used to find a target array with maximum size based on the row. Let m' be the number of logical rows and n' be the number of logical columns of the target array. The current logical array is $S = \{R_1, R_2, \cdots, R_{m'}\}$. Initially, all rows in the host array are selected for inclusion into the target array. Thus, each logical row in S is also a physical row. The formal description of *New_Row_First* is shown in Fig. 1 (right).

In *New_Row_First*, the code before the while-loop initializes the data structures, calculates all M_i and gets the initial target array. The running time needed by all these steps is bounded by $O((1-\rho)\cdot m \cdot n)$. In the while-loop, *New_LRE* runs in $O((1-\rho)\cdot n)$ and *GCR* runs in $O((1-\rho)\cdot m \cdot n)$. Hence, the while-loop runs in $O((m-r)\cdot(1-\rho)\cdot m\cdot n)$, *i.e.*, the time complexity of *New_Row_First* is $O((m-r)\cdot(1-\rho)\cdot m\cdot n)$.

Similarly, we can describe a procedure *New_Column_First* to find a target array with maximum size based on the column. Its time complexity is $O((n-c)\cdot(1-\rho)\cdot m\cdot n)$.

The structure of the main algorithm, denoted as *New_RCRoute*, is the same as that of *RCRoute*. but the subprocedures have been improved. The largest array derived from procedures *New_Row_First* and *New_Column_First* is taken as the target array for H. The time complexity of algorithm *New_RCRoute* is $O(\max\{(m-r),(n-c)\}\cdot(1-\rho)\cdot m\cdot n)$, which is the lower bound of the complexity for row and column rerouting[15].

4 Experimental Results

We report our experimental results in this section. We have implemented the algorithm *RCRoute* (corrected version) and the improved algorithm *New_RCRoute* in C on a personal computer—Intel Pentium-III 500 MHZ. The implementations of the two algorithms are modified accordingly to find maximal target arrays and maximal square target arrays. In our experiments, *harvest* and *degradation*, formulated in [13],[14] and [15], are calculated for each target array. In order to make a fair comparison between *New_RCRoute* and *RCRoute*, we keep the same assumptions as in [14],[15], *i.e.*, the faults in random host arrays were generated by a uniform random generator; The fault size in host array is from 0.1% to 10% for the experiments. Both algorithms are tested with the same random input instances. The running time and the size of each target array obtained by *New_RCRoute* is compared with the corresponding array obtained by *RCRoute*[15]. Table 1 summarizes the experimental results for the random host arrays with different sizes. The improvement in running time is calculated by

$$(1-\frac{running\_time\_of\_New\_RCRoute}{running\_time\_of\_RCRoute})\times 100\%.$$

The calculations required to arrive at solutions for maximal target array encompass the solution for maximal square target array. Hence, without loss of

Table 1. The comparison of running time for 20 random instances

			Performance		Running Time		
Host Array	Fault Size	(%)	Harvest (%)	Degrad. (%)	RCRoute (s)	New_RCRoute (s)	Improve (%)
64 × 64	4	0.1	98.53	1.56	0.296	0.150	49.3
64 × 64	40	1.0	96.29	4.65	0.296	0.149	49.7
64 × 64	409	10.0	84.52	23.91	0.286	0.143	50.0
128 × 128	16	0.1	98.85	1.24	2.313	1.191	48.5
128 × 128	163	1.0	97.15	3.82	2.307	1.189	48.5
128 × 128	1638	10.0	84.61	23.84	2.213	1.152	47.9
256 × 256	65	0.1	99.24	0.86	18.316	9.441	48.5
256 × 256	655	1.0	97.56	3.41	18.160	9.369	48.4
256 × 256	6553	10.0	84.37	24.07	17.340	9.067	47.7
512 × 512	262	0.1	99.41	0.69	147.291	76.148	48.3
512 × 512	2621	1.0	97.92	3.06	145.004	75.081	48.2
512 × 512	26214	10.0	84.89	23.60	137.259	72.033	47.5

generality, we collect the running time only in the case of finding maximal target array. Table 1 shows the running time comparisons for the maximal target array. For each random instance, the running time required by *New_RCRoute* is significantly less than that required by *RCRoute*. For example, for the host array of size 256 × 256 with 655 fault elements, the running time for *New_RCRoute* is 9.369 seconds, while it is 18.160 seconds for *RCRoute*. The improvement in running time is 48.4%, which is nearly equal to 50%. Except for small size instances such as 64 × 64, increase in fault density in the host array leads to less improvement in running time as more backtracking is needed in routing. For instance, for the 512 × 512 host array with the 26214 fault elements, the improvement in running time is 47.5%, which is a little less than 48.3%, the improvement running time for the 512 × 512 host array with the 262 fault elements.

We can conclude from the analysis above that our algorithm *New_RCRoute* reduced the running time by approximately 50%, especially, for low density of faults in the host arrays, without loss of harvest.

5 Conclusions

We have presented a degradation approach for the reconfiguration in VLSI/WSI arrays under the rerouting constraint *row and column rerouting*. We have proposed a new strategy for the row selection. The new strategy binds well for high-speed realizations. For different sized host arrays, our algorithm has maintained harvest and degradation while reducing the running time by approximately 50%. Method to overcome the flaw in one of the recent contributions in this area was also made. The improved algorithms have been implemented and experimental results have been collected. These running time results clearly reflect the underlying characteristics of the improved algorithm.

Acknowledgment. We wish to express our sincere thanks to the anonymous referees for their constructive suggestions. We are grateful to Ms Chandni R. Patel for pointing out oversights in an earlier draft of this paper.

References

1. T. E. Mangir and A. Avizienis, "Fault-tolerant Design for VLSI: Effect of Interconnection Requirements on Yield Improvement of VLSI design", *IEEE Trans. on Computers*, vol. 31, no. 7, pp. 609–615, July 1982.
2. J.W. Greene and A.E. Gamal, "Configuration of VLSI Array in the Presence of Defects", *J. ACM*, vol. 31, no. 4, pp. 694–717, Oct. 1984.
3. T. Leighton and A. E. Gamal. "Wafer-scal Integration of Systoric Arrays", *IEEE Trans. on Computer*, vol. 34, no. 5, pp. 448–461, May 1985.
4. C.W.H Lam, H.F. Li and R. Jakakumar, "A Study of Two Approaches for Reconfiguring Fault-tolerant Systoric Array", *IEEE Trans. on Computers*, vol. 38, no. 6, pp. 833–844, June 1989.
5. I. Koren and A.D. Singh, "Fault Tolerance in VLSI Circuits", *Computer*, vol. 23, no. 7, pp. 73–83, July 1990.
6. Y.Y. Chen, S.J. Upadhyaya and C. H. Cheng, "A Comprehensive Reconfiguration Scheme for Fault-tolerant VLSI/WSI Array Processors", *IEEE Trans. on Computers*, vol. 46, no. 12, pp. 1363–1371, Dec. 1997.
7. T. Horita and I. Takanami, "Fault-tolerant Processor Arrays Based on the 1.5-track Switches with Flexible Spare Distributions", *IEEE Trans. on Computers*, vol. 49, no. 6, pp. 542–552, June 2000.
8. S.Y. Kuo and W.K. Fuchs, "Efficient Spare Allocation for Reconfigurable Arrays", *IEEE Design and Test*, vol. 4, no. 7, pp. 24–31, Feb. 1987.
9. C.L. Wey and F. Lombardi, "On the Repair of Redundant RAM's", *IEEE Trans. on CAD of Integrated Circuits and Systems*, vol. 6, no. 2, pp. 222–231, Mar. 1987.
10. M.G. Sami and R. Stefabelli. "Reconfigurable Architectures for VLSI Processing Arrays", *Proc. IEEE*, vol. 74, no. 5, pp. 712–722, May 1986.
11. R. Negrini, M.G. Sami and R. Stefanelli, *Fault tolerance through reconfiguration in VLSI and WSI arrays*. The MIT Press, 1989
12. F. Distante, M.G. Sami and R. Stefanelli, "Harvesting through array partitioning: a solution to achieve defect tolerance Defect and Fault Tolerance in VLSI Systems", Proc. 1997 IEEE International Symp. *Defect and Fault Tolerance in VLSI Systems*, Paris, pp. 261–269, 1997
13. S.Y. Kuo and I.Y. Chen, "Efficient Reconfiguration Algorithms for Degradable VLSI/WSI Arrays." *IEEE Trans. on Computer-Aided Design*, vol. 11, no. 10, pp. 1289–1300, Oct. 1992.
14. C.P. Low and H.W. Leong, "On the Reconfiguration of Degradable VLSI/WSI Arrays", *IEEE Trans. on Computer-Aided Design of Integrated Circuits and Systems*, vol. 16, no. 10, pp. 1213–1221, Oct. 1997.
15. C.P. Low, "An Efficient Reconfiguration Algorithm for Degradable VLSI/WSI Arrays", *IEEE Trans. on Computers*, vol. 49, no. 6, pp. 553–559, June 2000.

Automated Generation of Kinetic Chemical Mechanisms Using Rewriting

Olivier Bournez[1], Guy-Marie Côme[2], Valérie Conraud[2], Hélène Kirchner[1], and Liliana Ibănescu[1]*

[1] LORIA & INRIA Lorraine
615, rue du Jardin Botanique, BP 101
F-54602 Villers-lès-Nancy Cedex, France
{Olivier.Bournez,Helene.Kirchner,Mariana-Liliana.Ibanescu}@loria.fr
[2] Département de Chimie Physique des Réactions (DCPR), INPL-ENSIC
1, rue Grandville, BP 451
F-54001, Nancy, France
{Guy-Marie.Come, Valerie.Conraud}@ensic.inpl-nancy.fr

Abstract. Several software systems have been developed recently for the automated generation of combustion reactions kinetic mechanisms using different representations of species and reactions and different generation algorithms. In parallel, several software systems based on rewriting have been developed for the easy modeling and prototyping of systems using rules controlled by strategies. This paper presents our current experience in using the rewrite system ELAN for prototyping the automatic generation of the combustion reactions mechanisms previously implemented in the EXGAS kinetic mechanism generator system. We describe how to express in ELAN acyclic and cyclic molecules, reactants, elementary reactions and the primary mechanism for acyclic species. Examples and generated outputs are given.

1 Introduction

Combustion reactions will still be the main source of energy in the 21st century. Understanding the fundamental mechanisms of these reactions is highly desirable for the optimal design and operation of efficient, safe and clean engines, burners, incinerators (see for e.g. [5]). Due to the complexity of mechanisms, several software systems have been developed for automatic generation of combustion reactions mechanisms using different representations of species and reactions, and different generation algorithms. Even if only simple molecules are used as reactants, the number of elementary reactions generated is very often big: for example 3662 reactions involving 470 species are generated by the software system EXGAS for explaining the combustion mechanism of n-hexane [7]. Traditional techniques for automatic generation of mechanism are limited by the computational power of computers, and by the difficulty to express chemists expertise into computational concepts.

* Work supported by Peugeot Citroën Automobiles.

P.M.A. Sloot et al. (Eds.): ICCS 2003, LNCS 2659, pp. 367–376, 2003.

Term and graph rewriting has been developed since the last thirty years, leading to a deep and solid corpus of knowledge about the rewrite relation induced by a set of equations (see for e.g. [2]). More recently, rule based languages focused on the use of rewriting as a modeling tool, which results in making the out-coming specification executable in a very efficient way (see for e.g. [8]).

The aim of our study is to use rewrite systems for automatic generation mechanisms. We plan to benefit from the elegance and expressiveness of rewriting as a computational paradigm implemented in the ELAN system [1,3] developed by the team of computer scientists at LORIA, and from the chemical expertise from the team of chemists at DCPR that developed EXGAS [5,15].

The paper is organized as follows: in Sect. 2 we describe the problem of generation of kinetic mechanisms, in Sect. 3 we introduce some notions about rewrite systems and in Sect. 4 we explain how to code in ELAN chemical objects like acyclic and cyclic molecules (in Sect. 4.1), reactants (in Sect. 4.2), elementary reactions (in Sect. 4.3) and the primary mechanism for acyclic species (in Sect. 4.5).

The results obtained so far and the positive feedback from chemists encourage us to continue our study, to set-up a prototype and to extend our technique to cyclic molecules.

2 Automatic Generation of Combustion Reactions Mechanisms

Combustion reactions take place in engines, burners and industrial chemical reactors to produce mechanical or thermal energy, and also to incinerate pollutants or to manufacture chemical substances. In order to understand the fundamental mechanism of combustion reactions in particular and gas-phase thermal reactions in general, different models are used for numerical simulations of the phenomena.

A methodology for the elaboration of fundamentals models for the gas-phase thermal reactions is given in [5] and the three main phases are described in Fig. 1:

- The elaboration of the initial reaction model consists of generating a mechanism for the reaction, by hand or using a computer program, of estimating the thermodynamic data of molecules and free radicals involved in the mechanism and the kinetic data of the elementary reactions constituting the mechanism.
- The numerical resolution of the balance equations of the laboratory reactor; the input data are the initial model of reaction, the operating conditions, as well as, possibly, transport data for the species.
- The validation of the reaction model by comparison of the results of simulations with experimental results and adjustment where necessary of certain thermodynamic, kinetic or transport parameters.

The final reaction model generated by this methodology can be incorporated into a calculation code of a reactor or a computer program of reacting fluid dynamics for simulations.

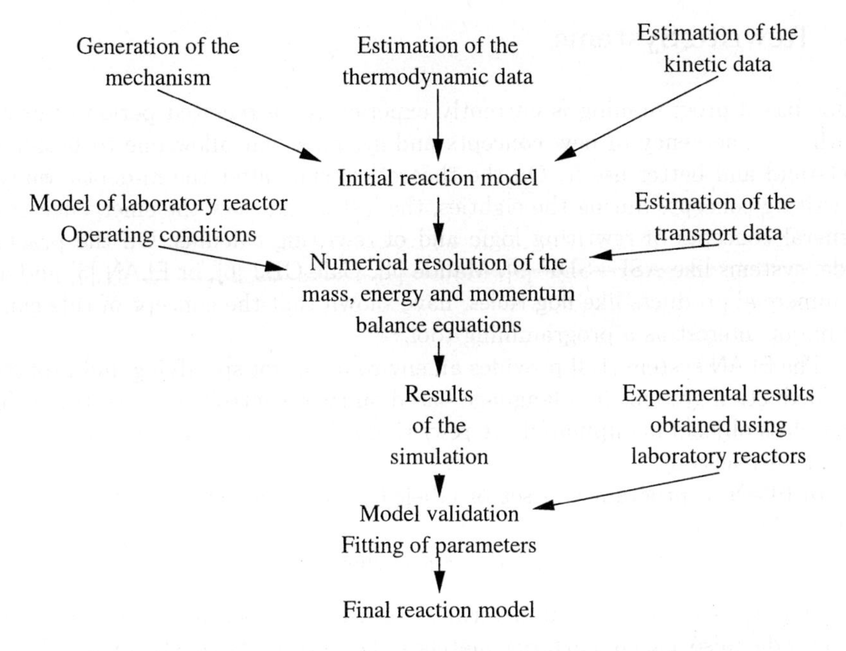

Fig. 1. Elaboration and validation of a model of a gas-phase thermal reaction

The generation of combustion mechanisms for a mixture of organic compounds in a large temperature field requires to consider several hundred chemical species and several thousands of elementary reactions. Therefore using computers is unavoidable. The generation of reactions mechanisms assisted by the computer has been rendered possible due to experimental and theoretical knowledge accumulated over the last three decades about the thermodynamic properties of molecules and free radicals, about elementary reactions and about the morphology of the reaction mechanism themselves (see for e.g. [5]).

A number of software systems have been developed for automatic generation of detailed mechanisms using both the expertise of researchers in thermodynamics and statistical mechanics, chemical kinetics, transport phenomena and theory of reactors and the expertise of computer scientists. A non exhaustive list of these software systems is the following: MAMOX [13], NetGen [14], EXGAS [5, 15], COMGEN [12].

Most of these systems are implemented using traditional imperative programming, using rather ad hoc data-structures and procedures for the representation of molecules and for the generation of the mechanisms. In particular, none of them benefits from flexibility and ease of modification of the rules that are used for the generation of mechanisms, even if their practical use requires to often modify the rules, according to chemist expertise or according to new experimental data.

3 Rewrite Systems

Rule-based programming is currently experiencing a renewed period of growth with the emergency of new concepts and systems that allow one to better understand and better use it. On the theoretical side, after the in-depth study of rewriting concepts during the eighties, the last decade saw the emergence of the general concepts of rewriting logic and of rewriting calculus. On the practical side, systems like ASF+SDF [9], Maude [4], Cafe-OBJ [6], or ELAN [3] and also commercial products like Ilog Rules, have shown that the concept of rule can be of major interest as a programming tool.

The ELAN system [1,3] provides an environment for specifying and prototyping deduction systems in a language based on rules controlled by strategies. The first class objects manipulated are rewriting rules, which may be used to model chemical reactions.

In ELAN, a program is a set of labeled conditional rewrite rules with local affectations

$$\ell : l \Rightarrow r \text{ if } c \text{ where } w$$

Informally, rewriting a ground term t consists of selecting a rule whose left-hand side (also called pattern) matches the current term (t), or a sub-term ($t_{|\omega}$), computing a substitution σ that gives the instantiation of rule variables ($l\sigma = t_{|\omega}$), and if the instantiated condition c is satisfied ($c\sigma$ reduces to *true*), applying substitution σ enriched by local affectation w to the right-hand side to build the reduced term.

In general, the normalization of a term may not terminate, or may terminate with different results corresponding to different selected rules, selected sub-terms or non-unicity of the substitution. So evaluation by rewriting is essentially non-deterministic and backtracking may be needed to generate all results. One of the main originalities of the ELAN language is to provide strategies as first class objects of the language. This allows the programmer to specify in a precise and natural way the control on the rule applications.

The full ELAN system includes a preprocessor, an interpreter, a compiler, and standard libraries available through the ELAN web page [1]. From the specific techniques developed for compiling strategy controlled rewrite systems [10,11], the ELAN compiler is able to generate code that applies up to 15 millions rewrite rules per second on typical examples where no non-determinism is involved and typically between 100 000 and one million controlled rewrite per second in presence of associative-commutative operators and non-determinism. We think that this may help to handle the combinatorial explosion of the number of reactions modeling a kinetic mechanism.

Note that among the possibilities offered by the ELAN system that we will use in this paper, we can notice that the ELAN system offers the power of mixfix signatures and matching, the possibility of defining and using associative-commutative operators.

4 Chemical Objects in ELAN

4.1 Molecule Representation

Molecules are represented in a simplified form as labeled graphs, describing the composition and the constitution of a molecule: vertices correspond to atoms, edges represent covalent bonds between atoms; bond lengths or bond angles are not taken into account.

In order to represent molecules in ELAN we use a *term* representation inspired from chemical linear SMILES notation [16] that code molecular graphs as trees:

1. Molecules are represented as hydrogen-suppressed molecular graphs.
2. If the hydrogen-suppressed molecular graph has cycles, we transform it into a tree applying the following rule to every cycle: choose one fresh digit and one single or aromatic bond of the cycle, break the bond and label the 2 atoms with the same digit.
3. Choose a root of the tree, and represent it like a concatenation of the root and the list of its sons.

For example, the term `C 1 (C) C C 1` represents methylcyclopropane and the term `C C (= O)O` represents acetic acid: see Fig. 2.

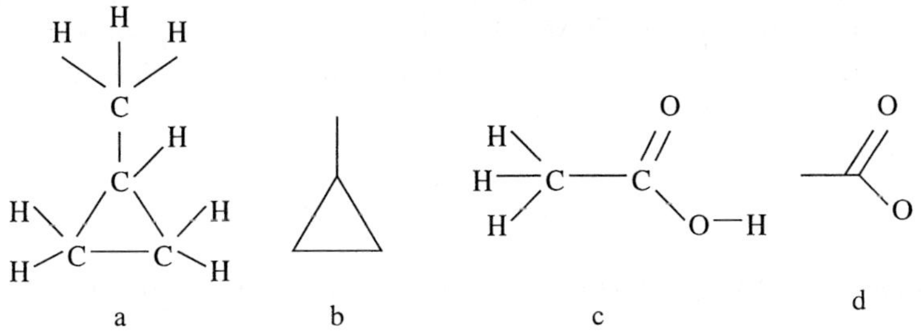

Fig. 2. Representations of molecules: a. molecular graph of methylcyclopropane; b. corresponding hydrogen-suppressed molecular graph; c. molecular graph of Acetic acid; d. corresponding hydrogen-suppressed molecular graph

In Fig. 3 we give the mix-fix signature in ELAN of our notation.

4.2 Reactant Representation

Reactants are represented in ELAN using an associative-commutative operator +. For example, `C C C C + O=O + C C (= O)O` is a term that represents a mixture of n-butane, oxygen and acetic acid.

```
C, c, O, o, H   : symbol;       /* Atom specification */
@ @      : (symbol int) symbol;/* Labels for cycle closure specification*/
 -, =, #, ':'   : link;         /* Link specification */
e        : radical;
@        : (symbol) radical;   /* Molecules and radicals specification */
@ @      : (symbol radical_list) radical;
@        : (radical) radical_list;     /* List of radicals specification */
@ @      : (link radical) radical_list;
(@)      : (radical_list) radical_list;
(@) @    : (radical_list radical_list) radical_list (AC);
```

Fig. 3. ELAN signature of our SMILES like notation for molecules

4.3 Elementary Reactions in ELAN

The ten main classes of generic reactions in EXGAS are: 1) unimolecular initiation (ui); 2) bimolecular initiation (bi); 3) addition of free radicals to oxygen (ad); 4) isomerisation of free radicals (is); 5) unimolecular decomposition of free radicals by beta-scission (bs); 6) unimolecular decomposition of hydroperoxyalkyl free radicals to cyclic ethers (cy); 7) oxidation of free radicals (ox); 8) metathesis (me); 9) combination of free radicals (co); 10) disproportionation of free radicals (di).

Generic reactions like addition of free radicals to oxygen, oxidation of free radicals, combination of free radicals and disproportionation of free radicals are expressed in a direct way in ELAN: the generic reaction is expressed into a term transformation which is described using ELAN code into rewrite rules.

For example, the generic reaction for addition of free radicals to oxygen is the following

$$O{=}O + \bullet R \longrightarrow \bullet OOR$$

It corresponds to the graph transformation given in Fig. 4 and is coded directly

```
O       X           O
||  +  / \   --->  / \
O     e   RL      e   O
                      |
                      X
                      |
                      RL
```

Fig. 4. Term transformation coding addition of free radicals to oxygen

by the following ELAN rewriting rule:

```
[ad] O=O + X (e) RL => O (e) O X RL end
```

where `X` is a variable of sort `symbol` and `RL` is a variable of sort `radical_list`.

The ELAN system generates the following output as the result for addition of butyl to oxygen:

```
Addition of C(e) C C C to O=O
 [ad] O=O + C(e)C C C  ----->  O(e)O C C C C
```

4.4 Generating All Visions

The generic reaction for unimolecular initiation is the following

$$\text{A–B} \longrightarrow \bullet\text{A} + \bullet\text{B}$$

Breaking the C–C bonds during unimolecular initiation corresponds to the term transformation given in Fig. 5 and can be implemented by the following ELAN

```
    X               X
   / \    --->     / \    + molec2Rad(Rad)
 Rad  RL          e   RL
```

Fig. 5. Term transformation coding unimolecular initiation

rewriting rule, a *named rule*

```
[ui] X (Rad) RL => X(e) RL + molec2Rad(Rad) end
```

where `X` is a variable of sort `symbol`, `Rad` is a variable of sort `radical` and `RL` is a variable of sort `radical_list`. The operator `molec2rad( )` transforms its argument into a radical and is defined by two ELAN rewriting rules, *no-named* rules:

```
[] molec2rad(X Rad)        => X (e) Rad  end
[] molec2rad(X (Rad) RL) => X (e) (Rad) RL  end
```

Note that the power of the associative-commutative matching allows us to give one generic ELAN rewriting rule for the unimolecular initiation and it will be applied to all sub-terms of root `X`.

The ELAN system applies a named rule to a term by matching the left-hand side of the rule to the term. In order to obtain all the results of the unimolecular initiation for the molecule `X (Rad) RL`, the rewriting rule `[ui]` has to be applied everywhere inside the term. To do this we suggest the following technique:

1. We define the *vision of a molecule* to be a representation of a molecule as a tree, given by the choice of the root (see Fig. 6).
2. We provide an operator `AllVis` that generates all the visions of a molecule by choosing every node of the tree representing the molecule to be the root.

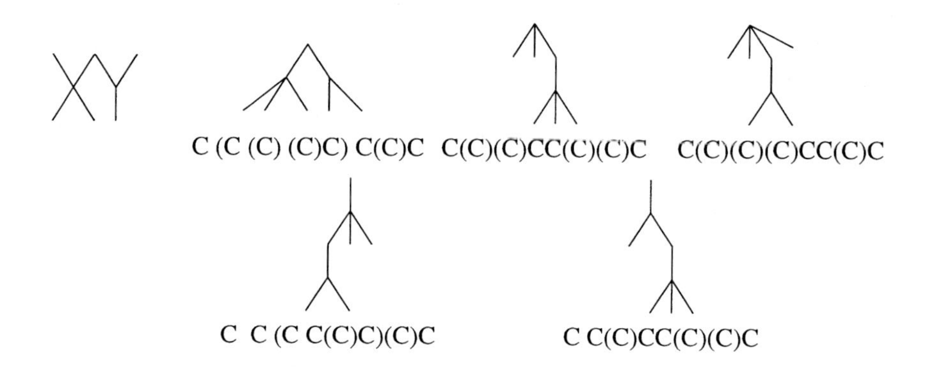

Fig. 6. The set of all distinct visions for ISO-octane

3. We apply the generic ELAN rewriting rule for unimolecular initiation to every vision of the molecule given by the `AllVis` operator.

The ELAN system generates the four first reactions from Fig. 7 as the result for the unimolecular initiation of ISO-octane by breaking the `C–C` bonds.

Equality Test for Two Molecules. Equality of two molecules M_1 and M_2, represented in the ELAN signature, means that M_1 and M_2 are visions of the same hydrogen-suppressed molecular graph. The `AllVis` operator is used for the equality test:

$$\texttt{is\_eq}(M_1,\ M_2)\ \texttt{=> true if}\ M_1 \in \texttt{AllVis}(M_2)$$

4.5 Primary Mechanism in ELAN

The primary mechanism in EXGAS is the following:

1. initiation reactions (ui and bi, reaction patterns 1 and 2 from Sect. 4.3) are applied to initial reactants (a mixture of molecules) and free radicals are generated;
2. the set of generic propagation reactions (reaction patterns 3—8 from Sect. 4.3) are applied to the free radicals until no new radical is generated;
3. the termination reactions (reaction patterns 9 and 10) are applied to free radicals and generate molecules.

In ELAN the primary mechanism is defined in a natural way as the concatenation of three strategies corresponding to each phase, `tryInit` for the initiation phase, `tryPropag` for the propagation phase and `tryTermin` for the termination phase:

```
[] mec_prim     => tryInit; tryPropag; tryTermin end
```

The user defined strategies tryInit and tryTermin are expressed using the ELAN choice strategy operator **dk** applied to the strategies (the ELAN rewriting rule) defining the generic reactions. The **dk** operator (*dont know choose*) takes all strategies given as arguments and returns, for each of them the set of all its results.

For example the following ELAN rule expresses a strategy that returns all the results of the unimolecular and bimolecular initiation:

```
[] tryInit        => dk(ui, bi) end
```

The output generated by ELAN after the initiation phase was applied to ISO-octane is given in Fig. 7.

```
Initiations for  C(C(C)C)C(C)(C)C
 [ui] C C(C)C C(C)(C)C -----> C e +  C(C)(C C(C)(C)C)e
 [ui] C C(C)(C)C C(C)C -----> C e +  C(C)(C)(C C(C)C)e
 [ui] C(C(C)C)C(C)(C)C -----> C(C(C)C)e +  C(C)(C)(C)e
 [ui] C(C(C)C)C(C)(C)C -----> C(C(C)(C)C)e + C(C)(C)e
 [bi] O=O + C(C)(C)C C(C)(C)C  -----> C(C)(C)(C C(C)(C)C)e +  O(e)O
 [bi] O=O + C C(C)C C(C)(C)C  -----> C(C(C)C C(C)(C)C)e +  O(e)O
 [bi] O=O + C C(C)(C)C C(C)C  -----> C(C(C)(C)C C(C)C)e +  O(e)O
 [bi] O=O + C(C(C)C)C(C)(C)C  -----> C(C(C)C)(C(C)(C)C)e +  O(e)O
```

Fig. 7. Initiation reactions of ISO-octane combustion in ELAN

Strategy tryPropag is defined as the iteration of one step of propagation using the ELAN strategy iterator **repeat***.

```
[] tryPropag => repeat*(propagOne) end
```

Strategy **repeat** iterates the strategy until it fails and returns the terms resulting from the last unfailing call of the strategy.

Strategy propagOne is defined in a similar way as tryInit using a **dk** operator applied to the generic reactions of the propagation phase.

5 Conclusions and Further Work

We described in this paper how to express in ELAN: acyclic and cyclic molecules, reactants, elementary reactions and the primary mechanism for acyclic species. This work is in a preliminary phase, since not all the functionalities of the EXGAS system have been implemented using the ELAN system. However, we think that at this time we have proved the feasibility of this approach.

The results obtained so far and the positive feedback from chemists encourage us to continue our study, to set-up a prototype and to extend it to cyclic molecules, which are not yet fully supported by any of the mentioned automatic kinetic mechanism generator systems.

References

1. Elan web site. http://elan.loria.fr.
2. Franz Baader and Tobias Nipkow. *Term Rewriting and all That*. Cambridge University Press, 1998.
3. Peter Borovanský, Hélène Kirchner, Pierre-Etienne Moreau, and Christophe Ringeissen. An overview of elan. In Claude Kirchner and Hélène Kirchner, editors, *Electronic Notes in Theoretical Computer Science*, volume 15. Elsevier Science Publishers, 2000.
4. M. Clavel, F. Duran, S. Eker, P. Lincoln, N. Marti-Oliet, J. Meseguer, and J.F. Quesada. Towards maude 2.0. In Kokichi Futatsugi, editor, *Electronic Notes in Theoretical Computer Science*, volume 36. Elsevier Science Publishers, 2001.
5. Guy-Marie Côme. *Gas-Phase Thermal Reactions. Chemical Engineering Kinetics.* Kluwer Academic Publishers, 2001.
6. K. Futatsugi and A. Nakagawa. An overview of CAFE specification environment – an algebraic approach for creating, verifying, and maintaining formal specifications over networks. In *Proceedings of the 1st IEEE Int. Conference on Formal Engineering Methods*, 1997.
7. Pierre-Alexandre Glaude, Frédérique Battin-Leclerc, René Fournet, Valérie Warth, Guy-Marie Côme, and Gérard Scacchi. Construction and simplification model for the oxidation of alkanes. *Combustion and Flame*, 122:451–462, 2000.
8. Hélène Kirchner and Pierre-Etienne Moreau. Promoting rewriting to a programming language: A compiler for non-deterministic rewrite programs in associative-commutative theories. *Journal of Functional Programming*, 11(2):207–251, 2001.
9. Paul Klint. A meta-environment for generating programming environments. *ACM Transactions on Software Engineering and Methodology (TOSEM)*, 2(2):176–201, 1993.
10. Pierre-Etienne Moreau. Rem (reduce elan machine): Core of the new elan compiler. In Leo Bachmair, editor, *Rewriting Techniques and Applications, 11th International Conference, RTA 2000, Norwich, UK, July 10-12, 2000, Proceedings*, volume 1833 of *Lecture Notes in Computer Science*, pages 265–269. Springer, 2000.
11. Pierre-Etienne Moreau and Hélène Kirchner. A compiler for rewrite programs in associative-commutative theories. In Catuscia Palamidessi, Hugh Glaser, and Karl Meinke, editors, *Principles of Declarative Programming, 10th International Symposium, PLILP'98 Held Jointly with the 7th International Conference, ALP'98, Pisa, Italy, September 16-18, 1998, Proceedings*, volume 1490 of *Lecture Notes in Computer Science*, pages 230–249. Springer, 1998.
12. Artur Rakiewicz and Thanh N. Truong. Application of chemical graph theory for automated mechanism generation. *accepted for publication*, 2002.
13. E. Ranzi, T. Faravelli, P. Gaffuri, and A. Sogaro. Low-temperature combustion: automatic generation of primary oxydation reaction and lumping procedures. *Combustion and Flame*, 102:179, 1995.
14. Roberta G. Susnow, Anthony M. Dean, William H. Green, P.Peczak, and Linda J. Broadbelt. Rate-based construction of kinetic models for complex systems. *The Journal of Physical Chemistry A*, 101:3731–3740, 1997.
15. Valérie Warth, Frédérique Battin-Leclerc, René Fournet, Pierre-Alexandre Glaude, Guy-Marie Côme, and Gérard Scacchi. Computer based generation of reaction mechanisms for gas-phase oxidation. *Computers and Chemistry*, 24:541–560, 2000.
16. David Weininger. Smiles, a chemical language and information system. 1. introduction to methodology and encoding rules. *Journal of Chemical Information and Computer Science*, 28:31–36, 1988.

Contraction versus Relaxation: A Comparison of Two Approaches for the Negative Cost Cycle Detection Problem

K. Subramani and L. Kovalchick

LCSEE,
West Virginia University,
Morgantown, WV
{ksmani, lynn}@csee.wvu.edu

Abstract. In this paper, we develop a greedy algorithm for the negative cost cycle detection problem and empirically contrast its performance with the "standard" Bellman-Ford (BF) algorithm for the same problem. Our experiments indicate that the greedy approach is superior to the dynamic programming approach of BF, on a wide variety of inputs.

1 Introduction

In this paper, we are concerned with the Negative Cost Cycle Detection problem (NEG): *Given a directed graph* $\mathbf{G} =< \mathbf{V}, \mathbf{E} >$, *where* $|\mathbf{V}| = n$ *and* $|\mathbf{E}| = m$, *and a cost function* $c : \mathbf{E} \rightarrow \Re$, *is there a negative cost cycle in* $\mathbf{G}$*?*

Our main contribution is the proposal of a greedy algorithm for NEG, based on vertex contraction. *All approaches to the negative cost cycle problem in the literature are based on dynamic programming; our approach is the first and only greedy approach to this problem, that we know of.* Scaling approaches have also been proposed for NEG ([Gol95]); however, these algorithms are efficient, only when the edge-weights are small integers. We do not place any restrictions on the edge costs. We note that the problem, as specified, is a decision problem, in that all that is asked of an algorithm is to *detect* the presence of a negative cycle. This problem finds application in a wide variety of areas such as Constraint Analysis [DMP91], Compiler Construction [Pug92], VLSI Design [WE94] and Scheduling [Sub02].

Our experiments indicate that Vertex Contraction is an effective alternative to the "standard" Bellman-Ford (BF) algorithm for the same problem; this is most surprising since in the case of sparse graphs, BF is provably superior to Vertex Contraction (from the perspective of asymptotic analysis).

2 The Vertex-Contraction Algorithm

The *vertex contraction* procedure consists of eliminating a vertex from the input graph, by merging all its incoming and outgoing edges. Consider a vertex v_i

P.M.A. Sloot et al. (Eds.): ICCS 2003, LNCS 2659, pp. 377–387, 2003.

with incoming edge e_{ki} and outgoing edge e_{ij}. When v_i is contracted, e_{ki} and e_{ij} are deleted and a single edge e'_{kj} is added with cost $c_{ki} + c_{ij}$. This process is repeated for each pair of incoming and outgoing edges. Consider the edge e'_{kj} that is created by the contraction; it falls into one of the following categories:

1. It is the first edge between vertex v_k and v_j. In this case, nothing more is to be done.
2. An edge e_{kj} already existed between v_k and v_j, prior to the contraction of v_i. In this case, if $c'_{kj} < c_{kj}$, keep the new edge and delete the previously existing edge (since it is redundant); otherwise delete the new edge (since it is redundant).

Algorithm (2.1) is a formal description of our technique.

Function NEGATIVE-COST-CYCLE($\mathbf{G}, n$)
1: **for** $(i = 1$ **to** $n)$ **do**
2: VERTEX-CONTRACT($\mathbf{G}, v_i$)
3: **end for**
4: **return(false)**

Algorithm 2.1: Negative cost cycle detection

We defer a formal proof of the correctness of the VC algorithm to the journal version of this paper. In the full version, an analysis of VC is also provided; we show that the algorithm runs in worst case time $O(n^3)$.

Thus, for dense graphs, Algorithm (2.1) is competitive with Bellman-Ford (BF); however for sparse graphs, the situation is not so sanguine. For instance, an adversary could provide the graph in Figure (1) as input.

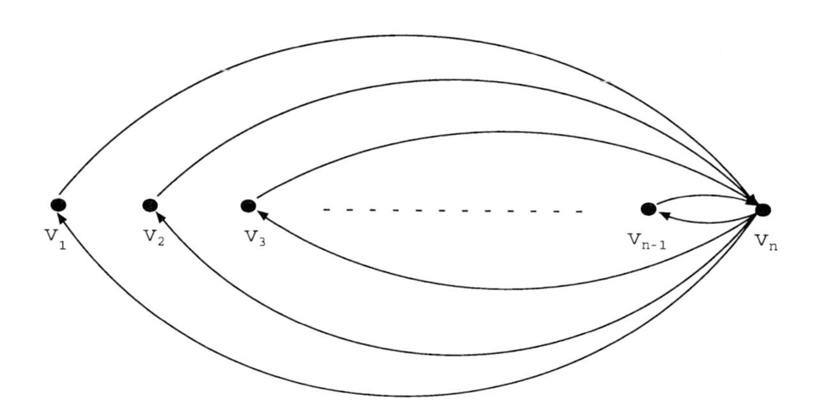

Fig. 1. Sparse graph that becomes dense after vertex contraction

```
Function VERTEX-CONTRACT(G, v_i)
 1: for (k = 1 to n) do
 2:   for (j = 1 to n) do
 3:     if (e_ki and e_ij exist) then
 4:       {Let c_kj denote the cost of the existing edge between v_k and v_j; note that
          c_kj = ∞ if there does not exist such an edge}
 5:       Create edge e'_kj with cost c'_kj = c_ki + c_ij
 6:       Delete edges e_ki and e_ij from G
 7:       if (j = k) then
 8:         {A cycle has been detected}
 9:         if (c'_jj < 0) then
10:           return(true)
11:         else
12:           Delete edge e_jj
13:         end if
14:       else
15:         if (c'_kj < c_kj) then
16:           Replace existing edge e_kj with e'_kj in G
17:         else
18:           Delete edge e'_kj
19:         end if
20:       end if
21:     end if
22:   end for
23: end for
```

Algorithm 2.2: Vertex Contraction

The above graph is sparse and has exactly $2 \cdot (n-1)$ edges. Observe that if vertex v_n is contracted first, the resultant graph is the complete graph on $n-1$ vertices and therefore dense. We call this graph *the cruel adversary*; in our experiments, we made it a point to contrast the performance of the vertex contraction algorithm with BF on this input. It is clear that any well-defined order of selecting the next vertex to be contracted is susceptible to attack by a malicious adversary; we could however choose the vertex to be contracted at random, without affecting the correctness of the algorithm. We have implemented Algorithm (2.1) in two different ways; in one implementation, the vertex to be contracted is chosen in a well-defined order, whereas in the second implementation, it is chosen at random. Algorithm (2.3) is a formal description of the random vertex contraction algorithm.

3 Implementation

Our experiments are classified into various categories, based on the following criteria:

1. Type of input graph - Sparse with many small negative cycles (Type A), Sparse with a few long negative cycles (Type B), Dense with many small

Function RANDOM-NEGATIVE-COST-CYCLE($\mathbf{G}, n$)
1: Generate a random permutation Π of the set $\{1, 2, 3 \ldots, n\}$.
2: **for** ($i = 1$ **to** n) **do**
3: VERTEX-CONTRACT($\mathbf{G}, v_{\Pi(i)}$)
4: **end for**
5: **return(false)**

Algorithm 2.3: Random negative cost cycle detection algorithm

negative cycles (Type C), Dense with a few long negative cycles (Type D), and the Cruel Adversary (Type E).
2. Type of Algorithm - Bellman-Ford (BF), Vertex-Contraction (VC) or Random Vertex-Contraction (RVC).
3. Type of Graph Data Structure - Simple Pointer or Array of Pointers.

All times recorded were averaged over 5 executions of each implementation.

3.1 Machine Characteristics

Table 1. Implementation System.

Machine Model	Silicon Graphics Onyx2
Processors	IR2/R10 250 Mhz
Cache	8 MB
Memory	2 GB
Operating System	IRIX 6.5.15
Language	C
Software	gcc

3.2 Graph Data Structures

Two different types of graph data structures were used for the experiments. We implemented BF, VC and RVC with an array of pointers structure and a simple pointer structure.

The array of pointers structure is a new representation. This representation makes use of an array of n pointers, one for each of the n vertices of the graph. Each pointer points to an n element array, which corresponds to the n vertices of the graph. Initially all entries of the array are assigned an undefined value. For a vertex v_i, if there exists an edge from v_i to another vertex v_j, position v_j of the array that v_i points to is assigned the cost of the edge between v_i and v_j. It should be noted that this representation is different from the adjacency-matrix representation [CLR92].

The simple pointer structure, also known as the adjacency-list representation [CLR92], requires only linear space. This representation makes use of an array of n lists, one for each of the n vertices of the graph.

3.3 Experimental Setup for Sparse Graphs

Sparse graphs were generated using the generator developed by Andrew Goldberg [CG96], which generates multiple edges between two vertices. Sparse graphs are defined as graphs with $o(n \cdot \log n)$ edges. We generated each graph 5 times using 5 different seeds for the random number generator.

Graphs of Type A and B were tested, with a number of vertices ranging from 500 to 5,500 in increments of 500.

We define a small negative cycle as one consisting of at most $\frac{n}{100}$ vertices. We define a long negative cycle as one consisting of $\Omega(\frac{n}{2})$ vertices. The number of long negative cycles in the input graphs was set to 4.

n	Array of Pointers (Time in Seconds)	
	VC	BF
500	0.15351	2.80657
750	0.50453	9.37442
1,000	1.58202	27.9044
1,250	2.23023	54.0744
1,500	4.74535	105.143
1,750	5.55235	156.953
2,000	12.7588	257.852
2,250	19.5588	337.136
2,500	13.9183	514.046
2,750	24.3229	624.652
3,000	30.4645	883.024
3,250	34.8372	1034.04
3,500	49.6497	1400.41
3,750	48.4852	1606.85
4,000	88.8478	2104.50
4,250	70.4305	2319.88
4,500	132.506	3094.30
4,750	82.0854	3180.42
5,000	108.178	4229.53
5,250	116.699	4377.80
5,500	133.606	5453.65

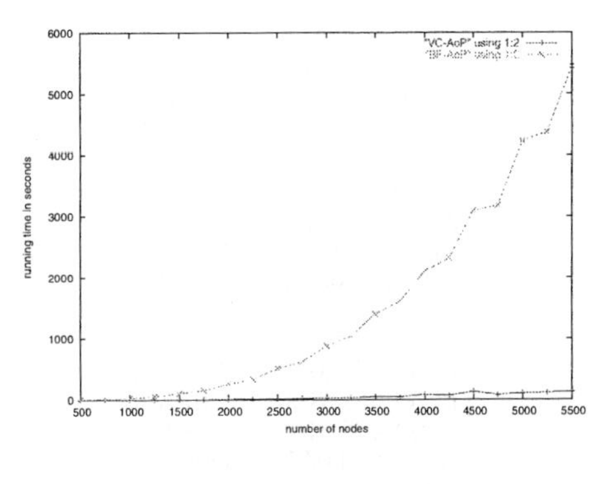

Fig. 2. Comparison of Vertex Contraction (VC), and Bellman-Ford (BF) Array of Pointer (AoP) implementation execution times (seconds) required to solve the Negative Cost Cycle problem for Type A graphs.

n	Simple Pointer (Time in Seconds)	
	VC	BF
500	0.003933	1.65399
750	0.007623	5.19749
1,000	0.009573	11.8637
1,250	0.023780	22.5836
1,500	0.013797	38.9001
1,750	0.013525	64.8949
2,000	0.022071	103.797
2,250	0.022178	155.955
2,500	0.025375	222.823
2,750	0.030861	304.137
3,000	0.040336	403.182
3,250	0.046731	519.489
3,500	0.047264	656.995
3,750	0.071233	814.206
4,000	0.063790	995.579
4,250	0.073681	1199.38
4,500	0.101693	1433.95
4,750	0.083590	1688.46
5,000	0.124874	1981.08
5,250	0.084357	2295.98
5,500	0.087477	2650.91

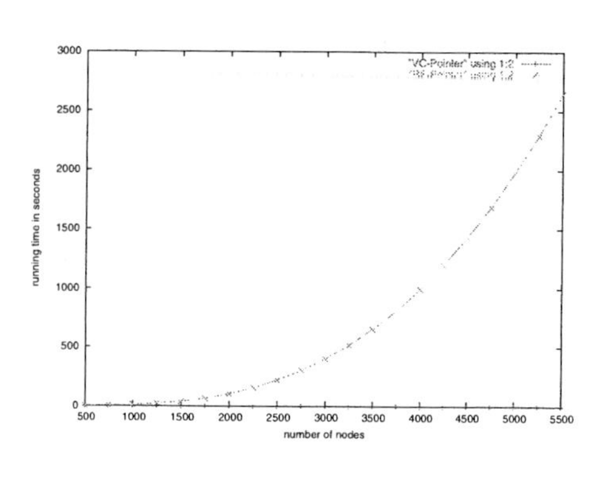

Fig. 3. Comparison of Vertex Contraction (VC), and Bellman-Ford (BF) Simple Pointer implementation execution times (seconds) required to solve the Negative Cost Cycle problem for Type A graphs.

n	Array of Pointers (Time in Seconds)	
	VC	BF
500	0.22578	2.80053
750	0.51732	9.40496
1,000	1.63598	25.2701
1,250	2.76916	52.7881
1,500	4.03085	103.765
1,750	4.58197	152.435
2,000	11.8345	253.484
2,250	20.4233	330.726
2,500	13.9014	502.027
2,750	23.9882	607.284
3,000	27.1102	875.921
3,250	35.6303	995.875
3,500	49.7201	1383.19
3,750	48.6051	1577.46
4,000	77.5183	2071.36
4,250	65.2763	2307.49
4,500	120.153	2978.97
4,750	83.6087	3209.21
5,000	92.0987	4076.75
5,250	151.066	4376.84
5,500	130.703	5408.41

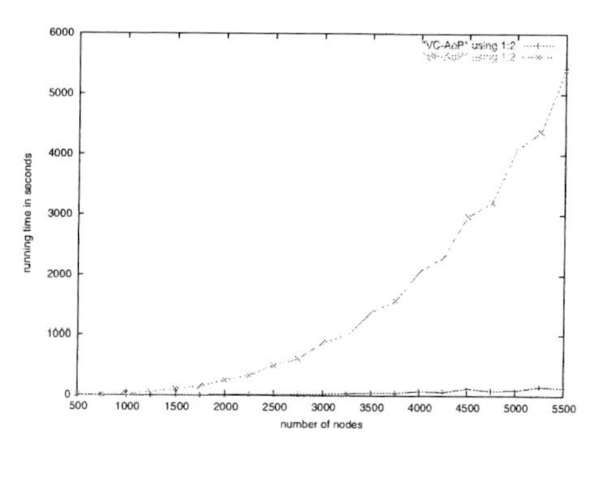

Fig. 4. Comparison of Vertex Contraction (VC), and Bellman-Ford (BF) Array of Pointer (AoP) implementation execution times (seconds) required to solve the Negative Cost Cycle problem for Type B graphs.

n	Simple Pointer (Time in Seconds)	
	VC	BF
500	0.004119	1.65394
750	0.008052	5.20284
1,000	0.011301	11.8367
1,250	0.022105	22.5755
1,500	0.099097	38.8326
1,750	0.022232	64.7921
2,000	0.021255	103.191
2,250	0.037886	154.749
2,500	0.026206	222.234
2,750	0.030613	303.657
3,000	0.037332	403.146
3,250	0.050565	518.724
3,500	0.047130	655.168
3,750	0.078139	813.916
4,000	0.188993	993.696
4,250	0.078959	1199.66
4,500	0.106838	1432.46
4,750	0.059203	1690.80
5,000	0.128170	1977.29
5,250	0.096562	2293.64
5,500	0.114865	2646.47

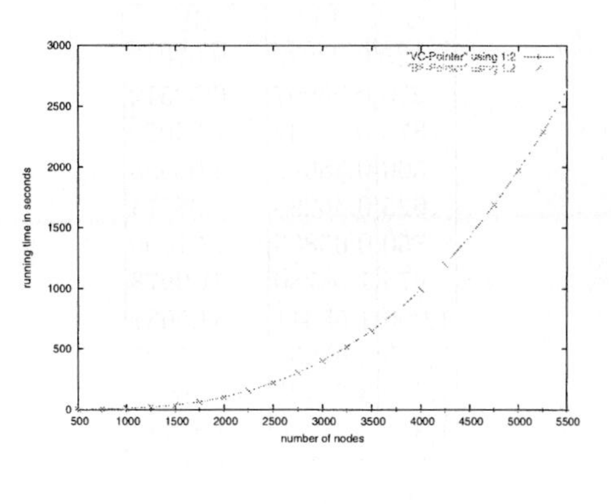

Fig. 5. Comparison of Vertex Contraction (VC), and Bellman-Ford (BF) Simple Pointer implementation execution times (seconds) required to solve the Negative Cost Cycle problem for Type B graphs.

3.4 Conclusions

It is easy to see from the tables and graphs in Figure (2) through Figure (5) that VC outperforms BF using either data structure; this is true for both types of sparse graphs that were tested. We conclude that VC is far superior to BF for sparse graphs.

An asymptotic analysis would indicate that BF is superior to VC for dense graphs, although, our experiments contradict this indication.

3.5 Experimental Setup for Dense Graphs

Dense graphs were generated using the generator developed by Andrew Goldberg [CG96]. Dense graphs were defined as those with $\Omega(\frac{n^2}{8})$ edges. We generated each graph 5 times using 5 different seeds for the random number generator.

Graphs of Type C and D were tested, with a number of vertices ranging from 125 to 1,875 in increments of 125, with small negative cycles and long negative cycles defined as in Section §3.3.

n	Array of Pointers (Time in Seconds) VC	BF
125	0.03830	0.07012
250	0.05997	0.52512
375	0.15247	1.74020
500	0.28095	4.08695
625	0.46288	7.98574
750	0.65885	13.7706
875	1.48780	21.9978
1,000	1.55311	34.5631
1,125	2.97252	51.5897
1,250	3.37425	74.6079
1,375	4.37236	100.182
1,500	7.30295	132.641
1,625	6.22661	168.864
1,750	8.63348	210.632
1,875	8.00939	260.838

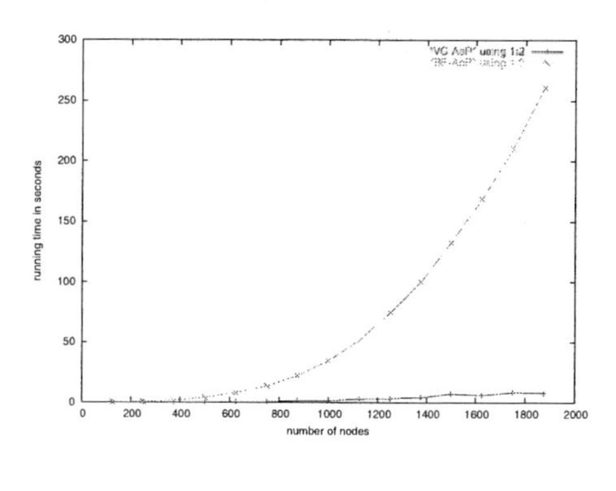

Fig. 6. Comparison of Vertex Contraction (VC), and Bellman-Ford (BF) Array of Pointer (AoP) implementation execution times (seconds) required to solve the Negative Cost Cycle problem for Type C graphs.

n	Simple Pointer (Time in Seconds) VC	BF
125	0.00048	0.09019
250	0.00194	1.02020
375	0.00303	4.50625
500	0.00675	13.2450
625	0.00750	30.9083
750	0.01562	62.0953
875	0.03498	123.824
1,000	0.09200	293.591
1,125	0.11334	672.256
1,250	0.19100	1350.14
1,375	0.25657	2447.76
1,500	0.42457	4079.35
1,625	0.41033	6346.27
1,750	0.65944	9445.69
1,875	0.99798	13480.9

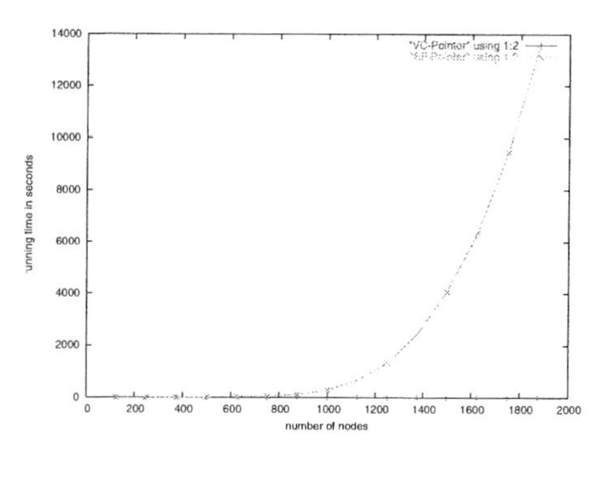

Fig. 7. Comparison of Vertex Contraction (VC), and Bellman-Ford (BF) Simple Pointer implementation execution times (seconds) required to solve the Negative Cost Cycle problem for Type C graphs.

n	Array of Pointers (Time in Seconds)	
	VC	BF
125	0.00146	0.06872
250	0.06294	0.52747
375	0.21069	1.73264
500	0.30008	4.09149
625	0.65476	7.98092
750	0.74009	13.7807
875	2.03858	21.9476
1,000	1.50686	35.0864
1,125	3.64647	51.5824
1,250	3.17255	72.8195
1,375	6.92824	101.460
1,500	6.88772	133.293
1,625	6.67336	167.244
1,750	7.48221	212.698
1,875	15.9974	263.763

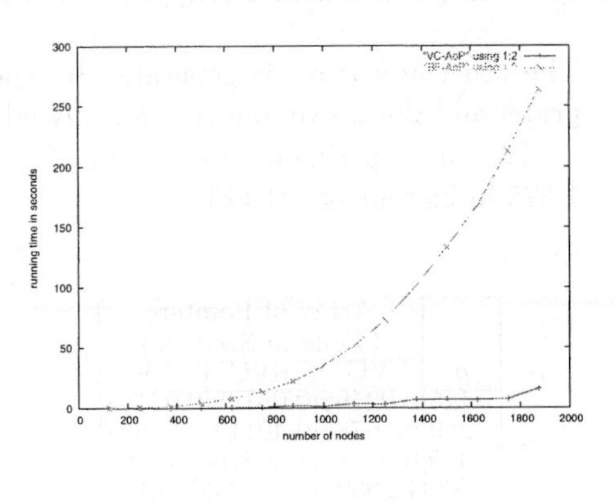

Fig. 8. Comparison of Vertex Contraction (VC), and Bellman-Ford (BF) Array of Pointer (AoP) implementation execution times (seconds) required to solve the Negative Cost Cycle problem for Type D graphs.

n	Simple Pointer (Time in Seconds)	
	VC	BF
125	0.00069	0.08752
250	0.00185	1.02023
375	0.00488	4.44321
500	0.00706	13.2395
625	0.01379	30.6152
750	0.01765	62.0956
875	0.05033	122.713
1,000	0.07456	293.617
1,125	0.20491	664.932
1,250	0.22301	1348.96
1,375	0.41339	2426.95
1,500	0.39452	4079.71
1,625	1.00850	6299.06
1,750	0.59050	9447.84
1,875	1.43688	13418.7

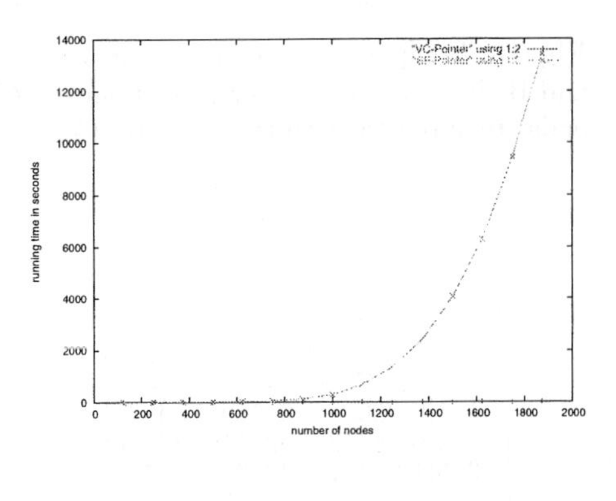

Fig. 9. Comparison of Vertex Contraction (VC), and Bellman-Ford (BF) Simple Pointer implementation execution times (seconds) required to solve the Negative Cost Cycle problem for Type D graphs.

3.6 Conclusions

It is easy to see from the tables and graphs in Figure (6) through Figure (9) that VC outperforms BF using either data structure; this is true with both types of dense graphs that were tested. We conclude that VC is far superior to BF for dense graphs.

3.7 Experimental Setup for Cruel Adversary Graphs

The cruel adversary is generated by specifying the number of vertices in the graph and the maximum cost for any edge.

For our experiments we generated graphs with vertices ranging from 125 to 1,875 in increments of 125.

n	Array of Pointers (Time in Seconds)		
	VC	RVC	BF
125	0.02168	0.05125	0.04738
250	0.16735	0.40114	0.34556
375	0.55377	1.35226	1.13898
500	1.29808	3.08334	2.66123
625	2.54713	6.22039	5.19088
750	4.38616	10.7309	9.02499
875	7.01832	15.6965	14.3931
1,000	10.9170	25.7463	23.6483
1,125	15.7832	37.1957	35.7989
1,250	21.9961	52.2092	54.5719
1,375	30.1900	71.4197	72.7322
1,500	41.3305	93.0058	94.8282
1,625	53.3580	66.8170	121.354
1,750	66.5882	147.186	152.360
1,875	83.2914	171.411	188.266

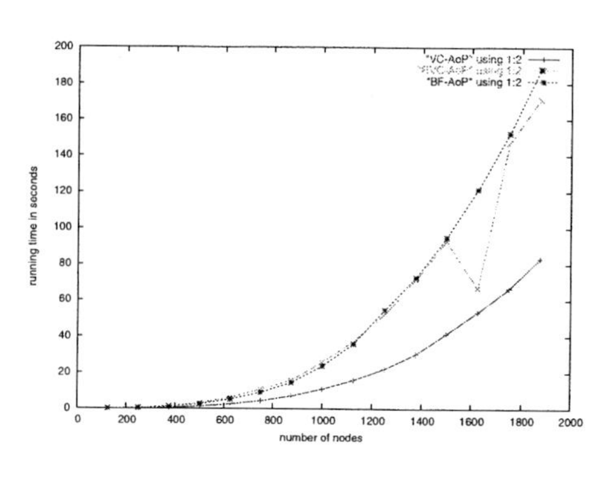

Fig. 10. Comparison of Vertex Contraction (VC), Random Vertex Contraction (RVC) and Bellman-Ford (BF) Array of Pointer (AoP) implementation execution times (seconds) required to solve the Negative Cost Cycle problem for Type E graphs.

n	Simple Pointer (Time in Seconds)		
	VC	RVC	BF
125	0.06657	0.01769	0.04572
250	0.48752	0.09771	0.33976
375	1.61851	0.47191	1.11752
500	4.55655	0.53125	2.61450
625	10.2743	2.45196	5.06600
750	19.7552	3.40885	8.70514
875	33.6204	5.76912	13.7683
1,000	52.5211	7.97386	20.4959
1,125	76.4936	25.6282	29.0852
1,250	106.248	33.1245	39.9211
1,375	144.869	9.84373	53.0680
1,500	187.058	67.5207	69.8548
1,625	244.296	66.2070	91.4537
1,750	304.338	78.6134	118.606
1,875	379.341	29.4672	151.796

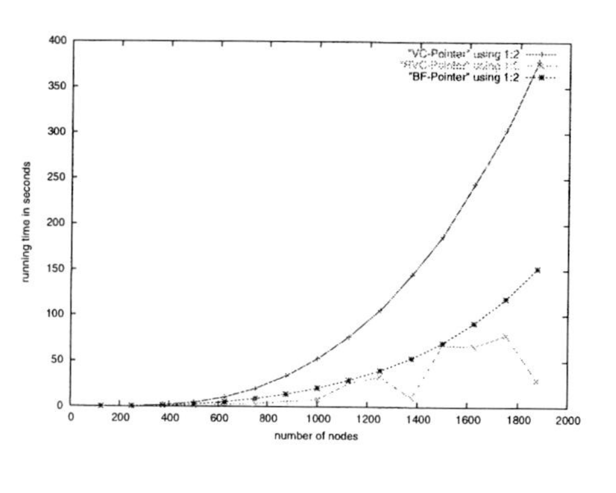

Fig. 11. Comparison of Vertex Contraction (VC), Random Vertex Contraction (RVC) and Bellman-Ford (BF) Simple Pointer implementation execution times (seconds) required to solve the Negative Cost Cycle problem for Type E graphs.

3.8 Conclusions

VC does considerably better than both RVC and BF, as observed from the table and graph in Figure (10) of the Array of Pointer implementation on Type E graphs. The results of RVC and BF are similar with RVC doing better in most instances.

VC does very poorly, as observed from the table and graph in Figure (11) of the Pointer implementation on Type E graphs. RVC does much better than VC and outperforms BF by a large margin on most instances. One conclusion that can be drawn from the data is that the time taken by RVC varies greatly depending on the random sequence of vertices chosen.

4 Conclusion

In this paper, we designed and analyzed a greedy algorithm called the vertex contraction algorithm (VC) for the negative cost cycle detection problem. Although vertex contraction is asymptotically inferior to the Bellman-Ford algorithm on sparse graphs, it is vastly superior from an empirical perspective.

We are currently working on two extensions: (a) Comparing our strategy with the Goldberg approach, (b) Combining the main idea of our approach, with heuristics such as contracting the vertex with the smallest degree-product.

References

[CG96] Boris V. Cherkassky and Andrew V. Goldberg. Negative-cycle detection algorithms. In Josep Díaz and Maria Serna, editors, *Algorithms—ESA '96, Fourth Annual European Symposium*, volume 1136 of *Lecture Notes in Computer Science*, pages 349–363, Barcelona, Spain, 25–27 September 1996. Springer.

[CLR92] T. H. Cormen, C. E. Leiserson, and R. L. Rivest. *Introduction to Algorithms.* MIT Press and McGraw-Hill Book Company, 6th edition, 1992.

[DMP91] R. Dechter, I. Meiri, and J. Pearl. Temporal constraint networks. *Artificial Intelligence*, 49:61–95, 1991.

[Gol95] Andrew V. Goldberg. Scaling algorithms for the shortest paths problem. *SIAM Journal on Computing*, 24(3):494–504, June 1995.

[Pug92] W. Pugh. The omega test: A fast and practical integer programming algorithm for dependence analysis. *Comm. of the ACM*, 35(8):102–114, August 1992.

[Sub02] K. Subramani. An analysis of zero-clairvoyant scheduling. In Joost-Pieter Katoen and Perdita Stevens, editors, *Proceedings of the 8th International Conference on Tools and Algorithms for the construction of Systems (TACAS)*, volume 2280 of *Lecture Notes in Computer Science*, pages 98–112. Springer-Verlag, April 2002.

[WE94] Neil H. Weste and Kamran Eshragian. *Principles of CMOS VLSI Design.* Addison Wesley, 1994.

Direct Method for Solving a Transmission Problem with a Discontinuous Coefficient and the Dirac Distribution

Hideyuki Koshigoe

Urban Environment System
Chiba University
1-33 Yayoi, Inage
263-8522, Japan
koshigoe@tu.chiba-u.ac.jp

Abstract. We construct finite difference solutions of a transmission problem with a discontinuous coefficient and the Dirac distribution by the direct method which we call the successive elimination of lines and then show that the limit function of them satisfies the transmission equation in the sense of distribution.

1 Introduction

This paper is devoted to the construction and the convergence of finite difference solutions based on the direct method coupled with the fictitious domain method ([2],[9])and distribution theoretical argument ([1]).
Let Ω be a rectangular domain in R^2, Ω_1 be an open subset of Ω and $\Omega_2 = \Omega \backslash \overline{\Omega_1}$, the interface of them be denoted by $\Gamma (= \overline{\Omega_1} \cap \overline{\Omega_2})$ and Γ be of class C^1. The transmission problem considered here is the followings.

Problem I. For $f \in L^2(\Omega)$, $\sigma \in L^2(\Gamma)$ and $g \subset H^{1/2}(\partial\Omega)$, find $u \in H^1(\Omega)$ such that

$$- \operatorname{div}(a(x,y)\,\nabla u) = f + \sigma\,\delta_\Gamma \quad \text{in} \quad D'(\Omega)\,, \tag{1}$$

$$u = g \quad \text{on} \quad \partial\Omega\,. \tag{2}$$

Here we assume that the discontinuous function a is given by

$$a(x,y) = \epsilon_1\,\chi_{\Omega_1}(x,y) + \epsilon_2\,\chi_{\Omega_2}(x,y),$$

where $\epsilon_i > 0$ is a parameter $(i = 1, 2)$ and χ_Π is defined by

$$\chi_\Pi(x,y) = \begin{cases} 1 & if\ (x,y) \in \Pi \\ 0 & if\ (x,y) \notin \Pi \end{cases}$$

for any subset Π of Ω.

Equations (1) of this type are arisen in various contexts. One of such examples can be found in the context of electricity and $\{\epsilon_1,\ \epsilon_2\}$ is corresponding to the

P.M.A. Sloot et al. (Eds.): ICCS 2003, LNCS 2659, pp. 388–398, 2003.

dielectric constant of the material $\{\Omega_1, \Omega_2\}$.
We now notice that Problem I is equivalent to the following problem II.

Problem II. Find $\lambda \in H^{1/2}(\Gamma)$ and $\{u_1(\lambda), u_2(\lambda)\} \in H^1(\Omega_1) \times H^1(\Omega_2)$ such that

$$-\,\epsilon_1\,\Delta u_1(\lambda) = f \quad \text{in} \quad \Omega_1\,, \tag{3}$$

$$-\,\epsilon_2\,\Delta u_2(\lambda) = f \quad \text{in} \quad \Omega_2\,, \tag{4}$$

$$u_1(\lambda) = u_2(\lambda) = \lambda \quad \text{on} \quad \Gamma\,, \tag{5}$$

$$u_2(\lambda) = g \quad \text{on} \quad \partial\Omega\,, \tag{6}$$

and

$$\epsilon_1\,\frac{\partial u_1(\lambda)}{\partial \nu} - \epsilon_2\,\frac{\partial u_2(\lambda)}{\partial \nu} = \sigma \quad \text{on} \quad \Gamma\,. \tag{7}$$

Here ν is the unit normal vector on Γ directed from Ω_1 to Ω_2 .
Hence introducing the Dirichlet-Neumann map T defined by

$$T : H^{1/2}(\Gamma) \ni \lambda \to \epsilon_1 \frac{\partial u_1(\lambda)}{\partial \nu} - \epsilon_2 \frac{\partial u_2(\lambda)}{\partial \nu} \in H^{-1/2}(\Gamma),$$

Problem I is reduced to find λ satisfying

$$T\lambda = \sigma\,. \tag{8}$$

From this point of view, the purpose of this paper is to show how to solve (8) directly.

This paper is organized as follows. Section 2 describes the finite difference approximation of Problem I. Section 3 is devoted to our numerical algorithm from the viewpoint of the successive elimination of lines coupled with the geometry of domains Ω_1 and Ω_2. In Sect. 4, we shall prove the justification of the finite difference scheme and finally discuss the convergence of approximate solutions constructed in Section 3.

2 Finite Difference Approximation of Problem I

Without loss of generality we assume that $g = 0$ and that Ω is the unit square in R^2, i.e., $\Omega = \{(x, y)|\ 0 < x, y < 1\ \}$. Let $h \in R$ be a mesh size such that $h = 1/n$ for an integer n and set $\Delta\, x = \Delta\, y = h$. We associate with it the set of the grid points:

$$\begin{aligned}\overline{\Omega}_h &= \{P_{i,j} \in R^2 \mid P_{i,j} = (i\ h,\ j\ h),\ 0 \le i, j \le n\},\\ \Omega_h &= \{P_{i,j} \in R^2 \mid P_{i,j} = (i\ h,\ j\ h),\ 1 \le i, j \le n-1\}.\end{aligned}$$

With each grid point $P_{i,j}$ of $\overline{\Omega}_h$, we associate the panel $\omega^0_{i,\ j}$ with center $P_{i,j}$:

$$\omega^0_{i,j} \equiv \Big(\ (i - 1/2)h,\ (i + 1/2)h\ \Big] \times \Big(\ (j - 1/2)h,\ (j + 1/2)h\ \Big], \tag{9}$$

and the cross $\omega^1_{i,j}$ with center $P_{i,j}$:

$$\omega^1_{i,j} = \omega^0_{i+1/2,j} \cup \omega^0_{i-1/2,j} \cup \omega^0_{i,j+1/2} \cup \omega^0_{i,j-1/2} \tag{10}$$

where e_i denotes the i th unit vector in R^2 and we set

$$\omega^0_{i\pm1/2,j} = \omega^0_{i,j} \pm \frac{h}{2}\, e_1, \quad \omega^0_{i,j\pm1/2} = \omega^0_{i,j} \pm \frac{h}{2}\, e_2. \tag{11}$$

Moreover using the datum in Problem I, we define

$$\begin{cases} a^E_{i,j} = \frac{1}{\Delta x \Delta y}\int_{\omega^0_{i+1/2,j}} a(x,y)\, dxdy, & a^W_{i,j} = \frac{1}{\Delta x \Delta y}\int_{\omega^0_{i-1/2,j}} a(x,y)\, dxdy, \\ a^N_{i,j} = \frac{1}{\Delta x \Delta y}\int_{\omega^0_{i,j+1/2}} a(x,y)\, dxdy, & a^S_{i,j} = \frac{1}{\Delta x \Delta y}\int_{\omega^0_{i,j-1/2}} a(x,y)\, dxdy, \\ f_{i,j} = \frac{1}{\Delta x \Delta y}\int_{\omega^0_{i,j}} f(x,y)\, dxdy, & \sigma_{i,j} = \frac{1}{\Delta\, l_{i,j}}\int_{\Gamma\, \cap\, \omega^0_{ij}} \sigma(s)\, ds, \\ \quad \Delta\, l_{i,j} = \int_{\Gamma\, \cap\, \omega^0_{i,j}} ds. \end{cases} \tag{12}$$

We then propose the discrete equation of Problem I as follows.

Problem F. Find $\{u_{i,j}\}$ $(1 \leq i,\ j \leq n-1)$ such that

$$\begin{aligned} &-\frac{1}{\Delta x}\left(a^E_{i,j}\, \frac{u_{i+1,j} - u_{ij}}{\Delta x} - a^W_{i,j}\, \frac{u_{ij} - u_{i-1,j}}{\Delta x} \right) \\ &-\frac{1}{\Delta y}\left(a^N_{i,j}\, \frac{u_{i,j+1} - u_{i,j}}{\Delta y} - a^S_{i,j}\, \frac{u_{i,j} - u_{i,j-1}}{\Delta y} \right) \\ &= f_{i,j} + \frac{\Delta l_{i,j}}{\Delta x\, \Delta y}\, \sigma_{i,j}, \quad 1 \leq i,j \leq n-1. \end{aligned} \tag{13}$$

Remark 1. The construction of solutions of Problem F will be discussed section 3. Then introducing the base function $\theta_{i,j}$:

$$\theta_{i,j}(x,y) = \begin{cases} 1, & (x,y) \in \omega^0_{i,j} \\ 0, & (x,y) \notin \omega^0_{i,j} \end{cases},$$

we define the piecewise functions σ_h and u_h by

$$\begin{aligned} \sigma_h &= \sum_{i,j=1}^{n-1} \frac{\Delta l_{i,j}}{\Delta x\, \Delta y}\, \sigma_{i,j}\, \theta_{i,j}(x,y), \\ u_h &= \sum_{i,j=1}^{n-1} u_{i,j}\, \theta_{i,j}(x,y) \end{aligned} \tag{14}$$

respectively. In section 4 we shall show that

(i) $\sigma_h \to \sigma \cdot \delta_\Gamma$ in $D'(\Omega)$,

(ii) $u_h \to u$ weakly in $L^2(\Omega)$, $u \in H^1(\Omega)$, and

(iii) u is the solution of Problem I in the sense of distrubution.

3 Construction of the Solution of (13)

3.1 Geometry of Domain and Principle of the Successive Elimination of Lines

In this subsection we deal with the $(n-1)$ vectors $\{U_i\}$ instead of the $(n-1)^2$ unknowns $u_{i,j}$. For each i, set $U_i = {}^t[u_{i,1},\ u_{i,2}, \cdots,\ u_{i,n-1}\ ]\ (1 \le i \le n-1)$. From the equations (13), it follows that

$$\begin{aligned} &(a_{i,j}^W + a_{i,j}^E + a_{i,j}^S + a_{i,j}^N)\ u_{i,j} - \ a_{i,j}^S u_{i,j-1} - \ a_{i,j}^N u_{i,j+1} \\ &= a_{i,j}^W u_{i-1,j} + a_{i,j}^E u_{i+1,j} + (\Delta x)^2\ f_{i,j} + \ \sigma_{i,j} \cdot \Delta l_{i,j} \end{aligned} \tag{15}$$

Now fix i $(1 \le i \le n-1)$. Paying attention to the vector U_i in (15) and setting $a_{i,j}^\epsilon = a_{i,j}^W + a_{i,j}^E + a_{i,j}^S + a_{i,j}^N$, Problem F *w.r.t.* $\{u_{i,j}\}$ is reduced to Problem M *w.r.t.* $\{U_i\}$.

Problem M. Find U_i $(1 \le i \le n-1)$ satisfying

$$A_i^\epsilon\, U_i = A_i^W\, U_{i-1} + A_i^E\, U_{i+1} + F_i \quad (1 \le i \le n-1) \tag{16}$$

where $U_0 = 0, U_n = 0,\ F_i$ is given by the data $\{f, \sigma\}$, A_i^ϵ is a tridiagonal matrix defined by

$$A_i^\epsilon = \begin{pmatrix} a_{i,1}^\epsilon & -a_{i,1}^N & 0 & \cdots & \cdots & \cdots & 0 \\ -a_{i,2}^S & a_{i,2}^\epsilon & -a_{i,2}^N & 0 & \vdots & \vdots & \vdots \\ 0 & \ddots & \ddots & \ddots & 0 & \vdots & \vdots \\ \vdots & 0 & \ddots & \ddots & \ddots & 0 & \vdots \\ \vdots & \vdots & 0 & \ddots & \ddots & \ddots & 0 \\ \vdots & \vdots & \vdots & 0 & -a_{i,n-2}^S & a_{i,n-2}^\epsilon & -a_{i,n-2}^N \\ 0 & \cdots & \cdots & \cdots & 0 & -a_{i,n-1}^S & a_{i,n-1}^\epsilon \end{pmatrix} \tag{17}$$

and A_i^W, A_i^E are the diagonal matrices given by

$$A_i^W = \mathrm{diag}[a_{i,j}^W]_{1 \le j \le n-1} \text{ and } A_i^E = \mathrm{diag}[a_{i,j}^E]_{1 \le j \le n-1} \tag{18}$$

Remark 2. For each $i (1 \le i \le n-1)$, A_i^ϵ is a symmetric matrix. In fact, $a_{i,j}^N = a_{i,j+1}^S$ holds from the definition (12).

Moreover in order to reduce the numbers of equations of Problem M, we separate unknown vector U_i into two parts considering the geometry of the domain Ω and the interface Γ. We first introduce the set of interface lattice points Γ_h and boundary lattice points $\partial\Omega_h$ as follows;

(i) $\Gamma_h = \{P_{i,j} = (ih, jh) \mid \Gamma \cap \omega_{i,j}^1 \ne \emptyset\}$,

(ii) $\partial\Omega_h = \overline{\Omega}_h \setminus \Omega_h$.

Division of the Unknown Vector $\{U_i\}$

For each $U_i = \{u_{i,j}\}_{1\leq j\leq n-1}$, we define $U_i' = \{u_{i,j}'\}_{1\leq j\leq n-1}$ and $W_i = \{w_{i,j}\}_{1\leq j\leq n-1}$ as follows;

$$u_{i,j}' = \begin{cases} 0 & \text{if } P_{i,j} \in \Gamma_h \\ u_{i,j} & \text{if } P_{i,j} \in \Omega_h \backslash \Gamma_h, \end{cases} \qquad w_{i,j} = \begin{cases} u_{i,j} & \text{if } P_{i,j} \in \Gamma_h \\ 0 & \text{if } P_{i,j} \in \Omega_h \backslash \Gamma_h \end{cases} \tag{19}$$

and devide U_i into two parts by

$$U_i = U_i' + W_i. \tag{20}$$

We then introduce the new vector $\{V_i\}$ defined by

$$V_i = A_i^W U_i' \; (= A_i^E U_i' \;) \qquad (1 \leq i \leq n-1). \tag{21}$$

From the definition of $\{U_i'\}$ and $\{V_i\}$, we get

Lemma 1. *$A_i^\epsilon U_i' = B\,V_i$, $A_i^w U_{i-1}' = V_{i-1}$ and $A_i^E U_{i+1}' = V_{i+1}$ hold $(1 \leq i \leq n-1)$. Here B is a block tridiagonal matrix in the discretization of the Laplace operator in Ω with homogeneous Dirichlet boundary conditions. i.e., $B = [b_{ij}]$ is an $(n-1)\times(n-1)$ tridiagonal matrix such that $B = \text{tridiag}[-1,\ 4,\ -1]$.*

Therefore the following equations are derived from Problem M, (17)-(21) and Lemma 1.

Problem PN. Find $\{V_i\ ,\ W_i\}$ such that for $i(1 \leq i \leq n-1)$,

$$B\,V_i = V_{i-1} + V_{i+1} + \; F_i \; + \Big(\; A_i^W\,W_{i-1} - A_i^\epsilon\,W_i + A_i^E\,W_{i+1} \; \Big) \tag{22}$$

where $V_0 = V_n = W_0 = W_n = 0$.

Moreover in order to deduce the equation of $\{W_i\}$ from Problem PN, we review the princple of the successive elimination of lines. The following proposition 1 was proved under two assumptions

Assumption 1. Let $B = \text{tridiag}[-1,\ 4,\ -1] \in \mathrm{R}^{(\mathrm{n}-1)\times(\mathrm{n}-1)}$.

Assumption 2. Let X_i and $Y_i \;\in\; R^{(n-1)}$ be satisfying the equations of the form : $B\,X_i \;=\; X_{i-1} \;+\; X_{i+1} \;+\; Y_i \quad (1 \leq i \leq n-1)$.

Proposition 1. *Under the above assumptions, X_k $(1 \leq k \leq n-1)$ is directly determined by*

$$Q\,X_k \;=\; \sum_{i=1}^{k-1} \; D_{\,n-k,\ i}\,Q\,Y_i \;\; + \; \sum_{i=k}^{n-1} \; D_{\,k,\ n-i}\,Q\,Y_i \tag{23}$$

where each $D_{\,l,\ i}$ $(1 \leq l, i \leq n-1)$ is a diagonal matrix defined by

$$\begin{aligned} D_{l,i} &= \text{diag}\left[\Big(\sinh(l\ \lambda_j)\ \sinh(i\ \lambda_j)\Big) / \Big(\sinh(n\ \lambda_j)\ \sinh(\lambda_j)\Big)\right]_{1\leq j\leq n-1} \\ \lambda_j &= \quad \text{arccosh}(2 - \cos(j\pi/n)), \end{aligned} \tag{24}$$

and $Q(=(q_{i,j})_{1\le i,j\le n-1})$ *is the othogonal matrix such that*

$$q_{i,j} = \sqrt{\frac{2}{n}}\ sin\left(\frac{i\,j\,\pi}{n}\right)\ (1 \le i,j \le n-1). \tag{25}$$

Remark 3. We call this proposition the princile of the successive elimination of lines (see also [6],[7],[11]).

Remark 4. Set $Q_i\ =^t\ (q_{i,1}, q_{i,2}, \cdots\cdots, q_{i,n-1})\quad (1 \le i \le n-1)$. Then $\{Q_i\}_{1\le i\le n-1}$ is the orthonormal system, which is used in the next subsection.

3.2 Numerical Algorithm

In this subsection, we show our numerical algorithm by use of the principle of the successive elimination of lines. First applying directly Proposition 1 to Problem PN, we have

Lemma 2. *Problem PN is equivalent to find* $\{V_k, W_k\}(1 \le k \le n-1)$ *satisfying*

$$\begin{aligned} Q\ V_k =\ & \textstyle\sum_{i=1}^{k-1}\ D_{n-k,i}\ Q\Big(\ A_i^W\,W_{i-1} - A_i^\epsilon\,W_i + A_i^E\,W_{i+1}\Big) \\ & + \textstyle\sum_{i=k}^{n-1}\ D_{k,n-i}\ Q\Big(\ A_i^W\,W_{i-1} - A_i^\epsilon\,W_i + A_i^E\,W_{i+1}\Big) \\ & + \Big(\ \textstyle\sum_{i=1}^{k-1}\ D_{n-k,i}\ Q\ F_i + \sum_{i=k}^{n-1}\ D_{k,n-i}\ Q\ F_i\Big). \end{aligned} \tag{26}$$

Using the orthogonal property of Q and the definitions of V_k and Γ_h, we get

$${}^tQ_l\ QV_k = 0 \text{ for any } l \text{ such that } P_{k,l} \in \Gamma_h,$$

from which it follows

Lemma 3. $\{W_i\}_{1\le i\le n-1}$ *in (26) satisfies the equations (27):*

$$\begin{aligned} & \textstyle\sum_{i=1}^{k-1}\ {}^tQ_l\ D_{n-k,i}\ Q\Big(\ -A_i^W\,W_{i-1} + A_i^\epsilon\,W_i - A_i^E\,W_{i+1}\Big) \\ +\ & \textstyle\sum_{i=k}^{n-1}\ {}^tQ_l\ D_{k,n-i}\ Q\Big(\ -A_i^W\,W_{i-1} + A_i^\epsilon\,W_i - A_i^E\,W_{i+1}\Big) \\ =\ & {}^tQ_l\Big(\ \textstyle\sum_{i=1}^{k-1}\ D_{n-k,i}\ Q\ F_i + \sum_{i=k}^{n-1}\ D_{k,n-i}\ Q\ F_i\ \Big) \end{aligned} \tag{27}$$

for (k,l) *such that* $P_{k,l} \in \Gamma_h$.

Conversely one may have a question whether it is possible to construct $\{V_k, W_k\}$ uniquely satisfying (26) from the equation (27). But the answer is yes and we shall prove it in the next section as the following theorem.

Theorem 1. *There exists a unique solution* $\{W_i\}_{1\le i\le n-1}$ *of the linear system (27).*

Hence the remainder part $\{V_k\}_{1\le k\le n-1}$ of $\{U_i\}_{1\le i\le n-1}$ is automatically computed by Theorem 1 and Lemma 2. i.e.,

Theorem 2. *V_k is determined by*

$$\begin{aligned} v_{k,l} = & \sum_{i=1}^{k-1} {}^tQ_l\ D_{n-k,i}\ Q\Big(\ A_i^W\ W_{i-1} - A_i^\epsilon\ W_i + A_i^E\ W_{i+1}\ \Big) \\ + & \sum_{i=k}^{n-1} {}^tQ_l\ D_{k,n-i}\ Q\Big(\ A_i^W\ W_{i-1} - A_i^\epsilon\ W_i + A_i^E\ W_{i+1}\ \Big) \\ + & {}^tQ_l\Big(\ \sum_{i=1}^{k-1}\ D_{n-k,i}\ Q\ F_i + \sum_{i=k}^{n-1}\ D_{k,n-i}\ Q\ F_i\Big). \end{aligned}$$

for (k,l) such that $P_{k,l} \in \Omega_h \backslash \Gamma_h$.

Therefore we summarize our numerical algorithm.

Numerical Algorithm
1st step: Calculate the solution $\{W_i\}$ on Γ_h of (27).
2nd step: Compute $\{V_k\}$ on $\Omega_h \backslash \Gamma_h$ by use of the formulation in Theorem 2.

4 Convergence of Approximate Solutions

4.1 Function Space V_h

In order to justify our numerical scheme(13), we first define the piecewise function $\theta_{\alpha,\beta}$ $(\ 0 \le \alpha, \beta \le n)$ as follows;

$\theta_{\alpha,\beta}(x,y) = \theta(x - \alpha\ h, y - \beta\ h)$ where $\theta(x,y) = \begin{cases} 1, & (x,y) \in \omega_{0,0}^0 \\ 0, & (x,y) \notin \omega_{0,0}^0\ , \end{cases}$

and $\theta_{0,j} = \theta_{n,j} = \theta_{i,0} = \theta_{i,n} = 0\ (i,j = 1,\cdots,n)$. We then introduce the function space V_h generated by $\theta_{i,j}$. i.e., $\phi \in V_h$, is of the form:

$$\phi(x,y) = \sum_{i,j=1}^{n-1}\ \phi_{i,j}\ \ \theta_{i,j}(x,y), \quad \phi_{i,j}\ \in\ R\ . \tag{28}$$

We now introduce the following approximation $\{\delta_h^1, \delta_h^2, \nabla_h, (\mathrm{div})_h\}$ of $\{\partial/\partial x, \partial/\partial y, \nabla,\ \mathrm{div}\ \}$.
(i) $\delta_h^1, \delta_h^2 : L^\infty(R^2) \to L^\infty(R^2)$ are defined by

$$\begin{aligned} (\delta_h^1\ u)(x,y) &= \tfrac{1}{h}\Big(\ u(x + \tfrac{1}{2}h, y) - u(x - \tfrac{1}{2}h, y)\Big), \\ (\delta_h^2\ u)(x,y) &= \tfrac{1}{h}\Big(\ u(x, y + \tfrac{1}{2}h) - u(x, y - \tfrac{1}{2}h)\Big). \end{aligned}$$

(ii) $\nabla_h : L^\infty(R^2) \to (L^\infty(R^2))^2$ is defined by

$$(\nabla_h\ u)(x,y) = \Big((\delta_h^1\ u)(x,y),\ (\delta_h^2\ u)(x,y)\Big). \tag{29}$$

(iii) $(\mathrm{div})_h : (L^\infty(R^2))^2 \to L^\infty(R^2)$ is defined by

$$(\mathrm{div})_h\ (u(x,y),\ v(x,y))\ =\ (\delta_h^1\ u)(x,y)\ +\ (\delta_h^2\ v)(x,y) \tag{30}$$

for $u,\ v\ \in\ L^\infty(R^2)$.

Then the norm $\|\cdot\|$ in V_h is equipped as follows;

$$\|u\| = \sqrt{\|\ u\|_{L^2(\Omega)}^2 + \|\nabla_h\ u\|_{L^2(\Omega)}^2}\ \text{ for } u \in V_h, \tag{31}$$

from which we get

Lemma 4. *(i) V_h is a Hilbert space.*

$$(ii)\ \left(\delta_h^i\ u, \phi\right)_{L^2(\Omega)} = -\left(u,\ \delta_h^i\ \phi\right)_{L^2(\Omega)} \quad \text{for } u,\ \phi \in V_h\ \ (i{=}1,2). \tag{32}$$

Furthermore using the notations $\{a^W_{i,j},\ a^S_{i,j},\ f_{i,j},\ \Delta l_{i,j}, \sigma_{i,j}\}$ in (12), we define approximate functions of a, f and σ respectively as follows:

$$\begin{aligned}
a_h^W(x,y) &= \textstyle\sum_{j=1}^{n-1} \sum_{i=1}^{n}\ a^W_{i,j}\ \theta_{i-1/2,\ j}(x,y),\\
a_h^S(x,y) &= \textstyle\sum_{i=1}^{n-1} \sum_{j=1}^{n}\ a^S_{i,j}\ \theta_{i,\ j-1/2}(x,y),\\
f_h(x,y) &= \textstyle\sum_{i,j=1}^{n-1}\ f_{i,j}\ \theta_{i,j}(x,y),\\
\sigma_h(x,y) &= \textstyle\sum_{i,j=1}^{n-1}\ \frac{\Delta\ l_{i,j}}{\Delta x \Delta y}\ \sigma_{i,j}\ \theta_{i,j}(x,y).
\end{aligned}$$

4.2 Approximate Solution in V_h of Problem I

In this subsection the approximate solution in V_h for Problem I is considered. We first propose the following approximation of Problem I in V_h.

Problem V. Find $u_h \in V_h$ such that

$$\begin{aligned}
&-\ (\mathrm{div})_h\ \left(a_h^W(x,y)\ (\delta_h^1\ u_h)\ ,\ a_h^S(x,y)\ (\delta_h^2\ u_h)\right)(x,y)\\
&\qquad = f_h(x,y) + \sigma_h(x,y)\ \ \text{for } (x,y) \in \textstyle\bigcup_{i,j=1}^{n-1} \omega^0_{i,j}.
\end{aligned} \tag{33}$$

We then get a following relation between Problem F and Problem V.

Lemma 5. *Problem F and Problem V are equivalent.*

Proof. Using the notations in 4.1 and the property of the support for piecewise functions, the equation (33) is of the form

$$\begin{aligned}
&-\ \textstyle\sum_{i,j=1}^{n-1}\ \Big(\ \frac{1}{\Delta x}\big(\ a^E_{i,j}\ \frac{u_{i+1,j}\ -\ u_{ij}}{\Delta x}\ -\ a^W_{i,j}\ \frac{u_{ij}\ -\ u_{i-1,j}}{\Delta x}\ \big)\\
&+\ \frac{1}{\Delta y}\big(\ a^N_{i,j}\ \frac{u_{i,j+1}\ -\ u_{ij}}{\Delta y}\ -\ a^S_{i,j}\ \frac{u_{ij}\ -\ u_{i,j-1}}{\Delta y}\ \big)\Big)\ \theta_{i,j}(x,y)\\
&=\ \textstyle\sum_{i,j=1}^{n-1}\ \Big(\ f_{i,j}\ +\ \frac{\Delta l_{i,j}}{\Delta x\,\Delta y}\ \sigma_{i,j}\ \Big)\ \theta_{i,j}(x,y)
\end{aligned}$$

for $(x,y) \in \displaystyle\bigcup_{i,j=1}^{n-1} \omega^0_{i,j}$. Hence this lemma holds.

Using the discrete Poincaré inequality and the trace theorem ([5]), we get

Proposition 2. *There exists a unique function $u_h \in V_h$ satisfying (33).*

The uniqueness of $\{W_i\}$ in (27) is now proved.

Proof of Theorem 1. Assume that there are two solutions $\{W_i\}$ and $\{\widetilde{W}_i\}$ satisfying the linear system (27). Then from Lemma 2, and (19)-(21), there are two solutions $\{U_i\}$ and $\{\widetilde{U}_i\}$ of Problem F. But this is contradictory to Proposition 2 by use of Lemma 5. Therefore the unique existence of the solution $\{W_i\}$ is ensured.

4.3 Convergence Theorem

We proceed to discuss the convergence of $\{u_h\}$.

Theorem 3. *(i) There exists $u \in H_0^1(\Omega)$ such that $u_h \to u$ weakly in $L^2(\Omega)$.*
(ii) u satisfies that for any $\phi \in D(\Omega)$,

$$\left\langle -\operatorname{div}\left(a \nabla u\right), \phi \right\rangle_{D'(\Omega)} = \left(f, \phi\right)_{L^2(\Omega)} + \left(\sigma, \phi\right)_{L^2(\Gamma)} \tag{34}$$

Proof. We divide the proof into four steps.

Step 1. There exists a subsequence u_h, also denoted by u_h, such that $u_h \to u$ weakly in $L^2(\Omega)$ and $\nabla_h u_h \to \nabla u$ weakly in $L^2(\Omega)$.
In fact, it follows from the bilinear form of (33) in V_h and the discrete Poincaré inequality.
Step 2. $f_h \to f$ *in* $L^2(\Omega)$ and $a_h^W \to a$ *a.e.* in Ω, $a_h^S \to a$ *a.e.* in Ω.
Because $f \in L^2(\Omega)$ and a is continuous in $\Omega \backslash \Gamma$.
Step 3. $\sigma_h \to \sigma \cdot \delta(\Gamma)$ in $D'(\Omega)$.
In fact, Set $I \equiv \left\langle \sigma_h - \sigma \cdot \delta(\Gamma), \phi \right\rangle$. Then

$$I = \sum_{i,j=1}^{n-1} \int_{\omega_{i,j}^0} \frac{\Delta l_{i,j}}{h^2} \sigma_{i,j}\, \phi(x,y)\, dxdy - \int_\Gamma \sigma(s)\phi(x(s),y(s))ds$$

$$= \sum_{i,j=1}^{n-1} \int_{\omega_{i,j}^0 \cap \Gamma} \sigma(s)\, \{\frac{1}{h^2} \int_{\omega_{i,j}^0} \phi(x,y)\, dxdy\}\, ds - \int_\Gamma \sigma(s)\phi(x(s),y(s))ds.$$

Since $\phi \in D(\Omega)$, there exists a point $(x_{i,j}, y_{i,j})$ in $\omega_{i,j}^0$ such that

$$\frac{1}{h^2} \int_{\omega_{i,j}^0} \phi(x,y)\, dxdy = \phi(x_{i,j}, y_{i,j}), \quad 1 \le i,j \le n-1.$$

Hence

$$\left|I\right| = \left| \sum_{i,j=1}^{n-1} \int_{\omega_{i,j}^0 \cap \Gamma} \sigma(s)\, \phi(x_{i,j}, y_{i,j})\, ds - \int_\Gamma \sigma(s)\phi(x(s),y(s))ds \right|$$

$$= \left| \int_\Gamma \sigma(s) \sum_{i,j=1}^{n-1} [(\phi(x_{i,j}, y_{i,j}) - \phi(x(s),y(s))\, \theta_{i,j}(x(s),y(s))]\, ds \right|$$

$$\le \left(\int_\Gamma |\sigma(s)|^2 \, ds \right)^{1/2} \left(\int_\Gamma \sum_{i,j=1}^{n-1} |\phi(x_{i,j}, y_{i,j}) - \phi(x(s),y(s))|^2\, \theta_{i,j}(x(s),y(s))\, ds \right)^{1/2}$$

$$\le \left(\int_\Gamma |\sigma(s)|^2 \, ds \right)^{1/2} \left(\sum_{i,j=1}^{n-1} \int_{\omega_{i,j}^0 \cap \Gamma} |(x_{i,j}, y_{i,j}) - (x(s),y(s))|^2 \cdot |\nabla\phi|_{L^\infty(\Omega)}^2\, ds \right)^{1/2}$$

$$\le \sqrt{2\,\mu(\Gamma)}\, h\, |\sigma|_{L^2(\Gamma)} \cdot |\nabla\phi|_{L^\infty(\Omega)} \quad \text{where } \mu(\Gamma) = \int_\Gamma ds.$$

This shows the statement of Step 3.

Step 4. For $\phi \in D(\Omega)$, the equation

$$-\left\langle \frac{\partial}{\partial x}\left(a \frac{\partial u}{\partial x} \right) + \frac{\partial}{\partial y}\left(a \frac{\partial u}{\partial y} \right), \phi \right\rangle_{D'(\Omega)} = \left(f, \phi\right)_{L^2(\Omega)} + \left(\sigma, \phi\right)_{L^2(\Gamma)} \tag{35}$$

holds.

In fact, it follows from Proposition 2 that for sufficiently small h,

$$\left(-\ (\mathrm{div})_h \left(a_h^W(x,y)\ (\delta_1\ u_h)\ ,\ a_h^S(x,y)\ (\delta_2\ u_h)\right)(x,y),\ \phi(x,y) \right)_{L^2(\Omega)}$$
$$= \left(\left(a_h^W(x,y)\ (\delta_1\ u_h)(x,y)\ ,\ a_h^S(x,y)\ (\delta_2\ u_h)(x,y)\right),\ \nabla_h\ \phi(x,y) \right)_{L^2(\Omega)}$$
$$= \left(\ f_h(x,y) + \sigma_h(x,y),\ \phi(x,y)\ \right)_{L^2(\Omega)}.$$

We then use the results from the Step1 to Step 3 and as $h \to 0$ in the above equation, we have

$$\left(a \frac{\partial u}{\partial x}, \frac{\partial \phi}{\partial x} \right)_{L^2(\Omega)} + \left(a \frac{\partial u}{\partial y}, \frac{\partial \phi}{\partial y} \right)_{L^2(\Omega)} = \left(f, \phi\right)_{L^2(\Omega)} + \left(\sigma, \phi\right)_{L^2(\Gamma)}. \tag{36}$$

Therefore combining it with the distribution formula, Step 4 is shown.
Finally we are able to conclude that the full sequence $\{u_h\}$ converges weakly to the solution u of Problem I since Problem I has a unique solution in $H^1(\Omega)$ as well known fact(cf. [8,10]).

Acknowledgment. The author is grateful to Prof. Kawarada of Chiba University for his significant comments and also wish to thank Prof. Kitahara of Kogakuin University for helpful discussions.

References

1. H. Fujita,H. Kawahara and H. Kawarada, Distribution theoretical approach to fictitious domain method for Neumann problems, East-West J. Math., vol. 3, no. 2 (1995), 111–126.
2. R. Glowinski, T.W. Pan and J. Periaux, A fictitious domain method for Dirichlet problem and applications, Computer Methods in Applied Mechanics and Engineering 111 (1994), 283–303.
3. H. Han and Z. Huang, The direct method of lines for the numerical solutions of interface problem, Computer methods in Applied Mechanics and Engineering, 171 (1999), 61–75.
4. F. John, Lectures on Advanced Numerical Analysis, Gordon and Breach Science Publishers, Inc., New York (1967).
5. H. Kawarada, Free boundary problem – theory and numerical method, Tokyo University Press (1989)(in Japanese) .
6. Koshigoe, H. and Kitahara, K., *"Method of lines coupled with fictitious domain for solving Poisson's equation"*, Mathematical Sciences and Applications (Gakuto International Series), Vol. 12 (1999), 233–241.
7. Koshigoe, H. and Kitahara, K., *"Numerical algorithm for finite difference solutions constructed by fictitious domain and successive eliminations of lines"*, Japan SIAM, Vol. 10, No. 3 (2000), 211–225 (in Japanese).

8. J.L. Lions, Optimal control of systems governed by partial differential equations, 170, Springer-Verlarg (1971).
9. G.I. Marchuk,Y.A. Kuznetsov and A.M. Matsokin, Fictitious domain and domain decomposition methods, Sov. J. Numer. Anal. Math. Modelling, Vol. 1, No. 1 (1986) 3–35.
10. S. Mizohata, The theory of partial differential equations, Cambridge at the University Press (1973).
11. K. Nakashima, Numerical computation of elliptic partial differential equations I, Method of Lines, Memoirs of the school of science & engineering, Waseda Univ., No. 29, 115–127,1965.

Track on
Web Engineering

CORBA Based Real-Time Object-Group Platform in Distributed Computing Environments

Su-Chong Joo[1], Sung-Kwun Oh[1], Chang-Sun Shin[1], and Jun Hwang[2]

[1] School of Electrical, Electronic and Information Engineering, Wonkwang University, Korea
{scjoo,ohsk,csshin}@wonkwang.ac.kr

[2] Department of Computer Science, Seoul Women University, Korea
hjun@swu.ac.kr

Abstract. Recently, the distributed real-time services are developing in distributed object computing environments in a way that can support a new programming paradigm of the distributed platform that requires interoperability among heterogeneous systems. These services are based on distributed middleware and object-oriented technologies. But we have the difficulties of managing of distributed objects and providing real-time objects with the timing constraints. For simultaneously solving these problems, we designed a new model, a real-time object group (RTOG) platform that can manage and group the distributed objects for reducing their own complicated managements and interfaces, and add distributed real-time requirements to real-time objects without modifying the ORB itself. The structure of the real-time object group we suggested contains several components reflecting the object grouping concepts and real-time service requirements analyzed by referring OMG CORBA specifications and real-time properties. To construct our RTOG platform, we designed the RTOG structure and the functional class diagram of components in a RTOG according to the analyzed requirements, and defined the timing constraints of real-time objects. And we explained the functional definitions and interactions among the components from the following points of view: distributed object management and real-time services. Finally, as results of the implemented our RTOG platform, we showed the real-time executing procedures via the Scheduler and the Real-Time Manager in RTOG platform charging with real-time services, and verified our findings by using the EDF algorithm to see whether real-time services can be applied to our platform.

1 Introduction and Related Works

The distributed object computing is becoming a widely accepted new programming paradigm for distributed applications that require seamless interoperability among heterogeneous environments. The Object Management Group (OMG) had developed the Common Object Request Broker Architecture (CORBA) [1,2] as a standard software specification for distributed environments. This standard software specification specifies an Interface Definition Language (IDL) for the interface descriptions of the functional behaviors of distributed components. It also specifies object services,

P.M.A. Sloot et al. (Eds.): ICCS 2003, LNCS 2659, pp. 401–411, 2003.

which facilitate standard interactions with a set of capabilities (i.e. Naming, Event, etc.), and an Object Request Broker (ORB), which is the middleware that allows for the seamless interactions between distributed objects with roles of clients or servers.

Many distributed real-time applications, such as automated factory controls, avionics navigation and military target tracking system, have benefited from a standard architecture like CORBA. Though many designers of distributed applications have been adapting the CORBA specification for developing their own middleware architectures, it is currently inadequate to support requirements for distributed real-time services. For example, the IDL describes the interfaces of functional behaviors of distributed components, but it is impossible to explicitly describe the timing constraints for their real-time behaviors. Furthermore, system services provided by distributed environments offer little support for end-to-end real-time scheduling [3,6] across the environment due to not providing services as synchronized clocks and bounded message latencies.

Recently, the Real-Time Special Interest Group (RT-SIG) [4,7] organized in the OMG has being propelled to examine the current CORBA standard and determine requirements for supporting distributed real-time applications. Especially, this RT-SIG is focusing on supporting the ability to express and enforce the timing constraints by extending the current CORBA standard. The RT-SIG has produced white papers that detail the desired capabilities for a distributed computing environment to support real-time applications. The desired real-time capabilities specified in the CORBA/RT SIG white paper are classified into the following three areas: desired capabilities for the operating environment; desired capabilities for the ORB architecture and desired capabilities for the object services and facilities. But lots of researchers have been done to only improve the system performance by using real-time CORBA itself, or modified a part of CORBA compliance. But, considering the integrated real-time CORBA environments, we still have problems with embedding real-time applications into CORBA implementation. For solving the above problems, it is necessary to extend the current OMG CORBA standard so that it can meet real-time requirements and the desired capabilities of the application without changing the CORBA standard itself. Also, we used the concept of an object group that we have already studied [8–12]. This object group concept was suggested for providing the efficient and convenient management of distributed objects, and for making the individual object independent and reusable. Based on the object group concept, we designed and implemented that a RTOG platform for not only supporting the distributed real-time properties, but also for managing the distributed objects based on the current CORBA standard.

Our paper is organized as follows; Section 2 describes the real-time object group model. Here, we designed the real-time object group (RTOG) and defined the timing constraints of real-time objects. In Sect. 3, we explain the functional definitions and interconnections of the components included in the RTOG from the following services; distributed object management service and real-time service. In Sct.4, we depict how to implement and execute components in RTOGs on our RTOG platform. As a result, we show the executing procedures via Scheduler object and Real-Time Manager in the RTOG charging with real-time services, and verified our findings by using the EDF algorithm to see whether real-time services can be applied to our platform. Finally, we discuss our conclusions and future works.

2 Real-Time Object Group

In a real-time system, first of all, the timing constraints must be met for given real-time applications. Real-time requirements are typically derived from the practical systems interacting with the physical environment. There are two kinds of the relating approaches suggested to adapt real-time services to CORBA. One approach has been producing a CORBA specification or coding implementations that support faster performance. The other approach taken by some vendors has been embedding their CORBA implementations to the real-time operating system [3,5]. But, because both approaches are insufficient for meeting the timing constraints, the supports for the expression and system-wide enforcement of the timing constraints are also necessary. For this reason, we suggested the real-time object group platform (called RTOG platform after this).

2.1 RTOG Structure

Figure 1 below shows the RTOG structure, The RTOG we suggested consists of the following objects; Group Manager (GM), Real-Time Manager (RTM), Object Factory, Security, Scheduler, Timer, Object Information Repository, Service Objects (SOs), and Sub-object groups. Here, the RTOG is logically presented as a single view system for object management service and real-time service of SOs existing in distributed systems connected over networks physically.

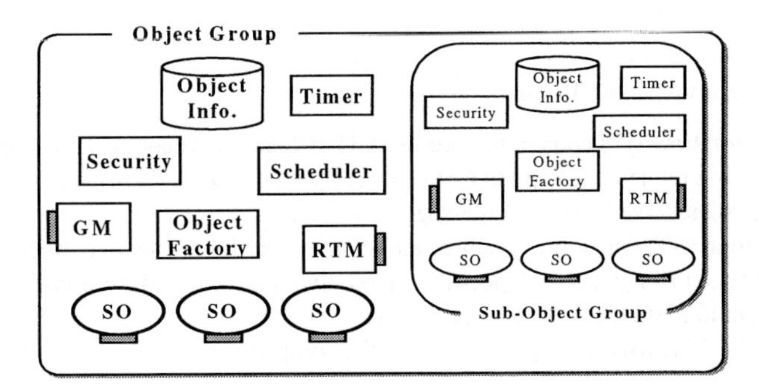

Fig. 1. RTOG structure

We suggest our RTOG for efficient management of distributed objects and the execution of services simply reflecting the real-time properties in distributed environments. And the major role of the components in the RTOG is categorized in two classes to reduce the side effect of interoperability between the object management and the real-time services. First, for supporting the object management services, our RTOG consists of the Group Manager, the Security, the Object Information Repository, and the Object Factory as components.

The Group Manager (GM) is responsible for managing all of distributed objects in group model, and interoperating between a client and target SOs. The Object Factory receives the object creation's request from the Group Manager, creates the requested new object using the information from the Object Information Repository, and then returns the reference of the created object to the Group Manager. The Security Object checks its own object group's access possibility for the request of a client in another group by searching the access control list (ACL) and the access rule. The Object Information Repository stores and maintains the information related to creation of SOs and the subgroup's Group Manager.

Secondly, the components in the RTOG taking care of client requests for real-time service are the Real-Time Manager, the Timer, and the Scheduler. The Real-Time Manager transparently supports the expression of real-time requirements and simply controls complex procedures for service calls binding between distributed objects by interoperating with the Scheduler and the Timer. This Timer has an alarm function for reporting the missing deadline to the requesting SO, without a reply message related to the missing deadline from the requested SO. Finally, the Scheduler executes a given scheduling policy. At this time, the developers can arbitrarily alter an appropriate scheduling mechanism into the Scheduler according to the real-time property of the special-purposed distributed application.

2.2 Timing Constraints

The basic flow of procedures to obtain the result returned from a client's request is broken into three steps; the service request step, the service process step and the result return step. Each step has the timing constraints and must guarantee the timeliness by explicitly outlining the timing constraint's definition. To do this support, we define the five timing constraints in our study as follows;

- An invocation time (IT) constraint specifies CIT (Client's Invocation Time), as a time that a client sends a request message to a SO and SIT(Service Object's Invoked Time), as a time that a SO received a request message of a client.
- A service time (ST) constraint specifies the relative time for the service execution of a SO.
- A transfer time (TT) constraint specifies the relative time required for transferring a request from a client to a SO (SIT-CIT), or inversely.
- A service deadline (SD) constraint specifies the completed service time of a SO as an absolute time.
- A request deadline (RD) constraint specifies the absolute time that a client received the result returning from a SO after a client requested a SO. Here, a RD is the timing constraint that must be guaranteed in a distributed real-time application.

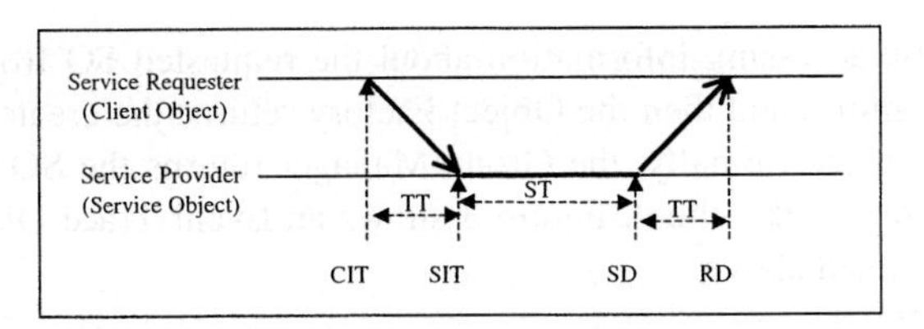

Fig. 2. The definition of timing constraints

Figure 2 shows the definition of the timing constraints. From these timing constraints, we can redefine RD and SD as following expressions;

$$RD \geq CIT+ST + (2 * TT) + \text{slack time},$$
$$SD \leq RD - TT, \textit{ where } TT = SIT\text{-}CIT \textit{ and } ST = SD - SIT.$$

The deadline for service execution (SD) is the sum of the client's invocation time(CIT), the service execution time (ST), transfer time (2*TT) and *slack time*. Here, the slack time is for a constant as a factor for deciding the time of the adequate RD. This service deadline (SD) must be smaller than and equal a time value which subtracts transfer time (TT) from a client's request deadline (RD) for guaranteeing real-time service. In our RTOG platform, we will use the timing constraints mentioned above for applying an Early Deadline First (EDF) algorithm.

3 Interacting with Components in RTOGs

In this section, we describe the interacting procedures between the components in a RTOG or RTOGs on distributed platform from the following points of view; distributed object management and real-time services. And we show interacting procedures and operations via an Event Trace Diagram (ETD).

3.1 Supporting Object Management Services in RTOGs

The Group Manager is responsible for the whole management of the components of the RTOG. The Group Manager can create and withdraw SOs or Sub-object groups in the RTOG itself, and also receive the reference of a SO requested by clients. In this paper, among these object management functions, we will discuss only the procedures for obtaining a SO's reference requested from a client in order to save space. A client must execute the object management procedures firstly before supporting real-time service. These procedures are consisted of following several steps. A client in an Object Group obtains the reference of the Group Manager in another Object Group involving the desired SO via a Name Server. A client requests the SO's reference from the Group Manager using the Object Group's reference obtained before. The Group Manager invokes the Security to decide whether a client can access a SO. After the Security checks the access control list, it returns the access right of the requested a SO to the Group manager. If a client has the access right for a target SO, the Group Manager requests the creation of it from the Object Factory. The Object Factory creates it by accessing information about the requested SO from the Object Information Repository, and then the Object Factory returns the created SO's reference to the Group Manager. Finally, the Group Manager returns the SO's reference from the Object Factory to the client. Figure 3 shows an Event Trace Diagram (ETD) for procedures explained above.

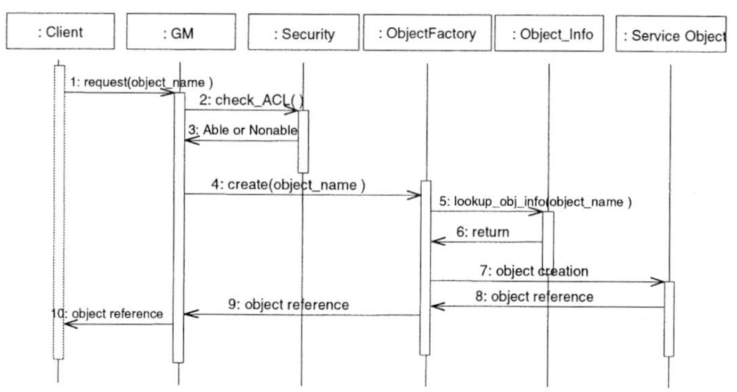

Fig. 3. The ETD for obtaining the Service Object (SO)'s reference

3.2 Supporting Real-Time Services in RTOGs

A client requests the real-time service to the SO with a reference obtained via the object management procedures. Here, as we previously mentioned, the service request of a client is necessary to involve the expression, that is, the RI (Real-time Information structure), about the timing constraints because of the service request with real-time property. In addition, we assume that the Real-Time Manager (RTM) in each object group maintains the global time information using global time service. But, we do not focus on the detailed global time service in our study.

When a client invokes real-time service to a SO in another object group, it passes the deadline information to the RTM in its own object group. This deadline is for an absolute time until the result message is returned. The RTM in a client's object group stores the current global time and the client's request deadline, that is, requested by a client, in RI. The RTM sets a client's deadline to the Timer in its own object group. If this deadline expires, the Timer notifies the missed deadline to a client. After a sequence of procedures mentioned above in an object group including a client is completed, a client requests the real-time service to SO in another object group. At this time, the RI information having the timing constraints stored by the RTM is simultaneously sent with the service call to a SO. A SO passes the RI received to the RTM in its own object group. The RTM calculates the service deadline with consideration to the transfer time, and passes the calculated service deadline to the Scheduler. The Scheduler schedules and decides the priority for the service execution of the client. According to the decided priority, the RTM sets the calculated service deadline on the Timer, and this Timer can also predict and notify the SO with the missed service deadline to a client. With setting the Timer, the RTM sends a client's information to a SO. After a SO performs the services, it returns the executed results to a client, inversely, and it reports the completion of the client's request to the RTM in its own object group. Then the RTM in a SO's object group disarms the Timer. Also, the RTM in a client's object group disarms Timer by getting the result from a SO. Figure 4 shows the Event Trace Diagram for the real-time service's procedures described above.

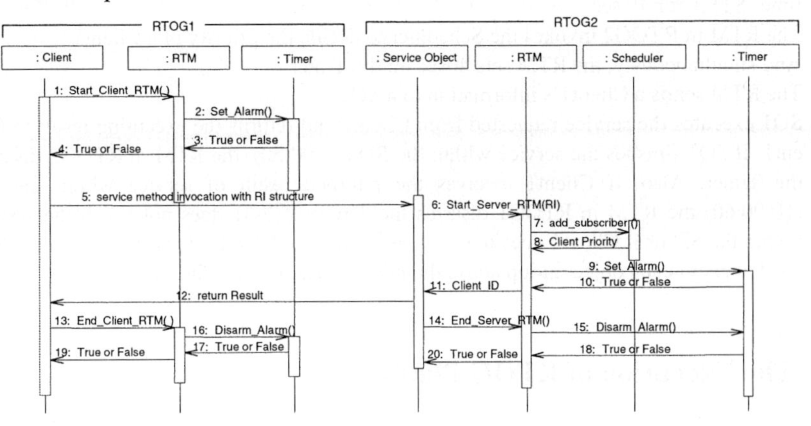

Fig. 4. The ETD for the support of real-time service

To illustrate how the components in our RTOG model work together for supporting a real-time service, the Fig. 5 is shown the expression and the real-time processing procedures along with given timing constraints in RI parameter, as an example of a typical interactions between Client1 and Service Object1 (SO1) by requesting real-time service.

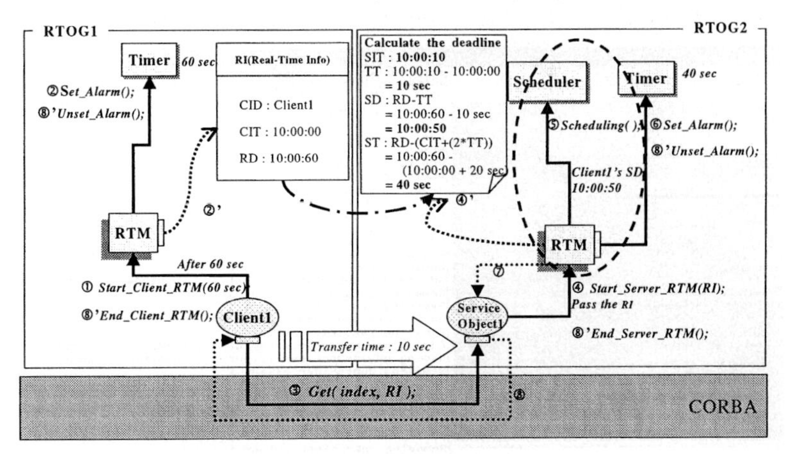

Fig. 5. The example of real-time service in RTOGs

We explain the procedure of real-time service shown in Fig. 5 and time-based numerical analysis from our model using the timing constraints described Sect. 2.2. Here, we assume that for convenient calculations, a unit of time is the second.

- Client1 sends the deadline information (CID, CIT, RD) to the RTM in RTOG1.
- The RTM in RTOG1 sets 60sec on the Timer and stores global time information(CID : Client1, CIT : 10:00:00, client's RD : 10:00:60)

- Client1 requests the real-time service with RI to Service Object1 (SO1) in RTOG2.
- SO1 passes the received RI to the RTM in RTOG2. And the RTM calculates the service deadline (SD) from following expression; SD = RD- TT; Here, when let the transfer time (TT) be 10sec, the calculated service deadline will be 10:00:50 sec. Also, the service time (ST) is for 40 sec.
- The RTM in RTOG2 invokes the Scheduler to decide the priority of a Client1,
- And simultaneously, the RTM sets 40sec on the Timer.
- The RTM sends a Client1's information to a SO1.
- SO1 executes the service requested from Client1 and returns the executing result to Client1. If SO1 finishes the service within the SD (10:00:50), the RTM in RTOG2 disarms the Timer. Also, if Client1 receives the returned result of service within the RD (10:00:60) the RTM in RTOG1 disarms the Timer. If SO1 does not finish the service within the SD or if Client1 does not receive the returned result of service within the RD, the Timers in each object group alarm the missing deadline to Client1.

4 The Execution of RTOG Platform

The components of RTOG were designed and implemented to IDL for independently supporting heterogeneous distributed computing environments, and implemented by using Visi-broker and Visual C++ 6.0 running on Windows. In this section, we will show only a procedure of real-time service. The supports of distributed object group management service have already published as researching papers [9,11,12]. Hence, we will not explain this part.

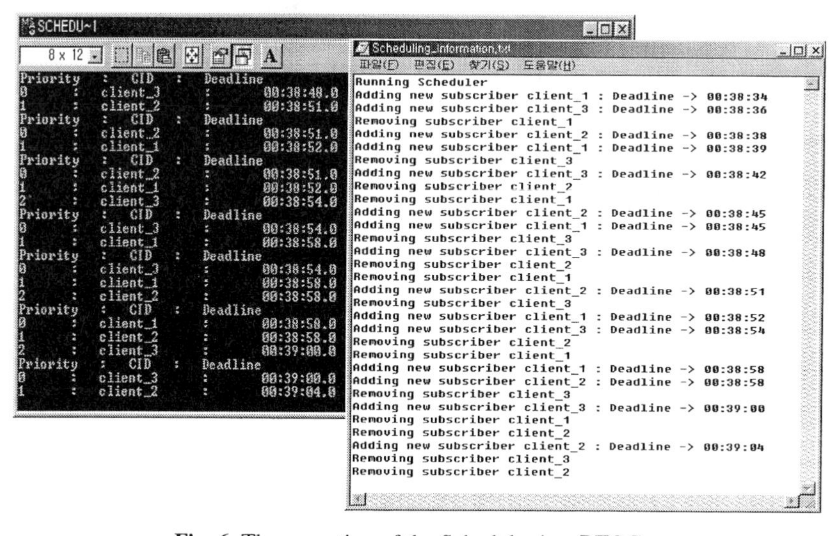

Fig. 6. The execution of the Scheduler in a RTOG

To verify the real-time execution procedures of the implemented RTOG platform, Figs. 6 and 7 below are shown the real-time scheduling procedures on the Scheduler and succeeding/missing procedures of a request deadline given from a client on the RTM, when an arbitrary SO is requested by a number of clients over our platform.

In Fig. 6, we display that the priorities of a client have been changed according to an arbitrary given scheduling policy of the Scheduler, as an example. Here we adopted the EDF algorithm as an example of the scheduling policy of the Scheduler. In this executing procedure, the service request of a client having the shorter deadline has the higher priority by EDF policy. We can show two cases deleting clients' requests from the scheduling list in the execution of the Scheduler. One is when a client's request does not complete within a given deadline from a SO's service; deadline missing. The other is when a client's request does complete normally; deadline success. In Fig. 7, we show as the executing procedures on the Real-Time Manager. The deadline success rates of each client's request are 2 per 5 (40%) in a client-1, 4 per 5 (80%) in a client-2, and 5 per 5 (100%) in a client-3, respectively. These results may differ according to each given client's deadline and scheduler policy. In this paper, we only used the EDF algorithm for verifying whether the Scheduler in our model could be operated well correctly, but not showing the improved performance, like decrease of deadline missing rate, of our model itself. Through this verification, we will develop a more flexible prototypical platform that can adopt various real-time scheduling policies to the Scheduler in our model according to given real-time application's properties.

Fig. 7. The execution of the real-time manager in a RTOG

5 Conclusions

The distributed object computing technology is becoming a new program paradigm for distributed applications requiring interoperability between heterogeneous systems. Especially, when developing a distributed real-time software, we have to consider the timing constraints for distributed real-time services. Therefore, for the needs of a new paradigms and trends, we suggest the real-time object group (RTOG) platform for providing object group concepts and real-time constraints. This platform supports real-time service at the distributed application level and object management service at the CORBA level, respectively. At the development of this platform, we first surveyed and analyzed studies relating to real-time service models for distributed object-oriented systems. From lots of researches based on the above environment, the existing researches have been attempting to only improve the performance of systems by using real-time CORBA, or modifying the part of CORBA compliance. Up to now, this kind of a new platform agreed to our goal has not ever tried to be developed yet.

Hence, we designed a new model of a real-time object group platform that can support the real-time requirement without modifying the ORB. In this paper, we defined the roles of the components of a RTOG, and described functions of each component and their relationships of interoperability between the components including in an object group. To verify the executing procedures of the implemented real-time object group platform, we simulated the processing procedures for the real-time service execution requested by several clients in our platform. As final results, we showed that our real-time object group platform is able not only to manage distributed objects and/or object groups, but also to guarantee the real-time requirements of distributed real-time applications. In this paper, we only used the EDF algorithm for verifying whether the Scheduler in our model could be scheduled correctly, but not showing the improved performance of our model itself.

In future, using this RTOG model, we will develop more flexible prototypical platform that can variously adopt new real-time scheduling policies to the Scheduler in our model according to given real-time application's properties and support general-purpose real-time services, not special-purpose real-time services. And then we want to verify the execution power of the distributed real-time application on our platform via various simulations for improving real-time service.

Acknowledgements. This work reported here was supported in part by Research funds of Wonkwang University and the Brain Korea 21-Project of KRF, 2003.

References

1. OMG Real-time Platform SIG: Real-time CORBA A White Paper-Issue 1.0. http://www.omg.org/realtime/real-time_whitepapers.html (1996)
2. OMG: The Common Object Request Broker: Architecture and Specification revision 2.2. http://www.omg.org/corba/corbaCB.htm (1998)
3. Victor Fay Wolfe. (eds.): Expressing and Enforcing Timing Constraints in a Dynamic Real-Time CORBA System. http://www.cs.uri.edu/rtsorac/publication.html (1997)
4. OMG TC: Realtime CORBA Extensions: Joint Initial Submission. OMG TC, Document orbos/98-01-09 (1998)
5. Nguyen Duy Hoa: Distributed Object Computing with TINA and CORBA. Technical Report Nr. 97/7. http://nenya.ms.mff.cuni.cz/thegroup (1997)
6. Scheduling in Real-Time System. http://www.realtime-os.com/sched_o3.html (1996)
7. Andrew S. Tanenbaum: Distributed Operating Systems. Prentice Hall (1995)
8. S.C. Joo: A Study on Object Group Modeling and Its Performance Evaluation in Distributed Processing Environments. Final Researching Report. ETRI (1997)

9. G.M. Shin, M.H. Kim, S.C. Joo: Distributed Objects' Grouping and Management for Supporting Real-Time in CORBA Environments. Journal of The Korea Information Processing Society, Vol. 6, No. 5 (1999) 1241–1252
10. B.T. Jun, M.H Kim, S.C. Joo: The Construction of QoS Integration Platform for Real-Time Negotiation and Adaptation Stream in Distributed Object Computing Environments. Journal of The Korea Information Processing Society, Vol. 7, No. 11S (2000) 3651–3667
11. M.H. Kim, S.C Joo: Construction of CORBA Object-Group Model for Distributed Real-Time Service. Journal of The Korea Information Science Society, Vol. 7, No. 6 (2001) 602–613
12. M.K. Kim, C.W Jeong, C.S Shin, S.C Joo: Design and Implementation of Distributed Object Management Model. In Proceedings of IASTED International Conference-Parallel and Distributed Computing and Systems. Anaheim CA, U.S.A (2001) 222–227

Structural Similarity between XML Documents and DTDs

Patrick K.L. Ng and Vincent T.Y. Ng

Department of Computing, the Hong Kong Polytechnic University, Hong Kong
{csklng,cstyng}@comp.polyu.edu.hk

Abstract. The use of XML documents in the Internet continues to grow. Need for the analysis of XML documents from heterogeneous sources is arisen, in which documents would conform to different DTDs. In this paper, we propose a measure on the structural similarity among XML documents and DTDs, which is natural to understand and fast to calculate. The measure is defined as a weighted sum of the local measures of document elements with a weighting scheme based on their subtree sizes. While the local measure of an element is defined as its edit distance against its declaration, viewed as regular expression, in the DTD. Based on our definition, an algorithm for edit distance calculation between a string and a regular expression is proposed, which is modified from the algorithm applied in the regular expression matching problem. The advantage of the measure comes with its natural definition and linear complexity.

1 Introduction

DTD is a formal description in XML declaration syntax of a particular type of XML documents. It specifies the structures for XML documents; what names are to be used for different types of elements, where they may occur, and how they all fit together. Any document, which conforms to the specification in DTD, is called *valid* with respect to the DTD.

XML documents conforming to a DTD will have similar structure. This structure itself can also contains the semantic information, for example, a document having the tags, <portfolio>, <deal>, <maturity>, etc., is probably describing an investment portfolio. Obviously, the documents describing related contents may not have the same structure, especially when they come from different sources. Instead, they are more likely to have similar structures. This arises the interest to quantify the structural similarity so that similar structured documents can be clustered. It can be applied in document indexing as well.

In this paper, we first define the local similarity of elements using the edit distance and then aggregate the values by a weighting scheme based on the subtree sizes of the document. This approach takes into consideration the element order and the element similarity irrespective of the level. The idea is natural, easy to understand and fast to calculate, which makes it useful in both document indexing and clustering.

P.M.A. Sloot et al. (Eds.): ICCS 2003, LNCS 2659, pp. 412–421, 2003.

2 Related Work

This problem was addressed recently in [5], in which metric was defined as the weighted sum of the aggregated *PMC* (**p**lus, **m**inus and **c**ommon tags) value against the closest instance DTD. However, the definition of *PMC* assumed that the elements were unordered but it is not true for both the XML documents and DTDs. Moreover, the *PMC* value was calculated by comparing elements from the same levels, this level restriction made all the substructures of an unmatched element ignored.

In [12], measure was defined between two XML documents based on edit distance with restricted edit operations. While the measure is target for document clustering, however, this metric cannot fully achieve our objective; namely, any two XML documents conforming to the same DTD with different degree of repetitions for an "*" operator will have distance of their degree difference, which is not zero.

In our definition, we preserve the element order in XML documents and are able to identify similarity in unmatched levels. Moreover, the more a document is conforming to a DTD, the higher the similarity value it is, in particular, highest similarity (one) is always returned when the document conforms to the DTD.

3 Local Similarity Consideration

Our idea comes from a very natural sense; if two composite objects are equivalent or similar, they should also be equivalent or similar in local sense. In the context of structural similarity, we consider the similarity at element level using edit distance, which is local in nature. This constitutes the basis of our metric.

3.1 Matching XML Documents with DTD with Errors Allowed

The main purpose of DTD is to ensure the validity of the XML documents. What's meant by a valid XML document? It means that every elements (and attributes) in the XML document are present and conform to the structure of the hierarchical specification in the DTD. In order to make the discussion concise and clear, we focus on the elements only. The treatment of attributes is very similar so that we leave it to the reader. Figure 1 below shows an example of DTD, the corresponding valid XML document and the tree representation of the XML document.

Pick the element *deal* with id "*BOND1016*" from the XML document as example, it has the following child element list *(id, customer, maturity, schedule, schedule)* which matches the corresponding declaration of element *deal* in the DTD: *(id, customer?, maturity, (amount|schedule*))*, in the sense of regular expression matching. Note that there is no ambiguity in which element declaration to refer to (if it ever exists) because each element is uniquely declared in the DTD. However, this does not imply that the child element, say *schedule*, will also conform to the DTD, which demands further validation against its declaration in DTD. It is necessary to go through the DTD validation process for every elements appearing in the XML document.

```
<!DOCTYPE portfolio
[
<!ELEMENT portfolio (name, owner+,
 deal*)>
<!ELEMENT name (#PCDATA)>
<!ELEMENT owner (#PCDATA)>
<!ELEMENT deal (id, customer?,
 maturity, (amount|schedule*))>
<!ELEMENT id (#PCDATA)>
<!ELEMENT customer (#PCDATA)>
<!ELEMENT maturity (#PCDATA)>
<!ELEMENT amount (#PCDATA)>
<!ELEMENT schedule (date, amount)>
<!ELEMENT date (#PCDATA)>
]>
```

```
<portfolio>
 <name>OTC_GS</name>
 <owner>GSCM</owner>
 <deal>
  <id>WK0612</id>
  <customer>WINPAT</customer>
  <maturity>27DEC2003</maturity>
  <schedule>
   <date>16FEB2003</date>
   <amount>150000</amount>
  </schedule>
  <schedule>
   <date>06JUN2003</date>
   <amount>180000</amount>
  </schedule>
 </deal>
 <deal>
  <id>PN1202</id>
  <maturity>02DEC2003</maturity>
  <amount>1000000</amount>
 </deal>
</portfolio>
```

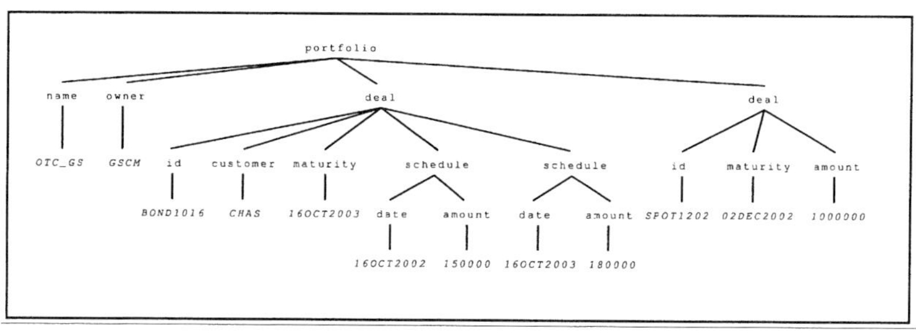

Fig. 1. Example of DTD (top left), XML document (top right) and the tree representation of the XML document (bottom)

Definition 1. If every elements of *X* (XML) conform to *D* (DTD), then we denote as "X matches D". Then it is obviously true if X is a valid with respect to D.

We can do label substitutions, say *a=deal, b=id, c=customer, d=maturity, e=schedule, f=amount*, then checking of the id "*BOND1016*" can be viewed as checking whether the string "*bcdee*" matches the regular expression "*bc?d(f|e*)*". With the setting above, matching the XML elements to the DTD can be viewed as matching the corresponding child element strings with their regular expressions in DTD.

In the previous discussion, we only consider whether an XML is valid or not. However, even a document is not valid with respect to DTD, we are still interested in how close to a valid document. For any element *e* in a document *X*, which is not necessarily valid with respect to a DTD *D*, it will fall into one of the three cases: (1) *e* matches D; (2) *e* exists in D but does not matches D; (3) *e* does not exist in D. If we can quantify the above cases, it can represent the local structural similarity between X and D, i.e. degree of matching at element level. In this section, we will only consider the cases (1) and (2) above while case (3) will be tackled at the end of Section 3.

In both cases (1) and (2), there exists regular expression *r* from *D* corresponding to each element *e* from *X*. We can make use of the edit distance to determine the distance between *s* and *r*. Edit distance is the metric to determine the distance between two strings; any string can be transformed into any other string by a combination of three basic operations; insertion, deletion and substitution. The edit distance is defined as the minimum number of basic operations for such a transformation. For the strings *abcdeh* and *aaacde*, it can be shown that the three operations: (1) substitute *b* into *a* (*aacdeh*); (2) insert *a* before *c* (*aaacdeh*); (3) delete *h* (*aaacdeh*), is a minimum transformation and hence the edit distance is 3. With this metric, we can easily extend the definition for strings and regular expressions.

Definition 2. Denote *EDIST*$[s_1, s_2]$ as the edit distance between two strings s_1 and s_2. Then define the edit distance between a string *s* and a regular expression *r*, *EDIST[s, r]*, as *min*{*EDIST*$[s, s_r]$: s_r matches *r*}. If *e* is an element in an XML document and there exists the corresponding declaration in the DTD, we denote *EDIST[e]* as *EDIST*$[s_e, r_e]$ where s_e is the child element string of *e* and r_e is the regular expression representing the declaration of *e* in the DTD. For example, *EDIST[aa, a*b]* = *min*{*EDIST[aa, b]*, *EDIST[aa, ab]*, *EDIST[aa, aab]*, *EDIST[aa, aaab]*, …}, which is equal to 1, from either *EDIST[aa, ab]* or *EDIST[aa, aab]*. Of course, the determination of edit distance becomes more complicated if the string and the regular expression are not simple.

3.2 Determination of Edit Distance

The problem of matching a string *s* with a regular expression *r* allowing errors has been widely studied, which is called the *Approximate Regular Expression Matching* problem [2,3,4,7,etc]; the problem statement is to find out all the substrings s_i of *s* such that *EDIST*$[s_i, r] <= d$, for some given error *d*. In [3], it defined a function *E[]*, which is very similar our *EDIST[]* function, as *min*{*EDIST*$[s_i, r]$: s_i is a substring of *s* ended at the last character of *s*}.

The difference is that our *EDIST[]* function considers only the whole string *s* while *E[]* took into account of all substrings of *s* ended at the last character of *s* in the string edit distance. In this section, we will present the idea and recurrence formula for *E[]*, following similar arguments in [3] (the proof was come from [2]) for completeness. Then we will show that modifying the initial condition in the formula will make it accounting for our *EDIST[]* function.

Definition 3. For a regular expression *r*, an automaton *M* can be constructed using the Thompson's construction [3]. Define

$\Pr e(i)$ as the set of predecessors of the node i in M

$\overline{\Pr e(i)}$ as the subset of Pre(i) excluding back edges

where *i* is the topological numbering of nodes in *M*. For any string $s = s_1 s_2 \ldots s_n$, *E[i, j]* is defined as the minimum edit distance between any string that can reach node *i* in *M* and any substring of *s* that ends at s_j. If *I* is a set of nodes in *M*, denote *E[I, j]* as *min*{*E[i, j]* for all *i* in *I*}.

Definition 4. *E[i, j]* must be equal to *E[k, j-1]* for some k ($\leq i$) plus the cost of editing a string w into s_j where w can range over all the strings on paths from $k+1$ to i in M. Define *E'[i, j]* as the best edit distance over all strings w that do not take across a back edge in M.

The idea is to use recurrence to calculate *E[]*. However, for any node i in M, the last word w may contain nodes coming from later node through some back edges, which is still not yet evaluated at that moment. Therefore, we by pass the problem by first calculating the *E'[]*, which assumes w will not pass any back edge. By considering different cases of recurrence, we arrive at the recurrence formula (stated in [3]).

$$\text{For } j \in [0,n],\ E[\theta, j] = 0$$

$$\text{For } i \neq \theta,\quad E[i,0] = \begin{cases} \min E[\overline{\text{Pre}(i)},0]+1 & \text{if } i \text{ is a non-}\varepsilon\text{ node} \\ \min E[\overline{\text{Pre}(i)},0] & \text{if } i \text{ is an } \varepsilon \text{ node} \end{cases}$$

$$\text{For } j \in [1,n] \text{ and } i \neq \theta,$$

$$E'[i,j] = \begin{cases} \min\left(E[i,j-1]+1,\ E[\text{Pre}(i),j-1] + \begin{cases} 0 & \text{if } r_i = s_j \\ 1 & \text{otherwise} \end{cases} \right) \\ E'[\overline{\text{Pre}(i)},j] \end{cases}$$

$$E[i,j] = \min\left(E'[i,j],\ \min\left(E'[\text{Pre}(i),j-1],\ E'[\overline{\text{Pre}(i)},j-1]\right) + \begin{cases} 1 & \text{if } i \text{ is a non-}\varepsilon\text{ node} \\ 0 & \text{if } i \text{ is an } \varepsilon \text{ node} \end{cases} \right)$$

Remember that the difference between *E[]* and *EDIST[]* is whether we consider substrings in the distance definition. However, the major arguments in [3] did not make any difference if we consider the whole string rather that all the substrings ended at the last character. At a closer look at the formula above, the only difference comes from the initial condition; in *E[]* case, $E[\theta, j] = 0$ for all j as arisen from empty substring. For our *EDIST[]* case, $EDIST[\theta, j] = j$ for j between 0 and n.

This constitutes the recurrence formula for our *EDIST[]* function, which runs at the same efficiency as the original algorithm.

3.3 Normalization of Edit Distance

With the distance function *EDIST[]*, we can determine how far an element in an XML document is from its declaration in the DTD. However, the edit distances for different elements in the same document may not be comparable; for example, we have two elements *a, b* having child strings "*c*" and "*efghijklmn*" and their corresponding regular expressions "*y**"; "*z**" respectively from a DTD.

Obviously, *EDIST[a]* = *EDIST[c, y*]* = *EDIST[c, empty string]* = 1 while *EDIST[b]* = *EDIST[efghijklmn, z*]* = *EDIST[efghijklmn, empty string]* = 10. Intuitively, both element *a* and *b* are simply totally different from the corresponding declarations in the DTD but the edit distances show very different numbers. That arises the need to normalize the edit distances to make them comparable.

For a string *s* and a regular expression *r*, we denote *len(s)* as the length of string s and *minlen(r)* as *min{len(s_r)*: s_r matches *r*}. Then the maximum possible *EDIST[s, r]* is *max{len(s), minlen(r)}* because *s* can always be converted into a string s_r matching *r* by inserting (or deleting) the extra length of the longer side and substituting the characters of the common length. We then normalize the edit distances by using this maximum possible distance as the denominator.

Definition 5. Define *Sim[s, r]*, the structural similarity function between a string s and a regular expression *r*, as 1 – *(EDIST[s, r] / max(len(s), minlen(r)))*.

If *s* presents a child element string of an element *e* in an XML document while *r* represent the corresponding declaration in DTD, then we define *Sim[e]* as *Sim[s, r]*. Obviously, *Sim[e]* is a number between 0 and 1 inclusively. Intuitively, value 1 means totally match while 0 means totally mismatch.

The *Sim[]* function provides a mean to quantify the local structural similarity for the elements in a XML document with a DTD. In Sect. 3.1, we have not yet tackled the case (3) of those elements *e* existing in the document while not in the DTD. Intuitively, it is simply a total mismatch with the DTD and we can assign a value 0 to the *Sim[]* function making the function well-defined in all cases. Then it is obvious that *Sim[e]* = 1 for all *e* if *X* is a valid document.

4 Global Similarity Consideration

Our goal is to consider the structural similarity among XML documents and DTDs as a whole instead of only locally. Therefore, we discuss in this section on how to promote the local consideration into global.

Artificial Root. In the previous section, it stated that "*Sim[e]* = 1 for all *e* if *X* is a valid document"; however, the sufficient condition statement is not true. Consider the following XML document against the DTD described in Fig. 1. All the elements in the document match the declarations in the DTD.

```
<schedule>
  <date>16OCT2002</date>
   <amount>150000</amount>
</schedule>
```

However, the document does not conform to the DTD because a valid document should have the root element *<portfolio>* instead of *<schedule>*. But why did not the matching check (by *Sim[]*) identify that? It is because the root element <schedule> has no parent element so that it did not fall into any matching check. In order to force the root element into the checking, we can add an artificial root, say *R*, to both the document and the DTD with the original roots as their child nodes. In other term, we effectively add an element R having the child element string *<schedule>*, which corresponds to the regular expression *<portfolio>*. Then we get *Sim[R]* = 0 because the <schedule> is totally different from *<portfolio>*, which successfully detected the invalidity.

We called such XML documents and the DTDs as rooted, it is obvious that the artificial root will not affect either the values of the *Sim[]* function or the validity of the original documents. In other words, a rooted XML document X is valid with respect to a rooted DTD D if and only if $Sim[e] = 1$ for all elements e in X. From now on, all the XML documents and DTDs assume rooted unless otherwise specified.

Weighting Scheme. Intuitively, different elements should have different level of importance in an XML document. In other words, the bigger the subtree of an element in the document tree, the greater the importance the element should be.

Definition 6. For any element e in an XML document, *Weight[e]* is defined as the size of the subtree rooted at e, excluding e itself. As a result, all the leave nodes (i.e. actual data) have zero weights. Figure 2 below shows the element weights for the XML example in figure 1, which has total weight (for all elements) of 99.

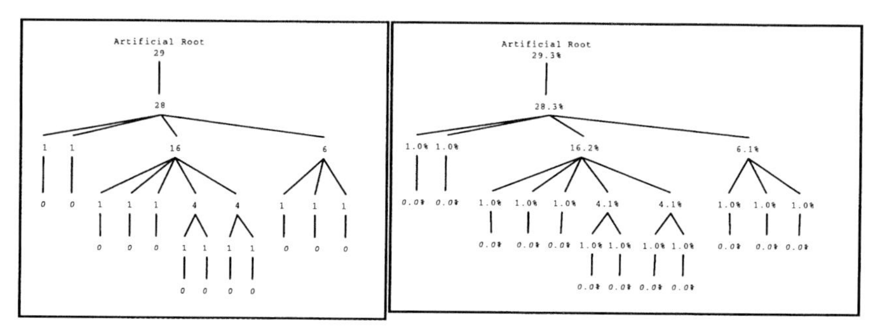

Fig. 2. Element Weights (left) and the weight distribution (right)

We can make use of the element weights to define a weighting scheme, as a consequence, the greater the substructure of an element, the greater the weighting percentage. Then we can gather local pieces of similarity information to give a global view, which takes care of the semantic meaning for XML documents.

5 Structural Similarity

Now, we can define the structural similarity as the weighted sum of local similarity (*Sim[]*) of the elements in an XML document X with respect to a DTD D.

$$Sim[X,D] = \frac{\sum_{e \in X} Weight[e] \times Sim[e]}{\sum_{e \in X} Weight[e]}$$

Hence X is valid with respect to D if and only if $Sim[X, D] = 1$. On the other extreme cases, say X has no common element with D, $Sim[X, D] = 0$. The figures below show an example of an XML document and a DTD, and illustrate the similarity function evaluation.

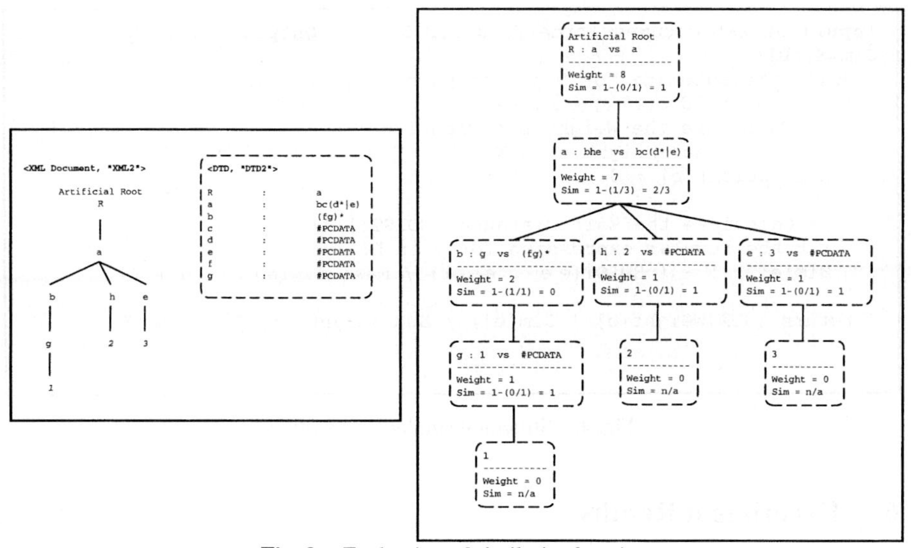

Fig. 3. Evaluation of similarity function

From the above figure, we can calculation the similarity as

$$Sim[XML2, DTD2] = \frac{8\times 1\big|_R + 7\times \frac{2}{3}\Big|_a + 2\times 0\big|_b + 1\times 1\big|_h + 1\times 1\big|_e + 1\times 1\big|_g}{8\big|_R + 7\big|_a + 2\big|_b + 1\big|_h + 1\big|_e + 1\big|_g} = 0.78$$

The definition of the similarity function is natural and simple, hence the evaluation is correspondingly straightforward. The following are the main steps for the calculation: (1) Calculate the weights of every elements in the document (hence the total weight); (2) Calculate edit distances and the maximum possible edit distances of every elements with their declarations in the DTD; (3) Calculate the weighted sum.

The part (1) above can be done by traversing the XML document tree once according to the postorder numbering. It is due to the property that the postorder numbering always goes through all the child nodes before parent nodes for every subtree in a tree. The weights can be obtained by accumulating the weights of child nodes in every subtrees.

For part (2), we can pre-build the automata for every declaration in the DTD indexed by the declaring elements, as the same automata maybe referred by many instances of the elements in the document. Then the edit distances can be calculated by the recurrence formula shown in the previous section. While the maximum possible edit distances can be much easier to calculate by counting both the lengths of the strings and the regular expressions.

By gathering the results in part (1) and (2), part (3) can be computed accordingly. The time complexity in part (2) dominates the other two parts. The time complexity of the overall algorithm is $O(R_D N_X)$, where R_D is longest *minlen*(*r*) among all *r* in *D* and N_X is the size of *X*, which is fast in practice. The algorithm is shown below.

```
Input: an XML document tree X, a DTD D          Output: Sim[X,D]
Sim[X, D]{
  Build the automata for D by Thompson's Construction
  for e (= child string e1e2...en) in X in postorder numbering{
    /* Calculate the Weight Distribution */
    if n > 0 then Weight[e] = Weight[e1] + ... + Weight[en] + n;
    else Weight[e] = 0;

    /* Calculate the Edit Distance, EDIST[] */
    r = the regular expression for e in D;
    Sim[e] = 1 - ( EDIST(e1e2...en, r) / Max( minlen(r), n ) )
  }
  return {SUM(Weight[e] * Sim[e]) / SUM(Weight[e]) for e in X}
}
```

Fig. 4. Structural similarity algorithm

6 Experiment Results

The proposed similarity measure has been implemented in Java using the DOM libraries [8, 9]. In order to evaluate the accurateness and effectiveness of the measure and the algorithm, both synthetic and real data are evaluated.

Synthetic Data: Synthetic data is generated by an XML document generator [13], which generates valid documents for a chosen DTD. An error rate can be configured so that generated documents will not fully conform but closely conform to the DTD, depending on the error rate specified. 3 sets of 500 documents were generated this way with different error rates: (A) No error, (B) 10%, (C) 50%. Similarity measures of the 3 datasets are run against the chosen DTD and shown below.

Similarity Range	Data Set A	Data Set B	Data Set C
Sim = 1.0	500	0	0
0.8 <= Sim < 1.0	0	166	0
0.6 <= Sim < 0.8	0	237	14
0.4 <= Sim < 0.6	0	25	114
0.2 <= Sim < 0.4	0	58	26
0.0 <= Sim < 0.2	0	14	346
TOTAL	500	500	500

It verified that the similarity measure always generates the value 1 for valid documents, while < 1 for invalid documents. Moreover, the majority of data sets B and C are considered as "good but not exact matched" (403/500 between 0.6 and 1.0) and "badly matched" (486/500 between 0.0 and 0.6) against the DTD respectively, hence it successfully measures the similarity against a DTD.

Real Data: Real data are downloaded from the Internet to evaluate the effectiveness of the measure. 727 XML documents are retrieved from a web site [6], generated by a common DTD. 2 documents are selected randomly from the population to generate a DTD based on the XTRACT system mechanism (Generalization, Factoring and MDL

Principle) [10], which appeared to be about 10% different from the original DTD. Then the similarity values of all the documents against both the original and generated DTD are calculated and listed below.

Similarity Range	Original DTD	Generated DTD
Sim = 1.0	727	339
0.8 <= Sim < 1.0	0	58
0.6 <= Sim < 0.8	0	335
0.0 <= Sim < 0.6	0	0
TOTAL	727	727

Even the generated DTD was based only on a small portion (2 out of 727) of the documents, it still showed "good" similarity (>= 0.6) against the whole population.

7 Conclusion

In the paper we propose a metric for the structural similarity between XML documents and DTD. The metric takes care of the semantic meaning of XML documents, which is easy to understand. We also present an algorithm, which is fast to run in practice. With carefully chosen DTD among an arbitrary set of XML documents, this metric can be used for XML clustering, classification and indexing, which constitutes the future direction of the project.

References

1. Zhang, K., Shasha, D.: Simple Fast Algorithm for the Editing Distance Between Trees and Related Problems, SIAM J. COMPUT 18(6) (1989) 1245–1262
2. Myers, E. W., Miller, W.: Approximate matching of regular expressions, Bull. Math. Biol., 51 (1989) 5–37
3. Wu, S., Manber, U., Myers, E.: A Subquadratic Algorithm for Approximate Regular Expression Matching, Journal of Algorithms, 19 (1995) 346–360
4. Myers, E. W.: A four-Russian algorithm for regular expression pattern matching, J. Assoc. Comput. Mach. 39(2) (1992) 430–448
5. Bertino, E., Guerrini, G., Mesiti, M., Rivara, I., Tavella, C.: Measuring the Structural Similarity among XML Documents and DTD, Technical Report DISI-TR-02-02, Dipartimento di Informatica e Scienze dell'Informazione, Universita` di Genova (Dec 2001)
6. ACM SIGMOD Record: XML Version, `http://www.acm.org/sigmod/record/xml/`
7. Navarro, G.: A guided tour to approximate string matching, ACM Computing Surveys 33(1), (2001) 31–88
8. W3C, Document Object Model (DOM) (1998)
9. The Apache XML Project, `http://xml.apache.org/`
10. Garofalakis, M., Gionis, A., Rastogi, R., Seshadri, S., Shim, K.: XTRACT: A System for Extracting Document Type Descriptors from XML Documents, Proceedings of ACM SIGMOD (2000)
11. Schlieder, T., Naumann, F.: Approximate Tree Embedding for Querying XML Data, Proceedings of the ACM SIGIR Workshop on XML and Information Retrieval (2000)
12. Nierman, A., Jagadish, H. V.: Evaluating Structural Similarity in XML Documents, Fifth International Workshop on the Web and Databases (WebDB 2002)
13. XML at Sun, `http://wwws.sun.com/software/xml/`

Web Personalisation with the Cover Coefficient Algorithm

Matthew Anderson, Irfan Altas, and Geoff Fellows

School of Information Studies
Charles Sturt University
Wagga Wagga, NSW, 2678
Australia
ialtas@csu.edu.au

Abstract. In this paper we discuss how to personalise web pages dynamically, based upon customer profiles generated from a click stream dataset using the cover coefficient algorithm. The personalisation model can be applied in an environment where there is a need to know the habits of customers that is beneficial to both the organisation and the web server administrator.

1 Introduction

Casselman [1] states that getting up close and personal with customers is also the goal of progressive companies operating in the online world. An online company attempts to imitate the closeness of a storeowner to a customer in the bricks and mortar world. An example of this is when you enter a store and you are recognised, the storeowner is aware that you have been there before. The storeowner knows your preferences and is then able to recommend products and a service based upon previous purchases and enquires.

Larger companies such as Amazon [2] and Yahoo [3] try to model this traditional retailer environment in an attempt to 'personalise the online experience'. The expectation is to have competitive advantages flowing from increased customer loyalty and retention rates by tailoring web site contents to suit their customers' desires.

They record click stream data from the hyper-text transfer protocol (HTTP) requests made to the web server. The web server log files store Internet Protocol (IP) address, cookies and the files requested [4,5]. The purpose of this exercise is to recognise patterns and models of navigational habits of users that are found in companies' click stream dataset. This information can be used to personalise web advertisements and web content and to promote new services that may be of interest to their customers. However, only few online companies are really effectively implementing personalisation today, mainly because of complexity and cost issues.

There are some common click stream models such as *click fact models* and *session fact models* [6]. These models are used to define profiles of web users' naviga-

P.M.A. Sloot et al. (Eds.): ICCS 2003, LNCS 2659, pp. 422–431, 2003.

tional habits, and these profiles are used to personalise the web documents. Although modelling the click stream data using the above models is possible, there is a need to try and gain an understanding of the customer's navigational habits in an automated way. This can be achieved by using data mining techniques on the click stream data-set.

Data mining systems have many techniques that help answer vague questions that would not normally be able to be answered by someone looking at the raw dataset. These techniques are broken up into studies that allow an implementation of a model to search the data in a specific way and output results. One technique includes a classification study that is a form of *supervised learning* and a clustering study that is *unsupervised learning* [7].

A classification technique classifies an item within a dataset into a predefined class [8]. The number of defined classes can be infinite and do not have to be in sequence. On the other hand a clustering technique processes rows of information that have similar trends and patterns in common and groups them together. These groups do not always contain unique values; it is often found that values can overlap across many clusters [9]. In this work, our aim is to implement a clustering model based on the cover coefficient technique (CC) [10]. The clusters obtained from this process can be used to recommend pages to a new user of a web site.

Since the early days of e-commerce, one of important goals is the development of recommendation based systems to automate the process of recommending products or services to web users. A successful recommendation system in an interactive environment is collaborative filtering that works in a comparative manner. It compares the similarity between users and their habits to make recommendations and predictions. Being able to create accurate predictions to be made from recommendation systems enables web personalisation to be more effective. One of such recommendation model is the top-N recommendation system [11]. In the top-N algorithm the main purpose is to identify a set of N items that will be of interest to a certain user. A top-N algorithm puts information into a binary matrix. From this matrix it can predicted whether a user will be interested in a particular item [12]. The benefits of using the top-N recommendation algorithm is that it shows that both cosine and probability-based schemes have a higher than average accurate recommendations compared with traditional collaborative filtering techniques [11]. Measuring similarities amongst documents are also extensively studied in the information retrieval context and referred to as resemblance coefficients [13]. Resemblance coefficients can be classified into four classes: distance, probabilistic, correlation and association coefficients.

The (probabilistic) CC technique can be used to measure similarities between documents in the information retrieval context. In this work, we implement the CC technique to measure similarities amongst web users by employing the click stream dataset collected from a web site. As a result of these similarity measurements it can be predicted whether a user will be interested in a particular web document and, hence, it can be used as a top-N recommendation model to personalise web documents. The CC technique can store input data from a click stream dataset in either a binary or numerical form. We discuss collecting click stream data in Section 2. The CC technique works from one particular item to calculate a similarity with all other items in the entire item collection where the entire collection includes the item itself.

Our aim in this work is to find out whether the CC concept can satisfactorily be implemented in web personalisation area. We discuss the CC technique in Section 3. Conclusions are presented in section 4.

2 Collecting Click Stream Data

We implemented the CC algorithm on a subset of web documents found on the Association of Societies for Growing Australian Plants (ASGAP) web site [14]. The web site contains information about the cultivation, propagation, conservation and appreciation of Australia's native flora. The structure of a subset of ASGAP web site is illustrated in Fig. 1. Each box in the Fig. 1 represents a web document while each arrow represents a link to the connected web document. Any page can be the entry point to the web site rather than just the homepage.

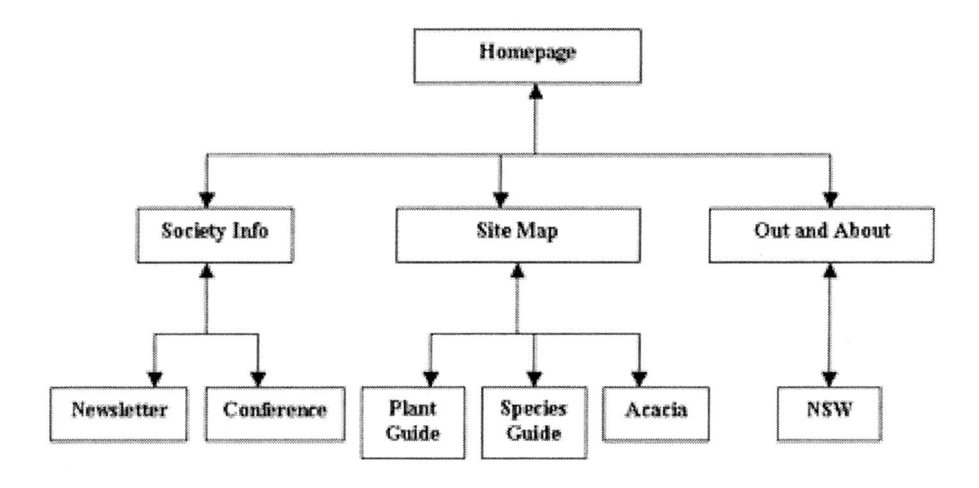

Fig. 1. A subset of the ASGAP web site [14]

The movement of a particular web user has been recorded in an access log file on the web server. An illustration from this log file for two users presented in the Fig. 2.

```
192.168.1.164.3000 GET /outAndAbout.html HTTP/1.0 [07Feb/2002:13:55:01 +100]
192.168.1.164.4000 GET /siteMap.html HTTP/1.0 [07/Feb/2002:13:40:08 +1100]
```

Fig. 2. Two user records from the click stream dataset

The meaning of the fields for the first record of Fig. 2 is as follows. The first field gives the host IP address rather than hostname to cut down the workload of the Domain Name Server (DNS) system. The second one contains the identification number of the web user that is stored in the cookie. The third field contains the HTTP request type made which includes the relative uniform resource locater (URL) where the file is located on the web site and the HTTP version. The last field is the date stamp for the file request event.

An important issue that can affect the click stream data is that some ISPs use proxy servers. Requests made to any resource on the World Wide Web are made via the proxy server. As a result all IP addresses logged in the web server access log file will be that of the proxy server rather than the individual web user, making multiple web users be viewed as one. This will decrease the value of the collected data. For example in Fig. 2 IP numbers are the same that may be assumed as a proxy server IP number.

To overcome the proxy problem above, cookies can be used. They store unique identifying number on the web user's hard disk drive, so upon return to a web site the web user's request will include this identification number. Figure 2 demonstrates this further. Even though both requests were made from the same IP address, 192.168.1.164, the cookie values, 3000 and 4000, identify individual web user.

3 The Cover Coefficient Technique and Web Personalisation

The CC identifies relationships between documents and users of a web site by use of a matrix. The CC technique works from one particular web user to calculate a probabilistic similarity measure with an entire web user collection including that particular user. The measure is the probability of randomly selecting a web document from one particular user, and from all the users containing that web document, the probability of randomly selecting a second particular user [10,3].

The CC technique creates a mxn matrix, say D, by using the click stream dataset. Its columns represent the web documents, $(d_1, d_2, \ldots d_n)$, and its rows represent users, $(u_1, u_2, \ldots u_m)$, who requested web documents. The entries of the matrix, D, simply indicate whether a user visited a particular document (1) or not (0) (see Fig. 3). It is possible to assign different meanings to the entries of the matrix such as the time spent on a particular web document by a particular user or how many times a user visited a particular web document. However, we work only binary values in this paper.

At the second stage of the CC algorithm, the D matrix is mapped into an m x m matrix, say C. The C matrix indicates the relationship between web users. Web user u_i contains n web documents $(d_1, d_2 \ldots d_n)$ and the probability of randomly selecting any one of these web documents, from all the web documents in the set, is s_i. Web document d_n is contained in m web users $(u_1, u_2, \ldots u_m)$ and the probability of randomly selecting any one of these web users, from all the web users in the set, is s'_n. The probability of selecting the same web documents from web users u_m and u_i, which is to say the extend to which web user u_i is covered by web user u_m, is therefore s_n multiplied by s'_n [13]. The entries of the C matrix in Fig. 4 can be calculated by using this definition as is illustrated for c_{12}

$$c_{12} = \sum_{k-1}^{10} s_k s_k^{'} = (\frac{1}{5}\frac{1}{6}) + (\frac{1}{5}\frac{1}{4}) + (0\frac{1}{2}) + (0\,0) + \ldots + (\frac{1}{5}0) + (0\,0) \tag{1}$$

User ID	home	Society Info	News	Conf.	Site Map	Plant Guide	Species Guide	Acacia	Out & About	NSW	Total
1000	1	1	0	0	1	1	0	0	1	0	5
2000	1	1	1	0	1	0	1	0	0	0	5
3000	1	0	0	0	0	0	0	0	1	1	3
4000	0	0	0	0	1	1	0	1	0	0	3
5000	0	1	0	0	1	0	0	0	1	0	3
6000	1	0	0	0	0	0	0	0	1	0	2
7000	1	0	0	0	0	0	0	0	0	0	1
8000	1	1	0	1	1	1	0	0	0	0	5
Total	6	4	1	1	5	3	1	1	4	1	

Fig. 3. D matrix associating web users to web documents

Each entry in the C matrix indicates how well web user u_i covers web user u_j, including when web user u_i and web user u_j are the same web user. The diagonal terms of the matrix C are known as the decoupling coefficients. The diagonal term, c_{ii}, is the dissimilarity of u_i to all other users found in the collection. The coupling coefficient, calculated as (1- c_{ii}), is the indication to how similar the user u_i is to all other web users in the collection.

The decoupling coefficient is used to estimate the number of clusters to be created. The number of clusters can be calculated as

$$n_c = \sum_{i=1}^{n} c_{ii} \tag{2}$$

where m denotes the number of users in the collection. Thus, for this example the number of clusters is

$$n_c = 0.240 + 0.523 + 0.472 + 0.511 + 0.233 + 0.208 + 0.167 + 0.390 = 2.744 \approx 3$$

U. ID	1000	2000	3000	4000	5000	6000	7000	8000	Total
1000	0.24	0.123	0.083	0.107	0.140	0.083	0.083	0.19	1.00
2000	0.123	0.523	0.033	0.04	0.09	0.033	0.033	0.123	1.00
3000	0.139	0.055	0.472	0.0	0.083	0.139	0.56	0.056	1.00
4000	0.178	0.067	0.00	0.511	0.067	0.0	0.0	0.178	1.0
5000	0.233	0.150	0.083	0.066	0.233	0.083	0.0	0.15	1.0
6000	0.208	0.083	0.208	0.0	0.125	0.208	0.083	0.083	1.0
7000	0.167	0.167	0.167	0.0	0.0	0.167	0.167	0.167	1.0
8000	0.19	0.123	0.033	0.107	0.090	0.033	0.033	0.390	1.0

Fig. 4. C Matrix representing similarities of web users

Once the number of clusters is calculated the next step is to select a single user that will represent a single cluster. Such a representative user is called a seed. Calculating the seed powers using the following formula identifies the seeds of the clusters

$$p_i = c_{ii}(1 - c_{ii})\sum_{j=1}^{n} d_{ij} \tag{3}$$

Where d_{ij} is the (i,j) entry of the matrix D. Thus, for the example the seed powers are

1000: 0.240 x 0.760 x 5 = 0.912

Similarly for 2000: 1.247; for 3000: 0.748; for 4000: 0.750; for 5000: 0.536; for 6000: 0.329; for 7000: 0.139; and for 8000: 1.190.

Then, three seed clusters would be the web user 2000 (1.247), the web user 8000 (1.190) and the web user 1000 (0.912) that are the first three web users with the largest seed powers. We refer them as the seed users. Using these seed users we can then distribute users to particular clusters. For example, the web user 3000 will be allocated to the seed user 1000 for the following reason. We identify the probabilistic cover coefficient numbers from the fourth row of the C matrix in Fig. 4 for the web user 3000 corresponding to each seed user. In this case, they are 0.139, 0.055 and 0.056 for the seed users 1000, 2000 and 8000, respectively. Amongst those numbers the largest probabilistic cover coefficient is 0.139 that corresponds to the seed user 1000. Therefore, the web user 3000 will be allocated to the cluster identified by the seed user 1000.

In some cases, the probabilistic cover coefficient values corresponding to the seed users may be equal. In that case, we allocate that user to the cluster with the least number of web users. For example, the user 7000 has the probabilistic cover coefficient value 0.167 corresponding to every seed user. Therefore, it is allocated to the cluster identified with the seed user 2000.

The three clusters for the example are given in Fig. 5. In implementations there is another cluster referred to as red bag cluster. Web users who request web documents that have not been requested by any other web users are allocated to the rag bag cluster due to unknown comparisons to any seed.

Cluster 1	Cluster 2	Cluster 3
1000	8000	2000
3000	4000	7000
5000		
6000		

Fig. 5. Cluster generation for the example

Once the clusters are generated they can be used for recommendations to the new web users entering to the web site. After a new user requests a few web documents they can be allocated to one of the clusters that have been previously created. Then, the system can dynamically present some recommended web documents (top-N recommendation) to the user according to the cluster, which the user was allocated to during their initial navigation of the web site.

The standard technique implemented on all recommendations is that the top-N most requested web documents by the users of that particular cluster be recommended to a new user allocated to that cluster. For example, if a new user is allocated to the cluster 1, then the new user may be recommended with the four most commonly requested web documents that were previously identified for this cluster. From Fig. 6 those will be "out & about", "home", "Society Info" and "Site Map".

User ID	home	Society Info	News	Conf.	Site Map	Plant Guide	Species Guide	Acacia	Out & About	NSW
1000	1	1	0	0	1	1	0	0	1	0
3000	1	0	0	0	0	0	0	0	1	1
5000	0	1	0	0	1	0	0	0	1	0
6000	1	0	0	0	0	0	0	0	1	0
Total	3	2	0	0	2	1	0	0	4	1

Fig. 6. Frequencies of the web document requests for the cluster 1

The system initially creates a set of clusters from existing click stream data. It uses these clusters to assign new web users to a particular cluster until the number of new users reaches for a predefined number. The system re-clusters all web users as a background job when the number of new users reaches to this predefined level. As soon as a new clustering of web users is available the system works with these new clusters. This is mainly to the time consuming nature of the clustering process. In our implementations we found little difference between real-time and delayed clustering approaches.

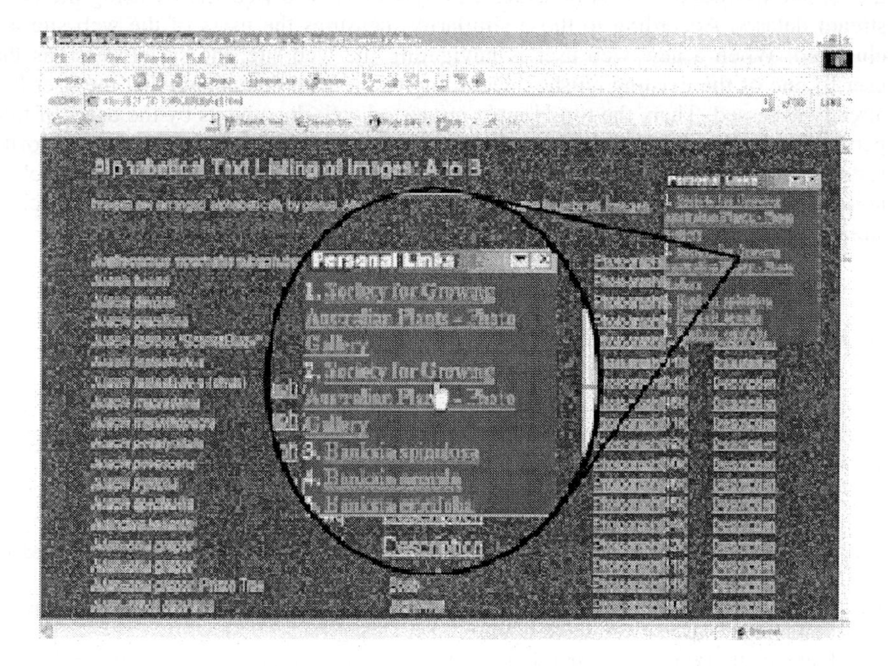

Fig. 7. Personalised web documents

This prototype model is implemented on the click stream data obtained from ASGAP web site. It is implemented on a Java Virtual Machine running on a Microsoft Windows 2000 platform. The web server used to handle the HTTP requests is Jakarta Tomcat version 4.1.12. The system runs as an application using Java to parse the web server log file. Web users and web documents are represented in memory as Java objects. The implementation details of the system can be found in [15].

Figure 7 is a screen snapshot from the system. It represents how the web document would be altered and presented to the web user. The windows application opens the document and parses it looking for the <TITLE> tag as this should be a very short description of the information within the web page. The title tags from the recommended web documents are then presented back to the user as links in a separate window as the top-5 recommendations.

4 Conclusions

In this work we developed a top-N recommendation system to dynamically introduce web documents to a web user of a site. The introduced web documents are predicted to be in the interest area of the user. The model uses the cover coefficient technique as the engine to measure the similarities amongst users of a web site by using the click stream dataset. According to those similarity measures the users of the web site are clustered. When a new web user is navigating the web site, the system assigns the user to one of the existing clusters according to the web documents that the user has already accessed. Then, the top-N most requested web documents by the users of that particular cluster will be recommended to this new user. The system is satisfactorily tested over a controlled dataset for which the number of clusters and member of clusters were manually created. Then, as a prototype it was implemented over a click stream dataset obtained from the ASGAP web site.

References

1. Casselman, G. 2001, *Web Personalization* (online). http://www.casselman.net/artlist/webpersonalization.htm [Accessed 22 Oct. 2002].
2. Amazon.com. 2002, Amazon.com Privacy Notice (online). http://www.amazon.com/exec/obidos/tg/browse/-/468496/104-2622634-4059946 [Accessed 16 May 20002].
3. Yahoo Inc. 2002, Yahoo! Privacy Policy (online). http://privacy.yahoo.com/ [Accessed 15 May 2002].
4. Keen, P. 1987, Information systems education: recommendations and implementation. Cambridge University Press, New York, USA.
5. Fielding, R., Gettys, J., Modul, J., Frystyk, H., Masinter, L., Leach, P. and Berner_Lee, T. 1999, Hypertext Transfer Protocol – HTTP/1.1 (online). http://www.w3.org/Protocols/rfc2616/rfc2616.txt [Accessed 24 June 2002].
6. Anderson, D. 2000, Personalizing: Port 80: Docks: Communities (online). http://riccistreet.net/port80/docks/personalizing.htm [Accessed 19 Sept. 2002].
7. Groth, R. 1998, Data Mining A Hands-on Approach for Business Professionals. Prentice Hall, New Jersey, USA.
8. Kurzeme, I. 1996, VICNET's users: a longitudinal market survey of the users of Victoria's network, VICNET. RMIT.
9. Hamilton, J., Gurak, E., Findlater, L. & Olive W. 2000, *The Virtuous Cycle of Data Mining* (online). http://www.cs.uregina.ca/~dbd/cs831/notes/virtuous_cycle/virtuous_cycle.html [Accessed 5 Apr. 2002].
10. Can, F. and Ozkarahan, E. 1990, Computation of Term/Document Discrimination Values by Use of the Cover Coefficient Concept, Journal of American Society for Information Science, 38(3), pp. 171–183.
11. Karypis, G. 2000, Analysis of Recommendation Algorithms for E-Commerce, in Proceedings of the 2nd ACM conference, ACM Press, pp. 158–167.
12. Yin, K. 1994, Case Study Research: Design and methods. 2nd edn. Sage, USA.
13. Lindley, D. 1997, Interactive Classification of Dynamic Document Collections. Phd Thesis, University of New South Wales, Australia.

14. ASGAP. 2002, The Society for Growing Australian Plants (online). http://farrer.riv.csu.edu.au/ASGAP [Accessed 30 June 2002].
15. Anderson, M. 2002, Effectiveness of Web Personalisation Using the Cover Coefficient Algorithm, Honours Thesis, School of Information Studies Charles Sturt University, Australia

Infrastructure of an Agile Supply Chain System: A Virtual Agent Modeling Approach

H.C.W. Lau and Christina W.Y. Wong

Dept. of Industrial & Systems Engineering
The Hong Kong Polytechnic University, Hong Kong

Abstract. This paper attempts to suggest an infrastructural framework for the design and development of an agile supply chain system which is characterized by its ability to cope with unprecedented changes related to the management of suppliers and the flow of parts within the value chain of the entire production network. In particular, a virtual agent modeling approach, which provides the adaptive and predictive capability to meet the requirements of this agile system, is proposed. The infrastructure of this agile supply chain system highlights the deployment of virtual agents, which are basically intelligent functional objects emulating the behavior of human beings with relatively high adaptability. The significance of contribution of this paper is concerned with the suggestion of an infrastructural framework, which is able to enhance the agility as well as adaptability of value-added activities in a supply chain, thereby providing more alternatives and ideas for those researchers who are interested in this field of study.

1 Introduction

In general, supply chain is a set of activities that cover enterprise functions from ordering and receiving of raw materials through the manufacturing of products in relation to the distribution and delivery to the customers. By flattening out the organizations and making use of "ad-hoc" workgroups or project teams, customer-focused business activities can be more responsive to ever-changing market conditions. To meet the relatively unpredictable customer demands, the agility of the supply chain is an important issue that is justifiable to be addressed. In general, agility is the ability of an enterprise to adapt to unforeseeable changes and this is different from flexibility which is more concerned with the ability of companies to respond to a variety of customer requirements that exist within the defined constraints [2]. Apart from the issue on customer requirements, the study on information flow within the supply chain entities is equally important in order to implement an effective network that is based on the formation of a value chain consisting of functional entities, providing timely and accurate information to achieve the efficient management of suppliers as well as the flow of parts.

From another point of view, there are activities taken place within a supply chain and this movement of activities is generally referred to as the *workflow* which encompasses various types of tasks that can be performed in one time interval. Evidences [18,23] have shown that the adoption of various workflow approaches can help regu

P.M.A. Sloot et al. (Eds.): ICCS 2003, LNCS 2659, pp. 432–441, 2003.

late and monitor the flow of information around the company. In general, in order to stay responsive towards market condition, both flexibility and agility are the important factors that need to be given appropriate consideration in the formulation of a company workflow [6,13]. As agility is more related to the "coping with the unpredictables", it is of utmost importance in a marketplace with growing uncertainty. The development of a supply chain system with such agility is the subject of study of this research.

The requirement of the capability of agility lends itself to the adoption of virtual agent modeling approach which focuses on the deployment of virtual agents composed of software programs designed to accomplish specific tasks like "real" human agents possessing specialized skills. With virtual agents that may equip themselves with certain level of intelligence, agility of the supply chain workflow can be enhanced. However, the technique to include the element of intelligence to the whole workflow needs to be carefully studied. In general, the virtual agent model is characterized by its communication between a set of autonomous, collaborative and independent agents which are designed to be business-oriented and try to accomplish business goals in a complex and ever-changing market [29,22]. In general, the technologies incorporated into the virtual agent model are mostly concerned with object technology, rule-based reasoning and even generic algorithm, all of which are not developed for speculating upcoming events. In this respect, it seems necessary that a certain "ingredient" of predictability such as projection of possible outcomes needs to be incorporated in order to enhance the "intelligence" of the agents.

The underlying technology of this agile supply chain system is the synergetic combination of neural networks and object technology, both of which are to be adopted to leverage the level of agility of the system. Neural networks, also called *learning automata*, are capable to learn relationships among complex sets of data through a learning process whereas object technology is featured by its encapsulation of both data and methods (functions) within a data object. The blend of these two technologies provide the perfect ingredient for the formation of *virtual agents* which are basically "intelligent functional objects" emulating the behavior of human beings albeit in a relatively adaptive way. These virtual agents will act as the "knowledge worker" of this supply chain, facilitating the efficient workflow of the entire supply chain network. With the presence of these agents, various issues such as process control, customer classification, business partner selection and demand prediction can be carried out in a way based on some adaptive reasoning mechanism. An infrastructural framework of an agile supply chain, capitalizing on the latest development in various emerging technologies, is proposed. This research is expected to have significant impact on the future development of supply chain workflow by virtue of its capabilities to progressively introduce machine self-learning and intelligence to the whole operating supply chain network over time.

2 Paradigms of Supply Chain Workflow

Recent years have seen significant changes made in terms of enterprise strategy and manufacturing paradigms particularly for those companies which are keen to remain world competitive in this ever-changing and volatile market [3, 14, 31]. As the manufacturing market is becoming more borderless by national borders, a number of global supply chain networks have been established, taking advantage of the fast-growing networking and information technologies [27]. Study indicates that while a number of infrastructural frameworks related to dispersed supply chain systems have been described in contemporary publications [14, 21, 30, 31], the detailed structure and formulation of a "smart" monitoring infrastructure for such a system has not received too much attention.

Review on contemporary publications indicates that whilst the research findings so far have contributed to the progressive introduction of intelligence level to the models in terms of tasks decomposition and optimization of process planning operations, the *machine self-learning* aspect and in particular, the ability to predict possible outcomes, has not achieved the anticipated research advances [4, 7, 11, 12, 25, 28, 30]. Machine self-learning enables the progressive addition of intelligence and corporate information to a system through a systematic knowledge creation process. This is automatic and self-tuned, thereby continually adding value to the involved system in terms of creation of a smart information management system for the monitoring of a dispersed manufacturing network. The reason of deficiency of this machine learning feature in the previous model is clear. Rule-based reasoning works by narrowing attention to a specific problem domain, then codifying human-developed rules of thumb and patterns of reasoning and the recommendations provided are normally similar to what a human expert would conclude given the same evidence whereas object technology is featured by its inheritance of data to related objects and its "properties plus behaviours" encapsulation. However, research findings based on the past years of study [16, 20, 30] show that the self-learning feature although exists yet is not remarkable due to the distinct inherent features of these technologies which are not specifically designed for the building of an autonomous knowledge creation mechanism with the ability to predict events - should they occur at a later stage.

This paper presents an agile supply chain workflow which incorporates a combination of technologies and can be deployed in a network of companies for converting complex business data into smart information, detecting opportunities and suggesting business strategies. The main feature of such a system is concerned with its self-learning capability which is favourable to the progressive introduction of intelligence and useful knowledge to the system database over time.

3 Tools and Technologies

There are various technologies and relevant tools which can be used for the development of a high performance supply chain system as described in the following context.

3.1 Distributed Object Technology

The efficiency of the supply chain is strongly related to the efficiency and accuracy of information exchange among the business partners and the frequency of information updating within the whole network [1]. With the emergence of the need to do business globally, it is crucial that information from various partners is able to be interchanged in a seamless, timely and accurate manner. However, the reality is that individual business partners are likely to be using dissimilar computing platforms with different formats and the efficient exchange of information is undoubtedly a concern that needs to be addressed.

To build a supply chain system as interoperable as possible, distributed object technology can provide the support of building a cross-platform information exchange network. The approach of deploying the Object Request Broker (ORB), which is built upon the distributed object technology, can be considered to increase the degree of integration and provide more useful features of the proposed supply chain system. The ORB can be regarded as a common interface over which requests are passed between objects (data), thereby ensuring interoperability between applications on different machines in a heterogeneous distributed environment. Common Object Request Broker Architecture (CORBA), which is built around an ORB, can be regarded as a standard architecture for distributing objects in networks, aiming to simplify the communication in a heterogeneous environment.

In general, the CORBA standard provides flexibility on the company's investment (to make the right choice for its particular environment and needs), improve the manageability (more effective information handling on business activities and relationships) and enhance the speed on application development (better communication between functional departments).

The need of a common standard in facilitating the realization of a cross-platform "information object" has prompted Microsoft to develop its DCOM (Distributed Component Object Model) architecture, based on which services to build distributed applications can be developed within the Windows environment. The detailed architecture of DCOM is covered in a number of articles and publications [5] and not to be covered here. In short, DCOM is an architecture that enables components (processes) to communicate across a network which may contain heterogeneous platforms, thereby providing services such as distributed messaging, object request brokering, distributed transaction services, and data connectivity services - over its own Remote Processing Control (RPC) mechanism [26].

In general, the benefits of using DCOM include (i) ability to enable a robust transaction processing based on Windows NT platform, (ii) virtually free to be bundled with Windows NT. Generally speaking, DCOM services are very closely linked with the Windows NT platform. In order to promote DCOM as a cross-platform product, Microsoft has broadened the support of DCOM to Unix with the collaboration with third-party companies. However, there is certain degree of limitation due to the inherent differences of various operating systems.

The purpose of using distributed object technology (CORBA and DCOM) is to ensure interoperability between applications on different machines in a heterogeneous distributed environment. This technology can simplify the communication between the heterogeneous objects and each business objects of the supply chain can remain unique with shared data and logic elements. This approach is able to improve the manageability of the company (more effective information handling on business activities and relationships) and enhance the speed on application development via better communication between functional departments. In brief, distributed object technology is characterized by its provision of a transparent information communication platform. This feature allows a wide range of organizations within the supply chain to have transparent access to information and data. In effect, the boundaries between applications disappear and each object in an enterprise-wide environment can locate any other object without having to know where the object is located. Once this distributed platform is established, a company can change any one application without having to worry that any other application will be affected. This feature is useful in a supply chain network where companies from various regions need to share information in a timely manner without worrying about the platforms they are with.

3.2 Neural Networks

The neural network serves to provide the recommended change of parameters based on what the network has been trained on. As such, it is important that enough data sample for the input and output layers are provided for training purpose. Neural network is a technology that has typically been used for prediction, clustering, classification and alerting of abnormal pattern. Parroting the operation of the human brain, neural network technology comprises many simple processing units connected by adaptive weights. They create predictive networks by considering a "training set" of actual records. In theory, the formation of neural network is similar to the formation of neural pathways in the brain as a task is practiced. Also a neural network refines its network with each new input it considers. To predict a future scenario, the neural network technology is able to work with a training set of data from the past history of records. It will use the training set to build a network, based on which the neural network is able to predict future scenario by supplying a set of attributes. As neural networks are meant to learn relationships between data sets by simply having sample data represented to their input and output layers, the training of the network with the layers mapped to relevant parameter values with the purpose to develop the correlation between these two groups of data will not, in principle, contradict the basic principle of neural network.

3.3 Rule-Based Reasoning

In general, rule-based reasoning works by narrowing attention to a specific problem domain, then codifying human-developed rules of thumb and patterns of reasoning, providing recommendations that are normally similar to what a human expert would conclude. Regarding the design of Rule-Based Reasoning (RBR) mechanisms, it can be seen that a number of contemporary publications in this area are available [10, 15, 19, 24]. Krishnamoorthy and Rajeev [15] describe in detail the operations of two main methods of inference or RBR, namely forward and backward chaining. Backward chaining is a goal-driven process, whereas forward chaining is data driven [9]. In order to illustrate the operation of the inference mechanism, Krishnamoorthy and Rajeev [15] describes a practical example for the selection of bearings with 16 rules and a complex "knowledge net" included The structure and the coding of the rules are explained in detail in the example. In another article, a Knowledge-Based Front End (KBFE) expert system has been developed to circumvent the restriction of knowledge-representation formalisms, including frames, classes, objects and list [10]. This KBFE system is characterized by the flexibility of its inference technique. Forward chaining is initially performed on base rules, and backward chaining is implemented, where appropriate, in order to prove individual antecedents of the base rules.

4 Neural Object Module

The proposed infrastructure of the agile supply chain system comprises a variety of virtual agents responsible for different tasks. There are production scheduling VA, purchasing VA, resource planning VA, transportation management VA and logistics planning VA. Agents of specific skills can be deployed to the system as deemed necessary and can be removed when their existence is no longer required.

Apart from the traditional job activities such as scheduling, these agents are also capable for discovering patterns and clusters in business data and trigger processes when data or events occur that are outside normal patterns. In brief, VAs are for self-learning systems [8] and they are particularly capable to handle real-life ambiguous situations in which the best of us have learned to extrapolate probable outcomes based on the pattern of events we see and then decide on the action to take. In general VAs are equipped with two main technologies including (i) neural networks which are featured by their adaptive statistical reasoning and the distinct ability to *learn* with proven applications in inventory optimization, factory control, customer classification, process planning optimization as well as material demand prediction, and (ii) object technology which has three main features including encapsulation, inheritance and polymorphism [17]. The synergetic combination of these two technologies is named Neural Object Technology (NOT) in this research.

With the introduction of the NOT, a Neural Object Module (NOM) can be formulated. NOM, which comprises a number of neural agents responsible for undertaking various tasks, includes a rule-based reasoning mechanism for tasks decomposition and allocation to various neural agents. The proposed NOM links with the existing business objects of the supply chain network. When confronted with a job request, the rule-based reasoning mechanism undergoes an inference process and deduces the associated solutions, resulting from the process of decomposition and allocation of tasks based on available information. The coordination and monitoring of the progress of the tasks are undertaken by appropriate VAs. As VAs possess the synergetic features of both neural networks and object technology, they are specialized in predicting probable outcomes by learning from the past experience and data (neural networks features) as well as encompassing their own attributes and methods (object technology features), all of which are essential for the operations that emulate the "smart" behaviour of human operators. In general, this NOM will provide critical business suggestions for running the whole supply chain network. In short, this NOM can be deployed on an existing operational network with minimal disruption in order to improve the performance and functionality of the running dispersed network and in particular, it greatly enhances the central-management role of the manufacturing network.

5 Infrastructure of an Agile Supply Chain System

The NOM is the "brain" of the whole agile supply chain system. However, apart from this monitoring and control module, there are three others to form the whole framework, including Outside Interface (OI), Business Objects (BO) and System Repository (SR). The OI module is meant to handle two main activities including (i) job request for capturing data of request from external sources and (ii) conversion of external data to match the format of rule-based reasoning mechanism which normally needs some input data as facts to trigger the "firing" of rules. The BO module comprises a number of business objects which contain relevant group of related data used in business processes such as document processing, task allocation and workflow planning. Different companies may have various business natures and subsequently the relevant activities are also dissimilar. In this respect, various business objects should be built in accordance with the related nature of work. Business object are created based on the underlying principle of object technology which has been discussed in the previous context. To ensure the exchange of data within various data objects, a distributed mechanism is devised in the proposed infrastructure, adopting the Distributed Component Object Model (DCOM) architecture which allows data communication across a network in a distributed way. DCOM provides distributed messaging services, object request broker services, distributed transaction services, and data connectivity services [26] and through the support of the DCOM architecture, information from business objects can be exchanged within the supply chain suppliers in a distributed way.

The information flow among the virtual agents and the various Information Systems (IS) in supplier chain partners. Basically the information flow can be categorized into two levels, i.e. strategic and operation. In the strategic level, various virtual agents such as order acquisition VA, scheduling VA, transportation management VA, resource management VA and logistics VA work together to share the required data and compute the needed data objects to the partner companies. The technique to be recommended is based on the deployment of a workflow VA for keeping the needed information. Data interchanges between the VAs and system repository are via ORB, in order to enable efficient data exchange among various information systems from business partners that may reside in distributed platforms over geographically-isolated regions. These agents are created based on corresponding functions performed in traditional supply chain. Concurrently, they aim to develop preliminary business activity plans and scheduling strategies, which are then sent to the transaction VA at the operational level.

Each transaction agent makes particular plans and scheduling strategies based on its resource capacity information that is collected from business partners. Then each transaction agent sends the corresponding task to the business partners. At the same time, information systems in the business partners interact with the system repository for updating information. Thus, the workflow agent executes the particular plans and scheduling strategies that are received from each transaction agent. It should be noted that the strategic level is more concerned with the information flow among the web server platform, the data format of which may not be fully compatible with the database platform adopted by individual partner companies of the operation level handled by the transaction VA. In this respect, a Data Communication Translator (DCT) is adopted to ensure the smooth data transfer between the strategic level and the operation level.

6 Testing and Evaluation

Based on the proposed system framework, a prototype system has been under development. A testing and evaluation plan, which aims to determine the feasibility of such system in a real industrial situation, has been scheduled into three phases as shown in the following.

1. The first phase involves the technical evaluation of the design and operations of the system prototype, by verifying that the object-oriented routines are doing the job as specified in the system framework.
2. The second phase is concerned with system evaluation. The prototype, with modifications made according to the problems encountered in the first phase, is deployed in a local area network, and the results are observed and recorded by the project team members.
3. The third phase is concerned with the site evaluation of the system. It is important that the agile supply chain system is to be field-tested by the real end-users, in order to determine the possible problems when operating in a practical environment.

At the present stage, the first phase has been completed, and phase two will commence once the modification of the system prototype and the setting up of a local area network have been finalized.

7 Conclusion

In order to attain business in the information age, companies are required to integrate business activities into a global response strategy. To cope with unpredictable changes, companies need to conduct their business activities in a flexible manner, and this task cannot be achieved without the deployment of an agile supply chain system. This paper suggests an infrastructural framework, involving various emerging technologies, for the design and development of an agent-based supply chain system with the distinct feature of the ability to cope with unexpected changes with the support of a machining learning mechanism. Further research on the infrastructural framework particularly relating to the synergetic combination of the two technologies (object technology and neural network) is needed in order to leverage the "intelligence" level of the virtual agents. In general, the proposed infrastructure paves the way for an approach to achieve agility of a supply chain system using a combination of tools and techniques. The result, if handled effectively, is essential to provide insights related to the strategic use of the supply chain concept, which can enhance company's competitiveness in a significant way.

References

1. ADC News & Solutions (magazine articles) (1998), "Driving the supply chain with better information", *ADC News & Solutions online,* News and Application Trends, May, 1998, `http://www.manufacturing.net/magazine/adc/`

2. Backhouse, CJ & Burns, ND (1999), "Agile value chains for manufacturing – implications for performance measures", *International Journal of Agile Management Systems*, Vol. 1, No. 2, 76–82.
3. Bidanda, B., Cleland, D.I., and Dharwadkar, S.R. (1993). *Shared Manufacturing: A Global Perspective*, McGrawHill, pp. 2–9, 15–16.
4. Bose, R. (1996), "Intelligent agents framework for developing knowledge-based decision support systems for collaborative organizational processes", *Expert Systems with Applications*, 11, 247–261.
5. Eddon, G., and Eddon, H. (1998), *Inside distributed COM*, Microsoft Press, April.
6. Ellis, S., Keddara, K. & Rozenberg, G. (1995), "Dynamic Change within Workflow Systems", *ACM Conference on Organizational Computing Systems (COOCS 95).*
7. Etzioni, O. and Weld, D.S. (1995) "Intelligent agents on the internet: Fact, fiction, and forecast", *IEEE Expert*, 10, 44–50.
8. Eunice, J. (1997), "Neural networks ninety nine", *Illuminata*, Nashua (`http://www.illuminata.com`)
9. Giarratano, J.C. and Riley, G.D. (1993) *Expert Systems: Principles and Programming*, International Thompson Publishing.
10. Hartle, S.L. and Jambunathan, K. (1996), "Knowledge Representation and Inferencing Techniques Developed for a Knowledge-based Front End", *Engineering Applications of Artificial Intelligence*, Vol. 9, No. 3, 245–259.
11. Hedberg, S.R. (1995), "Intelligent agents: The first harvest of softbots looks promising", *IEEE Expert*, 10, 6–9.
12. Jennings, N.R. and Wooldrige, M. (1995), "Intelligent agents: Theory and practice", *The Knowledge Engineering Review*, 10.
13. Kappel G., Lang P., Rausch-Schott S. and Retschitzegger W. (1995), "Workflow Management Based on Object, Rules and Roles", *IEEE Data Engineering*, March.
14. Karlsson, C. and Ahlstrom, P. (1997), "A lean and global smaller firm", *International Journal of Operations and Production Management*, Vol. 17, No. 10, pp. 940–952.
15. Krishnamoorthy, C.S. and Rajeev, S. (1996) *Artificial intelligence and expert systems for engineers*, CRC Press.
16. Lau, H., Tso, SK and Ho, J. (1998), "Development of An Intelligent Task Management System in a Manufacturing Information Network", *Expert System with Applications*, Vol. 15, pp. 165–179.
17. Lau, H., Wong, T.T. and Pun, F. (1999), "Neural-fuzzy modelling of plastic injection molding machine for intelligent control", *Expert Systems with Applications*, Vol. 17, pp. 33–43
18. Lawrence, P. (Editor) (1997), *Workflow handbook 1997*, John Wiley & Son, Chichester.
19. Lee, J.K. and Kwon, S.B. (1995). "ES: An Expert System Development Planner Using a Constraint and Rule-Based Approach", *Expert Systems with Applications*, Vol. 9, No. 1, 3–14.
20. Lee, W.B and Lau, H. (1999), "Multi-agent modeling of dispersed manufacturing network", *Expert System with Applications*,Vol. 16, No. 3, pp. 297–306.
21. Link, D., Darlow, N.R. and Baines, T.S. (1997), "A framework for the role of manufacturing strategy in global manufacturing", *Proceedings of CIRP International Symposium – Advanced Design and Manufacture in the Global Manufacturing Era*, pp. 417–422. Hong Kong.
22. Nwana, H. (1996), "Software Agents: An Overview", *Knowledge Engineering Review*, 11(3), pp. 205–244, October/November.
23. Poyssick, G. & Hannaford, S. (1996), *Workflow Reengineering*, Adobe Press, Mountain View, California.

24. Ragothaman, S., Carpenter, J., and Buttars, T. (1995). "Using Rule Induction for Knowledge Acquisition: An Expert Systems Approach to Evaluating Material Errors and Irregularities", *Expert Systems with Applications*, Vol. 9, No. 4, 483–490.
25. Ramsus, D.W. (1995) "Intelligent agents: PC", *AI*, 9, 27–32.
26. Rosemary, R. (1998), *DCOM Explained*, Digital Press, Boston.
27. Shi, Y., Gregory, M.J. and Naylor, M. (1997), "International manufacturing configurations map: a self assessment tool of international manufacturing capabilities", *Integrated Manufacturing Systems*, 8/5, pp. 273–282.
28. Sinha, A.P. and Popken, D. (1996), "Completeness and consistency checking of system requirements: An expert system approach", *Expert Systems with Applications*, 11, 263–273.
29. Sycara K., Decker, K., Pannu, A., Williamson, M. and Zeng, D. (1996), "Distributed Intelligent Agents", *IEEE Expert*, December.
30. Tso, SK, Lau, H., Ho, John & Zhang, WJ (1999), "A Framework for Developing Agent-based Collaborative Service Support System in a Manufacturing Information Network", *Engineering Applications of Artificial Intelligence*, Vol 12, pp. 43–57.
31. Vastag, G, Kasarda, J.D. and Bonne, T. (1994), "Logistical support for manufacturing agility in global markets", *International Journal of Operations and Production Management*, Vol. 14, No. 11, pp. 73–85.

Track on Networking

Performance Improvement of Deflection Routing in Optical Burst Switching Networks

SuKyoung Lee[1], Kotikalapudi Sriram[1], HyunSook Kim[2], and JooSeok Song[2]

[1] National Institute of Standards and Technology,
100 Bureau Drive, Stop 8920, Gaithersburg, MD 20899, USA.
[2] Dept. of Computer Science, Yonsei University, Seoul, Korea

Abstract. In Optical Burst Switching (OBS) networks, when contention occurs at an intermediate switch, two or more bursts that are in contention can be lost because a forwarding path reservation is not made for a burst until a control message for the burst arrives. While deflection routing protocol is proposed as one of the contention resolution techniques, there has been no appropriate deflection routing algorithm to find an alternate route. In this paper, we propose a novel deflection routing algorithm to compute alternate routes with better performance as compared to other known techniques. This algorithm deflects contending bursts using a path that is based on minimization of a performance measure that combines distance and blocking due to contention. We will show, through simulation results, that there is an improvement in terms of loss with increased network throughput.

1 Introduction

For building IP over WDM network, we have several alternatives. One promising approach is optical burst switching [1]. One of challenging issues in application of burst switching to optical domain is the resolution of contentions that results from multiple incoming bursts that are directed to the same output port. In an optical burst switch, various techniques designed to resolve contentions include optical buffering, wavelength conversion and deflection routing. Over the other techniques, deflection routing gains an advantage that buffer capacity is very limited to maintain a reasonable level of data losses while routing the contending bursts to an output port other than the intended output port. However, before attempting to implement deflection routing in OBS networks, the details of deflection routing have to be investigated.

Some most recent studies about deflection routing can be found in [2]-[5]. In [2], the performance of deflection routing is examined in OBS networks based on Just-Enough-Time (JET). The authors of [3] and [4] demonstrated via simulation test that the blocking probability is improved when deflection routing is used as a contention resolution. On the other hand, the authors of [5] introduced how deflection routing can be applied to the self-routing address scheme. However, they do not explain how to route an alternate path i.e. which constraints could be applied to the selection of an Alternate Route (AR).

P.M.A. Sloot et al. (Eds.): ICCS 2003, LNCS 2659, pp. 445–452, 2003.

In this paper, we introduce a Contention-based Deflection Routing (CDR) algorithm for computing Alternate Routes (ARs) in a way that avoids burst blocking due to contention. The objective of this paper is to formulate the deflection routing problems in mathematical form to motivate more work towards the development of efficient solution techniques. The main purpose of this algorithm is to recompute an optical path that is already established, so that the AR can be placed into the network being expected a burst would experience less blocking on the AR than on the shortest alternate path usually adopted in other known deflection routing works [2]-[5]. The proposed technique can also be applied to pre-configuration of primary routes with improved performance given that the burst intensity of each node is periodically measured by the network management system.

The rest of this paper is organized as follows. Section 2 describes CDR algorithm in detail. In Section 3, we present the simulation model used and performance of the proposed CDR is evaluated via computer simulation. Finally, we conclude the paper in Section 4.

2 Deflection Routing Algorithm

In OBS, a control packet is sent first to set up a connection by reserving an appropriate amount of bandwidth and configuring the switches along a path, followed by a data burst without waiting for an acknowledgement for the connection establishment. A delay referring to offset time or Fiber Delay Line (FDL) has been proposed to bring this form of reservation to fruition. The offset time allows for the control packet to be processed at each node while the burst is buffered electronically at the source; thus no FDLs are necessary at the intermediate nodes to delay the burst while the packet is being processed.

The proposed CDR can be applied to both styles of reservation now that CDR is not the mechanism of how to reserve the wavelength but how to select the alternative path. However, in consideration of the fact that the required FDL technology is not mature enough for it to be implemented in practice, it may be more beneficial to the current carriers that CDR is implemented on the basis of offset-based scheme since it maintains the most state information for the bursts and therefore allows for more flexible deflection routing.

In an OBS network, the deflection routing functions implemented in each switch automate the AR setups when a control packet encounters a congested node over the primary path. However, the switches have only local information about the network resources, especially the contention status. These local routing decisions for the ARs made in edge nodes may result in a degraded global network performance on the long run. As a solution for this, the proposed CDR algorithm is to perform periodical global re-optimization of ARs based on the most recent information of contention status. Even though this re-optimization is periodical, we mean that it is not performed not too frequently (i.e. daily or weekly). It is also possible for the re-optimization to be performed on demand.

To formulate the CDR problem, the network topology, a set of attributes pertaining to the resources and the constraints in the network are defined. The demands that are to be routed through ARs in the network are described by a set of attributes as well. Then, the problem is to find an optimal AR minimizing a cost function which explicitly accounts for the contention rate as well as the burst hop distance. The aforementioned CDR problem can be formulated as follows: Consider a physical network represented by a graph $G(N, L)$ where N is the set of nodes and L the set of links (*i.e.* fibers) connecting the nodes. It is assumed that each link between nodes i and j, has W_{ij} wavelengths with the same capacity of C. At each node n, $(n = 1, \ldots, |N|)$, the number of transmitters and receivers are defined as $P_n^{(t)}$ and $P_n^{(r)}$, respectively. If a node n has the number, P_n of ports, clearly, at most $\sum_n P_n$ wavelengths are needed to realize any possible virtual topology.

Let Λ be the set of traffic demands belonging to the loss-sensitive service class between a pair of edge nodes, where $\lambda_{ij}^{sd} \in \Lambda$ represents the arrival rate of bursts from source s to destination d that flows over an virtual link between node i and node j, and where $\lambda_{s_k d_k}$ denotes the average flow of bursts associated with the kth traffic demand requesting wavelength. Let $D = \{D_{ij}\}$ be the distance matrix from node i to node j that means a propagation delay from node i to node j $(i \neq j)$. As the cost of contention from node i to node j $(i \neq j)$, we introduce a constant b_{ij} to denote burst blocking rate that is collected periodically from the network.

In the CDR problem formulation, the variable that needs to be determined is x_{ij}. x_{ij} is a binary variable associated with link (i, j) to indicate whether $(x_{ij} = 1)$ or not $(x_{ij} = 0)$ the link (i, j) is established over AR. The problem is to find an AR from the congested node to the destination, for which the burst blocking rate and the distance are minimized. Then, this problem can be formulated for the kth traffic flow as follows:

$$Minimize\ g_d \sum_{i,j} x_{ij} D_{ij} + g_b[\log_{10}[1 - \prod\nolimits_{i,j}(1 - x_{ij} b_{ij})] \tag{1}$$

where g_d, and g_b denote the weights for delay and blocking, respectively.

such that

$$\sum_{\forall j \in N} x_{ij} \leq P_i^{(t)}, \sum_{\forall i \in N} x_{ij} \leq P_j^{(r)} \tag{2}$$

$$\lambda_{ij} = x_{ij} \sum_{s,d} \lambda_{ij}^{sd} + x_{ij} \lambda_{s_k d_k} \qquad \forall i, j \in N \tag{3}$$

$$\lambda_{ij}^{s_k d_k} \in \{0, \lambda_{s_k d_k}\} \qquad \forall i, j \in N \tag{4}$$

$$\sum_j x_{ij} - \sum_j x_{ji} = \begin{cases} 1, i = s_k \\ -1, i = d_k \\ 0, \text{otherwise} \end{cases} \quad \forall s, d, i \in N \tag{5}$$

$$\lambda_{ij} \leq W_{ij} C \qquad \forall i, j \in N \tag{6}$$

Eq. 1 defines the objective function. This objective function is a weighted sum of the end-to-end burst blocking rate and the distance for the route. To decrease the computational complexity, we can express the above objective function, Eq. 1 as

$$Miminize \qquad g_d \sum_{i,j} x_{ij} D_{ij} + g_b \sum_{i,j} x_{ij} \log_{10} b_{ij}, \tag{7}$$

where as the burst contention rate, b_{ij}, the real data can be used that has been collected into deflection routing information base. Constraint 2 ensures that the number of lightpaths originating from and terminating at a switch is not more than the switch's out-degree and in-degree, respectively. There are some constraints (3-6) related to the traffic flow on virtual topology for each link (i, j). Constraint 3 states that the total flow on the simplex link from node i to node j is expressed as the superposition of the existing traffic (*i.e.* bursts) and the new burst associated with the link. Since we are setting up an AR for the optical bursts coming from a specific traffic flow, the bursts of the demand $\lambda_{s_k d_k}$ are not segmented at any congested node in the network. Thus, constraint 4 defines that the traffic demand $\lambda_{s_k d_k}$ is routed from node i to j on a single AR. Constraint 5 formulates flow conservation at each node. The traffic flowing into a node should be equal to that flowing out of the node for any node other than the source and destinations for each traffic flow k. Constraint 6 assures that traffic flowing through a link cannot exceed the total link capacity.

We know that the objective function in Eq. 7 is more of a practical value than one involving distance alone. It includes Quality of Service (QoS) requirements regarding loss as well as distance. This objective function can be easily generalized to the case of multiple Classes of Service (CoS), where bursts of different CoS may have different QoS requirements regarding loss. The disparate CoS and their required QoS can be reflected into the routing decision by having different weights associated with each in our objective function. In the end, the AR would be set up according to the values of the x_{ij} determined from the above integer linear programming formulation.

As mentioned above, the optimization algorithm described in this section can be performed offline or online. In the former approach, multiple fixed ARs are considered when a contention occurs. Thus, each node in the network is required to maintain a deflection routing table that contains an ordered list of a number of fixed ARs to each destination node. In the on-demand CDR, the AR from a congested node to a destination node is chosen dynamically, depending on the current network state. The on-demand CDR method will require more computations and a longer response time than CDR based on pre-computed alternate routes and lookup-table. But the on-demand CDR approach is more flexible and would result in better resource utilization and performance than the latter approach.

For offset-based scheme [6], one of representative reservation styles in an OBS network, the burst will not arrive successfully to its destination over the AR computed by CDR without extra offset time or buffered delay. When deflection routing is performed due to a contention at an intermediate node, the offset time on the AR is different from(usually, longer than) that on the primary path. One solution to this problem is to render sufficient extra offset time to each burst while the other is making the control packet reserve FDL buffer to delay the burst. Even though the above problem is resolved, it may happen that the too much increased distance on AR cause longer delay than expected offset time or buffering time. Thus, if $t_{o,c}$ denotes a maximum limit of offset time for service class c, that consists of basic offset time and extra offset time, another constraint for offset time is defined as

$$\sum_{i,j} x_{ij} D_{ij} \leq t_{o,c} \qquad \forall i,j. \tag{8}$$

3 Performance Evaluation

In order to evaluate the effectiveness of CDR technique, simulation tests are conducted under two schemes: CDR algorithm and Shortest Path-based Deflection Routing (SPDR) algorithm which has been generally used in other deflection routing works [3]-[5]. Our simulation is run on JET [6] which is one of offset-based schemes.

In our simulation, we assume that each fiber link is composed of 6 wavelengths and its bandwidth is 1 Gbps. The burst sources were individually simulated with the on/off model, where the on and off periods were varied as (α^{-1} = 120 ms, β^{-1} = 880 ms) and (α^{-1} = 300 ms, β^{-1} = 700 ms)

The tests are carried out using two particular network scenarios: a simple network topology with five nodes as figure 1 in [3] and 14-node NSFNET in Fig. 1 as a larger topology. Over NSFNET topology, as soon as five ingress-egress node pairs are chosen randomly, optical bursts are generated at the ingress nodes of a optical burst network domain.

In Fig. 1, some bursts whose source and destination are CA1 and NJ, respectively, experience a contention at UT node on the primary path (CA1-UT-MI-NJ). As an example in our simulation, the deflection routing table at the node

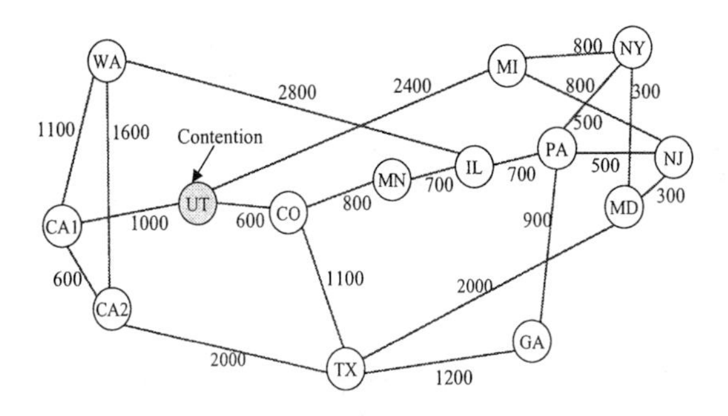

Fig. 1. Simulation network topology

Table 1. Burst drop rate and overall throughput improvement(Mbps) over five-nodes topology

Burst size (Bytes)	Burst drop rate		Throughput improvement
	CDR	SPDR	
576	1.14×10^{-5}	3.44×10^{-4}	2.39
2304	1.53×10^{-5}	3.36×10^{-4}	8.49
4608	1.62×10^{-5}	3.40×10^{-4}	17.27

lists the (UT-CO-MN-IL-PA-NJ) and (UT-CO-TX-MD-NJ) as alternate candidate paths. Of these, (UT-CO-MN-IL-PA-NJ) is the shortest alternate path from UT to NJ. However, the CDR scheme can very well select (UT-CO-TX-MD-NJ) as the preferred alternate path if that happens to be the only one that meets the requirement on burst blocking as well as distance. The focus of our performance evaluation is on data loss rate caused by contention. A burst will be dropped if both the primary and deflection paths are blocked. The data loss rate for the entire network is found by calculating the average of the burst drops for each source-destination pair. Thus, we first show that the data loss rate is indeed reduced by CDR technique in comparison to SPDR.

Table 1 shows the results of simulation performed with 3 different burst sizes as in [1] when offered traffic load is 0.6. As can be seen from this table, CDR scheme drops smaller number of bursts than SPDR with improved throughput.

Figs. 2 and 3 show simulation results comparing the burst blocking or loss rate for our CDR method with the SPDR method. The SPDR algorithm simply picks the shortest path alternate route available from the deflection routing table, whereas with the CDR scheme the alternate path selection is based on

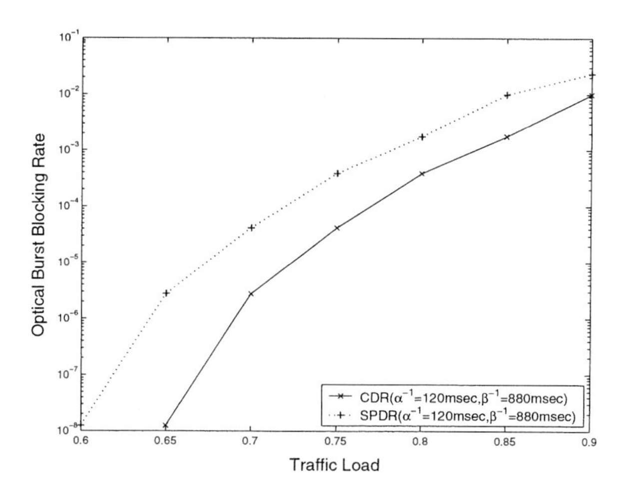

Fig. 2. Burst blocking rate for CDR and SPDR without FDL when activity is 0.12

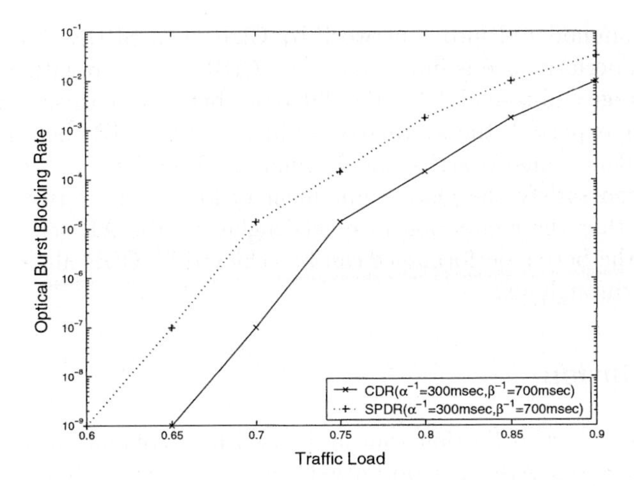

Fig. 3. Burst blocking rate for CDR and SPDR without FDL when activity is 0.3

minimizing a composite performance measure consisting of the alternate path distance as well as burst blocking along that path. For typical operating load values up to 0.75, the CDR algorithm improves burst blocking by more than an order of magnitude as compared to SPDR in the test cases that we have studied through simulation runs.

Fig. 4 shows how much the overall network throughput was improved by CDR algorithm, compared to SPDR algorithm. The throughput values shown are the values averaged during the entire simulation time. In this figure, as the traffic load increases and more bursts contend on the shortest AR, the more

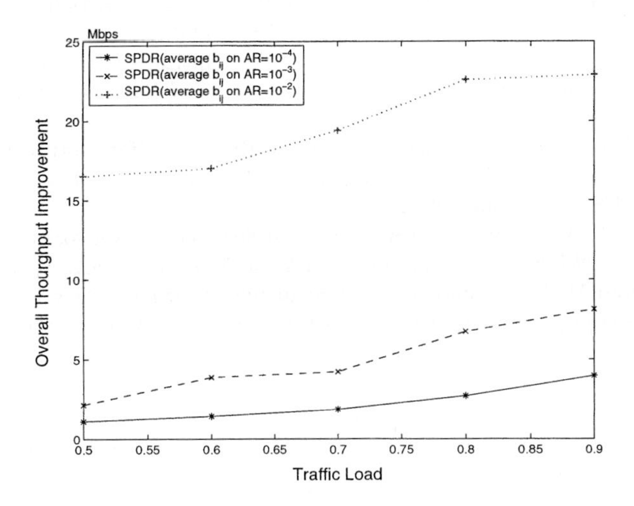

Fig. 4. Overall network throughput improvement by CDR when compared to SPDR

data is accommodated into a network by CDR than SPDR. We can see that throughput performance is improved under CDR compared with SPDR.

For average end-to-end delay the difference between both schemes is around 1.6%. As we expected, the average delay increases for CDR because the bursts have more deflections to arrive the destination. Therefore, we know that CDR algorithm can satisfy the QoS requirement of loss-sensitive traffic. Especially, in the case that the contention rate gets higher on the AR selected by SPDR algorithm, the better performance can be achieved by CDR algorithm in terms of loss and throughput.

4 Conclusion

We note that when deflection routing is used as a contention resolution, it is important to design ARs in an optimized fashion to minimize burst blocking. Our simulation results have shown that the proposed CDR scheme enables lower burst blocking rate and higher network throughput as compared to the SPDR scheme. This is because CDR selects ARs that are not necessarily shortest path but have the lowest burst blocking rate, based on periodically updated measurement data.

An area of future work can be to extend the CDR algorithm with scheduling policies that consider traffic types with different QoS requirements in OBS networks.

References

1. H.M. Chaskar, S. Verma, and R. Ravikanth, "Robust Transport of IP Traffic over WDM Using Optical Burst Switching", *Optical Network Magazine*, vol. 3, no. 4, pp. 47-60, Jul/Aug. 2002.
2. C-F. Hsu, T-L Liu and F-F Huang, "Performance Analysis of Deflection Routing in Optical Burst-Switched Networks", *IEEE INFOCOM'02*, 2002.
3. X. Wang, H. Morikawa, and T. Aoyama, "Burst Optical Deflection Routing Protocol for Wavelength Routing WDM Networks", *Optical Networks Magazine*, November/December 2002, pp. 12-19.
4. S. Kim, N. Kim, and M. Kang, "Contention Resolution for Optical Burst Switching Networks Using Alternative Routing", *IEEE ICC'02*, vol. 5, pp. 2678-2681, NewYork, USA, Apr/May. 2002.
5. C.Y. Li et al, "Deflection Routing in Slotted Self-routing Networks with Arbitrary Topology", *IEEE ICC'02*, vol.5, pp. 2781-2785, NewYork, USA, Apr/May. 2002.
6. C. Qiao, and M. Yoo, "Optical Burst Switching (OBS) - A New Paradigm for an Optical Internet," *Journal of High Speed Networks*, vol. 8, no. 1, pp. 69-84, 1999.

A Performance Evaluation on Qos-Supporting Multicast Protocol over Conventional Multicast Protocol

Won-Hyuck Choi, Tae-Seung Lee, and Jung-Sun Kim

School of Electronics, Telecommunication and Computer Engineering
Hankuk Aviation University, Seoul, Korea
rbooo@korea.com, thestaff@hitel.net,
jskim@mail.hangkong.ac.kr

Abstract. Recently there have been many requirements for reliability from the multicast communication services that have peer-to-multiple or multiple-to-multiple connecting ability unlike the previous peer-to-peer communications. Multicast protocols include the distance vector multicast routing protocol (DVMRP), multicast extension to open shortest path first (MOSPF), core based tree (CBT) and protocol independent multicast (PIM) structures. These protocols use their own local distance algorithms to utilize characteristics of network packets as routing costs in transferring data. However, multimedia applications require more efficient multicast protocols as well as broadband network connection, real-time transmission. To the requirement, the quality of service sensitive multicast Internet protocol (QosMIC) is urged to be used in Internet. The QosMIC supports several quality of service (Qos) properties: reliability, efficient resource management and flexibility of multicast networks. The QosMIC is in distinction from the existing protocols in that the QosMIC presents data in advance. This paper analyzes the QosMIC and CBT by using objective and practical bandwidths in transferring data packets, and measures their performances on a simulation to test such bandwidths.

1 Introduction

Recently multimedia workstations, distributed systems and high speed communication systems have had remarkable advances. The combination of these systems encourages the distributed multimedia systems to be developed, and not only workstations but also personal computers can transfer the non-text signals such as voice, image, audio and graphic. Such services can be considered as an aim of data transmission implemented on network and transport layers. The transmission protocols define and implement functions to support the atoms of the services. The up-to-date network communications require the higher layers than transport layer to take over appropriate roles. As a result, users will be ultimately able to enjoy several multimedia applications at a time, each application supplying its own characteristic service.

There are increasing requests on the current multicast communications to have reliability on peer-to-multiple or multiple-to-multiple communication services. In addition to it, broadband network connection, real-time transmission and more efficient multicast protocols are also needed to support multimedia applications. Nevertheless,

P.M.A. Sloot et al. (Eds.): ICCS 2003, LNCS 2659, pp. 453–462, 2003.

the multicast protocols such as the distance vector multicast routing protocol (DVMRP), multicast extension to open shortest path first (MOSPF), core based tree (CBT), protocol independent multicast (PIM) structures, are lack of communication processing performance, bandwidth and quality of service (Qos), which are fundamental requirements on viable multicasting. It is because such multicast protocols use only their own local distance algorithms as routing costs of packets.

This problem urges to introduce multicast group communication into the service layer newly added to Internet. The service layer provides services necessary to next generations of Internet. The multicasting has fundamental role in the service layer. Multicast must transfer similar information to multiple receivers. This fact results in that single multicast group might suffer from transmission delay which is a defect to unicast communication, since multiple packets are moved through network at a time. Moreover, the quality of user receiving has to be assured.

Network communication must be considered with respect to social exchange, value of the communication and dependency on service. Such properties are sustained by consistent management of Qos layer. The existing multicast routing protocols cannot implement applications to provide such Qos properties. Most of the protocols adopt the PIM, CBT and BGMP as the best efficient structures of Internet. The protocols present users with single path routing. If there is not sufficient resource, they cannot satisfy user's requirements.

Multicasts are divided into conventional methods and methods to improve network service. A representative method of conventional methods is the CBT. The CBT is the most effective method of the existing multicasts to enhance network speed, but it does not consider Qos. Qos enables resource of a source to be connected in shorter time when the source is disconnected from a receiver of network. Therefore, the multicast to present Qos can provide more reliability by preparing sufficient resource.

In this paper, the authors attempt to evaluate a Qos application by using a reliable and efficient multicast protocol, the quality of service sensitive multicast Internet protocol (QosMIC), on Internet. The multicast protocol does meet three requirements: (1) Qos support, (2) reliability, and (3) efficient resource management. The QosMIC is based on multiple routing and Qos supporting, so involves in multicast group connection and fully satisfies the multiple routing and Qos requirements. To compare the QosMIC and the CBT, the best popular one of the existing multicast protocols, a simulation is conducted [1,5].

The hereafter composition of this paper is as follows: section 2 describes the overview of multicast protocols, section 3 describes the QosMIC method, section 4 reports the result of the simulation to compare the QosMIC performance and CBT, and section 5 finally concludes our work.

2 Multicast Protocol

Of the previous multicast protocols, four have been commonly used: the source based tree (SBT) multicast protocol constructs trees for each source, the shared tree (ShT) shares trees with multiple sources, the Intra-Domain establishes trees in the same domain as to their locations, and the Inter-Domain establishes trees in different domains.

2.1 Source Based Tree Protocol

The SBT is also called the source distribution tree (SDT), or the shortest path tree (SPT) in that it is based on the shortest path algorithm. When trees of the SBT are represented, (S, G) is used to denote the source S. The SBT which has a good delay characteristic of traffic is efficient in multicast services like the internet broadcasting that has a few senders but many receivers. The distance vector multicast routing protocol (DVMRP), one derivation of the SBT, establishes the shortest path to the receiver to forward datagrams in multicasting mode. And the multicast extension to open shortest path first (MOSPF) protocol is the one that, as you can infer from its name, extends the second version of the open shortest path first (OSPF) (RFC1583). Its name signifies it is independent of unicast routing mechanism, though it has some similar characteristic to unicast protocols. The PIM uses only the resulting tables without associating with algorithms or protocols that put out an unicast routing table. When the density of a tree with a distribution of group members is considered, the PIM is classified into the dense mode (DM) if the density is high and the sparse mode (SM) otherwise [6].

2.2 Shared Tree Protocol

The ShT is denoted as (*, G), where * means the whole sources and G the group. The size of the real tree has the value of $O(|G|)$ independently of the number of sources since it is the shared tree. The cost in constructing trees is not expensive, but a serious traffic delay might be caused by the increasing number of sources. The ShT is appropriate when multicast services are supplied in the networks that have somewhat narrow bandwidth traffic but many senders. The core based tree (CBT) is one of the ShT and has a core router on the center of the shared tree. In contrast to the PIM-SM protocol that operates as an uni-direction tree and so has constraints on selecting the optimal routing path, the CBT tree operates as bi-direct tree and has more flexible extensibility of networks in comparison with the existing source-based multicast routing protocols.

The best advantage of the CBT protocol is that the change of overhead to manage trees is not very severe when the configuration of member or source is changed. It is due to single tree shared with all members of a multicast group. In addition to it, the CBT can utilize efficiently resource of network including bandwidth of a link because it does not require periodic information to manage groups as the IGMP does and use its bandwidth only for multicast traffic. On the other hand, the disadvantage of the CBT is that it might cause to increase link cost to establish the whole routing path of a tree on the basis of core router and make a bottleneck at core link to condense traffic to the core. That is, high link load needed for overloading the bandwidth of core link makes a bottleneck. When this situation gets more strained, it might result in the deadlock of core link and lead to the stop of multicast service.

The PIM-SM protocol uses the rendezvous point (RP) that is shared with the whole receiver routers for each multicast group, and it is used as an uni-direction tree [3].

3 Protocol Description: The QosMIC

3.1 Overview

The main parts of the QosMIC support functions to select a router. In other words, the QosMIC is defined as collection of Qos routing information about multiple paths. When a new node is selected from a multicast tree, Qos requires the exact path by gathering information. To the requirement, the QosMIC is classified into the most up-to-date router. The main parts of the router try to search the adjacent routers from tree. Figure 1 shows the node accessing method of the QosMIC. It has to be noted how the core router cannot be used. In the method, the group member of the near node is active and the router selects the Join, and this results in eliminating the necessity of the cores. The QosMIC is superior to the unicast protocol and core-based tree protocols when such advantage is born in mind [1].

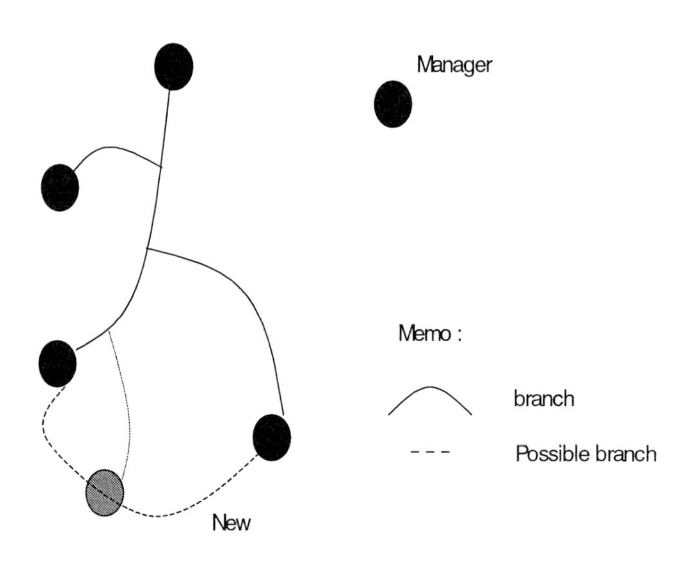

Fig. 1. Approach of the QosMIC

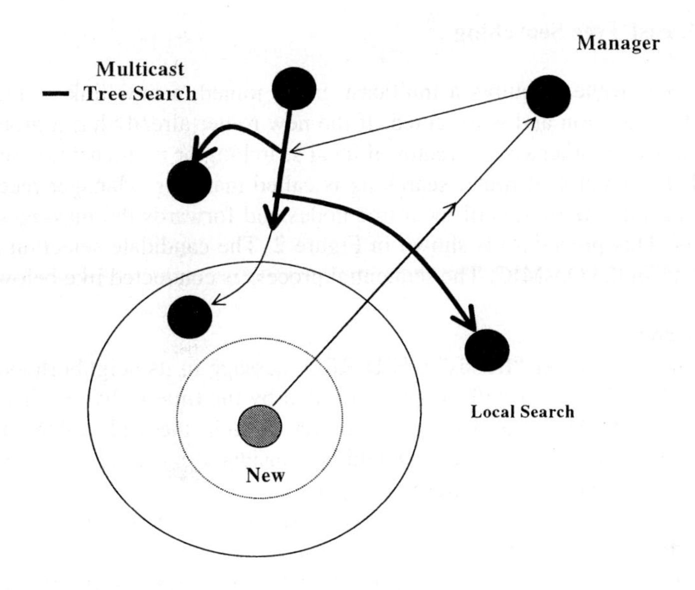

Fig. 2. Searching procedure of the QosMIC

3.2 Structure and Message Forwarding Method of the QosMIC

The transmission in the QosMIC structure is implemented as receiver sending on multicast tree and consists of three stages. In the first stage, a router is defined if a tree exists and the tree is able to join other trees through the interfaces of them. These trees are called candidates. In the second stage, the candidates inform the new router of their own information by command messages. In the final stage, the joined router selects nodes that need information. The QosMIC is a source-based tree that Qos characteristic is improved, and is switched such that sharing tree is avoided. The message handling of the QosMIC is implemented as follows. All the routers contain multicast routing information. The information forwards individual multicast packets and executes the appropriate members linked. Each link has entries of tree table in an isolated tree partition. If the packet transferred *matches* an entry of the routing table, it is linked. In unique tree method, when packets from all sources of a group are *matched*, the routing entry takes them from shared tree. In (S, G) of the source-based tree, G represents the group and S the source. The shared tree uses (*, G) notation, where * designates the source. The routing location in most multicast protocols of IETF takes lower level. It means that the routing location will be vanished. The adjoining messages correspond with each other periodically and the incoming links from them are newly combined [1], [4].

3.3 Multicast Tree Searching

When the new router requires a multicast group joined, a new link is added in the multicast tree location and is executed. If the new router already has a group entry, a link is established; otherwise a creator of local searching or multicast tree searching is employed. The creator of router searching is called manager. Manager receives messages from candidate routers of local tree nodes and forwards the messages to multicast routers. This procedure is shown in Figure 2. The candidate selection process is much serious in the QosMIC. The sequential process is conducted like below [8]:

1) Local Search

① The New router "floods" a BID-REQ message in its neighborhood. Reverse path multicasting with scope controlled by the time to live (TTL) field. Because QosMIC has the Multicast Tree Search, the TTL value can be kept very small. The advantage should be considered meaningful at the results of simulation conducted later in this paper.

② Every In-Tree router that receives a BID-REQ message, becomes a Candidate router, and unicast a BID message back to the New router. The BID message on its way collects information on the expected performance of the path, based on dynamic Qos metrics. The Candidate router considers the New router as a tentative dependent, and cannot leave the tree unless the tentative status is timed out.

③ The New router collects BID messages. The procedure termincates unsuccessfully, if the New router does not receive any replies before the expiration of a timer set for this purpose. Otherwise, we enter the phase of establishing the connection.

2) Multicast Tree Search

① New router sends an M-JOIN message to the Manager of the group.

② The Manager "orders" a bidding session with a BID-ORDER message. Some subset of the routers that receive the BID-ORDER become Candidates.

③ The Candidates unicast BID's to the New router. The BID's are identical to the BID's in the Local Search.

4 Simulation

We compare and analyze the multicast protocols, the QosMIC and CBT described in Sect. 3. The first experiment compares the average speed of traffic flows between terminals when the QosMIC and CBT are used alone. It is to measure the performances in speed, which is the most significant property in networks. The second experiment compares the average delays between terminals. It is to measure how much the Qos between terminals is satisfied when network information is transferred. To do this, a multicast simulation model is established first. The comparison of the QosMIC and CBT routings parameterizes the number and bandwidths of multicast groups, the delay time of packets and the size of data packets as to the characteristic of an application program. The simulation platform is a PC machine with 512MByte memory and 1.5GHz Intel Pentium Processor, the operating system of the machine the Linux RedHat 7.0 and the tool the Network Simulator – ns Version 2 which is used widely in PC-based environments. In the first experiment, packet sizes to test are 210byte and 512byte.

Figures 3 and 4 report the 4.53% improvement of the QosMIC with respect to the CBT for 210byte packets and the 10.25% for 512byte packets. The time measurements in the second experiment are the average delay times to the total traffic as with the increasing number of groups between terminals when all nodes linked with a router are considered.

Figures 5 and 6 present the measurements of the average delay times between terminals when the size of packets is 210byte and 512byte, respectively. It is noted that the delay characteristic of the QosMIC routing is slightly superior to that of the CBT. The delay characteristics of the QosMIC against the CBT are recorded as 12.23% for 210byte packets and 14.4% for 512byte packets.

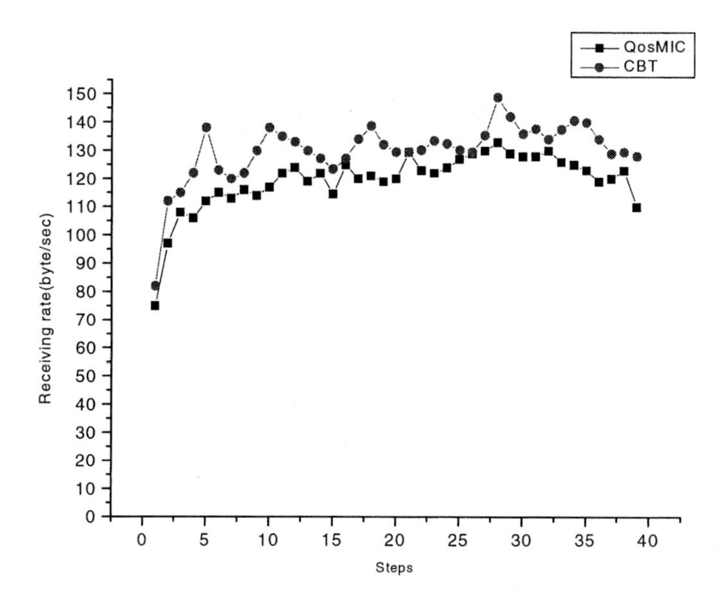

Fig. 3. Data receiving rates for 210byte packets

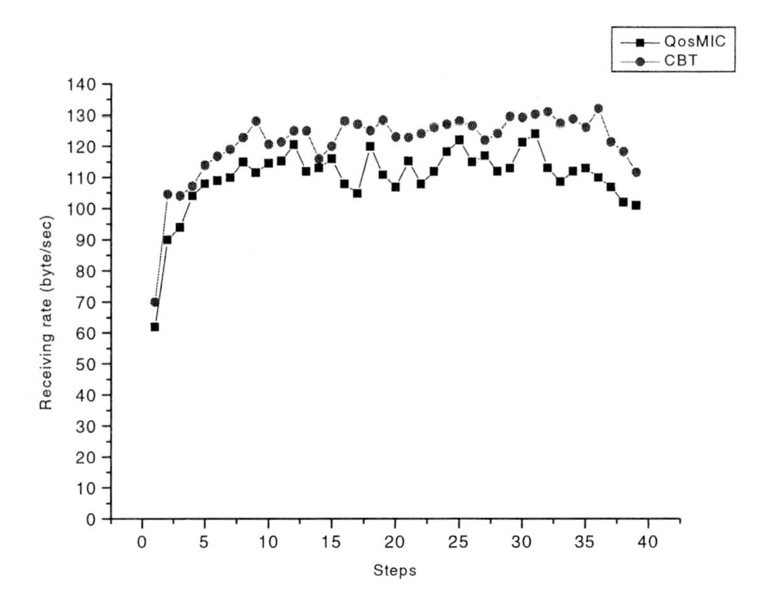

Fig. 4. Data receiving rates for 512byte packets

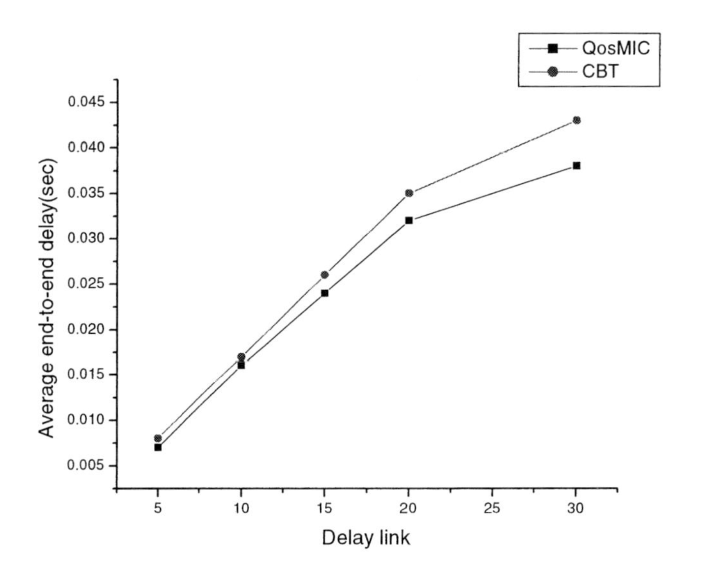

Fig. 5. The average delays between terminals for 210byte packets

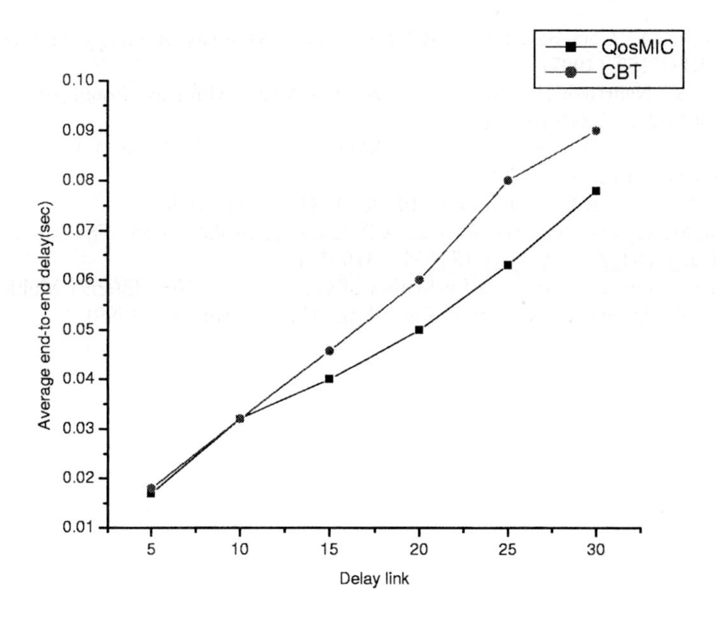

Fig. 6. The average delays between terminals for 512byte packets

5 Conclusion

In this paper, we measured the packet processing times and average delay times between terminals for 210byte packets and 512byte packets, respectively. From the first experiment results, the 4.53% and 10.25% improvements of the QosMIC against the CBT at the two packet sizes were obtained. And the average delay times were reduced by the 12.23% and 14.4%. The delay times of the CBT increase more than them of the QosMIC when the number of links of senders is high, the packet size is large and the number of groups of receivers is high. Those evidences strongly confirm the fact that the increasing number of CBT groups induces the increase of the number of groups in Core and the more delays of data packets, and the QosMIC remedies such defect. In the future work, it is required to measure the average delay time and data processing time between terminals as well as the reliabilities on higher levels. Additionally, other multicast protocols beside the CBT must be compared and analyzed in the process between the QosMIC and terminal, and the reliabilities of higher levels be analyzed.

References

1. The QoSMIC Implementation Web-Site. Available at `http://www.cs.ucr.edu/~michalis/qosmic.html`
2. Almerith, K.: A Long-Term Analysis of Growth and Usage Patterns in Multicast Backbone. IEEE INFOCOM. **2** (2000) 824–833

3. Ballardie, A.: Core Based Trees (CBT version 2) Multicast Routing – Protocol Specification. RFC2189. (1997)
4. Chen, S., Nahrstedt, K., Shavitt, Y.: A QoS-Aware Multicast Routing Protocols. IEEE INFOCOM. **3** (2000) 1594–1603
5. Fei, A., Gerla, M.: Receiver-Initiated Multicasting with Multiple QoS Constraints. IEEE INFOCOM. **1** (2000) 62–70
6. Moy, J.: Multicast Extensions to OSPF. IETF RFC 1584. (1994)
7. Parsa, M., Garcia-Luna-Aceves, J. J.: A Protocol for Scalable Loop-Free Multicast Routing. IEEE J. Select. Areas Comm. **15** (1997) 316–331
8. Yan, S.: Implementation and Evaluation of QoSMIC – A New Internet Multicast Routing Protocol. Master's thesis, Dept. of Elect. Eng., Univ. of Toronto. (1999)

Improving the Performance of Multistage Interconnection Networks under Nonuniform Traffic Pattern on Shorter Cycles

Hyunseung Choo[1] and Youngsong Mun[2]

[1] School of Information and Communication Engineering
Sungkyunkwan University, Suwon, Korea
choo@skku.edu
[2] School of Computing, Soongsil University, Seoul, Korea
mun@computing.ssu.ac.kr

Abstract. Multistage interconnection networks (MINs) have been recently identified as an efficient interconnection network for a switching fabric of communication structures such as gigabit ethernet switch, terabit router, and ATM switch. Even though there have been a number of studies about modeling MINs in the literature, almost all of them are for trends MINs under uniform traffic which dose not reflect the realistic. In this paper, we propose an analytical model to evaluate the performance of ATM switches based on MINs with the small clock cycle (SCC) scheme under nonuniform traffic. Here MINs of 6 and 10 stages with built-in buffer modules holding single or multiple cells are considered for the evaluation. Comprehensive computer simulation results present that the proposed model is effective for predicting the performance of ATM switches under the realistic nonuniform traffic. It also shows that the detrimental effect on the hot spot traffic which is typical in the Internet turns out to be more significant as the switch size increases.

1 Introduction

Multistage interconnection network (MIN) constructed by switching elements (SEs) with simple connections in multiple stages has been recognized as an efficient interconnection structure for super computer [1] and faster network devices [2]. Its regularity and modularity make MIN very appropriate for very large scale integration. It also has the self-routing capability, which is important for fast input to output connection. MINs thus have been considered for the switching fabric of new communication structures such as gigabit Ethernet switch [3], terabit router, and ATM switch.

ATM [4] is a high-speed connection-oriented packet switching technique which employs short fixed-length packets called cells. It allows much higher rate switching than typical packet switching due to its functional simplification. While numerous switching systems for ATM have been proposed, MIN is identified to be very effective for integrated service digital network (ISDN) and also it is expected to be an

P.M.A. Sloot et al. (Eds.): ICCS 2003, LNCS 2659, pp. 463–473, 2003.

outstanding candidate for the future networks supporting broadband services such as multimedia data and graphic oriented applications. Understanding and predicting the performance of the switches are very important in designing correct switch-based communication systems. There have been a number of studies investigating the performance of MINs in the literature [5–7]. However, almost all of these previous works are for studying the MINs under the uniform traffic pattern. Although the performance under uniform traffic is interesting to note, it does not provide the actual data on the real performance of MINs. The nonuniform traffic reflects the realistic traffic pattern of currently deployed integrated service networks where a wide range of bandwidths needs to be accommodated. Therefore, the performance of the MINs under nonuniform traffic must be studied for obtaining an efficient switch-based system. Even though there are few number of models investigated on nonuniform traffic patterns [6-7], they are not good enough on the performance of the models and even not reasonably verified.

In this paper, we propose an analytical modeling method to evaluate the performance of ATM switches based on MINs with small clock cycle (SCC) scheme under the nonuniform traffic. It is mainly achieved by properly implementing the nonuniform dispatch probability in modeling the operation for each switching element. To evaluate the accuracy of the proposed model, comprehensive computer simulation is performed on comparison purpose for two performance measures – throughput and delay. MINs with 6 and 10 stages with built-in buffer modules holding single or multiple cells are considered for the evaluation. As a typical nonuniform traffic pattern, hot spot traffic of 3.5% and 7% are investigated the comparison of the simulation data with the data obtained from the analytical model shows that the proposed method is very effective for predicting the performance of ATM switches under the realistic nonuniform traffic. The detrimental effect on the hot spot traffic on the network performance turns out to get more significant as the switch some increases.

2 The Proposed Model

The SCC consists of two phases in our models. In the first phase, the sending buffer modules check the empty space availability of the receiving buffer modules. Based on the availability (routing information) of the succeeding stages, each sending buffer module sends a packet to its destination or enters into the blocked state in the second phase. Other assumptions used in our model are as follows.

- 2×2 SEs with the buffer modules of size m are used, where the buffer modules are located at the input port; All packets have the same size (like in ATM); The probability that a packet arrives at each network input is the same for all inputs. Each inlet generates requests identically with a certain probability which are nonuniformly distributed over all outlets; Two buffer modules in a same 2×2 SE are statistically independent; Each packet has an equal probability to win the contention; A blocked packet is resubmitted to the original destination.

The number of possible states of a buffer module of size m is 2m such that *State*-0, *State*-n_k(1≤k≤m-1), and *State*-b_k(1≤k≤m). The state transition diagram of this model is shown in figure 1. Note that *State*-n_m does not exist. This is because if a buffer module is full, then it cannot receive a packet. If the packet at the head left the module, it would be in *State*-$n_{(m-1)}$. Otherwise, in *State*-b_m. Note also that the state of a buffer module depends only on the oldest packet (the packet at the head of the buffer module) whether it has participated in a contention or not. The buffers are assumed to be serviced on a first-come-first-served (FCFS) manner.

- *State*-0 : a buffer module is empty; *State*-n_k : a buffer module has *k* packets and the head packet moved into the current position in the previous network cycle; *State*-b_k : a buffer module has *k* packets and the head packet could not move forward due to the lack of the empty space of its destined buffer module in the previous network cycle.

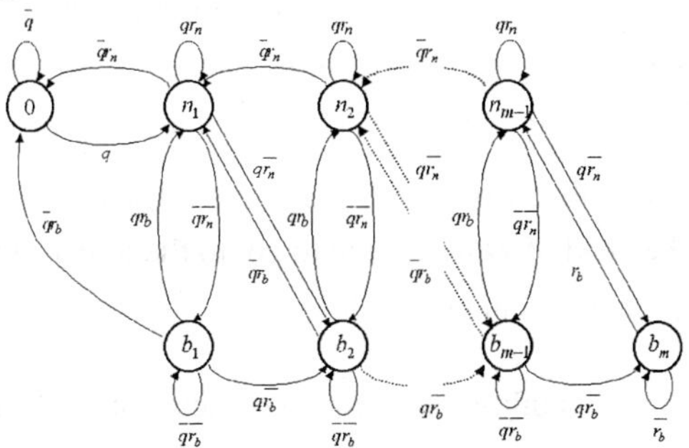

Fig. 1. The state transition diagram of the proposed model

The following variables are defined to develop our analytical model. Here *Q(ij)* denotes a *j*-th buffer module in stage *i* and two ports in the same SE are called as a conjugate port each other. We denote $Q(ij^c)$ as the conjugate buffer module of $Q(ij)$. Also t_b represents the time instance when a network cycle begins, while t_d represents the duration of a network cycle.

- *m* : number of buffers in a buffer module; *n* : number of switching stages. There are $n = \log_2 N$ stages for $N \times N$ MINs.

- $P_0(ij,t)$: probability that $Q(ij)$ is empty at t_b ; $\overline{P(ij,t)}$: probability that $Q(ij)$ is not full at t_b ; $P_{n_k}(ij,t)$ / $P_{n_k}(ij^c,t)$: probability that $Q(ij)$ / $Q(ij^c)$ is in *State* - n_k at t_b, where 1≤*k*≤*m*-1; $P_{b_k}(ij,t)$ / $P_{b_k}(ij^c,t)$: probability that $Q(ij)$ / $Q(ij^c)$ is in *State* - b_k at t_b, where 1≤*k*≤*m*; $P_b^u(ij,t)$ / $P_b^l(ij,t)$: probability that a head packet in $Q(ij)$ is a blocked one and it is destined to the upper / lower output port at t_b ; $SP_n(ij,t)$: $\sum_{k=1}^{m-1} P_{n_k}(ij,t)$, and $SP_b(ij,t)$: $\sum_{k=1}^{m} P_{b_k}(ij,t)$.

- $q(ij,t)$: probability that a packet is ready to come to the buffer module $Q(ij)$ during t_d .
- $r(ij)$ / $r_x(ij,t)$: probability that a normal/blocked head packet in $Q(ij)$ is destined to the upper output port; $r_n(ij,t)/r_b(ij,t)$: probability that a normal/blocked packet at the head of $Q(ij)$ is able to move forward during t_d ;
- $r_n^u(ij,t)$ / $r_n^l(ij,t)$: probability that a normal packet at the head of $Q(ij)$ can get to the upper / lower output port during t_d ; $r_{nn}^u(ij,t)$ / $r_{nn}^l(ij,t)$: probability that a normal packet at the head of $Q(ij)$ can get to the upper / lower output port during t_d by considering $Q(ij^c)$ in either $State\!-\!n_k$ or $State\!-\!b_k$. If $Q(ij^c)$ is in $State\!-\!b_k$, it is assumed that the blocked packet is destined to the lower / upper port (so no contention is necessary); $r_{nb}^u(ij,t)$ / $r_{nb}^l(ij,t)$: probability that a normal packet at the head of $Q(ij)$ is able to get to the upper / lower output port during t_d by winning the contention with a blocked packet at the head of $Q(ij^c)$;
- $r_b^u(ij,t)$ / $r_b^l(ij,t)$: probability that a blocked packet at the head of $Q(ij)$ can get to the upper / lower output port during t_d ; $r_{bn}^u(ij,t)$ / $r_{bn}^l(ij,t)$: probability that a blocked packet at the head of $Q(ij)$ is able to move forward to the upper / lower output port during t_d . Here it is assumed that $Q(ij^c)$ is empty or in the $State\!-\!n_k$; $r_{bb}^u(ij,t)$ / $r_{bb}^l(ij,t)$: probability that a blocked packet at the head of $Q(ij)$ is able to move forward to the upper / lower output port during t_d . Here it is assumed that $Q(ij^c)$ also has a blocked packet.
- $P^{na}(ij,t)$ / $P^{ba}(ij,t)$ / $P^{bba}(ij,t)$: probability that a buffer space in $Q(ij)$ is available (ready to accept packets) during t_d , given that no blocked packet / only one blocked packet / two blocked packets in the previous stage is destined to that buffer module.
- $X_n^u(ij,t)$ / $X_n^l(ij,t)$: probability that a normal packet destined to the upper / lower output port is blocked during t_d ; $X_b^u(ij,t)$ / $X_b^l(ij,t)$: probability that a blocked packet destined to the upper / lower output port is blocked during t_d .
- $T(ij,t)$: probability that an input port of $Q(ij)$ receives a packet.

2.1 Calculations for the Required Measures

[Obtaining $r_n(ij,t)$] A normal packet in an SE is always able to get to its desired output port when the other buffer is empty or it is destined to a different port from that of the normal packet in the other buffer. When two normal packets compete, each packet has the equal probability to win the contention. The probability that a normal packet in $Q(ij)$ does not compete with a blocked packet in the other buffer is $r(ij)\{1-r_x(ij^c,t)\}$ or $\{1-r(ij)\}r_x(ij^c,t)$. Therefore, probabilities $r_{nn}^u(ij,t)$ is as follows and $r_{nn}^l(ij,t)$ is obtained similarly.

$$r_{nn}^u(ij,t) = r(ij)P_0(ij^c,t) + [0.5r(ij)r(ij^c) + r(ij)\{1-r(ij^c)\}]SP_n(ij^c,t) \\ + r(ij)\{1-r_x(ij^c,t)\}SP_b(ij^c,t) \quad (1)$$

$r_{nb}^u(ij,t)$ and $r_{nb}^l(ij,t)$ are probabilities that a packet has the same destination as the blocked packet in the other buffer and it wins the contention. Thus they are as follows:

$$r_{nb}^u(ij,t) = 0.5r(ij)r_x(ij^c,t)SP_b(ij^c,t) \quad (2)$$

The probability that a buffer module is not full ($\overline{P(ij,t)}$) is simply

$$\overline{P(ij,t)} = 1 - P_{b_m}(ij,t)\,. \quad (3)$$

For $P^{ba}(ij,t)$, the originating buffer module of a packet is in $State$- b_k. If $Q(ij)$ received a packet in the previous network cycle with probability of $T(ij,t-1)$, it should not have been full. To make a space to accept a packet in network cycle t, a packet must have been moved if it had only one available space. If it did not receive a packet in the previous network cycle with the probability of $1-T(ij,t-1)$, it must have been full. To make a space to accept a packet in network cycle t, a packet must have been moved from the full buffer. Thus

$$P^{ba}(ij,t) = T(ij,t-1)\times A + \{1-T(ij,t-1)\}\frac{P_{b_m}(ij,t-1)r_b(ij,t-1)}{P_{b_m}(ij,t-1)}\,. \quad (4)$$

where

$$A = \frac{P_0(ij,t-1) + \sum_{k=1}^{m-2} P_{n_k}(ij,t-1) + \sum_{k=1}^{m-2} P_{b_k}(ij,t-1) + P_{n_{m-1}}(ij,t-1)r_n(ij,t-1) + P_{b_{m-1}}(ij,t-1)r_b(ij,t-1)}{P_0(ij,t-1) + \sum_{k=1}^{m-1} P_{n_k}(ij,t-1) + \sum_{k=1}^{m-1} P_{b_k}(ij,t-1)}$$

$P^{na}(ij,t)$ implies that the destined buffer can be in any state because no blocked packet is destined to it. If it received a packet in the previous network cycle with probability of $T(ij,t-1)$, it should have not been full. To make a space to accept a packet in network cycle t, a packet must have been moved if it had only one available space. If it did not receive a packet in the previous network cycle with probability of

1-*T*(*ij*,*t*-1), it could be in any state. To make a space to accept a packet in the network cycle *t*, a packet must have been if the buffer was full. $P^{na}(ij,t)$ is as follows.

$$P^{na}(ij,t) = T(ij,t-1) \times A + \{1-T(ij,t-1)\} \times B . \tag{5}$$

where $B = \dfrac{P_0(ij,t-1) + \sum_{k=1}^{m-1} P_{n_k}(ij,t-1) + \sum_{k=1}^{m-1} P_{b_k}(ij,t-1) + P_{b_{m-1}}(ij,t-1) r_b(ij,t-1)}{P_0(ij,t-1) + \sum_{k=1}^{m-1} P_{n_k}(ij,t-1) + \sum_{k=1}^{m} P_{b_k}(ij,t-1)}$.

For a packet to move to the succeeding stage, it should be able to get to the desired output port as well as the destined buffer module should be available. Thus $r_n^u(ij,t)$ is

as follows and $r_n^l(ij,t)$ is obtained similarly.

$$r_n^u(ij,t) = r_{nn}^u(ij,t) P^{na}((i+1)e,t) + r_{nb}^u(ij,t) P^{ba}((i+1)e,t) , \tag{6}$$

So $r_n(ij,t)$ is

$$r_n(ij,t) = r_n^u(ij,t) + r_n^l(ij,t) . \tag{7}$$

We can calculate $r_b(ij,t)$ using the similar method and we skip the detail due to the space constraint.

[Calculating $X_n^u(ij,t)$, $X_b^u(ij,t)$, $X_n^l(ij,t)$, $X_b^l(ij,t)$, $r_x(ij,t)$, $T(ij,t)$ **and** $q(ij,t)$ **]**

$X_n^u(ij,t)$ is the probability that a normal packet destined to the upper output port is blocked. Therefore $X_n^u(ij,t)$ is

$$X_n^u(ij,t) = r_{nn}^u(ij,t)\{1-P^{na}((i+1)e,t)\} + r_{nb}^u(ij,t)\{1-P^{ba}((i+1)e,t)\} + 0.5r(ij)r(ij^c)SP_n(ij^c,t) + 0.5r(ij)r_x(ij^c,t)SP_b(ij^c,t) . \tag{8}$$

The first two terms are probabilities for the destination unavailable. The last two terms are for the contention losing. $X_b^u(ij,t)$ is the probability that a blocked packet destined to the upper output port is blocked again. We can calculate this probability easily by the approach employed in $X_n^u(ij,t)$.

$$X_b^u(ij,t) = r_{bn}^u(ij,t)\{1-P^{ba}((i+1)e,t)\} + r_{bb}^u(ij,t)\{1-P^{bba}((i+1)e,t)\} + 0.5r_x(ij)r(ij^c)SP_n(ij^c,t) + 0.5r_x(ij)r_x(ij^c,t)SP_b(ij^c,t) . \tag{9}$$

$X_n^l(ij,t)$ and $X_b^l(ij,t)$ are obtained similarly.

Also $r_x(ij,t)$, which is the probability that a blocked head packet is destined to the upper output port is calculated as follows.

$$r_x(ij,t) = \frac{P_b^u(ij,t-1)}{P_b^u(ij,t-1) + P_b^l(ij,t-1)} \quad (P_b^u(ij,t-1) + P_b^l(ij,t-1) \neq 0) \tag{10}$$

Here $P_b^u(ij,t)$ and $P_b^l(ij,t)$ are calculated as follows.

$$P_b^u(ij,t) = X_n^u(ij,t)SP_n(ij,t) + X_b^u(ij,t)SP_b(ij,t)\ , \tag{11}$$

$$P_b^l(ij,t) = X_n^l(ij,t)SP_n(ij,t) + X_b^l(ij,t)SP_b(ij,t)\ . \tag{12}$$

If the buffer module $Q(ij)$ for *Stage-i* ($2 \le i \le n$) is connected into the upper output ports of the previous stage ($Q((i-1)g)$ and $Q((i-1)g^c)$), the throughput of $Q(ij)$ is as follows.

$$\begin{aligned} T(ij,t) = {} & SP_n((i-1)g,t)r_n^u((i-1)g,t) + SP_n((i-1)g^c,t)r_n^u((i-1)g^c,t) \\ & + SP_b((i-1)g,t)r_b^u((i-1)g,t) + P_b((i-1)g^c,t)r_b^u((i-1)g^c,t) \end{aligned} \tag{13}$$

The buffer modules which are connected to lower output ports of the previous stage are obtained similarly. $T(ij,t)$ ($1 \le i \le n$) also has the following relation with $T(ij,t)$.

$$T(ij,t) = q(ij,t)[\overline{P(ij,t)} + P_{n_m}(ij,t)r_n(ij,t) + P_{b_m}(ij,t)r_b(ij,t)] \tag{14}$$

Finally, $q(ij,t)$ ($2 \le i \le n$) is obtained.

$$q(ij,t) = \frac{T(ij,t)}{\overline{P(ij,t)} + P_{n_m}(ij,t)r_n(ij,t) + P_{b_m}(ij,t)r_b(ij,t)} \tag{15}$$

We need to separately consider the first and last stages because they have different conditions from other stages inside.

2.2 Calculating *r(ij)*, Throughput, and Delay

r(ij) is calculated by using the transformation method studied in [8]. It is a mapping scheme that transforms the given referencing pattern into a set of *r(ij)*'s which reflects the steady state traffic flow in the network. For a steady state referencing pattern, we represent it in terms of destination accessing probabilities A_j, the probability that a new packet generated by an inlet chooses the output port *j* as its destination. Then *r(ij)* can be represented as the conditional probability that the sum of A_j's which are connected by the upper output port of $Q(ij)$ under the given condition that the sum of A_j's of all possible destined output ports which are connected by the upper or lower output ports of $Q(ij)$. For example, *r(ij)*'s in three-stage MIN are as follows.

For the last stage: $r(31) = r(32) = \dfrac{A_1}{A_1 + A_2}$, $r(33) = r(34) = \dfrac{A_3}{A_3 + A_4}$,

For the second stage: $r(21) = r(22) = r(25) = r(26) = \dfrac{A_1 + A_2}{A_1 + A_2 + A_3 + A_4}$,

For the first stage, all $r(1j)$ ($1 \le j \le N$) are same:

$$r(11) = r(12) = \ldots = r(18) = \frac{A_1 + A_2 + A_3 + A_4}{A_1 + A_2 + A_3 + A_4 + A_5 + A_6 + A_7 + A_8} = A_1 + A_2 + A_3 + A_4$$

Normalized throughput of a MIN is defined to be the throughput of an output port of the last stage. If a port j is the upper output port of an SE, the normalized throughput in this port is as follows. If a port j is the lower output port of an SE, then it obtained similarly.

$$TNET(j,t) = SP_n(ng,t)r_n^u(ng,t) + SP_n(ng^c,t)r_n^u(ng^c,t) + SP_b(ng,t)r_b^u(ng,t) + SP_b(ng^c,t)r_b^u(ng^c,t) \quad (16)$$

The delay occurred for a packet at the buffer module $Q(ij)$ in the steady state is calculated by using Little's formula.

$$D(ij) = \lim_{t\to\infty} \frac{\sum_{k=1}^{m-1} k\{P_{n_k}(ij,t) + P_{b_k}(ij,t)\} + mP_{b_m}(ij,t)}{T(ij,t)} \quad (17)$$

As each port delay is different, the weight of it should be considered for the mean delay. Hence the mean delay is

$$D = \sum_{j=1}^{N} w_j D(j) \quad (18)$$

Here w_j - the weight of port j for the mean delay - is obtained by the rate of the normalized throughput of that port as follows.

$$w_j = \lim_{t\to\infty} \frac{TNET(j)}{\sum_{k=1}^{N} TNET(k,t)} \quad (19)$$

3 Performance Verification of the Proposed Model

Correctness of our model in terms of the network throughput and delay is verified by comparing them with the data obtained from computer simulation for various buffer sizes and traffic conditions. For the simulation, 95% confidence interval is used and following approaches are employed for the computer simulation.

- Each inlet generates requests at the rate of the offered input traffic load; The destination of each packet follows the given nonuniform traffic pattern – hot spot traffic pattern. Here each inlet makes a fraction h of their requests to a hot spot port, while the remaining (1-h) of their requests are distributed uniformly over all output ports including the hot spot port. Assume that usually the output port 1 is the hot spot port. If there is a contention between the packets in an SE, it is resolved randomly; The buffer operation is based on the FCFS principle.

Figure 2 shows the mean throughput and delay comparison of 6-stage single buffered MIN with the 7% hot spot traffic pattern using computer simulation. It plots the throughput and delay data for MIN's containing buffer modules with capacity to hold one cell (single buffered) of six stages (64×64) respectively. The offered traffic load varies form 0.1 to 1, and data are collected from the computer simulation. Here each simulation data is obtained by averaging 10 runs. In each run, 1,000,000 iterations are taken to collect reliable data, and the variations in the last 100,000 iterations are less than 0.1%. Figure 3 shows that the throughput comparison of the hot spot port and other ports in this case. The throughput of the hot spot port is much more than other ports since heavy traffics are concentrated on the hot spot port. Figures 4 and 5 show that comparison results of the more buffered MIN. In case of the uniform traffic pattern, more buffers can increase the performance of MINs (15~20%) in terms of throughput. In our model, the throughput of the multi-buffered MIN with the 3.5% hot spot traffic pattern is more increased (50%). But that with the 7% hot spot traffic pattern is not increased since severe blocking caused by the concentration of the traffic decreases the performance.

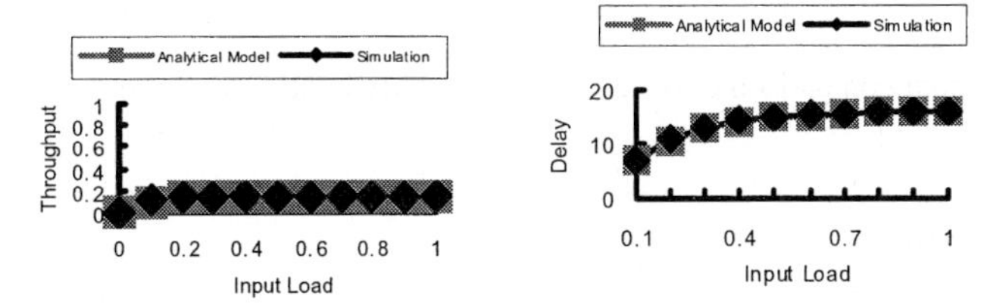

Fig. 2. Throughput and mean delay comparison of 6-stage single-buffered MIN with 7% the hot spot traffic pattern

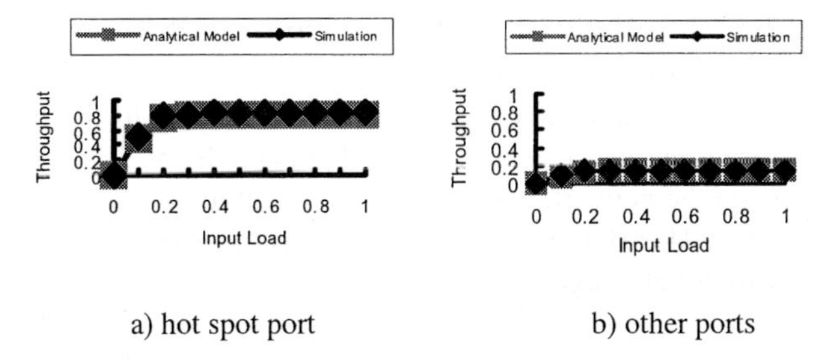

a) hot spot port b) other ports

Fig. 3. Throughput comparison of 6-stage single-buffered MIN with 7% the hot spot traffic pattern

These figures show that our models are exactly accurate in terms of the throughput under the practical size MIN. Delay is also very accurate in single buffered MIN but the accuracy of it is somewhat decreased in case of the multi-buffered MIN. For more sized MIN, the throughput and delay have the stable value from the time that the input load is very low (0.1) since heavy traffics are concentrated form a lot of (1024) input ports. The practical performance is very low (about 0.03) which means that switching structures are not useful when their size is very large with nonuniform traffics patterns. This phenomenon is observed similarly when we compare the multi-buffered cases.

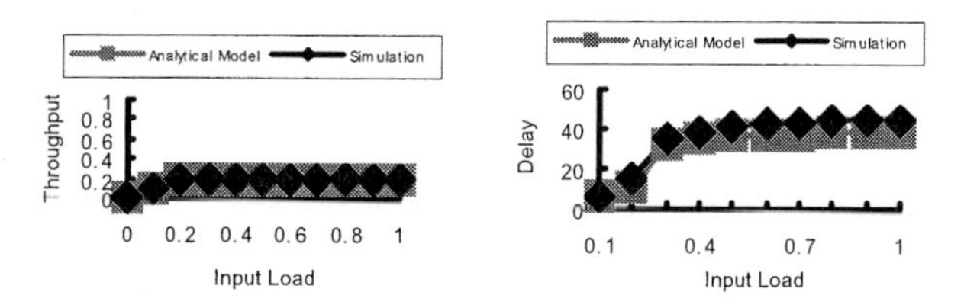

Fig. 4. Throughput and mean delay comparison of 6-stage 4-buffered MIN with 7% the hot spot traffic pattern

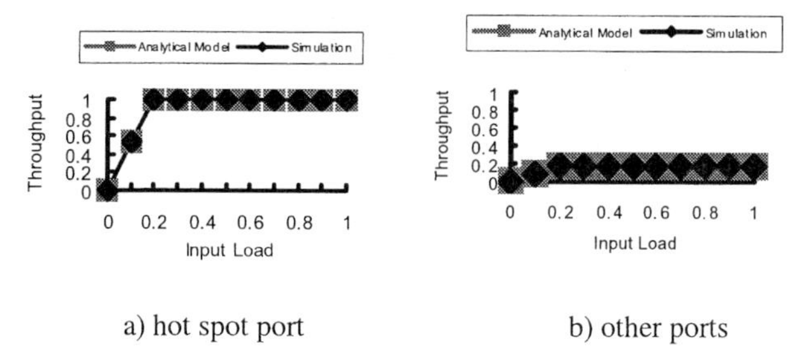

a) hot spot port b) other ports

Fig. 5. Throughput comparison of 6-stage 4-buffered MIN with 7% the hot spot traffic pattern

4 Conclusion

This paper has proposed an analytical modeling method for the performance evaluation of MINs with 2×2 switching elements using small clock cycles (SCC) under nonuniform traffic. The effectiveness of the proposed model was verified by computer simulation for various practical MINs; 6×6 and 10×10 switches, single and 4 buffered MIN with 3.5% and 7% hot spot traffic. According to the results, the proposed model is very accurate in terms of throughput and delay. The detrimental effect of hot spot traffic on the network performance turns out to get more significant as the switch size increases. Therefore hot spot traffic needs to be avoided as much as possible for especially relatively large size switches.

References

1. C. L. Wu and T. Y. Feng, " On a class of Multistage Interconnection Networks," IEEE Trans. on Computers, Vol. C-29, pp. 694–702, August 1980.
2. J.S. Turner, "Design of and integrated services packet networks," Ninth Data Commun. Symp., in ACM SigComm Comput. Commun. Rev., vol. 15, pp. 124–133, Sept. 1985.
3. Muh-rong Yang and GnoKou Ma, "BATMAN: A New Architectural Design of a Very Large Next Generation Gigabit Switch," IEEE International Conference on Communications, Vol.2/3, pp. 740–744, May 1997.
4. H. Rudin, "The ATM-Asynchronous Transfer Mode," Computer Networks and ISDN Systems, Vol. 24, pp. 277–278, 1992.

5. Y.C. Jenq, “Performance analysis of a packet switch based on single buffered Banyan network,” IEEE J. Select. Areas Commun, vol. SAC-3, pp. 1014–1021, Dec. 1983.
6. H. Kim and A. Leon-Garcia, “Performance of Buffered Banyan Networks Under Nonuniform Traffic Patterns,” IEEE Transaction on Communications, Vol. 38, No. 5, May 1990.
7. T. Lin and L. Kleinrock, “Performance Analysis of Finite-Buffered Multistage Interconnection Networks with a General Traffic Pattern”, ACM SIGMETRICS Conference on Measurement and Modeling of Computer Systems, San Diego, CA, pp. 68–78, May 21–24, 1991.
8. H.Y. Youn and H. Choo, “Performance Enhancement of Multistage Interconnection Networks with Unit Step Buffering,” IEEE Trans. on Commun. Vol. 47, No. 4, April 1999.

Symmetric Tree Replication Protocol for Efficient Distributed Storage System*

Sung Chune Choi[1], Hee Yong Youn[1], and Joong Sup Choi[2]

[1] School of Information and Communications Engineering
Sungkyunkwan University
440-746, Suwon, Korea +82-31-290-7952
{choisc,youn}@ece.skku.ac.kr

[2] Information Security Technology Division
Korea Information Security Agency
138-803, Seoul, Korea +82-2-405-5263
jschoi@kisa.or.kr

Abstract. In large distributed systems, replication of data and service is needed to decrease the communication cost, increase the overall availability, avoid single server bottleneck, and increase the reliability. Tree quorum protocol is one of the replication protocols allowing very low read cost in the best case but has some drawbacks such that the number of replicas grows rapidly as the level increases and root replica becomes a bottleneck. In this paper we propose a new replication protocol called symmetric tree protocol which efficiently solves the problems. The proposed symmetric tree protocol also requires much smaller read cost than the previous protocols. We conduct cost and availability analysis of the protocols, and the proposed protocol displays comparable read availability to the tree protocol using much smaller number of nodes. It is thus effective to be applied to survival storage system.

Keywords: replication protocol, tree quorum, availability, distributed system, symmetric tree

1 Introduction

In large distributed systems node and communication failures are likely to occur which can prevent the operations from being successfully carried out. This raises the need for introducing a sufficient level of fault tolerance into the distributed systems [1]. Other problems are increased communication cost as the system grows and bottleneck of a single server as many processes compete for the same service. Replication of data and services is one of the practical solutions to the problems. It can avoid the server bottleneck problem and increase overall availability [2]. However, the communication cost increases as the number of replicas increases. In order to minimize the

*This work was supported by Korea Research Foundation Grant (KRF-2002-041-D00421).
Corresponding author: Hee Yong Youn

P.M.A. Sloot et al. (Eds.): ICCS 2003, LNCS 2659, pp. 474–484, 2003.

communication cost, thus, the number of replicas should be kept as small as possible [3,4].

We consider a distributed system in which data are fully replicated and two types of operations are allowed on the replicated data, namely read and write. A read operation returns some value while write operation installs a new value. Proper synchronization is achieved if read operations return the value installed by the last write operation, and read and write operations or two write operations are not executed concurrently. Each node uses a centralized concurrency control scheme to synchronize the accesses to its local copy and a replication control protocol to coordinate the accesses to various replicas. The basic principle employed in maintaining consistency of replicated data is to require conflicting operations to lock at least one common copy.

A distributed system consists of a set of distinct sites that communicate with each other by sending messages over a communication network. No assumptions are made regarding the speed, connectivity, or reliability of the network. We assume that the sites are fail-stop. A distributed database consists of a set of objects stored at several sites. Users interact with the database by invoking transaction programs. Transactions are partially ordered sequence of read and write operations that are executed atomically. Execution of a transaction must appear atomic, i.e. a transaction either commits or aborts. In a replicated database, copies of an object may be stored at several sites in the network. Multiple copies of an object must appear as a single logical object to the transactions. This is termed as one-copy equivalence and enforced by replication control protocol.

In the literature there exist several replication algorithms aiming at different goals. The Read-One/Write-All algorithm [5] has minimum read cost and maximum read availability but has maximum communication cost for write operation. The quorum consensus [6] and dynamic voting [7] have good read and write availability but has a disadvantage of high read cost. For all these replication control protocols, making the read operation cheaper ends up with a more expensive write operation and vice versa. All the strategies have a cost of $O(n)$, which means the operation cost linearly depends on the number of replicas in the system [8].

By assigning a logical structure to a set of replicas, it is possible to reduce the communication cost. Multi-Level-Voting-Protocol [9], Weighted Voting [10], Tree quorum protocol [11] and Grid protocol [12] are such type of replication protocols, but they still have relatively high operation cost especially when the number of replicas is large and some failures exist. The tree quorum protocol has read cost of 1 in the absence of failures and provides graceful degradation when failures exist. Hence in this paper we develop a new replication protocol which has low operation cost. We compare the replication protocols which use a logical tree structure and the proposed symmetric tree structure. The proposed symmetric tree protocol allows comparable read availability to the tree protocol with much smaller number of nodes. It is thus effective to be applied to survival storage system.

The rest of the paper is organized as follows. In Sect. 2 we describe the earlier protocols, and in Sect. 3 we propose a new protocol. We compare the new symmetric tree protocol with earlier protocols in Sect. 4. Section 5 concludes the paper.

2 Review of Earlier Protocols

2.1 Logarithmic Replication Protocol

The class of replication protocols discussed here is a generalization of the Tree Quorum Protocol. These replication protocols use a logical tree structure. All the algorithms have read cost of 1 in the absence of failures and provide graceful degradation when failures exist. The replication protocols which use tree structure have varying costs and availabilities according to fault condition, whereas other replication protocols have constant costs and availabilities.

Let n be the number of replicas which can be organized as a tree of height h. A non-leaf replica R_i has S_{Ri} descendants. For each replica except the leaves we define read quorum rq_{Ri}, and a write quorum wq_{Ri}. A tree consisting of 13 replicas organized in three levels is shown in Fig. 1.

Algorithm:
The algorithm is recursive:
• A tree of height 0 consists only of a leaf replica and can be locked by simply locking the replica for either read or write operation.
Read operation:
• Locking a tree of height h for a read operation means to lock the root replica or rq_{ROOT} of its subtrees with the rq_{ROOT} descendants of the root serving as new root replicas of the subtrees of height h-1.
Write operation:
• Locking a tree of height h for a write operation means to lock the root replica and wq_{ROOT} of its subtrees. Here, the wq_{ROOT} descendants of the root serve as new root replicas for the subtrees of height h-1.

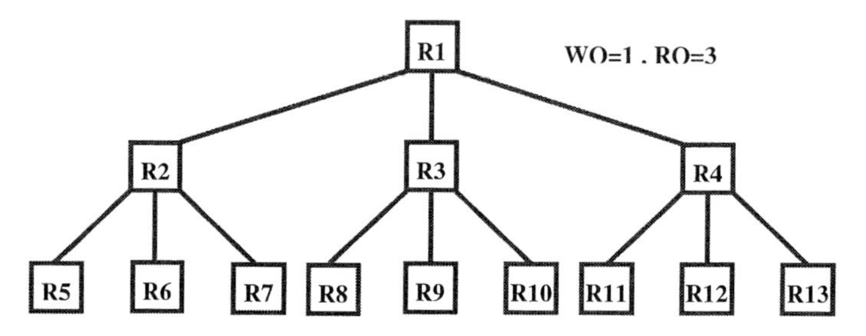

Fig. 1. A tree of height 3 with 13 replicas

Examples of valid read quorum sets (RQ) from Figure 1 are {R1}, {R2, R3, R4} when node-1 is not available, and {R5, R6, R7, R3, R11, R12, R13} when node-1, node-2, and node-4 are not available. Examples of valid write quorum sets (WQ) are {R1, R2, R5}, {R1, R3, R10}, and {R1, R4, R12}.

In order to maintain consistency of the replicated objects, the selected read and write quorums have to fulfill some requirements. To detect read/write conflict, any valid read quorum set must intersect with valid write quorum set. This requirement can be satisfied by $rq_{Ri} + wq_{Ri} > V_{Ri}$ where V_{Ri} is the number of descendants. To detect write/write conflicts, any two valid write quorum sets must intersect. This is always satisfied because every valid write quorum set includes the root. So there exist many cases of valid RQ and WQ in the tree quorum protocol. After reviewing all possible cases regarding the provided cost and availability, the strategy with RQ = S and WQ = 1 is selected as the best choice. The case is called *Logarithmic Protocol.*

2.2 Grid Protocol

Here the replicas are logically arranged in a grid topology, and read and write operations are required to lock the rows and columns of nodes such that conflicting operations request a common node. In this protocol read quorum consists of a node from every column and write quorum consists of a node from every column plus a full column of nodes. A grid network which consists of 4 rows and 4 columns of 16 replicas is shown in Fig. 2.

Algorithm:
Read operation:
• Select and lock any single node from each column. This set of nodes is called Column-cover (C-cover).
Write operation:
• Choose a C-cover and lock any whole column.

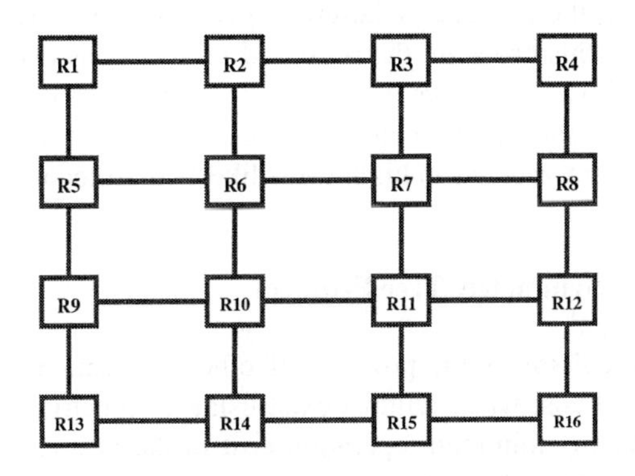

Fig. 2. A Grid network with 16 replicas

Examples of valid read quorum sets here are {R1, R6, R3, R12} and {R1, R6, R7, R8}, and examples of valid write quorum sets are {R1, R10, R3, R4, R2, R6, R14} and {R9, R6, R3, R8, R7, R11, R15}.

Here two write operations cannot be executed concurrently because each one locks a C-cover and all nodes in a column. If two operations have locked a C-cover, neither one can obtain write-locks from all the nodes in any column. Similar argument can be used to show that concurrent execution of read and write operations are not possible. The grid protocol has constant operation cost because the size of quorum set is always same. One disadvantage of this protocol is the operation cost is equally high regardless of failure condition. However, it has higher availability than tree quorum protocol. To solve the problems of the tree quorum and grid protocol, we next present the proposed symmetric tree protocol.

3 The Proposed Protocol

As mentioned in the previous section, tree quorum protocol and grid protocol have some drawbacks, respectively. In this section we propose a new replication protocol called symmetric tree protocol to solve the drawbacks of the previous protocols.

3.1 Motivation

In typical replication protocols almost all replicas need to be accessed for read and write operation. For example, in Quorum consensus protocol of 16 replicas, read quorum + write quorum need to be greater than 16. However, if we assign a logical structure to the replicas, then we can reduce the number of replicas because we can organize the structure in such a way that the tree and grid topology are used together.

As previously described, a major weakness of tree quorum protocol is that the number of replicas rapidly grows as the tree level grows. The tree quorum protocol has read cost of 1 in the absence of failures. As a result, root replica can become a bottleneck. To solve this problem, the proposed protocol has two root replicas using symmetric tree topology. Therefore, the proposed protocol can have the advantage of having minimum operation cost whether failures exist or not. Also it has the advantage of having not many replicas in high levels like grid protocol.

3.2 The Proposed Symmetric Tree Protocol

The proposed protocol solves the problem of other replication protocols of significantly increased read cost when failures exist using symmetric tree topology. It has the advantage of having minimum operation cost in the best case like tree structure protocol and also has the advantage of having not so many replicas in higher levels like grid protocol. In this protocol replicas are organized as shown in Figure 3, where two tree structures are joined whose leaf nodes are common. We call the two trees up-tree and down-tree, respectively.

Algorithm:

Read Operation:

• Read operation starts from either the root node of up-tree or that of down-tree. If the root replica is not available, all the descendants of it are read. Each the descendant serves as a new root replica for the subtree to the $h/2$-level of the symmetric tree, where h is the number of levels of the tree. After the $h/2$-level, the next level is read if any node of the parent nodes is not available.

Write Operation:

• Write operation starts from any root replica of the two trees. Only one replica of the descendants of the root must writes. A descendant of the root serves as a new root replica of the subtree. The descendant of the parents must write from the $h/2$-level to the last level.

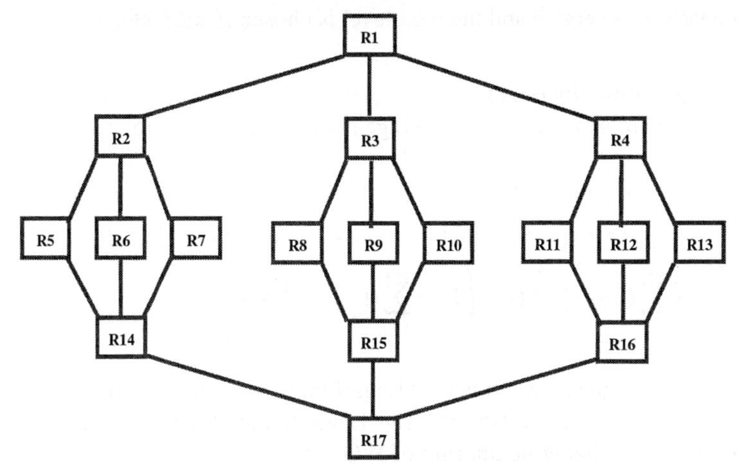

Fig. 3. A network for the proposed symmetric tree protocol with 17 replicas

From Fig. 3 examples of valid read quorum set are {R1}, {R2, R3, R4}, {R3, R4, R5, R6, R7}, {R3, R4, R14}, {R1, R8, R9, R10, R16}, {R14, R15, R16}, and {R17}. Examples of valid write quorum sets are {R1, R2, R5, R14, R17}, {R1, R3, R9, R15, R17}, and {R1, R3, R10, R15, R17}.

Consistency is maintained in selecting read, write quorum set to satisfy the requirement of detecting read/write conflict and write/write conflict. The Read/Write conflict can be detected because read operation should lock the whole descendants and write operation should lock at least one replica of whole descendants. The Write/Write conflict can also be detected because every write operation has to lock the root replica of the symmetric tree structure. Thus, if one replica is locked, then others cannot get the lock.

4 Performance Analysis

In this section cost and availability of the proposed protocol and tree protocol are analyzed.

4.1 Cost Analysis

For cost analysis, read cost and write cost are considered, which are computed by the number of sites involved in the operation. A read operation in both the protocols need to consult only the root replica which is very efficient if root replica is available. However, the root replica becomes a bottleneck if all operations are done on it. Therefore the algorithm is slightly modified as the level of the tree the operation will be performed is randomly selected. For this, uniformly distributed random variable X in the interval [0,1] and a parameter f in interval [0,1] are chosen. A random value x of X is generated and the top-level is chosen for performing read operation if $x \leq f$. Otherwise, a new x is generated and the next level is chosen if $x \leq f$, etc.

4.1.1 Logarithmic Protocol

Suppose there are h+1 levels. The average read cost computes to

$$\begin{aligned} C_{read} &= f + (1-f)\cdot f \cdot RQ + (1-f)^2 \cdot f \cdot RQ^2 + \cdots + (1-f)^{h-1} \cdot f \cdot RQ^{h-1} \\ &\quad + (1-(f + (1-f)\cdot f + (1-f)^2 \cdot f + \cdots + (1-f)^{h-1} \cdot f)) \cdot RQ^h \\ &= f\sum_{k=0}^{h-1}(1-f)^k \cdot RQ^k + \left(1 - f\sum_{k=0}^{h-1}(1-f)^k\right) \cdot RQ^h \end{aligned} \tag{1}$$

In the tree protocol, we consider Logarithmic protocol which is the best choice regarding the cost and availability. The Logarithmic Protocol is defined by a read quorum of RQ = S and write quorum of WQ = 1.

$$\min(C_{read}) = 1 \tag{2}$$

Average read cost is

$$C_{read} = f\frac{((1-f)S)^h - 1}{(1-f)S - 1} + ((1-f)S)^h \tag{3}$$

For write operation,

$$C_{write} = h + 1 \tag{4}$$

4.1.2 Symmetric Tree Protocol

For cost analysis, read cost and write cost are considered, which are computed by the number of sites involved in that operation. A read and write operation in both the protocols only need to consult the root replica which is very efficient if root replica is available.

Average read cost is computed by

$$C_{read} = f + (1-f)\cdot f\cdot S + (1-f)^2\cdot f\cdot S^2 + \cdots + (1-f)^{\frac{h}{2}}\cdot f\cdot S^{\frac{h}{2}}$$
$$+ (1-f)^{\frac{h}{2}+1}\cdot f\cdot S^{\frac{h}{2}-1} + \cdots + (1-f)^h\cdot f\cdot S$$
$$+ (1-(f + (1-f)\cdot f + \cdots + (1-f)^h\cdot f)) \quad (5)$$
$$= f\left(\left(\sum_{k=0}^{\frac{h}{2}}(1-f)^k\cdot S^k\right) + \left(\sum_{k=1}^{\frac{h}{2}-1}(1-f)^{\frac{h}{2}+k}\cdot S^{\frac{h}{2}-k}\right)\right) + \left(1 - f\sum_{k=0}^{h}(1-f)^k\right)$$

$$\min(C_{read}) = 1 \quad (6)$$

For write operation,

$$C_{write} = h+1 \quad (7)$$

Comparison of average read cost for both the protocols are shown in Fig. 4 for 121 nodes of tree quorum protocol and 161 nodes of symmetric tree protocol. Average write cost for them is depend on height of tree.

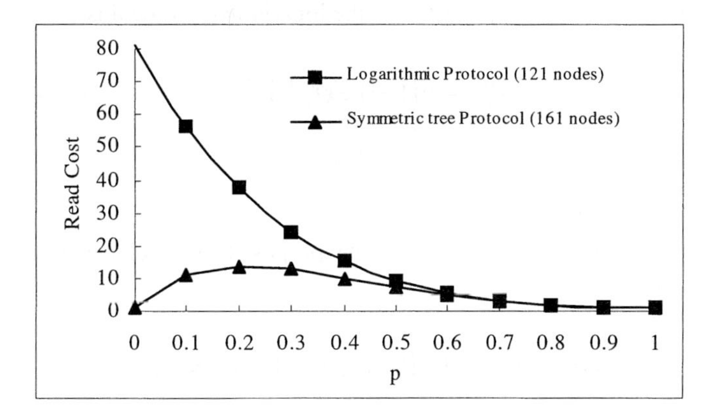

Fig. 4. Comparison of average read cost

The tree quorum protocol consists of 121 nodes of 5 levels and the last level has the maximal number of node. On the other hand, the symmetric tree protocol consists of 161 nodes of 9 levels and the fourth level has the maximal number of node. Also, the maximal number of node is the same. Observe from Figure 4 that read cost of the proposed protocol is always much smaller than logarithmic protocol in sprite of has more nodes. The difference between the two protocols gets more substantial when the failure of nodes is low.

4.2 Availability Analysis

4.2.1 Logarithmic Protocol

The availability of read operation for the tree quorum protocol performed on a tree of height $l > 0$ is

$$\wp_{read}^{(l)} = p + (1-p)\sum_{k=RQ}^{S}\binom{S}{k}\left(\wp_{read}^{(l-1)}\right)^{k}\left(1-\wp_{read}^{(l-1)}\right)^{S-k} \text{ with } \wp_{read}^{(0)} = p \quad (8)$$

So the availability of read operation for the logarithmic protocol is

$$\wp_{read}^{(l)} = p + (1-p)\left(\wp_{read}^{(l-1)}\right)^{S} \text{ since RQ} = \text{S} \quad (9)$$

The availability of write operation is

$$\wp_{write}^{(l)} = p\sum_{k=WQ}^{S}\binom{S}{k}\left(\wp_{write}^{(l-1)}\right)^{k}\left(1-\wp_{write}^{(l-1)}\right)^{S-k} \text{ with } \wp_{write}^{(0)} = p \quad (10)$$

So the availability of write operation for the logarithmic protocol is

$$\wp_{write}^{(l)} = p\left(1-\left(1-\wp_{write}^{(l-1)}\right)^{S}\right) \quad (11)$$

4.2.2 Symmetric Tree Protocol

The availability of read operation for the proposed protocol performed on height $h/2 \geq l > 0$ is

$$\wp_{read}^{(l)} = p + (1-p)\left(\left(\wp_{read}^{(l-1)}\right)^{S} + \left(1-\left(\wp_{read}^{(l-1)}\right)^{S}\right)p\right) \text{ with } \wp_{read}^{(0)} = p \quad (12)$$

The availability of write operation is

$$\wp_{write}^{(l)} = p^{2}\sum_{k=WQ}^{S}\binom{S}{k}\left(\wp_{write}^{(l-1)}\right)^{k}\cdot\left(1-\wp_{write}^{(l-1)}\right)^{S-k} \text{ with } \wp_{write}^{(0)} = p \quad (13)$$

Comparisons of read availabilities of the two protocols are shown in Fig. 5. For the comparison, we examine 3 cases, each one with a different number of replicas. They are 17 of level 5, 53 of level 7, and 161 replicas of level 9. As illustrated in Fig. 5, the read availability of the proposed protocol increases as the level of tree increases. Note that the read availability of symmetric tree protocol is higher than logarithmic protocol with much smaller number of nodes.

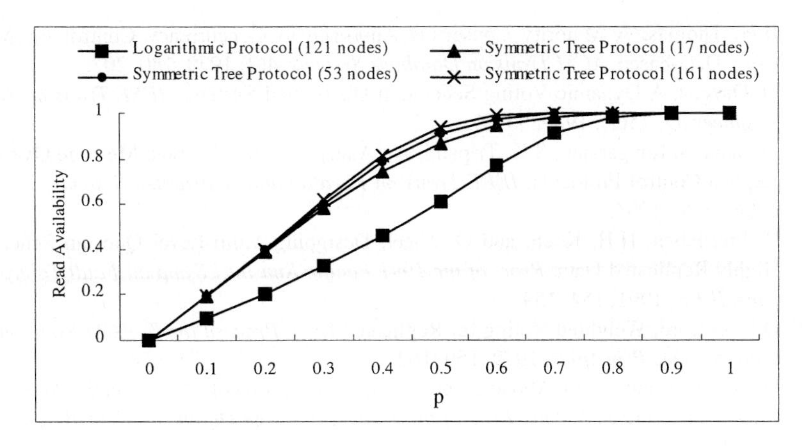

Fig. 5. Comparison of read availability

5 Conclusion

In this paper we have proposed a new symmetric tree replication protocol which is better than Logarithmic protocol in read cost and read availability using much smaller number of nodes. One of the main advantages of the proposed symmetric tree protocol is much smaller read cost than previous protocols even though the failure rate of the nodes increases. The choice of which replication protocol should be used depends on the primary performance criterion, and the new proposed symmetric tree protocol can be selected for low operation cost and high read availability.

As future work, we will analyze and compare the response time and throughput of the proposed protocol. Implementation of the protocols for survivable storage system will also be carried out.

Reference

[1] C. Amza,, A.L. Cox, W. Zwaenepoel, Data replication strategies for fault tolerance and availability on commodity clusters, *Proc. Int'l Conf on Dependable Systems and Networks (DSN)*, 2000, 459–467

[2] H.Y. Youn, B. Krishnamsetty, D. Lee, B. K. Lee, J.S. Choi, H.G. Kim, C.W. Park, and H.S. Lee, An Efficient Hybrid Replication Protocol for Highly Available Distributed System, *Proc. Int'l Conf on Communication and Computer Networks (CCN)*, Nov, 2002,

[3] K. Arai, K. Tanaka, M. Takizawa, Group protocol for quorum-based replication, *Proc. Seventh Int'l Conf on Parallel and Distributed Systems*, 2000, 57–64.

[4] G. Alonso, Partial Database Replication and Group Communication Primitives, *Proc. of the 2nd European Research Seminar on Advances in Distributed Systems (ERSADS'97)*, March 1997, 171–176.

[5] P.A. Bernstein and N. Goodman, An Algorithm for Concurrency Control and Recovery in Replicated Distributed Databases, *ACM Trans on Distributed Systems*, 9(4), 1984, 596–615.

[6] R.H. Thomas. A Majority Consensus Approach to Concurrency Control for Multiple Copy Data-based, *ACM Trans on Database Systems*, 4(2),1979, 180–207.
[7] D. Davcev, A Dynamic Voting Scheme in Distributed Systems. *IEEE Trans on Software Engineering*, 15(1), 1989, 93–97.
[8] D. Saha, S. Rangarajan, S.K. Tripathi, An Analysis of the Average Message Overhead in Replica Control Protocols, *IEEE Trans on Parallel and Distributed Systems*, 7(10), Oct. 1996, 1026–1034.
[9] B. Freisleben, H.H. Koch, and O. Theel, Designing Multi-Level Quorum Schemes for Highly Replicated Data. *Proc. of the 1991 Pacific Rim Int'l Symp on Fault Tolerant Systems, IEEE*, 1991, 154–159.
[10] D.K. Gifford, Weighted Voting for Replicated Data, *Proc. of the 7th ACM Symp on Operating Systems Principles*, 1979, 150–162.
[11] D. Agrawal and A. El Abbadi, The tree Quorum protocol: An Efficient Approach for Managing Replicated Data, *Proc of the 16th Very Large Databases (VLDB) Conf*, 1990, 243–254.
[12] S. Cheung, M. Ammar, and M. Ahamad, The Grid Protocol: A High Performance Scheme for Maintaining Replicated Data, *Proc of the 6th Int'l Conf on Data Engineering*, 1990, 438–445.

A Group Key Management Supporting Individual Batch Rekeying and Fault Tolerance for Secure Multicast*

Hojae Hyun, Keechon Kim, and Sunyoung Han

Konkuk University, 1 Hwayang-dong, Gwangjin-gu, Seoul, Korea
Phone: (02) 450-3537
{hjhyun,kckim,syhan}@kkucc.konkuk.ac.kr

Abstract. One of the major problem areas of secure multicast is group key management [2]. In this paper, we present a novel approach for scalable group key management for secure multicast. Unlike previous work, based on centralized structure, our group key management uses distributed group security manager (GSM). Different GSMs are used to manage each local group, reducing the problem of concentrating the overhead on a single group manager. Using individual batch rekeying, it has advantage of reducing the frequency of rekeying from the size and membership dynamics of the group. Also, GSM supports fault tolerance as well.

1 Introduction

Recently, such Internet applications as multimedia conferences, teleconferences, pay per view and computer-supported collaborative work are expected to be more important. These applications are based upon group communications over Internet.

Multicast communication [1] provides efficient delivery of data from one sender to multiple receivers. It reduces sender transmission overhead, network bandwidth and receiving delay. Also, it supplements drawback of unicast and broadcast. Therefore, multicast communication is an efficient solution to group communication and applied for many of applications. However, multicast communication does not support restricted communication among specified group of users. i.e., any user can receive all data of a multicast group as any host join or leave a multicast group by sending IGMP(Internet Group Management Protocol) [3] messages to their local router. Due to the openness of multicast, the potential for an attack may be greater than in unicast [8].

A typical solution to protect the privacy of group communication is known as encryption algorithm. Encryption algorithm takes messages and encrypts it with a key. If it is applied for group communication, senders encrypt messages using group key that was already distributed to members of the group. Only those who know the group key will be able to decrypt encrypted message. Group key management is an important issue of secure multicast communication [7].

* This research was supported by University IT Research Center Project.

P.M.A. Sloot et al. (Eds.): ICCS 2003, LNCS 2659, pp. 485–494, 2003.

It is required that the group key changes after a new member has joined (so that the new member will not be able to access past multicast data) or a member of group has left (so that the old member will not be able to continue to access multicast data) [9].

In this paper, we describe the design and architecture of a scalable group key management that uses periodic batch rekeying for secure multicast. Our key management is a solution targeted at MBone related secure applications that involve a large number of members and dynamic membership.

To support scalability without regarding group size, our group key management uses distributed Group Security Manager (GSM). Using periodic batch rekeying, it has advantage of reducing the frequency of rekeying from the size and membership dynamics of the group. To support scalable rekeying, we propose *individual batch rekeying*. It is a batch rekeying driven by time interval of each local group independently.

Fault tolerance of key management is the key manager's ability to maintain performance even if service is halted due to system failure or failure due to high traffic. If the member detects fault of GSM, he selects new GSM for fault tolerance..

The remainder of this paper is organized as follows. The next section describes the architecture of proposed system. Section 3 presents an operational overview of system. Section 4 describes performance analysis. Finally, in Sect. 5 we present the conclusion and future research.

2 System Architecture

In the rest of this paper, we use the notation described in Table 1.

Table 1. Notation

GSM : GSM manager mode
member : GSM member mode
MAA : Multicast Address Allocator
Km : individual key of member, Kr : random key
Klg : local group key , Kpg: parent group key
ACL : Access Control List
Klg-ver: Klg revision number
Kpg-ver: Kpg revision number
reg-no: registration number

The proposed group key management offers scalability using distributed hierarchical *GSM* based on *Iolus*[12] that manage dividing big group to several small group as shown in (Fig. 1) so that the changes in a local group does not influence in whole. *MAA* [4][5] allocates (group address, local group address) to group initiator. Group initiator becomes the highest-level GSM. MAA allocates local group address to child GSM. Parent GSM manages local members. Child GSM knows three multicast group addresses (group address, parent group address, local group address).

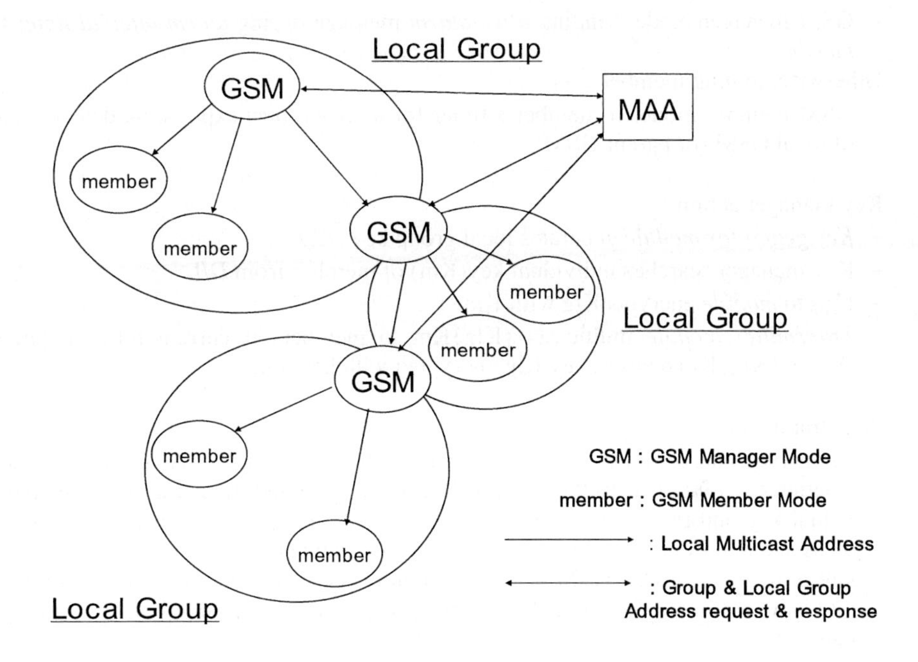

Fig. 1. GSM system architecture

2.1 GSM Component

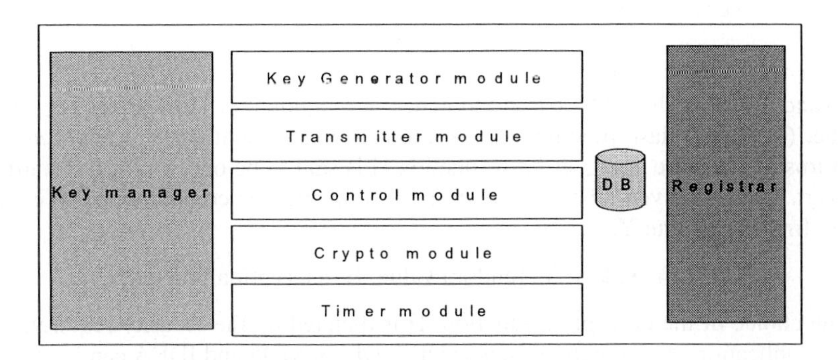

Fig. 2. GSM component

GSM is composed of *key manager*, *registrar*, *DB* and *5 function modules* (Fig. 2). One GSM in a local group collects member (or other GSM)'s requests during a *rekey interval* and processes the requests in a batch.
If a GSM does not exist in local group, joining member converts

- GSM manager mode: sending *alive alarm* message during *alarm interval timer* to *member.*

Otherwise, joining member acts

- GSM member mode: if member's timer for a *alive alarm* expires, he detects fault of local GSM (or parent GSM).

Key manager action

- *Key generator module* generates local group key (Klg).
- Key manager searches individual key(Km) of member from *DB.*
- *Crypto module* encrypts Klg with Km.
- *Transmitter module* multicasts {Klg}Km to members if current rekey expired. Where {Klg}Km denotes key Klg encrypted with key Km.

Registrar action

- Upon receiving request of each member, Registrar checks ACL and decides whether to grant or deny the request. If a request granted, registrar generates individual key and response message including local group address, parent group address and individual key through secure unicast channel. Registrar stores individual key of member in DB. GSM becomes a bottleneck when the member registration rate is high at once. To solve this problem, *registrar* can be added or removed dynamically.

GSM's DB is composed of *state, ACL, group ID, parent group ID, local group ID, group address, parent group address, local group address, parent group key, local group key, Kpg-ver, Klg-ver, PG-rekey interval, LG-rekey interval, member list, individual keys of local group, individual key for parent group, registration number of local group* and *registration number of parent group.*

When a member (or child GSM) is registered in local GSM (or parent GSM), *registration number* is stored in DB.

Key generator module action

- Random value should be hashed to generate individual or local group key of member (or child). Hash function [21] is a transformation that takes an input *m* and returns a fixed-size string. A hash function H is said to be one-way if it is hard to invert. Because given a hash value X hard to invert, it is computationally infeasible to find some input Y.

$$X = H(Y) \quad X\text{: key, } Y\text{: random value, } H\text{: encryption function} \tag{1}$$

The choice of the encryption function, H is dictated by the security requirements of the application. Any function such as DES, DES3, MD5 and IDEA can be used.

Transmitter module action

- *Parent group transmitter* multicast message to parent group encrypted with parent group key. Only child GSM use parent group transmitter.
- *Local group transmitter* multicasts messages (key update, alive alarm, data) to local group.

Control module generates a request message (*find_GSM, join, leave, NAK*) of member.

Crypto module encrypts (or decrypts) message (or encrypted message) of group with appropriate key.

Timer module cans startup (or reset) timers.
- *rekey interval timer* in GSM measures lifetime of valid key.
- *alive alarm timer* detects fault of GSM.
- *wait timer* in member detects loss of key update message.

3 Operational Overview of System

3.1 Group Initiation

Group initiator who wishes to create multicast group receives (*group address, local group address*) from MAA. The initiator can use this information to advertise the group session using the Session Announcement Protocol (SAP) [6]. Firstly, initiator acts as the highest-level GSM.

3.2 Member Join

A member can find local GSM using the Expanding Ring Search (ERS) mechanism [17]. He multicasts a *FIND_localGSM* packet with a small Time To Live (TTL) value to *group address*. TTL value defines a range for local group.

- GSM manager mode: After a moment, if there was no response, the member takes the place of local GSM. He multicasts a *FIND parentGSM* to group address. TTL value defines a range greater than local group. if the member receives *REPLY_FIND_parentGSM* packet, He checks it same email and get *GSM IP address* from it. He sends *join* request to parent GSM and receives message that is composed of *ACL, parent group address, individual key for parent group, registration number of parent group, PG-rekey interval* from parent GSM through secure unicast channel. He joins parent group. After current rekey interval, member receives *key update message* from parent GSM through multicast channel and decrypts it and stores *Kpg* in DB.

- GSM member mode: if the member receives *REPLY_FIND_localGSM* packet, He checks it same email and get *GSM IP address* from it. He sends *join* request to local GSM and receives message that is composed of (*parent group address:* if parent GSM exist), *local group address, individual key for local group, registration number of local group, LG-rekey interval* from local GSM through secure unicast channel. He joins local group. After current rekey interval, member receives *key update message* from local GSM through multicast channel and decrypts it and stores *Klg* in DB.

3.3 GSM Action

GSM generates new group. Upon receiving *find_GSM (local, parent)* request, he responses message including *GSM IP address* and *email of requester.*

He acts in a network that supports multicast and authenticates each member using an authentication protocol, such as SSL 3.0 [15]. He receives member's request and responses through secure unicast channel (SSL 3.0).

After member joins, he starts rekey interval timer. After current rekey interval, he multicasts key update message to member. If receiving leave request of member, registrar deletes member info from DB.

3.4 Key Update

GSM increases *Klg-ver* of key update message step 1. *Add-no* field includes the smallest *reg-no* (except for leaver's *reg-no*) of new member during current rekey interval. GSM updates the *local group key* after current rekey interval. He distributes the new *local group key* to members in local group. We use *key update message* (Fig. 3) by group-oriented rekeying strategy [11].

HDR	Klg-ver	Add-no	{Klg'} Klg	{Klg'} Km1	....	{Klg'} Kmn

Fig. 3. Key update message

If a member receives *key update message* then he checks updated key by *Klg-ver.* New member uses (*reg-no* – *add-no* + 5)th field to find *local group key* in the *key update message.* If no member leaves, only the remaining member that has *reg-no* smaller than *add-no* can update their *local group key* by decrypting *4th field* in the *key update message.* Otherwise, remaining (or new) members use (reg-no + 4)th field to find *local group key* in the *key update message.* In this case, two fields(*Add-no, {Klg'}Klg)* are not used.

3.5 Leave & Fault Tolerance

If a member wants to leave, he sends a *leave* request to the local GSM. If the GSM receives it, he deletes information concerning the member. When GSM receives a join request after member leaves, it sends leaver's *reg no* to new member for reusing empty spaces in DB, key update message.

GSM periodically sends *alive alarm* message including *parent group address, member list* for notifying own alive to local members.

(Fig 4) describes a pseudo-code about member's fault detection and tolerance.

```
while(alive alarm timer not expire){
 receive message from local GSM
if (message == alive alarm) reset alive alarm timer
time is increased automatically synchronized with system clock }
get remaining member's info of low reg-no on member list
if ( member's info of low reg-no == my info) set mode = GSM manager
```

Fig. 4. GSM's fault detection and tolerance

If the child GSM (or parent GSM) leaves from group then he checks remaining member of local group. If the member exists then they announce the member that has lower reg-no of member list as a new GSM. New GSM joins parent GSM through parent group address and takes place of fault GSM.

If GSM wants to expel a member, it processes *key update* and *leave* immediately to disallow the leaving member's continued participation in the local group. Member (or child GSM) resets *wait timer* to 0. GSM resets *rekey interval timer* to 0.

3.6 Retransmission Key Update Message

Each member starts up *wait timer* to recognize problem of lost *key update message*. Member decides *wait timer* by average time of receiving *key update message* from local GSM. If *wait timer* has expired, member sends *NAK* packet(Fig 5) with lost *Klg-ver* through multicast channel [13], [14]. If member receives *NAK packet*, he has two methods. One, if the member has lost same key update message then he resets *wait timer* and does not send NAK packet to avoid *NAK-implosion*. Two, if the member has receivd the *key update message* correctly then he multicasts the *key update message*.

If GSM receives NAK packet, he multicasts the key update message. otherwise, he deletes the *key update message* from buffer.

```
Struct Nak_hdr {
 double packetID;
 double Message ID;
 double Nak_no;
 double Length;
 double Klg-ver;} /* lost key update message */
```

Fig. 5. NAK header

3.7 Data Distribution

A member generates *Kr* and encrypts data with it. He encrypts it with *Klg* and multicasts *{data}Kr* , *{Kr}Klg* to local group. Upon receiving {Kr}Klg, child GSM decrypts it and encrypts it with *Kpg*. The child GSM multicasts *{data}Kr, {Kr}Kpg* to parent group through *parent group transmitter*.

Decrypting and re-encrypting the whole data is reduced to simply decrypting and re-encrypting *Kr* to reduce the overhead of GSM.

4 Performance Analysis

In order to analyze the performance of our system, we implemented a simple program that uses our key management. The program consists of the GSM and client application. We evaluated performance tradeoffs between our individual batch rekeying and individual rekeying of Iolus.

All tests were performed on a Linux kernel 2.4.18 running on a 500Mhz Pentium II. We assumed that interval of rekey was 20sec and number of remaining member was 30. We measured processing performance between GSA and GSM on member's join request increase from 0 to 60sec step 4 in a local group. We used DES3 as encryption algorithm and 128bit key as local group key.

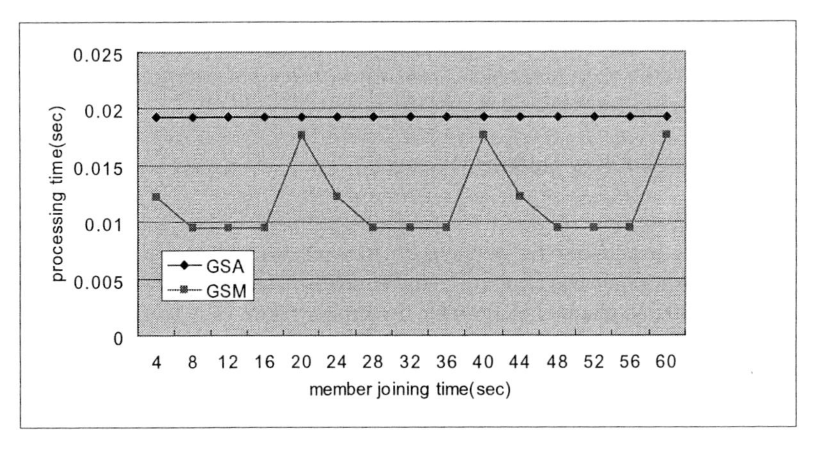

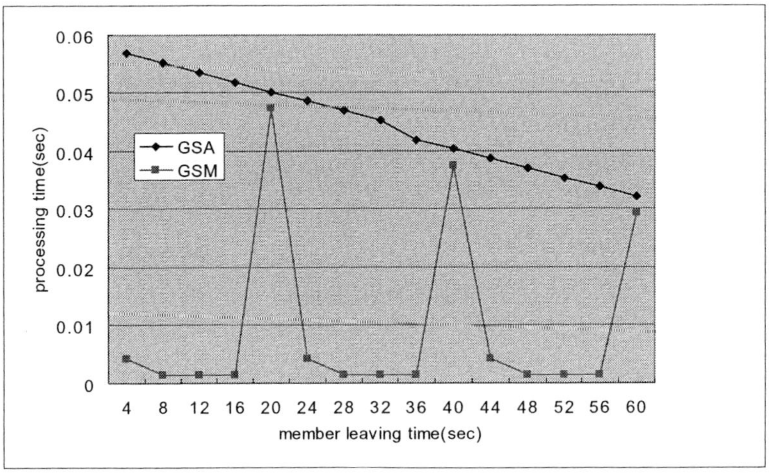

Fig. 6. Processing time of GSM and GSA for member join & leave

We show the processing time of local group manager per join (or leave) request for different rekeying policy (Fig. 6). Frequency of rekeying would become a bottleneck for local group manager. The advantage of our system is to collect requests during a rekey interval in a local group and update key in batch. The variable of performance is membership dynamics during a rekey interval.

5 Conclusions and Future Work

In this paper, we have proposed a novel approach that has distributed hierarchical GSM for secure multicast. A large multicast group is split into smaller local groups and each GSM is used to manage each local group. It minimized problem of concentrating the overhead on a GSM. Thus, exposure of key affects a local group are not reflected to other groups. Also, if a GSM fails, only its local group is affected.

Using individual batch rekeying, it has advantage of less number of rekeying from dynamic membership of the local group. In order to guarantee reliability of individual batch rekeying, our system supports retransmission mechanism.

We compare the performance of our system with Iolus. As a result, our system reduces the overhead of group manager through collects request during the rekey interval and batching key update.

Our key management increases the performance of data transmission through multithread of transmission and crypto module. Also, it supports fault tolerance of GSM for fully distributed structure.

Future works includes research about dynamic GSM for load balancing and IPv6 deployment.

References

1. S.E. Deering.: Host Extensions for IP Multicasting. RFC1112, August 1989
2. D. Wallner, E. Harder, R. Agee.: Key management for multicast: Issues and architectures. RFC 2627, June 1999
3. W. Fenner.: Internet Group Management Protocol, version2. RFC2236, November 1997.
4. D. Thaler, M. Handley, D. Estrin.: The Internet Multicast Address Allocation Architecture. RFC 2908, September 2000
5. P. Radoslavov, D. Estrin, R. Govindan, M. Handley, S. Kumar, D. Thaler: The Multicast Address-Set Claim (MASC) Protocol. RFC 2909, September 2000
6. M. Handley, C. Perkins, E. Whelan.: Session Announcement Protocol. RFC2974, October 2000
7. T. Hardjono, B. Cain, N. Doraswamy.: A Framework for Group Key Management for Multicast Security. Internet Draft, draft-ietf-ipsec-gkmframework-03.txt, August 2000
8. T. Hardjono, R. Canetti, M. Baugher, P. Dinsmore.: Secure IP Multicast: Problem Areas, Framework and Building Blocks. Internet Draft, draft-irtf-smug-framework-01.txt, September 2000
9. H. Harney, E. Harder.: Multicast Security Management Protocol (MSMP) Requirements and Policy. Internet Draft, draft-harney-sparta-msmp-sec-00.txt, March, 1999
10. T. Hardjono, B. Cain, I. Monga.: Intra-Domain Group Key Management Protocol. Internet Draft, draft-irtf-smug-intragkm-00.txt, September 2000

11. Chung Kei Wong, Mohamed Gouda, Simon S. Lam.: Secure group Communication Using Key Graphs. In Proceedings of ACM SIGCOMM '98, Vancover, B.C. September 1998
12. Mittra.: Iolus: A Framework for Scalable Secure Multicasting. In ACM SIGCOMM,volume 27,4 of Computer Communication Review, pages 277–288, New York, September 1997
13. S. Floyd, V. Jacobson, C. G. Liu, S. McCanne, L. Zhang.: A reliable multicast framework for light-weight session and application level framing. IEEE/ACM Transaction on Networking, 5(6):784–803, December 1997
14. D. Towsley, J. Kurose, S. Pingali.: A comparison of sender-initiated reliable multicast and receiver-initiated reliable multicast protocols. IEEE Journal on Selected Areas in Communications, 15(3):398–406, 1997
15. Alan O. Freier, Philip Karlton, Paul C. Kocher.: The SSL Protocol Version 3.0. Work in progress, Netscape Communications, November 1996
16. B. Preneel.: Analysis and Design of Cryptographic Hash Functions. Phd thesis, Katholieke University, Leuven, January 1993
17. D. Boggs.: Internet Broadcasting, XEROX Palo Alto Research Center, Technical Report CSL-83-3, 1983

Architecture for Internal Communication in Multi-gigabit IP Routers

Young-Cheol Bang[1], W.B. Lee[2], Hyunseung Choo[3] , and N.S.V. Rao[4]

[1] Department of Computer Engineering, Korea Polytechnic University
Kyunggi-Do, Korea
ybang@kpu.ac.kr

[2] Network Technology Laboratory, Electronics and Telecommunications Research Institute
161 Gajeong-dong, Yuseong-gu, Deajeon, 305-350 Korea
leewb@etri.re.kr

[3] School of Information and Communication Engineering
Sungkyunkwan University, Suwon 440-746, Korea
choo@ece.skku.ac.kr

[4] Computer Science and Mathematics Division, Oak Ridge National Laboratory
Oak Ridge, TN 37831 USA
raons@ornl.gov

Abstract. The IP packets from a source are suitably forwarded by the routers along the path(s) to the destination. The packets destined to the routers themselves are called non-forwarding packets, and their processing is crucial to the overall speed of IP routers. An architecture is proposed here for efficiently handling the non-forwarding packets for high-speed routers. This architecture includes the Inter-Processor Communication Message Protocol for internal communication needed within the router for distributed processing of the non-forwarded packets. Our implementation results show that this architecture improves the processing speed by 10% as compared to the existing mechanism based on UDP/IP.

1 Introduction

High-speed routers for fast packet forwarding are crucial to meeting the growing bandwidth demands of the Internet users and applications. High-speed routers are being designed with decentralized multiprocessing architecture for IP routing functions. Such a router consists of a number of host processors, line interfaces and switch fabrics [2,3]. Routing lookup algorithms, optimal buffer management algorithms and silicon techniques have been studied in order to increase performance of high-speed routers [4–6]. But the software architecture techniques still leave room for improvement. In particular for speeding up the high-speed routers, we need to efficiently implement various software modules such as Forwarding Engine (FE) in line interface, switch fabric module, FIB manipulation module and non-forwarding packet processing module.

P.M.A. Sloot et al. (Eds.): ICCS 2003, LNCS 2659, pp. 495–503, 2003.

A packet whose destination address is the router itself is termed *non-forwarded*, such as a routing protocol packet or an application packet. The processing time of the non-forwarding packets is a crucial factor in the overall performance of high-speed routers. In this paper, we present a software architecture for high-speed routers to efficiently process the non-forwarding packets. For communication between the processors within the router, we propose Inter-Processor Communication Message Protocol (IPCMP), which is a mechanism for Inter-Processor Communication (IPC). Our implementation results show that this architecture improves the processing speed by 10% as compared to the existing mechanism based on UDP/IP.

2 Architecture for High-Speed IP Router

The first generation routers are PC-based with a single shared bus. In second generation routers, the multiple bus architecture is used to support distributed processing. The third generation routers are based on multiple processor-space switching architecture, which employs switch fabric for fast forwarding [3][13]. This architecture contains line interfaces, a switch fabric, and a routing processor as shown in Fig. 1. Each line interface has its own processor and typically supports 4, 8, or 16 ports.

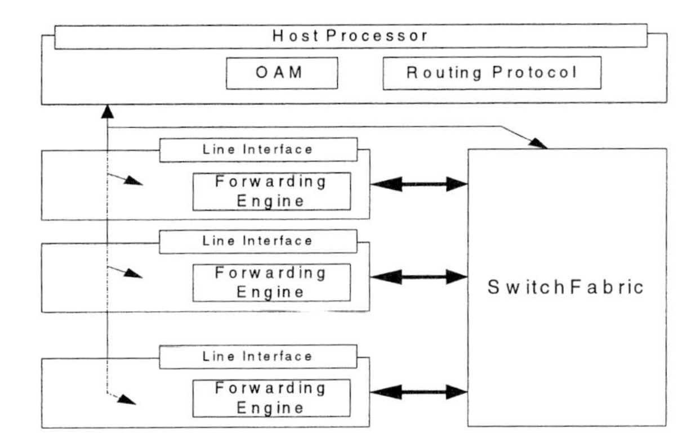

Fig. 1 High-speed IP router architecture

An IP (Internet Protocol) router examines the header of a packet and typically removes the link layer header (such as Ethernet header), modifies IP header, and replaces the link layer header for retransmission. Typically, packets are received at an inbound network interface and stored in the buffer. Then they are forwarded through the switch fabric to the outbound interface that transmits them on the next hop. Forwarding Engine provides the intelligence and processing power to analyze packet headers, lookup the routing table, classify the packets based on their destination and source addresses and other control information and rules, and also to provide queuing and policing of the packets. The switch fabric provides high-speed interconnection of the line interfaces. The basic functions of a switch fabric are the spatial transfer of packets from their incoming ports to destination ports and buffering of packets in case of contention.

The host processor chooses the next-hop for a packet based on the information in its routing database. It supports an interior gateway (IGP) to carry out distributed routing and an exterior gateway protocol (EGP) to exchange topology information with other autonomous systems. Generally, it OSPF (Open Shortest Packet First) and IS-IS (Intermediate System to Intermediate System) for IGP, and BGP (Border Gateway Protocol) as an EGP. The Operation and Maintenance (OAM) module provides network management and system support facilities, including logging, debugging, auditing, status reporting, exception reporting and control. It also provides means for diagnosing faults and measuring traffic.

3 Proposed Mechanism for Processing of Non-forwarded Packets

When a packet to be forwarded arrives at a line interface, the forwarding engine reads the header to determine how to forward the packet and then updates the header and sends the packet to the outbound line interface through the switch fabric for transmission. Processing speed of a forwarded packet is affected by forwarding engine's capacity. In Fig. 2, A and B represent the processing paths of the forwarded packets. The packet whose IP destination address is the router itself needs to be processed on the host processor. These packets include routing protocol packets, Simple Network Management Protocol (SNMP), application, and error reporting packets. In Fig 2, the C and D represent the processing paths of the non-forwarded packets.

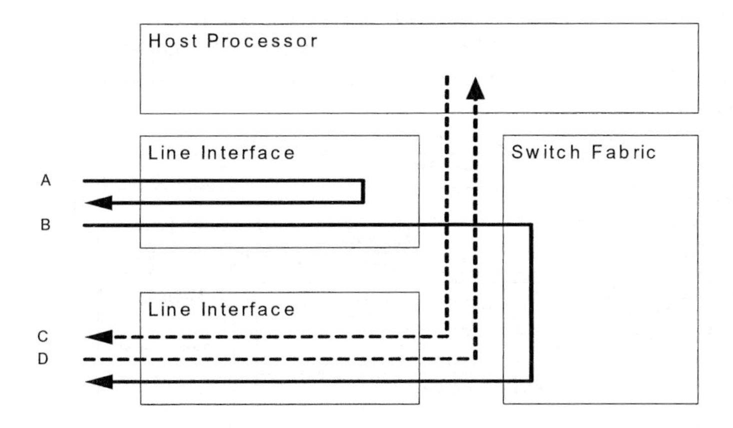

Fig. 2. Processing paths for forwarded and non-forwarded packets

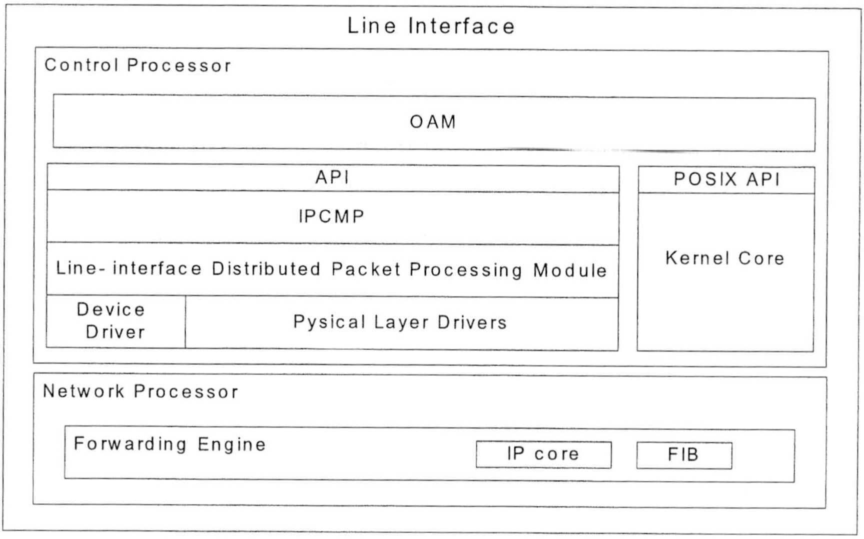

Fig. 3. Software architecture in line interface

The software architecture of the line interface can be divided into two parts as shown in Fig 3. The first part is the NP (Network Processor) that processes the incoming packets. It validates TOS (Type of Service), TTL (Time To Live), and checksum fields. The NP reads the header to determine how to forward the packet, updates the header and sends the entire packet to the outbound interface through the switch fabric for transmission. The second part is the CP (Control Processor) which processes the non-forwarded packets which include packets with: 1) router itself as the IP destination address, 2) broadcast MAC address such as ARP request packet 3) multicast destination address such as OSPF packet, RIP packet. 4) unknown IP destination address invoking ICMP error-report message. 5) invalid header invoking ICMP error-report message.

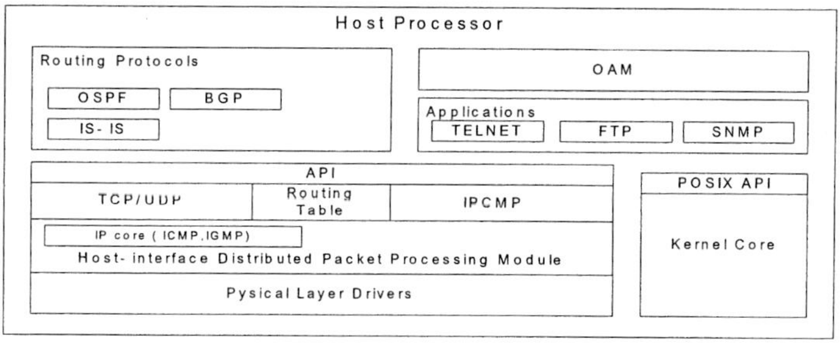

Fig. 4. Software architecture in host processor

LDPPM (Line-interface Distributed Packet Processing Module) incorporates the decision algorithm to decide whether to process a non-forwarded packet in line interface DPPM. The ARP request or response packet has to manipulate in host-processor

DPPM for table consistency while the ICMP error-report packet is handled by LDPPM. OAM module handles the control messages from host interface such as interface configuration, FIB update information and access list and so on. IPCMP gives the method of data passing between processors, host interface and line interface.

16- bit Source Port Number	16- bit Destination Port Number
16- bit IPCMP length	Undefine
data	

Fig. 5. IPCMP header

The software architecture of the host processor is shown in Fig 4. HDPPM (Host-interface Distributed Packet Processing Module) receives the non-forwarded packets from the device driver and performs packet assembly and IP option processing. The remaining packet processing is performed at the higher layer such as TCP (Transmission Control Protocol) or UDP (User Datagram Protocol). And ICMP messages (Echo reply message, Redirection message, Source Quench message, Time Stamp Reply message) are also managed by HDPPM

Since the most of IP packets generated by the host processor carry the IP address of the next hop, the host processor needs to access its individual routing table, which should be the master table of the entire router. The forwarding information base (FIB) in each line interface should match the master routing table of the host processor. The routing table in the host processor is updated either by the various routing protocol entities or by the network management action, which are assumed to happen infrequently. The FIB updates in the line interface can be done in two steps; first, any update information originated in the routing protocol has to be transferred to the line interface in the IPCMP format, and as a second step, pushed into the search database in the NP through API. Managing the routing table and FIB is up to the routing manager in OAM, but is not shown in Fig. 4.

This router architecture needs an IPC network to communicate between the routing processor and line interfaces, such as Cell Bus network, HDLC network, VME bus network, ATM Asynchronous Transfer Mode) switch bus network, and Ethernet switch bus network. Cell bus, VME and HDLC use a shared bus while ATM and Ethernet use point-to-point method. Throughput of HDLC and VME bus is low as compared to the point-to-point method [14,15]. In an Ethernet switch, we can implement IPC using TCP/IP or UDP/IP [16] using point-to-point or point-to-multipoint methods. The IPCMP in Figs. 3 and 4 is the substitute for the UDP/IP method, since it provides a direct Ethernet encapsulation to the application through the socket interface. In Fig. 5, Source Port Number is the port number of source application and Destination Port Number is the port number of destination application. IPCMP length field specifies the message size. Fig 6 shows the example code using IPCMP socket

API. IPCMP provides common sockets API which is also used for TCP/IDP sockets. We introduce a socket type called AF_IPCMP, which is similar to the AF_INET type used for TCP/IDP sockets. When a user calls "*socket (AF_IPCMP, SOCK_DGRAM, 0);*" to create a IPCMP type socket, we return the number as a new socket descriptor.

```
...
if ((sockfd = socket(AF_IPCMP, SOCK_DGRAM, 0)) < 0){
        printf("Can't open IPCMP socket\n");
        exit(1);
}
...
while (tmp_count < max_packet){
        slen = sendto(sockfd, (char *)&s_buf_person,

        sizeof(person),0,(struct sockadr *)&dst_addr,
      da_len);

        if (slen < 0){
                printf("SENDTO ERROR\n");
                close(sockfd);
                exit(1);
        }
        rlen = recvfrom (sockfd, r_buf, 1500, 0,
        (struct sockaddr*)&src_addr, &sa_len);
        if (rel < 0){
                printf("RECVFROM ERROR\n");
                close(sockfd);
                exit(1);
        }
...
}
...
```

Fig. 6. Example code using IPCMP socket API

4 Experimental Results

The hardware platform used for performance evaluation consists of two Linux systems with customized boards running Linux kernel 2.2.14. The host processor consists of a PowerPC MPC750 (300MHz) microprocessor with 512 Mbytes of main memory and the other system, the line interface, consists of a PowerPC 405GP (266MHz) microprocessor with 512 Mbytes of main memory [17,18].

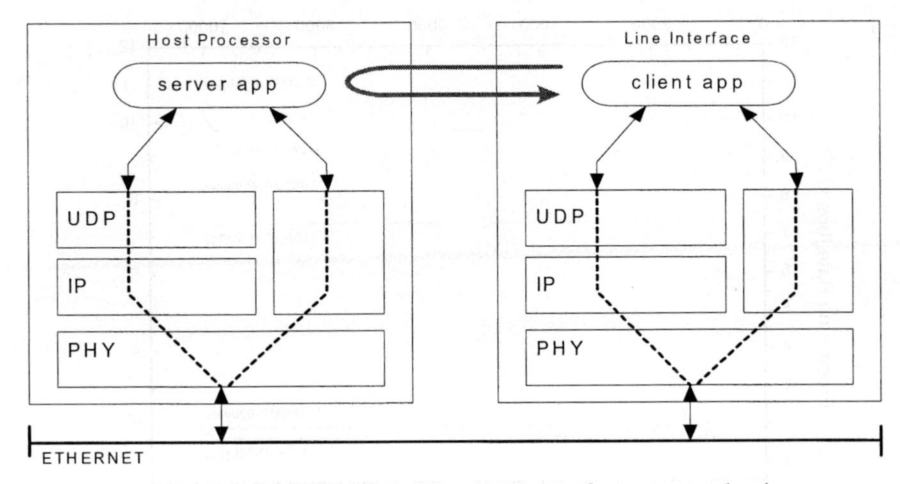

Fig. 7. The brief internal architecture for performance evaluation

In Fig. 7, the client application creates and sends variable packets to the server application and then server application replies them, like a ping-pong test. We measured the RTT of different sized packets as shown in Table 1.

Table 1. The round trip time of packets

Type	UDP/IP on Ethernet								
Bytes	60			500			1000		
Number of Packet	1000	5000	10000	1000	5000	10000	1000	5000	10000
Avg. RTT(msec)	289.83	1428.56	2845.32	548.52	2727.79	5459.54	804.63	4083.43	8250.13
Type	IPCMP on Ethernet								
Bytes	60			500			1000		
Number of Packet	1000	5000	10000	1000	5000	10000	1000	5000	10000
Avg. RTT(msec)	246.63	1225.71	2462.95	504.16	2541.01	5057.31	764.13	3820.52	7653.78

Based on RTT results, IPCMP resulted is faster by an average of 9.3% compared to the UDP/IP (60-byte packets: Avg. 14.2%, 500-byte packets: Avg.7.4%, 1000-byte packets: Avg.6.2%). This result is due to the one-layer packet processing instead of two layers, UDP and IP. Fig 8 compares the packet processing rate of IPCMP with that of UDP/IP. IPCMP shows the processing rate of Avg. 1.95 Mbps for 60-byte packets while UDP/IP shows the processing rate of Avg. 1.67 Mbps. Overall, we can increase packet processing rate of IPCMP by Avg. 10.3% compared to the UDP/IP.

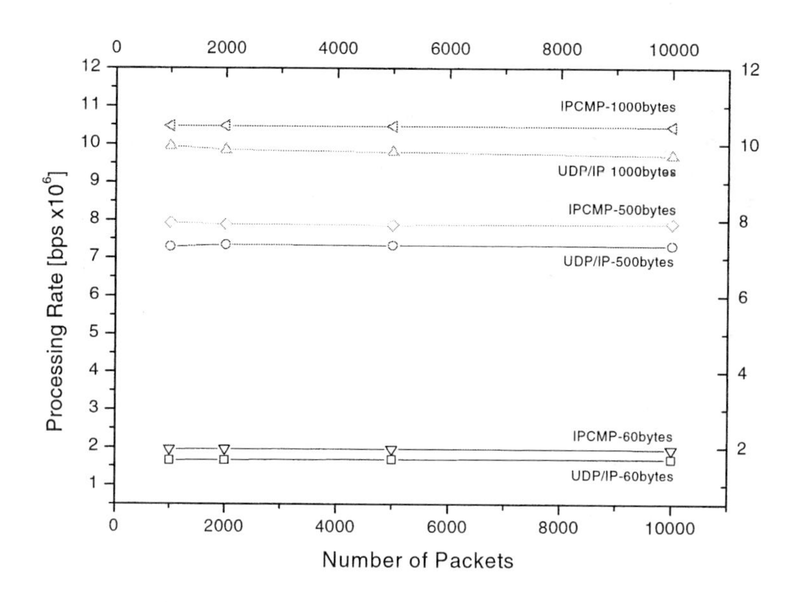

Fig. 8. Processing rate of packets

5 Conclusion

In this paper, we introduced a software architecture for efficient manipulations of high-speed non-forwarding packets such as routing protocol packets, application packets and so on. As shown in results, the non-forwarding packet processing is the crucial part in overall performance of the high-speed router for reliable routing. A proposed IPC mechanism called IPCMP outperforms the existing UDP/IP technique by 10% in terms of RTTs. Furthermore, we can reduce the complexity of software modules by removing TCP/UDP/IP stack in Line Interface.

Since non-forwarding packets affect in overall performances of high-capacity IP routers by drastic increment, the high-capacity IP router should be designed in taking into consideration non-forwarding packets.

Furthermore, we also plan to analyze non-forwarding traffics more in detail and design more efficient distribution method for the routing table information to forwarding table in line interfaces for high-speed IP router such as TeraBit-Router for the future study.

References

1. A. Bharagava and B. Bhargava, "Measurements and quality of service issues in electric commerce software," in Proc. Application-Specific System and Software and Technology, pp. 26–33, 1999.
2. S. Keshav, R. Sharma, "Issues and trends in router design," IEEE Communications Magazine, Vol. 36, pp. 144–151, May 1998.
3. V.P. Kumar, T.V. Lashman, D. Stiliadis, "Beyond Best Effort: Router Architectures for the Differentiated Services of Tomorrow's Internet," IEEE Communications Magazine, Vol.36, pp. 152–164, May 1998.
4. N. McKeown, P. Gupta, and S. Lin, "Routing lookups in hardware at memory access speeds," INFOCOM '98. Proceedings. IEEE , Vol. 3, pp. 1240–1247, 1998.
5. S. Nilsson, G.Karlsson, "IP-address lookup using LC-tries," Selected Areas in Communications, IEEE Journal on, Vol. 17, pp. 1083–1092, June 1999.
6. C. Donpaul, et al, "Implementing scheduling algorithms in High-Speed Networks," Selected Areas in Communications, IEEE Journal on , Vol. 17, pp. 1145–1158, June 1999.
7. J. Moy, OSPF Version2, RFC2328, April. 1998.
8. Christian Huitema, "Routing in the Internet," 2nd Edition, Prentice Hall, 1999.
9. D. Estrin, et al, "Protocol Independent Multicast-Sparse Mode (PIM-SM): Protocol Specification," Internet Request For Comments RFC 2362, June 1998.
10. J. Moy, "Multicast Extensions to OSPF," Internet Request For Comments RFC 1584, March. 1994.
11. F. Baker, "Requirements for IP Version 4 Routers," RFC 1812, June 1995.
12. Wang-Bong Lee, et al, "An Architecture of Distributed Multi-Gigabit IP Router," AIC 24th Conference, Seoul, Nov. 2000.
13. A. Asthana, C. Delph, H. Jagdish, and P. Kryzyanowski, "Towards a gigabit IP router," Journal of High-Speed Networks, 1(4), 1992.
14. J. Furnuas, et al, "A prototype for interprocess communication support, in hardware," Ninth Euromicro Workshop on Real-Time Systems, pp. 18–24, 1997.
15. Bup Joong Kim, et al, "Designed and Implementation of IPC Network in ATM Switching system," ICATM 2001, pp. 148–152
16. W.R. Stevens, TCP/IP Illustrated Volume 1, Addison-Wesley, New York, 1996.
17. The Linux kernel Archives. `http://www.kernel.org`
18. LinuxPPC. `http://www.linuxppc.org`

Low-Cost Fault-Tolerance Protocol for Large-Scale Network Monitoring*

JinHo Ahn[1], SungGi Min[2,**], YoungIl Choi[3], and ByungSun Lee[3]

[1] Dept. of Computer Science, College of Information Science, Kyonggi University
San 94-6 Yiuidong, Paldalgu, Suwonsi Kyonggido 442-760, Republic of Korea
jhahn91@hanmail.net

[2] Dept. of Computer Science & Engineering, Korea University
5-1 Anamdong, Sungbukgu, Seoul 136-701, Republic of Korea
sgmin@korea.ac.kr

[3] Network Technology Lab., Electronics and Telecommunications Research Institute
Yusong P.O. Box 106, Taejon 305-600, Republic of Korea
{yichoi,bslee}@etri.re.kr

Abstract. Distributed hierarchical network monitoring model has been proposed to solve scalability problem of centralized model. In this distributed model, a top-level monitoring manager, called main manager, obtains aggregate management information from mid-level managers, named domain managers, forming a hierarchical structure. However, if some of monitoring managers crash, network elements cannot be continuously and correctly monitored until the managers are repaired. To address this important, but previously unresolved issue, this paper presents a new fault-tolerance protocol for domain managers, named $DMFTP$, allowing the managers to efficiently utilize their organization structure. Therefore, this protocol can minimize failure detection overhead and the number of live managers affected by each manager node crash. Also, it tolerates concurrent manager failures and, after the failed managers have been repaired, ensures their immediate and consistent recovery.

1 Introduction

Network monitoring is an ability to collect traffic information from networked systems in a timely manner [4]. The traffic information is generally used in these systems for a variety of purposes such as quality of service, scheduling, capacity planning and traffic flow prediction and so on. As the systems scale up, the importance of the ability significantly increases because it is very difficult to control and manage the systems without any monitoring mechanism.

Monitoring systems should be designed to minimize network traffic incurred by them and latency in extracting the necessary information from the networked systems [5]. These systems can be classified into centralized and distributed(or

* This work was supported by Electronics and Telecommunications Research Institute Grant.

** Corresponding author. Tel.:+82-2-3290-3201; fax:+82-2-953-0771.

P.M.A. Sloot et al. (Eds.): ICCS 2003, LNCS 2659, pp. 504–513, 2003.

decentralized) based on their organization model. In the centralized monitoring model, a single manager collects information from all agents on monitored nodes and processes it. The model can be simply implemented, but become a performance bottleneck of the entire systems. Thus, the model is widely used to manage relatively small-scale networks. In order to pursue increased performance and scalability, the distributed monitoring model can be used for large-scale networked systems. In this model, monitoring managers are largely classified into main manager and domain manager depending on the role of each manager, and hierarchically organized. In other words, a hierarchical structure is formed with the main manager in top, several levels of domain managers in the middle and a collection of agents at the bottom. Thus, administrators can delegate monitoring tasks across a hierarchy of managers [2]. Each lowest-level domain manager collects management information from some agents. Similarly, a set of domain managers act as agents to a higher-level domain manager or to the main manager. The main manager obtains and processes aggregate and pre-filtered information from the domain managers. Several existing network monitoring systems based on this model were proposed to use more than one among some technologies such as SNMP, Distributed Object, Mobile Agent and so on [4].

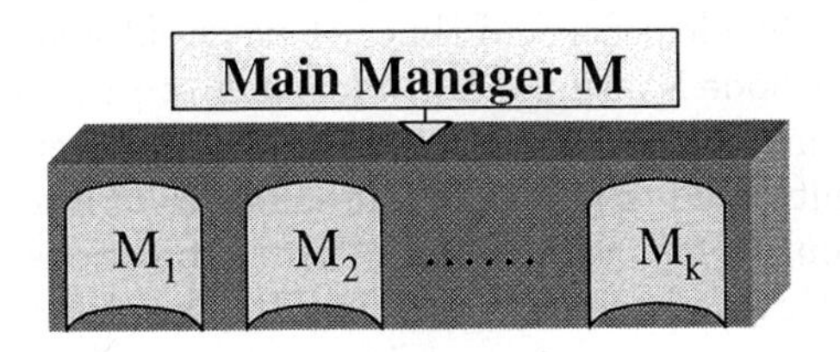

Fig. 1. A cluster of redundant main managers

However, suppose that some of managers crash in this model. In this case, network devices or nodes affected by the failures cannot be continuously and correctly monitored before the failed managers are recovered. To the best of our knowledge, few among previous works consider this important issue. Especially, if the main manager fails, the entire monitoring system may never play its role. To address the issue, our research group has performed developing a scalable and efficient fault-tolerance strategy for the model [1]. In this strategy, main manager fault tolerance is achieved by applying existing scalable replication-based protocols [3] to a cluster of redundant main managers like in Fig. 1 because the main manager should process very large volume of management information and perform significantly complex decision-making process for various applications. But, another fault-tolerance protocol is required for a hierarchy of domain managers due to their more lightweight role compared with that of the main manager. For this purpose, this paper proposes a scalable fault-tolerance protocol for domain managers, called $DMFTP$(*D*omain *M*anager *F*ault-*T*olerance *P*rotocol), to efficiently utilize their hierarchical structure based on the assumption that the main manager is failure-free. The protocol results in low failure detection overhead by each domain manager periodically sending a domain manager advertisement message only to its immediate super manager. Thus, when some of domain managers fail even concurrently, the protocol enables their immediate super managers to take over them. This behavior can achieve minimizing the number of live managers affected by the failures. Also, after failed managers have been recovered, it allows them to immediately play their pre-failure roles in order to improve entire monitoring system performance degraded by the failures.

The rest of the paper is organized as follows. In Sect. 2, we describe our designed protocol $DMFTP$ and in Sect. 3, prove its correctness. Finally, Sect. 4 concludes this paper.

2 Domain Manager Fault-Tolerance Protocol (DMFTP)

The proposed protocol $DMFTP$ is composed of three components, failure detection, takeover and recovery. Every manager i should maintain three variables, $MMngr_i$, $Parent_i$ and $SubMngrs_i$, for the protocol. $MMngr_i$ is the main manager's identifier needed when domain manager i recovers after repaired. $Parent_i$ is the immediate super manager's identifier of domain manager i. $SubMngrs_i$ is a tree for saving the identifier and timer of every sub-manager of main or domain manager i. Its node is a tuple (id, $timer$, sub). $timer$ for each sub-manager id is used so that manager i detects whether sub-manager id is alive or failed currently, and is initialized to a. sub for sub-manager id is a sub-tree for all sub-managers of the domain manager id. In the next subsections, basic concepts and algorithms of the three components of $DMFTP$ are described in details. Due to space limitation, there are no formal descriptions of the proposed protocol in this paper. The interested reader can find the descriptions in [1].

2.1 Failure Detection and Takeover

Every domain manager i periodically sends each domain manager advertisement message only to its immediate super manager $Parent_i$. Thus, monitoring the advertisement messages of its sub managers, the super manager $Parent_i$ can detect which managers fail or are alive among them. For example, Fig. 2(a) shows a hierarchy of managers consisting of a main manager M and twelve domain managers and Fig. 2(b), domain manager advertisement message interaction between four managers M, D_1, D_6 and D_{12}. In Fig. 2(b), domain managers D_1, D_6 and D_{12} periodically transmit each a domain manager advertisement message to their immediate super managers M, D_1 and D_6 by invoking procedure RCV_DMMNGRAM(). In this case, each super manager resets $timer$ for each corresponding sub manager to a like in Figs. 3(a)–3(c). Therefore, the protocol $DMFTP$ results in low failure-free overhead incurred by failure detection by efficiently utilizing the hierarchical structure of managers.

In $DMFTP$, each super manager sm decrements the timer for every sub manager by one every certain time interval. If there are some sub managers which cannot send each a domain manager advertisement message to sm until their timers become zero, sm suspects that the managers failed. In this case, it executes the takeover procedure of $DMFTP$. For example, if domain manager D_6 fails like in Fig. 4, manager D_1 cannot receive any advertisement message from D_6. Thus, the timer for D_6 eventually becomes zero like in Fig. 4(c) and so, D_1 invokes procedure TAKEOVER(). This procedure forces three sub managers of D_6, D_{10}, D_{11} and D_{12}, to call procedure UPDATE_PARENT(). In this case, like in Fig. 4(d), the sub managers set all $Parent_i$ to D_1 as their immediate super manager. Afterwards, they periodically send each an advertisement message to D_1 like in Fig. 4(b).

Also, the protocol $DMFTP$ performs the consistent takeover process even if domain managers fail concurrently. To illustrate this claim, Fig. 5 indicates an example that concurrent failures of two domain managers D_1 and D_6 occur. In this case, main manager M can receive no advertisement message from the domain manager D_1 like in Fig. 5(b). Thus, the timer for D_1 of M is finally expired like in Fig. 5(c) and then M executes procedure TAKEOVER(), which causes sub managers of D_1, D_4, D_5 and D_6, to invoke procedure UPDATE_PARENT(). However, M cannot receive any acknowledgement of the procedure from the failed manager D_6 like in Fig. 5(b). In this case, it sets the timer for D_6 to zero and then forces the sub managers of D_6 to perform procedure UPDATE_PARENT(). After having completed the takeover procedure, M can receive each advertisement message from D_{10}, D_{11} and D_{12} respectively every certain interval like in Fig. 5(b).

2.2 Recovery

This section describes the recovery algorithms of $DMFTP$ allowing repaired domain managers to play their pre-failure roles. For example, Fig. 6 shows that failed domain manager D_6 in Fig. 5 is recovered by executing $DMFTP$. In this example, after the manager D_6 has been repaired, it performs procedure RECOVERY(). This procedure first forces the main manager M to invoke procedure GET_PARENTS() like in Fig. 6(b) in order to obtain a list of super managers of D_6 known to be alive by M. In this case, the received list is $\{M\}$ because the manager D_1 previously failed. Thus, D_6 requires from M a tree of sub managers it managed before the failure by remotely calling procedure GET_SUBMNGRS() of M. At this point, M resets the timer for D_6 to a like in Fig. 6(c) and then causes three sub managers, D_{10}, D_{11} and D_{12}, to remotely invoke UPDATE_PARENT() so that the sub managers set all $Parent_i$ to D_6 as their immediate super manager. After having completed this procedure, D_6 obtains the tree from M like Fig. 6(d). Hereafter, they periodically send each an advertisement message to D_6 like in Fig. 6(b).

Next, $DMFTP$ can allow each repaired manager to be recovered consistently even if its immediate super manager fails concurrently when the recovery process is started. To illustrate this feature, Fig. 7 shows an example that D_1 crashes as soon as D_6 in Fig. 4 performs procedure RECOVERY(). In this case, D_6 receives a list of super managers of D_6, $\{M, D_1\}$, from M because M has not detected

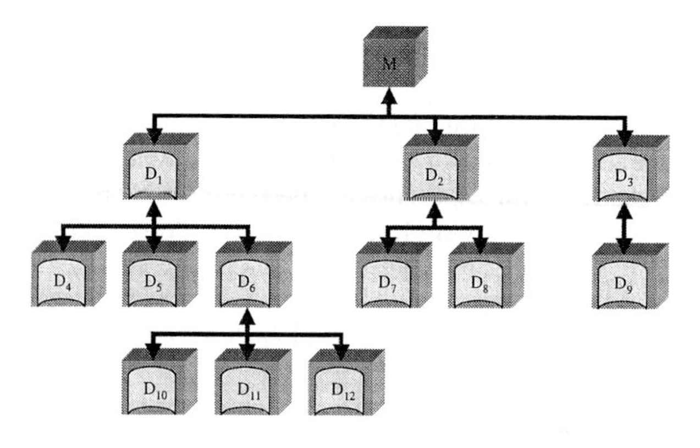

(a) a hierarchy of managers

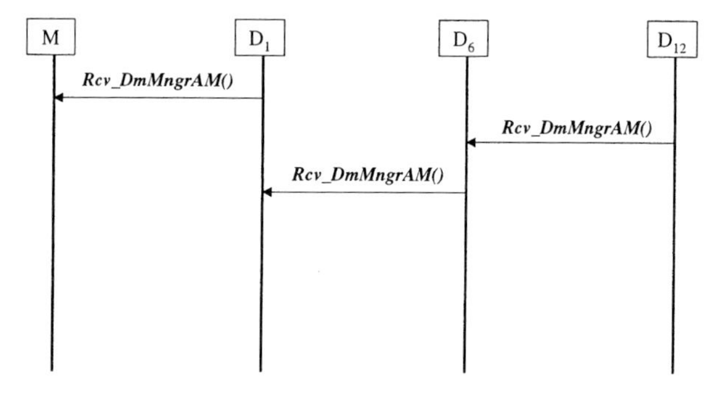

(b) Domain manager advertisement message interaction between four managers

Fig. 2. In case of no manager failure in $DMFTP$ (*continued*)

D_1's failure yet. Then, D_6 first invokes procedure GET_SUBMNGRS() of the lowest level super manager in the list, D_1, to obtain a tree of its monitored sub managers. However, like in Fig. 7, D_6 cannot get the tree from the currently crashed manager, D_1. Thus, it requires the tree from the secondly lowest level super manager, M, by calling procedure GET_SUBMNGRS() of M. At this point, M can detect D_1's failure by either D_6 or the timer for D_1. In the first case, it sets the timer for D_1 to zero and then forces the sub managers of D_1 to set all $Parent_i$ to M by invoking their procedure UPDATE_PARENT() respectively. Also, M resets the timer for D_6 to a and then causes D_6's sub managers to set their $Parent_i$ to D_6. Then, D_6 gets the tree from M. Afterwards, D_4, D_5 and D_6 periodically send each an advertisement message to M, and D_{10}, D_{11} and D_{12}, to D_6 like in Fig. 7.

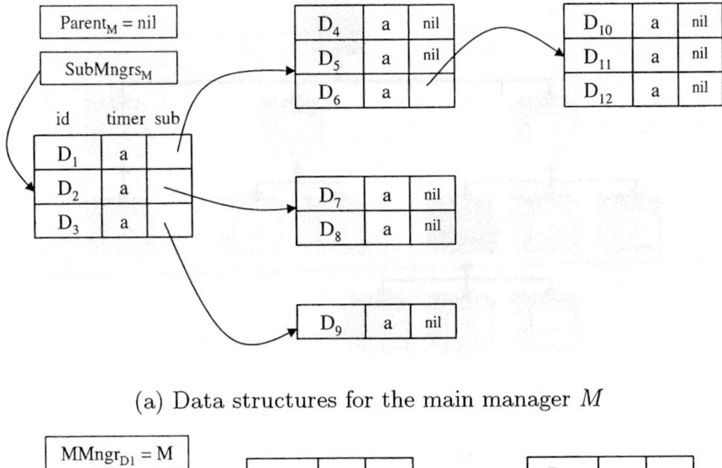

(a) Data structures for the main manager M

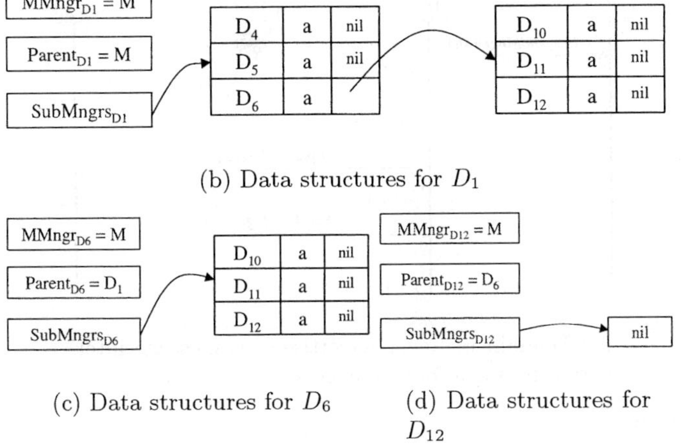

(b) Data structures for D_1

(c) Data structures for D_6

(d) Data structures for D_{12}

Fig. 3. In case of no manager failure in $DMFTP$

3 Correctness

This section shows theorem 1 and 2 to prove the correctness of our takeover algorithm and recovery algorithm in $DMFTP$. Due to space limitation, the proof of the two theorems is omitted. The interested reader can find them in [1].

Theorem 1. Even if multiple domain managers crash concurrently, our takeover algorithm enables another live managers to monitor all the network elements previously managed by the failed ones. □

Theorem 2. After every repaired domain manager has completed our recovery algorithm, it can manage its monitored network elements before it has failed. □

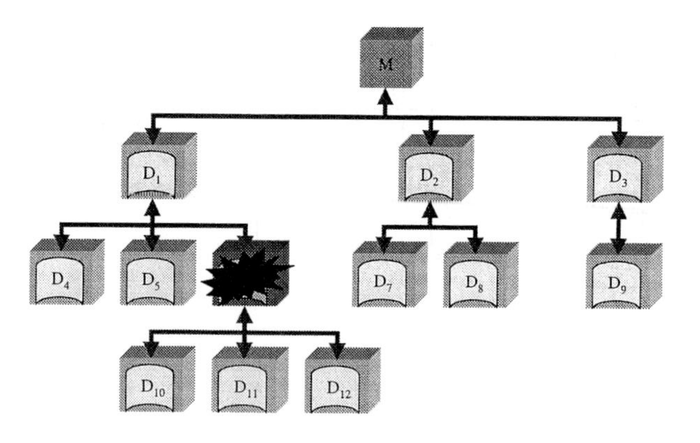

(a) a hierarchy of managers

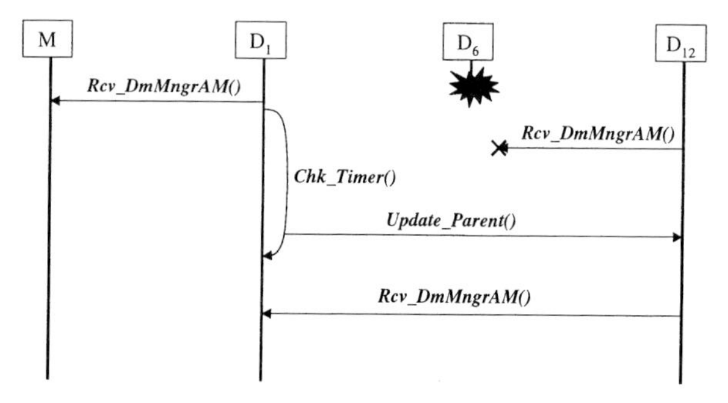

(b) Domain manager advertisement message interaction between four managers

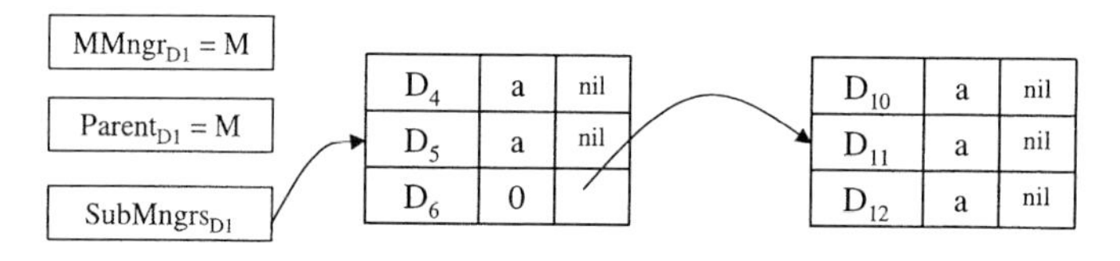

(c) Data structures for D_1

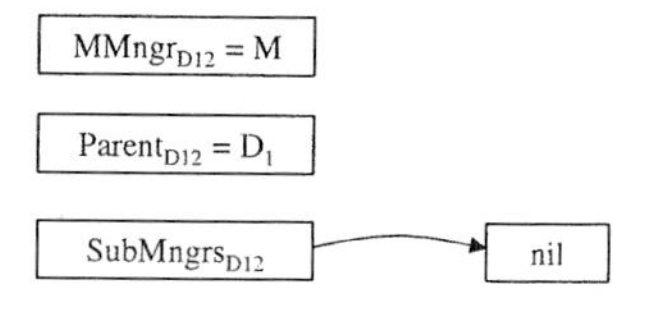

(d) Data structures for D_{12}

Fig. 4. In case of D_6's failure in $DMFTP$

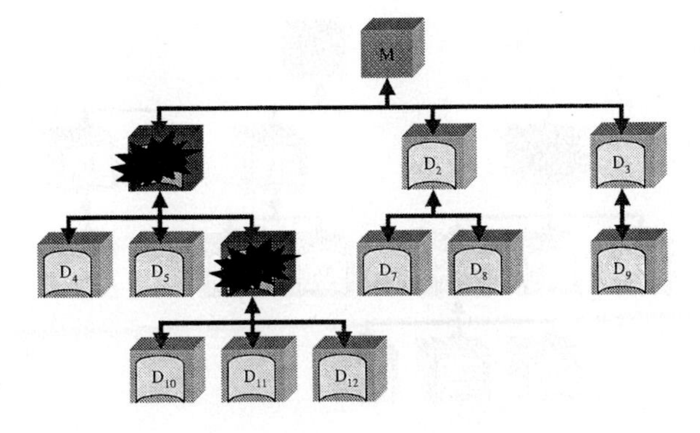

(a) a hierarchy of managers

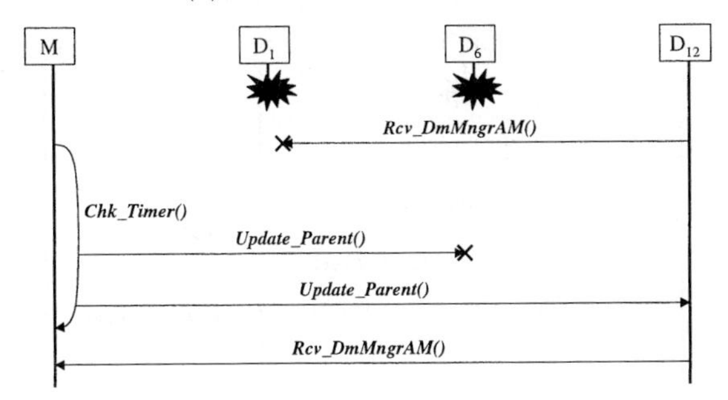

(b) Domain manager advertisement message interaction between four managers

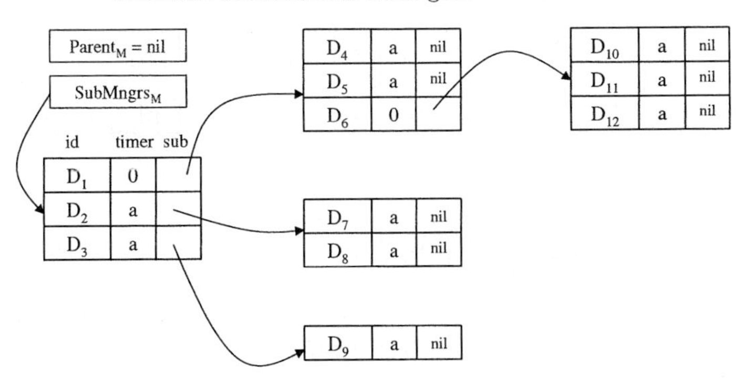

(c) Data structures for M

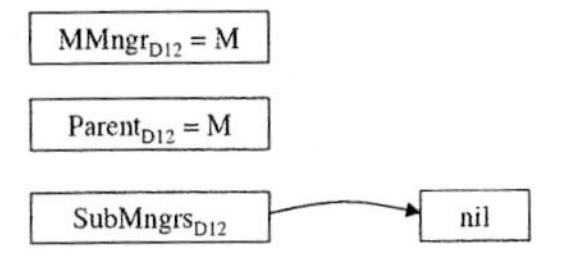

(d) Data structures for D_{12}

Fig. 5. In case of concurrent failures of two domain managers D_1 and D_6 in $DMFTP$

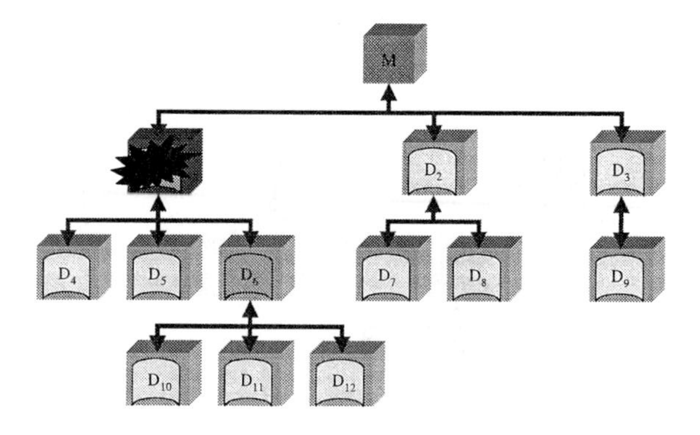

(a) a hierarchy of managers

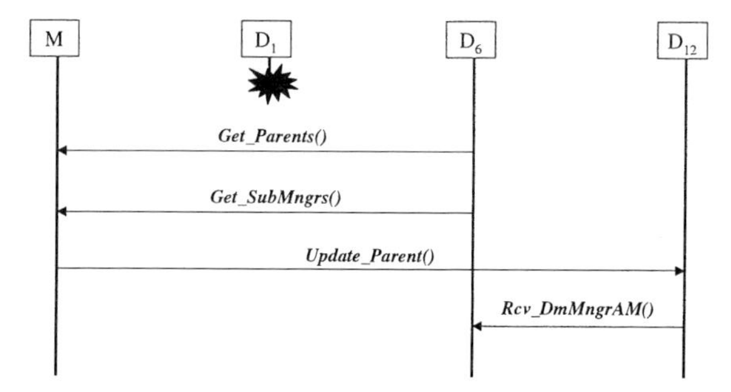

(b) Domain manager advertisement message interaction between four managers

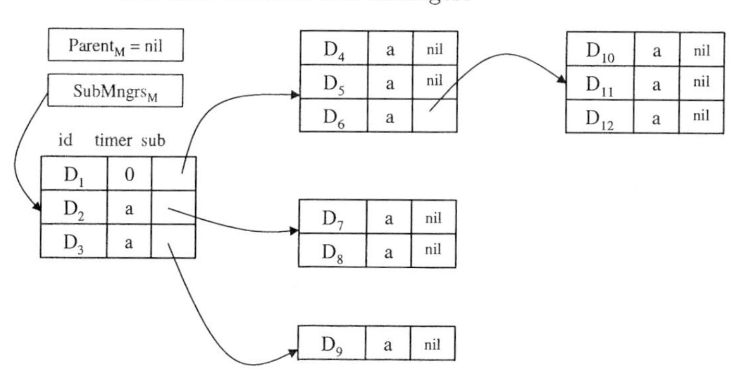

(c) Data structures for M

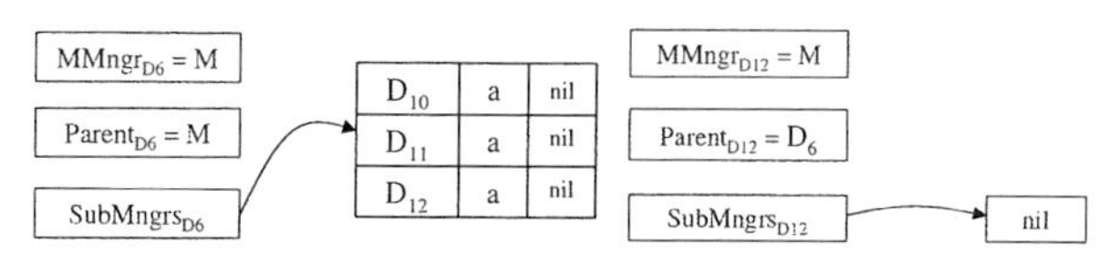

(d) Data structures for D_6

(e) Data structures for D_{12}

Fig. 6. An example of recovering domain manager D_6 in figure 5

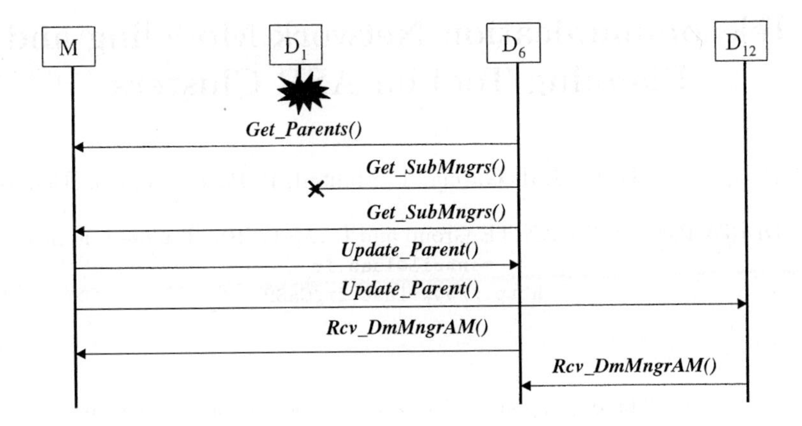

Fig. 7. In case of D_1's failure while recovering D_6 in figure 4

4 Conclusion

In this paper, a low cost fault-tolerance protocol for domain managers, $DMFTP$, was designed based on their hierarchical structure. This protocol can minimize failure detection overhead and the number of live managers affected by each manager node crash. Also, it tolerates concurrent manager failures and, after the failed managers have been repaired, ensures their immediate and consistent recovery.

Integrating $DMFTP$ with existing replication-based protocols [3] for the main manager is a research work requiring further investigation because the role of the main manager fault-tolerance protocols is very important for $DMFTP$ to work effectively [1]. Thus, we are currently examining which replication protocols are appropriate for $DMFTP$ through various experiments.

References

1. J. Ahn, S. Min, Y. Choi and B. Lee. A Novel Fault-Tolerance Strategy for Large-Scale Network Monitoring. *Technical Report KU-CSE-02-049*, Korea University, 2002.
2. G. Goldszmidt and Y. Yemini. Delegated Agents for Network Management. *IEEE Communication Magazine*, 36(3):66–70, March 1998.
3. R. Guerraoui and A. Schiper. Software-Based Replication for Fault Tolerance. *IEEE Computer*, 30(4):68–74, 1997.
4. J. Philippe, M. Flatin and S. Znaty. Two Taxonomies of Distributed Network and System Management Paradigms. *Emerging Trends and Challenges in Network Management*, 2000.
5. R. Subramanyan, J. Miguel-Alonso and J.A.B. Fortes. A scalable SNMP-based distributed monitoring system for heterogeneous network computing. *In Proc. of the 12nd ACM/IEEE International Supercomputing Conference.* Dallas, Texas, Nov. 2000.

Telecommunication Network Modeling and Planning Tool on ASP Clusters

P. Bacquet, O. Brun, J.M. Garcia, T. Monteil, P. Pascal, and S. Richard

DELTA Partners SA ANITE Group and LAAS-CNRS, Toulouse, France
monteil@laas.fr
http://www.laas.fr/CASP

Abstract. The development of Internet allows to consider new execution paradigm in industrial context: ASP model (Application Service Provider). An interesting execution support to implement this concept is clusters of PC. General principles of the ASP are described. The architecture for the resource manager and watcher is described. A set of existing or created tools to materialize a solution to this problem is explained. An industrial telecommunication modeling and planning software is used on the ASP. In particular, the analytical performance evaluation kernel of this software is parallelized on the cluster through the ASP. Several results on cluster are given.

1 Introduction

Many users around the world need a set of applications to administer the companies, to manage relations with customers or to solve scientific problems. Those kinds of software could be heavy, expensive and sometimes made with many specific components depending on utilization (plug in notion). They could require specific help during hard work time or for complex problems. They could also need powerful computers (processor, memory or data storage) and the purchase of a lot of licenses for punctual work.

For different reasons (financial, human resources, complexity of solution, ...) , the potential user of those software could have difficulties to efficiently use those applications on the machine in his office. Another aspect, is the response time requested for the application which can be very short.

A solution to those problems is to use remote applications on remote resources: ASP model (Application Service Provider) to save money and time. The most common way to offer all those services is to use Internet or Intranet. The machines that could run those remote applications should be multi-processors for parallel or sequential applications. The clusters of PC is, at this time, the best execution support due to its low price and its capacity of scalability, modularity and evolution.

Nevertheless, the utilization of clusters in ASP model requires specific software to administrate the clusters, to manage the resources, to communicate through the network and to insure security. An important aspect for industrial

P.M.A. Sloot et al. (Eds.): ICCS 2003, LNCS 2659, pp. 514–523, 2003.

utilization is to insure high availability, different classes of quality of service and exact accounting.

This article deals with an ASP solution developed in the CASP project (Clusters for Application Service Providers)[1] and its validation on the NetQUAD telecommunication modeling software (DELTA Partners SA, a company of Anite group) [1]. The main contributions of this article are to propose a set of tools to create ASP on clusters and to present the experiments done on an industrial application. In Sect. 2, an ASP model is proposed. Then, the NetQUAD software is described. A specific NetQUAD plug-in for ASP utilization and a new parallel kernel are detailed. Finally, some results are given on an ASP cluster.

2 ASP Model Proposed

2.1 Introduction

CASP (Clusters for Application Service Providers) project aims at creation of remote Internet services on a cluster for high performance applications. Two entities are defined: the customers and the providers. The providers own the resources (processors, disks, memory and applications servers) and the customers possess all the necessary tools to express their needs (quality of service, resources, software). The providers and the customers could be in the same network area, in this case the network used is an Intranet. The providers could also be outside the customers company. In this case the network between customers and providers could be dedicated or free like the Internet.

Cluster architecture is used because of its extension capacities and modularity capacities. A software infrastructure is needed to manage the cluster in order to make ASP. Resources management with quality of service and an Internet portal are also needed.

The goal of this project is to put on the market a generic product for ASP running on Linux with an end to end solution. It is composed of:

- Software managing reliable communications between a client and a remote application.
- Tools to guarantee a quality of service to make an industrial and commercial use (resources management, high availability).
- Software to manage the provider platform.

The project allows to use cluster in an industrial and commercial world, in the supercomputing and ASP domain.

---

[1] This project is a cooperation between Alinka SA, Delta Partners SA group Anite and LAAS-CNRS and is funded by the French Ministry of Industry and RNTL (Reseau National des Technologies Logicielles)

2.2 AROMA

AROMA (scAlable ResOurces Manager and wAtcher) is the part of CASP integrating all the necessary tools to manage the remote execution with a defined quality of service to produce an ASP. AROMA integrates a resource management system [2,3], an application launcher and an accounting system. This tool is an evolution of works done on Network-Analyser system [4].

It must have an easy-to-use graphical interface which must adapt to different classes of users: resource customers, administrators. They have access to different dynamic services: visualization of dynamic and static information on resources (CPU, memory, network, libraries) on different times (last seconds, minutes, hours, days, weeks, months, years) and mapping services. The customers can use different services depending on their permissions. The user interface is dynamic and depends on permissions and evolution of the services. A client defines a contract with the provider. According to this contract, the provider guarantees a quality of service and the client pays according to the account.

At first (Fig. 1), the customers build all the necessary information for the job to execute. The connection to the remote system is defined with a login, a group and a contract key. The contract will define utilization of resources and quality level for the remote execution. The first connection is used to define the different permissions on all the AROMA systems services and on the remote applications controlled by AROMA. The data base of AROMA is used to do this. It is also used to update or load services on the client host. It could be new services given to customers or services acquired with an evolution of their contract. It is dynamic and transparent for users.

A request is sent to AROMA to execute all or a part of the work. A negotiation is established between AROMA and the application depending on the quality of service. Finally the job is executed into AROMA environment.

AROMA is able to control the execution (real time resources consummation, end of application detection). This functionality is used by the accounting system for example to stop a job which has consumed too many resources. When the remote application is finished, customers are advertised and can manually or automatically get the results.

The AROMA resource manager has to face three problems: portability, dynamicity and efficiency.

- **Portability:** The Java programming language has been chosen to ensure portability. It guarantees code portability without re-writing.
- **Dynamicity:** Two levels of dynamicity can be distinguished. Dynamicity due to changes in the cluster topology and dynamicity due to users permissions. Both aspects are resolved using Jini technologies and Java Remote Method Invocation system [2]. Jini communication protocols are used to detect changes in the cluster architecture, while RMI dynamic class downloading is used to offer different services depending on the client permissions.

[2] Java, all Java-based marks, Jini and all Jini-based marks are trademarks or registered trademarks of Sun Microsystems, Inc. in the U.S. and other countries.

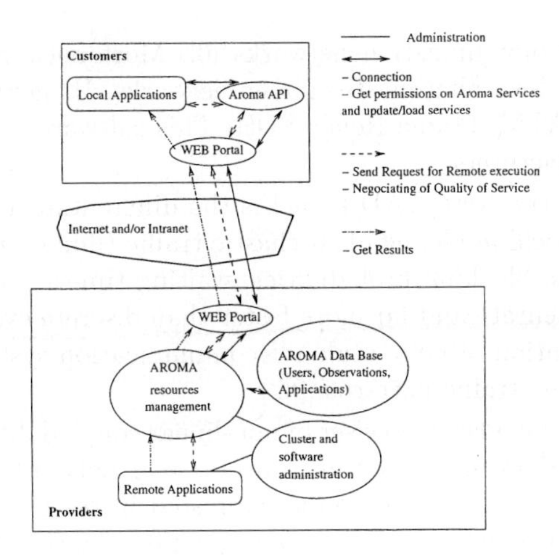

Fig. 1. Communication between customers and providers

- **Efficiency:** Mapping with Quality of Service implies watching a lot of resources. The mapping decision has to be as quick as possible and must not create a system overload. C language is used to collect system dependent information in a very efficient way. Communication between C modules and Java servers is ensured using the Java Native Interface.

2.3 WEB Portal

With the Internet portal an enterprise can have a personalized access to its applications. All dialogues between clients and providers use a WEB portal to insure security and transparent communications (Fig. 1).

2.4 Clusters and Application Administration

To insure a good quality of service of the whole system in a cluster utilization, the software Alinka raisin [5] is used. It manages quick installation and configuration of Linux kernel and useful software. Installation and management of clusters could be done using Alinka raisin. It helps providers to manage different clusters and operating system version. It could automatically create the architecture of AROMA. It is also used to link UNIX users on the provider clusters and remote users on the customer machine.

3 Telecom Modeling Application

3.1 Principle

The NetQUAD Network Planning Tool. NetQUAD is a software suite that provides a set of tools for the design, optimization and performance evaluation of large scale telecommunication networks [6]. Most network technologies are included in the various plug-in: Telephone networks, IP networks, IP-DiffServ-MPLS networks, ATM, Frame Relay, SDH. This software is used by many operators and manufacturers.

A key point in the NetQUAD kernel is the differential traffic theory that is used for the analytical modeling of stochastic traffic (link carried traffics, end to end carried traffics, blocking probabilities, waiting times, jitter ...) [7,8]. These models are very accurate and far more faster than discrete event simulations for performance evaluation of large scale telecommunication systems with complex routing and complex traffic patterns [9].

Nevertheless solving the whole set of non-linear coupled differential equations can take several hours of computing time for real networks. Computing time reduction is a major issue when one is interested in testing several changes in the network or implementing iterative optimization algorithms.

In the following we describe the circuit-switched models used in the NetQUAD Telephony plug-in and how these models have been parallelized in order to drastically reduce the computing times for very large scale networks (millions of flows, thousands of links, etc.).

Circuit-Switched Network Modeling in NetQUAD. Using classical exponential inter arrival time and service time laws, a rigorous description of the state of a circuit-switched network is given by a Markov chain. Though such hypothesis can be practically verified, they are not suited for modeling because of the huge number of states that need to be considered in a real network.

It is the aim of "traffic modeling" (queuing theory) to give ways to evaluate network performances such as the mean and variance of carried traffics as well as nodes and trunks blocking probabilities [10,11].

Theoretical studies have permitted to study the form of the mean differential equation associated with each network resource in the case of simple "cells" [12, 13]. These cells reflect classical "serial/parallel" structures that can be found within a network. This analysis has been done for call routing combining load-sharing and overflow. The first fundamental idea, developed within NetQUAD Telephony, is to model the traffic starting from the exact differential equation of each traffic flow. The second idea is to approximate blocking probabilities by means of fictitious offered traffics (equivalent traffics) linked to such set of differential equations.

Finally, the model is described by a set of ordinary differential equations whose structure (and number of equations) depends on the call routing rules. Within this model, call routing is generic; this allows any network topology associated to any call routing mechanism (multiple overflows and cross-overflows, load-sharing on primary or overflow routes, etc.)

These studies have led to the definition of a new efficient and generic tool targeted to circuit-switched networks performance analysis. NetQUAD Telephony includes an algorithm named VTAM (Voice Traffic Analytical

Modeling), the numerical translation of this model. The main advantages of VTAM are the following:

- good accuracy compared to other approaches,
- dynamical model avoiding heavy event-driven simulations when the need to study transient behavior arises,
- easy computation of the stationary solution due to an efficient fixed point algorithm (contraction property),
- differentiable with respect to network parameters (allowing gradients computation and parameter optimization).

In particular, the fixed point algorithm used to solve the nonlinear system of equations works as follows. At each iteration, once the blocking probabilities have been computed, new values of carried traffics are computed according to the traffic nonlinear equations. These equations are applied for each traffic flow independently. The blocking probabilities on trunks, set of trunks or path, are then updated by applying the Erlang-B formula to fictitious offered traffics. This process is repeated until the global convergence of the blocking probabilities.

VTAM Parallel Algorithm. The most straightforward way to take advantage of the inherent parallelism of the VTAM algorithm is to use a synchronous data parallel approach. It consists in partitioning the set of all flows into several independent separate subsets. Each subset is assigned to one processor.

The assignment of the N flows to the K parallel tasks is of course a major issue in achieving a good load-balancing. Assigning nearly N/K flows to each task does not always lead to a good load-balancing. Indeed the computational load of a parallel task greatly depends on the number of routing commands r_i of the flows i assigned to that task.

The assignment policy we have used can be described as follows. Let l_1^j and l_2^j be respectively the first and last flows assigned to task $j = 0, \ldots, K-1$. Obviously, we have the following relations (1):

$$l_1^0 = 0,\ l_1^1 = l_2^0 + 1,\ \ldots,\ l_1^K - 1 = l_2^{K-2} + 1 \quad \text{and} \quad l_2^{K-1} = N - 1 \tag{1}$$

For tasks $j = 0, \ldots, K-2$, the last flow l_2^j assigned to task j can be computed recursively using the following formula (2):

$$l_2^j = min\{k > l_1^{j-1} \text{ such that } \sum_{i=l_1^j}^{k} r_i > \sum_{i=0}^{N} r_i / K\} \tag{2}$$

Experimental results have shown that this assignment policy allows to obtain a very good load-balancing of the parallel tasks.

Once flows have been assigned to tasks, the associated nonlinear system of differential equations is solved. To this end, the K parallel tasks iteratively performs a fixed point algorithm. This parallel algorithm is based on the message passing paradigm and uses the MPI communication library. Iteration k of this algorithm proceeds as follows:

1. **Computation of Carried Traffics.** Each task $j = 0, \ldots, K-1$ computes new values of carried traffics $X_j^l(k)$ for each link l using its own subset of flows. This computation is done independently by the K tasks using the same algorithm than in the sequential algorithm.
2. **Global Reduction Operation.** Each task $j = 0, \ldots, K-1$ exchanges its vector $(X_j^0(k), \ldots, X_j^L(k))$ of carried traffics with all other tasks and computes the global carried traffic $X^l(k) = \sum_{j=0}^{K-1} X_j^l(k)$ for each link l.
3. **Update of Blocking Probabilities.** Each task $j = 0, \ldots, K-1$ independently updates the blocking probabilities of each link, set of links or path using the carried traffics $X^l(k)$, $l = 0, \ldots, L$.

This process is repeated until the global convergence of the blocking probabilities.

3.2 ASP Utilization

Interactions between NetQUAD and CASP servers are illustrated by (Fig. 2).

NetQUAD connection to AROMA is made without any change to the NetQUAD initial code. All the work is performed by a special plug-in called the Interactor.

The aim of the Interactor is to handle the communication work between the local NetQUAD and the CASP servers. The Interactor uses both NetQUAD and AROMA Application Programming Interfaces to perform authentication, authorization, submission, reporting and negotiation.

Authorization and authentication is done in two steps. First, a local verification is made in order to know if the user owns a NetQUAD license, then a remote verification is done to test if the user is known as an AROMA client. This remote verification is performed with each later communication between the Interactor and the CASP servers in order to be sure that the user has right to make the associated operation.

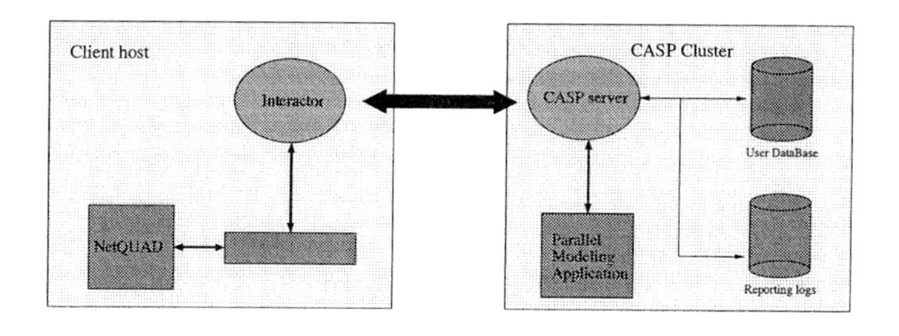

Fig. 2. Interaction between NetQUAD and CASP servers

Once the user has been identified as an AROMA client, he can submit its jobs. The Interactor then communicates with the CASP servers to make reporting or to collect results. Both operations can be done in-line or in batch mode.

An AROMA client is linked to a contract which specified the client limitations on resources utilization (globally: number of processors, maximum cpu time, memory size...). For every new submission, a verification is performed to test if the user respects its contract resource limitations. If it not the case, the Interactor is in charge of modifying and re-submitting the NetQUAD request so that it can suit the user contract limitations.

3.3 Validation on ASP Cluster

The results given in the following have been obtained on a linux PC cluster. The main features of the cluster nodes are given in Table 1. These nodes are interconnected by a 1 Gb/s myrinet network.

The testbed networks used to assess the performance of the parallel algorithm are described in Table 2. The AN2 topology approaches real topologies with very large trunk groups. The AN2-C/2 topology has been obtained from the AN2 topology by reducing the link capacities by a factor 2.

The parallel computing times obtained on the linux PC cluster with the 4 testbed networks are given in Fig. 3a while the corresponding parallel speedups are shown in Fig. 3b.

The best speedups are obtained for the large size networks AN2 and AN2-C/2. For the AN2-C/2 topology, the computing time falls from 138 seconds on 1

Table 1. Main features of the cluster nodes

Number of Processors	2
Model Name	Pentium III (Katmai)
CPU MHz	451.032
RAM memory	128 MB
Cache Size	512 KB

Table 2. Main features of the testbed networks

	EN2	AN1	AN2	AN2-C/2
Number of Nodes	113	208	312	312
Number of Links	287	388	606	606
Average capacity per link	136.64	380.94	532.83	266.41
Number of flows	12321	11200	25200	25200
Total Traffic (Erlangs)	12321	11200	25200	25200
Max Number of Hops	3	4	4	4
Max number of OverFlows	1	1	1	1
Number of Iterations	5	3	3	40
Size of the input file (MB)	19.9	42	100	100

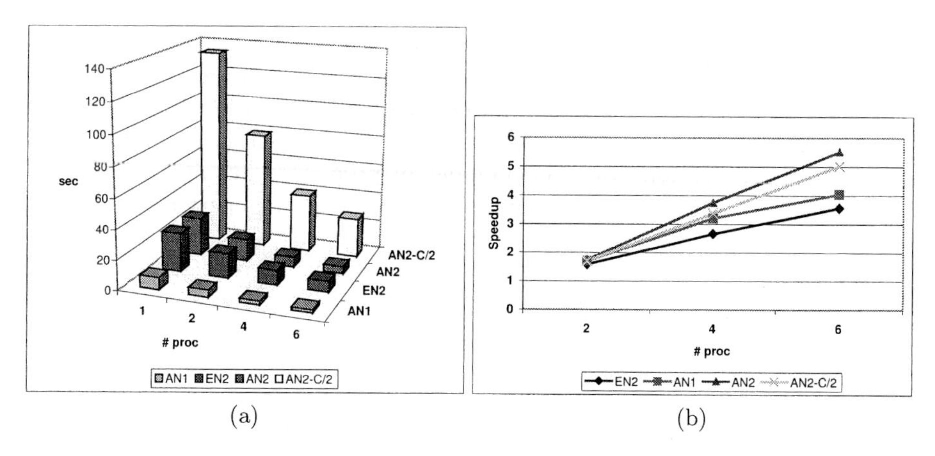

(a) (b)

Fig. 3. VTAM computing times and parallel speedups

processor to only 27.9 seconds on 6 processors. However, when the I/O processing time is a significant part of the computing time, worst speedups are obtained.

4 Conclusion

The ASP model developed in the CASP project will respond to industrial needs. Many high performance application will use remote Internet services on cluster with ASP. The Internet portal guarantees security and transparency to industrial customers. The client part is portable (use of Java Virtual Machine) and the API allows to convert different applications to the ASP model.

The ASP concept developed within the CASP project has been validated using the NetQUAD telecommunication modeling software, an industrial application used by many operators and manufacturers around the world. The new NetQUAD parallel plug-in specifically developed for ASP execution on cluster demonstrates the power of this concept. As can be seen from the experimental results, such analytic simulations are efficiently parallelized on clusters and can also support remote execution by using input and output files.

New researches are undertaken in order to supply several grades of service to applications running on the parallel cluster according to contracts between the customers and the provider.

References

1. `http://www.anite.com`
2. Sun Grid Engine. `http://www.gridengine.sunsource.net`
3. Karl Czajkowski, Ian Foster, Nick Karonis, Carl Kesselman and Stuart Martin. *A resource management architecture for metacomputing systems.* Proceedings 4th workshop on Job Scheduling Strategies for Parallel Processing,pp 62–82 Springer-Verlag LNCS 1459, 1998.

4. T. Monteil, J.M. Garcia *Task Allocation Strategies on Workstations Using Processor Load Prediction.* PDPTA97, International Conference on Parallel and Distributed Processing Techniques and Applications, pp. 416–421, Las Vegas USA, 1997
5. `http://www.alinka.com`
6. `http://www.delta.fr`
7. JM Garcia, D. Gauchard, O. Brun, P. Bacquet, J. Sexton and E. Lawless. *Modélisation Différentielle et Simulation Hybride Distribuée*, Réseaux et Systèmes Répartis, Volume 13, No 6, 2001.
8. Garcia J.M. *A new approach for analytical modelling of packet switched telecommunication networks* . LAAS Research Report N° 98443, 1998.
9. Misra J. *Distributed Discrete-Event Simulation.* ACM Computing Surveys, Vol.18, N° 1, 1986.
10. Roberts, J. et al Eds *Broadband Network Teletraffic, Final Report of Action Cost 242.* Springer Berlin, 1996.
11. Takacs, L. *Introduction to the Theory of Queues.* Oxford University Press, 1962.
12. Garcia J. M., Le Gall F. et Bernussou J. *A model for telephone networks and its use for routing optimization purposes*, newblock IEEE Journal on selected areas in communication, Special issue on communication network performance evaluation, Vol. 4, No 6, september 1986.
13. Garcia J.M. *Problèmes liés à la modélisation du traffic et à l'acheminement des appels dans un réseau téléphonique.* Thèse de doctorat, Université Paul sabatier (Toulouse), 1980.

Track on Parallel Methods and Systems

Fault-Tolerant Routing in Mesh-Connected 2D Tori

Eunseuk Oh[1], Jong-Seok Kim[2], and Hyeong-Ok Lee[3]

[1] Department of Computer Science
Texas A&M University, College Station, TX 77843-3112
eunseuko@cs.tamu.edu
[2] Department of Computer Science
[3] Department of Computer Education
Sunchon National University, Sunchon, Chonnam, 540-742, Korea
{rockhee,oklee}@sunchon.ac.kr

Abstract. In this paper, we study natural conditions for 2D tori with a large number of faulty nodes to remain connected. Under the suggested connectivity conditions, we develop efficient routing algorithms in 2D tori with a large number of faulty nodes. As long as a given input torus and the meshes within the torus satisfy the conditions, the routing algorithm successfully constructs a fault-free path by using only local information of the network. Also, our algorithms do not require faulty nodes and faulty blocks to be a special structure such as convex, rectangle, while each mesh in a given torus can sustain as many faulty nodes as possible, provided that non-faulty nodes of the mesh are connected and the mesh holds the connectivity conditions. Specifically, for a torus sustaining up to 22.2% faulty nodes, in linear time, our algorithm constructs a fault-free path of length bounded by six times the shortest path length between the two nodes.

1 Introduction

Due to rapid progress in hardware technology, designing a distributed parallel computing system connecting a large number of multiprocessors has become feasible. In such a system, multiprocessors communicate by message passing. As the size of the network interconnecting multiprocessors continuously increases, routing in networks with a large number of faults becomes unavoidable. Let G be a regular network of degree n. The node-connectivity of G is the minimum integer k such that removing any k nodes in G results in a disconnected or trivial graph. It is well-known that the node-connectivity of an n-regular graph is n. Thus, many fault-tolerant communication algorithms for n-regular graphs allow to contain up to $n-1$ faulty nodes. On the other hand, there has been much effort to try to allow n-regular networks to have more than $n-1$ faulty nodes [9,10,11].

Recently, a new connectivity measure, *local subcube-connectivity* has been introduced to identify conditions where hypercube networks with a very large

P.M.A. Sloot et al. (Eds.): ICCS 2003, LNCS 2659, pp. 527–536, 2003.

number of faulty nodes still remain connected [3,4]. A network is functional as long as every pair of non-faulty nodes in a network containing faulty nodes is still able to communicate. That is, any pair of two non-faulty nodes are connected by a fault-free path in such a network. Under the conditions that a hypercube network is still connected, Chen, et al. were able to contain up to 37.5% faulty nodes in hypercubes. Also, a formal theoretical analysis and experimental simulations have been provided to demonstrate a realistic and practical view of their connectivity measure [2,3,5]. They mentioned that their new connectivity measure is applicable to other hierarchical networks. In this paper, we establish the result for the applicability of their definitions to networks with a bounded degree such as tori and mesh.

A 2D torus is one of the most popular graph models used to represent interconnection networks [7]. A 2D torus represents parallel computers in which processors are connected by a two dimensional grid of communication links with wraparound connections. Due to a constant degree of tori and meshes, developing routing algorithms that can only tolerate constant faulty nodes would not be practical. In many real situations, fault-free routing paths in a torus or mesh can be constructed even though some nodes are disconnected from the network. Many faulty models on meshes that can contain faulty nodes greater than its tolerance has been introduced [1,6,12]. Under their faulty models, a fault-free path can be constructed on the mesh containing a large number of faulty nodes. However, most faulty models require that faulty blocks form special structures such as rectangle, convex, or nonconvex (L, T, or + shape), which regard all non-faulty nodes within these blocks as faulty. Our algorithms do not put such a restriction on faulty structures, and only assume that non-faulty nodes in the network make a connected graph and limit the number of faulty nodes on each mesh's boundary to be less than half.

For a given torus, our routing algorithm constructs a path, tracing meshes of similar sizes in the torus. As long as a given input torus and meshes within the torus satisfy the required conditions, the suggested algorithm successfully constructs the fault-free path even though the torus contains a large number of faulty nodes. Also, most of algorithms developed under faulty block models focused on wormhole routing, whose time complexity is more likely independent of the distance of the communication path if the message is long enough [8]. On the other hand, we consider our algorithm under the store-and-forward model, whose time complexity mostly depends on the distance of the communication path. Specifically, for a torus sustaining up to 22.2% faulty nodes, in linear time, our algorithm can construct a fault-free path of length bounded by six times the shortest path length between the two nodes. Further, our algorithm does not require global information about the network, and a fault-free path is constructed by only using information about directly connected nodes. Most of our discussions were focused on tori and their connectivity, but our results can easily be extended to meshes.

2 Preliminaries

A two-dimensional torus T(shortly, 2D torus or $n \times n$ torus) is an undirected graph with n^2 nodes whose label is identified by two coordinates (i, j) and with $2n^2$ edges. For a node (i, j) in a torus, it has four neighbors $(i-1, j)$, $(i+1, j)$, $(i, j-1)$, and $(i, j+1)$, where the "+" and "-" operations on indices are performed as modulo arithmetic. The distance between two nodes $u = (i_1, j_1)$ and $v = (i_2, j_2)$ in T is the shortest path length between them defined as

$$dist(u, v) = min(|i_1 - i_2|, n - |i_1 - i_2|) + min(|j_1 - j_2|, n - |j_1 - j_2|).$$

There are four simple and deterministic routing algorithms that connect the source node u and the destination node v: (1) the message is routed along the X dimension and then along the Y dimension, (2) the message is routed along the X dimension and then along the Y dimension using wraparound, (3) the message is routed along the X dimension using wraparound and then along the Y dimension, and (4) the message is routed along the X dimension using wraparound and then the Y dimension using wraparound. A minimal path is determined among them. The diameter, the maximum distance between any two nodes of an $n \times n$ torus, is $2\lfloor \frac{n}{2} \rfloor$.

2.1 Balanced 4-ary Partitioning Tree

A two-dimensional torus T can be divided into meshes in many different ways. To explain uniformly, we assume that T is divided into meshes of size at most h^2, where $h = \lceil \frac{n}{2^\alpha} \rceil$, and α is a natural number. Each node in a 2D mesh is labeled as (i, j) and $0 \leq i, j < n$. The lower-most West node is labeled as (0,0), and the upper-most East node is labeled as $(n-1, n-1)$. We first propose a labeling scheme to identify meshes in a torus when a torus is divided into meshes of similar sizes. The coordinates of the lower-most West nodes of the meshes are obtained by a *balanced 4-ary partitioning tree* which is defined below.

Definition 1. *A balanced 4-ary partitioning tree is constructed as follows.*

1. *the root node is labeled as $n \times n$.*
2. *for each node with a label $h_1 \times h_2$, generate the first, second, third, and fourth children of the node and label them as $h_{11} \times h_{21}$, $h_{12} \times h_{21}$, $h_{11} \times h_{22}$, and $h_{12} \times h_{22}$, respectively, where $h_{11} = \lfloor \frac{h_1}{2} \rfloor$, $h_{12} = \lceil \frac{h_1}{2} \rceil$, $h_{21} = \lfloor \frac{h_2}{2} \rfloor$, and $h_{22} = \lceil \frac{h_2}{2} \rceil$.*

For the balanced 4-ary partitioning tree with depth d, there are 4^d leaf nodes. The depth of the root is 0. Thus, an $n \times n$ torus is divided into 4^d meshes of size at most h^2, where $h = \lceil \frac{n}{2^d} \rceil$. The balanced 4-ary partitioning tree whose root is labeled as 13×13 is shown in Fig. 1. We assume that each node has pointers to nodes corresponding to neighboring meshes. The actual coordinates of the lower-most West node of a mesh in a torus can be calculated as follows. Let $\beta = \lfloor \frac{n}{2^d} \rfloor$.

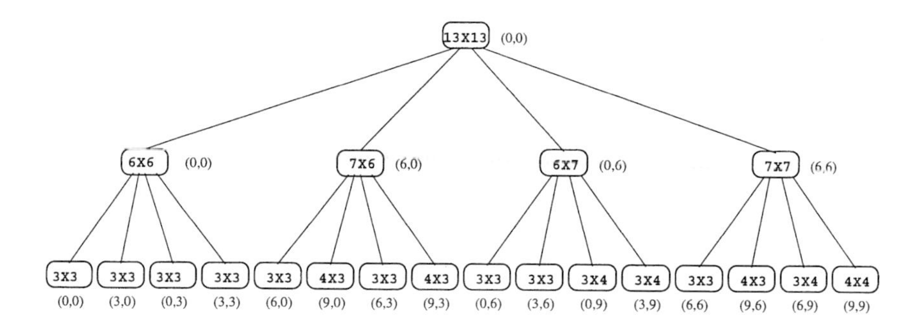

Fig. 1. Partition of a 13×13 torus by the balanced 4-ary partitioning tree with depth 2

Coordinates for Lower-Most West Nodes of Meshes

1. the lower-most West node of a mesh identified by the root node is (0,0).
2. for the lower-most West node (i, j) in a mesh identified by the node in a balanced 4-ary partitioning tree at depth d, the lower-most West nodes of four meshes corresponding to the first, second, third, and fourth children of the node (i, j) are (i, j), $(i + \beta, j)$, $(i, j + \beta)$, $(i + \beta, j + \beta)$.

Now, mesh labels in a torus are obtained through its respective coordinates, $[i, j]$ where a mesh can be viewed as a node in a two dimensional grid. For any two meshes in a torus divided by the balanced 4-ary partitioning tree, the number of nodes on two adjoining boundaries are the same.

3 The Local Mesh-Connectivity

After applying the balanced 4-ary partitioning tree with depth d to an $n \times n$ torus, a label $[i, j]$ corresponds to a mesh whose coordinate of the lower-most West node is $(\beta \cdot i, \beta \cdot j)$, where $\beta = \lfloor \frac{n}{2^d} \rfloor$. The size of the mesh is decided according to the definition of a balanced 4-ary partitioning tree. The set of meshes obtained by applying the balanced 4-ary partitioning tree with the depth d is called a *d-mesh set*, and a mesh in the d-mesh-set is called d-mesh. In addition, the distance between the two d-meshes $M_1 = [i_1, j_1]$ and $M_2 = [i_2, j_2]$ is called *d-distance* and defined as

$$\Delta(M_1, M_2) = min(|i_1 - i_2|, 2^d - |i_1 - i_2|) + min(|j_1 - j_2|, 2^d - |j_1 - j_2|).$$

Definition 2. *The 2D torus T is locally d-mesh-connected if in each d-mesh in T, the non-faulty nodes in the d-mesh are connected.*

For any given two non-faulty nodes u and v in a locally d-mesh-connected torus, a sufficient condition to exist a fault-free path between them is provided below. We denote it as $u \overset{p}{\leadsto} v$ if there is a path p between u and v, and as $u \rightarrow v$ if u is directly connected to v.

Theorem 1. *If, on any side of a boundary of size h in each d-mesh, the number of faulty nodes is at most $\lfloor \frac{h-1}{2} \rfloor$, then the non-faulty nodes in a locally d-mesh-connected 2D torus T make a connected graph.*

Proof. To prove that the non-faulty nodes in a locally d-mesh-connected 2D torus make a connected graph, it suffices to show that for any two given nodes u and v in a 2D torus, there is a fault-free path $u \sim v$. Since the non-faulty nodes in a d-mesh in T are connected, if u and v is in a d-mesh M, then $u \overset{p}{\sim} v$ where all nodes on p are in M. Suppose the node u is in a d-mesh M and the node v is in another d-mesh M'. Let a neighboring d-mesh of M be M_1. Then, there are h pairs of adjacent nodes such that one is on the boundary of M and another is on the boundary of M_1. Since we assume that on any boundary of a d-mesh, less than half of the nodes are faulty, there must be a path of adjacent non-faulty nodes $b_0 \rightarrow b_1$ such that b_0 is in M and b_1 is in M_1. By the definition of d-mesh connectivity, we have $u \overset{p}{\sim} b_0$, where all nodes on p are in M. Thus, we can construct a fault-free path $u \overset{p}{\sim} b_0 \rightarrow b_1$. We continue this construction until we have $u \sim v'$ such that v' is a non-faulty node on the boundary of M'. Inductively, the fault-free path $u \sim v'$ can be constructed by extending the path between them going through meshes between M and M'. Since non-faulty nodes in M' are connected, there is a fault-free path $v' \overset{p'}{\sim} v$ such that all nodes on p' are in M'. Therefore, the concatenation of the path $u \sim v'$ and the path between $v' \sim v$ gives a fault-free path $u \sim v$. □

Under the above boundary condition, the non-faulty nodes in a locally d-mesh-connected 2D torus T make a connected graph, though T may contain a very large number of faulty nodes. It shows that the local d-mesh connectivity leads to the global connectivity of the network with a large number of faulty nodes as long as the boundary condition is satisfied. Furthermore, as pointed out in [3], the local mesh-connectivity can also be detected in a distributed way by using a local managing program on a processor in each mesh.

Suppose that a d-mesh contains faulty nodes up to less than half of its total number of nodes. When d is small enough, the number of faulty nodes that can be contained in a locally d-mesh-connected torus T would be close to half of the total number of nodes in T. To give an intuition of the number of faulty nodes, consider a square $n \times n$ torus and set $d = log\frac{n}{h}$. Since a d-mesh has size of h^2 and $h = \frac{n}{2^d}$, a d-mesh may contain up to $4^d(\frac{1}{2}(\frac{n}{2^d})^2 - 1) = \frac{n^2}{2} - 4^d$ faulty nodes. According to the sufficient condition on the boundary of a mesh, a $log\frac{n}{3}$-mesh can contain up to 2 faulty nodes.

Corollary 1. *Under the boundary condition, if each d-mesh in a locally d-mesh connected torus T has at most 2 faulty nodes, then the non-faulty nodes in T make a connected graph, where $d = log\frac{n}{3}$.*

Corollary 2. *Under the boundary condition, a locally d-mesh-connected torus T is also locally k-mesh-connected for all $k < d$.*

So far, we assume that d is fixed in a locally d-mesh-connected torus and its non-faulty nodes are connected as long as the number of faulty nodes in any side of a boundary of d-mesh is less than half of the nodes on the boundary. We generalize the previous condition to allow a faulty d-mesh(the mesh whose nodes are all faulty), while the torus remains connected.

Definition 3. *The 2D torus T is locally mesh-connected if each d-mesh, $d \geq 1$, in T is contained in an h-mesh whose non-faulty nodes are connected, $h \leq d$.*

Let δ be the largest integer such that the locally mesh-connected 2D torus T is locally δ-mesh-connected. Such a number δ always exists. For example, $\delta = 0$. In this case, the locally mesh-connected 2D torus itself is locally 0-mesh-connected. If the boundary condition is satisfied, all non-faulty nodes in T are connected. The following theorem directly directly from the definition of δ.

Theorem 2. *If, on any side of a boundary of size h in each δ-mesh, the number of faulty nodes is at most $\lfloor \frac{h-1}{2} \rfloor$, then the non-faulty nodes in a locally mesh-connected 2D torus make a connected graph.*

4 Routing in Locally Mesh-Connected Tori

As described in Theorem 1, a path between two given non-faulty nodes in a locally mesh-connected torus can be constructed using only non-faulty nodes when all d-mesh satisfy the boundary condition. Before we present routing algorithms in a locally mesh-connected torus, we discuss the worst case to find two non-faulty nodes u and u_1, where u is in a mesh M and u_1 is in the adjoining boundary of a mesh M_1. Without loss of generality, we assume that the size of M is $h_1 h_2$ and the size of M_1 is $h_3 h_2$. The worst case would take time $O(h_1 h_2)$. We will focus more on routing algorithms whose time complexity is linear and the length of path. An example case that takes roughly half of the size of a d-mesh is given in Fig. 2.

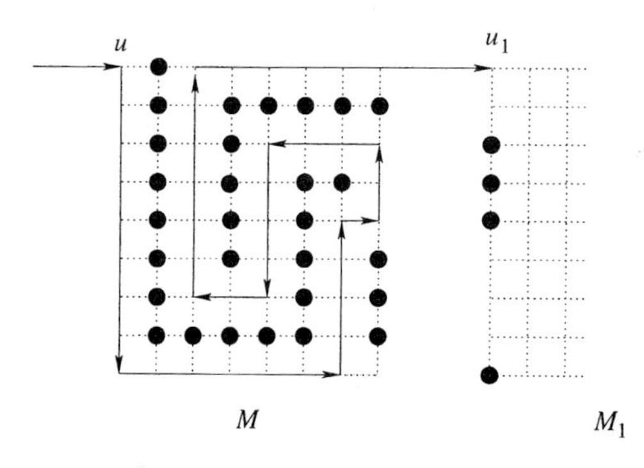

Fig. 2. An example for connecting two non-faulty nodes, one is in M and the other is in M_1

4.1 Routing in a Locally d-Mesh-Connected Torus

Assume that a d-mesh M contains the source node u and another d-mesh M' contains the destination node v. An efficient routing algorithm for a locally d-mesh-connected torus is given in Fig. 3.

Routing
Input: an $n \times n$ torus T and non-faulty nodes u and v in it
Output: a fault-free path $P = u \sim v$ or return FAIL if T is not locally d-mesh-connected or it does not satisfy the boundary condition.
{Assume that u is in a d-mesh M and v is in another d-mesh M'.}
Phase 1: find labels for d-meshes connecting M and M' whose d-distance is $\Delta(M, M')$.
  Let such d-meshes be $M(= M_0), M_1, \ldots, M' = (M_{\Delta(M,M')})$;
Phase 2: find a fault-free path from u in M to a non-faulty node in M';
  2.2 initialize the path $P = u$, and let $w = u$;
  2.3 **for** $k = 0$ **to** $\Delta(M, M') - 1$ **do**
    find a fault-free path $b_1 \to b_2$ such that b_1 is in M_k and b_2 is in M_{k+1};
    extend the path P to $w \overset{p}{\sim} b_1 \to b_2$, where all nodes of p are in M_k;
    let $w = b_2$;
Phase 3: find a fault-free path from a non-faulty node to v in M';
  {at this point, w is on a boundary of M'. }
  extend the path P to $w \overset{p'}{\sim} v$, where all nodes of p' are in M';
if, at any point, the algorithm can not proceed, return FAIL;

Fig. 3. Routing in a locally d-mesh-connected torus

Theorem 3. *For two given non-faulty nodes u in a d-mesh M and v in another d-mesh M' in a locally d-mesh-connected torus, the algorithm* **Routing** *constructs a fault-free path $u \sim v$ in time $O(hn)$ if all d-meshes satisfy the boundary condition, where $h = \lceil \frac{n}{2^d} \rceil$.*

Proof. The construction of a fault-free path P starts from the node u and traverses d-meshes detected in Phase 1. When u is in M and v is in M', d-meshes connecting M and M' whose length is $\Delta(M, M')$ can be found in time $2 \cdot 2^d = O(\frac{n}{h})$, which is chosen to have the minimum number of d-meshes among four possible routes from the label of M to the label of M' with or without wraparound. In the kth execution of the loop in Phase 2, a non-faulty node u in M_k is connected to a non-faulty node b_2 in M_{k+1} going through b_1 such that $b_1 \to b_2$, where b_1 is on a boundary of M_k and b_2 is on a boundary of M_{k+1}. Finding the path $b_1 \to b_2$ takes time $O(h)$. Since a non-faulty node u can be connected to these adjacent nodes in time $O(h^2)$, we can find a fault-free path $u \sim b_2$ in time $O(h^2)$. After $\Delta(M, M')$ executions of the loop in Phase 2, the node w is in a boundary of M'. Since the non-faulty nodes in M' make a connected graph, in time $O(h^2)$, we can find a fault-free path $w \sim v$, and can extend the path P to $w \sim v$. Therefore, if the $n \times n$ torus is locally d-mesh-connected and all d-meshes satisfy the boundary condition, then the algorithm **Routing** constructs a fault-free path $u \sim v$. The running time of the algorithm is bounded by $O(\frac{n}{h} + \Delta(M, M')h^2) = O(hn)$, since $\Delta(M, M') = O(\frac{n}{h})$. □

In the case where h is a small constant, for example $h \leq 4$, the algorithm **Routing** runs in linear time. As shown in Fig. 2, some structures of a d-mesh take time $O(h^2)$ to search non-faulty nodes. Thus, our time complexity is reasonably tight even for large h. The length of the path P constructed by **Routing** is bounded by $(\Delta(M, M') + 1)h^2 - 1$ because the length of P is increased by at most h^2 in each execution of the loop in Phase 2 and the length of the path constructed in Phase 3 is bound by $h^2 - 1$. Again, if h is small, the length of the path P becomes $O(n)$. Specifically, for a torus divided into 3×3 meshes by a 4-ary balanced partitioning tree, a locally d-mesh-connected torus may contain up to $\frac{2}{9}n^2$ faulty nodes, which is 22.2% of the total number of nodes in a torus. The number of nodes that can be contained in a locally mesh-connected torus would increase as the size of a d-mesh increases at the cost of the path length. For a locally d-mesh-connected torus consisting of 3×3 meshes, we have the following corollary.

Corollary 3. *For a locally d-mesh-connected torus consisting of $log\frac{n}{3}$-meshes, in linear time, the algorithm* **Routing** *constructs a fault-free path of length bounded by $6\Delta(M, M')$.*

4.2 Routing in a Locally Mesh-Connected Torus

Similarly a fault-free path $u \sim v$ in a locally mesh-connected torus T can be constructed when all δ-meshes satisfy the boundary condition. A routing algorithm for T is given in Fig. 4. The largest integer δ such that T is locally δ-mesh connected is not given in our algorithm. If the value δ is known, a path $u \sim v$ can be constructed by the algorithm **Routing**.

Theorem 4. *If the input torus T of the algorithm* **δ-Routing** *is locally mesh-connected, then for two non-faulty nodes u and v, and δ such that T is locally δ-mesh-connected, the algorithm* **δ-Routing** *constructs a fault-free path $u \sim v$ in time $O(hn)$, provided that all δ-meshes satisfy the boundary condition, where $h = \lceil \frac{n}{2^\delta} \rceil$.*

Proof. Suppose a path $P = u \sim w$, $w \neq v$ is constructed inductively. By the definition of the value δ, we must have an integer $k \geq \delta$ such that w is in a k-mesh M whose non-faulty nodes are connected, and a fault-free path $b_1 \to b_2$ can be constructed between b_1 in a boundary of M and b_2 in a boundary of k-mesh M_1. However, since we do not know the value δ, we will try all possible values from $k = \lceil log\frac{n}{2} \rceil$ to find a fault-free path $b_1 \to b_2$. In the worst case, $k = \delta$, we have $w \overset{p}{\sim} b_1 \to b_2$, where all nodes of p are in a δ-mesh. Using the balanced 4-ary

δ-Routing
Input: an $n \times n$ torus T and non-faulty nodes u and v in T
Output: a fault-free path $P = u \sim v$ or return FAIL if T is not locally mesh-connected or it does not satisfy the boundary condition.
1. let $w = u$;
2. initialize the path $P = w$;
3. **while** $w \neq v$ **do**
    $k = \lceil log\frac{n}{2} \rceil$; done = false;
    while done=false and w is not in M' **do**
3.1   find a k-mesh M_0 with w and its neighboring k-mesh M_1 such that $\Delta(M_1, M') = \Delta(M_0, M') - 1$, where M' is a k-mesh with v;
3.2   **if** there is $w \overset{p}{\sim} b_1 \rightarrow b_2$ such that b_1 is in M_0 and b_2 is in M_1;
        then
            extend the path P to the path $w \overset{p}{\sim} b_1 \rightarrow b_2$, where all nodes of p are in M_0; let $w = b_2$;
            done = true;
        else
            $k = k - 1$;
3.3 **if** w is in M'
        while there is no fault-free unused path $w \sim v$ **do**
            $k = k - 1$;
        let M'' be a k-mesh with v;
        extend the path P to $w \overset{p'}{\sim} v$,where all nodes of p' are in M'';
        return $P = u \sim v$;
if, at any point, the algorithm can not proceed, return FAIL;

Fig. 4. Routing in a locally mesh-connected torus

partitioning tree, Step 3.1 of the algorithm **δ-Routing** can be done in a constant time after identifying a k-mesh M_0 with u, which takes $O(log\frac{n}{2})$. Since we need to check whether non-faulty nodes in a k-mesh are connected and it satisfies the boundary condition for each value k, Step 3.2 of the algorithm **δ-Routing** takes $O(2^2) + O(4^2) + \cdots + O(h^2) = O(h^2)$. Thus, the total time of Step 3.1 and Step 3.2 is bounded by $log\frac{n}{2} + h^2\Delta(M, M') = O(hn)$, where $h = \lceil \frac{n}{2^\delta} \rceil$. If w is in a k-mesh M', then nodes w and v are in the same δ-mesh. We try to find a fault-free path $w \sim v$ in the smallest size mesh(thus, the largest k) from the value k found in Step 3.2. Thus, a fault-free path $w \sim v$ can be constructed in time $O(h^2)$. It shows that in time $O(hn)$, the algorithm **δ-Routing** constructs a fault-free path $u \sim v$ in the locally mesh-connected torus if all δ-meshes satisfy the boundary condition. □

In general, the running time of the algorithm **δ-Routing** would be much better than $O(hn)$. For example, if a torus T contains relatively fewer faulty nodes so that most small meshes in T satisfy the boundary condition and non-faulty nodes in them are connected, then the largest value k can be found in the earlier iteration of Step 3.2. Thus, the path P can be extended in meshes smaller than h^2 in most cases.

5 Conclusion

We have studied the connectivity measure for a 2D torus with a large number of faulty nodes to remain connected, and established the result for tori. Our study shows that the local subgraph-connectivity suggested in [3,4] also successfully extends to networks with a bounded degree. The algorithm developed for tori meeting these conditions is practical and efficient in most cases. Further, our algorithm does not require global information about the network, and a fault-free path is constructed by using only information about directly connected nodes. In this paper, we only consider the routing problem. Other communication problems in tori such as broadcasting and gossiping would be interesting to study under this new connectivity measure. In addition, our result can easily be extended to meshes. Without considering wraparound connections, a 2D torus has the same structure as a 2D mesh. Thus, our algorithm can be applied to a 2D mesh with a slight modification.

References

1. S. Chalasani and R.V. Boppana, "Communication in Multicomputers with Nonconvex faults," *Proc. Int'l Symp. Computer Architecture*, pp. 268–277, 1990.
2. J. Chen, I. A. Kanj, and G. Wang, "Hypercube Network Fault Tolerance: A Probabilistic Approach," *Proc. Int'l Conf. Parallel Processing*, pp. 65–72, 2002.
3. J. Chen, G. Wang, and S. Chen, "Locally Subcube-Connected Hypercube Networks: Theoretical Analysis and Experimental Results," *IEEE Trans. Computers*, vol. 51, no. 5, pp. 530–540, 2002.
4. J. Chen, G. Wang, and S. Chen, "Routing in Hypercube Networks with A Constant Fraction of Faulty Nodes," *Journal of Interconnection Networks*, vol. 2, pp. 283–294, 2001.
5. J. Chen, T. Wang, and E. Oh, "Adaptive Routing: Centralized versus Distributed," Technical Report, Texas A&M University, 2002.
6. A.A. Chien and J.H. Kim, "Planar-Adaptive Routing: Low-Cost Adaptive Networks for Multiprocessors," *Proc. Int'l Symp. Computer Architecture*, pp. 268–277, 1990.
7. W.J. Dally and C.L. Seitz, "The torus Routing Chip," *Distributed Computing*, pp. 187–196, 1986.
8. W.J. Dally and C.L. Seitz, "Deadlock-free message routing in multiprocessor interconnection network," *IEEE Trans Comput.* vol. 36, no. 5, pp. 547–553, 1987.
9. Q.-P. Gu and S. Peng, "k-Pairwise Cluster Fault Tolerant Routing in Hypercubes," *IEEE Trans. Computers*, vol. 46, pp. 1042–1049, 1997.
10. Q.-P. Gu and S. Peng, "Unicast in Hypercubes with Large Number of Faulty Nodes," *IEEE Trans. Parallel and Distributed Systems*, vol. 10, pp. 946–975, 1999.
11. S. Latifi, M. Hedge, and M. Naraghi-Pour, "Conditional Connectivity Measures for Large Multiprocessor Systems," *IEEE Trans. Computers*, vol. 43, pp. 218–222, 1994.
12. C.C. Su and K.G. Shin, "Adaptive Fault-Tolerant Deadlock-Free Routing in Meshes and Hypercubes," *IEEE Trans. Computers*, vol. 45, no. 6, pp. 666–683, 1996.

Network-Tree Model and Shortest Path Algorithm

Guozhen Tan[1], Xiaojun Han[1], and Wen Gao[2]

[1] Department of Computer Science and Engineering
Dalian University of Technology, 116024, China*
gztan@dlut.edu.cn
[2] Institute of Computing Technology
Chinese Academy of Sciences, 100080, China

Abstract. For the first time, this paper proposes the Network-Tree Model (NTM), route optimization theory and wholly new, high performance algorithm for shortest path calculation in large-scale network. We first decompose the network into a set of sub-networks by adopting the idea of multi-hierarchy partition and anomalistic regional partition, and then construct the NTM and the Expanded NTM. The Network-Tree Shortest Path (NTSP) algorithm presented in this paper narrows the searching space of the route optimization within a few sub-networks, so it greatly reduces the requirements for main memory and improves the computational efficiency compared to traditional algorithms. Experiment results based on grid network show that NTSP algorithm achieves much higher computational performance than Dijkstra algorithm and other hierarchical shortest path algorithms.

Keywords: nNetwork-tree model, shortest path algorithm, large-scale network, route optimization algorithm

1 Introduction

The shortest-path problem (SP) is a key problem of many fields [1] such as computer science, transportation engineering, communication engineering, system engineering, and operation research. Especially, the computing efficiency is the bottleneck problem of route optimization for large-scale network. Furthermore, the main memory requirement also is an important factor as to implement the algorithm [2].

As known, there are two kinds of algorithms for shortest path computing, that one is label-setting algorithm and the other is label-correcting algorithm, with the representation of the classic Dijkstra algorithm and Bellman-Ford algorithm respectively. All the other algorithms are the variations of these two kinds. People have widely realized that all these algorithms cannot meet the requirements of many applications when computing the shortest paths of large-scale network.

* This work was supported in part by Grand 99025 of Ministry of Education and Grand 9810200104 of Liaoning Science Foundation, China.

P.M.A. Sloot et al. (Eds.): ICCS 2003, LNCS 2659, pp. 537–547, 2003.

Partitioning the network and constructing new network model have become a fast and effective methods for improving computations in large-scale networks [3], which have been widely applied in many fields such as Intelligent Transportation System (ITS), Graphic Information System (GIS) and route protocols of network communication. Level graphs and associated approximate shortest path algorithm were presented in [4,5,6], which significantly improved the computing efficiency of path optimization. However, the level graph model cannot be abstracted from any given graph, and the papers only provide the approximate shortest path algorithm with the errors that cannot be overlooked. Hierarchical Encoded Path View (HEPV) model can be found in [2,7], and a path-query algorithm based on this model was proposed to precompute and store the shortest path of each sub-network for later on-line query. It did not provide significant improvement for route optimization of multi-level large-scale network. Furthermore, the method requires a large overhead main memory for storing the precomputed path information, which makes it impractical to use for calculating the shortest path in a large-scale network.

For the first time, we present the new network-tree model and the associated network-tree shortest path (NTSP) algorithm to solve effectively the storage requirement and efficiency problem of large-scale network optimization. The prominent advantages of network-tree model and NTSP algorithm compared to the previous ones can be founded in the following four aspects: (1) We can construct the network-tree model from any network; (2) The regional partition and hierarchical partition method of network-tree model is very useful to store the network date both hierarchically and distributively; (3) The concept of tree expresses clearly the internal relation among the sub-networks. (4) The route optimization can be constrained within a few sub-networks, which can greatly reduce the searching scope and significantly improve the computing efficiency. So the main memory requirements of our algorithm are very limited.

The remainder of this paper is organized as follows: In Sect. 2, we present the definitions of both the network-tree model and expanded network-tree model, and the optimization theorem. In Sect. 3, we present the preprocessing procedure and NTSP algorithm. The experiment result will be presented in Sect. 4. Section 5 is the conclusions.

2 Network-Tree Model

Network-Tree Model is the definitional combination and expansion of network and tree. Any network can be divided into a number of sub-networks, and each sub-network will be constructed to one node of network-tree. Specially, the node of network-tree is called macro-node to distinguish from the node of original network.

2.1 Definition of Network-Tree Model

Let the network-tree be $N\_Tree = (T, H)$, where T is the macro-node set of N. If only one macro-node exists in T, $H = \phi$, otherwise, H is a binary relation defined on T. There is only one macro-node m_r called the root of N_Tree. If $T - \{m_r\} \neq \phi$, there exists a division of $T_1, T_2, \cdots, T_n (n > 0)$, with $\forall j \neq k, 1 \leq j, k \leq n, T_j \cap T_k = \phi$. For $\forall i, 1 \leq i \leq n$, the only macro-node $m_i \in T_i$ satisfies $< m_r, m_i > \in H$. As the counterpart of the $T - \{m_r\}$ division, there exists the only division $H_1, H_2, \cdots, H_n$ of $H - \{< m_r, m_1 >, \cdots, < m_r, m_n >\}$, with $\forall j \neq k, 1 \leq j, k \leq n, H_j \cap H_k = \phi$. For $\forall i, 1 \leq i \leq n$, H_i is a binary relation defined on T_i, and (T_i, H_i) is also a network-tree defined above, and we call it the sub-tree of the root macro-node m_r.

Next, we will present the method of network division and network-tree model construction procedure from any network.

2.2 Construction of Network-Tree Model

Let the network be $N = (V, A, W)$, where V is a finite set of nodes, $A \in N \times N$ is a finite set of arcs, $W = \{w(a) | a \in A\}$, w is the mapping function of $A \rightarrow R^+$. First, we assign levels to all arcs according to properties in the original network, that is, we create a mapping function for arc set A.

Definition 1. $\forall a \in A, \exists h \in I, I = \{1, 2, \cdots, k\}$, there exists a mapping function $f : A \rightarrow I$, which makes $f(a) = h$, where h is the level of arc a. And $\forall h \in I$, which makes $f^{-1} = \{a | a \in A, f(a) = h\}$ as a non-null space. Let the arc set of level h be A_h, that is, $A_h = f^{-1}(h)$.

Generally, let A_1 be the highest level arc set. The more importance in rank in the network, the smaller value of arc level.

Based on the arc level, we create the network-tree $N\_Tree = (T, H)$ from the network $N = (V, A, W)$. Let the A_1 induced sub-network of N be the root macro-node m_1 of the network-tree, $m_1 \in T$. If $N - \{m_1\} \neq \phi$, we divide the sub-network $N - \{m_1\}$ into n_1 non-null sub-networks $T_{11}, T_{12}, \cdots, T_{1n_1}$ by A_1, and for $\forall p \neq q$, $1 \leq p, q \leq n_1$, we get $T_{1p} \cap T_{1q} = \phi$. Then $\forall i, 1 \leq i \leq n_1$, abstract the highest-level arcs from T_{1i} to make the arc set A^{1i}, let the A^{1i} induced sub-network of T_{1i} and the node intersection set of T_{1i} and m_1 be the ith son macro-node m_{1i} of m_1, that is $m_{1i} \in T$, $< m_1, m_{1i} > \in H$. If $T_{1i} - \{m_{1i}\} \neq \phi$, we continue to divide the sub-network $T_{1i} - \{m_{1i}\}$ into n_{1i} non-null sub-networks $T_{li1}, T_{1i2}, \cdots, T_{1in}$ by A^{1i}, and for $\forall p \neq q$, $1 \leq p, q \leq n_{1i}$, we get $T_{1ip} \cap T_{1iq} = \phi$. Then $\forall j$, $1 \leq j \leq n_{li}$, abstract the highest-level arcs from T_{1ij} to make the arc set A^{lij}, let the A^{lij} induced sub-network of T_{1ij} and the node intersection set of T_{1ij} and m_{1i} be the jth son macro-node m_{1ij} of m_{1i}, that is $m_{1ij} \in T$, $< m_{1i}, m_{1ij} > \in H$. Follow the step above to continue the construction until we get the network-tree model $N\_Tree = (T, H)$.

We get the following definitions and properties of NTM.

Definition 2. In network-tree model $N\_Tree = (T, H)$, any macro-node $m_c \in T$, if there exists the parent macro-node $m_p \in T$, that is, $< m_p, m_c > \in H$, the node set of $m_c \cap m_p$ is called the connecting node set of m_c, denoted by R_{m_c}. Obviously, the connecting node set of the root macro-node is null.

Definition 3. The mapping function $\sigma : V \to T$, for $\forall v \in V$, there exists the only macro-node $m_i \in T$, $v \in V(m_i)$ and $v \notin R_{m_i}$, then let $(v, m_i) \in \sigma$, denoted by $\sigma(v) = m_i$, and we call σ the mapping from node of original network to macro-node of network-tree.

Definition 4. The mapping function $\varphi : V \to I$, $\forall v \in V$, let $\varphi(v)$ denote the level of node v, with the level number of the macro-node $\sigma(v)$ in the network-tree being the node's level number. We assign the level number of root macro-node to 1.

Property 1. $\forall u, v \in V$, let $T_{\sigma(u)}$ be the sub-tree which root is $\sigma(u)$, and m_p be the parent macro-node of $\sigma(u)$, with m_p being non-null. If $\sigma(v) \notin T_{\sigma(u)}$, and there exists the path $\pi = \pi_1 = (v_0 = u, v_1, v_2, \cdots, v_n = v)$ or $\pi = \pi_2 = (v_0 = v, v_1, v_2, \cdots, v_n = u)$ between u and v in the original network, then $\exists v_i$, $0 < i \leq n$, with $v_i \in R_{\sigma(u)} \subseteq V(m_p)$, and if $\pi = \pi_1$, the path from u to v_i must be in $T_{\sigma(u)}$, and if $\pi = \pi_2$, the path from v_i to u must be in $T_{\sigma(u)}$.

Property 1 shows the special characteristics of path in network-tree, which can be directly extracted from the construction procedure of network-tree.

2.3 Theory of Path Optimization

Let $sp_{N_0}(u, v)$ denote the shortest path from u to v in network N_0, and $l(\pi)$ denote the distance of path π.

Definition 5. Network-tree model $N\_Tree = (T, H)$, if for each sub-tree T' of N_Tree, with macro-node m_r being the root of T', $\forall u, v \in V(m_r)$, $l(sp_{m_c}(u, v)) = l(sp_{T'}(u, v))$, thus N_Tree is called Expanded Network-Tree Model.

Definition 6. Network-tree model $N\_Tree = (T, H)$, let the transitive closure of macro-node m be $B_m = \{m_p | (m_p \in T) \wedge (m_p = m \vee (\exists m_q \in B_m)(< m_p, m_q > \in H))\}$.

In Expanded Network-Tree Model, we get the following optimal theorem.

Theorem 1. In the Expanded Network-Tree Model, $\forall s, t \in V$, let $B_{\sigma(s)}$, $B_{\sigma(t)}$ be the transitive closures of macro-nodes $\sigma(s)$ and $\sigma(t)$ respectively, and then there must be $l(sp_N(s, t)) = l(sp_{B_{\sigma(s)} \cup B_{\sigma(t)}}(s, t))$, that is, at least one shortest path from s to t in original network N exists in the sub-networks contained by $B_{\sigma(s)}) \cup B_{\sigma(t)}$.

Proof. Let one of shortest paths from s to t in original network N be $sp_N(s, t) = (v_0 = s, v_1, v_2, \cdots, v_n = t)$, so we only need to prove if $sp_N(s, t)$ does not exist in the sub-networks of $B_{\sigma(s)}) \cup B_{\sigma(t)}$, we always can find the path π from s to t in the $B_{\sigma(s)}) \cup B_{\sigma(t)}$, with $l(\pi) = l(sp_N(s, t))$.

We discuss the $sp_N(s, t)$ repeatedly under the following three cases:

Case 1: when $\varphi(s) > \varphi(t)$, let $T_{\sigma(s)}$ be the sub-tree which root is $\sigma(s)$, and m_p be the parent macro-node of $\sigma(s)$, obviously $m_p \neq \phi$ and $\sigma(t) \notin T_{\sigma(s)}$. According to the Property 1, there must be $\exists v_i$, $0 < i \leq n$, with $v_i \in R_{\sigma(s)} \subseteq V(m_p)$, and the path from s to v_i exists in $T_{\sigma(s)}$, so $l(sp_{T_{\sigma(s)}}(s,v_i)) = l(sp_N(s,v_i)))$. As $s,v \in \sigma(s)$, and according to the definition of expanded network-tree model, $l(sp_{\sigma(s)}(s,v_i)) = l(sp_{T_{\sigma(s)}}(s,v_i))$. Therefore, $l(sp_{\sigma(s)}(s,v_i)) = l(sp_N(s,v_i))$. Let $\pi = sp_{\sigma(s)}(s,v_i) + sp_N(v_i,t)$, and then $l(\pi) = l(sp_N(s,t))$. $\sigma(s) \in B_{\sigma(s)}$, so $sp_{\sigma(s)}(s,v_i)$ exists in $B_{\sigma(s)}$. Let $s = v_i$, and we continue to discuss the sub-path $sp_N(v_i,t)$ until all sub-paths of π exist in $B_{\sigma(s)} \cup B_{\sigma(t)}$.

Case 2: when $\varphi(s) < \varphi(t)$, Similar as the case 1, there must be $\exists v_i$, $0 < i \leq n$, with $v_i \in R_{\sigma(t)} \subseteq V(m_p)$, and the path from v_i to t exists in $T_{\sigma(t)}$. Let $\pi = sp_N(s,v_i) + sp_{\sigma(t)}(v_i,t)$, with $l(\pi) = l(sp_N(s,t))$, and $sp_{\sigma(t)}(v_i,t)$ exists in $B_{\sigma(t)}$. Let $t = v_i$, and we continue to discuss the sub-path $sp_N(s,v_i)$ until all sub-paths of π exist in $B_{\sigma(s)} \cup B_{\sigma(t)}$.

Case 3: when $\varphi(s) = \varphi(t)$, if $\sigma(s) \neq \sigma(t)$, discuss the path as case 1. And if $\sigma(s) = \sigma(t)$, that is, both s and t locate in the same macro-node, there exist two case: **A.**When $\sigma(s)$ is not the root macro-node of N_Tree, if $l(sp_{\sigma(s)}(s,t) = l(sp_N(s,t))$, let $\pi = sp_{\sigma(s)}(s,t)$, with π being in the $B_{\sigma(s)}$, satisfying the proof requirements, and quit the discussion. Otherwise, there must be $\exists v_i$, $o < i < n$, with $v_i \in R_{\sigma(s)}$, and $sp_N(s,v_i)$ exists in the $\sigma(s)$, that is, $l(sp_{\sigma(s)}(s,v_i)) = l(sp_N(s,v_i))$. Let $\pi = sp_{\sigma(s)}(s,v_i) + sp_N(v_i,t)$, and continue to discuss the rest path $sp_N(v_i,t)$ as case 2. **B.**When $\sigma(s)$ is the root macro-node of N_Tree, then $\pi = sp_{\sigma(s)}(s,t)$, obviously $l(\pi) = l(sp_N(s,t))$, and π exists in $B_{\sigma(s)}$, satisfying the proof requirements, and quit the discussion.

Stated as above, the path π will exist in the sub-networks contained by $B_{\sigma(s)}$ and $B_{\sigma(t)}$ after a finite number of discussions, with $l(\pi) = l(sp_N(s,t))$. The theorem of route optimization builds theoretical bases for high performance shortest path computing of large-scale network, and we only need to search a few macro-nodes when computing the shortest path of Expanded NTM.

3 Shortest Path Algorithm of Expanded Network-Tree Model

3.1 Preprocessing Procedure

We have presented the construction procedure from general network to network-tree in the previous section, but the constructed network-tree cannot be guaranteed to be an expanded network-tree. Therefore we need a preprocessing procedure to create the expanded network-tree from network-tree. First we present the definition of virtual arc.

Definition 7. In network model $N = (V, A, W)$, $\forall u, v \in V$, let the virtual arc from u to v be $\overline{a_{uv}} = (u, v, sp_N(u,v))$, with $w(\overline{a_{uv}}) = l(sp_N(u,v))$, that is, the virtual arc is constructed by the shortest path between the corresponding two nodes.

Preprocessing procedure will begin with the bottom level macro-nodes in the network-tree, and process all the macro-nodes from bottom to top. For any two connecting nodes in a macro-node m, if the shortest path between these two nodes in m is less than the corresponding shortest path in the parent macro-node of m, add the virtual arc between these two nodes in m to the parent macro-node of m. We present the following recursive procedure, where the function $Djik(u, v, N_0)$ denotes that we invoke the $Dijkstra$ algorithm to calculating the shortest path from u to v in the network N_0.

Preprocessing Algorithm:
/* Input: the macro-node m*/
[1] : **for** each son macro-node m_s of m **do**
[2] : Preprocess (m_s);
end for;
[3] : **if** the parent macro-node m_p of m not null **then**
[4] : **for** $\forall u, v \in R_m$**do**
[5] : **if** $Djik(u, v, m) < Djik(u, v, m_p)$, **then**
[6] : $new\_arc = (u, v, Djik(u, v, m))$, $m_p = m_p \cup \{new\_arc\}$;
end if; end for; end if; end.

So we can invoke the preprocessing procedure to convert any network-tree to the expanded network-tree, with the root macro-node of the network-tree being the input macro-node. Obviously, the preprocessed network-tree is sure to be the expanded network-tree.

3.2 Network-Tree Shortest Path ($NTSP$) Algorithm

There should be two steps in the $NTSP$ algorithm: computing the transitive closure B that the optimal path exists in and searching the optimal path in B.

We present the following $NTSP$ algorithm, where the network model is the expanded network-tree N_Tree, and $d[v]$ is the label of the shortest distance from starting node s to the node v, and $\pi(v_i)$ represents the shortest path from starting node s to node v_i, and *open* is the set of temporarily-labeled node during the searching procedure, and *close* is the set of permanently-labeled node.

NTSP Algorithm.
/* Input: Starting node s and destination node t; Output: The shortest path from s to t */
[1] Initialization:
$open = \{s\}$, $closed = \phi$; $\forall v \in N\_Tree$, $d[v] = \infty$, $d[s] = 0$;
$\forall i \in N\_Tree$, $\pi_i = Null$; $B = \phi$, $m_s = \sigma(s)$, $m_t = \sigma(t)$;
[2] **while** $m_s \neq \phi$ **do** $B = B + m_s$, $m_s = m_s.parent$;
[3] **while** $m_t \neq \phi$ **do** $B = B + m_t$, $m_t = m_t.parent$;
[4] **while** $open \neq \phi$ **do**
[5] $v_c = \min_{d[v]}\{v | v \in open\}$;
[6] $open = open - \{v_c\}$, $closed = closed + \{v_c\}$;
[7] **if** $v_c = t$ **then break**;
[8] **for each** arc $a_{v_c v}$ **in** B emanating from v_c **do**
[9] **if** $d[v] > d[v_c] + w(a_{v_c v})$ **then**
[10] $d[v] = d[v_c] + w(a_{v_c v})$;

[11] **if** $v \notin open$ **then** $open = open + \{v\}$;
end if; **end for**; **end while**;
[12] **if** $d[t] = \infty$ **then** $\pi(t) = null$;
[13] **return** $\pi(t)$, and algorithm terminates.

NTSP algorithm computes the transitive closure at step [2], [3], and B contains the transitive closures of both macro-node $\sigma(s)$ and $\sigma(t)$. The searching procedure always be constrained in B, satisfying the requirements of Theorem 1, so at least one shortest path from s to t is sure to exist in the B. As the label-marked searching technology is adopted, the algorithm will be able to find the shortest path from s to t in B that is also the shortest path in original network according to Theorem 1.

4 Experimental Results

4.1 Network-Tree Based on Lattice Structures

We present the following procedure to construct the network-tree *N_Tree* based on lattice structures, with $k(k \geq 2)$ being the level number of the network-tree, l_h and b_h are integers, $1 \leq h \leq k$, and $w_1 < w_2 < \cdots < w_{k-1} < w_k$.

I. Construct a $\prod_{h=1}^{k} l_h \times \prod_{h=1}^{k} b_h$ rectangular lattice, and place the lattice in the first quadrant of the Cartesian plane, with the lower left corner corresponding to $(0,0)$, and the lattice and Cartesian coordinates match;
II. The nodes of the lattice correspond nodes of the network; sides of the lattice correspond to the bi-directional arcs of the network. The following sub-procedure constructs the network-tree *N_Tree* and assigns weight value to the arcs:

Step1: Abstract the nodes which abscissa can be divided exactly by $\prod_{i=2}^{k} l_i$ or which ordinate can be divided by $\prod_{i=2}^{k} b_i$, and all the arcs between these nodes to construct the root macro-node m_r of *N_Tree*. Let the weight of all these arcs be w_1. The rest network is divided into $l_1 \times b_1$ sub-networks by m_r, and let the parent macro-node of these sub-networks be m_r, and process all these sub-networks as step2, with initially $h = 2$.

Step2: For each sub-network N_{sub}, if $h = k$, then let the whole sub-network N_{sub} be a macro-node m of *N_Tree*. Otherwise, in the sub-network N_{sub}, abstract the nodes which abscissa can be divided exactly by $\prod_{i=h+1}^{k} l_i$ or which ordinate can be divided by $\prod_{i=h+1}^{k} l_i$, and all the arcs between these nodes to construct a macro-node m. Let the parent macro-node of N_{sub} be the parent macro-node of m, and the weight of all the arcs in m be w_h. When $h \neq k$, the sub-network $N_{sub} - \{m\}$ is divided into $l_h \times b_h$ smaller sub-networks, and let the parent macro-node of all these new smaller sub-networks be m. Until all the sub-networks are processed, then continue to the step3.

Step3: Let $h = h + 1$, and if $h > k$, then go to the step4. Otherwise, go to step2 to process all the new smaller sub-networks.

Step4: The network-tree construction is completed.

For example, shown as Fig. 1, we create a network-tree based on lattice structures with $k = 3$, $l_1 = b_1 = 2$, $l_2 = b_2 = 3$, $l_3 = b_3 = 4$, $w_1 = 2$, $w_2 = 3$ and $w_3 = 5$.

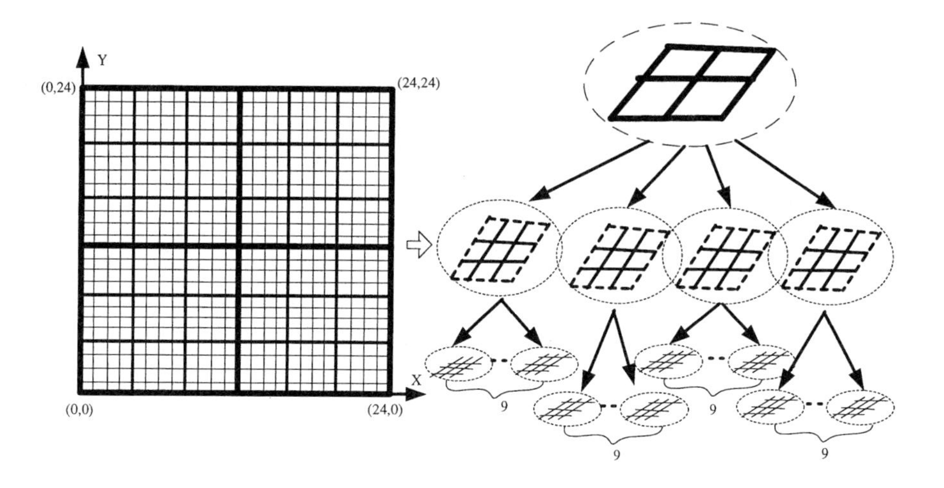

Fig. 1. The example of network-tree construction based on lattice structures

4.2 Experimental Results

We will create a large number of origin and destination nodes (OD pairs) randomly, and the computing efficiency will be measured by the average calculation time, that is $\overline{T} = \frac{\sum_{i=1}^{n} T_i}{n}$, with n being the number of OD pairs.

To better observe the algorithms' computing performance when calculating OD pairs of different distance, we define the test distance of OD pair: Let the original node be $s(x_s, y_s)$ and the destination node be $t(x_t, y_t)$, We adopt the Manhattan values as the test distance of this OD pair, that is $|x_t - x_s| + |y_t - y_s|$.

Our experiment uses a Pentium IV personal computer running at 1.7 GHz with 256 MB RAM. The operating system is Windows98; the programming language is Visual C++ 6.0.

We present the following computing efficiency comparisons of $NTSP$, hierarchical encoded path query ($HEPQ$), and $Dijkstra$ algorithm.

Experiment 1. Comparison of $NTSP$, $HEPQ$, and $Dijkstra$ algorithm with two-level expanded network-tree under different test distances. The parameters are set as $k = 2$, $l_1 = b_1 = 6$, $l_2 = b_2 = 10$, $w_1 = 2$ and $w_2 = 5$, thus the total node number is 3721, the total arc number is 14876, results are demonstrated in Fig. 2.

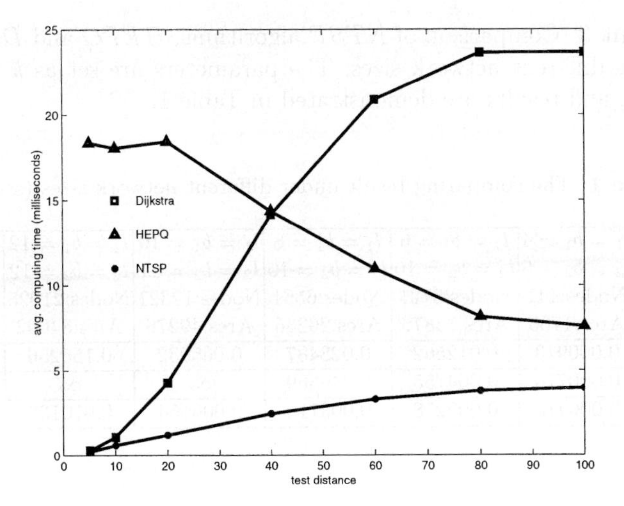

Fig. 2. Comparison of NTSP, HEPQ, and Dijkstra algorithm with two-level expanded network-tree

Experiment 2. Comparison of three *NTSP* algorithms and *Dijkstra* algorithm with three-level expanded network-tree under different test distances. The parameters are set as $k = 3$, $l_h = b_h = 6 (1 \leq h \leq 3)$, $w_1 = 2$, $w_2 = 4$ and $w_3 = 7$, thus the total node number is 47089, the total arc number is 188348, results are demonstrated in Fig. 3.

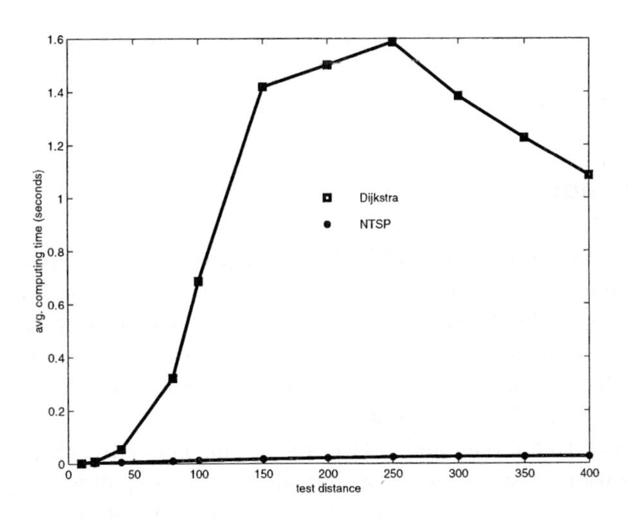

Fig. 3. Comparison of NTSP algorithms and Dijkstra algorithm with three-level expanded network-tree

Experiment 3. Comparison of *NTSP* algorithms, *HEPQ* and *Dijkstra* algorithm under different network sizes. The parameters are set as $k = 2$, $w_1 = 2$ and $w_2 = 5$, and results are demonstrated in Table 1.

Table 1. The comparing result under different network sizes (seconds)

Size	$l_1 = b_1 = 4$ $l_2 = b_2 = 5$ Nodes:441 Arcs:1756	$l_1 = b_1 = 6$ $l_2 = b_2 = 10$ Nodes:3721 Arcs:14876	$l_1 = b_1 = 8$ $l_2 = b_2 = 10$ Nodes:6561 Arcs:26236	$l_1 = b_1 = 10$ $l_2 = b_2 = 11$ Nodes:12321 Arcs:49276	$l_1 = b_1 = 12$ $l_2 = b_2 = 12$ Nodes:21025 Arcs:84092	$l_1 = b_1 = 16$ $l_2 = b_2 = 16$ Nodes:66049 Arcs:264188
Dijkstra	0.000913	0.012562	0.028487	0.065932	0.155256	1.371372
HEPQ	0.000776	0.004755	0.049369	∞	∞	∞
NTSP	0.000413	0.002278	0.003710	0.006154	0.010132	0.026215

4.3 Interpretation of Results

The test results show the different computing performances of *NTSP*, *HEPQ* and *Dijkstra* algorithm: Experiment 1 shows that the *NTSP* algorithm achieves higher computing efficiency than *HEPQ* and *Dijkstra* algorithm under all the test distances, especially with the network size increasing, such as Experiment 3, under the best case the *NTSP* algorithm improved computing time by a factor of 50 over the *Dijkstra* algorithm. The spatial complexity of the *HEPQ* algorithm is very high; therefore, the algorithm can accelerate the computing speed to a certain degree when applying to small or general network (see Experiment 1), but if the storage is constrained, the computing performance decreases when the network size increases to a certain size. As Experiment 3 shows, when the network size increases to 12,321 nodes, the *HEPQ* algorithm required more main memory than our test computer could provide, and the algorithm was unable to find a shortest path.

5 Conclusions

The network-tree model and NTSP algorithm provide a wholly new method to effectively implement the optimization of large-scale network. The advantages of network-tree model express not only at the great efficiency improvement of shortest path computing, but also at the network data partition according to the hierarchical characteristic and regional characteristic, which is very important for many definite application areas.

The network-tree model and NTSP algorithms not only can be apply to the path optimization of static network, but also can be expanded to the dynamic optimal path searching of time-dependent network. The algorithms have an excellent parallel structure, and are well suited to realizing parallel computations of the shortest path.

References

1. Cherkassky B.V., Goldberg A.V., Radzik T.: Shortest paths algorithms: Theory and experimental evaluation. Mathematical Programming, 1996, 73(2): 129–174
2. Ning Jing, Yun-Wu Wang, and Elke A. Rundensteiner: Hierarchical Encoded Path Views for Path Query Processing: An Optimal Model and Its Performance Evaluation. IEEE Transactions on Knowledge and Data Engineering, 1998, 10(3): 409–432
3. A.Car, G.Taylor, C.Brunsdon. An analysis of the performance of a hierarchical wayfinding computational model using synthetic graphs. Computers, Environment and Urban Systems, 2001, 25: 69–88
4. J. Shapiro, J. Waxman, D. Nir: Level Graphs and Approximate Shortest Path Algorithms, Networks, 1992, 22: 691–717
5. Yu-Li Chou, H. Edwin Romeijn and Robert L. Smith: Approximating Shortest Paths in Large-Scale Networks with an Application to Intelligent Transportation Systems, INFORMS journal on Computing, 1998, 10 (2): 163–179
6. Chan-Kyoo Park, Kiseok Sung, Seungyong Doh, Soondal Park: Finding a path in the hierarchical road networks. 2001 IEEE Intelligent Transportation Systems Conference Proceedings-Oakland (CA) USA, August 25–29, 2001, pp. 936–942
7. A. Fetterer, S. Shekhar: A Performance Analysis of Hierarchical Shortest Path Algorithms. IEEE Proceedings of the International Conference on Tools with Artificial Intelligence, Nov 3–8, 1997, pp. 84–92

Parallel Genetic Algorithm for a Flow-Shop Problem with Multiprocessor Tasks

C. Oguz[1], Yu-Fai Fung[2], M. Fikret Ercan[3], and X.T. Qi[1]

[1] Dept. of Management, [2] Dept. of Electrical Eng., The Hong Kong Polytechnic, University, Hong Kong SAR
{msceyda, eeyffung}@polyu.edu.hk
[3] School of Electrical and Electronic Eng., Singapore Polytechnic, Singapore
mfercan@sp.edu.sg

Abstract. Machine scheduling problems belong to the most difficult deterministic combinatorial optimization problems. Since most scheduling problems are NP-hard, it is impossible to find the optimal schedule in reasonable time. In this paper, we consider a flow-shop scheduling problem with multiprocessor tasks. A parallel genetic algorithm using multithreaded programming technique is developed to obtain a quick but good solution to the problem. The performance of the parallel genetic algorithm under various conditions and parameters are studied and presented.

Keywords: Genetic algorithms, parallel architectures, parallel computing

1 Introduction

Multiprocessor task scheduling is one of the challenging problems in computer and manufacturing processes. The general problem of multiprocessor task scheduling can be stated as scheduling a set of independent and simultaneously available tasks onto a set of parallel identical processors so that a given performance criterion is optimized. This type of scheduling problems is known to be intractable even with the simplest assumption [12]. An extensive survey on multiprocessor tasks scheduling can be found in [7]. This survey reveals that a single-stage setting for the processor environment is assumed in most of the multiprocessor task scheduling studies. Although this kind of assumption may be meaningful for some problems, there are many other practical problems that require jobs to go through more than one stage where each stage has several parallel processors. This type of environment is known as flow-shops in scheduling theory [9].

In this paper, we consider a multiprocessor task (MPT) scheduling problem in a flow-shop environment, which can be described formally as follows: There is a set J of n independent and simultaneously available MPTs to be processed in a two-stage flow-shop, where stage j consists of m_j identical parallel processors $(j = 1,2)$. Each $MPT_i \in J$ should be processed on p_{ij} identical processors simultaneously at stage j without interruption for a period of T_{ij} $(i = 1,2,\ldots,n \ and \ j = 1,2)$. Hence, each

P.M.A. Sloot et al. (Eds.): ICCS 2003, LNCS 2659, pp. 548–559, 2003.

$MPT_i \in J$ is characterized by its processing time, T_{ij}, and its processor requirement, p_{ij} $(i = 1,2,...,n$ *and* $j = 1,2)$. All the processors are continuously available from time 0 onwards and each processor can handle no more than one MPT at a time. Tasks flow from stage 1 to stage 2 by utilizing any of the processors while satisfying the flow-shop and the MPT constraints. The objective is to find an optimal schedule to minimize the maximum completion time of all tasks.

The motivation for this problem comes from machine vision systems developed to perform real-time image understanding [3,8]. These systems utilize multiple layers of multiprocessor computing platforms where data have to pass through from one layer to another. Algorithms on parallel identical processors of each layer process image data. These systems can be analyzed from a scheduling perspective since they resemble the multi-stage flow-shop environment with MPTs, where data represent the incoming MPTs and algorithms applied, such as, feature detection, feature grouping, or object recognition define their operations at respective stages. Another application where the above MPT scheduling problem encountered is in diagnosable microprocessor systems [11], where a number of processors must process a task to detect faults. Other applications arise when a task requires more than one processor, tool or workforce simultaneously (see for instance, power system transient stability computations [16]).

Despite many practical applications that may involve MPTs in multi-stage settings, majority of research in this area has focused on MPT scheduling in a single-stage; little attention has been given to MPT scheduling in multi-stage settings. Extensive surveys on scheduling MPTs can be found in [7]. Another limitation of previous studies is that they mainly addressed the computational complexity issues of the problems (see for instance [2]). Most recently, Oguz et al. [14] provided approximation algorithms for the MPT scheduling problems in a flow-shop environment.

In this paper, we combine two areas, namely scheduling and parallel computing by parallel implementing a scheduling algorithm for real-time machine vision systems. Since MPT scheduling is intractable even in its simplest forms [1,12], we focus on efficient approximate algorithms to find a near optimal solution. We also concentrate on the fact that the real-life problems, like machine vision systems, require a quick but a good solution. It is thus of interest how a fast approximation algorithm could be developed for machine vision systems, which can be modeled as a MPT flow-shop scheduling problem.

Considering the success of the genetic algorithms (GA) developed for scheduling problems [4, 6, 15], we choose to use this local search method to provide a good solution to our problem. GAs are introduced by Holland [10]. A typical GA starts with an initial population of possible solutions to the problem (chromosomes). Each chromosome is characterized by its fitness, which is determined by the associated value of the objective function. The fittest chromosomes in the population (parents) will be selected for the generation of new solutions (children). This generation will take place according to a genetic operation, such as mutation (introducing of variations into the chromosomes) and crossover (taking the best features of each parent and mixing the remaining features). Hence, the new chromosomes will be somewhat different than their parents. In the new generation, the fitness of the children is evaluated in a similar fashion as their parents, and the worst fitted chromosomes will die to maintain the desired population. The birth and death processes will define the population size but usually it remains constant from one generation to the next. This procedure is repeated until a desired termination criterion is reached. The output of this simulated evolution process will be the best chromosome in the final population, which can be a highly evolved solution to the problem.

Yet, excessive computation time in finding this highly evolved solution is a disadvantage of the GAs [5]. Hence, we present a parallel GA (PGA) in this paper. However, the benefit of the PGA is expected to be not only a speed-up in the computation time but also a better solution, that is shorter makespan for the scheduling problem, compared to a sequential GA)[5].

In the following section, we describe the design of the PGA. Next, the computational study is presented. We then report and discuss the computational results. Conclusions are given in the last section.

2 Parallel Genetic Algorithm

There are various kinds of implementation of PGAs that can be classified into three categories: global, migration and diffusion. These categories are mainly based on the structure of the population. The global PGA, often known as the worker/farmer model, treats the whole population as a single unit. Each chromosome can mate with any other chromosome in the entire population. The migration PGA, which is more similar to the natural evolution than the global PGA, divides the whole population into several subgroups. A chromosome can only mate within the subgroup and migrations may happen among the subgroups. The migration PGA is also called the coarse-grained model or island model. The diffusion PGA regards each chromosome as a separate unit. One chromosome can mate with another one in its neighborhood. The use of local neighbor leads to a continuous diffusion of chromosomes over the whole population. The diffusion PGA is also known as fine-grained, neighborhood, or cellular model.

In our PGA, the whole population is divided into G subgroups, each of which has s chromosomes and is processed by a sequential sub-GA. The sub-GAs run concurrently with some migrations among them. An epoch is the number of generations between two occurrences of migration. The effects of the number of subgroups, the value of s, the number of migrated chromosomes, and the length of an epoch on the performance of the PGA are analyzed in Section 3.

2.1 Hardware and Software Environment

The PGAs can be implemented on various parallel computing hardware and software environments, from networked PCs to mainframes. We implemented our PGA on SUN servers with multithreaded programming. With the emergence of the shared memory symmetric multiprocessor (SMP) computing systems, multithreaded programming provides the right programming paradigm to make maximum use of these new machines. The PGA algorithm is based on running several sequential GAs (SGA). Multithreaded programming is selected in the implementation so that each independent thread processes an SGA, since multithreading technique allows one program to execute multiple tasks concurrently. In the following, we will use the terms thread and subgroup interchangeably.

2.2 The Sequential Genetic Algorithm

Here, we will briefly describe the structure of the SGAs used in our PGA.

Chromosome Design: We define a chromosome as a string of $2n$ bytes. The first half of a chromosome is a permutation of $1,2,...,n$ representing the task list at stage 1. Similarly, the second half of a chromosome is a permutation of $1,2,...,n$, representing the task list at stage 2. A chromosome is decoded to a schedule by assigning the first unscheduled task in the task list to the machines at each stage.

Selection and Fitness: The fitness of chromosome x_k, i.e. the probability of chromosome x_k being selected to be a parent, is given by:

$$F(x_k) = \frac{f_{\max} - f(x_k)}{\sum_{i=1}^{N_{pop}} (f_{\max} - f(x_i))},$$

where $f(x_k)$ is the makespan of the schedule decoded from x_k, and $f_{\max}$ is the maximum $f(x_k)$ in the current generation, and N_{pop} is the population size. We use the well-known and commonly used roulette-wheel method as the selection operation [5]. Other selection methods, such as tournament selection, are also considered, but no improvement is observed from experimental results.

Crossover Operation: The crossover is an operation to generate two children chromosomes from two parent chromosomes selected. Three crossover operators are considered in this study: the one-point crossover (c1), the two-point crossover (c2) and the uniform crossover (c3). Since, in our study, a chromosome is composed of two parts, how to apply the crossover to these two parts is a problem. In our computational experiments, we find that it is better to crossover the two parts in the same position. The details are as follows:

(1) One-point crossover. For two parents, x_1 and x_2, a crossover position r, $r < n$, is randomly generated. The first r bits of the first child are the same as the first r bits of the parent x_1, and the bits from $r+1$ to n of the first child are in the same order as they are in parent x_2. The second half of the first child chromosome is generated in the same way. The child has the same absolute task sequence as parent x_1 and the same relative task sequence as parent x_2. Another child is generated with the same absolute sequence as parent x_2 and with the relative sequence as parent x_1. For example, if $x_1 = [(1,2,3,5,4,6)(1,2,5,3,4,6)]$, $x_2 = [(2,1,4,3,5,6)(2,4,1,3,5,6)]$ and the crossover position is 3, then the two children will be $C_1 = [(1,2,3,4,5,6)(1,2,5,4,3,6)]$, $C_2 = [(2,1,4,3,5,6)(2,4,1,5,3,6)]$.

(2) Two-point crossover. We randomly generate two positions r and s, $r < s$. One child will have the same absolute sequence as x_1 in the bits from 1 to r and from $s+1$ to n, and other bits have the same relative sequence as x_2. Another child is generated

correspondingly. For the above example, if $r = 2$ and $s = 4$, then the two children will be $C_1 = [(1,2,3,5,4,6)(1,2,3,5,4,6)]$, $C_2 = [(2,1,3,4,5,6)(2,4,1,3,5,6)]$.

(3) Uniform crossover. Uniform crossover can be regarded as a multiple-point crossover. First a 0-1 string, called the mask string, is randomly generated. Then one child is generated with the same absolute sequence as x_1 where the corresponding mask string bit is 1, and other bits have the same relative sequence as x_2. For the above example, if the mask string is $(0,1,1,0,1,1)$, then $C_1 = [(1,2,3,5,4,6)(1,2,5,3,4,6)]$, $C_2 = [(2,1,3,4,5,6)(2,4,1,3,5,6)]$.

The crossover rate, that is, the probability of applying the crossover operator to the parents, is often considered to be 1 in scheduling problems [4, 13]. We also found from our computational study that the crossover rate of 1 is better.

Mutation Operation: The mutation operation modifies a chromosome. The arbitrary two-bit exchange mechanism is applied in our algorithm. In the arbitrary two-bit exchange, two positions are randomly selected and exchanged. After each mutation operation, we compare the new chromosome with the original one. If the fitness of the new chromosome is greater than that of the original, the new chromosome replaces the original. Otherwise, the original chromosome is kept in the population. This procedure can be regarded as a GA combined with a stochastic neighborhood search [6].

The mutation rate, that is the probability of the mutation to be applied to a chromosome, is reported to be large for scheduling problems. For instance, an initial mutation rate of 0.8 is used in [15], which is decreased by 0.01% at each generation. In [13], it is reported that the mutation rate of 1 is the best. A large mutation rate in scheduling problems may be due to the combinatorial character of the scheduling problems. Large mutation rate might help the GA to search the neighborhood of a schedule. Based on our computational study, we chose a mutation rate of 1.

Other Factors: The SGA uses a population size, which will be determined based on computational results given in Sect. 3.1, and an elitist strategy for reproduction, which is to remove the worst chromosome from the current population and include the best one from the previous population. The termination criterion is set to a limited number of generations, which is 500. While many researchers generate the initial population randomly, others favor a good solution as the initial "seed". In our problem, it is reasonable to say that in an optimal schedule, if an MPT is processed early at stage 1, probably it will be processed early at stage 2, too. Hence, the task sequence in two stages will be almost identical. Therefore, we use a ratio of 3:1 for initial chromosomes with the same task list in both stages and initial chromosomes with random task list.

2.3 Design of Migration

The design of migration concerns two aspects: the route of migration and the communication method among subgroups or threads. For the route of migration, we generate a migration table $(r_1, r_2, \ldots, r_n)$, which is a permutation of $(1,2,\ldots,n)$. According to the migration table, the emigrants of a subgroup s will go to the destination subgroup r_s. Two kinds of migration routes are tested: fixed route and random route. In the fixed route, the migration table of each epoch is defined as $(2,3,\ldots,n,1)$. In the random route, the migration routes are randomly generated for each epoch. Computational results show that the random route is better and the details are presented in Section 3.2.

The communication method between the subgroups will influence the computation time. One easy implementation is to synchronize all the threads for each epoch. In this way, threads will pause after one epoch to wait for the completion of other threads. When all threads have completed for an epoch, another independent thread (we use the main thread to save computer resource) will be in charge of dealing with migration, and then each thread continues. For example, in the fixed migration route, the main thread will first save the emigrants of thread 1 in a buffer, and then copy the emigrants of thread $s+1$ to replace the emigrants of thread s, $s=1,2,...,n-1$. Finally it will copy the emigrants in the buffer to replace the emigrants of thread n.

Another approach is the asynchronous method which is more complex but efficient. For each thread, a buffer is allocated to hold the emigrants coming from a different subgroup. A thread can place its emigrants in the buffer of the destination thread without waiting for the destination thread to finish. After reading the emigrants in its buffer, a thread can proceed with its next epoch. Since the buffer is a shared memory block, which can be accessed by different threads, it must be protected by locks. The mutual exclusion lock (mutex) is applied in our PGA. Each buffer has two mutex locks, namely read-mutex-lock and write-mutex-lock, each of which can be locked only by one thread at any time. Only by locking its own read-mutex-lock, a thread can read from its buffer. Similarly, only by locking the write-mutex-lock of a buffer, a thread can write to the buffer of the destination thread. The procedure of communication for a thread can be explained by the following pseudo codes:

Pseudo code for migration of one thread.

```
// Procedure of writing emigration
Determine the destination thread from the migration table;
Lock the write-mutex-lock of the destination thread;                (O1)
Copy the emigrants to the buffer of the destination thread;         (O2)
Unlock read-mutex-lock of the destination thread;                   (O3)
// Procedure of reading immigration
Lock its own read-mutex-lock;                                       (O4)
Read from its buffer the immigrants and replace the emigrants;      (O5)
Unlock its own write-mutex-lock;                                    (O6)
```

Vector (r,w,b) represents the state of the buffer, where $r=0$ (or $w=0$) means the read-mutex-lock (or write-mutex-lock) is unlocked, $r=1$ (or $w=1$) means the read-mutex-lock (write-mutex-lock) is locked, $b=0$ means the buffer is empty and $b=1$ means the buffer is full. The following example explains the state of the buffer after each operation in the above code, where the symbol "?" represents an undetermined state: Consider operation (O1) at state $(1,?,?)$, which means that the read-mutex-lock is locked, and the write-mutex-lock is undetermined. If $w=1$, which means the buffer of the destination thread cannot be written to, then operation (O1) is blocked to wait for w to become 0; if $w=0$ (now b must be 0), then operation (O1) can continue. By operation (O1), the thread locks the buffer of the destination thread and changes w to 1 and the state of the destination thread becomes $(1,1,0)$. Similarly, before operation (O4), the state of a thread is undetermined and when (O4) finishes, it must become $(1,1,1)$. For the destination thread, we have $(1,?,?) \Rightarrow (1) \Rightarrow (1,1,0) \Rightarrow 2 \Rightarrow (1,1,1) \Rightarrow (0,1,1)$. For the thread itself, we have

$(?,?,?) \Rightarrow (4) \Rightarrow (1,1,1) \Rightarrow 5 \Rightarrow (1,1,0) \Rightarrow (1,0,0)$. The initial state of each thread is $(1,1,0)$, i.e., the read-mutex-lock is locked and the write-mutex-lock is unlocked. To avoid deadlock, a thread is not assigned as its own destination.

2.4 Property of SGA

Since each subgroup is processed with a sequential sub-GA, the final result of the PGA will be affected from the application of different crossover operations. As mentioned in Section 2.2, we consider three different crossover operations. By applying them to the sub-GAs, we obtain three PGAs, denoted by PGA-c1, PGA-c2, PGA-c3, respectively. In addition, we employ a combined structure of the sub-GAs such that the crossover operations for some sub-GAs are different from others. For example, half of the sub-GAs use the one-point crossover and half of the sub-GAs use the two-point crossover, which is denoted by PGA-c12. Similarly, we have PGA-c13 and PGA-c23. In PGA-c123, each of the three crossover operations is used in one third of all the sub-GAs. The combined structure simulates the natural evolution environment in which each subgroup evolves in different conditions.

3 Experimental Results

The performance of the PGA under different hardware platforms and parameters are studied. We focused on the properties of the parallel computation, including the speed-up of computation time, the number of sub-groups, the migration and different crossover operations. The program is coded in C++ and run on SUN servers. All the results are the means of running 50 problems, where the number of machine is 16 for both stages. Unless explicitly specified, the PGA is the PGA-c12 with random migration route and has a migration size of two individuals. The number of epoch is 25 and the epoch length is 20 generations, which means a total of 500 generations. The number of subgroups is 15 and the size of a subgroup is 16.

The speed-up ratio of computation time of the PGA under different hardware platforms and processor workloads is reported. The speed-up ratio is defined as the ratio of computation time with a single thread to computation time with multiple threads and different workload. Experimental results show that the speed-up depends on the hardware configuration and the workload of the computers. Hence, three different SUN servers, S1, S2 and S3, are tested. S1 is a SUN SPARCserver 1000E with 4 CPUs and the CPU usage is less than 50% in normal conditions. The other two machines, S2 and S3, are logically derived from a SUN Ultra Enterprise (UE 10000) server with 16 CPUs. S2 and S3 have 4 and 8 CPUs, respectively, while the CPU usage for both machines is almost 100% in normal conditions. We tested the problems with 50, 100, 150, 200 and 300 jobs and with different number of threads, $G = 1$, 4, 8, 10, 15, 20, 30, and 60. In each case, the population size of a subgroup is $s = 240/G$. Thus the total computation requirements are identical for different number of threads. The speed-up ratios obtained under different machines and conditions are listed in Tables 1, 2 and 3.

Table 1. Speed-up ratios for S1: 4 CPUs, not busy.

Threads	50 jobs	100 jobs	150 jobs	200 jobs	300 jobs
4	3.69	3.67	3.70	3.76	3.77
8	3.79	3.74	3.72	3.84	3.84
10	3.82	3.80	3.83	3.84	3.85
15	3.78	3.83	3.83	3.85	3.85
20	3.74	3.82	3.85	3.84	3.86
30	3.72	3.80	3.83	3.83	3.85
60	3.72	3.68	3.69	3.73	3.75

Table 2. Speed-up ratios for S1: 4 CPUs, busy.

Threads	50 jobs	100 jobs	150 jobs	200 jobs	300 jobs
4	2.26	2.58	2.62	2.63	3.10
8	2.83	3.42	3.50	3.51	3.50
10	2.87	3.61	3.68	3.71	3.72
15	3.10	4.01	4.32	4.41	4.43
20	3.46	4.09	4.55	4.45	4.43
30	3.52	4.30	4.70	4.69	4.70
60	3.18	4.20	4.57	4.57	4.58

Table 3. Speed-up ratios for S1: 8 CPUs, busy.

Threads	50 jobs	100 jobs	150 jobs	200 jobs	300 jobs
4	2.46	2.53	2.62	3.14	3.15
8	3.46	4.30	4.31	4.65	4.89
10	3.82	5.14	5.10	5.26	5.57
15	3.89	6.01	6.11	6.93	6.95
20	3.95	6.54	6.60	7.62	7.60
30	4.19	6.96	6.97	8.67	8.68
60	4.10	6.52	6.87	8.28	8.31

Table 4. Processing time (in sec.) by a single thread under the three different servers.

Computers	50 jobs	100 jobs	150 jobs	200 jobs	300 jobs
S1	238.4	563.5	278.4	1069	2859
S2	207.3	435.1	606.8	1001	1652
S3	99.30	292.4	486.5	922.5	1565

From the results, we can examine the relationship between the speed-up ratio and the number of jobs of the problem as well as the number of threads. The processing times of the PGA using a single thread are presented in Table 4 and these reflect the complexity of the PGA algorithm.

The parallel implementation of the GA will create certain overheads in the processing time such as the time required to create the threads and to synchronize two threads. These times are in the scale of microseconds and can be ignored when compared to the processing time demanded by the algorithm (see Table 4). Moreover, the synchronization between threads only takes place during migrations and these occur in a very limited number of times (25 in our tests) for the complete process.

As our algorithms are implemented on general-purpose servers, during run time, they are competing with other users' processes for the available system resources (CPUs). Usually, the Operating System, Solaris in our case, is responsible for the fair allocation of resources to users' processes. If there are more jobs than the available number of CPUs, then the jobs will form a queue and share the CPUs based on the round-robin mechanism. This is the major source of overhead induced in the computation time.

When a problem has more jobs, the computing requirement increases. Hence, the effect of the overhead is reduced and a better speed-up ratio is obtained. This is substantiated by the results presented in Tables 1, 2, and 3. Based on the results, we can observe that if the number of threads is fixed, the speed-up ratio has a tendency to increase for cases with more jobs. There are, however, some exceptional cases. These may be caused by workloads submitted by other users, or by tasks performed by the operating system while the PGA program is being executed.

The speed-up ratio increases with the increase in the number of threads because this increases the share of PGA on the system. However, the speed-up ratio begins to decrease if there are too many threads, for example, 60 threads. The total processing time of the PGA is equal to the total duration of all threads, i.e., it is determined by the last thread terminated. The duration of a thread is the combination of the total time when it is served by a CPU (t_s) and the time when it is waiting for available resources (t_w). When the number of thread increases, the term t_s decreases since the workload assigned to a thread is reduced. On the other hand, the term t_w increases as there are more threads waiting in the queue. A speed-up is obtained when the total duration ($t_s + t_w$) of the last terminating thread is reduced compared to the single thread case. When both terms are minimized then the speed-up ratio will be optimized. In our experiments, the optimal result is obtained when 30 threads are created. The speed-up ratios decrease in the 60-thread case because the total duration is increased due to the term t_w.

3.1 Number of Subgroups

The advantage of PGA is not only the speed-up of computation time, but also the improvement of the solution. We can find a better solution by dividing a large population into several subgroups. Table 5 depicts such a result. We have a whole population of 240 individuals, which means that the total computing requirements are fixed. With different number of subgroups G, the subgroup size is $s = 240/G$. To compare the solutions for different subgroups, the solution obtained from the single thread (group) case is used as a reference. The ratio of the solutions obtained from using different number of subgroups to the single group case are evaluated and listed in Table 5.

From Table 5, we can see that PGA can obtain better solutions than the sequential GA on a large whole population (the case of one thread). Computational results show that 15 subgroups each with 16 individuals produce the best result. If the subgroup size is too small, the improvement is less significant. In addition, PGA gives a better result if we have more jobs to schedule. For the problems with 50 jobs the maximum improvement is 0.010 while for the problems with 300 jobs the improvement is 0.025.

Generally, when the amount of computation increases, a better solution can be obtained. The size and the number of subgroups determine the amount of computation and we have shown that a small subgroup is better than a large one. Next we will study the effect of altering the number of subgroups by fixing the size of each subgroup. We concentrate on the 300-job case and we use 16 and 30 as the size of a subgroup for comparison. The results, which are given in Table 6, indicate that increasing the number of subgroups can improve the solutions, but the improvement becomes less significant when the number of subgroups exceeds 20. If the number of subgroups is fixed, the subgroup size of 30 gives better results than 16. However the improvement is not significant.

Table 5. Ratio of solutions found by different number of subgroups to the single group case.

Sub-groups	50 jobs	100 jobs	150 jobs	200 jobs	300 jobs
1	1.000	1.000	1.000	1.000	1.000
4	0.994	0.984	0.981	0.978	0.989
8	0.991	0.982	0.977	0.976	0.984
10	0.991	0.982	0.977	0.975	0.980
15	0.990	0.982	0.976	0.974	0.975
30	0.993	0.983	0.978	0.976	0.980
60	0.994	0.984	0.979	0.977	0.983

Table 6. Ratio of solutions found by different number of subgroups to the single group case.

Subgroup	1	4	8	10	15	20	30	40
16	1.000	0.987	0.983	0.977	0.973	0.971	0.970	0.970
32	0.998	0.98 6	0.982	0.975	0.972	0.970	0.969	0.969

3.2 Design of Migrations

In Table7, we compare the fixed route migration and random route migration mechanism with different number of migration. The problems with 300 jobs are used and the solution of the fixed route mechanism with zero migration is taken as a base. It is easy to observe that the random route migration performs better than the fixed route migration. For the number of individuals for migration, 2 or 3 should be a better choice. We also consider the complete island model in which no migration occurs among the subgroups, that is, when migration number is zero. The result of the complete island model is worse than the result of none-zero migration, which demonstrates the usefulness of migration.

Table 7. Ratio of solutions found by different numbers of migration to the zero migration.

No. of migra- tion	0	1	2	3	4	5
Fixed route	1.000	0.988	0.985	0.985	0.986	0.987
Random route	1.000	0.987	0.984	0.984	0.985	0.987

The frequency of migration, which is represented by the epoch length, also affects the performance of PGA. The short epoch length will weaken the outcome of the island model, which will become similar to the global PGA. The long epoch length will lead to the complete island model. To compare the different epoch lengths, we chose the epoch length to be 10, 20, 30, 40, 50 and 100, with the corresponding number of epochs of 50, 25, 17, 13, 10 and 5, so that the total number of generations will be almost identical, i.e. 500, or slightly more. The results are provided in Table 8 for problems with 300 jobs, where the problem with the epoch length of 100 is taken as a base. The results are consistent with the above analysis and the epoch length of 20 gives the best result.

Table 8. Ratio of solutions found by different length of epoch.

Length of Epoch	10	20	30	40	50	100
Relative error	0.989	0.987	0.988	0.990	0.995	1.000

3.3 PGAs with Different Crossover Operations

As mentioned in Section 2.3, seven PGAs, namely PGA-c1, PGA-c2, PGA-c3, PGA-c12, PGA-c13, PGA-c23 and PGA-c123, are analyzed. The performances of the seven PGAs for the 300-job case are compared and the results are depicted in Table 9, where PGA-c12 is taken as a base. Results show that PGA-c1 and PGA-c2 have similar performances and are better than PGA-c3. On the other hand, the performances of all these PGAs, with only one kind of crossover operation, are worse than that of the PGAs with combined crossover operations. Among different combinations of the crossover operations, the combination of two operations seems to be sufficient for obtaining reasonable performance since the combination of three operations, PGA-c123, does not produce a better result.

Table 9. Ratio of solutions found by different crossover operations.

PGA-c1	PGA-c2	PGA-c3	PGA-c12	PGA-c13	PGA-c23	PGA-c123
1.0010	1.0008	1.0147	1.0000	1.0003	1.0002	1.0005

4 Conclusions

The purpose of this study is to provide a quick but a good solution for MPT scheduling in flow-shops. To achieve this objective, we developed a PGA. In the paper, we introduced the design of a SGA, which is the basic element of the PGA, together with the different characteristics of the PGA. The algorithm was implemented by multi-threaded programming on SUN servers. It was observed that if the workload of the server is high, by creating more threads, a better speed-up ratio could be obtained. We found that a relatively small size of a subgroup, about 16, and medium number of subgroups, about 15, are suitable. We compared different methods of migration among subgroups and found that random migration with 2 or 3 individuals is a better choice. In conclusion, the parallel implementation of the genetic algorithm can achieve both speed-up of computation time and the improvement of the near optimal solutions. As a prototype, this parallel genetic algorithm can be used to solve other complex scheduling problems.

Acknowledgement.
The work described in this paper was partially supported by a grant from The Hong Kong Polytechnic University (Project no. G-S551).

References

1. Blazewicz J., Drabowski M., Weglarz J.: Scheduling Multiprocessor Tasks to Minimize Schedule Length, IEEE Trans. Computers **C-35/5** (1986) 389-393
2. Brucker P.: Scheduling Algorithms, Springer, Berlin (1995)
3. Cantoni V., Ferretti M.: Pyramidal Architectures for Computer Vision, Plennium Press, New York (1994)
4. Chen C.L., Vempati V.S., Aljaber N.: An Application of Genetic Algorithms for Flow Shop Problems, European Journal of Operational Research **80** (1995) 389-396
5. Chipperfield A., Fleming P.: Parallel Genetic Algorithms, In: Zomaya, A.Y. (ed.): Parallel and Distributed Computing Handbook, McGraw-Hill (1996)
6. Dorndorf U., Pesch E.: Evolution Based Learning in a Job Shop Scheduling Environment, Comp. Opns. Res. **22** (1995) 25-40
7. Drozdowski M.: Scheduling Multiprocessor Tasks - An Overview, European Journal of Operational Research **94** (1996) 215-230
8. Ercan M.F., Fung Y.F.: The Design and Evaluation of a Multiprocessor System for Computer Vision, Microprocessors and Microsystems **24** (2000) 365-377
9. Gupta J.N.D., Hariri A.M.A., Potts C.N.: Scheduling a Two-stage Hybrid Flow Shop with Parallel Machines at the First Stage, Ann. Oper. Res. **69** (1997) 171-191
10. Holland H.: Adaptation in Natural and Artificial Systems. Ann Arbor, The University of Michigan Press (1975)
11. Krawczyk H., Kubale M.: An Approximation Algorithm for Diagnostic Test Scheduling in Multicomputer Systems, IEEE Trans. Comput. **34/9** (1985) 869-872
12. Lloyd E.L.: Concurrent Task Systems. Opns Res. **29** (1981) 189-201
13. Murata T., Ishibuchi H., Tanaka H.: Genetic Algorithms for Flowshop Scheduling, Comp. Ind. Engng, **30** (1996) 1061-1071
14. Oguz C., Ercan M.F., Cheng T.C.E., Fung Y.F.: Multiprocessor Task Scheduling in Multi Layer Computer Systems, in print European Journal of Operations Research.
15. Reeves C.R.: A Genetic Algorithm for Flowshop Sequencing, Comp. Opns. Res. **22** (1995) 5-13
16. Scala M. L., Bose A., Tylavsky J., Chai J. S.: A Highly Parallel Method for Transient Stability Analysis, IEEE Trans. Power Systems **5** (1990) 1439-1446

Performance Variability of Highly Parallel Architectures

William T.C. Kramer[1] and Clint Ryan[2]

[1] Department of Computing Sciences, University of California at Berkeley and the National Energy Research Scientific Computing Center, Lawrence Berkeley National Laboratory

[2] Department of Computing Sciences, University of California at Berkeley

Abstract. The design and evaluation of high performance computers has concentrated on increasing computational speed for applications. This performance is often measured on a well configured dedicated system to show the best case. In the real environment, resources are not always dedicated to a single task, and systems run tasks that may influence each other, so run times vary, sometimes to an unreasonably large extent. This paper explores the amount of variation seen across four large distributed memory systems in a systematic manner. It then analyzes the causes for the variations seen and discusses what can be done to decrease the variation without impacting performance.

1 Motivation

Application performance for computer systems, including parallel system hardware and architectures, is well studied. Many architectural features are assessed with application benchmarks – be it on real or simulated systems – so the impact of the functions can be evaluated. To a lesser degree, system software is evaluated for performance of both the code itself and the applications that use it.

Despite the studies there is a crucial area of performance that has not received enough attention – specifically how much variation in performance exists in systems and what contributes to that variation, particularly when software and hardware are viewed as a single system. The variability of performance is a reliability issue as importance as availability and mean time between failures. The user's productivity is impacted at least as much when performance varies by 100% as when system availability is only 50%.

Large scale applications run on multiple CPUs. Many of the most challenging applications, including climate, combustion, material science and life science applications need hundreds to thousands of processors for the most demanding problems. Currently, the only technology that can address these large-scale problems is distributed memory systems consisting of clusters of Shared Memory Processing (SMP) nodes, connected with dedicated interconnect networks. The components of the system that can be adjusted for the application workload, typically fall into three major categories.

P.M.A. Sloot et al. (Eds.): ICCS 2003, LNCS 2659, pp. 560–569, 2003.

- Intensive computation with large memory and fast communication
- Intensive computation with limited memory and less communication
- Intensive computing with large data storage

Architectural features contribute to the performance and variation differently within these broad categories.

2 Variation Exists in Application Performance

Performance variation is caused by many factors. On multi-user systems with multiple jobs running within a shared memory processor, frequent causes are memory contention, overloading the system with work, and priorities of other users.

However, on large-scale distributed memory systems, it is rare the compute-intensive parallel applications share SMP nodes. The NERSC IBM SP system, "Seaborg," is a 3,328-processor system with Power 3+ CPUs. It uses an IBM "Colony" switch to connect the SMP nodes. The most efficient manner for applications to use this type of systems is to have nodes dedicated to a single parallel application that uses communication libraries such as MPI to communicate between tasks running in parallel. Thus, many factors normally contributing to variation are not present on these systems, yet, as shown below, application run times can still vary widely.

Figure 1 shows the variation present on the NERSC IBM SP system when it was first installed. Previous experience had shown that a number of software factors could cause variation, including slightly different system software installation on the nodes, system management event timing (daemons running at particular times) and application performance tuning. A typical code shows modest improvements in performance but high variability with a more aggressive setting. The goal is to recalibrate the timing interval to give the best performance with the least variation. Despite the challenges, some of which are outlined above, it is possible to make such a large system as the NERSC IBM-SP operate in a reliable manner. On a heavily used (85–95% utilization) system running the general scientific workload, the NPB benchmarks have consistently low performance variation over multiple runs averaging a CoV of 1.3%, with the maximum variation shown by the CG benchmark as 3.25%.

3 Hypothesis

As noted above, many things can be done to minimize performance variation, including strict system administration and management, nodes dedicated to single application, eliminating bugs, adding more resources and configuration tuning. Nonetheless, fundamental questions remain about how much variation is acceptable, how low variation can be on specific systems, and what most influences variation. This study is the initial attempt to explore the question *Do parallel system architectures/designs influence performance variation in addition to performance itself?* We believe they do.

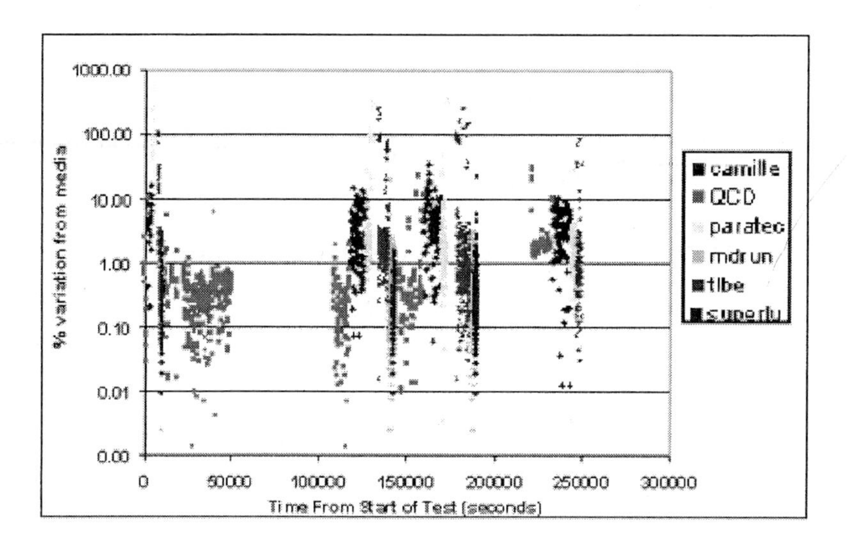

Fig. 1. The variation in performance of six full applications that were part of the NERSC IBM running with 256-way concurrency SP benchmark suite used for system acceptance. The codes were run over a three day period with very little else on the system. The run time variation shows that large-scale parallel systems exhibit significant variation unless carefully designed and configured

The approach of this study is to define experiments to understand factors that determine the relevant key features that influence variability of results. We assembled a set of portable codes that are representative of highly parallel applications and can be used to assess the variability of different system architectures. The codes were run on different architectures and the results were analyzed.

4 The Basic Tests

A number of codes were considered for testing architectures, including SPLASH[1] and Spec[2] benchmarks. Of the benchmark suites available the most effective for this purpose are the NAS Parallel Benchmarks[3] (NPBs), created at NASA Ames Research Center. The benchmarks have been heavily used on a wide range of architectures. They are portable and are known to give comparable results across systems. Since NPBs have been in use for over 10 years (evolving from Version 1 to Version 2), they are well understood. The codes were sensitive to other activities running on the systems. Finally the benchmarks have been correlated to real scientific workloads at a number of sites.

The NPB benchmarks are implemented primarily in FORTRAN, with some C. They use MPI as the message passing interface. The suite has different problem set sizes, three for parallel code and one for serial execution.

For the sake of simplicity, three NPB benchmarks were chosen for use.

• **LU** – The LU benchmark solves a finite difference discretization of the 3-D compressible Navier-Stokes equations through a block-lower-triangular block-upper- triangular approximate factorization of the original difference scheme. The Computation to Communication ratio is high so the code scales well. LU uses many small messages.

• **FT** – A 3-D FFT PDE with a 3-D array of data is distributed according to z-planes of the array. One or more planes are stored in each processor. The forward 3-D FFT is then performed as multiple 1-D FFTs in each dimension. An array transposition is performed, which amounts to an all-to-all exchange. Thus FT shows big, busty communication patterns amongst all nodes in between periods where all nodes are computing on their own data.

• **EP** – (Embarrassingly Parallel). Each processor independently generates pseudorandom numbers in batches and uses these to compute and tally pairs of normally distributed numbers. No communication is needed until the very end. This test was included to give a baseline for CPU performance.

5 Architectures Evaluated

Four systems with different architectural features were used in the evaluation. The complete features are listed in the full report[4], but a brief summary if provided here.

• **Cray T3E** – The oldest system is the Cray T3E at the NERSC, placed into service in 1997. The "mcurie" system consists of 696 CPUs, each with 256 MB of local memory. Together with interconnect hardware, the processor and local memory form a Processing Element (PE) or node. The PEs are connected by a network arranged in a 3-dimensional Torus with low latency and relatively high bandwidth. The processors are T3E Alpha EV-57 running at 450 MHz, capable of two floating point operations per cycle. The system uses a UNIX like operating system that has a Chorus derived microkernel on the 644 compute nodes and UNICOS/mk on the OS and command nodes. The T3E is the only system with static routing.

• **IBM SP** – The next system is a 3,328 processor IBM-RS/6000-SP at NERSC called "Seaborg." It is composed of a 184 compute nodes containing 16 Power 3+ processors connected to each other with a high bandwidth, switching network known as the "Colony" switch in a Omega topology. A full instance of AIX runs on every node. Each node has two switch adapters. Four nodes have 64 GB, 64 nodes have 32 GB and the rest have 16 GB.

• **Compaq SC** – The Lemieux Compaq SC system at the Pittsburgh Supercomputer Center (PSC) is composed of 750 Compaq Alphaserver ES45 nodes and a separate front end node. Each computational node contains four 1-GHz processors capable of two Flop/s per cycle and runs a full incidence of the Tru64 Unix operating system. A Quadrics Elan3 interconnection network connects the nodes in a Fat Tree topology. Each node is a four-processor SMP, with 4 GB of memory and two switch adapters.

• **Intel** – The final system is LBNL's "Alvarez" commodity cluster of 85 two-way SMP Pentium III nodes connected with Myrinet 2000, another Fat Tree. The CPUs are xSeries 330, running at 866 Mhz with 1 GB SDRAM each. Each node runs Linux RedHat distribution.

Thus, the systems studied show three types of network topology, four operating systems, and four types of processors.

6 Test Results

On each system, a number of runs were executed for each of the three NPB codes. All codes were run using the largest (Class C) problem sets with a 128-way concurrency using 128 MPI tasks. This was chosen because it used at least eight nodes and ran long enough to minimize the effects of start up events. The jobs were run in sets ranging from 10 to 30 runs of each code. Each system allocated dedicated nodes to the tasks. All runs used a one-to-one mapping of CPUs to tasks, which meant that the nodes were fully packed and all CPUs were used. Table 1 summarizes the results for the primary test runs. There was no significant variability due to time of day or which nodes were used by the scheduler to to run the jobs.

Table 1. The basic statistics for the test runs. Including some of the special tests discussed below, over 2,500 test runs were made

System		EP	LU	FT
Cray T3E	Number of Runs	70	119	118
	Mean Run Time (sec)	35.5	305.2	106.5
	Standard Dev (sec)	2.2	47.8	12.1
	Coefficient of Variance	6.11%	15.58%	11.33%
IBM SP	Number of Runs	424	165	210
	Mean Run Time (sec)	17.4	74.6	41.5
	Standard Dev (sec)	0.09	3.4	2.4
	Coefficient of Variance	0.52%	4.58%	5.70%
Compaq SC	Number of Runs	336	359	371
	Mean Run Time (sec)	5.03	42.8	30.6
	Standard Dev (sec)	0.35	1.9	1.0
	Coefficient of Variance	6.91%	4.53%	3.18%
Intel Cluster	Number of Runs	112	71	119
	Mean Run Time (sec)	17.6	408.7	90.7
	Standard Dev (sec)	0.03	10.7	1.0
	Coefficient of Variance	0.17%	2.62%	1.07%

7 EP Variation

Two machines, the T3E and PSC Compaq, showed an unexpectedly high variation for EP runs. As its name suggests, EP does very little communication.

However, because it has such a short run time, it is possible that individual cases of network congestion caused this variation. If a version of EP that does not use the network still shows significant variation, the individual node must be at fault. To test this, we ran the serial version on the T3E and a four-CPU (a single node) version on the Compaq. The coefficient of variation for the Compaq dropped to less than 1.6%, indicating that something on the node was causing variation. Variation dropped to 1.5% on the T3E, but we wished to determine the effect of migration. We measured only CPU time for the T3E and found less than 0.5% variation. From this we can conclude that the network and, in the case of the T3E, the NPB method of timing, were responsible for most of the variation. Only on the Compaq system do individual nodes contribute to the variation.

8 Changing the Number of Adapters

Two machines in the study, the IBM SP and the Compaq system, have a variable number of network ports (adapters) on each node. We expected the use of more adapters would mean run time and variation for the LU and FT benchmarks, but not for the EP benchmark. A set of test runs was made of all three benchmarks using both one and two adapters.

Changing from one to two adapters had a statistically significant effect on mean run time only for the EP and FT program runs on the SP machine ($p < .01$ in both cases). Using an F-test to compare changes in the variation, we found using two adapters increases variation for the FT benchmark on both systems ($p < .05$ in both cases). Variation for the LU benchmark decreased with two adapters on the Compaq ($p < .01$) and increased with two on the SP ($p < .01$). As expected, changing the adapter had no significant effect on the EP benchmarks.

These results agree with many, but not all, of our hypotheses. Increasing the number of adapters did not have much effect on mean run time, probably because the codes do not send enough data to benefit from an increase in bandwidth. We cannot explain why EP shows an increase in performance on the SP with two adapters. In addition, we do not yet know why LU performed more consistently on the Compaq with two adapters. These results do suggest two things. First, different factors influence variation; one cannot simply say that changing one particular aspect of a network will decrease variation for all programs that use the network. Second, when optimizing a program for a particular system, a programmer should consider things that do not necessarily increase megaflop rate or decrease memory usage. Analysis of Variance (ANOVA) suggest that changing adapters had the following effects on run time: a) no effect on PSC, b) no effect on SP for the LU code and c) an effect on the SP both the EP and SP codes.

9 High Variation on the Cray T3E

With the exception of the T3E, all of the machines studied had distributions such as the one shown in Fig. 2a. In essence, the distributions were tight bell curves with long right tails. Almost every job experienced some normally distributed slowdown, while a few suffered significantly more. A typical distribution for the T3E is shown in Fig. 2b. The majority of jobs experienced only a very small slowdown, while a significant portion suffered a far larger slowdown (40% or more in some cases). To test for this, we examined the system logs for some of the runs and measured only system time. This caused the histograms to collapse into ones similar to Fig. 2c.

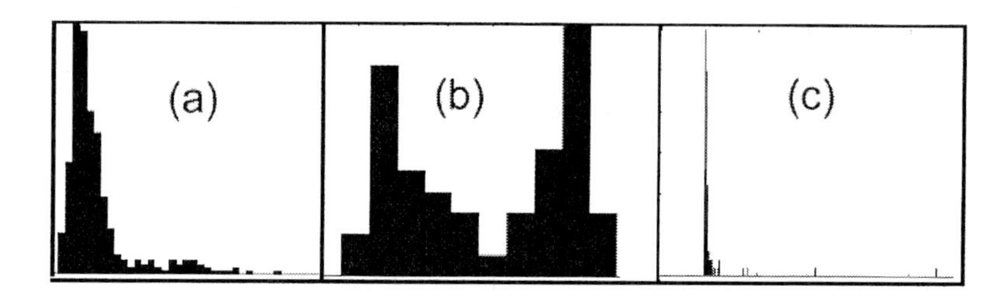

Fig. 2. Histograms of (a) LU times from the Compaq SC system. Note Gaussian distribution with a long fat tail. (b) LU runs from the T3E using the NPB report times. Note a much tighter distribution. (c) T3E distribution based on accounting data, not the run time as reported by LU. This eliminates the impact of migration and shows a much more tightly packed distribution

We were surprised at the variation indicated for the Cray T3E, which seemed unusually high, and investigated further. In order to make efficient use of the network, the T3E assigns logical node numbers to physical PEs at boot time.[5] Physical node numbers are based on how the node physically connects in the interconnect network. Logical numbers are assigned deterministically to minimize routing. The switch does direction order routing and special routing, but adaptive routing was never implemented, so the T3E only routes data through a predefined path using virtual channels.

Jobs are scheduled on logically contiguous nodes. This means that large contiguous blocks of PEs gradually become fragmented, making it increasingly difficult to run jobs requiring large numbers of CPUs. The T3E addresses this problem by periodically scanning all the PEs and identifying ones that have no work assigned. In a manner similar to memory shuffling, the system "migrates" jobs to pack all the running PEs together. This creates larger sets of contiguous PEs for new jobs to start. Jobs are assigned to PEs using a number of parameters, including an alignment measure that indicates how the starting point and/or ending point of the application aligns to power of two logical PE node number.[6]

In order to efficiently schedule new jobs, the T3E system software called the Global Resource Manager (GRM) scans all the nodes to look for opportunities to migrate. The frequency of scans is site selectable; on the T3E under study occur at five second intervals.

When jobs are migrated, system accounting is adjusted to compensate for the time the job is moving and not processing. However, the real time clock continues, which is what is used to report the NPB run times. System accounting logs have been correlated with the output of the NPB tests. Comparisons of T3E coefficients of variance using actual NPB Run Time reports and system accounting data are made. The NPB run time reports calculate the "wall clock" time for the test – and do not adjust for time lost due to migration or checkpoints. The wall CoV for the wall clock time of EP, LU, and FT were 6.11%, 15.58%, and 11.33% respectively, while the CPU time not including time spent during job migration, which most users pay attention to, was much lower at 0.8%, 0.6%, and 0.93%. When adjusted for time spent migrating, the variation of the T3E improves considerably. The situation caused by the T3E having to migrate to maintain a mapping of PEs to the location in the switch fabric has interesting trade offs. Much of the impact of variation discussed earlier is mitigated, since jobs will not abort due to exceed run times. Yet there are consequences, such as users waiting longer for results. Not migrating also has consequences, since certain work will not progress through the system as rapidly and system productivity will decrease.

10 Detecting and Reacting to Variation

Off-line detection and reaction to variation is possible and is done. Since minimal variation is not typically designed into architectures, most remediation is via system management and software. Making nodes strictly homogenous in hardware is key – but in differing amount so memory is not an issue unless the application tries to exploit virtual memory. Making nodes strictly homogeneous with respect to system software and timing of daemon task runs can have great impact. Indeed some parallel applications take less time if they do not use all the CPUs in a node to avoid the intrusive impact of system tasks that run periodically. Thus a coordinated, system wide timing "heartbeat" to coordinate the execution of system housekeeping tasks on all nodes may be beneficial.

Architecturally, variation may be impacted most by how the interconnect functions. Thus a clear evaluation of the likely variability of messages for different loaded conditions and patterns would seem called for as part of switch design. The effort to drive variation down to single digits after the fact is large and complicated.

The work discussed at NERSC required a team of 12 experts working together for six months, having skills in such areas as switch software and hardware, operating systems, MPI, compilers, mathematical libraries, applications and system administration. The improvements also involved major modifications to the switch micro code, lowest level software drivers and global file systems.

Detecting variation in real time so an application can respond is difficult. Dynamically detecting and responding in the proper manner is even more difficult. Some codes, such as the Gordon Bell Prize-winning LSM, are internally instrumented to report the performance of internal steps – such as reporting the overall performance or length of time taken for a time step of simulation. From there, it is feasible to consider monitoring the periods and identifying whether the past period is within an appropriate range.

Another way to instrument codes is with tools that acquire information from the hardware. While normally used for debugging and after the fact analysis of performance, such tools conceivably could be used to assess variation. Past systems such as the Cray YMP and C-90 could monitor and report hardware performance with virtually no overhead. In order to assess whether today's tools do the same, the benchmarks were built with the IBM performance tools[7], which monitor and report performance. Seven runs of the three instrumented benchmarks were made on the IBM SP and compared to the runs made without the monitoring software. The codes performed up to 250% slower when instrumented. Performance variation for the instrumented codes also was dramatically higher, with increases between 250% and 400%. This is in part because retrieving information counters in network adapters stops all traffic for a period of time. Unless more efficient methods are developed to decrease the overhead of performance monitoring, it is unlikely applications will be able to use these tools directly.

Even when variation is detectable, it is not clear what the proper action to take would be without a better model of what is going on. For example, it is unclear whether adding more CPUs to an application improves variation, because it changes both network traffic patterns and forces the application to scale more. Likewise, it may be decreasing CPUs would improve the variation and possibly performance. Several teams using large shared codes specifically run special tests on target systems to assess performance tradeoffs. However, it is uncommon for people to also test for tradeoffs in variation. Another concern is the fact that an application spending time monitoring and deciding what to do will have a longer run time and possibly a variation in performance.

11 Conclusions

Performance variation does exist on large distributed memory systems and can have a very significant impact on the useability and effectiveness of the system. While difficult, it is possible to constrain but not eliminate wide variation with good system management, tuning and design.

Performance variation still remains a problem and comes from complex tradeoffs of design and implementation. The T3E's need to migrate jobs in order to have contiguous nodes assigned to jobs due to switch routing increases variation. Yet, without migration, system scheduling would be much less efficient. If the architecture supported adaptive routing this conflict might be mitigated.

Architectural features do influence performance variation. The study of one or two adaptors shows architectural features impact performance and variation unexpected ways.

Some of the responsibility to control performance variation belongs in the domain of system managers. They have to assess how configuration and tuning changes will impact performance variation as well as absolute performance.

Responsibility for understanding variation also rests with the application user. The adapter study shows selecting the use of architectural features may not change performance but may positively or negatively influence the variation of performance.

Finally, the study of performance variation is important and can have impact in design of systems and applications. It is worth more resources to continue this understanding.

Acknowledgements. Thanks are given to the Pittsburgh Supercomputer Center and the National Energy Research Scientific Computing Center. In particular, David Skinner, Tina Butler, Nicholas Cardo, Adrian Wong Thomas Davis and Jonathon Carter contributed ideas to this paper. Special thanks are also owed to Prof John Kubitowicz for his suggestions and guidance.

References

1. Woo, Stephan, et al., The SPLASH-2 Programs: Characterization and Methological Considerations, Proc ISCA, June 1995, Santa Margherita Ligure, Italy, pp. 24–36.
2. Henning, John L., SPEC CPU 2000: Measuring CPU performance in the New Millenium, IEEE Computer, July 2000, pp. 28–35.
3. David H. Bailey, et al., The NAS Parallel Benchmarks, Intl. Journal of Supercomputer Applications, vol. 5, no. 3 (Fall 1991), pp. 66–73 http://www.nas.nasa.gov/Research/Reports/Techreports/1996/nas-96-010-abstract.html
4. William T.C. Kramer and Clint Ryan, Performance Varibility of Highly Parallel Architectures, National Energy Research Scientific Computing Center, Lawrence Berkeley National Laboratory Report LBNL-52591, March 2003. http://www.nersc.gov/ kramer/variation-systems.pdf
5. Cray T3E System Support Skills, Cray Research Technical documentation, Number R-T3ESSS, Revision R/H, May 1997
6. UNICOS/mk Resource Administration, SG-2602, version 2.0.2
7. IBM Parallel Environment for AIX Dynamic Probe Class Library Programming Guide, Version 3 Release 1, Document Number SA22-7420-00

JDOS: A Jini Based Distributed Operating System

M. Saywell and J.S. Reeve

The Department of Electronics and Computer Science
University of Southampton, Southampton SO17 1BJ, UK
jsr@ecs.soton.ac.uk

Abstract. J-DOS provides and integrated JAVA environment for the execution of a program across an interconnected network of heterogeneous computers. The system provides a file system, shared memory and a distributed execution scheme, all of which is transparent to the user. The framework used to provide these service is sufficiently general as to allow the provision of extra services by the user. We describe the client-server execution, remote execution and the shared file system, paying particular attention to the techniques used to distribute threads over many nodes. Distributed Mandelbrot set generation and rendering is used to benchmark and validate the the remote execution and load balancing aspects of the system.

1 Introduction

The purpose of a computer cluster is to present a number of physically distinct machines as a single networked virtual computer[2]. However such systems are typically only distributed at the processing level, requiring independently configured third party software to provide distributed functionality. For example NFS[6] is commonly used to share file systems and X11[3] provides support for remote GUIs. Our purpose is to provide the essential distributed services of remote processing, shared memory and a common storage medium and offer a framework which is readily extensible so that supplementary services can be added with minimal programming overhead. In contrast to high performance computing environments like Beowulf[5] which usually provide The Message Passing Interface[8], J-DOS provides the ability to adapt and scale in an un-managed way. Additionally the system doesn't attempt to disguise its distributed nature and instead allows the programmer to distribute processes when it is most suitable to do so.
J-DOS provides a dynamic extensible distributed computing environment in 100% pure Java. The system includes a file system, shared memory and a distributed processing environment. Provision is made for extra services to be added by the user without detailed knowledge of the system or the underlying Jini and RMI behaviour. The J-DOS environment is "zero configuration", in the sense that the introduction of a new service is made simply by running it to trigger the automatic registration of the service which then becomes available to

P.M.A. Sloot et al. (Eds.): ICCS 2003, LNCS 2659, pp. 570–580, 2003.

clients. Finally, J-DOS provides classes which automate the thread distribution and collection processes.

2 Other Systems

There are two systems, similar to our own, that aim at providing a distributed Java environment.
Javaparty[4] was developed by the Institute for Program Structures and Data Organisation at the University of Karlsruhe, and aims to provide transparent remote objects in a Java environment. An additional class modifier ,"remote", is introduced into the language which is interpreted by a pre-processor that generates the necessary RMI calls and interfaces and deals with network exception handling. Javaparty distributes instances of any "remote" objects among other machines, thereby spreading the load over all machines in the cluster. "While regular Java classes are limited to one Java virtual machine, remote classes and their instances are visible and accessible anywhere in the distributed JavaParty environment. As far as remote classes are concerned, the JavaParty environment can be viewed as a Java virtual machine that is distributed over several computers"[4].
Cluster Virtual Machine for Java[10, 11], cJVM, was developed at the IBM Haifa Laboratory and "provides a single system image of a traditional JVM while execution on a cluster"[10]. The underlying architecture of cJVM is similar to RMI, so when an object is located on a remote node, the local node communicates with it via a proxy object. From the applications perspective, the object and its proxy are indistinguishable. This allows multi-threaded programs to benefit from the cluster with no change to the code. Although threads are executed on different nodes, they cannot migrate between them. The load sharing function allocates a new thread to the best suited node when the thread was started, consequently load imbalance can arise when over time as the system load changes.

3 Description

In the following we describe the components of JDOS; a communications backend, an infrastructure to allow processes to run on remote hosts and a shared file system. Our system builds on Remote Method Invocation RMI[9], which allows a JVM running on one host to invoke methods on another host and the Jini[7] set of classes, which provide a client server architecture in which the client can locate and utilise services by name reference.

3.1 Communications

In order to understand the design decisions made here we must first give a more detailed account of the the communications process used between a typical Jini Service and a client. Initially the only knowledge a client has is that it requires

a service matching an interface which it possesses. This is used to query the lookup server and download a proxy class. The proxy then uses RMI to invoke methods on the service, so a Remote Interface must be declared containing the signature of each method which is to be invoked over the network.
For example:-

```
public interface Protocol extends Remote {
  public String getName() throws RemoteException;
  public void setName(String s) throws RemoteException;
}
```

The proxy class which is downloaded by the client contains a reference to this interface:-

```
public class Proxy
     implements Serializable, PublishedInterface {
  Protocol comms;
  int     age;
  public Proxy(Protocol p) {
    this.comms = p;
    this.age = 8;
  }
  public String getName() {
    return comms.getName();
  }
  public void setDetails(int i, String s) {
    this.age = i;
    comms.setName(s);
  }
}
```

Finally the service will either implement or contain a sub-class which implements this interface:-

```
public class Backend extends UnicastRemoteObject
                implements Protocol {
  String name = "Ben";
  Backend() throws RemoteException {}
  public String getName() throws RemoteException {
    return name;
  }
  public void setName(String s) throws RemoteException {
    name = s;
  }
}
```

The service is responsible for creating and publishing the proxy object. Notice that the proxy requires an object implementing the Protocol interface in its constructor - this is the Backend object which receives remote invocations and its proxy which is used to communicate with the backend:-

```
Backend backend = new Backend();
```

```
Proxy proxy = new Proxy(backend);
```

Having created the proxy it must be registered and uploaded to at least one lookup server before a client can download and execute it. The standard serialization process recursively converts each field of an object to a byte stream which can then be piped across a network or saved to disk.

The backend object extends UnicastRemoteObject, which implements the methods necessary to customise Serialization, which in short replaces the reference to the backend with a "stub" object. Stubs are complete Java objects, dynamically generated at compile time they implement the same interface as the service (in this example the "Protocol" interface) and relay invocations complete with parameters over the network to the service which created them. Since both stub and service implement the same interface, from the client's perspective there is no difference between the two. The overall resultant behaviour however, is that all invocations made by a client are realised on the same backend service object.

There is one final complication to this procedure which is the means by which the client obtains the class definition of the stub itself. This is required in order for the JVM to be able to instantiate a stub object once it has fully streamed from the server since one of the main design goals of Jini is that a client only needs an interface in order to fully communicate with a service. RMI solves this problem by providing a custom ClassLoader which retrieves class definitions from an HTTP server. This ClassLoader is then used automatically during the de-serialization process of the stub. The address of the server must be provided at run time, typically specified as a command line option to the client JVM.

The resulting framework preserves the strong typing provided by Java while simultaneously blurring the remote nature between client and service, however it does not provide a good base for a distributed system such as J-DOS in which clients are able to invoke any method of an arbitrary objects on a remote node and as such the interface used for communication must be as generic as possible. In fact it is possible to reduce it to a single method, namely:-

```
public Object invoke(String methodName, Object[] args)
                                   throws RemoteException;
```

As with traditional RMI this invokes the method named "invoke" on the Backend, however by using reflection this single method can be used to invoke any other method on the service.

```
public Object invoke(String methodName, Object[] args)
                                  throws RemoteException {
  Class[] cTypes = new Class[args.length];
  for(int i=0; i < args.length; i++) {
    cTypes[i] = args[i].getClass();
  }
  Method m=oService.getClass().getMethod(methodName,cTypes);
  Object foo = m.invoke(oService, args);
  return foo;
}
```

In the above code “oService” is a reference to the service which owns the object and performs no processing itself other than to invoke the specified method on its parent service object.
The following code is an example of how the proxy can utilise this universal communications protocol:

```
public Collection getMatching(Class c)
                            throws OFSException {
  try {
    Object[] args = {cwd, c};
    Object o = remoteComms.invoke("getMatching", args);
    return (Collection) o;
  }
  catch (RemoteException ex) {
    rethrow(ex);
  }
  return null;
}
```

The previous code was taken from the File System proxy and is used to retrieve all leaves in the current node which are an instance of the specified Class. The arguments passed to the communications backend are “cwd”, the path of the current node, and “c”, the class to match against. his call will invoke the “getMatching” method and the Collection of matching leaves is passed back to the client application.
The advantage of this implementation is that additional methods can be added to the service without re-compiling or even restarting existing clients. If existing method signatures are changed then corresponding changes are required only in the proxy. Additionally, the RemoteCommsBackend class can be re-used for every service. Its constructor takes a single argument which is the Service it belongs to, and therefore the object upon which it will invoke methods.
If an exception should be thrown by code executing on the server it is caught and passed back to the client encapsulated in a RemoteException. It is then rethrown, hence providing the client with meaningful Exceptions from the server.

3.2 Remote Execution

Our design of a remote execution system has three essential considerations:-

1. Remote methods executing of service nodes must not block. This is desirable from a processing perspective, to extract maximal gain in performance.
2. Methods often have side-effects, that is they may alter the internal state of the object in some way e.g. by setting an instance variable, producing output to the screen etc. Our design provides means for the client to retrieve the object after the method has been invoked and consequently these side effects are preserved.
3. The client must be able to upload the class definitions to the server so they can be accessed locally by the JVM.

In certain situations it may be advantageous to have side effects occurring on a remote machine, particularly with regards to resources available on that machine. For example, a method may read in or save to a file which is on the remote machine.

The implementation of our design is thread based, extending the concept of Runnable objects:-

```
public void remoteRun(RemotelyRunnable rThread)
```

RemotelyRunnable is an interface which extends both Serializeable and Runnable. When the client invokes this method the rThread object is serialized and sent to the service as with a remoteInvocation, however the method returns void. This is because the remote Service spawns a new thread which in turn executes the run() method of the RemotelyRunnable rThread object. Therefore the method will return before the rThread has completed execution, thus allowing the client to farm out many requests in a truly distributed fashion.

In a direct contrast to the remoteInvocation method, all results which are to be passed back to the client by some means must now be stored within the the rThread object itself. Once the run() method has completed the object must then be made available to the client by some means.

To do this we provide an RMI callback method within the client which is called by the service whenever a thread completes, the method itself could be integrated into the RemotelyRunnable class allowing the client to specify the action to be performed.

This approach is both the fast and efficient since the service sends the result directly to the client as soon as the thread had completed. The additional complexity required in the client is the result of it becoming an RMI server object, although once again much of this complexity will be common across all clients and so consequently can be encompassed into helper classes.

For a distributed system to be of practical use it should be possible for a user to run programs which they have compiled or otherwise obtained themselves, without providing each node in the system which is to run the program with the necessary class definitions. RMI addresses this issue by providing a mechanism to download these class definitions from an HTTP server, however this requires the downloading party to know the address of the HTTP server in advance. If the class definition is being held by the client (of which there may be many, all with different class definitions) then the server has no way of knowing where to look. Ideally the service should download the class definition from the client, which we provide by manually streaming the object which the client wishes to be run and instantiating it using a special ClassLoader which can load definitions from the client. In order to do this it is obvious that a form of 2 way communication is required, consequently the client must become an RMI server object.

With our approach the client sends the server an RMI stub, instead of the object to be run. The stub has several methods which the server must call in order to set up a connection over the network, through which the RemotelyRunnable object can be streamed. the RMI interface is as follows:

```
public interface ClientComms extends Remote {
```

```
  public InetAddress getAddress(Integer pID)
                                    throws RemoteException;
  public int getPort(Integer pID) throws RemoteException;
  public void receiveResult(RemotelyRunnable result)
                                    throws RemoteException;
  public ClassFile getClassDefn(String s)
                                    throws RemoteException;
}
```

On receiving the stub, the service immediately makes a network connection to the host and port returned by the first 2 methods, of the above interface. The client is waiting for a connection on this address.
Next the service instantiates a new instance of RemoteClassLoader. This is a customised ClassLoader which uses the getClassDefn method in the client stub to resolve class definitions. The client is only queried if the definition can not be found locally, hence it is not possible for the client to override local definitions and performance is much greater than if every definition was taken from the client. The method returns a ClassFile object, a simple wrapper for an array of bytes which constitute the class definition itself.
Following this a custom ObjectInputStream is created. This is a very simple extension of the ObjectInputStream provided by the Java API:-

```
class JDOSObjectInputStream extends ObjectInputStream {
  private ClassLoader loader;

  public JDOSObjectInputStream(InputStream in,ClassLoader loader)
                            throws IOException {
    super(in);
    this.loader = loader;
  }

  protected Class resolveClass(ObjectStreamClass v)
                            throws ClassNotFoundException {
    return Class.forName(v.getName(),true,this.loader);
  }
}
```

This class overrides the resolveClass method (used to load class definitions) forcing the use of a custom ClassLoader, in this case the RemoteClassLoader. The method is called when a complete object has been received from the client and is about to be instantiated, the RemoteClassLoader returns the Class necessary for the received object to be instantiated, having uploaded the definition from the client if necessary. This allows the remote host to run any program that the client can provide class definitions for, without restarting the service or manually uploading files to another host, hence greatly increasing the transparency of the system.
The Jini classes, contained in JavaSpaces, are used as a means of distributing the stubs: the client writes an object containing the stub and a unique process

ID to a JavaSpace which then notifies all interested parties (in this case the execution services) that a new object has been written. Each service can then choose whether or not to take the object. This process is done on a first-come first-served basis, for example if 5 objects are written to the JavaSpace and there are 10 executing services then only the first 5 services will successfully retrieve a matching object from the JavaSpace. The process ID contained in this object is used when communicating with the client over RMI (see the RMI protocol interface above), as all the services communicate with the same RMI backend on the client. This approach could be seen as limiting the scalability of the system, however the RMI stubs are very small so large amounts of processing would be needed before a bottleneck was realised. If this scenario were to arise it could be easily dealt with by having multiple JavaSpaces, services would receive notifications from all of them whereas clients would upload stubs to a single JavaSpace selected at random.
This approach makes the client's role in distributed processing much simpler. It need only to write the stubs to a JavaSpace and then wait for the results to be returned by means of another RMI call to the client. This call is relayed back to the client application, which should implement the RRCallback interface, this specifies a single method:-

```
public void receiveResult(RemotelyRunnable rr)
```

All the behaviour described above has been incorporated into a series of "helper" classes, the most important of which is RemoteThreadRunner, as it automates the entire process I have described. The client need only to instantiate this class and provide the above callBack method to be able to farm out and receive threads. New RemotelyRunnable objects can be distributed by simply invoking the remoteRun method of the RemoteThreadRunner object.

3.3 Remote File System

Another requirement of a distributed environment is some form of shared file system. While JavaSpaces partially meets this need, it does not provide any kind of structure and introduces unnecessary complications such as requiring objects be renewed with leases. As such, it was decided that an additional service be added to the system providing an interface which mimics that found in traditional file systems.
The suggested structure is a tree comprised of nodes which contains leaves and other nodes, with a single root node, named "/" by default. Objects are encapsulated into leaves which then may be placed into any node. We have made extensive use of Interfaces and abstract classes to allow different underlying storage mechanisms, with a RAM based backend for testing but potential for a persistent implementation.
At the lowest level of the file system is a series of interfaces which define the basic properties of leaves and nodes. Note that at this level serialization is not enforced, hence it is possible to create branches containing more fluid entries similar to /dev or /proc in Linux. For example a node could contain a list of currently

executing threads, or a list of currently known processing nodes. Building on top of this layer is the series of RemoteFileSystem interfaces, these are identical except that Serialization is enforced.

As the file system is entirely Java-based it seemed logical to support the reading and writing of objects directly, note however that if the object is written to a remote node it must still be Serializeable, so this does not offer any particular advantage except for convenience.

Like any other Jini service the Remote File System may be accessed by multiple clients at any one time. Care must taken concerning the serialization of objects. If a client is passed the root node, the entire tree will be serialized including all sub-nodes and leaves, and any changes the client makes to the tree will not be reflected at the server as the client now has a local copy.

A second issue is that each client must be responsible for remembering its current location in the file system, if this is not so when one client changes node this will be reflected in all other clients. This is clearly an undesirable behaviour.

In order to address these issues the concept of a "window" is used, this is an object used by the client which holds state (current directory) and is used to communicate with the server. Many of the methods, particularly those associated with navigation, take and return String objects, thus avoiding the serialization problem. Whenever the window communicates with the server it also passes a String denoting the directory in which the action should be performed, this addresses the state issue while maintaining a clean interface to the client.

```
public abstract class ObjectFSWindow extends ObjectFSNode
                                    implements Serializable {
  abstract public void moveTo(String s) throws OFSException;
  abstract public void openNode(String s) throws OFSException;
  abstract public void openParent() throws OFSException;
  ....
}

public abstract class ObjectFSNode
                                    implements ObjectFSMember {
  abstract public String[] peekNodes() throws OFSException;
  abstract public String[] peekLeaves() throws OFSException;
  abstract public void removeLeaf(String s) throws OFSException;
  abstract public void removeNode(String s) throws OFSException;
  abstract public String getName() throws OFSException;
  abstract public void setName(String s) throws OFSException;
  ....
}
```

In addition any implementation of this service must take care that operations are thread safe. In particular, care must be taken to handle the scenario which arises when when one client performs an action which affects a node currently in use by another client.

4 Performance Results

The algorithm chosen to demonstrate the distributed processing ability of the system was the generation of Mandelbrot fractals. The set is generated by the iteration formula:-

$$Z_n = Z_{n-1}^2$$

where $Z_n = X_n + iY_n$ is a complex number, with $X_n$and Y_n real. For each point on the X, Y plane Z is iterated until it converges and that point of the plane is coloured in proportion to the number of iterations that it takes to converge. If Z fails to converge in 256 iterations then the point is black. The region covered was the plane $-1 < X < 1$ and $-1 < Y < 1$. This was partitioned into square sub-regions and the colour map for each sub-region was computed independently on different processors.
The platform used to test J-DOS was a Beowulf[5] cluster running MOSIX[1] consisting of 30 dual processor, 1GHz Pentium3 machines each with 1Gbyte of ram and linked with 100Mbit Ethernet. MOSIX itself is capable of migrating processes so care taken to ensure that it didn't do so during our experiments. Table 1 shows that the problem does indeed benefit from distribution albeit in a way that isn't in proportion with the number of processors. Part of the problem is the bus architecture of the cluster and part is the overhead in distributing threads over several JVMs. The second contribution is clearly displayed in table 2. Where the problem only involves a single thread and we use more than one processor that single thread is not run on the host machine and the extra second that it takes to run can be attributed to the time it takes to transfer the 15Mbytes of data back to the host node. The benefits of parallelisation show up somewhat better with higher resolution images but the bottleneck associated with collecting data at a common node remains.

Table 1. Times to render a 2000×2000 image when the number of processors matches the number of threads

Threads	1	4	16	25
Time (secs)	8.4	3.4	3.9	3.9

5 Conclusions and Future Developments

The system is currently being developed by adding a remote GUI interface so that graphical applications could be displayed on one node and the J-DOS interface on another. Inter-thread communication is also being developed so that programmers are not restricted to the client server model. This would allow parallel programs with any logical communication patterns to be run under the

Table 2. Time to render a 2000×2000 image when the number of threads doesn't match the number of processors

	Nodes			
Threads	1	4	16	25
1	6.8	8.3	8.3	8.3
4	3.8	3.3	3.5	3.4
16	4.3	3.9	2.8	2.4
25	4.2	31.	2.6	2.7

system. The file system, as so far developed, is not distributed and would benefit from being so by allowing more overall storage capacity and allowing data searches to be distributed over all available nodes. Redundancy could also be built into such a file system making it more robust.
Overall J-DOS provides a dynamic extensible distributed computing environment in Java, with well defined interfaces by which programmers can write and distribute there own programs, without detailed knowledge of the cluster and using either client server or shared memory communication or both.

References

1. A.Barak and O.La'adan. The MOSIX multicomputer operating syatem for high performance cluster computing. *Journal of Future Generation Computer Systems*, 13:361–372, 1998.
2. C.Catlett and L.Smarr. Metacomputing. *Communications of the ACM*, 35:44–52, 1992.
3. D.Young. *X Window system programming and applications with XT*. Prentice Hall, 1990.
4. M.Philippesen and M.Zenger. JavaParty-transparent remote objects in java. *Concurrency:Practice and Experience*, 9:1225–1242, 1997.
5. T.Sterling nad G.Bell and J.Kowalik. *Beowulf Cluster Computing with Linux*. MIT Press, 2001.
6. R.Sandberg. The SUN network file system: Design, implementation and experience. Technical report, SUN Microsystems, Inc., 1985.
7. S.Oaks and H.Wong. *Jini in a nutshell*. O'reilly, 2000.
8. W.Gropp, M.Snir, B.Nitzberg, and E.Lusk. *MPI: The Complete Reference*. MIT Press, 1998.
9. W.Grosso. *Java RMI*. O'reilly, 2002.
10. Y.Aridor, M. Factor, and A.Teperman. cJVM: A single image of a JVM on a cluster. In *Proceedings of the International Conference on Parallel Processing, Aizu-Wakamatsu, Japan*, pages 4–11, 1999.
11. Y.Aridor, M. Factor, and A.Teperman. Implementing java on clusters. *Lecture Notes on Computer Science*, 2150:722–731, 2001.

Parallel Blocked Sparse Matrix-Vector Multiplication with Dynamic Parameter Selection Method

Makoto Kudo[1], Hisayasu Kuroda[2], and Yasumasa Kanada[2]

[1] Department of Computer Science
Graduate School of Information Science and Technology
The University of Tokyo
[2] Super Computing Division, Information Technology Center
The University of Tokyo
{mkudoh,kuroda,kanada}@pi.cc.u-tokyo.ac.jp

Abstract. A blocking method is a popular optimization technique for sparse matrix-vector multiplication (SpMxV). In this paper, a new blocking method which generalizes the conventional two blocking methods and its application to the parallel environment are proposed. This paper also proposes a dynamic parameter selection method for blocked parallel SpMxV which automatically selects the parameter set according to the characteristics of the target matrix and machine in order to achieve high performance on various computational environments. The performance with dynamically selected parameter set is compared with the performance with generally-used fixed parameter sets for 12 types of sparse matrices on four parallel machines: including PentiumIII, Sparc II, MIPS R12000 and Itanium. The result shows that the performance with dynamically selected parameter set is the best in most cases.

1 Introduction

Sparse matrix-vector multiplication (SpMxV), $y = y + Ax$ where A is a sparse $(N \times N)$-matrix, x and y are dense N-vector, is an important kernel which appears in many scientific fields. Since parallel computers have recently become popular, most SpMxV computations have been executed on parallel environments. The computation time of SpMxV often accounts for significant part of the total application time. Therefore, fast SpMxV routine is important for high performance computing field. This paper proposes a parallel SpMxV routine in which aggressive optimizations are exploited. The target machines range over distributed memory parallel machines including clusters. SMPs are also considered via SPMD programming model. Each node is assumed to have memory hierarchy of cache and memory.

In local computation part on each node, our routine uses blocking method in order to efficiently exploit higher level memory. In addition to conventional two blocking methods for SpMxV, a new blocking method which generalizes

P.M.A. Sloot et al. (Eds.): ICCS 2003, LNCS 2659, pp. 581–591, 2003.

those two methods is proposed. This proposed blocking method is applied to the parallel environment.

There are several parameters that affect the performance of blocked parallel SpMxV computation. The optimal parameter set highly depends on the characteristics of target matrix and machine. Therefore, selecting the optimal parameter set according to these characteristics is important. The optimal parameter set is automatically selected at the run time according to the characteristics of target matrix and machine. Therefore, our routine can achieve high performance on various computational environments.

2 Experimental Environments

The features of 4 parallel machines used in the experiment are shown in Table 1. C language is used as a programming language. For communication library, MPI library is used. The implementation of MPI library is MPICH [1] ver 1.2.1.

12 test sparse matrices are shown in Table 2. They are collected from Tim Davis' Sparse Matrix Collection [2] except the dense matrix of No. 12. Column 'Size' is the dimension of the matrix. All matrices are square matrix. Column 'Nonzeros' is the number of non-zero elements in the matrix. Column 'Density' is the percentage of existence of non-zero elements.

3 Local Computation Optimization

A blocking method is a popular optimization technique for SpMxV on machines that have memory hierarchy. Non-zero elements in sparse matrices from real-world scientific fields are not randomly scattered, but often exist in dense blocks [3]. By blocking such dense blocks, high performance of SpMxV can be achieved. This blocking method is called as 'Register Blocking' [4], because its main focus is efficient usage of the register memory. The blocking shape is generally rectangular. Since most of dense blocks in sparse matrices are small, a block size

Table 1. Machine's architectures

Processor	Intel Pentium III	SUN Ultra Sparc II	SGI MIPS R12000	Intel Itanium
Clock	800 MHz	333 MHz	350 MHz	800 MHz
# of PEs	8 (dual $\times$ 4)	8	8	8
Network	100 base-T Ethernet	SMP	DSM	SMP
Compiler	Portland Group's C compiler	SUN WorkShop Compiler	MIPSpro Compiler	Intel Itanium
Compiler Version	3.2-3	5.0	7.30	5.0.1
Compiler Option	-fast -tp p6 -Mdalign -Mvect= assoc,prefetch,nosse	-xO5 -dalign -xtarget=native -xarch=v8plusa -xrestrict=all	-Ofast=ip27 -TARG:proc=r10k -TARG:isa=mips4 -r12000 -64	-O3

Table 2. Test matrix set

No.	Name	Description	Size	Nonzeros	Density
1	pre2	Harmonic balance method	659033	5,959,282	0.00137
2	cfd2	CFD, symmetric pressure matrix	123440	3,087,898	0.0203
3	pwtk	Stiff matrix	217918	11,634,424	0.0245
4	gearbox	Aircraft flap actuator	153746	9,080,404	0.0384
5	venkat01	Flow simulation	62424	1,717,792	0.0441
6	nasasrb	Structural engineering	54870	2,677,324	0.0889
7	ct20stif	CT20 engine block	52329	2,698,463	0.0985
8	bcsstk32	Stiff matrix for automobile	44609	2,014,701	0.101
9	gupta2	Linear programming matrix	62064	4,248,286	0.110
10	3dtube	3D pressure tube	45330	3,213,618	0.156
11	gupta3	Linear programming matrix	16783	9,323,427	3.31
12	dense	Dense matrix	3162	9,998,244	100

should also be small, such as 2×2 or 3×3. By blocking in such a small size block, a usage of the highest level memory, register memory, can be optimized, and the number of access to the column index array is reduced.

The access order to matrix data in blocked SpMxV computation changes from non-blocked SpMxV. Therefore, the data format of sparse matrix is reconstructed from Compressed Row Storage (CRS) format to Block Compressed Row Storage (BCRS) format [5]. Array elements of non-zero values are reordered to the accessing order, and column index array is also reconstructed.

There have been used two types of register blocking method so far. The difference of these two methods is that whether zero elements in the original sparse matrix can enter the block of the blocked matrix or they can not.

One of the conventional methods which does not allow zero elements entering into the blocked matrix (we call this method as type 1 method) splits the original matrix into two matrices, blocked matrix A_{block} and remaining matrix A_{remain} [6,7,8] (See the left of Fig. 1). We start searching blocks from the top-left of the matrix to the right-bottom direction. If a dense block larger than the block size $r \times c$ is found, it is stored into the blocked matrix A_{block}. The dense block smaller than the block size is not blocked and stored into the remaining matrix A_{remain}. The data storage format of A_{remain} is CRS format. In this blocking method, the

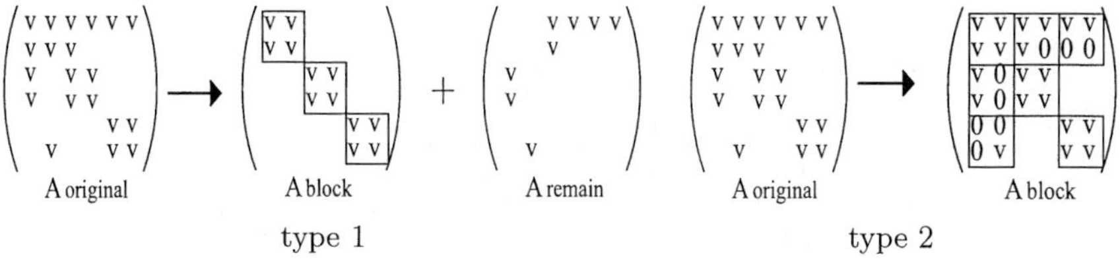

A original → A block + A remain (type 1); A original → A block (type 2)

Fig. 1. The example of conventional two blocking methods (2×2)

total number of the non-zero elements is the same before and after the blocking. Therefore, the number of floating-point number operations is also the same. However, since the original matrix is split into two matrices, SpMxV operation is computed twice. The performance may decline by overhead of two matrices computation. Since blocks can not contain zero elements, an opportunity of blocking many non-zero elements may be missed.

The other method of conventional blocking (type 2 blocking) is that even if there are dense blocks in the original matrix whose size is smaller than block size $r \times c$, the elements are stored in the blocked matrix with padding zero elements [4,9]. The original matrix is split into logical $r \times c$ blocks. If there is at least one element in a logical block, that block is stored into the blocked matrix with padding zero elements. If zero elements in the original matrix are blocked, these zero elements are treated as non-zero elements in the blocked matrix (See the right of Fig. 1). All the non-zero elements are blocked in this manner. The original matrix is not split unlike the type 1 blocking. However, the number of non-zero elements may increase. Therefore, the number of floating-point operations may also increase, and it can cause performance decline.

3.1 The New Blocking Method

Assume that the original matrix is blocked with block size $r \times c$. The type 1 blocking method makes block if there are at least rc elements in a block. And the type 2 blocking method makes block if there is at least 1 element in a block. We can easily find that the both methods are the parts of a general method that if there are at least t elements in a block, that block is stored into the blocked matrix. t is the threshold integer number. This generalized method is our proposal method. This method uses the threshold number t, if the number of non-zero elements in a block is larger than t, the block is stored into the blocked matrix with padding zero elements if necessary. And if the number of non-zero elements in a block is smaller than t, the block is stored into remaining matrix in CRS format. This method does not miss the non-zero elements to be blocked unlike the type 1 blocking, and does not increase a lot of non-zero elements unlike the type 2 blocking. The example of blocking by this method with block size 2×2 and threshold 2 is shown in Fig. 2.

Next, we show the effect of our new blocking method. The performance of non-blocked SpMxV, that of conventional two methods and that of our new

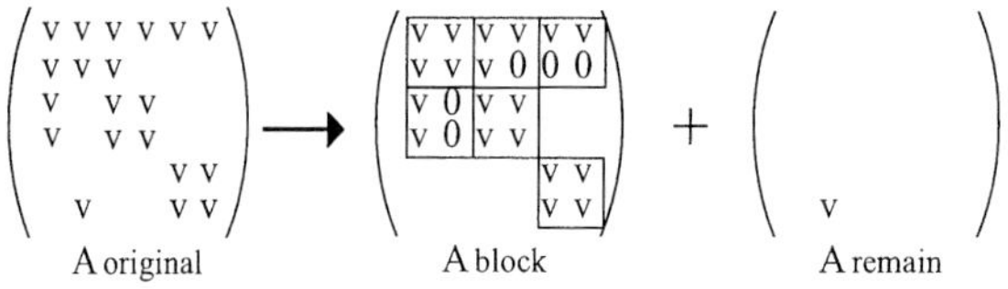

A original A block A remain

Fig. 2. The example of proposed blocking method $(2 \times 2, 2)$

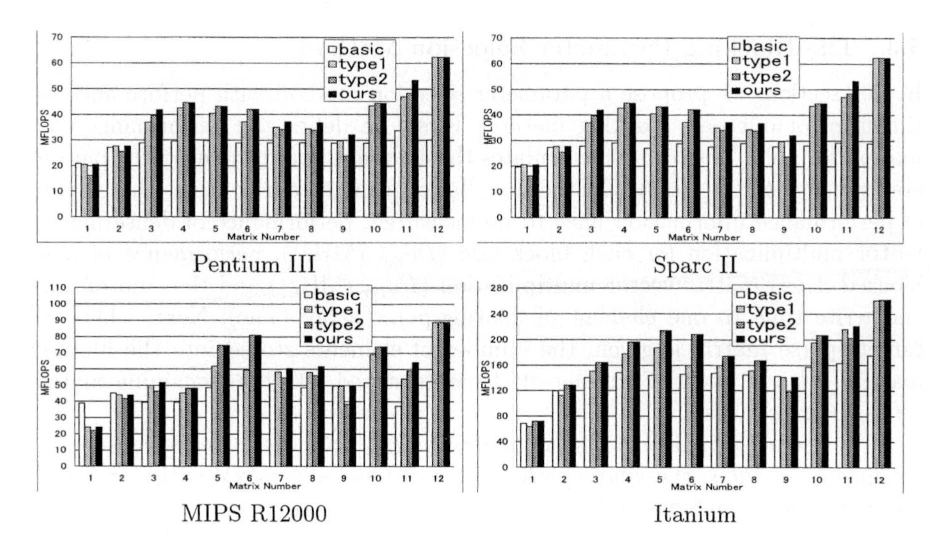

Fig. 3. The performance of four types of SpMxV

method are shown in Fig. 3. The block sizes used in the figure are the best block sizes that are selected by exhaustive search. The block sizes used in this paper are from 1×2 to 8×8. From the result, it can not be said that one of the conventional two methods are better for all cases. There are cases that the difference of performance of type 1 and type 2 is more than 20 %. Therefore, it can be said that fixing the blocking method to either of conventional two methods is not a good idea. On the other hand, since our new method includes type 1 and type 2, the performance of our new method is at least the same performance of conventional two methods. In some cases, its performance is better than both of conventional methods.

3.2 Selection of the Optimal Blocking Parameter Set

There are 3 parameters for our new blocking method, row size r, column size c and threshold t. To achieve high performance, the optimal parameter set (r, c, t) should be used. t can take from 1 to rc for the block size $r \times c$. The number of combinations which the parameter set (r, c, t) can take is 1926 for the block sizes from 1×2 to 8×8. One of the selecting methods to find the best parameter set is measuring the performance exhaustively for all parameter sets. As stated before, the matrix data have to be reconstructed from CRS format to BCRS format before the series of computation of blocked SpMxV.

The reconstruction operation takes about 10 times of non-blocked SpMxV computation time in this study. Therefore, it takes more than 10,000 times of non-blocked SpMxV computation time for exhaustive search. Since it is not practical, the performance of the blocked SpMxV for a given parameter has to be estimated without matrix data reconstruction.

3.3 The Blocking Parameter Selection Method

In this section, we propose a parameter selection method with performance estimation for our new blocking method. First, we define the performance estimation formula. The formula calculates the estimated performance with parameter set (r, c, t), $P(r, c, t)$. On the installation time of our routine, three types of performance information have to be measured, performance of dense matrix vector multiplication for each block size ($Pd_{r,c}$ (MHz)), performance of non-blocked dense matrix vector multiplication (Pd_{nb} (MHz)) and the sum of read and write time to one element of floating-point array (Tac). Next, when the target sparse matrix is given, the number of elements stored into the blocked matrix ($Nb_{r,c,t}$) and the number of elements stored into the remaining matrix ($Nr_{r,c,t}$) are counted.

If we use performance of blocked matrix as $Pd_{r,c}$ and performance of remaining matrix as Pd_{nb}, the whole performance of blocked SpMxV with parameter set (r, c, t) is estimated as follows,

$$P(r, c, t) = \begin{cases} \dfrac{nnz}{\frac{Nb_{r,c,t}}{Pd_{r,c}} + M \cdot Tac}, & t = 1 \\ \dfrac{nnz}{\frac{Nb_{r,c,t}}{Pd_{r,c}} + \frac{Nr_{r,c,t}}{Pd_{nb}} + 2M \cdot Tac}, & t > 1 \end{cases} \quad (1)$$

where M is row size of matrix and nnz is the number of non-zero elements in the original matrix. $M \cdot Tac$ means the access time to the destination vector, y. Three variables in (1), $Pd_{r,c}$, Pd_{nb} and Tac have to be measured only once at the installation time of the routine. However, $Nb_{r,c,t}$ and $Nr_{r,c,t}$ have to be counted at the run-time after the target sparse matrix is given. Therefore, it is important to quickly count these two variables. Next, we explain the fast counting method of these two variables.

The original matrix is logically split into the $r \times c$ size blocks. Then, the number of non-zero elements in each block is counted. Let $nnzb$ be this value. $nnzb$ is counted for all the logical blocks, and then the number of blocks for each $nnzb$ is counted. That is, we collect the information that the number of blocks whose $nnzb = 1$ is a and the number of blocks whose $nnzb = 2$ is b, etc. Let $b_{r,c,i}$ be the number of blocks whose $nnzb = i$ (block size is $(r \times c)$). The two variables $Nb_{r,c,t}$ and $Nr_{r,c,t}$ in (1) are calculated by the following equation, $Nb_{r,c,t} = \sum_{i=t}^{rc} b_{r,c,i} rc$, $Nr_{r,c,t} = \sum_{i=1}^{t-1} b_{r,c,i} i$. By using this counting method, if we count $b_{r,c,i}$ once for a block size, $Nb_{r,c,t}$ and $Nr_{r,c,t}$ of all the threshold values for the block size can be quickly calculated by small number of integer instruction.

$P(r, c, t)$ for all the combination of (r, c, t) is calculated in this manner, and the parameter set (r, c, t) with which $P(r, c, t)$ is best is selected as the optimal parameter set.

4 Data Communication Optimization

The matrix and vector data are distributed in the row block distribution manner so that each processor has nearly the equal number of non-zero elements. On parallel SpMxV computation, vector elements that reside on other processors may be required. In such case, communication of vector data between processors is necessary. Communication time becomes more dominant as the number of processors becomes larger. Therefore it is important for parallel SpMxV operation to reduce the communication time. In this paper, two communication methods are used; 'range limited communication' and 'minimum data size communication'. It is dependent on the non-zero structure of target matrix and machine's architecture which method is fast.

Range Limited Communication: By this communication method, processors send only minimum contiguous required block (See the left of Fig. 4). Communications between processors that are not required are not sent. Since unnecessary elements are often contained in the block, the communication data size is not minimum on most cases. However, since this communication method does not need local copy operation, the overhead time is small.

Minimum Data Size Communication: This method communicates only elements that have to be sent to other processors. (See the right of Fig. 4). Therefore the communication data size is minimum. In order to complete communication within two processors by one message passing, we need pack and unpack operations, because the elements to be sent are not placed consecutively in the source vector. These pack and unpack operations require a little overhead time.

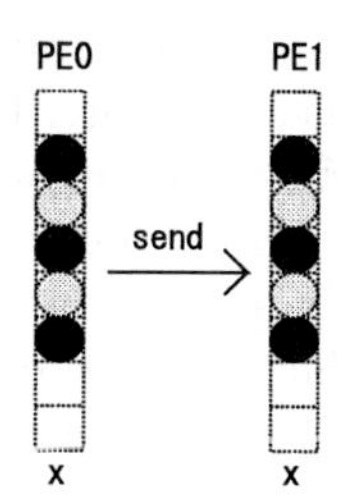

Range limited communication

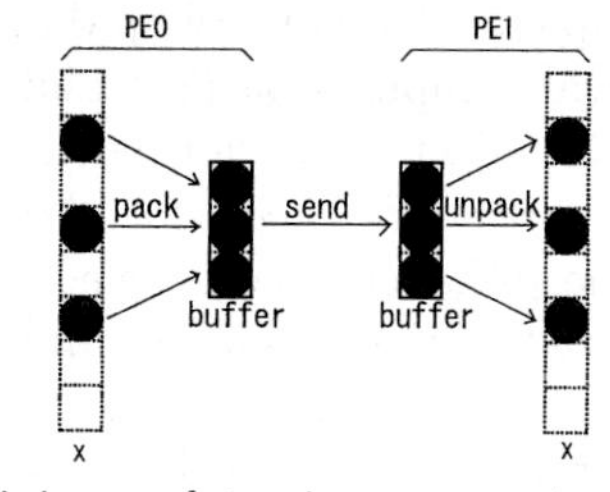

Minimum data size communication

Fig. 4. The range limited communication and the minimum data size communication The black circles in PE0 represents the source vector elements that must be sent to PE1. The gray circles are the dummy data that does not need to be sent to PE1 originally, however, these elements also have to be sent in range limited communication.

4.1 Implementation of Communication Method

The range limited communication and minimum data size communication use one to one communication. In this study, three implementation types of one to one communication are used, 'Send-Recv', 'Isend-Irecv' and 'Irecv-Isend'. 'Send-Recv' uses MPI_Send and MPI_Recv functions. 'Isend-Irecv' uses non-blocking communication (MPI_Isend and MPI_Irecv). All the communication types implemented in this study are summarized in Table 3.

Table 3. The communication Methods

Method	Explanation
sendrecv R	Range limited comm. with MPI_Send and MPI_Recv
isendrecv R	Range limited comm. with non-blocking communication in send-receive order
irecvsend R	Range limited comm. with non-blocking communication in receive-send order
sendrecv M	Minimum data size comm. with MPI_Send and MPI_Recv
isendrecv M	Minimum data size comm. with non-blocking communication in send-receive order
irecvsend M	Minimum data size comm. with non-blocking communication in receive-send order

4.2 Application of New Blocking to the Parallel Environment

On single processor environment, blocking parameter set is selected by performance estimation formula described in Sect. 3.3. This parameter selection method is applied to parallel environment. The block size is selected by performance estimation concurrently on each processor. That is, a processor selects the block size with only information of the matrix data that is distributed to the processor. Therefore, data communication is unnecessary on the blocking parameter selection time. Because the estimated blocking parameter can differ on each processor, each processor may select different blocking parameter set.

5 Numerical Experiment

5.1 The Dynamic Parameter Selection Method

We have implemented parallel SpMxV routine which has optimization techniques on both local computation part and data communication part. It has four parameters that affect the performance of parallel SpMxV, *the row size of the block, the column size of the block, the threshold number and the data communication method.* The parameter set is selected automatically at the run time according to the characteristics of matrix and machine in the following order, **1.** Select the communication method, **2.** Select the blocking parameter set. First, all the communication methods in Table 3 are executed and their time is measured. The fastest communication method is selected as the optimal method. Next, the blocking parameter set is selected by the selection method described in Sect. 3.3.

5.2 The Performance of Our Parallel SpMxV Library

To show the performance of our parallel SpMxV routine, the experimental results are shown in this section. The performance with dynamically selected parameter set is compared with the performance of fixed parameter sets that is generally used for all matrices and machines. The two fixed parameter sets used in this experiment are (No-blocking and isendrecv-M) and (2×2, 4 and isendrecv-M).

The computation time of parallel SpMxV is shown in Fig. 5.2. The time in the figure is an average time of 8 processors. There are 3 bars for each matrix. The left 2 bars are the time of fixed parameter sets. The right most bar is the time of our routine of which parameter set is dynamicaly selected. All bars are scaled so that the time of our routine becomes to 1.0. The light part of bar is the time

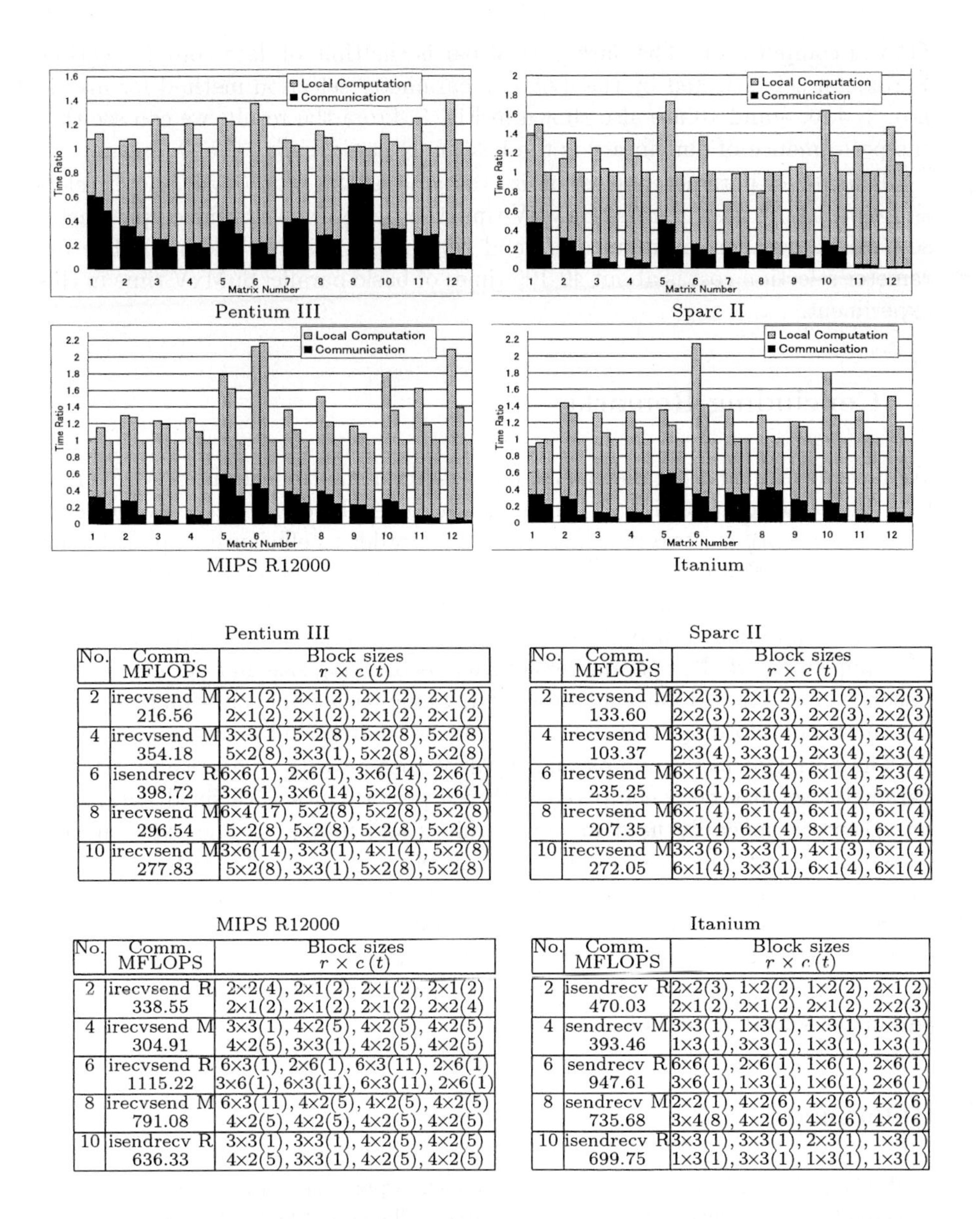

Pentium III

No.	Comm. MFLOPS	Block sizes $r \times c\,(t)$
2	irecvsend M 216.56	2×1(2), 2×1(2), 2×1(2), 2×1(2) 2×1(2), 2×1(2), 2×1(2), 2×1(2)
4	irecvsend M 354.18	3×3(1), 5×2(8), 5×2(8), 5×2(8) 5×2(8), 3×3(1), 5×2(8), 5×2(8)
6	isendrecv R 398.72	6×6(1), 2×6(1), 3×6(14), 2×6(1) 3×6(1), 3×6(14), 5×2(8), 2×6(1)
8	irecvsend M 296.54	6×4(17), 5×2(8), 5×2(8), 5×2(8) 5×2(8), 5×2(8), 5×2(8), 5×2(8)
10	irecvsend M 277.83	3×6(14), 3×3(1), 4×1(4), 5×2(8) 5×2(8), 3×3(1), 5×2(8), 5×2(8)

Sparc II

No.	Comm. MFLOPS	Block sizes $r \times c\,(t)$
2	irecvsend M 133.60	2×2(3), 2×1(2), 2×1(2), 2×2(3) 2×2(3), 2×2(3), 2×2(3), 2×2(3)
4	irecvsend M 103.37	3×3(1), 2×3(4), 2×3(4), 2×3(4) 2×3(4), 3×3(1), 2×3(4), 2×3(4)
6	irecvsend M 235.25	6×1(1), 2×3(4), 6×1(4), 2×3(4) 3×6(1), 6×1(4), 6×1(4), 5×2(6)
8	irecvsend M 207.35	6×1(4), 6×1(4), 6×1(4), 6×1(4) 8×1(4), 6×1(4), 8×1(4), 6×1(4)
10	irecvsend M 272.05	3×3(6), 3×3(1), 4×1(3), 6×1(4) 6×1(4), 3×3(1), 6×1(4), 6×1(4)

MIPS R12000

No.	Comm. MFLOPS	Block sizes $r \times c\,(t)$
2	irecvsend R 338.55	2×2(4), 2×1(2), 2×1(2), 2×1(2) 2×1(2), 2×1(2), 2×1(2), 2×2(4)
4	irecvsend M 304.91	3×3(1), 4×2(5), 4×2(5), 4×2(5) 4×2(5), 3×3(1), 4×2(5), 4×2(5)
6	irecvsend R 1115.22	6×3(1), 2×6(1), 6×3(11), 2×6(1) 3×6(1), 6×3(11), 6×3(11), 2×6(1)
8	irecvsend M 791.08	6×3(11), 4×2(5), 4×2(5), 4×2(5) 4×2(5), 4×2(5), 4×2(5), 4×2(5)
10	isendrecv R 636.33	3×3(1), 3×3(1), 4×2(5), 4×2(5) 4×2(5), 3×3(1), 4×2(5), 4×2(5)

Itanium

No.	Comm. MFLOPS	Block sizes $r \times c\,(t)$
2	isendrecv R 470.03	2×2(3), 1×2(2), 1×2(2), 2×1(2) 2×1(2), 2×1(2), 2×1(2), 2×2(3)
4	sendrecv M 393.46	3×3(1), 1×3(1), 1×3(1), 1×3(1) 1×3(1), 3×3(1), 1×3(1), 1×3(1)
6	sendrecv R 947.61	6×6(1), 2×6(1), 1×6(1), 2×6(1) 3×6(1), 1×3(1), 1×6(1), 2×6(1)
8	sendrecv M 735.68	2×2(1), 4×2(6), 4×2(6), 4×2(6) 3×4(8), 4×2(6), 4×2(6), 4×2(6)
10	isendrecv R 699.75	3×3(1), 3×3(1), 2×3(1), 1×3(1) 1×3(1), 3×3(1), 1×3(1), 1×3(1)

Fig. 5. The performance of parallel SpMxV and selected parameters. The graph shows time of parallel SpMxV with each parameter set. All bars are scaled so that the time with auto-tuned parameter set becomes to 1.0. There are 3 bars for each matrix. The left most bar shows time with parameter set (No-blocking and isendrecv-M). The middle bar shows time with parameter set (2×2, 4 and isendrecv-M). The right most bar shows time with auto-tuned parameter set. The table shows the selected parameter set which is selected by auto-tuning system

of local computation. The dark part of bar is the time of data communication. Parameter sets selected by the dynamic parameter selection method for matrix nos. 2, 4, 6, 8 and 10 are also shown in Fig. 5. From the result, we can see that the performance of our library is the best in 41 cases among the 48 cases.

The cost of parameter selection is evaluated as the ratio of parameter selection time to the basic parallel SpMxV time that uses a non-optimized parameter set (MPI_Allgather communication and Non-blocking SpMxV kernel). The parameter selection cost is about 40-100 times of basic parallel SpMxV time in this experiment.

6 Concluding Remarks

We have developed a parallel sparse matrix-vector multiplication (SpMxV) routine. The local computation on each node uses a new blocking method which generalizes conventional two blocking methods in order to efficiently exploit higher level memory. From the experimental result, it has been shown that our new blocking method is equal or better compared with the conventional two methods. On data communication part, six types of data communication methods have been implemented. The parameter set such as blocking size and data communication type is selected automatically by dynamic parameter selection method at the run time according to the characteristics of matrix and machine.

The performance of our parallel SpMxV routine with dynamically selected parameter set has been compared with the performance with fixed parameter sets that are generally used for all machines and matrices. The experimental result has shown that the performance of our routine is the fastest in most cases. Therefore, it can be said that the dynamic parameter selection method works well and our parallel SpMxV routine is high performance for various kinds of sparse matrices and on many types of computers.

As future work, it would be important to develop the faster and more accurate parameter selection method.

References

1. William, G., Lusk, E.: (User's guide for mpich, a portable implementation of mpi)
2. Davis, T.: University of florida sparse matrix collection (1997) NA Digest, vol. 97., Jun 1997. `http://www.cise.ufl.edu/~davis/sparse/`.
3. White, J., Sadayappan, P.: On improving the performance of parallel sparse matrix-vector multiplication. In: Proceedings of Supercomputing'97, San Jose, CA (1997)
4. Im, E.: Optimizing the Performance of Sparse Matrix - Vector Multiplication. PhD thesis, University of California at Berkeley (2000)
5. Barrett, R., Berry, M., Chan, T.F., Demmel, J., Donato, J., Dongarra, J., Eijkhout, V., Pozo, R., Romine, C., der Vorst, H.V.: Templates for the Solution of Linear Systems: Building Blocks for Iterative Methods. Society for Industrial and Applied Mathematics, Philadelphia, PA, USA (1994)

6. Geus, S.R.R.: Towards a fast parallel sparse matrix-vector multiplication. In: Parallel Computing: Fundamentals & Applications, Proceedings of the International Conference ParCo'99, 17–20 August 1999, Delft, The Netherlands. (2000) 308–315
7. Pinar, A., Heath, M.T.: Improving performance of sparse matrix-vector multiplication. In: Proceedings of Supercomputing'99. (1999)
8. Toledo, S.: Improving the memory-system performance of sparse-matrix vector multiplication. IBM Journal of Research and Development **41** (1997) 711–725
9. Vuduc, R., Demmel, J.W., Yelick, K.A., Kamil, S., Nishtala, R., Lee, B.: Performance optimizations and bounds for sparse matrix-vector multiply. In: Proceedings of Supercomputing 2002. (2002)

Parallelization of the Discrete Gradient Method of Non-smooth Optimization and Its Applications

G. Beliakov[1], J.E. Monsalve Tobon[2], and A.M. Bagirov[2]

[1] School of Information Technology, Deakin University, 221 Burwood Hwy. Burwood 3125, Australia, gleb@deakin.edu.au
[2] Centre for Informatics and Applied Optimization
School of Information Technology and Mathematical Sciences
The University of Ballarat, Victoria 3353, Australia
esteban@caribe.apana.org.au,a.bagirov@ballarat.edu.au

Abstract. We investigate parallelization and performance of the discrete gradient method of nonsmooth optimization. This derivative free method is shown to be an effective optimization tool, able to skip many shallow local minima of nonconvex nondifferentiable objective functions. Although this is a sequential iterative method, we were able to parallelize critical steps of the algorithm, and this lead to a significant improvement in performance on multiprocessor computer clusters. We applied this method to a difficult polyatomic clusters problem in computational chemistry, and found this method to outperform other algorithms.

1 Introduction

Numerical optimization is a generic task in the core of many models in science. One important example comes from theoretical chemistry, where a potential energy of a molecule as a function of individual atom positions, needs to be minimized to find the most stable conformation of this molecule. Instances of this problem appear in studies of molecular conformations, protein folding, polyatomic clusters, etc. [11,13,15]. One characteristic feature of many such models is very complicated objective function, which usually comes in the form of a black-box (i.e., a third party proprietary software), and very few assumptions can be made about it. Derivatives, even if they formally exist, are extremely difficult to express analytically, since most molecular descriptions involve some internal sets of coordinates, not readily expressed in terms of optimization variables [1, 7,11]. Moreover, not all models involve smooth functions. It is therefore quite important to have very robust, effective derivative free optimization algorithms, which do not rely on assumptions about certain properties of the objective function (such as differentiability), as these may turn out to be wrong. In this paper we study one such method, called Discrete Gradient (DG) [3,4].

The typical number of variables in molecular conformation problems ranges from several dozens to several hundreds, and the potential energy surface (PES)

P.M.A. Sloot et al. (Eds.): ICCS 2003, LNCS 2659, pp. 592–601, 2003.

is very rugged, involving multiple local minima and other stationary points. This makes optimization procedure computationally very expensive, which dictates the need for parallelization [9].

Multiprocessor clusters are a viable (and inexpensive) alternative to traditional supercomputers. The gain in performance is achieved by executing parts of the algorithm in parallel on different processors, and then merging the results. There are tools for automatic parallelization of serial program code, however given that optimization methods based on descent are serial in their very nature (i.e., iteratively moving from one point to a better one along the direction of descent), these tools are likely to fail in such cases. It is necessary to have a good understanding of the algorithm to identify those few opportunities that can help improve performance of serial algorithms.

After briefly describing the main ideas behind the Discrete Gradient method (for in-depth mathematical treatment refer to [3,4]), we discuss the ways in which parts of this method can be parallelized, and their limitations. We will then present results of some numerical experiments, which confirm the success of our strategies. In the last section we discuss application of DG to one benchmark problem in computational chemistry, that of polyatomic clusters. We will briefly introduce this application, show how it can be solved using DG, and compare our results to previous approaches.

2 Discrete Gradient Method

Non-smooth optimization is an important area of mathematical programming. As the name implies, no assumptions about differentiability of the objective function are made. This is frequently the case where the objective function f is piecewise continuously differentiable or given as a black-box, possibly as a solution of another complicated problem. We only assume Lipschitz continuity of f.

Many methods of smooth optimization, when all involved functions are assumed to be continuously differentiable or twice continuously differentiable, are based on gradient, Hessian or their various approximations. In nonsmooth optimization functions involved are no longer continuously differentiable. Nonsmooth analysis is a theoretical basis for nonsmooth optimization. Comprehensive description of nonsmooth analysis can be found, for example, in [5,6]. The notion of a subgradient is one of the key notions in nonsmooth analysis. The subgradient is a generalization of the notion of a gradient for Lipschitz continuous functions. The set of subgradients is called a subdifferential. Different generalizations of gradient were proposed and studied by many authors. Two of them: the Clarke subdifferential and Demyanov-Rubinov quasidifferential are widely used. In this paper we will consider a version of the discrete gradient method which based on the approximations to the quasidifferential. First we recall the definition of the quasidifferential.

Let f be a Lipschitz continuous function defined on an open set $X \subset \mathbb{R}^n$, where $\mathbb{R}^n$ is n-dimensional Euclidean space. This function is called directional differentiable at a point $x \in X$ if the limit

$$f'(x,g) = \lim_{\alpha \to +0} \alpha^{-1}[f(x+\alpha g) - f(x)]$$

exists for any $g \in \mathbb{R}^n$. The function f is called quasidifferentiable at a point $x \in X$ if it is directionally differentiable and there exist compact, convex sets $\underline{\partial} f(x)$ and $\overline{\partial} f(x)$ such that

$$f'(x,g) = \max_{v \in \underline{\partial} f(x)} \langle v, g \rangle + \min_{w \in \overline{\partial} f(x)} \langle w, g \rangle.$$

Here $\langle \cdot, \cdot \rangle$ stands for a scalar product in $\mathbb{R}^n$. The pair $Df(x) = [\underline{\partial} f(x), \overline{\partial} f(x)]$ is called a quasidifferential of the function f at a point x. The set $\underline{\partial} f(x)$ is said to be a subdifferential and the set $\overline{\partial} f(x)$ a superdifferential of the function f at a point x.

Let a function f be quasidifferentiable at $x \in \mathbb{R}^n$. For point x to be a minimum point of f on $\mathbb{R}^n$ it is necessary that

$$-\overline{\partial} f(x) \subset \underline{\partial} f(x). \tag{1}$$

A point $x \in \mathbb{R}^n$ satisfying (1) is called an inf-stationary point of the function f on $\mathbb{R}^n$. If $x \in \mathbb{R}^n$ is not an inf-stationary point then

$$-\overline{\partial} f(x) \not\subset \underline{\partial} f(x)$$

and

$$\max_{w \in \overline{\partial} f(x)} \min_{v \in \underline{\partial} f(x)} \|v + w\| > 0.$$

Then the direction $g^0 = -\|v^0 + w^0\|^{-1}(v^0 + w^0)$ where

$$\|v^0 + w^0\| = \max_{w \in \overline{\partial} f(x)} \min_{v \in \underline{\partial} f(x)} \|v + w\| > 0,$$

is a direction of steepest descent. Thus, we need to compute the entire quasidifferential of the function f to define the steepest descent direction. However, often it is impossible.

In [4] a method for minimizing quasidifferentiable functions based on the notion of a discrete gradient is developed. The discrete gradient is a finite difference estimate to a subgradient. Unlike many other finite difference estimates to a subgradient the discrete gradient is defined with respect to a given direction which allows one to get good approximation for the quasidifferential.

In [4] an algorithm for the calculation of the descent direction of a quasidifferentiable function by using discrete gradients is proposed. This is a terminating algorithm, i.e. it calculates discrete gradients step by step, and after a finite number of iterations either the descent direction is calculated, or it is found that the current point is an approximate stationary point. In the discrete gradient method Armijo's algorithm is used for a line search [2].

The discrete gradient method proceeds as follows. At a given approximation, it calculates the descent direction by calculating the discrete gradients step by step, and improving the approximation of the Demyanov-Rubinov quasidifferential. Once the descent direction is calculated, Armijo's algorithm is used for line search. The local minimum in this direction is chosen as the next approximation. Detailed treatment of this method is in [3,4].

A key step in the discrete gradient is the calculation of the descent direction. We calculate it using approximations to the subdifferential and superdifferential. This problem is reduced to a certain quadratic programming problem which is solved repeatedly at each iteration.

One of the characteristic features of DG method is its ability to "skip" many shallow local minima of the objective function, and converge if not to the global, to a deep local minimum. This feature is particularly important for difficult multiextrema optimization problems, such as those that arise in theoretical chemistry. The number of local minima in these problems can be of order 10^{20}, most of them shallow and not important from the chemical point of view. This feature has been predicted from theoretical point of view, but in this study we empirically demonstrated that this is actually the case. We compared DG method with another derivative free local method (which does converge to the nearest local minimum), and found that DG converges to a "better" solution (see Sect. 5).

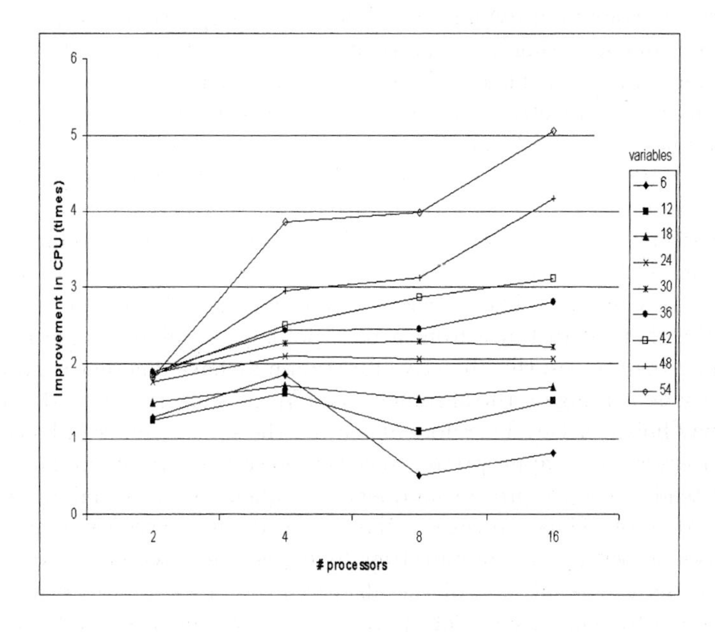

Fig. 1. Improvement in CPU time (with respect to run on a single processor) as a function of the number of processors and number of variables

3 Parallelization Strategies

The need for parallelization stems from the high computational cost of multivariate non-smooth optimization problems, and the need to solve these problems many times (e.g., as part of another algorithm). However, since the optimization strategies based on iterative descent are not parallel in their nature, there is only limited scope for parallelization. In the case of Discrete Gradient method, we identified two places, where the algorithm spends considerable time, that could be parallelized.

The way the algorithm computes the discrete gradient is by taking n values of the function (n is the number of variables), in the neighborhood of a current approximation, in the dynamically calculated directions. Then, two sets (approximations to the subdifferential and superdifferential) of points are built using these values. A subproblem of calculating the minimal distance from one set to the convex hull of the other is solved. The latter is a quadratic programming problem, and a terminating Wolfe's algorithm is used for this purpose. In degenerate cases, one of the sets may collapse into a point (the origin).

One obvious place to parallelize the algorithm is the Wolfe's algorithm [19], which solves a quadratic programming problem. A collection of such problems needs to be solved and this can be done in parallel. However, it is necessary to supply various processors with the coordinates of all the points from the two sets, and these are computed dynamically. The second opportunity to parallelize the algorithm is in computing $n+1$ values of the function in the neighborhood of the current approximation. This improvement would have a big impact on problems with complicated objective function, such as those coming from chemistry.

We implemented both parallelization strategies using the Message Passing Interface (MPI) standard [18]. Running DG algorithm on m processors, one of the processors takes the role of the main driver (master), and the rest take the role of workers (slaves). Master processor distributes the work to slaves and then merges the results. Since usually $m < n$, slaves will usually have more than one basic task to complete. These tasks (computing the value of the function and finding the shortest distance from a base point to a polytope) are not equivalent among themselves: the objective function may have different computational complexity depending on the argument, and Wolfe's algorithm for different base points may finish earlier. Therefore allocating the same number of basic tasks to slave processors is not appropriate, since the slowest processor will be the bottleneck. Instead, the slave processors query to a queue of tasks maintained by the master processor. Thus, processors that finished their tasks early are not idle, and processors with more computationally expensive tasks have fewer tasks.

Of course, the serial part of the algorithm becomes its bottleneck, and we did not hope to achieve complete parallelization of this essentially serial algorithm. However, parallelization of expensive parts of the serial algorithm yielded substantial improvements. The gain in computing time is illustrated in Fig. 1.

Of course, there is no much point in having more processors than the number of variables n, and for $m = n$, the slowest processor will dominate the speed of the algorithm. There is an optimal value for the ratio m/n, but it varies much with the objective function. This optimal ratio can be found empirically for a class of objective functions, and subsequently used in cases where multiple optimization tasks are performed (i.e., optimizations from various initial approximations).

4 Application: Polyatomic Clusters Problem

In this section we discuss one practical application of DG, which came from theoretical chemistry. This problem is well studied, it is easy to formulate (but extremely difficult to solve), and it serves as a benchmark for optimization algorithms. Its variations are also important on their own (e.g., in studies of noble gases, interstellar dust), and as subproblems in other models, such as protein folding problem [8,10,12,16,17].

The simplest instance of this problem is the Lennard-Jones cluster problem. It consists in finding the geometry of a cluster of N identical particles loosely bound by interatomic forces. The optimal geometry minimizes the potential energy of the cluster, expressed as a function of Cartesian coordinates

$$E(x,y,z) = \sum_{i=1}^{N} \sum_{j=i+1}^{N} v(r_{ij}) \tag{2}$$

The pairwise potential (the Lennard-Jones potential) is given by

$$v(r_{ij}) = \left(\frac{1}{r_{ij}^{12}} - \frac{1}{r_{ij}^{6}} \right) \tag{3}$$

where

$$r_{ij} = (x_i - x_j)^2 + (y_i - y_j)^2 + (z_i - z_j)^2.$$

Variations of this problem include Carbon C60 clusters, water molecule clusters, An -clusters, etc., these are presented in [17].

The objective function of the Lennard-Jones problem (2) is very simple to compute, but it has extremely complicated landscape, with huge number of local minima (for the 147 atoms problem). The number of variables for an N-atom problem is $3N-6$ (accounting for 3 translations and 3 rotations). There are at least $N!$ global minima (due to permutations of identical atoms).

Even though the objective function is smooth, the ability of DG to skip many local minima will be of great value (it is indeed the case as shown in Sect. 5). Traditional approach to this problem is to build clusters incrementally, by using the solution for $N-1$ atom problem as the initial approximation to the N-atom problem [10,16]. Most of the optimal geometries follow this rule, however in special cases (such as $N = 38, 75, 76, 77, \ldots$) this rule is broken, and solution, say at $N = 38$ is completely different from that of $N = 37$. Hence these special "difficult" clusters essentially require random starting points and serve as benchmarks for many optimization algorithms [10].

One approach taken in [10] is to modify the Lennard-Jones potential with a penalty, favoring compact clusters

$$\bar{v}(r) = \left(\frac{1}{r^{12}} - \frac{1}{r^6} + \mu r + \beta \max\{0, r^2 - D^2\}\right) \quad (4)$$

μ, β and D are penalty parameters. The authors of [10] reported that first minimizing the modified objective function, and then using its minimum as the initial approximation to the original problem yielded convergence to the global minimum with much greater success than starting local optimization with random points (Table 1). Thus, the expected number of runs for locating the global minimum is substantially smaller. This can be projected to other instances of polyatomic clusters problem [8,10,17].

In our study we followed the penalty approach from [10], but used DG rather than other local optimization technique. The ability of DG to skip local minima paid off, and the global minima were located with higher frequency than that of [10].

5 Results

In this section we present three types of results of numerical experiments with the Discrete Gradient method. The first result relates to parallelization of the algorithm. Figure 1 presents improvements in CPU time taken by our implementation of the DG method running on the specified number of processors. As the objective function, the Lennard-Jones clusters problem (2) was taken for the specified number of variables. Ideally the improvement should be linear in the number of processors, but interprocessor communications and serial parts of the algorithm reduce it. This value can become smaller than 1 (loss in efficiency). It is clear from the plot that running DG in parallel on a number of processors is warranted only for a relatively high number of variables.

The second result illustrates the ability of DG to skip shallow local minima. We used the Lennard-Jones clusters problem with $N = 10$ (24 variables) as the objective function. The true global minimum is at $f = -28.42$. We started DG and another local derivative free method UOBYQA by M.J.D. Powell [14] from the same randomly chosen initial points, and compared the local minima both methods converged to. Figure 2 presents the plot of values at local minima of one method against the other. Points on the diagonal correspond to both methods converging to the same local minimum, whereas points above the diagonal mean the DG has found a better minimum. Figure 3 shows the frequency of locating a minimum in a given interval starting from a random point. It is clear from both plots that DG systematically converges to substantially better local minima (thus skipping the shallow minima other local methods are trapped in).

Finally, we used the modification of the objective function (3) to improve the frequency of locating the global minimum, as proposed in [10]. We ran DG from a number of random starting points and calculated the relative frequency of locating the global minimum. Table 1 compares the success rate of DG with

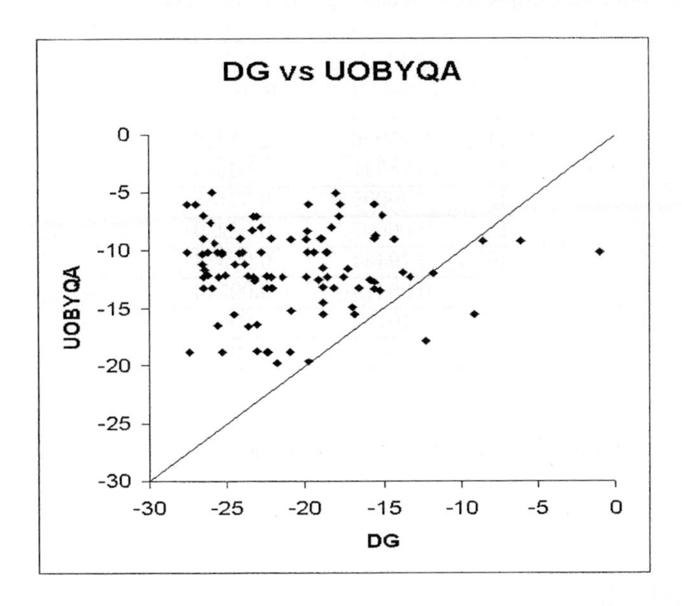

Fig. 2. Solution found by the Discrete Gradient vs. that of M.J.D. Powell's method

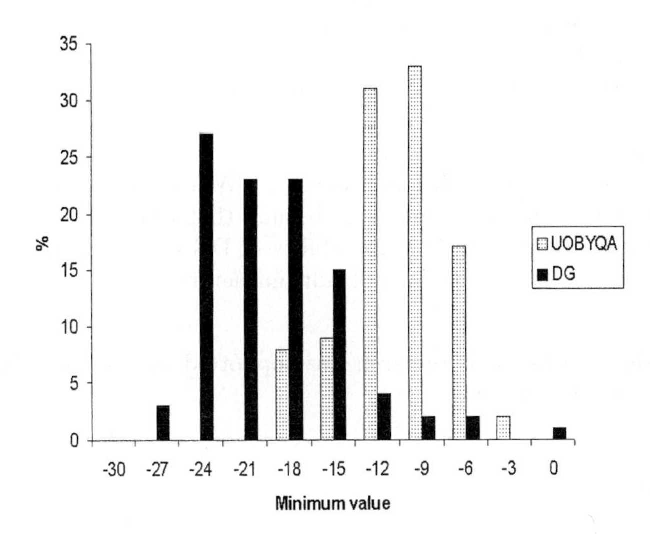

Fig. 3. Frequency of locating a local minimum from a randomly chosen initial point

Table 1. The frequency of locating the global minimum from a random initial point by DG and standard Conjugate Gradient method from [10]

N	DG Method Success Rate	CG Success Rate
11	0.25926	0.01488
12	0.78947	0.02254
13	0.20000	0.02185
15	0.06667	0.02230
16	0.19444	0.01338
17	0.05419	0.00540
23	0.10170	0.00428
38	0.41667	0.00887

that of the standard Conjugate Gradient method from [10]. Again, clearly DG is superior, since it is consistently more reliable in locating the global minimum. It is most noticeable for $N = 38$ "difficult" cluster.

6 Conclusion

Non-smooth optimization has many important applications in scientific computing. Computational chemistry provides a range of such applications related to molecular structure prediction. The objective function in such problems can have very complicated structure, or even given as a black-box, and no assumptions about differentiability can be made. The Discrete Gradient method is based on generalizations of the notion of gradient equipped with a full-scale calculus.

We have shown that parts of the DG method can be parallelized for its use on distributed memory computer clusters. Such parallelization only makes sense for a relatively large number of variables. This is the kind of problems that frequently come from computational chemistry. We examined one such problem, the Lennard-Jones clusters problem, and found that DG has an advantage over other optimization methods. It is the ability of DG to skip many shallow local minima that makes it suitable for difficult multiextrema problems.

Acknowledgements. This research was supported by the Victorian Partnership for Advanced Computing.

References

[1] M.P. Allen and D.J. Tildesley, Computer Simulations of Liquids, Oxford Science Publications, Oxford, 1990.

[2] L. Armijo, Minimization of functions having continuous partial derivatives, Pacific J. of Mathematics, 16 (1966), pp. 1–13.

[3] A. Bagirov, Derivative-free methods for unconstrained nonsmooth optimization and its numerical analysis, Journal Investigacao Operacional, 19 (1999), pp. 75–93.
[4] A. Bagirov, A method for minimization of quasidifferentiable functions, Optimization Methods and Software, 17 (2002), pp. 31–60.
[5] F.H. Clarke, Optimization and Nonsmooth Analysis, John Wiley, New York, 1983.
[6] V.F. Demyanov and A.M. Rubinov, Constructive Nonsmooth Analysis, Peter Lang, Frankfurt am Main, 1995.
[7] C.A. Floudas, Deterministic global optimization: Theory, methods, and applications, Kluwer Academic Publishers, Dordrecht, London, 2000.
[8] B. Hartke, Global geometry optimization of molecular clusters: TIP4P water, Zeitschrift fur Physikalische Chemie, 214 (2000), pp. 1251–1264.
[9] J.Y. Lee, J. Pillardy, C. Czaplewski, Y. Arnautova, D.R. Ripoll, A. Liwo, K.D. Gibson, R.J. Wawak and H.A. Scheraga, Efficient parallel algorithms in global optimization of potential energy functions for peptides, proteins, and crystals, Computer Physics Communications, 128 (2000), pp. 399–411.
[10] M. Locatelli and F. Schoen, Fast global optimization of difficult Lennard-Jones clusters, Computational Optimization and Applications, 21 (2002), pp. 55–70.
[11] A. Neumaier, Molecular modeling of proteins and mathematical prediction of protein structure, SIAM Review, 39 (1997), pp. 407–460.
[12] R.V. Pappu, R.K. Hart and J.W. Ponder, Analysis and application of potential energy smoothing and search methods for global optimization, Journal of Physical Chemistry B, 102 (1998), pp. 9725–9742.
[13] P.M. Pardalos and C.A. Floudas, Optimization in computational chemistry and molecular biology : local and global approaches, Kluwer Academic Publishers, Boston, 2000.
[14] M.J.D. Powell, UOBYQA: unconstrained optimization by quadratic approximation, Mathematical Programming, 92 (2002), pp. 555–582.
[15] H.A. Scheraga, The Multiple Minima Problem in Protein Folding, Polish Journal of Chemistry, 68 (1994), pp. 889–891.
[16] D.J. Wales and D.J., Global optimization by basin-hopping at the lowest energy structures of Lennard-Jones clusters containing up to 110 atoms, J. Phys. Chem., 101 (1997), pp. 5111–5116.
[17] D.J. Wales and H.A. Scheraga, Review: Chemistry – Global optimization of clusters, crystals, and biomolecules, Science, 285 (1999), pp. 1368–1372.
[18] B. Wilkinson and C.M. Allen, Parallel programming: techniques and applications using networked workstations and parallel computers, Prentice Hall, Upper Saddle River, N.J., 1999.
[19] P.H. Wolfe, Finding the nearest point in a polytope, Math. Progr., 11 (1976), pp. 128–149.

A Parallel Prime Edge-Length Crystallographic FFT

Jaime Seguel and Daniel Burbano

University of Puerto Rico at Mayaguez, Mayaguez PR 00680, USA
(Jaime.Seguel)(dburbano)@ece.uprm.edu

Abstract. Although other methods are available, computational X-ray crystallography is still the most accurate way of determining the atomic structure of crystals. For large scale problems such as protein or virus structure determination, a huge amount of three-dimensional discrete Fourier transforms (DFT) conform the core computation in these methods. Despite the fact that highly efficient fast Fourier transform (FFT) implementations are available, significant improvements can be obtained by using FFT variants tailored to crystal structure calculations. These variants, or crystallographic FFTs, use a-priori knowledge of the specimen's crystal symmetries to lower the operation count and storage requirement of a usual, asymmetric FFT. The design and implementation of crystallographic FFTs brings about several problems of its own. And, as is usually the case with prime length FFTs, prime edge-length crystallographic FFTs pose the hardest challenges among them. This paper develops and tests a parallel multidimensional crystallographic FFT of prime edge-length, whose performance is significantly better than that of the usual FFT.

1 Introduction

Let $Z^d/N = Z/N \times \cdots \times Z/N$, be the Cartesian product of d copies of Z/N, the set of integers modulo N. The *d-dimensional discrete Fourier transform* (DFT) of *edge length* N of a real- or complex-valued mapping f defined in Z^d/N is the linear operator

$$\hat{f}(\boldsymbol{k}) = \frac{1}{\sqrt{N}} \sum_{\boldsymbol{l} \in Z^d/N} f(\boldsymbol{l}) w_N^{\boldsymbol{k}\cdot\boldsymbol{l}}, \quad \boldsymbol{k} \in Z^d/N; \tag{1}$$

where $w_N = \exp(-2\pi i/N)$, $\cdot$ denotes the dot product, and $i = \sqrt{-1}$. By endowing Z^d/N with an order (for example, the *lexicographic order*), (1) can be thought as the product of a $N^d \times N^d$ complex matrix by a vector. If $d = 1$, we denote this matrix F_N. Thus, (1) is computed in $O(N^{2d})$ time. For $d = 1$, a *fast Fourier transform* [3] (FFT) computes (1) in $O(N \log N)$ operations. For $d \geq 2$, the usual FFT method consists of applying N^{d-1} one-dimensional FFTs along each of the d dimensions. Thus, (1) is computed in $O(N^d \log N)$ time. This complexity bound cannot be improved for general DFT computations. For

P.M.A. Sloot et al. (Eds.): ICCS 2003, LNCS 2659, pp. 602–611, 2003.

problems whose data is endowed with redundancies, such as x-ray crystal diffraction intensity data, attempts have been made to reduce the FFT run time. A starting point in these attempts is to eliminate all redundancies induced by the presence of a crystallographic symmetry both, from the input and output data sets. The resulting input and output data sets are usually called *fundamental data sets.* This data reduction is indeed a lossless compression, since each data set can be reconstructed from the corresponding fundamental data set by applying symmetry relations. In order to compute (1) within fundamental data sets, the DFT equations have to be modified. This modification, which consists mainly in factoring repeated input in the DFT sum (1) and eliminating redundant outputs, reduces the operation count of the DFT. However, the complexity order of computing with the reduced system of DFT equations remains within $O(N^{2d})$, and therefore, DFT reductions do not produce a fast Fourier transform method. The design of fast algorithms for computing the reduced DFT equations, collectively known as *crystallographic FFTs,* poses several mathematical and algorithmic challenges, which are summarized is Sect. 2. These challenges, most of them still open problems, are most probably the main reason behind the rather slow emergence of crystallographic FFT methods in the scientific computing community. A crystallographic FFT for composite edge-lengths was first introduced by Ten Eyck in [7]. This method is applicable to a few crystal symmetries. A *universal crystallographic FFT* of composite edge-length, this is a crystallographic FFT for all crystal symmetries, is proposed in [6]. In [6] it is also shown that composite edge-length crystallographic FFTs are computed in $O(M \log M)$ time, where M is the size of the subset of non-redundant input data. A prime edge-length crystallographic FFT method was first introduced by Auslander *et. al.* in [1]. Although this method is universal, its time complexity is on average, higher than that of the usual FFT [5]. A method whose time complexity varies with the symmetry from an optimal $O(M \log M)$ to a worst case bound smaller but close to Auslander's bound, is proposed in [5]. In the same article, parallel computing is suggested as a way of speeding up the execution time of problems that yield worst case complexity bounds. In this paper we explore such alternative by designing and testing a parallel version of the prime edge-length crystallographic FFT introduced in [5]. Our preliminary computer results show the validity of this approach.

This article is organized as follows: in Sect. 2 we review the main concepts and issues related to crystallographic FFT design. Section 3 provides a detailed description of the parallel crystallographic algorithm, and Sect. 4 summarizes the results of our computer experiments.

2 Background

For the rest of this paper, we assume that N is prime. A matrix or vector is said to be *over* Z/N if all its entries are in Z/N. All operations with matrices and vectors over Z/N are performed in modulo N. A square matrix over Z/N is nonsingular, if its determinant is different from zero in modulo N.

The atomic structure of a crystal can be determined from X-ray diffraction intensity data through a intensive computer procedure. Measured intensities provide only the magnitudes $|E|$ of the *normalized structure factors* $E = |E|\exp(i\phi)$ of the crystal specimen. The elucidation of the crystal structure requires a knowledge of the *phase* ϕ, which is unobtainable from any physical experiment. The problem of reconstructing phases from magnitudes, or *phase problem*, is solved by evaluating numerous trial structures. Each trial structure is refined through a sequence of steps which involve three-dimensional discrete Fourier transform (DFT) calculations. For instance, one of the most successful methods, the *shake-and-bake* method [8], [9], alternates optimization in Fourier space with filtering in real space, performing thus, a forward and a backward three-dimensional DFT at each refinement step. Since crystal structures consist of repeating symmetric unit cells, their intensity data is highly redundant. In mathematical terms, the observed intensities are represented by a real mapping f mapping defined on Z^d/N, and redundancies described by the action of S, a $d \times d$ nonsingular matrix over Z/N satisfying

$$f(\boldsymbol{l}) = f(S\boldsymbol{l}) \quad \text{for all } \boldsymbol{l} \in Z^d/N. \tag{2}$$

Matrix S is called a *symmetry* and f is said to be S-symmetric. For example, the mapping f defined in $Z^2/5$ by the two-dimensional data array $[f(k,l)]_{(k,l)}$

$$f(Z^2/5) = \begin{bmatrix} 8 & 2.2 & 5.9 & 5.9 & 2.2 \\ 2.2 & 4 & 1.2 & 6 & 4 \\ 5.9 & 6 & 7.7 & 7.7 & 1.2 \\ 5.9 & 1.2 & 7.7 & 7.7 & 6 \\ 2.2 & 4 & 6 & 1.2 & 4 \end{bmatrix} \tag{3}$$

is S-symmetric with

$$S = \begin{bmatrix} 0 & 1 \\ -1 & 0 \end{bmatrix} = \begin{bmatrix} 0 & 1 \\ 4 & 0 \end{bmatrix} \quad \text{modulo 5.} \tag{4}$$

Clearly, an S-symmetric mapping f satisfies $f(S^j\boldsymbol{a}) = f(\boldsymbol{a})$ for all j. Thus, the restriction of f to the subset of Z^d/N

$$O_S(\boldsymbol{a}) = \{S^j\boldsymbol{a} : \ j \text{ integer } \} \tag{5}$$

yields a constant function. The subset $O_S(\boldsymbol{a})$ is called S-*orbit* of $\boldsymbol{a}$. It follows that an S-symmetric function is completely determined by its values on a *fundamental set* $\mathcal{F}_S$ formed by one and only one element from each S-orbit. For example, the S-orbits and their images under f for symmetry (3) are

$$O_S(0,0) = \{(0,0)\}\ ,\ f(O_S(0,0)) = \{8\} \tag{6}$$

$$O_S(0,1) = \{(0,1),\ (1,0),\ (0,4),\ (4,0)\}\ ,\ f(O_S(0,1)) = \{2.2\} \tag{7}$$

$$O_S(0,2) = \{(0,2),\ (2,0),\ (0,3),\ (3,0)\}\ ,\ f(O_S(0,2)) = \{5.9\} \tag{8}$$

$$O_S(1,1) = \{(1,1),\ (1,4),\ (4,4),\ (4,1)\}\ ,\ f(O_S(1,1)) = \{4\} \tag{9}$$

$$O_S(1,2) = \{(1,2),\ (2,4),\ (4,3),\ (3,1)\}\ ,\ f(O_S(1,2)) = \{1.2\} \tag{10}$$
$$O_S(1,3) = \{(1,3),\ (3,4),\ (4,2),\ (2,1)\}\ ,\ f(O_S(1,3)) = \{6\} \tag{11}$$
$$O_S(2,2) = \{(2,2),\ (2,3),\ (3,3),\ (3,2)\}\ ,\ f(O_S(2,2)) = \{7.7\}. \tag{12}$$

A fundamental indexing set for this symmetry is then,

$$\mathcal{F}_S = \{(0,0),(0,1),(0,2),(1,1),(1,2),(1,3),(2,2)\}. \tag{13}$$

This choice of fundamental indexing set is not unique. It is easy to show that if f is S-symmetric, then $\hat{f}$ is S_*-symmetric, where S_* is the transpose of the inverse of S. Therefore, $\hat{f}$ is also determined by its values in a fundamental set $\mathcal{F}_{S_*}$. We call $f(\mathcal{F}_S)$ and $\hat{f}(\mathcal{F}_{S_*})$ *fundamental input data* and *fundamental output data*, respectively. In example (3) a fundamental input data set is

$$f(\mathcal{F}_S) = \{8,\ 2.2,\ 5.9,\ 4,\ 1.2,\ 6,\ 7.7\}; \tag{14}$$

while, as a straight computation shows,

$$\hat{f}(\mathcal{F}_{S_*}) = \{116,\ -11.1602,\ 13.6602,\ 6.1133,\ -13,\ 11,\ 14.3867\} \tag{15}$$

is a fundamental output data set. The reduction of the usual N^d-point input data set to a subset of fundamental input data induces a modification on Eq. (1). In fact, since $f(S^j\boldsymbol{a}) = f(\boldsymbol{a})$ for all j, the input datum $f(\boldsymbol{a})$ can be factored out of the terms of $\hat{f}(\boldsymbol{k})$ that are indexed in $O_S(\boldsymbol{a})$ in (1). This is,

$$\sum_{\boldsymbol{l}\in O_S(\boldsymbol{a})} f(\boldsymbol{l})w_N^{\boldsymbol{k}\cdot\boldsymbol{l}} = f(\boldsymbol{a})\left(\sum_{\boldsymbol{l}\in O_S(\boldsymbol{a})} w_N^{\boldsymbol{k}\cdot\boldsymbol{l}}\right) = f(\boldsymbol{a})K_N(\boldsymbol{k},\boldsymbol{a}). \tag{16}$$

The expression $K_N(\boldsymbol{k},\boldsymbol{a}) = \sum_{\boldsymbol{l}\in O_S(\boldsymbol{a})} w_N^{\boldsymbol{k}\cdot\boldsymbol{l}}$ is the so-called *symmetrized DFT kernel.* The linear transformation

$$\hat{f}(\boldsymbol{k}) = \sum_{\boldsymbol{a}\in\mathcal{F}_S} K_N(\boldsymbol{k},\boldsymbol{a})f(\boldsymbol{a}), \quad \boldsymbol{k}\in\mathcal{F}_{S_*}, \tag{17}$$

is called *symmetrized DFT.* It can be shown that of $\mathcal{F}_S$ and $\mathcal{F}_{S_*}$ have the same size M. Thus, Eq. (17) computes the symmetric DFT in $O(M^2)$ time. For instance, in example (3), the symmetrized DFT involves approximately $7^2 = 49$ complex multiplications, and $7\cdot 6 = 42$ complex additions. This is certainly less than the approximately $25^2 = 625$ complex multiplications and $25\cdot 24 = 600$ complex additions required by the usual DFT. However, since the size of the fundamental set is a constant fraction of N^d, in terms of asymptotic growth, the mere reduction of redundant data does not yield a superior FFT method. Auslander *et. al.* [1] propose an algorithm for computing the core of (17) with *fast cyclic convolutions.* Fast cyclic convolutions are $O(Q\log Q)$ methods for computing a matrix by vector product $Y = HX$, where H is a $Q\times Q$ *circulant.* The use of fast cyclic convolutions is expected to lower the method's asymptotic complexity growth. However, as pointed out in [5], even for some very common crystallographic symmetries Auslander's method involves such a large number of cyclic convolutions that its complexity is often higher than the FFT complexity boundary. In what follows, we examine basic ideas underlying the use of fast cyclic convolutions in crystallographic FFT computations. As in [1] and [5], we declare a $Q \times Q$ matrix H to be circulant if it satisfies $H(k,l) = H(k',l')$ whenever $k + l = k' + l'$ modulo Q. It is easy to show that if H is circulant,

$$\Delta(H) = F_Q H F_Q \tag{18}$$

is a $Q \times Q$ diagonal matrix. Thus, $Y = HX$ is computed in $O(Q \log Q)$ time through

$$Y = F_Q^{-1} \Delta(H) F_Q^{-1} X. \tag{19}$$

Equation (19) is often referred as *fast cyclic convolution* algorithm. Fast cyclic convolutions are also used for computing $Y = TX$. where T is a $P \times R$ matrix that satisfies $T(k,l) = T(k',l')$ whenever $k + l = k' + l'$. We call such matrices, *pre-Hankel*. A pre-Hankel can always be *embedded* in a circulant H. This means that T occupies the upper leftmost corner of H. In such a case, H is a $Q \times Q$ with $Q \geq P + R - 1$ [5]. Auslander *et. al.* turn the core computations in (17) into pre-Hankel blocks. Central to this aim is the fact that for each $n \neq 0$ in Z/N, there is an integer $0 \leq j \leq N - 2$, such that $n = g^j$. Auslander *et. al.* [1] use this property for partitioning both $\mathcal{F}_S$ and $\mathcal{F}_{S_*}$ into subsets of the form

$$O_g(\boldsymbol{a}) = \{g^j \boldsymbol{a} : j = 0, ..., N-2\}. \tag{20}$$

Each pair formed by one such subset in $\mathcal{F}_{S_*}$ and one in $\mathcal{F}_S$ corresponds to a pre-Hankel block in (17). Since $O_g(\boldsymbol{a})$ contains at most $N - 1$ elements, Auslander's crystallographic FFT involves $\Lambda = O\left((M/(N-1))^2\right)$ cyclic convolutions. But since M is $O(N^3)$, $\Lambda = O(N^4)$, which is higher than the usual FFT. A way around this problem is suggested in [5]. The main idea in [5] is using orbit segments generated by a $d \times d$ matrix C over Z/N, instead of (20). In [5], a significant reduction in the number of cyclic convolutions is shown, including reductions from $\Lambda = O(N^4)$ in Auslander's method to $\Lambda = 1$. In these cases, the proposed crystallographic FFT achieves an optimal lower time complexity of $O(M \log M)$. Unfortunately, such optimal reduction is not always achievable, since, Λ is bounded below by a constant that depends on the crystal symmetry. In [5], a method that parallelizes most of the cyclic convolution computations, is outlined. In the next section, we review the main ideas behind this method and provide a detailed description of its parallel version, which we call *prime crystallographic FFT*, (PCFFT). Our experiments confirm the potential of the PCFFT method for overcoming the execution time limit imposed by the lower bound of Λ.

3 Parallel Prime Edge-Length Crystallographic FFT (PCFFT)

The crystallographic FFT proposed in [5] is derived with the help of a $d \times d$ matrix over Z/N, which commutes with the symmetry S. This matrix is precomputed through an exhaustive, direct search, which is out of the scope of this article. We recall, however, that as shown in [5], the action of C partitions $\mathcal{F}_S$ into subsets of the form

$$\{C^j \boldsymbol{b} : \ 0 \leq j \leq L_{\boldsymbol{b}} - 1\}. \tag{21}$$

Here, $\boldsymbol{b}$ ranges over some $I_S \subset \mathcal{F}_S$, and $L_{\boldsymbol{b}}$ is the number of elements in (21). Similarly, $\mathcal{F}_{S_*}$ is partitioned into subsets generated by the action of C^t, the transpose of C, on a subset $I_{S_*} \subset \mathcal{F}_{S_*}$. For each pair $(\boldsymbol{a}, \boldsymbol{b})$ in $I_{S_*} \times I_S$, and k, l, k', and l' nonnegative integers such that $k + l = k' + l'$,

$$\begin{aligned} K_N({C^t}^k \boldsymbol{a}, (C)^l \boldsymbol{b}) &= K_N({C^t}^{k+l} \boldsymbol{a}, \boldsymbol{b}) \\ &= K_P({C^t}^{k'+l'} \boldsymbol{a}, \boldsymbol{b}) \\ &= K_N({C^t}^{k'} \boldsymbol{a}, C^{l'} \boldsymbol{b}); \end{aligned}$$

showing that the block $W_{(\boldsymbol{a},\boldsymbol{b})}$ whose entries are $W_{(\boldsymbol{a},\boldsymbol{b})}(k,l) = K_N(C^k \boldsymbol{a}, {C^t}^l \boldsymbol{b})$, $0 \leq k < L_{\boldsymbol{a}}$, and $0 \leq j < L_{\boldsymbol{b}}$; is pre-Hankel. The core computations in (17) are then expressed as

$$Y = \left[W_{(\boldsymbol{a},\boldsymbol{b})}\right]_{(\boldsymbol{a},\boldsymbol{b})} \left[f\left(\mathcal{F}_S\right)\right]. \tag{22}$$

These computations include neither the input datum $f(\mathbf{0})$ nor the output datum $\hat{f}(\mathbf{0})$. Computations with these data conform the so-called *border computations*. In order to turn core computations into cyclic convolutions, each $W_{(\boldsymbol{a},\boldsymbol{b})}$ is embedded into a circulant. Following [5], we propose a common size $Q \times Q$ for these circulant. Thus. $Q \geq 2L - 1$, where

$$L = \max\{L_{\boldsymbol{a}} : \ \boldsymbol{a} \in I_{S_*}\} = \max\{L_{\boldsymbol{b}} : \ \boldsymbol{b} \in I_S\}. \tag{23}$$

Since the choice of Q is also guided by the efficiency of the Q-point FFT, it is reasonable to assume that $2L - 1 \leq Q \leq 2^{\lceil \log_2(2L-1) \rceil}$, where $\lceil x \rceil$ is the smallest integer that is greater than or equal to x. Let $H_{(\boldsymbol{a},\boldsymbol{b})}$ be the $Q \times Q$ circulant in which $W_{(\boldsymbol{a},\boldsymbol{b})}$ is embedded. Then, the core computations are represented as $\boldsymbol{U} = H\boldsymbol{V}$ where

$$H = \left[H_{(\boldsymbol{a},\boldsymbol{b})}\right]_{(\boldsymbol{a},\boldsymbol{b})} \tag{24}$$

is a block matrix. The input vector $\boldsymbol{V}$ is composed of Q-point vector segments $\boldsymbol{V}_{\boldsymbol{b}}$, each consisting of the $L_{\boldsymbol{b}}$ values of $f(C^j \boldsymbol{b})$, $0 \leq j < L_{\boldsymbol{b}}$; followed by $Q - L_{\boldsymbol{b}}$ zeros. Vector $\boldsymbol{U}$, in turn, is composed by Q-point segments denoted $\boldsymbol{U}_{\boldsymbol{a}}$, $\boldsymbol{a} \in I_{S_*}$, that result from

$$\boldsymbol{U}_{\boldsymbol{a}} = \sum_{\boldsymbol{b} \in I_{S_*}} H_{(\boldsymbol{a},\boldsymbol{b})} \boldsymbol{V}_{\boldsymbol{b}} = \sum_{\boldsymbol{b} \in I_{S_*}} F_Q^{-1} \Delta\left(H_{(\boldsymbol{a},\boldsymbol{b})}\right) F_Q^{-1} \boldsymbol{V}_{\boldsymbol{b}} \tag{25}$$

$$= F_Q^{-1}\left(\sum_{b\in I_{S_*}} \Delta\left(H_{(a,b)}\right) F_Q^{-1} V_b\right). \tag{26}$$

Parallelism is readily seen in (26). We assume that P processors are available. For simplicity, P is assumed to be a power of two that divides Λ. We denote the ratio by $\rho = \Lambda/P$. We decompose V in P groups of size ρQ and denote the set of indices of the i-th group by P_i. Thus, processor i receives the collection of input vector segments V_b, with $b \in P_i$. A precomputed array of size ΛQ containing the diagonal entries of $\Delta(H_{(a,b)})$, is stored in processor i, as well. This array is denoted D_b. For a Q-point vector Y, we define $D_b \odot Y$ as the Hadamard product of the vector formed by concatenating Λ copies of Y and D_b. Using these conventions, we translate formula (26) into the next parallel crystallographic FFT (PCFFT) method. The first steps are:

Step 1. For each $i = 0$ to $P-1$, processor i computes $Y_a = F_Q^{-1} V_a$, for each $a \in P_i$.
Step 2. For each $i = 0$ to $P-1$, processor i computes $Z_a = D_a \odot Y_a$, for each $a \in P_i$.
Step 3. For each $i = 0$ to $P-1$, processor i computes $X_i = \sum_{a\in P_i} Z_a$.

Steps 1 to 3 require no interprocessor communication. However, in order to complete the sum in (26), all ΛQ-point data vectors X_i, $o \le i \le P-1$; must be added together. Interprocessor communications and vector additions overlap while computing this sum. Our *communication topology* is a hypercube. At each step, a different set consisting of all the *parallel edges* of the hypercube is selected. At each step, each pair of processors linked by by a hypercube edge, exchange data. There will be $\log P$ sets of parallel edges. Thus, for each $0 \le j \le \log P - 1$, we denote by E_j one such set of parallel hypercube edges.

Step 4. For each $j = 0, ..., \log P - 1$, for each pair $(i, l) \in E_j$, processor i sends the upper segment of X_i to processor l, and processor l sends the lower segment of X_l, to processor i. Then, processor i adds the lower segment of X_l to the lower segment of X_i, and stores the result a new, half size vector X_i; while processor l adds the upper segment of X_i with the upper segment of X_l, and stores the result in a new, half size vector X_l.

At the end of step 4, processor i holds in X_i, a ρQ-point segment of the sum in (26). The core computation is completed by performing
Step 5. For each $i = 0$ to $P-1$, processor i computes $U_a = F_Q^{-1} X_a^{(i)}$, for each $a \in P_i$.

The addition of datum $f(\mathbf{0})$, which is part of the border computation, is performed as follows

Step 6. For each $i = 0$ to $P-1$, processor i computes $\hat{f}(O_M^{l_a}(a) = \frac{1}{\sqrt{N}}\left[f(\mathbf{0})\right]_{l_a} + U_a^*$, for each $a \in P_i$. Here U_a^* is the column vector formed by the L_a first entries of the vector U_a computed in step 5.

Finally, the computation of $\hat{f}(\mathbf{0})$ is not worth parallelizing. We simply perform

Step 7. $\hat{f}(\mathbf{0}) = \frac{1}{\sqrt{N}} \sum_{\mathbf{b} \in I_{S_*}} f(\mathbf{b})|O_S(\mathbf{b})|$. Here $|O_S(\mathbf{b})|$ is the size of $O_S(\mathbf{b})$.

The contribution to the time complexity of the PCFFT of steps 6 and 7 is negligible. Steps 1 and 5 are performed in $O(\rho Q \log Q)$ time. Steps 2 and 3 are performed in $O(\rho \Lambda Q)$ time. Step 4 is performed in $\log P$ passes. The total number of data-points transmitted in parallel is ΛQ, and a similar number of parallel additions are performed. Thus, this step is $O(\Lambda Q)$.

4 Computer Experiments

Following [5], we consider an S-symmetric input array whose redundancies are described by the symmetry

$$S = \begin{bmatrix} 0 & 0 & 1 \\ 0 & 1 & 0 \\ -1 & 0 & 0 \end{bmatrix}. \tag{27}$$

We choose three-dimensional data arrays of several prime edge-lengths, and compute the the corresponding parameters Q and Λ. The results are shown in Table 1. We kept $Q = 2L - 1$ in all our computer runs. We implemented our algorithm in C/MPI and run it on a commodity cluster conformed by eight 650 MHz Pentium III dual processors, interconnected by a 16-port switch. The cluster runs under Linux/OSCAR. The one-dimensional, Q-point FFTs in steps 1 and 5 of the PCFFT are computed with the 1-D FFTW [4] package.

Table 2 compares the execution times in seconds of the 3-D FFTW [4] and our parallel crystallographic FFT. We also computed the speed up ratios between the 3-D FFTW and the PCFFT. In doing this we had adopted the classical speed up definition as the ratio between the execution times of the parallel method and the best serial algorithm, for the same problem size. The less classical but more common speed-up definition, which is the ratio between the execution times of the parallel method in P processors and the parallel method in one processor, is meaningless in our case, since the PCFFT in one processor is on average, far less efficient than the 3-D FFTW. We also compared the PCFFT run times with those of the parallel MPI-FFTW [4]. It turns out, however, that due to

Table 1. PCFFT parameters for different problem sizes

Prime Edge	Problem Size	Λ	Q
199	7880599	200	19799
223	11089576	224	24863
239	13651919	240	28599
263	18191447	264	34583
271	19902511	272	36719
311	30080231	312	48359

Table 2. 3-D FFTW vs. PCFFT

Problem Size	3-D FFTW time	PCFFT time	Speed up
7880599	19.887	3.018	6.59
11089576	52.428	3.572	14.68
13651919	61.019	5.184	11.77
18191447	254.969	10.336	24.67
19902511	269.612	5.803	46.46
30080231	619.351	77.259	8.02

Table 3. Performance of the Hypercube Reduce-Scatter vs. MPI Reduce-Scatter routine for computing step 4. Times are measured in seconds

Problem Size	Hypercube R-S	MPI R-S
7880599	.430	.972
11089576	.563	1.121
13651919	.67	1.404
18191447	1.142	1.735
19902511	.836	1.837
30080231	45.496	54.512

the load unbalancing induced by the irregularity of the problem, and above all, the large data array transpositions and communications that were required, the MPI-FFTW performed very poorly in our system. In fact, in most runs, the MPI-FFTW took longer than the 3-D FFTW in our cluster. We could not compare the PCFFT with a parallel symmetrized FFT since, to our knowledge, no other parallel prime edge-length crystallographic FFT is currently available.

It is worth remarking that in the usual, asymmetric multidimensional FFT no overlaps between communications and computations are possible. This overlapping is, thus, an additional attractive computational feature of the PCFFT. As a way of assessing the impact of a careful exploitation of this feature in the overall PCFFT performance, we compare the performance of two implementations of step 4 of the PCFFT. The first uses an MPI Reduce-Scatter function, while the second, a communication/addition method tailor made for the problem. Results are reported in Table 3. The tailor made routine, which we call *hypercube method*, has been crafted to maximize overlaps between communications and computations. The authors are still researching on further improvements on this step of the PCFFT which is crucial for the scaling of the method to massive distributed systems

Acknowledgments. This work was partially supported by NIH/NIGMS grant No. S06GM08103 and the NSF PRECISE project of the University of Puerto Rico at Mayagüez

References

1. L. Auslander, and M. Shenefelt, *Fourier Transforms that Respect Crystallographic Symmetries*, IBM J. Res. and Dev., 31, pp. 213–223, 1987.
2. M. An, J.W. Cooley, and R. Tolimeri, *Factorization Methods for Crystallographic Fourier Transforms*, Advances in Appl. Math. 11, pp. 358–371, 1990.
3. J. Cooley, and J. Tukey, *An Algorithm for the Machine Calculation of Complex Fourier Series*, Math. Comp., 19, pp. 297–301, 1965.
4. M. Frigo, S.G. Johnson, *An adaptive software architecture for the FFT* ICASSP Conference Proceedings, 3 (1998), pp 1381–1384.
5. J. Seguel, D. Bollman, E. Orozco, *A New Prime Edge-length Crystallographic FFT*, ELSEVIER Lecture Notes in Computer Science, Vol. 2330, pp. 548–557,2002.
6. J. Seguel, *A Unified Treatment of Compact Symmetric FFT Code Generation*, IEEE Transactions on Signal Processing, Vol 50, No. 11, pp. 2789–2797, 2002.
7. L.F. Ten Eyck, *Crystallographic Fast Fourier Transforms*, ACTA Crystallogr. A, vol. 29, pp. 183–191, 1973.
8. C.M. Weeks, G.T. DeTitta, H.A. Hauptman, H.A Thuman, R. Miller, *Structure Solution by Minimal Function Phase Refinement and Fourier Filtering: II Implementation and Applications*, Acta Crystallogr. A50, pp. 210–220, 1994.
9. C.M. Weeks, R. Miller *Optimizing Shake-and-Bake for Proteins*, Acta Crystallogr. D55, pp 492–500, 1999.

A Service-Oriented Framework for Parallel Medical Image Reconstruction*

S. Benkner, A. Dimitrov, G. Engelbrecht, R. Schmidt, and N. Terziev

Institute for Software Science, University of Vienna,
Liechtensteinstrasse 22, A-1090 Vienna, Austria
{sigi,dimitrov,gerry,rainer,terziev}@par.univie.ac.at

Abstract. This paper presents an overview of the eMERGE system, a service-oriented system for medical image reconstruction on parallel computers. Within the eMERGE system, parallel image reconstruction codes running on SMP clusters or other parallel machines may be exposed as Jini services, as Web Services, or as OGSA services and accessed over the Internet by means of a browser-based GUI. The system provides support for dynamic service management, service selection and monitoring. We describe the parallelization of a fully 3D iterative image reconstruction code using a combination of MPI and OpenMP, discuss the major design and implementation issues of the various service types supported by our system, and present experimental results comparing the performance of RMI-based to SOAP-based image data transfers.

1 Introduction

Novel algorithms enable very accurate 3D reconstruction of medical images from 2D scanner data by considering principal 3D effects of data acquisition. However, the high computational requirements restrict the deployment of these methods to dedicated research centers. In order to make advanced image reconstruction services available to hospitals that have no in-house high performance computing facilities, we are developing the eMERGE system, a service-oriented framework that supports near real-time 3D image reconstruction by providing transparent access to remote parallel computers over the Internet. Our current prototype consists of a browser-based GUI, a parallel fully 3D image reconstruction code written in C/MPI/OpenMP, and a framework for dynamic service management, service selection and monitoring. Parallel image reconstruction codes provided on PC clusters or other parallel machines may be exposed in our system as Jini services, as Web Services, or as OGSA-compliant Grid services.

This article is organized as follows: Section 2 provides an overview of the image reconstruction algorithm and its parallelization. Sections 3 and 4 describe the main components of the eMERGE system and the different service variants,

* This research was partially supported by the Austrian Science Fund as part of the AURORA project under contract SFB F1102 and by the European Commission as part of the IST Project GEMSS under contract IST-2001-37153.

P.M.A. Sloot et al. (Eds.): ICCS 2003, LNCS 2659, pp. 612–621, 2003.

respectively. Performance results of image transfers using different protocols are presented in Section 5, followed by conclusions and future plans in Section 6.

2 Parallel Fully 3D Image Reconstruction

In emission tomography (e.g. SPECT), the spatial distribution of a radioactive tracer is reconstructed from projection data. In fully 3D image reconstruction a solution for the whole image volume X is found simultaneously by using the whole set of projection values Y. The imaging process is characterized by the following equation:

$$Y = A\,X + E \tag{1}$$

The image volume X is represented by means of a vector of size J which stores the mean activity of each voxel. Vector Y of size I represents the whole set of projection values, counting the photons detected for each projection value during the whole acquisition. Element a_{ij} of the system matrix A represents the probability that a photon emitted in voxel j is detected at detector position i. The system matrix allows a flexible modeling of scanner geometry, detector efficiency, scatter and attenuation. In typical clinical applications J, the number of unknowns is 128^3 and I, the number of equations is $128^2 \times 120$ assuming 3° rotational increment over a range of 360°.

The ill-posedness of A and the error term E, in general, allow no exact solution of Eqn. 1. In our system we utilize an improved variant of the well-known ML-EM algorithm for emission tomography [12], which is an iterative method for finding a feasible solution of Eqn. 1. The ML-EM algorithm takes into account the Poisson statistics of photon emission and detection. During one iteration step, the voxel $x_j^{(n)}$ of the n-th iteration is updated by

$$x_j^{(n+1)} = x_j^{(n)} \sum_i a_{ij} \frac{y_i}{\sum_k a_{ik} x_k^{(n)}} \quad . \tag{2}$$

A more detailed description of our reconstruction algorithm, which implicitly corrects for scatter and attenuation, can be found in [2].

Experiments with a sequential implementation of the image reconstruction algorithm in C, have shown that the time required for the reconstruction of a 128^3 image volume on a single processor workstation can be in the range of several hours to a few days, depending on the configuration of the algorithm and the hardware.

2.1 Parallelization for SMP Clusters

In order to allow on-site clinical analysis the reconstruction code has been parallelized. Since our main target platforms are SMP clusters, we adopted a hybrid parallelization strategy relying on a combination of MPI [7] and OpenMP [11]. Distributed-memory parallelization, based on the *single-program multiple-data* (SPMD) paradigm, is achieved by parameterizing the code in such a way that it can be executed by multiple MPI processes, each computing its own part of the image volume. This was implemented by strip-mining the outer loop representing the computations defined by Eqn.2 across multiple processes. By utilizing appropriate OpenMP work sharing constructs, each MPI process employs multiple OpenMP threads. At the end of a full iteration cycle, the voxels updated by individual MPI processes are transferred to all other processes in a collective communication phase.

Since most of the elements of the system matrix are zero we utilize the compressed row storage format (CRS) in our implementation. Moreover the symmetries of the system matrix are exploited to further reduce the storage requirements. In order to minimize communication overheads, the system matrix and some other data structures are allocated on each node.

The parallelized source program is compiled with an OpenMP C compiler and linked with the MPI library. The resulting program is executed on an SMP cluster as an MPI program by a set of parallel processes, each usually running on a separate node of the cluster within its own local address space. Due to the use of OpenMP, each MPI process employs multiple threads (usually one thread per processor) which execute concurrently in the shared address space of a node.

The combination of MPI and OpenMP offers greater flexibility with respect to memory management than an approach based on MPI only. Moreover, due to the shared memory available within nodes, the number of communications can be reduced while message sizes can be increased. In a pure MPI implementation of our algorithm, not only more messages but also much more memory is required. On the other hand, by using OpenMP, certain synchronization overheads are introduced whenever multiple threads update data structures allocated in the shared memory of a node. Experimental results comparing hybrid-parallel versions of the reconstruction algorithm to a pure message passing version are presented next.

2.2 Performance Experiments

We present performance experiments with different versions of the parallel image reconstruction code on a Beowulf-type PC cluster consisting of 6 nodes connected by Fast Ethernet. Each node is equipped with 4 Pentium III Xeon processors (700 MHz) and 4 GB of memory. For the compilation of the MPI/OpenMP source program the C compiler from Portland Group Inc. was used together with the mpich library.

In Figure 1 the speed-ups (with respect to the sequential code executed on a single processor of the cluster) of three different code variants are compared for an image of size 128x128x128 using data from a clinical standard phantom. A pure MPI version is compared to hybrid-parallel versions which employed 2 and 4 threads within each MPI process, respectively. The hybrid-parallel versions have been executed in such a way that each thread is always running on its own processor. On up to two nodes (8 processors), all three versions yield approximately the same speed-ups. On more than two-nodes, the pure MPI-version

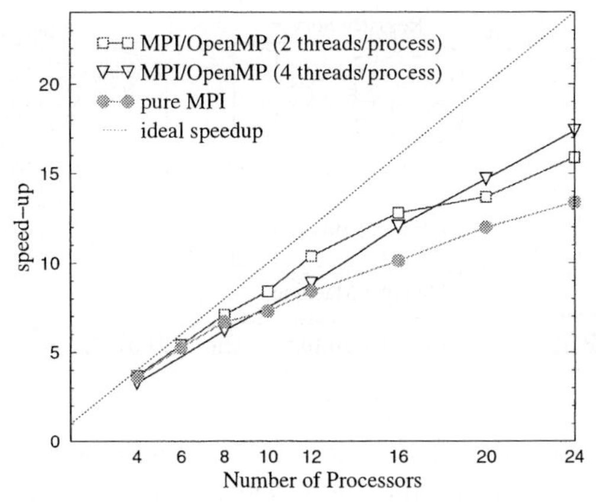

Fig. 1. Experimental comparison of a pure MPI version to hybrid-parallel MPI/OpenMP versions with 2 and 4 threads per MPI process, respectively.

is outperformed by the MPI/OpenMP versions. One reason for the worse scaling of the pure MPI version is the larger communication overhead (more and smaller messages), since communication is performed also within nodes. Moreover, the MPI version requires more memory because replicated data structures are allocated within each process, while in the MPI/OpenMP version all threads have access to the data allocated within their parent processes. As can be observed from Figure 1, the performance of the MPI/OpenMP version where only 2 threads were generated within each MPI process decreases when more than one MPI process is run on a node, i.e. if more than 12 processors are used. If all 24 processors of the cluster are utilized, a significant performance difference between the pure MPI version and the MPI/OpenMP versions can be seen.

Our reconstruction algorithm has been implemented in such a way that it is portable across distributed-memory architectures, shared-memory architectures and clusters. A performance evaluation on other parallel computing platforms, including a heterogeneous SMP cluster (2-processor nodes and 4-processor nodes), an SGI Origin, and an NEC SX-5 multi-node vector supercomputer is currently under way. In order to efficiently utilize the computing power offered by vector supercomputers such as the SX-5, modifications of our implementation will be required for exploiting all three levels of parallelism (i.e. distributed-memory, shared-memory, and vector parallelism).

3 The eMERGE System

Within the eMERGE system, parallel image reconstruction codes running on PC clusters may be exposed as Jini services, as Web Services, or as OGSA-compliant Grid services. The eMERGE system provides basic support for automatic service selection, dynamic service management, and service monitoring. Services may be accessed via a browser-based graphical user interface either in direct mode, in

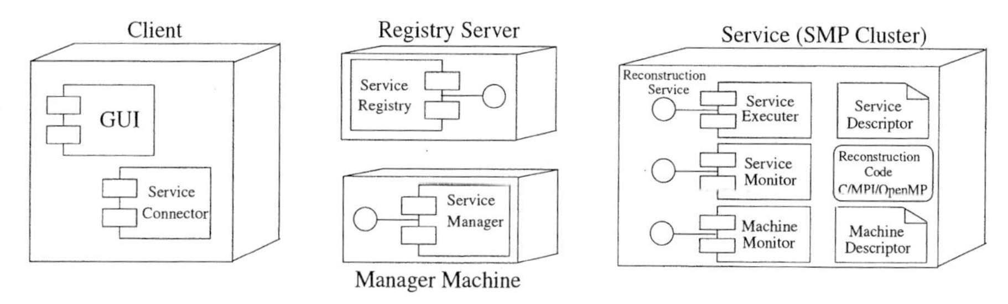

Fig. 2. Main components of the eMERGE system.

which case a predefined reconstruction service is accessed, or in managed mode, where a service manager automatically selects from the set of available parallel machines the one on which the reconstruction job is expected to be completed within the shortest period of time. The main components of the eMERGE system are shown in Figure 2.

In a typical use-case scenario, projection data from a CT scanner are loaded from the client's local file system and displayed in the GUI. The user selects the slices to be reconstructed and various other reconstruction parameters. When the user decides to start the reconstruction process, the raw image and an XML file, which contains meta data about the image and the reconstruction process, are transferred over the Internet to a remote parallel machine. On the target machine, an intermediate Java program handles the transfer of image data, and initiates the execution of the native image reconstruction code. By monitoring the reconstruction process, the user is provided with an estimated service completion time and corresponding progress information. After completion of the reconstruction process, the reconstructed 3D image is transferred back to the client and displayed in the GUI.

3.1 Reconstruction Services and Server-Side Components

The native image reconstruction code, written in C/MPI/OpenMP, may be installed on PC clusters and other parallel computers and exposed as a service through an intermediate server-side Java program, called *service executor*, which handles the data transfer between clients and services, and the execution of the reconstruction job on the parallel target machine. The service executor implements the interface `ReconstructionService`, which contains all methods a client may invoke on an image reconstruction service. The service executor relies on a *service monitor* for monitoring the state of a running service, and a *machine monitor* for obtaining an estimate of the available computational power of the target machine at a certain point in time. For this purpose the machine monitor needs access to the local batch system. Upon start-up, the machine monitor is configured with an XML machine descriptor, which currently contains the number of available processors as well as their estimated performance.

In order to support the concurrent execution of multiple reconstruction jobs on the same target machine, the service executor generates a unique id for each client, which is passed as an argument to all methods provided by the reconstruction service. As a consequence, a single instance of the service executor can be accessed by multiple clients at the same time.

3.2 Service Manager and Service Registry

Reconstruction services may be accessed in direct mode or in managed mode. In direct mode, the client has to know the location of the service. In managed mode, the client contacts a service manager, which dynamically selects a reconstruction service to be accessed by the client. Whenever a reconstruction service is added to the system, the service registers itself in the service registry. Upon an incoming client request, the service manager selects from the set of registered services the one that is expected to complete the reconstruction job within the shortest period of time. For this purpose, the service manager maintains state information of all registered service machines and chooses the machine that provides the maximum free computational capacity (currently measured as the accumulated performance of free processors) at the time a service is requested. Whenever the number of available free processors on a registered service machine changes, the service manager receives a corresponding notification from the machine monitor. After having selected a service, the service manager forwards an appropriate service proxy to the client which then accesses the chosen service directly.

3.3 User Interface and Client-Side Components

The image reconstruction service system may be accessed via a graphical user interface that can be provided to end users as a Java applet via a browser. Since the GUI needs to access the client's local file system for loading and saving medical images, we utilize a signed applet which presents a X.509 certificate upon download to the user for granting the required permissions.

The GUI utilizes a flexible *service connector* component which is capable of loading or generating different service proxies in order to access reconstruction services via RMI or Web Services (JAX-RPC, SOAP) protocols.

4 Service Technologies

Initially, the eMERGE system has been developed based on Jini. Support for Web Services and OGSA has been added recently and is a major objective of ongoing work.

4.1 Jini

Within the Jini-based implementation of the eMERGE system, reconstruction codes installed on parallel computers are exposed as services accessible via the Java RMI protocol. Using the basic mechanisms of Jini (discovery, join and lookup), a dynamic federation of multiple image reconstruction services on different parallel hosts may be established. In such a federation of services, at least one service registry (realized as a Jini lookup service) and one service manager (realized as a Jini service) must be instantiated. Whenever a reconstruction service is added to the system, the service registers itself with the service registry. The service manager obtains service proxies of all available reconstruction services from the service registry.

Upon startup of the client GUI, the service connector component discovers the service registry and obtains a proxy object of the service manager. At the time a client issues a request for an image reconstruction service, the service manager selects a suitable image reconstruction service (see below) and returns the proxy of the corresponding service executor to the client. The client's service connector then accesses the image reconstruction service directly. The raw image data and the XML-based meta-data are transferred over RMI to the target machine by calling appropriate methods of the service executor. Thereupon, the service executor launches the parallel reconstruction process on the target machine, usually utilizing all available free processors.

At the moment our service manager uses a fairly simple strategy for selecting a reconstruction service: the machine with the maximum available computing resources, currently measured as the number of free processors times the estimated processor performance in FLOPs, is selected. In order to provide up-to-date information about the actual free computational capacity of a machine, the monitoring component notifies the service manager whenever the number of free processors changes.

In order to support realistic settings, where server-side machines as well as clients are behind firewalls, we utilize HTTP tunneling of all RMI calls.

4.2 Web Services

Web Services are rapidly becoming the standard way of performing Enterprise Application Integration (EAI) within as well as across organizations. Web Services are based on XML vocabularies standardized by the W3C (SOAP, WSDL, UDDI, etc.) and provide an architecture for exposing applications as services that are platform and programming language independent. Web services are identified by an URI, expose their public interfaces and protocol bindings through a Web Services Description Language (WSDL) document, and are usually accessed via SOAP over HTTP, even across firewalls.

Our current Web Services implementation of image reconstruction services supports access in direct mode only. For the development we mainly used the Java Web Service Developer Pack (JWSDP 1.0 [8]) which provides a framework for developing and hosting web services based on servlets and Tomcat. Moreover, we have utilized other Web Services frameworks, mainly for the purpose of performance comparisons, including Glue [5] and Apache Axis [1].

The starting point for realizing image reconstruction services as Web Services is the Java interface `ReconstructionService`, which specifies all methods clients may invoke on a reconstruction service. In order to support access to reconstruction services based on Web Services technologies, a WSDL document that describes the public interfaces and protocol bindings is required. This WSDL document has been generated automatically from the Java interface provided for reconstruction services by using the `xrpcc` tool of the JWSDP. Besides generating a WSDL document from a Java interface, the `xrpcc` tool also generates stubs and ties for marshalling of arguments on the client-side and server-side, respectively.

On the client-side, we provide an additional service connector for accessing reconstruction services over JAX-RPC (which is SUN's implementation of SOAP RPC). The service connector loads the stub class of the reconstruction service over HTTP (by means of an URLClassLoader). Upon method invocation, the stub creates a SOAP request message that is conveyed via HTTP to a servlet on the server side. The servlet parses the request message, converts arguments from XML to Java and invokes the corresponding method of the service executor. The method's result object is transformed into an appropriate SOAP response message by the servlet, transferred back to the client, and transformed by the stub into a Java object.

Compared to a pure Java/RMI communication protocol, SOAP calls imply an additional overhead for marshalling arguments and for parsing SOAP messages both on the client side and the server side (see Section 5).

In addition to the stub-based service connector, we have implemented a connector that makes use of the Web Services Invocation Framework (WSIF 2.0 [13]) in order to access reconstruction services independently of the actual implementation and deployment as described in the next section.

4.3 Open Grid Service Architecture

We have implemented a variant of the medical image reconstruction service according to the Open Grid Service Architecture (OGSA) based on the Open Grid Service Infrastructure (OGSI) Application Framework (OGSI TP-5 [10]).

An OGSA service is a Web Service that provides standard mechanisms for creating, naming and discovering services [9]. These mechanisms are realized by an OGSI container which provides the runtime environment for Grid services. The OGSI container itself makes use of the Tomcat servlet engine and of Apache Axis. The OGSA specification requires that a compliant Grid service implements at least the interface (PortType) `GridService`, which is the base interface definition in OGSA.

As in the Jini and Web Services implementation, the OGSA implementation of the medical image reconstruction service is based on the Java interface `ReconstructionService` from which a WSDL document is generated. In order to provide an OGSA compliant implementation, our service executor component extends the class `PersistentServiceSkeleton` provided by the OGSI application framework implementing the interface (PortType) `GridService`. Moreover, a deployment descriptor has to be provided, specifying the URI of the service and other properties. Upon deployment, the service is automatically registered in the registry of the OGSI container.

At the current stage of implementation, the OGSA-compliant version of the service can be used in direct mode only. That is, the client application must know the service endpoint, which is an URI that designates the WSDL of the reconstruction service (also referred to as Grid Service Handle (GSH) in OGSA terminology). In order to access the OGSA service from our GUI, we have implemented a service connector based on the Web Service Invocation Framework (WSIF 2.0) [13]. Using WSIF, the service connector constructs a SOAP request based on the grid service reference (GSR) of the reconstruction service (which is the WSDL document of the service decorated by the container with runtime specific information) without the need of any OGSA specific classes or stubs. Another major advantage of WSIF is that besides SOAP other bindings and transport protocols may be supported (eg. EJB using RMI/IIOP).

5 Experimental Results

In this section we present performance results of image data transfer based on Java RMI in comparison to various implementations of SOAP (XML-RPC).

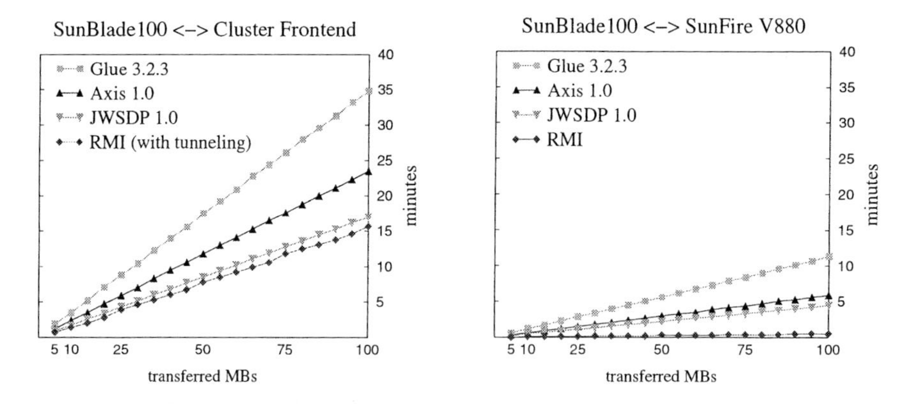

Fig. 3. Performance of data transfers: RMI vs. different SOAP implementations.

Although in our current reconstruction service the size of medical images is only approximately 5 MBs, we have measured transfers of up to 100 MBs on a Fast Ethernet back and forth (ping-pong) between a client and a reconstruction service executor. The client application was executed on a Sun Blade 100 (sparcv9, 502MHz, 640MB RAM). The service executor was hosted on the front-end PC (dual Pentium II, 400MHz, 1GB total RAM) of our 64 processor SMP cluster (see left hand side of Fig. 3) and on a Sun Fire V880 (four sparcv9, 750MHz, 8GB RAM), respectively. Apparently for both settings, RMI data transfers are the fastest, followed by JAX-RPC with JWSDP [8], SOAP XML-RPC with Axis [1], and finally Glue's implementation [5] of SOAP-RPC. Note that in the figure on the left hand side we utilized HTTP tunneling of RMI calls in order to get access to the cluster frontend across the firewall. The overhead of SOAP based data transfers compared to RMI is mainly caused by the generation and parsing of SOAP messages and to a much lesser extent by the conversion of image data represented as a Java byte array to/from XML.

6 Conclusions

In this paper we gave an overview of a service-oriented system that provides access to parallel image reconstruction codes on different parallel computers via the Internet based on RMI and SOAP. We are currently extending our system to provide support for automatic service selection for the Web Services and OGSA-based service implementations. Moreover our Jini-based service manager is being extended in order to allow transparent access to a local Jini federation through Web Services or OGSA protocols similar to the ICENI system [6].

Major directions of future work will include the extension of the service manager with more sophisticated methods for automatic service selection based on user-specified criteria, support for authorization and authentication, and for secure data transfers. Our services will be integrated into a larger Grid-enabled Medical Simulation System which is currently being developed within the EU project GEMSS. Within GEMSS, negotiable QoS support for medical services and security of patient data under the constraints of EU legislation will be of major importance.

References

1. Apache Axis. `http://ws.apache.org/axis`
2. W. Backfrieder, S. Benkner, G. Engelbrecht. Web-based parallel ML-EM reconstruction for SPECT on SMP clusters, Proc. METMBS'01, CSREA Press, 2001.
3. I. Foster, C. Kesselman, J. Nick, S. Tuecke. The Physiology of the Grid: An Open Grid Services Architecture for Distributed Systems Integration, Open Grid Service Infrastructure WG, Global Grid Forum, June 22, 2002.
4. I. Foster, C. Kesselman, S. Tuecke. The Anatomy of the Grid: Enabling Scalable Virtual Organizations, International J. Supercomputer Applications, 15(3), 2001.
5. Glue. `http://www.themindelectric.com/glue`
6. N. Furmento, W. Lee, A. Mayer, S. Newhouse, J. Darlington. ICENI: An Open Grid Service Architecture Implemented with Jini. Proceedings of the IEEE/ACM SC2002 Conference, November 2002.
7. Message Passing Interface Forum. MPI: A Message-Passing Interface Standard. Vers. 1.1, June 1995. MPI-2: Extensions to the Message-Passing Interface, 1997.
8. Java Web Services Developer Pack. `http://java.sun.com/webservices`
9. S. Tuecke, K. Czajkowski, I. Foster, J. Frey, S. Graham, C. Kesselman. Grid Service Specification: Open Grid Service Infrastructure WG, GGF, Draft 2, 7/17/2002.
10. T. Sandholm, R. Seed, J. Gawor. OGSI Technology Preview Core - A Grid Service Container Framework, OGSI Technology Preview 5, DRAFT 11/08/2002. `http://www.globus.org/ogsa/releases/TechPreview`
11. OpenMP Architecture Review Board. OpenMP C and C++ Application Program Interface. Version 2.0, March 2002. `http://www.openmp.org`.
12. L. A. Shepp and Y. Vardi, Maximum likelihood reconstruction for emission tomography. IEEE Trans. Med. Imaging, vol. 1, no. 2, pp. 113-122, 1982.
13. Web Services Invocation Framework. `http://ws.apache.org/wsif`

Parallel Implicit Solution of Full Navier-Stokes Equations

V.Ü. Ünal[1] and Ü. Gülçat[2]

[1]Physics Department
Yeditepe University
81120, Kayışdağı, İstanbul, Turkey
vunal@yeditepe.edu.tr
[2]Faculty of Aeronautics and Astronautics
Istanbul Technical University
80626, Maslak, İstanbul, Turkey
gulcat@itu.edu.tr

Abstract. Parallel implicit solution of the incompressible Navier-Stokes equations based on two fractional steps in time and Finite Element discretization in space is presented. The accuracy of the scheme is second order in both time and space domains. Large time step sizes, with CFL numbers much larger than unity, are taken. The Domain Decomposition Technique is implemented for parallel solution of the problem with matching and non-overlapping sub domains. The segragate solution to tempereature field is obtained for the flow case where the forced convection is one order of magnitude higher than the free convection.

1 Introduction

As is well known, explicit schemes impose severe restrictions on the time step size for analyzing complex viscous flow fields, which are resolved with sufficiently fine grids. To remedy this, implicit flow solvers are used in analyzing such flows. Accuracy of the scheme is also a major issue in the numerical study of complex flows. Naturally, the higher order accurate schemes allow one to resolve the flow field with less number of grid points. Resolving the flow field with less number of points gives a great advantage to implicit schemes since the size of the matrix to be inverted becomes small.

In this study a second order accurate implicit scheme for solution of full Navier-Stokes equations is developed and implemented. The space is discretized with brick elements while modified version of the two-step fractional method is used in time discretization of the momentum equation. At each time step, the momentum equation is solved only once to get the half time step velocity field. The pressure, on the other hand, is obtained via an auxiliary scalar potential which satisfies the Poisson's equation. For the parallel implicit solution of the matrix equations, modified version of the Domain Decomposition Method, [1,2], is utilized, and direct inversion in each domain is performed. Super linear speed ups were achieved [3].

P.M.A. Sloot et al. (Eds.): ICCS 2003, LNCS 2659, pp. 622–631, 2003.

The implicit formulation for the momentum equation and the parallel implementation of the momentum, the pressure and the energy equation will be given in the following sections. The parallel solutions results obtained for the forced cooling of a room with chilled ceiling will be given at the end.

The scheme here is made to run on SGI Origin 3000 utilized with 8 processors running Unix operating system. Public version of the Parallel Virtual Machine, PVM 3.3, is used as the communication library.

2 Formulation

2.1 Navier-Stokes Equations

The flow of unsteady incompressible viscous fluid is governed with the continuity equation, [4],

$$\nabla.\mathbf{u} = 0 \tag{1}$$

the momentum (Navier-Stokes) equation

$$\frac{D\mathbf{u}}{Dt} = -\nabla p + \frac{1}{Re}\nabla^2\mathbf{u} + \frac{Gr}{Re^2}T\mathbf{k} \tag{2}$$

and the energy equation

$$\frac{DT}{Dt} = \frac{1}{RePr}\nabla^2 T. \tag{3}$$

The equations are written in vector form (here on, boldface type symbols denote vector or matrix quantities). The velocity vector, pressure, temperature and time are denoted by $\mathbf{u}=\mathbf{u}(u,v.w)$, p, T and t, respectively. The variables are non-dimensionalized using a reference velocity and a characteristic length. Re is the Reynolds number, $Re = UH/\nu$ where U is the reference velocity, H is the characteristic length and ν is the kinematic viscosity of the fluid. Gr represents the Grashof number, $Gr = g\beta\Delta TH^3/\nu^2$, where β is the coefficient of volumetric expansion. Pr denotes Prandtl number and **k** denotes the unit vector in z direction.

2.1 FEM Formulation

The integral form of Eq. (2) over the space-time domain reads as

$$\int_{\Omega}\int_{t}\frac{\partial \mathbf{u}}{\partial t}\mathbf{N}\, d\Omega dt = \int_{\Omega}\int_{t}(-\mathbf{u}.\nabla\mathbf{u} - \nabla p + \frac{1}{Re}\nabla^2\mathbf{u} + \frac{Gr}{Re^2}T^n\mathbf{k})\mathbf{N}\, d\Omega dt \tag{4}$$

where **N** is an arbitrary weighting function. The time integration of both sides of Eq. (4) for half a time step, $\Delta t/2$, from time step n to n +1/2 gives

$$\int_\Omega (\mathbf{u}^{n+1/2} - \mathbf{u}^n)\,\mathbf{N}\,d\Omega = \frac{\Delta t}{2}\int_\Omega (-\mathbf{u}.\nabla\mathbf{u}^{n+1/2} - \nabla p^n + \frac{1}{Re}\nabla^2\mathbf{u}^{n+1/2} + \frac{Gr}{Re^2}T^n\,\mathbf{k})\,\mathbf{N}\,d\Omega. \tag{5}$$

At the intermediate time step the time integration of Eq. (4), where the convective and viscous terms are taken at n +1/2 and pressure term at time level n, yields

$$\int_\Omega (\mathbf{u}^* - \mathbf{u}^n)\mathbf{N}\,d\Omega = \Delta t\int_\Omega (-\mathbf{u}.\nabla\mathbf{u}^{n+1/2} - \nabla p^n + \frac{1}{Re}\nabla^2\mathbf{u}^{n+1/2} + \frac{Gr}{Re^2}T^n\,\mathbf{k})\,\mathbf{N}\,d\Omega. \tag{6}$$

For the full time step, the averaged value of pressure at time levels n and n+1 is used to give

$$\int_\Omega (\mathbf{u}^{n+1} - \mathbf{u}^n)\,\mathbf{N}\,d\Omega = \Delta t\int_\Omega (-\mathbf{u}.\nabla\mathbf{u}^{n+1/2} + \frac{1}{Re}\nabla^2\mathbf{u}^{n+1/2} - \nabla\frac{p^n + p^{n+1}}{2} + \frac{Gr}{Re^2}\frac{T^{n+1} + T^n}{2}\mathbf{k})\,\mathbf{N}\,d\Omega. \tag{7}$$

Subtracting (6) from (7) results in

$$\int_\Omega (\mathbf{u}^{n+1} - \mathbf{u}^*)\mathbf{N}\,d\Omega = -\frac{\Delta t}{2}\int_\Omega \nabla(p^{n+1} - p^n)\mathbf{N}\,d\Omega - \frac{\Delta t}{2}\int_\Omega (T^{n+1} - T^n)\mathbf{k}\mathbf{N}\,d\Omega. \tag{8}$$

If one takes the divergence of Eq. (8), the following is obtained;

$$\int_\Omega \nabla.\mathbf{u}^*\,\mathbf{N}\,d\Omega = -\frac{\Delta t}{2}\int_\Omega \nabla^2(p^{n+1} - p^n)\mathbf{N}\,d\Omega - \frac{\Delta t}{2}\int_\Omega \frac{\partial}{\partial z}(T^{n+1} - T^n)\mathbf{N}\,d\Omega. \tag{9}$$

Subtracting (4) from (5) yields

$$\mathbf{u}^* = 2\mathbf{u}^{n+1/2} - \mathbf{u}^n\,. \tag{10}$$

2.3 Numerical Formulation

Defining the auxiliary potential function $\varphi = -\Delta t(p^{n+1} - p^{n})$ and choosing N as tri-linear shape functions, discretization of Eq. (5) gives

$$\left(\frac{2\mathbf{M}}{\Delta t} + \mathbf{D} + \frac{\mathbf{A}}{\mathrm{Re}}\right)\mathbf{u}_{\alpha}^{n+1/2} = \mathbf{B}_{\alpha} + p_e\,\mathbf{C}_{\alpha} + \frac{2\mathbf{M}}{\Delta t}\mathbf{u}_{\alpha}^{n} + \frac{\mathrm{Gr}}{\mathrm{Re}^2}\mathbf{M}\,\mathrm{T}^{n}\,\mathbf{k} \tag{11}$$

where α indicates the Cartesian coordinate components x, y and z, **M** is the lumped element mass matrix, **D** is the advection matrix, A is the stiffness matrix, **C** is the coefficient matrix for pressure, **B** is the vector due to boundary conditions and **E** is the matrix which arises due to incompressibility.

Equation for the temperature, Eq. (3) is solved explicitly using:

$$\mathbf{M}\,\mathrm{T}^{n+1} = \mathbf{M}\,\mathrm{T}^{n} - \frac{\Delta t}{2}\left(\frac{\mathbf{A}}{\mathrm{Pr.Re}} + \mathbf{D}\right)\mathrm{T}^{n} \tag{12}$$

The discretized form of Eq. (9) reads as

$$\frac{1}{2}\mathbf{A}\,\varphi = -\frac{1}{2}\mathbf{A}\left(p^{n+1} - p^{n}\right)\Delta t = 2\mathbf{E}_{\alpha}\mathbf{u}_{\alpha}^{n+1/2} - \frac{\Delta t}{2}\frac{\mathrm{Gr}}{\mathrm{Re}^2}(\mathrm{T}^{n+1} - \mathrm{T}^{n})_e C_z \tag{13}$$

Subtracting Eq. (6) from Eq. (7) and introducing the auxiliary potential function φ, one obtains the following;

$$\begin{aligned}\mathbf{M}\mathbf{u}_{\alpha}^{n+1} &= \mathbf{M}\mathbf{u}_{\alpha}^{*} + \frac{1}{2}\mathbf{E}_{\alpha}\,\varphi + \frac{\Delta t}{2}\frac{\mathrm{Gr}}{\mathrm{Re}^2}(\mathrm{T}^{n+1} - \mathrm{T}^{n}) \\ &= 2\,\mathbf{M}\mathbf{u}_{\alpha}^{n+1/2} - \mathbf{M}\mathbf{u}_{\alpha}^{n} + \frac{1}{2}\mathbf{E}_{\alpha}\,\varphi + \frac{\Delta t}{2}\frac{\mathrm{Gr}}{\mathrm{Re}^2}(\mathrm{T}^{n+1} - \mathrm{T}^{n})\end{aligned} \tag{14}$$

The element auxiliary potential φ_e is defined as

$$\varphi_e = \frac{1}{\mathrm{vol}(\Omega_e)}\int_{\Omega_e}\mathbf{N}_i\,\varphi_i\,d\Omega_e, \qquad i = 1,\ldots\ldots\ldots,8, \tag{15}$$

where Ω is the flow domain and $\mathbf{N}_i$ are the shape functions.

The following steps are performed to advance the solution one time-step.

i. Eq. (11) is solved to find the velocity field at time level n+1/2 with domain de composition,
ii. Eq. (12) is solved explicitly and temperature field T^{n+1} is obtained,
iii. Knowing the half step velocity field and the temperature, Eq. (13) is solved with domain decomposition to obtain the auxiliary potential φ.
iv. With this φ, the new time level velocity field $\mathbf{u}^{n+1}$ is calculated via Eq.(14).

v. The associated pressure field p^{n+1} is determined from the old time level pressure field p^n and φ obtained in the second step.

The above procedure is repeated until the desired time level. In all computations lumped form of the mass matrix is used.

3 Domain Decomposition

The domain decomposition technique, [3,5,6,7], is modified and applied for the efficient parallel solution of the momentum, Eq. (11) and the Poisson's Equation for the auxiliary potential function, Eq. (13).

3.1 For the Momentum Equation

Initialization: The momentum, Eq. (11) is solved with a direct solution method, separately in each domain Ω_i with boundary of $\partial\Omega_i$ and interface S_j, with vanishing Neumann boundary condition on the domain interfaces,

At the beginning of each time level: $\mu^k=0$ and $aw^k=0$,

$$\overline{\mathbf{A}}\,\mathrm{y_i} = \mathrm{f_i} \qquad \text{in } \Omega_i,$$

$$\mathrm{y_i} = \mathrm{g_i} \qquad \text{on } \partial\Omega_i,$$

$$\frac{\partial \mathrm{y_i}}{\partial \mathrm{n_i}} = (-1)^{i-1}\mu^k \quad \text{on } S_j,$$

$$\mu^o = \mu^k,\ w^o = \mathrm{g}^o \text{ and } \mathrm{g}^o = aw^k - (\mathrm{y}_2^o - \mathrm{y}_1^o)S_j.$$

where subscript k denotes the interface iteration level at which convergence is obtained, subscript o stands for the initial interface iteration level at the each time level and $\overline{\mathbf{A}} = \frac{2\mathbf{M}}{\Delta t} + \mathbf{D} + \frac{\mathbf{A}}{\mathrm{Re}}$ in Eq. (11) and $\mathrm{y_i} = \mathbf{u}_\alpha^{n+1/2}$.

Unit Problem: Inside the interface iteration cycle, each slave starts to solve the unit problem using the direct solution, together with the initialized Neumann condition received from the master at the end of the initialization step.

$$\overline{\mathbf{A}}\,\mathrm{x}_i^n = 0 \qquad \text{in } \Omega_i,$$

$$\mathrm{x}_i^n = 0 \qquad \text{on } \partial\Omega_i,$$

$$\frac{\partial x_i^n}{\partial n_i} = (-1)^{i-1} w^n \quad \text{on } S_j$$

Each slave sends its new computed interface values to the master processor at the end of the unit problem and master starts post-processing of the received data from the slaves. The interface iteration cycle starts up at the master processor. Master optimizes the interface nodal values of the unknowns using the following '*steepest descent*' procedure.

Steepest Descent: Interface iteration cycle goes on until the master processor finds correct Neumann conditions at the interfaces, μ^k, at the end of its post processing of the interface data received from the slave processors,

$$aw^n = \left(x_1^n - x_2^n\right)S_j$$

$$g^{n+1} = g^n - \beta^n aw^n \qquad w^{n+1} = g^{n+1} + s^n w^n$$

$$\beta^n = \frac{\sum_j \int_{S_j} \left|g^n\right|^2 ds}{\sum_j \int_{S_j} (aw^n) w^n ds} \qquad s^n = \frac{\sum_j \int_{S_j} \left|g^{n+1}\right|^2 ds}{\sum_j \int_{S_j} (g^n)^2 ds}$$

$$\mu^{n+1} = \mu^n - \beta^n w^n$$

where n denotes the interface iteration level.

Convergence check: If $\left|\mu^{n+1} - \mu^n\right| < \varepsilon_\varphi = 10^{-4}$ then $\mu^{n+1} = \mu^k$

Finalization: Having obtained the correct Neumann boundary condition for each interface, the momentum equation, Eq. (11), is solved via the direct solution method at each sub-domain, together with the correct interface conditions,

$$\overline{\mathbf{A}}\, y_i = f_i \quad \text{in } \Omega_i,$$

$$y_i = g_i \quad \text{on } \partial\Omega_i,$$

$$\frac{\partial y_i}{\partial n_i} = (-1)^{i-1} \mu^k \quad \text{on } S_j$$

3.2 Parallel Implementation

During parallel implementation, in order to advance the solution single time step, the momentum equation is solved implicitly with domain decomposition. Solving Eq. (11)

gives the velocity field at half time level, which is used at the right hand sides of Poisson's Eq. (13), in obtaining the auxiliary potential. The solution of the auxiliary potential is obtained with again domain decomposition where an iterative solution is also necessary at the interface. Therefore, the computations involving an inner iterative cycle and outer time step advancements have to be performed in a parallel manner on each processor communicating with the neighboring one.

4 Results and Discussion

The flow in a room, given in Fig. 1, with a Reynolds numbers of 1000, based on 1m. length, inflow speed and the kinematic viscosity, is studied. The computational domain has the following dimensions: sides x = 4m, y = 2m (symmetry plane) and maximum height z = 2.6m. The solution is advanced up to the dimensionless time level of 75, where the steady state is reached, with time step of 0.025. Computations are carried with the 6-domain partitioning grids with total number of 61x31x35 nodes. The following dimensionless numbers are employed. Re = 1000, Pr = 0.7, $Gr/Re^2 = 0.01$. The elapsed time per time step per grid point is 0.0002 seconds.

$$0<x_c<0.2 : \quad y_c = \frac{m}{p^2}(2px_c - x_c^2)$$

$$0.2<x_c<1 : \quad y_c = \frac{m}{(1-p)^2}((1-2p)+2px_c - x_c^2))$$

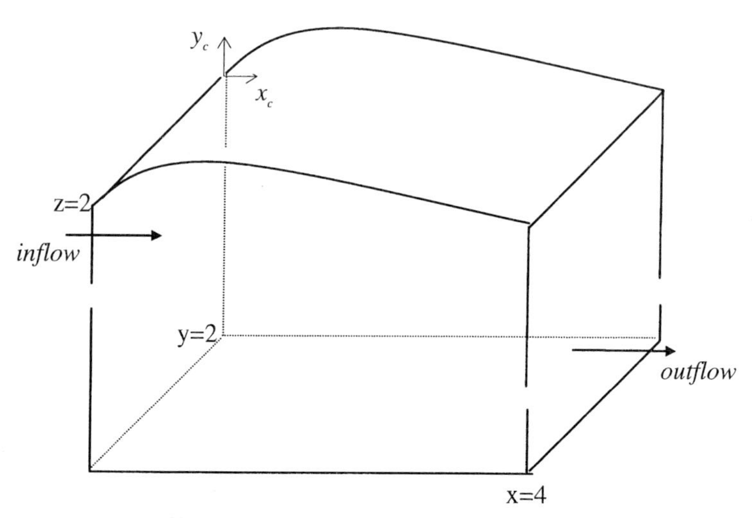

Fig. 1. Definition of the real problem

Initial & Boundary Conditions:

$u=\sqrt{2}/2$, $v=0$, $w=\sqrt{2}/2$

$T = -0.7$	at the top wall
$T=T(x)=(x-2)/4$	inside the cavity
$T_c = -0.5$	cold left wall
$T_h = 0.5$	hot right wall

Computational grid with 6-domain partitioning (along the y-axis) is shown in Fig. 2. Velocity and temperature fields at the symmetry are presented in Fig. 3. The cooling effect of the chilled curved ceiling is obvious on the isotherm plots. The initially linear temperature field changes, because of the circulation in the room, so that the cool front moves towards the hot wall. Shown in Fig. 4 is the isotherms at x = 2 plane where initially temperature is zero. Because of secondary currents in the room cool air packages are observed around the x = 2 plane.

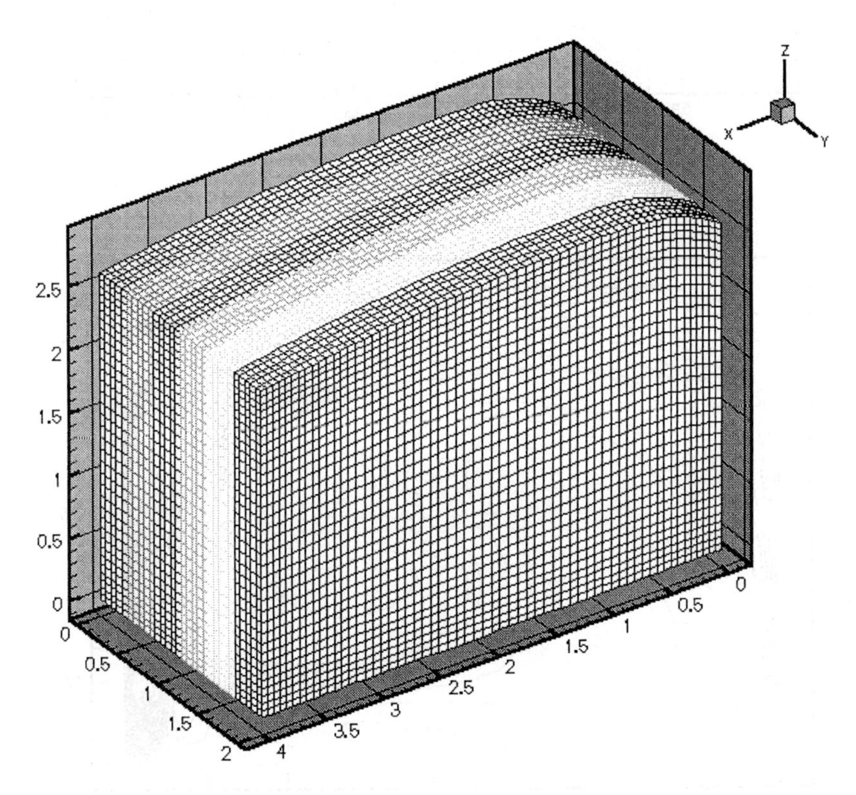

Fig. 2. 6-domain partitioning with total number of 61x31x35 nodes

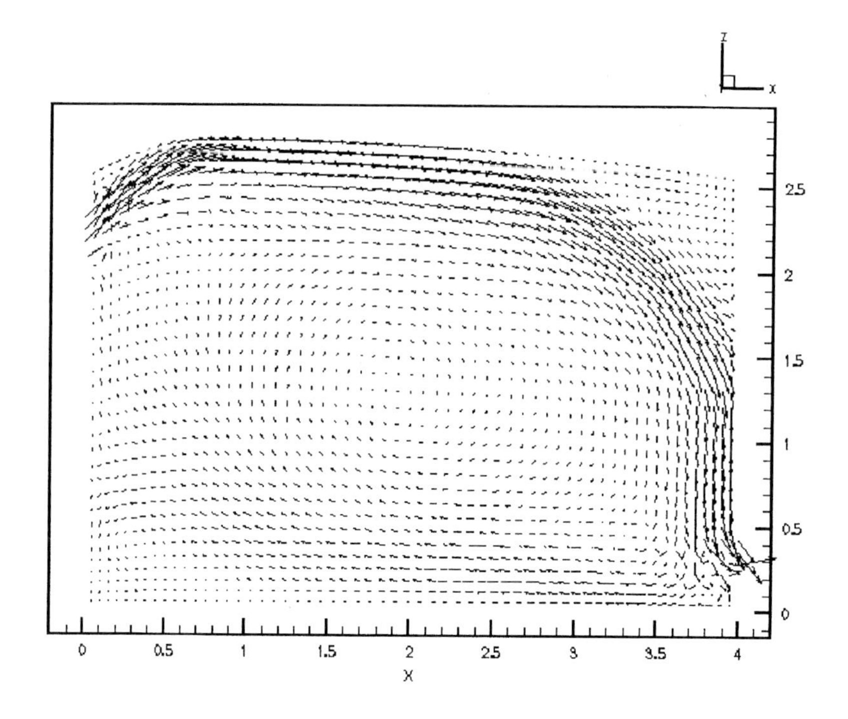

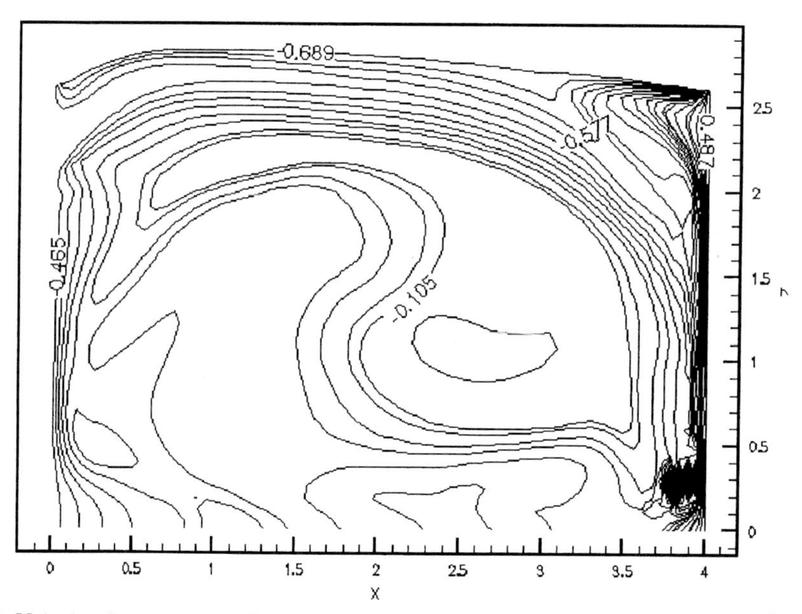

Fig. 3. Velocity & temperature fields at y=2 symmetry plane, Re=1000, Pr=0.7, Gr/Re^2=0.01, $\Delta t = 0.025$

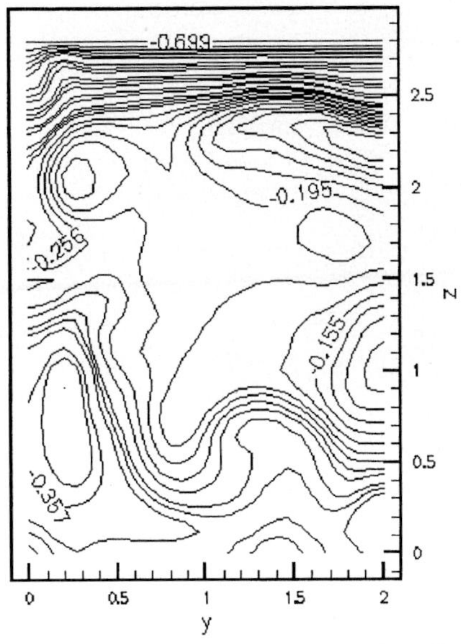

Fig. 4. Temperature field at x=2 plane, Re=1000, Pr=0.7, Gr/Re^2=0.01, $\Delta t = 0.025$

References

1. Gülçat, Ü. and Aslan, A.R., 1997. Accurate 3D Viscous Incompressible Flow Calculations with the FEM, *International Journal for Numerical Methods in Fluids,* 25, 985–1001.
2. Gülçat, Ü. and Üstoglu Ünal V., 2000. Accurate Implicit Solution of 3D Navier-Stokes Equations on Cluster of Workstations, *Parallel CFD 2000 Conference,* Trodheim, Norway, May 22–25.
3. Gülçat, Ü. and Üstoglu Ünal V., 2002. Accurate Implicit Parallel Solution of 3D Navier-Stokes Equations, *Parallel CFD 2002 Conference,* Japan, May 20–22.
4. Schlichting H., Boundary Layer Theory, McGraw-Hill, 1968.
5. Glowinski, R., and Periaux, J., 1982. Domain Decomposition Methods for Nonlinear Problems in Fluid Dynamics, Research Report 147, Inria, FRANCE.
6. Dinh, Q.V., Ecer, A., Gülçat, Ü., Glowinski, R., and Periaux, J., 1992. Applications to Fluid Dynamics, *Parallel CFD 92,* Rutgers University, May 18–20.
7. Glowinski, R., Pan, T.W., and Periaux, J., 1995. *Parallel CFD 95.*

Track on Data Mining

Application of the Confidence Measure in Knowledge Acquisition Process

Michal Wozniak

Chair of Systems and Computer Networks, Wroclaw University of Technology, Poland
The Higher Vocational State School in Legnica, Poland
wozniak@zssk.pwr.wroc.pl

Abstract. Paper deals with the knowledge acquisition process. Different experts formulate the rules for decision support systems. We assume they have different knowledge about the problem and therefore obtained rules have different qualities. We will formulate the proposition of the confidence measure and its application to the decision process based on Bayes formulae. We will propose how calculate the value of measure under consideration for typical statistical learning process. On the base on the proposed measure of the knowledge quality we propose the procedure of the contradictions elimination for the set of logical rules.

1 Introduction

Machine learning is the attractive approach for building decision support systems [9]. For this type of software, the quality of the knowledge base plays the key-role. In many cases we can meet following problems:

- the experts can not formulate the rules for decision problem, because they might not have the knowledge needed to develop effective algorithms, but we can get the learning data from databases and experts only classify each record to the correct class,
- the knowledge given by experts is uncompleted or qualities of experts (knowledge sources) are different for each of them.

For the first problem machine learning algorithms can give us the set of rules (hypothesis) on the base on the learning set, but we have to answer following questions:

1. Who made the object descriptions? (Can we trust the operator?)
2. Who confirmed the diagnosis and what was the expert quality?

In the second problem we get the rules from different experts and their qualities are different. This problems was partly described for the induction learning [2,3,7] and for the concept description [1]. The following paper concerns on the quality of rule for the probabilistic reasoning, but the proposed measure can be modified to acquisition process for another form of rule.

The content of the work is as follow: Section 2 introduces necessary background and provides the probabilistic decision problem statement. Next section presents the

P.M.A. Sloot et al. (Eds.): ICCS 2003, LNCS 2659, pp. 635–643, 2003.

form of rule for the probabilistic expert systems and proposes the rule-based algorithm. Section 4 defines statistical confidence measure of the knowledge and shows how modify knowledge base according to confidence measure of rules. In this section we also presents the interpretation of proposed measure for the estimation process based on the typical statistical model. Section 5 proposes the procedure of contradictions elimination for the set of rules obtained from different sources. The last section concluded the paper.

2 Decision Problem Statement

Among the different concepts and methods of using "uncertain" information in pattern recognition, an attractive from the theoretical point of view and efficient approach is through the Bayes decision theory. This approach consists of assumption [5] that the feature vector $x = (x^{(1)}, x^{(2)}, ..., x^{(d)})$ (describing the object being under recognition) and number of class $j \in \{1,2..., M\}$ (the object belonged to) are the realization of the pair of the random variables X, J. For example in medical diagnosis X describes the result of patient examinations and J denotes the patient state. Random variable J is described by the prior probability p_j, where

$$p_j = P(J = j). \tag{1}$$

X has probability density function

$$f(X = x|J = j) = f_j(x) \tag{2}$$

for each j which is named conditional density function. These parameters can be used to enumerating *posterior* probability according to Bayes formulae:

$$p(j|x) = \frac{p_j f_j(x)}{\sum_{k=1}^{M} p_j f_j(x)} \tag{3}$$

The formalisation of the recognition in the case under consideration implies the setting of a optimal Bayes decision algorithm $\Psi(x)$, which minimizes probability of misclassification for 0-1 loss function[4]:

$$\Psi(x) = i \text{ if } p(i|x) = \max_{k \in \{1, ..., M\}} p(k|x). \tag{4}$$

In the real situation the prior probabilities and the conditional density functions are usually unknown. Furthermore we often have no reason to decide that the prior probability is different for each of the decisions. Instead of them we can used the rules and/or the learning set for the constructing decision algorithms [12,13].

3 Rule-Based Decision Algorithm

Rules as the form of learning information is the most popular model for the logical decision support systems. For systems we consider the rules given by experts have rather the statistical interpretation than logical one.
The form of rule for the probabilistic decision support system[6] is usually as follow

`if A then B with the probability` β,

where β is interpreted as the estimator of the *posterior* probability, given by the following formulae:

$$\beta = P(B|A) \tag{5}$$

More precisely, in the case of the human knowledge acquisition process, experts are not disposed to formulate the exact value of the β, but he (or she) rather prefers to give the interval for its value

$$\underline{\beta} \leq \beta \leq \overline{\beta} \,. \tag{6}$$

The analysis of different practical examples leads to the following form of rule $r_i^{(k)}$:

IF $x \in D_i^{(k)}$
THEN `state of object is` i
WITH `posterior probability` $\beta_i^{(k)}$ `greater than` $\underline{\beta}_i^{(k)}$ `and less than` $\overline{\beta}_i^{(k)}$,

where

$$\beta_j^{(k)} = \int_{D_i^{(k)}} p(i|x)dx \tag{7}$$

The graphical interpretation of the *posterior* probability estimator for the decision area given by rule is depicted on Fig. 1.

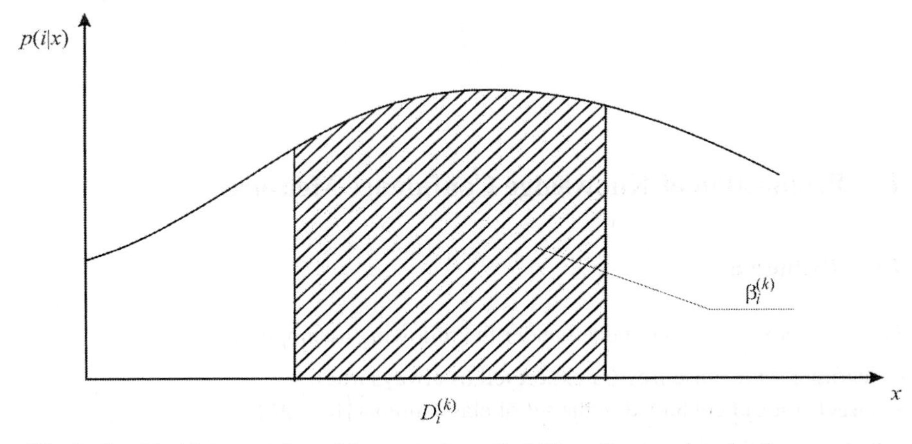

Fig. 1. Graphical interpretation of the *posterior* probability estimator given by the expert rule

For that form of knowledge we can formulate the decision algorithm $\Psi_R(x)$

$$\Psi_R(x) = i \text{ if } \hat{p}(i|x) = \max_k \hat{p}_k(i|x), \tag{8}$$

where $\hat{p}(i|x)$ is the *posterior* probability estimator obtained from the rule set.

The knowledge about probabilities given by expert estimates the average *posterior* probability for the whole decision area. As we see for decision making we are interested in the exact value of the *posterior* probability for given observation.

Lets note the rule estimator will be more precise if:

- rule decision region will be smaller,
- differences between upper and lower bound of the probability given by expert will be smaller.

For the problem under consideration definition of the relation "more specific" between the probabilistic rules pointed at the same classes is very useful.

Definition

Rule $r_i^{(k)}$ is "more specific" than rule $r_i^{(l)}$ if

$$\left(\overline{\beta}_i^{(k)} - \underline{\beta}_i^{(k)}\right)\left(\frac{\int_{D_i^{(k)}} dx}{\int_X dx}\right) < \left(\overline{\beta}_i^{(l)} - \underline{\beta}_i^{(l)}\right)\left(\frac{\int_{D_i^{(l)}} dx}{\int_X dx}\right). \tag{9}$$

Hence the proposition of the *posterior* probability estimator $\hat{p}(i|x)$ is as follow.

From subset of rules $R_i(x) = \left\{r_i^{(k)} : x \in D_i^{(k)}\right\}$ choose the "most specific" rule $r_i^{(m)}$.

$$\hat{p}(i|k) = \frac{\left(\overline{\beta}_i^{(m)} - \underline{\beta}_i^{(m)}\right)}{\int_{D_i^{(m)}} dx} \tag{10}$$

4 Proposition of Knowledge Confidence Measure

4.1 Definition

This estimator $\hat{p}(i|x)$ is obtained under the following assumption:

- learning set is noise free (or expert tell us always true),
- target concept contained in the set of class number $\{1, ..., M\}$,
- the prior probabilities of classes are unknown.

We consider decision under the first assumption given by the following formulae:

$$P(\text{If A then B with probability } \beta) = 1 .$$

During the expert system designing process the rules are obtained from different sources and the sources have the different confidence. For the knowledge given by experts we can not assume that expert tell us true or/and if the rule set is generated by the machine learning algorithms we can not assume the learning set is noise free. Therefore we postulate that we have not to trust all information we got or the believe on it only with the γ factor, proposed as the confidence (quality) measure. It can be formulated as [14]

$$P(\text{If A then B with probability } \beta) = \gamma \leq 1 .$$

Lets as $\gamma_i^{(k)}$ denote the value of the confidence measure of rule $r_i^{(k)}$.

4.2 Using Confidence Measure for Rule Base Modification

We defined confidence measure. Now, lets us show how it can be utilized for the modification the set of rules. For the form of rule we described in section 3 we propose the following procedure, which should be started after the acquisition process:

for i:=1 to M

for each rule $r_i^{(k)}$ pointed at class i:

$\overline{\beta}_i^{(k)} = 1 - \left(1 - \overline{\beta}_i^{(k)}\right)\gamma_i^{(k)}$;

$\underline{\beta}_i^{(k)} = \underline{\beta}_i^{(k)}\, \gamma_i^{(k)}$

endfor.

endfor.

The proposition of the application of the confidence measure for the set of the logical rules will be shown in the next section.

4.3 Confidence Measure for the Statistical Estimation

The central problem of our proposition is how to calculate the confidence measure. For human experts the values for their rules is fixed arbitrary according to the quality of creator. The presented problem we can also find in the typical statistical estimation of unknown parameter β, where we assume the significant level[12]. The significant level can be interpreted as the confidence measure. Each rule gives the index of the class. If the feature vector value belongs to the decision area given by the rule, the decision depends on the previous state and on the applied therapy. While constructing the artificial rule set, we have to define somehow the decision areas for the new rule set. For example we can want to obtain for each rule *posterior* probability estimator, which is not less than a fixed value or in the practice we can use the one of very well known machine learning algorithms like AQ,CN2 [9].

For each of the given intervals k we have to obtain the estimator of the *posterior* probability.

We use the following statistical model [11]:

- the learning set is selected randomly from a population and there exists two class of points: marked (point at the class $i \in \{1, \ldots, M\}$) and unmarked (point at the class l, where $l \in \{1, \ldots, M\}$ and $l \neq i$),
- the expected value for the population is p,
- the best estimator of p is

$$\hat{p} = \frac{m}{n}, \tag{11}$$

where n means the sample size and m - the number of the marked elements.

Let us concentrate on two cases

Small Sample ($n \leq 100$)

$$t = \frac{\hat{p} - p}{s/\sqrt{n}} \tag{12}$$

has the Student's t-distribution. We want to estimate one parameter - the expected value, therefore we use t-distribution with n-1 degree of freedom.
For the fixed significance level α we get

$$P(|t| < \mu_\alpha) = 1 - \alpha \tag{13}$$

using the short-cut formula

$$s^2 = \frac{\sum (x_i - \hat{p})^2}{(n-1)} \approx \sum x_i^2 + \frac{\left(\sum x_i\right)^2}{n} \tag{14}$$

we obtain

$$P\left(\frac{m}{n} - \mu_\alpha \sqrt{\frac{\frac{m}{n}\left(1 - \frac{m}{n}\right)}{n}} < p < \frac{m}{n} + \mu_\alpha \sqrt{\frac{\frac{m}{n}\left(1 - \frac{m}{n}\right)}{n}} \right) \leq 1 - \alpha \tag{15}$$

The μ_α is the value of the t-distribution on n-1 degrees of freedom and for the significance level α. In this case we get rule $r_i^{(k)}$, for which confidence measure of rule $\gamma_i^{(k)}$ is given by the following equation

$$\gamma_i^{(k)} = 1 - \alpha, \tag{16}$$

$$\underline{\beta}_i^{(k)} = \frac{m}{n} - \mu_\alpha \sqrt{\frac{\frac{m}{n}\left(1-\frac{m}{n}\right)}{n}} \text{ and } \overline{\beta}_i^{(k)} = \frac{m}{n} + \mu_\alpha \sqrt{\frac{\frac{m}{n}\left(1-\frac{m}{n}\right)}{n}} \quad (17)$$

Big Sample (*n*> 100)
For the big sample, the distribution is similar to the normal distribution. We thus have the equation of the same for as (15), but in this case μ_α is the value of normal standardized N(0, 1) distribution for the significance level α.

5 Contradictions Elimination Algorithm

As we have mentioned the proposed method of the quality management can be applied to the logical knowledge representation (where "if-then" means logical implication). E.g. for the unordered set [2,9] of logical rules acquisition process we can attribute the value of confidence to each of rule. It could be used in the case if the contradiction in the set of rules would be detected.
First we note the set of rule *R* consists of the *M* subsets

$$R = R_1 \cup R_2 \cup ... \cup R_M , \quad (18)$$

where R_i denote subset of rule pointed at the *i*-th class.
For this form of rule the two of them contradict each other if

$$\begin{gathered} \exists\, x \in X \wedge \exists\, k, l \in \{1, 2, ..., M\}, k \neq l \wedge \exists\, i, j \\ x \in D_i^{(k)} \wedge x \in D_j^{(l)} \end{gathered} \quad (19)$$

where *i*, j denote the number of rule.
The equation (19) means that we can find observation, which belongs to the decision area of the rule pointed at class *i* ($R_i^{(k)}$) and decision areas of the rule pointed at different class *j* ($R_j^{(l)}$).
Let us propose the idea of the contradictions elimination algorithm.

```
//for each class number
for i:= 1 to M:
//for each rule in R_i
 for k:=1 to |R_i|
  //for each class number bigger than i
  for j:= i to M:
   //for each rule in R_j
   for l:= 1 to |R_j| :
    //if r_i^(k) and r_j^(l) contradict each other
```

```
            if D_i^(k) ∩ D_j^(l) ≠ ∅
              then
                //if confidence of r_i^(k) is higher then
                //confidence of r_j^(l)
                if γ_i^(k) ≥ γ_j^(l)
                  then
                    //remove the common part from decision
                    //area of rule r_j^(l)
                    D_j^(l) := D_j^(l) \ (D_j^(l) ∩ D_i^(k))
                  else
                    //remove the common part from decision
                    //area of rule r_j^(l)
                    D_i^(k) := D_i^(k) \ (D_j^(l) ∩ D_i^(k))
                  fi
            fi
          endfor
        endfor
      endfor
    endfor
```

6 Conclusion

The paper concerned probabilistic reasoning and the proposition of the quality measure for that formulated decision problems. We proposed the idea of the contradiction detection and elimination method for the logical representation of experts' knowledge too. We hope this idea of confidence management can be helpful for other problems whose can be met during the knowledge acquisition process from different sources. Presented ideas need the analytical and simulation researches. Let us draw some future works under the concept of the information quality:

1. developing the method how to judge the expert quality (we formulated only the method of counting the confidence measure for rules obtained via machine learning algorithms but we propose arbitrary judgment for rule given by experts),
2. applying proposed method to the real medical decision problems (work in progress),
3. analytical researches into proposed method properties,
4. performing simulation experiments on computer generated data to estimate the dependencies between the size of the decision area and the data quality versus correctness of classification,
5. developing the software for the contradictions elimination algorithm.

References

1. Bergadano F., Matwin S., Michalski R.S., Zhang J., Measuring of Quality Concept Descriptions, *Proc. Of the 3rd European Working Session on Learning*, Aberdeen, Scotland, 1988.
2. Bruha I., Kockova S., Quality of decision rules: Empirical and statistical approaches, *Informatica*, no 17.
3. Dean P., Famili A., Comparative Performance of Rule Quality Measures in an Inductive Systems, *Applied Intelligence*, no 7, 1997.
4. Devijver P.A., Kittler J., *Pattern Recognition: A Statistical Approach*, Prentice Hall, London 1982.
5. Duda R.O., Hart P.E., *Pattern Classification and Scene Analysis*, John Wiley and Sons, New York, 1973
6. Giakoumakis E., Papakonstantiou G., Skordalakis E., Rule-based systems and pattern recognition, *Pattern Recognition Letters*, No 5, 1987.
7. Gur-Ali O., Wallance W.A., Induction of rules subject to a quality constraint: probabilistic inductive learning, *IEEE Transaction on Knowledge and Data Engeineering*, vol. 5, no 3, 1993.
8. Kurzynski M., Wozniak M.: Rule-based algorithms with learning for sequential recognition problem, *Proceedings of the Third International Conference on Information Fusion "Fusion 2000"*. Paris, July 10–13, 2000.
9. Mitchell T., *Machine Learning*, McGraw Hill, 1997.
10. Pearl J., *Probabilistic Reasoning in Intelligent Systems: Networks of Plausible Inference*, Morgan Kaufmann Pub. Inc., San Francisco, California, 1991.
11. Sachs L., *Applied Statistic. A Handbook of Techniques*, Springer-Verlag, New York Berlin Heideberg Tokyo, 1984.
12. Wozniak M., Blinowska A., Unification of the information as the way of recognition the controlled Markov chains, *Proc. of the Congress on Information Processing and Management of Uncertainty in Knowledge Based Systems,* Granada, Spain 1996.
13. Wozniak M., Jackowski K., Rule-based contextual pattern recognition algorithm – parametric case. *Proceedings of International Symposium on Pattern Recognition*, Royal Military Academy. Brussels, February 12, 1999.
14. Wozniak M., Idea of Knowledge Acquisition for the Probabilistic Expert Systems, *Proceedings of the International Conference on Computational Intelligence for Modeling, Control and Automation*, Vienna (Austria), 12–14 February 2003 (be published).

Computational Science and Data Mining

Flaviu Adrian Mărginean

Department of Computer Science, The University of York
Heslington, York YO10 5DD, United Kingdom
flav@cs.york.ac.uk

Abstract. The last decade has witnessed an impressive growth of Data Mining through algorithms and applications. Despite the advances, a computational theory of Data Mining is still largely outstanding. This paper discusses some aspects relevant to computation in Data Mining from the point of view of the Machine Learning theoretician. Computational techniques used in other fields that deal with learning from data, such as Statistics and Machine Learning, are potentially very relevant. However, the specifics of Data Mining are such that most often those techniques are not directly applicable but require to be re-cast and re-analysed within Data Mining starting from first principles. We illustrate this with a PAC-learnability analysis for a Data Mining-like task. We show that accounting for Data Mining specific requirements, such as inference of weak predictors and agnosticity assumptions, requires the generalisation of the classical PAC framework in novel ways.

1 Introduction

Data Mining is a relatively recent research field, formed over the last ten years at the intersection of Database Theory, Statistics and Machine Learning [12]. It is perhaps surprising, given the three parent fields, that the theoretical foundations of Data Mining are still in the incipient development stages. Databases have been studied since the '60s, while Statistics is century-old. Machine Learning has existed as a problem since Turing's seminal work in the early '50s. All three fields have by now acquired solid theoretical foundations and can be regarded as mature fields. By way of contrast, not enough theoretical understanding exists about the nature and purposes of Data Mining, and the mere fact that ONE mythical theory of Data Mining is still being searched for is, perhaps, a sign of the immaturity of the field. The efforts to develop full-fledged foundations for Data Mining might have been hampered by the so-called reductionist approaches [14], i.e. those aiming to reduce Data Mining to one or another of its three parents mentioned above. The difficulty with these approaches seems to lie with the fact that, while some aspects of Data Mining fit neatly in either Database Theory, Statistics or Machine Learning, neither of the parents alone can provide a fully satisfactory account of all the more important characteristics of Data Mining. More problematically, even when the general methodology can be traced to one

P.M.A. Sloot et al. (Eds.): ICCS 2003, LNCS 2659, pp. 644–651, 2003.

of the parent fields, it may be the case that the relevant techniques or tools are not directly applicable, but require to be re-invented instead, within Data Mining, starting from first principles.

One such problem is the issue of learnability in Data Mining. An early preoccupation of Machine Learning theoreticians has been to assign meaning to the learning process, i.e. to build computational models of learning. A mathematical model of learning should be able to tell us something about the learning target, as well as about the computational resources needed in order to learn successfully (information, time, space). In contrast, in Data Mining one usually searches for the unknown and unexpected: *"unknown and unexpected patterns of information" (Parsaye), "previously unknown relationships and patterns within data" (Ferruzza), "valid, novel, potentially useful, and ultimately understandable patterns in data" (Fayyad), "previously unknown, comprehensible, and actionable information" (Zekulin), "unsuspected relationships in observational data sets" (Hand-Mannila-Smyth, [6]).* A direct application of computational models developed for Machine Learning is therefore not possible, for at least two reasons. First, a Data Mining task may consist of finding many weak predictors in a hypothesis space whereas in Machine Learning one strong predictor is normally sought [4]. This is the so-called *local-global* problem [13]. Secondly, Machine Learning algorithms usually make an assumption of representability of the data-generating mechanism within the hypothesis space of the learner. This is certainly not the case within Data Mining, wherein the lack of assumptions regarding the phenomenon generating the data is a matter of principle as described above. This is the so-called *agnosticity* problem [17].

The purpose of this paper is to give a computational treatment of learning in Data Mining and a PAC-learnability analysis. Such analyses have not so far been done, to our knowledge. We wish to emphasise that this is not the only possible computational treatment of learning in Data Mining. Many computational models of learning exist in the theoretical literature on Machine Learning (PAC-learning, U-learning, etc.) as well as many forms of learning (inductive, analytical, Bayesian, reinforcement, etc.). There is no a priori reason to expect the situation to be any different in Data Mining. Rather we wish to showcase a type of computational analysis, which takes account of the specifics of Data Mining as mentioned above. The paper is organised as follows. In Section 2 we provide some technical preliminaries needed in the rest of the work: a description of Valiant's PAC-learnability framework in Section 2.1; a brief description of Mitchell's agnostic learning in Section 2.2; and a presentation of the local-global problem in Section 2.3. Section 3 is the main section of the paper, containing the definitions necessary for our extended PAC framework and the main result. Finally, Section 4 concludes with a discussion and gives some directions for further work.

2 Background

2.1 PAC-Learnability Framework

We first present the PAC model of learning, generally following Mitchell [17].

In the PAC model of learning (whereof many variants exist), we are customarily given an instance space X and two subsets of the powerset of X, the concept space $\mathcal{C}$ and the hypothesis space $\mathcal{H}$: $\mathcal{C}, \mathcal{H} \subseteq 2^X$; they can equally well be thought of as spaces of Boolean-valued functions over X. These two spaces are usually implicitly defined by some representational schemes, for instance DNF or CNF formulae, first-order representations etc. However, this will not be important in our analysis. It is assumed in the classical model that $\mathcal{C} \subseteq \mathcal{H}$, i.e. any possible target is representable in the hypothesis space of the learner. This entails the possibility of finding a unique strong predictor in the hypothesis space.

In Data Mining, we no longer make this rather strong assumption. Models of learning that do not make this assumption have been described before under the name agnostic [17] or robust learning [8], however those models differ in a number of ways from the treatment that we propose here.

It is further assumed in the PAC model that a distribution $\mathcal{D}$ over X is given, unknown to the learner but fixed in advance. The purpose will be to probably approximately learn $c \in \mathcal{C}$ by querying an oracle that makes random independent draws from the distribution $\mathcal{D}$. Every time the oracle is queried by the learner, it draws an instance $x \in X$ at random according to $\mathcal{D}$ and returns the pair $\langle x, c(x) \rangle$ to the learner. An approximation $h \in H$ to $c \in \mathcal{C}$ is evaluated with respect to the distribution $\mathcal{D}$ over X: the error of the approximation is the probability that an instance from X randomly drawn according to $\mathcal{D}$ will be misclassified by h. It is required that a learner, using reasonable amounts of computational and informational resources (time, space, queries to the random oracle), output a hypothesis that with high confidence approximates the target well-enough. The use of resources such as time and space define the computational complexity of PAC learning, while the number of queries to the random oracle needed to probably approximately infer the unknown target defines the sample complexity or information-theoretic complexity of the PAC learning. It is the latter we will be concerned with in this paper.

2.2 Agnostic Learning

In agnostic learning [17, 8], no prior assumption is made to the effect that $\mathcal{C} \subseteq \mathcal{H}$. Since the process generating the data is not assumed to be representable in the hypothesis space of the learner, in agnostic learning one seeks the hypothesis with the smallest error over the training data, h^D_{best}. It is then shown using Hoeffding bounds that the true error of this hypothesis will not exceed an ϵ overhead compared to the error over the training data. This framework is not fully satisfactory and cannot be applied directly to the study of Data Mining. It does not appear clear in agnostic learning why the hypothesis h^D_{best} would be interesting for us from the point of view of learning the target c. Whilst it is guaranteed that its true error will be not much bigger than its error over the training data, it is not clear why this would be enough justification for singling out h^D_{best}. Its error over the training data might be smallest of all hypotheses, but it can still be too big for h^D_{best} to be a reasonable predictor for c. If we however accept that we are also interested in weak predictors, and not necessarily in strong predictors, it is not clear why we should pick just one of them. Rather, from Data Mining we know that we are usually interested in any number of weak predictors satisfying some quality criteria. This is the object of Section 2.3.

2.3 Local and Global Issues in Data Mining

Data Mining may be concerned [13] with one of two problems. The probabilistic modelling research tradition views Data Mining as the task of approximating a global model underlying the data in the form of a joint distribution.

The other approach can be seen as as *"the essence of Data Mining — an attempt to locate nuggets of value amongst the dross" ([5, Hand])*. The typical example is the discovery of frequently occurring patterns, wherein patterns and their associated frequencies give local properties of the data that can be understood without having information about the global mass of the data. The collection of such patterns may be used as a first step towards the global analysis of the data, whereby the collection of patterns collected in the mining phase undergoes various aggregation operations aimed at building a global description of the data. However, global approaches more genuinely belong to Machine Learning, where the assumption of a global generative process behind the data is better supported. Most prominent example of work in the local tradition is the research on association rule mining [13, 4, 1, 15]. We shall be concerned in this paper with the local problem. The trouble with building weak predictors (models) of the data, based on local information, is that of sampling. How can we make inductive leaps from training data to weak predictors when *"selecting only a sample may discard just those few cases one had hoped to detect" ([5, Hand])*? This problem, to our knowledge, has not been given a theoretical treatment so far.

3 PAC-Learnability Analysis for Mining

In this section we present our PAC framework for mining analysis, paying attention to the necessity to model both requirements specific to the mining process: *agnosticity* as described in Section 2.2 and *locality* as described in Section 2.3. Given that we no longer can rely on the assumption $\mathcal{C} \subseteq \mathcal{H}$ which would allow us to define the version space of the hypotheses consistent with the data, we have to find another way of defining interesting hypotheses from the point of view of consistency with the data. We do so by introducing the following quasi-order on $\mathcal{H}$, which is relative to the training data D and target c : $h_1 \succeq_{\mathcal{OD}(c,D)} h_2$ *iff* $\forall \langle x, c(x) \rangle \in D : h_2(x) = c(x) \Rightarrow h_1(x) = c(x)$. The order is subscripted $\mathcal{OD}$, indicating *desirability* of hypotheses in $\mathcal{H}$ (or Order of Desire).

It is trivial to verify that the quasi-order axioms (transitivity, reflexivity) are satisfied. The notion of Version Space $VS_{\mathcal{H},D}$ [17] generalises as follows:

$$VS_{\mathcal{H},D} \stackrel{\text{def}}{=} \{h \in H \mid h|_D = c|_D\}$$

$$SVS_{\mathcal{H},D} \stackrel{\text{def}}{=} \{h \in H \mid \nexists h' \in \mathcal{H} \textit{ such that } h' \succ_{\mathcal{OD}(c,D)} h\}$$

We call the generalised Version Space, Soft Version Space $SVS_{\mathcal{H},D}$. The word *soft* we use to indicate the graceful degradation of hypotheses with respect to consistency. With classical Version Spaces, hypotheses are classified in a crisp manner: hypotheses are either in (consistent) or out (inconsistent) of the Version Space. Soft Version Spaces $SVS_{\mathcal{H},D}$ retain the hypotheses maximally consistent with the data D, but the gap between them and the hypotheses left out of the version space is not as dramatic as in the classical case. Rather, consistency comes in degrees and there may be all sorts of shades, i.e. hypotheses that are more or less consistent with the data D according to the order $\mathcal{OD}(c, D)$. When $D = X$ we denote the Soft Version Space by $SVS_{\mathcal{H},c}$, the set of hypotheses in $\mathcal{H}$ maximally consistent with c over the entire instance space X. Let $\sim_{\mathcal{OD}(c,D)}$ be the equivalence relationship on $\mathcal{H}$ canonically induced by the quasi-order $\mathcal{OD}(c, D)$:

$$h_1 \sim_{\mathcal{OD}(c,D)} h_2 \;\; \textit{iff} \;\; h_1 \succeq_{\mathcal{OD}(c,D)} h_2 \textit{ and } h_2 \succeq_{\mathcal{OD}(c,D)} h_1$$

For the case $c \in \mathcal{H}$, the partial order $\mathcal{OD}(c, D)/\sim_{\mathcal{OD}(c,D)}$ on $\mathcal{H}/\sim_{\mathcal{OD}(c,D)}$ becomes the boolean order $B_1 \stackrel{\text{def}}{=} (0 < 1)$, with the subclass of consistent hypotheses corresponding to 1 and the subclass of inconsistent hypotheses corresponding to 0. Therefore, in this case the Soft Version Space becomes the set of consistent hypotheses, i.e. it reduces to the classical black-and-white definition of the Version Space $VS_{\mathcal{H},D}$. The following theorem establishes the sample complexity of "soft learning" for maximally consistent learners[1].

Theorem 1 (Soft Version Spaces are $\epsilon, \delta-$Exhaustible). *Let $\mathcal{C}, \mathcal{H} \subseteq 2^X$ be a concept space and a hypothesis space respectively, and let $VC(\mathcal{H}) < \infty$ be the finite Vapnik-Chervonenkis dimension of $\mathcal{H}$. For all $0 < \epsilon, \delta < 1$, for all $D \subseteq (X \times \{0,1\})^m$ training data such that $m \geq \frac{1}{2\epsilon^2}(4\log_2(2/\delta) + 8\,VC(\mathcal{H})\log_2(13/\epsilon))$ and for all $h \in SVS_{\mathcal{H},c}$ there is $h' \in SVS_{\mathcal{H},c}$ and $h'' \in SVS_{\mathcal{H},D}$ such that, with probability at least $1 - \delta$, $error(h', c) \leq error(h, c) + \epsilon$ and $error(h'', h') < \epsilon$.*

Proof Outline. For reasons of space we only indicate the main steps of the proof's theorem. The first step is to show that for all $h \in SVS_{\mathcal{H},c}$ there is $h' \in SVS_{\mathcal{H},c}$ such that, with probability at least $1 - \delta$, $error(h', c) \leq error(h, c) + \epsilon$. This is done by restricting h to D and choosing h' such that $h' \succeq_{\mathcal{OD}(c,D)} h$ (one such maximal element with respect to $\mathcal{OD}(c, D)$ is bound to exist). It is also possible to show that h' can be so chosen that not only does $h' \in SVS_{\mathcal{H},D}$ but also $h' \in SVS_{\mathcal{H},c}$. However such a choice will necessarily be non-constructive. h'

[1] Compare with similar results in [17] regarding the complexity of PAC learning for consistent learners.

will do at least as well as h on the training data; therefore, by using Hoeffding bounds, we can bound with high confidence the true error of h' versus the true error of h. The second step involves showing that h' is probably approximable by an $h'' \in SVS_{\mathcal{H},D}$ with high confidence. This is essentially done by taking h' as a target, showing that there are elements in $SVS_{\mathcal{H},D}$ that are consistent with h' on the training data D, and applying the classical result of PAC-learnability from [17]. Unlike in the case of h', such elements can be chosen effectively, in effect as any $h'' \succeq_{\mathcal{OD}(c,D)} h$. Fewer examples would be needed for the second step, only $\frac{1}{\epsilon}(4\log_2(2/\delta) + 8\,VC(\mathcal{H})\log_2(13/\epsilon))$ according to the classical result for Version Spaces. However, this would only guarantee that elements in $SVS_{\mathcal{H},D}$ probably approximate elements in $SVS_{\mathcal{H},c}$, rather than ensuring that all elements in $SVS_{\mathcal{H},c}$ are probably approximately learned in the reasonable sense described by the theorem's statement. ∎

4 Discussion and Further Work

A learnability analysis for Data Mining has been presented within the general methodology of Valiant's PAC-learning framework. The analysis shows that this type of analysis is feasible and that the process of mining is meaningful: the weak predictors $SVS_{\mathcal{H},D}$ inferred by maximal consistency from a polynomial sample have predictive power in a well-defined way. Moreover, the collection of all these predictors approximates the true target $SVS_{\mathcal{H},c}$, thereby collectively giving some global information about the data — just as practitioners of Data Mining would expect [13]. As far as the author is aware, this is the first learnability analysis for a Data Mining type of task.

There are various ways in which this work can be extended. First, we have investigated learnability with respect to only one resource, i.e. sample size. A mining algorithm will need to behave well not only in respect of the informational resources it requires (sample complexity), but also from the point of view of the hardware and time requirements (computational complexity). In other words, we have proved in this paper that any maximally consistent learner is effective, but there remains to be proved that there are such learners for given hypotheses spaces that are also efficient. Secondly, other models of learning may also be considered, as the PAC learning, although most common, is far from being the only learning model. It is, however, the simplest and best understood and it is the model of choice for first-time analyses of an inductive learning process [2, 10, 7].

We see the importance of this paper in conveying a principle: computational analyses of Data Mining problems and algorithms are possible, provided one re-develops the relevant techniques in the specific context of Data Mining rather than attempting a blind translation of techniques from other fields that deal with learning from data. Furthermore, the computational setting in this paper suggests new algorithms based on maximal consistency computation.

Traditionally, the APRIORI algorithm and its variants [1, 15] have handled the boolean inductive query evaluation problem with respect to single monotonic constraints (e.g. minimum frequency). More recently, extensions have been proposed [3] that combine APRIORI with data structures based on VERSION SPACES in order to evaluate boolean inductive queries defined as conjunctions of both monotonic and anti-monotonic constraints. Those formalisms work well when the boolean inductive query evaluates to a non-empty set. However, natural computational settings exist wherein this is not a reasonable expectation for most of the queries. It is clear, in such settings, that the database still possesses some structure and some answers are better than others; for instance, one answer may satisfy more constraints in the inductive query than another. In cases where the classical formalism returns an empty answer set, we would instead be interested in computing those answers that most closely satisfy the inductive query. This problem requires the extension of the classical APRIORI formalism at both the conceptual and algorithmic level. In this paper we defined a framework entitled SOFT VERSION SPACES that can be viewed as describing the optimal 'soft match' $SVS_{\mathcal{H},D}$ between a language of patterns $\mathcal{H}$, herein more generally regarded as a set of hypotheses, and an inductive query D consisting of a conjunction of monotonic and anti-monotonic constraints over $\mathcal{H}$, herein more restrictedly viewed as the conjunction of the *h-membership* relations for positive and negative *c*-examples, respectively. This can be shown to generalise the classical APRIORI-based formalism in a natural way. The development of this idea and of a suitable SOFT-APRIORI algorithm for computing $SVS_{\mathcal{H},D}$ are topics for a future paper.

Acknowledgements.
The author thanks the anonymous reviewers for their comments. My mother, Maria Adriana (Puşi), and brother, Emil Raul, have provided moral support and encouragement during the writing of this paper. In loving memory of my departed father, Emil.

References

1. R. Agrawal, H. Mannila, R. Srikant, H. Toivonen, and A. Verkamo. Fast Discovery of Association Rules. In U. Fayyad, G. Piatetsky-Shapiro, P. Smyth, and R. Uthurusamy, editors, *Advance in Knowledge Discovery and Data Mining*, pages 307–328. AAAI Press/MIT Press, 1996.
2. M. Anthony and N. Biggs. *Computational Learning Theory*. Cambridge University Press, 1997.
3. L. De Raedt, M. Jaeger, S. Lee, and H. Mannila. A Theory of Inductive Query Answering. In *Proceedings of the 2002 IEEE International Conference on Data Mining (ICDM'02)*, Maebashi, Japan, December 9–12 2002. Extended abstract.
4. D. Gunopulos, H. Mannila, R. Khardon, and H. Toivonen. Data Mining, Hypergraph Transversals, and Machine Learning (extended abstract). In *Proceedings of the 16th ACM SIGACT-SIGMOD-SIGART Symposium on Principles of Database Systems*, pages 209–216, Tucson, Arizona, USA, 1997. ACM Press.

5. D.J. Hand. Statistics and Data Mining: Intersecting Disciplines. *ACM SIGKDD Explorations*, 1(1):16–19, 1999.
6. D. Hand, H. Mannila, and P. Smyth. *Principles of Data Mining.* MIT Press, 2001.
7. D. Haussler. Quantifying inductive bias: AI learning algorithms and Valiant's learning framework. *Artificial intelligence*, 36:177 – 221, 1988.
8. D. Haussler, S. Ben-David, N. Cesa-Bianchi, and P. Long. Characterizations of Learnability for Classes of {0,...,n}-valued Functions. *J. Comp. Sys. Sci.*, 50(1):74–86, 1995.
9. H. Hirsh. *Incremental Version Space Merging: A General Framework for Concept Learning.* Kluwer, 1990.
10. Michael J. Kearns and Umesh V. Vazirani. *An Introduction to Computational Learning Theory.* The MIT Press, 1994.
11. N. Lavrač, D. Gamberger, and V. Jovanoski. A Study of Relevance for Learning in Deductive Databases. *Journal of Logic Programming*, 40(2/3):215–249, 1999.
12. H. Mannila. Data mining: machine learning, statistics, and databases. In *Proceedings of the Eighth International Conference on Scientific and Statistical Database Management*, pages 1–8, Stockholm, June 18–20 1996.
13. H. Mannila. Local and Global Methods in Data Mining: Basic Techniques and Open Problems. In *Proceedings of ICALP 2002, 29th International Colloquium on Automata, Languages, and Programming*, Malaga, Spain, July 2002. Springer.
14. H. Mannila. Theoretical Frameworks for Data Mining. *ACM SIGKDD Explorations*, 1(2):30–32, January 2000.
15. H. Mannila and H. Toivonen. Levelwise Search and Borders of Theories in Knowledge Discovery. *Data Mining and Knowledge Discovery*, 1(3):241–258, 1997.
16. T.M. Mitchell. *Version Spaces: An Approach to Concept Learning.* PhD thesis, Electrical Engineering Department, Stanford University, 1979.
17. T.M. Mitchell. *Machine Learning.* McGraw-Hill, New York, 1997.
18. L. Pitt and L. G. Valiant. Computational limitations on learning from examples. *Journal of the ACM*, 35(4):965–984, 1988.

Data Sharing Model for Sequence Alignment to Reduce Database Retrieve*

Min Jun Kim[1], Jai-Hoon Kim[1], Jin-Won Jung[2], and Weontae Lee[2]

[1] The Graduate School of Information and Communication, Ajou University
Suwon, 442-749, Republic of Korea
{xwind,jaikim}@ajou.ac.kr

[2] HTSD-NMR Laboratory, Department of Biochemistry
Yonsei University, Seoul 120-749, Republic of Korea
{solwind,wlee}@spin.yonsei.ac.kr

Abstract. Many programs of bioinformatics provide biochemists and biologists retrieve and analysis services of gene and protein databases. These services access databases for each arrival of user's request, which takes a long time, increases server's overload and response time. In this paper, two data sharing models are presented to reduce database retrieve for sequence alignments. (1) Grouping model is proposed to share database access between many requests arrived in a certain period of time by utilizing database access patterns in bioinformatics. (2) Carpool model is also proposed to reduce response time as well as to increase system throughput by servicing new arriving request immediately together with the previous on going requests to share database accesses without waiting time. Simulation results show that two data sharing models can reduce the number of database access by sharing it among many requests and increase system throughput.

1 Introduction

In the early 21st century, success in human gene project accelerates development in bio-science technology. In the Postgenome era spread by completion of the human genetic map, researches about structures and functions of proteins and discovery of genes will be performed a lot. These researches produce huge quantity of gene data. Gene data is expressed by four characters (A, T, G, C). Data more than three billion DNA bases are accumulated in bio-database, many of which can be accessed via Web. There are many popular bio-databases such as SwissProt [6], GenBank [7], EMBL [8] etc.

Whole records in database are retrieved in many bioinformatics applications such as sequence alignments and structure comparisons. The order of data retrieve does not affect on the results in many applications. We propose new data sharing models to

* This study was supported by the Brain Korea 21 Project in 2003, a grant of the International Mobile Telecommunications 2000 R&D Project, Ministry of Information & Communication, Republic of Korea, and the Ministry of Science and Technology of Korea / the Korea Science and Engineering Foundation through the NRL program of MOST NRDP (M1-0203-00-0020).

P.M.A. Sloot et al. (Eds.): ICCS 2003, LNCS 2659, pp. 652–661, 2003.

increase performance of applications that access bio-databases in this characteristic. First, grouping model is proposed, which accesses database records once and process several requests together by gathering many users' requests arrived in a scheduled period. Grouping model can decrease the number of database access and reduce total system cost. Second, carpool model is proposed to avoid waiting time until the end of period for grouping users' requests unlike the grouping model. New arrival of request is combined with the previous on going requests at any time and shares the database accesses in the carpool model, while new request has to wait until the end of grouping period in the grouping model. Carpool model can reduce an average response time for each user as well as the number of total database accesses.

2 Traditional Model

Many applications of bioinformatics perform gene sequence alignments and protein structure comparisons between genes and proteins. These applications and gene and protein databases are provided on the Web. In such an environment, users send gene or protein sequences through the Web. Then, programs on server such as FastA and Blast read known sequences in databases and compare them with sequences contained in users' requests.

2.1 Traditional Architecture

Many programs used for sequence alignments or structure comparisons retrieve the whole records in gene databases such as SwissProt [6], GenBank [7], EMBL [8], etc. Such programs as FastA and Blast must respond to each user's request after retrieving whole records in database and comparing each of records with user's sequence.

In traditional model, new arrival of request is immediately processed without waiting time when system is idle. However, when the previous request is processing, new request is registered to a queue in the order of arrival, then processed only after all the previous requests are serviced.

2.2 Cost Analysis

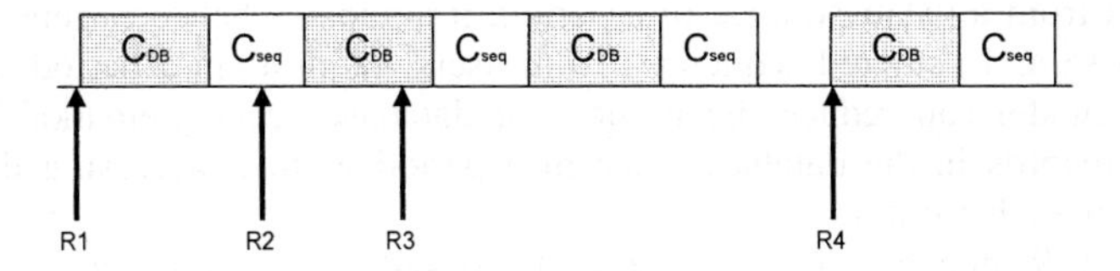

Fig. 1. Traditional program model

Figure 1 shows the timing diagram to demonstrate arrival and servicing the users' requests, R1 to R4. C_{DB} denotes an assumed cost for database retrieve for all records. Also, C_{seq} denotes an assumed cost for comparison between gene sequence from a user request and all gene sequences from a database.

An average cost to process a user request for a gene sequence alignment is the sum of database retrieve cost (C_{DB}) and comparison cost (C_{seq}) between gene sequences. Thus, the average cost for a request, C^{org}_{avg}, can be computed as follows:

$$C^{org}_{avg} = C_{DB} + C_{seq} \tag{1}$$

It is assumed that the arrival of user request is Poisson process with an arrival rate λ. User's request is registered to a queue by the order of arrival, and processed sequentially. It is M/G/1 queuing model. The service rate is $1/C^{org}_{arg}$. Using M/G/1 queuing model, response time can be derived as follows:

$$w_O = C^{org}_{avg} + \frac{\lambda {C^{org}_{avg}}^2}{2(1-\lambda \cdot C^{org}_{avg})} \tag{2}$$

3 Grouping Model

Problem in the traditional model is retrieving the whole records in a database for a sequence alignment on each arrival of user's request. Bioinformatics programs like FastA and Blast have to access whole records in the database repeatedly to service many users' requests. Grouping model reduces the number of database access by grouping users' requests and sharing the database access. Thus, it can reduce a total system cost and increase system throughput.

3.1 Grouping Architecture

Grouping model gathers users' requests and processes together periodically without accessing a database every time on each user's request. Grouping model retrieves sequences from a database once in a period, it compares between gene sequences in the database and gathered sequences from users' requests in a period. By doing so, grouping model can reduce the number of database access remarkably because it retrieves records in the database once in a period without accessing database individually for each request.

In Fig. 2, Rj denotes j-th users' requests arrived in the i-th period. Users' requests are processed after the end of i-th period. For grouping model, user's request may be delayed up to the length of a period (D) even if there is no other request in the period for grouping. But, the period can be adjusted according to the arrival rate of user request. When the arrival rate of user request is high, grouping model can reduce the number of database access remarkably.

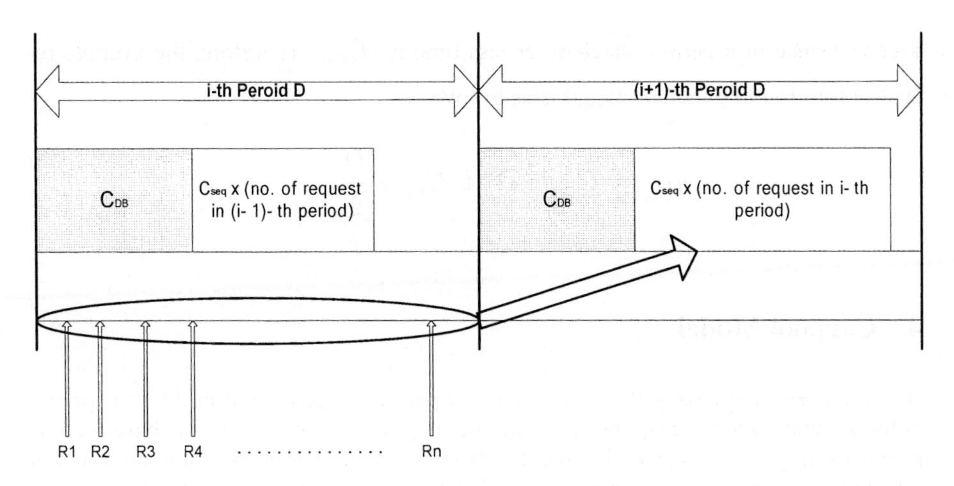

Fig. 2. Processing in grouping model

3.2 Cost Analysis

When the arrival of user request is governed by Poisson process with an arrival rate λ, the average number of user request arrived during a period (D) is $D \cdot \lambda$. Cost for comparison between gene sequences gathered from users' requests in a period and gene sequences in a database is $D \cdot \lambda \cdot C_{seq}$. Cost for retrieving whole records in a database is assumed to C_{DB}. Thus, total system cost per period is as follows:

$$C^{grp}_{total} = C_{DB} + D \cdot \lambda \cdot C_{seq} \tag{3}$$

An average cost to service a request is as follows:

$$C^{grp}_{\arg} = \frac{C_{DB}}{D \cdot \lambda} + C_{seq} \tag{4}$$

The probability of the arrival of at least one user's request during a period (D) is $1 - e^{-\lambda D}$. Because the database is accessed only when there is a request during a period (D), an average cost to service a user for grouping model is as follows:

$$C^{grp}_{avg} = \frac{C_{DB}}{D \cdot \lambda} \cdot \left(1 - e^{-\lambda D}\right) + C_{seq} \tag{5}$$

Also, the average response time is the sum of retrieving all records in database and comparing them with a user request, and an average waiting time ($D/2$) until the end of period. The average number of user requests arrived during a period is $D \cdot \lambda$. Thus, time for processing data is $D \cdot \lambda \cdot C_{seq}$ in a period (D). Because database is accessed once in a period, database access time is C_{DB}. Therefore, the average response for grouping model (w_{grp}) is as follows:

$$w_{grp} = C_{DB} + D \cdot \lambda \cdot C_{seq} + \frac{D}{2} \tag{6}$$

4 Carpool Model

Because users' requests gathered in a current period are processed in the next period in the grouping model, response time can be long. Retrieving whole database records is general in gene sequence alignment. Also, the result of sequence alignment is not affected on the order of sequence analyzed. Thus, we propose a model to process each user's request immediately by combining with the previously arrived requests to share and reduce database accesses without waiting until the end of period unlike the grouping model.

4.1 Carpool Architecture

In a carpool model, user's request is processed immediately with other requests being processed. When a server retrieves one record (sequence) from a sequence database, new request is processed together with the previously arrived requests.

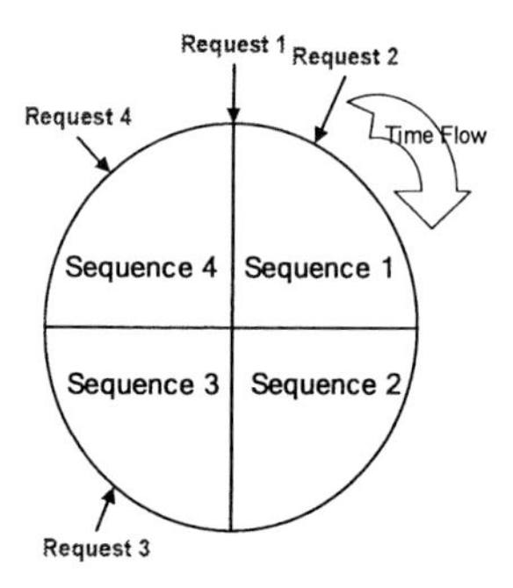

Fig. 3. Database access

Figure 3 shows an example. In the example, we assume a sequence database has four sequences (records), Sequence 1 to 4. Sequence alignment program, such as FastA or Blast, compares the gene sequence from a request with all gene sequences in a database in the order of Sequence 1, 2, 3, and 4. Figure 3 shows that other users' requests, Request 2, Request 3, and Request 4, are arrived during the service for Request 1.

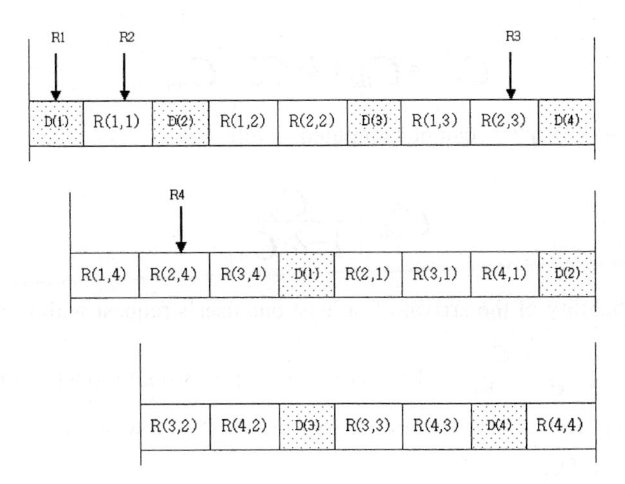

Fig. 4. Processing order in carpool mechanism

Figure 4 shows service procedures for the carpool model for four users' requests, Request 1 to 4. Database has four sequences, as shown in Fig.3. D(i) denotes a cost for retrieve the i-th sequence from a database, and R(i, j) denotes comparing between Sequence j and a sequence in user Request i.

Sequence 1 is retrieved (D(1)) and compared with Request 1 (R(1, 1)). Because Request 2 is arrived while processing Request 1, Sequence 2 is retrieved (D(2)) and compared with both requests, Request 1 (R(1,2)) and Request 2 (R2(2,2)). Database access can be reduced because sequence is accessed once and Request 1 and Request 2 are processed together. Request 1 is finished after retrieve the Sequence 4 (D(4)) and process (R(1,4)). After Sequence 4 is read (D(4)) from a database and processed for three requests, Request 1 (R(1,4)), Request 2 (R(2,4)), Request 3 (R(3,4)), Sequence 1 is retrieve (D(1)) to service Request 2, Request 3 and Request 4 because three requests have not been compared with Sequence 1. User's request of carpool model is processed immediately from the next retrieval of database record until all database records are retrieved and compared with the request sequence. Thus, in a carpool model, a new request does not have to wait unlike the grouping mechanism.

4.2 Cost Analysis

Total cost for sequence analysis adopting carpool model can be computed from the database retrieve time and processing time. The arrival rate of user's request is assumed to λ governed by Poisson process. Time to retrieve all records in a database and process (compare) them with a sequence of user's request is assumed to C^{cp}_{total}. The average number of request is $\lambda \cdot C^{cp}_{total}$ during interval C^{cp}_{total}. All records in database is retrieved once (C_{DB}) during the interval. Thus, total cost during the interval is computed as follows: (C_{seq} is the same as the cost analysis in Sect. 2.2)

$$C_{total}^{cp} = C_{DB} + \lambda \cdot C_{total}^{cp} \cdot C_{seq} \tag{7}$$

From formula (7), C_{total}^{cp} can be computed as follows:

$$C_{total}^{cp} = \frac{C_{DB}}{1 - \lambda \cdot C_{seq}} \tag{8}$$

The probability of the arrival of at least one user's request within interval C_{total}^{cp} is $1 - e^{-\lambda C_{total}^{cp}} = 1 - e^{-\lambda \frac{C_{DB}}{1-\lambda \cdot C_{seq}}}$. The average cost of a carpool model can be computed as follows: (We derive analysis similar to formula (5), however use interval C_{total}^{cp} instead of period D.)

$$C_{total}^{cp} = \frac{C_{DB}}{C_{total}^{cp} \cdot \lambda} \cdot \left(1 - e^{-\lambda \cdot C_{total}^{cp}}\right) + C_{seq} = \left(\frac{1}{\lambda} - C_{seq}\right) \cdot \left(1 - e^{-\lambda \cdot \frac{C_{DB}}{1-\lambda C_{seq}}}\right) + C_{seq} \tag{9}$$

Response time (w_{cp}) of carpool model is the same as the sum of time to process user's request ($C_{DB} + C_{seq}$) and time to process other users' requests ($\lambda \cdot w_{cp} \cdot C_{seq}$) arrived during the response time (w_{cp}).

$$w_{cp} = C_{DB} + C_{seq} + \lambda \cdot w_{cp} \cdot C_{seq} \tag{10}$$

From Eq. (10), w_{cp} can be computed as follows:

$$w_{cp} = \frac{C_{DB} + C_{seq}}{1 - \lambda \cdot C_{seq}} \tag{11}$$

5 Performance Evaluation

For performance evaluation, C_{DB} and C_{seq} is measured by executing a real sequence alignment program FastA(version 3) accessing Genbank(Release 72.02) database, and cytochrome as sequence of user request on Pentium 4 system (1.6GHz, 256 Mbyte Ram (pc2700)). C_{DB} and C_{seq} is 3.99 sec. and 19.98 sec., respectively, from the experiment. Cytochrome is a gene acting in cellular redox of human protein.

5.1 Cost Analysis

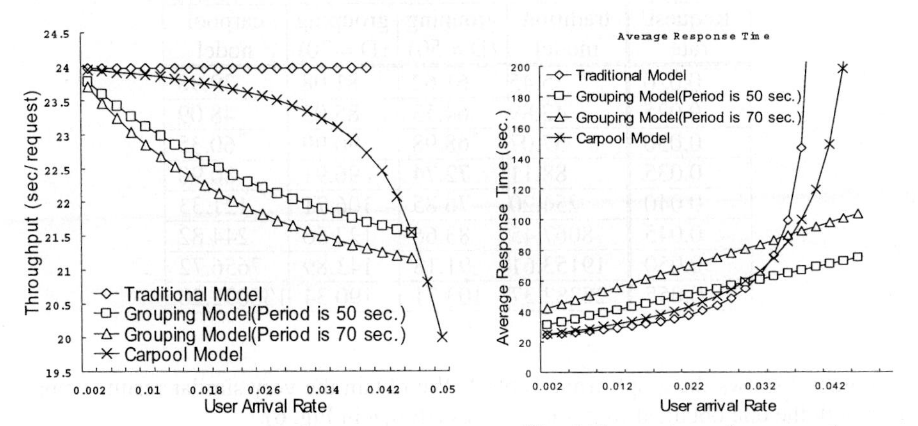

Fig. 5. Comparison of throughput (Average time per request)

Fig. 6. Comparison of response time

Figure 5 depicts the average cost comparisons between traditional model, grouping model and carpool models. Period is 50 sec. and 70 sec. for grouping model. X axis is the rate of user request (λ) and y axis is system cost per user request. For grouping model and carpool model, system cost is decreased when λ value increases by sharing a database retrieve. Also, for grouping model, system cost is reduced as D increases because more requests are sharing the database access. We draw curves up to the value of λ where the system can admit the rate of arrival. Carpool model can admit the highest request rate (up to 0.05) while the traditional model accommodates the lowest (up to 0.04).

Figure 6 shows response time for different models. X axis is request rate (λ) and y axis is an average response time for user's request. Because a request of traditional model is accumulated in a queue, response time is increased rapidly as request rate is increased. Response time of carpool model is shorter than that of grouping model at the low request rate because each request is immediately processed without grouping time unlike the grouping model.

5.2 Simulation

We simulated the three models to verify the analysis in Sects. 2.2, 3.2, and 4.2. C_{DB} and C_{seq} are set to 3.99 sec. and 19.98 sec., respectively. We simulated grouping model with 2000 requests, and obtain an average response time. Also, carpool model is simulated with 60000 requests. The results of simulation are shown in Table 1.

Table 1. Response time (seconds)

Request rate	tradition model	grouping (D = 50)	grouping (D = 70)	carpool nodel
0.020	35.45	61.62	81.08	39.98
0.025	42.89	64.53	85.05	48.09
0.030	57.07	68.98	90.99	60.35
0.035	88.11	72.74	96.91	80.90
0.040	254.20	76.85	106.71	121.33
0.045	8067.45	83.66	122.56	244.82
0.050	19153.61	91.18	143.89	7656.72
0.055	28282.37	103.71	190.34	130092.20

Figure 7 shows the graph from Table 1. We obtain the very similar results comparing with the one obtained from the analysis (shown in Fig. 6).

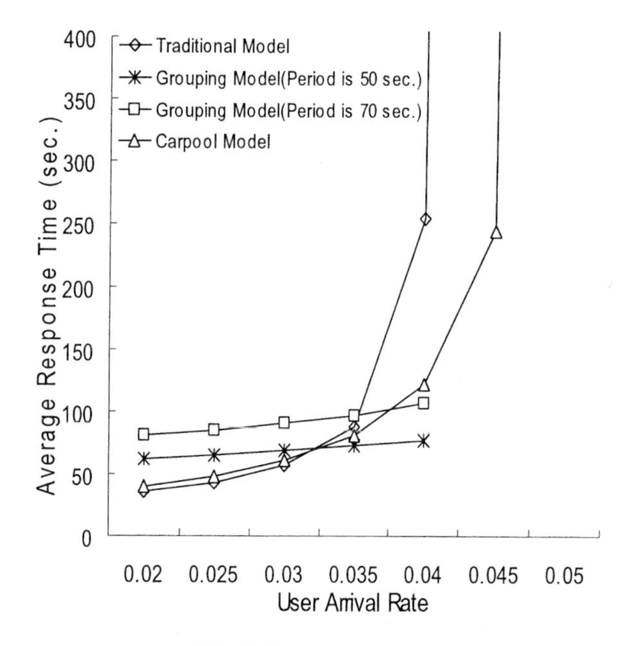

Fig. 7. Response time

We can know that response time of traditional model increases rapidly as the request rate increases. Grouping model shows the stable response time even if request rate increases. Carpool model has shorter response time than traditional model because each user's request is processed immediately and by sharing the database access.

6 Conclusion

There are huge amount of data in gene or protein databases. To access such database efficiently, we propose two data sharing models, grouping and carpool models. Grouping model processes set of users' requests gathered in a period. By processing many users' requests together, database access can be shared, thus the number of database access can be reduced. Carpool model shares database access between requests and processes user's request immediately without waiting unlike the grouping model.

By these data sharing models, bioinformatics programs can retrieve database efficiently, reduce database access and admit higher rate of user request. Results from analysis and simulation show that these data sharing models can service more users as well as commit a shorter response time than a traditonal model.

References

[1] Ozden, B., R. Rastogi, and A. Silberschatz, and C. Martin "Demand Paging for Video on Demand Servers," IEEE International Conference on Multimedia Computing and Systems, May 1995.

[2] Stephen F. Altschul, Warren Gish, Webb Miller, Eugene W. Myers and David J. Lipman, "Basic Local Alignment Search Tool," J. Mol. Biol., pp. 403–410, 1990.

[3] Tieng K. Yap, Ophir Frieder, and Robert L. Martino, "Parallel Homologous Sequence Searching in Large Databases," IEEE Proceedings of the Fifth Symposium on the Frontiers of Massively Parallel Computation, 1995.

[4] Pearson, W.R., and Lipman, D. "Improved Tools for Biological Sequence Comparison," Proc. National Academic Science, USA, pp. 85, 1988.

[5] Jerry Banks, John S. Carson, II, Barry L. Nelson "Discrete-Event System Simulation," Prentice Hall, 2000.

[6] SWISSPROT: http://www.expasy.org/

[7] GENBANK: http://www.ncbi.nlm.nih.gov/Genbank/

[8] EMBL: http://www.ncbi.nlm.nih.gov/Genbank/

Workshop on Parallel Linear Algebra (WoPLA03)

Self-Adapting Software for Numerical Linear Algebra Library Routines on Clusters*

Zizhong Chen[1], Jack Dongarra[1], Piotr Luszczek[1], and Kenneth Roche[1]

Computer Science Department, University of Tennessee Knoxville, 1122 Volunteer Blvd., Suite 203, Knoxville, TN 37996-3450, U.S.A. dongarra@cs.utk.edu

Abstract. This article describes the context, design, and recent development of the LAPACK for Clusters (LFC) project. It has been developed in the framework of Self-Adapting Numerical Software (SANS) since we believe such an approach can deliver the convenience and ease of use of existing sequential environments bundled with the power and versatility of highly-tuned parallel codes that execute on clusters. Accomplishing this task is far from trivial as we argue in the paper by presenting pertinent case studies and possible usage scenarios.

1 Introduction

Driven by the desire of scientists for ever higher levels of detail and accuracy in their simulations, the size and complexity of required computations is growing at least as fast as the improvements in processor technology. Scientific applications need to be tuned to extract near peak performance even as hardware platforms change underneath them. Unfortunately, tuning even the simplest real-world operations for high performance usually requires an intense and sustained effort, stretching over a period of weeks or months, from the most technically advanced programmers, who are inevitably in very scarce supply. While access to necessary computing and information technology has improved dramatically over the past decade, the efficient application of scientific computing techniques still requires levels of specialized knowledge in numerical analysis, mathematical software, computer architectures, and programming languages that many working researchers do not have the time, the energy, or the inclination to acquire. With good reason scientists expect their computing tools to serve them and not the other way around. And unfortunately, the growing desire to tackle highly interdisciplinary problems using more and more realistic simulations on increasingly complex computing platforms will only exacerbate the problem. The challenge for the development of next generation software is the successful management of the complex computing environment while delivering to the scientist the full power of flexible compositions of the available algorithmic alternatives and candidate hardware resources.

* This work is partially supported by the DOE LACSI - Subcontract #R71700J-29200099 from Rice University and by the NSF NPACI – P.O. 10181408-002 from University of California Board of Regents via Prime Contract #ASC-96-19020.

P.M.A. Sloot et al. (Eds.): ICCS 2003, LNCS 2659, pp. 665–672, 2003.

With this paper we develop the concept of Self-Adapting Numerical Software (SANS) for numerical libraries that execute in the cluster computing setting. The central focus is the LAPACK For Clusters (LFC) software which supports a serial, single processor user interface, but delivers the computing power achievable by an expert user working on the same problem who optimally utilizes the resources of a cluster. The basic premise is to design numerical library software that addresses both computational time and space complexity issues on the user's behalf and in a manner as transparent to the user as possible. The software intends to allow users to either link against an archived library of executable routines or benefit from the convenience of prebuilt executable programs without the hassle of properly having to resolve linker dependencies. The user is assumed to call one of the LFC routines from a serial environment while working on a single processor of the cluster. The software executes the application. If it is possible to finish executing the problem faster by mapping the problem into a parallel environment, then this is the thread of execution taken. Otherwise, the application is executed locally with the best choice of a serial algorithm. The details for parallelizing the user's problem such as resource discovery, selection, and allocation, mapping the data onto (and off of) the working cluster of processors, executing the user's application in parallel, freeing the allocated resources, and returning control to the user's process in the serial environment from which the procedure began are all handled by the software. Whether the application was executed in a parallel or serial environment is presumed not to be of interest to the user but may be explicitely queried. All the user knows is that the application executed successfully and, hopefully, in a timely manner.

2 Related Efforts

Since the concept of self-adaptation appeared in the literature [27] it has been successfully applied in a wide range of projects. The ATLAS [31] project started as a "`DGEMM()` optimizer" [13] but continues to successfully evolve by including tuning for all levels of Basic Linear Algebra Subprograms (BLAS) [7, 8, 10, 9] and LAPACK [2] as well by making decisions at compilation and execution time. Functionality similar to ATLAS, but much more limited, was included in the PHiPAC [6] project. Iterative methods and sparse linear algebra operations are the main focus of numerous efforts. Some of them [30, 3] target convergence properties of iterative solvers in a parallel setting while others [1, 21, 20, 28, 26] optimize the most common numerical kernels or provide intelligent algorithmic choices for the entire problem solving process [5, 24]. In the area of parallel computing, researchers are offering automatic tuning of generic collective communication routines [29] or specific collectives as in the HPL project [12]. Automatic optimization of the Fast Fourier Transform (FFT) kernel has also been under investigation by many scientists [19, 18, 23]. In grid computing environments [17], holistic approaches to software libraries and problem solving environments such as defined in the GrADS project [4] are actively being tested. Proof of concept efforts on the grid employing SANS components exist [25] and have helped in forming the approach followed in LFC.

3 LAPACK for Clusters Overview

The LFC software addresses the motivating factors from the previous section in a self-adapting fashion. LFC assumes that only a C compiler, an MPI [14–16] implementation , and some variant of the BLAS routines, be it ATLAS or a vendor supplied implementation, is installed on the target system. Target systems are intended to be "Beowulf like". There are essentially three components to the software: data collection routines, data movement routines, and application routines.

4 Typical Usage Scenario

The steps involved in a typical LFC run start with a user's problem that may be stated in terms of linear algebra. The problem statement is addressable with one of the LAPACK routines supported in LFC. For instance, suppose that the user has a system of n linear equations with n unknowns, $Ax = b$. There is a parallel computing environment that has LFC installed. The user is, for now, assumed to have access to at least a single node of said parallel computing environment. This is not a necessary constraint - rather a simplifying one. The user compiles the application code (that calls LFC routines) linking with the LFC library and executes the application from a sequential environment. The LFC routine executes the application returning an error code denoting success or failure. The user interprets this information and proceeds accordingly.

On the LFC software side, a decision is made upon how to solve the user's problem by coupling the cluster state information with a knowledge of the particular application. Specifically, a decision is based upon the scheduler's ability to successfully predict that a particular subset of the available processors on the cluster will enable a reduction of the total time to solution when compared to serial expectations for the specific application and user parameters. The relevant times are the time that is spent handling the user's data before and after the parallel application plus the amount of time required to execute the parallel application. If the decision is to solve the user's problem locally (sequentially) then the relevant LAPACK routine is executed. On the contrary, if the decision is to solve the user's problem in parallel then a process is forked that will be responsible for spawning the parallel job and the parent process waits for its return in the sequential environment. The selected processors are allocated (in MPI), the user's data is mapped (block cyclically decomposed) onto the processors (the data may be in memory or on disk), the parallel application is executed (e.g. ScaLAPACK), the data is reverse mapped, the parallel process group is freed, and the solution and control are returned to the user's process.

5 Performance Results

Figure 1 demonstrates the strength of the self-adapting approach of the LFC software. The problem sizes tested were 512, 1024, 2048, 4096, 8192, 12288, 14000. LFC chose 2, 3, 6, 8, 12, 16, 16 processes for these problems respectively. The oracle, which in theory knows the best parameters, utilized 4, 4, 8, 10, 14, 16, 16 processes respectively. The runs were conducted on a cluster of eight Intel Pentium III, 933 MHz dual processors, connected with a 100 Mb/s switch. In each run the data was assumed to start on disk and was written back to disk after the factorization. In the parallel environment, both the oracle and LFC utilized the I/O routines from ROMIO to load/store the data and the ScaLAPACK routine PDGESV() for the application code.

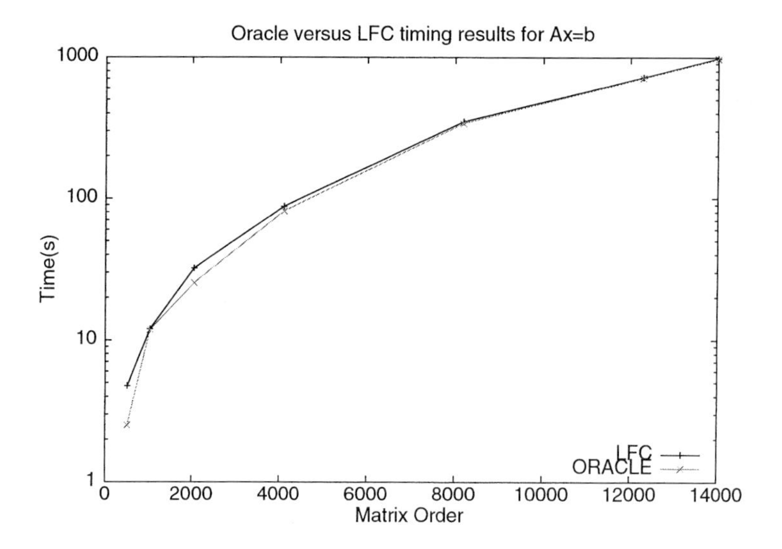

Fig. 1. Parellel execution times of the linear solver run by the oracle and LFC on a cluster.

Figure 2 illustrates the fact that the situation is more complicated than just selecting the right grid aspect ratio (e.g. the number of process rows divided by the number of process columns). Sometimes it might be beneficial to use a smaller number of processors. This is especially true if the number of processors is a prime number which leads to a flat process grid and thus very poor performance on many systems. It is unrealistic to expect that non-expert users will correctly make the right decisions here. It is either a matter of having expertise or experimental data to guide the choice and our experiences suggest that perhaps a combination of both is required to make good decisions consistently. Another point stressed by the figure is the widening gap (excluding merge points at prime processor numbers) between the worst and the optimal resource

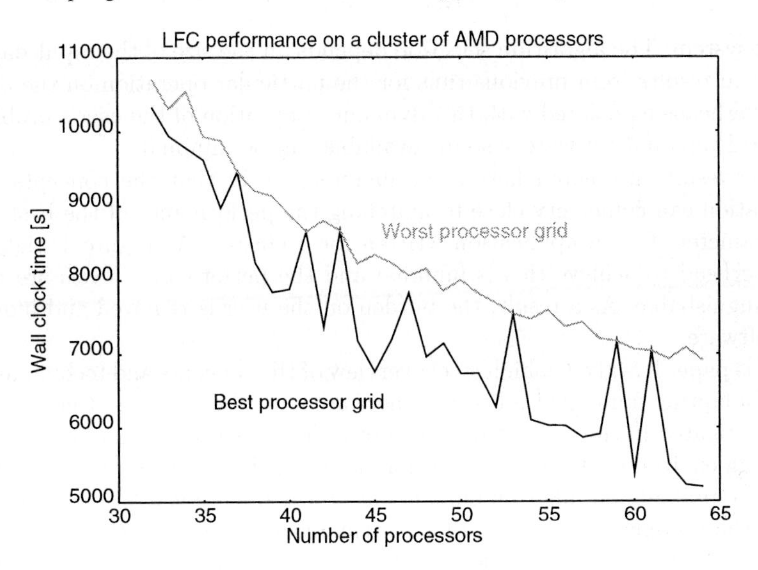

Fig. 2. Timing results for solving a linear system of order 70000 with the best and worst possible rectangular processor grid topologies reported.

choices for increasing number of processors. This shows the increasing chance for a user to use the hardware and software in a suboptimal way as more powerful computers become available.

As a side note, with respect to experimental data, it is worth mentioning that the collection of data for Figure 2 required a number of floating point operations that would compute the LU factorization of a square dense matrix of order almost three hundred thousand. Matrices of that size are usually suitable for supercomputers (the slowest supercomputer on the Top500 [22] list that factored such a matrix was on position 16 in November 2002) – an unlikely target machine for majority of users.

6 Conclusions and Future Work

As computing systems become more powerful and complex it becomes a major challenge to tune applications for high performance. We have described a concept and outlined a plan to develop numerical library software for systems of linear equations which adapts to the user's problem and the computational environment in an attempt to extract near optimum performance. This approach has applications beyond solving systems of linear equations and can be applied to most other areas where users turn to a library of numerical software for their solution.

At runtime our software makes choices at the software and hardware levels for obtaining a best parameter set for the selected algorithm by applying expertise from the literature and empirical investigations of the core kernels on the target system. The algorithm selection depends on the size of the input data and empirical results from previous runs for the particular operation on the cluster. The overheads associated with this dynamic adaptation of the user's problem to the hardware and software systems available can be minimal.

The results presented here show unambiguously that the concepts of self adaptation can come very close to matching the performance of the best choice in parameters for an application written for a cluster. As Figure 1 highlights, the overhead to achieve this is minimal and the performance levels are almost indistinguishable. As a result, the burden on the user is removed and hidden in the software.

This paper has given a high level overview of the concepts and techniques used in self adapting numerical software. There are a number of issues that remain to be investigated in the context of this approach [11]. Issues such as adapting to a changing environment during execution, reproducibility of results when solving the same problem on differing numbers of processors, fault tolerance, rescheduling in the presence of additional load, dynamically migrating the computation, etc all present additional challenges which are ripe for further investigation. In addition, with Grid computing becoming mainstream, these concepts will find added importance [4].

Acknowledgements.
We wish to thank the Ohio Supercomputing Center (OSC), the Computational Science and Mathematics Division at Oak Ridge National Laboratory (XTORC cluster), the Center for Computational Sciences at Oak Ridge National Laboratory (Cheetah, Eagle), the Dolphin donation cluster (part of the SinRG program at the University of Tennessee Knoxville), the San Diego Supercomputing Center (SDSC), and the National Energy Research Scientific Computing Center (NERSC) for research conducted on their resources. We also wish to thank NPACI, the National Partnership for the Advancement of Computational Infrastrucure, for including LFC in its NPACkage.

References

1. R. Agarwal, Fred Gustavson, and M. Zubair. A high-performacne algorithm using preprocessing for the sparse matrix-vector multiplication. In *Proceedings of International Conference on Supercomputing*, 1992.
2. E. Anderson, Z. Bai, C. Bischof, Suzan L. Blackford, James W. Demmel, Jack J. Dongarra, J. Du Croz, A. Greenbaum, S. Hammarling, A. McKenney, and Danny C. Sorensen. *LAPACK User's Guide.* Society for Industrial and Applied Mathematics, Philadelphia, Third edition, 1999.
3. Richard Barrett, Michael Berry, Jack Dongarra, Victor Eijkhout, and Charles Romine. Algorithmic bombardment for the iterative solution of linear systems: A poly-iterative approach. *Journal of Computational and Applied Mathematics*, 74(1-2):91–109, 1996.

4. F. Berman. The GrADS project: Software support for high level grid application development. *International Journal of High Performance Computing Applications*, 15:327–344, 2001.
5. A. Bik and H. Wijshoff. Advanced compiler optimizations for sparse computations. *Journal of Parallel and Distributing Computing*, 31:14–24, 1995.
6. J. Bilmes et al. Optimizing matrix multiply using PHiPAC: a portable, high-performance, ANSI C coding methodology. In *Proceedings of International Conference on Supercomputing*, Vienna, Austria, 1997. ACM SIGARC.
7. Jack J. Dongarra, J. Du Croz, Iain S. Duff, and S. Hammarling. Algorithm 679: A set of Level 3 Basic Linear Algebra Subprograms. *ACM Transactions on Mathematical Software*, 16:1–17, March 1990.
8. Jack J. Dongarra, J. Du Croz, Iain S. Duff, and S. Hammarling. A set of Level 3 Basic Linear Algebra Subprograms. *ACM Transactions on Mathematical Software*, 16:18–28, March 1990.
9. Jack J. Dongarra, J. Du Croz, S. Hammarling, and R. Hanson. Algorithm 656: An extended set of FORTRAN Basic Linear Algebra Subprograms. *ACM Transactions on Mathematical Software*, 14:18–32, March 1988.
10. Jack J. Dongarra, J. Du Croz, S. Hammarling, and R. Hanson. An extended set of FORTRAN Basic Linear Algebra Subprograms. *ACM Transactions on Mathematical Software*, 14:1–17, March 1988.
11. Jack J. Dongarra and Victor Eijkhout. Self-adapting numerical software for next generation applications. Technical report, Innovative Computing Laboratory, University of Tennessee, August 2002. `http://icl.cs.utk.edu/iclprojects/pages/sans.html`.
12. Jack J. Dongarra, Piotr Luszczek, and Antione Petitet. The LINPACK benchmark: Past, present, and future. *Concurrency and Computation: Practice and Experience*, 15:1–18, 2003.
13. Jack J. Dongarra and Clint R. Whaley. Automatically tuned linear algebra software (ATLAS). In *Proceedings of SC'98 Conference*. IEEE, 1998.
14. Message Passing Interface Forum. MPI: A Message-Passing Interface Standard. *The International Journal of Supercomputer Applications and High Performance Computing*, 8, 1994.
15. Message Passing Interface Forum. MPI: A Message-Passing Interface Standard (version 1.1), 1995. Available at: `http://www.mpi-forum.org/`.
16. Message Passing Interface Forum. MPI-2: Extensions to the Message-Passing Interface, 18 July 1997. Available at `http://www.mpi-forum.org/docs/mpi-20.ps`.
17. Ian Foster and Carl Kesselman. *The Grid: Blueprint for a New Computing Infrastructure*. Morgan Kaufmann, San Francisco, 1999.
18. M. Frigo. A fast Fourier transform compiler. In *Proceedings of ACM SIGPLAN Conference on Programming Language Design and Implementation*, Atlanta, Georgia, USA, 1999.
19. M. Frigo and S. G. Johnson. FFTW: An adaptive software architecture for the FFT. In *Proceedings International Conference on Acoustics, Speech, and Signal Processing*, Seattle, Washington, USA, 1998.
20. E.-J. Im. *Automatic optimization of sparse matrix-vector multiplication*. PhD thesis, University of California, Berkeley, California, 2000.
21. E.-J. Im and Kathy Yelick. Optimizing sparse matrix-vector multiplication on SMPs. In *Ninth SIAM Conference on Parallel Processing for Scientific Computing*, San Antonio, Texas, 1999.

22. Hans W. Meuer, Erik Strohmaier, Jack J. Dongarra, and Horst D. Simon. *Top500 Supercomputer Sites*, 20th edition edition, November 2002. (The report can be downloaded from `http://www.netlib.org/benchmark/top500.html`).
23. D. Mirkovic and S. L. Johnsson. Automatic performance tuning in the UHFFT library. In *2001 International Conference on Computational Science*, San Francisco, California, USA, 2001.
24. Jakob Ostergaard. *OptimQR – A software package to create near-optimal solvers for sparse systems of linear equations.* `http://ostenfeld.dk/~jakob/OptimQR/`.
25. Antoine Petitet et al. Numerical libraries and the grid. *International Journal of High Performance Computing Applications*, 15:359–374, 2001.
26. Ali Pinar and Michael T. Heath. Improving performance of sparse matrix-vector multiplication. In *Proceddings of SC'99*, 1999.
27. J. R. Rice. On the construction of poly-algorithms for automatic numerical analysis. In M. Klerer and J. Reinfelds, editors, *Interactive Systems for Experimental Applied Mathematics*, pages 31–313. Academic Press, 1968.
28. Sivan Toledo. Improving the memory-system performance of sparse matrix-vector multiplication. *IBM Journal of Research and Development*, 41(6), November 1997.
29. Sathish Vadhiyar, Graham Fagg, and Jack J. Dongarra. Performance modeling for self adapting collective communications for MPI. In *Los Alamos Computer Science Institute Symposium (LACSI 2001)*, Sante Fe, New Mexico, 2001.
30. R. Weiss, H. Haefner, and W. Schoenauer. *LINSOL (LINear SOLver) – Description and User's Guide for the Parallelized Version.* University of Karlsruhe Computing Center, 1995.
31. R. Clint Whaley, Antoine Petitet, and Jack J. Dongarra. Automated empirical optimizations of software and the ATLAS project. *Parallel Computing*, 27(1-2):3–35, 2001.

V-Invariant Methods for Generalised Least Squares Problems

M.R. Osborne

Mathematical Sciences Institute, Australian National University, ACT 0200, Australia

Abstract. An important consideration in solving generalised least squares problems is the dimension of the covariance matrix V. This has the dimension of the data set and is large when the data set is large. In addition the problem can be formulated to have a well determined solution in cases where V is illconditioned or singular, a class of problems that includes the case of equality constrained least squares. This paper considers a class of methods which factorize the design matrix A while leaving V invariant, and which can be expected to be well behaved exactly when the original problem solution is well behaved. Implementation is most satisfactory when V is diagonal. This can be achieved by a preprocessing step in which V is replaced by the diagonal matrix D which results from the modified Cholesky factorization $PVP^T \to LDL^T$ where L is unit lower triangular and P is the permutation matrix associated with diagonal pivoting. Conditions under which this replacement is satisfactory are investigated.

1 Introduction

The generalised least squares problem is

$$\min_{\mathbf{x},\mathbf{r}} \mathbf{r}^T V^{-1} \mathbf{r}; \ \mathbf{r} = A\mathbf{x} - \mathbf{b} \tag{1}$$

where $A \in R^p \to R^n$, rank $(A) = p < n$, $\mathbf{x} \in R^p$, $\mathbf{r} \in R^n$. In this form it requires V to be positive definite and hence invertible. However, the necessary conditions for this problem to have a solution can be cast in the form

$$\begin{bmatrix} V & -A \\ -A^T & 0 \end{bmatrix} \begin{bmatrix} \boldsymbol{\lambda} \\ \mathbf{x} \end{bmatrix} = \begin{bmatrix} -\mathbf{b} \\ 0 \end{bmatrix} \tag{2}$$

where $\boldsymbol{\lambda}$ is the vector of Lagrange multipliers and is related to the residual vector by $\mathbf{r} = V\boldsymbol{\lambda}$. This proves to have solutions under weaker conditions on V.

In (2) the solution of the generalised least squares problem requires the solution of a linear system with symmetric, indefinite matrix. This poses problems in the data analytic context because n may well need to be large as a consequence of the generic $n^{-1/2}$ rate of convergence in stochastic estimation problems, while p has to be small in typical regression situations. Thus the size of V is potentially the main source of difficulty in otherwise well behaved problems. This would also suggest that if structure is present in V then advantage should be taken of this fact. One possible approach is considered here based on the paper [2]. The generalisation embodied in the formulation

P.M.A. Sloot et al. (Eds.): ICCS 2003, LNCS 2659, pp. 673–682, 2003.

(2) has the interesting advantage that it permits least squares problems subject to linear equality constraints to be considered in the same framework as the generalised least squares problem. Let the constrained problem be written

$$\min_{\mathbf{x},\mathbf{r}_2} \mathbf{r}_2^T V^{-1} \mathbf{r}_2;\ \mathbf{r}_2 = A\mathbf{x} - \mathbf{b},\ 0 = C\mathbf{x} - \mathbf{d},$$

where $C \in R^p \to R^m$, $m < p$, is assumed to have full rank. This ensures that the constraints are consistent for arbitrary problem data $\mathbf{d}$. One possible approach involves reformulating the problem using penalised least squares:

$$\min_{\mathbf{x},\mathbf{r}} \left\{ \mathbf{r}_2^T V^{-1} \mathbf{r}_2 + \sum_{i=1}^{m} \sigma^2 W_i^{-2} (\mathbf{r}_1)_i^2 \right\} \tag{3}$$

where $\mathbf{r}_1 = C\mathbf{x} - \mathbf{d}$, and $W = \operatorname{diag}\{W_i\}$ is some bounded positive scaling matrix. The idea is that as σ gets large the penalised term will dominate unless $\|\mathbf{r}_1\| \to 0$. It is straightforward to show a convergence rate of $O\left(\sigma^{-2}\right)$. But (3) is just a generalised least squares problem and can be reformulated as

$$\min_{\mathbf{x},\mathbf{s}} \mathbf{s}^T \mathbf{s};\ \begin{bmatrix} \sigma^{-1} W & \\ & L \end{bmatrix} \mathbf{s} = \begin{bmatrix} C \\ A \end{bmatrix} \mathbf{x} - \begin{bmatrix} \mathbf{d} \\ \mathbf{b} \end{bmatrix}.$$

This problem has the limiting form as $\sigma \to \infty$

$$\min_{\mathbf{x},\mathbf{s}} \mathbf{s}^T \mathbf{s};\ \begin{bmatrix} 0 & \\ & L \end{bmatrix} \mathbf{s} = \begin{bmatrix} C \\ A \end{bmatrix} \mathbf{x} - \begin{bmatrix} \mathbf{d} \\ \mathbf{b} \end{bmatrix}.$$

The solution of this problem is well determined provided

$$M = \begin{bmatrix} \begin{bmatrix} 0 & 0 \\ 0 & V \end{bmatrix} & -\begin{bmatrix} C \\ A \end{bmatrix} \\ -\begin{bmatrix} C^T & A^T \end{bmatrix} & 0 \end{bmatrix}$$

is nonsingular. The ordering of terms here proves important.

The development of this paper is as follows. The next two sections consider V-invariance and its application to the factorization of the design matrix A with only minor differences from [2]. Practical implementation consideration puts an emphasis on diagonal V with elements ordered in increasing size, and it is suggested [4] that the LDL^T, L lower triangular with unit diagonal, modification of the basic Cholesky factorization of V using diagonal pivoting [3] could prove useful in weakening this requirement. The major contribution in this paper is made in the concluding section where it is argued that the error that occurs when small elements in the computed D are set to zero is, with some qualification, of the same size as these small elements. In fact this argument extends to the case where D supports multiple scales switching between these in discrete steps. Also, it is not required that all steps of the Cholesky factorization be completed successfully in the case that V is (almost) semi-definite. It is shown that one consequence of the ability to support multiple scales is a requirement for column pivoting in the factorization of the design matrix A, especially when equality constraints are present.

2 V-Invariance

The motivation for introducing algorithms for the solution of generalised least squares problems based on transformations with special invariance properties derives from the special case of the methods for the linear least squares problem based on orthogonal factorization of the design matrix A where an orthogonal matrix Q is constructed to reduce A to upper triangular form while preserving column lengths corresponding to $V = I$. The condition for V-invariance generalises this metric condition to

$$JVJ^T = V.$$

Now assume that a V-invariant factorization of A is given by

$$JA = \begin{bmatrix} U \\ 0 \end{bmatrix}, \tag{4}$$

and that V has the form

$$V = \begin{bmatrix} O_k & 0 \\ 0 & V_2 \end{bmatrix} \tag{5}$$

where O_k is the $k \times k$ zero matrix and V_2 has the form

$$V_2 = \begin{bmatrix} V_{11} & V_{12} \\ V_{21} & V_{22} \end{bmatrix} \tag{6}$$

where $V_{11} \in R^{p-k} \to R^{p-k}$. Then the transformation applied to equation (2) gives

$$\begin{bmatrix} \begin{bmatrix} 0 & 0 \\ 0 & V_{11} \\ 0 & V_{21} \end{bmatrix} & \begin{bmatrix} 0 \\ V_{12} \\ V_{22} \end{bmatrix} & -\begin{bmatrix} U \\ 0 \end{bmatrix} \\ -\begin{bmatrix} U^T & 0 \end{bmatrix} & & 0 \end{bmatrix} \begin{bmatrix} \left(J^{-T}\boldsymbol{\lambda}\right)_1 \\ \left(J^{-T}\boldsymbol{\lambda}\right)_2 \\ \mathbf{x} \end{bmatrix} = -\begin{bmatrix} (J\mathbf{b})_1 \\ (J\mathbf{b})_2 \\ 0 \end{bmatrix}.$$

This system reduces to the equations

$$U^T \left(J^{-T}\boldsymbol{\lambda}\right)_1 = 0,$$
$$V_{22} \left(J^{-T}\boldsymbol{\lambda}\right)_2 = -(J\mathbf{b})_2,$$
$$\begin{bmatrix} 0 \\ V_{12} \end{bmatrix} \left(J^{-T}\boldsymbol{\lambda}\right)_2 - U\mathbf{x} = -(J\mathbf{b})_1.$$

Remark 1. If U in (4) is nonsingular, and V_2 is ordered so that V_{22} is nonsingular, then the generalised least squares problem has a well determined solution.

To construct factorizations it is convenient to develop elementary V-invariant transformations. Let

$$J = I - 2\mathbf{u}\mathbf{v}^T.$$

If $\mathbf{v}^T V \mathbf{v} > 0$ then the condition of V-invariance determines $\mathbf{u}$:

$$\mathbf{u} = \frac{1}{\mathbf{v}^T V \mathbf{v}} V \mathbf{v}, \; J = I - \frac{2}{\mathbf{v}^T V \mathbf{v}} V \mathbf{v} \mathbf{v}^T. \tag{7}$$

In this case $\mathbf{u}^T \mathbf{v} = 1$ so that J is also an elementary reflector ($J^2 = I$). If $\mathbf{v} = \begin{bmatrix} \mathbf{v}_1 \\ 0 \end{bmatrix}$, $\mathbf{v}_1 \in R^k$ then $\mathbf{v}^T V = 0$ and the above derivation of J breaks down. A V-invariant transformation is given by

$$J = I - 2 \begin{bmatrix} \mathbf{u}_1 \\ \mathbf{u}_2 \end{bmatrix} \begin{bmatrix} \mathbf{v}_1^T & 0 \end{bmatrix}, \tag{8}$$

and $J^2 = I$ if $\mathbf{v}_1^T \mathbf{u}_1 = 1$. There is a useful connection between the two forms of transformations. Let

$$V_\varepsilon = \begin{bmatrix} \varepsilon W & \\ & V_2 \end{bmatrix}$$

where $V_\varepsilon \in R^k \to R^k$ is a positive diagonal matrix. Then the V_ε-invariant transformation derived from (7) but written in terms of $\mathbf{u}$ is

$$\begin{aligned} J_\varepsilon &= I - \frac{2}{\mathbf{u}^T V_\varepsilon^{-1} \mathbf{u}} \mathbf{u} \mathbf{u}^T V_\varepsilon^{-1}, \\ &= I - \frac{2}{\mathbf{u}^T \varepsilon V_\varepsilon^{-1} \mathbf{u}} \mathbf{u} \mathbf{u}^T \varepsilon V_\varepsilon^{-1}, \\ &\to I - \frac{2}{\mathbf{u}_1^T W^{-1} \mathbf{u}_1} \begin{bmatrix} \mathbf{u}_1 \\ \mathbf{u}_2 \end{bmatrix} \begin{bmatrix} \mathbf{u}_1^T W^{-1} & 0 \end{bmatrix}, \end{aligned} \tag{9}$$

as $\varepsilon \to 0$. Note that the action of this transformation can be considered to be on the scale defined by ε.

Numerical stability of the elementary V-invariant transformations depends on the size of the elements of J and hence on its norm. In this connection the following result [2] is important.

Lemma 1. *Let J be an elementary reflector, and set $\eta = \|\mathbf{u}\| \, \|\mathbf{v}\| > 1$. Then the spectral norm of J is given by*

$$\|J\| = \eta + \sqrt{\{\eta^2 - 1\}}.$$

In the case of the V-invariant transformation (7)

$$\eta = \frac{\|\mathbf{v}\| \, \|V \mathbf{v}\|}{\mathbf{v}^T V \mathbf{v}}. \tag{10}$$

Thus the relative size of $\mathbf{v}^T V \mathbf{v}$ is important.

3 V-Invariant Factorization

The basic idea is to build up a V-invariant transformation J taking A to upper triangular form using elementary V-invariant transformations. Assume that a partial factorization J_{i-1} has been constructed:

$$J_{i-1}A = \begin{bmatrix} U_{i-1} & U_{12}^i \\ & A_i \end{bmatrix}.$$

The aim now is to construct $\mathbf{v}$ such that the elementary V-invariant transformation (7)

$$J^i = I - \frac{2}{\mathbf{v}^T V \mathbf{v}} V \mathbf{v}\mathbf{v}^T$$

has the action

$$J^i \left(J_{i-1} A \mathbf{e}_i \right) = J^i \begin{bmatrix} \mathbf{u}_i \\ \mathbf{a}_i \end{bmatrix} = \begin{bmatrix} \mathbf{u}_i \\ \gamma \mathbf{e}_1 \end{bmatrix}$$

giving

$$J_i = J^i J_{i-1},\ U_i = \begin{bmatrix} U_{i-1} & \mathbf{u}_i \\ & \gamma \end{bmatrix}.$$

Note that the scale of $\mathbf{v}$ is disposable. Thus the choice

$$V\mathbf{v} = \begin{bmatrix} 0 \\ \mathbf{a}_i - \gamma \mathbf{e}_1 \end{bmatrix} \tag{11}$$

is allowed. This would be suitable except that computation of $\mathbf{v}$ requires the solution of a system of linear equations with matrix V, and this would seem to reintroduce exactly the kind of complication that we seek to avoid. There are considerable simplifications if V_2 is diagonal. In this case transformation of the design matrix to upper triangular form is made easier by some structural assumptions on V. These are that V has the reduced form (5) where $V_2 \in R^{n-k} \to R^{n-k}$, and

$$V_2 = \text{diag}\left\{ \nu_{k+1}, \nu_{k+2}, \cdots, \nu_n \right\},\ 0 < \nu_{k+1} \le \nu_{k+2} \le \cdots \le \nu_n. \tag{12}$$

If V is a positive semidefinite matrix in general form then this could be achieved by

1. Scaling V so that the diagonal elements of the transformed matrix are unity. Here $V \to S\overline{V}S$ where $S = \text{diag}\left\{ V_{11}^{1/2}, \cdots, V_{nn}^{1/2} \right\}$.
2. A diagonal pivoting (rank revealing) LDL^T factorization [3] applied to $\overline{V}$ to construct a matrix of known factorization and rank which closely approximates $\overline{V}$

$$P\overline{V}P = LDL^T.$$

Note that the ordering achieved by this factorization gives the elements of D in decreasing order of magnitude which is the inverse of that assumed in (5).

3. Transformation of the problem to one with covariance matrix D

$$\mathbf{r} \to L^{-1}PS^{-1}\mathbf{r} \Rightarrow A \to L^{-1}PS^{-1}A,\ \mathbf{b} \to L^{-1}PS^{-1}\mathbf{b}.$$

The conditioning of the forward substitution is aided here by the use of diagonal pivoting which ensures that $|L_{ij}| \le 1,\ j < i$. Thus illconditioning in V tends to be concentrated in D by the rank revealing factorization.

4. Permutation of the elements of the new covariance into increasing order with permutation matrix Q

$$D \to QDQ^T,\ \mathbf{r} \to Q\mathbf{r}.$$

To compute γ let

$$V_2 = \begin{bmatrix} V_{11} & \\ & V_{22} \end{bmatrix},\ V_{11} = \operatorname{diag}\{\nu_{k+1}, \cdots, \nu_{i-1}\},\ V_{22} = \operatorname{diag}\{\nu_i, \cdots, \nu_n\},$$

where the partitioning is chosen to correspond to the i'th factorization step. Then

$$J^i = \begin{bmatrix} I_k & & \\ & I_1 & \\ & & I_2 - \frac{2}{\mathbf{v}^T V_{22}\mathbf{v}} V_{22}\mathbf{v}\mathbf{v}^T \end{bmatrix}$$

is V-invariant where I_k is the $k \times k$ unit matrix and I_1 and I_2 are unit matrices conformable with V_{11} and V_{22} respectively. Then (11) gives

$$\mathbf{v} = \begin{bmatrix} 0 \\ V_{22}^{-1}\mathbf{a}_i \end{bmatrix}.$$

The properties of the elementary reflector to give

$$\begin{aligned} \begin{bmatrix} \mathbf{u}_i \\ \mathbf{a}_i \end{bmatrix} &= J^i \begin{bmatrix} \mathbf{u}_i \\ \gamma \mathbf{e}_1 \end{bmatrix}, \\ &= \left(I - \frac{2}{\mathbf{v}^T V \mathbf{v}} V \mathbf{v}\mathbf{v}^T \right) \begin{bmatrix} \mathbf{u}_i \\ \gamma \mathbf{e}_1 \end{bmatrix}. \end{aligned}$$

Taking the scalar product with $\mathbf{v}$ gives a quadratic equation for γ

$$(\mathbf{a}_i - \gamma \mathbf{e}_1)^T V_{22}^{-1} (\mathbf{a}_i + \gamma \mathbf{e}_1) = 0.$$

This has the solution

$$\gamma = \theta \sqrt{\left\{\nu_i \mathbf{a}_i^T V_{22}^{-1} \mathbf{a}_i\right\}}.$$

where the standard argument which seeks to minimize cancellation suggests that $\theta = -\operatorname{sgn}((\mathbf{a}_i)_1)$ is appropriate. An interesting feature is the appearance of the term $N^i = \nu_i V_{22}^{-1}$ which means that γ is independent of the scale of V. [2] give the stability criterion

$$\eta = \frac{\left\|V_{22}^{-1}(\mathbf{a}_i - \gamma \mathbf{e}_1)\right\| \left\|\mathbf{a}_i - \gamma \mathbf{e}_1\right\|}{(\mathbf{a}_i - \gamma \mathbf{e}_1)^T V_{22}^{-1} (\mathbf{a}_i - \gamma \mathbf{e}_1)} \ge \frac{\|\mathbf{a}_i\|}{2|\gamma|}.$$

Remark 2. Column pivoting can be used to exchange any column $\begin{bmatrix} \mathbf{u}_j \\ \mathbf{a}_j \end{bmatrix}$, $j = i, i+1, \cdots, p$ of $\begin{bmatrix} U_{12}^i \\ A_i \end{bmatrix}$ into the pivotal position. Let $\rho_i(A,j) = \mathbf{a}_j^T N^i \mathbf{a}_j$. The above analysis suggests that j be chosen to maximize $\rho_i(A,j) / \|\mathbf{a}_j\|^2$ in order to make the lower bound for η as small as possible. However, if $V \neq I$, lengths are not preserved by the V-invariant transformations and a consequence is that denominators cannot be updated economically in general. Thus the selection of the pivotal column would typically be made on the basis of the size of the $\rho_i(A,j)$ alone.

If $1 \leq l \leq k$ then the key mapping

$$J\mathbf{a}_l = \gamma \mathbf{e}_l, \; l \leq k,$$

needs to make use of the second family of V-invariant transformations (8). Let $\mathbf{a}_1 = \begin{bmatrix} \mathbf{a}_1^1 \\ \mathbf{a}_1^2 \end{bmatrix}$ in conformity with the partitioning of V. Here the limiting argument leading to (9) provides insight into the reason why V-invariant transformations work well in the multiscaled situations. If $l = 1$ the resulting transformation matrix $J = I - 2\mathbf{c}\mathbf{d}^T$ has the generic form (8) with

$$\sqrt{2}\mathbf{c} = \left(\mathbf{a}_1 + \operatorname{sgn}\left(\left(\mathbf{a}_1^1\right)_1\right) \left\|\mathbf{a}_1^1\right\|_2 \mathbf{e}_1\right) / \left\|\mathbf{a}_1^1\right\|_2,$$

$$\sqrt{2}\mathbf{d} = \begin{bmatrix} \mathbf{a}_1^1 + \operatorname{sgn}\left(\left(\mathbf{a}_1^1\right)_1\right) \left\|\mathbf{a}_1^1\right\|_2 \mathbf{e}_1 \\ 0 \end{bmatrix} / \left(\left\|\mathbf{a}_1^1\right\|_2 + \left|\left(\mathbf{a}_1^1\right)_1\right|\right).$$

This transformation will have large elements if

$$\left\|\mathbf{a}_1^1\right\|_2 \ll \left\|\mathbf{a}_1\right\|_2. \tag{13}$$

This is the limiting case of (10) which characterizes illconditioning in the other class of transformations. The extreme case corresponds to $\left\|\mathbf{a}_1^1\right\|_2 = 0$, and this can certainly occur in the case $k > 0$ in (5), (12). This is illustrated in the following example which provides a justification for the use of column pivoting in implementing the V-invariant factorization.

Example 1. Column pivoting is necessary for V-invariant solution methods if the generalised least squares problem is subject to equality constraints. Here the pivotal column at the i'th step is chosen as the one that maximizes $\rho_i(A,j) = \sum_{q=i}^{n+m} N_q^i A_{qj}^2$, $j = i, i+1, \cdots, p$, where m is the number of equality constraints and N^i is the scaled diagonal weighting matrix. To illustrate the requirement consider the design matrix $A_2 \in R^p \to R^n$ given by

$$\begin{aligned} (A_2)_{*1} &= S\mathbf{e}, \\ (A_2)_{i(2j)} &= S_i \cos\left(2\pi j\,(i-1)\,h\right), \\ (A_2)_{i(2j+1)} &= S_i \sin\left(2\pi j\,(i-1)\,h\right), \\ i = 1, 2, \cdots, n,\; j &= 1, 2, \cdots, k, \end{aligned}$$

where $p = 2k+1,\ S = \text{diag}\left\{1/\sqrt{2}, 1, \cdots, 1, 1/\sqrt{2}\right\},\ h = 1/(n-1)$. The columns of A_2 are orthogonal and similarly scaled so a least squares problem with A_2 as design is very well conditioned. Let the rows of the constraint matrix $A_1 \in R^p \rightarrow R^2$ be given by

$$(A_1)_{1*} = \mathbf{e}^T,\ (A_1)_{2*} = \mathbf{e}^T - p\mathbf{e}_{k+1}^T.$$

Then the constrained least squares problem

$$\min_{\mathbf{s},\mathbf{x}} \mathbf{s}^T\mathbf{s};\quad \begin{bmatrix} 0 & \\ & I \end{bmatrix} \mathbf{s} = \begin{bmatrix} A_1 \\ A_2 \end{bmatrix} \mathbf{x} - \begin{bmatrix} \mathbf{b}_1 \\ \mathbf{b}_2 \end{bmatrix},$$

with

$$\begin{bmatrix} \mathbf{b}_1 \\ \mathbf{b}_2 \end{bmatrix} = \begin{bmatrix} A_1\mathbf{e} \\ A_2\mathbf{e} + \varepsilon \end{bmatrix},\ \varepsilon_i = \sin\left(2\pi\left(k+1\right)\left(i-3\right)h\right),\ i = 3, 4, \cdots, n+2,$$

has the exact solution $\mathbf{x} = \mathbf{e}$. However, the leading 2×2 submatrix of A_1 is singular if $k \geq 2$. This causes a breakdown of the V-invariant factorization at the second step (13) as a consequence of the form of N^2 if column interchanges are not used. In the particular case corresponding to $n = 8,\ p = 5$, the ordering resulting from column pivoting is $\{3, 2, 1, 4, 5\}$. The first interchange is a consequence of the largest coefficient in the second constraint, and it succeeds in forcing the leading 2×2 submatrix to be nonsingular.

4 Does the Use of the LDL^T Factorization of V Make Sense?

In this section it is assumed that V has full rank, but is illconditioned as a result of a cluster of small eigenvalues. The results extend immediately to the case where V has a known nullspace plus a cluster of small eigenvalues. Here the rank-revealing Cholesky factorization of V, performed in exact arithmetic, has the form

$$PVP^T \rightarrow L\,\text{diag}\left\{D_n, D_{n-1}, \cdots, D_1\right\} L^T$$

where P is the permutation matrix corresponding to the diagonal pivoting, and the diagonal pivoting ensures that

$$D_n \geq D_{n-1} \geq \cdots \geq D_1.$$

This order is the reverse of that required here, and it must be inverted in order to construct the factorization of the design matrix based on V-invariant transformations. Conditions which guarantee that $n-k$ steps of the Cholesky factorization with diagonal pivoting can be computed leaving a small remainder are given in [3]. One form compatible with the above assumption is

$$\begin{aligned} &\Delta_1 = \text{diag}\left\{D_1, D_2, \cdots, D_k\right\} > 0, \text{ small}, \\ &D_k \ll D_{k+1}, \\ &\Delta_2 = \text{diag}\left\{D_{k+1}, \cdots, D_n\right\} \text{ not small}, \\ &\text{where } k \leq p. \end{aligned}$$

It would be expected that if the computed factorization does go to completion then the computed values $\{D_1, D_2, \cdots, D_k\}$ could have high relative error as a result of cancellation. Does this matter? The following argument suggests strongly that it does not. It assumes that the factorization is stopped after $n-k$ steps. This yields the incomplete transformation

$$PVP^T = L\begin{bmatrix} \Delta_2 & \\ & V^k \end{bmatrix} L^T,$$

where use has been made of the result that the unit lower triangular matrix L has the particular structure

$$L = \begin{bmatrix} L_{n-k} & \\ L_2^{(n-k)} & I \end{bmatrix},$$

and

$$L^{-1}\begin{bmatrix} 0 & 0 \\ 0 & V^k \end{bmatrix} L^{-T} = \begin{bmatrix} 0 & 0 \\ 0 & V^k \end{bmatrix}.$$

First note that the case $D = \text{diag}\{0, \cdots, 0, D_{k+1}, \cdots, D_n\}$ corresponds to the equality constrained problem

$$\min_{\mathbf{x}} \mathbf{s}^T\mathbf{s};\ \begin{bmatrix} 0 & \\ & \Delta_2^{1/2} \end{bmatrix} \mathbf{s} = \begin{bmatrix} A_1 \\ A_2 \end{bmatrix} \mathbf{x} - \begin{bmatrix} \mathbf{b}_1 \\ \mathbf{b}_2 \end{bmatrix}.$$

Here the Cholesky L has been absorbed into the design to simplify notation ($\mathbf{r} \to L^{-1}\mathbf{r}$), and it is assumed that necessary permutations to order D have also been applied. This is the limiting problem as $\lambda \to \infty$ associated with the penalised objective

$$\min_{\mathbf{x}} \left\{\mathbf{r}_2^T \Delta_2^{-1} \mathbf{r}_2 + \lambda \mathbf{r}_1^T E^{-1} \mathbf{r}_1\right\};\ \mathbf{r} = \begin{bmatrix} A_1 \\ A_2 \end{bmatrix} \mathbf{x} - \begin{bmatrix} \mathbf{b}_1 \\ \mathbf{b}_2 \end{bmatrix} \tag{14}$$

where $\lambda^{-1}E = V^k$ where $V^k \in R^k \to R^k$ is small as the remainder after $n-k$ steps of the rank-revealing Cholesky factorization [3] p. 212, E is positive definite, and $\lambda = 1/\left\|V^k\right\|$. The problem (14) has the alternative form

$$\min_{\mathbf{x}} \mathbf{s}^T\mathbf{s};\ \begin{bmatrix} \lambda^{-1/2}E^{1/2} & \\ & \Delta_2^{1/2} \end{bmatrix} \mathbf{s} = \begin{bmatrix} A_1 \\ A_2 \end{bmatrix} \mathbf{x} - \begin{bmatrix} \mathbf{b}_1 \\ \mathbf{b}_2 \end{bmatrix}.$$

Now penalty function theory can be used to show $\left\|\mathbf{x}(\lambda) - \widehat{\mathbf{x}}\right\| = O(1/\lambda),\ \lambda \to \infty$ provided

$$M = \begin{bmatrix} A_2^T \Delta_2^{-1} A_2 & A_1^T E^{-T/2} \\ E^{-1/2} A_1 & 0 \end{bmatrix}$$

has full rank [1]. This is a weaker condition than both A_1 and A_2 having full rank. Here the conditioning of the scaled perturbation matrix E could also be important. Assuming that E is reasonably well conditioned then the equality constrained problem obtained by setting $\Delta_1 = 0$ has a well defined solution which differs from the exact solution of that based on the computed LDL^T factorization by $O(\|\Delta_1\|)$.

The above argument identifies a class of problems where V is illconditioned for inversion, but where solution by V-invariant methods is satisfactory after preliminary problem transformation based on the LDL^T factorization of V with diagonal pivoting . One relevant case is the fourth example in [4]. This is defined to have an illconditioned L, and unit diagonal. However, if the LDL^T factorization is recomputed using diagonal pivoting then the illconditioning is moved to D and does not cause problems.

References

1. A.V. Fiacco and G.P. McCormick, *Nonlinear programming: Sequential unconstrained minimization techniques*, John Wiley and Sons, Inc., 1968.
2. M. Gulliksson and P. Wedin, *Modifying the QR-decomposition to constrained and weighted linear least squares*, SIAM J. Matrix Anal. Appl. **13** (1992), no. 4, 1298-–1313.
3. N.J. Higham, *Accuracy and stability of numerical algorithms*, SIAM, 1996.
4. I. Søderkvist, *On algorithms for generalized least squares problems with ill-conditioned covariance matrices*, Computational Statistics 11 (1996), 303—313.

Parallelisation of Sparse Grids for Large Scale Data Analysis

Jochen Garcke[1], Markus Hegland[2], and Ole Nielsen[2]

[1] Institut für Angewandte Mathematik,
Rheinische Friedrich-Wilhelms-Universität Bonn
Wegelerstr. 6
D-53115 Bonn
garcke@iam.uni-bonn.de*
[2] Mathematical Sciences Institute, Australian National University

Abstract. Sparse Grids (SG), due to Zenger, are the basis for efficient high dimensional approximation and have recently been applied successfully to predictive modelling. They are spanned by a collection of simpler function spaces represented by regular grids. The combination technique prescribes how approximations on simple grids can be combined to approximate the high dimensional functions. It can be improved by iterative refinement.
Fitting sparse grids admits the exploitation of parallelism at various stages. The fit can be done entirely by fitting partial models on regular grids. This allows parallelism over the partial grids. In addition, each of the partial grid fits can be parallelised as well, both in the assembly phase where parallelism is done over the data and in the solution stage using traditional parallel solvers for the resulting PDEs. A simple timing model confirms that the most effective methods are obtained when both types of parallelism are used.

Keywords: predictive modelling, sparse grids, parallelism, numerical linear algebra

1 Introduction

Data mining algorithms have to address two major computational challenges. First, they have to be able to handle large and growing datasets and secondly, they need to be able to process complex data. Datasets used in data mining studies have been doubling in size every year and many are now in the terabyte range. The second challenge is sometimes referred to as the *curse of dimensionality* as the algorithmic complexity grows exponentially in the number of features or dimension of the data. Data mining aims to find patterns or structure in the data. Parallel processing is a major tool in addressing the large computational requirements of data mining algorithms.

* Part of the work was supported by the German Bundesministerium für Bildung und Forschung (BMB+F) within the project 03GRM6BN.

P.M.A. Sloot et al. (Eds.): ICCS 2003, LNCS 2659, pp. 683–692, 2003.

One important type of pattern discovered in data mining algorithms is represented by functions among selected dataset features. In data mining, the discovery of such functions is referred to as *predictive modelling* and includes both classification and regression. Here we will consider regression but the same algorithms are used in classification as well. Different types of models are obtained from different functional classes. The classical methods of least squares uses linear and nonlinear functions with relatively few parameters. Modern nonparametric methods are characterised by large numbers of parameters and can flexibly approximate general function sets. They include artificial neural nets, Bayesian nets, classification and regression trees (CART) [1], Multivariate Adaptive Regression Splines (MARS) [2], Support Vector Machines, ANOVA splines and additive models.

All approaches are able to characterise a large class of behaviours and involve training or fitting the model to the dataset. For example, one may wish to predict the vegetation cover of a particular region based on cartographic measurements such as elevation, slope, distance to water, etc. Other examples are prediction of the likelihood of a car insurance customer making a claim, a business customer to purchase a product or a resident to commit taxation fraud.

For a given response variable y and predictor variables $x_1, \ldots, x_d$ a predictive model is described by a function

$$y = f(x_1, \ldots, x_d).$$

We will only consider the case where the function f is an element of a linear space and we'll discuss methods to compute the its representation from data in then following.

2 Sparse Grids for Predictive Modelling

Recently a technique called sparse grids [3], based on a hierarchical basis approach, has generated considerable interest as a vehicle for reducing dimensionality problems where approximations of high dimensional functions are sought. Generalised sparse grids functions $f(\boldsymbol{x})$ are approximations which can be described by additive models of the form

$$f(\boldsymbol{x}) = \sum_{\alpha} f_\alpha(\boldsymbol{x}) \tag{1}$$

where the partial functions f_α are simpler than f in some sense. Typically, the partial functions only depend on a subset of the variables or the dependence has a coarser scale as discussed below.

Sparse grids for the solution of partial differential equations, numerical integration, and approximation problems have been studied for more than a decade by Griebel, Zenger et al. They also developed an algorithm known as the combination technique [4] prescribing how the collection of standard grids can be combined to approximate the high dimensional function. More recently, Garcke, Griebel and Thess [5,6] demonstrated the feasibility of sparse grids in data mining by using the combination technique in predictive modelling.

Additive models of the form of equation (1) generalise linear models and thus form a core technique in nonparametric regression. They include the Multivariate Adaptive Regression Splines (MARS) [2], and the Additive Models by Hastie and Tibshirani [7,8].

Challenges include the selection of function spaces, variable selection etc. Here we will discuss algorithms for the determination of the function f and hence the partial functions f_α given observed function values when the function spaces are given. For observed data points $\boldsymbol{x}^1, \ldots, \boldsymbol{x}^n$ and function values $y^1, \ldots, y^n$ we define the function f from some finite dimensional function space V to be the solution of a penalised least squares problem of the form

$$J(f) = \frac{1}{n}\sum_{i=1}^{n}\left(f(\boldsymbol{x}^{(i)}) - y^i\right)^2 + \beta\,\|Lf\|^2 \tag{2}$$

for some (differential) operator L. The solution of this problem can be viewed as a projection of a generalised thin plate spline function, see [9]. If the partial functions f_α are known to be orthogonal with respect to the corresponding norm (here the standard 2-norm), then they can be computed independently as minima of J. A slightly more general case is considered in the case of the combination technique where the projections into the spaces of the partial functions are known to commute. In this case, if g^α are the projections into the partial spaces, then the partial functions are known to be multiples of these projections with known (integer) coefficients, the combination coefficients c_α and thus

$$f = \sum_\alpha c_\alpha g_\alpha.$$

However, these approximations can break down when the projections do not commute. As a generalisation of this approach, an approximation has been suggested in [9] where the partial functions are again multiples of the projections g_α but the coefficients are this time determined by minimising the functional J. Thus one gets

$$f = \sum_\alpha \gamma_\alpha g_\alpha.$$

This also generalises the approximations obtained from the additive Schwarz method which is

$$f = \gamma \sum_\alpha g_\alpha$$

and experimental evidence shows that the performance is in many cases close to that of the multiplicative Schwarz method, which in statistics is known under the term of backfitting [7]. The approaches above can be further improved by iterative refinement.

The interesting aspect of these problems which we will discuss here is the opportunities for parallel processing and the tradeoff between parallel processing and the performance of the solvers. We use a two level iterative solver and parallelism is exploited at both levels.

2.1 Multiresolution Analysis and Sparse Grids

The sparse grid idea stems from a hierarchical subspace splitting [3]. Consider the space $V_{\mathbf{l}}$, with $\mathbf{l} = (l_1, ..., l_d) \in \mathbf{N}_0^d$, of piecewise d-linear functions which is spanned by the usual d-linear 'hat' functions

$$\varphi_{\mathbf{l},\mathbf{j}}(\mathbf{x}) \; := \; \prod_{t=1}^{d} \varphi_{l_t,j_t}(x_t), \qquad j_t = 0, ..., 2^{l_t}.$$

Here, the 1 D functions $\varphi_{l,j}(x)$ are

$$\varphi_{l,j}(x) = \begin{cases} 1 - |2^l \cdot x - j|, & x \in [\frac{j-1}{2^l}, \frac{j+1}{2^l}]; \\ 0, & \text{otherwise.} \end{cases}$$

The number of basis functions needed to resolve any $f \in V_l := V_{l,\dots,l}$ is now larger than 2^{ld}. With a resolution of just 17 points in each dimension ($l = 4$), say, a ten dimensional problem would require computation and storage of about 2×10^{12} coefficients which is more than one can expect on computers available today – the curse of dimensionality.

Now we define the difference spaces $W_{\mathbf{l}}$, with $\mathbf{e}_t$ denoting the t-th unit vector,

$$W_{\mathbf{l}} \; := \; V_{\mathbf{l}} - \sum_{t=1}^{d} V_{\mathbf{l}-\mathbf{e}_t}.$$

These hierarchical difference spaces lead to the definition of a multilevel subspace splitting, i.e., the definition of the space V_l as a direct sum of subspaces,

$$V_l \; := \; \sum_{l_1=0}^{l} \cdots \sum_{l_d=0}^{l} W_{(l_1,..l_d)} \; = \; \bigoplus_{|\mathbf{l}|_\infty \le l} W_{\mathbf{l}}, \tag{3}$$

Figure 1, showing the norms of the errors for the reconstruction of a two-dimensional function on a logarithmic scale, indicates that spaces $W_{\mathbf{l}}$ with large $|\mathbf{l}|_1$ contribute very little. In fact, it can be shown [10,3] that the size of the error committed by removing the space $W_{\mathbf{l}}$ is proportional to $2^{-r|\mathbf{l}|_1}$, where $r = 2$ in the case of piecewise linear functions. This suggests removing all spaces where the sum of resolutions is 'large'. The choice of $|\mathbf{l}|_1 \le l$ in (3) results in the sparse grid of [3] (see Fig. 1 for an example in three dimensions) but the grids can be chosen more generally.

Sparse grid spaces can also be achieved with the so-called combination technique [4] through the combination of certain spaces $V_{\mathbf{l}}$ instead of the difference

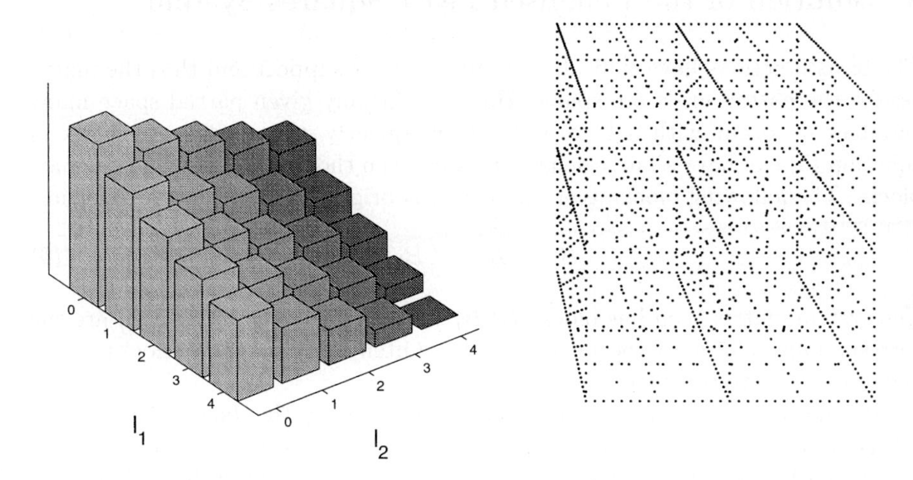

Fig. 1. Left: Norms of errors of the difference spaces $W_{\mathbf{l}}$ on a logarithmic scale. The example function is $u(x_1, x_2) = e^{-(x_1^2+x_2^2)}$ with $(x_1, x_2) \in [0,1] \times [0,1]$. Right: Sparse grid of refinement level $l = 5$ in three dimensions

spaces $W_{\mathbf{l}}$. As mentioned the grids can be chosen more generally, so we will now let the indices belong to an unspecified index set I, which leads to the generalised sparse grid space

$$S_I^d = \bigcup_{\mathbf{l} \in I} V_{\mathbf{l}}. \tag{4}$$

Each term in this sum is a tensor product of one dimensional spaces, but they are now restricted by the index set I and will generally have a much lower complexity.

See [4,6,9,10] for details and further references on this subsection.

2.2 Penalised Least Squares on Sparse Grids

To compute the partial functions $f_\alpha = f_{\mathbf{l}} \in V_{\mathbf{l}}$ on each grid, the functional $J(f)$ in equation (2) has to be minimised. Substituting the representation of $f_{\mathbf{l}} = \sum_i^m \alpha_i \varphi_i$, with $\{\varphi_i\}_{i=1}^m$ a basis of $V_{\mathbf{l}}$, into (2) and differentiation with respect to α_i results in the linear system, in matrix notation,

$$(\beta C + B^T B)\alpha = B^T y. \tag{5}$$

Here C is a square $m \times m$ matrix with entries $C_{j,k} = n\,(L\varphi_j, L\varphi_k)_{L_2}$ ($(\cdot,\cdot)_{L_2}$ denotes the standard scalarproduct in L_2) and B is a rectangular $n \times m$ matrix with entries $B_{i,j} = \varphi_j(\boldsymbol{x}^{(i)}), i = 1, \ldots n, j = 1, \ldots m$. Since $n \gg m$ one stores $B^T \cdot B$ and not B, this also allows to only use one matrix structure for both C and $B^T \cdot B$. See [6] for further details.

3 Solution of the Penalised Least Squares System

The basis functions have typically a small compact support and thus the matrices for the partial grids are sparse. However, for any given partial space many variables do occur with a large scale. Consequently, the coefficient matrix of equation (5) can have a fairly dense structure even though it has sparse diagonal blocks. The following blocking of the matrix B originates from the partial grids:

$$B = [B_1, \dots, B_k]. \tag{6}$$

The system matrix then has blocks $B_i^T B_j + \beta C_{i,j}$ where the blocks $C_{i,j}$ are the corresponding blocks of the penalty terms. Often, only diagonal blocks of the penalty terms are considered.

Solving for the partial grids amounts to solving the (sparse) equations $(B_i^T B_i + \beta C_{i,i})x_i = B_i^T y$. In general, we cannot assume that the x_i are good approximations for the components of the solution to the full problem. However, they are a good starting point. Note that in terms of the linear algebra, the large system of linear equations is usually singular, as the spaces V_j have non-trivial intersections. In fact, we will assume that the collection of partial spaces V_j is *relatively large*, and, in particular, for each pair V_i and V_j the intersection $V_i \cap V_j$ is also a space V_k. Solving the partial problems amounts to a block Jacobi preconditioning which does not directly lead to a good approximation. Instead one uses the combination formula where the partial solutions are scaled with a (combination) factor c_i. The combination formula becomes necessary as the same part of the solution may be contained in several different partial functions and thus needs to be subtracted again. This only needs to be taken into account for the block Jacobi variant; in the case of block Gauss-Seidel, the residuals are subtracted from the right-hand side after every partial fitting step. This, however, comes at a cost as the parallelism over the partial grids is lost as one needs to update the right-hand-side between fitting partial grids. As we solve our systems in parallel the block Gauss-Seidel method is less attractive.

Using the combination formula in the way described may lead to good solutions, however, these are in general an approximation to the solution of the full fitting problem. The solution can be improved with iterative refinement. This provides a block Jacobi-like variant for the solution of the full problem and thus we introduce an outer iteration to correct the approximation obtained from the combination formula. This outer iteration should also deal with errors introduced through the randomness of the data, which may be amplified by the combination method. In terms of the outer iteration, the solution of the partial systems together with the combination formula forms a preconditioner. We further accelerate the convergence of this outer iteration by using conjugate gradients. Experiments show that these outer iterations converge within a couple of steps.

The partial problems themselves are solved iteratively as well using conjugate gradients. While fitting to a regular grid is a standard procedure, it may converge slowly. Thus the question arises if one can get away with only a few iteration steps for the inner iterations as well. We have conducted initial experiments which show that the number of inner iterations can have a substantial influence on the solution and we observed saturation effects, where the outer iterations did not seem to be able to improve beyond a certain limit for low numbers of inner iterations. This behaviour did seem to be dependent on the type of problem and further investigations will be done to understand this better.

While this interaction between inner and outer iterations is an interesting topic for further investigation we will not further pursue it here but discuss the levels of parallelism which can be exploited both over the grids (corresponding to the other iterations) and within the grids (corresponding to the inner iterations).

4 Strategies for Exploiting Parallelism

In general, coarse grain parallelism is preferred, in particular for the application of distributed memory computing [11] as it typically has less (communication) overhead. However, in many cases the amount of available coarse grain parallelism of the algorithm is limited. If one would like to further parallelise the computations (if sufficient parallel resources are available) one would need to look at utilising finer grain parallelism as well. Such fine grain parallelism is also well suited to some shared memory and vector computations, in which case it can be competitive with the coarse grain parallelism.

The combination technique is straightforwardly parallel on a coarse grain level [12]. A second level of parallelisation on a fine grain level for each problem in the collection of grids can be achieved through the use of threads on shared-memory multi-processor machines. Both parallelisation strategies, i.e., the direct coarse grain parallel treatment of the different grids and the fine grain approach via threads, can also be combined and used simultaneously. This leads to a parallel method which is well suited for a cluster of multi-processor machines. See [13] for first results concerning speedups and efficiency.

4.1 Parallelisation across Grids

The linear systems 5) for the partial functions f_α of the collection of grids can be computed independently of each other, therefore their computation in each outer iteration step can be simply done completely in parallel. Each process computes the solution on a certain number of grids. If as many processors are available as there are grids in the collection of grids then each processor computes the solution for only one grid. The control process collects the results and computes the final function f on the sparse grid. Just a short setup or gather phase, respectively, is necessary. Since the cost of computation is roughly known *a-priori*, a simple but effective static load balancing strategy is available; see [14].

4.2 Parallelisation across Data

To compute $B^T \cdot B$ in (5) for each data instance $\boldsymbol{x}^{(i)}$, the product of the values of all basis function, which are non-zero at $\boldsymbol{x}^{(i)}$, has to be calculated, i.e.,

$\sum_{j,k} \varphi_j(\boldsymbol{x}^{(i)}) \cdot \varphi_k(\boldsymbol{x}^{(i)})$, and the results have to be written into the matrix structure at $\left(B^T \cdot B\right)_{j,k}$. These computations only depend on the data and therefore can be done independently for all instances. Therefore the $d \times n$ array of the training set can be separated in p parts, where p is the number of processors available in the shared-memory environment. Each processor now computes the matrix entries for n/p instances. Some overhead is introduced to avoid memory conflicts when writing into the matrix structure. In a similar way the evaluation of the classifier on the data points can be threaded in the evaluation phase.

4.3 Parallelisation of Solvers

After the matrix is built, threading can be used on SMP architectures in the solution phase as fine grain parallelism. Since we are using an iterative solver most of the computing time is used for the matrix-vector-multiplication. Here the vector α in (5) of size m can be split into p parts and each processor now computes the action of the matrix on a vector of size m/p.

4.4 Combination of Coarse and Fine Grain Parallelism

We have seen in [13] that coarse grain parallelism yields the highest speedups and the better efficiency. However, the number of grids may be such that the parallel resources are not fully utilised. Here we show how much additional speedup to expect when using fine grain parallelism in addition to the coarse grain parallelism.

If p processors are used to parallelise a computation with k grids one can expect a maximal speedup of $k/\lceil k/p \rceil$. This is displayed for the case of $p = 30$ and $k = 1, \ldots, 300$ in Fig. 2. In order to use the fine grain parallelism we first use the coarse grain approach for a first stage where $p\lfloor k/p \rfloor$ grids are processed and then, in a second stage the processors are distributed evenly among the remaining tasks. After the first step there are

$$p_x = k - p\lfloor k/p \rfloor < p$$

tasks remaining. Thus one has $p_l = \lfloor p/p_x \rfloor$ processors per remaining task. These are then used to parallelise all the remaining tasks and one thus gets a total speedup of

$$S_{p,k} = k/(\lfloor k/p \rfloor + \text{sign}(p_x) * (0.1 + 0.9/p_l).$$

This is again displayed in Fig. 2. We assume here that the fine grain parallelism has ten percent overhead which cannot be parallelised.

This approach can be refined further through the concurrent use of fine and coarse grain parallelism, e.g., parallelising $p/2$ grids with the parallelism across grids and a 2 processor shared memory parallelism for each grid.

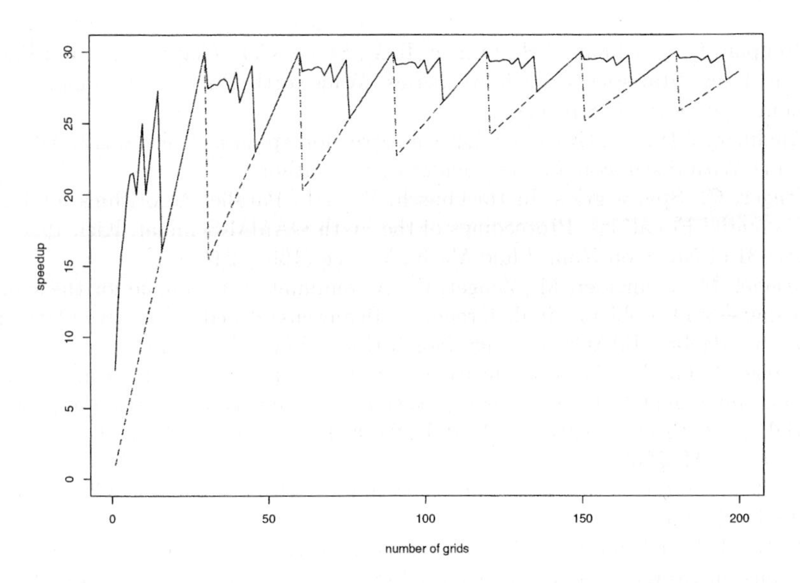

Fig. 2. Theoretical speedups for coarse grain parallelism (dashed line) and coarse grain with added fine grain

5 Conclusion

In this paper we extend results from [13], where two parallelisation strategies for the sparse grid combination technique where shown. Through the combination of both strategies the speedup results can be further improved. This leads to a parallel method which is well suited for a cluster of multi-processor machines.

The performance of the outer iteration using new combination coefficients does depend on the commutation properties of the projection operators into the partial spaces. In particular, if the projection operators commute, the outer iteration gives the correct result after one step. In addition, the performance depends on the partial spaces in a different way: If all the partial spaces are one-dimensional, then the outer iteration again gives the exact result after one step. Alternatively, if the spaces form an ordered sequence $V_1 \subset V_2 \subset \ldots V_m$ and the dimensions are $\dim(V_k) = k$ then again the new combination method gives an exact result after one step (which can be seen by using orthogonalisation). It would thus seem that in general more spaces would provide better convergence of the outer iteration. In the case of the solution of the Laplace equation with finite elements, it is known that the classical combination method acts like extrapolation and larger error terms cancel [4]. Note that these classical combination methods do also use a large number of subspaces. Thus larger numbers of subspaces would appear to provide better performance. The extrapolation properties of the general combination method for fitting problems will be further investigated in the future.

References

1. Breiman, L., Friedman, J.H., Olshen, R.A., Stone, C.J.: Classi.cation and Regression Trees. Statistics/Probability Series. Wadsworth Publishing Company, Belmont, California, U.S.A. (1984)
2. Friedman, J.H.: Multivariate adaptive regression splines. Ann. Statist. **19** (1991) 1–141 With discussion and a rejoinder by the author.
3. Zenger, C.: Sparse grids. In Hackbusch, W., ed.: Parallel Algorithms for Partial Di.erential Equations, Proceedings of the Sixth GAMM-Seminar, Kiel, 1990. Volume 31 of Notes on Num. Fluid Mech., Vieweg (1991) 241–251
4. Griebel, M., Schneider, M., Zenger, C.: A combination technique for the solution of sparse grid problems. In de Groen, P., Beauwens, R., eds.: Iterative Methods in Linear Algebra, IMACS, Elsevier, North Holland (1992) 263–281
5. Garcke, J., Griebel, M.: Classification with sparse grids using simplicial basis functions. Intelligent Data Analysis 6 (2002) 483–502 (shortened version appeared in KDD 2001, Proc. Seventh ACM SIGKDD, F. Provost and R. Srikant (eds.), pages 87–96, ACM, 2001).
6. Garcke, J., Griebel, M., Thess, M.: Data mining with sparse grids. Computing **67** (2001) 225–253
7. Hastie, T.J., Tibshirani, R.J.: Generalized additive models. Volume 43 of Monographs on Statistics and Applied Probability. Chapman and Hall Ltd., London (1990)
8. Hastie, T., Tibshirani, R.: Generalized additive models. Statist. Sci. 1 (1986) 297–318 With discussion.
9. Hegland, M.: Additive sparse grid .tting. In: Proceedings of the Fifth International Conference on Curves and Surfaces, Saint-Malo, France 2002. (2002) submitted.
10. Hegland, M., Nielsen, O.M., Shen, Z.: High dimensional smoothing based on multilevel analysis. Submitted (2000) Available at `http://datamining.anu.edu.au/publications/2000/hisurf2000.ps.gz`.
11. Blackford, L.S., Choi, J., Cleary, A., D'Azevedo, E., Demmel, J., Dhillon, I., Dongarra, J., Hammarling, S., Henry, G., Petitet, A., Stanley, K., Walker, D., Whaley, R.C.: ScaLAPACK Users' Guide. Society for Industrial and Applied Mathematics, Philadelphia, PA (1997)
12. Griebel, M.: The combination technique for the sparse grid solution of PDEs on multiprocessor machines. Parallel Processing Letters 2 (1992) 61–70
13. Garcke, J., Griebel, M.: On the parallelization of the sparse grid approach for data mining. In Margenov, S., Wasniewski, J., Yalamov, P., eds.: Large-Scale Scienti.c Computations, Third International Conference, Sozopol, Bulgaria. Volume 2179 of Lecture Notes in Computer Science. (2001) 22–32
14. Griebel, M., Huber, W., Störtkuhl, T., Zenger, C.: On the parallel solution of 3D PDEs on a network of workstations and on vector computers. In Bode, A., Cin, M.D., eds.: Lecture Notes in Computer Science 732, Parallel Computer Architectures: Theory, Hardware, Software, Applications, Springer Verlag (1993) 276–291

An Augmented Lanczos Algorithm for the Efficient Computation of a Dot-Product of a Function of a Large Sparse Symmetric Matrix

Roger B. Sidje[1], Kevin Burrage[1], and B. Philippe[2]

[1] Department of Mathematics
University of Queensland
Brisbane QLD 4072, Australia
{rbs,kb}@maths.uq.edu.au
[2] INRIA/IRISA
Campus de Beaulieu
35042 Rennes Cedex, France
philippe@irisa.fr

Abstract. The Lanczos algorithm is appreciated in many situations due to its speed and economy of storage. However, the advantage that the Lanczos basis vectors need not be kept is lost when the algorithm is used to compute the action of a matrix function on a vector. Either the basis vectors need to be kept, or the Lanczos process needs to be applied twice. In this study we describe an augmented Lanczos algorithm to compute a dot product relative to a function of a large sparse symmetric matrix, without keeping the basis vectors.

1 Introduction

To compute the action of a matrix-function on a vector, $f(A)v$, where A is a large sparse symmetric matrix of order n, and f is a function for which A is admissible (i.e., $f(A)$ is defined), the general principle of the Lanczos method relies on projecting the operator A onto the Krylov subspace $\mathcal{K}_m(A, v) = \text{Span}\{v, Av, \ldots, A^{m-1}v\}$ and approximating the action of the function in the subspace. Here, m is the prescribed dimension of the Krylov subspace and in practice $m \ll n$. The classical Lanczos Algorithm 1 will start with the vector v and perform a three-term recurrence scheme to construct a symmetric tridiagonal matrix T_m and an orthonormal basis V_m of $\mathcal{K}_m(A, v)$:

$$T_m = \begin{bmatrix} \alpha_1 & \beta_2 & & \\ \beta_2 & \alpha_2 & \ddots & \\ & \ddots & \ddots & \beta_m \\ & & \beta_m & \alpha_m \end{bmatrix}, \qquad V_m = [v_1, v_2, \ldots, v_m]$$

and for any $j = 1, \ldots, m$ we have the fundamental relations

P.M.A. Sloot et al. (Eds.): ICCS 2003, LNCS 2659, pp. 693–704, 2003.

$$\begin{aligned} V_j^T V_j &= I \\ AV_j &= V_j T_j + \beta_{j+1} v_{j+1} e_j^T \\ V_j^T A V_j &= T_j \end{aligned} \quad (1)$$

where $V_j = [v_1, \ldots, v_j]$, T_j is the jth leading submatrix of T_m, e_j is the jth column of the identity matrix of appropriate size. The pseudo-code for the Lanczos process is indicated below.

ALGORITHM 1: Lanczos(m, A, v)

$\beta = \|v\|_2$; $v_1 := v/\beta$; $\beta_1 := 0$;
for $j := 1 : m$ **do**
  $p := Av_j - \beta_j v_{j-1}$;
  $\alpha_j := v_j^T p$; $p := p - \alpha_j v_j$;
  $\beta_{j+1} := \|p\|_2$; $v_{j+1} := p/\beta_{j+1}$;
endfor

From this, $v = \beta V_m e_1$ where $\beta = \|v\|_2$. To compute $w = f(A)v$, the underlying principle is to use the approximation $w \approx \widetilde{w}$ where

$$\widetilde{w} = \beta V_m f(T_m) e_1. \quad (2)$$

Thus, the distinctive feature is that the original large problem is converted to the smaller problem (2) which is more desirable. Moreover the special structure of T_m (symmetric tridiagonal) means that, when possible, the smaller matrix-function action $f(T_m)e_1$ can be handled in an efficient, tailor-made manner. Theory and practice over the years have confirmed that this simplified process is amazingly effective for a wide class of problems. In general however, the basis $V_m = [v_1, \ldots, v_m]$ has to be kept for the final evaluation of $\widetilde{w} = \beta V_m f(T_m) e_1$. The basis V_m is a dense matrix of size $n \times m$, and when n is very large, such a storage is seen as a drawback in certain configurations. In specific contexts such as linear systems (conjugate gradient-type methods) or eigenvalue problems (where eigenvectors are not needed), it is possible to avoid storing the basis vectors and this is considered as a significant strength of the Lanczos scheme. In principle however, it is also possible to avoid this storage by applying the Lanczos process twice. But the perception that Lanczos storage savings are lost when evaluating $f(A)v$ has prompted some authors to offer alternative methods outside the Krylov-based paradigm. For instance, Bergamaschi and Vianello [2] used Chebyshev series expansions to compute the action of the exponential operator in the case of large sparse symmetric matrices arising after space discretization of 2D and 3D parabolic PDEs by finite differences and finite elements. The configuration used for their experiments was a 600Mhz Alpha station with 128MB of RAM. Although the Chebyshev series method requires an *a priori* estimation of the leftmost and rightmost eigenvalues of A, and is known to converge slower than the Krylov approach, they showed that when memory is at premium, some advantages of the Krylov method may need to be traded.

In this study, we present an augmented Lanczos algorithm which alleviates the drawback, allowing the Krylov framework to be fully retained when the issue at hand is really to compute $u^T f(A)v$ for given vectors u and v, i.e., when we are interested in a weighted sum of the action of matrix-function on a vector. As a particular example, taking $u = e_i$ and $v = e_j$ allows retrieving the ij-th entry of $f(A)$. Assuming the matrix is positive definite, this was used to bound entries of a matrix function [1,5,6]. In this study however, we do not assume that the matrix is positive definite. Another example worth mentioning is the evaluation of the norm $\|f(A)v\|_2$ which can be handled as a special case. Thus, compared to a method such as the Chebyshev method, the benefits are two-fold:

- Economy of storage: the Lanczos basis vectors need not be saved. This alleviates the primary drawback that motivated the Chebyshev method, and
- Potential re-usability: the computed data can be re-used to get $\{u_p^T f(A)v\}$ for a family of vectors $\{u_p\}$. Note that because the matrix is symmetric, computing $\{u_p^T f(A)v\}$ is algorithmically equivalent to computing $\{u^T f(A)v_q\}$ (but not equivalent to computing $\{u_p^T f(A)v_q\}$ which would need several sweeps). The computed data can also be re-used to get $u^T f_p(A)v$ for a family of functions $\{f_p\}$. This can be useful when the argument matrix depends on another variable, e.g., the matrix may belong to the parameterized family $A(t) = tA$ where t is the 'time' variable. As a concrete example, evaluating $\int_0^T u^T f(tA)v$ may involve discretizing the integration domain and computing the matrix function at several knots. Sometimes also, the matrix may vary in terms of rank-updates – such problems can be seen in Bernstein and Van Loan [3] though in the unsymmetric context. A careful implementation of our scheme can offer a suitable re-usable framework where appropriate.

2 The Augmented Lanczos Algorithm

The principle of augmenting a Krylov subspace is well-known in the literature and has been used in several other contexts. Differences between approaches usually lie in the strategy by which the augmentation is made. Probably, the most popular Krylov algorithm obtained via augmentation is the GMRES algorithm of Saad and Schultz [11] which resulted from taking full advantage of the information computed by the Arnoldi procedure (or FOM – Full Orthogonalization Method). Our proposed augmented scheme falls in a class of similar approaches. To get $u^T f(A)v$, the idea is to augment the Lanczos basis V_m as follows. Define the orthogonal projector onto V_m as $P_m = V_m V_m^T$, and let

$$\hat{V}_{m+1} = [V_m \mid \hat{v}_{m+1}]$$

where

$$\hat{v}_{m+1} = \frac{(I - P_m)u}{\|(I - P_m)u\|_2},$$

and take an approximation in this augmented subspace as $u^T w \approx u^T \hat{w}$ where

$$\hat{w} = \beta \hat{V}_{m+1} f(\hat{V}_{m+1}^T A \hat{V}_{m+1}) e_1.$$

Now, we have the symmetric matrix

$$\hat{V}_{m+1}^T A \hat{V}_{m+1} = \begin{bmatrix} V_m^T \\ \hat{v}_{m+1}^T \end{bmatrix} A[V_m \mid \hat{v}_{m+1}] = \begin{bmatrix} T_m & V_m^T A \hat{v}_{m+1} \\ \hat{v}_{m+1}^T A V_m & \hat{v}_{m+1}^T A \hat{v}_{m+1} \end{bmatrix}$$

and expanding further, we get

$$\hat{v}_{m+1}^T A V_m = \hat{v}_{m+1}^T (V_m T_m + \beta_{m+1} v_{m+1} e_m^T) = \beta_{m+1} \hat{v}_{m+1}^T v_{m+1} e_m^T$$

and letting

$$\hat{\beta}_{m+1} \equiv \beta_{m+1} \hat{v}_{m+1}^T v_{m+1} \tag{3}$$

and

$$\hat{\alpha}_{m+1} \equiv \hat{v}_{m+1}^T A \hat{v}_{m+1} \tag{4}$$

we see that

$$\hat{V}_{m+1}^T A \hat{V}_{m+1} \equiv \hat{T}_{m+1} = \left[\begin{array}{cccc|c} & & & & 0 \\ & & T_m & & \vdots \\ & & & & 0 \\ & & & & \hat{\beta}_{m+1} \\ \hline 0 & \cdots & 0 & \hat{\beta}_{m+1} & \hat{\alpha}_{m+1} \end{array}\right].$$

Hence $\hat{T}_{m+1}$ remains *tridiagonal*. Furthermore, just as with the standard Lanczos process, the work at one step is a dovetail update from the work at the previous step, without the need to store the entire set of basis vectors. Indeed $V_m^T u = [v_1^T u, \ldots, v_m^T u]^T = [(V_{m-1}^T u)^T, v_m^T u]^T$ and $(I - P_m)u = (I - V_m V_m^T)u = (I - v_m v_m^T) \cdots (I - v_1 v_1^T)u$ can be computed incrementally as basis vectors become available. We now give additional details on how $\hat{\alpha}_{m+1}$ can be efficiently updated instead of using just (4). Direct calculation shows that

$$\begin{aligned} \hat{\alpha}_{m+1} \equiv \hat{v}_{m+1}^T A \hat{v}_{m+1} &= \frac{u^T (I - P_m) A (I - P_m) u}{\|(I - P_m)u\|_2^2} \\ &= \frac{u^T A u - u^T V_m T_m V_m^T u - 2\beta_{m+1} (v_{m+1}^T u)(v_m^T u)}{\|(I - P_m)u\|_2^2}. \end{aligned}$$

It is therefore clear that we also have

$$\hat{\alpha}_m = \frac{u^T A u - u^T V_{m-1} T_{m-1} V_{m-1}^T u - 2\beta_m (v_m^T u)(v_{m-1}^T u)}{\|(I - P_{m-1})u\|_2^2}.$$

But further calculation shows that

$$u^T V_m T_m V_m^T u = u^T V_{m-1} T_{m-1} V_{m-1}^T u + 2\beta_m (v_{m+1}^T u)(v_m^T u) + \alpha_m (v_m^T u)^2,$$

and hence we end up with the updating formula

$$\hat{\alpha}_{m+1} \|(I - P_m)u\|_2^2 = \hat{\alpha}_m \|(I - P_{m-1})u\|_2^2 - 2\beta_{m+1} (v_{m+1}^T u)(v_m^T u) - \alpha_m (v_m^T u)^2$$

in which we also have

$$\|(I - P_m)u\|_2^2 = \|(I - P_{m-1})u\|_2^2 + (v_m^T u)^2.$$

Therefore, the vector u interacts with A only through one extra matrix-vector product needed to form the initial $\hat{\alpha}_2$. And from there subsequent updates are made by scalar recurrences using quantities readily available.

Remarks

Rather than adding the augmented vector u at the end, we could have put it at the beginning. And then, we could either 1) orthogonalize u against each incoming vector as before, or 2) orthogonalize each incoming vector against u.

1. In the first case, we would get $\tilde{V}_{m+1} = [\hat{v}_{m+1} \mid V_m]$, with V_m *unchanged* and $\hat{v}_{m+1}$ resulting from orthogonalizing u against each incoming vector as before. This is therefore equivalent to a permutation $\tilde{V}_{m+1} = \hat{V}_{m+1}\widetilde{P}_{m+1}$ where $\widetilde{P}_{m+1} = [e_{m+1}, e_1, \ldots, e_m]$ is the $(m+1)$-by-$(m+1)$ permutation matrix that produces the effect. It follows that

$$\tilde{V}_{m+1}^T A \tilde{V}_{m+1} = \widetilde{P}_{m+1}^T \hat{V}_{m+1}^T A \hat{V}_{m+1} \widetilde{P}_{m+1} = \widetilde{P}_{m+1}^T \hat{T}_{m+1} \widetilde{P}_{m+1}$$

   and this shows that the tridiagonal form is destroyed. If we leave out $\hat{v}_{m+1}$ we recover the classical Lanczos quantities, whereas our primary interest in this study is to explore an augmented approach and summarize our findings.
2. If u is added at the beginning and each incoming vector is orthogonalized against it, the resulting matrix from iterating on $\{u, v, Av, \ldots\}$ is not tridiagonal (although the restriction onwards from v is). Augmenting at the beginning is also what is often termed as *locking* or *deflation* when used as a means to accelerate the convergence of eigenvalue problems or linear systems.

It is worth noting that when dealing with the same augmented subspace $\mathcal{K}_m(A, v) + \{u\}$, there exists transformation matrices between the computed bases. Our proposed approximation scheme offers the advantage of retaining the standard Lanczos elements. Existing Lanczos implementations would therefore provide a good starting code base and as we shall see, the scheme is also well suited for establishing further results. It should also be clear such an augmented scheme can be applied to the Arnoldi procedure. But, by itself, the Arnoldi procedure already needs all basis vectors to perform the Gram-Schmidt sweeps.

3 Alternative Approaches

To compute $u^T f(A) v$, our motivation was to avoid keeping the entire set of Lanczos basis vectors. Nor did we want to apply the process twice and incur the associated cost. Aside from leaving out u and embedding $u^T V_m$ in the classical Lanczos scheme as noted earlier, these goals could be achieved by other approaches such as the block Lanczos algorithm and the bi-orthogonal Lanczos algorithm in the ways outlined below.

If started with $[v, u]$, the block Lanczos computes a block tridiagonal matrix and an orthonormal basis V_{2m} of $\mathcal{K}_m(A, v) + \mathcal{K}_m(A, u)$. Since u and v can be expressed in the computed basis as $u = V_{2m} V_{2m}^T u$ and $v = V_{2m} V_{2m}^T v$, an approximation to $u^T f(A) v$ could be retrieved without storing all of V_{2m}. However, doing a block Lanczos in this context is twice the cost of the augmented scheme, although the approach satisfies a particular theoretical characterization as we shall see below.

Another approach is the bi-orthogonal Lanczos algorithm which is usually meant for unsymmetric problems. But it is possible to deliberately apply it here to simultaneously compute a basis of $\mathcal{K}_m(A, v)$ and a basis of $\mathcal{K}_m(A, u)$ from where to approximate $u^T f(A)v$. Below we briefly summarize this well-known algorithm using its general, unsymmetric notation. We do not dwell here on the obvious notational simplifications that can be made since the matrix is symmetric.

<u>ALGORITHM 2</u>: Bi-orthogonal Lanczos

$\beta := \|v\|_2\,; \quad \alpha := (u^T v)/\beta\,;$
$v_1 := v/\beta\,; \quad u_1 := u/\alpha\,;$
for $j := 1 : m$ **do**
  $p := Av_j\,; \quad q := A^T u_j\,;$
  if $j > 1$ **then**
    $p := p - \gamma_j v_{j-1}\,; \quad q := q - \beta_j u_{j-1}\,;$
  endif
  $\alpha_j := u_j^T p\,;$
  $p := p - \alpha_j v_j\,; \quad q := q - \alpha_j u_j\,; \quad s := q^T p\,;$
  $\beta_{j+1} := \sqrt{|s|}\,;$
  $\gamma_{j+1} := \mathrm{sign}(s)\,\beta_{j+1}\,;$
  $v_{j+1} := p/\beta_{j+1}\,; \quad u_{j+1} := q/\gamma_{j+1}\,;$
endfor

Upon completion of this algorithm, we get a tridiagonal matrix

$$T_m = \begin{bmatrix} \alpha_1 & \gamma_2 & & \\ \beta_2 & \alpha_2 & \ddots & \\ & \ddots & \ddots & \gamma_m \\ & & \beta_m & \alpha_m \end{bmatrix},$$

and $V_m \equiv [v_1, \ldots, v_m]$ is a basis of $\mathcal{K}_m(A, v)$, $U_m \equiv [u_1, \ldots, u_m]$ is a basis of $\mathcal{K}_m(A^T, u)$. The bases are bi-orthogonal, i.e.,

$$U_m^T V_m = V_m^T U_m = I \tag{5}$$

and furthermore we have the relations

$$AV_m = V_m T_m + \beta_{m+1} v_{m+1} e_m^T \tag{6}$$
$$U_m^T A V_m = T_m \tag{7}$$
$$A^T U_m = U_m T_m^T + \gamma_{m+1} u_{m+1} e_m^T. \tag{8}$$

From these, we see that $v = \beta V_m e_1$ and $u = \alpha U_m e_1$ and the approximation is computed as $u^T f(A) v = \alpha\beta e_1^T U_m^T f(A) V_m e_1 \approx \alpha\beta e_1^T f(U_m^T A V_m) e_1 = \alpha\beta e_1^T f(T_m) e_1$. And this relationship allows us to avoid storing the bases explicitly. In our symmetric context, we replace A^T with A. Each step requires two matrix-vector products, so it is twice as expensive as of the augmented scheme, but as we shall see below the bi-orthogonal Lanczos algorithm turns out to have certain theoretical characteristics similar to that of the block Lanczos.

4 Some Properties

The notation used here is that introduced earlier for each relevant method. We first start with the proposed augmented process. In analogy with (1) the augmented process satisfies the following relations

$$\begin{aligned}\hat{V}_{m+1}^T\hat{V}_{m+1} &= I\\ \hat{V}_{m+1}^T A\hat{V}_{m+1} &= \hat{T}_{m+1}\\ A\hat{V}_{m+1} &= \hat{V}_{m+1}\hat{T}_{m+1} + (\beta_{m+1}v_{m+1} - \hat{\beta}_{m+1}\hat{v}_{m+1})e_m^T - \hat{\beta}_{m+1}v_m e_{m+1}^T +\\ &\quad + (A - \hat{\alpha}_{m+1}I)\hat{v}_{m+1}e_{m+1}^T\\ &= \hat{V}_{m+1}\hat{T}_{m+1} + (I - \hat{v}_{m+1}\hat{v}_{m+1}^T)\Big(\beta_{m+1}v_{m+1}e_m^T +\\ &\quad + (I - v_m v_m^T)A\hat{v}_{m+1}e_{m+1}^T\Big)\end{aligned}$$

where e_m and e_{m+1} are of length $m+1$. We have already seen that the first and second relations hold by construction. To get the third relation, we write

$$A\hat{V}_{m+1} = A[V_m \mid \hat{v}_{m+1}] = [V_m T_m + \beta_{m+1}v_{m+1}e_m^T \mid A\hat{v}_{m+1}]$$

and

$$\begin{aligned}\hat{V}_{m+1}\hat{T}_{m+1} &= [V_m \mid \hat{v}_{m+1}]\begin{bmatrix} T_m & \hat{\beta}_{m+1}e_m \\ \hat{\beta}_{m+1}e_m^T & \hat{\alpha}_{m+1}\end{bmatrix}\\ &= [V_m T_m + \hat{\beta}_{m+1}\hat{v}_{m+1}e_m^T \mid \hat{\beta}_{m+1}v_m + \hat{\alpha}_{m+1}\hat{v}_{m+1}]\end{aligned}$$

and the relation comes after some algebraic manipulation. The last follows from the fact that $\hat{\alpha}_{m+1} = \hat{v}_{m+1}^T A\hat{v}_{m+1}$ and $\beta_{m+1}\hat{v}_{m+1}^T v_{m+1} = \hat{\beta}_{m+1} = v_m^T A\hat{v}_{m+1}$.

The basis V_m and the tridiagonal matrix T_m computed by the original Lanczos process are known to satisfy $p(A)v_1 = V_m p(T_m)e_1$ for any polynomial p of degree $\leq m-1$ (see, e.g., [9, Lemma 3.1]). This is still valid with the augmented process since it fully retains the original Lanczos elements. Thus we have $p(A)v_1 = \hat{V}_{m+1}p(\hat{T}_{m+1})e_1$ for any polynomial p of degree $\leq m-1$. In fact, we now show that we can extend this property to a higher degree in our context.

Proposition 1. *At the m-th step of the augmented Lanczos process, we have* $u^T p(A)v_1 = u^T\hat{V}_{m+1}p(\hat{T}_{m+1})e_1$ *for any polynomial p of degree* $\leq m$.

Proof. Since the result is already true for a degree $\leq m-1$, it remains to establish the degree m. Let $p_m(z) = a_0 + zq_{m-1}(z)$ be a polynomial of degree m in which q_{m-1} is a polynomial of degree $m-1$ and $a_0 = p_m(0)$ is a scalar. We have

$$\begin{aligned}p_m(A)v_1 &= a_0v_1 + Aq_{m-1}(A)v_1 = a_0v_1 + A\hat{V}_{m+1}q_{m-1}(\hat{T}_{m+1})e_1\\ &= a_0v_1 + \Big(\hat{V}_{m+1}\hat{T}_{m+1} + (I - \hat{v}_{m+1}\hat{v}_{m+1}^T)\Big(\beta_{m+1}v_{m+1}e_m^T +\\ &\quad + (I - v_m v_m^T)A\hat{v}_{m+1}e_{m+1}^T\Big)\Big)q_{m-1}(\hat{T}_{m+1})e_1\\ &= \hat{V}_{m+1}p_m(\hat{T}_{m+1})e_1 + (I - \hat{v}_{m+1}\hat{v}_{m+1}^T)\Big(\beta_{m+1}v_{m+1}e_m^T +\\ &\quad + (I - v_m v_m^T)A\hat{v}_{m+1}e_{m+1}^T\Big)q_{m-1}(\hat{T}_{m+1})e_1\end{aligned}$$

Since $\hat{T}_{m+1}$ is a tridiagonal matrix of order $m+1$ and q_{m-1} is a polynomial of degree $m-1$, we have $e_{m+1}^T q_{m-1}(\hat{T}_{m+1})e_1 = 0$. Furthermore $u^T(I - \hat{v}_{m+1}\hat{v}_{m+1}^T)v_{m+1} = 0$.

Proposition 1 means that the approximation is *exact* if the minimal degree of v with respect to the matrix A is $\leq m$. In general, the *a priori* error bounds obtained using functional calculus (as in [7]) for the usual Lanczos scheme for approximating $f(A)v$ can be transposed here. In particular, it is known from the theory of matrix functions [8, Chap.6] that

$$f(A)v = p_{n-1}(A)v \tag{9}$$

where p_{n-1} is the Hermite interpolation polynomial of f at $\lambda(A)$, the spectrum of A. Hence,

$$u^T\hat{w} = \beta u^T\hat{V}_{m+1}f(\hat{T}_{m+1})e_1 = \beta u^T\hat{V}_{m+1}p_m(\hat{T}_{m+1})e_1 = u^Tp_m(A)v$$

with p_m being the Hermite interpolation polynomial (of degree m) at the modified 'Ritz' values [1]. The error can be written in a polynomial form as

$$|u^Tf(A)v - \beta u^T\hat{V}_{m+1}f(\hat{T}_{m+1})e_1| = |u^Tp_{n-1}(A)v - u^Tp_m(\hat{T}_{m+1})v|$$

The theory of Krylov subspaces [10] shows that $\lambda(\hat{T}_{m+1})$ not only approximates a larger subset of $\lambda(A)$ as m increases but also approximates it more accurately. The next results shed some light on the behavior of the bi-orthogonal Lanczos process and the block Lanczos process.

Proposition 2. *Let T_m be the resulting tridiagonal matrix at the m-th step of the bi-orthogonal Lanczos algorithm, then $u_1^Tp(A)v_1 = e_1^Tp(T_m)e_1$ for any polynomial p of degree $\leq 2(m-1)$.*

Proof. Any polynomial p of degree $\leq 2(m-1)$ can be factored as $p(z) = q(z)r(z) + s(z)$ where q, r, s are polynomials of degree $<= m-1$. From this, if follows that $u_1^Tp(A)v_1 = (q(A)u_1)^Tr(A)v_1 + u_1^Ts(A)v_1 = (U_mq(T_m)e_1)^TV_mr(T_m)e_1 + e_1^TU_m^TV_ms_m(T_m)e_1$ and the bi-orthogonality property (5) yields the result.

Proposition 3. *Let T_{2m} be the resulting block-tridiagonal matrix at the m-th step of the block Lanczos algorithm started with MGS(v,u) where MGS stands for the Modified Gram-Schmidt procedure, then $u^Tp(A)v = \beta(\omega_1e_1^T + \omega_2e_2^T)p(T_{2m})e_1$ for any polynomial p of degree $\leq 2(m-1)$ where ω_1 and ω_2 are the coefficients relative to u from the MGS(v,u) procedure.*

Proof. The block Lanczos procedure will generate an orthonormal basis for $\mathcal{K}_m(A,v) + \mathcal{K}_m(A,u)$, hence the assertion is already true for any polynomial of degree $\leq m-1$. The rest of the proof proceeds as in Proposition 2 using the basis of dimension $2m$ generated by the block Lanczos algorithm.

[1] These are not really the Ritz values since the tridiagonal matrix is augmented in a special way. But since the eigenvalues of $\hat{T}_{m+1}$ embed the Ritz values according to Cauchy's interleaving theorem, they remain close.

We should draw the attention of the reader on the fact that an apparent higher polynomial order of an approach is not synonymous with higher accuracy. A higher degree polynomial does not necessarily convey a better approximation to a *matrix function.* An illustrative case in point is the Taylor series expansion which can still converge slowly even after summing a large number of terms. As relation (9) showed, the underlying approximations are going to depend on how good the spectrum of A is approximated in the respective situations. Since the three approaches usually yield comparable approximations to the spectrum, the augmented approach stands out due to its cheaper cost.

5 Numerical Experiments

There are a number of matrix functions that arise frequently in the literature. In our first experiment, we consider the sine function $f_1(x) = \sin x$. In the second experiment, we consider an example from Druskin and Knizhnerman [4]: the exponential function with square-root $f_2(x) = e^{-\theta\sqrt{x}}$, $\theta \geq 0$. This example results from the boundary value problem

$$Aw - \frac{d^2w}{d\theta^2} = g(\theta)v.$$

The function f_2 is the solution of the boundary value problem $w(0) = v$, $w(+\infty) = 0$ with $g = 0$. References in Druskin and Knizhnerman [4] include applications where this problem arises, for example in geophysical computed tomography. In our experiments, we used the functions on symmetric matrices from the Harwell-Boeing collection.

Let f_{exact} be the computed value of $u^T f(tA)v$. For the purpose of the experiments, it was computed by diagonalizing the matrix. Each figure includes three error curves corresponding to $\|f_{\text{exact}} - f_{m^{\text{th}}\text{approx}}\|_2 / \|f_{\text{exact}}\|$, where the m-th approximation, $f_{m^{\text{th}}\text{approx}}$, is either the approximation at the m-th iteration with the proposed augmented scheme (plain curve with circles), the approximation with the bi-orthogonal Lanczos (dashed curved with triangles), or the classical approximation $u^T\tilde{w} = \beta u^T V_m f(tT_m)e_1$ (dotted curve with crosses).

On the reported examples, all methods tend to converge very well as generally observed with Krylov-based approximations to matrix functions. The first example has some interesting observations. The bi-orthogonal Lanczos scheme exhibits faster convergence at the beginning but is subsequently caught up by the other methods. What is even more noticeable is that the convergence of the bi-orthogonal scheme stalls at some point. Further investigation showed us that a marked lost of orthogonality started in the bi-orthogonal process and so the process was contaminated from that point onwards (since we were not dealing with a production code, our testings did not involve re-orthogonalization techniques that may be needed with Lanczos-based algorithms). Since in general, the bi-orthogonal scheme is more sensitive to lost of orthogonality, that is yet another unfavorable argument against using it without cross-checking.

5.1 Example 1

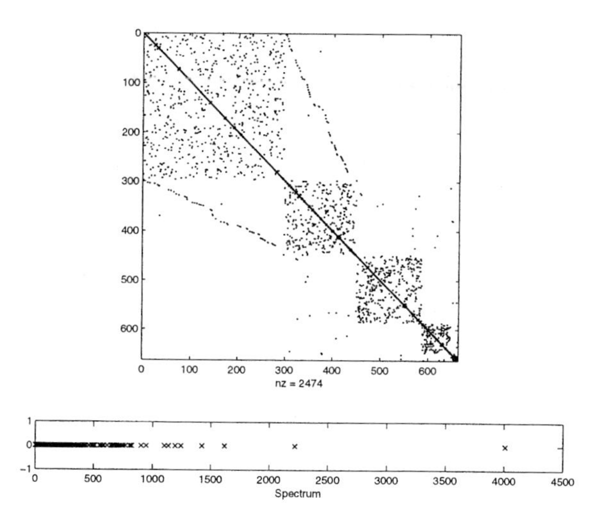

Sparsity pattern and spectrum

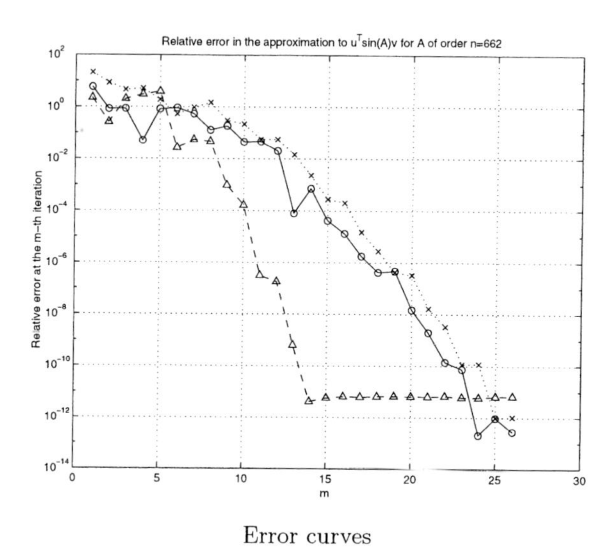

Error curves

Fig. 1. We consider $f_1(x) = \sin x$ and compute $u^T f_1(tA)v$ for random u and v, and a 662-by-662 matrix A known as 662bus in the Harwell-Boeing collection. The sparsity pattern and spectrum are shown on the left side (the spectrum lies on the real line since A is symmetric). We took $t = 0.01$, for A is of large norm.

5.2 Example 2

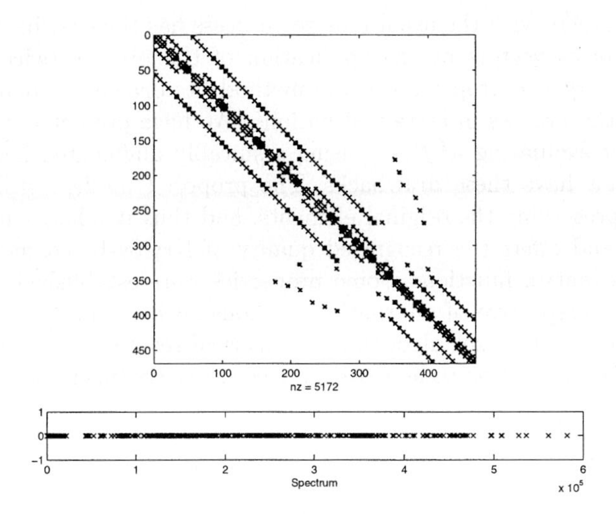

Sparsity pattern and spectrum

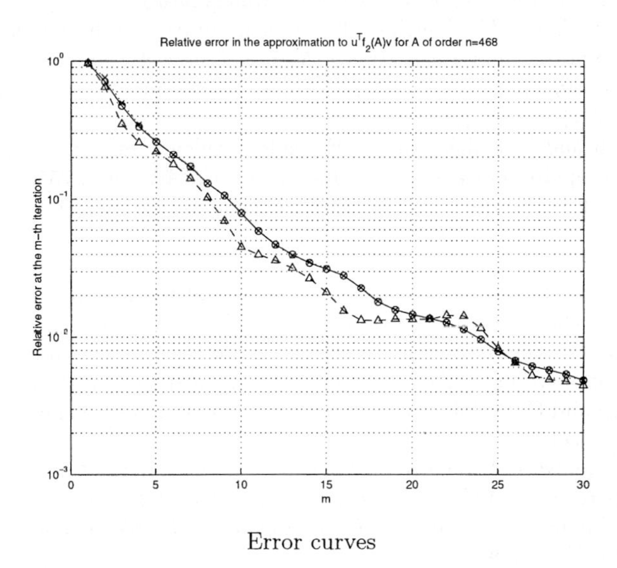

Error curves

Fig. 2. We consider $f_2(x) = e^{-\theta\sqrt{x}}$, $\theta = 0.01$ and compute $u^T f_2(A)v$ where u and v are randomly generated, and A is the 468-by-468 matrix known as `nos5` in the Harwell-Boeing collection

6 Conclusion

Evaluating $f(A)v$ with the usual Lanczos process has the drawbacks of the storage of the basis vectors or the application of the process twice. Despite the reputed convergence properties of the method, its drawbacks deter some users from using the process in certain situations. We have presented a cost-effective approach for evaluating $u^T f(A)v$ using a suitably augmented Lanczos process that does not have these drawbacks. The proposed method still has the advantage of preserving the original elements, and thus it inherits from the same foundation and offers the recognized quality of Krylov-based methods for approximating matrix functions. Some properties were established and placed in the broader perspective of alternative methods such as the block Lanczos and the bi-orthogonal Lanczos algorithms. Numerical results were made to test the quality of the method on some frequently used matrix functions.

References

1. M. Benzi and G. Golub. Bounds for the entries of matrix functions with applications to preconditioning. *BIT Numerical Mathematics*, 39(3):417–438, 1999.
2. L. Bergamaschi and M. Vianello. Efficient computation of the exponential operator for large, sparse, symmetric matrices. *Numer. Linear Algebra Appl.*, 7:27–45, 2000.
3. D. S. Bernstein and C. F. Van Loan. Rational matrix functions and rank-1 updates. *SIMAX*, 22(1):145–154, 2000.
4. V. Druskin and L. Knizhnerman. Extended krylov subspaces: Approximation of the matrix square root and related functions. *SIMAX*, 19(3):755–771, 1998.
5. G. Golub and G. Meurant. Matrices, moments and quadrature. In D. F. Griffiths and G. A Waston, editors, *Numerical Analysis 93.* Pitman Research Notes in Mathematics 303, Longman Scientific and Technical, 1993.
6. G. Golub and G. Meurant. Matrices, moments and quadrature. ii. how to compute the norm of the error in iterative methods. *BIT*, 37(3):687–705, 1997.
7. M. Hochbruck and Ch. Lubich. On Krylov subspace approximations ot the matrix exponential operator. *SIAM J. Numer. Anal.*, 34(5):1911–1925, October 1997.
8. R. A. Horn and C. R. Johnson. *Topics in Matrix Aanalysis.* Cambridge University Press, Cambridge, 1991.
9. Y. Saad. Analysis of some Krylov subspace approximations to the matrix exponential operator. *SIAM J. Numer. Anal.*, 29(1):208–227, 1992.
10. Y. Saad. *Numerical Methods for Large Eigenvalue Problems.* John Wiley & Sons, Manchester Univ. Press, 1992.
11. Y. Saad and M. H. Schultz. GMRES: A generalized minimal residual algorithm for solving nonsymmetric linear systems. *SIAM J. Sci. and Stat. Comp.*, 3(7):856–869, July 1986.

Memory Hierarchy Optimizations and Performance Bounds for Sparse $A^T Ax$

Richard Vuduc, Attila Gyulassy, James W. Demmel, and Katherine A. Yelick

Computer Science Division, University of California, Berkeley
richie,jediati,demmel,yelick@cs.berkeley.edu

Abstract. This paper presents uniprocessor performance optimizations, automatic tuning techniques, and an experimental analysis of the sparse matrix operation, $y = A^T Ax$, where A is a sparse matrix and x, y are dense vectors. We describe an implementation of this computational kernel which brings A through the memory hierarchy only once, and which can be combined naturally with the register blocking optimization previously proposed in the SPARSITY tuning system for sparse matrix-vector multiply. We evaluate these optimizations on a benchmark set of 44 matrices and 4 platforms, showing speedups of up to 4.2×. We also develop platform-specific upper-bounds on the performance of these implementations. We analyze how closely we can approach these bounds, and show when low-level tuning techniques (e.g., better instruction scheduling) are likely to yield a significant pay-off. Finally, we propose a hybrid off-line/run-time heuristic which in practice automatically selects near-optimal values of the key tuning parameters, the register block sizes.

1 Introduction

This paper considers automatic performance tuning of the sparse matrix operation, $y \leftarrow y + A^T Ax$, where A is a sparse matrix and x, y are dense vectors. This computational kernel—Sp$A^T A$ hereafter—is the inner-loop of interior point methods for mathematical programming [17], algorithms for computing the singular value decomposition [5], and Kleinberg's HITS algorithm for finding hubs and authorities in graphs [10], among others. The challenge in tuning sparse kernels is choosing a data structure and algorithm that best exploits the non-zero structure of the matrix for a given memory hierarchy and microarchitecture: this task can be daunting and time-consuming because the best implementation will vary across machines, compilers, and matrices. Purely static solutions are limited since the matrix may not be known until run-time.

Our approach to automatic tuning of Sp$A^T A$ builds on experience with dense linear algebra [2,18], sparse matrix-vector multiply (SpM×V) [9,14], and sparse triangular solve (SpTS) [16]. Specifically, we apply the tuning methodology first proposed in the SPARSITY system for SpM×V [9]. We show how Sp$A^T A$ can be algorithmically cache-blocked to reuse A in a way that also allows register-level blocking to exploit dense subblocks (Section 2). The set of these implementations, parameterized by block size, defines an *implementation space*. We search

P.M.A. Sloot et al. (Eds.): ICCS 2003, LNCS 2659, pp. 705–714, 2003.

Table 1. Evaluation platforms. Dense BLAS data reported using ATLAS 3.4.1 [18] on the Ultra 2i and Pentium III, ESSL v3.1 on the Power3, and MKL v5.2 on Itanium.

Property	Sun Ultra 2i	Intel Pentium III	IBM Power3	Intel Itanium
Clock rate (MHz)	333	500	375	800
Peak Main Memory Bandwidth (MB/s)	500	680	1600	2100
Peak Flop Rate (Mflop/s)	667	500	1500	3200
DGEMM, $n = 1000$ (Mflop/s)	425	331	1300	2200
DGEMV, $n = 2000$ (Mflop/s)	58	96	260	315
STREAM Triad Bandwidth (MB/s)	250	350	715	1100
No. of FP regs (double)	16	8	32	128
L1 size (KB), line size (B), latency (cy)	16,16,1	16,32,1	64,128,.5	16,32,1 (int)
L2 size (KB), line size (B), latency (cy)	2048,64,7	512,32,18	8192,128,9	96,64,6-9
L3 size (KB), line size (B), latency (cy)	n/a	n/a	n/a	2048,64,21-24
Memory latency (cycles, $\approx$)	36-66 cy	26–60	35-139 cy	36-85 cy
Vendor C compiler version	v6.1	v6.0	v5.0	v6.0

this space by (1) benchmarking the routines on a synethetic matrix *off-line* (*i.e.*, *once* per platform), and (2) predicting the best block size using estimated properties of the matrix non-zero structure and the benchmark data. Our experiments on four hardware platforms (Table 1) and 44 matrices show that we can obtain speedups between 1.5×–4.2× over a reference implementation which computes $t = Ax$ and $y = A^T t$ as separate steps.

We evaluate our SpA^TA performance relative to upper bounds on performance (Section 3). We used similar bounds for SpM×V to show that the performance (Mflop/s) of SPARSITY-generated code is often within 20% of the upper bound, implying that more low-level tuning (*e.g.*, better instruction scheduling) will be limited [14]. Here, we derive upper bounds for our SpA^TA implementations, and show that we typically achieve between 20%–40% of this bound. Since we rely on the compiler to schedule our fully unrolled code, this finding suggests that future work can fruitfully apply automatic low-level tuning methods, in the spirit of ATLAS/PHiPAC, to improve further the SpA^TA performance.

This paper summarizes the key findings of a recent technical report [15]. We refer the reader there for details omitted due to space constraints.

2 Memory Hierarchy Optimizations for Sparse A^TAx

We assume a baseline implementation of $y = A^TAx$ that first computes $t = Ax$ followed by $y = A^Tt$. For large matrices A, this implementation brings A through the memory hierarchy twice. However, we can compute A^TAx by reading A from main memory only once. Denote the rows of A by $a_1^T, a_2^T, \ldots, a_m^T$. Then,

$$y = A^TAx = (a_1 \ldots a_m) \begin{pmatrix} a_1^T \\ \ldots \\ a_m^T \end{pmatrix} x = \sum_{i=1}^{m} a_i(a_i^T x). \quad (1)$$

$$A = \begin{pmatrix} a_{00} & a_{01} & 0 & 0 & a_{04} & a_{05} \\ a_{10} & a_{11} & 0 & 0 & a_{14} & a_{15} \\ 0 & 0 & a_{22} & 0 & a_{24} & a_{25} \\ 0 & 0 & a_{32} & a_{33} & a_{34} & a_{35} \end{pmatrix}$$

$$\texttt{b\_row\_ptr} = \begin{pmatrix} 0 & 2 & 4 \end{pmatrix}, \texttt{b\_col\_idx} = \begin{pmatrix} 0 & 4 & 2 & 4 \end{pmatrix}$$

$$\texttt{b\_value} = \begin{pmatrix} a_{00} & a_{01} & a_{10} & a_{11} & a_{04} & a_{05} & a_{14} & a_{15} & a_{22} & 0 & a_{32} & a_{33} & a_{24} & a_{25} & a_{34} & a_{35} \end{pmatrix}$$

Fig. 1. 2×2 BCSR storage. Elements are stored in the `b_value` array. The column index of the (0,0) entry of each block is stored in `b_col_idx`. The `b_row_ptr` array points to block row starting positions in the `b_col_idx` array. (Figure taken from Im [9].)

Assuming sufficient cache capacity, each row a_i^T is read from memory into cache to compute the dot product $t_i = a_i^T x$, and reused for the vector scaling $t_i a_i$. We can also take each a_i^T to be a *block of rows* instead of just a single row, allowing us to combine the cache optimization of Equation (1) with a previously proposed *register blocking* optimization [9]. Below, we review register blocking, and describe our heuristic to choose the key tuning parameter, the block size.

Register blocking improves register reuse by reorganizing the matrix data structure into a sequence of "small" dense blocks, keeping small blocks of x and y in registers. An $m \times n$ sparse matrix in $r \times c$ register blocked format is divided logically into $\frac{m}{r} \times \frac{n}{c}$ submatrices, each of size $r \times c$. Only blocks containing at least one non-zero are stored. Multiplying by A proceeds block-by-block: for each block, we reuse the corresponding r elements of y and c elements of x. For simplicity, assume that r divides m and c divides n. We use the blocked compressed sparse row (BCSR) storage format [11], a 2×2 example of which is shown in Figure 1. When $r = c = 1$, BCSR reduces to compressed sparse row (CSR) storage. BCSR can store fewer column indices than CSR (one per block instead of one per non-zero). We fully unroll the $r \times c$ submatrix computation to reduce loop overhead and expose scheduling opportunities to the compiler.

Figure 1 also shows that creating blocks may require filling in explicit zeros. We define the *fill ratio* to be the number of stored values (*i.e.*, including zeros) divided by the number of true non-zeros. We may trade-off extra computation (*i.e.*, fill ratio > 1) for improved efficiency from uniform code and memory access.

To select r and c, we adapt the SPARSITY v2.0 heuristic for SpM×V [14] to Sp$A^T A$. First, we measure the speed (Mflop/s) of the blocked Sp$A^T A$ code for all $r \times c$ up to 8×8 for a dense matrix stored in sparse BCSR format. These measurements are made only once per architecture. Second, when the matrix is known at run-time, we estimate the fill ratio for all block sizes using a recently described sampling scheme [14]. Third, we choose the r and c that maximize

$$\text{Estimated Mflop/s} = \frac{\text{Mflop/s on dense matrix in } r \times c \text{ BCSR}}{\text{Estimated fill ratio for } r \times c \text{ blocking}} \quad . \tag{2}$$

The run-time overhead of picking a register block size and converting into our data structure is typically between 5–20 executions of naïve Sp$A^T A$ [14]. Thus, the optimizations we propose are most suitable when SpA^TA must be performed many times (*e.g.*, in sparse iterative methods).

3 Upper Bounds on Performance

We use performance upper bounds to estimate the best possible performance given a matrix and data structure but *independent* of any particular instruction mix or ordering. In our work on sparse kernels, we have thus far focused on data structure transformations, relying on the compiler to produce good schedules. An upper bound allows us to estimate the likely pay-off from low-level tuning.

Our bounds for the SpA^TA implementations described in Section 2 are based on bounds we developed for SpM×V [14]. Our key assumptions are as follows:

1. We bound time from below by considering only the cost of *memory* operations. We also assume write-back caches (true of the evaluation platforms) and sufficient store buffer capacity so that we can consider only loads and ignore the cost of stores.
2. Our model of execution time charges for cache and memory *latency*, as opposed to assuming that data can be retrieved from memory at the manufacturer's reported peak main memory bandwidth. We also assume accesses to the L1 cache commit at the maximum load/store commit rate.
3. As shown below, we can get a lower bound on memory costs by computing a lower bound on cache misses. Therefore, we consider only compulsory and capacity misses, *i.e.*, we ignore conflict misses. Also, we account for cache capacity and cache line size, but assume full associativity.

We refer the reader to our prior paper [14] and our full SpA^TA technical report [15] for a more careful justification of these assumptions.

Let T be total time of SpA^TA in seconds, and P the performance in Mflop/s:

$$P = \frac{4k}{T} \times 10^{-6} \tag{3}$$

where k is the number of non-zeros in the $m \times n$ sparse matrix A, *excluding* any fill. To get an *upper bound on performance*, we need a *lower bound* on T. We present our lower bound on T, based on Assumptions 1 and 2, in Section 3.1. Our lower bounds on cache misses (Assumption 3) are described in Section 3.2.

3.1 A Latency-Based Execution Time Model

Consider a machine with κ cache levels, where the access latency at cache level i is α_i seconds, and the memory access latency is α_{mem}. Suppose SpA^TA executes H_i cache accesses (or cache hits) and M_i cache misses at each level i, and that the total number of loads is L. We take the execution time T to be

$$T = \sum_{i=1}^{\kappa} \alpha_i H_i + \alpha_{\text{mem}} M_\kappa = \alpha_1 L + \sum_{i=1}^{\kappa-1} (\alpha_{i+1} - \alpha_i) M_i + \alpha_{\text{mem}} M_\kappa \tag{4}$$

where we use $H_1 = L - M_1$ and $H_i = M_{i-1} - M_i$ for $2 \leq i \leq \kappa$. According to Equation (4), we can minimize T by minimizing M_i, assuming $\alpha_{i+1} \geq \alpha_i$.

3.2 A Lower Bound on Cache Misses

Following Equation (4), we obtain a lower bound on M_i for SpA^TA by counting compulsory and capacity misses but ignoring conflict misses. The bound is a function of the cache configuration and matrix data structure.

Let C_i be the size of each cache i in double-precision words, and let l_i be the line size, in doubles, with $C_1 \leq \ldots \leq C_\kappa$, and $l_1 \leq \ldots \leq l_\kappa$. Suppose γ integer indices use the same storage as 1 double. (We used 32-bit integers; thus, $\gamma = 2$.) Assume full associativity and complete user-control over cache data placement.

We describe the BCSR data structure as follows. Let $\hat{k} = \hat{k}(r, c)$ be the number of stored values, so the fill ratio is $\hat{k}/k$, and the number of stored blocks is $\frac{\hat{k}}{rc}$. Then, the total number of loads L is $L = L_A + L_x + L_y$, where

$$L_A = 2\left(\hat{k} + \frac{\hat{k}}{rc}\right) + \frac{m}{r} \qquad L_x = \frac{\hat{k}}{r} \qquad L_y = \frac{\hat{k}}{r}. \tag{5}$$

L_A contains terms for the values, block column indices, and row pointers; the factor of 2 accounts for reading A twice (once to compute Ax, and once for A^T times the result). L_x and L_y are the total number of loads required to read x and y, where we load c elements of each vector for each of the $\frac{\hat{k}}{rc}$ blocks.

To correctly model capacity misses, we compute the amount of data, or *working set*, required to multiply by a block row and its transpose. For the moment, assume that all block rows have the same number of $r \times c$ blocks; then, each block row has $\frac{\hat{k}}{rc} \times \frac{r}{m} = \frac{\hat{k}}{cm}$ blocks, or $\frac{\hat{k}}{m}$ non-zeros per row. Define the *matrix working set*, $\hat{W}$, to be the size of matrix data for a block row, and the *vector working set*, $\hat{V}$, to be the size of the corresponding vector elements for x and y:

$$\hat{W} = \frac{\hat{k}}{m}r + \frac{1}{\gamma}\frac{\hat{k}}{cm} + \frac{1}{\gamma} \quad , \quad \hat{V} = 2\hat{k}/m \tag{6}$$

For each cache i, we compute a lower bound on the misses M_i according to one of the following 2 cases, depending on the relative values of C_i, $\hat{W}$, and $\hat{V}$.

1. $\hat{W} + \hat{V} \leq C_i$: *Entire working set fits in cache.* Since there is sufficient cache capacity, we incur only compulsory misses on the matrix and vector elements:

$$M_i \geq \frac{1}{l_i}\left(\frac{m}{r}\hat{W} + 2n\right) \quad . \tag{7}$$

2. $\hat{W} + \hat{V} > C_i$: *The working set exceeds the maximum cache capacity.* In addition to the compulsory misses shown in Equation (7), we incur capacity misses for each element of the total working set that exceeds the capacity:

$$M_i \geq \frac{1}{l_i}\left[\frac{m}{r}\hat{W} + 2n + \frac{m}{r}(\hat{W} + \hat{V} - C_i)\right] \quad . \tag{8}$$

The factor of $\frac{1}{l_i}$ optimistically counts only 1 miss per cache line. We refer the reader to our full report for detailed derivations of Equations (7)–(8) [15].

4 Experimental Results and Analysis

We measured the performance of various Sp$A^T A$ implementations on 4 platforms (Table 1) and the 44 matrices of the original SPARSITY benchmark suite. Matrix 1 is a dense matrix in sparse format. Matrices 2–17 come from finite element method (FEM) applications: 2–9 have a structure dominated by a single block size aligned uniformly, while 10–17 have irregular block structure. Matrices 18–39 come from non-FEM applications, including chemical process simulation and financial modeling, among others. Matrices 40–44 arise in linear programming.

We use the PAPI hardware counter library (v2.1) to measure cycles and cache misses [4]. Figures 2–5 summarize the results, comparing performance (Mflop/s; y-axis) of the following 7 bounds and implementations for each matrix (x-axis):

- **Upper bound** (shown as a solid line): The fastest (highest value) of our performance upper bound, Equation (3), over all $r \times c$ block sizes up to 8×8.
- **PAPI upper bound** (shown by triangles): An upper bound in which we set L and M_i to the true load and miss counts measured by PAPI. The gap between the two bounds indicates how well Equations (5)–(8) reflect reality.
- **Best cache optimized, register blocked** implementation (squares): Performance of the best code over all $r \times c$, based on an exhaustive search.
- **Heuristically predicted implementation** (solid circles): Performance of the implementation predicted by our heuristic.
- **Register blocking only** (diamonds): Performance without algorithmic cache blocking, where the block size is chosen by exhaustive search.
- **Cache optimization only** (shown by asterisks): Performance of the code with only the algorithmic cache optimization, (*i.e.*, with $r = c = 1$).
- **Reference** ($\times$'s): No cache or register-level optimizations have been applied.

Matrices which are small relative to the cache size have been omitted.

We draw the following 5 high-level conclusions based on Figures 2–5. More complete and detailed analyses appear in the full report [15].

1. *The cache optimization leads to uniformly good performance improvements.* On all platforms, applying the cache optimization, even without register blocking, leads to speedups ranging from up to 1.2× on the Itanium and Power3 platforms, to just over 1.6× on the Ultra 2i and Pentium III platforms.

2. *Register blocking and the cache optimization can be combined to good effect.* When combined, we observe speedups from 1.2× up to 4.2×. Moreover, the speedup relative to the register blocking only code is still up to 1.8×.

3. *Our heuristic always chooses a near-optimal block size.* Indeed, the performance of heuristic's block size is within 10% of the exhaustive best in all but four instances—Matrices 17, 21, and 27 on the Ultra 2i, and Matrix 2 on the Pentium III. There, the heuristic performance is within 15% of the best.

4. *Our implementations are within 20–30% of the PAPI upper bound for FEM matrices, but within only about 40–50% on other matrices.* The gap between actual performance and the upper bound is larger than what we observed previously for SpM×V and SpTS [14,16]. This result suggests that a larger payoff is expected from low-level tuning by using ATLAS/PHiPAC techniques.

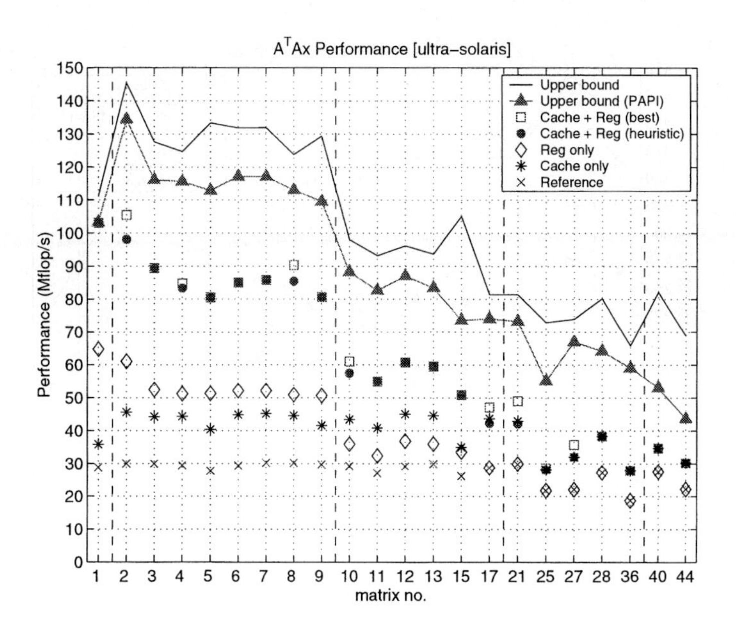

Fig. 2. Sp$A^T A$ performance on the Sun Ultra 2i platform.

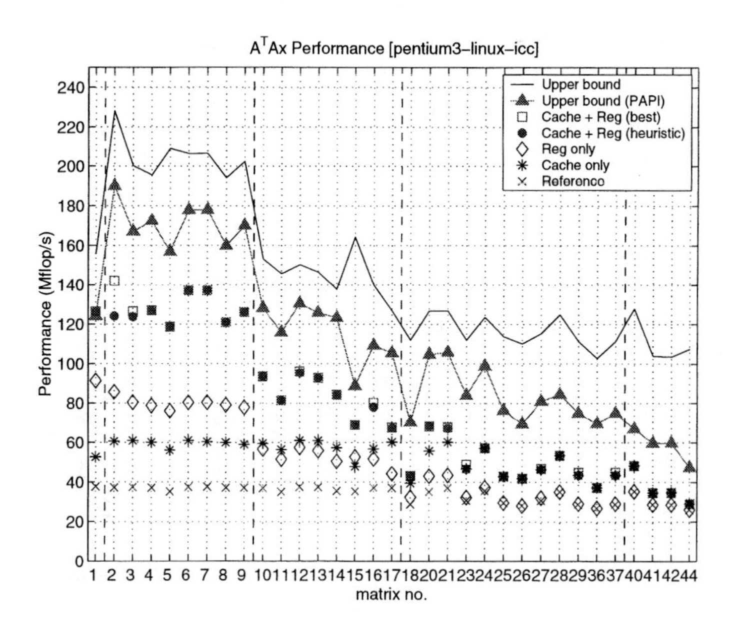

Fig. 3. Sp$A^T A$ performance on the Intel Pentium III platform.

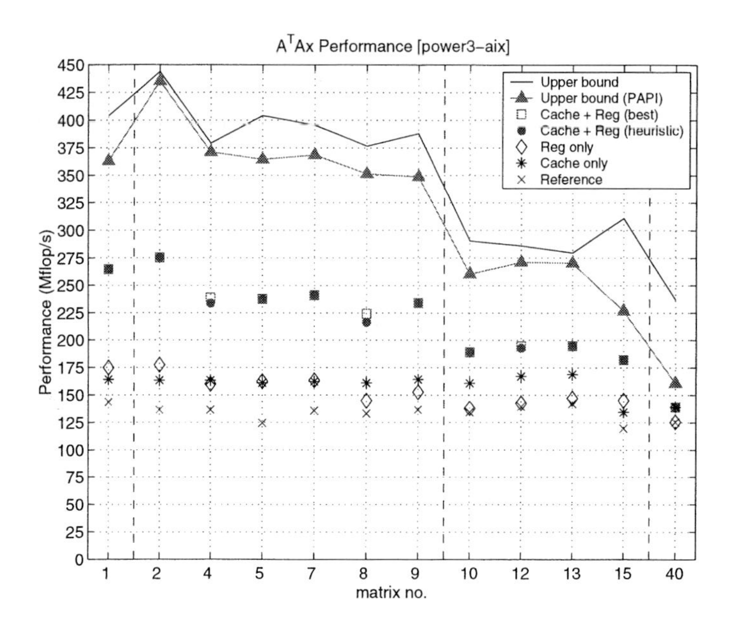

Fig. 4. Sp$A^T A$ performance on the IBM Power3 platform.

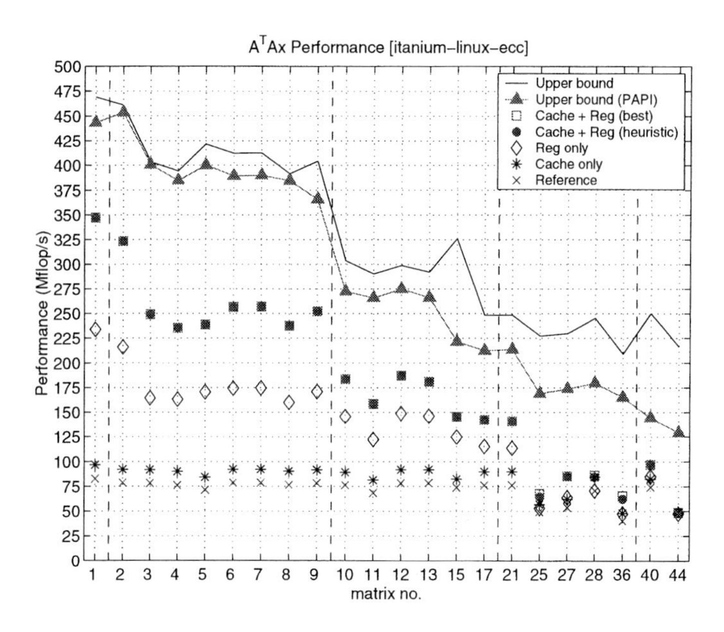

Fig. 5. Sp$A^T A$ performance on the Intel Itanium platform.

5. *Our analytic model of misses is accurate for FEM matrices, but less accurate for the others.* For the FEM matrices 1–17, the PAPI upper bound is typically within 10–15% of the analytic upper bound, indicating that our analytic model of misses is accurate in these cases. Matrices 18–44 have more random non-zero structure, so the gap between the analytic and PAPI upper bounds is larger due to our assumption of only 1 miss per cache line.

5 Related Work

There have been a number of sophisticated models proposed for analyzing memory behavior of SpM×V. Temam and Jalby [13], and Fraguela, *et al.* [6] have developed sophisticated probabilistic cache miss models for SpM×V, but assume uniform distribution of non-zero entries. To obtain lower bounds, we account only for conflict and capacity misses. Gropp, *et al.*, use bounds like the ones we develop to analyze and tune a computational fluid dynamics code [7]; Heber, *et al.*, analyze a sparse fracture mechanics code [8] on Itanium. We address matrices from a variety of applications, and furthermore, have proposed an explicit model execution time for sparse, streaming applications.

Work in sparse compilers, *e.g.*, Bik *et al.* [1] and the Bernoulli compiler [12], complements our own work. These projects address the expression of sparse kernels and data structures for code generation. One distinction of our work is our use of a hybrid off-line, on-line model for selecting transformations.

6 Conclusions and Future Directions

The speedups of up to 4.2× that we have observed indicate that there is tremendous potential to boost performance in applications dominated by SpA^TA. We are incorporating this kernel and these optimizations in an automatically tuned sparse library based on the Sparse BLAS standard [3].

Our upper bounds indicate that there is more room for improvement using low-level tuning techniques than with prior work on SpM×V and SpTS. Applying automated search-scheduling techniques developed in ATLAS and PHiPAC is therefore a natural extension. In addition, refined versions of our bounds could be used to study how performance varies with architectural parameters, in the spirit of SpM×V modeling work by Temam and Jalby [13].

Additional reuse is possible when multiplying by multiple vectors. Preliminary results on Itanium for sparse matrix-multiple-vector mltiplication show speedups of 6.5 to 9 [14]. This is a natural opportunity for future work with SpA^TA as well. We are exploring this optimization and other higher-level kernels with matrix reuse (*e.g.*, A^kx, matrix triple products).

Acknowledgements. We thank Michael de Lorimier for the SpATA initial implementations. This research was supported in part by the National Science Foundation under NSF Cooperative Agreement No. ACI-9813362, the Department of Energy under DOE Grant No. DE-FC02-01ER25478, and a gift from

Intel. The information presented here does not necessarily re.ect the position or the policy of the Government and no o.cial endorsement should be inferred.

References

1. A.J.C. Bik and H.A.G. Wijsho.. Automatic nonzero structure analysis. *SIAM Journal on Computing*, 28(5):1576–1587, 1999.
2. J. Bilmes, K. Asanović, C. Chin, and J. Demmel. Optimizing matrix multiply using PHiPAC: a portable, high-performance, ANSI C coding methodology. In Proceedings of the International Conference on Supercomputing, July 1997.
3. S. Blackford et al. Document for the Basic Linear Algebra Subprograms (BLAS) standard: BLAS Technical Forum, 2001. Chapter 3: `http://www.netlib.org/blast`.
4. S. Browne, J. Dongarra, N. Garner, K. London, and P. Mucci. A scalable crossplatform infrastructure for application performance tuning using hardware counters. In *Proceedings of Supercomputing*, November 2000.
5. J.W. Demmel. Applied Numerical Linear Algebra. *SIAM*, 1997.
6. B.B. Fraguela, R. Doallo, and E.L. Zapata. Memory hierarchy performance prediction for sparse blocked algorithms. *Parallel Processing Letters*, 9(3), 1999.
7. W.D. Gropp, D.K. Kasushik, D.E. Keyes, and B.F. Smith. Towards realistic bounds for implicit CFD codes. In *Proceedings of Parallel Computational Fluid Dynamics*, pages 241–248, 1999.
8. G. Heber, A.J. Dolgert, M. Alt, K.A. Mazurkiewicz, and L. Stringer. Fracture mechanics on the intel itanium architecture: A case study. In *Workshop on EPIC Architectures and Compiler Technology* (ACM MICRO 34), Austin, TX, 2001.
9. E.-J. Im and K.A. Yelick. Optimizing sparse matrix computations for register reuse in SPARSITY. In *Proceedings of ICCS*, pages 127–136, May 2001.
10. J.M. Kleinberg. Authoritative sources in a hyperlinked environment. *Journal of the ACM*, 46(5):604–632, 1999.
11. Y. Saad. SPARSKIT: A basic toolkit for sparse matrix computations, 1994. `http://www.cs.umn.edu/Research/arpa/SPARSKIT/sparskit.html`.
12. P. Stodghill. A Relational Approach to the Automatic Generation of Sequential Sparse Matrix Codes. PhD thesis, Cornell University, August 1997.
13. O. Temam and W. Jalby. Characterizing the behavior of sparse algorithms on caches. In *Proceedings of Supercomputing '92*, 1992.
14. R. Vuduc, J.W. Demmel, K.A. Yelick, S. Kamil, R. Nishtala, and B. Lee. Performance optimizations and bounds for sparse matrix-vector multiply. In *Proceedings of Supercomputing*, Baltimore, MD, USA, November 2002.
15. R. Vuduc, A. Gyulassy, J.W. Demmel, and K.A. Yelick. Memory hierarchy optimizations and performance bounds for sparse ATAx. Technical Report UCB/CS-03-1232, University of California, Berkeley, February 2003.
16. R. Vuduc, S. Kamil, J. Hsu, R. Nishtala, J.W. Demmel, and K.A. Yelick. Automatic performance tuning and analysis of sparse triangular solve. In *ICS 2002: POHLL Workshop*, New York, USA, June 2002.
17. W. Wang and D.P. O'Leary. Adaptive use of iterative methods in interior point methods for linear programming. Technical Report UMIACS-95-111, University of Maryland at College Park, College Park, MD, USA, 1995.
18. C. Whaley and J. Dongarra. Automatically tuned linear algebra software. In *Proc. of Supercomp.*, Orlando, FL, 1998.

Issues in the Design of Scalable Out-of-Core Dense Symmetric Indefinite Factorization Algorithms

Peter E. Strazdins

Australian National University, Acton, ACT 0200. Australia
Peter.Strazdins@cs.anu.edu.au, http://cs.anu.edu.au/~Peter.Strazdins

Abstract. In the factorization of indefinite symmetric linear systems, symmetric pivoting is required to maintain numerical stability, while attaining a reduced floating point operation count.
However, symmetric pivoting presents many challenges in the design of efficient algorithms, and especially in the context of a parallel out-of-core solver for dense systems. Here, the search for a candidate pivot in order to eliminate a single column potentially requires a large number of messages and accesses of disk blocks.
In this paper, we look at the problems of scalability in terms of number of processors and the ratio of data size relative to aggregate memory capacity for these solvers. We find that diagonal pivoting methods which exploit locality of pivots offer the best potential to meet these demands.
A left-looking algorithm based on an exhaustive block-search strategy for dense matrices is described and analysed; its scalability in terms of parallel I/O is dependent on being able to find stable pivots near or within the current elimination block.

1 Introduction

Large symmetric indefinite systems of equations arise in many applications, including incompressible flow computations, optimization of linear and non-linear programs, electromagnetic field analysis and data mining. An important special case, occurring in the latter two applications, are semi-definite and *weakly-indefinite* (i.e. most columns can be stably eliminated without pivoting) systems [1,2].

For the largest of such systems, even the aggregate memory of a distributed memory processor may not be sufficient [3]. Thus, out-of-core methods, scalable both with respect to matrix size and the number of processors, are needed in order to factor (and hence solve) these systems.

Algorithms for factoring $N \times N$ symmetric indefinite systems that are stable and yet exploit symmetry to have only $\frac{N^3}{3} + O(N^2)$ floating point operations are well known (see [4] and the references within). The diagonal pivoting methods are dominant in the literature, and several high-performance implementations of algorithms based on this method have been given [5,6,7], with some more recent papers considering distributed memory implementation [8,9].

P.M.A. Sloot et al. (Eds.): ICCS 2003, LNCS 2659, pp. 715–724, 2003.

Work so far on parallel out-of-core dense factorization algorithms has covered algorithms based on LU, LL^T and QR factorizations [10,11,12,13]. In all cases, left-looking versions of blocked algorithms are preferred; this is because these reduce the number of matrix elements written to disk from $O(N^3)$ to $O(N^2)$ [11] and permit easier checkpointing [13]. Within this, two approaches have been employed: a *slab-based* approach, where the block of columns currently being eliminated are kept in core, and a *block-based* approach, where only a part of the column block need be kept in core. The latter permits a wider column width to be used, which in turn permits greater re-use of data fetched from disk; thus it is more scalable with respect to linear system size relative to aggregate memory capacity [10,13]. However, it is not easily applied to LU factorization with partial (row) pivoting, as this requires a full column block to be in core [11].

While the difficulties of a block-based approach seem even greater for algorithms requiring symmetric pivoting (involving interchange of both row and columns), we shall see that diagonal pivoting methods offer a solution that can preserve stability guarantees.

This paper is organized as follows. Diagonal pivoting algorithms, with an outline of issues when applied to the parallel out-of-core case, are described in Section 2. Block- and slab-based algorithms based on locality of interchanges are described in Section 3, with some analysis of their potential performance. Conclusions are given in Section 4, together with a description of future work.

2 Diagonal Pivoting Algorithms

In the diagonal pivoting method, the decomposition $A = PLDL^TP^T$ is performed, where P is a permutation matrix, L is an $N \times N$ lower triangular matrix with a unit diagonal, and D is a block diagonal matrix with either 1×1 or 2×2 sub-blocks [4].

Figure 1 indicates the partial factorization of a left-looking blocked algorithm (see [8] for a description of the right-looking algorithm, and for more details on step 4). As A is symmetric, it is represented by a lower triangular matrix[1]. Here, A_2 represents the current elimination block, and the first j_1 columns of L and D (D_{11}) have already been factored and stored in $\left(\frac{A_{11}}{\frac{A_{21}}{A_{31}}}\right)$.

As this algorithm is left-looking, interchanges within A_2 resulting from the column eliminations in A_{11} are performed only when A_2 is ready for elimination (step 1).

In a left-looking LU factorization, it is possible to defer the row interchanges in A_1 (cf. step 5) until the factorization is complete; this can be useful as it

[1] Most implementations in the literature store a dense lower triangular matrix in a rectangular matrix, with memory locations corresponding to the upper triangular matrix being unused. With the advent of virtual memory, this incurs only an $O(N)$ storage overhead, as pages containing no elements of the lower matrix are never allocated.

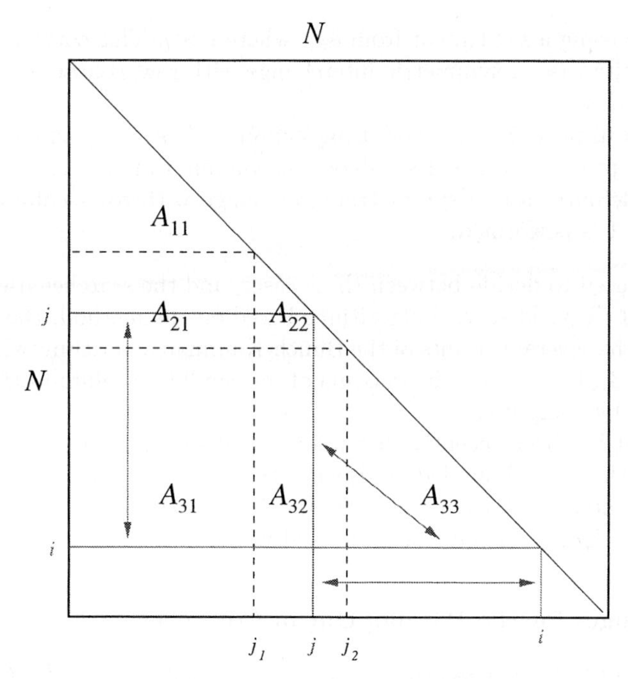

1. apply pivots from A_{11} to $A_2 = \begin{pmatrix} A_{22} \\ \hline A_{32} \end{pmatrix}$
2. A_{22} -= $A_{21} W_{21}^T$, $W_{21} = D_{11} A_{21}$
3. A_{32} -= $A_{31} W_{21}^T$
4. eliminate block column A_2 (applying pivots within A_2)
5. apply pivots (row interchanges) from A_2 to $A_1 = \begin{pmatrix} A_{21} \\ \hline A_{31} \end{pmatrix}$

Fig. 1. Partial left-looking LDL^{T} factorization algorithm, showing a symmetric interchange between columns j and i

avoids having to repeatedly modify the already factored (and hence out-of-core) part of the matrix [11]. However, this is problematic when one tries to exploit symmetry as here the row interchanges operate on only part of the row and furthermore the current elimination block A_2 might only include one of the rows involved in the interchange.

The elimination of A_2 (step 4) proceeds column by column; in the elimination of column j, 3 cases can arise:

1. Eliminate using a 1×1 pivot from $a_{j,j}$. This corresponds to the definite case, and will be used when $a_{j,j}$ is sufficiently large, i.e. $|a_{j,j}| \geq \alpha\gamma_j$. Here $\gamma_j = \max(|a_{j+1:N,j}|)$ and $\alpha \in (0,1)$ is a constant determining the growth rate (typically $\alpha = 0.5$).

2. Eliminate using a 1×1 pivot from $a_{i,i}$, where $i > j$. This corresponds to the semi-definite case; a symmetric interchange with row/columns i and j must be performed.
3. Eliminate using a 2×2 pivot using columns i', $i' \geq j$, and i, $i \geq j+1$ (this case produces a 2×2 sub-block at column j of D). This corresponds to the indefinite case; a symmetric interchange with rows/columns i',i and and $j, j+1$ is performed.

The tests used to decide between these cases, and the searches used to select column i (and i'), yield several algorithms based on the method, the most well-known being the several variants of the Bunch-Kaufman algorithm (where always $i' = j$ in Case 3). In all cases, the maxima of the candidate column i (i'), denoted γ_i ($\gamma_{i'}$), must be computed.

However, it has been recently shown for this algorithm there is no guarantee that the growth of L is bounded, and hence the algorithm has no overall strong stability guarantee [7]. Variants such as the *bounded Bunch-Kaufman* and *fast Bunch-Parlett* algorithms have been devised which overcomes this problem [7].

2.1 Challenges for the Parallel Out-of-Core Case

In the parallel out-of-core case, the matrix A is distributed over a $P \times Q$ processor grid with an $r \times r$ block-cyclic matrix distribution, i.e. (storage) block (i, j) of A will be on processor (i mod P, j mod Q). Assume that this distribution applies to both disk and memory; thus, loading a portion of A into memory involves only an access to the processor's local disk. Thus, the disk can be viewed simply as another level of the (distributed) memory hierarchy. Assume, for the moment, that the storage is column-oriented, both on disk and in memory.

In a left-looking blocked $\mathrm{LDL}^{\mathrm{T}}$ factorization of A, as indicated in Figure 1, consider the elimination of column j where the candidate pivot i lies outside the current elimination block A_2. For the sake of simplicity, assume a slab-based algorithm, i.e. all of the A_2 is in core.

To determine whether i is a suitable pivot, $a_i = (a_{i,\,j:i}, a_{i+1:N,\,i})$ must be aligned with $a_j = a_{j:N,\,j}$, and, for a left-looking algorithm, *all* updates from the $j-1$ eliminations so far must be applied to it. The alignment involves reading a potentially large number of disk blocks from remote processors, since parts of a row and a column must be accessed. The updates require all of A_1 to be read.

At this point, it may of course be discovered that i is not a suitable pivot and this process has to be repeated! In the out-of-core situation, it may be efficient to amortize this large cost by fetching a small block of pivots, and seeing if any are suitable.

If pivot i proves suitable, the interchange must then be completed. Firstly, a_i must be over-written by a_j (note that the portion in A_{33} must have the values of a_j *before* any updates were applied, for a left-looking algorithm). Secondly, at some later stage, the interchanges of rows i and j in A_1 need to be performed (step 5). However. with a slab-based algorithm, these can performed in-core whenever A_1 is loaded into memory [11].

Thus, if a significant number of searches for pivots from outside the current elimination block must be made, algorithm performance will be severely degraded. From the point of view of having to update a_i, it has been argued right-looking algorithms are strongly favoured for LDL^T [14], and may even be preferable for the out-of-core case.

However, if i is within the current elimination block, the only overhead will be in the actual message exchanges, and even this will only occur if column i is outside the storage block of column j [8].

2.2 Elimination Block Search Strategies

A method searching for (stable) pivots in the current elimination block used for sparse matrices is the *exhaustive pivoting strategy* [7], which originates from the earlier MA27 codes by Duff and Reid [15]. This technique is used in multi-frontal methods, where this has a large payoff in preserving the sparsity of the matrices [7].

Here, a block of ω_a columns currently under consideration for elimination are arranged in a queue. Each iteration of the search selects column i (initially from the head of the queue, working inwards). After computing γ_i, it is then matched with each column i' preceding it in the queue. Note that at this point, it must be ensured that all updates from previous eliminations have been applied to columns i and i'.

A simple test involving matrix elements at positions (i, i), (i, i') and (i', i'), and γ_i and γ_i' determines whether columns i and i' can form a stable 2×2 pivot [7]. If it fails, column i is then considered for a 1×1 pivot.

One remaining point is what to do if no stable pivots are found in the current block. A number of fresh columns (from the trailing sub-matrix A_{33}) will be then brought to the head of the queue, and the whole process will then be repeated. While the queue size ω_a may thus grow to be $O(N)$, eventually the whole factorization will complete [7].

This strategy has been adapted for the dense parallel case [9]. The elimination block is the current storage block and so is of fixed size; thus all columns (rows) of the block are held within a column (row) of the processor grid (cf. Figure 1). Thus interchanges within the block can be performed without communication. If no stable pivot from within the block could be found for a column, it is eliminated using a pivot from outside using the bounded Bunch-Kaufman algorithm. Here the storage block size is determined by optimal matrix-matrix multiply (cf. steps 2–3 in Figure 1) performance, which is in turn determined by the cache size.

Such a more rigid scheme is a good tradeoff in the dense parallel (in-core) case: it permits the BLAS computational routines to operate a peak efficiency, whereas the cost for searching for a pivot outside the elimination block is low enough to be tolerated provided it occurs in the minority of cases. Indeed, it was found that best performance was obtained by limiting the search to $\omega_s \leq \omega_a$ columns in advance (e.g. $\omega_s = 16$ with $r = \omega_a = 64$ being optimal on a Fujitsu AP3000) [9], as the overhead of computing a number of column maxima will at some point exceed the communication costs of a symmetric interchange.

With this limited strategy, it was found that in the worst case (for highly indefinite matrices) only $0.15N$ searches outside the elimination block were required. This number reduced to $< 0.05N$ for the 'weakly indefinite' matrices typically generated to electromagnetic compatibility analysis [9].

However, as we have seen in Section 2.1 for the out-of-core situation, the cost of searching outside the elimination block is much higher, and more exhaustive strategies will be favored.

3 Diagonal Pivoting Methods for a Parallel Out-of-Core Algorithm

In this section, we develop both slab- and block- based algorithms exploiting locality of interchanges, and give some analysis of their behaviors.

3.1 Blocking and Data Layout Issues

To obtain high performance, such an algorithm must consider all levels of the (parallel) memory hierarchy, and exploit locality in searching for pivots as much as possible. Correspondingly, the following blocking factors can be identified.

- ω_a, the blocking factor of the outer algorithm ($\omega_a = j_2 - j_1 + 1$ in Figure 1). This should be as large as possible to maximize data-reuse at the disk level. Let M denote the amount memory available for the algorithm; for a slab-based algorithm, ω_a is constrained by:

$$\frac{N\omega_a}{PQ} \leq M \tag{1}$$

- ω_c, the optimal blocking factor for matrix-matrix multiply (as performed in step 3 in Figure 1), determined by maximizing data-reuse at the cache level. An optimal algorithm will require $\omega_a \geq \omega_c$; with $\omega_c < \omega_a$, two levels of blocking may be incorporated [11].
- r, the storage block size, primarily determined by concerns of minimizing communication latencies [14]. The situations $r = \omega_c$ (storage blocking) and $r \approx 1$ (algorithmic blocking) are typical [14]. Note that to fully utilize aggregate memory capacity, we will require:

$$\omega_a \geq \max(P, Q)r \tag{2}$$

- ω_s, the number of columns to be searched in the elimination block search strategy (see Section 2.2); this should be sufficiently large to ensure a successful search most of the time. However, as all of the columns must be kept up-to-date after each elimination, so that their maxima can then be computed, all of these columns must be in core, i.e.:

$$\frac{N\omega_s}{PQ} \leq M \tag{3}$$

  There may be a reduction in communication costs if $\omega_s \leq r$ holds [8].

– ω_d, where ω_d^2 equals the disk block size. For a block-based algorithm, $\omega_a^2 >> \omega_d^2$ will be required to amortize disk latencies.

A second issue is the organization of A upon disk: while in Section 2.1 we assumed it conformed to the block-cyclic data distribution, there is still freedom over whether disk blocks are column or row-oriented, or oriented towards block rows and columns (with ω_s and r being candidates for the sizes of these blocks). The choice depends on the dominant access patterns of the algorithm.

3.2 A Slab-Based Algorithm Exploiting Locality in Interchanges

We will now describe how the partial left-looking factorization algorithm of Figure 1 can be combined with an exhaustive elimination block search strategy, as described in Section 2.2, for a slab-based algorithm using a fixed width of ω_a.

Assume that there are u_1 uneliminated columns left over from the previous stage, where $0 \leq u_1 < \omega_a$, which are still in core. These columns will become the first u_1 columns of A_2 for this stage of the algorithm. The last $\omega_a - u_1$ columns of A_2 are now loaded into core, and steps 1–3 in Figure 1 are applied to only these columns.

Insert a new step 3a: interchange the first u_1' columns of A_2 with the last u_1' columns, where $u_1' = \min(u_1, \omega_a - u_1)$. The motivation for this is to move the 'difficult to eliminate' columns to the back of A_2, so that updates from the fresh columns have a chance of changing these columns.

Step 4 can use an exhaustive block search strategy ($\omega_s = \omega_a$); after completion, the $\omega_a - u_2$, $0 \leq u_2 < \omega_a$, successfully eliminated columns from this stage are written back to disk. To ensure progress, if the elimination block search strategy resulted in no eliminations, at least one column must be eliminated using non-local interchanges, e.g. by the bounded Bunch-Kaufman method.

Thus, this algorithm proceeds operating on a series of overlapping slabs; provided the elimination block search strategy always can eliminate one column, all interchanges will be within these slabs. In this case, no interchanges could have yet been applied to columns $j_1 + u_1 : N$ from previous stages, so step 1 is empty and A_{21} and A_{31} will have their rows aligned with A_2 for the updates of steps 2–3. Furthermore, step 5 will not require any updates to A_{31}.

Note that A_{21} and A_{31} can be read in column blocks of width much less than ω_a (but the width should be at least ω_c for good matrix multiply performance), which are then applied to A_2; provided $\omega_c << \omega_a$, only a negligible amount of extra memory need be reserved for them [12,13].

The performance of this algorithm thus depends on the u_i's being small, e.g. $u_i \leq 0.2\omega_a$, i.e. the (exhaustive) block search strategy is highly successful at finding stable pivots. The pivoting behavior of the limited strategy of [9] mentioned in Section 2.2 provide encouragement that this will hold in practice.

In this algorithm, ω_a is limited by Equation 1. Achieving good load balance in terms of both computation and memory utilization (Equation 2) favours a low value of r. This, plus the fact that the slabs can overlap, means that the elimination blocks cannot be confined within storage blocks. In other words, an algorithmic blocking regime must be used here, even though this means interchanges within a slab will require communication.

Assuming $u_i \approx 0$, the number of matrix elements read from disk are dominated by steps 2–3:

$$\Sigma_{i=0}^{N/\omega_a} i\omega_a(N - i\omega_a) = \frac{N^3}{6\omega_a} + O(N^2) \tag{4}$$

and approximately $2 \cdot \frac{N^2}{2}$ matrix elements are written (including step 5).

As the dominant disk accesses are column oriented, organizing disk blocks in a column-based fashion is desirable. However, placing $r' > 1$ columns per disk block will reduce the number of blocks read in step 5.

3.3 The Need for Improved Memory Scalability

Considering Equation 4 and 1, the question arises of whether, on contemporary parallel machines, a slab-based algorithm is sufficient for good performance for problem sizes that can be solved in 'reasonable time'.

For a processor with 0.5 GB (equivalent to 6.7×10^7 double precision words) of memory available to matrix data and can sustain 1 GFLOPS over a factorization, a matrix of size $N_1 = 10^4$ will (just) fit in core, and take $t_1 \approx 5$min to factorize. For a matrix of size $N_2 = 10^5$, a slab of width $\omega_a \approx 670$ would fit within core and is likely to be wide enough for high re-use of disk accesses; this computation would require $(\frac{N_2}{N_1})^3 t_1 = 10t_1$ time, somewhat under 100 hours.

For a $P \times P$ grid of these processors, where $P = 10$, a matrix of size $N_2 = 10^5$ would just fit in core and, assuming an ideal speedup, factorize in time $(\frac{N_2}{N_1})^3/P^2 t_1 = 10t_1$. For a matrix of size $N_3 = 10^6$, a slab of (ample) width $\omega_a \approx 660 \times 10$ would fit in core, and would factor in time of $(\frac{N_3}{N_1})^3/P^2 t_1 = 10^4 t_1$ (over one month).

Thus, unless one wishes to factor systems requiring much more time than this, a slab-based algorithm will be adequate for moderate-sized grids of contemporary machines.

3.4 A Block-Based Algorithm

The algorithm of Section 3.2 can be converted to require a slab of width of only $\omega_s < \omega_a$, where instead ω_s is constrained by Equation 3. As indicated in Section 3.3, the resulting value of ω_s should be ample on contemporary parallel machines to permit a successful block search strategy and good load balance.

The idea is to simply to apply the left-looking factorization algorithm internally to step 4 with a blocking factor of ω_s. $\omega_s \geq \omega_c$ is desirable for good cache performance (and, as we have seen in Section 3.3, likely to be obtainable for situations of interest).

This implies that columns of A_2 must be read repeatedly as they were for A_1 in the slab algorithm, with the total number of such elements being given by:

$$\Sigma_{i=0}^{N/\omega_a} \Sigma_{j=0}^{\omega_a/\omega_s} i\omega_a \cdot j\omega_s = \frac{N^2\omega_a}{4\omega_s} + O(N) \tag{5}$$

which are dominated by the number of reads in Equation 4.

Steps 2–3 can proceed with a block-based algorithm as is done for LL^T in [13], using $\omega_a \times \omega_a$ size blocks, and similarly for step 5. As these account for the dominant number of disk reads, storing the factored columns L in a row-block oriented fashion is desirable.

4 Conclusions and Future Work

Factorizing and hence solving symmetric indefinite systems is an interesting computation where there is a tradeoff in accuracy and performance.

Especially for the parallel out-of-core case, we have shown that there are potentially very large overheads in terms of disk accesses and messages in this computation, due with (long-distance) symmetric interchanges.

We have developed both slab and a (limited)-based algorithms which avoid this by having overlapping elimination blocks and search exhaustively for stable pivots within each block. The flexibility permitted in pivot searches by the diagonal pivoting methods was an important ingredient.

In either case, balanced memory utilization is important, favoring a small storage block size (i.e. the algorithmic blocking technique), even if this results in some extra communications.

However, the algorithms' performance is highly dependent on finding stable pivots within (the next ω_s columns of) the current elimination block in the majority of cases.

While the block-based algorithm has a limited scalability in that the next ω_s columns must be kept in core, this seems not to be crucial given the memory capacity of contemporary processors. Indeed, it is unclear that for medium-to-large sized contemporary parallel processors whether the greater scalability of the block-based algorithms is really necessary for problems that can be solved in reasonable time, given their extra complexity in implementation.

Future work will firstly involve investigating the success of block search strategy for large in-core algorithms. If proved successful, as expected from earlier studies, the out-of-core algorithms will be implemented. Their performance with respect to disk and communication overheads need then to be evaluated, and the latter also compared with the performance in the in-core case with existing implementations [8,9].

References

1. Homma, K., Nagase, K., Noro, M., Strazdins, P., Yamagajo, T.: Frequency Interpolation Methods for Accelerating Parallel EMC Analysis. In: Proceedings of the 15th International Parallel and Distributed Processing Symposium (for the 2nd Workshop on Parallel and Distributed Scientific and Engineering Computing with Applications), San Francisco, IEEE Press (2001) 10
2. Christen, P., Hegland, M., Nielsen, O., Roberts, S., Strazdins, P., Altas, I.: Scalable Parallel Algorithms for Surface Fitting and Data Mining. Parallel Computing **27** (2001) 941–961

3. Cwik, T., Patterson, J., Scott, D.: Electromagnetic Scattering Calculations on the Intel Touchstone Delta. In: Proceedings of the Supercomuting 92, IEEE Press (1992) 538–542
4. Golub, G., Loan, C.V.: Matrix Computations. second edn. John Hopkins University Press, Baltimore (1989)
5. Anderson, C., Dongarra, J.: Evaluating Block Algorithm Variants in LAPACK. In: Fourth SIAM Conference for Parallel Processing for Scientific Computing, Chicago (1989) 6 pages.
6. Jones, M.T., Patrick, M.L.: Factoring Symmetric Indefinite Matrices on High-Performance Architectures. SIAM Journal on Matrix Analysis and Applications **12** (1991) 273–283
7. Ashcraft, C., Grimes, R.G., Lewis, J.G.: Accurate Symmetric Indefinite Linear Equation Solvers. SIMAX **20** (1998) 513–561
8. Strazdins, P.E.: Accelerated methods for performing the LDLT decomposition. ANZIAM **42** (2000) C1328–C1355
9. Strazdins, P., Lewis, J.: An Efficient and Stable Method for Parallel Factorization of Symmetric Indefinite Matrices. In: Proceedings of The 5th International Conference and Exhibition on High-Performance Computing in the Asia-Pacific Region, Gold Coast (2001) 13
10. Toledo, S., Gustavson, F.: The design and implementation of solar, a portable library for scalable out-of-core linear algebra computations. In: Proceedings of the 4th Annual Workshop on I/O in Parallel and Distributed Systems, Philadelphia (1996) 28–40
11. Dongarra, J.J., Hammarling, S., Walker, D.W.: Key Concepts for Parallel Out-of-Core LU Factorizations. Technical Report CS-96-324, LAPACK Working Note 110, Computer Science Dept, University of Tennessee, Knoxville (1996)
12. D'Azevedo, E.F., Dongarra, J.J.: Key Concepts for Parallel Out-of-Core LU Factorizations. Technical Report CS-97-347, LAPACK Working Note 118, Computer Science Dept, University of Tennessee, Knoxville (1997)
13. Gunter, B.C., Riley, W., van de Geigin, R.A.: Parallel Out-of-core Cholesky and QR Factorizations with POOCLAPACK. In: Proceedings of the 15th International Parallel and Distributed Processing Symposium (for the 2nd Workshop on Parallel and Distributed Scientific and Engineering Computing with Applications), San Francisco, IEEE Press (2001) 10
14. Strazdins, P.E.: Parallelizing Dense Symmetric Indefinite Solvers. In: PART'99: The 6th Annual Australasian Conference on Parallel And Real-Time Systems, Melbourne, Springer-Verlag (1999) 398–410
15. Duff, I., Reid, J.: MA27: A Set of Fortran Subroutines for Solving Sparse Symmetric Sets of Linear Equations. Technical Report AERE R 10533, Harwell (1983)

Application of the Multi-level Parallelism (MLP) Software to a Finite Element Groundwater Program Using Iterative Solvers with Comparison to MPI

Fred Tracy

Engineer Research and Development Center
Information Technology Laboratory
Major Shared Resource Center
Vicksburg, MS, USA 39180

Abstract. The purpose of this paper is to give the results of the performance evaluation of the Multi-Level Parallelism (MLP) software versus MPI using the groundwater program FEMWATER on a remediation of a military site where iterative solvers are employed to solve the system of nonlinear equations. A one-to-one correspondence in functionality between MPI and MLP was maintained for all parallel operations so the performance comparisons would be consistent. An unstructured mesh application is one of the most difficult parallel applications, so this represents a good test. In this study, MLP did better in general on 64 PEs or less, but MPI proved more scalable as it did as good or better when using 128 PEs.

1 Introduction

Users of large shared-memory architectures on high performance computers are often challenged to find a method of parallel programming that is both efficient and easy to use. On shared-memory machines with large numbers of processors, the question of the "best" parallel method for a given problem is a difficult one. Recently, the Message Passing Interface (MPI) has been combined with OpenMP threads to make a dual-level or mixed-mode parallel programming paradigm [1]. While OpenMP is rather simple to apply to an existing code, it has inherent limitations with respect to efficient parallelization for larger numbers of processors. If programmed properly, MPI can create codes that scale to very large numbers of processors, but it is typically rather difficult to implement.

The Multi-Level Parallelism (MLP) software [2] utilizes a new method that takes advantage of large shared-memory SGI architectures, and excellent performance on NASA Ames applications has been reported [3]. An independent evaluation of MLP [4] gave inconclusive results on the performance of MLP and the NASA Ames computational fluid dynamics code Overflow because the MPI version and the MLP version of Overflow that were used in the evaluation had different algorithms for partitioning the work. Therefore, a more detailed study is warranted.

P.M.A. Sloot et al. (Eds.): ICCS 2003, LNCS 2659, pp. 725–735, 2003.

1.1 Purpose

The purpose of this paper is to give the results of the performance evaluation of MLP versus MPI using the groundwater program FEMWATER [5] on a remediation of a military site [6] where iterative solvers are employed to solve the system of nonlinear equations. A one-to-one correspondence in functionality between MPI and MLP was maintained for all parallel operations so the performance comparisons would be consistent. This was done, for instance, by replacing a call to an MPI reduction routine to get the global maximum value by a call to an equivalent MLP subroutine (described below). Another example also described below is the updating of ghost nodes where an equivalent subroutine was written in MLP. An unstructured mesh application is one of the most difficult parallel applications, so this represents a good test.

1.2 The MLP Parallel Programming Paradigm

Using MLP, the problem is explicitly partitioned among processing elements (PEs) as in MPI. However, rather than using sends, receives, broadcasts, reductions, etc., as in MPI, MLP communicates data among PEs by using shared variables as in OpenMP. As in MPI, combining OpenMP threads with forked processes is very natural with MLP, which, in fact, is indicated in its name.

2 Description of the Application

Figure 1 illustrates a typical top view of a 3-D finite element mesh for a remediation study. Several layers are used to model the soil layers underneath this surface. The mesh used for this comparison has 102,996 nodes and 187,902 3-D prism elements. The number of PEs used in this study starts at eight and is repeatedly doubled to 16, 32, 64, and 128 with the number of elements also being doubled each time the number of PEs is doubled.

3 Flow Equations

Pressure head in FEMWATER is modeled by applying conservation of mass to obtain,

$$\frac{\rho}{\rho_0} F \frac{\partial h}{\partial t} = \nabla \cdot \left[\mathbf{K} \cdot \left(\nabla \mathbf{h} + \frac{\rho}{\rho_0} \nabla \mathbf{z} \right) \right] + \sum_{m=1}^{N_{ss}} \frac{\rho_m^*}{\rho_0} Q_m \delta \left(\mathbf{r} - \mathbf{r_m} \right) , \tag{1}$$

$$h = \frac{p}{\rho_0 g} , \tag{2}$$

$$F = n \frac{dS}{dh} + S\alpha' + \theta\beta' , \tag{3}$$

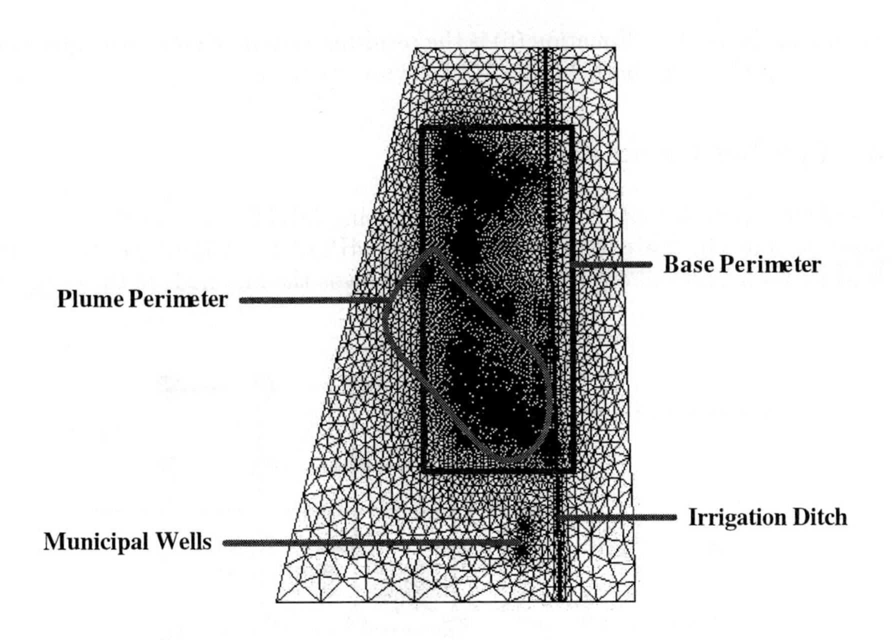

Fig. 1. This figure illustrates a typical top view of a 3-D finite mesh for a remediation study. Several layers are used to model the soil layers underneath the surface. The mesh used for this comparison has 102,996 nodes and 187,902 3-D prism elements

$$\alpha' = \rho_0 g \alpha \ , \tag{4}$$

$$\beta' = \rho_0 g \beta \ . \tag{5}$$

where α is the compressibility of the soil medium, β is the compressibility of water, δ is the Dirac delta function, g is the acceleration due to gravity, h is the pressure head, $\mathbf{K}$ is the hydraulic conductivity tensor, n is the porosity, N_{ss} is the number of source/sink nodes for flow, Q_m is the quantity of flow at the mth source/sink node, p is the pressure, $\mathbf{r}$ is a vector from the origin to an (x, y, z) point in space, ρ is the density with contaminant, ρ_0 is the density without contaminant, ρ_m^* is the density of the mth source/sink, $\mathbf{r_m}$ is the location of the mth source/sink node, S is the saturation, t is the time, and θ is the moisture content.

The Galerkin finite element method is then applied to obtain

$$\mathbf{M^{n+1}} \left(\mathbf{h^{n+1}} - \mathbf{h^n} \right) + \triangle \mathbf{t} \mathbf{K^{n+1}} \mathbf{h^{n+1}} = \triangle \mathbf{t} \mathbf{Q'^n} \ . \tag{6}$$

for the $(n+1)^{th}$ time-step. Here $\mathbf{M}$ is the mass matrix, $\mathbf{K}$ is the stiffness matrix, and $\mathbf{Q'}$ is a collection of flow type terms for the right-hand side, and $\triangle t$

is the time increment. Equation (6) is the resulting system of nonlinear equations that is solved, where both $\mathbf{M}$ and $\mathbf{K}$ are symmetric.

4 Parallel Paradigm

The finite element mesh is first partitioned using METIS [7]. The ghost cells as illustrated in Fig. 2 are updated using either MPI or MLP. Border elements are kept by both PEs and owned by the PE that owns the first node of the element.

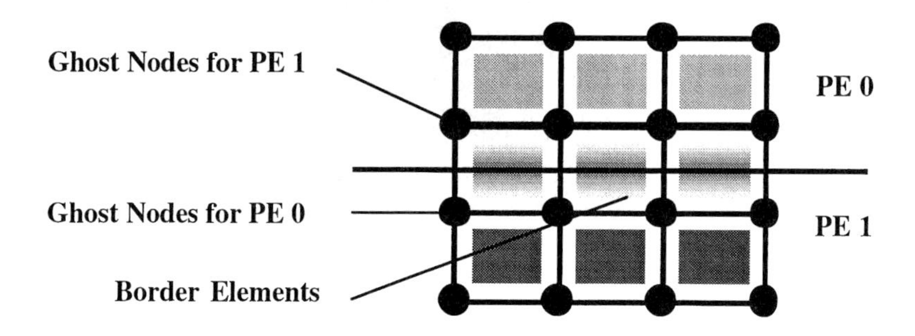

Fig. 2. The ghost cells are updated using either MPI or MLP. Border elements are kept by both PEs and owned by the PE that owns the first node of the element

5 Solvers Tested

Equation (6) is solved using a Picard iteration resulting in a symmetric, positive-definite linear system of equations of the typical form,

$$\mathbf{A}\mathbf{x} = \mathbf{b} \,. \tag{7}$$

The first solver tested consists of a forward relaxation step described by

$$\mathbf{x}^{\mathrm{new}} = \mathbf{x}^{\mathrm{old}} + \omega \left(\mathbf{D} + \mathbf{L}\right)^{-1} \left(\mathbf{b} - \mathbf{A}\mathbf{x}^{\mathrm{old}}\right) \quad 0 < \omega \leq 2 \,, \tag{8}$$

followed by a backward relaxation step,

$$\mathbf{x}^{\mathrm{new}} = \mathbf{x}^{\mathrm{old}} + \omega \left(\mathbf{D} + \mathbf{U}\right)^{-1} \left(\mathbf{b} - \mathbf{A}\mathbf{x}^{\mathrm{old}}\right) \quad 0 < \omega \leq 2 \,. \tag{9}$$

where $\mathbf{D}$ is the diagonal of $\mathbf{A}$, $\mathbf{L}$ is the lower part of A, $\mathbf{U}$ is the upper part of $\mathbf{A}$, and ω is the relaxation factor. In the parallel version, the owned nodes do the relaxation in parallel with matrix elements in $\mathbf{L}$ (forward step) and $\mathbf{U}$ (backward step) set to zeroes anywhere beyond the owned nodes and the ghost nodes. The ghost nodes are updated after the forward step and again after the backward step.

The second solver tested is a conjugate gradient solver with diagonal and incomplete LU preconditioners. The ILU preconditioner [8] is of the form,

$$\tilde{\mathbf{A}} = (\mathbf{D} + \omega\mathbf{L})\,\mathbf{D}^{-1}\,(\mathbf{D} + \omega\mathbf{U}) \quad \mathbf{0} \leq \omega \leq \mathbf{1}\,, \tag{10}$$

It was implemented to minimize communication by setting to zero all terms in **L** and **U** that were beyond owned and ghost nodes of the PE. Note that with $\omega = 0$, the preconditioner becomes **D**.

6 MLP and MPI Programming Details

The MLP and MPI programming details will now be presented in detail.

6.1 Initialization

While MPI is an extensive library, MLP has relatively few routines. Therefore, more is required when using MLP to set up. With MLP, rather than using sends, receives, broadcasts, reductions, etc., as in MPI, shared variables are used to communicate information as in OpenMP. The computer code given below shows the setup process as was done in FEMWATER.

MLP Setup Process for FEMWATER

```
 implicit real * 8 (a-h, o-z)
 parameter (ighnmx = 6113, npmx = 16)
 dimension bufv8(ighnmx, npmx), bufs8(npmx), ibufs(npmx)
 common / mlp_dat / ipt1, ipt2, ipt3
 dimension numcpu(npmx)
c   Establish pointers.
 pointer (ipt1, bufv8)
 pointer (ipt2, bufs8)
 pointer (ipt3, ibufs)
 integer * 8 isizes(3), ipoint(3), numvar
c   Get memory and link with pointers.
 numvar = 3
 isizes(1) = ighnmx * npmx * 8
 isizes(2) = npmx * 8
 isizes(3) = npmx * 4
 call mlp_getmem (numvar, isizes, ipoint)
 ipt1 = ipoint(1)
 ipt2 = ipoint(2)
 ipt3 = ipoint(3)
c   Set the number of processes.
 noproc = npmx
c   Pin-to-node option.
 npinit = 1
```

```
c   Choose one thread per process.
 do i = 1, noproc
   numcpu(i) = 1
 end do
c   Fork processes.
 call mlp_forkit (noproc, myid, numcpu, npinit)
```

First, the general-purpose shared variables of `bufv8`, `bufs8`, and `ibufs` are defined to keep a real vector for each PE, a real scalar for each PE, and an integer scalar for each PE, respectively. The pointer variables are then used to associate the value of the pointer to where the variable resides in memory as determined by the call to `mlp_getmem`. The pointers can be placed in a common block, but the actual variables cannot, as their memory locations are already set in `mlp_getmem`.

In the example above, the number of processes is set to 16, and the number of threads per process is set to 1 (OpenMP directives are not inserted). Note that it is easy to set the number of threads to a different value for each process. Threads can also migrate from their assigned node (the SGI O3K has four PEs per node), but experience indicates that performance improves when using multiple threads per process if the threads are pinned to their respective nodes as done by `ipinit` = 1. The call to `mlp_forkit` finishes the initialization process by forking the processes.

6.2 Reduction

A typical reduction in FEMWATER using MPI is given by

Reduction Using MPI

```
 call MPI_ALLREDUCE (res, resg, 1, MPI_REAL8, MPI_MAX,
&  MPI_COMM_WORLD, ieror)
```

where the maximum value of `res` over all the PEs is placed into `resg`. The program code given below shows how this is done using MLP.

Reduction Using MLP

```
c   Put data in the shared variable.
 bufs8(myid) = res
c   Make sure all PEs are through.
 call mlp_barrier (myid, noproc)
c   Have all PEs compute the maximum value.
 gmax = -1.0d30
 do i = 1, noproc
   gmax = dmax1 (bufs8(i), gmax)
 end do
 resg = gmax
```

Here, myid goes from 1 to 16, and each PE places its value of res into the bufs8 variable. Adding the mlp_barrier call ensures that all the PEs have finished. Finally, the maximum value is manually computed and stored in resg.

6.3 Updating Ghost Nodes

Most of the communication in FEMWATER is done in the solver when updating the ghost nodes. The send loop using MPI is as follows:

MPI Send Loop

```
 do i = 1, noproc
   num = nodgh(1, i)
   if (num .ne. 0) then
     jfn = num + 1
     do j = 2, jfn
       jloc = nodgh(j, i)
       buff(j - 1) = v(jloc)
     end do
     itag = 100
     call MPI_SEND (buff, num, MPI_REAL8, i - 1, itag,
&      MPI_COMM_WORLD, ierror)
   end if
 end do
```

The store data loop using MLP is

MLP Store Data Loop

```
 ibufs(myid) = 0
 iplace = 0
 do i = 1, noproc
   num = nodgh(1, i)
   if (num .ne. 0) then
     ibufs(myid) = ibufs(myid) + 1
     iplace = iplace + 1
     bufv8(iplace, myid) = i
     iplace = iplace + 1
     bufv8(iplace, myid) = num
     jfn = num + 1
     do j = 2, jfn
       iplace = iplace + 1
       jloc = nodgh(j, i)
       bufv8(iplace, myid) = v(jloc)
     end do
   end if
 end do
```

nodgh is an array that contains the number of nodes and the local node numbers whose values are to be sent to different PEs. The MLP version stores the number of messages in ibufs and the actual data of the messages in the bufv8 array. The receive loop (actually done before the send loop) using MPI is the following:

MPI Receive Loop

```
 do i = 1, noproc
   num = numngh(i)
   if (num .ne. 0) then
     itag = 100
     nst = nstngh(i)
     call MPI_IRECV (v(nst), num, MPI_REAL8, i - 1,
&      itag, MPI_COMM_WORLD, ireq(i), ierror)
   end if
 end do
```

The wait loop for MPI to complete the ghost cell update is done by

MPI Wait Loop

```
 do i = 1, noproc
   if (numngh(i) .ne. 0) then
     call MPI_WAIT (ireq(i), istat, ierror)
   end if
 end do
```

numgh is the number of ghost cells to be updated from the various PEs, and nst is where the ghost cells start in the local v vector. Finally, the retrieve data loop for MLP is the following:

MLP Retrieve Data Loop

```
 do i = 1, noproc
   nummes = ibufs(i)
   if (nummes .ne. 0) then
     iplace = 0
     do k = 1, nummes
       iplace = iplace + 1
       idest = bufv8(iplace, i)
       iplace = iplace + 1
       num = bufv8(iplace, i)
       if (idest .eq. myid) then
         nst = nstngh(i)
         do j = 1, num
           v(j + nst - 1) = bufv8(iplace + j, i)
         end do
```

```
            end if
            iplace = iplace + num
          end do
        end if
      end do
```

Synchronization is achieved by placing a call to `mlp_barrier` between the store and retrieve loops.

7 Performance Results

Performance tests using both the MPI and MLP version of FEMWATER were conducted on the SGI Origin 3800 located at the U.S. Army Engineer Research and Development Center (ERDC) Major Shared Resource Center (MSRC), Vicksburg, MS, for the remediation test problem using three different iterative solvers. Table 1 summarizes the results. The iterative linear solvers used were (1) the relaxation solver with one linear iteration being a forward loop, a ghost node update, a backward loop, and a second ghost node update, (2) a CG solver with the diagonal preconditioner, and (3) a CG solver using the ILU preconditioner with $\omega = 1$. Five hundred nonlinear iterations with 20 linear iterations each were done in each case. Both CG solves converged much quicker than the relaxation solver, but since this is a performance test of MPI versus MLP, the same number of linear iterations (10,000) was done in each case. In this series of tests, no OpenMP directives were added. The original problem was first run with 8 PEs. Next, the number of PEs was doubled along with the number of elements until 128 PE runs were done. These tests were run in a very busy production environment on the O3K, so running times varied as much as 10%. The best times from two runs in each case were used for Table 1.

8 Conclusions

The MLP results were generally better than the MPI results for PEs 8-64, and the MPI results were as good or better when 128 PEs were used. The percentage differences were always less than 10%, which means that no significant advantage was achieved with either parallel paradigm. The number of source lines of code (SLOC) not counting comments for the single subroutine that does the ghost cell update is 35 for MPI and 52 for MLP, so each is rather simple to implement, especially since the tedious work of determining where the data are to be sent and received has already been done. However, the edge goes to MPI for simplicity. More SLOC are required in the initialization and reduction for MLP, but the MPI libraries are much more extensive and thus already have initialization and reduction routines supplied.

For FEMWATER, a one-to-one correspondence implementation from MPI to MLP does not yield much benefit. However, if the shared variables could be used in a clever way to avoid all the partitioning and preparation, MLP still

Table 1. FREMWATER running times (sec)

PEs	8	16	32	64	128
Nodes	102,996	197,409	386,235	763,887	1,519,191
Elements	187,902	375,804	751,608	1,503,216	3,006,432
Solver	Relaxation	Relaxation	Relaxation	Relaxation	Relaxation
MPI	506	533	569	615	612
MLP	495	531	558	590	675
Solver	CG	CG	CG	CG	CG
Preconditioner	Diagonal	Diagonal	Diagonal	Diagonal	Diagonal
MPI	464	484	505	514	582
MLP	437	473	492	524	619
Solver	CG	CG	CG	CG	CG
Preconditioner	ILU	ILU	ILU	ILU	ILU
MPI	601	637	639	668	708
MLP	581	613	639	656	708

could have merit. Further research is needed here, as well as investigating the use of multiple OpenMP threads for each process.

Acknowledgment. This work was supported in part by a grant of computer time from the DoD High Performance Computing Modernization Program at the ERDC MSRC.

References

1. Fahey, R, and Smith, J.: STWAVE: A Case Study in Dual-Level Parallelism. ERSC MSRC Technical Report 01-28, Vicksburg, MS (2001)
2. Taft, J.R.: MLP Version 2.1 (computer program). Sienna Software, Inc., 1105 Terminal Way, Ste 202, Reno, NV (2002)
3. Taft, J.R.: Overflow Gets Excellent Results on SGI Origin 2000. NAS News, Vol. 3, No. 1, NASA Ames Research Center, Moffett Field, CA (1998)
4. Wornom, S.F., Tracy, F.T., Duffy, D.Q., and Alter, R.W.: A Performance Evaluation of the Multi-Level Parallelism (MLP) Software with MPI and OpenMP. DoD HPC Users Group Conference Proceedings, Austin, TX (2002)
5. Lin, H.J., Richards, D.R., Talbot, C.A., Yeh, G.T., Cheng, J.R., Cheng, H.P., and Jones, N.L.: FEMWATER: A Three-Dimensional Finite Element Computer Model for Simulating Density-Dependent Flow and Transport in Variably Saturated Media. Technical Report CHL-97-12, U.S. Army Engineer Research and Development Center (ERDC), Vicksburg, MS (1997)

6. Tracy, F.T., Talbot, C.A., Holland, J.P., Turnbull, S.J., McGehee, T.L., and Donnell, B.P.: The Application of the Parallelized Groundwater Model FEMWATER to a Deep Mine Project and the Remediation of a Large Military Site. DoD HPC Users Group Conference Proceedings, Monterey, CA (1999)
7. Karypis, G.: METIS (computer program). `http://www.users.cs.umn.edu/~karypis/metis/`, University of Minnesota, Minneapolis, MN (2002)
8. Dongara, J.J., Sorensen, D.C., and van der Vorst, H.A.: Numerical Linear Algebra for High-Performance Computers, SIAM, Philadelphia (1998), 203

Workshop on Java in Computational Science

Visual Parameteric Modeler for Rapid Composition of Parameter-Sweep Applications for Processing on Global Grids

Shoaib Burq[1], Steve Melnikoff[1], Kim Branson[2], and Rajkumar Buyya[1,*]

[1] Grid Computing and Distributed Systems Lab
Dept. of Computer Science and Software Engg.
The University of Melbourne, Australia
[2] Structural Biology
Walter and Eliza Hall Institute
Parkville, Melbourne, Australia

Abstract. Grids are emerging as a platform for the next-generation parallel and distributed computing. Large-scale parametric studies and parameter sweep applications find a natural place in the Grid's distribution model. There is little or no communication between jobs. The task of parallelising and distributing existing applications is conceptually trivial. These properties of parametric studies make it an ideal place to start developing integrated development environments (IDEs) for rapidly Grid-enabling applications. However, there is a lack of the availability of IDEs for scientists to Grid-enable their applications, without the need of developing them as parallel applications explicitly. This paper presents a Java based IDE called Visual Parameteric Modeler (VPM), developed as part of the Gridbus project, for rapid creation of parameter sweep applications. It supports automatic creation of parameter script and parameterisation of input data files, which is compatible with the Nimrod-G parameter specification language. The usefulness of VPM is demonstrated by a case study on a composition of molecular docking application as a parameter sweep application. Such applications can be deployed on clusters using the Nimrod/enFuzion system and on global Grids using the Nimrod-G grid resource broker.

1 Introduction

As high-speed networks become ubiquitous and research in middleware technologies matures, new windows of opportunity for application scientists to run their applications on parallel and distributed computing environments, such as clusters and Grids [3], are increasing. The underlying infrastructure, providing the low-level facilities to run applications in a heterogeneous and distributed environment; and high-level tools that facilitate the creation of Grid applications and their deployment on distributed resources, makes up the Grid.

There exist a number of models for the construction of parallel and distributed applications. Parameter sweep is one of the simplest and most practical of the models that can yield powerful results. Parameter sweep applications consist of programs that

* Correspondence to: Rajkumar Buyya, email: raj@cs.mu.oz.au

P.M.A. Sloot et al. (Eds.): ICCS 2003, LNCS 2659, pp. 739–749, 2003.

are run independently on different nodes with different input parameters or data sets. There are numerous application areas where parametric studies find a use. Some application scenarios include:

- molecular biologist (drug designer) looking for compounds, in a large chemical data sets , that best dock with a particular protein [8];
- geologist looking at the change in the density and depth of ore-body and the overlying rock's density to optimize cost and production;
- aerospace engineer understanding the role of geometry parameters in the aerodynamic design and optimization process [11];
- high energy physicist investigating on the origin of mass by analysing petabytes of data generated by high-energy accelerators such as the LHC (Large Hadron Collider) [14]; and
- neuroscientist performing brain activity analysis by conducting pair-wise cross co-relation analysis of MEG (Magneto-EncephaloGraphy) sensors data [13].

The practical implications of performing parametric studies make it difficult for an application scientist, who has little or no knowledge of distributed computing, to use it effectively. The vision of the Grid is precisely to bridge this gap by providing a seamless access to compute and other scientific resources without the need of users concerning about the lower-level details of the computing infrastructure or the resource management issues [1]. High-level tools for creation of distributed applications and their deployment on the Grid make up an essential part of this vision. Currently, there is still lack of the availability of integrated development environments (IDEs) with visual interface for scientists to rapidly Grid-enable their existing applications.

This paper presents a Java based IDE called Visual Parameteric Modeler (VPM), developed as part of the Gridbus project, for rapid creation of parameter sweep applications. VPM provides a simple visual interface for the manipulation of scripts or input files of existing applications. Users can assign parameters to certain values by highlighting them. They can select from a number of different data types and domains to describe their parameters. VPM also incorporates a task editor for creating the tasks carried out by different jobs during different stages of a distributed execution. The parameters and tasks together provide the basis of each run. VPM allows the rapid creation and manipulation of the parameters. While being flexible, it is also simple enough for a non-expert to create a parameter script, known as a plan file. The parameter sweep applications composed using VPM can be deployed on global Grids using the Nimrod-G resource broker that supports scheduling based on the user's quality of service (QoS) requirements – such as the deadline, budget, and optimization preference – and the access price of resources.

The rest of this paper is organised as follows. Section 2 presents related tools and their capabilities including differences. The VPM architecture is discussed in Sect. 3 and the design and implementation is discussed in Sect. 4. The use of VPM for composing molecular docking application as a parameter sweep application is presented in Sect. 5, followed by a conclusion in Sect. 6.

2 Related Work

VPM draws inspiration from or builds on the concepts developed in Nimod [2] and its commercial version (Enfuzion [1]); and its Grid-enabled version (Nimrod-G [9]) that support the creation and execution of parametric applications on clusters and Grids respectively. A declarative language, called parameter specification language, supported by Nimrod describes the parameters and the tasks that make up the plans.

For the creation of plans, Enfuzion takes a wizard approach. Enfuzion will take a user through the operation of creating a job specification file step-by-step, because it is too complex for novice users to create parameter script on their own. In the input file to the application, the user must change the value assigned to a parameter to a place marker. Although simple and less prone to error, this approach is too rigid, slow and cumbersome for someone working on several input files at the same time. As the parameter script and parameterized input data files generated by VPM confirm to the Nimrod parameter specification language, it serves as a complimentary tool. This ensures that VPM can be used by EnFuzion and Nimrod-G users.

Using VPM, the users can select all application input data/configuration files and parameterise easily. The users can drag and select the value in the input file that they wish to assign a parameter to, or they can create parameters independent of an input file. This gives the user a great deal of flexibility and control. By giving the user fields to input their parameter configuration and then generating the plan specification automatically we can prevent errors. Even if the users create parameter script in their favorite editor, VPM allows them to import and make use of its capabilities. Once the plan specification is created, the users proceed to execution phase during which they have an option of changing values assigned parameters. Like enFuzion, the VPM will automatically create application jobs each with different parameter values will be created. Such jobs can be analysed on clusters or Grids using enFuzion or Nimrod-G respectively.

Other related works include, APST (AppLeS Parameter Sweep Template) [10] and NASA IPG (Information Power Grid) parameter process specification tool [11]. APST expects application scientists to explicitly create jobs and assign parameter values to them. IPG provides graphical environment for parameterising the data files. Both the APST and IPG schedulers use traditional system centric policies for resource allocation. As VPM confirms to the Nimrod-G parameter specification language, it enables the users to harness Grid resources using the Nimrod-G resource broker depending on their QoS requirements and the access price of resources. Thus, it supports the Grid economy, which is essential for management and allocation of resources based on the supply and demand.

3 Architecture

The visual parameteric modeler architecture and parameter sweep application creation flow model is shown in Fig. 1. VPM supports the creation of a new parameter sweep applications from scratch or the utilisation of the existing parameterised application plans with further update. In the first case, the users can add all those files to be parameterised and use VPM to parameterise data items of interest. In the second case, the users can import the existing parametric plans and pass through the VPM scanner and parser that identify parameters and make them available for further update. The users can use the VPM task editor to create a task to be associated with jobs. Based on parameter types and their values a number of jobs, each representing a different parameter scenario, are generated automatically.

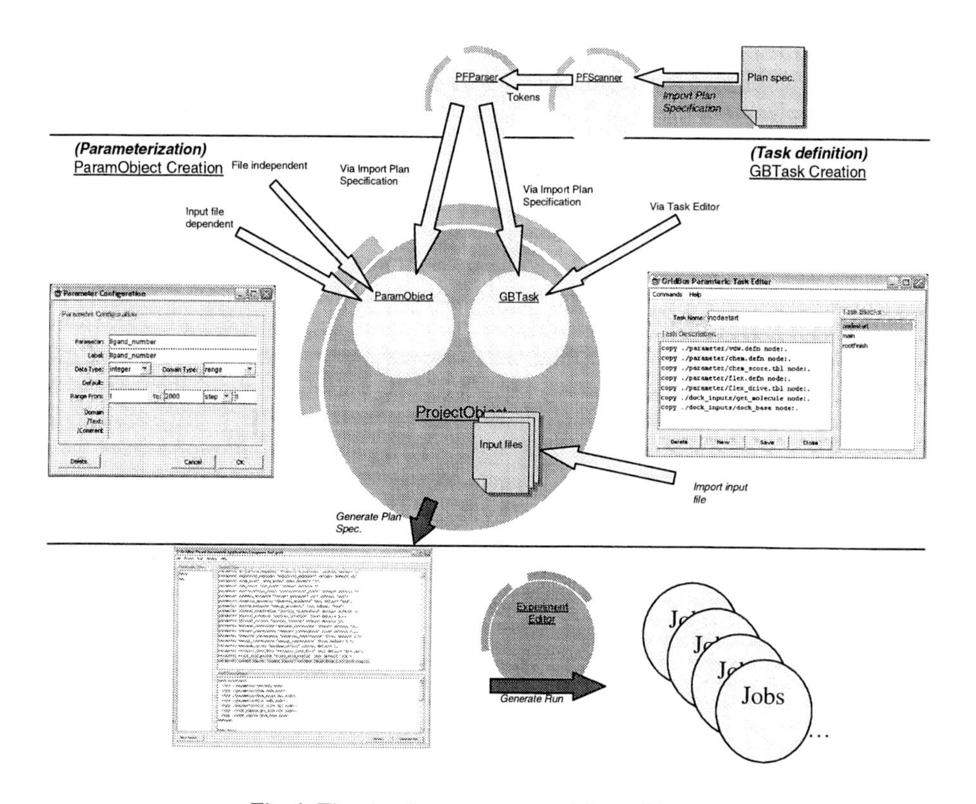

Fig. 1. The visual parameteric modeler architecture

VPM consists of three major visual components: Project, Input Files and Tasks. These components are represented as Project Window, Input File Window and Task Editor, respectively. The design of VPM, shown in Fig. 2, allows a single project to have several input data files and tasks.

These visual components provide the user access to the objects that encapsulate the plan's information-model, namely to ParamObject and GBTask. ParamObject is created and manipulated from the project window or input file window, while the GBTask is created and manipulated using the TaskEditor.

A plan consists of parameters and task. In VPM, parameters are internally represented as ParamObjects and tasks as GBTasks. ParamObjects are created by any of the following three methods.

- File dependent parameterization
- File independent parameterization
- Via imported plan specification

File Dependent Parameterization

Once an input file or a script file is imported into VPM, values that have to be assigned parameters are highlighted and the parameter defined and assigned by a simple click of mouse.

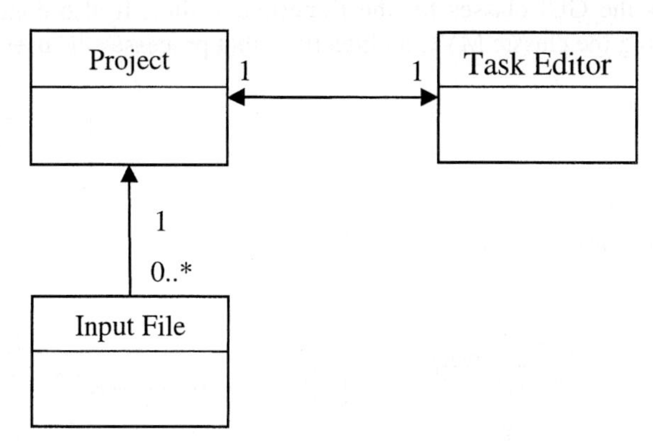

Fig. 2. Basic visual components of VPM

File Independent Parameterization

New parameters may also be created by simply defining its properties.

Via Imported Plan Specification

VPM contains a LALR (Look Ahead, Left to Right) parser for plan specification that confirms to the Nimrod parameter specification language. This allows the reuse of an existing plan file (parameter script). The parser translates each parameter definition into a ParamObject and each task description into a GBTask (see Figure 1).

Experiment Editor and Job Generation

Once a plan specification is completed, VPM can generate a run specification. This enumerates every value lying within the range of the parameters described by the plan, and a description of the jobs in terms of the values assigned to them. Hence, the run specification describes the distribution model of the application parameterized using VPM.

4 Design and Implementation

VPM is coded in Java and MVC (Model-View-Controller) architecture [12] design pattern that decouples the data model from the component that represents it on the screen. The graphical user interfaces are created using the Java Swing component set that uses MVC architecture consistently.

Besides the above-mentioned objects, VPM has various components that facilitate the creation of a plan specification (parameter script) and parameterisation on input data files. VPM consists of many packages and associations and reverse-associations between them are shown in Fig. 3. The arrow heads point at the dependent packages. Notice, a single class, Jobs, in GBJobs package, is responsible for the production of Grid enabled jobs. This can be extended to support creation of job specification for different scheduling systems.

ExperimentEditor

This contains the GUI classes for the ExperimentEdior. It also contains a controller class (following the classic MVC architecture) that processes the user input.

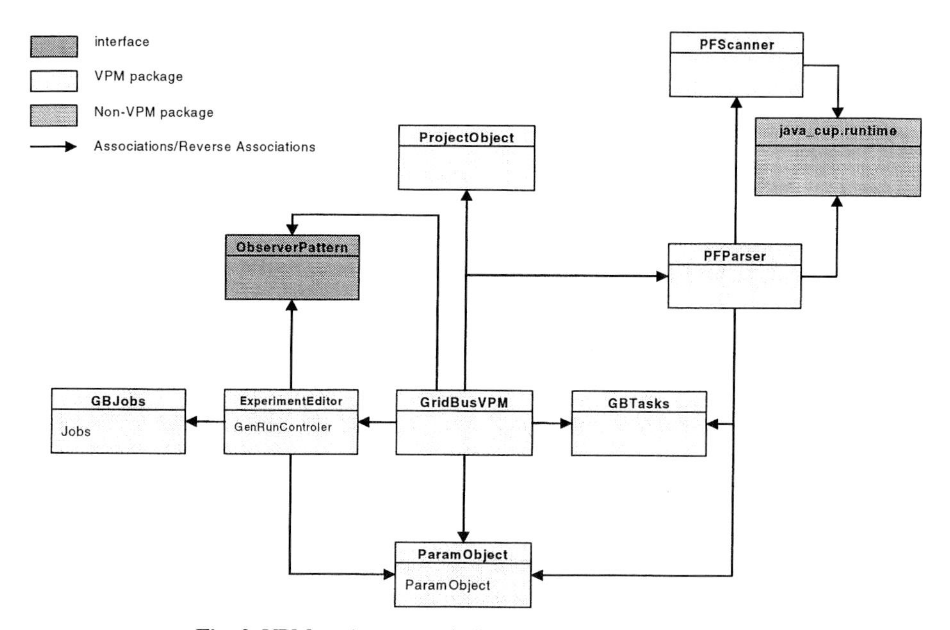

Fig. 3. VPM package associations and reverse-associations

GBJobs

This contains a single class, `Jobs`. "`Jobs`" takes as its input a count (N) of those parameters that have a range of values and an array of integers of size N containing the maximum value taken by each of these parameters.

GBTask

This package contains a single class, `GBTask`. It is a serializable object. It encapsulates the commands that execute during different phases of the distributed run.

GridBus

This is the largest package containing mostly the GUI classes for VPM. Following the MVC architecture, it contains all the "view" components. It also includes a utility class, called `GBFileManager`, for handling all file operations within VPM. In addition, this package contains the class that has VPM's main method, named `Project`.

ObserverPattern

This package contains two interfaces `Observer` and `Subject`. This facilitates the implementation of MVC architecture, by decoupling related objects [4]. A subject may have a number of observers. All observers are notified when the subject undergoes a state change. In response, the observer may query the subject to synchronize its state with the subject. The observer implements the `update()` method while the subject implements the `addObserver()` and `removeObserver()` method. On a state change, the subject calls each observer's update method.

ParamObject

This package contains a single class: `ParamObject`. The `ParamObject` is the heart of VPM. It is a serializable object encapsulating the state of a parameter, it contains two key methods: `makePlanStep()` and `makeRunStep()`. These methods are responsible for automating the process of plan and run specification creation. The `makePlanStep` method converts the fields of the ParamObject into a line of the simple declarative language following the grammar of Fig. 4. `makeRunStep` converts a parameter's declaration into a statement of a run specification. This declaration identifies the possible value(s) taken by the parameter. Currently `makeRunStep` generates a Nimrod-G readable statement.

plan → *plan rest* | *error* | ε
rest → *planStep* | *taskBlock* | *newline*
planStep → *PARAMETER ID label type domain SEMI*
label → *LABEL QUOTE* | *QUOTE* | ε
type → *INTEGER* | *FLOAT* | *TEXT* | *FILE*
domain → *DEFAULT value_opt*
| *RANGE range_values* domain2
| *SELECTANY value_list default_opts*
| *SELECTONE value_list default_opt*
| *RANDOM range_values* points_opt
| COMPUTE expr
| JITP jitp_expr
points_opt → POINTS value_opt | ε
default_opts → value_opt value_list | value_opt
domain2 → POINTS value_opt | STEP value_opt | ε
value_opt → ID | QUOTE | NUM
expr → expr PLUS term | expr MINUS term | term
term → term TIMES factor | factor
factor → NUMBER | LPAREN expr RPAREN

Fig. 4. Context free grammar for plan specification

PFScanner
PFScanner, (plan file scanner) created using an open source tool called JLex [5], performs lexical analysis of the plan specification. It comes into play when the user wishes to import an existing plan specification into VPM. It interfaces with the PFParser (discussed below) providing it with a stream of identified tokens.

PFParser
PFParser, (plan file parser) written using an open source tool called CUP [6], interfaces with the PFScanner and attempts to match the stream of tokens to a complete parameter or task definition as described by the A context-free grammar shown in grammar in Fig. 4. All caps denote the terminals. In doing so, it generates new `ParamObjects` or `GBTasks`. It contains two public methods for the retrieval of ParamObject and GBTasks: `getParams()` and `getTasks()`.

ProjectObject
`ProjectObject` encapsulates all the attributes necessary to describe a VPM project. It contains the `ParamObjects`, `GBTasks`, paths to input files and other attributes that uniquely identify a project.

5 Use Case Study – Molecular Docking Application

Molecular modeling for drug design involves screening millions of ligand records or molecules of compounds in a chemical database (CDB) to identify those that are potential drugs. This process is called molecular *docking* [7]. It helps scientists explore how two molecules, such as a drug and an enzyme or protein receptor, fit together. Docking each molecule in the target chemical database is both a compute and data intensive task. In [8], a virtual laboratory environment has been developed and demonstrated distributed execution of molecular docking application on Global Grids. The application has been formulated as a parameter sweep application using a simple parameter specification language and deployed on global Grids using the Nimrod-G resource broker.

We now discuss how the application has been parameterized (i.e., the creation of parameter script and parameterisation of data files) using the VPM. In [8], the creation of parameter script and parameterisation of data/configuration files has been carried out manually using a text editor. Although this task is simple, it becomes cumbersome when an application contains multiple data files and has a large number of data entries to be parameterised. This approach is also prone to creating parameter script with syntax errors. The use of visual modeler helps overcome these limitations and aids in the rapid parameterisation of the molecular docking application such as the "Dock" [7] software package.

Fig. 5 shows the parameterisation of docking application configuration input file using VPM. First, the configuration input file is imported into the VPM. When the value of a data item to be parameterised is selected (see the highlighted text "S_1" in Fig. 5), it appears in the dialogue box where the parameter name can be defined along with the attributes (data type and values). In this example, the name of a data item, "ligand_atom_file", indicates the molecule to be screened. As the aim of parameterisation is to screen multiple molecules, this parameter need to be defined as the "range" data type and then assign values for index start, end, step. For example, to screen the first 2000 molecules in the chemical data base, the initial values to be assigned are 1, 2000, and 1 respectively. VPM will automatically create a parameter statement and add to the script (see the highlighted statement in Figure 6). A task specification creation module provides dialogue facility selection of appropriate commands associated with the execution of a parametric job (see a small window in Fig. 6).

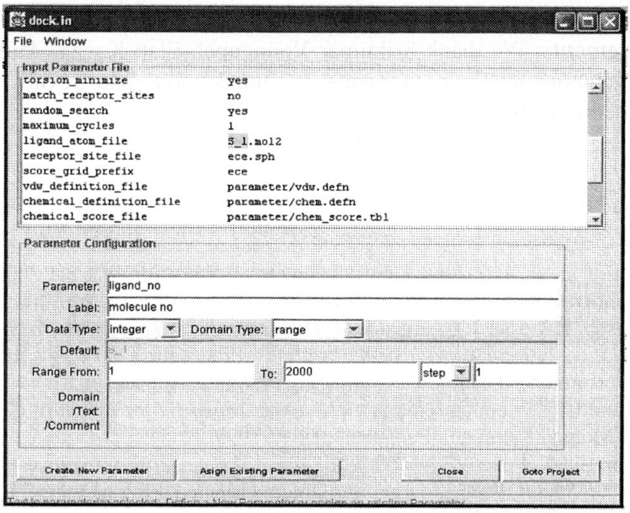

Fig. 5. The parameterisation of docking configuration input file

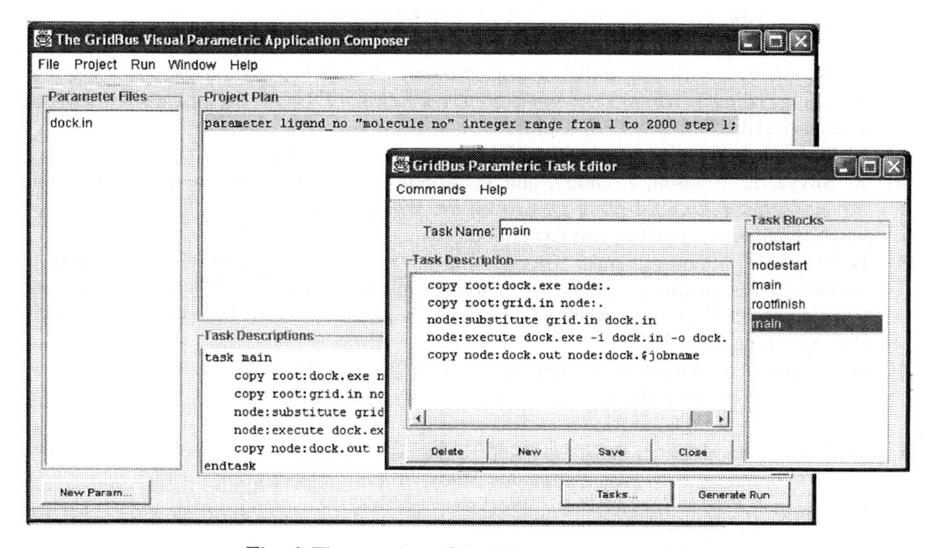

Fig. 6. The creation of docking parameter script

6 Conclusion

In this paper, we outlined the need for the development of IDEs and other applications and tools in order to provide the applications scientist with user-friendly environments to run their code on the Grid. We introduced VPM developed to provide one such environment for parameter sweep applications. We identified its key features, while giving some of its implementation details. Finally, we showed how application scientists can use VPM to parameterise their applications. Such parameterised applications can be deployed on Global Grids using the Nimrod-G resource broker.

Acknowledgements. We thank Srikumar Venugopal, Elan Kovan, Anthony Sulistio, and Sarana Nutanong for their comments. We thank anonymous reviewers for providing excellent comments.

References

[1] D. Abramson et. al., EnFuzion Tutorial, Chapter 4, EnFuzion Manual, 2002. Available at: `http://www.csse.monash.edu.au/cluster/enFuzion/tutorial.htm`

[2] D. Abramson, R. Sosic, J. Giddy, and B. Hall, *Nimrod: A Tool for Performing Parameterised Simulations using Distributed Workstations*, Proceedings of the 4th IEEE Symposium on High Performance Distributed Computing, Virginia, August 1995.

[3] I. Foster and C. Kesselman (editors), *The Grid: Blueprint for a Future Computing Infrastructure*, Morgan Kaufmann Publishers, USA, 1999.

[4] G. Krasner and S. Pope, *A cookbook for using the model-view controller user interface paradigm in Smalltalk-80,* Journal of Object-Oriented Programming, 1(3):26–49, August/September 1988.

[5] E. Berk, *JLex: A lexical analyzer generator for Java(TM)*, Department of Computer Science, Princeton University Version 1.2.5, September 6, 2000 `http://www.cs.princeton.edu/~appel/modern/java/JLex/`

[6] S. E. Hudson, *CUP: LALR Parser Generator for Java(TM)*, GVU Center, Georgia Tech. Version 0.10, July 1999 http://www.cs.princeton.edu/~appel/modern/java/CUP

[7] T. Ewing (editor), *DOCK Version 4.0 Reference Manual*, University of California at San Francisco (UCSF), USA, 1998. Online version: `http://www.cmpharm.ucsf.edu/kuntz/dock.html`

[8] R. Buyya, K. Branson, J. Giddy, and D. Abramson, *The Virtual Laboratory: Enabling Molecular Modelling for Drug Design on the World Wide Grid*, Journal of Concurrency and Computations: Practice and Experience, Wiley, USA, Jan 2003.

[9] R. Buyya, D. Abramson, and J. Giddy, *Nimrod-G: An Architecture for a Resource Management and Scheduling System in a Global Computational Grid*, The 4th International Conference on High Performance Computing in Asia-Pacific Region (HPC Asia 2000), Beijing, China, May 2002.

[10] H. Casanova, G. Obertelli, F. Berman, and R. Wolski, *The AppLeS Parameter Sweep Template: User-Level Middleware for the Grid*, Proceedings of the Super Computing (SC 2002) Conference, Dallas, USA.

[11] M. Yarrow, K. McCann, R. Biswas, and R. Van der Wijngaart, *An Advanced User Interface Approach for Complex Parameter Study Process Specification on the Information Power Grid*, Proceedings of the 1st Workshop on Grid Computing (GRID 2002), Bangalore, India, Dec. 2000.

[12] Java and MVC architecture,
`http://javanook.tripod.com/patterns/java-mvc.html`
[13] R. Buyya, S. Date, Y. Mizuno-Matsumoto, S. Venugopal, and D. Abramson, *Economic and On Demand Brain Activity Analysis on Global Grids*, Technical Report, Grid Computing and Distributed Systems (GRIDS) Lab, The University of Melbourne, Australia, Jan. 2002.
[14] CERN, the LHC Grid Project, `http://lcg.web.cern.ch/LCG/`

Method Call Acceleration in Embedded Java Virtual Machines

M. Debbabi[1,2], M. Erhioui[2], L. Ketari[2], N. Tawbi[2], H. Yahyaoui[2], and S. Zhioua[2]

[1] Panasonic Information and Networking Technologies Laboratory
Princeton, NJ, USA
debbabim@research.panasonic.com
[2] Computer Science Department, Laval University, Quebec, Canada
{erhioui,lamia,tawbi,hamdi,zhioua}@ift.ulaval.ca

Abstract. Object oriented languages, in particular Java, use a frequent dynamic dispatch mechanism to search for the definition of an invoked method. A method could be defined in more than one class. The search for the appropriate method definition is performed dynamically. This induces an execution time overhead that is significant. Many static and dynamic techniques have been proposed to minimize the cost of such an overhead. Generally, these techniques are not adequate for embedded Java platforms with resource constraints because they require a relatively big memory space. The paper proposes a dynamic, flexible and efficient technique for accelerating method calls mechanism in embedded systems. This acceleration technique spans over 3 aspects of the method call: (1) lookup, (2) caching, and (3) synchronized methods.

1 Motivations

The main intent of this research is to accelerate Java-based embedded systems. Nowadays, there is growing interest and potential for embedded systems to make a contribution in scientific computing. For instance, a Java-based sensor that collects (often massive) data for scientific experiments needs to be accelerated for a shorter pre-processing time. This acceleration is highly relevant especially in the context of Java-based distributed systems supporting HPC activities. An increased performance will be perceived as both useful and interesting by members of the community.

In this paper, we focus on accelerating embedded Java virtual machines. More accurately, we put the emphasis on one particular aspect that is the method call. Method invocation is a crucial aspect in the design and implementation of the runtime of any object oriented language. This aspect is even more important when it comes to dynamically typed languages such as Java. In fact, Java offers a dynamic dispatch mechanism that allows to invoke the appropriate method depending on the runtime class of the receiving object. Usually, this process is inevitable because the actual invoked method can not be known statically. In fact, the method could be defined in more than one class. Hence, to compute the

P.M.A. Sloot et al. (Eds.): ICCS 2003, LNCS 2659, pp. 750–759, 2003.

invoked method, a dynamic search (or a dynamic dispatch) procedure should be performed. This search procedure starts from the call receiver class. It spans over the hierarchy to know if a superclass of the call receiver class defines a method having the same signature as the statically computed method for this call. Unfortunately, method invocations involving dynamic dispatch are slower in an interpreter and incur an additional overhead which is significant. So the frequent use of this mechanism hurts badly the execution performance of Java applications. Many static and dynamic techniques have been proposed to minimize the cost of such an overhead. Generally, these techniques are not adequate for embedded Java platforms with resource constraints because they require an important memory space. Consequently, lightweight and tiny optimizations, with minimum space and time strategy, are required to accelerate Java calls in embedded virtual machines.

The remainder of this paper is organized as follows. Section 2 presents the Related Work. Section 3 outlines our approach. Section 4 presents the results. Finally, Sect. 5 concludes the paper.

2 Related Work

Object-oriented languages support inheritance and polymorphism to allow the development of flexible and reusable software. The type of a specific object would usually be determined at run time. This feature is called the dynamic binding. In this context, the selection of the appropriate method to execute is based on a lookup mechanism which means that the actual method to be executed after an invocation is determined dynamically based on the type of the method's receiver, the class hierarchy and the method inheritance or overloading schema. The lookup mechanism consists of determining the actual method to be executed when an invocation occurs. If this class implements a method that has the same signature as the called one, the found method will be executed. Otherwise, the parent classes will be checked recursively until the searched method is found. If no method is found, an error is signaled (*MsgNotUnderstood*). Unfortunately, this operation is too frequent and is very expensive. The principal dynamic binding algorithm is called the Dispatch Table Search [1] (DTS). It proceeds as mentioned above. The DTS is good in terms of memory cost, however the search overhead makes the mechanism too slow. Many techniques have been proposed to minimize the overhead associated to DTS: static techniques which pre-compute a part of the lookup and dynamic techniques which cache the results of previous lookup, thus avoiding other lookups. In the following, we present one of these techniques: the Selector Table Indexing technique (STI) [2]. Given a class hierarchy of C classes and S selectors, a two-dimensional array of $C * S$ entries is built. Classes and selectors are given consecutive numbers and the array is filled by pre-computing the lookup for each class and selector. An array entry contains a reference to the corresponding method or to an error routine. These tables are computed for a complete system. The main drawback of STI is that space requirements are huge for a large system. Hence, many dis-

patch table compression techniques have been proposed (Selector coloring [7], Row displacement [3], etc) to minimize space overhead. Another drawback of this technique is that the compiled code is very sensitive to changes in the class hierarchy. However, this technique delivers fast and constant time lookup.
Devirtualization technique with code patching mechanism [4] is another optimization technique that converts a virtual method call to a direct call. Given a dynamic site call, the current class hierarchy is analyzed by the compiler to determine if the call can be devirtualized. If it is true and if the method size is small, the compiler generates the inlined code of the method with the backup code of making the dynamic call. When devirtualization becomes invalid, the compiler performs code patching to make the backup code executed later. Otherwise (devirtualization is valid), only the inlined code is actually executed. The main drawback of this technique is that it relies on heavy analysis (flow-sensitive type analysis, preexistence analysis, dynamic class hierarchy analysis, etc.) so it is not convenient for embedded systems due to the fact that it can be too expensive in both time and space.
A static method call resolution technique [8] has been proposed to solve dynamic method calls. A variable-type analysis and a declared-type analysis use the whole class hierarchy program to compute a set of a method call receivers. This technique is limited because it does not deal with the dynamic class loading problem. In fact, the class hierarchy could change while the program is executing. This could change a method call receivers set and make the static performed optimizations inaccurate.
Dynamic techniques consist in caching results of previous lookups. Cache-based techniques eliminate the requirements to create huge dispatch tables, so memory overhead and table creation time are reduced. There are two main approaches to caching: global caches [1] and small inline caches [5]. Global cache technique stores the previous lookup results. In the global cache table, each entry consists of triplets (receiver class, selector and method address). The receiver class and the selector are used to compute an index in the cache. If the current class and the method name match those in the cached entry at the computed index, the code at the method address is executed. Hence, method lookup is avoided. Otherwise, a default dispatching technique (usually DTS) is used and at the end of this search, a new triplet is added to the cache table and control is transferred to the found method. The run-time memory required by this algorithm is small, usually a fixed amount of the cache and the overhead of the DTS technique. The main disadvantage of this technique is that a frequent change of the receiver class slows the execution. The inline cache technique consists in caching the result of the previous lookup (method address) in the code itself at each call site. Inline cache changes the call instruction by overwriting it with a direct invocation of the method found by the default method lookup. Inline cache assumes that the receiver's class changes infrequently otherwise, the inline cache technique delivers slow execution time. Polymorphic inline cache [9] is an extension of the inline cache technique. The compiler generates a call to a special stub routine. Each call site jumps to a specific stub function. The function is initially a call

to a method lookup. Each time the method lookup is called, the stub function is extended. This technique has the cost of a test and a direct jump in the best case. Moreover, the executable code could expand dramatically when the class hierarchy is huge and the receiver class changes frequently.

3 Approach

We propose a dynamic, flexible and efficient technique for accelerating method calls in embedded systems. This acceleration technique spans over 3 aspects of the method call: (1) lookup, (2) caching, and (3) synchronized methods.

3.1 Lookup

The lookup is accelerated by the application of a direct access to the method tables. This is achieved by using an appropriate hashing technique. Actually, we build a hash table for each virtual method table. Each index of the hash table is a hashing result of the method signature. The size of the hash table should be carefully chosen so as to have a low footprint while minimizing the collisions between method signatures. By doing so, we get a more efficient and flexible lookup mechanism. Efficiency stems from the fact that we have direct access to the method tables. Flexibility stems from the fact that we can tune the size of the hash table so as to have the best ratio for speed versus footprint.
In what follows we explain how this lookup acceleration could be implemented within a conventional embedded Java virtual machine such as KVM [6] (Kilobyte Virtual Machine). The method lookup mechanism in KVM is linear i.e. it uses a sequential access. A hash-based lookup will definitely yield a better performance. The implementation of such a mechanism will affect two components of the virtual machine: the loader and the interpreter. The loader is modified to construct hashed method tables for each loaded class. The interpreter is modified to take advantage of the new method tables to perform fast and direct-access lookups. During the loading process of a class, a hash method table is built. A hash is computed from the method signature. Each entry in the hash table consists of two components. The first component is a flag indicating whether the class contains a method with such a definition. The second component is a pointer to the method definition. In the case of a collision, this second component is a pointer to a list of method definitions. The method hash table construction algorithm is depicted in Fig. 1.

The original lookup algorithm is linear. It tests in each method table, by iteration over its elements, of a class if it has a method that has the same signature as that invoked. In Fig. 2 we give the original lookup mechanism.

The new lookup algorithm uses the hash obtained from the method signature to access the corresponding entry in the hash table. If the flag associated with this entry is ON, it accesses the method definition thanks to the second component of the entry. If the flag is OFF, this means that the class does not implement

```
HashTableBuild() {
   for each method of the class method table {
     h = compute_hash(method);
     element = get_element(hashTable, h);
     if (element->flag == ON) {
       allocate_space(method);
       register_method_in_collision_list();
     }
    else {
       register_method_in_hashTable();
       method->flag = ON;
     }
    }
}
```

Fig. 1. Method hash table construction algorithm

```
lookupMethod(class, key) {
    while (class) {
       table = get_method_table(class);
       for each method of table {
        if (method->signature == key) {
          return method;
         }
        class = class->superclass;
       }
    }
}
```

Fig. 2. Original lookup algorithm

```
lookupMethod(class, key) {
   while(class) {
     hashTable = get_hash_table(class);
     h  = compute_method_hash(key);
     entry = get_element(hashTable, h);
     if (entry->flag == ON) { /*Test the flag field of the hashing table entry*/
       for each element in collision list {
         if (key == element->signature)
             return element;
       }
     }
    class = class->superclass;
   }
}
```

Fig. 3. New lookup algorithm

such a method and the search is directed to the super-class. In Fig. 3, we give the new lookup algorithm.

The new lookup algorithm performs fewer iterations than the original one. In fact, in the worst case, the whole collision list has to be visited. The lookup method acceleration depends on the hash table size. In fact, a big size requires a high memory space but it minimizes the collisions. On the other hand, it might induce an additional cost in terms of memory management (allocation, garbage collection, compaction, etc.).

3.2 Caching

Another factor in the acceleration of a method call is the caching approach. We claim in this paper that the inline cache technique could achieve a significant speed-up of a program execution by slightly modifying the cache layout. Actually, the modification consists in adding a pointer to the receiver object in the cache entry. We explain hereafter why such a simple modification will result in a significant speed-up. In the conventional inline cache technique (such as the one implemented in KVM), only a pointer to the method definition is stored in the cache structure. When a method is invoked, the class of the receiver is compared to the class of the invoked method. If there is an equality between these two classes, the method definition is retrieved thanks to the cache entry. If there is no equality, a dynamic lookup is performed to search for the method definition in the class hierarchy. This inline cache technique could be significantly improved by avoiding many of the dynamic lookups when there is a mismatch between the class of the receiver and the class of the invoked method. Actually, when there is such a mismatch, if we can detect that the receiver has not changed, we can retrieve the method definition from the cache. This is done by:

- Adding a pointer to the receiver in the cache structure,
- Modifying the condition that guards cache retrieval. Actually, when a method is invoked, the condition to get a method definition from the cache is:
  - The class of the receiver is equal to the class of the invoked method, or,
  - The current receiver is equal to the cached receiver (the receiver has not changed).

Here is an illustration when this inline cache technique yields a significant speed-up. Assume that we have a class B that inherits a non-static method m from a class A. Assume also that we have a loop that will be executed very often in which we have the following method invocation: the method m is invoked on an object say o_B that is instance of the class B. In our inline caching technique the object o_B is going to be cached after the first invocation of the method m. In the subsequent invocations of the method m, since the class of the receiver (B) is different from the class of the invoked method (A), the behavior of the conventional inline cache technique will be very different from the one of the proposed inline technique. The conventional technique will perform a dynamic lookup for each subsequent invocation of the method m resulting in a significant overhead. The inline technique with the modified cache structure will simply test if the current receiver equals the one that is stored in the cache. Accordingly, the method definition will be retrieved from the cache for all subsequent invocations of the method m resulting in a significant speed-up.

3.3 Synchronization

Before executing a synchronized method, a thread must acquire the lock associated with the receiver object. This is necessary if two threads are trying to execute this method on the same object. Locking an object slows the execution.

The proposed technique is related to the acceleration of multithreaded applications. It improves the synchronization technique used in many virtual machine interpreters. The synchronization technique used in conventional embedded Java virtual machines (especially those based on the KVM) associate to an object one of four states: unlocked, simple locked, extended locked and associated to a monitor.

The unlocked state is for an object that is not synchronized yet. The object passes form the unlocked state to the simple lock when a thread tries to lock it for the first time. The object state passes from the simple lock state to the extended lock state when a thread tries to lock the object for the second time. The extended lock state will be the state of the object until a second thread tries to lock it. From any of these states, a creation of a monitor can happen when a second thread tries to lock an object while another is the owner of the lock. A transition from any state to the monitor state could happen. Exiting a synchronized method triggers the transition from a monitor state or extended lock to a simple lock or an unlocked state. Note that a transition to a bad state can occurs when an exception is raised. We present in the following the automaton that represents the object state. Figure 4 depicts the original automaton where the states are:

- A: unlocked.
- B: simple lock state.
- C: extended state.
- D: monitor state.
- E: exception state (bad state).

And actions are:

- u: set_unlocked.
- s: set_simple_lock.
- e: set_extended_lock.
- m: set_monitor.
- x: raise_exception

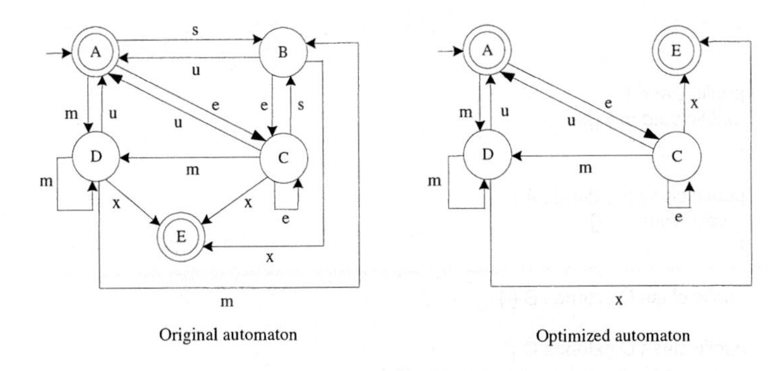

Fig. 4. Original and optimized automaton

By removing the state B, we can avoid transitions from and to the simple lock state, therefore going directly from the unlocked state to the extended lock state. Figure 4 depicts the optimized automaton.

An object passes from the unlocked state to the simple lock when a thread tries to lock it for the first time. In this case, the depth (number of time the object is locked) is equal to one. Since, an object passes from the simple lock state to the extended lock state when a thread tries to lock the object for the second time incrementing the depth, we can remove the simple lock state. In fact, it can be considered as an extended lock state which depth can be greater or equal to one. The elimination of the simple lock state improves thread performance because the execution of some portion of code is avoided. In fact, the transition from the simple lock to the extended lock is avoided.

4 Results

We implemented our technique in the original KVM 1.0.3. We did some testing on the embedded *CaffeineMark* benchmark without floats. Figure 6 presents a performance histogram that illustrates the comparison between the original and the optimized VMs. The comparison proves a reasonable speedup. For some typical examples (e.g. Java programs that frequently call inherited methods), our technique is capable to reach a speedup of more than 27%. Figure 5 shows a typical example that contains a class hierarchy and an invocation of an inherited method m. Figure 7 shows the execution time acceleration for this example. This time is given with respect to the hash table size.

5 Conclusion and Future Work

In this paper, we addressed the acceleration of method calls in an embedded Java virtual machine such as the KVM. We have proposed techniques for the acceleration of dynamic lookups, caching and synchronized method calls. The

```
public class A {
 public void m() {};
}

public class B extends A {
 public void m() {};
}

public class C extends B { }

public class D extends C {
    public static void main(String args[]) {
      A o;
      o = new D();
      int i = 0;
      while (i < 1000000) {
           o.m();
           i++;
      }
    }
}
```

Fig. 5. Typical example for our optimization technique

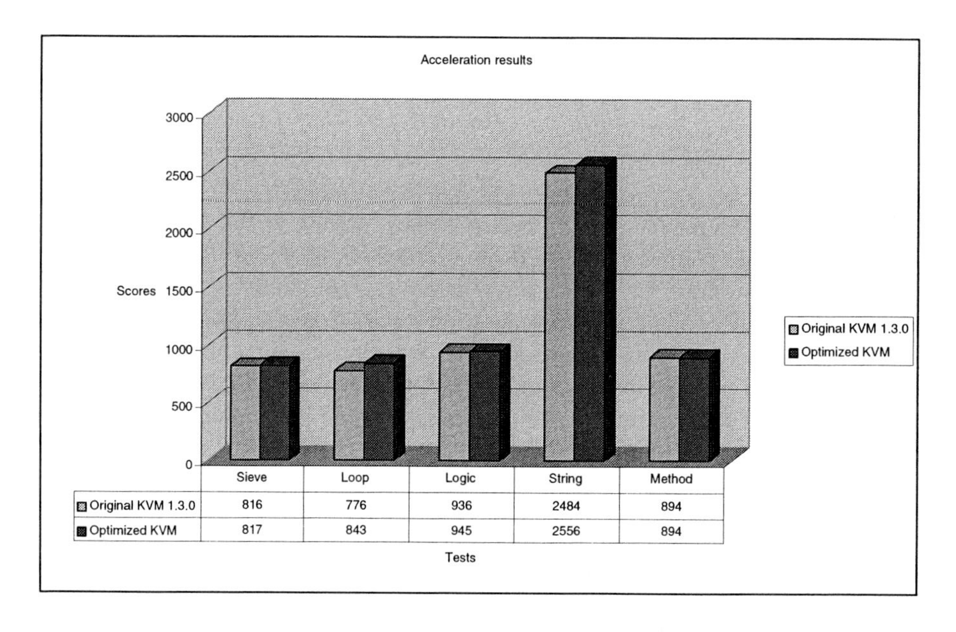

	Sieve	Loop	Logic	String	Method
Original KVM 1.3.0	816	776	936	2484	894
Optimized KVM	817	843	945	2556	894

Fig. 6. Acceleration results for the embedded *CaffeineMark* benchmark

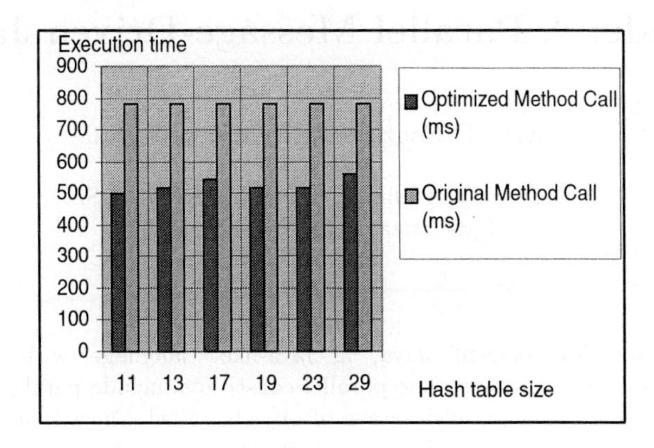

Fig. 7. Execution time acceleration for a typical example

results show that our optimizations are efficient and not expensive from the footprint standpoint. Moreover, these techniques are very generic and could be successfully applied to other virtual machines.

References

1. Goldberg A. and Robson D. *Smalltalk-80: The Language and its Implementation.* Addison–Wesley, 1985.
2. Cox B. *Object-Oriented Programming, An Evolutionary Approach.* Addison-Wesley, 1987.
3. Driesen K. Selector Table Indexing and Sparse Arrays. In *Proceedings of OOP-SLA'93*, Washington, DC, 1993.
4. Ishizaki K., Yasue T., Kawahito M., and Komatsu H. A Study of Devirtualization Techniques for a Java Just-In-Time Compiler. In *Proceedings of the ACM-SIGPLAN Conference on Object-Oriented Programmimg Systems, Languages and Applications*, pages 294–310, Minneapolis, Minnesota, USA, October 2000.
5. Deutsch L.P. and Schiffman A. Efficient Implementation of the Smalltalk-80 System. In *Proceedings of the 11th Symposium on Principles of Programming Languages*, Salt Lake City, UT, 1984.
6. Sun MicroSystems. KVM Porting Guide. Technical report, Sun MicroSystems, California, USA, September 2001.
7. Dencker P., Durre K., and Heuft J. Optimization of Parser Tables for Portable Compilers. *ACM TOPLAS*, 6(4):546–572, 1984.
8. Vijay S., Laurie H., Chrislain R.and Raja V., Patrick L., Etienne G., and Charles G. Practical Virtual Method Call Resolution for Java. In *Proceedings of the ACM-SIGPLAN Conference on Object-Oriented Programmimg Systems, Languages and Applications*, pages 264–280, Minneapolis, Minnesota, USA, October 2000.
9. Holzle U., Chambers C., and Ungar D. Optimizing Dynamically-Typed Object-Oriented Languages With Polymorphic Inline Caches. In *Proceedings of ECOOP'93*, 1993.

Jade: A Parallel Message-Driven Java

Jayant DeSouza and Laxmikant V. Kalé

University of Illinois, Urbana IL 61801, USA
{jdesouza,kale}@cs.uiuc.edu

Abstract. We present Java, a Java-like[1] language with parallel message-driven features. The parallel constructs include parallel classes called *Chares* and parallel arrays of objects called *ChareArrays*. Communication with a parallel object occurs through asynchronous method invocation. In the message-driven paradigm, when a method of a Chare or ChareArray is invoked, it continues to completion before any other method of the same parallel object can run. In contrast to Java's run-time compilation, the Java source code is translated to Charm++ source code, which is then *compiled* and executed on the target machine. The resulting code supports object migration and load-balancing and scales well to large number of processors. Java's standard libraries are not supported.

1 Introduction

A promising approach to parallel programming that we have been exploring for the past decade is to effect an "optimal" division of labor between the application programmer and the "programming system". Specifically, this approach requires the programmer to decompose their application into a large number of interacting objects, and passes to the runtime system the responsibility of mapping the objects to processors. The "objects" may be true C++ objects, or they may be user-level threads running MPI[3]. In either case, the model requires encapsulation: each object is allowed to access its variables, along with read-only data structures.

This approach leads to several benefits: powerful scalability, efficient execution on both shared and distributed architectures, a simple programmer cost model for the various features of the language, object migration and automatic dynamic measurement-based load balancing [8,4], automatic checkpointing, support for out-of-core execution, the ability to dynamically shrink or expand the set of processors used by the job.[12]

We have implemented this approach in C++, in the form of the Charm++ system.[6,9,11] Charm++ has been very successful in parallelizing several applications effectively. As an example, NAMD[7], a molecular dynamics application written in Charm++ is very scalable and runs on thousands of processors.[21] Recently, NAMD won the Gordon Bell award at SC2002.[22] Another example

[1] Java is a trademark of Sun Microsystems, Inc.

P.M.A. Sloot et al. (Eds.): ICCS 2003, LNCS 2659, pp. 760–769, 2003.

is the FEM framework[2] written in Charm++ and AMPI, which is being used to develop very large applications at the Center for the Simulation of Advanced Rockets.

The work reported here describes our preliminary attempts at bringing the benefits of the Charm++ approach to Java. Some of the benefits of combining Java's serial programming model with the parallel model of Charm++ are:

- The absence of pointers and pointer arithmetic in Java enables better compiler analysis because the aliasing issue is removed. We plan to do compiler analyses that will take advantage of this feature.
- Java's memory model and garbage collection make memory management easier and this removes a major source of difficult bugs, thereby improving programmer productivity.
- C++ is a hard language to parse. In fact, Charm++ does not parse user C++ code, but requires the user to write an interface file to identify parallel objects and other special Charm++ entities. Java is a simpler language than C++, thus enabling a Jade parser.
- In Charm++ when parameters are packed for transmission, the user needs to make some simple but essential modifications when passing an array as a parameter. The user also has to write a pack-unpack (PUP) method containing pointer-following codes if the object being passed contains dynamically allocated data and/or has pointers. All this is done automatically in Jade. This is not done automatically in Charm++ because of the lack of a parser and due to the aliasing issues that arises when working with pointers.
- To support migration and load-balancing in Charm++, the user needs to write pack-unpack (PUP) methods for all classes that might be migrated, whether or not the class has pointers. Without this step, which some users of Charm++ find cumbersome, the advantages of automatic load-balancing will not be realized. Since these methods can be auto-generated based on the data members of the class and since Java does not have pointers like Charm++, this is done automatically for the user in Jade.
- Java has limited support for parallel programming. It is run-time compiled and usually slower than compiled code. Java provides threads and simple synchronization, but assumes a shared-memory model in which threads run, making it hard to run Java codes in parallel on multiple nodes. Distributed parallelism could be achieved using RMI, but Java RMI is very inefficient and puts the burden of writing message-passing code on the user. There are no facilities for global synchronization and communication such as reductions and broadcasts. Java lacks higher level parallel constructs such as parallel arrays. And so on. By introducing message-driven features into Java, adding global communication and synchronization, providing higher-level language constructs for parallel objects, and making it compiled rather than run-time compiled, Jade offers a sturdy base for parallel programming in Java.

Our approach uses Java's syntax and basic semantics, extending them with Charm++ parallel constructs, and translating Jade to C++ /Charm++. This retains the performance benefits of C++ while providing the user with a clean programming model.

The paper surveys some of the past work on parallel Java in Sect. 2. Section 3 describes the execution model, language constructs and other features of Jade, and then compares and contrasts Jade with both Charm++ and Java. We then discuss some implementation details and translation issues in Sect. 4. Finally, we summarize the work presented and, keeping in mind that this is just a start to the work we plan to achieve with Jade, we discuss our future plans.

2 Related Work

Java has built-in features to support parallelism: typically, Java threads are used for shared memory applications and either Java RMI or the standard socket interface is used for distributed applications. Java RMI is slow, interpreted, requires explicit name binding, and provides no optimization for local objects.

The JavaParty[20] project takes an approach to parallel Java that is based on improving Java RMI. The execution time of a Java RMI call is analyzed and divided into time for low-level communication, serialization of parameters, and RMI overhead. Each of these issues is addressed by them with the ParaStation, UKA-Serialization[19], and KaRMI[17] projects, respectively. JavaParty code is interpreted, whereas Jade is compiled.

The Manta[1] project also works by improving Java RMI. It uses a compiler approach to generate specialized serializers, implements its own runtime and protocol to speed up RMI, and can use a fast low-level communication layer (Panda/LFC) to improve speed. Manta supports runtime generation of marshaller code, distributed garbage collection, and compiler optimizations such as execution of non-blocking remote methods without a thread context switch. (This last is an automatic feature of Jade since blocking methods are not supported in the parallel constructs.) Manta interoperates with standard JVM and supports the Sun RMI protocol.[14,13]

Titanium[23] translates Java into C. It adds a global address space and multidimensional titanium arrays to Java. However, on distributed memory architectures, the authors admit that accessing global data is significantly slower than accessing local data. Several compiler analyses are performed. Barriers and data exchange operations are used for global synchronization.

Our approach differs from JavaParty, Manta, and Titanium in that we introduce a different parallel programming style, the message-driven style, into Java. We also support object migration and load-balancing.

One of the authors has also implemented a previous version of parallel Java.[10] In that version, parallel constructs such as *chares* were added to Java. A JVM was executed on each processor of a parallel machine, and messages were injected into them using JNI and Java Reflection, i.e. the Java code was interpreted on a JVM. Parallel arrays, migration of objects, and global communication and synchronization were not supported.

3 Jade Parallel Constructs

We first describe the programming model and parallel constructs of Jade and then compare it with Charm++ and Java.

3.1 Jade

A Jade program consists of a number of parallel objects distributed across the available processors. The Jade runtime system manages the location, migration, message-forwarding, and load-balancing of these objects. Sequential objects in the system are owned by parallel objects and are tightly encapsulated within them, as explained below.

Figure 1 shows an example Jade program. Line 4 declares the Hello package which contains the subsequent entities. If no package is declared, the entities belong to a default null package.

Chares: A *chare* (line 6) is essentially a Java object without an explicit thread of control. When a method of the chare is invoked, it executes to completion without interruption. This implies that only one method of a chare can execute at a time, i.e. in the Java sense it is as if all methods are `synchronized`. A chare is declared by specifying the `synchronized` attribute for the class, and by extending from the `Chare` class, which implements Serializable.

```
 1 /** Simple parallel hello program using ChareArrays.
 2  main sends a message to hi[0], which sends it on to the
 3    next element, and so on.*/
 4 package Hello;
 5
 6 public synchronized class TheMain extends Chare {
 7     public int main(String []args){
 8         ChareArray hi1 = new Hello[5];
 9         hi1[0].sayHello(21);
10     }
11
12     public void memberMethod(int parameter) {...}
13 }
14
15 public synchronized class Hello extends ChareArray1D {
16     public Hello(){}
17
18     public void sayHello(int number) {
19         if (thisIndex < 4)
20             thisProxy[thisIndex+1].sayHello(number+1);
21         else
22             CkExit();
23     }
24 }
```

Fig. 1. Example Jade program: Array Hello

Jade does not assume a shared memory model. A chare can have member variables, which can be instances of sequential classes or other chares. However, because a chare is a parallel object, its member variables are not permitted to be public. Member functions (which, of course, can include access methods) can be public. They need not be `void`, but when invoked asynchronously, the return value is discarded.

Jade execution model: Execution of a Jade program begins by instantiating one copy of every chare that has a `main` (see line 7 of Fig. 1), and then invoking the `main` function of the chare on processor 0. This is different from the standard Java behavior, in which `main` is `static`, and only one `main` is invoked, without instantiating the class containing it. Typically, the `main` functions create parallel objects and invoke methods on them to get the work started. The reason we instantiate the class containing `main` is because most programs need an object to send back results to, print them out, etc. Usually one `main` is sufficient for a program, but multiple `main`'s are supported for Jade module initialization. Note that `ChareArrays` (described below) cannot have `main` functions.

ChareArrays: Jade supports 1D, 2D, and 3D arrays of parallel objects as its primary parallel construct. Support for user-defined index types is part of the language, but is not yet implemented. The arrays are sparse and created lazily, i.e. each element of an array is only created when first accessed. Each element of the array is essentially a chare. The elements of the array are mapped by the system to the available processors.

A parallel array is declared by making the class `synchronized` and extending `ChareArray1D`, `2D` or `3D`. Line 15 of Fig. 1 shows the declaration of a 1D array.

Line 8 of Fig. 1 shows the instantiation of a ChareArray containing five elements. (The language supports ChareArray constructors with arguments but the discussion of the syntax is omitted due to space constraints.) The next line invokes the sayHello method of the first element (index value 0) of the ChareArray. The `int thisIndex` member of ChareArray1D (line 19) returns the index of the current element, and the `CProxy thisProxy` member returns a proxy to the array itself. To broadcast a method invocation to all elements of an array, leave out the index, e.g.`A.sayHello()`.

Asynchronous method invocation(AMI): Invocation of methods on chares and ChareArray's is usually asynchronous, i.e. the caller of such a method does not block and wait for a response, e.g. see line 9 of Fig. 1. A feature of asynchronous invocation is that it makes (invocation-related) deadlock impossible. AMI is just like sending a message; but the associated features of the message-passing programming style popularized by MPI are not present, e.g. in Jade there is no corresponding blocking receive for messages.

Note that AMI does not directly permit traditional blocking function calls (possibly with return values). This can be achieved by splitting up the calling method into two methods, the first of which contains the code up to and including the AMI call. The called function must then send back a message to the second method to resume the computation.

Parameter passing: Method invocation can be either local or remote, depending on the location of the called chare (i.e. on which processor of the parallel system it is running). Parameters are passed between parallel objects by copying them over via a message: this involves packing and unpacking of parameters. Parameters are automatically packed and unpacked without user intervention by the Jade translator-generated code. Chares and ChareArrays can also be sent as parameters; however, they are passed by reference.

Since a Jade parallel object cannot have public data members, and since parameters are passed by copy between parallel objects, we can conclude that the address space of a parallel object is private to it and cannot be directly accessed by any other parallel object. This tight encapsulation of data facilitates migration of objects.

Unlike traditional languages, (but similar to Java with threads) `main` can terminate without causing the program to end. Thus, on line 9 of Fig. 1, `main` does not block waiting for the call to `sayHello()` to return. Control moves on to the next statement, and, in this case, since there is none, `main` terminates. Therefore, in Jade the end of computation is signaled by calling `CkExit()`, as shown in line 22 of the example program. CkExit gracefully terminates the program on all processors, ensuring that all messages have been delivered, flushing buffers, shutting down processes, etc.

3.2 Comparison with Charm++

The current implementation of Jade supports only asynchronous method invocation. Charm++ also supports *Synchronous method invocation(SMI)* through `threaded` and `sync` methods. `threaded` methods have an explicit thread of control and can therefore block. The Charm++ system still ensures that only one method of the chare containing the `threaded` method executes at a time, i.e. if a thread is blocked, no other method of the chare is invoked. This protects data accesses, i.e. no locking is required when accessing member variables. Calls to `sync` methods will block, and so they can only be called from `threaded` methods. There is no fundamental obstacle to implementing SMI in Jade and we plan to do this in the next version.

Charm++ only automates packing and unpacking of basic parameter types, but requires user intervention for passing arrays, nested objects, and pointers. Jade completely automates parameter passing.

3.3 Comparison with Java

Given the large amount of research into translation of (sequential) Java to C, we decided to focus our efforts on the parallel features rather than on implementing all Java features.

Java import statements are translated into #include's with a hierarchy of directories. Java `packages` and package-private classes are supported.

Differences: As mentioned above, parameter passing between parallel objects in Jade is by copy (except that parallel objects are passed by reference), whereas

Java parameter passing (which is retained in Jade for non-parallel objects) is by copy for primitive types and by reference for objects.

Java supports dynamic loading of classes based on the `CLASSPATH`. However, our compilation product is a single executable and dynamic loading is not supported. It is possible to implement this capability in C++, and if applications require it, we will add this feature to Jade.

Multiple classes can be defined in one Jade source file, unlike Java which has a very tight tying of classes to the naming structure to simplify dynamic class loading.

Jade treats `main` differently from Java as described in Sect. 3.1.

Java bytecode is portable across heterogeneous systems, while Jade code is compiled to a particular architecture. Java primitive types are standard across all platforms, but Jade types depend on the machine and C++ compiler used.

Restrictions: Java names can be Unicode, whereas we support only ASCII. We do not currently support the standard Java runtime and libraries, nor do we support exceptions. We plan to implement non-preemptive threads in the near future, as well as garbage collection.[2]

4 Translation Issues

Our translator is implemented using the ANTLR parsing system which supports Java as an implementation language. For our grammar, we used a public domain ANTLR Java grammar.[16]

The steps and byproducts involved in compiling a Jade source file are shown in Fig. 2. The Jade source file (`.pjs`) is translated into four files: the Charm++ interface file (`.ci`), a `.h` header file and a `.C` C++ source file

The `.ci` file is then fed to the Charm++ translator. It contains information about the parallel constructs in the program. The Charm++ translator gen-

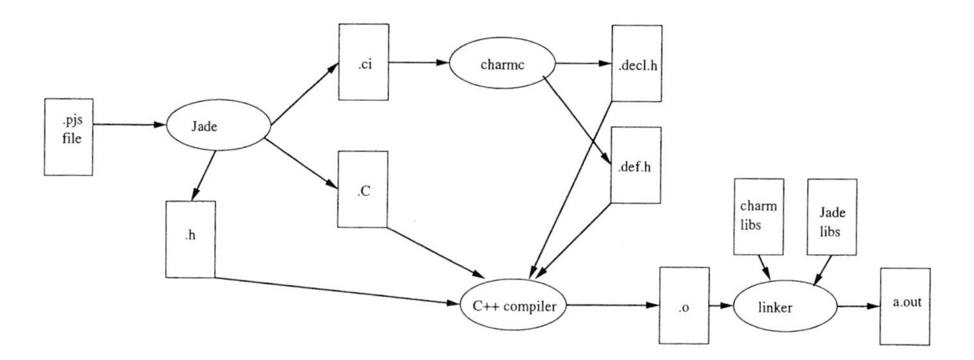

Fig. 2. Compilation of a Jade source file

[2] **`delete`** is supported in the interim.

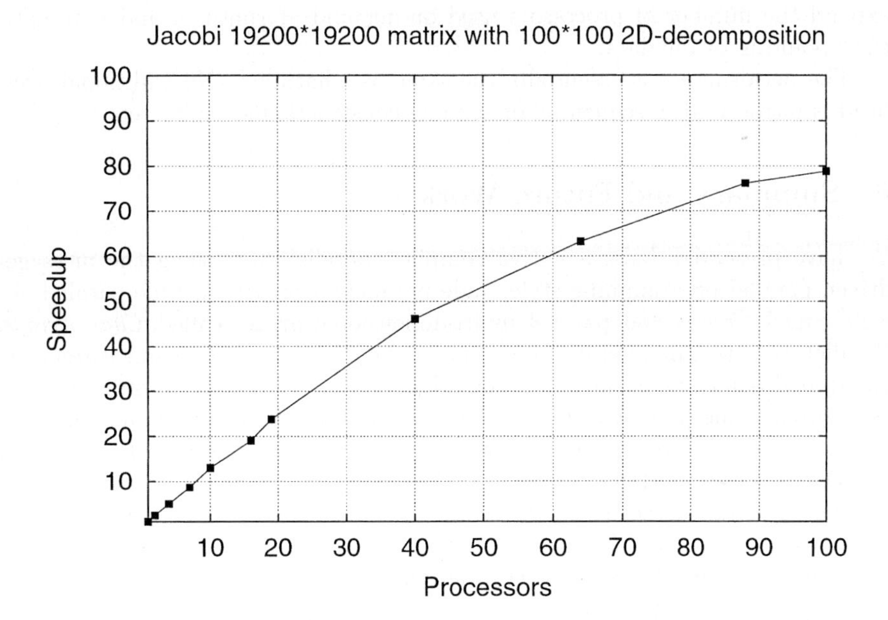

Fig. 3. Performance of Jacobi 2D written in Jade

erates `.decl.h` and `.def.h` files, which are `#include`'d in the `.h` and `.C` files respectively.

All the above generated files apart from the `.ci` file are then compiled with a C++ compiler and linked with the Charm++ and the Jade libraries to generate a single executable `a.out` file.

5 Implementation

It is important to remember that the objective of Jade is to retain the performance advantages of Charm++ while gaining the simplicity and ease of debugging of Java. It is not intended to outperform Charm++, but to bring the benefits of the message-driven programming style to Java.

We implemented a Jacobi 2D decomposition example in Jade. The resulting translated code runs on all the platforms supported by Charm++, such as the Intel ASCI-Red, SGI Origin2000, IBM SP-2, Cray T3E, Convex Exemplar, clusters of Unix (Linux, Solaris, etc) workstations, and even single-processor UNIX machines. Figure 3 shows the speedup on the LeMieux cluster at Pittsburgh Supercomputing Center. (Each node of LeMieux contains four 1 GHz Compaq Alpha CPU's; the nodes are connected by a high-speed Quadrix Elan interconnect.)

The implementation has demonstrated automatic parameter marshalling, automatic object migration and measurement-based load-balancing, efficient execution on both shared and distributed architectures, the ability to shrink and expand the number of processors used on-demand at runtime, and automatic checkpointing of the data.

The performance is essentially the same as Charm++. No additional overhead is introduced in sequential or parallel constructs, messaging, etc.

6 Summary and Future Work

We have presented Jade, a native compiled parallel Java using the message-driven parallel programming style. Jade parallel constructs include parallel objects called *Chares* and parallel multi-dimensional arrays called *ChareArrays*. Parallel objects can migrate across processors, and automatic measurement-based load balancing ensures even load distribution. Communication is through asynchronous method invocation and includes automatic parameter marshalling. Messages are forwarded to the target object even if it has migrated. Barriers and reductions are supported.

Our plans for further research include the enhancement of Jade with parallel garbage collection, threads and synchronous method invocation, and the addition of compiler optimizations in the translation process.

References

1. H. Bal et al. Manta: Fast parallel Java. URL: `http://www.cs.vu.nl/manta/`.
2. M. Bhandarkar and L. V. Kalé. A Parallel Framework for Explicit FEM. In M. Valero, V. K. Prasanna, and S. Vajpeyam, editors, *Proceedings of the International Conference on High Performance Computing (HiPC 2000), Lecture Notes in Computer Science*, volume 1970, pages 385–395. Springer Verlag, December 2000.
3. M. Bhandarkar, L.V. Kale, E. de Sturler, and J. Hoeflinger. Object-Based Adaptive Load Balancing for MPI Programs. In *Proceedings of the International Conference on Computational Science, San Francisco, CA, LNCS 2074*, pages 108–117, May 2001.
4. R.K. Brunner and L.V. Kalé. Handling application-induced load imbalance using parallel objects. Technical Report 99-03, Parallel Programming Laboratory, Department of Computer Science, University of Illinois at Urbana-Champaign, May 1999. Submitted for publication.
5. J. Gosling, B. Joy, and G. Steele. *The Java Language Specification.* Addison-Wesley, 1996.
6. L. Kalé and S. Krishnan. CHARM++: A Portable Concurrent Object Oriented System Based on C++. In A. Paepcke, editor, *Proceedings of OOPSLA'93*, pages 91–108. ACM Press, September 1993.
7. L. Kalé, R. Skeel, M. Bhandarkar, R. Brunner, A. Gursoy, N. Krawetz, J. Phillips, A. Shinozaki, K. Varadarajan, and K. Schulten. NAMD2: Greater scalability for parallel molecular dynamics. *Journal of Computational Physics*, 151:283–312, 1999.
8. L.V. Kale, M. Bhandarkar, and R. Brunner. Run-time Support for Adaptive Load Balancing. In J. Rolim, editor, *Lecture Notes in Computer Science, Proceedings of 4th Workshop on Runtime Systems for Parallel Programming (RTSPP) Cancun - Mexico*, volume 1800, pages 1152–1159, March 2000.

9. L.V. Kale, M. Bhandarkar, N. Jagathesan, S. Krishnan, and J. Yelon. Converse: An Interoperable Framework for Parallel Programming. In *Proceedings of the 10th International Parallel Processing Symposium*, pages 212–217, April 1996.
10. L.V. Kalé, M. Bhandarkar, and T. Wilmarth. Design and implementation of parallel java with a global object space. In *Proc. Conf. on Parallel and Distributed Processing Technology and Applications*, pages 235–244, Las Vegas, Nevada, July 1997.
11. L.V. Kale and S. Krishnan. Charm++: Parallel Programming with Message-Driven Objects. In G. V. Wilson and P. Lu, editors, *Parallel Programming using C++*, pages 175–213. MIT Press, 1996.
12. L.V. Kalé, S. Kumar, and J. DeSouza. A malleable-job system for timeshared parallel machines. In *2nd IEEE/ACM International Symposium on Cluster Computing and the Grid (CCGrid 2002)*, May 2002.
13. J. Maassen, R. van Nieuwpoort, R. Veldema, H. Bal, T. Kielmann, C. Jacobs, and R. Hofman. Efficient Java RMI for parallel programming. *ACM Transactions on Programming Languages and Systems (TOPLAS)*, 23(6):747–775, November 2001.
14. J. Maassen, R. van Nieuwpoort, R. Veldema, H. Bal, and A. Plaat. An efficient implementation of Java's remote method invocation. In *Proc. Seventh ACM SIGPLAN Symposium on Principles and Practice of Parallel Programming (PPoPP'99)*, pages 173–182, Atlanta, GA, May 1999.
15. M.W. Macbeth, K.A. McGuigan, and P.J. Hatcher. Executing Java threads in parallel in a distributed-memory environment. In *Proc. CASCON 98*, pages 40–54, Missisauga, ON, Canada, 1998.
16. J. Mitchell, T. Parr, et al. Java grammar for ANTLR. URL: `http://www.antlr.org/grammars/java`.
17. C. Nester, M. Philippsen, and B. Haumacher. A more efficient RMI for Java. In *Proc. ACM 1999 Java Grande Conference*, San Francisco, CA, June 1999.
18. T. Parr et al. Website for the ANTLR translator generator. URL: `http://www.antlr.org/`.
19. M. Philippsen and B. Haumacher. More efficient object serialization. In *Parallel and Distributed Processing, LNCS, International Workshop on Java for Parallel and Distributed Computing*, volume 1586, pages 718–732, San Juan, Puerto Rico, April 1999.
20. M. Philippsen and M. Zenger. JavaParty – transparent remote objects in Java. *Concurrency: Practice and Experience*, 9(11):1125–1242, November 1997.
21. J. Phillips, G. Zheng, and L.V. Kalé. Namd: Biomolecular simulation on thousands of processors. In *Workshop: Scaling to New Heights*, Pittsburgh, PA, May 2002.
22. J.C. Phillips, G. Zheng, S. Kumar, and L.V. Kalé. Namd: Biomolecular simulation on thousands of processors. In *Proceedings of SC 2002*, Baltimore, MD, September 2002.
23. K.A. Yelick, L. Semenzato, G. Pike, C. Miyamoto, B. Liblit, A. Krishnamurthy, P.N. Hilfinger, S.L. Graham, D. Gay, P. Colella, and A. Aiken. Titanium: A high-performance java dialect. *Concurrency: Practice and Experience*, 10(11–13), September–November 1998.

Workshop on Computational Earthquake Physics and Solid Earth System Simulation

Parallel Finite Element Analysis Platform for the Earth Simulator: GeoFEM

Hiroshi Okuda[1], Kengo Nakajima[2], Mikio Iizuka[2], Li Chen[2], and Hisashi Nakamura[2]

[1] Department of Quantum Engineering and Systems Science, University of Tokyo
7-3-1 Hongo, Bunkyo, Tokyo 113-8656, Japan
okuda@q.t.u-tokyo.ac.jp

[2] Research Organization for Information Science & Technology (RIST)
2-2-54 Naka-Meguro, Meguro, Tokyo 153-0061, Japan
{nakajima,iizuka,chen,nakamura}@tokyo.rist.or.jp

Abstract. GeoFEM has been developed as a finite element solid earth simulator using the Earth Simulator (ES) (35.61 Tflops/peak according to the Linpack benchmark test). It is composed of a platform and some pluggable 'analysis modules' for structural, electromagnetic thermal fluid, and wave propagation simulations. The platform includes three parts: parallel I/O interface, iterative equation solvers and visualizers. Parallel solvers have got very high performance on the ES. When using up to 176 nodes of the ES, the computational speed of the static linear analysis by the optimized ICCG (Conjugate Gradient method with Incomplete Cholesky Preconditioning) solver, has reached 3.8 TFLOPS (33.7% of peak performance of 176 nodes). Parallel visualizer can provide many visualization methods for analysis modules covering scalar, vector and tensor datasets, and have been optimized in parallel performance. The analysis modules have also been vectorized and parallelized suitably for the ES and coupled on memory with the parallel visualizer.

1 Introduction

In order to solve global environmental problems and to take measures against the natural disasters, Japanese Ministry of Education, Culture, Sports, Science and Technology (formerly, the Science and Technology Agency) has begun a five-year project called 'Earth Simulator project' from the fiscal year of 1997, which tries to forecast various earth phenomena through the simulation of virtual earth placed in a supercomputer. It is also expected to be a breakthrough in bridging the geoscience and information science fields. The specific research topics of the project are as follows:

(1) Development of the Earth Simulator (ES)[1], a high performance massively parallel processing computer;
(2) Modeling of atmospheric and oceanic field phenomena and carrying out high-resolution simulations;
(3) Modeling and simulation of solid earth field phenomena;
(4) Development of large-scale parallel software for the 'Earth Simulator'.

P.M.A. Sloot et al. (Eds.): ICCS 2003, LNCS 2659, pp. 773–780, 2003.

The ES started its operation in March, 2002. It has shared memory symmetric multi-processor (SMP) cluster architecture, and consists of 640 SMP nodes connected by a high-speed network (data transfer speed: 12.3 GB). Each node contains eight vector processors with a peak performance of 8 GFLOPS and a high-speed memory of 2 GB for each processor. According to the Linpack benchmark test, the ES was listed as the fastest machine on Top500 as of June 2002, having a peak performance of 35.61 TFLOPS [1]. The GeoFEM deals with the topics (3) and (4) of the above list.

The solid earth is a coupled system of various complicated phenomena extending over a wide range of spectrum of space and time. To simulate this, it needs large-scale parallel technologies, combining the sophisticated models of solid earth processes and the large amount of observation data [2]. GeoFEM [3, 4] challenges for long-term prediction of the activities of the plate and mantle near the Japanese islands through the modeling and calculation of the solid earth problems, including the dynamics and heat transfer inside the earth.

In the followings, some features of GeoFEM are described, i.e. system overview, parallel iterative solvers and parallel visualization techniques.

2 Software System Configuration

GeoFEM consists of the components '*platform*', '*analysis modules*' and '*utilities*', as shown in Fig. 1. *Platform* includes device-independent I/O interface, iterative equation solvers and visualizers. *Analysis modules* are linked to *platform on memory* by device-independent interface such that the copies of data should be avoided as much as possible for large scaled-data and high-speed calculations. To deal with various problems of solid earth, analysis modules plugged-in are elastic, plastic, visco-elastic, and contact structural analysis, wave propagation analysis, thermal analysis and incompressible thermal fluid analysis modules. *Utilities* are for domain partitioning and pre/post viewers.

As for the I/O support, device independent communicator, which is also a module, is supplied by the platform. For data, which has standardized format, e.g. mesh data, reader programs in the same interface are supported by the platform. On the other hand, generic data input is supported for analyzer-specific data, e.g. various control data like the time increment or the convergence criteria. To do this, GDL (GeoFEM Data Description Language) and GDL compiler are supplied by the platform. GDL, just like Fortran 90 language, defines the data structures and generates the reader program. A simple example of the GDL script is shown as follows:

```
begin structure_static_contact
  begin solver
     method BICGSTAB

     precondition   BLOCK
       iteration   real_parameter  1.0e-5
   end
   begin nonlinear_method
        ...
   end
end
```

Since the ES is an SMP cluster machine which has a hybrid-parallel architecture that consists of a number of nodes which are connected by a fast network. Each node consists of multiple processors which have access to a shared memory, while the data on other nodes may be accessed by means of explicit message-passing.

In order to get high performance on this type of machine, the three-level hybrid parallelization was adopted in GeoFEM. That means:

- Inter-SMP node: MPI;
- Intra-SMP node: OpenMP for parallelization;
- Individual PE: Compiler directives for vectorization / pseudo vectorization.

Except parallel visualizers, GeoFEM software is implemented by Fortran 90 because the compilers of Fortran 90 on the ES are highly optimized and Fortran 90 is very popular for most researchers in computing areas. However, in the visualization part, Fortran 90 is difficult to realize the specific functions. C language and a good interface between C and Fortran are used.

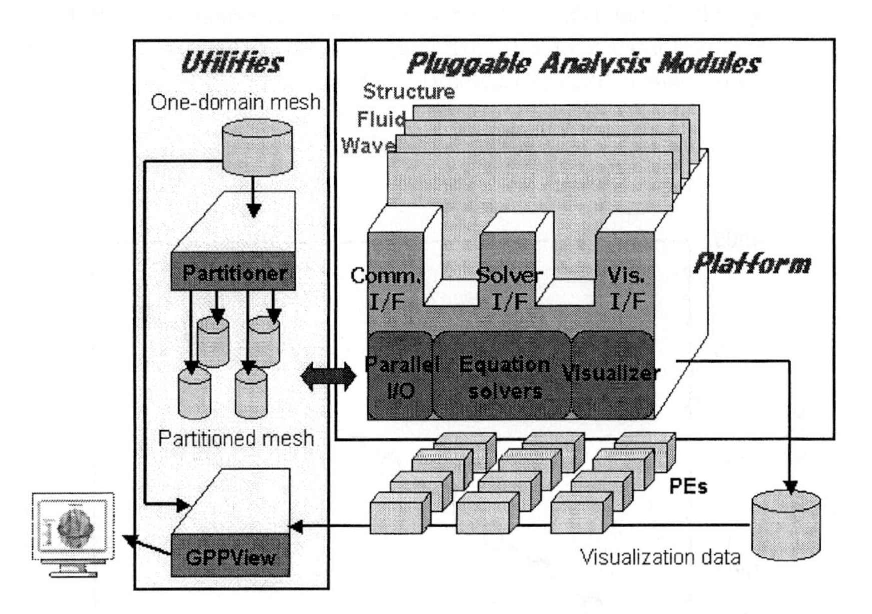

Fig. 1. System configuration of GeoFEM

3 Hybrid Parallel Iterative Solver on the ES [5–7]

When iterative solvers with preconditioning are used, a partitioning in a node-based manner at the defining point of degree of freedom is favorable. GeoFEM assumes that the finite element mesh is partitioned into subdomains before the computation starts. The partitioning is done by RCB (Recursive Coordinate Bisection), RSB (Recursive Spectral Bisection) or the Greedy algorithm. The communication table is generated during the partitioning process.

Each subdomain is assigned to one SMP node. In order to achieve the efficient parallel and vector computation for applications with unstructured grids, the following three issues are critical:

- Local operation and no global dependency;
- Continuous memory access;
- Sufficiently long loops.

Three-level hybrid parallel was implemented in the solvers. Compared with flat-MPI (only MPI is used in communication among PEs), the hybrid method can obtain better performance with the increase of data size. Simple 3D elastic linear problems with more than 2.2×109 DOF have been solved by localized 3×3 block ICCG(0) with the additive Schwarz domain decomposition. PDJDS/CM-RCM (Parallel Descending-order Jagged Diagonal Storage/Cyclic Multicolor-Reverse Cuthil McKee) [8, 9] reordering is applied. Using 176 SMP nodes of the ES, performance of 3.8 TFLOPS (33.7% of peak performance of 176 SMP nodes) has been achieved. Here, the problem size per SMP node is fixed as 12,582,912 DOF.

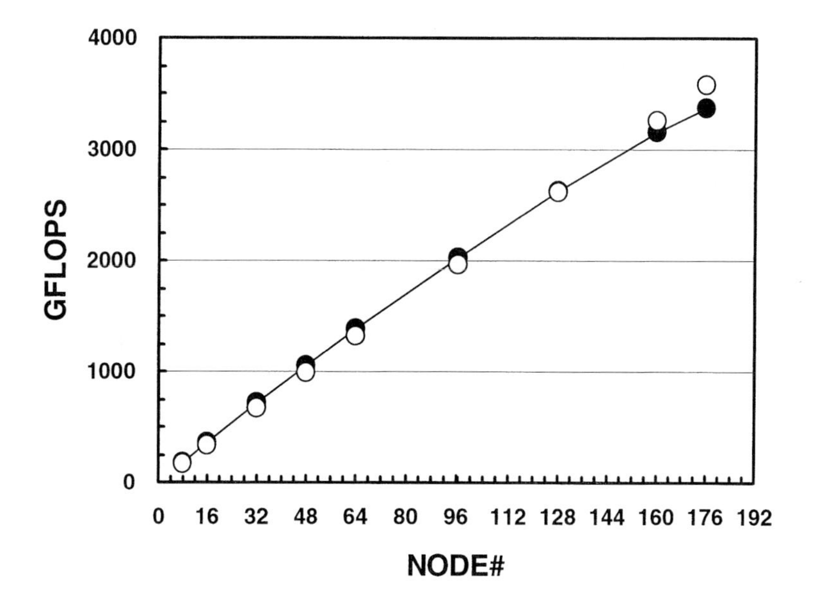

Fig. 2. Problem size and parallel performance (GFLOPS rate) on the ES for the 3D linear elastic problem, using up to 176 SMP nodes (Black circle: Flat MPI, White circle: Hybrid)

4 Parallel Visualization

The target of data size in GeoFEM simulation on the ES will reach one to ten petabytes. In order to help researchers understand the complicated relationship among different physical attributes in extremely large datasets well, it is very important to develop a parallel visualization system which can transform the huge data into appealing images in a high performance and quality. However, it is very challenging due to the complicated grids including unstructured, hierarchical and hybrid grids, extremely large data size with time-varying large timesteps, many kinds of physical attributes covering scalars, vectors and tensors, and no graphics hardware on the ES. Facing up to these challenges, we have developed a parallel visualization subsystem in GeoFEM for the ES, which has the following features [10, 11]:

(1) Concurrent with analysis modules on the ES
The concurrent visualization with analysis modules on the ES has been developed in GeoFEM. Thus, there is no need to save the results generated by analysis modules on the hard disk or transfer huge data by network, which can avoid the limitations of storage capacity for large-scale data. Meanwhile, full use of supercomputers' huge memory can be made to complete visualization. Moreover, at each time step, multiple visualization methods can be performed to generate visualization results by different visualization methods and different parameters. Two kinds of output styles are provided. One is to output the simplified geometric visual elements to clients. On each client, the users can set viewing, illumination, shading parameter values, and so on, and display the graphic primitives by the GPPView viewing software, which is also developed by the GeoFEM group [3]. On the computational server, the users only specify in the batch files the visualization methods such as cross-sectioning, streamlines, and related parameters. The second style is to output an image or a sequence of animation images directly to clients in case that even the simplified geometric primitives are still too large to transfer.
(2) Applicable for complicated unstructured datasets
Most current commercial visualization software can get good performance for regular datasets, but fail for unstructured datasets. The present visualization techniques are well designed for complicated unstructured datasets.
(3) A number of parallel visualization techniques available in GeoFEM
As shown in Table 1, a number of parallel visualization techniques are provided in GeoFEM for scalar, vector and tensor data fields, to reveal the features of datasets from many aspects.

Table 1. Available parallel visualization techniques in GeoFEM

Scalar Field	Vector Field	Tensor Field
Surface Rendering	Streamline	Hyperstreamline
Interval Volume Fitting	Particle Tracking	
Volume Rendering	Line Integral Convolution	
Hybrid Surface and Volume Rendering	Volume Rendering	

(4) High parallel performance
Optimization by three-level hybrid parallelization has been performed on our *Parallel Surface Rendering* and *Parallel Volume Rendering* modules on the ES, including the message passing for inter-node communication, p-thread or microtasking for intra-node parallelization, and the vectorization for each PE. Furthermore, the dynamic load balancing is considered for achieving good speedup performance.

Visualization images have been obtained by applying the above techniques to some unstructured earthquake simulation datasets, stress analysis of Pin Grid Array (PGA) and flow analysis of outer core. Also, speedup performances tested on Hitachi SR8000 and the ES demonstrate the high parallel performance of our visualization modules. Figure 3 is the volume rendered images to show the equivalent stress by the linear elastostatic analysis for PGA dataset with 5,886,640 elements and 6,263,201 grid nodes. On the ES using 8 SMP nodes, it took about 0.977 seconds to generate the right image in Fig. 3 (image resolution 460*300 pixels), and the speedup of 20.6 was attained by virtue of the three-level of parallelization compared with the result using flat MPI without vectorization. Figure 4 shows the parallel volume rendering image for a large earthquake simulation dataset of the Southwest Japan in 1944, with 2,027,520 elements. The magnitude of the displacement data attribute is mapped to color. On the ES, with 8 nodes, it took about 1.09 seconds to generate a single image (image resolution: 400*190 pixels), and about 5.00 seconds to generate 16 animation images with a rotation of the viewpoint along the x axis.

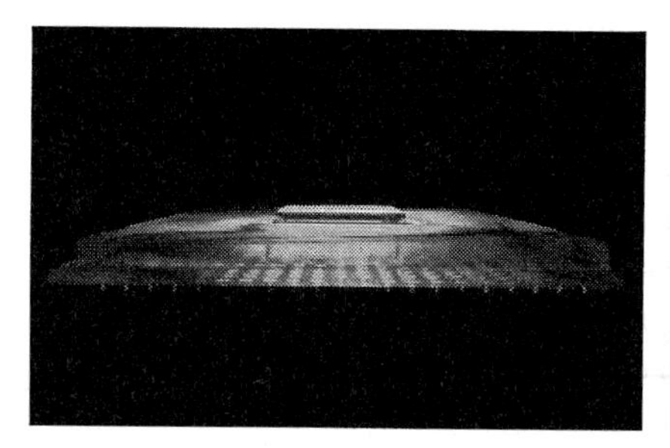
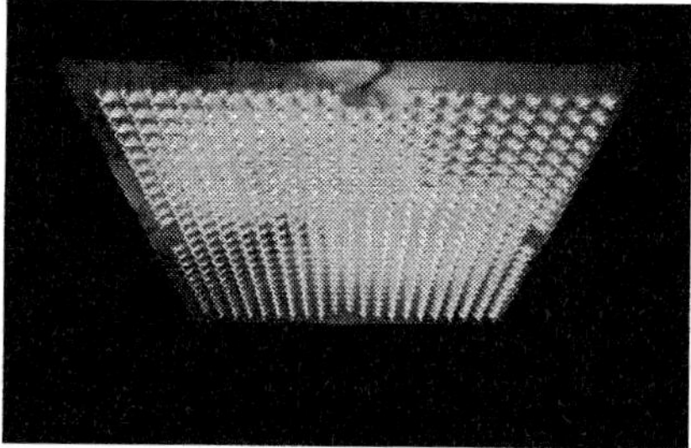

Fig. 3. Parallel volume rendered images from different viewpoints showing the equivalent stress value by the linear elastostatic analysis for a PGA dataset with 5,886,640 mesh elements

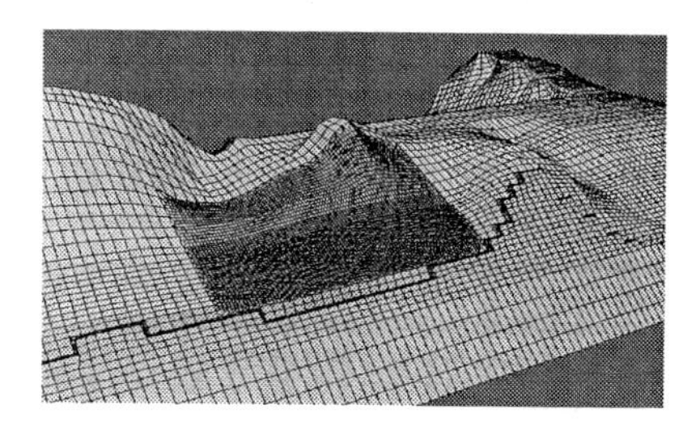
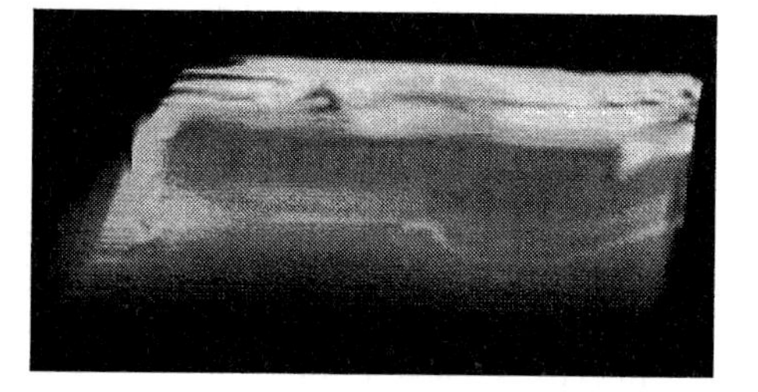

Fig. 4. Parallel volume rendering a large earthquake simulation dataset of the southwest Japan in 1944, with 2,027,520 elements. Left: complicated mesh of the dataset; Right: a volume rendering image in which the magnitude of the displacement data attribute is mapped to color

5 Concluding Remark

The characteristics of GeoFEM as a parallel FEM analysis platform are:

- High performance parallel solver;
- Pluggable design;
- Visualization for large-scale data.

By virtue of the pluggable design, the sequential finite element codes, developed by the geoscientists, can be easily converted to the parallel codes, which can enjoy the high performance of GeoFEM.

GeoFEM ported to the ES is expected to be a powerful tool to handle larger size of engineering problems as well as tackle more realistic solid earth problems. Meanwhile, except the ES, it can achieve high parallel performance on all the SMP cluster machines, and its flat MPI version can work efficiently on all distributed-memory machines.

Acknowledgements. This work is supported by MEXT's (Ministry of Education, Culture, Sports, Science and Technology) Science and Technology Promotion Fund entitled 'Study on Parallel Software for Precise Prediction of Global Dynamics'.

References

1. http://www.es.jamstec.go.jp
2. 1st ACES (APEC Cooperation for Earthquake Simulation) Workshop Proceedings, 1999
3. http://geofem.tokyo.rist.or.jp
4. Okuda, H., Yagawa, G., Nakajima, K., Nakamura, H.: Parallel Finite Element Solid Earth Simulator: GeoFEM. Proc. Fifth World Congress on Computational Mechanics (WCCM V), July 7-12 (2002) Vienna, Austria, ISBN 3-9501554-0-6, http://wccm.tuwien.ac.at, Vol.1, 160–166
5. Garatani, K., Nakajima, K., Okuda, H., Yagawa, G.: Three-Dimensional Elasto-Static Analysis of 100 Million Degrees of Freedom. Advances in Engineering Software, Vol. 32, No. 7, (2001) 511–518
6. Nakajima, K., Okuda, H.: Parallel Iterative Solvers with Localized ILU Preconditioning for Unstructured Grids on Workstation Clusters. International Journal for Computational Fluid Dynamics, Vol. 12 (1999) 315–322
7. Nakajima, K., Okuda, H.: Parallel Iterative Solvers for Unstructured Grids Using an OpenMP/MPI Hybrid Programming Model for the GeoFEM Platform on SMP Cluster Architectures. WOMPEI 2002 (International Workshop on OpenMP: Experiences and Implementations), Kansai Science City, JAPAN, May 15 (2002) 437–448
8. Washio, T., Maruyama, K., Osoda, T., Shimizu, F., Doi, S.: Blocking and Reordering to Achieve Highly Parallel Robust ILU Preconditioners. RIKEN Symposium on Linear Algebra and its Applications, The Institute of Physical and Chemical Research (1999) 42–49

9. Washio, T., Maruyama, K., Osoda, T., Shimizu, F., Doi, S.: Efficient Implementations of Block Sparse Matrix Operations on Shared Memory Vector Machines. SNA2000: CD Proccedings of the Fourth International Conference on Supercomputing in Nuclear Applications (2000)
10. Fujishiro, I., Chen, L., Takeshima, Y., Nakamura, H., Suzuki, Y.: Parallel Visualization of Gigabyte Datasets in GeoFEM. Journal of Concurrency and Computation: Practice and Experience, Vol. 14, No. 6-7 (2001) 521–530
11. Chen, L., Fujishiro, I., Nakajima, K.: Parallel Performance Optimization for Large-Scale Unstructured Data Visualization for the Earth Simulator. In: Reinhard, E., Pueyo, X., Bartz, D. (eds.): Proceedings of Fourth Eurographics Workshop on Parallel Graphics and Visualization, Germany (2002) 133–140

Mantle Convection Modeling with Viscoelastic/Brittle Lithosphere: Numerical and Computational Methodology

Louis Moresi[1], David May[2], Justin Freeman[1], and Bill Appelbe[2]

[1] Monash University, Dept. of Mathematical Science
Clayton, Victoria 3186, Australia
{Louis.Moresi,Justin.Freeman}@sci.monash.edu.au
http://wasabi.maths.monash.edu/MonashClusterComputing

[2] Victorian Partnership for Advanced Computing
Carlton South, Victoria 3053, Australia
{davidm,bill}@vpac.org
http://www.vpac.org/snark

Abstract. The Earth's tectonic plates are strong, viscoelastic shells which make up the outermost part of a thermally convecting, predominantly viscous layer; at the boundaries between plates the rheology is thought to be dominated by brittle processes. Brittle failure of the lithosphere occurs when stresses are high. In order to build a realistic simulation of the planet's evolution, the complete viscoelastic / brittle convection system needs to be considered. A Lagrangian Integration point finite element method is discussed which can simulate very large deformation viscoelasticity with a strain-dependent yield stress. We also describe the general, parallel implementation of this method (SNARK) and compare the performance to a highly optimized, serial prototype code (ELLIPSIS). The specialized code shows better scaling for a single processor. The parallel scaling of the general code is very flat for "realistic" problem sizes indicating efficient use of multiple processors.

1 Introduction

Solid state convection in the Earth's mantle drives the surface motion of a cool thermal boundary layer comprising a number of distinct lithospheric plates. Motions in the mantle are described by the equations of fluid dynamics. The rheology needed to describe deformation in the lithosphere is highly non-linear, and, in addition, near the surface where temperatures are less than $\approx 600°C$ it also becomes necessary to consider the role of elasticity (Watts et al, 1980). The strong correlation between seismicity and plate boundaries (e.g. Barazangi & Dorman, 1969) makes it seem likely that plate motions are associated with localization of deformation occuring when stresses reach the yield strength of the lithosphere. This suggests that the fundamental process is thermal convection; plate tectonics is the manner in which the system organizes. Having identified the need for efficient, large-scale convection simulations with elastic effects in an evolving cool lithosphere, we present a method for simulating viscoelastic/brittle materials during very large deformation, and describe the practical implementation of this method into software for parallel cluster architectures.

P.M.A. Sloot et al. (Eds.): ICCS 2003, LNCS 2659, pp. 781–787, 2003.

2 Mathematical Model

We begin with the classical momentum conservation equation:

$$\nabla.\boldsymbol{\sigma} = \boldsymbol{f} \tag{1}$$

where $\boldsymbol{\sigma}$ is the stress tensor and $\boldsymbol{f}$ a force term. As we are interested only in very slow deformations of highly viscous materials, (infinite Prandlt number) we have neglected all inertial terms in (1), and assume incompressible flow.

We employ a Maxwell viscoelastic model which assumes that the strain rate tensor, $\boldsymbol{D}$, defined as:

$$D_{ij} = \frac{1}{2}\left(\frac{\partial V_i}{\partial x_j} + \frac{\partial V_j}{\partial x_i}\right) \tag{2}$$

is the sum of an elastic strain rate tensor $\boldsymbol{D}_e$ and a viscous strain rate tensor $\boldsymbol{D}_v$. The velocity vector, $\boldsymbol{V}$, is the fundamental unknown of our problem and all these entities are expressed in the fixed reference frame x_i.

To account for the low-temperature lithospheric rheology, brittle behaviour has been parameterized using a non-linear effective viscosity. To determine the effective viscosity we introduce a Prandtl-Reuss flow rule for the plastic part of the stretching, and introducing another additive term in the stretching $\boldsymbol{D}_p$:

$$\frac{\overset{\triangledown}{\boldsymbol{\tau}}}{2\mu} + \frac{\boldsymbol{\tau}}{2\eta} + \lambda\frac{\boldsymbol{\tau}}{2\left|\boldsymbol{\tau}\right|} = \boldsymbol{D}_e + \boldsymbol{D}_v + \boldsymbol{D}_p \tag{3}$$

where λ is a parameter to be determined such that the stress remains on the yield surface, and $|\boldsymbol{\tau}| \equiv (\tau_{ij}\tau_{ij}/2)^{(1/2)}$. The plastic flow rule introduces a non-linearity into the constitutive law which, in general, requires iteration to determine the equilibrium state.

$\overset{\triangledown}{\boldsymbol{\tau}}$ is the Jaumann corotatonal stress rate for an element of the continuum, μ is the shear modulus and η is shear viscosity. The Jaumann stress rate is the objective (frame invariant) quantity given by

$$\overset{\triangledown}{\boldsymbol{\tau}} = \frac{D\boldsymbol{\tau}}{Dt} + \boldsymbol{\tau}\mathbf{W} - \mathbf{W}\boldsymbol{\tau} \tag{4}$$

where $\mathbf{W}$ is the material spin tensor.

In the mantle, the force term from equation (1) is a gravitational body force due to density changes. We assume that these arise, for any given material, through temperature effects:

$$\nabla \cdot \boldsymbol{\tau} - \nabla p = g\rho_0(1 - \alpha_T T)\hat{\mathbf{z}} \tag{5}$$

where g is the acceleration due to gravity, ρ_0 is material density at a reference temperature, α_T is the coefficient of thermal expansivity, and T is temperature. $\hat{\mathbf{z}}$ is a unit vector in the vertical direction. We have also assumed that the variation in density only needs to be considered in the driving term (the Boussinesq approximation).

The energy equation governs the evolution of the temperature in response to advection and diffusion of heat through the fluid:

$$\frac{DT}{Dt} = -\kappa \nabla^2 T \tag{6}$$

where κ is the thermal diffusivity of the material.

The viscosity of the mantle at long timescale is known to be a complicated function of temperature, pressure, stress, grain-size, composition (particularly water content) etc (Karato & Wu, 1993). Despite this complexity, the dominant effect on the viscosity from the point of view of the large-scale dynamics of the system is the effect of temperature (e.g. Solomatov, 1995).

2.1 Numerical Implementation

As we are interested in solutions where very large deformations may occur – including thermally driven fluid convection, we would like to work with a fluid-like system of equations. Hence we obtain a stress / strain-rate relation by expressing the Jaumann stress-rate in a difference form:

$$\overset{\triangledown}{\boldsymbol{\tau}} \approx \frac{\boldsymbol{\tau}^{t+\Delta t^e} - \boldsymbol{\tau}^t}{\Delta t^e} - \mathbf{W}^t \boldsymbol{\tau}^t + \boldsymbol{\tau}^t \mathbf{W}^t \tag{7}$$

where the superscripts $t, t + \Delta t^e$ indicate values at the current and future timestep respectively.

The equation of motion is then

$$\nabla(\eta_{\text{eff}} \mathbf{D}^{t+\Delta t^e}) - \nabla p = g\rho_0(1 - \alpha_T T)\hat{\mathbf{z}} - \nabla\left(\eta_{\text{eff}}\left[\frac{\boldsymbol{\tau}^t}{\mu \Delta t^e} + \frac{\mathbf{W}^t \boldsymbol{\tau}^t}{\mu} - \frac{\boldsymbol{\tau}^t \mathbf{W}^t}{\mu}\right]\right) \tag{8}$$

The velocity field $\mathbf{u}$ and pressure at $t + \Delta t^e$ can be solved for a given temperature distribution and the stress history from the previous step. Our system of equations is thus composed of a quasi-viscous part with modified material parameters and a right-hand-side term depending on values from the previous timestep. This approach minimizes the modification to the viscous flow code. Instead of using physical values for viscosity we use effective material properties (8) to take into account elasticity through a softening of the viscous term.

3 Computational Method

Having devised a suitable mathematical representation of the class of problems we wish to model, we need to choose a numerical algorithm which can obtain an accurate solution for a wide range of conditions. Our method is based closely on the standard finite element method, and is a direct development of the Material Point Method of Sulsky et al. (1995). Our particular formulation could best be described as a finite element representation of the equations of fluid dynamics with moving integration points.

A mesh is used to discretize the domain into elements, and the shape functions interpolate node points in the mesh in the usual fashion. Material points embedded in the fluid are advected using the nodal point velocity field interpolated by the shape functions.

The problem is formulated in a weak form to give an integral equation, and the shape function expansion produces a discrete (matrix) equation. Equation (1) in weak form becomes

$$\int_{\Omega} N_{(i,j)} \tau_{ij} d\Omega - \int_{\Omega} N_{,i} p d\Omega = \int_{\Omega} N_i f_i d\Omega \tag{9}$$

where Ω is the problem domain, and the trial functions, N, are the shape functions defined by the mesh; we have assumed no non-zero traction boundary conditions are present. For the discretized problem, these integrals occur over subdomains (elements) and are calculated by summation over a finite number of sample points within each element. For example, in order to integrate a quantity, ϕ over the element domain Ω^e we replace the continuous integral by a summation

$$\int_{\Omega^e} \phi d\Omega \leftarrow \sum_p w_p \phi(\mathbf{x}_p) \tag{10}$$

In standard finite elements, the positions of the sample points, $\mathbf{x}_p$, and the weighting, w_p are optimized in advance. In our scheme, the $\mathbf{x}_p$'s correspond precisely to the Lagrangian points embedded in the fluid, and w_p must be recalculated at the end of a timestep for the new configuration of particles.

4 Efficient Parallel Implementation

One advantage of the Lagrangian integration point FEM is that it permits the use of a grid which is significantly more regular than the geometrical complexity developed by the model. In a parallel code this has the advantage that the mesh connectivity and domain decomposition is either static or evolves much more slowly than the Lagrangian particle distribution. This fact leads to considerable efficiencies in the allocation and access of distributed global matrices.

In the original, serial implementation of the method (ELLIPSIS , as described in Moresi et al, 2003) a specialized multigrid velocity solver specifically optimized for geoscience problems (Moresi and Solomatov, 1995) was used. This multigrid solver was coupled to a preconditioned Uzawa iteration for pressure (Moresi and Solomatov, 1995) which was found to be robust for strong variations in material properties.

In the parallel code (SNARK), we sought a more general implementation which would be suitable for a broader range of problems, with a more flexible code design (See the paper by Appelbe et al in this volume). The PETSc framework was chosen as the platform since it provides both the interface for domain-decomposition, parallel vector / matrix storage, and a wide range of solution methods. PETSc makes it possible to switch between the many different included solvers very easily. For the velocity solver we found the included GMRES solver to be a reliable starting point for most problems, coupled to the same preconditioned Uzawa scheme developed for the specialized problem.

In this section we demonstrate the impact on solver efficiency of choosing a general approach over a highly specialized one.

4.1 Typical Problem

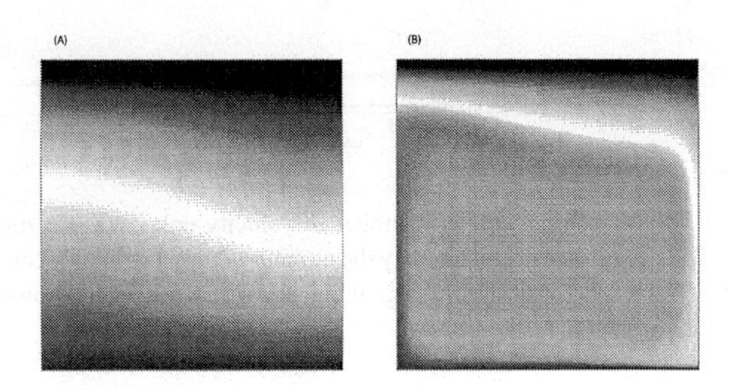

Fig. 1. Convection problem used for timings. The initial condition is on the left (A), and the final, steady-state temperature solution is on the right (B). Blue (dark) is cool, Red (mid-gray) is warm

We benchmark the ELLIPSIS and SNARK codes using a two dimensional convection problem in a 1×1 box of 64×64 elements, temperature fixed top ($T = 0$) and bottom ($T = 1$), free slip boundaries, Rayleigh number 10^6, and initial condition

$$T(x, y) = (1 - y) + (\cos \pi x \cdot \sin \pi y)/100 \tag{11}$$

and using a viscosity dependent on temperature:

$$\eta = \eta_0 \exp\left(\frac{E}{T + T_0}\right) \tag{12}$$

In which T_0 was chosen to be 1, and E, η_0 were chosen to give a basal viscosity of 1, and a viscosity contrast of 10^9 ($E = 41.44, \eta_0 = 10^{-9}$)

A thermal convection cell develops as shown in figure 1. The temperature dependence of viscosity leads to a thick, cool upper thermal boundary layer, with a vigorously convecting layer beneath it. Strong viscosity gradients develop in the thermal boundary layers which may influence the convergence rate of the solver. (See Moresi et al, 1996) In the case of the ELLIPSIS multigrid and the PETSc GMRES solvers, there was no appreciable difference in the solution time for overall viscosity variations of 10^9.

4.2 Solver Scaling with Problem Size

Multigrid, properly implemented, is particularly well suited for large scale problems because the solution time scales linearly with the number of unknowns. Figure 2 shows the scaling of the solution time for each of the methods, normalized by the number of unknowns and the solution time for the smallest problem. The times are the result of

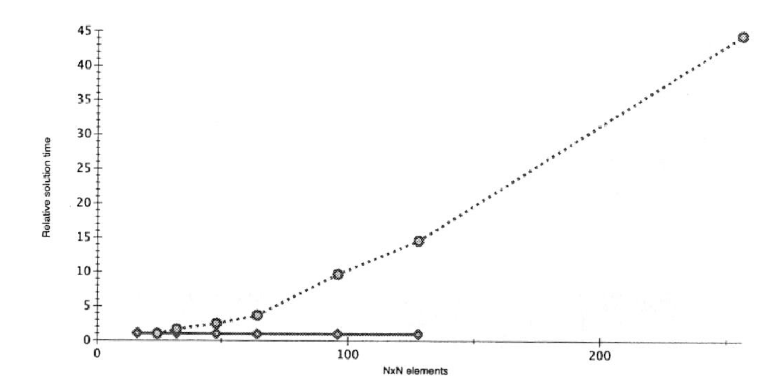

Fig. 2. Scaling of problem time with the number of velocity unknowns for multigrid (red) and GMRES (blue). The timings are normalized by the number of unknowns and scaled by the coarsest grid solution time. A flat curve is desirable since it indicates the work per unknown does not change with the size of the problem

averaging several runs with different viscosity laws. The multigrid scaling is very close to linear for the entire range of problem-sizes explored, whereas GMRES performance degrades quickly as the problem size increases.

4.3 Parallel Efficiency

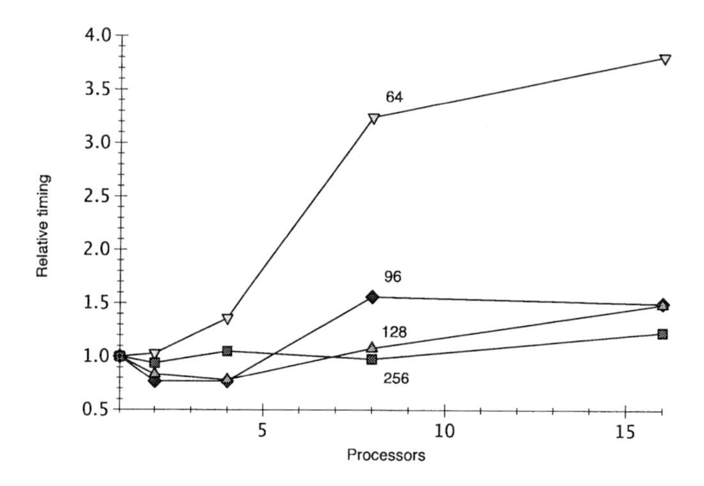

Fig. 3. Efficiency of the parallelization inherited from PETSc as a function of the number of processors. The timings are scaled to the solution time on a single processor in each case and then multiplied by the number of processors. Values below 1 indicate that the problem size is too large for efficient execution on a single processor, values significantly above 1 indicate inefficiencies in the parallel code – e.g. due to over-decomposition

The parallel benchmarks were run on a distributed-memory cluster machine with low-latency communication. In figure 3 we plot the solution time scaled by the number of processors for parallel runs on different grid sizes. In this case it is clear that the efficiency of the parallelization using PETSc is extremely good, provided that the problem is sufficiently large to warrant decomposition. The timings are averaged over several runs, but the fact that the cluster was heavily loaded during these benchmarks introduced some scatter into the results which is evident in the figure.

5 Discussion

The abstraction of parallelism provided by building a code upon the PETSc libraries provides significant benefits in code readability, compactness, flexibility and maintainability. The parallel performance of the resulting code scales very well on a typical supercomputing cluster. The only issue we have found to date is that the PETSc solvers need to be chosen carefully to reproduce the efficiency of a solver written especially for the problem in hand. Further exploration of the issue of general PETSc multigrid solvers for this problem is needed.

References

Barazangi, M., and Dorman, J., (1969), World seismicity maps compiled from ESSA, Coast and Geodetic Survey, epicenter data 1961–1967, Bull. Seism. Soc. Am. 59, 369–380.

Karato, S.-I. and Wu, P., (1993), Rheology of the upper mantle: A synthesis, Science, 260, 771–778.

McKenzie, D.P., (1977), Surface deformation, gravity anomalies, and convection, Geophys. J. R. Astr. Soc., 48, 211–238.

Moresi, L., Mühlhaus, H.-B., Dufour, F., Viscoelastic formulation for modelling of plate tectonics, In Bifurcation and localization in soils and rocks 99. Mühlhaus, Dyskin, A. and Pasternak, E. (ed), (Balkema, Rotterdam, 2001)

Moresi, L., Solomatov, V.S., (1995), Numerical investigations of 2D convection in a fluid with extremely large viscosity variations, Phys. Fluids, 7, 2154–2162

Moresi, L., Solomatov, V.S., (1998), Mantle convection with a brittle lithosphere: thoughts on the global tectonics styles of the Earth and Venus, Geophys. J. Int., 133, 669–682.

Moresi, L., Dufour, F., Mühlhaus, H.-B. (2002) Mantle convection models with viscoelastic/brittle lithosphere: Numerical methodology and plate tectonic modeling. Pure Appl. Geophys., 159, 2335–2356.

Moresi, L., Dufour, F., Mühlhaus, H.-B. (2003) A lagrangian integration point finite element method for large deformation modeling of viscoelastic geomaterials. J. Comput. Phys., 184, 476–497.

Moresi, L., Zhong, S.J. , Gurnis, M. (1996), The accuracy of finite element solutions of Stokes' flow with strongly varying viscosity. Phys. Earth Planet. Inter., 97, 83–94.

Solomatov, V.S., (1995), Scaling of temperature- and stress-dependent viscosity convection, Phys. Fluids, 7, 266–274.

Sulsky, D., Zhou, S.-J., Schreyer, H.L., (1995), Application of a particle-in-cell method to solid mechanics, Comput. Phys. Commun. 87, 236–252.

Watts, A.B., Bodine, J.H., Ribe, N.M., (1980), Observations of flexure and the geological evolution of the Pacific Basin. Nature, 283, 532–537.

Anisotropic Convection Model for the Earth's Mantle

H.-B. Mühlhaus[1,2], M. Čada[3], and L. Moresi[4]

[1] The University of Queensland
St Lucia, QLD 4072, Australia
muhlhaus@quakes.uq.edu.au
[2] CSIRO Division of Exploration and Mining
26 Dick Perry Ave., Kensington WA 6051, Australia
[3] Ludwig-Maximilians University, Institute of Geophysics
Theresienstr. 41, 80333 Munich, Germany
miro.cada@addcom.de
[4] School of Mathematical Sciences, Building 28, Monash University
Victoria 3800, Australia
louis.moresi@sci.monash.edu

Abstract. The paper presents a theory for modeling flow in anisotropic, viscous rock. This theory has originally been developed for the simulation of large deformation processes including the folding and kinking of multi-layered visco-elastic rock (Mühlhaus et al. [1,2]). The orientation of slip planes in the context of crystallographic slip is determined by the normal vector – the director – of these surfaces. The model is applied to simulate anisotropic mantle convection. We compare the evolution of flow patterns, Nusselt number and director orientations for isotropic and anisotropic rheologies. In the simulations we utilize two different finite element methodologies: The Lagrangian Integration Point Method Moresi et al [8] and an Eulerian formulation, which we implemented into the finite element based pde solver Fastflo (www.cmis.csiro.au/Fastflo/). The reason for utilizing two different finite element codes was firstly to study the influence of an anisotropic power law rheology which currently is not implemented into the Lagrangian Integration point scheme [8]and secondly to study the numerical performance of Eulerian (Fastflo)- and Lagrangian integration schemes [8]. It turned out that whereas in the Lagrangian method the Nusselt number vs time plot reached only a quasi steady state where the Nusselt number oscillates around a steady state value the Eulerian scheme reaches exact steady states and produces a high degree of alignment (director orientation locally orthogonal to velocity vector almost everywhere in the computational domain). In the simulations emergent anisotropy was strongest in terms of modulus contrast in the up and down-welling plumes. Mechanisms for anisotropic material behavior in the mantle dynamics context are discussed by Christensen [3]. The dominant mineral phases in the mantle generally do not exhibit strong elastic anisotropy but they still may be oriented by the convective flow. Thus viscous anisotropy (the main focus of this paper) may or may not correlate with elastic or seismic anisotropy.

1 Introduction

Layered rock structures typically exhibit spectacular deformation patterns, illustrations of buckling phenomena on a massive scale. Layered or, more generally, transversely

P.M.A. Sloot et al. (Eds.): ICCS 2003, LNCS 2659, pp. 788–797, 2003.

isotropic materials are indeed ubiquitous in the lithosphere ("the plate"). There is also mounting evidence (mainly from seismic measurements) that at least the upper part of the mantle exhibits acoustic wave anisotropy. A model for a layered mantle was proposed recently e.g. by Aki [5] and Takeuchi et al. [8]. Physical explanations for the presence of material anisotropy in the mantle may be based on flow alignment of the crystallographic slip planes of olivine (the dominant mineral in the mantle). Indeed, laboratory studies by Karato [6] in the context of flow alignment and seismic anisotropy have revealed two classes of anisotropy namely lattice preferred orientation and shape preferred orientation. Hence the proper consideration of the spatial orientation of the dominant slip plane as well as its kinematic and dynamic properties are crucial for the simulation of anisotropic mantle convection.

So far direct simulations of anisotropic mantle flows have been a highly specialized area (e.g. Christensen [3]; the paper also contains concise summary of possible mechanisms for anisotropic behavior). A possible reason may be that conclusive seismic anisotropy data have became available only relatively recently. In the following we give brief outlines of the constitutive theory [1,2] and the Lagrangian Integration Point finite element scheme (LIP), which we used for the solution of the example problems. Our LIP based code allows only linear strain rate dependency at present. In order to study the influence of a power law rheology on the evolution of viscous anisotropy we have implemented the governing equations into a finite element based partial differential equation solver package (Sect. 4). Results including comparisons of anisotropic and isotropic natural convection in a unit cell are presented in the Sects. 3 and 4. The question remains if there is a relationship between elastic (seismic) and viscous anisotropy. If the anisotropy is due to an alternating sequence of mineralocically and mechanically distinct constituents (Allegre and Turcotte, 1986) then the answer is yes; although the elastic strength of the anisotropy in terms of moduli contrast may differ from the viscosity contrast. If the seismic anisotropy is related to the elastic anisotropy in single crystals then the question is whether the reorientation of the crystallographic lattice is described sufficiently accurately by the evolution of the single director (normal vector) of the dominant slip system (Eq. (3)). In single slip and if elastic deformations are negligible this should be the case.

2 Mathematical Formulation

In the case of a material with layering or preferred crystallographic slip directions, the orientation of the director is normal to the layer or slip surfaces. Transverse-isotropic relations are characterized by two effective viscosities. We designate normal viscosity as η ($\sigma_{11} = -p + 2\eta D'_{11}$) and η_s ($\sigma_{12} = 2\eta_S D'_{12}$) is the shear viscosity. In the following simple model for a layered viscous material we correct the isotropic part $2\eta D'_{ij}$ of the model by means of the Λ_{ijkl} tensor (Mühlhaus et al. [1,2]) to consider the mechanical effect of the layering; thus

$$\sigma_{ij} = 2\eta D'_{ij} - 2(\eta - \eta_s)\Lambda_{ijlm} D'_{lm} - p\delta_{ij} \tag{1}$$

where a prime designates the deviator of the respective quantity, p is the pressure, D_{ij} is the stretching, σ_{ij} is the Cauchy or true stress and

$$\Lambda_{ijkl} = \frac{1}{2}(n_i n_k \delta_{lj} + n_j n_k \delta_{il} + n_i n_l \delta_{kj} + n_j n_l \delta_{ik}) - 2n_i n_j n_k n_l. \tag{2}$$

is the anisotropy tensor. In (1) and (2) the vector n_i is the unit orientation-vector of the director N_i. In the present applications we assume that the director transforms as a material surface element; in continuum mechanics theory the evolution of the director of the layers can be derived through the time derivative of Nanson's equation, which relates the current director to the reference director:

$$\dot{N}_i = N_{i,t} + v_j N_{i,j} = -v_{j,i} N_j \tag{3}$$

where v_i is the velocity vector. In 2D it is plausible that the planes of anisotropy or slip planes are aligned with the velocity vectors in steady states, which is equivalent to normality of the directors to the velocity vectors.

2.1 Numerical Method

The Lagrangian Integration Point finite element method ELLIPSIS uses a standard mesh to discretize the domain into elements, and the shape functions interpolate node points in the mesh in the usual fashion. Derivatives are computed on the mesh using the values of nodal variables but material property variations like the "director" are measured by the particles. The problem is formulated in a weak form to give an integral equation, and the shape function expansion produces a discrete (matrix) equation. For the discretized problem, these integrals occur over sub-domains (elements) and are calculated by summation over a finite number of sample points (tracers) within each element. For example, in order to integrate the components of the element stiffness matrix $\mathbf{K}^E$ over the element domain Ω^E:

$$\mathbf{K}^E = \int_{\Omega_E} \mathbf{B}^T(\mathrm{x})\mathbf{C}(\mathrm{x})\mathbf{B}(\mathrm{x})d\Omega \tag{4}$$

we replace the continuous integral by a summation

$$\mathbf{K}^E = \sum_p w_p \mathbf{B}_p^T(\mathrm{x}_p)\mathbf{C}_p(\mathrm{x}_p)\mathbf{B}_p(\mathrm{x}_p) \tag{5}$$

Here the matrix $\mathbf{B}$ consists of the appropriate gradients of interpolation functions which transform nodal point velocity components to strain-rate pseudo-vectors at any points in the element domain. $\mathbf{C}$ the constitutive operator corresponding to (1) is composed of two parts $\mathbf{C} = \mathbf{C}_{iso} + \mathbf{C}_{aniso}$.

In standard finite elements, the positions of the sample points, x_p, and the weighting, w_p are optimized in advance. In our scheme, the x_p's correspond precisely to the Lagrangian points embedded in the fluid, and w_p must be recalculated at the end of a time-step for the new configuration of particles. The Lagrangian points carry the history variables (in this case director orientation) which are therefore directly available for the element integrals without the need to interpolate from nodal points to fixed integration points. Moresi et al. [9] give a full discussion of the implementation and integration scheme.

In the Fastflo based simulations (www.cmis.csiro.au/Fastflo/) the advection terms in the heat equation and in the purely hyperbolic director evolution equation are stabilized using artificial diffusion terms a method which is also called tensor upwinding (Zienkiewicz and Taylor, [10]). See Sect. 4 for more details.

3 Application

We have simulated a basally heated convection problem in a box of aspect ratio 2×1, with free slip-boundary conditions where the temperature is fixed at the top and bottom and there is no heat flux at the sides. We assume a Rayleigh-number of $Ra = 5.64 \times 10^5$ and a constant ratio of $\frac{\eta}{\eta_S} = 10$ (Fig. 1). Subsequently the influence of the viscosity ratio on the time history of the Nusselt number is also investigated (Fig. 2). In the definition of the Rayleigh number for anisotropic viscous materials we follow Christensen [3] and

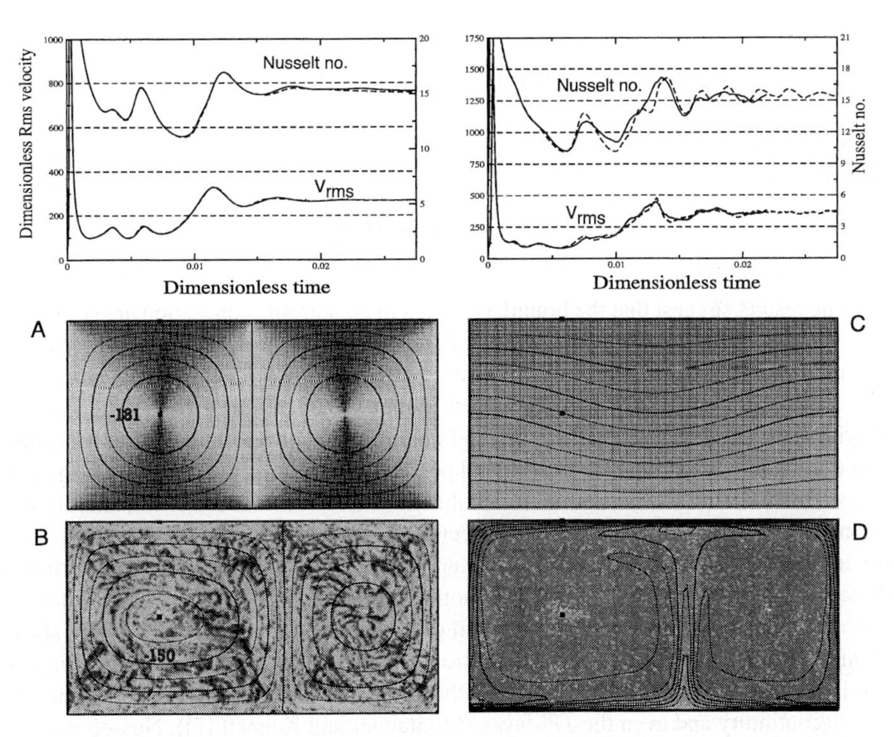

Fig. 1. Time dependent convection in a 2×1 box. $Ra = 5.6 \times 10^5$. Time series plots of velocity and Nusselt number isotropic convection (top left) and anisotropic convection, $\frac{\eta_S}{\eta}$ (top right). Dashed lines and lines are the results of the 32×32 and 64×64 elements simulations respectively. The initial state of alignment is shown in (a) and the initial thermal structure in (b), and after 4000 time-steps ($t = 0.04$) in (c) and (d)

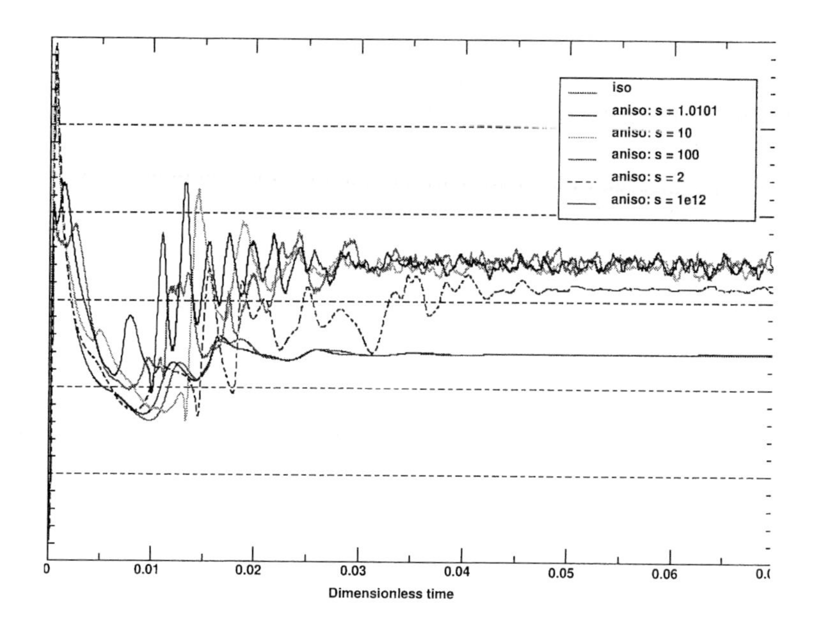

Fig. 2. Nusselt no. plots for different viscosity ratios: s = η/η_S

define:

$$Ra = \frac{2\alpha\rho_0^2 c_p g \Delta T H^3}{\kappa(\eta + \eta_s)} \tag{6}$$

The results suggest that the boundary layers in anisotropic convection are somewhat more stable than the equivalent isotropic boundary layers in medium to high Ra convection, leading to a reduced tendency for convection cells to break down and reform. The isotropic simulation passes through different stages of convective evolution, displaying plumes of hot material raising and cold material descending into its core, until two dominant convection cell emerge and persist in the steady state shown in Fig. 1. In the orthotropic simulation the director evolution follows the induced shear, and director alignment in rising plumes as well as director alignment in the boundary layers is visible in steady state. These aligned regions encircle a core of roughly randomly oriented directors. This suggests that seismic anisotropy is likely to be mostly dominant in up and down welling plumes and the hot and cold thermal boundary layers where shear-strain rates are high. This fits the observational evidence that deep mantle anisotropy is significant smaller, than in the shallow lithosphere (upper boundary layer) or the 660 km discontinuity and even the D'' layer (Montagner and Kennett [7]). Nusselt number histories for varying viscosity ratios are depicted in Fig. 2. It turns out that the graphs don't differ much for $\frac{\eta}{\eta_S} > 10$ i.e. there is little difference between the histories for the ratios 10, 100 and 10^{12}.

4 Power Law Induced Anisotropy

In this section we investigate the influence of a power law rheology on the spatial distribution of the strength of the anisotropy as expressed by the local value of the ratio η/η_S. The behavior is locally isotropic if the viscosity ratio is equal to 1; the most extreme case of anisotropy would correspond to a ratio of ∞ (see Fig. 2), where we have plotted the Nusselt-number of different viscosities ratios). In the study we use the simple relationship

$$\frac{\eta}{\eta_S} = (1 + \dot{\gamma}/\dot{\gamma}_Y)^{1-\frac{1}{n}}, \tag{7}$$

where in general $\dot{\gamma}$ is defined as:

$$\dot{\gamma} = \sqrt{D_{ij}\Lambda_{ijkl}D_{kl}} \tag{8}$$

In 2D there is a simpler way to calculate $\dot{\gamma}$: The shear strain vector on the Nslip surface is:

$$\dot{\gamma}_i = D_{ij}n_j \tag{9}$$

Let $\mathbf{m}$ be a vector in the slip plane so that $\mathbf{m} \cdot \mathbf{n} = 0$. In 2D the components of $\mathbf{m}$ read:

$$\mathbf{m} = \begin{pmatrix} n_2 \\ -n_1 \end{pmatrix} \tag{10}$$

The magnitude of the shear stress on the n-surface is then defined as:

$$\dot{\gamma} = |D_{ij}n_j m_i| = |n_1 n_2(D_{11} - D_{22}) + D_{12}(n_2^2 - n_1^2)| \tag{11}$$

The parameter $\dot{\gamma}_Y$ is a strain rate characterizing the transition from predominantly linear creep to power law creep and the superscript n is the power law exponent. We find the following limit values for the viscosity ratio:

$$\lim_{\dot{\gamma}\to 0} \left(\frac{\eta}{\eta_S}\right) = 1 \tag{12}$$

$$\lim_{\dot{\gamma}\to \infty} \left(\frac{\eta}{\eta_S}\right) = \infty \tag{13}$$

We consider a quadratic convection cell with ideal frictionless boundaries and zero normal velocities. The temperature difference between the top and the bottom boundary is fixed and on the sides we assume periodic boundaries (zero temperature fluxes). The domain is discretized into 710 six noded triangular elements with element concentrations along the sides of the model. Since the power law rheology is not yet implemented into our material point advection code ELLIPSIS (Moresi et al., 2001 [9]) we have used the finite element code Fastflo (see http://www.cmis.csiro.au/Fastflo/ for more information). In Fastflo the user programs the mathematical model and algorithms by means of the high level language Fastalk. In the Fastflo macro developed for this application the momentum equilibrium equation based on the rheology (1) with the viscosity ratio as

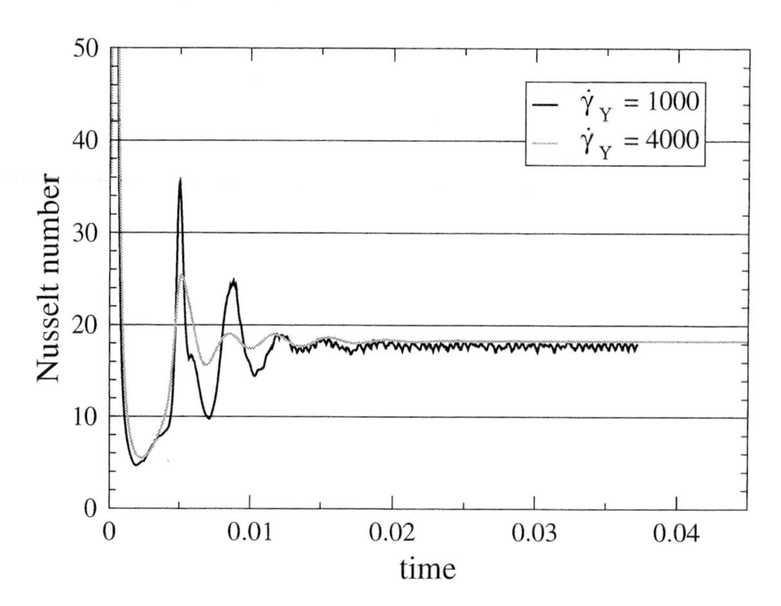

Fig. 3. Nusselt-number vs. time for a power law exponent of n=5 and dimensionless strain rates $\dot{\gamma}_Y$ = 1000 (dark) and 4000 (light) respectively with steady state values of 17.8 and 18.5 respectively

defined by (7) is solved iteratively by successively substituting $\dot{\gamma}$ for a given temperature at each time step. An average of 4 iterations were required for four digit accuracy in the r.m.s. of the nodal point velocity vector.

Subsequently the heat equation is solved using a backward Euler differencing scheme and tensor upwinding. The latter amounts to the inclusion of an additional diffusion term with the diffusivity tensor $v_i v_j \Delta t/2$ (Zienkiewicz and Taylor, [10]). Finally the director advection equation (3) is solved – again using an Euler backward scheme and tensor upwinding. Tensor upwinding is very important here, because of the hyperbolic nature of the director equation (see Fig. 5). The time-step is determined from the Courant condition-like criterion

$$\Delta t = \alpha \frac{\sqrt{\text{area}/2 \times \text{num. of elem.}}}{v_{max}}. \tag{14}$$

In the calculations we have assumed that $\alpha = 1/2$; The factor 2 in the numerator considers the fact that the elements are triangular. As in the other sections of this paper we assume an isotropic Rayleigh number of $Ra = 5.64 \times 10^5$, an initial temperature perturbation of

$$T = 1 - x_2 + \frac{1}{10} \cos(\pi x_1) \sin(\pi x_2), \tag{15}$$

and an initial director orientation of $\mathbf{n}^T = (0, 1)$ everywhere. Figure 3 shows the evolution of the Nusselt-number with time for a power law exponent of n=5 and dimensionless strain rates $\dot{\gamma}_Y$ = 1000 and 4000 respectively. The steady state Nusselt numbers are 17.8

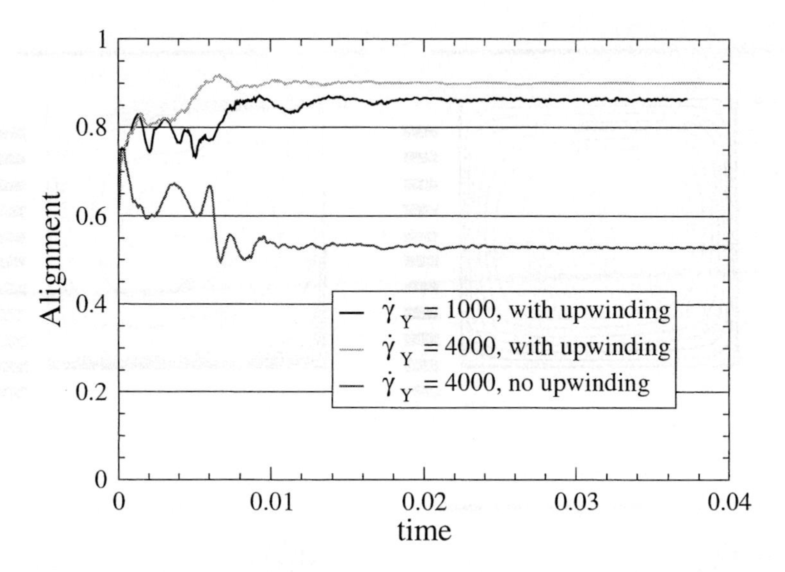

Fig. 4. Alignment for $\dot\gamma_Y$ = 1000 (dark graph; with upwind) and $\dot\gamma_Y$ = 4000 (medium; no upwind and light with upwind) respectively. The shear viscosity contrast and hence the anisotropy is largest in the up and down welling plumes. A milder degree of anisotropy is also observed in the cold and the hot boundary layers. The core of the convection cells is virtually isotropic

and 18.5 for $\dot\gamma_Y$ = 100 and 4000 respectively. In the calculations we have used the definition

$$Nu = 1 + \frac{1}{Ra}\int_V \sigma_{ij} D_{ij} dV, \tag{16}$$

of the Nusselt number which is also valid during the transient phase of convection (see eg. Christensen, 1984 [4]). In Eq. (16) all quantities are non dimensional.

A global measure for the degree of alignment of the director in dependency of the time is represented in Fig. 4. For a quadratic domain of unit area the degree of alignment is measured in terms of the sin of the mean value of the angle enclosed by the velocity vector and the director:

$$\int_V |\sin(\alpha)| dV = \int_V \left(\frac{|v_1 n_2 - v_2 n_1|}{|\mathbf{v}||\mathbf{n}|}\right) dV \tag{17}$$

If the velocity vector is everywhere orthogonal to the director (full alignment) then $\sin(\alpha) = 1$. In the simulations we achieve an average alignment of 0.85 and 0.91 for $\dot\gamma_Y$ = 1000 and $\dot\gamma_Y$ = 4000 respectively with tensor upwinding. Also shown is the alignment evolution for $\dot\gamma_Y$ = 1000 without upwinding (purpel graph in Fig. 4). In this case alignment does not take place because of the numerical ill-posedness of the unstabilised director equations.

Figure 5 (A) and (B) show the contours of the streamlines (determined in the usual way from the velocity potential ϕ where $v_1 = \phi_{,2}$ and $v_2 = -\phi_{,1}$) and the isotherms respectively; both plots are for $\dot\gamma_Y$ = 1000. The streamlines lines are more angular than

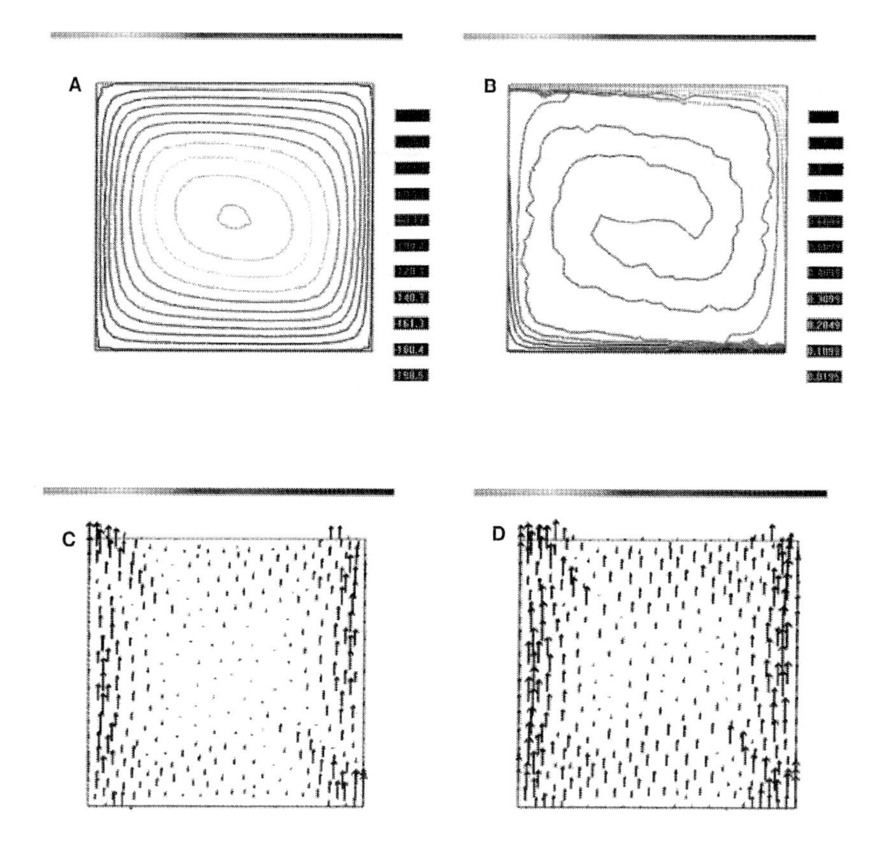

Fig. 5. Streamlines (A), isotherms (B) n=5 and $\dot{\gamma}_Y$ = 1000 in (A) and (B). (C) and (D): pseudu vectors $(0, 1 - \eta_S/\eta)$; The arrows are normalized with respect to the maximum arrow length. Actual max arrow length: 0.57 (C), 0.93 (D); Parameters: n=5, $\dot{\gamma}_Y$ = 4000 (C), n=5, $\dot{\gamma}_Y$ = 1000 (D)

in isotropic convection. Also characteristic for anisotropic convection is the spiral like shape of the isotherms. An important measure for the spatial distribution of anisotropy is the distribution of ratio of η_S/η; if $\eta_S/\eta = 1$ the material behaves locally isotropic; strong anisotropy is obtained for $\eta_S/\eta = 0$, i.e. in the limit $\dot{\gamma}/\dot{\gamma}_Y \to \infty$. In Fig. 5 the viscosity ratio is represented for visualization purposes by means of the distribution of the pseudo vector field $(0, 1 - \eta_S/\eta)$ anisotropy is strongest where the arrow is longest and *vice versa.* In the plots the vector are normalized such that the magnitudes range between 0 and 1. The actual magnitude ranges between 0.57 for $\dot{\gamma}_Y = 4000$ and 0.93 for $\dot{\gamma}_Y$ = 1000. The anisotropy is strongest in both cases in the hot and cold plumes on the sides of the cell. The core of the convection cells are more or less isotropic however in the case of the lower transition strain rate we observe significant anisotropy (in terms of the viscosity ratio) in the vicinity of the hot and cold thermal boundary layers as well.

Acknowledgement. The authors gratefully acknowledge the support of the supercomputing facilities of the Australian Earth System Simulator (ACcESS), a Major National Research Facility.

References

1. Mühlhaus, H.-B., Dufour, F., Moresi, L., Hobbs, B., (2002) *A director theory for viscoelastic folding instabilities in multilayered rock*, Int. J. Solids and Structures. Vol. 39; 3675–3691.
2. Mühlhaus, H.-B., Moresi. L., Hobbs. B., Dufour. F., (2002) *Large Amplitude Folding in Finely Layered Viscoelastic Rock Structures*, Pure. Appl. Geophys., 159, 2311–2333
3. Christensen, U.C., (1987) *Some geodynamical effects of anisotropic viscosity*, Geophys.J.R.astr. Soc. 91;711-736.
4. U. Christensen (1984) Convection with pressure and temperature dependent non-Newtonian rheology. Geophys. J.R. Astr. Soc.,77, 343–384
5. Aki, K., (1968) *Seismological evidence for the existence of soft thin layers in the upper mantle under Japan*, J. Geophys. Res., 73, 585–596.
6. Karato, S.-I., (1998) *Seismic Anisotropy in the Deep Mantle, Boundary Layers and the Geometry of Mantle Convection*, Pure appl. geophys., 151, 565–587.
7. Montagner, J.-P., and Kennett, B.L.N., (1996) *How to Reconcile Body-wave and Normal-mode Reference Earth Model*, Geophys. J.. Int. 125, 229–248.
8. Takeuchi, H., Y. Hamano, and Y. Hasegawa, (1968) *Rayleigh- and Lowe-wave discrepancy and the existence of magma pockets in the upper mantle*, J. Geophys. Res., 73, 3349–3350.
9. Moresi, L., Mühlhaus, H.-B., Dufour, F., (2001) *Particle-in-Cell Solutions for Creeping Viscous Flows with Internal Interfaces*, Proceedings of the 5th International Workshop on Bifurcation and Localisation (IWBL'99), Perth, H.-B. Mühlhaus, A. Dyskin, E. Pasternak Australia, Balkema, Rotterdam.
10. O.C. Zienkiewicz and R.L. Taylor (2000), *Finite Element Method: Volume 3, Fluid Dynamics*; ISBN 0750650508, Butterworth-Heinemann).
11. C.J. Allegre and D.L. Turcotte (1986) Implications of two component marble cake mantle. Nature, 323, 123–127

Finite Element Simulation of Stress Evolution in a Frictional Contact System

H.L. Xing[1], P. Mora[1], and A. Makinouchi[2]

[1]QUAKES, The University of Queensland, St. Lucia, Brisbane, QLD 4072, Australia
{xing,mora}@quakes.uq.edu.au
[2]Integrated V-CAD Research Program
The Institute of Physical and Chemical Research, Japan

Abstract. A 3-dimensional finite element algorithm for modeling nonlinear frictional contact behaviours between deformable bodies with the node-to-point contact element strategy has been proposed and applied here to investigate stress evolution processes of a nonlinear frictional contact system. The numerical results of a typical intra-plate fault bend model demonstrate the efficiency and usefulness of this algorithm.

1 Introduction

The understanding, simulation and prediction of behaviors of frictional contact systems are very important in both theory and practical applications. Although the important progresses have been achieved in the computational mechanics, how to describe the complex phenomena in a frictional contact system with reasonable accuracy still remains a problem to be overcome. Up to now, it is difficult for most of the existing finite element codes to get a stable solution even with the assumption that the interface parameters are constant during the frictional process. An arbitrarily shaped contact element strategy, named as node-to-point contact element strategy, was proposed based on the static-explicit algorithm and applied to handle the frictional contact between deformable bodies with stick and finite frictional slip [1–4]. This paper will focus on how to extend the algorithm to the simulation of stress evolution phenomena in a frictional contact system.

2 Finite Element Formulation

The updated Lagrangian rate formulation is employed to describe the non-linear contact problem. The rate type equilibrium equation and the boundary at the current configuration are equivalently expressed by a principle of virtual velocity of the form

$$\int_V \left(\sigma_{ij}^J - D_{ik}\sigma_{kj} + \sigma_{ik}L_{jk} - \sigma_{ik}D_{kj}\right)\delta L_{ij}\,dV = \int_{S_F} \dot{F}_i \delta v_i\,dS + \int_{S_c^1} \dot{f}_i^1 \delta v_i\,dS + \int_{S_c^2} \dot{f}_i^2 \delta v_i\,dS \quad (1)$$

P.M.A. Sloot et al. (Eds.): ICCS 2003, LNCS 2659, pp. 798–806, 2003.

where V and S denote respectively the domains occupied by the total deformable body B and its boundary at time t; S_F is a part of the boundary of S on which the rate of traction $\dot{F}_i$ is prescribed; $\boldsymbol{\delta v}$ is the virtual velocity field which satisfies the condition $\boldsymbol{\delta v} = \boldsymbol{0}$ on the velocity boundary; σ_{ij}^J is the Jaumann rate of Cauchy stress σ_{ij}; $\boldsymbol{L}$ is the velocity gradient tensor, $\boldsymbol{L} = \partial \boldsymbol{v} / \partial \boldsymbol{x}$; $\boldsymbol{D}$ is the symmetric part of $\boldsymbol{L}$; $\dot{\boldsymbol{f}}^{\alpha}$ is the rate of contact stress on contact interface S_c^{α} of the body α and calculated as follows.

Friction is by nature a path-dependent dissipative phenomenon that requires the integration along the slip path. In this study, a standard Coulomb friction model is applied in an analogous way to the flow plasticity rule, which governs the slipping behaviour. The basic formulations are summarized below [1]-[3] (The tiled (~) above a variable denotes a relative component between slave and master bodies, and *l, m=1,2; i, j, k = 1, 3* respectively in this paper).

Based on experimental observations, an increment decomposition is assumed

$$\Delta\tilde{u}_m = \Delta\tilde{u}_m^e + \Delta\tilde{u}_m^p, \tag{2}$$

where $\Delta\tilde{u}_m^e$ and $\Delta\tilde{u}_m^p$ respectively represent the stick (reversible) and the slip (irreversible) part of $\Delta\tilde{u}_m$, the relative displacement increment component along direction m on the contact interface. The slip is governed by the yield condition

$$F = \sqrt{f_m f_m} - \overline{F}, \tag{3}$$

where f_m ($m=1,2$) is the frictional stress component along the tangential direction m; $\overline{F}$, the critical frictional stress, $\overline{F} = \mu f_n$; μ is the friction coefficient, it may depend on the normal contact pressure f_n, the equivalent slip velocity $\dot{\tilde{u}}_{eq}\left(= \sqrt{\dot{\tilde{u}}_m \dot{\tilde{u}}_m}\right)$ and the state variable φ, *i.e.* $\mu = \mu(f_n, \dot{\tilde{u}}_{eq}, \varphi)$ (e.g. [5]).

If $F<0$, contact is in the sticking state and treated as a linear elasticity, *i.e.*

$$f_m = E_t \tilde{u}_m^e = E_t \sum \Delta\tilde{u}_m^e, \tag{4}$$

where E_t is a constant in the tangential direction.

When $F=0$, the friction changes its character from the stick to the slip. The frictional stress can be described as

$$f_m = \eta_m \overline{F} \text{ and } \eta_m = f_m^e \Big/ \sqrt{f_l^e f_l^e} \tag{5}$$

where $f_m^e = E_t(\tilde{u}_m - \tilde{u}_m^p\big|_0)$, and $\tilde{u}_m^p\big|_0$ is the value of $\tilde{u}_m^p$ at the beginning of this step.

The linearized form of Eq. (5) can be rewritten as

$$df_l = \frac{\overline{F}E_t}{\sqrt{f_m^e f_m^e}}(\delta_{lm} - \eta_l\eta_m)d\tilde{u}_m + \eta_l\mu\left(df_n + \frac{\partial\mu}{\partial f_n}df_n\right) + \eta_l f_n\left(\frac{\partial\mu}{\partial\dot{\tilde{u}}_{eq}}d\dot{\tilde{u}}_{eq} + \frac{\partial\mu}{\partial\varphi}d\varphi\right) \tag{6}$$

In addition, the penalty parameter method is chosen to satisfy the normal impenetrability condition. Finally, the frictional contact stress acting on a slave node can be described as (denote $\dot{f}_3 = \dot{f}_n$)

$$\dot{f}_i = G_{ij}\dot{\tilde{u}}_j + \dot{f}_{\varphi i}\,, \tag{7}$$

where $\boldsymbol{G}$ is the frictional contact matrix; $\dot{f}_{\varphi i}$ is from the contribution of the terms related with φ.

A node-to-point contact element strategy was proposed to handle the frictional contact problems between deformable bodies [1]-[4]. Assume a slave node s has contacted with point c on a surface element (master segment) E', and the surface element E' consists of γ nodes ($\gamma = 4$ in this paper if without special notation), thus the term related with contact in Eq. (1) can be described as $(\alpha = 1,(\gamma+1), \beta = 1,(\gamma+1))$

$$\dot{f}_i(\delta u_{si} - \delta u_{ci}) = \delta\tilde{u}_{sci\beta}\left(\left[\overline{K}_{fik}\right]_{\beta\alpha}\dot{u}_{sck\alpha} + R_\beta\dot{f}_{\varphi i}\right), \tag{8}$$

where

$$\left[\overline{K}_{fik}\right]_{\beta\alpha} = R_\beta\,\boldsymbol{e}_i\cdot\left\{G_{hk}R_\alpha\,\boldsymbol{e}_h + \left(H_{jm}\hat{\boldsymbol{e}}_j\left(\left(\overline{C}_{ll}R_{\alpha,m} - \overline{C}_{ml}R_{\alpha,l}\right)\boldsymbol{e}_k\cdot\tilde{\boldsymbol{x}} + R_\alpha\left(\overline{C}_{ll}\,\hat{\boldsymbol{e}}_m - \overline{C}_{ml}\,\hat{\boldsymbol{e}}_l\right)\cdot\boldsymbol{e}_k\right)\right)\right\} \tag{9}$$

$(h = 1,\ 3, l \neq m$ and no sum on l),

here $\overline{C}_{ml} = C_{ml} - g_n\,\boldsymbol{n}\cdot\hat{\boldsymbol{e}}_{m,l}$, $C_{ml} = \hat{\boldsymbol{e}}_m\cdot\hat{\boldsymbol{e}}_l$, $\wp = \overline{C}_{11}\overline{C}_{22} - \overline{C}_{12}\overline{C}_{21}$, $\tilde{\boldsymbol{x}} = \boldsymbol{x}_s - \boldsymbol{x}_c$, $E_{ijm} = \hat{\boldsymbol{e}}_{i,m}\cdot\hat{\boldsymbol{e}}_j$, $H_{jm} = \hat{f}_i E_{ijm}/\wp$, $\dot{\boldsymbol{u}}_{sc} = \lfloor\dot{\boldsymbol{u}}_s \ \ \dot{\boldsymbol{u}}_1 \ \ \dot{\boldsymbol{u}}_2 \ \ \ldots \ \ \dot{\boldsymbol{u}}_\gamma\rfloor$, $\boldsymbol{R} = \begin{bmatrix}1 & -N_1 & -N_2 & \ldots & -N_\gamma\end{bmatrix}^T$, N_p $(p = 1,\ \gamma)$ is the shape function value of the point c on the surface element E', $\hat{\boldsymbol{e}}_i$ and $\boldsymbol{e}_i$ are respectively the base vectors of the local natural and the local Cartesian coordinate systems on the master segment; $g_n = \boldsymbol{n}\cdot(\boldsymbol{x}_s - \boldsymbol{x}_c)$, while $\boldsymbol{x}_s$ and $\boldsymbol{x}_c$ are the position coordinates of a slave node and its corresponding contact point on a master surface, respectively.

The explicit time integration algorithm is applied with the R-minimum to limit the step size for avoiding a drastic change in states within an incremental step [e.g. 4]. In combination with the above equations, Eq. (1) can be rewritten as

$$(\boldsymbol{K} + \boldsymbol{K}_f)\Delta\boldsymbol{u} = \Delta\boldsymbol{F} + \Delta\boldsymbol{F}_f, \tag{10}$$

here $\boldsymbol{K}$ is the standard stiffness matrix corresponding to the total body B; $\Delta \boldsymbol{F}$ is the external force increment subjected to body B on S_F; $\Delta \boldsymbol{u}$ is the nodal displacement increment; $\boldsymbol{K}_f$ and $\Delta \boldsymbol{F}_f$ are the stiffness matrices and the force increments of all the node-to-point contact elements.

3 Numerical Investigation of Stress Evolution

A typical intra-plate fault bend model (300x300x50 mm^3) is analyzed here, which has a pre-cut fault that is artificially bent by an angle of 5.6^0 at the center E of the fault. The details of the geometry, the boundary conditions are shown in Figure 1. There exists no relative motion along the interface at the both ends (as depicted using the thick black lines CD and FG in Fig. 1). This can be easily achieved using the 'stick' algorithm in the code. While, for the other part of the fault interface (i.e. segments DE and EF in Figure 1), the widely applied rate- and state-dependent friction law [5] is used here with the following parameters: $\mu = 0.60 + (0.010 - 0.025)\boldsymbol{ln}(V / 0.001)$, $d\varphi/dt = 0$. Thus, the total fault consists of four fault segments: CD, DE, EF and FG in Fig. 1. Here all the materials have the same properties: density $\rho = 2.60\, g/cm^3$, Young's modulus $E=44.8$ GPa and Poisson's ratio $\gamma = 0.12$. As for the loading conditions, two loading stages are applied here: firstly, the pressure on the surfaces A and B are loaded up to 10 MPa, then sustaining this pressure on surface A, while all the nodes on surface B are moved in the x-direction at the velocity $Vx=-0.001 mm/sec$.

The stress evolution of the total system during the loading processes is calculated and some results at the different loading time are shown in Figs. 2 and 3. At the beginning of the second loading stage (i.e. at the end of the first loading stage), the stress distribution is rather smooth except that at the both ends as shown in Fig. 2, and the same situation remains until the nodes start to change their stick states to the slip (see Figs. 3: time=228.7s). However, with the increase of the prescribed displacement, the stress on the segment DE begins to decrease and to redistribute locally due to the local energy release induced by the state changes from the stick to the slip on the segment DE. This also results in the stress increase for the nodes around the segments CD and EF (see Fig. 3: Time=292.75s – 296.15s). Afterwards, the stress increases much for most of the nodes except the local energy release zone around the segment DE (see, e.g. Fig. 3: Time=302.98s). With the transition of the stick-slip instability and the occurrence of energy release at the node on the fault segment EF, the stress deceases and redistributes locally for the corresponding nodes which have just entered the slip state, while the stress of neighboring node that is in the stick state increases (see, e.g. Figs. 3: Time=311.08s – 369.78s). Finally, this also causes the obvious stress increase at both end zones (i.e. segments CD and FG in Fig. 1) (as shown in Figs. 3: Time=329.93s – 369.78s).

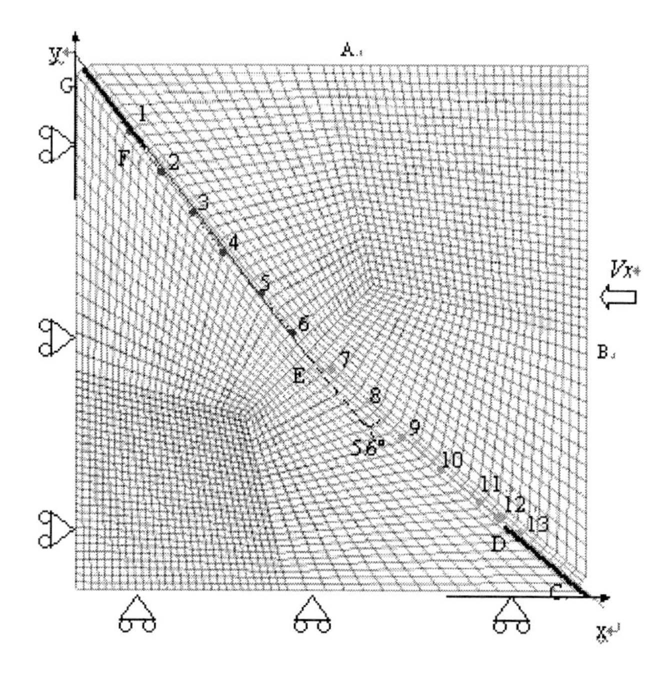

Fig. 1. The mesh, the boundary conditions and prescribed nodal positions used for the fault bend model in the x-y cross section. Here all the nodes at the surface marked by a triangular with two circles ( ) are free along the direction of two circles but fixed at the other direction

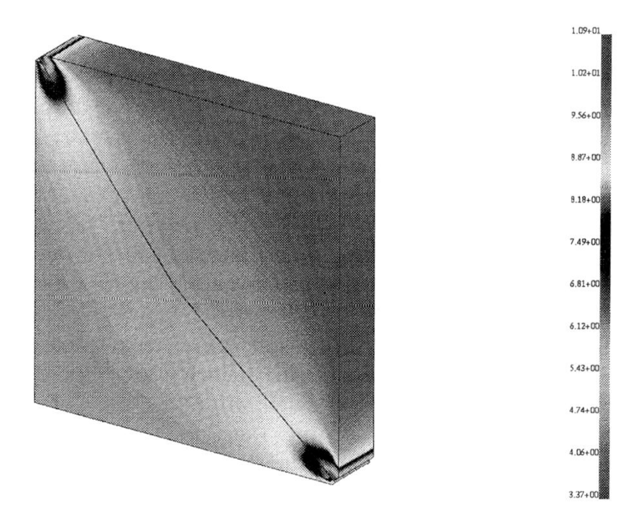

Fig. 2. Mises equivalent stress $\overline{\sigma}$ distribution at the end of the first loading stage(here $\overline{\sigma} = \sqrt{3\sigma'_{ij}\sigma'_{ij}/2}$, σ'_{ij} is the deviator stress tensor, unit: MPa)

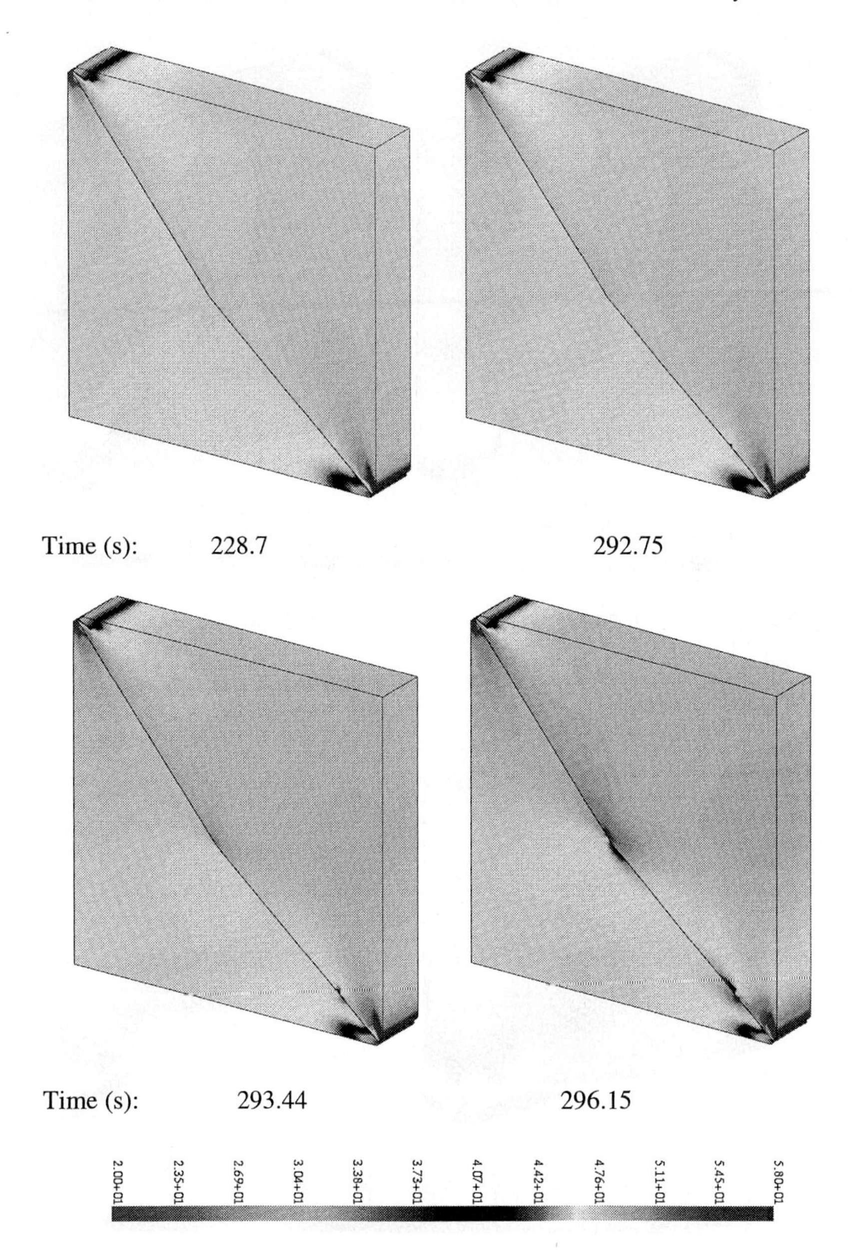

Fig. 3. Mises equivalent stress $\overline{\sigma}$ variation during the second loading stage (here $\overline{\sigma} = \sqrt{3\sigma'_{ij}\sigma'_{ij}/2}$, σ'_{ij} is the deviator stress tensor, unit: MPa)

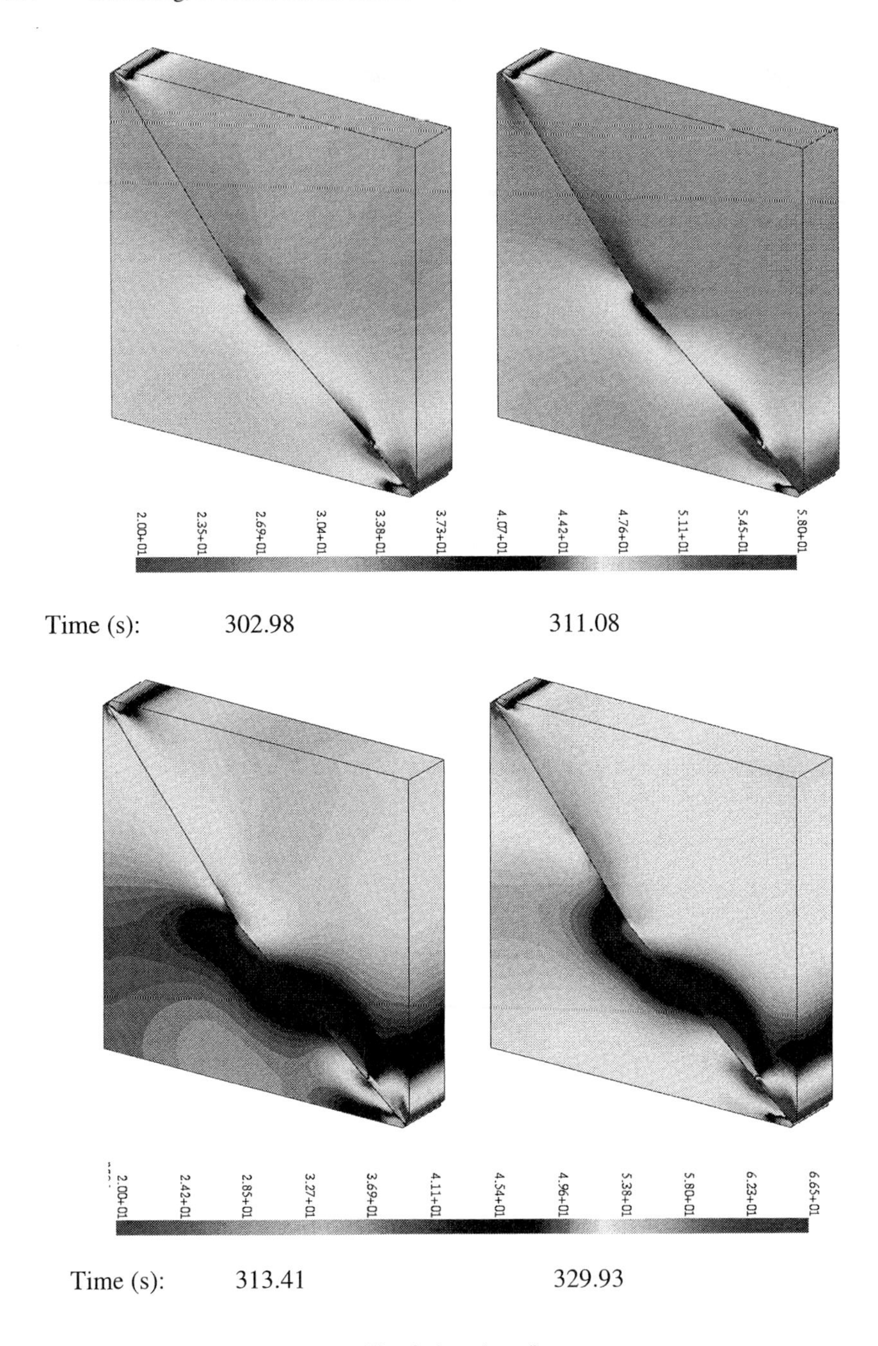

Fig. 3. (continued)

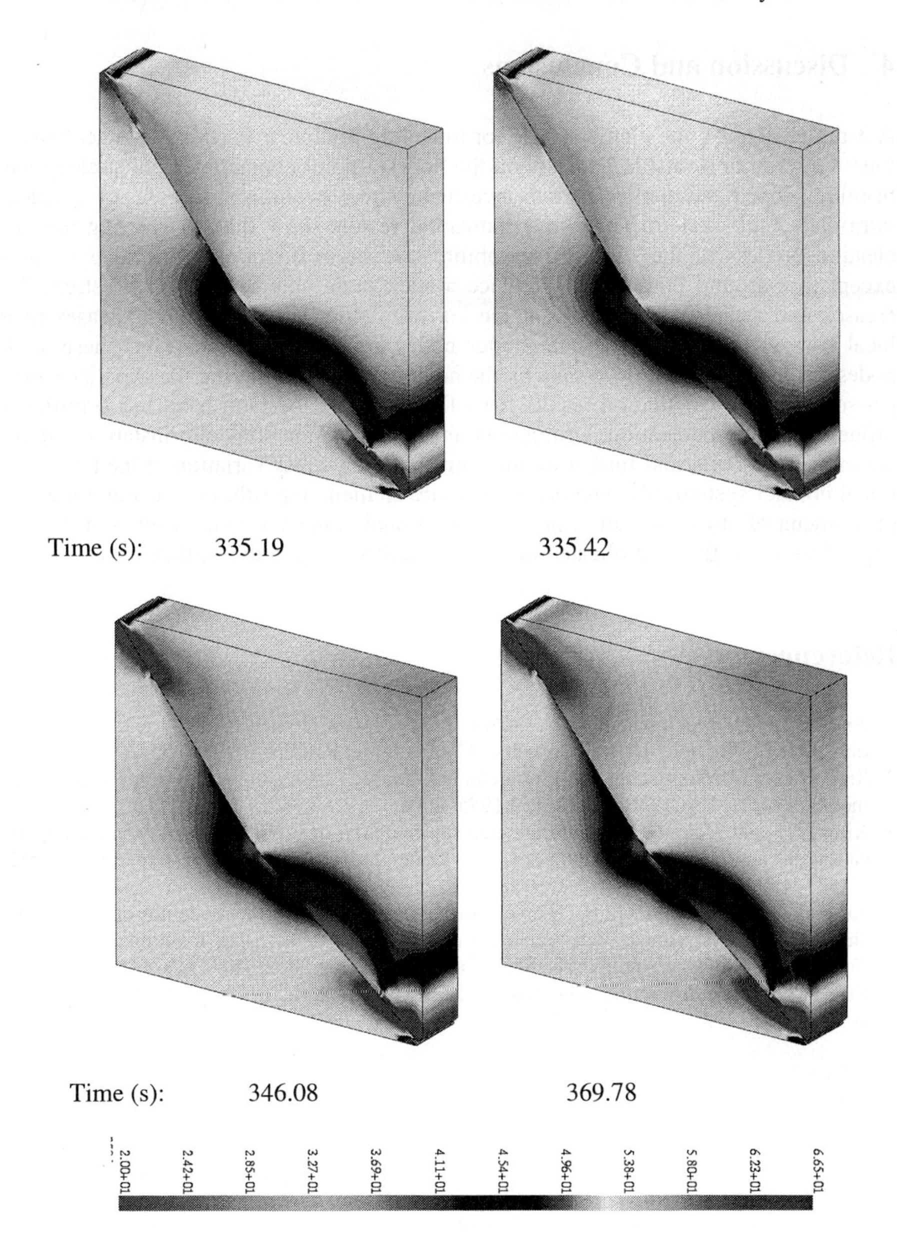

Fig. 3. (continued)

4 Discussion and Conclusions

A 3-dimensional finite element code for modeling nonlinear frictional contact behaviours between deformable bodies with the node-to-point contact element strategy has been developed and applied here to investigate stress evolution processes of a typical intra-plate fault system. The above numerical results show that: (1) During the nucleation process of the stick-slip instability, the stress distribution is rather smooth except that around both ends; (2) Once a stick node enters the slip, the stress decreases and redistributes locally at the corresponding slip zone, but increases in a local narrow zone around its neighbour nodes being in the stick. Thus these stick nodes are 'pushed' to the slip state by the neighbour nodes entering the slip. (3). Comparison with the calculation results for a flat fault [2], the fault bend has significant influence on the nucleation, termination and restart of the stick-slip instability along the intra-plate fault, and further on the corresponding stress variation of the total frictional contact system. (4). The proposed finite element algorithm can capture the key phenomena of stress evolution in such a frictional contact system easily and will be applied to the further complicated interacting fault system in the earthquake research.

References

1. Xing H.L. and Makinouchi A.: A node-to-point contact element strategy and its applications. RIKEN Review: High Performance Computing in RIKEN 2000; 30:35–39
2. Xing H.L. and Makinouchi A.: Finite element analysis of a sandwich friction experiment model of rocks, PAGEOPH 2002; 159:1985–2009
3. Xing H.L. and Makinouchi A.: Finite element modeling of multiboby contact and its application to active faults, Concurrency and Computation: Practice and Experience 2002; 14:431–450
4. Xing H.L. and Makinouchi A.: Three dimensional finite element modeling of thermomechanical frictional contact between finite deformation bodies using R-minimum strategy, Computer Methods in Applied Mechanics and Engineering 2002; 191:4193–4214.
5. Ruina A.L.: Slip instability and state variable friction laws, J. Geophys. Res. 1983; 88: 10359–10370

Transparent Boundary Conditions for Wave Propagation on Unbounded Domains

Dorin-Cezar Ionescu[1] and Heiner Igel[2]

[1] Queensland University Advanced Centre for Earthquake Studies (QUAKES)
The University of Queensland, Brisbane, QLD 4072, Australia
[2] Department für Geo- und Umweltwissenschaften, Ludwig-Maximilians-Universität
Theresienstr. 41, 80333 München, Germany

Abstract. The numerical solution of the time dependent wave equation in an unbounded domain generally leads to a truncation of this domain, which requires the introduction of an artificial boundary with associated boundary conditions. Such nonreflecting conditions ensure the equivalence between the solution of the original problem in the unbounded region and the solution inside the artificial boundary. We consider the acoustic wave equation and derive exact transparent boundary conditions that are local in time and can be directly used in explicit methods. These conditions annihilate wave harmonics up to a given order on a spherical artificial boundary, and we show how to combine the derived boundary condition with a finite difference method. The analysis is complemented by a numerical example in two spatial dimensions that illustrates the usefulness and accuracy of transparent boundary conditions.

1 Introduction

Modern trends in the development of numerical methods lead to higher and higher requirements for computational accuracy. The numerical solution of the wave equation on unbounded domains requires a truncation to fit the infinite region on a finite computer. Minimizing the amount of spurious reflections requires in many cases the introduction of an artificial boundary and of associated boundary conditions. The critical importance of these techniques becomes particularly evident when one considers that the gains made in the computational domain by using sophisticated high-order numerical approaches may vanish to a large extent as result of violating the conditions at the artificial boundary.

Despite the computational speed of finite difference schemes and the robustness of finite elements in handling complex geometries the resulting numerical error consists of two independent contributions: the discretization error of the numerical method used and the spurious reflection generated at the artificial boundary. This spurious contribution travels back and substantially degrades the accuracy of the solution everywhere in the computational domain. Unless both error components are reduced systematically, the numerical solution does not converge to the solution of the original problem in the infinite region.

There are various techniques for the approximate handling of boundary conditions at the external boundary of a finite domain constructed from the original

P.M.A. Sloot et al. (Eds.): ICCS 2003, LNCS 2659, pp. 807–816, 2003.

unbounded domain by means of truncation. One class of conditions is given by local differential operators [2], [3], including conditions that perfectly annihilate impinging waves at a finite number of *selected* angles of incidence [4], whilst a different approach is based on absorbing layers [5]. There are cases, where some of the difficulties with boundary conditions may be avoided partially by using a momentum space approach [6]. In contrast to grid methods in coordinate space where continuum waves spread over the entire space, in momentum space the waves are confined to a small finite volume and the dynamics stays localized around the origin at all times.

Exact nonreflecting boundary conditions that are *nonlocal* in both space and time have been investigated numerically in [7]. Numerical methods based on these exact conditions display a long-time instability, and a major disadvantage is related to the nonlocal character of the boundary condition in time. Due to the temporal nonlocality that requires information from previous time steps, these methods require a considerably longer computational time than explicit schemes for the wave equation. Recently, very important progress has been made by Grote and Keller [8] by deriving exact nonreflecting boundary conditions that are *local* in time.

In the present work we study exact transparent boundary conditions that are local in time for the scalar wave equation for the general case in two and three spatial dimensions. In contrast to [8] where an integral equation is used, our approach is based on the separation of variables combined with recurrence relations that provide a very direct derivation of the boundary condition. Since the derived condition is local in time and is equivalent to the result obtained in [8], our formulation complements the integral transform approach and is expected to have applications to cases where the latter method is difficult to apply.

The local character in time of the boundary condition and its explicit representation that requires only first order derivatives of the solution makes it relatively easy to apply in calculations based on explicit schemes. In contrast to earlier studies based on exact but nonlocal boundary conditions, the present finite differences implementation does not require a significant increase in the computational time. This highlights a key point in practical applications of such conditions in that the related implementations should not become computationally too expensive. In Sec. 2, we illustrate the fundamental ideas underlying the derivation of nonreflecting boundary conditions for the one-dimensional case and present the extension to higher dimensions. In Sec. 3 we discuss the implementation in a finite differences scheme and present a simple two dimensional numerical example that illustrates the usefulness of transparent boundary conditions. The conclusions of the present study are given in Sec. 4.

2 Theoretical Approach

2.1 The One-Dimensional Wave Equation

We consider the one-dimensional wave equation describing the propagation of perturbations along the positive real axis ($x \geq 0, t \geq 0$) with velocity $c = 1$ that are induced by a general and possibly nonlinear forcing term $f = f(x, t, \Phi, \partial_x \Phi)$

$$(\partial_t^2 - \partial_x^2)\,\Phi(x,t) = f \ , \tag{1}$$

where $\Phi(x,t)$ represents the displacement of an infinitely long string and $\partial_t = \partial/\partial t$. Upon requiring $\Phi(0,t) = 0$ for the state at rest we assume $\Phi(x,t)$ to describe the position of a string fixed at the origin. We define the initial conditions by the string position and velocity at $t = 0$ by $\Phi(x,0) = U_0$ and $\partial_t \Phi(x,t)\mid_{t=0} = V_0$.

The local character of the problem is defined by assuming $f = 0$ for $x \geq L$, $\forall\ t \geq 0$. Thus, the positive x-axis separates into two distinctly different regions: the bounded (interior) domain $x \leq L$, and the unbounded (exterior) region $x \geq L$ where the forcing term f vanishes. The two regions are separated by the artificial boundary at $x = L$.
To find the exact absorbing boundary condition at $x = L$ it is useful to separate outgoing from incoming waves by defining

$$v = \partial_t \Phi + \partial_x \Phi, \quad \text{and} \quad w = \partial_t \Phi - \partial_x \Phi \ . \tag{2}$$

Since $\Phi(x,t)$ is a solution of Eq. (1) for $x \geq L$, i.e. $(\partial_t^2 - \partial_x^2)\Phi = 0$, in the exterior region from (2)

$$\begin{aligned} 0 &= (\partial_t - \partial_x)\,[(\partial_t + \partial_x)\Phi] = (\partial_t - \partial_x)v \ , \\ 0 &= (\partial_t + \partial_x)\,[(\partial_t - \partial_x)\Phi] = (\partial_t + \partial_x)w \ . \end{aligned} \tag{3}$$

This system of first-order equations has the general solution

$$v(x,t) = \psi(x+t), \quad \text{and} \quad w(x,t) = \varphi(x-t) \ , \tag{4}$$

where ψ and φ are arbitrary functions that are determined by initial and boundary conditions. Thus, incoming (v) and outgoing (w) waves are defined as

$$\begin{aligned} v(x,t) &= \text{const., for } x + t = \text{const.} \quad \text{(incoming)} \\ w(x,t) &= \text{const., for } x - t = \text{const.} \quad \text{(outgoing)} \ . \end{aligned} \tag{5}$$

Since there are no incoming waves in the exterior region, it follows $v(L,t) = 0$. The requirement of a purely outgoing wave for $x \geq L$ combined with the definition for incoming waves v from eq. (2) yields the exact nonreflecting boundary condition for the displacement $\Phi(x,t)$

$$(\partial_t + \partial_x)\,\Phi(x,t)\mid_{x=L} = 0 \ . \tag{6}$$

This expression which is local in time guarantees that the artificial boundary at $x = L$ is perfectly transparent to both incoming and outgoing waves as they leave the interior region without any spurious reflection. Note that since the derivation of the exact boundary condition (6) depends solely on properties in the exterior domain, $x \geq L$, the problem inside the computational volume can be arbitrarily complex.

2.2 Transparent Boundary Conditions in Higher Dimensions

The derivation of exact absorbing boundary conditions in higher dimensions is considerably more challenging as compared to the one dimensional case discussed previously. Distinctly different from the one-dimensional case where waves can propagate in two directions only, in two and more dimensions waves propagate in infinitely many directions.

In the following we consider wave propagation in an unbounded region $\mathbb{R}^3$ and surround the computational region $\tilde{\Omega}$ containing the forcing term f by an artificial boundary S that is assumed to be the surface of a sphere with radius R. In the exterior domain one has $f = 0$, and the wave function $\psi(\mathbf{r}, t)$ satisfies the homogeneous wave equation with propagation velocity $c > 0$, i.e.

$$\left(\frac{1}{c^2}\,\partial_t^2 \;-\; \Delta\right)\psi(\mathbf{r},t) \;=\; 0 \;\text{ in }\; \mathbb{R}^3 \setminus \tilde{\Omega} \;, \tag{7}$$

with initial conditions $\psi(\mathbf{r}, 0) = 0$ and $\partial_t \psi(\mathbf{r}, 0) = 0$. It is useful to separate the variables by expanding the solution in spherical coordinates r, ϑ, φ

$$\psi(\mathbf{r},t) \;=\; \sum_{l=0}^{\infty} \sum_{m=-l}^{l} \psi_{l,m}(r,t)\, Y_{l,m}(\vartheta,\varphi) \;, \tag{8}$$

where the spherical harmonics

$$Y_{l,m}(\vartheta,\varphi) \;=\; \sqrt{\frac{(2l\,+\,1)\,(l\,-\,|m|)!}{4\pi\,(l\,+\,|m|)!}}\; P_l^{|m|}(\cos\vartheta)\; e^{im\varphi} \;, \tag{9}$$

are orthonormal and the functions $P_l^{|m|}$ are Legendre polynomials. Using the orthonormality properties of the $Y_{l,m}$ the radial time-dependent functions $\psi_{l,m}(r, t)$ may be written as

$$\psi_{l,m}(r,t) \;=\; \int_0^{\pi} d\vartheta\, \sin\vartheta \int_0^{2\pi} d\varphi\; Y_{l,m}^*(\vartheta,\varphi)\; \psi(r,\vartheta,\varphi,t) \;. \tag{10}$$

By inserting expression (8) in eq. (7) one obtains the radial equation

$$\left[\frac{1}{c^2}\,\partial_t^2 \;-\; \partial_r^2 \;-\; \frac{2}{r}\,\partial_r \;+\; \frac{l(l+1)}{r^2}\right]\psi_{l,m}(r,t) \;=\; 0 \;, \tag{11}$$

with the initial conditions $\psi_{l,m}(r, 0) = 0$ and $\partial_t \psi_{l,m}(r, 0) = 0$. The differential operator in the square brackets which we denote by R_l, satisfies a remarkable commutation-like relation $R_l\,[\partial_r - (l-1)/r] = [\partial_r - (l-1)/r]\,R_{l-1}$, or

$$R_l\left(\partial_r \;-\; \frac{l-1}{r}\right)\phi_{l-1}(r,t) \;=\; \left(\partial_r \;-\; \frac{l-1}{r}\right) R_{l-1}\,\phi_{l-1}(r,t) \;=\; 0\,, \tag{12}$$

whenever $R_{l-1}\,\phi_{l-1} = 0$. Thus, one obtains a solution of the l-th order equation $R_l\,\phi_l = 0$ from a $(l-1)$th order solution according to the recurrence relation

$$\phi_l(r,t) \;=\; \left[\partial_r \;-\; \frac{l\,-\,1}{r}\right]\phi_{l-1}(r,t) \;, \tag{13}$$

that can be also obtained by use of properties of spherical Bessel functions [10]. Recursive use of (13) yields a representation for the radial functions $\psi_{l,m}$

$$\psi_{l,m} = \left[\partial_r - \frac{l-1}{r}\right]\left[\partial_r - \frac{l-2}{r}\right]\phi_{l-2,m} = \ldots = \prod_{i=1}^{l}\left[\partial_r - \frac{i-1}{r}\right]\phi_{l,m} \quad (14)$$

where $\phi_{l,m}$ is a solution to the $l = 0$ version of eq. (11). It follows that the modified radial function $\Phi_{l,m}(r,t) = r\phi_{l,m}$ satisfies a simple one-dimensional wave equation, i.e.

$$\left(\frac{1}{c^2}\partial_t^2 - \partial_r^2\right)\Phi_{l,m}(r,t) = 0 \ . \quad (15)$$

As shown in Sec. 2.1, a general solution for *outgoing* waves is written as $\Phi(r-ct)$, such that for $l \geq 1$ the radial functions $\psi_{l,m}$ are expressed as

$$\psi_{l,m}(r,t) = \prod_{i=1}^{l}\left(\partial_r - \frac{i-1}{r}\right)\frac{1}{r}\Phi_{l,m}(r-ct) \ , \quad (16)$$

where the index denotes a radial wave function of order l. Recursive use of this relations enables one after some rearrangement to rewrite the l-th order radial function as a sum over l, i.e.

$$\begin{aligned}\psi_{l,m}(r,t) &= \sum_{i=0}^{l}\frac{(-)^i}{r^{i+1}}\rho_{l,i}\frac{\partial^{l-i}}{\partial r^{l-i}}\Phi_{l,m}(r-ct)\\ &= (-)^l\sum_{i=0}^{l}\frac{1}{c^{l-i}r^{i+1}}\rho_{l,i}\frac{\partial^{l-i}}{\partial t^{l-i}}\Phi_{l,m}(r-ct) \ , \end{aligned} \quad (17)$$

where $\rho_{l,i} = (l+i)!/[2^i\, i!\,(l-i)!]$. These coefficients can be obtained by using induction in eq. (17) to obtain recurrence relations for $\rho_{l,i}$, and we replaced the spatial derivative with a time derivative using

$$(-)^k\frac{\partial^k}{\partial r^k}\Phi_{l,m}(r-ct) = \frac{1}{c^k}\frac{\partial^k}{\partial t^k}\Phi_{l,m}(r-ct) \ . \quad (18)$$

In analogy with ref. [8] we now replace the radial derivative with a time derivative by applying the operator B_1 on the radial function

$$B_1\,\psi_{l,m} = \left[\partial_r + \frac{1}{c}\partial_t + \frac{1}{r}\right]\psi_{l,m} = \frac{(-)^{l+1}}{r}\sum_{i=0}^{l}\frac{i\,\rho_{l,i}}{c^{l-i}r^{i+1}}\frac{\partial^{l-i}}{\partial t^{l-i}}\Phi_{l,m}(r-ct), \quad (19)$$

where we used eq. (17). Finally, using expansion (8) and multiplying the last expression by Y_{lm} with summation over l and m, we obtain

$$B_1\,\psi(R,\vartheta,\varphi,t) = -\frac{1}{R}\sum_{l,m}Y_{l,m}(\vartheta,\varphi)\sum_{i=1}^{l}\frac{(-)^l i\rho_{li}}{c^{l-i}R^{i+1}}\frac{\partial^{l-i}}{\partial t^{l-i}}\Phi_{l,m}(R-ct) \ . \quad (20)$$

In the general case, the functions $\Phi_{l,m}$ are obtained by evaluating Eq. (17) at $r = R$. Since $\rho_{l0} = 1$ this leads to the solution of a linear differential equation

$$\frac{1}{c^l}\frac{d^l}{dt^l}\,\tilde{\Phi}_{l,m}(t) \;=\; -\sum_{i=1}^{l} \frac{\rho_{li}}{c^{l-i}R^i}\,\frac{d^{l-i}}{dt^{l-i}}\,\tilde{\Phi}_{l,m}(R-ct) \;+\; \psi_{l,m}(R,t)\,, \qquad (21)$$

where we have substituted $\tilde{\Phi}_{l,m} = (-)^l \Phi_{l,m}/R$ and the inhomogeneous term $\psi_{l,m}(R,t)$ is given by Eq. (10) evaluated at $r = R$.

Expression (20) represents the exact nonreflecting boundary condition in the form obtained in [8] where integral transforms formed the basis of the approach. Note that in practical calculations truncation of the summation over l at a finite value $l = L$ leads to an exact representation of modes with $l \leq L$. Thus, the boundary condition reduces to $B_1\psi|_{r=R} = 0$ for harmonic modes with $l' > L$. In particular, $B_1\psi|_{r=R} = 0$ is an exact boundary condition for spherically symmetric modes ($l = 0$).

3 Numerical Results Using Finite Differences

Using the results of the previous section, we illustrate the use of transparent boundary conditions and their numerical implementation using a finite difference formulation. The wave equation is discretized both in space and time using centred finite differences. At time $t_k = k\,\Delta t$, we denote by $\Psi^k(n)$ the numerical approximation to the time dependent wave function $\psi(\mathbf{r},t)$ and by $f^k(n)$ the forcing term at the n-th grid point r_n in radial direction [8]. The numerical solution is advanced in time using

$$\Psi^{k+1}(n) \;=\; 2\,\Psi^k(n) \;-\; \Psi^{k-1}(n) \;+\; (\Delta t)^2\left[\mathcal{D}\,\Psi^k(n) \;+\; f^k(n)\right]\;, \qquad (22)$$

where $\mathcal{D}$ represents a finite difference approximation to the Laplace operator Δ.

An apparent complication occurs when the Laplace operator is to be calculated at the outer most radial grid point, $r_n = R$, since this calculation uses values of $\Psi^k(n+1)$ belonging to the exterior region. However this problem can be solved by using the explicit representation of the derived boundary condition. More precisely, one obtains an additional relation between the quantities $\Psi^{k+1}(n)$ and $\Psi^k(n+1)$ by using a finite difference representation of the boundary condition equation (20) at $r_n = R$. Consequently, the problem is solved by coupling the two equations for $\Psi^{k+1}(n)$ and $\Psi^k(n+1)$, allowing one to solve for $\Psi^{k+1}(n)$. Due to the local character in time of the boundary condition, both the differential equation and the boundary condition are discretized in time about $t = t_k$, and only function values at $t = t_k$ are needed in a given time step [8].

This procedure becomes particularly simple and is best illustrated using a cartesian grid $x_n = n\Delta x$ in one dimension. The second space derivative ∂_x^2 is written as

$$\mathcal{D}\,\Psi^k(n) \;=\; \frac{\Psi^k(n+1) \;-\; 2\,\Psi^k(n) \;+\Psi^k(n-1)}{(\Delta x)^2}\;, \qquad (23)$$

and using the finite difference representation of the boundary condition we arrive after some algebra at

$$\Psi^k(n+1) \quad = \quad -\frac{1}{c}\frac{\Delta x}{\Delta t}\left[\Psi^{k+1}(n) \; - \; \Psi^{k-1}(n)\right] \; + \; \Psi^k(n-1) \; , \qquad (24)$$

which represents the additional equation for $\Psi^{k+1}(n)$. By combining the two equations for $\Psi^{k+1}(n)$ and $\Psi^k(n+1)$ one finds

$$\Psi^{k+1}(n) \quad = \quad \frac{2\,\Psi^k(n) \; + \; (\alpha-1)\Psi^{k-1}(n) \; + \; 2\,\alpha^2\left[\Psi^k(n-1) \; - \; \Psi^k(n)\right]}{1 \; + \; \alpha} \qquad (25)$$

where $\alpha = c\Delta t/\Delta x$. This expression clearly shows that the evaluation of the time extrapolated function $\Psi^{k+1}(n)$ at the boundary does not depend on function values located outside the computational domain.

Using the approach described above we are now in a position to analyse the time evolution of perturbations using transparent boundary conditions. In Figures 1 and 2 we display snapshots of the two-dimensional wave function at different times for $t = 82, 128, 270, 450$, and 900, respectively. The contour plots on the right hand side of the figures display the position of time evolved waves in the xy-plane, where the computational domain extends from an inner radius $r_< = 1000$ km to the outer radius $r_> = 6371$ km, and we used $\Delta t = 2.5$ s, $dr \simeq 3.5$ km, and $c = 5$ km/s. The plots on the left show the dependence of the associated wave function on the radial coordinate r. While the starting point of the calculation is $t = 0$, the perturbing source is assumed to be proportional to a Gaussian $\exp(t-t_0)^2$ and is located on a circle with $r_0 \simeq 4700$ km. This may be associated with an explosive point source that is located at the origin $r_< = r_0 \rightarrow 0$ in the far past. For $t = 82$ we observe a strong peak located in the vicinity of the source. In the contour plot on the right hand side, one sees that the wave is located in the area between the two bright circles.

For larger times, at $t = 128$, the wave separates in two independent contributions propagating in opposite directions along the radial coordinate r. While one of the waves moves outwards towards increasing r-values the other wave propagates inwards, as indicated by the two arrows in the left figure. Note that the amplitude decreases in magnitude with increasing r-values as a result of larger surface elements $r\,dr\,d\varphi$. At t $= 270$, it is seen that only the ingoing wave can be found inside the computational region as the outgoing one passes the artificial boundary ar $r = R$ without any reflection. On the other hand, the ingoing wave propagating towards smaller r-values changes its sign and the direction of propagation after encountering the margin $r_<$ of the inner circle. As a result, for larger times ($t = 450$) this wave propagates towards the margin of the computational domain but with opposite sign. Finally, for even larger times this wave passes the artificial boundary at $r = R$ without any reflection. Thus, for $t = 900$ the computational region is seen to be completely unperturbed and, as a result of the boundary condition, the artificial boundary appears perfectly transparent to the wave as there is no spurious reflection.

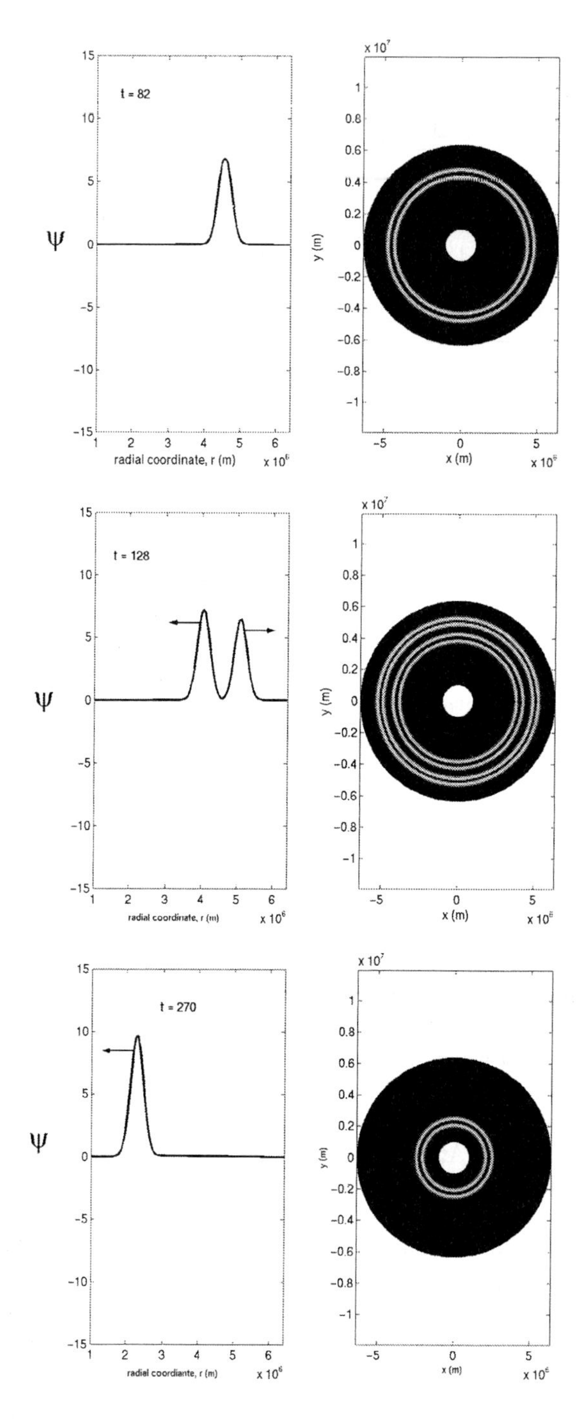

Fig. 1. Snapshots of the time evolved wave function obtained by the numerical solution of the wave equation incorporating the nonreflecting boundary condition for $t = 82, 128$, and $t = 270$. The initial wave separates in two parts propagating in opposite directions.

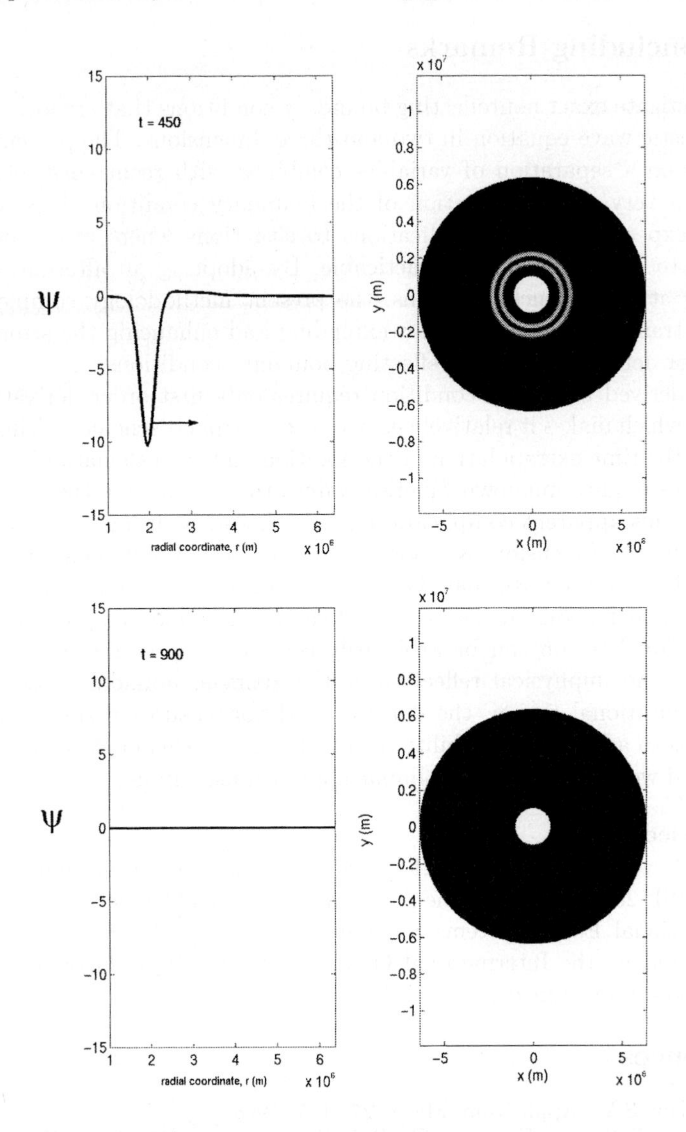

Fig. 2. The same as in Fig. 1 for larger times $t = 450$, and 900. At large asymptotic times both components leave the computational domain and no reflection at the artificial boundary is observed.

4 Concluding Remarks

We investigate exact nonreflecting boundary conditions that are local in time for the acoustic wave equation in two and three dimensions. The present approach is based on a separation of variables combined with recurrence relations that provide a very direct derivation of the boundary condition. This formulation can be expected to have applications to situations where other methods are difficult to apply or even impracticable. By adopting an alternative point of view for attacking such problems, the present methodology complements the integral transform approach, thus extending and enhancing the strength of the theory for deriving exact nonreflecting boundary conditions.

The derived boundary condition requires only first order derivatives of the solution which makes it relatively easy to use in explicit schemes. Using finite differences the time extrapolation of the solution and the calculation of the spatial derivatives require unknown function values that lie outside the computational domain. This apparent complication is solved, and by using a simple numerical example in two dimensions we show how these exterior values can be eliminated.

Finally, we emphasize that the derivation of the boundary condition depends only on the behaviour in the exterior domain, such that the problem inside the computational region can be arbitrarily complex, e.g. nonlinear. Furthermore, as there is no unphysical reflection at the artificial boundary associated with the computational region, the derived condition ensures perfect transparency that leads to a long-time stability of the numerical scheme. After this work was completed we learned about a similar approach used in [9].

Acknowledgement.
This work was supported in part by the University of Queensland under Grant No. UQRSF 2002001336, by the Australian Research Council, by the Australian Computational Earth Systems Simulator (ACcESS) Major National Research Facility, and by the International Quality Network (IQN) Georisk Program at the University of München.

References

1. Tsynkov, S.V.: Appl. Num. Math. **27**, 465 (1998).
2. Clayton, R.W. and Engquist, B.: Bull. Sei. Soc. Am. **67**, 1529 (1977).
3. Engquist, B. and Majda, A.: Commun. Pure Appl. Math. **32**, 313 (1979).
4. Higdon, R.L.: SIAM J. Num. Anal. **27**, 831 (1990).
5. Israeli, M. and Orszag, S.A.: J. Comput. Phys. **41**, 115 (1981).
6. Ionescu, D.C. and Belkacem, A.: Eur. Phys. Journal D **18**, 301 (2002); see also Phys. Scripta **T80**, 128 (1999).
7. Givoli, D. and Cohen, D.: J. Comput. Phys. **117**, 102 (1995).
8. Grote, M.J. and Keller, J.B.: SIAM J. Appl. Math. **55**, 280 (1995); and **60**, 803 (2000); see also J. Comput. Phys. **139**, 327 (1998).
9. Thompson, L. and Huan, R.: submitted to Comp. Meth. Appl. Mech. Eng. (2003).
10. Abramowitz, M. and Stegun, I.A.: *Handbook of Mathematical Functions* (Harri Deutsch, Thun, Frankfurt am Main, 1984).

A 2D Numerical Model for Simulating the Physics of Fault Systems

Peter Mora[1,2] and Dion Weatherley[1]

[1]Queensland University Advanced Centre for Earthquake Studies (QUAKES)
The University of Queensland, Brisbane, QLD 4072, Australia
[2]The Australian Computational Earth Systems Simulator
Major National Research Facility (ACcESS MNRF)
The University of Queensland, Brisbane, QLD 4072, Australia

Abstract. Simulations provide a powerful means to help gain the understanding of crustal fault system physics required to progress towards the goal of earthquake forecasting. Cellular Automata are efficient enough to probe system dynamics but their simplifications render interpretations questionable. In contrast, sophisticated elasto-dynamic models yield more convincing results but are too computationally demanding to explore phase space. To help bridge this gap, we develop a simple 2D elasto-dynamic model of parallel fault systems. The model is discretised onto a triangular lattice and faults are specified as split nodes along horizontal rows in the lattice. A simple numerical approach is presented for calculating the forces at medium and split nodes such that general nonlinear frictional constitutive relations can be modeled along faults. Single and multi-fault simulation examples are presented using a nonlinear frictional relation that is slip and slip-rate dependent in order to illustrate the model.

1 Introduction

Numerical shear experiments of granular regions have exhibited accelerating energy release in the lead-up to large events[3] and a growth in correlation lengths in the stress field[4]. While these results suggest a Critical Point-like mechanism in elasto-dynamic systems and the possibility of earthquake forecasting, they do not prove that such a mechanism occurs in the crust. Cellular Automaton (CA) models exhibit accelerating energy release prior to large events or unpredictable behaviour in which large events may occur at any time, depending on tuning parameters such as the dissipation factor and stress transfer ratio[6]. The mean stress plots of the granular simulations are most similar to the CA mean stress plots near the boundary of the predictable and unpredictable regimes suggesting that elasto-dynamic systems may be close to the borderline between the predictable and unpredictable. To progress in resolving the question of whether more realistic fault system models exhibit predictable behaviour and to determine whether they also have an unpredictable and predictable regime depending on tuning parameters as seen in CA simulations, we develop a 2D elasto-dynamic model of parallel interacting faults.

P.M.A. Sloot et al. (Eds.): ICCS 2003, LNCS 2659, pp. 817–826, 2003.

Preliminary work using the model[5] to simulate the dynamics of multiple parallel interacting faults have been performed and indicate, through calculations of the so called "inverse metric" (see [1]), that the system is non-ergodic. This has major implications to the analysis of crustal fault systems within a statistical physics framework. This work has also shown that such fault models exhibit so called "glassy" behaviour which implies that mean field theoretical analysis such as [1] require revision to introduce a memory kernel.

The elasto-dynamic parallel interacting fault model developed here may help provide a crucial link between CA maps of phase space and the behaviour of more realistic elasto-dynamic interacting fault system models, and thus, a means to improve understanding of the dynamics and predictability of real fault systems.

2 Numerical Model

The numerical model consists of a 2D triangular lattice of masses each of which is connected to its six nearest neighbours by a linear spring. This discretisation yields isotropic elasticity with compressional and shear wave speeds related by $V_s = V_p/\sqrt{3}$[2] which is a typical value for rocks. This simple discretisation allows elasto-dynamics to be simulated relatively efficiently, albeit with the restriction of only one Poisson's ratio. Furthermore, horizontal faults can be easily specified in the model by splitting masses in half along a horizontal row, and simulating the frictional interaction of split masses with one another. Henceforth, we will refer to the masses as nodes of the discrete lattice. In the following, $\alpha = 0, \ldots, 5$ is used to index the six lattice directions.

3 Calculation of Forces

The force on masses at medium nodes and at fault (split) nodes is calculated as

$$\mathbf{F}_i = \begin{cases} \mathbf{F}_i^e + \mathbf{F}_i^\nu + \mathbf{F}_i^T & , \quad i \in M \\ \mathbf{F}_i^f + \mathbf{F}_i^\nu + \mathbf{F}_i^T & , \quad i \in F^+ \ or \ i \in F^- \end{cases} , \tag{1}$$

where subscript i is used to denote mass (node) number i, $\mathbf{F}_i^T$ is a term representing "thermal noise" in the system, $\mathbf{F}_i^e$ is the elastic force due to springs connected to node i, and $\mathbf{F}_i^f$ is the sum of the elastic and frictional forces acting on the split nodes along faults.

3.1 Elastic Forces

The elastic forces are calculated by summing the elastic forces due to connected springs

$$\mathbf{F}_i^e = \begin{cases} \sum_{\alpha=0}^{5} k(d^\beta - d_0)\mathbf{e}^\beta & , \quad i \in M \\ (\mathbf{F}_i^e)^+ & , \quad i \in F^+ \\ (\mathbf{F}_i^e)^- & , \quad i \in F^- \end{cases} , \tag{2}$$

where $\beta = mod(\alpha, 6)$ is the lattice direction number, k is the spring constant, d^β is the distance to the mass linked in direction β, d_0 is the equilibrium separation or lattice spacing, M denotes the set of medium nodes, F^+ denotes the set of upper fault surface split nodes, and F^- denotes the set of lower fault surface split nodes. In the above equation, $(\mathbf{F}_i^e)^+$ is the force on the upper split nodes due to linked nodes above the fault, and $(\mathbf{F}_i^e)^-$ is the force of lower split nodes due to linked nodes below the fault, namely

$$(\mathbf{F}_i^e)^+ = \sum_{\alpha=0}^{3} k^\beta (d^\beta - d_0)\mathbf{e}^\beta \quad i \in F^+$$

$$(\mathbf{F}_i^e)^- = \sum_{\alpha=3}^{6} k^\beta (d^\beta - d_0)\mathbf{e}^\beta \quad i \in F^- \quad , \tag{3}$$

where

$$k^\alpha = \begin{cases} k^\alpha = k & , \quad \alpha = 1, 2, 4 \text{ } or \text{ } 5 \\ k^\alpha = k/2 & , \quad \alpha = 0, 3 \end{cases} \quad , \tag{4}$$

specifies spring constants connected to split nodes (i.e. spring constants in the medium are k whereas the horizontal springs along the fault are split in two so $k^\alpha = k/2$ for $\alpha = 0, 3$). As such, the total elastic force of split node pairs moving as a single mass m in unison (i.e. when split nodes are in static frictional contact) is the sum of the elastic forces on the split node pair $\mathbf{F}_i^e = (\mathbf{F}_i^e)^+ + (\mathbf{F}_i^e)^-$ and yields the same expression for force as for the medium nodes specified by Equation (2).

3.2 Viscous Forces

In order to damp energy in the closed elastic system, an artificial viscosity is introduced that attenuates elastic waves as $\exp(-\gamma t)$ where γ is the frequency independent attenuation coefficient which is related to the viscosity coefficient by $\gamma = \nu/2m$ (e.g. see [2]). The viscous force is given by

$$\mathbf{F}_i^\nu = -\nu(v_i - v_0) \quad , \tag{5}$$

where ν is the viscosity coefficient, v_i is the velocity of node i, and v_0 is a specified reference velocity (e.g. in constant strain rate shear experiments, we set v_0 to the velocity of a homogeneous elastic system undergoing shear). To have an attenuation coefficient γ that is uniform in space, the viscosity ν must be set to $\nu = 2m\gamma$. Hence, for a homogeneous medium, the viscosity coefficients at split nodes are half as large as at medium nodes.

3.3 Thermal Noise

Recent research[1, 4–6] suggests that fault system models may be understood using concepts developed in statistical physics and that their dynamics may have similarities to classical Critical-Point systems. A key parameter in such systems is the temperature. As such, we introduce a thermal noise term to provide a means to study the statistical physics of the system. This is achieved by adding a random forcing term at each time step

$$\mathbf{F}_i^T = \begin{cases} N_i & , \quad i \in M \\ N_i/2 & , \quad i \in F^+ \text{ or } i \in F^- \end{cases} , \tag{6}$$

where the magnitude of N_i relates to the effective temperature of the system. The factor of a half for the second case is because the noise – which is assumed to have uniform statistics in space – is shared equally by each of the split nodes on faults. This term models random time dependent fluctuations in stress within the earth due to seismic background noise (distant earthquakes, earth tides, human noise, wind and ocean noise, etc).

3.4 Vertical Component of Elastic Forces on Split Nodes

The term $\mathbf{F}^f$ in Equation (1) represents the force on the split nodes due to medium elasticity and the friction. When split nodes are in contact, half of the vertical elastic force due to linked springs is applied to the split node itself and half is applied to the touching split node

$$F_y^f = \left[(F_y^e)^+ + (F_y^e)^-\right]/2 \quad . \tag{7}$$

Hence, considering the mass of split nodes is $m/2$, split nodes in contact accelerate in unison with vertical acceleration $\left[(F_y^e)^+ + (F_y^e)\right]/m$ (i.e. the same vertical acceleration as a medium node at the split node location linked to the six neighboring nodes). When split nodes move out of contact (i.e. when $(F_y^e)^+ - (F_y^e)^- > 0$, there is no interaction between the split nodes so the vertical force on each split node is the elastic force due to its linked springs. Summarising the above, the vertical elastic force on split nodes is given by

$$F_y^f = \begin{cases} \left[(F_y^e)^+ + (F_y^e)^-\right]/2 & , \quad \left[(F_y^e)^+ - (F_y^e)^-\right] \le 0 \\ F_y^e & , \quad \left[(F_y^e)^+ - (F_y^e)^-\right] > 0 \end{cases} . \tag{8}$$

3.5 Horizontal Component of Frictional and Elastic Forces on Split Nodes

The horizontal force on a split node is the sum of the horizontal elastic force due to linked springs and a frictional force f

$$F_x^f = F_x^e + f \quad , \tag{9}$$

where F_x^e is the horizontal elastic force on the split node being considered given by Equation (3). In the case of a fault in static frictional contact, the friction is such that split nodes accelerate horizontally in unison so

$$f = \left[(F_x^e)' - F_x^e\right]/2 \quad , \tag{10}$$

where the $'$ is used to denote the other split node (ie. if we are calculating f for the upper split node, then the $'$ signifies the lower split node and vice versa). Hence, for the case of static frictional contact, the split nodes both accelerate with a horizontal acceleration $[(F_x^e)^+ + (F_x^e)^-]/m$ which is equal to the acceleration that would be calculated for a medium node of mass m that replaces the split node pair (c.f. case for vertical forces). When the elastic force is sufficiently great to overcome static friction, the split nodes will begin to slip. Therefore, we can write

$$|f| = \min(|[(F_x^e)' - F_x^e]/2|, |\tau|) \quad , \tag{11}$$

where τ is a function that prescribes the frictional constitutive relation with $\tau^+ = -\tau^-$ (i.e. the friction is equal in magnitude and opposite in direction on the upper and lower fault surfaces). Hence, the friction can be written as

$$f = \begin{cases} [(F_x^e)' - F_x^e]/2 & , \quad |[(F_x^e)' - F_x^e]/2| < |\tau| \\ \tau & , \quad \textit{otherwise} \end{cases} \tag{12}$$

The sign of τ is such that it opposes slip between the nodes. As such, we can write

$$sgn(\tau) = \begin{cases} [(F_x^e)' - F_x^e]/|(F_x^e)' - F_x^e| & , \quad S(t-\Delta t) = \textit{stick} \\ (V_x' - V_x)/|V_x' - V_x| & , \quad S(t-\Delta t) = \textit{slip} \end{cases} \quad , \tag{13}$$

where $S(t-\Delta t)$ specifies the state of the split node at the previous time step $t-\Delta t$ and may be either *stick* or *slip*. The first case ensures the friction will oppose the new slip velocity of a split node pair that is beginning to slip whereas the second case opposes the slip velocity of an already slipping split node pair. Hence, in the upper case, the state changes from $S(t-\Delta t) = \textit{stick}$ to $S(t) = \textit{slip}$ whereas in the lower case, the state remains unchanged (i.e. $S(t) = S(t-\Delta t) = \textit{slip}$).

The magnitude of the friction is given by

$$|\tau| = \mu[(F_y^e)^- - (F_y^e)^+]/2 \quad , \tag{14}$$

(i.e. Coulomb friction) where the friction coefficient μ may be a function of dynamic variables such as slip or slip-rate. In the following examples, we will use a friction that is slip and slip-rate dependent as follows

$$\mu(t) = \begin{cases} \mu_d + (\mu_s - \mu_d)(1 - s(t)/D_c)^{p_1} & , \quad s < D_c \\ \mu_d + (\mu_s - \mu_d)(1 - \dot{s}(t - \Delta t/2)/V_c)^{p_2} & , \quad s \geq D_c \end{cases} \quad , \tag{15}$$

where μ_s and μ_d are respectively the static and dynamic friction coefficients, p_1 and p_2 are exponents that control the functional form of friction with slip and slip-rate, s is the amount of slip during a rupture event, $\dot{s}$ is the slip-rate, D_c is the "critical slip weakening distance" over which friction weakens to the dynamic value, and V_c is the slip-rate when slip reaches D_c. The slip-rate is calculated using centred finite differences from the positions of the nodes at the current and past time steps which is the reason why $\dot{s}$ is at time $(t - \Delta t/2)$. This frictional relation is slip weakening until the slip equals D_c and then remains at the dynamic value until the slip-rate drops below the value it attained when slip first exceeded D_c. As the slip rate drops further, the fault re-strengthens as a function of velocity and reaches the static friction once the slip-rate drops to zero. The second term allows the fault to heal after passage of a rupture front and yields slip-pulse behaviour of simulated earthquake ruptures consistent with observations, rather than less realistic crack-like ruptures which result from simple slip weakening relationships. Exponent p_1 controls the sharpness of the leading edge of the pulse and p_2 controls the sharpness of the trailing edge of the pulse. In the following examples, we will set $p_1 = 2$ and $p_2 = 1$ which yields a relatively symmetrical slip pulse.

4 Time Integration Scheme

A second order finite difference scheme is used to extrapolate the positions in time as follows

$$(u_i)_\ell(t + \Delta t) \;=\; 2(u_i)_\ell(t) - (u_i)_\ell(t - \Delta t) \;+\; \Delta t^2 (a_i)_\ell(t) \quad , \tag{16}$$

where $(u_i)_\ell$ is the ℓ-th component of the displacement at the i-th node and the acceleration $\mathbf{a}_i$ is calculated from the force given by Equation (1) as

$$\mathbf{a}_i \;=\; \mathbf{F}_i/m_i \quad , \tag{17}$$

where m_i is the mass of the i-th node (e.g. for a homogeneous medium, m would be constant except for split nodes which would have $m_i = m/2$). Once the new displacements are calculated, we can evaluate the new slip-rate at time $(t+\Delta t/2)$. When the slip-rate changes sign, i.e. when $sgn(\dot{s}(t-\Delta t/2)) = -sgn(\dot{s}(t+\Delta t/2)$, it is assumed that slip is stopping and the change in sign is caused by numerical overshoot due to the finiteness of the time step Δt. In this case, we set $S(t+\Delta t/2)$ to *stick*.

5 Numerical Examples

In all examples, we set $\Delta t = 0.4$, $d_0 = \Delta x = 1$, $k = 1$ and $m = 1$ and use lattice dimensions of $N_x = 100$ and $N_y = 101$. These parameters yield a P-wave speed of $V_p \approx 1$ (see [2]). In the single fault examples, a horizontal fault was centered in the model at $y = y_{50}$ where $y_n = n\Delta y$ denotes the y-ordinates of the n-th row of lattice nodes and row indices have the range $n = 0, \ldots N_y - 1$. In all cases, the vertical strain was fixed at $\varepsilon_{yy} = 0.02$ and viscosity was set to $\nu = 0.04$. Boundary conditions are circular in x and rigid in y, and the thermal noise N_i is set to 0.

5.1 Waves Traveling through a Locked Fault

A point source was excited below the fault with the aim of verifying that split nodes in static frictional contact behave identically as medium nodes. The snapshot shown in Figure 1 shows circular compressional and shear waves propagating through the locked fault. There are no artificial reflections, thus verifying the implementation for static frictional contact.

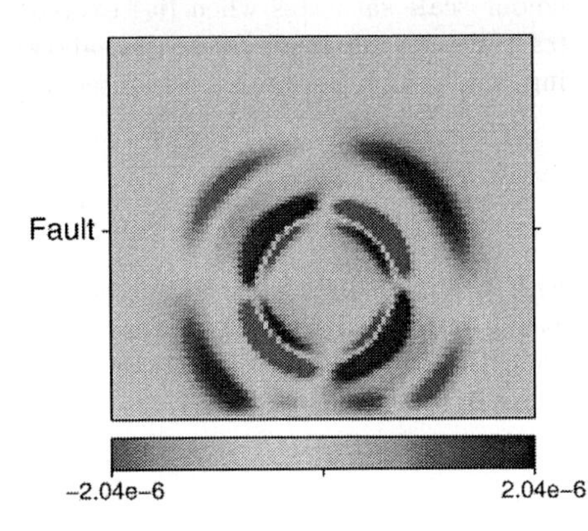

Fig. 1. Snapshot showing the vertical component of velocity at time $t = 25$ due to a point source below the fault located at $\mathbf{x} = (50, 34)$. The y-ordinate of the horizontal fault is indicated by tic marks on the frame. The source was excited by adding a Gaussian perturbation $Ke^{-\kappa(t-t_0)}$ to the horizontal component of displacement. Source parameters were $K = 0.0001$, $\kappa = 0.05$ and $t_0 = 25$. The colour scale saturates when $|v_y|$ exceeds 0.3 times its maximum value.

5.2 Rupture of a Homogeneous Fault

A numerical shear experiment was conducted by driving the upper and lower rows of nodes, which represent rigid driving plates, in opposite directions at a constant rate of 0.0005 (i.e. approx 0.0005 times the P-wave speed). The fault static friction was initialised to $\mu_s = 0.8$ everywhere except at the middle node located at $x = 50\Delta x$, which was set was set to $\mu_s = 0.75$. This provides a weak point or seed for the rupture to nucleate. The dynamic friction was set to $\mu_d = 0.7$ everywhere and the critical slip weakening distance was set to $D_c = 0.02$. Initially, there was no slip along the fault followed by quasi-static slip at the central weak point when the system became sufficiently stressed. As the stress builds up, a small region of slip grows quasi-statically around the weak node until a dynamic rupture is initiated which then propagates outwards at the compressional wave speed (Figures 2 and 3).

5.3 Rupture of a Heterogeneous Fault

A second shear experiment was conducted with the same parameters as the previous example except that both the static and dynamic friction were assigned values from a power law distribution, specifically $\mu(x_n) = \mathit{fft}^{-1}\{k_n^{-p} N_n\}$ where exponent $p = 0.3$, N_n is white noise, and x_n and k_n respectively denote the discrete locations and wavenumbers. The range of fluctuations for the static friction was $\mu_s \in [0.7, 0.8]$ and the range for the dynamic friction was $[0.55, 0.65]$. After several slip events, the stress becomes highly heterogeneous along the fault (Figure 4) and once this happens, ruptures typically propagate at approximately

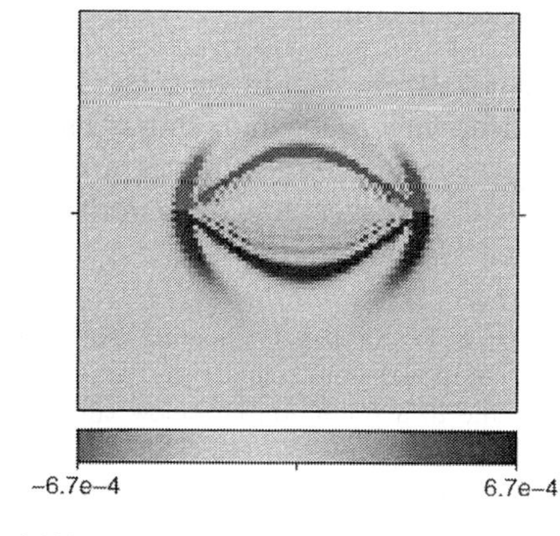

Fig. 2. Snapshot of the horizontal component of velocity showing a rupture propagating bi-directionally along the fault at the compressional wave speed. The y-ordinate of the horizontal fault is indicated by tic marks on the frame. The colour scale saturates when $|v_x|$ exceeds 0.1 times its maximum value thus allowing small amplitudes to be visualised.

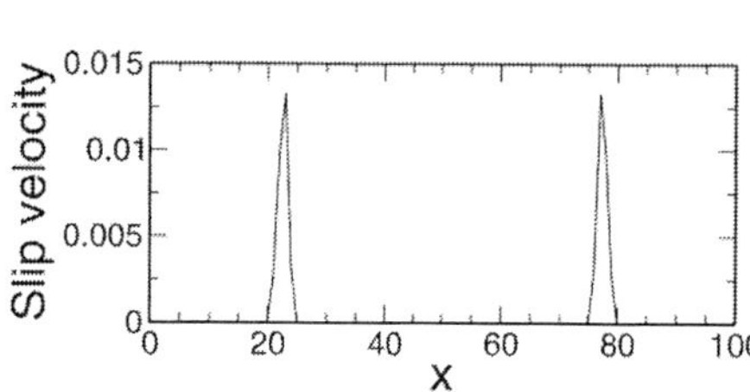

Fig. 3. Slip velocity on the fault defined as the horizontal velocity of upper split nodes subtracted from the horizontal velocity of lower split nodes.

the shear or Rayleigh wave speed (Figures 4 & 5). Because of the heterogeneity, it is typical that the rupture front will propagate in one direction only and leave a complex wave train following the rupture front as shown in Figure 4 (right).

5.4 Multi-fault Simulation

As the motivation to develop the numerical model is to study the physics of interacting fault systems, we present a shear experiment with a number of parallel faults to illustrate this capability. The same parameters were used as in the single heterogeneous fault case except that 11 faults were initialised at $y = 25 + 5j\Delta y$ where $j = 0, \ldots 11$ denotes the fault number. In this example, a lower shear rate was used than in previous examples. Namely, the speeds of the upper and lower rows were set to 0.0002 instead of 0.0005. Figure 6 shows the shear stress as a function of time. The saw-tooth shapes are characteristic of stick-slip behaviour, with each drop being caused by a dynamic rupture event on a fault.

Figure 7 shows snapshots of the shear stress σ_{xy} and horizontal component of velocity in the model at the 136050-th time step when a dynamic rupture event is occurring. The stress field in the fault region is complex and heterogeneous although coherent high stress bands can be seen running diagonally across the model. These are analogous to grain bridges which support stress in granular models. The numerical model allows the evolution of these complex stress patterns to be studied. One goal of such studies would be to determine whether there is a consistent growth in correlation lengths in the lead-up to large events in accord with the Critical Point Hypothesis for earthquakes and as seen in granular numerical models[4]. The horizontal component of velocity shows a rupture

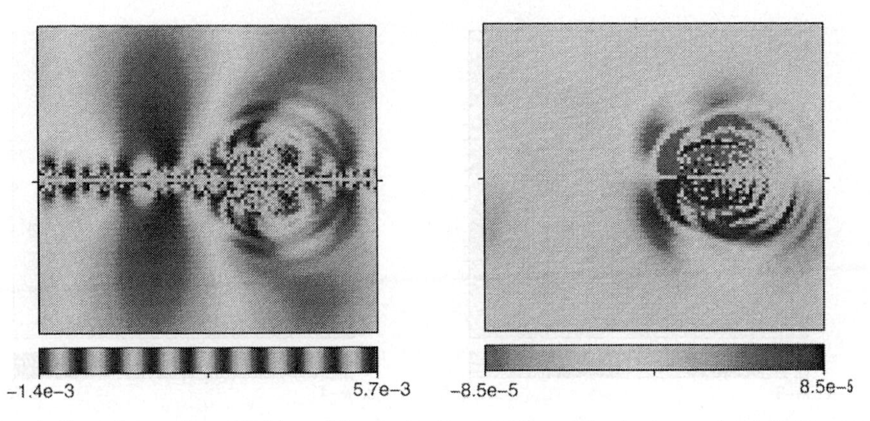

Fig. 4. Snapshots of σ_{yy} (left) and horizontal component of velocity v_x (right) showing a rupture propagating to the left at around the Rayleigh wave speed. The colour scale of v_x saturates when $|v_x|$ exceeds 0.015 times its maximum value.

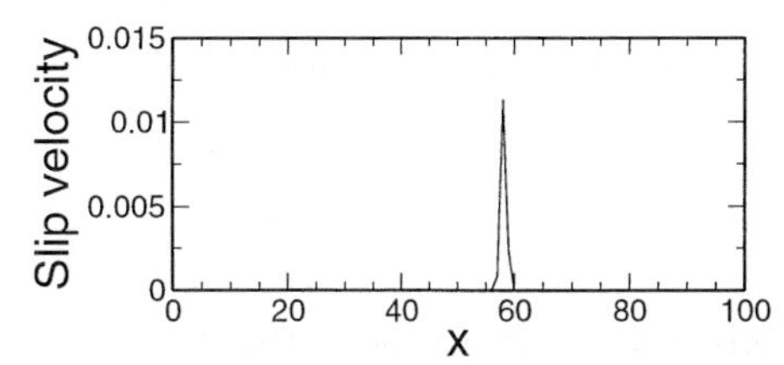

Fig. 5. Slip velocity on the fault at the same time as the snapshot shown in Figure 4.

propagating to the right along the 4-th fault from the bottom. In some cases, a rupture on one fault will trigger rupture on another fault. In most cases, ruptures propagate at around the Rayleigh wave speed although ruptures frequently accelerate to the P-wave speed, probably as a consequence of the high driving rate. Both unidirectional and bidirectional ruptures were observed.

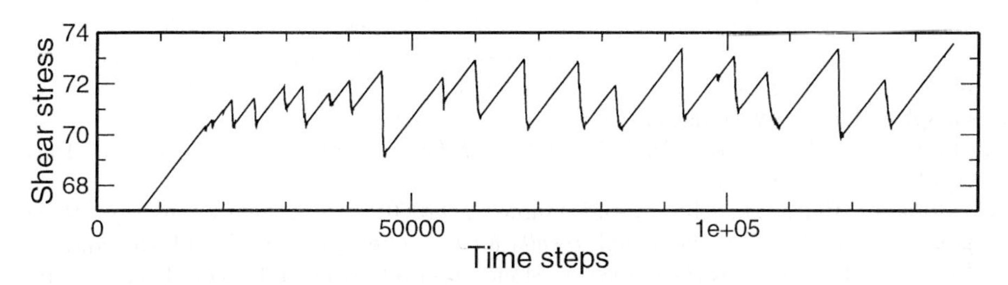

Fig. 6. Shear stress in the multi-fault model measured on the upper & lower edges of the lattice. The plot shows characteristic saw-tooth shapes associated with stick-slip behaviour. The initial shear stress was non-zero to minimise the first loading time.

6 Conclusions

A simple and relatively efficient numerical model is presented that provides a means to simulate the physics of parallel fault systems, and hence, a means to

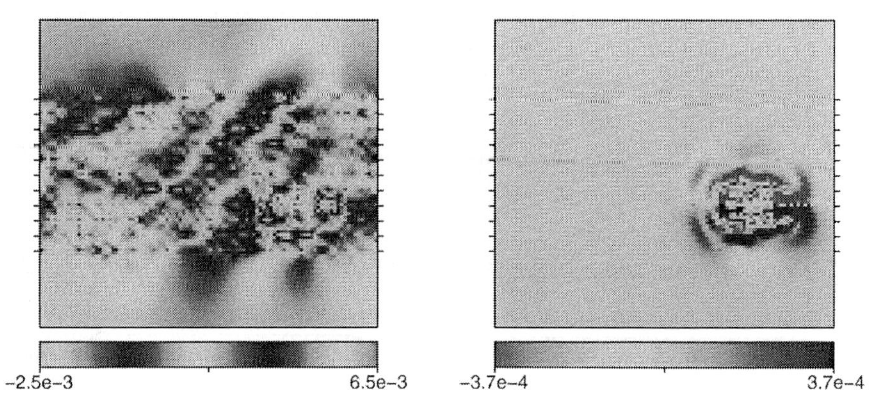

Fig. 7. Snapshots of shear stress (left) and the horizontal component of velocity (right). The colour scale of v_x saturates when $|v_x|$ exceeds 0.05 times its maximum value. The y-ordinates of the 11 horizontal faults are indicated by tic marks on the frame.

study whether elasto-dynamic crustal fault models may exhibit a dynamics that enables earthquakes to be forecast.

Acknowledgement.

This work has been supported by The University of Queensland, the Australian Research Council, and the Australian Computational Earth Systems Simulator Major National Research Facility.

References

1. Klein W., and Anghel M., Ferguson C.D., Rundle J.B., and Martins, J.S.Sá (2000) *Statistical analysis of a model for earthquake faults with long-range stress transfer*, in: Geocomplexity and the Physics of Earthquakes (Geophysical Monograph series; no. 120), eds. Rundle, J.B. and Turcotte, D.L., and Klein, W., pp 43-71 (American Geophys. Union, Washington, DC).
2. Mora, P., and Place, D. (1994) *Simulation of the frictional stick-slip instability*, Pure Appl. Geophys., **143**, 61-87.
3. Mora, P., Place, D., Abe, S. and Jaumé, S. (2000) *Lattice solid simulation of the physics of earthquakes: the model, results and directions*, in: GeoComplexity and the Physics of Earthquakes (Geophysical Monograph series; no. 120), eds. Rundle, J.B., Turcotte, D.L. & Klein, W., pp 105-125 (American Geophys. Union, Washington, DC).
4. Mora, P., and Place, D. (2002) *Stress correlation function evolution in lattice solid elasto-dynamic models of shear and fracture zones, and earthquake prediction*, Pure Appl. Geophys, **159**, 2413-2427.
5. Mora, P., Weatherley, D., and Klein, W. (2003) *Simulation of parallel interacting faults and earthquake predictability*, Proc. European Geophysical Society Annual Meeting.
6. Weatherley, D. and Mora, P. (2003) *Accelerating precursory activity within a class of earthquake analog automata*, Pure Appl. Geophysics, accepted.

Strategies for the Detection and Analysis of Space-Time Patterns of Earthquakes on Complex Fault Systems

John B. Rundle[1], William Klein[2], Kristy Tiampo[3], Andrea Donnellan[4], and Geoffrey Fox[5]

[1]Departments of Physics, Geology, Civil & Environmental Engineering, and Center for Computational Science and Engineering, University of California, Davis, Davis, CA 95616; and Distinguished Visiting Scientist, Earth & Space Sciences Division Jet Propulsion Laboratory, Pasadena, CA 91125
rundle@physics.ucdavis.edu; http://naniloa.ucdavis.edu/~rundle/
[2]Department of Physics and Center for Computational Science, Boston University Boston, MA 02215; and X-7, LANL, Los Alamos, NM
[3]Cooperative Institute for Research in Environmental Science, University of Colorado Boulder, CO 80309; University of Western Ontario, London, Ontario, Canada; and Universidad Complutense Madrid, Madrid, Spain
[4]Earth & Space Sciences Division, Jet Propulsion Laboratory, Pasadena, CA
[5]Department of Computer Science, Indiana University, Bloomington, IN

Abstract. Our research focuses on computational techniques to understand the dynamics of space-time patterns in driven threshold systems, particularly on earthquake fault systems. We discuss the scientific and computational formulation of strategies for understanding such space-time patterns, leading optimistically to earthquake forecasting and prediction. In particular, we describe pattern dynamics techniques that can be used to relate the observable processes on earthquake fault systems to the fundamentally unobservable dynamical processes. To illustrate our results, we study the emergent modes of the earthquake fault system in southern California, both with models (*Virtual California*) and with data.

1 Introduction

Earthquake faults occur in topologically complex, multi-scale networks that are driven to failure by external forces arising from plate tectonic motions. Such networks are prominent examples of driven threshold systems, other instances of which are the human brain, convective circulations in the atmosphere, one-dimensional electron waves in solids, driven foams, and magnetic de-pinning transitions in high temperature superconductors. The basic problem in this class of systems is that the true (force-displacement or current-voltage) dynamics is usually *unobservable*. Since to define the dynamics, one needs to know values of the variables in which the dynamics is formulated, lack of such knowledge precludes a deterministic approach to

P.M.A. Sloot et al. (Eds.): ICCS 2003, LNCS 2659, pp. 827–836, 2003.

forecasting the future evolution of the system. With respect to earthquakes, the space time patterns associated with the time, location, and magnitude of the sudden events (earthquakes) from the force threshold are *observable*. Our scientific focus is therefore on understanding the *observable space-time earthquake patterns* that arise from fundamentally *unobservable dynamics*, using new data-mining, pattern recognition, and ensemble forecasting techniques appropriate for these multi-scale systems.

2 Earthquakes

Earthquakes are a complex nonlinear dynamical system, so that techniques appropriate for the study of linear systems have not been of much utility. For example, earthquake forecasting based on linear fits to historic data have not been successful [1,2]. Moreover, the nonlinear dynamics, combined with our lack of microscopic knowledge about the system means that earthquake forecasting and predictability cannot be understood using deterministic dynamical methods, but must instead be approached as a stochastic problem. In addition, as Edward Lorenz showed [3], for nonlinear dynamical systems such as weather and earthquakes, *the past is not necessarily the key to the future*. Therefore, one must be extremely wary of simply extrapolating past behavior into the future, as the Parkfield prediction experience has shown [1,2].

2.1 Numerical Simulations

As elaborated below, there are two serious drawbacks to a purely observational approach to the problem of earthquake forecasting: 1) *Inaccessible* and *unobservable* stress-strain dynamics, and 2) *Multiscale dynamics* that cover a vast range of space and time scales. Because of these fundamental problems, the use of numerical simulations, together with theory and analysis, is mandatory if we are to discover answers to the questions above. Correspondingly, all types of earthquake-related data, including seismic, geodetic, paleoseismic, and laboratory rock mechanics experiments must be employed. The data are used both to determine physical properties of the models we simulate, a process of data assimilation, as well as to critically test the results of our simulation-derived hypotheses, so that future hypotheses can be developed. Moreover, as the computational complexity of the simulation process increases, we are necessarily led to the adoption of modern Information Technology (IT) approaches, that allow rapid prototyping of candidate models within a flexible, web-based computing framework. While advances in IT are thus an integral part of a systematic approach to numerical simulation of complex dynamical systems, our work is focused primarily on the use of simulations to advance scientific understanding of the fault system problem.

2.2 Unobservable Dynamics

Earthquake faults occur in topologically complex, multi-scale networks that are driven to failure by external forces arising from plate tectonic motions [4–7]. The basic problem in this class of systems is that the true stress-strain dynamics is *inaccessible* to direct observations, or *unobservable*. For example, the best current compendium of stress magnitudes and directions in the earth's crust is the World Stress Map [8], entries on which represent point static time-averaged estimates of maximum and minimum principal stresses in space. Since to define the fault dynamics, one needs dynamic stresses and strains for all space and time, the WSM data will not be sufficient for this purpose.

Conversely, the space time patterns associated with the time, location, and magnitude of the earthquakes are easily *observable.* Our scientific focus is therefore on understanding how the *observable space-time earthquake patterns* are related to the fundamentally *inaccessible* and *unobservable dynamics*, thus we are developing new data-mining, pattern recognition, theoretical analysis and ensemble forecasting techniques. In view of the lack of direct observational data, any new techniques that use space-time patterns of earthquakes to interpret underlying dynamics and forecast future activity must be developed via knowledge acquisition and knowledge reasoning techniques derived from the integration of diverse and indirect observations, combined with a spectrum of increasingly detailed and realistic numerical simulations of candidate models.

2.3 Multiscale Dynamics

The second problem, equally serious, is that earthquake dynamics is strongly coupled across a vast range of space and time scales that are both much smaller and much larger than "human" dimensions [9-13]. The important spatial scales span the range from the *grain scale*, of 1 nm to 1 cm; the *fault zone scale*, at 1 cm to 100 m; the *fault segment scale*, at 100 m to 10 km; the *fault system or network scale*, at 10 km to 1000 km; finally to the *Tectonic plate boundary scale* in excess of 1000 km. Important time scales span the range from the *source process time scale* of fractions of seconds to seconds; to the *stress transfer scale* of seconds to years; to *event recurrence time scales* of years to many thousands of years; finally to the *fault topology evolution scale*, in excess of many thousands of years up to millions of years. There is considerable evidence that many/most/all of these spatial and temporal scales are strongly coupled by the dynamics. Consider, as evidence, the Gutenberg-Richter relation, which is a power law for frequency of events in terms of cumulative event sizes. Power laws are a fundamental property of scale-invariant, self-organizing systems [14,15] whose dynamics and structures are strongly coupled and correlated across many scales in space and time. If the dynamics were were instead unconnected or random, one would expect to see Gaussian or Poisson statistics.

Simulations can help us to understand how processes operating on time scales of seconds and spatial scales of meters, such as source process times in fault zones, influence processes that are observed to occur over time scales of hundreds of years and spatial scales of hundreds of kilometers, such as recurrence of great earthquakes. Numerical simulations also allow us to connect observable surface data to underlying unobservable stress-strain dynamics, so we can determine how these are related. Thus we conclude that numerical simulations are mandatory if we are to understand the physics of earthquake fault systems. However, simulations of such complex systems are, at present, impossible to understand without some parallel theoretical investigation which gives us a framework to both interpret the vast amount of data generated and to ask the proper questions.

3 The Virtual California Model

Although all scales are important, we place more emphasis on the *fault system* or *fault network* scale, since this is the scale of most current and planned observational data networks. It is also the scale upon which the data we are interested in understanding, large and great earthquakes, occur. Furthermore, since it is not possible to uniquely determine the stress distribution on the southern California fault system, and since the friction laws and elastic stress transfer moduli are not known, it makes little sense to pursue a deterministic computation to model the space-time evolution of stress on the fault system. We therefore coarse-grain over times shorter than the source process time, which means we either neglect wave-mediated stress transfer, or we represent it in simple ways.

The Virtual_California model [5,6] is a stochastic, cellular automata instantiation of an earthquake *backslip model*, in that loading of each fault segment occurs via the accumulation of slip deficit S(t)-Vt, where S(t) is slip, V is long term slip rate, and t is time. At the present time, faults used in the model are exclusively vertical strike slip faults, the most active faults in California, and upon which most of the seismic moment release is localized. Thrust earthquakes, such as the 1994 Northridge and 1971 San Fernando faults, are certainly damaging, but they occur infrequently and are therefore regarded as perturbations on the primary strike slip fault structures. The Virtual_California model also has the following additional characteristics:

1. Surfaces of discontinuity (faults) across which slip is discontinuous at the time of an earthquake, and which are subject to frictional resistance. Surfaces analyzed in current models range from infinitely long, vertically dipping faults to topologically complex systems of vertically dipping faults mirroring the complexity found on natural fault networks.

2. Persistent increase of stresses on the fault surfaces arising from plate tectonic forcing parameterized via the backslip method.

3. Linear elastic stress transfer or interactions between fault surfaces. In some model implementations, elastic waves and inertial effects may be included in simple ways. In other implementations, details of the rupture and stress transfer process are absorbed into stochastic terms in the dynamical equations, and quasistatic stress Green's functions are used. Other types of linear stress transfer are also possible, including linear viscoelastic and linear poroelastic physics [16].

4. Parameters for friction laws and fault topology that are determined by assimilating seismic, paleoseismic, geodetic, and other geophysical data from events occurring over the last ~200 years in California.

5. Frictional resistance laws that range from the simplest Amontons-Coulomb stick-slip friction [17], to heuristic laws such as slip- or velocity-weakening laws [4,17,18], to laws based on recent laboratory friction experiments including rate-and-state [19–28] and leaky threshold laws [18,25,26,28], to other types of rupture and healing laws characterized by inner- and outer- physical scales [7].

In general, any of the friction laws described in bullet 5 can be written in either of the following representative, equivalent forms on an element of fault surface:

$$\frac{\partial\sigma(t)}{\partial t} = K_L V - f[\sigma(t), V(t)] \tag{1}$$

or

$$K_L \frac{dS(t)}{dt} = f[\sigma(t), V(t)]$$

Here σ is the shear stress, and K_L is the self-interaction or "stress drop stiffness" and $f[\sigma,V]$ is the *stress dissipation function* [29]. For example, the "Amontons" or Coulomb friction law, having a sharp failure threshold, can be written in the form (1) using a Dirac delta function [28].

4 Software Technology

Virtual California simulates fault interaction to determine correlated patterns in the nonlinear complex system of an entire plate boundary region. The evolution of these patterns enables forecasts of future large events. Capturing the nonlinear pattern dynamics of the fault system along a plate boundary implies the realization of a digital laboratory, which allows understanding of the mechanisms behind the observations and patterns. Our software technology development is based on the principle of scalability. When it is fully deployed, researchers will be able to create and verify patterns down to ever smaller spatial scales, which will enable cross-scale parameterization and validations, thus in turn enabling plate-boundary system analysis. The possibility of forecasting large earthquakes will therefore be greatly enhanced.

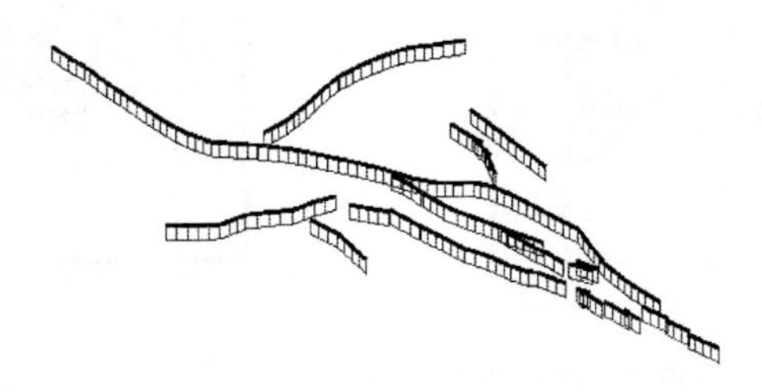

Fig. 1. Model fault segment network for southern California

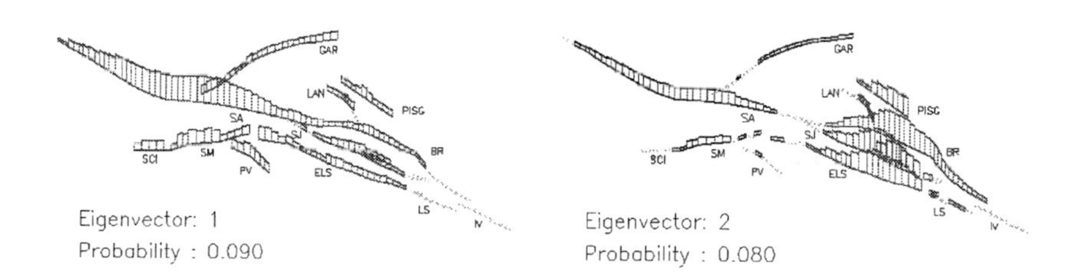

Fig. 2. First two correlation eigenpatterns for model fault network shown in Fig. 1

Pattern analysis methods are another type of recently developed tool. One method bins many decades of seismic activity on a gridded representation of California. Eigensystem analysis reveals clusters of correlated space and time activity. These eigenpatterns can be used to develop forecast methods [28]. When these methods attain parallel speedup we will produce better forecasts and enable speedy tests of earthquake correlations and predictions. This will be due to the ability to use much smaller geographic cell sizes and so forecast the frequent magnitude 4 earthquakes, not just the rare magnitude 6 events.

Examples of recent results using Virtual California simulations are shown in Figs. 1–2, while results from observed data in southern California are shown in Fig. 3. In Fig. 1 we have defined a network of fault segments based upon the actual network of faults in southern California. Figure 2 shows two of the eigenpatterns produced by activity in the model fault system. Red segments are positively correlated with red segments, negatively correlated with blue segments, and uncorrelated with other segments. Figure 3 shows similar eigenpatterns for the actual earthquake data in southern California.

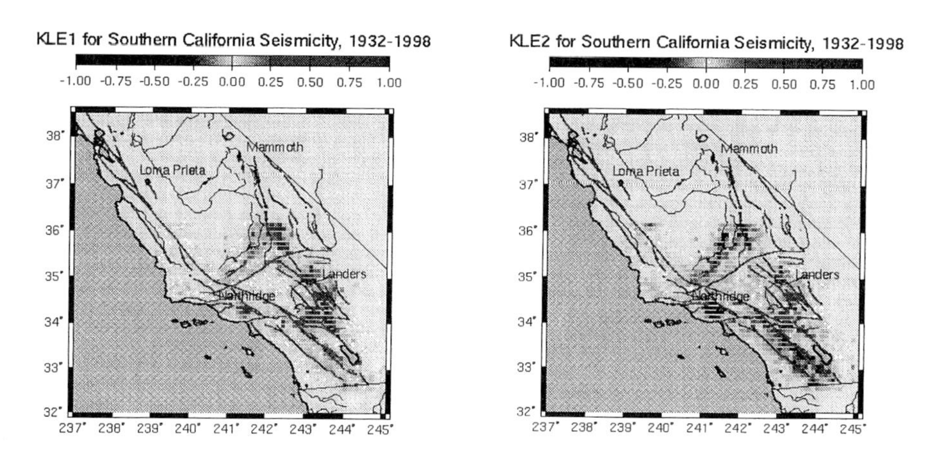

Fig. 3. First two correlation eigenpatterns for the actual earthquake data in southern California since 1932.

5 Community Grid

As part of our work, we are also leveraging the Grid-based Community Modeling Environment (CME) under development by the Southern California Earthquake Center (SCEC) to produce an emergent computational framework for a broad class of *stress-evolution simulations* (SES) for the dynamics of *earthquake fault systems*. The purpose of SES models is to enhance basic understanding of the phenomena of earthquakes, as well as to build the computational and scientific foundations that will be needed to forecast large and damaging events. The scientific and information technology requirements for this class of physics-based simulations demand that our emergent Grid-computational framework address the following requirements of modern IT systems:

1. Space-time pattern analysis using both simulation-based approaches to computational forecasts of catastrophic and damaging earthquakes.
2. Prototyping computational methods to separate stochastic and nonlinear dynamical factors in earthquake fault networks.
3. Multi-scale analysis and modeling of the physical processes and patterns.
4. Knowledge discovery, visualization and interpretation of highly heterogeneous simulation data sets.
5. Integration of data into the simulations arising from multiple sources including surface deformation, historic and recent seismic activity, geological field investigations, and laboratory-derived friction data.
6. Integrating advanced methods for information storage and retrieval to enhance the interoperability and linkage of fixed and streaming data from observations and simulations.

We term the international collection of scientists and Grid [30-38] or web resources in a particular field as a Community Grid. Our IT research is oriented at enabling and enhancing the interaction of *all* components of the Community Semantic Grid for Earthquake Science (CSGES) for the SES effort, leading to an emergent multi-scale Grid. We are using the ideas and artifacts produced by other relevant activities from both the computer science and earth science areas, including [42-52]:

1. The major SCEC/CME-led effort to build Grid resources for the earthquake field and an information system to describe them.
2. Ontologies and XML data structures being built by SCEC, ACES (with GEM) and related fields as illustrated by OpenGIS and XMML for the mining community.
3. Parallel simulation codes with a portal interface (Gateway) [39-41] developed by GEM as part of a NASA HPCC project. The Gateway Grid Computational Environment will be a key part of the initial CSGES deployment.
4. International ACES capabilities such as the GeoFEM software and Earth Simulator hardware from Japan and the Australian ACcESS simulation software.
5. Basic collaboration capabilities among some participants including audio/video conferencing (Access Grid) and the ability for simple sharing of web pages and other documents as implemented in our system Garnet.
6. Initial designs of use of Web service technologies (WSDL, UDDI) to provide a component model for some relevant web resources.

7. Development of new dynamic models for collaborative environments coming from use of JXTA and other Peer-to-Peer technologies, integrating these with existing event-based collaborative environments.
8. Server or peer-to-peer discovery mechanisms (as in UDDI and JXTA) combining with a component model (WSDL) to lay the groundwork for emergent information systems. Note all components (objects) in the system will have in the spirit of the semantic web, meta-data defined for them which will enable the linkage of components in our proposed Semantic Grid.

6 Final Remarks

The results of our research are critical prerequisites for the analysis and understanding of space-time patterns of earthquakes on arbitrary complex earthquake fault systems. In turn, such understanding is mandatory if earthquake forecasting and prediction is to become a reality. The instantiation of any and all simulations of the dynamics of complex earthquake fault networks must allow use of the full range of methods, attributes, and parameters represented in the current physics-based earthquake simulation literature. Our work is complementary to, but does not duplicate research underway within the SCEC/ITR effort, which is focused primarily on integrating conventional and elastic-wave- based seismic hazard analysis within an emergent Grid-computational environment. Our methods emphasize the process of *optimizing, executing, integrating* and *analyzing* SES. Indeed, we argue that it is precisely the *lack* of an adequate multi-scale computational framework allowing rapid prototyping and analysis of candidate SES models that has seriously retarded the development of these types of simulations.

Acknowledgements. This work has been supported by CIRES and NASA student fellowships (KFT and SM); by US DOE grant DE-FG03-95ER14499 (JBR); by NASA grant NAG5-5168 (SJG). WK was supported by USDOE/OBES grant DE-FG02-95ER14498 and W-7405-ENG-6 at LANL. WK would also like to acknowledge the hospitality and support of CNLS at LANL.

References

1. Bakun, W and Lindh, A., The Parkfield, California earthquake prediction experiment, Science, 89, (1985) 3051–3058
2. Jones, L.M., Earthquake prediction: The interaction of public policy and science, Proc. Nat. Acad. USA, 93, (1996) 3721–3725
3. Lorenz, E.N., Deterministic nonperiodic flow, J. Atmos. Sci., 20, (1963) 130–141
4. Scholz, C.H., The Mechanics of Earthquakes and Faulting, Cambridge University Press, Cambridge, UK (1990)
5. Rundle, J.B., Turcotte, D.L., and Klein, W., eds., Geocomplexity and the Physics of Earthquakes, Geophysical Monograph 120, American Geophysical Union, Washington, DC (2000).

6. Rundle, P.B., Rundle, J.B., Tiampo, K.F., Martins, J.S.S., McGinnis, S., and Klein, W., Nonlinear network dynamics on earthquake fault systems, Phys. Rev. Lett., 87, (2001) 148501(1–4) (2001)
7. Ward, S.N., San Francisco bay area earthquake simulations, a step towards a standard physical model, Bull. Seism. Soc. Am., 90, (2000) 370–386
8. Zoback, M.L., 1st-order and 2nd-order patterns of stress in the lithosphere – The World Stress Map project, J. Geophys. Res., 97, (1992) 11703–11728
9. GEM home page, http://geodynamics.jpl.nasa.gov/gem/
10. ACES home page, http://quakes.earth.uq.edu.au/ACES/
11. Southern California Earthquake Center, http://www.scec.org/
12. Mora, P., ed., Proceedings of the 1st ACES Workshop, published by APEC Cooperation for Earthquake Simulation, Brisbane, Queensland, AU (1999).
13. Matsu'ura, M., Nakajima, K., and Mora, P., eds., Proceedings of the 2nd ACES Workshop, published by APEC Cooperation for Earthquake Simulation, Brisbane, Queensland, AU (2001).
14. Vicsek, T., Fractal Growth Phenomena, World Scientific, Singapore (1989).
15. Gouyet, J.-F., Physics and Fractal Structures, Springer-Verlag, Berlin (1996).
16. Miller, S.A., Ben-Zion, Y., and Burg, J.P., A three-dimensional fluid-controlled earthquake model: Behavior and implications, J. Geophys. Res., 104, (1999) 10621–10638
17. Persson, B.N.J., Sliding Friction, Physical Principles and Applications (Springer-Verlag, Berlin (1998).
18. Rundle, J.B., Klein, W., Gross, S., and Ferguson, C.D., The traveling density wave model for earthquakes and driven threshold systems, Phys. Rev. E, 56, (1997) 293–307
19. Dieterich J.H., and Kilgore, B., Implications for fault constitutive properties for earthquake prediction, Proc. Nat. Acad. Sci. USA, 93, 3787 (1996).
20. Dieterich, J.H., and Kilgore, B., Direct observation for frictional contacts – New insights for state-dependent properties, PAGEOPH, 143, 283–302 (1994).
21. Dieterich, J.H., Modeling of rock friction 1, Experimental results and constitutive equations, J. Geophys. Res., 84, (1979) 2161–2175
22. Rice, J.R., Lapusta, N., and Ranjith, K., Rate and state dependent friction and the stability of sliding between elastically deformable solids, J. Mech. Phys. Sol., 49, (200) 1865–1898
23. Ranjith, K. and J.R. Rice, J.R., Slip dynamics between at an interface between dissimilar materials, J. Mech. Phys. Sol., 49, (2001) 341–361
24. Ruina, A., Slip instability and state-variable friction laws, J. Geophys. Res., 88, (1983) 359–370
25. Tullis, T.E., Rock friction and its implications for earthquake prediction examined via models of Parkfield earthquakes, Proc. Nat. Acad. Sci USA, 93, (1996) 3803–3810
26. Karner, S.L., and Marone, C., Effects of loading rate and normal stress on stress drop and stick slip recurrence interval, pp. 187–198 in Rundle, J.B., Turcotte, D.L., and Klein, W., eds., Geocomplexity and the Physics of Earthquakes, Geophysical Monograph 120, American Geophysical Union, Washington, DC (2000).
27. Beeler, N.M., Tullis, T.E., and Blanpied, M.L., and Weeks, J.D., Frictional behavior of large displacement experimental faults, J. Geophys. Res., 101, (1996) 8697–8715
28. Rundle, J.B., Tiampo, K.F., Klein, W., Martins, J.S.S., Self-organization in leaky threshold systems: The influence of near mean field dynamics and its implications for earthquakes, neurobiology and forecasting, in press, Proc. Nat. Acad. Sci. USA, (2001) 2463–2465
29. Klein, W., Rundle, J.B., and Ferguson, C.D., Scaling and nucleation in models of earthquake faults, Phys. Rev. Lett., 78, (1997) 3793–3796
30. Particle Physics Data Grid, `http://www.cacr.caltech.edu/ppdg/`
31. GriPhyN Project Site, `http://www.griphyn.org`
32. NEES Earthquake Engineering Grid, `http://www-meesgrod.org/`
33. Resource Description Framework (RDF), `http://www.w3.org/TR/REC-rdf-syntax/`

34. Ninja Project, http://ninja.cs.berkeley.edu/
35. Semantic Web, http://www.w3.org/2001/sw/
36. The Grid Forum http://www.gridforum.org
37. Grid Forum Computing Environment Working Group, http://www.computingportals.org/cbp.html
38. Grid Message System, http://aspen.csit.fsu.edu/users/shrideep/mspaces/
39. General Review of Portals by G. Fox, http://aspen.csit.fsu.edu/collabtools/CollabReviewfeb25-01.html
40. Portals and Frameworks for Web-based Education and Computational Science, http://www.new-npac.org/users/fox/documents/pajavaapril00/
41. Gateway Computational Portal, http://www.gatewayportal.org
42. See talk by CCA lead Rob Armstrong of Sandia at LLNL meeting on Software Components July 2001 http://www.llnl.gov/CASC/workshops/components_2001/viewgraphs/RobArmstrong.ppt
43. Web Services Description Language (WSDL) 1.1 http://www.w3.org/TR/wsdl.
44. Presentation on Web Services by Francesco Curbera of IBM at DoE Components Workshop July 23–25, 2001. Livermore, California. http://www.llnl.gov/CASC/workshops/components_2001/viewgraphs/FranciscoCurbera.ppt
45. Doe Babel project http://www.llnl.gov/CASC/components/babel.html
46. IBM Web Services Invocation Framework http://www.alphaworks.ibm.com/tech/wsif
47. XML based messaging and protocol specifications SOAP. http://www.w3.org/2000/xp/.
48. W3C Resource Description Framework http://www.w3.org/RDF/
49. Sun Microsystems JXTA Peer to Peer technology. http://www.jxta.org.
50. Universal Description, Discovery and Integration Project http://www.uddi.org/
51. Semantic Web from W3C to describe self organizing Intelligence from enhanced web resources. http://www.w3.org/2001/sw/.
52. W3C description of Naming and Addressing http://www.w3.org/Addressing/

Texture Alignment in Simple Shear

Frédéric Dufour[1], Hans Mühlhaus[2], and Louis Moresi[3]

[1] Laboratoire de Génie Civil de Nantes – Saint-Nazaire
1 rue de la Noë, 44321 Nantes Cedex, France
frederic.dufour@ec-nantes.fr
[2] Queensland University, Advanced Centre for Earthquakes Studies
St Lucia, QLD 4072, Australia
[3] MONASH University
PO Box 28M, Vic 3800, Australia

Abstract. We illustrate the flow behaviour of fluids with isotropic and anisotropic microstructure (internal length, layering with bending stiffness) by means of numerical simulations of silo discharge and flow alignment in simple shear. The Cosserat theory is used to provide an internal length in the constitutive model through bending stiffness to describe isotropic microstructure and this theory is coupled to a director theory to add specific orientation of grains to describe anisotropic microstructure. The numerical solution is based on an implicit form of the Material Point Method developed by Moresi et al. [1].

Keywords: director theory, Cosserat theory, grain flow, Lagrangian integration points

1 Introduction

The mechanics of granular materials has intrigued physicists and engineers for well over two centuries. At low strain rates, particulates such as sand or cereals behave like solids, but at high strain rates, the behaviour is fluid or gas like.

Computer simulations and specifically discrete element simulations [2] are an important tool for exploring the fundamental behaviours of granular flows, flow regimes, phase transitions, fluctuations etc. However computational requirements set strong limitations to the size of discrete element models. The much simpler and (where applicable) much more efficient continuum models for granular flows are valid when the typical grain size is much smaller than characteristic structural dimensions e.g. the outlet size in silo flows. Cosserat continuum theory [3] considers some of the salient features of the discrete microstructure (e.g. grain size, relative rotation between microstructure and the continuum) within the framework of a continuum theory. Such a theory fits between detailed discrete theories and the usual continuum theory.

The Cosserat- or micropolar theory may be employed for a variety of applications involving the need to describe the heterogenous microstructure of the material such as granular materials [4], layered materials [5] or crystals [6] within the framework of a continuum theory.

P.M.A. Sloot et al. (Eds.): ICCS 2003, LNCS 2659, pp. 837–844, 2003.

This class of theories can be implemented in the context of a classical finite element method (FEM). However, very large deformations are sometimes difficult to handle elegantly within the FEM because mesh distortion and remeshing can quickly present severe difficulties. The version of the Material Point Method (MPM, [7] and [8]) applied here combines the versatility of the standard FEM with the geometrical flexibility of pure particle schemes such as the Smooth Particle Hydrodynanmics (SPH, [9]).

In MPM Lagrangian integration points move through a spatially fixed Eulerian mesh. MPM is inspired by particle-in-cell (PIC) finite difference methods, originally designed for fluid mechanics, in which fluid velocities are solved on the mesh, and material strains and material history variables are recorded by Lagrangian particles. These particles serve as integration points for the stiffness matrix and force vectors. The MPM method applied here is implicit [1] as opposed to the explicit form proposed by Sulsky et al. [10].

We show simple simulations to illustrate the performance of the numerical scheme and the constitutive theory.

2 Mathematical Formulation

2.1 Cosserat Deformation Measures

We assign a local rigid cross to every material point (x_1, x_2, x_3) of the body in a Cartesian coordinate system (X_1, X_2, X_3). During deformation, the rigid crosses rotate at a rate ω_i^c about their axis i and are translated according to the conventional linear velocity vector $\mathbf{u}$. The angular velocity ω_i^c is considered to be independent of $\mathbf{u}$ and differs from the angular velocity of an infinitesimal volume element of the continuum ω_i,

$$\omega_i = -\frac{1}{2}\epsilon_{ijk} W_{jk} \tag{1}$$

where

$$W_{jk} = \frac{1}{2}(u_{j,k} - u_{k,j}) \tag{2}$$

In (1) ϵ_{ijk} designates the permutation symbol, and in (2) $(.)_{,k} \equiv \partial(.)/\partial x_k$ are partial derivatives.

In the classical theory the stretching tensor is given by:

$$D_{ij} = \frac{1}{2}(u_{i,j} + u_{j,i}) \tag{3}$$

In the Cosserat theory, as a rotational parameter has been added, in addition to the classical strain rate tensor $\boldsymbol{D}$, there is an additional rate measure,

$$\omega_i^{rel} = \omega_i - \omega_i^c \tag{4}$$

which represents the relative angular velocity between the material element and the associated rigid coordinate cross. In this case, the rate of the deformation tensor can be expressed by the rate of the distortion tensor, γ,

$$\gamma_{ij} = u_{i,j} - W^c_{ij} \tag{5}$$

where

$$W^c_{ij} = -\epsilon_{kij}\omega^c_k \tag{6}$$

and by the tensor representing the measure of relative angular velocity between the neighbouring rigid crosses,

$$\kappa_{ij} = \omega^c_{i,j} \tag{7}$$

The conventional strain rate tensor can be expressed as the symmetrical part of the rate of the distortion tensor,

$$D_{ij} = \frac{1}{2}(\gamma_{ij} + \gamma_{ji}) \tag{8}$$

and the relative angular velocity as the antisymmetrical part

$$W^{rel}_{ij} = \frac{1}{2}(\gamma_{ij} - \gamma_{ji}) \tag{9}$$

We have 2 deformation rate measures i.e. γ and κ. Both measures are objective. In a rotating observer frame γ and κ are obtained as $Q\gamma Q^T$ and $Q\kappa Q^T$ where $QQ^T = 1$ describes the rotation of the moving -with respect to the fixed-observer frame.

2.2 Constitutive Relationships for Granular Materials

In a 2D conventional continuum an isotropic material is characterised by a bulk viscosity B and a shear viscosity η, for a Cosserat continuum we also have a Cosserat shear viscosity η^c and a bending viscosity M. The constitutive relation for a generalised Newtonian fluid can be written in the usual pseudo-vector form:

$$\boldsymbol{\sigma} = \boldsymbol{\Lambda}\boldsymbol{D} \tag{10}$$

where the stress vector components are:

$$\boldsymbol{\sigma}^T = \{\sigma_{xx}, \sigma_{yy}, \sigma_{xy}, \sigma_{yx}, \mu_{zx}, \mu_{zy}\} \tag{11}$$

the deformation vector components are:

$$D^T = \{\gamma_{xx}, \gamma_{yy}, \gamma_{xy}, \gamma_{yx}, \kappa_{zx}, \kappa_{zy}\} \tag{12}$$

and the matrix $\boldsymbol{\Lambda}$ is expressed as:

$$\left\{\begin{array}{cccccc} B+\eta & B-\eta & 0 & 0 & 0 & 0 \\ & B+\eta & 0 & 0 & 0 & 0 \\ & & \eta+\eta^c & \eta-\eta^c & 0 & 0 \\ & & & \eta+\eta^c & 0 & 0 \\ & symm. & & & M & 0 \\ & & & & & M \end{array}\right\} \tag{13}$$

For benchmarking purposes, we use the simplest possible realisation of a granular, viscous medium. In the granular-elasticity model of Choi and Mühlhaus [11] we replace the contact stiffnesses K_n and K_m and relative displacements by contact dashpots η_n and η_m and relative velocities and relative rotation rates and obtain the relationships

$$\eta = \frac{1-n}{4\pi}k(\eta_n + \eta_m) \tag{14}$$

$$\eta^c = \frac{1-n}{2\pi}k\eta_m \tag{15}$$

$$B = \frac{1-n}{2\pi}k\eta_n \tag{16}$$

$$M = 2\eta^c R^2 \tag{17}$$

2.3 Director Theory

The salient mechanical feature of layered geomaterials is the existence of a characteristic orientation given by the normal vector field $n_i(x_k, t)$ of the layer planes, where (x_1, x_2, x_3) are Cartesian coordinates, and t is the time. We assume linear viscous behaviour and designate with η the normal viscosity and η_s the shear viscosity in the layer planes normal to n_i. The orientation of the normal vector, or director as it is sometimes called in the literature on oriented materials, changes with deformation. Using a standard result of continuum mechanics, the evolution of the director of the layers is described by

$$\dot{n}_i = W^n_{ij} n_j \text{ where } W^n_{ij} = W_{ij} - (D_{ki}\lambda_{kj} - D_{kj}\lambda_{ki}) \quad \text{and} \quad \lambda_{ij} = n_i n_j \tag{18}$$

where $\boldsymbol{L} = \boldsymbol{D} + \boldsymbol{W}$ is the velocity gradient, $\boldsymbol{D}$ is the stretching and $\boldsymbol{W}$ is the spin. The superscript n distinguishes the spin $\boldsymbol{W}^n$ of the director $\boldsymbol{N}$ from the spin $\boldsymbol{W}$ of an infinitesimal volume element dV of the continuum.

2.4 Specific Viscous Relationships

We consider layered materials. The layering may be in the form of an alternating sequence of hard and soft materials or in the form of a superposition of layers of equal width of one and the same material, which are weakly bonded along the interface. In the following simple model for a layered viscous material we correct the isotropic part $2\eta D'_{ij}$ of the model by means of the tensor $\boldsymbol{\Lambda}$ to consider the mechanical effect of the layering; thus

$$\sigma_{ij} = 2\eta D'_{ij} - 2(\eta - \eta_s)\Lambda_{ijlm} D'_{lm} - p\delta_{ij} \tag{19}$$

where p is the pressure, $\boldsymbol{D}'$ designates the deviator of $\boldsymbol{D}$ (i.e. $\boldsymbol{D}' = \boldsymbol{D} - tr(\boldsymbol{D})$), and

$$\Lambda_{ijkl} = \left(\frac{1}{2}(n_i n_k \delta_{lj} + n_j n_k \delta_{il} + n_i n_l \delta_{kj} + n_j n_l \delta_{ik}) - 2n_i n_j n_k n_l\right) \tag{20}$$

3 Applications

3.1 Silo Discharge

We now consider the discharge of a Cosserat material from a model silo. In this case we are interested in the influence of the internal length parameter on the discharge velocity of a Cosserat fluid as described by (10–13)

Figure 1 shows the geometry of the model as well as different snapshots along the computations. Heavy lines are free-slip boundaries. Flow is only due to the downward gravity field. The grid drawn on the flowing material is a "dye" to record deformation – it does not affect the material properties. Corridors along the edges provide space where upward flow of the passive background material can take place to equilibrate the pressure due to the downward flow of the Cosserat viscous material. The mechanical characteristics of each material are summarized in Fig. 2.

In Fig. 3 we plot the volume flowing out the reservoir versus time and for different values of the ratio $\alpha = R/a$ where a is the silo aperture. The flow rate is almost identical for all values of α (Fig. 3d and 3e) larger than the aperture (1.0) which corresponds, in the elastic case, to a situation in which no flow can occur. Note that in purely viscous materials static equilibrium states do not exist, ensuring that stable arches do not form. For $0 \leq \alpha \leq 1.0$, the smaller the internal length the faster the outflow. As for the elastic case the internal length provides a bending stiffness which slows down the flow.

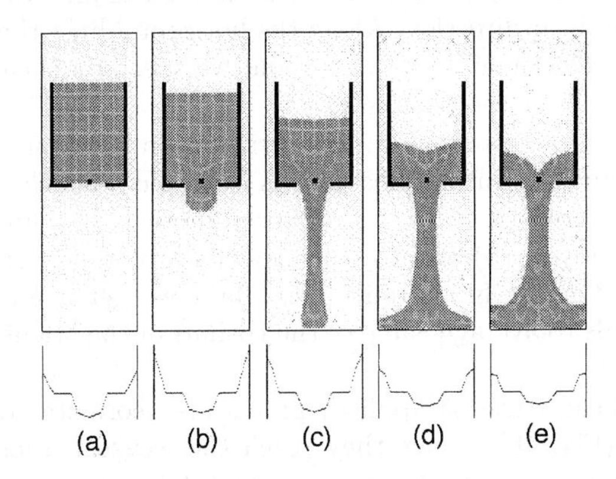

Fig. 1. Snapshots of the model. (a) Initial conditions, (b) $t = 4.75\ 10^{-4}$ sec., (c) $t = 1.42\ 10^{-3}$ sec., (d) $t = 2.85\ 10^{-3}$ sec. and (e) $t = 3.8\ 10^{-3} sec.$

	Granular material	Background
Internal length	R	0
Shear viscosity	1000	1
Bulk viscosity	$+\infty$	$+\infty$
Density	10^6	0

Fig. 2. Constant values.

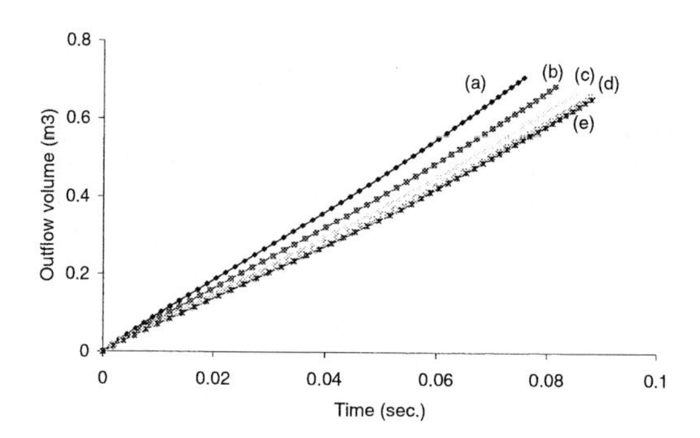

Fig. 3. Outgoing volume for (a) α=0, (b) α=1/3, (c) α=2/3, (d) α=5/3 and (e) α=10/3

3.2 Self-Alignment

In this example the flow of anisotropic particles (pencil-shaped) is modelled in an infinite shearlayer. The normal viscosity η is set to 1000 Pa.s and the shear viscosity η_s to 500 Pa.s. Each particle has an internal length R of 0.2 m. We specify periodic boundary conditions along the vertical lines to model the infinite dimension in the shear direction. Along the horizontal lines the cosserat rotation and the normal velocity are set to zero and we specify a shear stress of 10^4 Pa on top and -10^4 Pa at the bottom.

Initially (Fig. 4a), a random director orientation between $-\frac{\pi}{2}$ and $\frac{\pi}{2}$ is set to each particle. Thus the initial behaviour is isotropic. Particles along the central vertical line are "dyed" to track the material motion through time (Fig. 4). To each configuration (a), (b) and (c) corresponds a plot (a'), (b') and (c') of the isovalues of n_x. On the first profile (Fig. 4a'), we get a unique isovalue 0.64 which corresponds to average value of the cosinus of the orientation angle of the director.

While the shear stress is applied, grains are reorientated parallely to the shear direction (Fig. 4b'). Once they reach the weakest orientation ($n_x = 0$), they remain in that position. As shown on plot 4c', the material is now strongly anisotropic due to the preferential orientation of interpolation points.

4 Conclusions

The Cosserat model shows the internal length effect on flow velocity. The Cosserat theory coupled with the director theory in the shear model can explain qualitatively the anisotropy induced during strong shearing, for example in a silo discharge of non-spherical grains.

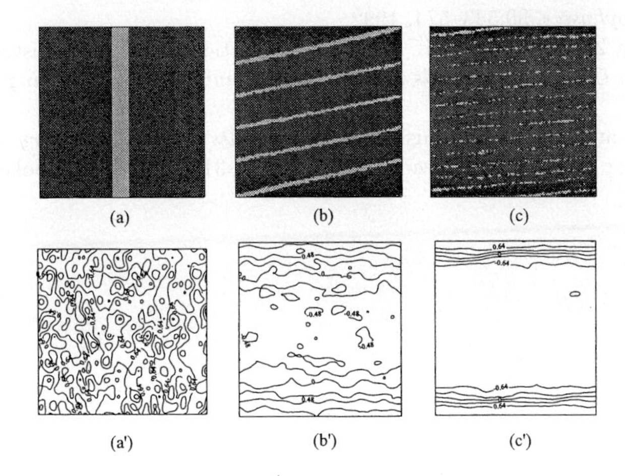

Fig. 4. (a) Initial conditions with random particle orientation, (b) beginning of the anisotropic behaviour and (c) the material is oriented. Plots (a'), (b') and (c') are the isovalues of the director orientation respectively corresponding to configuration (a), (b) and (c)

References

1. L. Moresi, F. Dufour, and H.-B. Mühlhaus. A lagrangian integration point finite element method for large deformation modeling of viscoelastic geomaterials. *Journal of Computational Physics*, In press, 2002.
2. H. Sakaguchi and H.-B. Mühlhaus. Mesh free modelling of failure and localization in brittle materials. In *Deformation and Progressive Failure in Geomechanics*, pages 15–21. Pergamon, 1997.
3. E. Cosserat and F. Cosserat. *Théorie des corps déformables.* Hermann, Paris, 1909.
4. H.-B. Mühlhaus, R. de Borst, and E.C. Aifantis. Constitutive models and numerical analysis for inelastic materials with microstrucutre. In Booker Eds. Beer and Carter, editors, *Computer Methods and Advances in Geomechanics*, pages 377–385, Rotterdam, 1991. Balkema.
5. D.P. Adhikary, H.-B. Mühlhaus, and A.V. Dyskin. Modeling the large deformations in stratified media-the cosserat continuum approach. *Mechanics of Cohesive-Frictional Materials*, 4:195–213, 1999.
6. S. Forrest, G. Cailletaud, and R. Sievert. A cosserat theory for elastoviscoplastic single crytals at finite deformation. *Archives of Mechanics*, 49(4):705–736, 1997.
7. L. Moresi, H.-B. Mühlhaus, and F. Dufour. Particle-in-cell solutions for creeping viscous flows with internal interfaces. In H.-B. Mühlhaus, A. Dyskin, and E. Pasternak, editors, *In Bifurcation and Localization in Soils and Rocks*, pages 345–353, Rotterdam, 2001. Balkema.
8. L. Moresi, H.-B. Mühlhaus, and F. Dufour. Viscoelastic formulation for modelling of plate tectonics. In H.-B. Mühlhaus, A. Dyskin, and E. Pasternak, editors, *In Bifurcation and Localization in Soils and Rocks*, pages 337–343, Rotterdam, 2001. Balkema.

9. J.J. Monaghan. Smoothed particle hydrodynamics. *Annual Review of Astronomy and Astrophysics*, 30:543–574, 1992.
10. D. Sulsky, Z. Chen, and H.L. Schreyer. A particle method for history-dependent materials. *Computer Methods in Applied Mechanics and Engineering*, 118:179–196, 1994.
11. S.K. Choi and H.-B. Mühlhaus. *Distinct elements vs Cosserat theory: A comparison for the case of an infinite shearlayer*, pages 315–319. Eds. Beer, Booker and Carter, 1991.

Mechanical Properties of the Earth's Crust with Self-Similar Distribution of Faults

Arcady V. Dyskin

School of Civil and Resource Engineering, The University of Western Australia
35 Stirling Hwy, Crawley, WA, 6009, Australia
phone: +618 9380 3987, fax: +618 9380 1044
adyskin@cyllene.uwa.edu.au

Abstract. The mechanical behaviour of the Earth's curst with a self-similar structure of faults is modelled using a continuous sequence of continua each with its own size of the averaging volume element. The overall tensorial properties scale isotropically, i.e. according to a power law with common exponent. In fault systems modelled by sets of 2-D cracks, the scaling is isotropic for both isotropically oriented and mutually perpendicular cracks. In the extreme anisotropic case of one set of parallel cracks, all effective compliances but one become zero, the non-vanishing component being scale independent. For isotropically oriented or mutually orthogonal cracks the wave propagation is characterised by extreme scattering, since any wave path is almost certainly intersected by a crack. The waves are not intersected only in the case of one set of parallel cracks. However, in this case due to its extreme anisotropy only extremely long waves (longer than all faults) can propagate.

1 Introduction

The multiscale structure of the Earth's crust makes it discontinuous in the sense that usually the continuum modelling can be applied only at very large scales. This makes the modelling quite challenging since the existing methods of discontinuous mechanics (eg. molecular dynamics type methods such as the discrete element method) are still computationally prohibitive in multi-scale situations. The situation becomes considerably more tractable if there are reasons to believe that the region of the Earth's crust being modelled is characterised by self-similar structure since then the concepts of fractal modelling can be used (eg, [1–5]). Such a modelling is based on a non-traditional philosophy according to which detailed local modelling, which is still not possible, is sacrificed in favour of a general description of scaling laws.

The fractal modelling can be enhanced by the adaptation of traditional notions of continuum mechanics to intrinsically discontinuous fractal objects. This can be accomplished by introducing a continuous set of continua of different scales [6–8]. These continua provide a combined description of the fractal object, which in the spirit of fractal modelling concentrates on scaling laws of properties or state variables which are integral to each continuum. The following reasoning underpins the method.

P.M.A. Sloot et al. (Eds.): ICCS 2003, LNCS 2659, pp. 845–854, 2003.

Traditional continuum mechanics is based on the introduction of length H which is considered to be infinitesimal. In practical terms, this length must be much smaller than the problem scale, L. Since real materials are discontinuous, at least at some scale, the length H must be considerably greater than the characteristic dimension of the discontinuities, l. Thus continuum modelling is only applicable to the situations when $l \ll H \ll L$. It is the left part of this inequality which is impossible to satisfy for geomaterials usually possessing discontinuities with dimensions covering a wide range scales (eg, [1]). If the Earth's crust structure is fractal or self-similar, the continuum modelling can be achieved by the introduction of a continuous set of modelling continua, each continuum being characterised by its own length H which is treated as infinitesimal within the continuum.

Formally this can be done as follows [8]. Consider a volume of the Earth's crust with a fractal structure, chose a scale, H, and remove all discontinuities of size H and greater. Thus we obtain a material with truncated structure and simultaneously set up a scale at which the truncated material can be modelled as a continuum. Therefore, the actual continuum modelling can be conducted for an object that is not fractal but rather a continuum that models, in some averaging manner, material with a certain structure of discontinuities of sizes smaller than H. By varying H one can model self-similar fractal objects by a continuous set of continua each of them being characterised by its own yardstick, H, specifying the scale. Each continuum models the fractal material at the scale H in the sense that the volume elements of size H in both the original material and the continuum show similar response to uniform loading. The yardstick H determines the resolution: no features with dimensions smaller than H are viewable in the H-continuum, [6]. Thus the H-continuum replaces the original material with the one possessing modified microstructure in which only those microstructural elements present that have characteristic sizes less than H. This leads to a continuum fractal modelling (Fig. 1) in which a model is produced that possesses an additional dimension, the scale [9], on top of conventional dimensions (four in the general case - three spatial dimensions and time).

When a fractal (self-similar) object is modelled by a continuum of continua, the overall characteristics of each continuum (eg, effective moduli) become functions of H. Usually these characteristics have the same sign for all continua, so according to the general theorem (eg, [10]) they must be represented by power functions, $f(H)=f^{*}H^{\alpha}$, where f^{*} is a prefactor. This is essentially the consequence of the fact that the fractal (self-similar) objects have no characteristic length. It should also be noted that if a certain characteristic is bounded (for instance Poisson's ratio) the exponent in its scaling law must be zero. Therefore bounded integral quantities are scale-independent.

Many integral quantities associated with a continuum are tensorial (overall properties such as effective moduli or compliances, integral state variables such as average stress, etc.). The main feature of tensorial properties is that their components undergo linear transformations when the coordinate frame is rotated. Since the transformed components must also scale as power functions and because the power functions with different exponents are linearly independent (e.g. [11]), the components must be either zero or infinite or scale with the same exponent. In other words for any tensor

$$f_{ijk\cdots}(H) = f^*_{ijk\cdots} H^{\alpha_{ijk\cdots}}, \quad 0 < f^*_{ijk\cdots} < \infty, \quad \alpha_{ijk\cdots} = \alpha = const \quad (1)$$

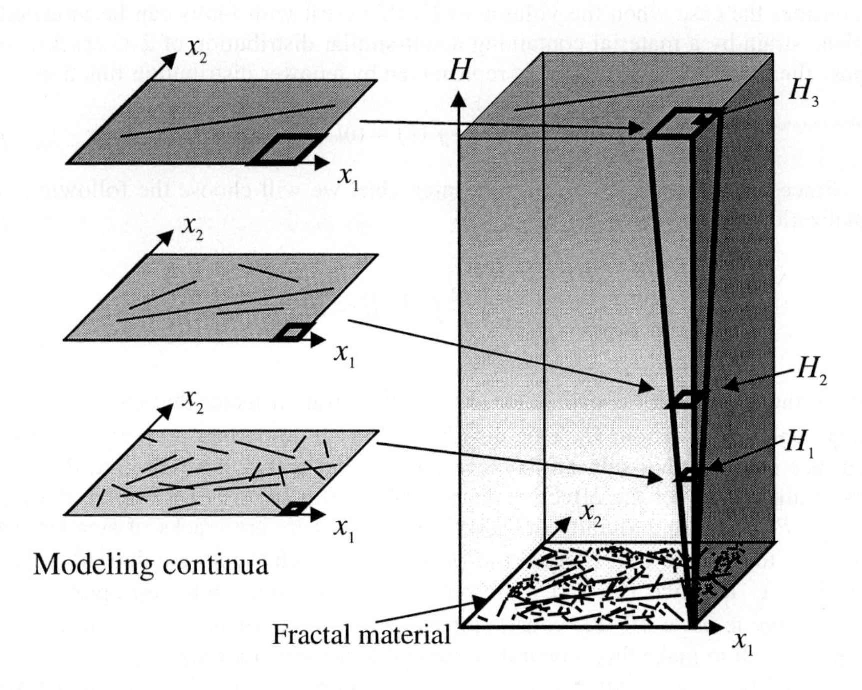

Fig. 1. Continuous fractal modeling. Only three continua out of a continuum of continua are shown

In particular, if the modelling continua are linearly elastic, the case considered hereafter, the tensors of general anisotropic moduli, C_{ijkl}, and compliances, A_{ijkl}, should scale as

$$C_{ijkl}(H) = c_{ijkl} H^{\alpha}, \quad A_{ijkl}(H) = a_{ijkl} H^{\beta}, \quad i, j, k, l = 1,2,3 \quad (2)$$

Therefore, the scaling of elastic characteristics must be isotropic. The anisotropy is only captured by the prefactors, c_{ijkl} and a_{ijkl}. The particular values of α, β, c_{ijkl} and a_{ijkl} depend on the material structure.

The present paper uses this approach to consider effective properties and wave propagation in the Earth's crust with self-similar faulting in the case where the faults can be represented as 2-D cracks.

2 Elastic Moduli of Earth's Crust with Self-Similar Distributions of Faults

Consider the case when the volume of Earth's crust with faults can be modeled in plane strain by a material containing a self-similar distribution of 2-D cracks. Suppose the crack size distribution is represented by a power distribution function

$$f(l) = \omega l^{-3} \tag{3}$$

Since the power functions are not integrable, we will choose the following normalization

$$\int_{l_{\min}}^{l_{\max}} l^2 f(l) dl = \Omega_t \tag{4}$$

Here, the dimensionless parameter ω is a concentration factor that ensures a specified total concentration, Ω_t. This distribution has an important property that distinguishes it from other self-similar distributions and makes it possible to determine the scaling laws for the effective characteristics [8]. In this distribution, the probability $P(n)$ that in a vicinity of a crack of size l, there are cracks of smaller sizes, from l/n to l, where $n>1$, is equal to $P(n)$~$\omega(n^2-1)$ which it does not depend upon the crack size, l. Since (3), (4) represent real distributions only asymptotically as $l_{max}/l_{min} \to \infty$, ie as $\omega \to 0$ (Ω_t=const), for any n the value of ω can be chosen sufficiently small to make this probability negligible for any crack size.

Therefore, in a vicinity of any crack only cracks of considerably smaller sizes can be found. Mechanically it means that the interaction between the cracks of similar sizes can be neglected; only the interaction between cracks of very different sizes is to be taken into account. Thus the differential self-consistent method, Salganik [12] can be used for calculating the effective characteristics.

According to the method, the compliance or modulus increments ΔA_{ij}, ΔC_{ij}, i,j=1, ..., 6, at each scale are determined by the contribution of non-interacting cracks of the corresponding scale considered in an effective continuum representing all cracks of smaller scales. The contribution is proportional to the concentration of this group of cracks, $\omega dH/H$. Therefore

$$A_{ij}(H + dH) = A_{ij}(H) + \omega S_{ij}(A_{11}, \ldots, A_{66}) dH/H \tag{5}$$

$$C_{ij}(H + dH) = C_{ij}(H) + \omega \Lambda_{ij}(C_{11}, \ldots, C_{66}) dH/H \tag{6}$$

where S_{ij}, Λ_{ij} are homogeneous functions of the first degree specific for the geometry and distribution of parameters of the cracks.

Substitution of (2) into (5) or (6) gives the scaling equations, which are equations to determine the exponent and prefactors:

$$\beta a_{ij} = \omega S_{ij}(a_{11}, a_{12}, \ldots a_{66}), \ \alpha c_{ij} = \omega \Lambda_{ij}(c_{11}, c_{12}, \ldots c_{66}) \quad (7)$$

Isotropically Oriented Cracks

In the case of randomly oriented cracks, the material is isotropic. Then, the expressions for effective Young's modulus, E, and Poisson's ratio, ν, written for the case of non-interacting cracks are (e.g., [12]):

$$E = E_m\left[1 - \frac{\pi}{4}\Omega\right], \quad \nu = \nu_m\left[1 - \frac{\pi}{4}\Omega\right] \quad (8)$$

where Ω is the crack concentration, E_m and ν_m are the Young's modulus and Poisson's ratio of the material. From here the corresponding components of tensor Λ_{ij} are $\Lambda_E = -E\pi/4$, $\Lambda_\nu = -\nu\pi/4$. Then, from the second equation (7) one gets the following equations for the exponent, α, and the prefactors e and ν:

$$\alpha e = -\frac{\pi}{4}\omega e, \quad \alpha\nu = -\frac{\pi}{4}\omega\nu \quad (9)$$

Since the Poisson's ratio is bounded, the second equation of (9) offers two solutions: either $\alpha=0$ or $\nu=0$. The first solution being substituted into the first equation of (8) leads to a trivial case of e=0. The second solution after substituting into the first equation of (8) gives the following scaling law

$$\nu = 0, \quad E = eH^{\alpha}, \quad \alpha = -\pi\omega/4 \quad (10)$$

In this scaling law the prefactor e is undeterminable and should be found independently, for instance from measurements at a certain scale. Then the scaling law (10) will determine the modulus for other scales. This scaling exponent is not related to the fractal dimension being D=2 for the case of cracks, because the cracks are considered as ideal cuts with no internal volume (or area in 2-D).

Plane with Two Mutually Orthogonal Sets of Cracks

Consider a plane with two mutually orthogonal sets of cracks, Fig. 2. It is assumed that the set of cracks perpendicular to the x_i axis is characterised by the distribution $\omega_i l^{-3}$ such that the total distribution has the concentration factor $\omega=\omega_1 + \omega_2$.

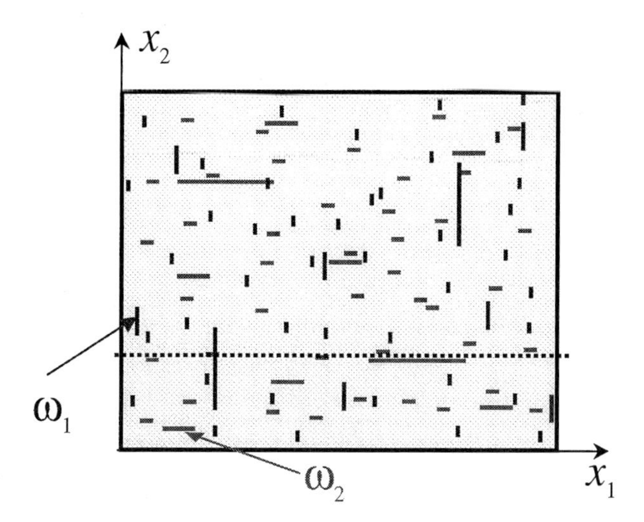

Fig. 2. Two mutually orthogonal sets of cracks with the concentration factors ω_1 and ω_2 respectively. Dotted line shows a possible wave path. It is almost certainly crosses a crack that is parallel to x_2 axis, however in the limiting case of a single set of parallel cracks (ω_1=0) the wave can propagate along the path shown without being interrupted

The analysis is based on Vavakin and Salganik's [13] solution for the effective compliances for an orthotropic plate with a set of non-interacting cracks aligned to one of the symmetry axes of the material. Generalizing their formula one can obtain the effective compliances for an orthotropic plate with two sets of non-interacting cracks of concentrations Ω_1 and Ω_2 normal to axes x_1 and x_2 (in a coordinate system x_1, x_2 aligned to the symmetry axes of the material):

$$\begin{cases} A_{11} = A_{11}{}^m + \dfrac{\pi}{4}\Omega_1\sqrt{A_{11}{}^m\left(2A_{12}{}^m + A_{66}{}^m + 2\sqrt{A_{11}{}^m A_{22}{}^m}\right)} \\ A_{22} = A_{22}{}^m + \dfrac{\pi}{4}\Omega_2\sqrt{A_{22}{}^m\left(2A_{12}{}^m + A_{66}{}^m + 2\sqrt{A_{11}{}^m A_{22}{}^m}\right)} \\ A_{66} = A_{66}{}^m + \dfrac{\pi}{4}\Omega_1\sqrt{A_{22}{}^m\left(2A_{12}{}^m + A_{66}{}^m + 2\sqrt{A_{11}{}^m A_{22}{}^m}\right)} \\ \qquad + \dfrac{\pi}{4}\Omega_2\sqrt{A_{11}{}^m\left(2A_{12}{}^m + A_{66}{}^m + 2\sqrt{A_{11}{}^m A_{22}{}^m}\right)} \\ A_{12} = A_{12}{}^m \end{cases} \tag{11}$$

Here $A_{11}{}^{m}$, $A_{22}{}^{m}$, $A_{12}{}^{m}$, $A_{66}{}^{m}$ are the compliances of the material, such that the Hook's law has a form: $\varepsilon_{11}=A_{11}{}^{m}\sigma_{11}+A_{12}{}^{m}\sigma_{22}$, $\varepsilon_{22}=A_{12}{}^{m}\sigma_{11}+A_{22}{}^{m}\sigma_{22}$, $\varepsilon_{12}=½A_{66}{}^{m}\sigma_{12}$.
Using the method outlined above the scaling equations can be obtained:

$$\begin{cases} \beta a_{ii} = \frac{\pi}{4}\omega_i\sqrt{a_{ii}\left(a_{66}+2\sqrt{a_{11}a_{22}}\right)}, \quad i=1,2 \\ \beta a_{66} = \frac{\pi}{4}\omega_2\sqrt{a_{11}\left(a_{66}+2\sqrt{a_{11}a_{22}}\right)}+\frac{\pi}{4}\omega_1\sqrt{a_{22}\left(a_{66}+2\sqrt{a_{11}a_{22}}\right)} \end{cases} \tag{12}$$

The solution represents the scaling laws:

$$\begin{aligned} &A_{ij} = a_{ij}H^{\beta}, \quad \beta = (\pi/2)\sqrt{\omega_1\omega_2} \\ &a_{22} = a_{11}(\omega_2/\omega_1)^2, \quad a_{66} = 2a_{11}(\omega_2/\omega_1), \quad a_{12}=0 \end{aligned} \tag{13}$$

For the case of a single set of cracks, i.e. when the concentration of the other set vanishes, say $\omega_2\to 0$, the exponent and all compliances except a_{11} vanish. The material becomes completely rigid in the direction x_2. This can easily be understood if one considers that in fractal materials with finite ω the total crack concentration is infinite implying the infinite total compliancy. What the scaling law determines is the way the moduli change in transition from one scale to another. Relative to such moduli the materials without cracks as well as the material in the directions not affected by cracks (the direction x_2 in this case) become infinitely rigid. In real situations we do not have true fractal materials as there are always lower and upper cut-offs. Then the complete rigidity simply means a very high modulus as compared to the values of other moduli.

3 On the Possibility of Wave Propagation

When a wave is sent through an Earth's crust with self-similar fault system its path may be intersected by the faults (as shown in Fig. 2). In order to determine the probability p of intersection we will use the renormalization technique. We introduce a scale, H and consider all faults with dimensions smaller than H. Take an arbitrary path of length $L>>H$. The path length L should be large enough to make the probability p_H of faults intersecting the path independent of particular realizations of fault positions. Divide the path into $m>>1$ segments. Each segment can be intersected either by faults of dimension less than H/m – this would happen with probability $p_{H/m}$ – or by faults of dimensions between H/m and H. This last probability will be denoted by q. Probability p_H as any property should scale according to the power law. On the other hand, $0\le p_H \le 1$. This implies that the exponent must be zero such that p_H =const. In particular, $p_{H/m}=p_H=p$. The path will not be intersected if all of its segments are not intersected. Therefore

$$1-p=(1-p)^m(1-q) \tag{14}$$

If q>0, i.e. faults of dimensions between H/m and H can intersect the path, the only solution of equation (14) is p=1. This for example corresponds to the cases of isotropically oriented faults or two sets of mutually orthogonal faults. When q=0, i.e. faults of dimensions between H/m and H cannot intersect the path, p=0. This is, for instance, the case of one set of cracks parallel to the x_1 axis (ω_1=0in Fig. 2) with wave path being parallel to the cracks.

This result suggests that only in the case of special crack arrangements, like parallel cracks one can expect the wave transmission. In other cases there should be extreme scattering. (In principle, the fact that the probability of intersection is one does not mean that there are no paths without intersections, just there are too few of them. So, the question of wave propagation in these cases would require further study.)

We now analyze the wave propagation parallel to faults in the case of one set. In order to do this we will formally write the wave equations for the case of two mutually orthogonal sets of cracks (Fig. 2) and then set ω_1=0.

Using (13) the Hook's law can be expressed in the following form

$$\varepsilon_{11} = a_{11}\sigma_{11}, \quad \varepsilon_{22} = a_{11}\left(\frac{\omega_1}{\omega_2}\right)^2 H^\beta \sigma_{22}, \quad \varepsilon_{12} = \frac{1}{2}a_{11}\left(\frac{\omega_1}{\omega_2}\right) H^\beta \sigma_{12} \tag{15}$$

Expressing σ_{ij} from (15) through strains and then through displacements and substituting the result into 2-D equations of motion $\partial^2\sigma_{ij}/\partial x_j^2=\rho\partial^2 u_i/\partial t^2$, where i=1,2, ρ is the density (scale independent in the case of cracks), one gets

$$\begin{aligned} &\frac{\partial^2 u_1}{\partial x_1^2} + \frac{1}{2}\frac{\omega_1}{\omega_2}\left(\frac{\partial^2 u_1}{\partial x_2^2} + \frac{\partial^2 u_2}{\partial x_1 \partial x_2}\right) = \rho a_{11}\ddot{u}_1 H^\beta \\ &\frac{1}{2}\frac{\omega_1}{\omega_2}\left(\frac{\partial^2 u_2}{\partial x_1^2} + \frac{\partial^2 u_1}{\partial x_1 \partial x_2}\right) + \left(\frac{\omega_1}{\omega_2}\right)^2 \frac{\partial^2 u_2}{\partial x_2^2} = \rho a_{11}\ddot{u}_2 H^\beta \end{aligned} \tag{16}$$

After the limiting transition $\omega_1 \to 0$ one has β=0 and system (16) reduces to a single equation

$$\frac{\partial^2 u_1}{\partial x_1^2} = \rho a_{11}\ddot{u}_1 \tag{17}$$

Solution of wave equation (17) will be sought in the form of the longitudinal wave

$$u_1(x_1, x_2, t) = \Phi(x_2)\exp(ik(x_1 - \nu t)) \tag{18}$$

where k is the wave number, ν is the frequency. Function Φ is determined after substitution of (18) into (17). It assumes the form

$$\Phi(x_2)=A\sinh(\nu/v\, x_2)+B\cosh(\nu/v\, x_2), \text{ where } v=(\rho a_{11})^{-1/2} \tag{19}$$

To determine constants A and B consider a case when the wave is between two large faults (much longer than the wave length). Let these faults be at a distance $2h$ apart. Then, recalling that the faults are modeled as cracks at which surfaces $\sigma_{12}=0$ one obtains $A=0$ and $\nu=0$. Therefore the waves that can propagate have vanishing frequency and, if the velocity ν/k is finite, the infinite length. This is the consequence of the extreme anisotropy associated with the fault distribution of this type. Recalling that the fractal medium is only an approximation to the Earth's crust one can conclude that even in the case of parallel faults only very long (very low frequency) longitudinal waves can propagate and they propagate with very low velocity.

4 Conclusions

The continuum fractal modelling of mechanical behaviour of the Earth's crust with self-similar structure is based on representing the object as a continuum of continua of different scales. It concentrates on scaling of overall mechanical properties and integral state variables, which is described by power laws. The tensorial quantities scale by power laws with exponents common for all components of the tensors.

Thus, effective elastic characteristics of the Earth's crust with self-similar faulting structure always scale isotropically. Even in the extreme case of one set of parallel faults when all compliances but one vanish, since the non-vanishing component is scale independent, the exponents are formally zero, so the scaling is still isotropic.

Any wave path will be intersected by a fault with probability one if the faults are not all parallel. Only in the case of a single set of parallel faults the waves are not intersected, but in this case due to the extreme anisotropy of the medium, only extremely long waves can propagate.

Acknowledgment. The author is grateful to the Australian Research Council for the financial support (Large grant A00104937).

References

1. Sadovskiy, M.A.: Distribution of Preferential Sizes in Solids. Transactions USSR Academy of Sciences. Earth Science Series, 269 (1983) 8–11
2. Scholz, C.H.: The Mechanics of Earthquakes and Faulting. Cambridge University Press, Cambridge, New York, Port Chester, Melbourne, Sydney (1990)
3. Barton, C.A., Zoback, M.D.: Self-Similar Distribution and Properties of Macroscopic Fractures at Depth in Crystalline Rock in the Cajon Pass Scientific Drill Hole. J. Geophys. Res. 97B (1992) 5181–5200
4. Turcotte, D.L.: Fractals and Chaos in Geology and Geophysics. Cambridge University Press (1993)
5. Dubois, J.: Non-Linear Dynamics in Geophysics. John Wiley and Sons, Chichester, New York, Weinheim, Brisbane, Singapore, Toronto (1998)
6. Dyskin, A.V.: Stress-Strain Calculations in Materials with Self-Similar or Fractal Microstructure. In: Valiapan, S., Khalili, N. (eds.) Computational Mechanics – New Frontiers for New Millennium. Vol. 2 (2001) 1173–1178

7. Dyskin, A.V.: Mechanics of Fractal Materials. In: Karihaloo, B. (ed.) Proc. of the IUTAM Symposium on Analytical and Computational Fracture Mechanics of Non-Homogeneous Materials. Kluwer Academic Press (2002) 73–82
8. Dyskin, A.V.: Continuum Fractal Mechanics of Earth's Crust. Pure and Applied Geophysics (PAGEOPH) (2003) (Accepted)
9. Rodionov, V.N., Sizov, I.A., Kocharyan, G.G.: The Discrete Properties of Geophysical Medium. In: Modelling of Natural Objects in Geomechanics, Nauka, Moscow (1989) 14–18 (in Russian)
10. Barenblatt, G.I., Botvina, L.R.: Application of the Similarity Method to Damage Calculation and Fatigue Crack Growth Studies, In: Sih, G.C., Zorski, H. (eds.) Defects and Fracture, Martinus Nijhoff Publishers (1980) 71–79
11. Achieser, N.I.: Theory of Approximation. Frederic Unger Publishing Co, New York (1956)
12. Salganik, R.L.: Mechanics of Bodies with Many Cracks. Mech. of Solids, 8 (1973) 135–143
13. Vavakin, A.S., Salganik, R.L.: Effective Elastic Characteristics of Bodies with Isolated Cracks, Cavities, and Rigid Nonhomogeneities. Mech. of Solids, 13 (1978) 87–97

Performance Modeling Codes for the QuakeSim Problem Solving Environment

Jay Parker[1], Andrea Donnellan[1], Gregory Lyzenga[1,2],
John Rundle[3], and Terry Tullis[4]

[1] Jet Propulsion Laboratory/California Institute of Technology,
Pasadena, California
Jay.W.Parker@jpl.nasa.gov
http://www-aig.jpl.nasa.gov/public/dus/quakesim

[2] Department of Physics, Harvey Mudd College
Claremont, California
lyzenga@hmc.edu

[3] Department of Physics,
University of California Davis,
Davis, California
rundle@physics.ucdavis.edu

[4] Department of Geological Sciences
Brown University, Providence, Rhode Island
terry_tullis@brown.edu

Abstract. The QuakeSim Problem Solving Environment uses a web-services approach to unify and deploy diverse remote data sources and processing services within a browser environment. Here we focus on the high-performance crustal modelling applications that will be included in this set of remote but interoperable applications. PARK is a model for unstable slip on a single earthquake fault represented as discrete patches, able to cover a very wide range of temporal and spatial scales. GeoFEST simulates stress evolution, fault slip and visco-elastic processes in realistic materials. Virtual California simulates fault interaction to determine correlated patterns in the nonlinear complex system of an entire plate boundary region. Pattern recognition tools extract Karhunen-Loeve modes and Hidden Markov state models from physical and virtual data streams. Sequential code benchmarking demonstrates PARK computes 15,000 patches for 500 time steps in under 8 hours (SGI Origin 3000), GeoFEST computes 50,000 tetrahedral elements for 1000 steps in under 14 hours (Sun Workstation), and Virtual California computes 215 fault segments for 10,000 time steps in under 0.5 hours (Pentium III). QuakeSim goals for June 2004 are to deploy MPI parallel codes that compute 400,000 patches (PARK), 16,000,000 tetrahedra (GeoFEST) and 700 segments (Virtual California) in essentially the same wallclock time, incorporating powerful tools such as stress field multipoles and the ESTO/PYRAMID mesh partitioning and refinement tools.

P.M.A. Sloot et al. (Eds.): ICCS 2003, LNCS 2659, pp. 855–862, 2003.

1 Introduction

The full objective over this three-year program (begun April 2002) is to produce a system to fully model earthquake-related data. Components of this system include:

1. A database system for handling both real and simulated data
2. Fully three-dimensional finite element code (FEM) with adaptive mesh generator capable of running on workstations and supercomputers for carrying out earthquake simulations
3. Inversion algorithms and assimilation codes for constraining the models and simulations with data
4. A collaborative portal (object broker) for allowing for seamless communication between codes, reference models, and data
5. Visualization codes for interpretation of data and models
6. Pattern recognizers capable of running on workstations and supercomputers for analyzing data and simulations

In order to develop a solid earth science framework for understanding and studying of active tectonic and earthquake processes, this task develops simulation and analysis tools to study the physics of earthquakes using state-of-the-art modeling, data manipulation, and pattern recognition technologies. We develop clearly defined accessible data formats and code protocols as inputs to the simulations. These are adapted to high-performance computers because the solid earth system is extremely complex and nonlinear resulting in computationally intensive problems with millions of unknowns. With these tools it will be possible to construct the more complex models and simulations necessary to develop hazard assessment systems critical for reducing future losses from major earthquakes. Use of multiple data types in a coherent modeling effort is illustrated in [1].

The system for unifying the data sources, modeling codes, pattern analysis and visualization is being constructed on a web-services model, and is described in a companion paper by Pierce *et al.* ("Interacting Data Services for Distributed Earthquake Modeling") and [2-5]. Some of the applications and their data linkages are shown in Fig. 1. Within our web services framework, these applications may be running multiple instances on separate machines located anywhere on the internet. Data sources (as in [1], where GPS and inSAR data is used) and fault databases such as the one under development [6-8] are also available through this portal system as distributed objects with clear interface definitions.

Three of these codes are targeted for improved performance, chiefly through design changes that make them efficient high-performance parallel codes. These codes are PARK, a boundary-element based code for studying unstable slip at the Parkfield segment of the San Andreas fault, Virtual California, which simulates the dynamic interaction of hundreds of fault segments comprising the active tectonics of California, and GeoFEST, a fully three-dimensional finite element code to model active tectonics and earthquake processes. Together with an adaptive mesh generator that constructs a mesh based on geometric and mechanical properties of the crustal structure, the GeoFEST system makes it possible to efficiently model time-dependent deformation of interacting fault systems embedded in a heterogeneous earth structure.

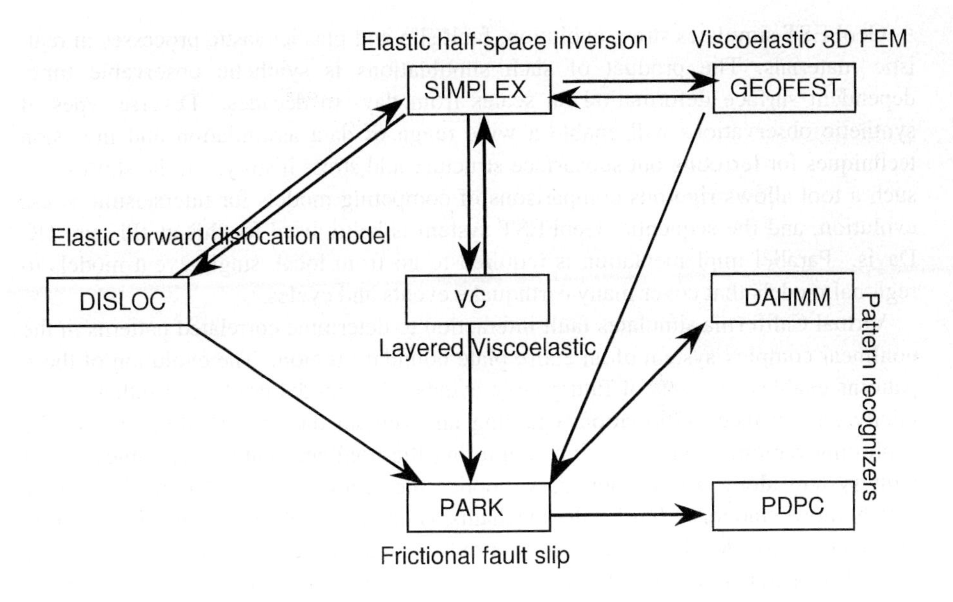

Fig. 1. Linkages between programs that maybe run separately or coupled to other programs. Low-level movement of actual data occurs between programs. There is also communication of metadata and information about the files being transferred

2 Scientific Significance

The full interoperable system will allow users from many environments to discover and exploit a very wide range of applications, models, and physical measurements in distributed databases. This is seen as of crucial importance for gaining full advantage of massive new national programs for gathering new regional and global data sets, such as the Plate Boundary Observatory, Earthscope, and NASA orbiting inSAR missions.

PARK is a model for unstable slip on a single earthquake fault. Because it aims to capture the instability it is designed to represent the slip on a fault at many scales, and to capture the developing seismic slip details over an extraordinary range of time scales (subseconds to decades). Its simulation of the evolution of fault rupture is the most realistic of the tools considered here. When transformed into an efficient parallel simulation, it will be the tool of choice for researchers seeking to determine the nature and detectability of earthquake warning signals such as surface strains and patterns of microseismicity. This is the first earthquake simulation code to seek enhanced scalibility and speed by employing a multipole technique. The multipole experience gained here will also be transferrable to the Virtual California code and other boundary element simulations. The power of massive parallel computing is required for this problem in order to support many small slip patch elements in order to cover the nucleation scale that initiates the instability.

GeoFEST simulates stress evolution, fault slip and plastic/elastic processes in realistic materials. The product of such simuluations is synthetic observable time-dependent surface deformation on scales from days to decades. Diverse types of synthetic observations will enable a wide range of data assimilation and inversion techniques for ferreting out subsurface structure and stress history. In the short term, such a tool allows rigorous comparisons of competing models for interseismic stress evolution, and the sequential GeoFEST system is being used for this at JPL and UC Davis. Parallel implementation is required to go from local, single-event models to regional models that cover many earthquake events and cycles.

Virtual California simulates fault interaction to determine correlated patterns in the nonlinear complex system of an entire plate boundary region. The evolution of these patterns enables forecasts of future large events. The model produces synthetic seismicity and surface deformation, enabling an eventual data assimilation system for exploiting regional data collection. Capturing the nonlinear pattern dynamics of the fault system along a plate boundary implies the realization of a digital laboratory which allows understanding of the mechanisms behind the observations and patterns. Our technology development aims to produce and demonstrate a scalable cluster code. When that is deployed researchers will be able to create and verify patterns down to smaller spatial scales, which will enable cross-scale parameterization and validations which will in turn enable plate-boundary system analysis and greatly enhanced forcasts of large earthquakes.

Pattern analysis methods are another tool type we are developing. Hidden Markov methods are one method incorporated into this framework [9]. Another method [10,11] bins many decades of seismic activity on a gridded representation of Califorinia. Eigensystem analysis reveals clusters of correlated space and time activity, which have been subjected to a phase dynamics forecast. When this method attains parallel speedup we will produce better forecasts and enable speedy tests of earthquake correlations and predictions. This will be due to the ability to use much smaller geographic cell sizes and so forecast the frequent magnitude 4 earthquakes, not just the rare magnitude 6 events.

3 Technology Accomplishments

The development of the three performance codes is at an early stage. Our planned milestones (including code parallel performance goals) are posted at the web site http://www-aig.jpl.nasa.gov/public/dus/quakesim/milestones.html.

The Milestone E code baseline benchmark marks the beginning of moving the three codes, PARK, geoFEST, and Virtual California to high-performance parallel computers, enabling the simulation of interacting fault systems leading to earthquakes. The scientific significance and performance results have been reported in http://www-aig.jpl.nasa.gov/public/dus/quakesim/MilestoneE_Baselines.pdf and are summarized in the following table:

Table 1. Summary of performance codes sequential computing time on baseline problem with resolution, time steps indicated

Code	Resolution/ Time Steps	Platform (processors)	Wallclock time
PARK	15,000 patches/ 500 steps	SGI Origin 3000, 600 Mhz (1)	7h53m
geoFEST	50,000 elements/ 1000 steps	Sparc, 450 Mhz (1)	13h43m
Virtual California	215 segments/ 10,000 steps	Pentium III, 1 Ghz (1)	5m38s (stress GF) 15m 48s (steps)

Performance for PARK on a single processor can handle a fault region with 15,000 fault elements over 500 time steps in just under 8 hours. But the problem of determining possible earthquake precursors requires a finer sampling. This project is committed to demonstrating the computation of 400,000 elements on 1024 processors, enabling 50,000 time steps in 40 hours by June, 2004. Such performance will be attained by combining efficient parallel problem decomposition and by exploiting an NlogN multipole technique [12] to reduce the currently N^2 element interaction compuations. This multipole method has been demonstrated in parallel astrophysics simulations with considerable success.

Performance for Virtual California [13] has been demonstrated at 215 interacting fault segments for 10,000 time steps in 20 minutes on a 1 GHz Linux workstation. This type of simulation allows determination of correlated patterns of activity that can be used to forecast seismic hazard, with relevance for earthquakes of Magnitude 6. Finer sampling may allow verification of such forecasts by including much smaller, more frequent earthquakes. An initial parallel demonstration appears in [14] The aims of this task are to validate parallel scaling of the Virtual California code, and evaluate the potential for the multipole technique pioneered in the PARK demonstration.

GeoFEST [15] is a direct implementation of stress-displacement volumetric finite elements. The mechanics include elasticity and visco-elastic deformation. Current workstation capacity can solve for the deformation and stress evolution due to a single-fault rupture, such as the Northridge event, using ~100,000 finite element equations. We seek to analyze the modes of interaction of the entire Southern California system of interacting faults, covering a portion of the crust ~1000 km on a side. Such a simulation would require ~5M equations to determine first-order effects, and certainly higher density for faithful representation of the time-dependent stress field. Current techniques require running such a model through thousands of time steps to attain a stable background stress field and assess the patterns of fault interactions. These considerations motivate tailoring the code toward hundreds of processors to attain solutions in reasonable turnaround time.

Our team has created a meshing tool with mesh generation and adaptation capabilities, while the ESS project has developed a parallel adaptive meshing library (the

Pyramid library) for supporting parallel dynamic adaptive meshing of unstructured meshes. Techniques of parallel adaptive meshing, combined with local error estimates, can be effectively used to compute an initial state of the displacement and stress fields, and to significantly reduce the computational load of computing the evolving stress field in the finite element modeling.

We are therefore in an excellent position to evaluate the ESS Adaptive Mesh Refinement (AMR) library with our finite element code, and to combine the strengths of both meshing tools into an integrated adaptive meshing library for a scalable parallel computing environment. The integrated adaptive meshing library will support the entire process of an unstructured parallel application, including initial mesh generation, dynamic parallel mesh adaptation with efficient mesh quality control, and dynamic load balancing. This will be a valuable tool for improving the computational efficiency of our finite element modeling.

Current GeoFEST performance computes 1000 time steps, and 50,000 volumetric finite elements with an unstructured mesh with large range of element sizes, in just under 14 hours on a 450 Mhz Sun Solaris computer. We plan to demonstrate problems with a complex embedded system of faults with 16,000,000 finite elements with the same number of time steps and solution time on 880 processors. We will use PYRAMID for parallel decomposition and adaptive mesh refinement, and a parallel iterative Conjugate Gradient method for the sparse matrix solution.

4 Status and Plans

In this fractional first year we have

- Agreed upon a software engineering plan,
- Selected a review board of regarded earthquake and computational science experts,
- Established a public web page area (http://www-aig.jpl.nasa.gov/public/dus/quakesim/) to allow documentation and access to our products,
- Demonstrated and documented the serial performance of three important codes believed to be scalable to high performance computing.
- Established requirements and design policy for a framework that will make community earthquake codes interoperable and deliverable to the scientific community.

In the coming year we will develop the three main codes for efficient parallel performance. Virtual California and geoFEST will be adapted for Riva and Parvox parallel rendering. Also, geoFEST will be linked with the PYRAMID mesh refinement library. We will develop the interoperability framework and tie in a fault database, mesh generator and the geoFEST code and demonstrate the prototype framework with a combined simulation. The pattern analysis software and data inversion and assimilation techniques will also be included within the framework.

This program differs from many past efforts in that the source code for the three performance codes is publicly available through the web site listed (geoFEST is subject to a no-cost license arrangement to US citizens through the Open Channel Foundation). The applications will also be available through the QuakeSim interoperability framework, including a supported ability to invoke the applications through the web-based portal for large problem solutions for authorized investigators. These modes of access will be supported as the applications are extended to efficient parallel software.

Acknowlegments. This work was performed at the Jet Propulsion Laboratory, California Institute of Technology, under a contract with the National Aeronautics and Space Administration.

References

1. Donnellan, A., J. Parker, and G. Peltzer, Combined GPS and InSAR models of postseismic deformation from the Northridge earthquake, *Pure and Appl. Geophys.*, 2261–2270, 2002.
2. Donnellan, Andrea, Geoffrey Fox, John Rundle, Dennis McLeod, Terry Tullis, Lisa Grant, Jay Parker, Marlon Pierce, Gregory Lyzenga, Anne Chen, John Lou, The Solid Earth Research Virtual Observatory: A web-based system for modeling multi-scale earthquake processes, *EOS Trans, AGU, 83*(47), Fall Meeting Suppl., Abstract NG528-08, Invited, 2002.
3. Fox, Geoffrey, Hasan Bulut, Kangseok Kim, Sung-Hoon Ko, Sangmi Lee, Sangyoon Oh, Shrideep Pallickara, Xiaohong Qiu, Ahmet Uyar, Minjun Wang, Wenjun Wu, Collaborative Web Services and Peer-to-Peer Grids to be presented at 2003 Collaborative Technologies Symposium (CTS'03).
http://grids.ucs.indiana.edu/ptliupages/publications/foxwmc03keynote.pdf
4. Fox, Geoffrey, Hasan Bulut, Kangseok Kim, Sung-Hoon Ko, Sangmi Lee, Sangyoon Oh, Xi Rao, Shrideep Pallickara, Quinlin Pei, Marlon Pierce, Ahmet Uyar, Wenjun Wu, Choonhan Youn, Dennis Gannon, and Aleksander Slominski, "An Architecture for e-Science and its Implications" in Proceedings of the 2002 International Symposium on *Performance Evaluation of Computer and Telecommunications Systems*, edited by Mohammed S.Obaidat, Franco Davoli, Ibrahim Onyuksel and Raffaele Bolla, Society for Modeling and Simulation International, pp 14–24 (2002).
http://grids.ucs.indiana.edu/ptliupages/publications/spectsescience.pdf
5. Fox, Geoffrey, Sung-Hoon Ko, Marlon Pierce, Ozgur Balsoy, Jake Kim, Sangmi Lee, Kangseok Kim, Sangyoon Oh, Xi Rao, Mustafa Varank, Hasan Bulut, Gurhan Gunduz, Xiaohong Qiu, Shrideep Pallickara, Ahmet Uyar, Choonhan Youn, *Grid Services for Earthquake Science, Concurrency and Computation: Practice and Experience in ACES Special Issue*, 14, 371–393, 2002.
http://aspen.ucs.indiana.edu/gemmauisummer2001/resources/gemandit7.doc
6. Gould, M., Grant, L., Donnellan, A., and D. McLeod. The GEM Fault Database: A Preliminary report on design and approach. *Proceedings and Abstracts*, SCEC Annual Meeting, p. 75–76, 2002.

7. Grant, L.B. and Gould, M.M. Paleoseismic and geologic data for earthquake simulation. Computational Science, Data Assimilation, and Information Technology for Understanding Earthquake Physics and Dynamics, 3rd ACES International Workshop, Maui, Hawaii, USA, p. 30, 2002.
8. Grant, L.B. and M.M. Gould. Assimilation of paleoseismic data for earthquake simulation. Submitted to Pure and Applied Geophysics, 2002.
9. Granat, R., and A. Donnellan, A hidden Markov model tool for geophysical data exploration, *Pure and Appl. Geophys*, 2271–2284, 2002.
10. Rundle, J.B., W. Klein, K.F. Tiampo, Andrea Donnellan, and G.C. Fox, Detection and Analysis of Space-Time Patterns in Complex Driven Threshold Systems, paper submitted to the 2003 International Conference on Computational Science, Melbourne, AU June 2–4, 2003.
11. Tiampo, K.F., J.B. Rundle, S. McGinnis and W. Klein, Pattern dynamics and forecast methods in seismically active regions, *Pure and Appl. Geophys.*, 159, 2429–2467, 2002.
12. Tullis, T.E., J. Salmon, N. Kato, and M. Warren, The application of fast multipole methods to increas the efficiency of a single fault numerical earthquake model, *Eos. Trans. Am. Geophys, Union, Fall Meeting Suppl.*, 80, F924 1999.
13. Rundle, J.B., P.B. Rundle, W. Klein, J. Martins, K.F. Tiampo, A. Donnellan and L.H. Kellogg, GEM plate boundary simulations for the Plate Boundary Observatory: Understanding the physics of earthquakes on complex fault systems, *Pure and Appl. Geophys.*, 159, 2357–2381 2002.
14. Tiampo, K.F., J.B. Rundle, S. Gross and S. McGinnis, Parallelization of a large-scale computational earthquake simulation program, *Concurrency & Computation, Practice and Experience*, 14, 531-550, ACES Special Issue, 2002.
15. Parker, Jay, Gregory Lyzenga, Jin-Fa Lee, John Lou, and Andrea Donnellan, Technologies for larger crustal studies by finite element simulation, Southern California Earthquake Center Annual Meeting, Oxnard, CA, September 8–11 2002.

Interacting Data Services for Distributed Earthquake Modeling

Marlon Pierce[1], Choonhan Youn[1,2], and Geoffrey Fox[1]

[1]Community Grid Labs, Indiana University
501 N. Morton Street, Suite 224
Bloomington, IN 47404-3730
{marpierc,gcf}@indiana.edu
[2]Department of Electrical Engineering and Computer Science, Syracuse University
cyoun@ecs.syr.edu

Abstract. We present XML schemas and our design for related data services for describing faults and surface displacements, which we use within earthquake modeling codes. These data services are implemented using a Web services approach and are incorporated in a portal architecture with other, general purpose services for application and file management. We make use of many Web services standards, including WSDL and SOAP, with specific implementations in Java. We illustrate how these data models and services may be used to build distributed, interacting applications through data flow.

1 Introduction

This paper describes our designs and initial efforts for building interacting data services for our earthquake modeling Web portal (QuakeSim). The QuakeSim portal targets several codes, including the following.: *Disloc* produces surface displacements based on multiple arbitrary dipping dislocations in an elastic half-space; *Simplex* inverts surface geodetic displacements to produce fault parameters; *GeoFEST* is a three-dimensional viscoelastic finite element model for calculating nodal displacements and tractions; *Virtual California* simulates interactions between vertical strike-slip faults; *PARK* is a boundary element program to calculate fault slip velocity history based on fault frictional properties. A complete code list and detailed descriptions may be found at [1].

The QuakeSim portal provides the unifying hosting environment for managing the various applications listed above. We briefly describe the portal here for reference, and include a more detailed description in the following section. QuakeSim is built out of portlet containers that access distributed resources via Web services, as described in [2]. The applications are wrapped as XML objects that can provide simple interactions with the hosting environments of the codes, allowing the codes to be executed, submitted to batch queuing systems, and monitored by users through the

P.M.A. Sloot et al. (Eds.): ICCS 2003, LNCS 2659, pp. 863–872, 2003.

browser. Web services are also used to manage remote files and archive user sessions. A general view of the portal architecture is shown in Fig. 1.

Going beyond simple submission and management of jobs, we must also support interactions between the portal's applications. The following motivating scenario [3] illustrates the value of code integration: InSAR satellite data produces a map of surface deformations. Simplex takes this surface data and produces models with errors of the underlying fault systems that produce the deformation. By using synthetic data, one may test the potential value of additional InSAR data. In particular, one may compare the improvement on errors in the fault models when one or two additional "look" angles are added. One additional angle may be obtained from the same satellite, collecting data in both the ascending and descending portions of the satellite's path. Two additional look angles, however, require an additional satellite. Disloc may be used to generate the necessary synthetic InSAR data. Simulation results indicate in fact that an additional data stream generates a significant improvement on the estimated error parameters of the modeled fault. However, adding a third data stream produces a much less noticeable improvement over two data streams.

Scenarios such as the above require some familiarity with the simulation codes in order to manage the code linkage. When we encounter the fully interacting system, we face the additional problem of scaling. Linking any two codes may be done in a one-time fashion, but linking multiple codes is a difficult problem greatly simplified by common data formats. Also, from the portal architecture point of view, it becomes possible to develop both general purpose tools for manipulating the common data elements and also a well-defined framework for adding new applications that will share data.

2 QuakeSim Portal Architecture Overview

QuakeSim is based around a Web Services model, illustrated in Fig. 1. The user interacts with the portal through a web browser, which accesses a central user interface server. This server maintains several client stubs that provide local interfaces to various remote services. These remote services are described in the Web Service Description Language (WSDL) [4] and are invoked via Simple Object Access Protocol (SOAP) [5] invocations over HTTP. These services are invoked on various service-providing hosts, which may in turn interact with local or remote databases (through JDBC, or Java Database Connectivity), as shown for Hosts 1 and 3; or with local queuing environments, as shown for Host 2. WSDL and SOAP are particularly useful when dealing with XML data: WSDL method (function) declarations can take both simple (string, integer, or float) and custom XML types as arguments and return types, and SOAP messages can be used as envelopes to carry arbitrary XML payloads. Readers interested in a general Web service overview are referred to [6].

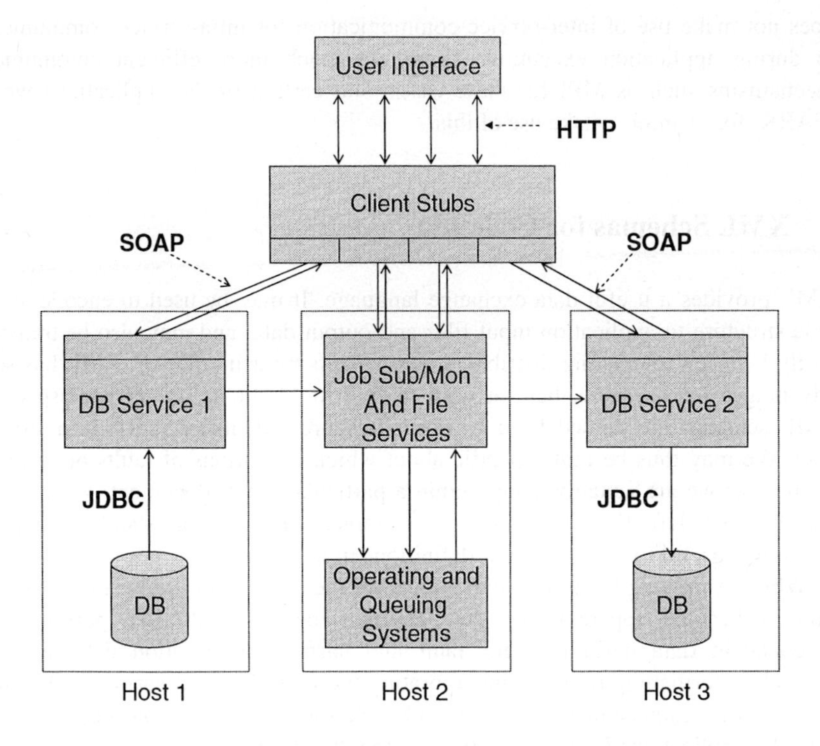

Fig. 1. QuakeSim portal architecture with Web service invocations of remote services. Arrows indicate remote invocations with indicated protocols

The service architecture allows the portal user to browse the database on Host 1, determine the interesting data file and transfer it to Host 2. Host 2 maintains application executables, and the application may be run with the appropriate input data on a queuing system (Host 2 may be a cluster or parallel computer). Following completion of the job execution, the user may transfer the output back to another database or file system, or she may download the files to her desktop.

More detailed descriptions of the portal architecture are available from [2]. The essential pieces for file transfer and job submission and monitoring have been developed. We must now address services for managing input and output data formats for specific applications. Before proceeding, however, we wish to clarify the communication infrastructure and performance. The system we present here enables loosely couple distributions of applications and data. The primary feature of this system is the use of WSDL to describe interfaces to services. These services may be bound to more than one protocol, so if high performance file transfer (for example) is needed, we may make use of a non-SOAP implementation of a higher performance protocol. In the current system, the data sizes are sufficiently small and the inter-service communication sufficiently infrequent, that communication performance is not an issue. System time is dominated by user interactions with the system, queuing system wait times, and (in the case of Simplex) execution time. The system we have described does not make use of inter-service communication (or intra-service communication) to during application execution. There are much more efficient communication mechanisms such as MPI for these cases, and certain of the applications we wrap (PARK, for example) make use of this.

3 XML Schemas for Code I/O

XML provides a useful data exchange language. It may be used to encode and provide structure to application input files and output data, and may also be transmitted easily between interacting distributed components by using SOAP. XML has several advantages for use here, but we wish to first highlight its use of namespaces. All XML schemas are named by a Uniform Resource Identifier (URI) in a structured way. We may thus be quite specific about which definitions of faults or surface deformations we are actually using within a particular XML document. We do not expect that our definitions for faults, for example, will be a final standard adopted by all, so it is useful to qualify all our definitions in this manner.

While examining the inputs and outputs for the applications to be added to QuakeSim, it became apparent that the data may be split into two portions: code-independent data definitions for fault and surface deformation data, and code-dependent formatting rules for incorporating the fault data and various code parameters, such as number of iterations and observation points. We consider as a starting case the applications Disloc and Simplex, together with fault characterization needed by all applications.

Our schema for faults is available from [8]. We highlight the major elements here. We structure our fault definitions as being composed of the following items. The *Map View* includes elements for longitude, latitude, and strike angle for the fault. The *Cartesian View* describes the location and dimensions of the fault in Cartesian space. Parameters include depth of the fault, width, length, and dip angle. *Material Properties* include various parameters needed to characterize the fault such as Lame parameters. *Slip* includes the strike slip, dip slip, and tensile components of the fault slip. Finally, *Rate* includes strike, dip, and tensile rates. Surface displacements may also be expressed in XML, as shown in [9]. Displacements may be characterized by their locations on a two-dimensional observation plane and the values of the three dimensional displacements (or rates of displacements) and errors.

Disloc and Simplex schemas may be viewed as [10] and [11], respectively. These may be compared to the actual input instructions for the codes as described in [12] and [13]. Note that we are not modifying the codes to take directly the XML input. Rather, we use XML descriptions as an independent format that may be used to generate input files for the codes in the proper legacy format. Both codes' schemas defer the definitions of Faults and Displacements to the appropriate external schema definitions, which may be included by the use of XML namespaces. The application schemas simply define the information necessary to implement the input files. The Disloc schema, for example, defines the optional format for observation points, which may be either on a regular grid or at a group of specified points.

4 Implementing Services for the Data Model

After defining the data models, we must next do two things: a) bind the schemas to a particular language, and b) implement services for interacting with the data through the language bindings. We first describe the general process and then look at the specific details for the Disloc, Simplex, Displacement and Fault schemas.

XML schemas map naturally to constructs in object oriented languages. For example, we may naturally map the schemas used to describe the data and code interfaces to Java data classes: each element of the schema has corresponding accessor (get and set) methods for obtaining and manipulating values. We do this (as a matter of course) with tools such as Castor [17], which automates the Java-to-XML translations.

The generated Java language bindings for the Fault, Disloc, and Simplex schemas are manipulated through service implementation files. This follows the same procedure described in [2] for Application Web Service schema. Essentially the service developer defines the programming interface she wants to expose and wraps the corresponding Java data class method invocations. This interface serves essentially as a façade to simplify interactions with the generated Castor classes. The service provider then translates the service interface into WSDL and "publishes" this interface. Publication may be done informally or through information services such as WSIL [14].

Client user interfaces for creating and manipulating the remote service may be built in the following manner: the developer downloads the WSDL interface and generates client-side stubs. The stubs are proxy objects that may be used as if they were local objects, but which invoke remote methods on the server. The client developer creates a particular interface based around the general purpose client stubs.

We now follow the above procedures for creating services for managing Disloc input and output. We describe the service implementation using Java interfaces and abstract classes, which define contracts that let an implementing class know what methods it must define for compatibility.

As described above, the Fault, Displacement, Disloc, and Simplex schemas may be converted automatically into Java classes using Castor, but we still must produce a developer-friendly façade over the data bindings. We must go a step further and define two generic interfaces that must be implemented by all applications: methods for handling Fault data and Displacement data. We also define an abstract (unimplemented) parent for code files which requires that its children implement methods for importing and exporting code data into legacy formats.

The FaultData interface fragment in Java has the following methods:

```
public interface FaultData {
    public Fault[] getAllFaults();
    public Fault getFault(String faultName);
    public void setFault(Fault sampleFault, String faultName);
...
}
```

This interface defines general methods possessed by all implementing classes. The variable Fault and the array of Faults, Fault[], are just Java representations of the XML Fault definition. Similarly, the DisplaceData interface defines general methods for manipulating Displacements, with corresponding method names and arguments.

Finally, we require an abstract parent that, when extended by a particular application, will implement the translations between XML legacy input and output data formats of the code. For example, we may express the input file for Disloc using XML, but to generate the input file for actually running the code, we must export the XML to the legacy format. Similarly, when Disloc has finished, we must import the output data and convert its legacy format into XML, where we may for example exchange Fault or Displacement data with another application. The reverse operations must also be implemented. The GEMCode abstract parent captures these requirements:

```
public abstract class GEMCode {
    pubic abstract void exportInputData(File f, GEMCode gc);
    public abstract GEMCode importInputData(File filename);
    public abstract void exportOutputData(File f, GEMCode gc);
    public abstract GEMCode importOutputData(File filename);
}
```

This defines the method names for general important and export methods, which must be fleshed out by the implementation.

Finally, the application implementation must extend its abstract parent and implement the FaultData and DisplaceData methods. It will also need to define relevant applications methods. For example, a partial listing (in Java) for Disloc would be

```
public class DislocData extends GEMCode implements
FaultData, DisplaceData {
    public void createInputFile() { ; }
    public void setObservationStyle(String obsvStyle) { ; }
    public void setGridObsvPoints(XSpan x, YSpan y) { ; }
    public void setFreeObsvPoints(PointSpec[] points) { ; }
}
```

DislocData thus is required to implement functions for manipulating Fault and Displacement data, import and export methods that translate between XML and Disloc legacy formats, and finally Disloc-specific methods for setting observation points, etc. Note the variables XSpan, YSpan, and PointSpec (or their arrays) are just Java classes automatically generated by the data bindings from the XML schema descriptions.

The class listed above provides methods for setting the observation style and observation points for outputs, as well as material properties of the fault. The observation points for Disloc output may be either in a regular grid or on specified surface points. The last two methods may be used to access the appropriate output of the surface deformations.

The above code fragment is next converted into WSDL (which is language-independent but too verbose to list here) and may be used by client developers to create methods for remote invocation. The value of WSDL and SOAP here is again evident: the Fault class, for example, may be directly cast back into its XML form and sent in a SOAP message to the remote service.

5 Data Service Architecture

The final step is to define the architecture that describes how the various services must interact with each other. We first consider the case for Disloc running by itself, illustrated in Fig. 2. Services are described in WSDL and SOAP is used for inter-component communication.

The browser interface gathers user code selections ("Disloc") and a desired host for executing Disloc. The user then fills out HTML forms, providing information needed to construct the Fault and Disloc schemas. These pages are created and managed by the User Interface Server, which acts as a control point (through client stubs) for managing the remote services. These HTTP requests parameters are translated into XML by the Disloc data service, which implements the interfaces described previously. This file is then exported to the legacy format and transferred to the execution host. When the code exits, the legacy data format may be transferred back to the Data Service and imported into XML.

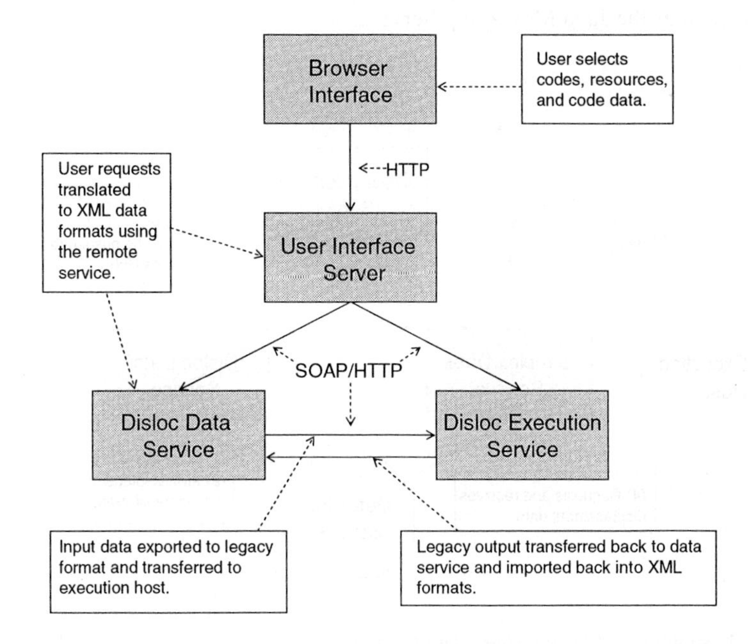

Fig. 2. Interactions of the Disloc data service. Shaded boxes indicate distributed components. Solid arrow lines indicate over-the-wire transmissions with the indicated protocols

The value of the last step is that it now allows the output data to be shared in a common format with other applications. See for example [15] as a similar approach. Visualization and analysis services may acquire the displacement data from Disloc, as may an application such as Simplex in our motivating scenario. Such data sharing is enabled by the common XML data format, but to be used requires an additional Data Hub service, as illustrated in Fig. 3.

The Data Hub service is responsible for extracting the Fault and Displacement data from the formatted XML output of applications. The Data Hub service may interoperate with other services such as the Database Service illustrated in Fig. 1. Fig. 3 illustrates the interaction of the Simplex and Disloc Data services with the hub.

Note that Fig. 3 assumes the interactions of Fig. 2 have already taken place. The User Interface server again acts as the central control point and issues commands (at either user request or through automating events) initially to the Disloc data service to transfer its output data to the hub (Step 1). The service accomplishes this in Step 2. In Step 3 the User Interface server requests that Simplex should import selected displacements from the Data Hub. The data (in XML) is then imported in Step 4. The Simplex Data Service may then (with additional Fault data) generate a Simplex legacy input file and execute the Simplex application as shown in Fig. 2.

We note that other architectures are possible here. In particular, the Data Hub service is shown to act as a "push" client when accepting Disloc data (Step 2), but we may also implement it using a publish/subscribe model based around messaging systems such as the Java Messaging Service.

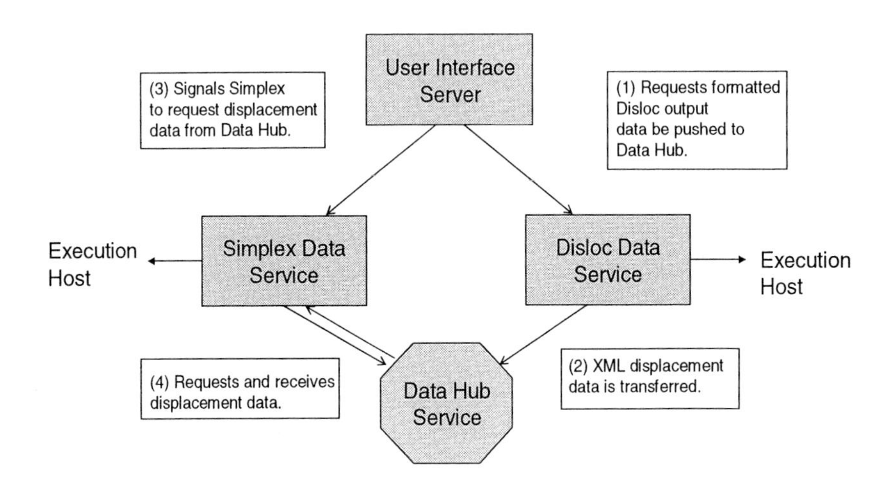

Fig. 3. Simplex and Disloc share data through the Data Hub service. Shaded objects represent distributed components. Closed arrows represent connections with execution services such as shown in Fig. 2. The open arrows represent SOAP over HTTP invocations

6 Summary and Conclusions

We have presented our initial architecture for implementing code specific data services. This consisted of three major steps. First, we devised XML schemas to express the application input data and code parameters. We gave examples of this process for Disloc and Simplex. Next we wrap these data models in Web services that can be plugged into our portal. These models must implement a set of specified interfaces for manipulating Faults and Displacements, as well as a parent interface that requires the service to implement import and export functions for converting between XML and legacy formats. These services may then be deployed and clients built following normal Web service development patterns. Finally, we provide a means for connecting two code-specific data services. A central data hub imports and exports XML-encoded fault and displacement data. This can be used to share, for example, synthetic Disloc displacement data with Simplex.

There are two possible revisions in our architecture. The first is in data encoding. XML is a verbose way for marking up data, and an alternate approach may be to encode only metadata in XML and use an alternative data format, such as HDF [16], for large data sets. We will need to base this on network and host performance tests for a large number of use cases. The second possible modification is in the nature of the Data Hub. As shown in Fig. 3, we have designed this to be a Web service component and assume point-to-point messaging. However, we must also explore alternative publish/subscribe mechanisms that will allow subscribed hosts to be notified when interesting data has been published. This mechanism would remove some of the low-level control capabilities from the User Interface server.

References

1. QuakeSim-System Documentation. Accessed from http://www-aig.jpl.nasa.gov/public/dus/quakesim/documentation.html
2. Pierce, M. et al.: Interoperable Web Services for Computational Portals. Proceedings of Supercomputing 2002. Available from http://sc-2002.org/paperpdfs/pap.pap284.pdf (2002)
3. Donnellan, A., et al: Final Report on the GESS Study – Inversion of Earthquake Fault Parameters Using Multiple Look Angles. NASA JPL internal report (2002).
4. Christensen, E., Curbera, F., Meredith, G., Weerawarana, S.: Web Service Description Language (WSDL) version 1.1. W3C Note 15 March 2001. Available from http://www.w3c.org/TR/wsdl.
5. Box, D., Ehnebuske, D., Kakivaya, G., Layman, A., Mendelsohn, N., Nielsen, H.F., Thatte, S., Winer, D.: Simple Object Access Protocol (SOAP) 1.1. W3C Note 08 May 2000. Available from http://www.w3.org/TR/SOAP/
6. Graham, S. et al.: Building Web Services with Java. SAMS, Indianapolis, 2002.

7. Thompson, H.S., Beech, D., Maloney, M., Mendelsohn, N. XML Schema Part 1: Structures. W3C Recommendation 2 May 2001. Available from http://www.w3.org/TR/xmlschema-1/;
Biron, P.V., Malhotra, A.: XML Schema Part 2: Datatypes. W3C Recommendation 02 May 2001. Available from http://www.w3.org/TR/xmlschema-2/.
8. QuakeSim Fault Schema. Available from http://www.servogrid.org/GCWS/Schema/GEMCodes/Faults.xsd.
9. QuakeSim Surface Displacement Schema: Available from http://www.servogrid.org/GCWS/Schema/GEMCodes/Displacement.xsd.
10. QuakeSim Disloc Schema. Available from http://www.servogrid.org/GCWS/Schema/GEMCodes/Disloc.xsd.
11. QuakeSim Simplex Schema. Available from http://www.servogrid.org/GCWS/Schema/GEMCodes/Simplex.xsd.
12. Donnellan, A, et al: Elastic Dislocation Fault Parameters. Available from http://www.servogrid.org/GCWS/Schema/GEMCodes/disloc.ps.
13. Lyzenga, G. Simplex Version 4.0 Input File Format. Available from http://www.servogrid.org/GCWS/Schema/GEMCodes/simplex4.ps.
14. Ballinger, K., Brittenham, P., Malhotra, A., Nagy, W. A., Pharies, S.: Web Service Inspection Language (WS-Inspection) 1.0. IBM and Microsoft November 2001. Available from http://www-106.ibm.com/developerworks/webservices/library/ws-wsilspec.html
15. Clarke, J., Namburu, R.R.: A Distributed Computing Environment for Interdisciplinary Applications. Currency and Computation: Practice and Experience. Vol. 14, No. 13–15, pp. 1161–1174 (2002).
16. NCSA HDF Home Page. Available from http://hdf.ncsa.uiuc.edu/.
17. The Castor Project. Available from http://castor.exolab.org/.

Apparent Strain Localization and Shear Wave Dispersion in Elastic Fault Gouge with Microrotations

E. Pasternak[1,2], H.-B. Mühlhaus[3] and A.V. Dyskin[1]

[1] School of Civil and Resource Engineering, The University of Western Australia
35 Stirling Highway, Crawley, WA, 6009, Australia
Elena@civil.uwa.edu.au

[2] Institut für Werkstoffkunde und Werkstofftechnik, Technische Universität Clausthal
Agricolastr. 6, D-38678, Clausthal-Zellerfeld, Germany

[3] CSIRO Division of Exploration and Mining, Australian Resource Research Centre
PoBox 1130, Bentley, WA 6102, Australia
and Department of Earth Sciences, The University of Queensland
St Lucia, QLD 4072, Australia
hans.muhlhaus@csiro.au

Abstract. Shear deformation of fault gouge or other particulate materials often results in observed strain localization, or more precisely, the localization of measured deformation gradients. In conventional elastic materials the strain localization cannot take place therefore this phenomenon is attributed to special types of non-elastic constitutive behaviour. For particulate materials however the Cosserat continuum which takes care of microrotations independent of displacements is a more appropriate model. In elastic Cosserat continuum the localization in displacement gradients is possible under some combinations of the generalized Cosserat elastic moduli. The same combinations of parameters also correspond to a considerable dispersion in shear wave propagation which can be used for independent experimental verification of the proposed mechanism of apparent strain localization in fault gouge.

1 Introduction

Strain localization under shear loading is often observed in granulate materials such as the material of fault gouge. If however one models a fault gouge as a homogeneous elastic layer of thickness $2h$, occupying the strip $-h<x_2<h$ in a Cartesian coordinate frame $(x_1x_2x_3)$ with specified displacement at the boundaries $u_1(x_1,\pm h,x_3)=\pm u_1^0$, $u_2(x_1,\pm h,x_3)= u_3(x_1,\pm h,x_3)=0$, then the shear strain and stress will be uniform i.e. $\varepsilon_{12}=u_1^0/2h$, $\sigma_{12}=\mu u_1^0/h$, where μ is the shear modulus of the gouge. Thus the strain localization is perceived as a manifestation of material instability which can only be achieved under either large deformations or certain inelastic constitutive laws, such as non-associated plasticity (eg, [1]) or post-peak softening.

An important feature of granulate materials overlooked by the conventional continuum mechanics is the presence of microrotations arising from the ability of particles to rotate independently, i.e. not in a chord with the rotations associated with the displacement field. This feature can be modelled by Cosserat continuum in which strain and displacement gradient become different. Observations of localization usu-

P.M.A. Sloot et al. (Eds.): ICCS 2003, LNCS 2659, pp. 873–882, 2003.

ally refer to the measurements of displacement gradients. In the present paper we show that in a Cosserat continuum localization of the displacement gradient can be observed even for elastic gouge. We show that the combinations of parameters producing the localization also lead to dispersion in shear wave propagation.

2 Three-Dimensional Continuum Elastic Model of Granulate Materials

Cosserat continuum models of a particulate material can be obtained by homogenisation of discrete equations of motion of the particles. Mühlhaus and Oka [2] suggested a continuum model for assemblies of identical spheres without resistance to relative rotations. The equations of motion were homogenised by expanding the difference expressions of the discrete model into Taylor series and retaining terms up to the second order. Since rotational degrees of freedom were introduced and higher order terms kept in the corresponding Taylor expansions, the resulted continuum description was a combination of a Cosserat theory and a strain gradient theory. An analysis of simple 1D particle arrangements ([3, 4]) showed that the resistance to relative rotations at particle contacts is important: its neglect leads to the loss of positive definiteness of the energy [2]. In the following we consider the resistance of the spheres to relative rotations at the contact points (see also [5]). The non-symmetry of the stress can then be balanced by moment stresses caused by relative rotations at the contacts alone.

We consider a three-dimensional assembly of identical spherical grains. The diameter D of the grain is assumed as much smaller than the problem dimension L, so that $D<<L$. The spheres are in permanent contact; and the orientation of the contact points is assumed as random. In our idealised model, every point of the equivalent continuum corresponds to the centroid of the reference sphere in the discrete material.

Interaction between each pair of neighbouring particles is represented by the total contact force $\vec{F}$ and contact moment $\vec{M}$. The contact moment reflects the fact that the real particles are neither absolutely rigid nor perfectly spherical and hence contact over a certain area. Then the relative rotation of one particle with respect to the other results in a non-symmetric distribution of contact forces, which in the first approximation can be described by the contact moment. It is supposed that the contact force and the moment are linearly dependent upon the relative displacement $\Delta\vec{u}$ and rotation $\Delta\vec{\varphi}$ between the neighbouring particles respectively:

$$\vec{F} = K\Delta\vec{u}\,, \quad \vec{M} = L\Delta\vec{\varphi}\,, \tag{1}$$

$$K = [K_{ij}],\ K_{ij} = (k_n - k_s)n_i n_j + k_s\delta_{ij},\ L = [L_{ij}],$$

$$L_{ij} = (k_{\varphi_n} - k_{\varphi_s})n_i n_j + k_{\varphi_s}\delta_{ij}\,. \tag{2}$$

Here K and L are the matrixes of the translational and rotational spring stiffnesses, k_n, k_s and $k_{\varphi n}$, $k_{\varphi s}$ are the normal and shear (tangential) contact stiffnesses of the translational and rotational springs and the indices in (2) refer to a spatially fixed Cartesian coordinate system.

Assuming that the particle arrangements are statistically homogeneous and applying the method of homogenisation by differential expansions [4] one obtains the following state and constitutive equations

$$\sigma_{ji,j} = \rho \ddot{u}_i, \quad \mu_{ji,j} + \varepsilon_{ijk}\sigma_{jk} = \rho \frac{D^2}{10} \ddot{\varphi}_i. \tag{3}$$

$$\sigma_{ji} = C_{ijlm}\gamma_{lm} + C_{lj}\gamma_{li}, \quad \mu_{ji} = D_{ijlm}\kappa_{lm} + D_{lj}\kappa_{li}, \tag{4}$$

Here σ_{ij} and μ_{ij} are non-symmetric stresses and moment stresses respectively, γ_{ij}, κ_{ij} are the classical Cosserat continuum deformation measures (eg, [6])

$$\gamma_{ji} = u_{i,j} - \varepsilon_{kji}\varphi_k, \quad \kappa_{ji} = \varphi_{i,j}, \tag{5}$$

where φ_i is the Cosserat rotation, γ_{ji} and κ_{ji} are strains and curvature twists.

The parameters of the constitutive relationships (4), the elastic moduli C_{ijlm}, C_{lj}, D_{ijlm}, D_{lj} have the form

$$C_{ijlm} = \frac{6\nu_s}{\pi D}(k_n - k_s)A_{ijlm}, \; C_{lj} = \frac{6\nu_s}{\pi D}k_s A_{lj},$$

$$D_{ijlm} = \frac{6\nu_s}{\pi D}(k_{\varphi_n} - k_{\varphi_s})A_{ijlm}, \; D_{lj} = \frac{6\nu_s}{\pi D}k_{\varphi_s} A_{lj}, \tag{6}$$

$$A_{lj} = \int_{\alpha/2} A n_l n_j dn = \frac{k}{6}\delta_{lj},$$

$$A_{ijlm} = \int_{\alpha/2} A n_i n_j n_l n_m dn = \frac{k}{30}\{\delta_{ij}\delta_{lm} + \delta_{il}\delta_{jm} + \delta_{im}\delta_{jl}\}. \tag{7}$$

Here $A(\mathbf{r},\mathbf{n})=k/4\pi$ for isotropic distribution of particle contacts, α is the spherical angle, $dn=\sin\theta d\theta d\phi$ in a spherical coordinate frame with the origin at the sphere centre (r,ϕ,θ).

3 Simple Shear of Elastic Gouge with Microrotations

We consider a model of fault gouge as an infinite Cosserat elastic layer occupying the area $|x_1|<\infty$, $|x_2|<h$ under plain strain conditions (Fig. 1) subjected to the following boundary conditions:

$$u_1(x_1,\pm h,x_3)=\pm u_1^0,\ u_2(x_1,\pm h,x_3)=0,\ \varphi_3(x_1,\pm h,x_3)=0. \tag{8}$$

The Cosserat continuum Lamé equations for the plain strain with no body forces and moments can be obtained by substituting (4), (5) into (3):

$$(2\mu+\lambda)\partial^2 u_1/\partial x_1^2+\lambda\partial^2 u_2/\partial x_1\partial x_2+(\mu+\alpha)\partial^2 u_1/\partial x_2^2+(\mu-\alpha)\partial^2 u_2/\partial x_1\partial x_2+2\alpha\partial\varphi_3/\partial x_2=0, \tag{9}$$

$$(\mu+\alpha)\partial^2 u_2/\partial x_1^2+(\mu-\alpha)\partial^2 u_1/\partial x_1\partial x_2+(2\mu+\lambda)\partial^2 u_2/\partial x_2^2+\lambda\partial^2 u_1/\partial x_1\partial x_2-2\alpha\partial\varphi_3/\partial x_1=0, \tag{10}$$

$$B(\partial^2\varphi_3/\partial x_1^2+\partial^2\varphi_3/\partial x_2^2)+2\alpha(\partial u_2/\partial x_1-\partial u_1/\partial x_2-2\varphi_3)=0, \tag{11}$$

where λ and μ are Lamé coefficients, α and B are the Cosserat elastic moduli, the latter is bending stiffness. The Lamé coefficients λ, μ and the Cosserat parameters α and B can be expressed through the micromechanical Cosserat model parameters, namely the solid volume fraction ν_s, coordination number k, the sphere diameter D and respective contact stiffnesses k_n, k_s, $k_{\varphi n}$, $k_{\varphi s}$ introduced in the previous section. This can be achieved by comparing micromechanical Cosserat elastic moduli (6), (7) in the constitutive equations (4) with corresponding terms in continuum Cosserat theory taken, for example from [6] that reads

$$\sigma_{11}=(2\mu+\lambda)\gamma_{11}+\lambda\gamma_{22},\ \sigma_{12}=(\mu+\alpha)\gamma_{12}+(\mu-\alpha)\gamma_{21},\ \sigma_{21}=(\mu+\alpha)\gamma_{21}+(\mu-\alpha)\gamma_{12},$$

$$\sigma_{22}=\lambda\gamma_{11}+(2\mu+\lambda)\gamma_{22}, \tag{12}$$

$$\mu_{13}=B\kappa_{13},\ \mu_{23}=B\kappa_{23}, \tag{13}$$

where the Cosserat deformation measures, strains γ_{ji} and curvatures κ_{ji}, are given by (5).

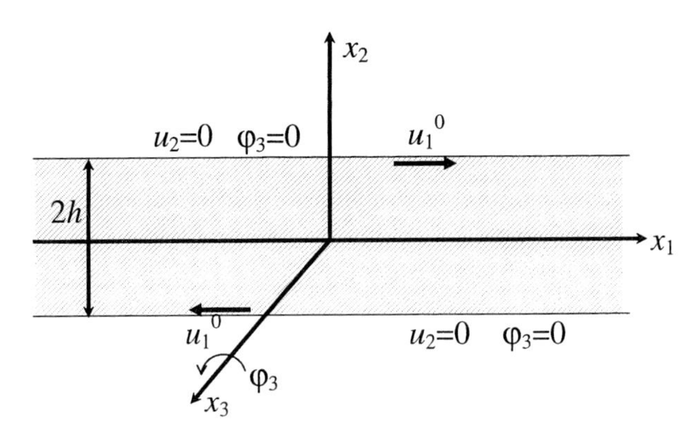

Fig. 1. A 2D model of fault with an elastic gouge with microrotations

By doing so one obtains

$$\lambda=\nu_s k(k_n-k_s)/(5\pi D),\ \mu=\nu_s k(2k_n+3k_s)/(10\pi D),\ \alpha=\nu_s k k_s/(2\pi D),$$

$$B=\nu_s k(k_{\varphi n}+4k_{\varphi s})/(5\pi D). \tag{14}$$

Noting that the boundary conditions (8) are homogeneous along x_1, we are seeking a homogeneous along x_1 solution of the system of the Lamé equations (9-11) satisfying these boundary conditions. If we find such a solution according to the uniqueness theorem this solution will be unique.

Such a solution can be written in the normalized form (h=1, u_1^0=1) as follows:

$$u_1=\left(x_2-\frac{\sinh(x_2/l)}{\sinh(1/l)}\Lambda^2 l\tanh(1/l)\right)(1-\Lambda^2 l\tanh(1/l))^{-1},\ u_2=0, \tag{15}$$

$$\varphi_3=-\frac{1}{2}\left(1-\frac{\cosh(x_2/l)}{\cosh(1/l)}\right)(1-\Lambda^2 l\tanh(1/l))^{-1}, \tag{16}$$

where

$$\Lambda=l_1/l,\quad l=(l_1^2+l_2^2)^{1/2},\quad l_1=(B/4\mu)^{1/2},\quad l_2=(B/4\alpha)^{1/2}. \tag{17}$$

Parameters l_1 and l_2 are two independent length scale parameters (normalized by h) reflecting the presence of a microstructure in the Cosserat continuum, with l acting as a "hypotenuse" of the Cosserat continuum length scale parameters. Their expressions through the microstructural parameters by using (14) can be written as follows:

$$l_1=\frac{1}{2}\sqrt{(k_{\varphi_n}+4k_{\varphi_s})\Big/(k_n+\frac{3}{2}k_s)},\quad l_2=\frac{\sqrt{2}}{2\sqrt{5}}\sqrt{(k_{\varphi_n}+4k_{\varphi_s})/k_s} \tag{18}$$

$$l=\sqrt{\frac{k_{\varphi_n}+4k_{\varphi_s}}{5k_s}\cdot\frac{k_n+4k_s}{2k_n+3k_s}}.$$

Consequently, after further normalisation (μ=1) of stress and moment stress fields one obtains:

$$\sigma_{12}=\left(1-2\Lambda^2\frac{\cosh(x_2/l)}{\cosh(1/l)}\right)(1-\Lambda^2 l\tanh(1/l))^{-1},$$

$$\sigma_{21}=(1-\Lambda^2 l\tanh(1/l))^{-1}, \tag{19}$$

$$\mu_{23} = 2\Lambda^2 l \frac{\sinh(x_2 / l)}{\cosh(1/l)} (1 - \Lambda^2 l \tanh(1/l))^{-1} . \tag{20}$$

Figures 2–4 show the distributions of displacement and displacement gradient for various ratios of microstructural parameters Λ, l. The displacements (Fig. 2) lie between the one for the standard elastic solution (Λ=0, l=0), and the Cosserat solution (Λ=1, l=1) that exhibits maximum deviation from the standard continuum solution. The displacement gradient for relatively small values of the parameter l (Fig. 3) exhibits a kind of plateau, which is insensitive to the values of the dimensionless parameter $0<\Lambda<1$. There is however a range of parameters Λ and l, for which the displacement gradient is highly non-homogeneous and displays localisation reaching its peak for the combination of parameters Λ=1, l=1 (Fig. 4). Indeed, as the displacement gradient $u_{1,2}$ reaches its maximum at x_2=0 for any values of Λ, l, the parametric analysis of the solution performed for the middle layer of the gouge

$$u_{1,2}(\Lambda, l)\Big|_{x_2=0} = \left(1 - \frac{\Lambda^2}{\cosh(1/l)}\right)(1 - \Lambda^2 l \tanh(1/l))^{-1} \tag{21}$$

shows that displacement gradient takes its max at Λ=1, l=1. This combination of parameters corresponds to l_2=0, l_1=l. This means that the bending stiffness $B<<(\mu+\alpha)l^2$ the quantity in brackets being the effective shear modulus.

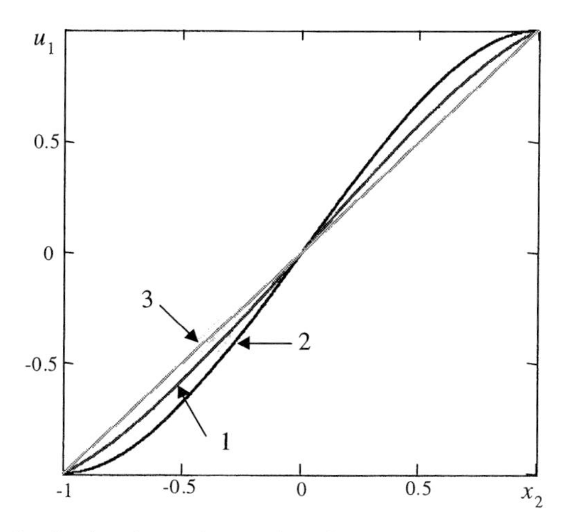

Fig. 2. Displacement distribution for various ratios of microstructural parameters Λ, l. Curve 1 corresponds to Λ=0.8, l=0.5, curve 2: Λ=1, l=1, curve 3: conventional elastic solution Λ=0, l=0

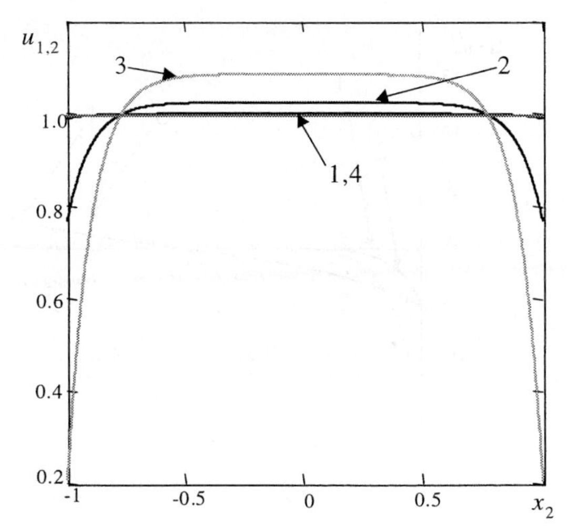

Fig. 3. Distribution of displacement gradients for small value of the parameter l and various values of microstructural parameter Λ. Curve 1 corresponds to Λ=0.1, l=0.1, curve 2: Λ=0.5, l=0.1, curve 3: Λ=0.9, l=0.1, curve 4: conventional elastic solution Λ=0, l=0

Figures 5 and 6 show the distributions of normalized moment stress μ_{23} and antisymmetric $\sigma_{[21]}$ and symmetric $\sigma_{(21)}$ parts of the shear stress. It is seen that the maximum values of the moment stress and the antisymmetric shear stress are attained at the gouge boundaries, while the maximum (localization) of the symmetric stress is achieved at the middle layer of the gouge the latter being in accordance with the localization of the displacement gradient.

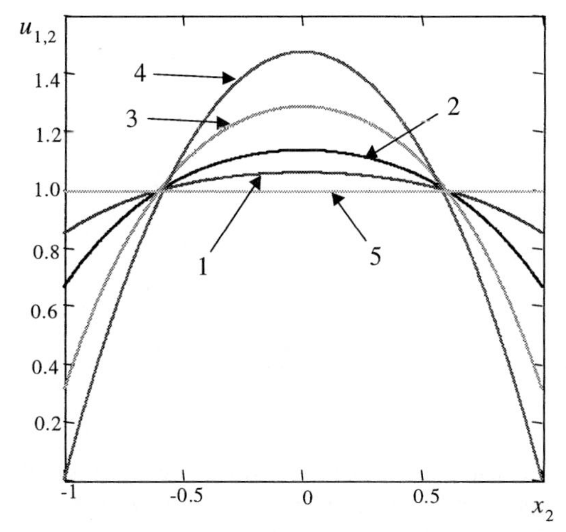

Fig. 4. Distribution of displacement gradients for large values of the parameter l and various values of microstructural parameter Λ. Curve 1 corresponds to Λ=0.5, l=0.5, curve 2: Λ=0.7, l=0.5, curve 3: Λ=0.9, l=0.5, curve 4: Λ=1, l=1, curve 5: conventional elastic solution Λ=0, l=0

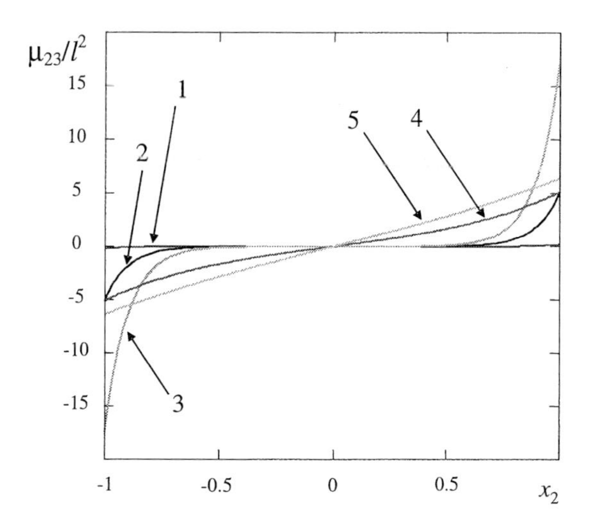

Fig. 5. Distribution of normalized moment stress for various values of the parameters l and Λ. Curve 1 corresponds to Λ=0.1, l=0.1, curve 2: Λ=0.5, l=0.1, curve 3: Λ=0.9, l=0.1, curve 4: Λ=0.9, l=0.5, curve 5: Λ=1, l=1

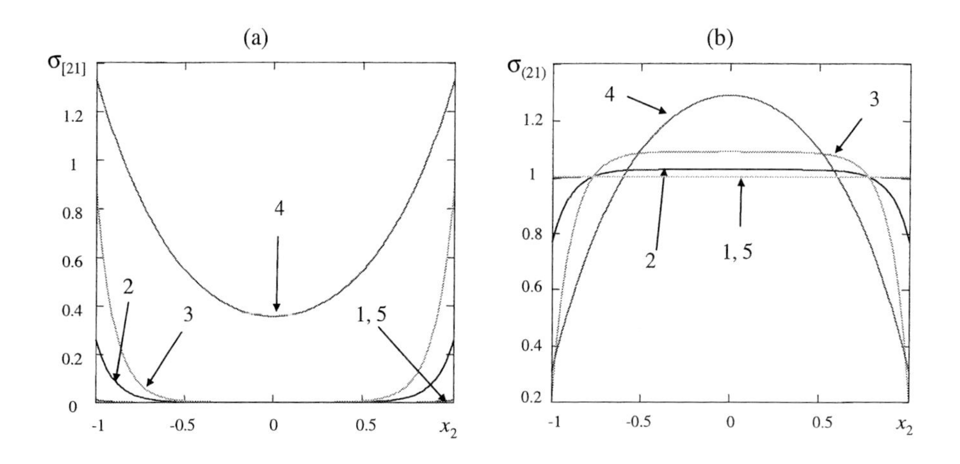

Fig. 6. Distribution of antisymmetric (a) and symmetric (b) parts of the shear stresses for various values of the parameters l and Λ. Curves 1 correspond to Λ=0.1, l=0.1, curves 2: Λ=0.5, l=0.1, curves 3: Λ=0.9, l=0.1, curves 4: Λ=0.9, l=0.5, curve 5: conventional elastic solution Λ=0, l=0

4 On the Possibility of Determining the Localization Regime from Wave Velocity Measurements

Prior determination of whether the fault gouge can show localization requires the measurements of calculations of the Cosserat parameters, which is somewhat involved. It would be attractive to use the measurements of velocities of wave propagation through the gouge material for this purpose.

Consider propagation of planar waves in an infinite Cosserat material. According to [6] there are the following types of planar waves: (1) a conventional longitudinal wave, which is insensitive to the microrotations; (2) a twist wave which reflects the Cosserat properties of the material, but difficult to measure and; (3) one or two (depending upon frequency) shear waves. These shear waves show dispersion which is a result of the Cosserat properties and could therefore be used for identifying possible localization. The wave number ξ for these shear waves satisfies the following characteristic equation (eg, [6])

$$c_2^2 c_4^2 \xi^4 + \left[\omega_*^2 c_5^2 - \omega^2\left(c_2^2 + c_4^2\right)\right]\xi^2 - \omega^2\left(\omega_*^2 - \omega^2\right) = 0 \tag{22}$$

where ω is the frequency and

$$\omega_*^2 = \frac{4\alpha}{J}, \quad c_2^2 = \frac{\mu + \alpha}{\rho}, \quad c_4^2 = \frac{B}{J}, \quad c_5^2 = \frac{\mu}{\rho} \tag{23}$$

where ρ is the material density, J is the rotational inertia which in the case of material consisted of spherical grains of diameter D is equal to $J=\rho D^2/10$.

When $\omega \ll \omega_*$ equation (22) gives only one solution which corresponds to the shear wave propagating with the velocity $\omega/\xi \cong c_5$ which is the conventional shear wave velocity. When $\omega=\omega_*$ equation (22) again gives only one solution corresponding to a shear wave traveling with velocity c_*, where

$$c_*^2 = \frac{(\mu + \alpha)B}{\alpha J + B\rho} \tag{24}$$

Comparison of these two velocities with the aid of (23) and (17) leads to

$$\frac{c_*^2}{c_5^2} = \frac{l^2}{l_2^2 + 0.025D^2} = \frac{1}{1 - \Lambda^2 + 0.025\,D^2/l^2} \tag{25}$$

If one assumes that the total characteristic length l is at least not smaller than the particle diameter D, then the localization case, $\Lambda \to 1$ corresponds to the case of $c_*/c_5 \gg 1$. Thus the localization case should correspond to considerable dispersion in the shear wave: its velocity should markedly increase with the increase in frequency. It should be noted that the waves at frequencies close to ω_* may considerably be attenuated by scattering since the length of the shear wave of frequency ω_* is

λ_*^2=0.025$D^2\Lambda^{-2}$/(1+0.025D^2/l_2^2) which, in the localization range $\Lambda\rightarrow 1$, is already below the particle diameter.

5 Conclusion

The presence of independent microrotations in particulate materials (e.g. the gouge) can lead to considerable localization of deformation gradients which when measured and interpreted in a sense of conventional medium create an impression of strain localization and associated material instability. Such a phenomenon, if experimentally confirmed will give a strong evidence of importance of the microrotational and Cosserat effects. The same phenomenon also leads to a considerable dispersion in shear wave propagation which can be used for independent experimental verification of the proposed mechanism of apparent strain localization in fault gouge.

Acknowledgment. The authors acknowledge the support of the UWA Small Research Grant 2001. The first author acknowledges the support of the ARC Discovery Grant DP0346148 and the Alexander von Humboldt Research Fellowship 2002-2003.

References

1. Rudnicki, J.W., Rice, J.R.: Conditions for the Localization of Deformation in Pressure-Sensitive Dilatant Materials. J. Mech. Phys. Solids. 23 (1975) 371–394
2. Mühlhaus, H-B., Oka, F.: Dispersion and Wave Propagation in Discrete and Continuous Models for Granular Materials. Int. J. Solids Structures. 33 (1996) 2841–2858
3. Mühlhaus, H.-B., Dyskin, A.V., Pasternak, E., Adhikary D.: Non-Standard Continuum Theories in Geomechanics: Theory, Experiments and Analysis. In: Picu, R.C., Krempl, E. (eds.): Proc. Fourth International Conference on Constitutive Laws for Engineering Materials. Troy, New York (1999) 321–324
4. Pasternak, E., Mühlhaus, H.-B.: Cosserat and Non-Local Continuum Models for Problems of Wave Propagation in Fractured Materials, In: Zhao, X.L., Grzebieta, R.H. (eds.): Structural Failure and Plasticity (IMPLAST 2000). Pergamon, Amsterdam (2000) 741–746
5. Pasternak, E., Mühlhaus, H.-B.: Cosserat Continuum Modelling of Granulate Materials. In: Valliappan. S. and Khalili. N. (eds.): Computational Mechanics – New Frontiers for New Millennium. Elsevier Science, Amsterdam (2001) 1189–1194
6. Nowacki, W.: Theory of Micropolar Elasticity. Springer, Wien (1970)

Efficient Implementation of Complex Particle Shapes in the Lattice Solid Model

Steffen Abe and Peter Mora

ACcESS
The University of Queensland
St Lucia, QLD 4072, Australia
{steffen, mora}quakes.uq.edu.au

Abstract. The lattice solid model is a particle based simulation model for the study of earthquake micro-physics and rock mechanics. It consists of particles interacting by various types of mechanisms such as elastic-brittle forces and friction. Results of laboratory experiments have shown that the grain shape has a major influence on the frictional properties of fault gouge. In order to enable realistic simulations it is thus important to include the capability to model non-spherical particles into the simulation software. To achieve this goal a new class of particles with variable shapes have been implemented in the lattice solid model. The shape of the particles is described by an arbitrary number of piecewise spherical patches. This leads to a good balance between the computational cost of the contact detection and calculation of interactions between particles and the range of particle shapes available.

1 Introduction

The lattice solid model (Mora and Place, 1994 [10], Place and Mora, 1999 [13]) is a particle based simulation model for the study of earthquake micro-physics and rock mechanics similar to the discrete element model (Cundall and Strack, 1979 [3]). It consists of particles interacting by various types of mechanisms such as brittle-elastic forces and friction. Results of laboratory experiments comparing the frictional behavior of fault gouge consisting of spherical grains with a gouge containing angular grains (Mair et al., 2002 [9]) have shown that grain shape has a major influence on the frictional properties of granular gouge. In order to enable realistic simulations it is thus important to include the capability to model non-spherical particles into the simulation software.

Place and Mora, 2000 [14] have successfully used strongly bonded groups of spherical particles to model angular grains in a 2D simulation of shear of a granular fault gouge. A similar approach has been by used by Pöschel and Buchholtz, 1993 [15] and Mahboubi et al., 1997 [8] using aggregates of variable sized spheres. While enabling the simulation of arbitrarily shaped grains in a simple and efficient way, the surface roughness of the grains is determined by the size of the spherical particles used to construct the grain. Thus grains formed by a low number of particles will always have a high surface roughness and

P.M.A. Sloot et al. (Eds.): ICCS 2003, LNCS 2659, pp. 883–891, 2003.

the modeling of grains with low surface roughness requires the grains to be constructed from a large number of spherical particles and thus leads to high computational cost.

Other approaches for the modeling of non-spherical grains in DEM simulations include elliptical or super-ellipsoidal particles (Ting et al., 1993 [17], Mustoe and DePoorter, 1993 [12], Cleary et al., 1997 [2]), polyhedra (Cundall, 1988 [4], Hogue and Newland, 1993 [6], Hogue and Newland, 1994 [7]) and discrete function representation (Williams and O'Connor, 1995 [18], Hogue, 1998 [5]). The use of elliptical or super-ellipsoidal particles strongly restricts the types of grain shapes which can be modeled, in particular with respect to the symmetry of the grains. While polyhedra and discrete function representations allow arbitrary shapes, a good approximation of smooth grain shapes requires a large number of vertices, leading to computationally expensive contact detection algorithms.

In order to achieve a balance between a high flexibility in the choice of grain shapes and computational efficiency a new class of grain shape representations for the use in lattice solid or DEM simulation has been developed based on grains constructed from overlapping spherical patches. Each grain consists of an arbitrary number of sphere segments forming the surface of the grain (Figure 1). The positions and orientations of the spherical patches are fixed relative to each other. A somewhat similar approach using overlapping spheres to construct non-spherical particles has been presented by Zhang and Vu-Quoc [19], however they used only four identical full spheres whereas the approach presented here uses an arbitrary number of spherical patches.

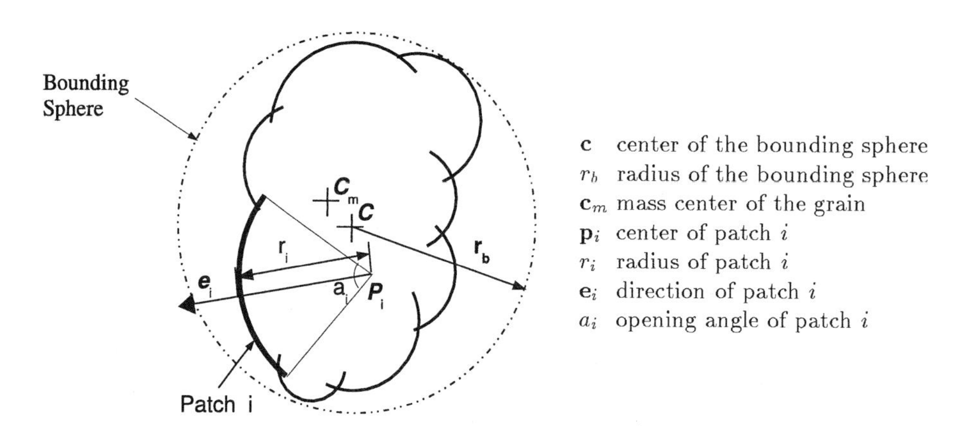

Fig. 1. Structure of a grain constructed from overlapping spherical patches. The figure is drawn in 2D for simplicity but the same structure applies to 3D particles.

The advantages of using spherical patches are that the detection of a contact between them can be implemented at a low computational cost (34Flops worst case) and that while arbitrary grain shapes can be constructed using a sufficient number of patches, some particle shapes can be approximated by a very small number of patches. However, some classes of grain shapes, in particular those with large flat surfaces and sharp edges would require a large number of patches for a good approximation and may be more efficiently simulated using other, for example polyhedral, approaches. The approach presented here is more suited to rounded grains.

2 The Algorithm

The Algorithm for the calculation of the motion of the grains can be separated into two parts: detection of the contacts between the grains including the calculation of the contact parameters and the calculation of the contact forces and the resulting grain movement.

2.1 Contact Detection

As a first step to find contact between two grains k and l the intersection between the bounding spheres of the grains is tested. If the bounding spheres intersect, i.e if the distance between their centers $\mathbf{c}_k$ and $\mathbf{c}_l$ is smaller that the sum of the radii r_k and r_l

$$|\mathbf{c}_k - \mathbf{c}_l| < r_k + r_l \quad , \tag{1}$$

contact between the two grains is possible. If the bounding spheres do not intersect, the contact detection algorithm for the pair of grain terminates and the next pair of grains is considered.

If the bounding spheres intersect, each pair of spherical patches i and j contained in grains k and l respectively is tested for intersection. This test is performed in two steps. First the distance between the patch centers $\mathbf{p}_i$ and $\mathbf{p}_j$ is compared with the sum of the patch radii r_i and r_j. If the distance between the patch centers is small enough, i.e.

$$|\mathbf{p}_i - \mathbf{p}_j| < r_i + r_j \quad . \tag{2}$$

it is then tested if the direction $\mathbf{e}^d_{ij}$ which connects the patch centers is within both patches. If the distance is too large no contact between the patches is possible and the next pair of patches is considered.

The direction of the vector $\mathbf{e}^d_{ij}$ between the patch centers can be calculated as

$$\mathbf{e}^d_{ij} = \frac{\mathbf{p}_i - \mathbf{p}_j}{|\mathbf{p}_i - \mathbf{p}_j|} \quad . \tag{3}$$

The direction $\mathbf{e}^d_{ij}$ is within a patch i if the angle between $\mathbf{e}^d_{ij}$ and the direction $\mathbf{e}_i$ is smaller that the bounding angle α_i of the patch i. Thus if the conditions

$$\mathbf{e}^d_{ij} \cdot \mathbf{e}_i < \cos\alpha_i \tag{4}$$

$$-\mathbf{e}^d_{ij} \cdot \mathbf{e}_j < \cos\alpha_j \tag{5}$$

are both fulfilled, a contact between the patches i and j has been found. The actual contact point is assumed to be in the center of the overlapping region between the patches along the line connecting the patch centers (Figure 2). The location x_{ij} of the contact can thus be computed as

$$\begin{aligned} \mathbf{x}_{ij} &= \mathbf{p}_i + \mathbf{e}^d_{ij} \left(\frac{\mathbf{p}_i - \mathbf{p}_j}{r_i + r_j} \right) \\ &= \mathbf{p}_j - \mathbf{e}^d_{ij} \left(\frac{\mathbf{p}_j - \mathbf{p}_i}{r_i + r_j} \right) \quad . \end{aligned} \tag{6}$$

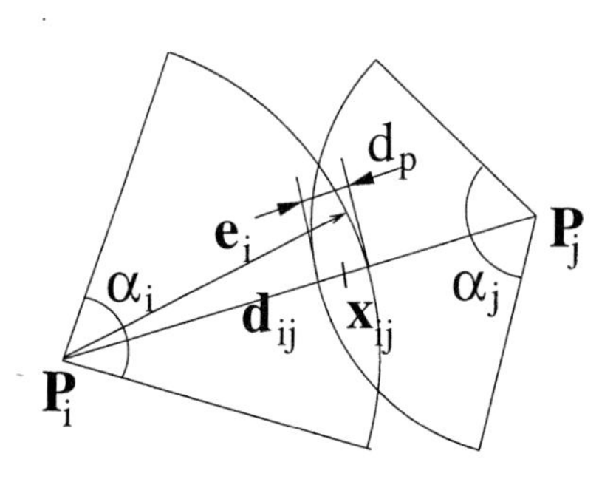

Fig. 2. Intersection between two patches.The figure is drawn in 2D for simplicity but the same structure applies to 3D particles. The point $\mathbf{x}_{ij}$ is the contact point assumed for force and motion calculations.

The contact normals $\mathbf{n}_i$ and $\mathbf{n}_j$ are assumed to be the surface normals of the patch at the contact point which, given that the patch surfaces are parts of spheres, can be calculated from the direction of the connecting line between the patch centers,

$$\mathbf{n}_i = \frac{\mathbf{d}_{ij}}{|\mathbf{d}_{ij}|} \tag{7}$$

$$\mathbf{n}_j = \frac{\mathbf{d}_{ji}}{|\mathbf{d}_{ji}|} = -\mathbf{n}_i \quad . \tag{8}$$

The penetration depth d_p (Figure 2) of the two patches at the contact point, which is needed for the subsequent calculation of the contact forces, can be calculated as the difference between the sum of the radii r_i and r_j and the distance between the patch centers $\mathbf{p}_i$ and $\mathbf{p}_j$

$$d_p = |(\mathbf{p}_i - \mathbf{p}_j)| - (r_i + r_j) \quad . \tag{9}$$

2.2 Calculation of Forces and Particle Motions

As a first step in the calculation of the motion of grains the contact forces are calculated from the contact parameters obtained by the contact detection algorithm. In lattice solid and DEM simulations a wide variety of contact forces have been implemented, including linear elastic forces (Mora and Place, 1994 [10]), Hertz-Mindlin contact laws (Morgan and Boettcher, 1999 [11]) and both linear and nonlinear friction forces (Cundall, 1979 [3], Place and Mora, 1999 [13], Abe et al., 2002 [1]). All those contact forces can be implemented here in the same way because the locally spherical shape of the grains at the point of contact. The total force at each contact $\mathbf{f}_{ij}$ can then be calculated as the sum of the forces due to each of the interactions at this contact

$$\mathbf{f}_{ij} = \mathbf{f}_{ij}^{elast} + \mathbf{f}_{ij}^{fric} + \dots \tag{10}$$

and the total force $\mathbf{f}_{grain}$ on each grain can be calculated as the sum of all contact forces applied to this grain

$$\mathbf{f}_{grain} = \sum_{patches} \mathbf{f}_{ij} \tag{11}$$

The calculation of the particle motions follows the approach derived in (Shabana, 1994 [16]. The linear acceleration $\mathbf{a}_i$ of grain i can be calculated directly from the total force $\mathbf{f}_i$ applied to this grain

$$\mathbf{a}_i = \frac{\mathbf{f}_i}{m_i} \tag{12}$$

where m_i is the mass of grain i. The velocity $\mathbf{v}_i$ and position $\mathbf{x}_i$ of the grain can then be updated using an appropriate integration scheme. Currently a simple first order method is used:

$$\mathbf{v}_i(t + \Delta t) = \mathbf{v}_i(t) + \Delta t \mathbf{a}_i \tag{13}$$

$$\mathbf{x}_i(t + \Delta t) = \mathbf{x}_i(t) + \Delta t \mathbf{v}_i \tag{14}$$

For the grain rotation it is necessary to calculate the torque τ_i applied to the grain from the contact forces $\mathbf{f}_{ij}$ and the contact positions $\mathbf{x}_{ij}$ relative to the center of mass $\mathbf{c}_m$ of the grain

$$\tau_i = \sum_{patches} (\mathbf{x}_{ij} - \mathbf{c}_m) \times \mathbf{f}_{ij} \quad . \tag{15}$$

Using the torque the change in angular momentum can be calculated

$$\dot{\mathbf{l}}_i = \frac{\tau_i}{\mathbf{I}_i} \quad , \tag{16}$$

where $\mathbf{I}_i$ is the inertia tensor of grain and from this the angular velocity ω and orientation $\mathbf{R}_i$ can be updated similar to the linear movement (Eq. 13)

$$\omega_i(t + \Delta t) = \omega_i(t) + \Delta t \dot{\mathbf{L}}_i \tag{17}$$

$$R_i(t + \Delta t) = \mathbf{R}_i(t) + \omega \star \mathbf{R}_i \Delta t \tag{18}$$

After updating the position $\mathbf{x}_i$ and orientation $\mathbf{R}_i$ of the grain the positions of the patch centers $\mathbf{p}_i$ and orientations $\mathbf{e}_i$ need to be updated

$$\mathbf{p}_i^{new} - \mathbf{c}_m = \mathbf{R}_i(\mathbf{p}_i^0 - \mathbf{c}_m) \tag{19}$$

$$\mathbf{e}_i^{new} = \mathbf{R}_i\mathbf{e}_i^0 \quad , \tag{20}$$

where $\mathbf{p}_i^0$ and $\mathbf{e}_i^0$ are the initial values for the patch center positions and patch directions. Also, the inertia tensor $\mathbf{I}_i^0$ of the grain needs to be updated

$$\mathbf{I}_i^{new} = \mathbf{R}_i\mathbf{I}_i^0\mathbf{R}_i^T \quad . \tag{21}$$

Changes to the inertia tensor $\mathbf{I}_i^0$ due to deformation of the grain are ignored.

3 Verification Test

In order to verify the correctness of the implementation of the complex particle shapes in the lattice solid model tests have been performed. If there is no energy input into a simulation model and no energy dissipated in the model, the total energy contained in the model should remain constant if the implementation is correct. While the conservation of energy is not proof of the correctness of the implementation, a violation of the conservation of energy would show that the implementation is incorrect. Thus the conservation of energy can provide a useful consistency check.

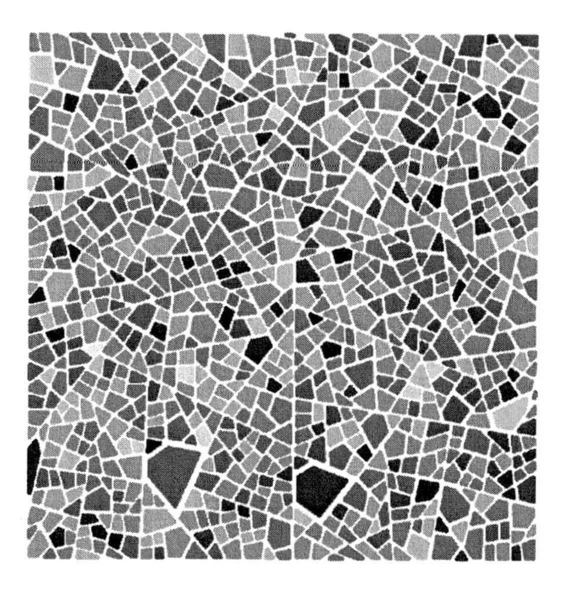

Fig. 3. Initial configuration of the model used for the verification test.

If the particles are interacting by elastic forces only, the only types of energy in the model are the kinetic energy of the grains E_{kin} and the elastic energy stored in the contacts between the grains $E_{elastic}$.

$$E_{total} = \sum_{particles} E^i_{kin} + \sum_{contacts} E^j_{elastic} \tag{22}$$

The kinetic energy of each grain E^i_{kin} can be calculated as the sum of translational part E^i_{trans} and a rotational part E^i_{rot}

$$\begin{aligned} E^i_{kin} &= E^i_{trans} + E^i_{rot} \\ &= \frac{1}{2} m_i v^2 + \frac{1}{2} \omega_i \mathbf{I}_i \omega_i^T \quad . \end{aligned} \tag{23}$$

The elastic energy stored in a linear elastic contact j can be calculated from the spring constant k^j for the normal elastic interactions and the penetration depth $\mathbf{d}^j_p$ (Eq. 9)

$$E^j_{elastic} = \frac{1}{2} k^j (d^j_p)^2 \quad . \tag{24}$$

The elastic interactions between the spherical patches used in this model are the same as the elastic interactions used between the spherical particles in the original lattice solid model (Mora and Place, 1994 [10]).

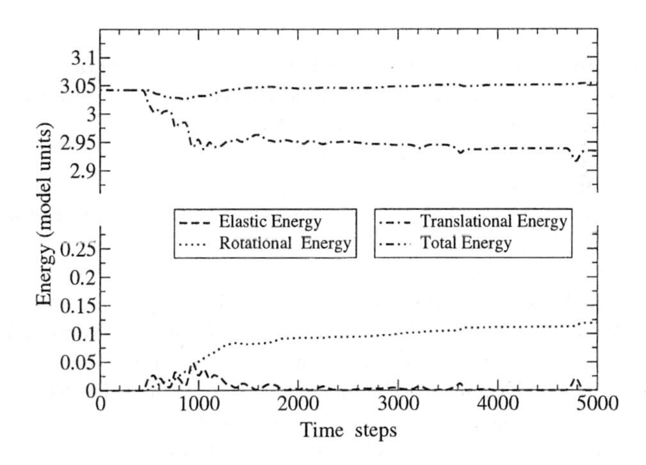

Fig. 4. Energy in the model during the simulation. The total energy is the sum of translational, rotational and elastic energy.

A 2D model is set up consisting of ≈ 900 random grains consisting of around 30 patches each (Figure 3) interacting only by linear elastic forces. The grains have been created by randomly splitting the model area into polygonal tiles and than generating each particle so that it best fits one of the tiles while maintaining a prescribed minimum patch radius. The grains have a random initial linear velocity and no initial angular velocity.

The simulation is then run for 5000 time steps and the translational, rotational and elastic energy in the system are recorded. The total energy is then calculated according to Equation 22. The results (Figure 4) show that the total energy in the system remains constant within about 2.5%.

4 Conclusions

A new class of particles allowing the representation of arbitrary grain shapes based on composite grains constructed from spherical patches has been implemented in the lattice solid model. This will enable the more realistic simulation of processes in granular media which are sensitive to grain shape, such as the frictional behavior of fault gouge.

Acknowledgments.
This work was supported by the Australian Computational Earth Systems Simulator (ACcESS) Major National Research Facility.

References

1. Abe, S., Dieterich, J.H., Mora.,P., Place,D.: Simulation of the Influence of Rate and State dependent Friction on the Macroscopic Behavior of Complex Fault Zones with the Lattice Solid Model Pure Appl. Geoph.,2002
2. Cleary, P.W., Stokes, N., Hurley, J.: Efficient Collision Detection of Three Dimensional Super-ellipsoidal Particles Comp. Techniques and Applications, 1997
3. Cundall, P. A., Strack, O. D. A.: A discrete numerical model for granular assemblies Geótechnique **29** 1979
4. Cundall, P. A.: Formulation of a Three-dimensional Distinct Element Model – Part I. A Scheme to Detect and Represent Contacts in a System Composed of Many Polyhedral Blocks Int. J. Rock. Mech. Min. Sci. & Geomech. Abstr. **25 No. 3** 1988
5. Hogue, C.: Shape Representations and Contact Detection for Discrete Element Simulations of Arbitrary Geometries Engineering Computations **15 No.3** 1998
6. Hogue, C., Newland, D.,: Efficient computer modelling of the motion of arbitrary grains in: Powders and Grains 93, Thornton (ed.), 1993
7. Hogue, C., Newland, D.,: Efficient computer simulation of moving granular particles Powder Technology, **78** 1994
8. Mahboubi, A., Cambrou, B., Fry, J.J.: Numerical modeling of the mechanical behavior of non-spherical, crushable particles in: Powders and Grains 97, Behringer and Jenkins (eds.) 1997
9. Mair, K., Frye, K.M., Marone, C.: Influence of Grain Characteristics on the Friction of Granular Shear Zones J. Geophys. Res. **107 B10** 2002
10. Mora, P., Place, D.: Simulation of the Stick-Slip Instability Pure Appl. Geophys. **143** (1994)

11. Morgan, J.K., Boettcher, M.S.: Numerical simulations of granular shear zones using the distinct element Method: 1. Shear zone kinematics and the micromechnics of localization J. Geoph. Res. **104 B2**, 1999
12. Mustoe, G.G.W., DePoorter, G.: A numerical model for the mechanical behavior of particulate media containing non-circular shaped particles in: Powders and Grains 93, Thornton (ed.), 1993
13. Place, D.,Mora, P.: The Lattice Solid Model to Simulate the Physics of Rocks and Earthquakes: Incorporation of Friction J. Comp. Phys. **150** (1999)
14. Place, D.,Mora, P.: Numerical Simulation of Localization Phenomena in a Fault Gouge Pure Appl. Geophys. **157** 2000
15. Pöschel, T., Buchholtz, V.: Static Friction Phenomena in Granular Materials: Coulomb Law versus Particle Geometry Phys. Rev. Lett. **71 n. 24** 1993
16. Shabana, A.A.: Computational Dynamics J. Wiley & Sons, 1994
17. Ting, J.M., Khwaja, M., Meachum, L.R., Rowell, J.D.: An Ellipse-Based Discrete Element Model For Granular Materials Int. J. Num. Ana. Meth. Geomech. **17** 1993
18. Williams, J. R., O'Connor, R. A Linear Complexity Intersection Algorithm for the Discrete Element Simulation of Arbitrary Geometries Engineering Computations **12** 1995
19. Zhang, X., Vu-Quoc, L,: Simulation of chute flow of soybeans using an improved tangential force-displacement model Mechanics of Materials **32**, 2000

A Method of Hidden Markov Model Optimization for Use with Geophysical Data Sets

Robert A. Granat

Jet Propulsion Laboratory Pasadena CA 91109, USA

Abstract. Geophysics research has been faced with a growing need for automated techniques with which to process large quantities of data. A successful tool must meet a number of requirements: it should be consistent, require minimal parameter tuning, and produce scientifically meaningful results in reasonable time. We introduce a hidden Markov model (HMM)-based method for analysis of geophysical data sets that attempts to address these issues. Our method improves on standard HMM methods and is based on the systematic analysis of structural local maxima of the HMM objective function. Preliminary results of the method as applied to geodetic and seismic records are presented.

1 Introduction

In recent years, geophysics research has been faced with a growing need for automated techniques by which to process the ever-increasing quantities of geophysical data being collected. Global positioning system (GPS) networks for measurement of surface displacement are expanding, seismic sensor sensitivity is increasing, synthetic aperture radar missions are planned to measure surface changes worldwide, and increasingly complex simulations are producing vast amounts of data. Automated techniques are necessary to assist in coping with the deluge of information. These techniques are useful in a number of ways: they can analyze quantities of data that would overwhelm human analysts, they can find subtle changes in the data that might evade a human expert, and they assist in objective decision making in cases where even experts disagree (for example, identifying aftershock sequences, or modes in GPS time series). These techniques are not expected to replace human analysis, but rather to be tools for human experts to use as part of the research cycle.

The field of geophysics poses particular challenges for automated analysis. The data is often noisy or of poor quality, due to the nature of the sensor equipment; for similar reasons it is also often sparse or incomplete. Furthermore, the underlying system is unobservable, highly complex, and still poorly understood by theory. Automated analysis is a useful tool only if it can satisfy several criteria. The results produced must be consistent across experiments on the same or similar data. Only minimal parameter tuning can be required, lest the results be

P.M.A. Sloot et al. (Eds.): ICCS 2003, LNCS 2659, pp. 892–901, 2003.

considered arbitrary. As well, the method must be computationally tractable, so results can be returned to the user in reasonable time.

In this work, we investigate the use of hidden Markov models (HMMs) [1–5] as the basis for an automated tool for analysis of geophysical data. We begin by giving a brief overview of hidden Markov models and introducing our notation. We then present the standard method for solving for the optimal HMM parameters and discuss the inherent local maxima issue associated with the HMM optimization problem. In answer to this we introduce our modified robust HMM optimization method and present some preliminary results produced by this method.

2 Hidden Markov Models

A hidden Markov model (HMM) is a statistical model for ordered data. The observed data is assumed to have been generated by a unobservable statistical process of a particular form. This process is such that each observation is coincident with the system being in a particular state. Furthermore it is a first order Markov process: the next state is dependent only the current state. The model is completely described by the initial state probabilities, the first order Markov chain state-to-state transition probabilities, and the probability distributions of observable outputs associated with each state.

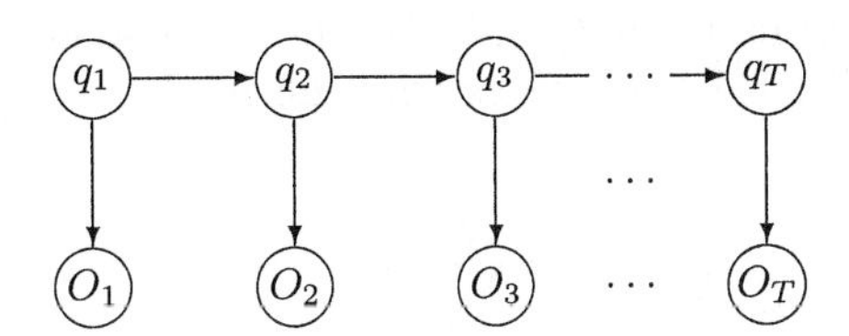

Fig. 1. A representation of the hidden Markov model, with hidden nodes in underlying system states q, and observable variables O.

Our notation is as follows: a hidden Markov model λ with N states is composed of initial state probabilities $\pi = (\pi_1, \ldots \pi_N)$, state-to-state transition probabilities $A = (a_{11}, \ldots, a_{ij}, \ldots, a_{NN})$, and the observable output probability distributions $B = (b_1, \ldots, b_N)$. The observable outputs can be either discrete or continuous. In the discrete case, the output probability distributions are denoted by $b_i(m)$, where m is one of M discrete output symbols. In the continuous case, the output probability distributions are denoted by $b_i(y, \theta_{i1}, \ldots, \theta_{ij}, \ldots, \theta_{iM})$ where y is the real-valued observable output (scalar or vector) and the θ_{ij}s are the parameters describing the output probability distribution. For the normal distribution we have $b_i(y, \mu_i, \Sigma_i)$.

3 Optimization Procedure for the HMM

For the series of observations $O = O_1 O_2 \cdots O_T$, we consider the possible model state sequences $Q = q_1 q_2 \cdots q_T$ to which this series of observations could be assigned. The probability of O given the model is obtained by summing the joint probability over all possible state sequences Q:

$$P(O|\lambda) = \sum_{\text{all } Q=q_1 q_2 \cdots q_T} \pi_{q_1} b_{q_1}(O_1) a_{q_1 q_2} b_{q_2}(O_2) \cdots a_{q_{T-1} q_T} b_{q_T}(O_T). \quad (1)$$

Although other optimization criteria are possible, most commonly we wish to optimize the model parameters so as to maximize this likelihood $P(O|\lambda)$. We can pose this as non-convex, non-linear optimization problem with constraints on π, A, and B that reflect the fact that they are probabilities. Often this problem is presented as the equivalent problem of maximizing the *log likelihood*, $\log P(O|\lambda)$.

The most common optimization technique employed to solve this problem is the Expectation-Maximization (EM) algorithm [6]. Briefly, using this method a likelihood $P(\lambda)$ with model parameters λ is optimized by repeated calculation and optimization of a so-called *Q-function*, which is defined as the expectation of the log of a certain underlying positive real-valued function $p(x, \lambda)$.

This method is inherently sensitive to the initial conditions and only guarantees eventual convergence to a local maxima of the objective function, not the global maximum. Nevertheless, it is widely used in practice and often achieves good results.

For the hidden Markov model, we employ the EM method in following manner. We have the Q-function $Q(\lambda, \lambda^{(k)})$ which we wish to maximize over λ at each iteration. We can view Q as the sum of three separable components, $Q = Q_1 + Q_2 + Q_3$:

$$Q_1(\lambda, \lambda^{(k)}) = \sum_{i=1}^{N} \tau_{i1}^{(k)} \log \pi_i, \quad (2)$$

$$Q_2(\lambda, \lambda^{(k)}) = \sum_{i=1}^{N} \sum_{j=1}^{N} \sum_{t=1}^{T-1} \tau_{ijt}^{(k)} \log a_{ij}, \quad (3)$$

$$Q_3(\lambda, \lambda^{(k)}) = \sum_{i=1}^{N} \sum_{t=1}^{T} \tau_{it}^{(k)} \log b_i(O_t). \quad (4)$$

Maximization of each component may be pursued separately. We have

$$\pi_i = \frac{\pi_i^{(k)} b_i^{(k)}(O_1)}{\sum_{j=1}^{N} \pi_j^{(k)} b_j^{(k)}(O_1)}, \quad (5)$$

as the maximizing solution for Q_1 and

$$a_{ij} = \frac{\sum_{t=1}^{T-1} \tau_{ijt}^{(k)}}{\sum_{t=1}^{T-1} \tau_{it}^{(k)}}, \quad (6)$$

as the maximizing solution for Q_2. If the outputs of the model are discrete, the maximizing solution for Q_3 is

$$b_i(m) = \frac{\sum_{t=1}^{T} \tau_{it}^{(k)} \delta(O_t - m)}{\sum_{t=1}^{T} \tau_{it}^{(k)}} \tag{7}$$

where m is a possible output symbol. If the outputs of the model are continuous, then there is no general explicit formula for the maximum value of the output distribution parameters. However, for certain special forms of the output distribution, the maximizing values can be calculated analytically. For example, in the case of multivariate Gaussian output distributions ($b_i(y) = n(\det(\Sigma_i))^{-1/2} \exp(-(y-\mu_i)^T \Sigma_i^{-1}(y-\mu_i)/2)$, where n is a normalizing factor), we have

$$\mu_i = \frac{\sum_{t=1}^{T} \tau_{it}^{(k)} O_t}{\sum_{t=1}^{T} \tau_{it}^{(k)}}, \tag{8}$$

and

$$\Sigma_i = \frac{\sum_{t=1}^{T} \tau_{it}^{(k)} (O_t - \mu_i^{(k+1)})(O_t - \mu_i^{(k+1)})^T}{\sum_{t=1}^{T} \tau_{it}^{(k)}}. \tag{9}$$

What remains is to calculate the probabilities τ_{it} and τ_{ijt}. To do so, it is possible to make use of the lattice structure of the HMM to perform an iterative calculation, known as the *forward-backward* procedure. For a more detailed explanation of this procedure, see [7].

4 Multimodality of the HMM Objective Function

As noted, the EM algorithm only guarantees convergence to a local maximum. Since the algorithm is deterministic, the initial model parameter selection controls which local maxima is eventually reached. In many cases, the EM algorithm functions well; this is one reason for its popularity. However, the likelihood function of an HMM potentially has an exponential number of local maxima; this makes the optimization problem much more difficult.

Suppose we consider a particular data set, one composed of S distinct segments s, each starting at $t_1(s)$ and ending at $t_T(s)$. For each segment the outputs $O_s = O_{t_1(s)} \cdots O_{t_T(s)}$ are all a single unique value, m_s. For this data set, the local maxima are solutions in which the possible output values for each state are unique, so that if $b_i(m) \neq 0$, then $b_j(m) = 0$ for all $i \neq j$, and are contiguous in the time sequence. More specifically, let the N_{s_i} segments $s_i(1), \ldots, s_i(N_{s_i})$ be associated with the state i; that is, let $b_i(m_{s_i(k)}) \neq 0, k = 1, \ldots, N_{s_i}$. Furthermore, let $L_{s_i(k)}$ be the length of the segment $s_i(k)$. Then a locally maximum model λ^* is such that

$$\pi_i^* = \begin{cases} 1 \text{ if } O_1 = m_{s_i(k)} \text{ for some } k \\ 0 \text{ otherwise} \end{cases},$$

$$a_{ij}^* = \begin{cases} \frac{\sum_{k=1}^{N_{s_i}} L_{s_i(k)} - 1}{\sum_{k=1}^{N_{s_i}} L_{s_i(k)}} & \text{if } i = j \\ \frac{1}{\sum_{k=1}^{N_{s_i}} L_{s_i(k)}} & \text{if } t_T(s_i(k)) + 1 = t_1(s_j(l)) \text{ for some } k, l \\ 0 & \text{otherwise} \end{cases},$$

$$b_i(m)^* = \begin{cases} \frac{L_{s_i(k)}}{\sum_{k=1}^{N_{s_i}} L_{s_i(k)}} & \text{if } m = m_{s_i(k)} \text{ for some } m_{s_i(k)} \\ 0 & \text{otherwise} \end{cases}. \tag{10}$$

We first present a simple illustrative example. Consider the sequence $O = 112233$ of length $T = 6$, on which we train a model of size $N = 2$. Consider

$$\lambda_1 = \left\{ \pi = \begin{pmatrix} 1 \\ 0 \end{pmatrix}, A = \begin{pmatrix} 0 & 1 \\ 0 & 1 \end{pmatrix}, b_1 = \begin{pmatrix} 1 \\ 0 \\ 0 \end{pmatrix}, b_2 = \begin{pmatrix} 1/5 \\ 2/5 \\ 2/5 \end{pmatrix} \right\},$$

$$\lambda_2 = \left\{ \pi = \begin{pmatrix} 1 \\ 0 \end{pmatrix}, A = \begin{pmatrix} 1/2 & 1/2 \\ 0 & 1 \end{pmatrix}, b_1 = \begin{pmatrix} 1 \\ 0 \\ 0 \end{pmatrix}, b_2 = \begin{pmatrix} 0 \\ 1/2 \\ 1/2 \end{pmatrix} \right\},$$

$$\lambda_3 = \left\{ \pi = \begin{pmatrix} 1 \\ 0 \end{pmatrix}, A = \begin{pmatrix} 2/3 & 1/3 \\ 0 & 1 \end{pmatrix}, b_1 = \begin{pmatrix} 2/3 \\ 1/3 \\ 0 \end{pmatrix}, b_2 = \begin{pmatrix} 0 \\ 1/3 \\ 2/3 \end{pmatrix} \right\}.$$

Then $P(O|\lambda_1) = 0.00512, P(O|\lambda_2) = 0.015625, P(O|\lambda_3) = 0.01$, so λ_2 is a local maximum. A second local maximum exists for which $q_1 \cdots q_4 = 1, q_5 q_6 = 2$; a third maximum is one for which the entire sequence is in the same state.

Now we present the general case and demonstrate that λ^* of the form described in (10) is in fact a local maximum. For ease of notation, we assume without loss of generality that $t_1(s_{i+1}(1)) > t_T(s_i(1))$, that is, the segment labels increase monotonically with t. We furthermore define $\mathcal{L}_i = \sum_{k=1}^{N_{s_i}} L_{s_i(k)}$. Then we have

$$P(O|\lambda^*) = \left(\frac{\mathcal{L}_1 - 1}{\mathcal{L}_1} \right)^{\mathcal{L}_1 - 1} \left(\frac{1}{\mathcal{L}_1} \right) \cdots$$
$$\left(\frac{\mathcal{L}_N - 1}{\mathcal{L}_N} \right)^{\mathcal{L}_N - 1} \cdot \prod_{k=1}^{N_{s_1}} \left(\frac{L_{s_1(k)}}{\mathcal{L}_1} \right)^{L_{s_1(k)}} \cdots \prod_{k=1}^{N_{s_N}} \left(\frac{L_{s_N(k)}}{\mathcal{L}_N} \right)^{L_{s_N(k)}} \tag{11}$$

Now consider a model λ which is slightly perturbed from λ^* so that

$$b_1(m_{s_2(1)}) = \frac{1}{\mathcal{L}_1 + 1}, \; b_1(m_{s_1(k)}) = \frac{L_{s_1(k)}}{\mathcal{L}_1 + 1}, \; k = 1, \ldots, N_{s_1}$$
$$b_2(m_{s_2(1)}) = \frac{L_{s_2(1)} - 1}{\mathcal{L}_2 - 1}, \; b_2(m_{s_2(k)}) = \frac{L_{s_2(k)}}{\mathcal{L}_2 - 1}, \; k = 2, \ldots, N_{s_2}, \tag{12}$$

and

$$a_{11} = \frac{\mathcal{L}_1}{\mathcal{L}_1 + 1}, \; a_{12} = \frac{1}{\mathcal{L}_1 + 1}, \tag{13}$$

$$a_{22} = \frac{\mathcal{L}_2 - 2}{\mathcal{L}_2 - 1}, \; a_{23} = \frac{1}{\mathcal{L}_2 - 1}. \tag{14}$$

In other words, this model λ corresponds to a state sequence Q such that $q_t = 1$ for $t = 1, \ldots, \mathcal{L}_1 + 1$. We have

$$P(O|\lambda) = \sum_{n=0}^{L_{s_2(1)}-1} \left\{ \left(\frac{\mathcal{L}_1}{\mathcal{L}_1 + 1} \right)^{(\mathcal{L}_1 - 1 + n)} \left(\frac{1}{\mathcal{L}_1 + 1} \right) \right.$$
$$\left(\frac{\mathcal{L}_2 - 2}{\mathcal{L}_2 - 1} \right)^{(\mathcal{L}_2 - 1 - n)} \left(\frac{1}{\mathcal{L}_2 - 1} \right) \cdots \left(\frac{\mathcal{L}_N - 1}{\mathcal{L}_N} \right)^{(\mathcal{L}_N - 1)} \cdot$$
$$\prod_{k=1}^{N_{s_1}} \left(\frac{L_{s_1(k)}}{\mathcal{L}_1 + 1} \right)^{L_{s_1(k)}} \left(\frac{1}{\mathcal{L}_1 + 1} \right)^n \left(\frac{L_{s_2(1)} - 1}{\mathcal{L}_2 - 1} \right)^{L_{s_2(1)} - n} \cdot$$
$$\left. \prod_{k=2}^{N_{s_2}} \left(\frac{L_{s_2(k)}}{\mathcal{L}_2 - 1} \right)^{L_{s_2(k)}} \prod_{k=1}^{N_{s_3}} \left(\frac{L_{s_3(k)}}{\mathcal{L}_3} \right)^{L_{s_3(k)}} \cdots \prod_{k=1}^{N_{s_N}} \left(\frac{L_{s_N(k)}}{\mathcal{L}_N} \right)^{L_{s_N(k)}} \right\}, \tag{15}$$

from which we can see that $P(O|\lambda^*) > P(O|\lambda)$. A similar analysis follows for the model λ perturbed from λ^* corresponding to the state sequence Q such that $q_t = 1$ for $t = 1, \ldots, \mathcal{L}_1 - 1$. We can extend this analysis to all such models λ such that A and B are perturbed in a like manner around the segment boundaries from A^* and B^*, so that $P(O|\lambda^*) > P(O|\lambda)$. From this we can conclude that λ^* is in fact a local maximum.

We note that for S unique segments there are $\binom{S-1}{N-1}$ local maxima λ^* of this form utilizing all N states, since we choose $N-1$ of the $S-1$ possible transitions between segments as our state transition points. We further note that this same analysis holds true for all models for which less than the full number of states are utilized. So in total there are $\sum_{n=1}^{N} \binom{S-1}{n-1}$ local maxima for this data set and model size N. If $S \geq N$, then $\sum_{n=1}^{N} \binom{S-1}{n-1} \geq 2^{N-1}$, so the lower bound on the number of local maxima is exponential in the model size.

An additional problem arises for certain forms of the output distribution B. For these forms there are values of the parameters θ_{im} such that the likelihood achieves an unfavorable global maximum. By unfavorable, we mean that these globally maximum model parameters are less informative about the values of the hidden variables than models with merely local maxima. For example, in the case of Gaussian output probability distributions, the likelihood goes approaches infinity as the eigenvalues of the variances approach zero. We can identify $\sum_{n=1}^{N} \binom{N}{n} \sum_{d=1}^{D} \binom{D}{d}$ such unfavorable global maxima, where D is the dimension of the observations, since the likelihood will approach infinity if even one eigenvalue of the variance of a single state approaches zero. This implies that the number of such global maxima is exponential in both the number of states and in the dimension of the observable data.

5 Q-Function Penalty Terms

The analysis of the previous section indicates that many fixed points of the EM transformation and sub-optimal local maxima are located in the model parameter space at predictable points where $b_i = b_j$. It would therefore appear to be advantageous to augment to the standard optimization procedure so as to avoid these parts of the parameter space. One way to do this is to add penalty terms to the Q-function.

No general penalty term exists to assist in avoiding the condition where $b_i = b_j$. However, for particular forms of the output distribution penalty terms can be devised. For example, for discrete output distributions, we can add a penalty term based on the inner product:

$$Q_3' = \sum_{i=1}^{N}\sum_{t=1}^{T}\sum_{m=1}^{M} \tau_{it}^{(k)}\delta(O_t - m)\log b_i(m) - \omega_{Q_3}\sum_{i=1}^{N}\sum_{j=1}^{N}\sum_{m=1}^{M} b_i(m)b_j(m) \quad (16)$$

where $\omega_{Q_3} > 0$ is a small weighting factor. As a second example, we consider the case of Gaussian output distributions. We add a penalty term based on the squared Euclidean distance:

$$Q_3' = \sum_{i=1}^{N}\sum_{t=1}^{T} \tau_{it}^{(k)}\Bigg(\log n - \frac{1}{2}\log\det(\Sigma_i) - \frac{1}{2}(m_i - \mu_i)^T\Sigma_i^{-1}(m_i - \mu_i)$$

$$- \frac{1}{2}(O_t - m_i)^T\Sigma_i^{-1}(O_t - m_i) + \frac{\omega_{Q_3}}{2}\sum_{j=1}^{N}(\mu_i - \mu_j)^T(\mu_i - \mu_j)\Bigg). \quad (17)$$

In both these cases conditions on the weighting terms ω_{Q_3} can be found such that the function Q_3 remains concave and thus has a single local maxima. Computing the solution to either maximization problem requires an iterative procedure with a computational cost per iteration which is cubic in the dimension of the observations. As the basic HMM optimization method requires inversion of the covariance matrices at each EM iteration, the modified method merely introduces a constant factor for bounded iterations in the inner loop. In practice, solutions to Q_3' can be found in very small (< 10) numbers of iterations.

We note in that these penalty terms do not help to escape from local maxima when the model parameters are already at a point where $b_i = b_j$. Although random initialization of the model parameters makes this unlikely, alternate initialization methods can make this more problematic. In such cases, one way to escape from the local maximum is to perturb the distributions by some small amount when the case $b_i = b_j$ is detected.

In the case of Gaussian output distributions we can impose an additional penalty term in order to deal with unfavorable global maxima located where the covariance matrices become singular. Our penalty term is based on the trace of the inverse of the covariance matrix, since

$$\operatorname{Tr}\Sigma_i^{-1} = \sum_{d=1}^{D}\frac{1}{\lambda_{id}} \quad (18)$$

where D is the dimension of the observations and $\lambda_{i1}, \ldots, \lambda_{iD}$ are the eigenvalues of the ith covariance matrix. The modified Q-function is

$$Q_3' = \sum_{t=1}^{T} \sum_{i=1}^{N} \tau_{it}^{(k)} \Big(\log n + \frac{1}{2} \log \det(\Sigma_i^{-1}) - \frac{1}{2}(m_i - \mu_i)^T \Sigma_i^{-1} (m_i - \mu_i) - \frac{1}{2} \operatorname{Tr} \Sigma_i^{-1} S_i - \frac{\omega_\Sigma}{2} \operatorname{Tr} \Sigma_i^{-1} \Big), \quad (19)$$

where ω_Σ is a weighting factor. This leads us to an optimum solution in which we add a diagonal matrix $\omega_\Sigma I$ to each covariance matrix.

Incorporating all of the above, our modified EM algorithm is then:

1. Start with $k = 0$ and pick a starting $\lambda^{(k)}$.
2. Calculate $Q'(\lambda^{(k)}, \lambda)$ (expectation step).
3. Maximize $Q'(\lambda^{(k)}, \lambda)$ over λ (maximization step). This gives us the transformation $\mathcal{F}$.
4. Set $\lambda^{(k+1)} = \mathcal{F}(\lambda^{(k)})$. If $Q'(\lambda^{(k+1)}, \lambda) - Q'(\lambda^{(k)}, \lambda)$ is below some threshold, stop. Otherwise, go to step 2.
5. Check to see if $b_i = b_j$ for any $i \neq j$. If so, then perturb the current model so that $\theta_i = \theta_i + \epsilon_\theta$, and go to step 2. Otherwise, stop.

6 Experimental Results

We applied our robust HMM method to GPS and seismicity data collected in the southern California region. In our implementation we assume Gaussian output probability distributions for both FMM and HMM for simplicity and ease of computation. Presented here are some preliminary experimental results.

The GPS data consists of surface displacement signals collected from a number of sites scattered around the southern California region. The data was three dimensional, consisting of east-west displacement, north-south displacement, and vertical displacement measurements, collected daily. Figure 2 shows a representative example of the results of the method applied to GPS data collected in the city of Claremont, California. The method determined that a five state model was optimal for this data set. Using a five state model, the HMM was able to separate the data into distinct classes that correspond to physical events. These classes are indicated in the figure by different shades and vertical lines. There is one instance of class 2 in the midst of class 3, corresponding to sharp north-south and vertical movements at that time sample, but otherwise the classes are sequential. The states before and after the Hector Mine quake of October 1999 are clearly separated, and distinct in turn from a period in 1998 in which well ground water drainage caused displacement in the vertical direction. Sharp movements in the north-south direction (as yet unattributed) were also isolated as a separate class.

The seismicity data was taken from the Southern California Earthquake Center (SCEC) catalog. For this experiment, the original data set was processed to

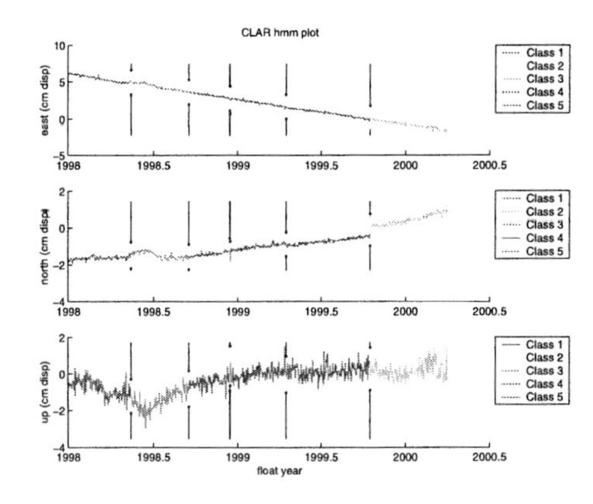

Fig. 2. HMM analysis results of global positioning system (GPS) relative displacement data collected from a receiver located in Claremont, California. Classes associated with different regimes are indicated by line coloration and vertical indicator lines.

produce six components for each observed seismic event between January 1st, 1960 and December 31st, 1999: latitude, longitude, depth, magnitude, time to next event, and time to previous event. Events of less than magnitude four were removed. The method determined that a model with 17 states would be optimal for this data sets. The data was grouped into scientifically meaningful classes, including clusters of aftershocks for the Hector Mine, Landers, and Northridge earthquakes, Transverse Range events, and swarm events in the Salten Sea area. Furthermore, relationships between the classes as indicated by the transition probabilities reveal evidence of scientifically meaningful phenomenon such as stress waves. Figure 3 show examples of the classifications produced by the method. Circles indicate the location of earthquakes; circle size corresponds to magnitude. Lines represent the major faults.

7 Conclusions and Future Work

We have presented a tool for geophysical data analysis that is based around the use of hidden Markov models (HMMs). The tool employs a method for estimating the optimal HMM parameters that is based on the analytical analysis of certain local maxima of the HMM objective function that originate in the model structure itself rather than the data. This analysis is then used to modify the standard optimization procedure through the application of penalty functions which enable the solution to avoid many local maxima. This improves both the quality and consistency of results. Preliminary experiments employing this method in the analysis of geodetic and seismic record data have yielded results which can be verified with respect to a priori scientific knowledge.

As part of our continued work on this method we are performing systematic analysis of the effect of our modified method on the quality and stability of the optimizated solution. In addition we are applying the method to a diverse assortment of geophysical data sets.

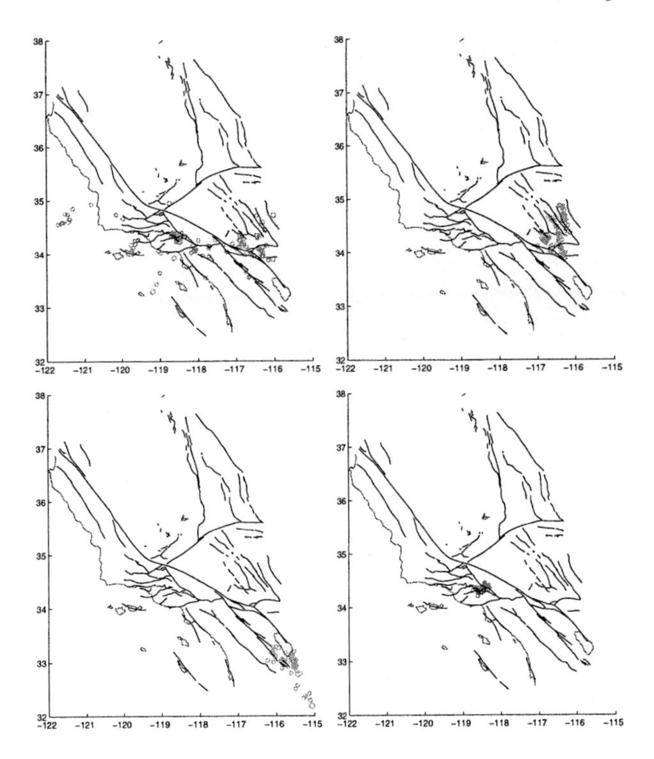

Fig. 3. HMM analysis result for SCEC catalog seismicity data. Upper left: the class of Transverse Range events; upper right: the class of Hector Mine and Landers earthquake aftershocks; bottom left: the class of Salten Sea swarm events; bottom right: the class of Northridge earthquake aftershocks.

References

1. L. E. Baum. An inequality and associated maximization technique in statistical estimation for probabilistic functions of markov processes. *Inequalities*, 3:1–8, 1972.
2. L. E. Baum and J. A. Egon. An inequality with applications to statistical estimation for probabilistic functions of a markov process and to a model for ecology. *Bull Amer Meteorol Soc*, 73:360–363, 1967.
3. L. E. Baum and T. Petric. Statistical inference for probabilistic functions of finite state markov chains. *Ann Math Stat*, 37:1554–1563, 1966.
4. L. E. Baum, T. Petrie, G. Soules, and H. Weiss. A maximization technique occuring in the statistical analysis of probabilistic functions of markov chains. *Ann Math Soc*, 41(1):164–171, 1970.
5. L. E. Baum and G. R. Sell. Growth functions for transformations on manifolds. *Pac J Math*, 27(2):211–227, 1968.
6. A. P. Dempster, N. M. Laird, and D. B. Rubin. Maximum likelihood from incomplete data via the em algorith. *J Roy Stat Soc*, 39(1):1–38, 1977.
7. L. R. Rabiner. A tutorial on hidden markov models and selected applications in speech recognition. *P IEEE*, 77(2):257–286, 1989.

Workshop on Performance Evaluation, Modeling, and Analysis of Scientific Applications on Large-Scale Systems

A Performance Model of Non-deterministic Particle Transport on Large-Scale Systems

Mark M. Mathis[1,*], Darren J. Kerbyson[2], and Adolfy Hoisie[2]

[1] Dept. of Computer Science, Texas A&M University
mmathis@cs.tamu.edu
[2] Performance and Architectures Lab, CCS-3, Los Alamos National Laboratory
{djk,hoisie}@lanl.gov

Abstract. In this work we present a predictive analytical model that encompasses the performance and scaling characteristics of a non-deterministic particle transport application, MCNP. Previous studies on the scalability of parallel Monte Carlo eigenvalue calculations have been rather general in nature [1]. It can be used for the simulation of neutron, photon, electron, or coupled transport, and has found uses in many problem areas. The performance model is validated against measurements on an AlphaServer ES40 system showing high accuracy across many processor / problem combinations. It is parametric with both application characteristics (e.g. problem size), and system characteristics (e.g. communication latency, bandwidth, achieved processing rate) serving as input. The model is used to provide insight into the achievable performance that should be possible on systems containing thousands of processors and to quantify the impact that possible improvements in sub-system performance may have. In addition, the impact on performance of modifying the communication structure of the code is also quantified.

1 Introduction

MCNP is a general purpose Monte-Carlo N-Particle code that represents part of the Accelerated Strategic Computing Initiative (ASCI) workload. It can be used for the simulation of neutron, photon, electron, or coupled transport [2]. Particle transport simulation has found uses in many problem areas including nuclear reactors, radiation shielding, and medical physics. There is great interest in the use of non-deterministic particle simulation on large-scale systems - both those currently in existence as well as future advanced systems being proposed.

A model of MCNP is required to assess the performance that can be obtained on current and future large-scale systems. In particular, a model can provide information to users on what size problem can be processed given a time allocation or what size problem needs to be processed in order to achieve a desired quality of results. The model can also be used in the procurement, and consequent

* Mathis is supported in part by a DOE High-Performance Computer Science Fellowship. Los Alamos National Laboratory is operated by the University of California for the National Nuclear Security Administration of the US Department of Energy.

P.M.A. Sloot et al. (Eds.): ICCS 2003, LNCS 2659, pp. 905–915, 2003.

installation, of future systems by providing information on what performance should be achievable prior to actual system availability (see for example [3]). The model can also be used to identify bottlenecks in the code and to make recommendations for its future development.

In this work we develop a detailed analytical performance model of MCNP. The model consists of two fundamental parts: an application model and a system model. The application model is based on a static analysis of the key portions of the code but is parameterized in terms of the data specific to the problem being simulated. The application model is combined with the system model in order to evaluate performance predictions for a specific system. The system model encapsulates key system characteristics such as communication (e.g. latency and bandwidth), and computational performance (e.g. processor speed). The two parts of the model are kept separate so the model can be re-used without alteration to explore a multitude of performance scenarios. For instance, one may evaluate a different problem by setting the appropriate input parameters to the application model, or evaluate a new machine by changing the input values to the system model. A similar modeling approach has been used to model other large-scale applications including deterministic transport on structured and unstructured meshes [4,5,6], and adaptive mesh refinement [7,8].

Motivation and Previous Work. Criticality safety is a vital part of the storage, transportation, and processing of fissionable materials. Criticality may be defined as that state of a nuclear chain-reacting medium when the nuclear fission chain reaction just becomes self-sustaining (critical). MCNP includes the capability to calculate eigenvalues for critical systems and forms the particular input studied in this work. The example geometry consists of an insulated barrel containing a number of hollow rods of fissionable material. Horizontal and vertical cross-sections of the geometry are shown in Fig. 1. The shading is used

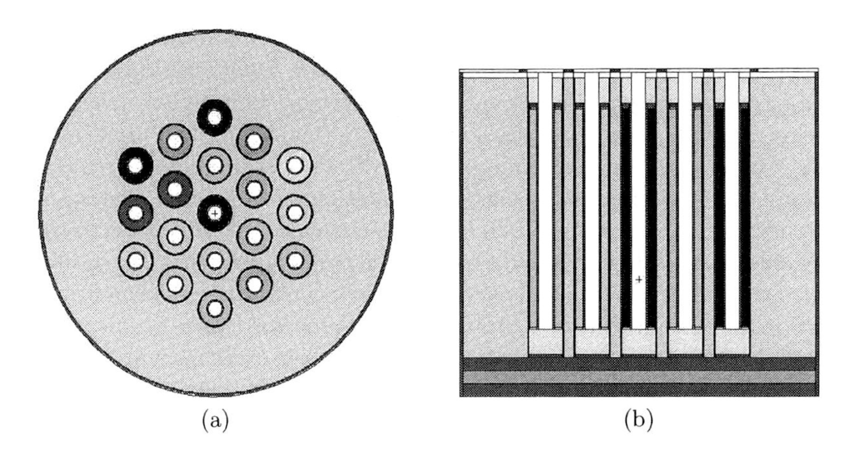

Fig. 1. Horizontal (a) and vertical (b) cross-sections of example geometry

to indicate the different material properties of each rod and also of the insulated barrel. The hollow portion of the rods are indicated as white. The goal of the simulation is to determine if the arrangement of rods is safe, i.e. non-critical.

2 MCNP

MCNP can trace its roots back to the invention of the Monte Carlo method at Los Alamos during World War II. The Monte Carlo method is generally attributed to Fermi, Metropolis, von Neumann, Richtmyer and Ulam [9]. It was one of the first application programs run on early computers in the 1950's. MCNP is the successor to their work and represents over 450 person-years of development. Version 4C of MCNP was used in this analysis.

Monte Carlo methods in general and MCNP specifically do not solve an explicit equation, but rather obtain answers by simulating the interactions between individual particles and a predefined geometry. The accuracy of the calculation increases in proportion to the number of particles used in the simulation. In general, the error in the calculation reduces as the square root of the number of particles. This is in contrast to deterministic transport methods, the most common of which is the discrete ordinates method, that actually solve the transport equation directly for the average particle behavior [10].

The input geometry for MCNP consists of a collection of cells defined as combinations of primitive shapes such as planes, cylinders and spheres. The material properties are retrieved from an external library . The behavior of each simulated particle and its interaction with the materials travelled through, as defined by the geometry, are recorded to produce a particle *history*. During this process, statistical information about certain events is gathered in histograms or *tallies*. The interaction of particle and geometry can result in several events such as neutron/photon scatter, capture, and leakage.

The current parallelization strategy of MCNP requires the geometry to be copied to all processors and thus the complexity of the geometry is constrained by the memory available in a single processing node [11]. Parallelism can be utilized to either solve the same problem faster by sub-dividing the simulated particles across all processors (strong scaling), or to give a more accurate simulation by simulating more particles in proportion to the number of processors (weak scaling). During each cycle of MCNP, each processor simulates a designated set of particles. At the end of each cycle, a single processor merges the results from all other processors during a *rendezvous*. This communication pattern requires several steps and a fairly high degree of coordination. Note that the achievable performance of MCNP is both input sensitive (the cost to simulate a particle depends on the complexity of the geometry and materials used) and output sensitive (the complexity of the output depends on the requested tallies).

3 Performance Model

MCNP is representative of a general parallel application paradigm known as master-slave. In this paradigm, a master process is responsible for dividing the work to be done across a number of slave processes. Work assignments and state information are distributed from master to slaves during a "scatter" phase at the start of each cycle. Once the slaves receive their assignments they may begin their local computation. Once the slaves have completed their work, they report their results to the master during a "gather" phase. The master then aggregates the results from all the slaves and a new cycle begins. The "scatter" and "gather" phases may actually consist of a sequence of messages. For MCNP the scatter phase consists of 2 communications, and the gather phase consists of 5 communications (Fig. 2).

MCNP is particularly well-suited to the master-slave paradigm due to the independent nature of particle simulation . During the work phase of MCNP there is no communication between slaves (i.e. any particle is simulated independently of any other particle). Unfortunately, this apparent strength reveals a hidden weakness of MCNP and the master-slave approach in general. First, all communication must go through the master and second, the entire geometry must be replicated. The amount of data transmitted from the master to each slave in the scatter phase, and from each slave to the master in the gather phase results in a scaling limitation. The performance of MCNP scales well until communication costs dominate the execution time. The first limitation could be potentially avoided by altering the manner in which data is reported to the master. The second could be overcome by allowing communication between slaves, to relay geometry or particle information.

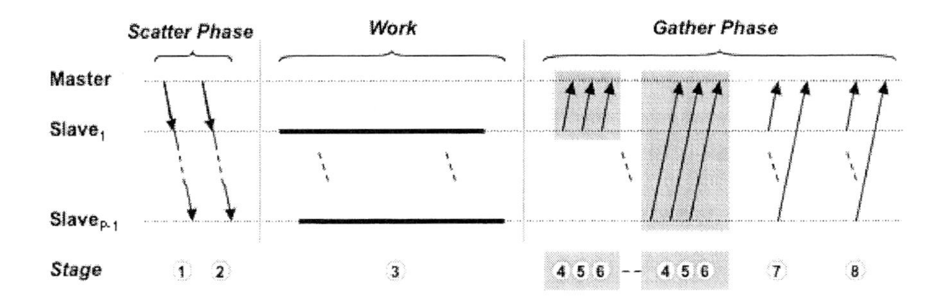

Fig. 2. Communication pattern for a single cycle of MCNP

3.1 Application Model

The application model includes only those portions of the parallel activity that significantly contribute to the overall execution time. The main stages of a cycle of MCNP corresponding to those depicted in Fig. 2 are listed in Table 1. The

Table 1. Summary of parallel activity for one cycle of MCNP. M=Master, S=Slave(s)

Stage	Source	Action	Size (bytes)	Description
1	M	bcast	$P * 8$	particle range to be computed by slaves
2	M	bcast	229240	update to current history
3	S	work	$T_{hist} * \lceil N_{ph}/(P-1) \rceil$	T_{hist} times the number of histories
4	S	pt2pt	5512	task common
5	S	pt2pt	320	tally data
6	S	pt2pt	204920	task array 1
7	S	pt2pt	$48 * \lceil N_{ph}/(P-1) \rceil$	task array 2
8	S	pt2pt	32	timing data

table summarizes the event source (either master or slave), its type (either collective broadcast, point-to-point communication, or computation), and also the weight associated with the event for each stage of the cycle. The weight is in bytes for all communication events (message sizes), and in terms of the number of particle histories for the computation events. The sizes of some of the messages are dependent upon the actual problem being solved. These sizes must be measured prior to the use of the model.

The first task array message (stage 6 in Table 1) is constant for each input geometry and the requested tallies. The tally data message (stage 5) can be calculated simply as the word size (8) times 2 plus the number of requested tallies (38 for this problem). The constant 48 involved in the task array 2 message (stage 7) is obtained by run-time measurement. This is related to the average number of collisions experienced by each particle.

The execution time for a single cycle of MCNP can be modeled as:

$$T_{total} = T_{scatter} + T_{slave} + T_{gather} \tag{1}$$

where the cycle time, T_{total}, is a summation of the scatter, work, and gather phases - $T_{scatter}$, T_{slave}, and T_{gather}, respectively. The form of this model is additive since the gather and scatter stages are in general synchronized by the bottleneck caused by the master, and the serialization of the communication from the slaves to the master. Each cycle begins with a scatter phase:

$$T_{scatter} = T_{bcast}(P * 8, P) + T_{bcast}(229420, P) \tag{2}$$

where the time to perform the collective broadcast operation, $T_{bcast}(S, P)$, is the time taken to broadcast S bytes across P processors on the target system. The scatter phase corresponds to the first two stages in Table 1.

The computation phase, performed on each slave, can be modeled as:

$$T_{slave}(P, N_{ph}, T_{hist}) = \left\lceil \frac{N_{ph}}{(P-1)} \right\rceil * T_{hist} \tag{3}$$

where N_{ph} is the number of particle histories per cycle which are divided amongst the $P-1$ slave processors. In general, it is more accurate to take the computation

Table 2. Summary of system model parameters (S in bytes)

$L\ (S), B\ (S)$	$T\quad(S)$	T
$\begin{cases} 5.05\mu s, 0.0MB/s & S < 64, \\ 5.47\mu s, 78MB/s & 64 \leq S < 512, \\ 10.3\mu s, 294MB/s & S \geq 512 \end{cases}$	$\begin{cases} 0.12ns & S < 32K, \\ 0.16ns & 32K \leq S \leq 4M, \\ 0.67ns & S > 4M \end{cases}$	$798\mu s$

time for the slowest slave. However, since each slave is responsible for an equal number of particles, we assume that all slaves will take the same time. The time to perform a single particle history, T_{hist}, can be measured on a single processor for the problem being solved. The gather phase can be modeled as:

$$T_{gather}(P, N_{ph}) = \sum_{i=1}^{P-1} \Bigg(T_{pt2pt}(5512, i, 0) + T_{pt2pt}(320, i, 0) + T_{pt2pt}(204920, i, 0) + T_{pt2pt}(48 * \left\lceil \frac{N_{ph}}{(P-1)} \right\rceil, i, 0) + T_{pt2pt}(32, i, 0) \Bigg) \quad (4)$$

where the five point-to-point communications, listed as stages 4-8 in Table 1, are effectively performed in a serialized way due to the master bottleneck. However, an examination of the current messaging within MCNP indicates that some of the data transfered between all slaves and the master (specifically stages 4, 5, possibly part of 6, and 8 in Table 1) can be at least partially implemented as collective reductions. If we assume that all of stages 4, 5, and 8 as well as half of stage 6 can be reduced, equation 4 can be re-written as:

$$T_{gather}(P, N_{ph}) = \sum_{i=1}^{P-1} \Bigg(T_{pt2pt}(102460, i, 0) + T_{pt2pt}(48 * \left\lceil \frac{N_{ph}}{(P-1)} \right\rceil, i, 0) \Bigg) + T_{reduce}(5512, P) + T_{reduce}(320, P) + T_{reduce}(102460, P) + T_{reduce}(32, P) \quad (5)$$

3.2 System Model

For the application model as formulated in Sect. 3.1, the required components of the system model are point-to-point communication times, collective broadcast times, the time required to perform a single particle history , and also the memory performance of a single node (for packing). MCNP actually uses the UPS messaging library [12] for communication between processors. UPS provides a generic interface with a retargetable backend. It allows a message of arbitrary length to be built from many smaller variables using packing functions in a similar way to that of PVM - a feature that is heavily utilized in MCNP. In the analysis that follows a 32 node AlphaServer ES40 cluster is used as the experimental testbed. This machine has four processors per node interconnected using the Quadrics QsNet high-performance network [13]. This network boasts high-performance communication with a typical MPI latency of $5\mu sec$ and a throughput of up to $340MB/s$ in one direction.

Measured MPI latency and bandwidth for inter-node unidirectional communication (point-to-point) were obtained using in-house benchmarks. The collective broadcast and reduction operations for P processors can be assumed to take $log_2(P)$ times that of a single point-to-point communication. The communication costs also include packing operations, implemented in UPS. The point-to-point, broadcast, and reduction communication operations are modeled as:

$$T_{pt2pt}(S, src, dest) = T_{pack}(S) + L_c(S) + S/B_c(S) \quad (6)$$

$$T_{bcast}(S, P) = T_{pack}(S) + T_{pt2pt}(S) * log_2(P) \quad (7)$$

$$T_{reduce}(S, P) = T_{pack}(S) + T_{pt2pt}(S) * log_2(P) \quad (8)$$

where S is the size of the message in Bytes, $T_{pack}(S)$ is the time to pack a single byte, $L_c(S)$ and $B_c(S)$ are the latency and bandwidth of a message of size S bytes. The parameters used in the system model are summarized in Table 2.

Model Validation. The model is validated against measurements made on our testbed system showing high accuracy – typically to within a 10% error. Measurements and predictions are shown in Fig. 3 for 7 sets of particle histories per cycle (100, 500, 1000, 5000, 10000, 50000, and 100000) on a range of processor counts (a strong-scaling analysis). The geometry is the same for all runs. Note that for small N_{ph}, the communication costs soon dominate the processing time, resulting in poor scalability. For large N_{ph}, the scalability is better up to a higher processor count.

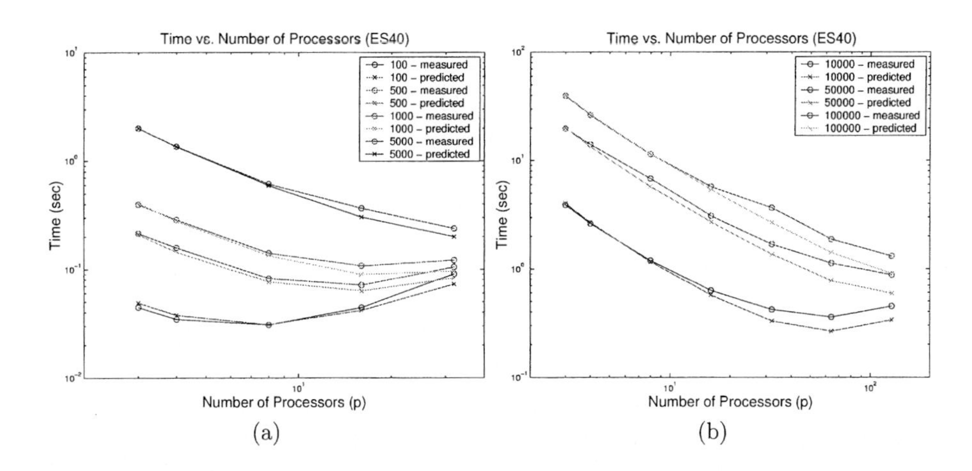

Fig. 3. Measured and predicted times for small (a) and large (b) inputs

4 Performance Study of MCNP

In this section the model is used to explore the performance of MCNP on larger configurations of current systems. It is also used to explore the possible improvements resulting from refining the code to make use of reduction operations as suggested in Sect. 3.1. The model is further utilized to investigate the performance of future systems assuming performance improvements in individual sub-system characteristics such as latency, bandwidth, and processing speed.

Scaling Behavior of Larger Systems. The MCNP performance model is used to explore the expected performance on larger AlphaServer ES40 systems in Fig. 4 for both strong and weak scaling models. As the processor count increases in the strong scaling mode, the amount of work per slave decreases and hence communication costs soon become a significant percentage of the overall run-time. In weak-scaling, as the processor count increases the amount of computation per processor is constant and thus the overall run-time increases more gradually due to increased communication costs. Overall it can be seen that in a strong-scaling mode, MCNP scales up to 512 processors on the problem being studied due to communication costs soon becoming significant. In weak-scaling, the performance of MCNP is much better and actually scales up to 8192 processors.

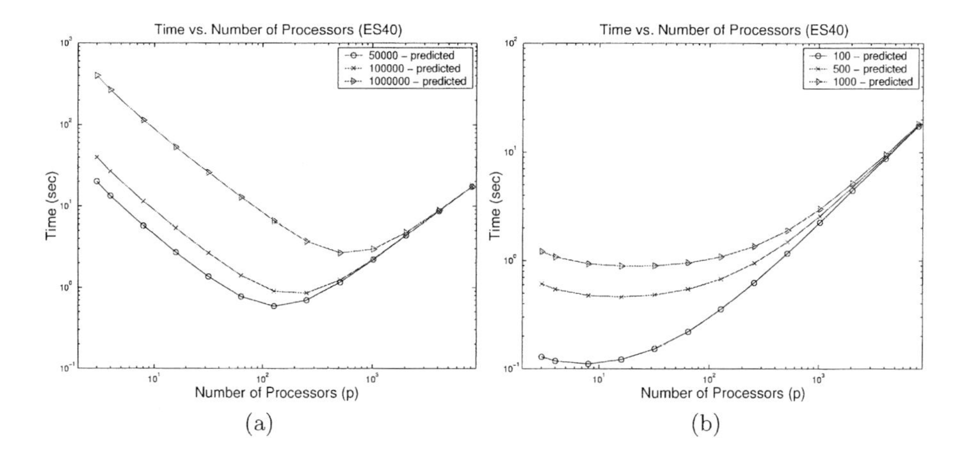

Fig. 4. Expected cycle times on larger ES40 systems for strong (a) and weak (b) scaling

Performance Predictions on Faster Systems. The impact of system performance improvements on the run-time of MCNP can also be quantified in advance of such systems being available. Here we examine a number of what-if scenarios by considering the performance of the communication and computation sub-systems to be improved by a factor of 8 individually. The factor of 8 was chosen to be indicative of what may happen to these sub-system performances

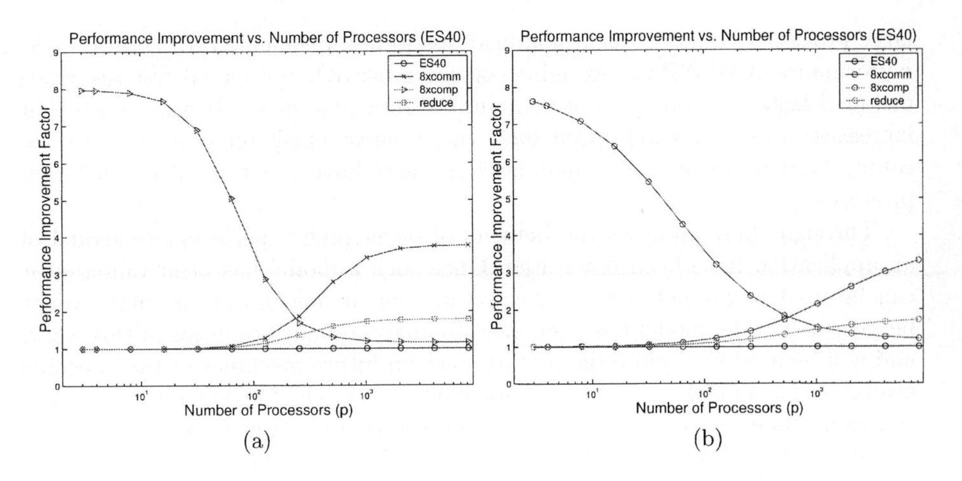

Fig. 5. Relative impact of improved computation and communication capabilities for strong (a) and weak (b) scaling

over the next 5 years.. The relative improvement over the existing ES40 system is shown in Fig. 5. Also included in Fig. 5 is the relative performance improvement that could be obtained if just the modifications to the code as described in Sect. 3.1 above were implemented.

It can be seen from Fig. 5 that an increase in computation capability has a much greater impact on performance for small numbers of processors, but rapidly declines as the number of processors is increased. Similarly on larger processor counts, the increase in communication capability will have a larger impact (due to communication constituting a larger percentage of the execution time as the number of processors increases).

5 Conclusion

In this work we have developed a detailed analytical performance model for MCNP. The model includes the main code characteristics and separates out the application and system characteristics. The model is based on a static analysis of the application but is parameterized in terms of its dynamic behavior. Through a validation process on a 32 node AlphaServer ES40 cluster, we have shown the model to be accurate with a typical error of 10%.

The model has been used to explore a number of performance scenarios. In a scalability analysis, the model was used to give expected performance on larger ES40 systems. This analysis showed that in a weak-scaling mode the application will scale to thousands of processors whereas in a strong-scaling mode the application scales to only hundreds of processors.

The performance of MCNP was also examined for the case of modifying the communication structure in the application to include the use of collective reductions. This analysis indicated that the performance could be improved on large processor counts if such modifications were implemented. In addition, the performance of MCNP was examined on a number of hypothetical systems which included faster computation or communication sub-systems. It was shown that increases in computation speed have the greatest effect on smaller processor counts, and increases in communication speed have greatest effect on larger processor counts.

Through these analyses the benefits of developing a performance model of an application have been illustrated. Once such a model has been validated it can be used to predict performance on systems or configurations that cannot be measured. The model has been used to analyze many scenarios in this work, and will be used to explore the performance on future machines as they become available. The model is part of an ongoing effort to model the ASCI workload and complements existing models for deterministic particle transport on structured and unstructured meshes and for adaptive mesh refinement applications.

References

1. Matsuura, S., Blomquist, R.N., Brown, F.B.: Parallel Monte Carlo Eigenvalue Calculations. Transactions of the American Nuclear Society **71** (1994) 199–202
2. Briesmeister, J.F.: MCNP™- A General Purpose Monte Carlo N-Particle Transport Code, Version 4C. Los Alamos National Laboratory. (2000)
3. Kerbyson, D.J., Hoisie, A., Wasserman, H.J.: Use of Predictive Performance Modeling During Large-Scale System Installation. In: 1st Int. Workshop on Hardware/Software Support for Parallel and Distributed Scientific and Engineering Computing, SPDSEC02, Charlottesville (2002)
4. Hoisie, A., Lubeck, O., Wasserman, H.J.: Performance and Scalability Analysis of Teraflop-Scale Parallel Architectures Using Multidimensional Wavefront Applications. Int. J. of High Performance Computing Applications **14** (2000) 330–346
5. Kerbyson, D.J., Hoisie, A., Pautz, S.D.: Performance Modeling of Deterministic Transport Computations. In: Performance Analysis and Grid Computing, Kluwer (2003)
6. Mathis, M.M., Amato, N.M., Adams, M.L.: A General Performance Model for Parallel Sweeps on Orthogonal Grids for Particle Transport Calculations. In: ICS, Santa Fe (2000) 255–263
7. Kerbyson, D.J., Alme, H.J., Hoisie, A., Petrini, F., Wasserman, H.J., Gittings, M.L.: Predictive Performance and Scalability Modeling of a Large-scale Application. In: Supercomputing, Denver (2001)
8. Kerbyson, D.J., Wasserman, H.J., Hoisie, A.: Exploring Advanced Architectures using Performance Prediction. In: Innovative Architecture for Future Generation High-Performance Processors and Systems, IEEE CS Press (2002) 27–37
9. Metropolis, N., Ulam, S.: The Monte Carlo Method. J. Amer. Statist. Assoc. **44** (1949) 335–341
10. Koch, K.R., Baker, R.S., Alcouffe, R.E.: Solution of the first-order form of the 3D discrete ordinates equation on a massively parallel processor. Transactions of the American Nuclear Society **65** (1992) 198–199
11. Cox, L.J.: DMMP Upgrade for MCNP4C™. Los Alamos National Laboratory Research Note (2001)

12. Barrett, R., McKay, M.: UPS: Unified Parallel Software User's Guide and Reference Manual. Los Alamos National Laboratory. (2002)
13. Petrini, F., Feng, W.C., Hoisie, A., Coll, S., Frachtenberg, E.: The Quadrics Network: High-Performance Clustering Technology. IEEE Micro **22** (2002) 46–57

A Compiler Approach to Performance Prediction Using Empirical-Based Modeling

Pedro C. Diniz

University of Southern California / Information Sciences Institute
4676 Admiralty Way, Suite 1001
Marina del Rey, California 90292, USA
pedro@isi.edu

Abstract. Performance understanding and prediction are extremely important goals for guiding the application of program optimizations or in helping programmers focus their efforts when tuning their applications. In this paper we survey current approaches in performance understanding and modeling for high-performance scientific applications. We also describe a performance modeling and prediction approach that relies on the synergistic collaboration of compiler analysis, compiler-generated instrumentation (to observe relevant run-time input values) and multi-model performance modeling. A compiler analyzes the source code to derive a discrete set of parameterizable performance models. The models use run-time data to define the values of their parameters. This approach, we believe, will allow for higher performance modeling accuracy and more importantly to more precise identification of what the causes of performance problems are.

1 Introduction

Despite the tremendous peak performance of high-end computing architectures, they deliver abysmally poor performance for current scientific and engineering applications. As these applications have millions of lines of source code and manipulate vast amounts of data, manual instrumentation and program understanding are infeasible. The standard approach to the problem of performance understanding relies on tools that profile the code execution and provide aggregate measures of performance. While it is useful to know that a given do loop has substantial *L1 cache misses* or *TLB misses*, that information gives little insight on how to remedy the performance problems.

The effects of compiler optimizations exacerbate this problem. Compilers apply a wide variety of transformations making it very difficult to map the effects of the code into high-level programming abstractions developers can reason about. We believe the key to address the problems of performance understanding and prediction is to develop techniques that take into account the effects of compilers at the instruction level and reason about the mapping of the instructions to the target architecture. To be useful such tools must be automated and must retain as much information about the high-level programming abstractions as possible.

P.M.A. Sloot et al. (Eds.): ICCS 2003, LNCS 2659, pp. 916–925, 2003.

This paper describes a performance modeling and prediction approach that uses traditional compiler analysis techniques, both at the source code level and assembly level, in collaboration with empirical performance modeling techniques. The compiler isolates sequences of instructions (called basic blocks in the compiler parlance) and maps them to high-level programming abstractions. Associated with each basic block the compiler builds a set of discrete performance models tailored for specific run-time execution scenarios. For example a *cache miss* or a *TLB miss* may lead to severe pipelining execution problems. In order to determine which of the set of models to apply the compiler generates and executes a skeleton of the application. The execution of the skeleton, will allow the compiler to extract the relevant run-time data which was identified statically. The compiler then feeds the information gathered by the skeleton and derives the actual model parameters and frequency of application of each model to predict the overall performance of the original application with its real data.

If successful this approach would provide, we believe, not only more accurate performance prediction but as a by-product, an understanding of what the cause-effect relations of program constructs on the performance are. The proposed performance modeling techniques can also be used as part of a fully automated program optimization tool. Such a tool would iterate over a given section of the code trying out several transformations observing which sequences of transformations would lead to better predicted performance before committing to the application of such transformations.

The remainder of this paper is organized as follows. In the next section we survey current approaches to performance modeling and understanding in high-performance scientific applications. Section 3 describes in more detail the modeling approach proposed in this paper. Section 4 describes three applications of the proposed approach followed by a brief discussion of the research challenges in Sect. 5. Finally we summarize this presentation in Sect. 6.

2 State-of-the-Art

We now describe current approaches in the area compiler optimizations specifically by addressing performance-aware compilation systems and feedback-directed optimizations. We also describe various efforts in performance modeling and prediction for large parallel codes on current and future processor architectures.

2.1 Performance-Aware Compilation

In may instances the lack of statically available information may prevent the compiler from applying program transformations more aggressively or simply from applying them all together. Several systems address this problem by a combination of static information and run-time testing. The inspector/executor approach dynamically analyzes the values in index arrays to automatically parallelize computations that access irregular meshes [1]. Speculative approaches optimistically execute loops in parallel, rolling back the computation if the parallel execution violates the data dependences [2]. Dynamic compilation systems enable code generation at run time [3,4] allowing the compiler to exploit knowledge about input values and hence generate more efficient code. The Dynamo system [5] continuously optimizes code based on performance data gathered incrementally at run-time.

As an alternative, researchers have develop approaches in which the compiler generates for selected computation sections a limited set of compiled versions each of which corresponds to the application of a particular set of program transformations. At run-time the generated code selects which version of the code to choose based on a set of compiler generated predicates or even by evaluating the performance of each alternative implementation and choosing the one with highest performance. This approach can be done entirely automatic as in [6] or with the involvement of the programmer to specify application specify adaptation and optimization strategies as in [7].

There are several on-going research projects in empirical optimization of scientific libraries through basic performance driven selection of multiple code variants (*e.g.*, ATLAS [8] or PhiPAC[9]). In these projects the compiler generates many implementation of the same computation, *e.g.*, *matrix-multiply*, for different optimization strategies in a purely off-line fashion and then selects, based on previous executions, which version performs best for each target architecture.

The GrADS[10] project aims at extending the notion of compile-time and run-time by creating a malleable object code that can be configured to a wide variety of resource availability scenarios. A configurable object program contains, in addition to the application code, strategies for mapping the application to different collections of resources and a resource selection model that provides an estimate of the performance of the application on a specific collection of resources. The GrADS approach project also relies on notions of performance contract to specify when reconfiguration of application code or resources should be triggered to maintain acceptable performance levels.

2.2 Performance Prediction

Other researchers have developed static estimators to guide the application of program transformations with the ultimate goal of improving performance.

For example in [11] the authors have developed a series of empirical models for the impact of data distribution in the performance of parallel applications on distributed memory machines. The system first "trains" the estimator with known communication and data partition patterns found in kernel routines and use the resulting estimator models for complete applications.

In the context of the POLARIS system researchers have also developed a methodology for statically predicting the performance of applications using a combination of static analysis and profiling information[12]. The approach uses source code analysis information to derive analytical expressions defining the number of expected *L1 cache misses* and basic arithmetic operators as a function of loop bounds and uses architecture characteristics *e.g.* number of functional units and pipelining depth). Using these analytical expression and real run-time data such as loop bounds the compiler can predict the overall execution time.

2.3 Performance Modeling & Understanding

Other researchers have also modeled application performance based on the high-level definition of the problem rather than directly on the code implementation. Kerbyson *et.al* [13] describe a series of modeling case studies where they use the data and computation partitions as well as their own empirical model for communications to define analytical expressions that track very well the observed performance on multiprocessor machines.

Other approaches have focused on modeling program behavior for large-scale parallel and distributed applications (*e.g.*, [14,15]). This work aims at understanding the sources of the applications' performance. In [16] the authors developed a set of simple models to capture the memory and communication behavior of each application. These metrics are capture include the cache-miss ratios and memory bandwidth via simulation and then convolved with the target architecture parameters to determine the expected performance. This approach aims at attaining better accuracy then *back-of-the-envelope* calculations without the extreme cost of cycle-level accuracy.

Other researchers have also studied the scalability of parallel scientific computations by empirical measurements using statistical instrumentation with currently available performance monitoring tools regarding computation and communication. In [17] the authors use the observed metrics to refine the explanations of the factors that influence application performance and scalability.

3 Empirical Modeling

Performance modeling techniques offer an alternative way of enabling the compiler to derive and possibly select a set of program transformations. This modeling approach is particularly valuable in scenarios in which the application can take an extremely long time to execute making profiling impractical or when attempting to predict the performance on future architectures.

3.1 Overview

The modeling approach described here and depicted in Fig. 1 relies on collaboration between static compiler analysis and run-time data, but unlike post-mortem profiling uses cues from the execution of a skeleton of the original code to derive better performance prediction models.

During a first phase the compiler identifies the application's control-flow-graph (CFG) isolating the basic blocks of instructions and retaining the mapping between relevant instructions (such as loads and stores) to the high-level programming constructs such as array accesses and arithmetic expressions. While

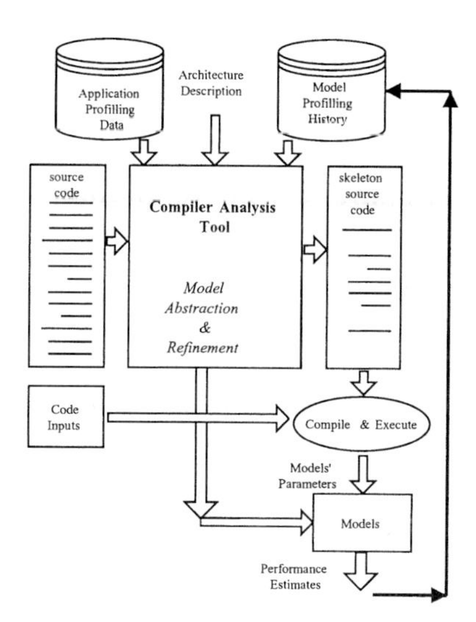

Fig. 1. Performance modeling approach

in general this seems a daunting problem given the various internal and target-specific compiler transformations, we believe it is possible to develop effective techniques that derive a mapping that will allow the compiler to provide useful information to the programmer. The recent experience with the HPCView[18] tool supports this claim. During this phase the compiler collects static information for each basic block such as the number of memory references and address calculations along with floating-point and integer operations[1]

For each basic block the compiler also derives a set of possible execution scenarios. These scenarios are based on the values of particular run-time variables that are bound to significantly affect the performance. For example a given loop might exhibit very distinct performance if the arrays it accesses are not in cache and/or if the stride of the accesses is larger than the cache line size. While in some cases the compiler can understand statically which of the array references are bound to create performance problems in other cases the stride of the access depends on the values of other variables. In other cases, such as conflicting misses the compiler can only determine this information once the array have been laid out in the code's virtual address space. Based on the assumptions for each scenario the compiler then derives a performance model. This model is parameterizable by run-time values such as loop bounds; array access stride or even the layout of the various arrays and their relative position in memory.

In order to determine the exact values of the various parameter models the compiler must determine the exact run-time values of the program variables that can affect the parameters of the various performance models for each pro-

[1] Collecting such static metrics by looking at the source is bound to lead to terribly inflated metrics as compilers apply program transformations such as common code elimination enabled by similarities in array indexing.

gram section. To derive these values the compiler generates a *skeleton* program based on the original code and runs it with the original code inputs. This *skeleton* program bypasses some of the code of the original application as the exact outputs are not relevant but retains the values of a set of variables (*e.g.*, loop bounds) that are important for modeling purposes. The exact set of variables whose values need to be captured by the *skeleton* program are derived statically by compiler analysis. The compiler then executes the *skeleton* program and collects the actual data values. Using the run-time values the compiler selects which of the pre-analyzed models to apply for each code section. Using the resulting output metrics for each selected model (by evaluation analytical performance expressions associated with that model using actual run-time values) the compiler derives execution estimates for the performance contributions of each region.

3.2 Program Skeletons

In many of these situations analyzing the structure of the computation and run-time variable values such as loop bounds and array access strides provides valuable information for the purpose of performance modeling and prediction of run-time execution. This data can be obtained, in many cases, by simple inspection of a subset of the variable the program manipulates. The example in Fig. 2 illustrates a case where the out of 4 array accesses, only 1 has variable stride and is bound to create substantial *cache* or *TLB misses.*

In the *skeleton* code the `do` loop is eliminated and instead the code extracts the relevant information to apply to an execution the model. It saves in an internal data structure the values of `n`, `mstride` and records which arrays have long and small access strides. The arithmetic functions `abs` and `sqrt` as well as the loop have been removed and the corresponding execution time can be accounted by using target architecture dependent constants. When encountering control-flow that is dependent on computed values, the compiler must use profile-based data about the frequency with which each of the branches was taken or retain in the skeleton code the computation that conditional statements depend on. While in the worst case scenario a given conditional statement would force the execution of the entire application, we believe that many computations will not elicit this behavior. As a fallback position the compiler can rely on accurate profiling [19] information regarding the frequency of the execution of each branch and bypass the information the execution of a skeleton code provides.

3.3 How to Derive Models

The model for each basic block of instructions in the code is derived by looking at the data dependences between the data required by each instruction. Internally the compiler builds a data-flow graph for each basic block and determines the set of data that is generated in register for each block as well as the set of data values generated to be used by other blocks.

Using the target architecture description in terms of the number of functional units, pipelines and their depth the compiler can derive a set of performance

```
mstride = ...
do ip=1,n
 b(ip) = two*vx(ip) + abs(...)
 b(ip) = sqrt(...)*face(mstride*ip)
enddo
```

```
mstride = ...
call save(address,mstride,n)
if(mstride .gt.cache_line) then
  call record_long_stride(address,array,"face")
else
  call record_unit_stride(address,array,"face")
endif
call record_unit_stride(address,array,"a")
call record_unit_stride(address,array,"b")
call record_unit_stride(address,array,"vx")
```

Fig. 2. Skeleton Extraction Example (original code on the left, skeleton on the right)

analytical expressions for a set of scenarios and determine for each of these scenarios what the expected performance is in terms of consumed clock cycles and peak performance the execution of the code section would take. For example, if one of the various memory references in the basic block causes a *TLB miss* that leads to a pipeline stall (due to data dependences) this leads to a substantial decrease in overall performance. Another scenario could explore the performance consequences of the references to the sparse array not being in cache. For each of the scenarios, the corresponding model can be obtained either empirically and/or by using target architecture cycle level accurate simulations.

The compiler analyzes a discrete set of such scenarios and enumerates the corresponding models. When generating the skeleton code the compiler also generates code to abstract, and keep track at run time, of the portion of the processor state that is relevant to each model. While in general this approach can lead to a full blown functional-level processor architecture simulation we believe it is possible to develop simple models for selected components of the architecture whose poor performance provide important insight into the overall program behavior. Tracking the values of consecutive array accesses by inspection of the corresponding indices, for example, allows compilers to signal which of the array references will clearly lead to potential performance problems.

4 Applications

4.1 Compiler Optimizations

Compiler writers have a wide range of program transformations at their disposal to attempt to improve the quality of the generated executable code. Unfortunately it is not easy to determine statically what the best sequence of program transformations is.

A compilation and program optimization system could use the approach outline above by using performance profiling information of previous runs to refine the performance models. This knowledge would help the compiler to select which optimization strategies are likely to produce better results. Because the performance of generated codes can varies widely with distinct data input characteristics, the compiler could use the performance models to select a set optimization strategies geared towards different data settings. The compiler would generate multiple code version of the same computation and select at run-time which of the selected code versions to use based on the assumptions for each code version.

4.2 Generating Performance Assertions

In the quest for understanding the performance bottlenecks of their applications researchers have developed the mechanism of performance assertions [20]. The current practice calls for the programmers to manually specify what the performance assertions would be.

Using this mechanism programmer must examine selected portions of their code and determine manually what a reasonable performance expectation a given segment of the code should deliver. When violated, the corresponding performance violation handling code (typically a write statement) will indicate the location and nature of the violation. Besides being tedious and error prone this process is highly non-portable. A given performance assertion might be acceptable to one target architecture but very unrealistic in another. This leads to an excessive number of spurious performance assertions violations which detracts the programmer from its main purpose find the real performance bottleneck problems. The automated approach proposed in this paper would aims at deriving the performance assertion directly from architecture specifications only raising performance assertion exceptions when a given threshold metrics say 10% of peak performance were predicted and providing additional information about why the performance model is reaching that particular performance level.

4.3 Interactive Performance Understanding

Ultimately it is the programmer who can profoundly impact the performance of its application. We foresee the application of the techniques proposed in this paper as part of interactive performance understanding systems that allow programmer to understand which data structure are substantially impacting performance and provide insight why and what to do about it. For example, programmers often use pointer variables for extreme flexibility. In the contexts of tight numeric intensive loop dereferencing pointers to retrieve/store data from/to memory might lead to severe pipeline stalling. Based on poor performance estimates suggested by its pipeline models the compiler could suggest the programmer to restructure its code by converting pointer access in the tight numeric loop to array accesses to a temporary array variable which the programmer loads (using a pointer-based only loop) before the numeric loop is executed.

5 Research Challenges

While appealing this approach raises several implementation and design challenges in building an automated and effective performance modeling and prediction systems for large scientific codes. First, how accurate can this approach be?

If the models are too simple the effort might not warrant the benefits; if too expensive the quantity and quality of the parameters might be as hard to extract as examining the impact of each instruction. Is there a meaningful middle-ground? Second, can basic blocks of instructions be meaningfully mapped to high-level programming constructs so as to provide good high-level program information about performance problems? Third, what is the precision of this approach in the presence of more sophisticated architectural features that are so hard to model? Forth, is it feasible for a static compiler analysis to generate a code skeleton to extract a set of meaningful parameters for each model? Can the implementation effectively capture a limited set of context representative of a wide set of execution scenarios? Finally, and given the large scale nature of the target applications, how does this approach scale?

6 Summary

In this paper we described an approach that relies on the synergistic collaboration of static compiler analysis, compiler-generated instrumentation (to observe relevant real run-time input values) and multi-model performance modeling of sequences of instruction for the target architecture (derived empirically and calibrated and validated off-line by cycle-accurate simulations). While there are many challenges to a practical implementation of the proposed approach we believe it is possible to build a program analysis tool that can deliver realistic performance estimates that are useful to programmers in understanding the source of their applications' performance issues.

References

1. Saltz, J., Berryman, H., Wu, J.: Multiprocessors and run-time compilation. Concurrency: Practice & Experience **3** (1991) 573–592
2. Rauchwerger, L., Padua, D.: The LRPD test: speculative run-time parallelization of loop with privatization and reduction parallelization. In: Proc. of the ACM Conference on Programming Language Design and Implementation (PLDI'95), ACM Press (1995)
3. Auslander, J., Philipose, M., Chambers, C., Eggers, S., Bershad, B.: Fast, Effective Dynamic Compilation. In: Proc. ACM Conference on Programming Language Design and Implementation (PLDI'96), ACM Press (1996)
4. Engler, D.: VCODE: A retargetable, extensible, very fast dynamic code generation system. In: Proc. of the ACM Conference on Program Language Design and Implementation (PLDI'96), ACM Press (1996)
5. Bala, V., Duestervald, E., Banerjia, S.: Dynamo: A Transparent Run-time Optimization System. In: Proc. of the ACM Conference on Programming Languages Design and Implementation (PLDI'00), ACM Press (2000)
6. Diniz, P., Rinard, R.: Dynamic feedback: An effective technique for adaptive computing. In: Proc. of the ACM Conference on Programming Language Design and Implementation (PLDI'97), ACM Press (1997)

7. Voss, M., Eigenmann, R.: High-Level Adaptive Program Optimization with ADAPT. In: Proc. of the ACM Conference on Principles and Practice of Parallel Processing (PPoPP'01), ACM Press (2001)
8. Whaley, C., Dongarra, J.: Automatically tuned linear algebra software. In: Proc. of Supercomputing (SC'98). (1998)
9. Bilmes, J., Asanovic, K., Chen, C.W., Demmel, J.: Optimizing matrix multiply using phipac: a portable high-performance ansi-c coding methodology. In: Proc. of the ACM International Conference on Supercomputing (ICS'97). (1997)
10. Kennedy, K., Mazina, M., Mellor-Crummey, J., Cooper, K., Torczon, L., Berman, F., Chien, A., Dail, H., Sievert, O., Angulo, D., Foster, I., Gannon, D., Johnsson, L., Kesselman, C., Aydt, R., Reed, D., Dongarra, J., Vadhiyar, S., Wolski, R.: Toward a framework for preparing and executing adaptive grid programs. In: Proc. of NSF Next Generation Systems Program Workshop (International Parallel and Distributed Processing Symposium 2002). (2002)
11. Bala, V., Fox, G., Kennedy, K., Kremer, U.: A static performance estimator to guide data partitioning decisions. In: Proc. of the third ACM Symposium on Principles and Practice of Parallel Programming (PPoPP'91), ACM Press (1991)
12. Cascaval, C., DeRose, L., Padua, D., Reed, D.: Compile-time based performance prediction. In: Proceedings of the 12th International Workshop on Languages and Compilers for Parallel Computing (LCPC'99), Springer-Verlag (1999) 365–379
13. Kerbyson, D., Alme, H., Hoisie, A., Petrini, F., Wasserman, H., Gittings, M.: Predictive Performance and Scalability Modeling of a large-scale Application. In: Proc. of the Supercomputing Conference (SC'01). (2001)
14. Snavely, A., Wolter, N., Carrington, L.: Modeling application performance by convolving machine signatures with application profiles. In: Proc. of the IEEE 4th Annual Workshop on Workload Characterization, Austin (2001)
15. da Lu, C., Reed, D.: Compact application signatures for parallel and distributed scientific codes. In: Proc. of the Supercomputing Conference (SC'02). (2002)
16. Snavely, A., Carrington, L., Wolter, N.: A framework for performance modeling and prediction. In: Proc. of the Supercomputing Conference (SC'02). (2002)
17. Vetter, J., Yoo, A.: An empirical performance evaluation of scalable scientific applications. In: Proc. of the Supercomputing Conference (SC'02). (2002)
18. Mellor-Crummey, J., Fowler, R., Marin, G., Tallent, N.: HPCView: A Tool for Top-Down Analysis of Node Performance. Journal of Supercomputing **23** (2002) 81–104
19. Ammons, G., Ball, T., Larus, J.: Exploiting hardware performance counters with flow and context sensitive profiling. In: Proc. of the ACM Conference on Programming Language Design and Implementation (PLDI'97), ACM Press (1997)
20. Vetter, J., Wooley, J.: Performance assertions. In: Proc. of the Supercomputing Conference (SC'02). (2002)

A Performance Prediction Framework for Scientific Applications

Laura Carrington, Allan Snavely, Xiaofeng Gao, and Nicole Wolter

San Diego Supercomputer Center, University of California, USA
{lnett,allans,xgao,wolter@sdsc.edu}

Abstract. This work presents a performance modeling framework, developed by the Performance Modeling and Characterization (PMaC) Lab at the San Diego Supercomputer Center, that is faster than traditional cycle-accurate simulation, more sophisticated than performance estimation based on system peak-performance metrics, and is shown to be effective on the LINPACK benchmark and a synthetic version of an ocean modeling application (NLOM). The LINPACK benchmark is further used to investigate methods to reduce the time required to make accurate performance predictions with the framework. These methods are applied to the predictions of the synthetic NLOM application.

1 Introduction

In this work, we report our ongoing progress to develop a general performance prediction framework to predict and explain the performance of scientific applications on current and future HPC platforms. The framework is not designed for a specific application or architecture but is designed to work for an arbitrary application on an arbitrary machine. In previous work we introduced our convolution method [4-6] that is a computational mapping of an application's signature (a representation of an applications fundamental operations) onto a machine profile (a characterization of a machine's ability to perform fundamental operations) to arrive at a performance prediction. We introduced Memory Access Patter Signature (MAPS), a benchmark probe tool for collecting machine profiles. We introduced MetaSim Tracer, a tool for gathering application signatures. See `http://www.sdsc.edu/PMaC` for previous papers and access to these tools. Finally, we showed that the framework could model and improve understanding of the performance of small parallel scientific kernels and applications on several different HPC architectures. Here we provide an update on the ongoing work to make full applications modeling tractable via the convolution method.

2 The Convolution Method

To create a model for the performance of a parallel application's serial sections (between communication events) we map the memory trace component of an application signature to the corresponding information in a machine profile in order to model this (presumed to be dominant) factor in performance. Next, we map a communication trace to its corresponding information in a machine profile to get a model of the

P.M.A. Sloot et al. (Eds.): ICCS 2003, LNCS 2659, pp. 926–935, 2003.

communication events. Then we combine the single-processor model (possibly supplemented with model terms for floating-point work, and other kinds of work), along with the communication model, to arrive at a performance model of a full parallel application.

Details of the convolution method for serial sections involve mapping each basic-block's expected dataset location onto the benchmark-probe curves from MAPS. The process is illustrated in Table 1 and Fig.1. Table 1 is the output of MetaSim Tracer on a Portable, Extensible Toolkit for Scientific Computation (PETSc) [15] application convolved with machine memory-hierarchy parameters from Pittsburgh Supercomputer Center's TCSini Compaq machine.

Table 1. Application signature example via MetaSim tracer

Block #	% Mem. Ref.	Ratio Rand	L1 hit Rate	L2 hit Rate	Data Set Location	Memory Bandwidth
55	0.9198	0.07	93.47	93.48	L1 Cache	4166.0
53	0.0271	0.00	90.33	90.39	Main Memory	1809.2
60	0.0232	0.00	94.81	99.89	L2 Cache	5561.3
5885	0.0125	0.20	77.32	90.00	L1/L2 Cache	1522.6

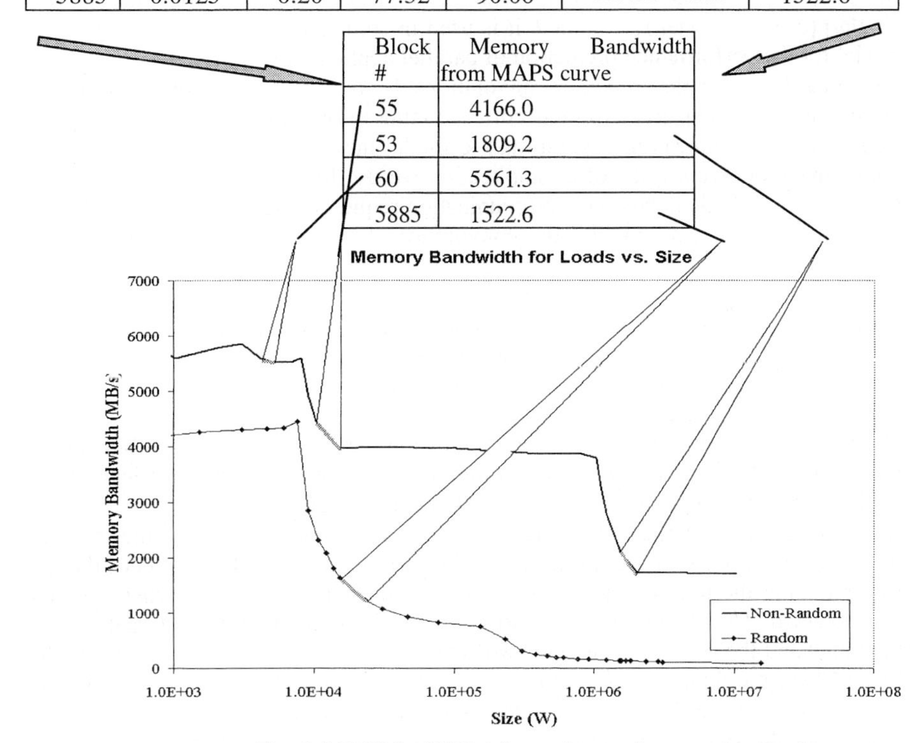

Block #	Memory Bandwidth from MAPS curve
55	4166.0
53	1809.2
60	5561.3
5885	1522.6

Fig. 1. MAPS for TCSini for random and non-random loads

The convolution represented between Figure 1 and Table 1 is carried out automatically by the MetaSim Convolver [16] and can be written as:

$$1)\quad \text{Memory Execution Time} = \sum_{i=1}^{n} \left(\text{MemOps BB}_i \,/\, \text{MemRate BB}_i\right)$$

Equation 1 predicts that the Memory Execution Time for an application between communication events is the sum, over all the basic-blocks in the application, of the expected time required to carry out the loads and stores in each basic-block. The expected execution time depends on the rates at which the machine can carry out loads and stores based on instruction type, access pattern, and where the references fall in the memory hierarchy. MemOps BB_i is the total number of dynamic memory references in basic block i. MemRate BB_i is the rate at which the machine can sustain these operations. MemOps BB_i subcomponents (random loads to main memory, stride 1 accesses to L2 cache etc.) are determined by the convolver. MemRate BB_i has subcomponent rates taken from the MAPS curves. This simple example illustrates only predictions involving memory operations, but more complex convolutions can deal with other kinds of operations, and the interactions and overlap between those operations. If an application is heavily memory bound, Memory Execution Time may anyway be a large percentage of total execution time. Otherwise, additional model terms are added to the convolution to account for cycles spent doing non-overlapped floating-point work, branches, file I/O, communications etc. Once the convolution for the single-processor model is complete it is used in conjunction with the communication model for the performance prediction of parallel applications.

Similar to the single-processor convolutions, for the communication model we map communication event trace to their corresponding information in the machine profile (speeds of communications operations) to model the communication events. For the communication event trace of an application we use MPIDtrace, which contains the sequence of CPU demands and communication requests launched by the CPU during an application's execution. For the convolution step we use the network simulator, Dimemas [14], which using this trace can model the communication requests of the application on an arbitrary machine provided a user inputted parameterized network (machine profile) of that machine. So a modeler inputs the performance parameters of an arbitrary machine they desire to model along with the communications trace. Dimemas will then calculate the expected performance of the application's communication events on that machine.

Dimemas not only models the communication part of the application but can also model the CPU demands between communication events of the application. The MPIDtrace of the application contain CPU demand information about the application specified in terms of the CPU time consumed on the machine where the trace was obtained. Dimemas uses a parameter (CPU ratio) to scale the CPU bursts of the trace for predicting the performance of a machine other than the machine where the trace was obtained. The ratio being the ratio of processor speeds between a target machine to predict and the machine where the MPIDtrace was collected. A naïve way of calculating CPU ratio might be to use the ratio of clock speed or the ratio of peak floating-point issues between the two processors. Our framework improves upon this idea by calculating the ratio from the single-processor models we develop using MetaSim Tracer and MAPS data. This new ratio will be the expected single-processor performance of the target machine being predicted (taking into account especially the performance of its memory hierarchy) to the single-processor performance of the machine where the MPIDtrace was collected. So given the single-processor model of an application along with the application signature and machine profile parts for the communication model, Dimemas can predict the performance of a parallel application.

Using detailed single-processor models with a special emphasis on the memory hierarchy and the network simulator Dimemas, the framework was shown in [4-6] to model the performance of NPB kernels, a PETSc kernel, and PETSc small applications with an error range of 1% to 16%. These performance predictions were consistent across a range of compute platforms (SDSC Power3 system Blue Horizon, PSC Compaq Lemiuex, a Cray T3E-600, TACC Power4 Longhorn), and across a range of processors from 2 to 128 for both weak and strong scaling. Some of the predictions used more complicated convolutions than Equation 1 and took into account (in addition to memory work) floating-point work and instruction level parallelism with overlap of memory and floating-point work.

The collection of an application signature via the MetaSim Tracer is needed only once per application-prediction series, allowing the convolution to run multiple times with the same application signature. Unfortunately MetaSim Tracer can require orders of magnitude slowdown of the instrumented code.

3 Speeding Up Tracing

In a quest to supply modelers with a general performance prediction framework that could work in a relatively short amount of time even on long-running applications, several methods of reducing trace time were investigated. The LINPACK Benchmark and the synthetic version of the Navy Land Ocean Model (synNLOM) [8] application were used in the investigation of trace time reduction and the results of predictions with these new methods is discussed in section 4. The application synNLOM was of particular interest because it exhibited many traits of a scientific application plus, the run time of the application took over an hour, making the reduction of trace time critical in its prediction.

The idea of trace sampling is to turn the trace collection of an application on and off at certain intervals while running the application. MetaSim Tracer processes all of the memory addresses of an application on-the-fly. Sampling would allow the tracer to process only a percentage of the addresses in an attempt to reduce the trace time. In our implementation the user can specify the size of the interval and the number of sequential addresses to process per interval (sample size). If interval size is set for 1,000,000 and sample size is set for 10,000 then as the application runs, every 1,000,000 addresses the first 10,000 will be processed. This would result in a 10% sampling of the full trace. The idea is that as each basic-block is traversed and processed the cache-hit rates may not change significantly. So calculating the hit rates based on traversing and processing the basic-block only 10% of the time may yield similar hit rates to those based on traversing the basic-block 100% of the time. It is the hit rate values that are used in the convolution to determine the bandwidth for that basic-block. This bandwidth is then used in the convolution Equation 1 to determine the Memory Execution Time.

Another issue with trace sampling is that the total number of memory references estimated for a basic-block may differ depending on the percent sampled, this is the MemOps BB value from Equation 1. This can be remedied by collecting two traces.

The first, uses no sampling but only counts (no processing of) memory references, instructions, and floating-point operations, thus it collects the correct value for MemOps BB for each basic-block. The second is the detailed trace collection for memory accesses with sampling. The first trace, because it does no processing, only takes a minimal amount of time, but is significant in that the data collected in this trace used in conjunction with the second trace ensures correct values for number of memory references and floating point operations for each basic-block. The accuracy of predictions using different sampling sizes and their respective trace times are discussed in Sect. 4.

Another way to reduce the trace time is to put an upper limit on the number of times a basic-block is traversed and processed. This means that if the user sets an upper limit as for example 1,000 then after a basic-block is traversed more than 1,000 times, the tracer no longer processes that information. The reasoning is that each time a basic-block is traversed it behaves similarly so the hit rates for a basic-block averaged over 1,000,000 traverses is going to be similar to the hit rates for a basic-block averaged over 1,000 traverses. This basic-block trace limit along with sampling can be used together to reduce the total trace time. Results of predictions for a series of traces using different basic-block limits are discussed in section 4.

Instead of using sampling to reduce trace time, tracing only certain sections of the application is also an option. The idea is that, in a lot of applications, there are only a small number of basic-blocks that account for most of the wall-clock time in the application. The reduction in trace time comes from the fact that instead of tracing all 100,000 basic-block in the application you only trace 100. The trick is determining which basic-blocks to trace in order to capture most of the applications performance attributes. This method also requires two traces. The first uses only counts (no processing of) memory references, instructions, and floating-point operations for the entire application. Using this trace, one can determine those basic-blocks that are contributing to the majority of memory references of the application. In the second trace, only those basic-blocks determined from the first trace are traced and processed. This results in only a fraction of the total number of basic-blocks from the application being traced. Section 4 discusses the results of tracing different numbers of basic-blocks.

Another approach enabled of this tracing method is that the trace of an application can be collected in phases, where a phase represents a certain number of basic-blocks. For example, an application containing 100 basic-blocks could be traced in two runs. The first run would trace basic-blocks 1-50 and the second run would trace basic-blocks 51-100. The has the advantage of being able to reduce trace time of a phase in order to fit into the queuing limits of the machine used to collect the traces. Also, this allows each phase to be collected simultaneously (multiple jobs in the queue). So although this method does not effect the cumulative trace collection time, it makes trace collection more parallel and flexible. Results of using this phase collection are discussed in Sect. 4.

The methods of trace time reduction using sampling, basic-block trace limits and tracing only a fraction of the basic-blocks can be combined to further reduce tracing time and enable a more flexible trace collection process. The results of both prediction accuracy and the trace time reductions of all methods, including combinations of them are discussed next in Sect. 4.

4 Results and Discussion

The framework was first used to confirm the accuracy of using it in the prediction of the performance of the LINPACK benchmark. The LINPACK benchmark was predicted on four different machines at different number of processors. The size of the problem solved by the code was scaled with the number of processors (i.e. weak scaling). The results of these predictions are shown in Tables 2-5. Since Blue Horizon was used to collect the MPIDtraces, eliminating the single-processor ratio, predictions are for that machine are viewed as an accuracy check of the network simulator rather than validation of the entire framework. Tables 2-5 have the real run time for each machine-processor pair, the predicted run time by the framework, and the relative error. The results show that for varying numbers of processors and different machines, the framework is accurate in its predictions.

Tables 2–5. Real and predicted time for LINPACK benchmark

Number of Processors	**PSC Lemieux**[1]		
	Real Time (s)	Predicted Time (s)	% Error[5]
4	9.3	8.9	4.3
16	21.9	20.9	4.6
64	23.9	22.8	4.6
256	22.2	21.4	3.6

1 Pittsburgh Supercomputer Center Compaq Alphaserver ES45 with 1-GHz processors and Quadrics interconnect.

Number of Processors	**TACC Longhorn**[2]		
	Real Time (s)	Predicted Time (s)	% Error[5]
4	8.5	8.8	-3.5
16	21.1	20.4	3.3
64	25.4	22.8	10.2
256	NA	21.1	NA

2 Texas Advanced Computing Center IBM Regatta-HPC with an IBM high-speed switch (SP Switch2).

Number of Processors	**NERSC Seaborg**[3]		
	Real Time (s)	Predicted Time (s)	% Error[5]
4	18.8	17.7	5.9
16	41.7	39.0	6.5
64	45.4	44.9	1.1
256	51.7	45.2	12.6

3 National Energy Research Scientific Computing Center IBM SP RS/6000

Number of Processors	SDSC Blue Horizon[4]		
	Real Time (s)	Predicted Timc (s)	% Error[5]
4	18.8	19.2	-2.1
16	41.4	40.9	1.2
64	45.4	43.1	5.1
256	43.0	41.3	4.0

4 San Diego Supercomputer Center IBM SP RS/6000
5 Percent relative error: (Real Time – Predicted Time)/(Real Time) *100.

As Tables 2–5 confirm, the framework is an accurate predictor of the LINPACK benchmark, therefore this code was then used to investigate the viability of using the trace time reduction methods discussed in section 2.4. The LINPACK Benchmark run on 64 processors was used to compare the accuracy of using each of the different trace time reduction methods as well as their overall trace time reduction. Then the methods were combined to predict the synNLOM application run on 28 processors (the usual size for the Navy's production runs). Trace time reduction method was essential to predict the synNLOM application due to the long runtime of the application and the queuing limits on the machine used to collect the traces.

The first investigation was into the use of different sampling percentages in tracing. Sampling percentages of 100% (no sampling), 10%, 5%, and 1% are shown in tables 6 and 7. Table 6 shows the trace time slowdown for each sampling size on the application. Remember that a cycle-accurate simulation typically results in a 1,000,000 times slow down so tracing, while slow, is not *that* bad. Table 6 shows that initially the full trace slowed down the application by a factor of 859 times, whereas the sampling trace can reduce this time by a factor of 8. Table 7 shows the results of predictions using the different sampling size traces. It illustrates that sampling not only reduces the trace collection time to one ten thousandths of that of a typical cycle-accurate simulation, but it is able to predict the performance of the application with only a maximum error of 8.4%.

Table 6. MetaSim Trace collection time comparison using trace sampling

Sampling size[1]	Slowdown factor[2]
NO Sampling	859
10%	152
5%	141
1%	132

1 The sampling size used for the trace, where 100% means no sampling
2 The slowdown factor is the number of times longer the trace takes to collect than the application

Table 7. Predictions for PSC's Lemieux results for different sampling sizes

Sampling size	Predicted Time (s)	Real Time (s)	% Error[3]
NO Sampling	22.8	23.9	4.6
10%	24.8	23.9	-3.8
5%	25.0	23.9	-4.6
1%	25.9	23.9	-8.4

3 The percent error is calculated: (Real time – Predicted time)/(Real time)*100

The second investigation was into reducing trace time by using the basic-block trace limit discussed in section 3 with no sampling. Three different limits were compared to a case with no limit. Table 8 displays the trace slowdown factors for each case. This table shows that there is a trace time-reduction benefit depending on the basic-block limit but that the accuracy of the prediction is sacrificed a bit for this method of trace time reduction.

Table 8. MetaSim Trace collection time using basic-block trace limits

Basic-block limit[1]	**Slowdown factor**
NO Sampling/limit	859
10,000	134
1,000	130
100	130

1 The basic-block limit is the limit of the number of times a basic-block is traced.

Table 9. Predictions for PSC's Lemieux for different basic-block trace limits

Basic-block limit	Predicted Time (s)	Real Time (s)	% Error[3]
All BB	22.8	23.9	4.6
10,000	28.7	23.9	-20.1
1,000	29.2	23.9	-22.2
100	29.5	23.9	-23.4

The third investigation was into the trace reduction time by tracing only a small number of basic-blocks as discussed in section 3. Three different basic-block numbers were compared both in reduction of trace time and accuracy of prediction. Table 10 displays the trace slowdown factors for each case. This table shows that this method does have trace reduction benefits and Table 11 confirms that the accuracy is still relatively good.

Table 10. MetaSim Trace collection time tracing only certain basic-blocks

Num. Basic-block	**Slowdown factor**
All	859
20	154
10	151

Table 11. Predictions for PSC's Lemieux results for different basic-block groups

Num. Basic-block	Predicted Time (s)	Real Time (s)	% Error[3]
All	22.8	23.9	4.6
20	26.8	23.9	-12.1
10	26.2	23.9	-9.6

In the prediction of synNLOM a combination of all the trace reduction methods were used. This application, run on 28 processors of SDSC's Blue Horizon takes over 1 hour to complete. To simulate 1 hour of an application on a cycle-accurate simulator could take around 114 years of CPU time. Using no trace reduction method, the collection of the MetaSim trace would take around 850 hours. To reduce this to a more manageable time the trace was collected using 1% sampling, collecting the top 100 basic-blocks of the application, a basic-block limit of 200, and the trace collected in 10 phases. This allowed each phase to be collected in 2-6 hours, easily fitting into the queuing limits of most HPC machines. This also allowed 10 different jobs to be run (simultaneously) on the machine ranging from 2-6 hours, quite a reduction from the years required for cycle-accurate simulation. Table 12 shows the results of predicting the synNLOM on PSC's Lemieux, NERSC's Seaborg, TACC's Longhorn, and SDSC's Blue Horizon. The percent error of the prediction, >9%, shows that not only is the framework accurate in its prediction, but with trace reduction methods it is capable of predicting the entire scientific application's run for long periods of time and doing these predictions in a reasonable amount of time, something not feasible by cycle-accurate simulation.

Table 12. Prediction of synNLOM for different machines

Machine	Real Time (s)	Predicted Time (s)	% Error
PSC's Lemieux	1818	1816	0.1
SDSC's Blue Horizon	4462	4594	-3.0
NERSC's Seaborg	4375	4756	-8.7
TACC's Longhorn	1944	1872	3.7

The results of performance predictions shown in Tables 5 through 12 illustrate the accuracy of using the performance prediction framework to predict scientific applications. Such predictions can be completed nearly 10,000 times faster than using a cycle-accurate simulator. In addition, the framework is flexible enough to be applied to many different architectures and applications.

This work was sponsored in part by the Department of Energy Office of Science through SciDAC award "High-End Computer System Performance: Science and Engineering". This work was sponsored in part by a grant from the Department of Defense High Performance Computing Modernization Program (HPCMP) and the National Security Agency. This research was supported in part by NSF cooperative agreement ACI-9619020 through computing resources provided by the National Partnership for Advanced Computational Infrastructure at the San Diego Supercomputer Center. Computer time was provided by the Pittsburgh Supercomputer Center, the Texas Advanced Computing Center, and the National Energy Research Scientific Computing Center.

References

1. D.J. Kerbyson, H. Alme, A. Hoisie, F. Petrini, H. Wasserman, M. Gittings, "Predictive Performance and Scalability Modeling of a Large-Scale Application", *Supercomputing 2001.*
2. J. Lo, S. Egger, J. Emer, H. Levy, R. Stamm, and D. Tullsen, "Converting Thread-Level Parallelism to Instruction-Level Parallelism via Simultaneous Multithreading", *ACM Transactions on Computer Systems*, August 1997.

3. J. Gibson, R. Kunz, D. Ofelt, M. Horowitz, J. Hennessy, and M. Heinrich, "FLASH vs. (Simulated) FLASH: Closing the Simulation Loop", *Proceedings of the 9th International Conference on Architectural Support for Programming Languages and Operating Systems (ASPLOS)*, pages 49–58, November 2000.
4. A. Snavely, N. Wolter, L. Carrington, R. Badia, J. Labarta, A. Purkasthaya, "A Framework to Enable Performance Modeling and Prediction", *Supercomputing 2002.*
5. L. Carrington, N. Wolter, and A. Snavely, "A Framework for Application Performance Prediction to Enable Scalability Understanding", Scaling to New Heights Workshop, Pittsburgh, May 2002
6. A. Snavely, N. Wolter, and L. Carrington, "Modeling Application Performance by Convolving Machine Signatures with Application Profiles", *IEEE 4th Annual Workshop on Workload Characterization*, Austin, Dec. 2, 2001.
7. LINPACK
8. See http://www7320.nrlssc.navy.mil/html/lsm-home.html
9. J. Simon, J.-M. Wierum, "Accurate performance prediction for massively parallel systems and its applications", proceedings, *Proceedings of European Conference on Parallel Processing EURO-PAR '96*, Lyon, France, v2, pages 675-688, Aug. 26–29, 1996.
10. See http://www.cepba.upc.es/tools_i.html
11. See http://www.sdsc.edu/PMaC/MAPS/
12. See http://www.cs.virginia.edu/stream/
13. See http://www.sdsc.edu/PMaC/Benchmark/
14. See http://www.cepba.upc.es/
15. See http://www-fp.msc.anl.gov/petsc/
16. See http://www.sdsc.edu/PMaC/MetaSim/

Identification of Performance Characteristics from Multi-view Trace Analysis

Daniel Spooner[1] and Darren Kerbyson[2]

[1] High Performance Systems Group, Dept. Of Computer Science
University of Warwick, Coventry, UK
dps@dcs.warwick.ac.uk

[2] Performance and Architectures Laboratory (PAL), CCS-3
Los Alamos National Laboratory, Los Alamos, USA
djk@lanl.gov

Abstract. In this paper, we introduce an instrumentation and visualisation tool that can be used to assist in analytical performance model generation. It is intended to provide a means of focusing the interest of the performance specialist, rather than automating the entire formulation process. The key motivation for this work was that while analytical models provide a firm basis for conducting performance studies, they can be time-consuming to generate for large, complex applications. The tool described in this paper allows trace files from different runs of an application to be compared and contrasted in order to determine the relative performance characteristics for critical regions of code. It is envisaged that the tool will develop to identify and summarise specific performance issues such as communication strategies through the use of novel visualisation techniques.

1 Introduction

The design and implementation of high-performance systems is a highly complex problem requiring knowledge of many factors. The peak performance of a system is a result of the underlying hardware architecture including the processor design, memory hierarchy, inter-processor and communication system, and their interaction. Moreover, the achievable performance is dependent upon the workload applied to the system, and how this workload utilises the resources within the system.

Performance modelling is a key approach that can provide information on the expected performance of a workload, given a particular architectural configuration. It is useful throughout the entire system life-cycle: starting at the design stage where no system is available for measurement, through comparison of systems and procurement, to implementation, installation and verification, and finally to examine the effects of system updates over time. At each stage, a performance model can provide an expectation of the achievable performance of the workload with reasonable fidelity.

P.M.A. Sloot et al. (Eds.): ICCS 2003, LNCS 2659, pp. 936–945, 2003.

Performance models are widely used: from large-scale, tightly-coupled systems through to dynamic and distributed Grid based systems. For instance, performance modelling is being used to validate the performance during the installation of ASCI Q at Los Alamos National Laboratory (LANL) [1], to compare the performance of large-scale systems such as the Earth Simulator [2], and has been used in the procurement of ASCI purple (expected to be a 100Tflop system). Performance models have also been applied in dynamic and distributed 'Grid type' environments to consider service-orientated metrics in the provision of resource management services [3] and in the mapping of business applications to resources [4].

The accuracy of a model, and hence, its effectiveness lie in its ability to capture an application's performance behaviour. It is considered advantageous to parameterise a model in terms of system configuration (e.g. for scalability analysis), and calculation behaviour (e.g. input data-set size). This allows for the exploration of the performance space without being specific to a particular 'performance point'.

It is, however, generally acknowledged that the formation of a performance model is a complex task. It may involve a thorough code analysis, inspection of important data structures and analysis of profile and trace data. It can therefore be time-consuming to generate a detailed model given the large size of many scientific applications and the relative complexities of advanced data structures and optimised communication strategies. Several semi-automated approaches have been proposed that aim to make the formation of a performance model a simpler task using 'black-box' techniques in which individual performance aspects are observed but not necessarily understood. Examples include modelling the scaling behaviour of basic-block performance [5] and modelling the memory behaviour of basic-blocks and extrapolating to other systems [6]. These approaches tend to be specific to a particular processor configuration and/or problem size.

In this work we consider an approach that aims to simplify the process of generating a performance model, but not to automate it entirely. The purpose is not to simplify the resultant performance model, nor detract from the skill-set required by the modeller, but rather remove unnecessary steps during formulation. While the answer to this question lies, in part, with the experience of the performance-modeller; we believe that tools can be developed (or adapted) to help locate and *focus* on the performance critical regions. Such regions are typically those whose execution behaviour changes when the system configuration or application input-data is varied.

While there are a number of post-analysis and diagnosis tools that can assist with identifying performance constraints such as bottlenecks [7] and communication patterns [8], many are aimed at resolving *problems* with the application rather than trying to *characterise* the application's behaviour. In this case, it is useful to identify the differences in particular idiosyncratic behaviours as well as to detect problems.

A tool is introduced here that uses a combination of static and dynamic call-graph analysis to attempt to identity regions of code that are sensitive to data-set and scalability variations in order to reduce the time-to-model. It provides a compact view of multiple executions of an application using colour cues to draw attention to areas of interest. Although the current implementation is a prototype; it is envisaged that other methods of visualisation could be employed to summarise large-scale performance characteristics "at a glance", such as those used already in code maintenance [9].

The paper is organised as follows: Section 2 describes the approach taken that can lead to a performance model and identifies areas where tools can assist. Section 3 introduces a tracing tool that can create call-graph trace files for post-analysis using source-code instrumentation. In Section 4, we describe a further tool that uses the trace files to create multi-view visualisations. Conclusions and future work is discussed in Section 5.

2 Identification of Performance Characteristics from Multiple Executions

The performance modelling work at Los Alamos National Laboratory to date has primarily focussed on applications representative of the ASCI (Accelerated Strategic Computing Initiative) workload where analytical techniques are employed to develop entire-application models for the large-scale ASCI computing resources. This differs from a number of other performance activities that tend to focus on smaller applications in distributed computing environments such as [10]. The Los Alamos models are used most prominently to explore the scaling behaviour of applications on existing and speculative future architectures.

The approach to developing a performance model is based upon a detailed understanding of the performance effects that occur when changes to the system and application configuration are applied. In the initial stages of formulation, the application is typically executed with fixed input data sizes and a varying number of processing elements (PE) to observe changes in the overall execution-time and resource use. This can reveal basic information, such as whether the program scales weakly or strongly. Likewise, observations can be taken by fixing the PEs and varying the data-set sizes.

Instrumentation and profiling are subsequently used to obtain an understanding of the code. Highlighting the changes in communication patterns between processors as the PE count changes can, for example, provide insight into the type and method of domain decomposition. Instrumentation of this type can reveal message size changes in SAGE [11], an adaptive mesh hydro code, due to its 1D decomposition, and the difference in neighbouring processors in Tycho [12], radiation transport code, due to its use of an unstructured grid.

The problem partitioning, and related messaging, revealed through instrumentation typically provides a strong indication of the arrangement of the application's data structures, which can be confirmed by thorough code analysis of the relevant regions. Generating static call-graphs can assist with identifying the functional dependencies of the code sections and dynamic call-graphs (through the use of traces collected at run-time) can illustrate the flow of execution. In addition, comparing the relative number of instructions issued for a given subroutine in different runs/iterations can highlight the impact of configuration change on calculation and computational areas.

The overall objective of these activities is to obtain a model based on timings of sequential elements parameterised by expressions that are subject to input parameters and differing levels of parallelism. By identifying messaging, data placement and computational sensitivity to configuration and data-set sizes, it is possible to locate regions of the code that can be described by a single timing (for a given architecture) or by an analytical expression that captures the computation/communication characteristics with respect to the input parameters. While the approach is based on understanding the application's behaviour, it is apparent that much of the initial work is based upon observing the performance effects of input and configuration variation (the dotted region in Fig. 1). It is when these effects are compared across configurations that areas of interest can be brought to the attention of the performance specialist.

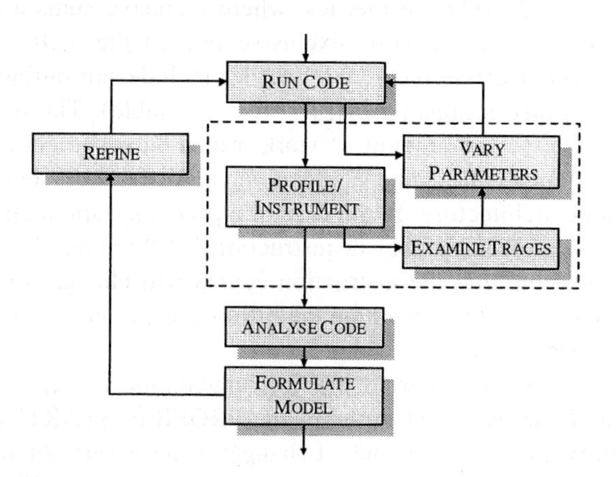

Fig. 1. The key stages of model development. Initially the code is run with a set of different input parameters (e.g. number of assigned processors, data-set sizes) to get a view of its scaling behaviour. Typically instrumentation or profiling is used to produce trace files that identity performance critical regions in the code. This is followed by detailed code analysis to formulate an initial performance model. This can then be refined until a satisfactory level of fidelity (predicted vs. measured) is obtained

Without losing the depth of detail that the models provide, a suitable tool can assist with instrumentation and call-graph generation producing a visualisation of the key differences to indicate "areas of interest". This can limit the performance modelling process, allowing a performance specialist to focus on a subset of the application rather then the entire program.

3 Call-Graph Collection

In order to assist the instrumentation, an automatic source-code level modification tool is used to indicate where and when a subroutine is entered and exited. The tool currently supports Fortran-77/90 and C. A lightweight profiling library is also used and linked with the application, which stores subroutine "events" into a page-based list whenever a subroutine is called. To minimise the overhead, each event records a limited amount of information including a source-file identifier, source-line number, and a field to denote that a 'context' has been entered into or exited from. As with nested subroutines, contexts are linked so that when subroutine *main* calls subroutine *init*, the instrumentation library is in the context of both *main* and *init* and any events that occur are associated with both of these contexts. This property is used to reduce the storage space of the event list as it is constructed dynamically in memory. The event-list is written to file when the application exits, although it is possible to allow on-line paging. The trace-file can then be processed by a subsequent utility.

In a similar manner to Paradyn [13], the instrumentation library utilises the concept of inclusive and exclusive metrics, where inclusive sums a metric for a subroutine and all its children, while exclusive returns the metric for a subroutine without its children. Current recordable metrics include subroutine duration and the number of issued instructions (if PAPI [14] is available). The instruction count is not used to determine the amount of work, rather to relate code density between different processors in the same application, or differences between application runs for the same architecture. In the case of a particular application run, each subroutine will issue a given number of instructions; if the input data or configuration is varied it is possible that the instruction counts will change for a particular subroutine. It is then possible to examine the differences, or error, to ascertain how the application was effected by the change.

When the application is started, the instrumentation is initially disabled which results in virtually no overhead. An explicit "PROFILE_START" call is required to enable the library and store the context changes as necessary. In addition to context changes, MPI calls are logged as events through the profiling MPI (PMPI) interface which is connected to the capture library and can assist with identifying the communication patterns that occur. The relevant parameters (source, destination, collective, size, type) are stored as part of an MPI event. Currently, only a limited number of MPI calls are wrapped which covered our test cases; it is reasonably straightforward to include further commands or to employ a third party tool such as VampirTrace.

The process allows rapid instrumentation of the source code with a subsequent compile, link and run sequence to obtain the required tracing information. The events that appear are essentially a call-graph of the program and can be subjected to a wide variety of analysis tools.

The instrumentation method is lightweight, consisting of a few array operations to store events in order to minimize the performance impact on the application. Depending upon the page size, periodic memory allocation is required which will incur a slight performance penalty. However, the principal purpose of this instrumentation utility is to illustrate the functional operation of the application which can be achieved by examination of the call-graph and the number of instructions issued, as opposed to performance-timings which are usually sensitive to experimental error.

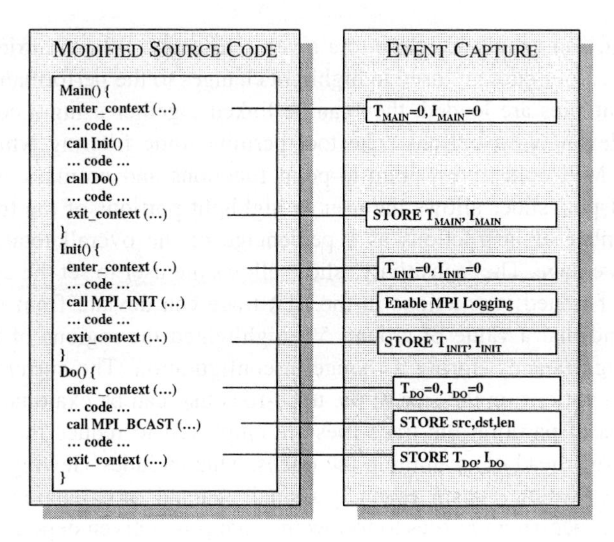

Fig. 2. The instrumented source consists of calls to a backend event capture library. This library links events together based on context and writes out a text trace file when the application quits. On entering a context, a timer and instruction counter is reset. These are then stored on exit of the context for post-analysis

4 Multi-view Visualisation

There are a number of visualisation tools that can assist with understanding an application's runtime behaviour. For MPI applications, Vampir provides a large suite of views that provide a good level of detail. It is essentially possible to "playback" the application and determine where communication occurs (and which processors are involved) and where periods of computation occur. However, these tools tend to produce views of a single application, albeit across many processors. In order to develop analytical performance models, it is useful to visualise the application behaviour over various iterations, processors or different runs.

Placing two traces "side-by-side" and allowing a tool to highlight the difference between the code densities (number of instructions in a subroutine) for two application-runs (presumably with different input-data sizes) is useful when attempting to identify critical regions of code. Using this approach, a performance modeller can rapidly determine if a code region is worth analysing or whether a single timing can be taken.

The tool developed in this work provides a multi-view visualisation of the trace files. Loading an event file generated by the instrumentation library described in Section 3, it reconstructs a complete call-graph of the application and compares it with subsequently loaded event files. The tool utilises an algorithm that searches for the largest groups of call-chain (or grouped call-chain) entries to locate similar code-regions. Entries in the trace files are deemed similar if the call-graph nodes match and that the instruction counts are close to a given sensitivity. Where entries differ or are non-existent, colour cues are used to highlight changes to the performance modeller.

When event-files are loaded, they can be linked together using a control panel to associates relevant view-options. The tool permits some filtering which allows the user to view MPI collectives, point-to-point functions and normal context (subroutines). A 'hotspot' slider allows the user to highlight portions of the trace where the inclusive number of instructions as a percentage of the overall total, is above the threshold percentage. The 'sensitivity' slider allows the user to set the extent to which the number of issued instructions in the first trace can deviate from the second. In tests, we found that a value of around 5% highlighted the regions of the call graph that altered significantly during a change in configuration. The current implementation places limitations on the size of the trace-files that can be examined at any given time. Schemes to partition the trace-files are employed to reduce the computational and memory overhead of examining the traces. This includes viewing a single iteration of an application (which requires a small manual modification to the source code) or post-processing the files to follow the chains to a given depth.

The screenshots shown in Figs. 3 and 4 illustrate traces that were loaded into the tool and visualised. Figure 3 demonstrates the comparative algorithm attempting to identify the largest similar run within the call-graph and creating links to the next trace. Where calls have not been made, they are highlighted in red and the link cursor between the traces points to where they should exist in the second trace. Where calls exist that are not in the second trace, they are highlighted in yellow and again link to where they are expected. The common colour, blue, indicates sections that exist in both traces but are different (i.e. instructions issued differ by a given threshold). The tool allows a chain of comparisons to be established, so that trace 1 is compared with 2, 2 with 3 and so on. It is assumed that the call-chains are broadly similar (such as differences in loop counts, conditionals, and communication patterns). Where multiple traces are entirely different, the tool effectiveness would be limited.

In addition to viewing the call graph, the tool is able to group call points together to obtain a frequency view of the application's trace. This is useful for grouping similar communications together and fits well with the combined analytical expressions that describe the overall application. Figure 4 illustrates this effect with traces from the Sweep3D ASCI demonstrator application. These are also subject to the same highlighting, so it is immediately obvious where one application has spent more time in a particular subroutine or exhibited different communication behaviour.

Together the tools allow a performance modeller to rapidly profile a code, run it under different configurations (such as on a 4x4 processor network, and then on 8x8 network) and then load in the traces to analyse the result for particular performance characteristics (see Table 1). In the case of moving from one processor arrangement to another, it is likely that the communication patterns will be different and, if the application scales weakly, large differences in the computation sections and message sizes. These visual cues aim to provide an effective summarised view of the program's dynamic operation. Once alerted to a particular region of code that appears sensitive to a particular change it is possible to direct attention to that part of the program which can ultimately assist in constructing the performance model.

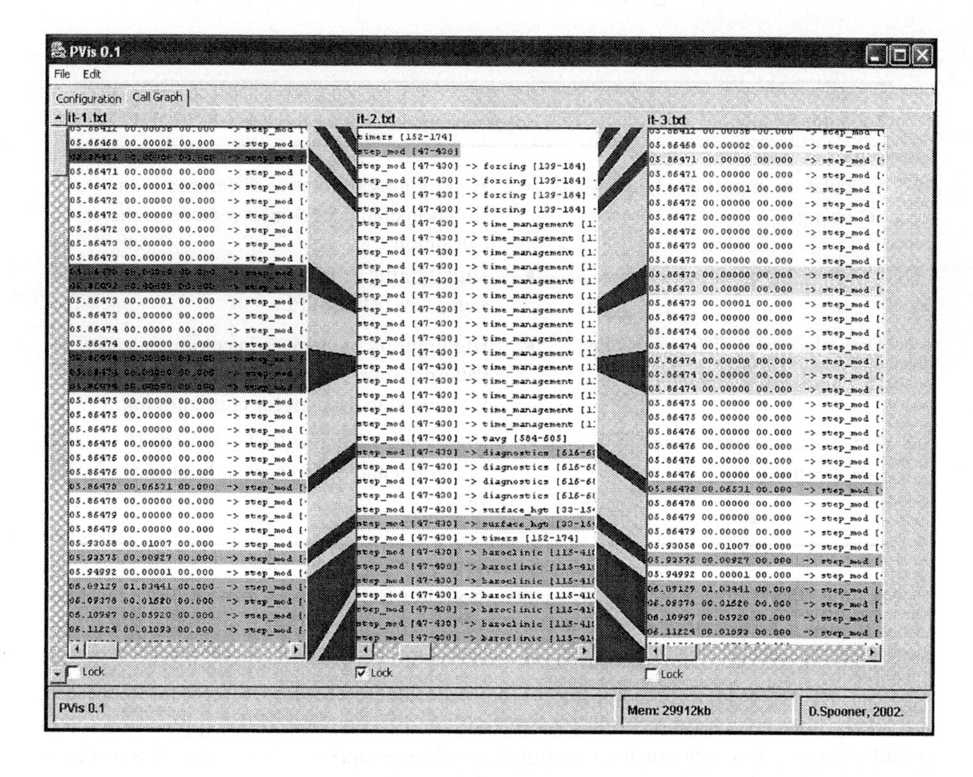

Fig. 3. Visualisation screenshot comparing three iterations of the same application on the same processor. Where regions match, but have differences (such as the number of issued instructions), the tool connects and highlights the sections using colour cues. Regions that have been removed or added are also identified

rank1

```
001 00.01528 00.889 .source.f [2-47]
001 01.50463 98.828 .sweep.f [2-625]
008 00.00012 00.000 ..octant.f [1-114]
080 00.67697 00.024 ..mpi_stuff.f [85-102]
040 00.00000 00.000 ...MPI_RECV from 3 to 1 (750 ele
080 00.21291 00.010 ..mpi_stuff.f [73-83]
040 00.00000 00.000 ...MPI_SEND from 1 to 0 (510 ele
040 00.00000 00.000 ...MPI_SEND from 1 to 3 (750 ele
040 00.00000 00.000 ...MPI_RECV from 0 to 1 (510 ele
001 00.18991 00.001 .mpi_stuff.f [187-200]
001 00.00000 00.000 ..MPI_ALLREDUCE (1 element)
001 00.00405 00.283 .flux_err.f [2-36]
001 00.02339 00.000 ..mpi_stuff.f [161-172]
001 00.00000 00.000 ...MPI_ALLREDUCE (1 element)
```

rank2

```
001 00.01527 00.890 .source.f [2-47]
001 01.55163 98.825 .sweep.f [2-625]
008 00.00010 00.000 ..octant.f [1-114]
120 00.64003 00.037 ..mpi_stuff.f [85-102]
040 00.00000 00.000 ...MPI_RECV from 3 to 2 (510 ele
040 00.00000 00.000 ...MPI_RECV from 4 to 2 (750 ele
120 00.29338 00.015 ..mpi_stuff.f [73-83]
040 00.00000 00.000 ...MPI_SEND from 2 to 0 (750 ele
040 00.00000 00.000 ...MPI_RECV from 0 to 2 (750 ele
040 00.00000 00.000 ...MPI_SEND from 2 to 4 (750 ele
040 00.00000 00.000 ...MPI_SEND from 2 to 3 (510 ele
001 00.09949 00.001 .mpi_stuff.f [187-200]
001 00.00000 00.000 ..MPI_ALLREDUCE (1 element)
001 00.00826 00.284 .flux_err.f [2-36]
001 00.00944 00.001 ..mpi_stuff.f [161-172]
001 00.00000 00.000 ...MPI_ALLREDUCE (1 element)
```

Fig. 4. Visualisation screenshot comparing consolidated traces for two processes in a single iteration of Sweep3D. The tool has highlighted an area where change is identified: in this case, the communications patterns are different due to rank placement in a 2x6 processor grid

Table 1. Example identifiable performance features from visualization tool

Example Characteristic	Tool Assistance
Calculation impact due to data-set change.	• Highlighting of relative difference in instruction counts in similar regions of the call-graph. • Highlighting of relative difference in message sizes between PEs.
Data-placement differences due to variation in PE count.	• Highlighting of changes in communication patterns (source, destination) as topology is changed to account for configuration.
Functional differences in application iterations.	• Highlighting of new/deleted regions in call-graph and differences in frequency views.

Together the tools allow a performance modeller to rapidly profile a code, run it under different configurations (such as on a 4x4 processor network, and then on 8x8 network) and then load in the traces to analyse the result for particular performance characteristics (see Table 1). In the case of moving from one processor arrangement to another, it is likely that the communication patterns will be different and, if the application scales weakly, large differences in the computation sections and message sizes. These visual cues aim to provide an effective summarised view of the program's dynamic operation. Once alerted to a particular region of code that appears sensitive to a particular change it is possible to direct attention to that part of the program which can ultimately assist in constructing the performance model.

The problem with viewing traces in this format is that screen estate is typically limited and it only takes a few traces to become impractical. In order to see more detail, the traces must be reduced into a more compact form and we envisage using individual pixels to visualise events in the trace-files. Displaying traces at this density should an observer to rapidly distinguish areas of interest and then "drill down" to view the relevant code.

5 Future Work and Conclusions

The approach described in this paper utilises dynamic trace files and a multi-view visualisation tool to highlight areas of interest when input parameters, data-sets and resource configurations are modified. By *focusing* on the areas of a code that are sensitive to configuration and input data, some of the drudgery is removed in terms of isolating the critical regions that govern the performance characteristic of an application.

Future developments of this approach include developing the visualisation tool to utilise the Vampir trace facility which is widely utilised in parallel performance studies. Additional work will focus on utilising exotic forms of visualisation to obtain compact views of multiple trace files relating to different performance scenarios.

Acknowledgements. Los Alamos National Laboratory is operated by the University of California for the National Nuclear Security Administration of the US Department of Energy.

References

1. Kerbyson, D.J., Hoisie, A., Wasserman, H.J.: Use of Predictive Performance Modelling During Large-Scale System Installation, in Proc. PACT-SPDSEC02, Charlottesville, VA. (September 2002)
2. Kerbyson, D.J., Hoisie, A., Wasserman, H.J.: A Comparison Between the Earth Simulator and AlphaServer Systems using Predictive Application Performance Models, Computer Architecture News, ACM (December 2002)
3. Spooner, D.P., Jarvis, S.A., Cao, J., Saini, S., Nudd, G.R.: Local Grid Scheduling Techniques using Performance Prediction, IEE Proc. – Computers and Digital Techniques (2003)
4. Perry, S.C., Grimwood, R.H., Kerbyson, D.J., Papaefstathiou, E., Nudd, G.R.: Performance Optimisation of Financial Option Calculations, Parallel Computing, **26**(5), Elsvier (2000) 623–639
5. Mellor-Crummey, J., Marin, G.: Building Parameterized Models for Black-Box Applications, in Proc. Los Alamos Computer Science Institute Symposium (LACSI), Santa Fe (October 2002)
6. Snavely, A, Carrington, L., Purkayastha, A., et. al.: A Framework for Application Performance Modeling and Prediction, in Proc. SC2002, Baltimore (2002)
7. Cain, H.W., Miller, B.P., Wylie, B.J.N.: A Callgraph-based Search Strategy for Automated Performance Diagnosis, Concurrency and Computation: Practice and Experience **14** (2002) 203–217.
8. Vampir, Pallas. `http://www.pallas.com/e/products/vampir/index.htm`
9. Eick, S.C., Steffen, J.L., Sumner, E.E.: Seesoft – A Tool for Visualizing Line Oriented Software Statistics, IEEE Trans. on Software Eng., **18**(11) (1992) 957–968
10. Nudd, G.R., Kerbyson, D.J., et.al.: PACE: A Toolset for the Performance Prediction of Parallel and Distributed Systems, Int. J. of High Performance Computing Applications **14** (2000) 228–251
11. Kerbyson, D.J., Alme, H.J., Hoisie, A., Petrini, F., Wasserman, H.J., Gittings, M.L.: Predictive Performance and Scalability Modeling of a Large-scale Application, in Proc. SC2001, Denver (2001)
12. Kerbyson, D.J., Hoisie, A., Pautz, S.D.: Performance Modeling of Deterministic Transport Computations, in Performance Analysis and Distributed Computing, Kluwer (2003)
13. Miller, B.P., Callaghan, M.D., Cargille, J.M., Hollingsworth, J.K., Irvin, R.E., Karavanic, K.L., Kunchithapadam, K., Newhall, T.: The Paradyn Parallel Performance Measurement Tool, IEEE Computing **28**(11) (1995) 37–46
14. Browne, S., Dongarra, J., Garner, N., Ho, G., Mucci, P.: A Portable Programming Interface for Performance Evaluation on Modern Processors, Int. J. of High Performance Computing Applications **14**(3) (2000) 189–204

Compiler Directed Parallelization of Loops in Scale for Shared-Memory Multiprocessors

Gregory S. Johnson[1] and Simha Sethumadhavan[2]

[1] Department of Computer Sciences &
Texas Advanced Computing Center
The University of Texas at Austin, Austin TX 78712, USA
johnson@tacc.utexas.edu
[2] Department of Computer Sciences
The University of Texas at Austin, Austin TX 78712, USA
simha@cs.utexas.edu

Abstract. Effective utilization of symmetric shared-memory multiprocessors (SMPs) is predicated on the development of efficient parallel code. Unfortunately, efficient parallelism is not always easy for the programmer to identify. Worse, exploiting such parallelism may directly conflict with optimizations affecting per-processor utilization (i.e. loop reordering to improve data locality). Here, we present our experience with a loop-level parallel compiler optimization for SMPs proposed by McKinley [6]. The algorithm uses dependence analysis and a simple model of the target machine, to transform nested loops. The goal of the approach is to promote efficient execution of parallel loops by exposing sources of large-grain parallel work *while* maintaining per-processor locality. We implement the optimization within the Scale compiler framework, and analyze the performance of multiprocessor code produced for three microbenchmarks.

1 Introduction

Effective exploitation of multiprocessor systems is hampered by the complexity of developing efficient parallel code. It is not always intuitively clear to the programmer which regions of a code are parallel and how each might be tuned to achieve high per-processor performance.

Consider the simple loop nest in Fig. 1a. Spatial and temporal locality considerations favor ordering the loops as shown. However, achieving the maximal granularity of parallelism favors moving the i loop to the outer position (since the i loop is parallel while the j loop is not). Doing so distributes neighboring iterations of the i loop across processors resulting in shared cache lines. Though processors update distinct values of a, a given $a[i]$ is present in the cache on multiple processors. Thus a write to an $a[i]$ by one processor incurs the expense of invalidating the previous value of that $a[i]$ in the caches of the other processors. Advanced compiler techniques capable of addressing contradictions involving locality and the granularity of parallelism are required.

McKinley proposes a compiler optimization algorithm which promotes efficient execution of loops on SMP machines [6]. It does so by exposing sources of large-grain parallel work $while$ maintaining per-processor locality. The algorithm computes the cost

P.M.A. Sloot et al. (Eds.): ICCS 2003, LNCS 2659, pp. 946–955, 2003.

```
for j ← 1 to m do
  for i ← 1 to n do
    a[i] ← a[i] * b[i,j]
  endfor
endfor
```

(a)

```
for k ← 1 to n by strip do
  for j ← 1 to m do
    for i ← k to min(k + strip - 1, n) do
      a[i] ← a[i] * b[i,j]
    endfor
  endfor
endfor
```

(b)

Fig. 1. A simple loop nest is shown in (a). The same loop nest is shown in (b) following the application of the optimization algorithm. The loop nest has been transformed by strip-mining and interchange, such that the k loop features large-grained parallelism, while the i loop maintains good locality

of distinct orderings of the loops in a nest in terms of cache lines used. It then orders the loops such that those with the most reuse (over the fewest cache lines used) are placed innermost. Finally, the algorithm selects the outermost legally parallelizable loop, applies strip-mining, and moves any resulting iterator loop to the outermost position.

Figure 1b is the result of this algorithm applied to the loop nest in 1(a). The i loop has been strip-mined and the resulting iterator k moved to the outermost position. The iteration space of the i loop is broken into $strip$ sized contiguous regions. The k loop is parallel and results in one strip of i per processor (with $strip$ set to the trip count of i / the number of processors). Locality is maintained within each strip, and the granularity of the work per processor is maximized (as the parallel loop is outermost).

We develop an implementation of this algorithm within the Scale compiler framework. We then examine the execution time and cache performance of multiprocessor code produced for several microbenchmarks by the augmented Scale compiler. Our results indicate that significant performance gains are achievable using a straightforward implementation of this algorithm. However, we also find that this optimization is sensitive to loop structure and the availability of robust dependence testing within the compiler.

We present this work as follows. In the next section we briefly introduce the reader to the Scale compiler framework, and the specific features which support our optimization algorithm. The algorithm itself is described in Sect. 3. We illustrate its design by way of pseudocode and an example showing the transformation of a simple loop nest. In Sect. 4, we detail the key components of our implementation, showing how each affects the structure of a loop nest which performs matrix multiply. We examine the performance of this implementation in Sect. 5. Finally, we relate our work to previous efforts in parallelizing compilers for SMPs.

2 Scale

The Scalable Compiler for Analytical Experiments (Scale) was developed at the University of Massachusetts, for the purpose of enabling research in compiler optimizations. Scale includes a modular framework specifically designed to permit new optimizations to be rapidly prototyped and tested.

Our selection of Scale is driven by pragmatic considerations as well as the availability of low-level features which directly support our implementation. Scale includes frontends for the high-level languages in which many benchmarks are written (C and Fortran). Additionally, Scale includes key low-level facilities such as dependence analysis, loop abstractions, reference groups, and a simple model of the host machine. The relationship between the latter two facilities and our implementation is as follows.

2.1 Reference Groups

To accurately quantify the locality available in a loop, it is necessary to determine which array references access the same set of cache lines. Scale provides this functionality, organizing references into groups. Each group corresponds to a set of related references which exhibit one of the following: spatial locality (references refer to neighboring data elements), temporal locality (references are loop invariant), or no locality (a single reference is assigned to its own group). If the size of a cache line is also known, reference groups can be used to estimate the cost (in cache lines accessed) of a specific ordering of the loops in a nest.

2.2 Machine Model

Scale implements a simple model of the target machine which includes cache characteristics such as L1 line size. Given reference groups, cache line size cls, and the trip count of a loop C, the cost of the loop nest (with the target loop placed innermost) is estimated as follows. A reference R, representative of a group which appears in the target loop, is selected. If R is loop invariant, its cost (in cache lines required) in the context of that loop is 1. Such a reference is likely to be stored in a register. If R varies as a function of the loop index in the first array subscript dimension, its cost is taken to be C / cls cache lines (adjusted appropriately for non-unit strides). If R carries no reuse, its cost is estimated to be cls (a new cache line is required for this reference on each iteration). The total cost of the nest is the cost of the target loop multiplied by the trip counts of the outer loops. *MemoryOrder* and *NearbyPermutation* (introduced in the next section) reorder the loops in a nest by cost, such that loops with the most reuse (lowest cost) are innermost. In practice, this placement very often promotes the best overall reuse [7]. An example of this approach, applied to a triply-nested loop, is illustrated in Sect. 4.1.

Additionally, we extend the Scale machine model to include a processor count. During strip-mining, this value is used to divide the iteration count of the target loop into exactly P strips of roughly equal size. Note that McKinley's optimization assumes that the processors are homogeneous, and that each is equipped with a local L1 cache. In tandem with locality-driven loop ordering, strip size computation based on processor count insures that the potential for false sharing is minimized. Having set the stage, we now examine the optimization algorithm itself in greater detail.

3 Optimization Algorithm

Our work is based on an algorithm proposed by McKinley [6]. It utilizes dependence analysis and a simple model of the target machine, to strip-mine and interchange loops

```
INPUT: A loop nest L = {l_1, ..., l_k}
OUTPUT: An optimized loop nest P
ALGORITHM:
  procedure LoopParallelize(L)
    MO = MemoryOrder(L)
    P = NearbyPermutation(L, MO)
    for j = 1, m {outermost to innermost loop of P}
      if (isParallel(p_j) == true)
        r_j = StripMine(p_j)  where r_j is the resulting outer loop
        markParallel(r_j)
        if (j != 1) permute r_j into the outermost legal position in P
        break
      endif
    endfor
```

Fig. 2. *LoopParallelize* reorders, strip-mines, and parallelizes loops in a given nest. The resulting nest features both large-grain parallel work and good per-processor locality.

in a nest. The goal of the algorithm is to promote efficient execution of parallel loops by exposing sources of large-grain parallel work while maintaining per-processor locality.

Our implementation utilizes the Scale dependence machinery to identify perfectly nested loops containing no function calls with unknown side-effects, within a target procedure. Each such nest is passed to our main optimization routine *LoopParallelize*, which may interchange, strip-mine, and / or mark parallel a member loop.

Pseudocode for *LoopParallelize* is shown in Fig. 2, and is very similar to the corresponding routine in [6]. *MemoryOrder* reorders the loops in a nest such that those with the most reuse are innermost and those with the least reuse outermost. This routine employs the Scale reference group data and cache line sizes for the target machine to compute the cost of each loop in terms of cache lines accessed. In the example in Fig. 1a, the i loop accesses fewer cache lines than the j loop, and is thus assigned to the innermost position by *MemoryOrder*. *NearbyPermutation* utilizes the dependence vectors produced by Scale to determine if the loop order proposed by *MemoryOrder* is legal (i.e. the vector for the reordered loop is lexicographically positive). If the proposed loop order is not legal, *NearbyPermutation* computes a close variation on this ordering which is legal. Refer to [6] for a full description of *NearbyPermutation*.

LoopParallelize now examines the newly reordered loop nest for parallelism. Working from the outermost loop to the innermost, and using the dependence information provided by Scale, *LoopParallelize* finds the first loop which is parallelizable. *LoopParallelize* strip-mines this loop. Strip-mining breaks the iteration space of the loop into contiguous "strips". The process converts a single loop into a doubly nested loop. The inner loop operates as usual, but only over "strip" iterations. The new outer "iterator" loop iterates over the strips. Strip-mining the i loop in Fig. 1a, results in the new i and k loops in Fig. 1b. *LoopParallelize* picks the strip size to be C / P where C is the trip count of i, and P is the number of processors in the target machine. Kennedy and McKinley have shown that this approach works well in the case where the iteration space is not smaller than the number of processors times the size of a cache line [3]. We assume this to always be the case. If the resulting iterator loop is not in the outermost position, *LoopParallelize* moves it to the outermost legal position (to maximize the granularity of parallelism), and marks it parallel. This step moves the k loop in Fig. 1b to the outermost position as shown.

The result of this effort is a set of optimized (where applicable) loop nests, each of which effectively balances large-grained parallel work with per-processor locality.

4 Implementation

Our augmented Scale compiler implements the McKinley optimization. We illustrate its behavior by way of application to the matrix multiply C code in Fig. 3a. The subsequent transformations are detailed step-by-step, through the following subsections. For clarity, the effect of each transformation is represented in classical C, rather than in the lower-level form produced by Scale.

4.1 MemoryOrder and NearbyPermutation

Recall that *MemoryOrder* computes an ordering of the loops in a nest such that those with the greatest locality over the fewest cache lines are placed innermost. If this loop order is illegal, *NearbyPermutation* finds a close variation that is legal, and performs the actual reordering. Consider the matrix multiply code in Fig. 3a, and assume row-major storage order. Given the original loop order as $<i, j, k>$, *MemoryOrder* computes that the best locality is achieved with the ordering $<i, k, j>$. Figure 3b illustrates how this order is computed, given a cache line size of four array elements, and the cost rules in Sect. 2.2. Observe that $c[i][j]$ and $b[k][j]$ exhibit spatial locality with loop j placed innermost, and $a[i][k]$ with k innermost. Also, $c[i][j]$ is loop invariant (temporal locality) if k is innermost, $a[i][k]$ if j is innermost, and $b[k][j]$ if i is innermost. Clearly the three reference groups benefit most (in terms of reuse), if loop j is placed innermost, followed by k and finally i. As this loop order is legal, *NearbyPermutation* interchanges loops j and k, resulting in the code seen in Fig. 4a.

```
for (i = 0; i < 100; i++) {
  for (j = 0; j < 100; j++) {
    for (k = 0; k < 100; k++) {
      c[i][j] += a[i][k] * b[k][j];
    }
  }
}
```

(a)

reference group	loop i innermost	loop j innermost	loop k innermos
c[i][j]	100 * 100 * 100	25 * 100 * 100	1 * 100 * 100
a[i][k]	100 * 100 * 100	1 * 100 * 100	25 * 100 * 100
b[k][j]	1 * 100 * 100	25 * 100 * 100	100 * 100 * 100
total cost	2,010,000	510,000	1,260,000

(b)

Fig. 3. A straightforward implementation of matrix multiply is shown in (a). The total cost of the loop nest in terms of cache lines required is shown in (b). A cost is computed for each loop in the nest to estimate the effect of placing it innermost. The cache line size in this example is 4 array elements. The table indicates that placing j innermost promotes the most reuse, followed by k and i

```
for (i = 0; i < 100; i++) {
  for (k = 0; k < 100; k++) {
    for (j = 0; j < 100; j++) {
      c[i][j] += a[i][k] * b[k][j];
    }
  }
}
```

(a)

```
for (ii = 0; ii < 100; ii += strip) {
  for (i = ii; i < min(100, ii + strip); i++) {
    for (k = 0; k < 100; k++) {
      for (j = 0; j < 100; j++) {
        c[i][j] += a[i][k] * b[k][j];
      }
    }
  }
}
```

(b)

Fig. 4. The loop nest with loops j and k interchanged by *MemoryOrder* and *NearbyPermutation* to exploit greater locality, is shown in (a). The same nest is shown in (b) after applying *LoopStripMine* to loop i. The resulting loop ii breaks i into "strips" of size *trip count / processor count*

4.2 LoopStripMine

Following *NearbyPermutation*, *LoopParallelize* strip-mines the outermost parallel loop (no loop-carried dependencies at the level of that loop, or procedure calls with unknown side-effects at that level or below), via *LoopStripMine*. *LoopStripMine* divides the iteration space of the loop into contiguous "strips", such that each processor of the target SMP runs roughly equal iterations (one strip). It does so by adjusting the lower and upper loop bounds to match the "width" of a strip, and encloses the loop in an outer loop which iterates over the strips. As we allocate one strip per processor, this iterator loop is used only to demarcate the bounds of the parallel region which is later marked as such by *markParallel*. Figure 4b shows the permuted matrix multiply loop nest (Fig. 4a) following the application of *LoopStripMine*.

4.3 markParallel

LoopParallelize moves the iterator loop created by *LoopStripMine* to the outermost legal position (if it is not already there), and marks it as parallel using *markParallel*. *markParallel* marks the expression node representing the initialization of the loop index variable, and the loop exit node in the Scale control flow graph representation of the iterator loop. These nodes demarcate the bounds of the instructions which compose the target loop, and thus the bounds of the desired parallel region.

The Scale SPARC backend does not currently support multithreaded code. The complexity of modifying it to do so is significant. Instead, we modify the source-to-C emission routines to replace marked loops with OpenMP parallel regions, as shown in Fig. 5.

5 Results

We examine the performance of our optimizing compiler by analyzing the runtime behavior and cache performance of multiprocessor code produced for several microbenchmarks. Specifically, we compare the execution times and cache hit rates of the binaries over a range of thread counts on a 14-processor Sun Enterprise 5500 SMP machine. Issues with the C and Fortran frontends used by Scale (neither accepts all legal ANSI programs), and with the Scale dependence infrastructure, focus our results on three

```
#pragma omp parallel firstprivate(ii, i, k, j)
{
  ii = omp_get_thread_num() * strip;
  for (i = ii; i < min(100, ii + strip); i++) {
    for (k = 0; k < 100; k++) {
      for (j = 0; j < 100; j++) {
        c[i][j] += a[i][k] * b[k][j];
      }
    }
  }
}
```

Fig. 5. The strip-mined matrix multiply code, after the application of *markParallel* and code emission. Loop ii has been replaced by an *OpenMP* parallel region

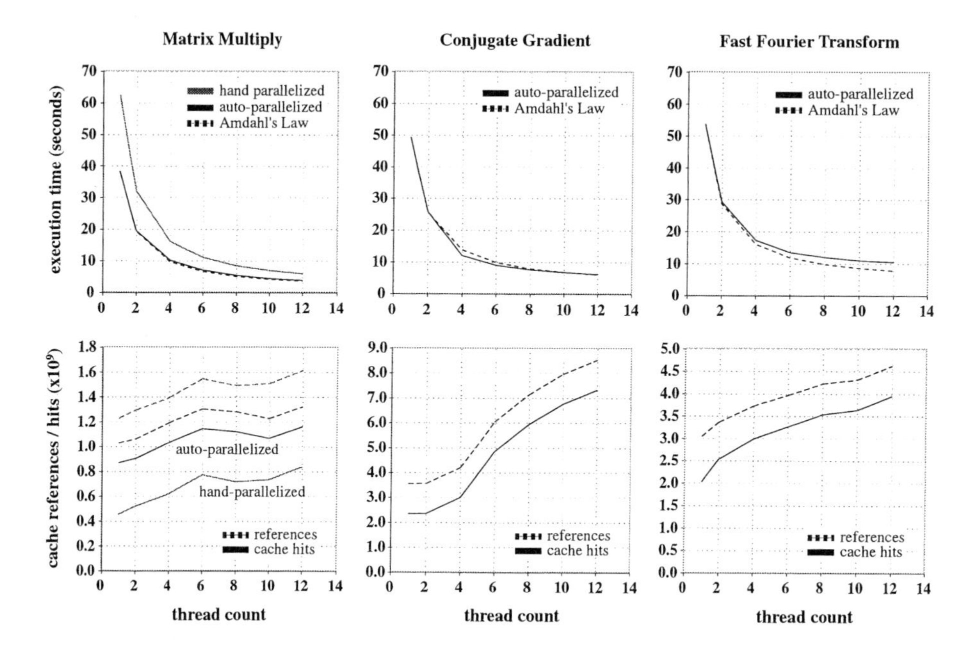

Fig. 6. Execution and cache performance of three benchmarks over multiple threads

microbenchmarks: Matrix Multiply (MM), Conjugate Gradient (CG), and Fast Fourier Transform (FFT). We perform source-to-source conversion via Scale with loop permutation (*MemoryOrder* and *NearbyPermutation*) and loop parallelization optimizations enabled. The resulting sources are compiled "-O3" with the Sun Forte 6.2 compiler. Our results are presented in Fig. 6. All measurements are averaged over five runs.

MM multiplies two 800 x 800 integer matrices. CG is a distillation of the kernel loop from the NAS Parallel Benchmarks (NPB) CG. Our CG loop nest performs an identical calculation, over the same iteration count as the Class-A version of the NPB CG. FFT computes the 2D FFT of a matrix with 4096 x 4096 elements. Our implementation correctly strip-mines and parallelizes the outermost (legally parallelizable) loop in the kernel nests of each benchmark. Additionally, Scale reorders the loops in MM as shown in Sect. 4, and (correctly) does not reorder the loops in either CG or FFT. A straightforward coding order for the outer levels of the CG and FFT loops matches the locality-driven ordering recommended by *MemoryOrder*, and imperfect nesting of the inner levels prevents more aggressive tuning.

5.1 Execution Time

The top half of Fig. 6 shows the execution performance of the benchmarks on 1, 2, 4, 6, 8, 10, and 12 threads (processor availability prevented collection of 14-thread timings). The graphs include results for the auto-parallelized code, and predictions based on Amdahl's Law. Additionally, the performance of a hand-parallelized version of MM (straightforward parallelization of loop i, with loop order $<i, j, k>$ as in Fig. 3a) is shown. Recall that our optimizer also parallelizes i, but reorders the loops as $<i, k, j>$.

Amdahl's Law states that the performance of a parallel code is limited by the presence of even a small fraction of serial code. More precisely: $speedup = (s + p) / (s + p / P)$, where s is the fraction of the uniprocessor code that must execute serially, p is the fraction that may execute in parallel, and P is the number of processors. We compute Amdahl's Law for our codes on T threads by measuring the execution time on one thread (the thread creation event demarcates the time required for the serial setup code s_s from the parallel computation time p). We also measure the time s_t required to generate T-1 threads, and compute Amdahl's Law with $s = s_s + s_t$.

Amdahl's Law might be thought to predict the best possible performance of a given code on a given number of processors. However, Amdahl's Law does not consider cache effects. As a result, we see our auto-parallelized CG outperforming the prediction on 4 and 6 threads! This is likely due to a slight performance boost resulting from a larger aggregate cache. The cache performance graph for CG supports this. Note that the ratio of hits to references increases slightly between 4 and 6 threads. In all other cases, our codes perform nearly (and in the case of MM, *very* nearly) as well as the predictions.

5.2 Cache Performance

The lower half of Fig. 6 shows the cache behavior of the benchmarks. Total memory references (dashed lines) and cache hits (solid lines) are shown for the auto-parallelized codes, and in the case of MM, for the hand-parallelized code as well.

Consider the sizable differences in the cache performance of MM, as a result of merely reordering the constituent loops. Not only is the total reference count reduced, but the ratio of hits to references is dramatically improved. Notice too that the improvement is relatively stable across thread counts. The consistently high hit ratio for the auto-parallelized code suggests that cache performance was not a significant inhibitor of parallel efficiency. The overlapping lines in the graph of execution time, for the auto-parallelized code and that predicted by Amdahl's Law, supports this.

The sharp increase in references seen on CG and less so on FFT (neither benefits from loop permutation due to loop structure) as the number of threads increases, underscores the need for locality optimizations for SMPs. In the next section, we examine other efforts in this area, including a class of approaches which seek to improve locality around loop structure which inhibits optimizations such as this one, though at the expense of portability and complexity.

6 Related Work

Here, we detail the relationship between our implementation and prior art in locality optimizations and parallelizing compilers.

6.1 Locality Optimizations

Data Layout Restructuring (DLR) approaches to locality optimization improve the *spatial* locality of references to datum (typically array elements), increasing cache utilization and subsequently performance. DLR algorithms are advantageous where complex loop structure prevents reordering to improve locality. However, the analysis required for DLR is extremely complex for arrays referenced by more than one loop. Li et al. [5] propose a generalized framework for configurable DLR. They argue for affine application-specific arrays instead of a conventional row / column format. Leung [4] proposes an analysis for static array restructuring, and Chandramouli et al. [2] propose an analysis for performing dynamic array restructuring with hardware support for memory management. The latter work is complimentary in nature to the algorithm implemented here.

Program Control Restructuring (PCR) methods alter the control flow of the program to enhance *spatial* and *temporal* locality, thereby improving cache performance. A key feature of PCR algorithms is that they are less architecture-dependent than their DLR analogs. The *MemoryOrder* and *NearbyPermutation* components of the algorithm we implement, are PCR transformations. In a related effort, Wolf and Lam [8] propose a mathematical basis for quantifying reuse, and propose a unified framework for locality-improving loop transformations including interchange, reversal, skewing and tiling.

6.2 Parallelizing Compilers

Information on commercial parallelizing compilers is only sparsely available. We therefore refrain from qualitative comparisons and instead summarize the known [1] general characteristics of auto-parallelizing compilers. In particular, we focus on the Sun SPARC compiler suite, as it is closely related to our work.

The SPARC compilers attempt to parallelize *do-all* loops in Fortran and *for* loops in C. Parallel code is generated for loops with integer indices and iteration counts known at compile time. Serial and parallel code is emitted for loops with iteration counts that are not known at compile time. The serial code is executed at runtime if the iteration count is less than that required to overcome the overhead (due to thread creation and synchronization) of parallel execution. Our implementation attempts to parallelize all Fortran and C loops, irrespective of profitability. The SPARC compilers are also capable of performing loop interchange and strip-mine, but it is unclear under what conditions these are used in tandem with parallelization to reduce false-sharing on SMP machines.

7 Conclusion

We present an implementation of a parallelizing compiler optimization proposed by McKinley [6]. This optimization restructures loop nests to promote high reuse, while enabling maximally-grained parallel work. We analyze the performance of our implementation by applying it to several microbenchmarks and executing the resulting binaries on a 14-way SPARC-based SMP. Our results indicate that while the relationship between cache performance and parallel efficiency is complex, each clearly interferes with the other. Good locality can promote high parallel efficiency (MM), but as parallelism increases, it inhibits aggregate cache reuse (CG, FFT, and to a lesser degree MM). Strengthening the former effect, and reducing the latter, are key goals of this optimization. It, and others like it, offer the potential of increased utilization of large parallel machines and reduced time-to-solution for multiprocessor codes.

References

1. C. Aoki, P. Damron, K. Goebel, V. Grover, X. Kong, M. Lai, K. Subramanian, P. Tirumalai, and J. Wang. A parallelizing compiler for UltraSPARC. 1996.
2. B. Chandramouli, J.B. Carter, W.C. Hsieh, and S.A. McKee. A cost framework for evaluating integrated restructuring optimizations. In *Proceedings of the International Conference on Parallel Architectures and Compilation Techniques*, pages 131–141, Spain, September 2001.
3. K. Kennedy and K.S. McKinley. Optimizing for parallelism and data locality. In *Proceedings of the ACM International Conference on Supercomputing*, pages 323–334, Washington, DC, July 1992.
4. S. Leung. Array restructuring for cache locality. Technical Report UW-CSE-96-08-01, University of Washington, Department of Computer Science, August 1996.
5. W. Li and K. Pingali. Access normalization: Loop restructuring for NUMA compilers. *ACM Transactions on Computer Systems*, 11(4):353–375, November 1993.
6. K.S. McKinley. A compiler optimization algorithm for shared-memory multiprocessors. *IEEE Transactions on Parallel and Distributed Systems*, 9(8):769–787, August 1998.
7. K.S. McKinley, S. Carr, and C. Tseng. Improving data locality with loop transformations. *ACM Transactions on Programming Languages and Systems*, 18(4):424–453, July 1996.
8. M.E. Wolf and M.S. Lam. A data locality optimizing algorithm. In *Proceedings of the ACM SIGPLAN Conference on Programming Language Design and Implementation*, New York, NY, 1991.

A New Data Compression Technique for Event Based Program Traces

Andreas Knüpfer

Center for High Performance Computing
Dresden University of Technology, Germany
knuepfer@zhr.tu-dresden.de

Abstract. The paper presents an innovative solution to the problem of the very huge data sets that are regularly produced by performance tracing techniques – especially on HPC programs. It designs an adapted data compression scheme that takes advantage of regularities frequently found in program traces. Algorithms to reveal repetition patterns in a programs call structure and run time behavior are discussed in detail, solutions to some problems arising on practical application are addressed as well. Two examples demonstrate the capabilities of the approach and document its behavior. Finally, some thoughts are given regarding how the patterns revealed in the process of data compression may assist the automatic analysis of traces.

1 Introduction

While tracing is a well established method of program performance analysis, there is one critical part regularly encountered: the amounts of data generated. Of course, this problem is already addressed by instrumentation procedures. Furthermore, trace libraries do pay close attention on efficient encoding of trace data. In any case fact is that trace data sets grow bigger and bigger. On the one hand, this is caused by faster processors that produce more trace events per time. On the other hand, the amount of trace data is multiplied by the number of processes that work together to solve a problem. Especially in the High Performance Computing (HPC) field the trend to massive parallel computation causes immense trace data sizes.

At the same time the extractable information of a trace does not necessarily increase along with its size. So the question arises what the actual information is inside a trace. The answer to that question is tricky. But it is easy to see what non-information is included: ever repeating sequences produced by program loops over some essential parts of a program.

Such regularities are exploited to construct a data compression method specially suited for program traces. Of course, general data compression methods are being utilized for program traces, e.g. gzip and others, but compression ratios[1] for traces achieved by them grow rarely over a factor of five to ten.

[1] Compression ratio is defined as $\mathcal{K} :=$ original Size / compressed Size.

P.M.A. Sloot et al. (Eds.): ICCS 2003, LNCS 2659, pp. 956–965, 2003.

The new approach was motivated by the work [2,3,4] about analysis and visualization of very large and highly parallel program traces of the local research group – especially in conjunction with the adaption of VAMPIR towards the ASCI project.

2 Construction of Meta-events

The given approach notices only events that denote function calls (enter events) or function returns (leave events) and ignores any other events like sending and receiving messages. Furthermore, the method takes only the current thread of a parallel program into account – thus for now it can be assumed that it has to deal with sequential programs only.

2.1 Model

By the restriction to enter and leave events – the most common classes of events in general – additional properties are assured: Following the procedural programming paradigm for every enter event e there is exactly one associated leave event l. This relation is denoted as $e \sim l$. Given that the relation $<$ expresses the order of events in terms of time, for two pairs of associated events $e \sim l$ and $f \sim g$ the equivalence

$$e < f < l \Leftrightarrow e < g < l \tag{1}$$

is implied. This is called the stack property since it is caused by the LIFO scheme of the function calls. Now meta-events are defined as follows:

Definition 1. *A collection of* $n \in \mathbb{N}$ *consecutive events* $m := \{e_0, e_1, ...e_{n-1}\}$ *is called a meta-event of order* n *if the implication* $e_i \in m \Rightarrow \exists^* \, e_j \in m : e_i \sim e_j$ *is satisfied for every* $i, 0 \leq i < n$.

With this definition in mind it is easy to put up the three rules for constructing meta-events – see also Fig. 1:

1. The pair of an enter event e and a leave event $l \sim e$ immediately following makes up a meta-event of order 2.
2. A meta-event m of order n, an enter event e immediately before m and a leave event $l \sim e$ immediately following m make up a meta-event m' of order $n + 2$.
3. Two meta-events m and m' of order n and n' make up a new meta-event of order $n + n'$ if there are no events or meta-events between both.

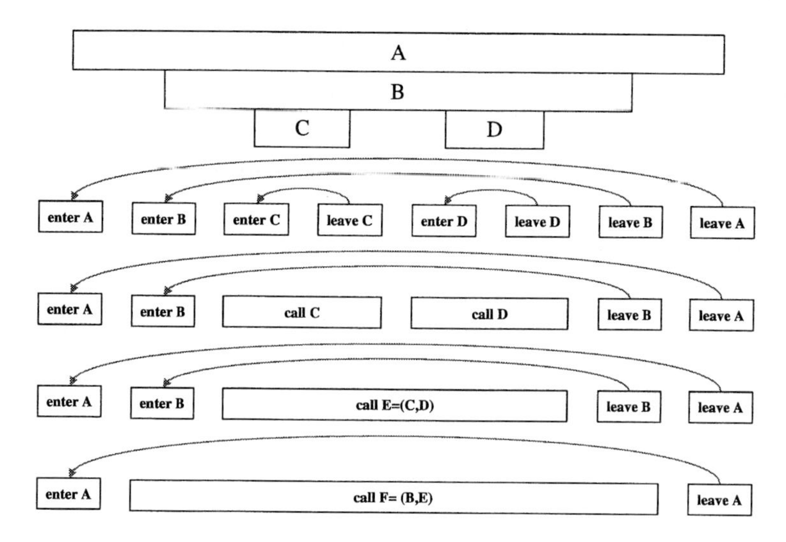

Fig. 1. Transformation from events to meta-events

2.2 Implementation

The first step for an application of the above scheme is to reveal the association $\sim$ between the single events. In fact, it is sufficient to be able to refer to the enter event $e \sim l$ that is associated with any given leave event l. This can be achieved very easily by extending the call stack representation of a trace library.

The second step is the application of the three construction rules for meta-events to the trace buffer – this works quite straight forward and it is done best by in-place substitution. The transformation algorithm has to traverse the event records in the order of creation and has to act only on leave records. For every possible situation exactly one of the three construction rules can be applied, compare Fig. 1:

1. If the current leave record references to the previous record as its associated one, then apply construction rule (1) – compare Fig. 1, (enter C, leave C).
2. If between the current leave record and its associated enter record there is exactly one meta-event record, apply construction rule (2) – compare Fig. 1, (enter B, call E, leave B). Note that by that time all leave records prior to the current one have already been processed.
3. If there are $l \geq 2$ meta-events between associated events, apply construction rule (3) exactly $l-1$ times – compare Fig. 1, (enter B, call C, call D, leave B). After that branch 2) will be applicable.

Out of practical considerations it is suggested that the algorithm shall raise the order of the created meta-events only up to a certain order N but not further. This implies that the sequence of events is transformed into a sequence of events and meta-events but in general not meta-events only!

2.3 Naming

It is customary in tracing to refer to functions by tokens. A token is an integer number that can be stored with a fixed amount of bits. There is a bijective relation between the function set and the tokens set.

This naming scheme can be extended to meta-events as well, it might even share the same token set. The definition of a meta-event token must take the way of construction (rule 1, 2, 3) into account and the thereby referenced other events/meta-events resp. their tokens.

3 Extending Meta-events to Event-Patterns

Up to now the algorithm recognizes repetitions in a function call structure. In a second stage similarities in terms of execution times are to be detected: that is a meta-event that appears multiple times with "similar" vectors of timestamps.

3.1 Comparison of Timestamp Vectors

The timestamp vector of a meta-event of order n is the vector

$$(t_0, t_1, ...t_{n-1}) \in \mathbb{R}_+^n,\ t_i \leq t_{i+1}\, \forall i \in [0, n-2] \tag{2}$$

of dimension n that simply contains the timestamps of the atomic events. In order to reveal similarity[2] it is convenient to transform a timestamp vector into the equivalent representation of a time differences vector together with the very first time stamp value t_0:

$$(t_0; d_0, d_1, ...d_{n-2}) \in \mathbb{R}_+ \times \mathbb{R}_+^{n-1},\ d_i := t_{i+1} - t_i \geq 0\, \forall i \in [0, n-2]. \tag{3}$$

Now it is very simple to recognize two time differences vectors $d := (d_0, ...d_{n-2})$ and $e := (e_0, ...e_{n-2})$ as similar if all the corresponding vector components are similar. This can be tested by an absolute measure $a_i := |d_i - e_i|$ or by a relative measure $r_i := |1 - \frac{e_i}{d_i}|$. The measure for the whole vectors is taken as maximum over all components $m_{\text{abs}} := \max_{0 \leq i < n-1}(a_i)$ and $m_{\text{rel}} := \max_{0 \leq i < n-1}(r_i)$. The suggestion is to combine the two ways of measuring: the two vectors are rated

$$\text{similar} :\Leftrightarrow m_{\text{abs}} \leq T_{\text{abs}} \vee m_{\text{rel}} \leq T_{\text{rel}} \tag{4}$$

with two thresholds T_{abs} and T_{rel}.

3.2 Event-Patterns

Where a meta-event represents several successive atomic events an event-pattern also respects the time behavior of the underlying part of a program.

[2] Testing for equality would be to strict.

Definition 2. *An event-pattern is defined as a 4-tuple of a meta-event of order n, a time differences vector (the representative vector) of dimension $n-1$ and the the absolute and relative thresholds $t_{abs} \leq T_{abs}$ and $t_{rel} \leq T_{rel}$ for time differences vectors.*

The transformation of meta-events into event-patterns is performed subsequent to the transformation of events into meta-events, i.e. it is only performed for the meta-events of maximal order. For each of the maximum-order meta-events the according time differences vector is compared with the representative vectors of all (so far defined) event-patterns based on the same meta-event. If similar event-patterns are found choose the best fit, otherwise define a new event-pattern with the given time differences vector as the new representative vector.

3.3 Over-All Data Compression Ratio

In order to estimate the over-all data compression ratio trace data sizes are to be compared: First there is always a (relative small) number of general definition records that shall be ignored here: $l_D := 0$ bytes. For classic enter or leave events there is a fixed amount of $l_E := 18$ bytes storage space required[3]. So the total size for a trace file with e events results in

$$L_0 := l_d + l_E \cdot e \,\text{bytes}. \tag{5}$$

An event-pattern is stored with $l_P := 30$ bytes regardless of its order n. For every event-pattern a definition record is required that needs $l_{EP} := 26 + 8 \cdot n$ bytes. That event-pattern references a tree of meta-event definitions. A single meta-event definition record requires $l_{ME} := 23$ bytes. The number of meta-event definitions k depends on the order n of the meta-event that forms the root of the tree and is

$$k := \begin{cases} (\log_2 n) - 1 \approx \log_2 n & \text{minimum} \\ n - 1 \approx n & \text{maximum/average} \end{cases} . \tag{6}$$

The total trace file size of a new-style trace file is then

$$L_1 := l_d + l_{ME} \cdot k + l_{EP} + l_P \cdot \frac{e}{n}. \tag{7}$$

With the sizes $L_0(e)$ and $L_1(e, n)$ the compression ratio achieved by the data compression approach can be predicted as

$$\mathcal{K} := \frac{L_0}{L_1} = \frac{l_E \cdot e}{l_{ME} \cdot k + l_{EP} + l_P \cdot \frac{e}{n}}, \tag{8}$$

see Fig. 2. For arbitrary high number of events and $k = \text{const}$ that results in

$$\mathcal{K} := \frac{L_0}{L_1} \longrightarrow \frac{l_E}{l_P} \cdot n = \frac{3}{5} \cdot n \; (e \to \infty, n = \text{const}), \tag{9}$$

[3] All record sizes refer to the VTF3 binary trace file format [16].

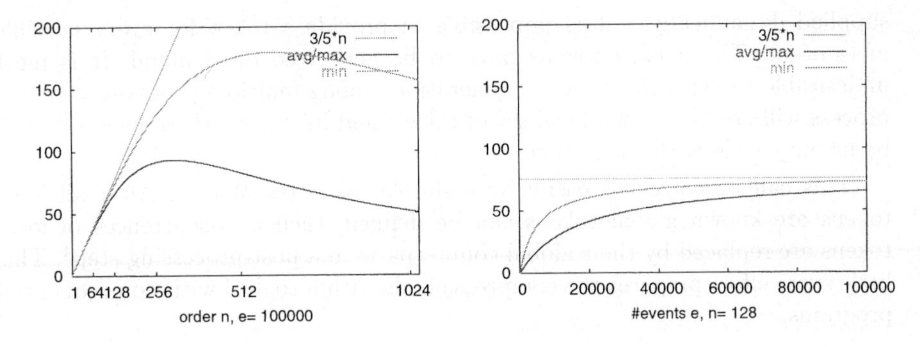

Fig. 2. Impact of the event count e and the maximum order n on the compression ratio $\mathcal{K}$. This shows idealistic behavior for a single repeated event-pattern

see Fig. 2b. The influence of n (Fig. 2a) is more interesting: it occurs that $\mathcal{K}$ grows with n up to a certain level. For higher values of n and e = const the compression ratio declines. The optimal n grows with the number of events e. Thus $\mathcal{K}$ can be increased arbitrary by increasing n as long as e is big enough!

4 Further Details

There are two issues that are necessary to analyze at this stage. Both deal with problems that arise only in certain situations but which can destroy the usability of the whole approach, if not handled adequately.

4.1 Correction Pairs

The first problem is related to the limited regularity of execution times: Even if a program shows the most regular execution times there is always a source of irregularity present: the operating systems task scheduler. Without special precautions the algorithm would produce a set of event-patterns for an over and over repeated meta-event: one event-pattern shows the undisturbed execution of the meta-event, all the others have one or more components of the time differences vector influenced by the disturber. The data compression algorithm works nevertheless but suffers a higher definition overhead.

The solution to this problem is quite simple since a disturbed time differences vector does not differ much from the undisturbed version but in one (or few) components: the solution is to store only index and values of the disturbed components (called a correction pair) and refer to an existing event-pattern for all remaining ones.

4.2 Name Spaces

The second problem for general usability is connected with parallel programs. Unlike the classic approach tokens for meta-events and event-patterns must be supplied dynamically – it is impossible to provide a token for every possible meta-event. This means tokens have to be generated on demand. It is most undesirable to synchronize token generation among multiple processes so every process will create an own local set of tokens and in general these local sets will be incompatible with each other.

This can be corrected easily by a simple token translation: After all local tokens are known global tokens can be defined, then all occurrences of local tokens are replaced by their global counterparts in a post-processing step[4]. This little extra effort prepares the compression algorithm to deal with non-sequential programs.

5 Examples

Now, two examples are presented in order to show the real world behavior of the algorithm. The examples shall address very regular and very irregular cases. For both the well known quick sort algorithm is utilized.

For the very regular case quick sort is called for sorting a rather small array as the body of a large loop, where the data to be sorted is always identical. This example can be understood as a general loop with anything in the body, where 'anything' is the same for all iterations. Figure 3 shows the compression ratios achieved for the regular example: $\mathcal{K}$ goes up to 45 where actually 4.5 MB trace data is reduced to 100 KB! The charts compare nicely to the theoretical result shown in Fig. 2. This plot is less smoothly because here the the given value for N is an upper bound and not necessarily reached.

The second example uses quick sort to have a very large array of random data sorted once. This produces a huge and complex recursive call structure. Such a behavior might arise and is a rather bad case but after all there is still

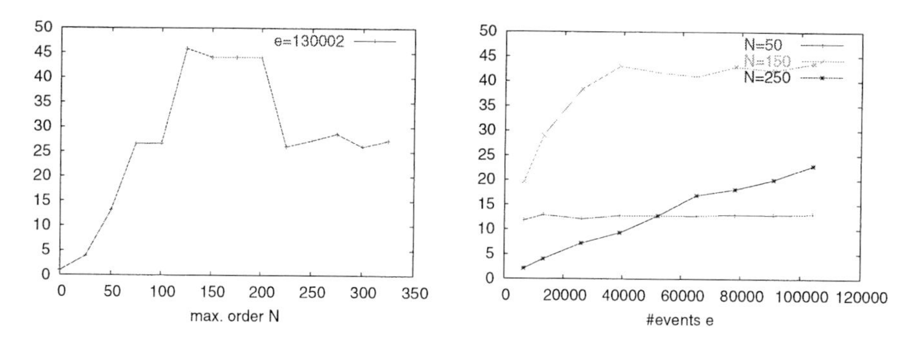

Fig. 3. Regular example: Influence of maximum order N and event count e on $\mathcal{K}$

[4] This can be combined with the merging of multiple process traces.

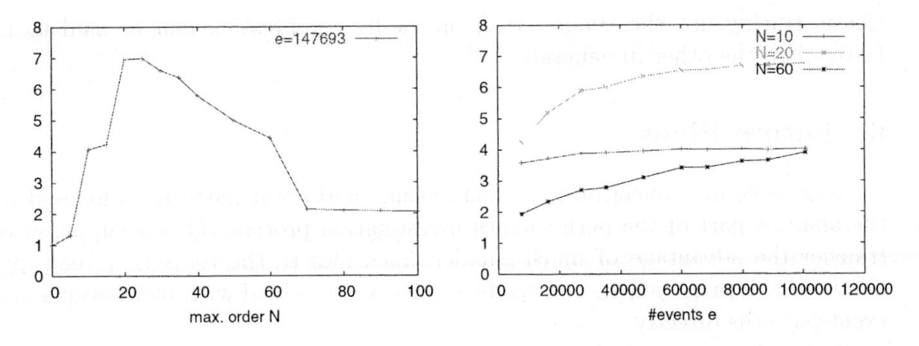

Fig. 4. Irregular example: Influence of maximum order N and event count e on $\mathcal{K}$

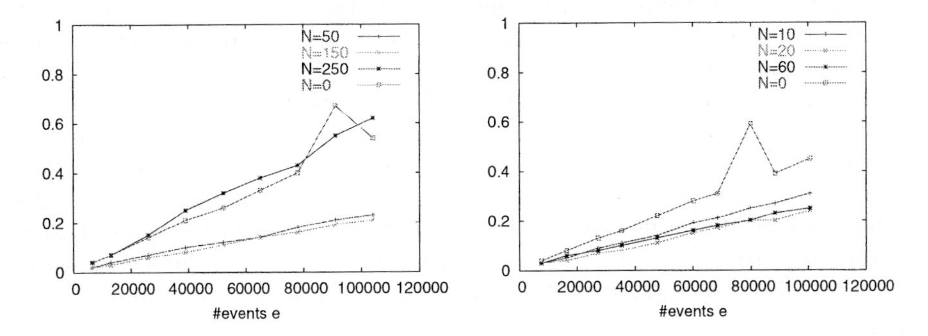

Fig. 5. Run times of regular and irregular experiments as measured on a PC AMD Athlon 1.4 GHz, disk writing performance $\approx 20\,\text{MB/s}$

some kind of regularity left[5]. In Fig. 4 the charts of $\mathcal{K}$ depending on N and e are given for the irregular case. Of course, the result is not as good as in Fig. 3 but it reveals the ability of the algorithm to cope well with non-regular situations! And most important there are never values of $\mathcal{K} < 1.0$!

The run time consumed by the trace data compression algorithm might not be as important if it is performed post mortem. But the following figures shall proof that the algorithm can be used at trace time as well, i.e. trace data is compressed on-the-fly, no intermediate non-compressed output is generated at all. In Fig. 5a the run time durations are given for the experiments shown in Fig. 3a, Fig. 5b resp. for all experiments of Fig. 4a. The case $N = 0$ which means classical tracing without any compression is shown for comparison.

Obviously, the compression algorithm does not necessarily suffer a penalty for the increased computational effort. In fact, in the actual test cases it is faster than the classic tracing most of the time! This can be explained by the smaller amount of data to be written to disk after on-the-fly compression. Since this effect is heavily dependent on processor speed and disk I/O performance neither

[5] The worst case would be something like choosing the next function to call at random which is assumed to never ever happen in real life.

classic tracing nor the version with on-the-fly compression can be said to be faster than the other in general.

6 Future Plans

Current work in connection with meta-events and event-patterns is focused on the analysis part of the performance investigation process. This is supposed to transfer the advantage of much smaller trace files to the tools that read and handle the data as well as to introduce the option to deal with meta-events and event-patterns directly.

So far there has only been discussion about compression but not about the decompression. Though, with a basic understanding of the compression scheme the according decompression is simple and works for every meta-event/event-pattern by itself. This is especially useful when accessing portions of a trace. A most elegant way to examine a compressed trace file would be to extract only the (few) event-patters that relate to the trace portion of interest! Thus the compression provides not only an advantage in terms of storage space but also in terms of access speed and response time during user interaction.

Another very interesting approach is using meta-events like more abstract functions, i.e. do not use functions as smallest entities but meta-events of certain (variable) sizes. This provides additional abstraction layers for trace analysis where so far there are merely two: statistics and fully detailed reproduction.

7 Related Work

Everything that is presented here is covered much more comprehensively in the German language diploma thesis [9]. There connections are drawn to string pattern matching [1,6,10,12], information encoding and general data compression algorithms [5,11,15] especially the Lempel-Ziv family.

Furthermore, for requirements and demands of trace analysis software tools [7,8,13,17,18] are referenced.

[14] presents another very interesting data compression scheme for parallel program traces that is more or less orthogonal to the one presented here.

References

1. Jun-ichi Aoe. *Computer Algorithms: String Pattern Matching Strategies.* IEEE Computer Society Press, Los Alamitos, 1994.
2. H. Brunst, H.-Ch. Hoppe, and W.E. Nagel. Group Based Performance Analysis for Multithreaded SMP Cluster Applications. In *Proceedings of Euro-Par2001, Manchester, UK*, volume 2150 of *LNCS*, page 148ff. Springer-Verlag Berlin Heidelberg New York, August 2001.
3. H. Brunst, H.-Ch. Hoppe, W.E. Nagel, and M. Winkler. Performance Otimization for Large Scale Computing: The Scalable VAMPIR Approach. In *Proceedings of ICCS2001, San Francisco, USA*, volume 2074 of *LNCS*, page 751ff. Springer-Verlag Berlin Heidelberg New York, May 2001.

4. H. Brunst, W.E. Nagel, and S. Seidl. Performance Tuning on Parallel Systems: All Problems Solved? In *Proceedings of PARA2000 - Workshop on Applied Parallel Computing*, volume 1947 of *LNCS*, pages 279–287. Springer-Verlag Berlin Heidelberg New York, June 2000.
5. T.H. Cormen, C.E. Leiserson, and R. L. Rivest. *Introduction to Algorithms.* MIT Press, 1990.
6. G. Davies and S. Bowsher. Algorithms for Pattern Matching. *Software - Practice and Experience*, June 1986.
7. R. Jain. *The Art of Computer Systems Performance Analysis.* John Wiley & Sons, Inc., 1991.
8. R. Klar et al. *Messung und Modellierung paralleler und verteilter Rechnersysteme.* B.G.Teubner, Stuttgart, 1995.
9. A. Knuepfer. Analyse von Programmspuren: Entwurf und Implementierung eines effizienten Algorithmus zur Datenreduktion. Technical Report ZHR-R-0202, Zentrum für Hochleistungsrechnen, TU-Dresden, June 2002.
10. D.E. Knuth. *The Art of Computer Programming: Fundamendal Algorithms*, volume 1. Addison Wesley Longman, zweite edition, 1997.
11. D.E. Knuth. *The Art of Computer Programming: Sorting and Searching*, volume 3. Addison Wesley Longman, zweite edition, 1998.
12. D.E. Knuth, J.H. Morris, and V.R. Pratt. Fast Pattern Matching in Strings. *SIAM Journal of Computing*, 06 1977.
13. B. Mohr. *Ereignisbasierte Rechneranalysesysteme zur Bewertung paralleler und verteilter Systeme.* VDI Verlag, 1992.
14. O.Y. Nickolayev, P.C. Roth, and D.A. Reed. Real-time statistical clustering for event trace reduction. *The International Journal of Supercomputer Applications and High Performance Computing*, 11(2):144–159, Summer 1997.
15. R. Sedgewick. *Algorithmen.* Addison Wesley, 1991.
16. S. Seidl. VTF3 - A Fast Vampir Trace File Low-Level Library. personal communications, May 2002.
17. F. Wolf. EARL – Eine programmierbare Umgebung zur Bewertung paralleler Prozesse auf Message-Passing-Systemen. Technical report, Forschungszentrum Jülich GmbH, June 1998. JÜL-3551.
18. F. Wolf and B. Mohr. Automatic Performance Analysis of SMP Cluster Applications. Technical report, Forschungszentrum Jülich GmbH, August 2001. FZJ-ZAM-IB-2001-05.

Exploiting Stability to Reduce Time-Space Cost for Memory Tracing

Xiaofeng Gao and Allan Snavely

San Diego Supercomputer Center, University of California, USA
xgao@cs.ucsd.edu, allans@sdsc.edu

Abstract. Memory traces record the addresses touched by a program during its execution, enabling many useful investigations for understanding and predicting program performance. But complete address traces are time-consuming to acquire and too large to practically store except in the case of short-running programs. Also, memory traces have to be re-acquired each time the input data (and thus the dynamic behavior of the program) changes. We observe that individual load and store instructions typically have stable memory access patterns. Changes in dynamic control-flow of programs, rather than variation in memory access patterns of individual instructions, appear to be the primary cause of overall memory behavior varying both during one execution of a program and during re-execution of the same program on different input data. We are leveraging this observation to enable approximate memory traces that are smaller than full traces, faster to acquire via sampling, much faster to re-acquire for new input data, and have a high degree of verisimilitude relative to full traces. This paper presents an update on our progress.

1 Introduction

Research in performance modeling and prediction relies heavily on application traces, especially memory traces. Previous researches have shown that interactions between a program and the memory-hierarchy of the machine on which it executes can largely determine its performance [1,4]. In our own previous work [4], we have shown that summarized memory traces are, to a first approximation, machine-independent; we used memory traces in performance models to predict and explain the performance of scientific applications with different problem sizes using both strong and weak scaling and across several modern High Performance Computing (HPC) platforms. We further used the models to predict the performance of future hardware upgrades and new machines. Thus, given a memory trace it is possible to explain observed cache hit-rates (for example) and to predict cache-hit rates for future machines. It is further possible to guide the tuning of an application, to match that application with machines well suited for its memory demands, and to design future machines towards the needs of the application.

Unfortunately, complete memory traces that are the most perfect representation of a program's memory behavior require Gigabytes of storage. And we found that methods of trading time for space to summarize memory traces (including our own methods for recording predicted cache-hit rates by processing the address stream on-the-

P.M.A. Sloot et al. (Eds.): ICCS 2003, LNCS 2659, pp. 966–975, 2003.

fly) are not fully satisfactory for three reasons: 1) the time required for summarizing may increase the already severe slowdown for tracing, 2) a summary may reduce the ability to use the same trace to predict performance on a different machine, and 3) this approach generally does not keep continuity of the dynamic execution, so any other analysis, or even the same analysis with minor changes in the parameters requires one to repeat the trace.

Full traces can be compressed rather than summarized to save space. Several researchers [2,8] have successfully used run-of-length, SEQUITUR [10], and other traditional lossless compression techniques such as LZV [16] to reduce trace size. However these compression techniques neither consider nor preserve control flow structure in the trace. Thus the compressed traces do not reflect the structure of the application. It is hard or impossible to get a hint of what will come out from the decompression pipe next, so it is nearly impossible to use techniques such as fast forwarding to reduce analysis cost. Also the compression ratio varies widely depending on the nature of the trace. For a memory trace with mostly random accesses or unpredictable major branches, the compression ratio is quite low and file-size savings minimal [13].

We are developing approximate memory traces that are reasonably small and preserve dynamic execution information. The behaviors of the stable memory instructions are represented by patterns similar to regular expressions. We approximate the behaviors of random instructions with synthetically generated random numbers. Infrequent paths in the control flow may be pruned off to further save space. This approximating method stores smaller traces than methods for memory trace compression when a significant amount of accesses are random. It allows fast-forwarding and preserves a high degree of verisimilitude relative to full traces.

Rubin et.al [8] used the SEQUITUR[10] algorithm to compress memory traces and used the resulting compressed trace to study data layout optimizations. Such lossless compression schemes work well for streams of memory accesses which are regular and have a lot of *exact* repetitions. But there are cases where lossless approaches do not work well. Since many scientific applications touch a lot of memory addresses and contain substantial interspersed randomness, there is not much space saved by using lossless compression techniques on their memory traces. The SIGMA tool [13] from IBM may also adopt similar approaches but with an undisclosed compression algorithm. Judging from their published results, the quality of compression largely depends upon the nature of the application and the input, and the user has little power to control the output size.

In point of fact, trace analysis tools, such as various simulators for the memory sub-system, may not be very sensitive to minor changes in the trace and therefore lossy traces can have satisfactory results. Several researchers have shown that cache simulators can still provide reasonable results when using sampled memory traces as the input [18–20]. However, in these works the sampling rate is a rule-of-thumb. There is no universal rule proposed as to where and how the memory trace should be sampled. Different applications may require quite different sampling rates to remain close to the original trace. Even for the same application, different input may also require quite different sampling rates.

Several other lossy compression schemes have been proposed for particular analysis. Kaplan suggested a lossy reduction scheme for virtual memory simulations [17]. It drops addresses guaranteed to be not visible to virtual memory. This scheme also makes certain assumptions about the hardware. Agarwal and Huffman

[21] suggested a lossy trace compression scheme by exploiting spatial locality in conjunction with temporal locality. All these lossy compression schemes are less than satisfactory when a significant amount of memory accesses are random.

We are looking for a scheme that can find high-level memory access patterns from the trace, yet preserves continuity and enough details for accurate performance prediction. We also want to be able to control the size and accuracy of the trace.

In our previous work, we found it critical to distinguish regular and stable behaviors (constant or clear patterns for memory accesses) from irregular and random ones. Currently, we have found that for the random access areas the trace can be approximated without having noticeably bad effects on subsequent analysis and modeling. Absolute values from the random parts, where generic compression schemes and other lossy schemes fail to have a satisfying compression ratio, turn out to be not too important from a performance standpoint and can be replaced by generated random values. We also observed that several very similar yet unequal sequences in memory traces puzzle generic compression schemes and cause them to fail to have satisfactory compression ratios. But when minor variations in sequences are ignored and lumped together, the compression ratio can be improved by orders of magnitude again without discernable impact on analysis and modeling steps. So when trace size is a concern, similar sequences and random sequences are the best candidates to be approximated. For regular and stable sequences, any generic compression scheme may be used without loss. Based on these observations, we here propose a framework to detect stability in the trace, to classify sequences by similarity and randomness, and to use that information to compress and approximate the trace using various approaches appropriate to each. We show that the tradeoff between the size and the quality of the trace can be controlled by definitions of randomness and similarity.

2 Memory Trace Break-Down

The order of addresses accessed by a program during execution is a function of its dynamic traversal of the control-flow graph *and* the memory access patterns of its individual load and store instructions. As an elucidating example, consider the code fragment in Figure 1.1. We assume the content of array **i** is nearly random. It is difficult for an encoding scheme based on exact pattern detection to summarize the memory access pattern of this fragment. The difficulty arises from two factors. First there is no particular pattern in the effective addresses of array **X**. Although array **i**, **Y** and **Z** are accessed with fixed stride their patterns are defaced in the address stream by **X** due to its random nature. Discernable order in the address stream is further mangled by the branch instruction since the two paths in the loop have each different numbers of memory references. Thus if an encoder simply observes the generated address stream and attempts to detect and encode patterns, it will have difficulty achieving much compression.

However if we focus on individual instructions and study the stride patterns of each instruction the hidden patterns in the stream suddenly becomes clear. Instruction 1, 4, 5 all have fixed stride, while 2,3,6,7 are random. Also, there is a pattern in the order of *instructions* that can be given by the regular expression
(1,4,5,6,7,1,2,3,1,2,3)*. If one random address in the same range is as good (or as bad) as another from a performance standpoint, then reproducing the fixed strides of instructions 1,4,5 and any random values for the addresses touched by 2,3,6,7 along

with the order these instructions are encountered, will serve well enough to represent this fragment's memory behavior.

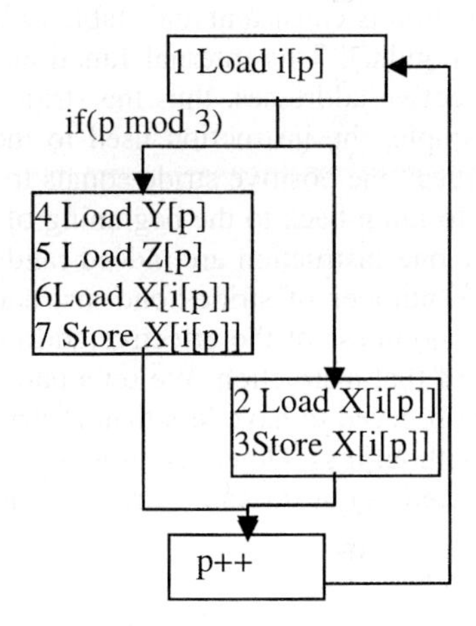

Fig. 1.1. Code Fragment

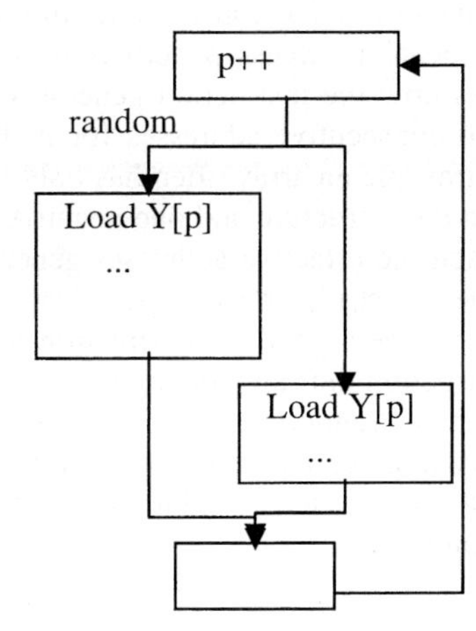

Fig. 1.2. Code Fragment with Randomness

We gather an approximate memory trace by collecting two primary kinds of information. One keeps track of memory instruction ids in the trace and tries to find patterns in order-of-instructions. The other keeps track addresses for individual instructions, and detects stride patterns in these addresses. When no clear pattern is found, these instructions are classified as random—this is a *good* thing from the compression standpoint as we assume their behavior can be usefully mimicked by some random sequence of addresses in the same range. At this time, we should mention that the stride patterns of each instruction typically do not change too much over different input (section 4). It is the order-of-instructions that usually can change a lot depending on input.

When two memory instructions are in the same basic block, their relative order in the trace is fixed: they always appear together with the same instruction distance between them. There is a one-to-one mapping from a sequence of memory instruction in a trace to a path in the control flow graph (we assume for each function call, there is an edge from the calling point to the entry pointer of the callee). So we can use a stream of basic block indices to replace a stream of memory instruction ids. This is particularly important for storing small traces. The number of basic blocks in an application is usually significantly smaller than the number of memory instructions.

3 Detecting Memory Stride Patterns

In our framework, we use different compression approaches for different memory instructions. If an instruction generates effective addresses randomly, we discard the real effective addresses and use some random numbers in the same range to replace them. When an instruction shows clear patterns in the effective addresses, we record the patterns without loss. One immediate question is how to efficiently detect pattern and randomness of the effective addresses one instruction generates.

Common sense suggests that when a compiler generates a memory instruction, it has a particular functionality; either it accesses a temporary variable, or a data structure in some order. The functionality of the instruction is consistent and stable (which does not mean its access pattern is necessarily regular). This internal functionality determines how the instruction generates the effective addresses, thus the stride between two consecutive addresses for it. For example, the instruction used to incrementally traverse an array often has only two strides: one positive stride equals to the size of the data structure and one negative stride to jump back to the beginning of the array. When the effective addresses generated by one instruction are indeed random, the number of strides must appear large. So the number of strides one instruction generates can be a good approximation of the randomness of the effective addresses of the instruction thus an indication of the nature of that instruction. We set a parameter R to define "randomness". If an instruction is observed to have less than R strides, we classify it as regular and record the patterns without loss. Otherwise it is regarded as random and its effective addresses will be replaced in our traces by a random number generator.

Table 1. Stride pattern categories

	1	2	3	4	5-16	>16
1	1369/ 54.95%	33/ 0.00%	80/ 0.13%	0/ 0.00%	12/ 0.02%	6/ 0.01%
2		82/ 8.18%	30/ 0.06%	9/ 0.00%	2/ 0.10%	4/ 0.00%
3			125/ 3.29%	20/ 0.00%	8/ 0.00%	9/ 0.00%
4				26/ 0.02%	6/ 0.00%	8/ 0.00%
5-16					53/ 7.87%	104/ 0.83%
>16						149/ 24.65%

Because we study stride patterns independent if dynamic control flow, these patterns will be particularly useful if they do not change much over different inputs. Table 1 shows how many instructions change their number-of-strides categories in gzip from SPEC2000 over 5 different reference runs[1].

Entry (i,i) shows the number of instructions that always have i strides for all five runs. Entry (i,j) shows the number of instructions have either i or j strides in the 5 different runs. There are only 20 static instructions that have more than 2 strides, these are counted multiple times in the table. The percentage entries show how much these instructions contribute to the dynamic instruction mix. (The total dynamic instruction count is the *average* of the 5 inputs.)

As can be seen, most of the instructions (static or dynamic) have the same number of strides regardless of input. There are only a few instructions that have different numbers of stride patterns depending on input (entries off the diagonal). In this im-

[1] The inputs of the five reference runs are input.graph, input.source, input.program, input.random, input.log

plementation, we choose R between 4 and 16. The shaded area highlights the maximum number of instructions affected by these choices.

Errors are introduced if we regard an instruction as random when in reality it is not. Figure 1.2 shows an unpredictable branch. On either path, array **Y** is accessed. Because the branch takes the two paths randomly, the stride pattern of the two memory instructions are random in our definition. However, if it is recognized that both load instructions access the same array then it is clear that **Y** is not really accessed randomly. In our implementation two random functions are used to approximate them independently. Although we do not record the absolute values of the random addresses, it will be necessary to study the correlations of the effective addresses generated by the memory instructions in the same loop to reduce these errors. It is also necessary to record statistical properties such as the range one random instruction can access. Using these statistical properties, we can generate more "accurate" random numbers.

4 Application Signatures

An application signature is a compressed dynamic control flow of the application for a given input. It summarizes and approximates the time-varying behaviors of the application on the given input. The pertinent features of signatures include the number of iterations for loops, what paths are taken and how they are taken and how the functions calls behave etc. Application signature provides valuable information about how the program's control flow changes with different inputs. It is can also be used to study correlation between dynamic behaviors and the inputs.

Just as we approximate memory access pattern information, we also approximate dynamic control flows based on stability. We instrument all the basic blocks in the application with DyninstAPI [6]. An online analysis breaks the stream of basic block ids into regions and studies the dynamic behaviors of those regions. For stable regions (with no or very few variations), we keep the exact dynamic behaviors. For unstable regions, we keep only approximated behaviors. The signature of the entire application is composed of a set of signatures of regions and the transition patterns among these signatures. Less significant signatures and similar signatures can be merged to simplify the dynamic control flow, thus reducing the size of the application signatures.

In order to be able to tell the boundaries and study the dynamic behaviors of each individual region, we have an extensive static analysis of the control flow graph before instrumentation. We first find all the loop heads and procedure entry blocks. These basic blocks are called region leaders and are used to mark the boundaries of the regions. The rest of the basic blocks are assigned to the inner-most containing region. Figure 2.2 gives an example of the basic blocks contained in the three regions in Fig. 2.1. The purpose of this assignment is to enable on-line analysis to study the regions' dynamic behaviors independently. For example, the loop headed by block 1 may have very complicated behaviors, but it does not affect the stable behavior of the inner loop headed by 6. Notice if a region is a loop, we do not regard the loop head as part of region. The loop head is included in the outer region as a place-holder to mark some variation of the inner loop happens on that path.

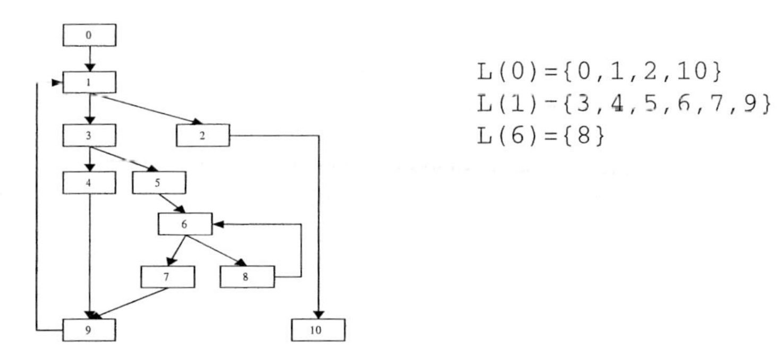

Fig. 2.1. Control Flow with three regions

Fig. 2.2. Three regions

A static path in a region is a longest non-repetitive sequence of basic blocks contained in that region. It also must start from the region's leader. For example, the region led by basic block 1 in Fig. 2 has two static paths: 135679 and 1349.

If a static path is set, the number and order of the events that can happen along the path are fixed. The events can be procedure calls, nested loops or memory instructions. This does not mean that the actual events that happen along this path are fixed. A static path just acts as a template and indicates some variations of the events will happen in a deterministic manner along this static path. We use dynamic path to summarize the actual events that happened along this static path. We define a dynamic path as an instance of static path with nesting structures having a signature.

After the static analysis, we instrument all the basic blocks in the application with their region ids. An on-line inspector will then capture a trace of the basic blocks and uses a stack to simulate the entering and exiting of the regions. Upon the completion of each static path, a dynamic path is generated to summarize all the events that happened along the just-passed path. This dynamic path is compared to the previously visited dynamic paths of this static path. If this is a new one, it will be inserted in the dynamic list for later usage. Upon the exit of each region, the inspector generates a signature to summarize the dynamic behaviors of the region in the just-past instance. For example, the summary of a loop structure includes loop iterations, the number of iterations the loop executed, the dynamic paths visited and the transition patterns and distributions among the dynamic paths. The newly generated signature is then compared to previous signatures of this structure and the signature list is updated if there are no same-signatures recorded before. Otherwise the signature index is passed to the outer region to be part of dynamic path information. Finally the application signature is the signature of the outmost region, for example the main() function.

Although it is possible to have the inspector connected to a database and store all the recorded signatures and dynamic paths on disk without any loss, we want to approximate unstable regions using the expected behaviors. In our implementation, each static path has a maximum number of dynamic paths. If there are too many different dynamic paths, the less significant ones will be merged and treated as one dynamic path. Each region also has a maximum number of signatures. Whenever this cap is exceeded, the inspector will choose two similar signatures and merge them. Useful definitions of *similar* are an area of our ongoing investigation. One open investigation is in how to define and compute the similarity distance between two signatures of a structure. In our implementation, we assign different weights to each structure field ad hoc. For example, the exact transition sequence of dynamic paths within a loop may be less important from a performance standpoint than the distribution of these dynamic paths. So we may call two invocations of the same loop similar if they contain similar distributions of dynamic paths. The distance between two signatures is computed by summing the distances of each field times the specified weights. Another intuition is that it may be useful to only merge the less visited signatures. In future work we plan to describe and explore the utility of various formal definitions of structure similarity.

5 Generating Memory Traces: Analysis of Results

An approximate memory trace is generated by embedding memory stride patterns into the applications signature.

```
(2001,<A4,S4,B0,ST>,(4000:4):1048576),

(1000,(1001,<A5,S4,B0,ST>:65536),

(1003<A4,S4,B0,LD><A7,S4,B0,ST>:1048575),

(1005<A7,S4,B0,LD><A5,RAN,*,LD><A5,IMM0,*,ST>:1048575),

(1007<A5,S4,B8,LD><A5,I-4,*,LD><A5,IMM0,*,ST> :65534), (1009:5)

(1,1000,(1001,<A5,S4,B0,ST>:65536),(1003<A4,S4,B0,LD><A7,S4,B0,ST>:10
48575),(1005 <A7,S4,B0,LD><A5,RAN,*,LD><A5,IMM0,*,ST>:1048575),(1007
<A5,S4,B8,LD><A5,I-4,*,LD><A5,IMM0,*,ST> :65534),(1009:5):10),

7000,10000,11000,6000,3000,
(3001<A5,RAN,*,LD><A5,IMM0,*,ST> <A4,RAN,*,ST><A7,S4,B4,LD>:1048575),

(3008<A4,S4,B8,LD><A4,IMM0,*,LD>:1048573)
```

Fig. 3. Generated full memory trace for IS.W with default input

The simplified approximate memory trace for IS.W is shown in Figure 3. Memory accesses are embedded in the applications signature and demarcated with <>. Each memory access has four fields to specify the data structure to which it belongs, the stride, the base address and the type of the operation. For example <A4,S4,B0,ST> means the memory instruction accesses the array 4 with stride of 4 bytes from the index 0, and it is a store operation. <A5,RAN,*,LD><A5,IMM0,*ST> shows two connected memory instructions. The first one is a random load from Array 5. The second is a store operation, it stores to the exact same address as the previous load. The application signature is demarcated with (). The number after semicolon is the expected iterations of the loop. For conciseness, constant memory accesses, and the jump-back negative strides for loops are removed from this trace.

We used this framework to obtain memory traces of the NPB benchmark suite. Table 3 compares the size of our application signatures and strides pattern information to the size of full memory traces for these benchmarks and gives the error in dynamic instruction count for our approximations. The sizes of *compressed* traces by traditional means would not be much less than full traces for CG and IS because of their significant amounts of random accesses. Using approximate traces, we can get consistent high "compression" ratios for all these applications.

Table 2. Size comparison and errors

	Estimated memory trace size[2]	Signature Size	Size of the stride patterns	Error in Instruction count	Error in MOPs	Error in FLOPs
IS.S	98MB	5KB	13KB	-0.001%	0.000%	-0.001%
CG.S	898MB	32KB	368KB	0.006%	0.009%	0.006%
FT.S	1718MB	54KB	129KB	0.203%	0.230%	-0.010%
MG.S	125MB	287KB	554KB	2.247%	2.247%	-3.905%
EP.S	1007MB	1KB	5KB	-0.183%	-0.185%	-0.198%
SP.S	3216B	272B	1KB	0.298%	-3.511%	0.000%

Error in total instruction count, as given in Table 2, is only a rough indicator of the verisimilitude of an approximate trace. The real question of interest is how well it mimics an application's true performance behaviors. We used the fast computation methods described in [11] to calculate cache hit rates from the generated IS.W memory trace *without* simulation on three different processors: Power 4, Alpha EV67 and McKinley. For loops with regular stride patterns, such as 1003 and 1007, the calculated cache miss rates and measured ones are indiscernible. For loops having random access patterns, such as 1005 and 3001, the average error in cache hit rates is 2.5%.

6 Conclusion and Future Work

In this paper, we presented a framework to generate approximate memory traces. By dividing the memory trace into stride pattern and application signature, we can differentiate instructions with simple and regular access patterns from complicated or nearly random ones. Different compression or approximation schemes are used for different cases. In this framework, we also proposed methods to make tradeoffs between the trace size and the accuracy of the trace. Initial results have shown that this framework with multiple compression and approximation schemes works fairly well. The memory trace can be approximated and recorded with a small file, yet remain fairly close to the original trace from the perspective of performance estimation. There are still several open questions such as how to compute the distance between two signatures, and the effect of false random instructions on the quality of the overall trace need to be further investigated. The entire scheme needs to be scaled up to an investigation of multiple full scientific applications and we are undertaking that now. This work was sponsored in part by the Department of Energy Office of Science through SciDAC award "High-End Computer System Performance: Science and Engineering". This work was sponsored in part by a grant from the Department of Defense High Performance Computing Modernization Program (HPCMP) and the National Security Agency.

[2] The full trace size is estimated by the multiplication of the dynamic memory instruction count and the length of the each address.

References

1. Chen Ding and Ken Kennedy "Bandwidth-Based Performance Tuning and Prediction" IASTED, Cambridge, MA November, 1999
2. Richard A. Uhlig, Trevor N. Mudge "Trace-driven Memory Simulation: A Survey" ACM Computing Surveys, V29 No. 2, 1997
3. D.H. Bailey, T. Harris et.al The NAS Parallel Benchmarks 2.0 The International Journal of Supercomputer Applications 1995
4. A. Snavely, N. Wolter, L. Carrington, R. Badia, J. Labarta, A. Purkasthaya, A Framework to Enable Performance Modeling and Prediction Supercomputing 2002
5. ATOM, see http://www.tru64unix.compaq.com/developerstoolkit/#atom
6. B. Buck, J. Hollingsworth An API for Runtime Code Patching The International Journal of High Performance Computing Application, 2000
7. S. P. Reiss, M. Renieris Encoding Program Executions SIGSOFT 2001
8. S. Rubin, R. Bodik, T. Chilimbi An Efficient Profile-Analysis framework for data-layout optimization POPL 2002
9. G. Ammons, J.R. Larus Improving Data-flow Analysis with Path Profiles SIGPLAN conference on programming language design and implementation. 1998
10. C.G. Nevill-Manning, I.H. Witten Compression and explanation using hierarchical grammars. The Computer Journal 1997
11. R.E. Lander, J.D. Fix, and A. LaMarca Cache Performance Analysis of Traversals and Random Accesses SODA 99
12. T. Ball The Concept of Dynamic Analysis ESEC/SIGSOFT FSE 1999
13. Luiz DeRose, K.Ekanadham, Jeffery K.Hollingsworth SIGMA: A Simulator Infrastructure to Guide Memory Analysis, SuperComputing 2002
14. T. Sherwood, E. Perelman, G. Hamerly and B. Calder "Automatically Characterizing Large Scale Program Behaviors" ASPLOS 2002
15. Steve Muchnick, Advanced Compiler Design & Implementation, Morgan Kaufmann, 1997
16. J. Ziv and A. Lempel, “A universal algorithm for sequential data compression” IEEE Transactions on information theory, pp. 337–343, 1977.
17. Scott F. Kaplan, Yannis Smaragdakis, and Paul R. Wilson, "Trace Reduction for Virtual Memory Simulations" SIGMETRICS '99
18. R.E. Kessler, Mark D. Hill, David A. Wood "A Comparison of Trace -Sampling Techniques for Multi-Megabyte Caches " IEEE Transactions on Computers(1994)
19. D.A Wood, M.D. Hill, R.E. Kessier, "A model for Estimating Trace-sampling Miss Ratios" ACM SIGMETRICS Performance Evaluation Review 1991
20. Thomas M. Conte, Mary Ann Hirsch, Wen-Mei W. Hwu "Combining Trace Sampling with Single Pass Methods for Efficient Cache Simulation" IEEE Transaction on Computers 1998
21. Anant Agarwal, Minor Huffman, "Blocking: exploiting spatial locality for trace compaction", Proceedings of the ACM SIGMETRICS 1990

Workshop on Scientific Visualization and Human-Machine Interaction in a Problem Solving Environment

Oh Behave! Agents-Based Behavioral Representations in Problem Solving Environments

M. North[1], C. Macal, and P. Campbell

[1]Argonne National Laboratory, 9700 S. Cass Avenue, Argonne, IL 60439
{north,macal,campbell}@anl.gov

Abstract. The development of deregulated electricity systems around the world has produced the need for simulation systems that are capable of addressing the complexities that arise in the new markets. Agent-based models allow the use of complex adaptive systems approaches that are capable of producing tools or problem solving environments that can address the behavior of each of the participants within the electricity market. The agents in the tools are allowed to establish their own objectives and apply their own decision rules. They can be developed to learn from their previous experiences and change their behavior when future opportunities arise. In this paper, we will argue that the same type of agent-based technology that is used to produce "realistic" agent behavior in agent-based simulation tools at Argonne National Laboratory can also be used to embed these tools in problem solving environments.

1 Introduction

The development of deregulated electricity systems around the world has produced the need for simulation systems that are capable of addressing the complexities that arise in the new markets. As these electric utility systems continue to evolve from regulated, vertically integrated monopoly structures to open markets that promote competition among suppliers and provide consumers with a choice of services, the unbundling of the generation, transmission, and distribution functions that is part of this evolution creates opportunities for many new players, or agents, to enter the market. It even creates new types of industries, including power brokers, marketers, and load aggregators or consolidators. As a result, fully functioning markets are distinguished by the presence of a large number of companies and players that are in direct competition. Economic theory holds that this will lead to increased economic efficiency expressed in higher quality services and products at lower retail prices. Each market participant has its own unique business strategy, risk preference, and decision model. Decentralized decision-making is one of the key features of the new deregulated markets.

Agent-based models (ABMs) allow the use of complex adaptive systems approaches that are capable of producing tools or problem solving environments (PSE) that can address the behavior of each of the participants within the electricity market. The agents in the tools are allowed to establish their own objectives and apply their own decision rules. They can be developed to learn from their previous experiences and change their behavior when future opportunities arise.

P.M.A. Sloot et al. (Eds.): ICCS 2003, LNCS 2659, pp. 979–984, 2003.

> A PSE is a computer system that provides all the computational facilities needed to solve a target class of problems. These features include advanced solution methods, automatic and semiautomatic selection of solution methods, and ways to easily incorporate novel solution methods. Moreover, PSEs use the language of the target class of problems, so users can run them without specialized knowledge of the underlying computer hardware or software. By exploiting modern technologies such as interactive color graphics, powerful processors, and networks of specialized services, PSEs can track extended problem solving tasks and allow users to review them easily. Overall, they create a framework that is all things to all people: they solve simple or complex problems, support rapid prototyping or detailed analysis, and can be used in introductory education or at the frontiers of science [1].

An agent is a software representation of a decision-making unit. Agents are self-directed software objects with specific traits and typically exhibit bounded rationality, meaning that they make decisions using limited internal decision rules that depend only on imperfect local information. Emergent behavior is a key feature of ABMs. Emergent behavior occurs when the behavior of a system is more complicated than the simple sum of the behaviors of its components [2].

In this paper, we will argue that the same type of agent-based technology that is used to produce "realistic" agent behavior in the Electricity Market Complex Adaptive Systems model (EMCAS), and other agent-based simulation tools at Argonne National Laboratory (ANL), can also be used to embed these tools in a PSE-type environment – i.e. one in which all the intricacies of the underlying computer hardware and software are hidden from the user, who is then free to focus on modeling meaningful solutions.

2 EMCAS

EMCAS is an electricity market model related to several earlier models [3,4]. EMCAS includes a large number of different agents to model the full range of time scales – from hours to decades – that are needed to understand the domain [5]. The focus of agent rules in EMCAS varies to match the time continuum, as shown in Fig. 1. Over longer time scales, human economic decisions dominate. Over shorter time scales, physical laws dominate. Many EMCAS agents are relatively complex, or "thick," compared to typical agents. EMCAS agents are highly specialized to perform diverse tasks, ranging from acting as generation companies to modeling transmission lines, as shown in Fig. 2. To support specialization, EMCAS agents include large numbers of highly specific rules. EMCAS agent strategies are highly programmable. Users can easily define new strategies to be used for EMCAS agents and then examine the marketplace consequences of these strategies. EMCAS and its component agents are currently being subjected to rigorous quantitative validation and calibration.

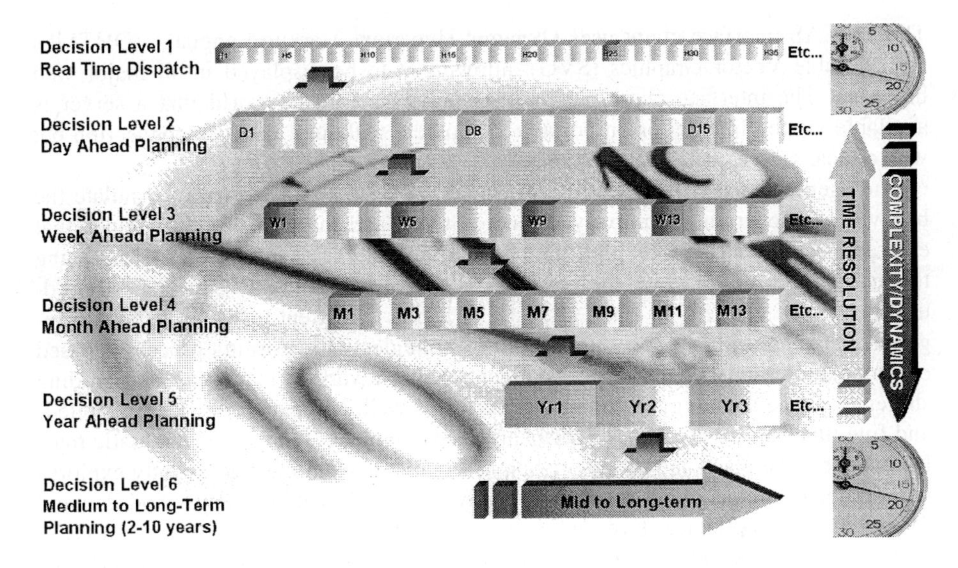

Fig. 1. EMCAS time scales and decision levels

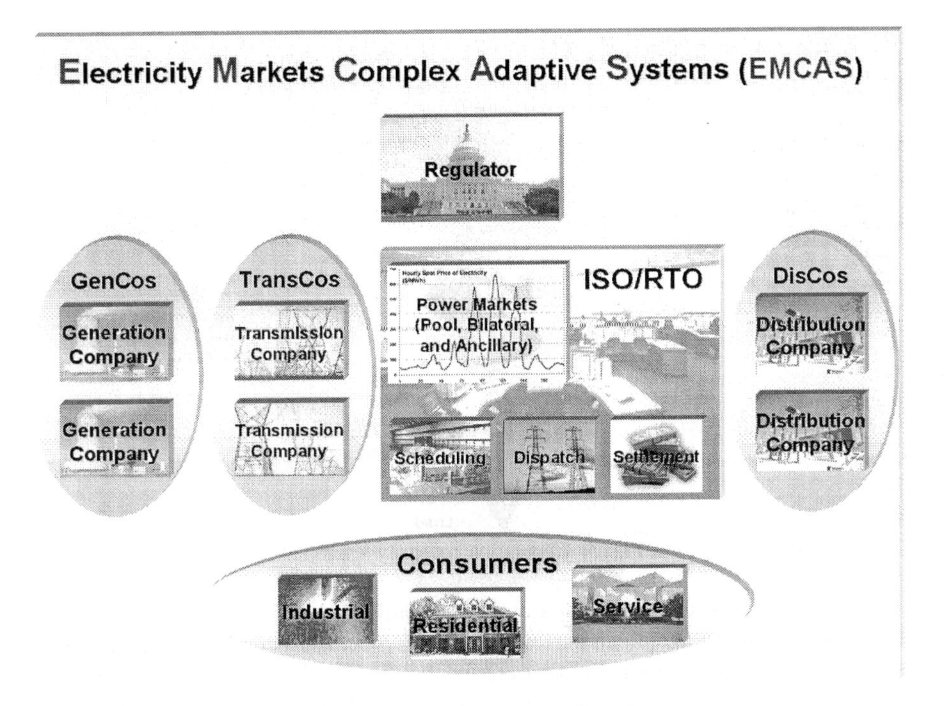

Fig. 2. EMCAS structure and agents

The EMCAS interface client uses Dynamic Hypertext Markup Language (DHTML) and Scalable Vector Graphics (SVG), allowing it to be displayed in all major web browsers. The interface client can be used anywhere in the world that a server is available via the Internet or on portable computers without a network connection but with a local server.

One agent class used in the EMCAS modeling system is designed to simulate the behavior of Generating Company Agents (GCAs) and the marketing strategies that emerge as GCAs strive to exploit the physical limitations of the power system using the market rules under which they operate, as shown in Figure 3. GCAs can sell products in various markets. In EMCAS, a GCA learns the extent to which local and regional prices are influenced by its marketing strategies. This learning process is based on an "explore and exploit" process. Agents explore various marketing and bidding strategies. Once a strategy is found that performs well, it is exercised (i.e., exploited) and fine-tuned as subtle changes occur in the marketplace. When more dramatic market changes take place and a strategy begins to fail, an agent more frequently explores new strategies in an attempt to adapt to the dynamic and evolving supply-and-demand forces in the marketplace. Even when a strategy continues to perform well, a GCA periodically explores and evaluates other strategies in its search for one that performs better. However, the exploration rate tends to be significantly lower than under stressful conditions.

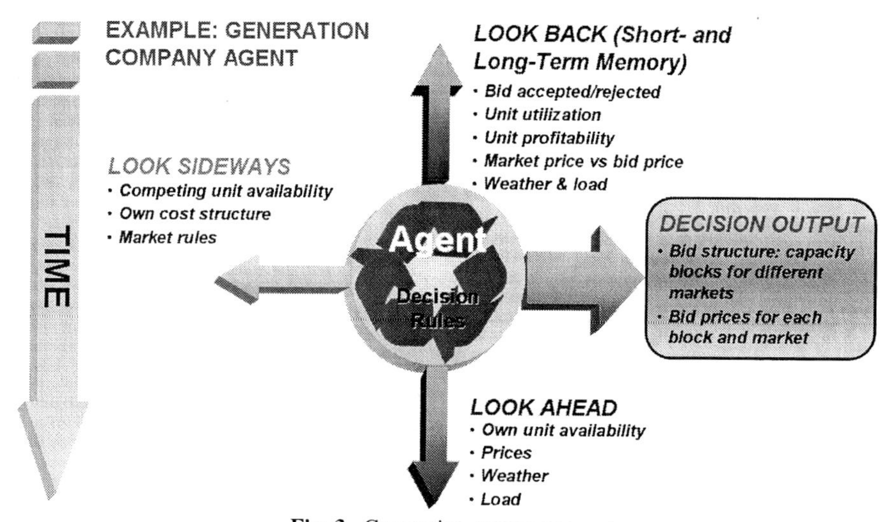

Fig. 3. Generation company agent

In EMCAS, a GCA is comprised of a number of building blocks that represent different tasks or actions an agent can perform. Each GCA seeks to arrange and parameterize these building blocks in a way that allows the market player to maximize its corporate utility. A building block consists of a set of one or more relatively simple rules. For example, one very simple agent rule may be if the GCAs sets the marketing clearing price in the last bidding period, then the GCA bid price in the next period will be fractionally higher." One parameter in this building block specifies the rate of change in the bid price. Although the basic building blocks are available to all GCAs, an exploited strategy may not utilize a building block if it is discovered that it is not beneficial. However, if market conditions change or if the GCA discovers a new way to combine the building block with another one, it can be used to develop a new strategy. When a GCA owns and operates more than one generating unit, an integrated strategy is formulated, and the combined effects of unit-level actions are important. This may entail losing money at one facility to gain more profits at another one.

We propose that this same agent building block approach can be used to develop elements of a PSE. In particular, the approach can be adapted to address automatic ontology construction/extension; personalization; and real-time visual representation of both the program state and the "object of interest" state within a PSE.

3 Automatic Ontology Construction and Extension

An EMCAS agent makes decisions based on past experiences and anticipated conditions in the future and in the context of current market rules and the potential impact that other players will have on the markets. In the same way, a PSE environment can be described within a particular domain. The possible/reasonable paths through the PSE can then be broken down into directed graphs of discrete steps, or building blocks, each corresponding to a function/action that allows the user to progress towards the goal of "solving" the problem. Analytical agents can then be constructed from the discrete steps in response to goals set by the user via the user interface. As the problem or analysis is worked through, the agent reevaluates its context at each step, assembling the necessary blocks as required. An ontology constructed this way can then be saved for further use in later PSE applications. Since, in use, each step is accompanied by an evaluation of the blocks needed for the next step(s), the ontology is extended by simply using the PSE. We have found it possible to build arbitrarily complex behavior paths using this approach.

3.1 Personalization

By providing a user interface that allows each user to enter their preferences, this initial personalization becomes part of the PSE environment that the agents automatically use to assemble the problem solving paths that are to be used. As the user continues to use the PSE, the agents learn preferred paths, or tool use, by the continual assessment of the internal PSE environment.

3.2 Real-Time Visual Representations within PSE

True flexibility in a user input-and-display environment can be achieved by having functions delegate these functions to other services. This can be achieved for most functions, but is perhaps most easily discussed/illustrated for the case of real-time visual representation. Meta-protocols have been developed that transparently link domain objects. This allows domain objects to publish available data and functionality at run time, and allows changes in the state of domain objects to be displayed as they occur.

Domain objects include both those responsible for functional behavior in the PSE and those objects that are the subject of the analysis – usually data objects of some kind. The user can therefore watch a display of the PSE elements evolving during use, improving their understanding of the analysis process, e.g. which solver, lookup table, data set, etc., is being used. At the same time, the change in state of each actual agent in use and the values that the agent represents can also be displayed.

4 Conclusion

ABMs allow the use of complex adaptive systems approaches that are capable of producing tools or PSEs that can address the behavior of each of the participants within complex systems. In this paper, we argued that the same type of agent-based technology that is used to produce "realistic" agent behavior in EMCAS and other agent-based simulation tools at ANL can also be used to embed these tools in a PSE-type environment.

References

1. Gallopoulos, S., Houstis, E., Rice, J., Computer as Thinker/Doer: Problem-Solving Environments for Computational Science, IEEE Computational Science and Engineering, IEEE, 11-23 Vol. 1, No. 2: Summer 1994.
2. Bonabeau, E., Dorigo, M., Theraulaz, G., Swarm Intelligence: From Natural to Artificial Systems, Oxford University Press: 1999.
3. VanKuiken, J.C., Veselka, T.D., Guziel, K.A., Blodgett, D.W., Hamilton, S., Kavicky, J.A., Koritarov, V.S., North, M.J., Novickas, A.A., Paprockas, K.R., Portante, E.C., Willing, D.L., APEX User's Guide (Argonne Production, Expansion, and Exchange Model for Electrical Systems) Version 3.0, Argonne National Laboratory: 1994.
4. Veselka, T.D., Portante, E.C., Koritarov, V.S., Hamilton, S., VanKuiken, J.C., Paprockas, K.R., North, M.J., Kavicky, J.A., Guziel, K.A.., Poch, L.A., Folga, S., Tompkins, M.M., Novickas, A.A., Impacts of Western Area Power Administration's Power Marketing Alternatives on Electric Utility Systems, Argonne National Laboratory: 1994
5. North, M., Koritarov, V., Boyd, G., Veselka, T.D., Macal, C.M., Conzelmann, G.C. Thimmapuram, P.R., E-Laboratories: Agent-Based Modeling of Electricity Markets," American Power Conference, Electronic Proceedings, PennWell Corporation: April 2002

JBeanStudio: A Component-Oriented Visual Software Authoring System for a Problem Solving Environment – Supporting Exploratory Visualization –

Masahiro Takatsuka

School of IT, The University of Sydney, NSW 2006 AUSTRALIA
masa@it.usyd.edu.au, http://www.it.usyd.edu.au/~masa/

Abstract. This paper discusses the benefits of a Component-Oriented visual software authoring system that provides the seamless integration of various software tools in a unified environment. It employs a visual component assembly paradigm for ease of construction, Java™and JavaBeans™component architecture for the open environment, and recursive development methods, all of which allow us to rapidly construct and share applications. Moreover, it is highly interactive and fully configurable in order to support exploratory visualization. This versatility has the potential to improve the integration of independently developed analysis tools and the dissemination of research findings.

1 Introduction

We have all witnessed the phenomenal technological progress in the fields of electronics engineering, and computer science[1][2]. This dramatic improvement has led many computer scientists and engineers to ever challenging problems involving supercomputing, parallel and distributed technologies, and much more complex algorithms that use these high-performance computing technologies. Moreover, the advances in other engineering fields such as aerospace and mechanical [3][4], has changed how massive scientific data are collected and processed. Computer scientists and Software Engineers are now in great demand for producing various types of computational tools to process and analyze these data.

1.1 Diversity of Computational Tools

Diverse groups of researchers have been building computational tools in order to support such datasets and computational demands for more sophisticated data analyzes and problem solving.

These tools are written in a variety of programming languages (C, C++, Visual Basic, FORTRAN, and Java), and also developed on different platforms (UNIX, Windows/Dos, Mac, etc.). The form of their deployment also varies.

P.M.A. Sloot et al. (Eds.): ICCS 2003, LNCS 2659, pp. 985–994, 2003.

Some tools are distributed as programming libraries and some are deployed as standalone applications. This often constitutes a burden for the end users. If a user is fortunate enough to have all his/her tools in library form and in the same environment, the user can write his/her own code to integrate various functions provided by the tools. If some of the tools are standalone applications, which require different platforms, a user has to move in and out of different environments to execute each tool (application). Some tool producers have chosen to use cross platform programming languages like Java^{TM}to avoid the intertwist of heterogeneous environments. However, these cross platform tools do still demand that a user integrates them at the coding level.

1.2 Need for Better Integration

Some generic mathematical and computational programming environments (such as Mathematica[1]and Matlab[2]) offer smooth integration by providing an infrastructure into which independently developed modules can easily be incorporated. Those environments, however, still require manual coding to integrate tools.

Problem Solving Environments (PSEs) have been used to address the issue presented above by providing application development systems that support the flexible deployment of analysis modules, often with a visual programming interface [5]. For instance, National Instruments' LabView[6], HP's visual programming tool: VEE[7]), OpenDX[3], IRIX Explorer[8] and AVS[9] provide a visualization-centered PSEs via visual programming interfaces. They also allow users to create their own computational/visualization modules using an application's framework. Most of the current solutions to seamless integration rely on the use of proprietary interfaces or frameworks. Independent development and deployment of tools are possible in those infrastructures but only within the particular application environment. So, for example, some analysis tools developed for AVS cannot be smoothly integrated with tools from OpenDX without modifications.

Moreover, many such environments are less flexible in configuring how data and pieces of information are exchanged. Many PSEs are interactive and configurable in the sense that a user can combine various tools in accordance with their needs. In visualization-oriented PSE, however, more flexibility should be provided at a lower level than at the level of just selecting tools and their configuration. At the lower level, a user would be able to interactively configure how data and pieces of information should be exchanged between tools. In many PSEs, tools are combined by strongly-typed functions/methods. Tools usually exchange numerical/nominal values and more complex structured data of the same type. In order to encourage interactive exploratory visualization, the process of configuring how each piece of data is mapped onto a visual variable needs

[1] http://www.wolfram.com
[2] http://www.mathworks.com
[3] http://www.opendx.org

to be provided. This is because users often do not know how best to visually represent the given data.

This project aims to provide an interactive and fully configurable visual software authoring system based on open-standards so that independently developed and deployable analysis and visualization components can be seamlessly integrated in a non-proprietary environment.

2 JBeanStudio

JBeanStudio is a component-oriented software authoring system. Users utilize its visual environment to rapidly wire components together into a useful application. It is called an *authoring* system because it does not generate code but rather generates running software on the fly. The application is always "live" while it is designed, so a design can always be changed as a consequence of using the constructed application and evaluating the results produced. Such seamless integration of design and runtime environments enhances productivity since it allows rapid adaptation of the program in accordance with new analysis requirements. It fully utilizes the unique characteristic of a component:

> "...a unit of composition with contractually specified interfaces and explicit context dependencies only. A software component can be deployed independently and is subject to composition by third parties." Szyperski and Pfister, 1997[10]

The feature *independent deployment*, or more specifically, *runtime deployment* is the key feature which needs to be exploited. If components could be imported into a system (which an end user is using) at the runtime, the user would not need to go through the coding and compiling processes in order to enjoy the advantages of the components. Consequently, it allows highly interactive and customizable program development. This is important since there are often as yet no rules as to how best to connect or configure components in order to perform the best problem solving tasks.

JBeanStudio offers:

1. Ease of use, but with the capability to construct complex analysis applications,
2. Rapid development and modification of applications, minimizing programming requirements,
3. Support for sharing and exchanging of developed applications.

The first two features allow many computational and other scientists to focus on actual problems and not on the underlying programming logic that provides these tools and facilitates their dynamic interactions. They are therefore free to fully utilize their domain knowledge. The third feature is even more important for peer review (replication of research results) and dissemination of research findings[11].

2.1 System Architecture

JBeanStudio is built around the well-established, open standard Java™programming language[12] and JavaBeans™component architecture[13] forming a layered model as shown in Figure 1. There are a few technologies which can be used to build component-oriented systems such as OMG's CORBA and Microsoft's .NET. However, Java has been accepted in various fields due

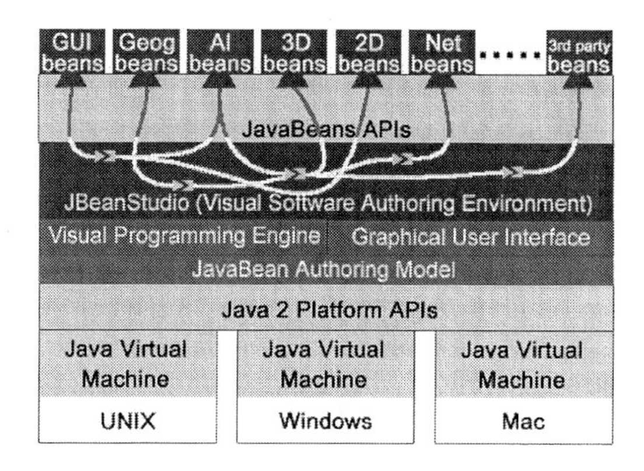

Fig. 1. System Architecture

to its many benefits, such as cross platform support (the bottom layer in Figure 1) and ease of use, as evidenced by the growing transition from C++ to Java. JavaBeans technology provides a standard Application Programming Interface(API) to make reusable components in the Java programming language. The combination of Java and JavaBeans provides better support for Component-Oriented Programming (COP). It is known that C++ does not directly support the concept of COP[14].

A program in JBeanStudio is constructed from building blocks - beans - (see the connections made at the layer of the visual authoring environment in Figure 1). By combining many components, more complex programs can be developed. As long as beans are created according to the JavaBeans standard, one can assume that they will work together with other beans created by different software vendors or individuals (see the top layer of Figure 1).

There are few applications similar to JBeanStudio. Sun Microsystems' BeanBox and BeanBuilder [15] are examples utilizing standard Java and JavaBeans component architecture. However, they are only a reference implementation to illustrate the possibilities of visual programming using Java and JavaBeans component architecture. They present various shortcomings to be addressed in order to fully support exploratory visualization and probably data analyses. Some of those problems are:

1. information/data exchanges between components are strictly via Java's event model. This means a component has to pass some pieces of information in the form of an event object. This implies that a special event needs to be created for each information which needs to be passed.
2. only subclasses of a particular Java component (java.awt.Component) can fully utilize the Java's event model in such an environment. In other words, if an event originates a Java software component, which is not a subclass of the java.awt.Component, the system fails to properly recognize the event dispatcher and listener relationships.
3. when a pieces of information is passed onto another component, the data type has to match between a sender and a receiver. For example, an integer value cannot be sent to a method (function) which takes a real number as its argument.
4. all events are handled by a single thread. It is almost impossible to create an application with loops.

Other visualization-oriented visual programming environments like AVS, IRIS Explorer and OpenDX provide excellent Problem Solving Environments and programming-based tools like Visualization Tool Kit (VTK)[16], but each of them forms its own proprietary environment making it difficult for components from different environments to work together.

3 Component Wiring and Dataflow

There are two types of dataflow models common to the above mentioned systems: 1) event-driven and 2) demand-driven. Most visual programming based systems use event-driven dataflow. This approach naturally follows the cause-effect paradigm and provides a very natural component wiring mechanism to users [17]. The demand-driven approach is often used when precise program control of how to execute branching and conditional execution is needed[18].

JBeanStudio aims to provide a more complete visual software authoring system using a hybrid approach (event-driven demand execution). This approach addresses the previously mentioned shortcomings by providing:

1. five types of information passing mechanisms - using 1) a standard Java event object, 2) event object's accessor method, 3) event dispatcher's accessor methods, 4) accessor methods of another object, which is not an event dispatcher, and 5) a constant value. This means that the event receiver uses the event as a simple trigger to obtain extra pieces of information from various sources.
2. a generic event handling - any JavaBean component can dispatch and receive events. A component does not need to be a subclass of a particular object or to use proprietary interfaces. This means any component can send and receive a message via Java's event model.
3. a data converter manager - it runs in the background and automatically converts one data type to another when an appropriate data converting

module is present in the system. It provides a basic set of data converter modules but a user can register any kind of customized data converters at the run-time in order to support custom data types.
4. multi-threading - each event dispatch-receive connection is executed in a separate thread. Hence, the execution of a loop and branching control will not halt or delay executions of other parts of an application.

4 Interactively Building a Visualization Application

JBeanStudio was designed to unify component-assembly and runtime environments. When JBeanStudio is first launched, it presents its main window with various menus and bundled components (see Figure 2(a)) along with an empty design and GUI windows.

Each component is stored in a folder of a component palette. A user can customize and create folders and palettes as well as load palettes from XML files. This feature is useful when customized components and/or groups of components need to be shared among colleagues. It also allows a user to import any JavaBeans components developed by third parties.

In JBeanStudio, a user designs an application by dragging components from the palettes and dropping them onto either the design or GUI window. Once components are placed in the design they can be customized and wired together as shown in Figure 2(b). Due to the visual assembly mechanism, no programming is required; in fact code writing is not supported (the user can write the code for beans outside of JBeanStudio). This makes it different from other development tools in which a user has to write some code to produce software components, applets or applications; hence our users (especially non-programmers) should find development less burdensome. The design of the application is saved using XML. This application is then ready to be distributed over the net. Since the whole design information is included in the saved design file, the other users can improve and/or customize the application further to suit their own needs.

As described before, JBeanStudio integrates the environments for building and running a program and an application is always running while it is being designed (see Figure 2(c)). Hence, a user can adopt the design in it as a final working program. However, the designer of the program might need to package the design into an applet or a standalone application as a customized product for other users. Alternatively, the user might want to package the design into a JavaBeans component for inclusion in a larger application. To support creation of Java beans, applets and applications, it provides functionality that does all of the source code generation and compiling work. This is the only function in JBeanStudio involving code generation. It automatically generates Java source code for the designed application, compiles it and packages it up into an archived file for distribution.

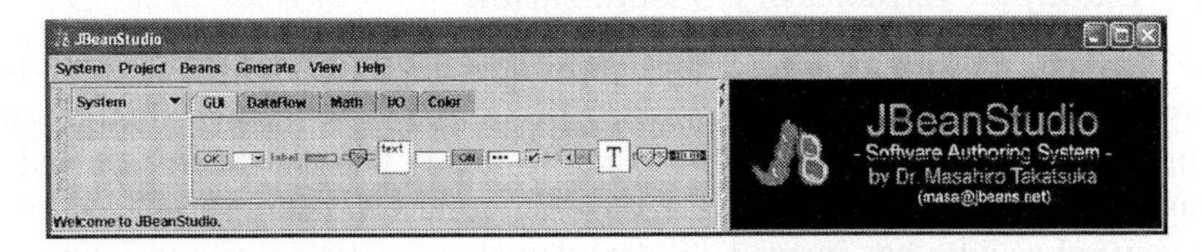

(a) Components are organized in the palettes and folders.

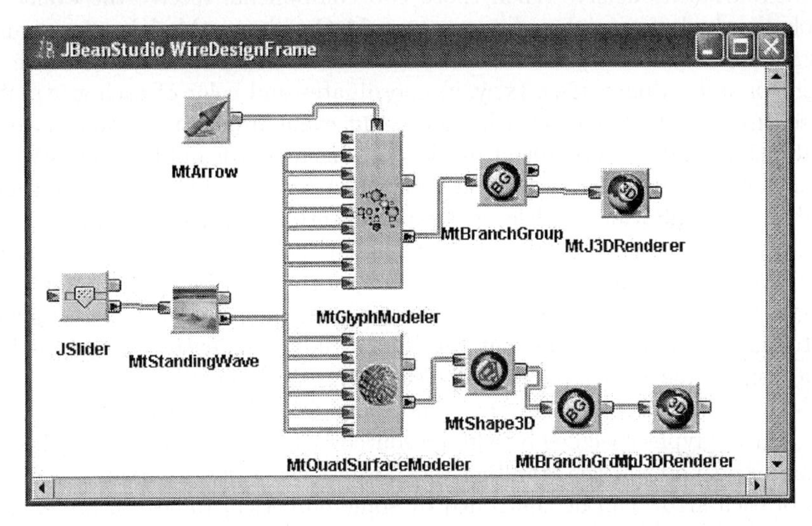

(b) Components are wired together in the design panel.

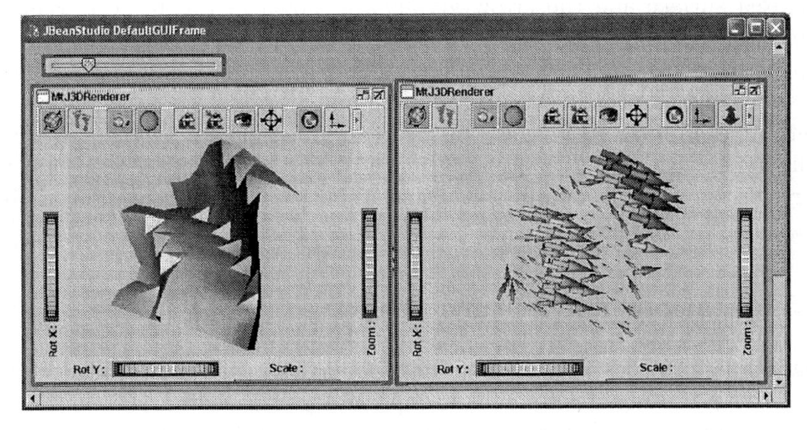

(c) GUIs of components are displayed in this panel while an application is designed and executed.

Fig. 2. JBeanStudio's Main, Design and GUI panels.

4.1 Example: Exploratory Visualization

The example shown in Figure 2 illustrates how JBeanStudio provides an interactive exploratory visualization system. The application in the example simply visualizes a dynamically changing standing wave. The component named MtStandingWave generates a standing wave in an $M \times N$ grid format according to the value of a slider bar (named JSlider). Upon generating a new standing wave, MtStandingWave sends an event to other components (MtGlyphModeler and MtGridSurfaceModeler). When those two components receive the event, they invoke methods from other data sources. MtGridSurfaceModeler, for example, has six input methods, each of which is used to compute a grid surface with colors (x and y dimensions, (x, y, z) coordinates and color of each grid). When those input methods receive a branched out event from MtStandingWave, MtGridSurfaceModeler component invokes various methods in MtStandingWave to retrieve various data for each input method. After obtaining all the information needed, MtGridSurfaceModeler generates a synthetic grid surface and send it to a 3-D renderer component. It should be noted that the source of the demanded information does not have to be the event source. In other words, MtGridSurfaceModeler can invoke a method to obtain, say, the color information for each grid from a different component. In JBeanStudio, the event can be used as a data transfer mechanism as well as just a triggering signal.

MtGlyphModeler receives the same event from MtStandingWave. It, however, places glyphs specified by a glyph source (MtArrow component in this case) at each grid point of the standing wave. Moreover, the size, color and orientation of each arrow can be controlled by some numerical values. In the example, color and z (height) values of the wave were fed into MtGlyphModeler's method from MtStandingWave in order to compute the orientation of arrow glyphs. The generated surface and the arrow glyphs are shown in Figure 2(c). While this simple visualization system is displaying the dynamically created surface and a set of arrow glyphs, one can re-configure how events are passed around and how demanded pieces of information are obtained. In other words, while a user is observing the produced images, the user can re-assign various numerical values to different visual variables (orientation of arrow, size of cone head and body, color, etc.) in order to obtain the best possible visual representation of his/her data.

5 Limitations and Future Works

JBeanStudio provides a visual programming environment and the flexibility and interactivity of being able to build an exploratory visualization system. However, since all data/information exchanges need to be visually specified, a novice user, who might have no experience in programming, might find JBeanStudio still difficult to use (program). However, the aim of the very first stage of JBeanStudio development is to provide a core visual authoring environment, which can be used to build a further higher-level framework. Hence, a team of users can develop a framework which defines a basic information exchange mechanism so that the connections that need to be made between multiple software components can be minimized. The approach of building an extra framework on top of JBeanStudio would be useful if a set of components needs to be closely coordinated.

Moreover, it currently does not have any component layout algorithms. Therefore, the design can be very cluttered and hard to read once it gets large and complicated. The use of an automatic clustered graph layout algorithm [19] is now under investigation in order to improve the user interface of JBeanStudio.

6 Conclusion

This paper describes the use of JBeanStudio as one example utilizing an open component-oriented infrastructure that provides seamless integration of various computational tools. It offers an easy-to-use, visual assembly environment that supports rapid construction of sophisticated software. This allows scientists and engineers to concentrate on solving their domain problems rather than dealing with programming.

JBeanStudio employs Java-based component architecture. Any JavaBeans components can be wired together in this environment to form a larger application. Those components can be independently developed without being concerned with the integration environment (JBeanStudio). Moreover, it is capable of automatically producing a JavaBeans component (which can also be a Java applet or an application) from a design. This allows a user to create a reusable and scalable system as well as to distribute and share the created design on the Web. The flexibility of being able to customize an application and the seamless transition from a desktop data analysis/visualization program to a Web application allows scientists to freely apply and experiment with their knowledge in scientific analyses and share results instantly and accurately.

References

1. Ceruzzi, P.E.: A History of Modern Computing. MIT Press, Cambridge, Mass. (1998)
2. Geppert, L.: The 100-million transistor IC. IEEE Spectrum **36** (1999) 22–24
3. NASA: TERRA: The EOS flagship. http://terra.nasa.gov (1999)
4. European Space Agency (ESA): Envisat. http://envisat.esa.int (2002)
5. Rice, J.R., Boisvert, R.F.: From scientific software libraries to problem-solving environments. IEEE Computational Science and Engineering 3 (1996) 44–53
6. Baroth, E.C., Hartsough, C., Johnsen, L., McGregor, J., Powell-Meeks, M., Walsh, A., Wells, G., Chazanoff, S., Brunzie, T.: A survey of data acquisition and analysis software tools, part. Evaluation Engineering Magazine (1993) 128–140
7. Hewlett Packard: VEE: Visual engineering environment. technical data 5091–1142EN, Hewlett Packard (1991)
8. Numerical Algorithms Group (NAG): Iris explorer user's guide. http://www.nag.co.uk/visual/IE/iecbb/DOC/html/nt-ieug5-0.htm (2000)
9. Systems, A.V.: Information visualization : Visual interfaces for decision support systems. http://www.avs.com/ (2002)

10. Szyperski, C., Pfister, C.: Workshop on component-oriented programming, summary. In Mühlhäuser, M., ed.: Special Issues in Object-Oriented Programming – ECOOP96 Workshop Reader. dpunkt Verlag, Heidelberg (1997)
11. Takatsuka, M., Gahegan, M.: Sharing exploratory geospatial analysis and decision making using GeoVISTA studio: From a desctop to the web. the Journal of Geographical Information and Decision Analysis (JGIDA) 5 (2001) 129–139
12. Joy, B., Steele, G., Gosling, J., Bracha, G.: The Java^{TM} Language Specification. Second edition edn. The Java Series. Addison-Wesley Pub Co. (2000)
13. Sun Microsystems Inc.: The javabeans^{TM} 1.01 specification (1997)
14. Szyperski, C.: Component Software: Beyond Object-Oriented Programming. ACM Press and Addison-Wesley, New York (1999)
15. Microsystems, S.: BeanBox and BeanBuilder. http://java.sun.com/products/javabeans/software/ (2002)
16. Schroeder, W., Martin, K., Lorensen, B.: The Visualization Toolkit: An Object-Oriented Approach to 3-D Graphics. 2nd edn. Prentice Hall PTR (1998)
17. Abram, G., Treinish, L.A.: An extended data-flow architecture for data analysis and visualization. In: Proceedings of the 1996 IBM Visualization Data Explorer Symposium, CA, IBM, IBM Corporation (1996) http://www.research.ibm.com/dx/proceedings/proc.htm.
18. Schroeder, W.J., Martin, K.M., Lorensen, W.E.: The design and implementation of an object-oriented toolkit for 3d graphics and visualization. In: Proceedigns of Seventh Annual IEEE Visualization '96, IEEE Computer Society, IEEE Computer Society (1996) 93–100
19. Pulo, K., Takatsuka, M.: Inclusion tree layout convention: An empirical investigation. In Pattison, T., Thomas, B., eds.: Volume 24 – Information Visualization 2003, Australian Computer Society, Australian Computer Society (2003) 27–35

Multi-agent Approach for Visualisation of Fuzzy Systems

Binh Pham and Ross Brown

Faculty of IT, Queensland University of Technology,
GPO Box 2434 Brisbane AUSTRALIA
{b.pham,r.brown}@qut.edu.au

Abstract. Complex fuzzy systems exist in many applications and effective visualisation is required to gain insights in the nature and working of these systems, especially in the implication of impreciseness, its propagation and impacts on the quality and reliability of the outcomes. This paper presents a design of a visualisation system based on multi-agent approach with the aim to facilitate the organisation and flow of complex tasks, their inter-relationships and their interactions with users. This design extends our previous work on the analysis of the fundamental ontologies which underpin the structure and requirements of fuzzy systems.

1 Introduction

Many real world problems can be represented as complex fuzzy systems which may involve a large amount of fuzzy data, fuzzy variables and fuzzy relationships. Fuzzy logic has been used extensively to model these systems in many application areas, ranging from engineering, science, medicine to environmental planning and social sciences [13]. While mathematical models are based on algebraic operations (equations, integrals), logic models rely on logic-type connectives (*and*, *or*, *if-then*), often with linguistic parameters, which give rise to rule-based and knowledge-based systems. Fuzzy logic models can combine both of these types of modelling via the fuzzification of algebraic and logical operations [1]. There are three common classes of fuzzy logic models: *information processing models* which describes probabilistic relationship between sets of inputs and outputs; *control models* which control the operations of systems governed by many fuzzy parameters; and *decision models* which model human behaviour by incorporating subjective knowledge and needs, using decision variables [6].

For some applications, fuzzy systems often perform better than traditional systems because of their capability to deal with non-linearity and uncertainty. Another advantage is that linguistic rules, when used in fuzzy systems, would not only make tools more intuitive, but also provide better understanding and appreciation of the outcomes. However, the complexity arisen from information uncertainty makes it more difficult for humans to understand the way these systems work, especially how

P.M.A. Sloot et al. (Eds.): ICCS 2003, LNCS 2659, pp. 995–1004, 2003.

to interpret the implication of the impreciseness of each variable on its interaction with other variables, and how the propagation of such impreciseness affects the level of confidence in the outcomes at every stage.

Visualisation has been used extensively during the last decade, but the bulk of research work has been focused on those systems which involve crisp data and crisp relationships. A few current approaches have some limitations due to either their ad hoc nature, or their ability to deal with only a specific aspect of the problem of visualisation of fuzzy systems [2, 4, 5, 8, 10, 11]. In addition, visualisation methods are often focused on data sets and only loosely coupled with the analytical process. It is left to users to decide how they deploy those visualisation tools provided.

The usefulness of a visualisation system would therefore be enhanced if it is driven primarily by those tasks that need to be performed, and not by data sets because such a system would link more tightly with the analytical process which underpins human understanding and decision making. Another aspect that needs to be considered is how to cater for different types of users.

In a previous paper [3], we presented a thorough analysis on visualisation requirements for fuzzy systems by investigating the fundamental ontologies which underpin the structure and requirements of these systems. This paper focuses on the design of a multi-agent based visualisation framework with the aim to facilitate the organisation and flow of complex tasks, their inter-relationships and their interactions with users.

Section 2 discusses briefly the characteristics of fuzzy systems and their visualisation requirements. Section 3 provides the motivations behind the multi-agent approach and an overview of the system. Subsequent sections present the structure and activities of each of these agent classes, and plan for their implementation.

2 Fuzzy Systems and Their Visualisation Requirements

To design an effective generic framework for visualization of fuzzy systems, we need to understand their essence: what are they composed of, how things are related to each other, what activities are being performed, and who are the main users of these systems. To facilitate the understanding of the rest of this paper, we briefly describe these requirements here. More detailed analysis on these requirements was presented in [3].

A typical fuzzy system consists of 6 main components: entities, data objects, relationships, events, tasks and outcomes. The *entities* include both physical (e.g. machines, workers) and abstract (e.g. returns of investment). *Data objects* may be represented in different forms: numerical, symbolic (e.g. rules), visual (e.g. diagrams, images) or audio. *Relationships* which underpin the working of a fuzzy system can be classified into five categories: data-data, data-task, data-user, task-task, and user-user. Each of these categories needs to be examined carefully in order to find appropriate visualization methods to facilitate the understanding of these relationships. *Events* change the system state and exert influence on the system performance, hence it is crucial to note and record them. To distinguish the level of complexity of *tasks*, they can be grouped into low-level and high-level tasks. The former includes the computation of numerical data, degree of fuzziness and the operation of fuzzy rules. The latter covers the detection of unusual patterns, data mining, learning process, optimization and prediction. The *outcomes* of a fuzzy system include not only the values of state variables, but also the level of acceptance of quality, degree of confidence, and degree of impreciseness of the outcomes.

We wish to examine the visualization requirements for fuzzy systems from user- and task-oriented points of view, so that a user is allowed to interact and select what to visualize and how to do it on the fly. Thus, it is also necessary to distinguish three main types of users and their different needs. The *users of fuzzy systems* wish to be able to interpret data and its salient characteristics, to understand the implication of each decision by setting up 'what-if' scenarios, and to adapt the system to their individual needs and preferences. The *designers of fuzzy systems*, on the other hand, require information on the internal structures of these systems for planning, verification and analysis. They also seek for conditions under which optimal solutions are obtained at each stage. The *designers of visualization systems* wish to understand how users make use of visualization techniques and the effectiveness of these techniques with the intention to identify drawbacks and to find ways to continuously improve the systems. We categorise four main types of visualization tasks:

- *Interactive exploration* to provide insights into: the degree of uncertainty of each variable and its effects on each task; the inter-dependency of two or more fuzzy variables or fuzzy rules; and the effects of different operations performed on fuzzy rules.
- *Automatic computer-supported exploration* to automatically highlight salient characteristics and unusual results; to display and compare alternatives (e.g. using statistical analysis); to optimize tasks under specified constraints; and to support batch processing of tasks via scripting or visual languages.
- *Capturing feedback from users* such as instructions on tasks; input parameters, variables, constraints; users' preferences, judgements and desired degree of fulfillment of outcomes in qualitative forms.
- *Capturing users' profiles and adaptation* in order to re-organise data and re-prioritise tasks to suit; and to automatically provide tasks and data according to detected patterns.

3 A Multi-agent Visualisation Framework

Our aim is to design a systematic framework based on a high level of abstraction, where visualisation is driven by users' needs which in turn are driven by application tasks and personal view points. Search and navigation methods and tools should be context-sensitive and should operate only on relevant information space. Thus, data should be organised according to task requirements to ensure efficiency.

To this end, we propose a visualisation framework based on 5 classes of agents: control agent, computation agent, symbolic agent, visualisation agent and profile agent. Fig. 1 shows a schematic diagram of the system architecture.

The *control agent* receives users' input which includes specifications, queries and parameters. Based on such input, this agent distributes tasks to appropriate agents. It also receives results and demands from other agents when a task is completed or when further information is needed. Another duty for this agent is to generate new tasks if required based on the results sent by other agents. The control agent may be viewed as a representative of the user in an automatic mode. In our model, the user can be included in the loop and allowed to intercept the control agent in order to give different instructions if desired. The user can also intercept other agents to select different methods for performing an operation instead of the default ones built into the system. The *computation agent* performs all numerical computation required by the system (e.g. statistics, probabilistic calculus, rough set operations, fuzzy set operations). It receives instructions from both the control agent and the visualisation agent. The *symbolic agent* makes use of the knowledge base to performs rule inferencing. It receives instructions from both the control agent and the visualisation agent. The *visualisation agent* receives instructions from the control agent and request information from the computation agent and rule agent in order to select appropriate visualisation techniques to provide displays. The results of the display then trigger the control agent or the user to issue another task. Another cycle then continues.

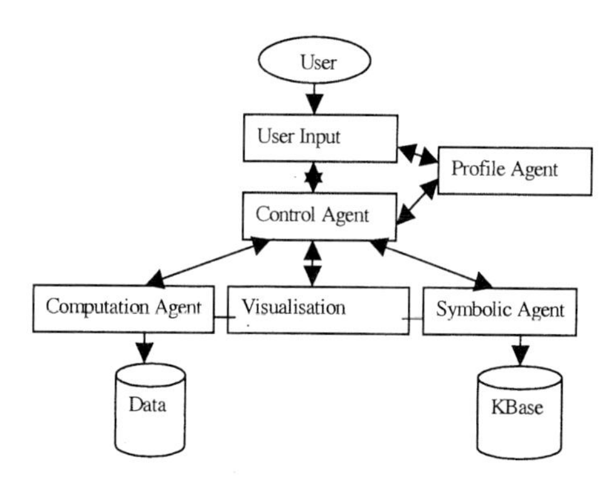

Fig. 1. System overview diagram.

The *profile agent* records the pattern of the user's behaviour in terms of the selection of tasks, visualisation techniques, numerical methods or inference rules. Based on this information, the profile agent then modifies the instructions issued by the control agent (e.g. re-prioritise tasks, change preferences, modes of display, etc.).

The following subsections show, using an object-based paradigm, the general methods within the agents as a specification of their functionality.

3.1 Control Agent

The control agent maintains a list of the other agent instantiations within the system and thus is able to control the flow of data around the agent-based visualisation system. The main tasks of the control agent are to process user events, distribute tasks to other agents and to process agent results and demands. Therefore it communicates with the Profile, Computation, Visualisation and Symbolic Agents. It is a form of automatic user within the system. However, the user can override its tasks through the *ProcessUserEvent* method.

Description	Parameters	Outputs
ProcessUserEvent-processes events generated by the user and other agents in the system.	*Specifications*–Choice of actions by the user. *Parameters*–for the specified action by the user.	
DistributeTask-takes any tasks as defined by the user and other processes and distributes them to other agents within the system.	*AgentID*–ID number of the agent to be notified. *TaskParameters*–parameters for the task to be sent to the agent.	
ProcessAgentResults-takes the results from agents and decides on next step in visualisation sequence. Involves distributing further tasks to other agents.	*AgentID*–ID number of the agent returning results. *TaskResults*–results returned by the agent.	*AgentID*–ID number of agent to pass on new task. *TaskParameters*–parameters for the new task.
ProcessAgentDemands-receives demands from other agents in the system to perform a task. Involves distributing further tasks to other agents.	*AgentID*–ID number of the agent making a demand. *DemandParameters*—parameters for the demand by the agent.	*AgentID*–ID number of the agent to accept the new task. *TaskParameters*–parameters for the new task.

3.2 Computation Agent

The main items stored by this agent includes the data to be visualised within the system, and a complete list of operations that can be performed by the computation agent. It is not an autonomous agent, as it is entirely controlled by the visualisation and control agents.

The computation agent processes the data upon requests from the control and or visualisation agent–for example, statistics, probabilistic calculus, rough set operations, fuzzy set operations, etc.

Description	Inputs	Outputs
ProcessData-processes the data according to user specifications or agent specifications in the parameters.	*DataOpID*–ID number of computation to enforce on data. *ProcessParameters*–parameters for computation, including data to load.	*ProcessResults*–results of computations.

3.3 Symbolic Agent

This agent is the interface to the knowledge database for the visualization system. It is used by the control agent and the visualization agent. It is not autonomous, as it simply provides a front end query interface to the knowledge database.

The main function of the symbolic agent is to process queries directed at the knowledge database. This database contains knowledge of appropriate visualisation techniques for the fuzzy data. Thus, the agent returns the appropriate information about techniques, parameters to use etc., as responses to queries from the control and visualisation agents.

Description	Inputs	Outputs
ProcessQuery-a query is made upon the knowledge base within the symbolic agent and it then returns inferences for the other agent to enact within the visualisation task.	*AgentID*–ID number of agent seeking inference from knowledge base. *QueryParameters*–parameters for the knowledge base query.	*QueryResults*–results returned from knowledge base query.

3.4 Visualization Agent

The visualization agent handles the graphical rendering of data to the output device. This agent draws direction from the symbolic agent, and is able to thus recommend a visualisation technique automatically, based upon the qualities inherent in the data. The data is received from querying the computational agent, which is directed by the visualisation agent to provide the data in a valid format for the visualisation technique.

There is only one major function listed, as the visualisation agent is fairly autonomous in its ability to organise a visualisation of data. Any information required for the visualisation is queried from the computational and symbolic agents, by using their querying methods. The control agent can, at the behest of the user, override the visualisation agent by making a call to the VisualiseData method.

Description	Inputs	Outputs
VisualiseData-produces a visualization of the data passed in to this function.	*VisOpID*–ID number of visualization technique to use. *VisParameters*–parameters for chosen visualization technique including the data to visualise.	*VisResults*–results of visualisation in form of image or movie.

As an example, we trace the execution of the visualization of the Iris data shown in a previous paper, which is a commonly used test data set for classification algorithms. The classification is done based on a training set of 75 plants, 11 fuzzy rules, 4 features (sepal length, sepal width, petal length and petal width) and 3 classes. In this case the user commences the system and inputs the Iris data file as the beginning of the visualization. The control agent then commences the dialog by noting the multidimensional nature of the data to be visualized. Information about the nature of the visualization is elicited from the user, who specifies a visualization for rule culling purposes. The control agent then passes this information onto the visualization agent.

The visualization agent then queries the symbolic agent for suggested visualization techniques. The symbolic agent replies with a suggestion of using the parallel coordinate visualization technique as shown in Fig. 3 (Left) [2]. In this method, n Cartesian coordinates are mapped into n parallel coordinates, and an n-dimensional point becomes a series of (n-1) lines connecting the values on n parallel axes. The visualization agent then requests the fuzzy data in an appropriate format for the parallel coordinate technique. Therefore, the visualization agent commences the rendering of the 2D parallel coordinate visualisation. However, the user requires a 3D version of the visualization, and chooses this using the appropriate menu options.

The control agent at this stage interrupts the visualization agent thread, and then enforces a 3D parallel coordinate visualization, as shown in Fig. 3 (Right). This is then rendered by the visualisation agent, whereupon control is given back to the user to interact with the visualization. This process is then repeated until termination by the user.

4 Visual Features

In previous work we have analysed in detail the mapping of various visual features to visualisation of fuzzy logic information [3]. In this section, we summarise some of the most importance features and show examples of their mappings to visualisation tasks.

Hue is heavily used to highlight data that is different, or to represent gradients in the data [9][12][8]. For fuzzy data, it can also be used to categorise the membership of a particular data point. In the example in Fig. 2 (Left), we see that the membership of different fuzzy terms can be illustrated by different hues. The region where the colours overlap indicates intuitively the location where these membership functions share areas of the domain.

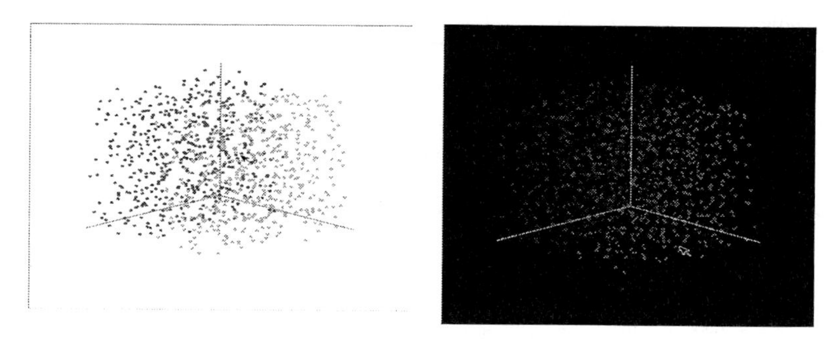

Fig. 2. (Left) Different colours used to indicate membership of different membership function terms. In this case, the red indicates slow motion, while the green indicates fast motion. (Right) Intensity difference as an indication of the membership degree for a fuzzy membership function Hot.

Luminance may be used to signify categories and highlight differences within scalar data. Luminance can also be used to directly indicate the Degree of Fulfilment (DOF) value for a single membership function. In the example in Fig. 2 (Right), the intensity represents the DOF value for a fuzzy function of the term Hot. The regions with the highest intensity indicate where the term Hot has its highest DOF value.

A number of other visual features can be utilised to indicate the precision of fuzzy data used in a visualisation, when applied to the visualisation objects:

- *size*-can be used to indicate the imprecision of data [12];
- *transparency*-can show the possibility of the fuzzy variable;
- *texture*-can indicate the level of precision, ambiguity or fuzziness;
- *glyphs* and *icons*-for example data points with error bars indicate imprecision;
- *particles*-represent the fuzziness of a region by varying the space between them;
- *blurring*-can be used to show the indistinct nature of data points [7].

5 Higher Spatial Representations

The visual features listed above are usually spatially arranged to form a coherent display in graphic forms, which enable the perception of various patterns in the data. We can combine the use of such visual features for denoting the imprecision in data with a number of common representation methods employed to display spatial data in higher dimensions: 2D, 3D, parametric, dynamic, metaphors and multimedia sensors. These methods have already been discussed in [3] and are not repeated here. Instead, we now discuss ways by which we can improve the visualization method provide by [2] by using parallel coordinates. The authors used the thickness or grey intensity of lines to indicate the fuzziness of points. One drawback is that it is difficult to visually distinguish fine grades of grey level on single lines. Another drawback is that it is not possible to perceive the core and support of a fuzzy set simultaneously.

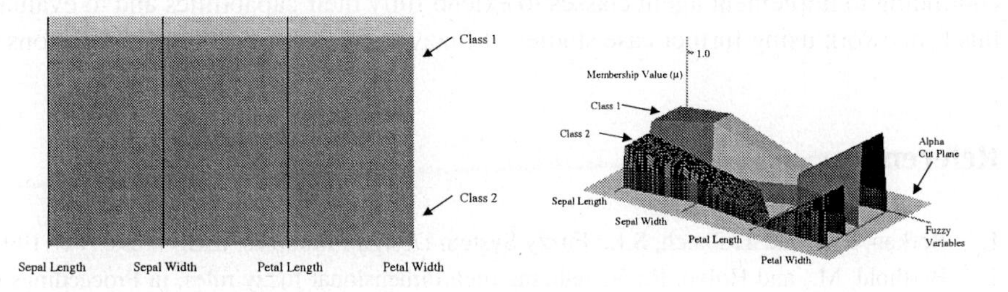

Fig. 3. (Left) Illustration of 2D method developed by Holve and Berthold, for representing multidimensional fuzzy rules using fuzzy parallel coordinates [2]. Two rules are illustrated from their Iris data example. (Right) Visualisation of the Iris data with lit and textured surfaces showing the same Iris data. Note how the alpha cutting of the membership function for Rule 2 on the Petal Length dimension is easily perceived.

One way of addressing these drawbacks is to use a 3D representation where the parallel coordinates are displayed on the x-y plane, and the fuzzy set membership functions are displayed in the z-coordinate (Fig. 3 Right). Different alpha-cuts of fuzzy rules can be identified by applying horizontal cutting planes. The separation of classes based on the confidence of a decision can be highlighted by using filled polygons with texture (Fig.3 Right) or using colour.

6 Profile Agent

This agent records a user's profile in various ways: patterns of tasks performed; patterns of usage of data and operations applied on fuzzy rules; specific types of constraints; desired degree of fulfilment of outcomes; and choice of visualization techniques. By communicating with both the user and the Control Agent, the Profile Agent uses these detected patterns to issue instructions to re-organise data and re-prioritise tasks. It also automatically offers choice of operations on fuzzy rules and degree of fulfilment of outcomes. Such adaptation would allow a user to gradually customize the visualization system to own application and subjective preferences.

7 Conclusion and Future Work

We have discussed the motivations behind the use of a multi-agent approach to develop a framework for visualization of fuzzy systems which is user- and task-oriented. This framework is based on our design of fundamental ontologies underlying the structure and requirements of these systems. We have also presented the structure and activities of these agent classes and how they would be implemented. A case study has been investigated to provide visualization support for a fuzzy model which had been developed to predict electricity spot prices [3]. We are continuing to implement agent classes to extend fully their capabilities and to evaluate this framework using further case studies of fuzzy systems for different applications.

References

1. Berkan, R.C. and Trubatch, S.L.: Fuzzy System Design Principles, IEEE Press, NY (1997)
2. Berthold, M., and Holve, R.: Visualizing high dimensional fuzzy rules, in Proceedings of the 19th International Conference of the North American Fuzzy Information Processing Society: NAFIPS, Dept. of Electr. Eng. & Comput. Sci., California Univ., Berkeley, CA, USA (2000) 64–68
3. Brown, R. and Pham, B., Visualisation of Fuzzy Decision Support Information: A Case Study, IEEE International Conference on Fuzzy Systems, St Louis, USA, May (2003) in print
4. Cox, Z., Dickerson, J.A., and Cook, D.: Visualizing Membership in Multiple Clusters after Fuzzy C-means Clustering, in Proceedings of Visual Data Exploration and Analysis VIII (2001) 60–68.
5. Dickerson, J.A., Cox, Z., Wurtele, E.S., and Fulmer, A. W.: Creating metabolic and regulatory network models using fuzzy cognitive maps, in Proceedings of IFSA World Congress and 20th NAFIPS International Conference, Joint 9th, Dept. of Electr. Eng, Iowa State Univ., Ames, IA, USA, 4 (2001) 2171–2176.
6. Farwowski, W., and Mita, A. (Eds.) Applications of fuzzy set theory in human factors (1986)
7. Gershon, N. D.: Visualization of fuzzy data using generalized animation, Visualization '92, Proceedings, Mitre Corp., McLean, VA, USA, (1992) 268–273
8. Jiang, B.: Visualisation of Fuzzy Boundaries of Geographic Objects, Cartography: Journal of Mapping Sciences Institute, Australia, 27, (1998) 31–36
9. Keller, P., and Keller, M.: Visual Cues. Piscataway, USA: IEEE Press (1993)
10. Nurnberger, A., Klose, A., and Kruse, R.: Discussing cluster shapes of fuzzy classifiers, in Fuzzy Information Processing Society, 1999. NAFIPS. 18th International Conference of the North American, Fac. of Comput. Sci., Magdeburg Univ., Germany, (1999) 546–550.
11. Nurnberger, A., Klose, A., and Kruse, R.: Analyzing borders between partially contradicting fuzzy classification rules, in Fuzzy Information Processing Society, NAFIPS. 19th International Conference of the North American, Fac. of Comput. Sci., Magdeburg Univ., Germany, (2000) 59–63
12. Tufte, E.: The Visual Display of Quantitative Information. Cheshire, USA: Graphics Press (1983)
13. Zadeh, L. A.: Toward a Theory of Fuzzy Information Granulation and its Centrality in Human Reasoning and Fuzzy Logic, Fuzzy Sets and Systems, 90, 2, (1997) 111–127.

Towards Biomedical Problem Solving in a Game Environment

Yang Cai[1], Ingo Snel[2], B. Suman Bharathi[2,3], Clementine Klein[4], and Judith Klein-Seetharaman[1-3,5]

[1] School of Computer Science, Carnegie Mellon University, Pittsburgh, PA15213, USA
ycai@andrew.cmu.edu

[2] Institute for Organic Chemistry, University of Frankfurt, 60439 Frankfurt, Germany
ingosnel@aol.com

[3] Institute for Biological Information Processing, Research Institute Jülich, 52425 Jülich, Germany
bsuman_1979@yahoo.com

[4] Berufskolleg Kartäuserwall, Abteilung Medien, Kartäuserwall 30, 50678 Köln, Germany
clekle@hotmail.com

[5] Department of Pharmacology, University of Pittsburgh Medical School, Pittsburgh, PA15261, USA
judithks@cs.cmu.edu

Abstract. Biomedical systems involve complex interactions between diverse components. Problem solving in such systems requires insight, i.e. the capability to make non-obvious connections. In this paper, we present a game-based problem solving environment, where users can explore biological interactions with navigation on atomic to macroscopic scales, role-play, and networked collaboration. The study investigates the system architecture of the biological game, bio-morphing characters, and bio-interactions with biosensing and biodynamics. The prototype has been implemented on PC and tested in a preschool environment where users have little knowledge in biology. The experiment shows that the game greatly inspired users both in concept learning and entertainment.

1 Introduction

Rapid advances in the convergent technologies "nano-bio-info-cogno", referring to nanoscience and -technology, biotechnology and biomedicine, information technology and the cognitive sciences, are believed to have the potential to result in a "comprehensive understanding of the structure and behavior of matter from the nanoscale up to the most complex system" if these advances are exploited in a synergistic fashion [1]. In biomedicine, high-throughput methodology now allows the accumulation of unprecedented amounts of scientific data, such as genome sequences, gene expression profiles, structural and functional proteomic data. These advances have stirred great hopes for understanding and curing diseases, but the quantity of data requires convergence with information technology to interpret and utilize these data to advance human performance and quality of life [1]. This requires an understanding of

P.M.A. Sloot et al. (Eds.): ICCS 2003, LNCS 2659, pp. 1005–1014, 2003.

the complex interactions between the components of biomedical systems by both domain and non-domain experts. This is particularly challenging in the biomedical domain because of the massive data and knowledge accumulation. Facilitating convergence of nano-bio-info-cogno technologies therefore requires a novel Problem Solving Environment (PSE) for the biomedical domain.

A biomedical PSE should be a computer system that provides all the computational facilities needed to immerse into a biomedical problem. It should hide the intricacies of computer modeling of physical phenomena so that the user can concentrate on developing an approach to cure a disease for example. The PSE has to be scientifically accurate and include access to the state-of-the-art in available data, knowledge and technology without requiring the user to bring domain expertise and extensive experience with the technical intricacies of the PSE that would present the user with tiresome activation barriers. Novel approaches are being developed, for example a storytelling system has been presented to fertilize multidisciplinary biomedical problem solving [2]. Furthermore, modern biomedical education makes extensive use of visualization of biomedical processes and concepts, for example the publication of the human genome sequence was accompanied by a CD-ROM that presented genome background as well as DNA sequencing techniques in animations [3]. However, these visualization tools are mostly designed to complement traditional teaching techniques and are not interactive. In contrast, interactive virtual laboratories have been developed for chemical and biomedical laboratory experiments (e.g. [4]), but these are targeted to students and researchers for solving very specific problems. Such systems provide too much domain specific information and too little insight for solving discovery problems for inexperienced users to serve as an integrated PSE.

Problem solving in biomedical systems requires insight, i.e. the capability to make non-obvious connections between the complex interactions of the components of these systems [5]. Such insightful solutions can often be found in an interactive and visual PSE, as demonstrated for example by the fact that despite the modern numerical computing technologies, biophysicists today still use Gedanken experiments for concept development [6]. Although there are many virtual reality three-dimensional molecular models available, biochemists still use hand-made models for intuitive reasoning. It is striking that simple intuitive simulation is still one of the most powerful approaches to creative problem solving.

Since the early days of artificial intelligence, issues of modeling scientific reasoning and its representation, in particular for those connected with everyday knowledge of the behavior of the physical world, have been studied [7]. At least two aspects have been explored: multiple representation and qualitative reasoning. Computation with Multiple Representations (CaMeRa) is a model that simulates human problem solving with multiple representations, including pictures and words [8]. CaMeRa combines a parallel network, used to process the low-level pictorial information, with rule-based processes in higher-level pictorial and verbal reasoning. Furthermore, many AI systems have been developed to simulate the cognition about physical and biological knowledge. What will happen if we spill a glass of milk on the floor? For humans, the answer is common sense, but understanding this process is non-trivial for computers. To arrive at an exact solution, the computer has to solve a set of non-linear partial differential equations of hydrodynamics that are computationally intractable even for simple boundary conditions [9]. A few studies have focused on the qualitative simulation of physical phenomena. Thus, Gardin uses two-dimensional diagrams to represent physical objects and their interaction [10] and Forbus uses the fuzzy language of "qualitative physics" to model the physical variables [11]. Lower-resolution qualitative models have made significant impact in many fields, including biology. A typical example is the Game of Life, a "Cellular Automaton" [12]. A cellular automaton is an array of identically programmed automata, or "cells", which interact with one another. The state of each cell changes from one generation to the next depending on the state of its immediate neighbors. By building appropriate rules, complex behavior can be simulated, ranging from the motion of fluids to outbreaks of starfish on a coral reef. Even if the line of cells starts with a random arrangement of states, the rules force patterns to emerge in life-like behavior. Empirical studies by Steven Wolfram [13] and others show that even the simple linear automata behave in ways reminiscent of complex biological systems. In light of this discovery, we intend to use simple biological characters to generate dynamic interactions.

Here we present a computer game as a novel environment for biological problem solving, where it provides a real-time interactive platform for users. The goal of this study is to develop a game-based PSE for users to explore multi-modal interactions inside a biological system. It includes essential biological simulation models for the immune system and the blood system. It allows users to manipulate and to participate in the interactions of the components of the system. The biological characters are simulated by software agents in the game.

2 System Architecture

As a test bed for the development of a game-based PSE for biomedical science, we designed a scientific problem that is derived from our ongoing research projects. By applying computational language technologies to the large amounts of whole genome sequence data publicly available, we have identified “genome signatures” that may provide new the development of vaccines against pathogenic microorganisms such as Neisseria [14]. The biomedical problem to be explored here is to find treatment for fatal meningitis. The new idea for the biomedical PSE is to develop an interactive interface modeled after traditional game engines, that teaches users without background in biology the understanding of this research problem ranging in hierarchy from atomic to macroscopic scales. In this hierarchy, the macroscopic level is that of infection of a human body with Neisseria. Fighting the infection at this level, however, requires a molecular level understanding of the processes involved. The goal is to provide the user with the necessary insight to creatively generate and test possible approaches to solving this problem using the PSE. The PSE contains three interaction modes: *role-play, voyage and networked problem solving.* Users can select a mode from the main menu.

- *Role-Play.* The system allows the user to be a biological character in the game. Cognition Science shows that role-play is an important way to stimulate creative ideas. It enables the user to have an intimate connection to the character. Also, personalization of a biological character makes a game more interactive.
- *Voyage.* The user can navigate through the biological system in the game. This gives the user an opportunity to look at the interactive components from different aspects, for example, travelling through capillaries and tissues. The voyage allows exploration at the user's chosen leisure, accommodating users with various backgrounds.
- *Distributed Problem Solving.* The game engine allows users to play the game over the Internet so that users can solve large problems in a collaborative way. For example, some users can play macrophages and others can play bacteria. The distributed problem solving enables diverse game strategies and more excitement of the game.

BioSim version 1.0 is a rapid prototype for this PSE. It is a two-stage game that simulates the journey of red blood cells and white blood cells (macrophages). The goal is to introduce the basic concepts of cellular interaction and the human immune system. The game begins with an animated scene of a blood stream with red and white cells moving passively with the heartbeat. Using a mouse and the arrow keys, the player can take the role of a biological character, for example a macrophage, and navigate inside the blood stream. The user can also actively squeeze out in capillary regions to access tissue that is infected by bacterial cells, which multiply at a certain speed. Screen shots of these processes are shown in Fig. 1.

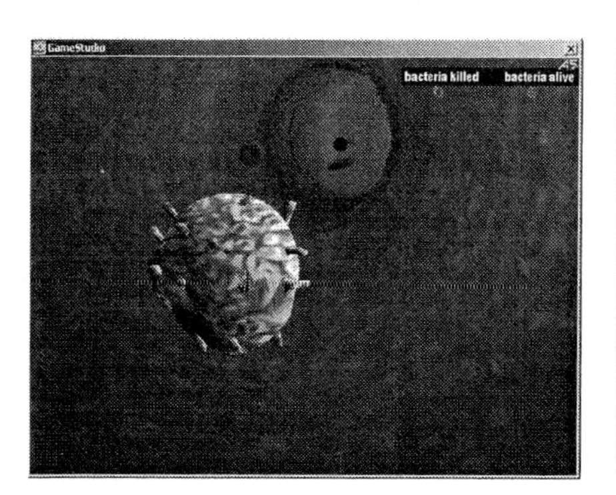

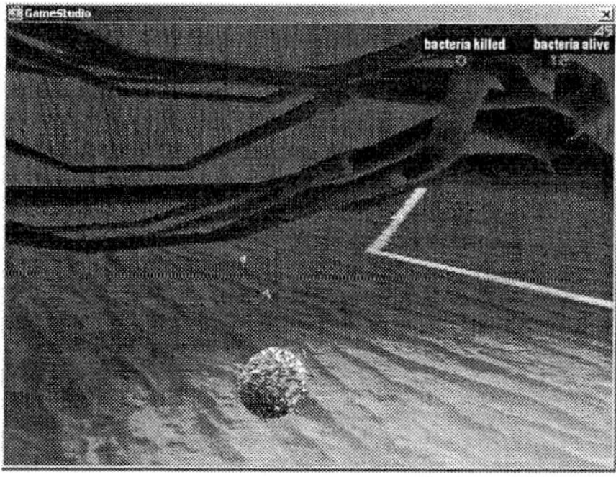

Fig. 1. The Macrophage (white blood cell) is inside the blood stream with red blood cells (left). After actively moving out of the blood stream, the macrophage approaches bacteria that infected human body tissue (right)

3 Biological "World Model"

In game design, "world models" are similar to the theatre stage or film scene with which actors and characters interact. A world model is often static and large in size. In this project, we developed a comprehensive world model that includes the vascular system with artery, veins and capillaries, as well as tissues (Fig. 2). In this world, the user can fly, walk or run through as one of the biological characters. In the prototype BioSim 1.0, we developed two scenes: inside the capillary and outside of the capillary. The transition of the scenes is possible by "squeezing" a character actively from the capillary to the tissue.

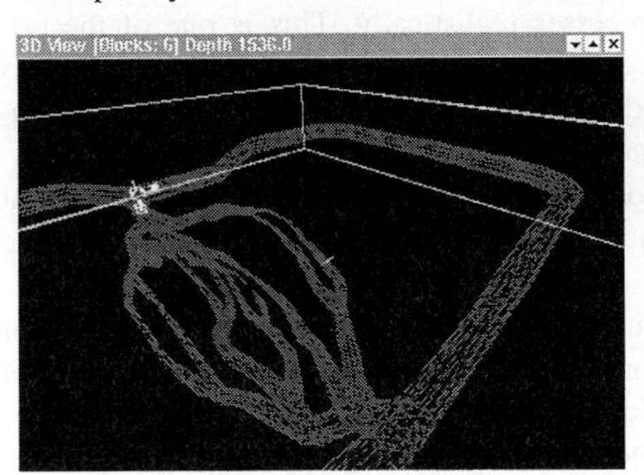

Fig. 2. Vascular system represented in the game. The 3-D wireframe model (left) and the 3-D photorealistic model (right) include arteries and capillaries

4 Biological Characters

So far, we have defined the following 3D animated characters that simulate biological behavior: bacteria, macrophages, and red blood cells. For the stand-alone characters, we *apply bio-morphing* to assign key frames to them. Bio-morphing is accomplished by digitizing deformed shapes from microscopic images of organisms, building wire frames and attaching texture and color skins. The transitions of each character are represented by a state machine (Fig. 3, left). For example a macrophage's states include the transitions to deform, shrink, eat, walk and die.

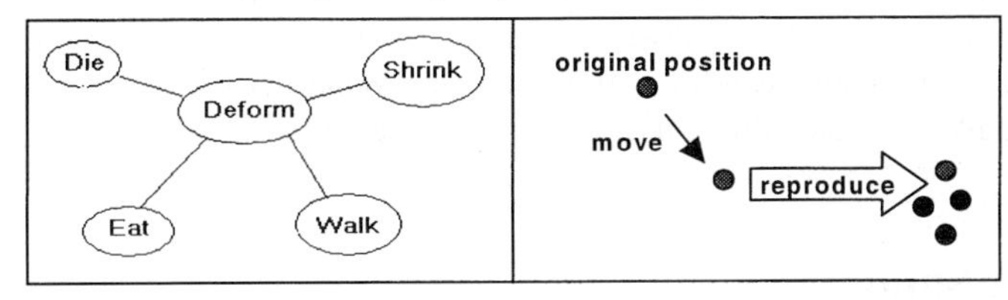

Fig. 3. State machine for a macrophage (left) and dynamics of an organism modeled by cellular automata (right)

5 Biological Interactions

Interaction is the key to computer games, and we believe similarly to an efficient biomedical PSE. We therefore allow the user to control the behavior of biological characters by realistic and scientifically accurate bio-interactions. These consist of bio-sensing and bio-dynamics.

Bio-sensing describes the ability of biological systems to sense environmental stimuli. For example, white blood cells can "smell" bacteria and move toward them by chemotaxis. To simulate such sensing capabilities, we define a circular envelope around the character as a sensing region. When the target is inside the region, the character will move towards the target and engage in interactions with it.

Bio-dynamics simulate biological processes realistically. This is one of the primary components adding excitement to the game. For each character, we define its interactive modes, such as motion, reproduction and death. Taking bacteria for example, we use the following rules (Fig. 3, right):

- *Autonomous motion.* Given a fixed duration, each bacteria moves a distance x at angle y, where x and y are random values. The distance should not exceed the maximal distance.
- *Reproduction.* Given a predefined duration, each bacterium reproduces a copy of its own which is placed beside its original position. The *Logistic growth model* [15] adequately describes the reproduction process of simple organisms over limited time periods by equation (1), where M is the carrying capacity of the population.

$$B_{n+1} = B_n + rB_n(1 - B_n/M) . \qquad (1)$$

- *Death.* If a bacterium's life cycle is over or if other cells eat it, it is removed from the scene.

6 Implementation

The prototype of BioSim 1.0 is implemented on PC. Photorealistic 3D models of components of the system were created with 3D Studio Max and imported into Game Studio 3D. 3D Modeler of GameStudio 3D was used to create the game scenes, biomorphed characters and the integration of the world/character dynamics and interactions. C-script, a C-style language, was used to encode the bio-dynamics and biosensing behaviors. Game Studio is run under the Windows operation system. It provides capability for either single user or multiple users across the Internet.

7 A Case Study

We conducted experiments in the effectiveness of the game to raise an awareness of the important issues in biomedical research on users with no background. The ideal group at this stage of implementation of the game is young children, for two reasons. One, children are unbiased and without background. Second, children learn optimally when the material to be learned is presented to them in an accurate way to avoid the build-up of incorrect models by implicit learning [16]. Implicit learning of correct biomedical concepts by children therefore requires the same fundamental issue of scientific accuracy as other users will require once the game reaches the stage of providing a PSE for users with any background. We tested BioSim 1.0 on 14 children at KinderCare, Cranberry, PA, on August 9th, 2002 and February 25, 2003. We let four- and five-year-old children play with the game on a laptop and focused our attention on strategic aspects and active questioning in the children's behavior.

Two strategies were quantified, the speed of macrophage movement towards the bacteria (Fig. 4, left) and the use of antibiotics in aiding the killing of the bacteria (Fig. 4, right). All children learnt quickly to shift from fast pace chasing to slow pace chasing so that their capture rate was improved. We then tested a more challenging concept, that of usage of antibiotics to aid the killing of the bacteria. We included the ability of bacteria to develop resistance in our growth model. Thus, the children had to discover that antibiotics at some stage in the game no longer inhibit bacterial growth. This was only observed by a single 5-year old, all other children kept on administering antibiotics despite energy consumption and lack of effect (Fig. 4, right). These types of quantitative assessment of strategic behavior of users open novel ways to analyze learning of problem solving skills that would not be possible with conventional teaching methods.

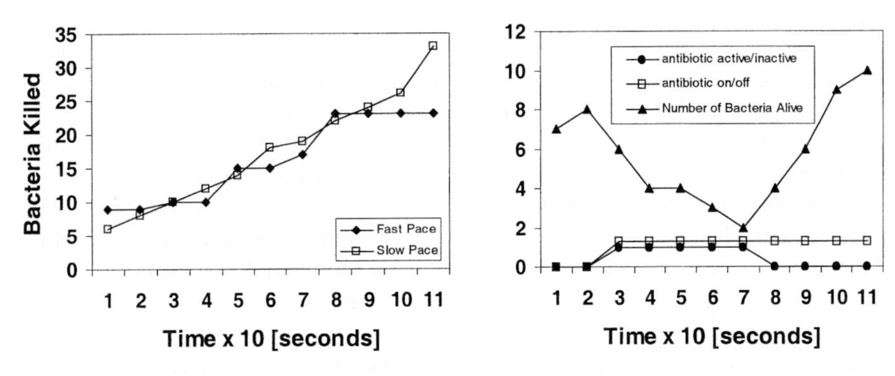

Fig. 4. Strategies against bacterial infection that can be explored in the game, speed of macrophage movement (left) and use of antibiotics (right). The fast pace strategy often leads to missing targets. The slow pace strategy gains steady capture rate. The use of antibiotics can be limited by gradual development of bacterial resistance. At that point, administration of drug does not inhibit bacterial growth.

Next, we asked learning assessment questions, such as "How does the macrophage get out of the capillary?" or "How do you kill bacteria?" The children used intuitive metaphors, for example the analogy of "vacuum" and "crash into" to describe how the macrophage attacks bacteria. This shows that the players are very sensitive to the intimate design details of the game, which opens a window for game developers to *encode* very subtle knowledge about complex biological interactions.

Finally, we tested the game-induced stimulation of questioning in the children. The results are summarized in Table 1. The five-year-old children asked several meaningful questions, for example: "Are bacteria germs?" "Where do the white cells go?" "What's a red cell?" "Where do the bacteria live?" "Is the macrophage good or bad?" Overall, four-year-old children asked fewer questions, and most of their questions were not relevant, for example, "Do you have other games?" "I don't want my head eaten off." These observations suggest that there may be a turning point between ages 4 and 5 where a PSE can become effective.

Table 1. Comparison of the reactions of two groups of children to the game. The children in the first group were 4 years old, those in the other were 5 years old. Each group consisted of 7 children, and the total number of children tested was 14.

Observation	4-year old	5-year old
Asked relevant questions	0	4
Controlled the game successfully	2	5
Described bacterial growth behavior	1	5
Described macrophage behavior	2	6

8 Further Developments in Game Design

We are currently in the process of adding more capabilities for the user to interact with different components of the game, and also with multiple players connected to the same game via the Internet *(Distributed Problem Solving)*. In the next version of the game, the user will choose between two aims, rather than playing the role of a single biological component. This modification will allow the user to assume the roles of multiple biological characters, thus studying their individual influence on a particular aim. These aims will be either to induce an infection with Neisseria and ensure its successful propagation in the human body or to fight the Neisseria infection. To enable more complex means of interaction with the biological world, each user will be equipped with a ship that allows for effective immersion into the environment (Fig. 5, upper left). The ship provides a means of transportation *(Voyage)* and action *(Role Playing)*. Each activity is determined by availability of "energy points", which have to be carefully balanced to minimize consumption and maximize effectiveness. The user knows the status of energy points via a control panel, which also provides for the various possibilities of action (Fig. 5, upper right). For example, in a state of high energy, the user can afford to travel actively with the ship to a point of infection. However, in a state of low energy, the user would choose to travel passively with the blood stream. This will allow the user to further develop decision-making skills in a biological PSE.

There will also be additional biosensing capabilities available through the control panel, for example a histamine sensor (Fig. 5 lower left) and mechanisms of the immune system to distinguish self from non-self (Fig. 5, lower right). This will allow introduction of molecular level information, for example the user will need to use molecular docking of the immune system's antibody structures to those of the bacterial surface structures. This will train users to view protein structures and understand the mechanisms of complementarities of two structures. The player seeking to evade the immune system would need to develop strategies to evade antibody marking, e.g. through surface mutation. Thinking about possible strategies from each point of view will allow the user to gain deep insight into the factors controlling the health of the organism, from the molecular to the macroscopic level, ultimately aiding in the development of novel solutions for biomedical problems such as the Neisseria infection.

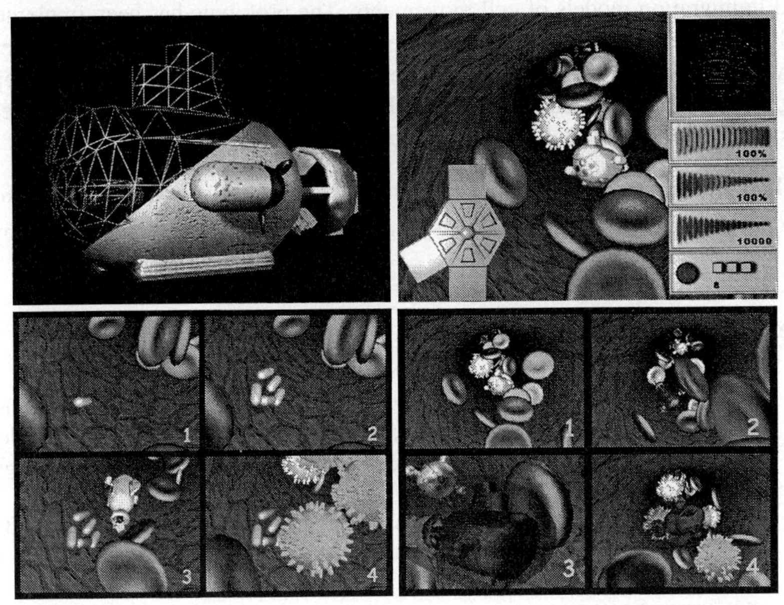

Fig. 5. Immersion of the user in the PSE. A ship provides transportation through the body (upper left). A control panel inside the ship has sensing and action capabilities (upper right). Marking of bacteria with histamines. Bacteria that have not been marked can divide undisturbed (gray bacteria in 1,2). When the ship (yellow) approaches the bacteria (3), the histamine sensor identifies the bacterial infection and the user can mark them (color change). After marking, the user can attract macrophages that will "eat" the bacteria (4) (Lower left). Increased complexity in future game. Internet connectivity will allow multiple users to participate in collaboration or as antagonists in a single game (yellow ship, 1, and blue ship, 2). Since ships are not-self, the human immune system would identify them as such. Therefore, one ship can mark another ship using immune system tools, i.e. antibodies. This is represented by the halo of the blue ship. Once labeled with antibodies, the user can again direct macrophages to "eat" the blue ship (lower right).

9 Summary

Future biomedical problem solving is beyond traditional means because of the existing challenges in cross-disciplinary communication and interpretation and utilization of vast quantities of available biomedical data. We want to build a virtual PSE that combines advanced computer graphics, computer vision, artificial intelligence technologies and creative instruction technologies. In this PSE, cross-disciplinary education will be on-demand, entertaining and interactive. This will allow focus on discovery and creativity rather than one-way tutoring. Towards this long-term goal, here, we have presented a game-based PSE, where users can explore complex biological interactions with navigation, role-play, and networked collaboration. The study investigates the system architecture of the biological game, bio-morphing characters, and bio-interactions with bio-sensing and bio-dynamics. The game is based on realistic biological models, such as logistic growth models of simple organism reproduction and immigration models of cell movements. The prototype has been implemented on PC and tested in a preschool environment where users have little knowledge in biology. The experiment shows that the game greatly inspired users both in concept learning and entertainment suggesting that the game-based PSE helps users to learn bio-system dynamics and multiple object interactions.

Acknowledgements. This work was supported by the Alexander von Humboldt-Foundation and Zukunftsinvestitionsprogramm der Bundesregierung, Germany.

References

1. Roco, M.C. and W.S. Bainbridge: Overview. In: Converging Technologies for Improving Human Performance: Nanotechnology, Biotechnology, Information Technology, and Cognitive Science, Bainbridge, W.S. (ed.) National Science Foundation: Arlington, Virginia. (2002) 1–23
2. Kuchinsky, A., K. Graham, D. Moh, A. Adler, K. Babaria, and M.L. Creech: Biological storytelling: A software tool for biological information organization based upon narrative structure. AVI (2002)
3. http://www.sciencemag.org
4. http://ir.chem.cmu.edu/irproject/applets/virtuallab/
5. Cai, Y.: Pictorial Thinking, in Journal of Cognition Science.: Taiyuan, China (1986)
6. Hayden, T.: The inner Einstein, in US News (2002)
7. McCarthy, J. and P.J. Hayes: Some philosophical problems from the standpoint of artificial intelligence, in Machine Intelligence, Michie, D. (ed.) Edinburgh University Press: Edinburgh (1969)
8. Simon, H.: Models of Thoughts. Vol. II. Yale Press (1989)
9. Feynman, R.: The Feynman Lectures on Physics. Reading, Massachusetts: Addison-Wesley (1963)
10. Gardin, F. and B. Meltzer: Analogical representations of naive physics. Artificial Intelligence **38** (1989) 139–159
11. Forbus, K.D.: Qualitative Process Theory. Artificial Intelligence **24** (1994) 85–168
12. Gardner, M.: Mathematical games: The fantastic combinations of John Conway's new solitaire game "life". Scientific American **223** (1970) 120–123
13. Wolfram, S.: A new kind of science Wolfram Publishing (2002)
14. Ganapathiraju, M., D. Weisser, R. Rosenfeld, J. Carbonell, R. Reddy, and J. Klein-Seetharaman: Comparative n-gram analysis of whole-genome protein sequences. in Proc. Human Language Technologies San Diego (2002)
15. Guyton, A.C. and Hall, J.E.: Textbook of Medical Physiology Philadelphia: W.B. Saunders Co (1997)
16. Perrig, P. and W.J. Perrig: Implicit and explicit memory in mentally retarded, learning disabled, and normal children. Swiss J of Psychology **54**(2) (1995) 77–86

Learning Surgical Interventions by Navigating in Virtual Reality Case Spaces

Piet Kommers, Steffan Rödel, Jan-Maarten Luursema, Bob Geelkerken, and Eelco Kunst

University of Twente, Faculty of Behavioral Sciences
Medical Spectrum Hospital
Kunst en v Leerdam
Enschede, The Netherlands

Abstract. Virtual Reality is becoming a serious candidate for a learning environment for complex skills like vascular interventions. The diagnostics, dimensioning and insertion of the endograft stent has been modeled as a decision making process and now faces its implementation in a VR learning space.

1 Introduction

The DIME project (Distributed Interactive Medical Exploratorium for 3D Medical Images) aims at conceptualizing, implementing and researching the effectiveness of a VR-based pre-surgical planning and teaching environment. The DIME explorations aim at a pre-surgical planning and teaching applications, which will most likely result in better post-surgical results, lower health care costs and increased efficiency in the training of fellow-surgeons. The project is a collaboration of a 'computer science' group from the University of Amsterdam (UvA)[1], a 'medically oriented computer science' group from the Leiden University Medical Center (LUMC)[2], a 'cognition' group from the University of Twente (UT) and medical specialists from LUMC) and The Medical Spectrum Twente.

2 Surgical Training in Virtual Reality

The surplus of surgical training skills as well as being oriented in medical disciplines like anatomy, histology, physiology etc is evident. The optimization of its professional training elements needs to most complete repertoire of learning technology like the media spectrum and the full repertoire of new learning paradigms.

[1] Prof Dr. P. Sloot (UvA) and Dr. E. Zudilova

[2] Prof Dr. JHC Reiber (RUL) and Dr. J. Schaap

P.M.A. Sloot et al. (Eds.): ICCS 2003, LNCS 2659, pp. 1015–1024, 2003.

The aim of this paper is to show the compatibility between the most advanced visualization methods currently feasible for the average desktop work stations [2]. The overall conception is that the clinical training will gradually be extended with VR learning systems, in order to make the supervised real operations more effective and safe.

2.1 Relevance of Virtual Reality

1. Surgical techniques have become increasingly complex, thus making the learning curve to master these techniques steeper and longer.
2. More complex intervention techniques are rapidly developed and introduced in the daily practice.
3. The conventional surgical teaching method is a close daily working relation between the experienced teacher (trainer) and the unskilled pupil (trainee).
4. In traditional teaching the steep learning curve takes place during the interaction with real patients.
5. The modern patient does not accept any mutilation attributed neither to the disease nor to the intervention.
6. It is clear that a perfect preoperative visualization and planning, and rehearsals of these interventions are essential.
7. This means that while there is an increased demand for surgical training, experienced surgeons have increasingly less time and opportunity to cope with this demand. A dedicated medical VE is badly needed to lift this burden from their shoulders.
8. Of importance is also the possibility to allow trainees to explore critical situations and to let them experiment with an underlying model of the phenomena and processes in the human body, without the stress of having to deal with an actual patient.
9. Virtual surgical tools should be available for life-long medical education and assessment of the surgical consultant.
10. Based upon the disappointing experiences with "Intelligent Tutoring Systems" in the '80ies we do not want to undertake the paradigm of "training dummy mannequins" as it lacks the notions of "continuous learning" and the "surgeon as active problem solver".

2.2 The Urgency for Laparoscopic Interventions

Nowadays the laparoscopic cholecystectomy is the preferred technique in many hospitals. However, the majority of the surgeons performing laparoscopic cholecystectomies are autodidactic. They heard about the technique on congresses. They visited clinical demonstrations in centres of excellence. Thereafter they planned the first procedures in their own hospital. It is not surprising that the results are not as good as reported in the literature in the early periods. The steep learning curve was moved on patients. A sufficient training and formal assessment of the surgical team before introducing the new technique into the hospital is not available. Moreover, more complex intervention techniques are rapidly developed and introduced in the daily practice. An example of this is the endovascular exclusion of infernal aortic aneurysms with an endograft. Cuijper recently reported in his thesis that only after an endovascular experience of 30 electively treated triple A the complication ratio is sloping down to acceptable levels. In the Netherlands only a few hospitals have such an experience. Also the first ruptured aortic aneurysms are treated in the endovascular way with a very good outcome.

However, this emergency procedure demands a large endovascular experienced team available during day and night. Gaining enough experience with this procedure is not possible in most of the hospitals in the Netherlands. Unfortunately, it is not possible to transport patient with a ruptured triple A to centres of excellence because of hemodynamical instability. In other words, the patients do not survive delay of treatment due to transportation. The next generation, more complex endografts with the possibility of perirenal sealing is underway. The results of the first clinical experiments came from "down under". It is clear that a perfect preoperative visualisation and planning and a dummy operation of the whole procedure is essential.

2.3 The Urgency for Training in Virtual Reality

The conventional and still actual surgical teaching method, introduced more than a century ago, is a close daily working relation between the practised teacher (a consultant surgeon) and the unskilled pupil (the surgical resident). Working-weeks of 70 to 90 hours were accepted and after 6 years of gaining theoretical and especially practical skills under direct supervision of the consultant the resident becomes a surgeon. Nowadays our society does not accept such long periods of formal learning and the working week is shortened to a maximum of 48 hours. This results in a 40% decline of directly supervised practical experience of residents in their first years of surgical training. Moreover, the government asked the surgical society to offer the basic surgical training in only 5 years!

On the other side the same government makes laws as the BIG and the WGBO. The surgeons are obliged to be qualified (formal licensed) and to be skilled and properly trained to offer and execute an intervention to or on a patient. Due to the "Schengen convention" there is a right of free establishing of the citizens of the European community in the participating countries. The formal training of resident and consultants in the European countries showed large diversities. Objective and proper methods for assessment and comparison of the outcome of the surgical training in the European countries are not available. It is clear that the conventional surgical teaching methods do not fulfil the demands of patients, the society, the government and the surgical profession. New training methods have to be developed.

The development of virtual reality surgical tools for theoretical and practical training and assessment of the resident is urgent [4]. Moreover, virtual surgical tools should be available for life-long medical education and assessment of the surgical consultant maintaining a high level of expertise and skills in the profession. This project intends to contribute to the steep increasing need for practical training and objective assessment for the surgeon in training.

2.4 Goals of the DIME Project

1. To create a VR learning environment that allows surgical trainees to both practice their skills in the 'Operating Room' Virtual Environment (VE) and enhance their understanding of the procedure under study by using the 'Library' VE or the peer-to-peer chat function [5].
2. DIME aims at identifying the more 'objective' training elements that need to be conveyed before the constructionistic learning starts. This is the reason that the first stage arranges teams with the various expertises that play a role for the definition of the anchoring points in the training of the future surgeon [7].
3. Most of the VR projects have invested in the actual building of the models and have no didactic interface yet. The DIME project sees this need and aims at defining a generic instructional method that intermediates between a VR medical model and a novice who needs to understand and optimize its functioning.
4. To specify and evaluate VR elements for the pre-clinical training phase of novices in artery surgery.

3 A-priory Expert System for Risk-Evaluation of Endovascular Stent Prosthesis Placement

The VREST[3] group has undertaken the formalization of dimensioning the stent orthesis for the AAA patients. In order to obtain an uncomplicated passage and a lasting exclusion of an infrarenal abdominal aorta aneurysm (AAA) through endovascular placement of a stent-prosthesis, one has to take into account many unique anatomical properties of the aortic-iliac-femoral trajectory and many unique properties of the stent prosthesis.

Because of this large amount of relevant anatomical and stent-prosthestic variables it is not easy, even for the experienced clinician to make the correct assessment. A validated stent-prosthesis expert system can offer support to the clinician in choosing between endovascular and transabdominal exclusion of the AAA. Such a system can also offer support in choosing the optimal type of stent-prosthesis, and in planning the procedure.

---

[3] The members of the VREST team (VREST: Virtual Reality Educational Surgical Tools) (Rödel SGJ, Kunst EE, Teijink JAW, Herwaarden JA van, Berg JC van den, Oude Groothuis P, Moll F, Palen J van der, Huisman A, Det RJ van, Geelkerken RH)

3.1 Goal

The goal of this prior study was the validation of a custom developed stent-prosthesis expert system. The anatomical AAA criteria were measured in 202 patients from two endovascular centers. Every AAA was divided in ten segments; suprarenal aorta, infrarenal aorta, aneurysm, aorta bifuration, right, left and common iliac artery and common left and right artery. For each segment, the following characteristics were recorded: length, thrombus, sclerosis, angulation and configuration. These 202 AAA's were then judged by five endovascular trained clinicians on anatomical fitness for placing the stent prosthesis [6].

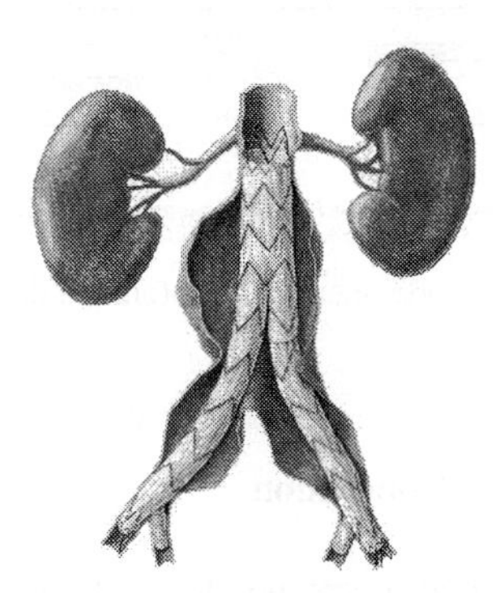

Fig. 1. The Stent Orthesis for the AAA Patients

Table 1. Correspondence Between the Clinicians and the Expert System

	Agreement	*False positive*	*False negative*	*Discussion*
Surgeon 1	*73.4*	*0.0*	*1.7*	*24.9*
Surgeon 2	*75.5*	*0.7*	*2.5*	*21.3*
Surgeon 3	*73.7*	*0.0*	*1.8*	*24.7*
Radiologist 1	*70.6*	*0.5*	*4.2*	*24.8*
Radiologist 2	*64.8*	*0.0*	*8.1*	*27.1*
Total	***71.6***	***0.2***	***3.7***	***24.6***

Agreement: *Same advice clinician and expert system.*
False positive: *Advice clinician = no intervention, advice expert system = intervention.*
False negative: *Advice clinician = intervention, advice expert system = no intervention.*
Discussion: *Advice clinician = high risk intervention, expert system = no intervention*

The chances for successful sealing were independently expressed in 'complication rating' for each AAA by five clinicians and the expert program. The complication rating was divided in 0% to 59% (low to intermediate risk), 60% to 94% (intermediate to high risk), 95% to 98% (very high risk), 99% (practically impossible) and 100% (impossible) to obtain sealing. It was allowed to choose between all configurations of three commercially available stent-prostheses. A total of 3030 AAA assessments were given by the five clinicians. These were compared to the assessments of the expert system.

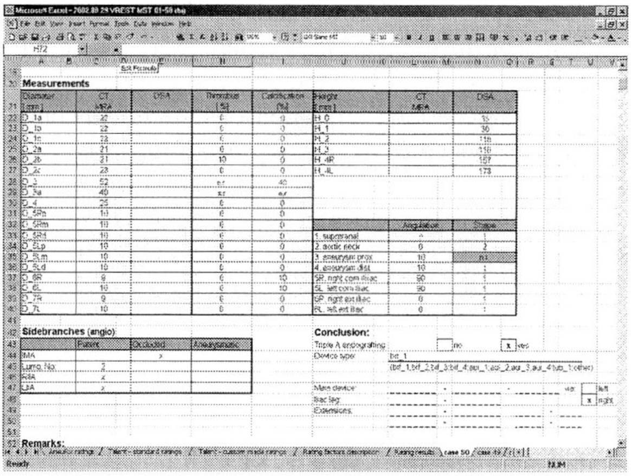

Fig. 2. Abdominal aorta aneurysm anatomical data input screen

3.2 Results and Preliminary Conclusion

There appeared to be a good correspondence between the assessments of the clinicians and the expert system as finally visualized as in Figure 2. Specifically the chances for an incorrect positive advice from the expert system are minimal (1:500 assessments). In case of a high complication rating the expert system tends to advise negative on stent placement in more cases then the experienced clinician does.

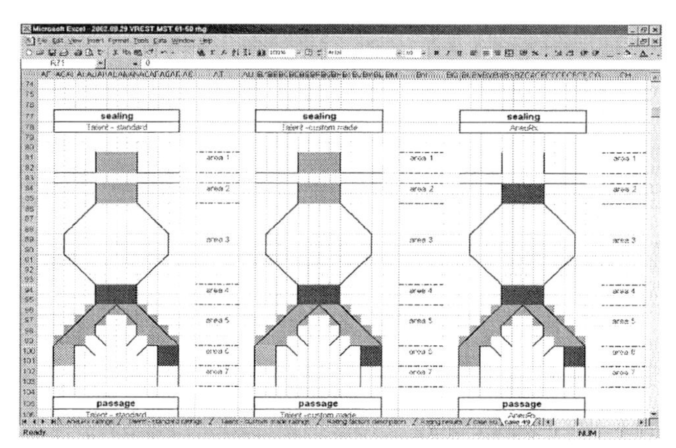

Fig. 3. Discrete Modeling Rational for the go-no go and the Dimensioning of the Stent Artifact

The basic underlying rational for the go-no go and the dimensioning of the stent artifact has been certified in this way. The envisaged DIME project plans the support of the 3D aspects of the particular patient in concern. Both the experienced and the novice vascular surgeon will increase effectiveness and task efficiency as the MRA images are transformed into three-dimensional models that can be inspected for critical morphologies and to anticipate better to the actual medical intervention.

3.3 Initial Architecture of the Proposed VR Learning Context

The surgical skills and its continuous sophistication; how should it benefit ideally from the virtual Operating Room? In this practice "space" (at the right side of Figure 2) the cycle goes around diagnosis, prepare and execute the intervention. But at unforeseen moments we expect that students consult a library of domain expertise. Based on the log data generated by this VE,

- the coach and trainee can evaluate his/her progress and set out an appropriate personal learning path
- the coach can extract general trends in learning of his/her trainees, as well as differences in learning styles of individual trainees.

Though the overall of this architecture looks adequate and robust, we decided not to adopt it basically, because of the suggested antagonism between the two [2]. More adequate and fair seemed the model where the learning between the various approaches due to variations in patients and the subsequent surgeon experts and the various trainees was the central core of the learning.

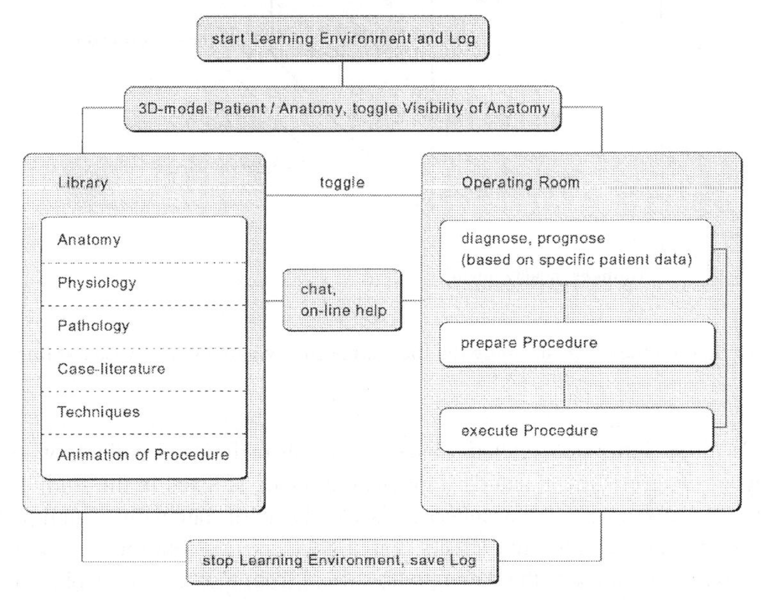

Fig. 4. The interplay between the students' theory versus skill-driven learning

The alternative to the theory/practice model above was left behind and the key idea for the VR training space was called "Interconnected Expertise". It aims at bring the novice in a quasi continuum of surgical interventions. Learning in this space allows a fluent transition between patients, surgeons and various stages in the intervention. The more natural one is by tracing the treatment by one surgeon in one patient through the natural chronology of the operation. But there are good reasons to switch between patients as it shows typical morphologies that clarify the reason why the initial treatment was needed there [1].

4 Imagining VR as Interconnected Expertise

VR as representation of the targeted object world may be propagated sufficiently. More intriguing is the question how learners may benefit from the prior experts and successful peer-learners. This question was recently addressed and came to the idea that three main dimensions need to be articulated before a meaningful navigation by the learner may take place.

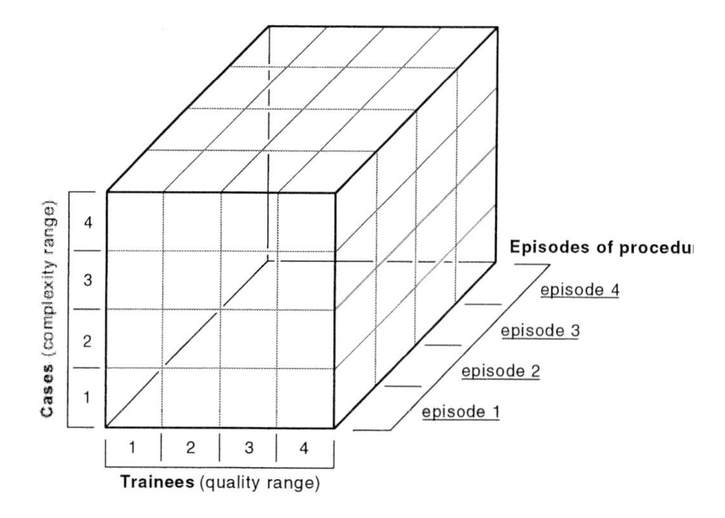

Fig. 5. Three main dimensions for allowing the learner to navigate in prior vascular interventions

1. The first dimension (called **Cases**) is the set of indexed patients who vary from obeying the prototypical standard medical problem as listed in the study books; in this case the abdominal aorta aneurism is clearly there, but with a minimum of complicated side effects and the patients is in a good overall condition. At the other end of the continuum there is the patient who suffers from a severe AAA phenomena, but at the same time a large number of constraining factors can be discerned like stenosis, aneurisms before and after the bifurcation and a complex artery morphology

2. The second dimension (called **Trainees**) varies between the highly-experienced vascular surgeons who performed in the Endograft stent VR model quite punctual and highly correct AAA interventions, to most of the embedded cases. At the other end of this dimension one can find recorded AAA interventions by freshmen demonstrating the many thinkable flaws. At each of the suboptimal interventions the ultimate surgeon marked the reason of labeling the flaw as being suboptimal and activates links to the real good solutions that should have been made.
3. The third dimension (called **Episode** 1..n) captures the subsequent stages in the AAA intervention. Two criteria for ordering them can be taken: The chronology of sub actions from early to later, versus the ordering from easy to complex. Analytical and experimental validation still needs to be performed in order to make a sensible choice here. In any case following the strict order of this dimension allows the learner to follow the prior interventions in it chronological order.

5 Conclusion

Though virtual reality is one of the prime candidates in vitalizing learning by its realism and direct appeal to the students' natural affordance to act upon urgencies rather than to "know" what experts are saying; VR in itself is not enough to make the learning more effective. By logic: Realism in VR does not suffice to exceed the real situation itself. We know from for instance link trainers for airplane pilots that the simulation can be more effective, once it elicits the novice to go into critically complex situations; exactly those situations that we never hope to meet in reality. Its added value is not only that the learner's reflexes are trained to survive in the panic of preciously decisive seconds. Its value is also that after all the fundamental understanding of complex mechanisms can best be understood if the learner is allowed to walk on the edge of what is a success versus a failure.

References

1. Beck K. and Cunningham W. (1989), A laboratory for teaching object-oriented thinking. In *Proc. OOPSLA'89*, ACM Sigplan Notices 17(4), pp. 1-6.
2. Kommers, P.A.M & Zhiming, Z.; (1998) Conceptual Support with Virtual Reality in Web-based Learning. (Co-author Zhao Zhiming). In: International Journal of Continuing Engineering Education and Life-Long Learning. ISSN 0957-4344. Volume 8, No 1/2. pp 184-204.
3. Kommers, P.A.M.; (2003). Experiential Learning through Constructivist Learning Tools. In: International Journal of Computers and Applications, Vol. 25, No 1, 2003. (pp 1-12). ACTA Press. ISSN: 1206-212X (202).
4. Lanier, Jaron; (1992). Virtual Reality: The Promise of the Future. In: Interactive Learning International, v8 n4 p275-79 Oct-Dec 1992 ISSN: 0748-5743.

5. McLellan, Hilary (1995) 6p.;Magical Stories: Blending Virtual Reality and Artificial Intelligence In: Imagery and Visual Literacy: Selected Readings from the Annual Conference of the International Visual Literacy Association (26th, Tempe, Arizona, October 12- 16, 1994).
6. Moshell, J.M., and Hughes, C.E. (1994, January). Shared Virtual Worlds for Education. Virtual Reality World, 2 (1), 63-74.
7. Psotka, Joseph; Immersive Training Systems: Virtual Reality and Education and Training. In: Instructional Science, v23 n5-6 p405-31 Nov 1995. ISSN: 0020-4277.

Virtual Reality and Desktop as a Combined Interaction-Visualisation Medium for a Problem-Solving Environment

E.V. Zudilova and P.M.A. Sloot

Section Computational Science
Faculty of Science, University of Amsterdam
Kruislaan 403, 1098 SJ Amsterdam, The Netherlands
[elenaz|sloot]@science.uva.nl

Abstract. The paper addresses the problem of how to make a human-machine interaction user-friendlier within a problem-solving environment. Two different projection modalities – virtual reality and desktop solution - are compared in respect to interaction capabilities provided by the Virtual Radiology Explorer (VRE) – case study of this research. The VRE is a problem-solving environment for vascular reconstruction, developed by the Section Computational Science of the University of Amsterdam. The potential users of the VRE are physicians, whose attitudes and motivations vary. The combination of virtual and desktop interaction modes within the same environment may help to satisfy the wider range of VRE users, in comparison to the case when only one projection modality is used. A Personal Space Station is considered as a possible solution for deploying this concept.

1 Introduction

A problem-solving environment (PSE) provides a primary (or end) user with a set of hardware and software resources for building a specific framework to solve a target class of problems. Ideally this framework is to be built in such a way that a user can exploit modern technologies without specialised knowledge of underlying hardware and software.

In reality the situation is far from ideal. It is supposed that a primary user knows how to use simulation and visualisation programs (libraries, modules, software components, etc.) and that he is able to characterise a problem to be solved using a specific definition language. Most of developers do not take into account the fact that a primary user of a PSE is a scientist, who focuses mostly on his research area, and he is not very experienced as a computer user and needs intuitive interaction capabilities and a feedback adapted to his skills and knowledge.

PSEs' developers focus today more and more on modern advanced technologies concerning also the projection equipment used as an interaction-visualisation

P.M.A. Sloot et al. (Eds.): ICCS 2003, LNCS 2659, pp. 1025–1034, 2003.

medium. The choice of projection equipment today is mostly task-related or even spontaneous. As a result, a human-machine interaction provided by PSEs is far from intuitive. Existing projection modalities have not been investigated yet in respect to usability factors. Meanwhile, the selection of an appropriate projection modality in accordance with user's tasks, preferences and personal features has to be used as a basis for building a motivated PSE.

The PSE introduced in this paper is a framework for rapid prototyping of an exploration environment that permits a user to explore interactively the visualised results of a simulation and manipulate the simulation parameters in near real-time, where a pre-operative planning of a vascular reconstruction procedure is a test case for making experiments.

Section 2 of the paper contains the description of the Virtual Radiology Explorer (VRE): its functionality, architecture and main interaction capabilities. Section 3 classifies users of the VRE and their needs in respect to different projection modalities. In section 4 two different approaches of deploying interaction-visualisation capabilities within a PSE are presented. The Virtual Operating Theatre and the Individual Desktop Environment are differentiated by functional and interaction capabilities needed by different groups of users of the VRE. The notion of a Personal Space Station is introduced as a possible solution to combine two these mediums within the same environment.

2 The Simulated Environment for Vascular Reconstruction

2.1 Introduction to the Vascular Reconstruction

Vascular diseases affect arteries and veins. Vascular disorders in general fall into two categories: aneurisms and stenosis. An aneurysmal disease is a balloon like swelling in the artery. Stenosis is a narrowing or blockage of the artery. The purpose of the vascular reconstruction is to redirect and increase blood flow or repair a weakened or aneurysmal artery if necessary.

There are several imaging techniques that can be used to detect vascular disorders. 3D data acquired by computed tomography (CT) or magnetic resonance imaging (MRI) is converted into a set of 2D slices that can be displayed and evaluated from various perspectives and at different levels. Magnetic resonance angiography (MRA) is a technique for imaging blood vessels that contain flowing blood. It is very popular among cardiovascular specialists because of its ability to non-invasively visualise a vascular disease.

The verification of the operation plan is one of the most complicated tasks in vascular surgery. Different treatments for vascular diseases exist today. They include adding shunts and bypasses in the case of aneurysms and applying thrombolysis techniques, balloon angioplasty, bypasses and stent placement for a stenosis. The best treatment is not always obvious because of the complexity of a vascular disease and because of another diseases that a patient may have.

2.2 A Design Concept

The Virtual Radiology Explorer (VRE) gives a possibility to verify whether the selected treatment is the best in the current circumstances.

The VRE is a PSE, that puts a user into an experimental cycle simulated by a computer and let him apply his expertise to find better solution for the treatment of a vascular disease. The criterion of the success of a treatment is the normalisation of a blood flow in the affected area. The procedure of adding a bypass is of the most interest to us as it can be used both for treatment of aneurysms and stenosis.

The design concept of the VRE is represented on Fig.1.

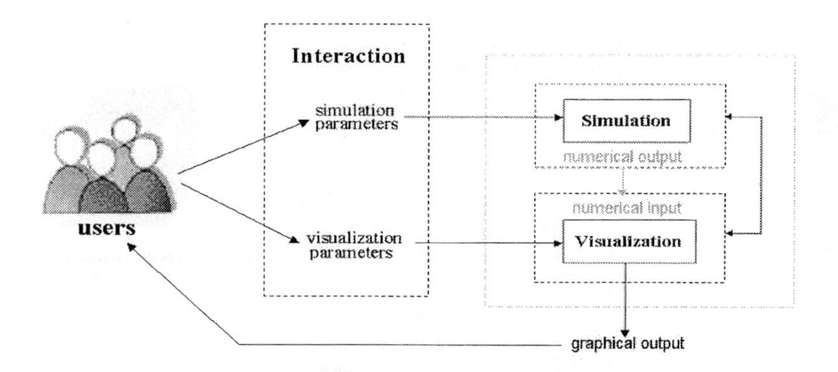

Fig. 1. A design concept

A simulation part of the VRE simulates the parameters of a patient's blood flow, i.e. velocity, pressure and shear stress. A user monitors the simulation process and controls the blood flow parameters. He has a possibility to add a bypass and to check whether a blood flow was normalised or not. He can also change the visualisation parameters, such as sample rate, scale, colour palette, light effects, etc.

The input for conducting experiments is scanned data of a patient stored in a database, which also contains an archive of interesting cases and interesting vascular images. This information is useful for both diagnostics and planning a treatment.

Fig.2 represents the current architecture of the VRE system, where the starting point is a scanner, and the front-end to the system is the Distributed Real-time Interactive Virtual Environment (DRIVE) system – a PC based environment, including a single-wall projection device developed by our group in the University of Amsterdam [1, 12].

The concept of interaction in virtual reality was selected as a basis for building the DRIVE system. Stereovision is the normal way almost everyone sees in the real world [4]. The virtual reality permits to build an environment where users interact freely with a 3D space and entities within it.

The working prototype of the VRE is provided with a multi-modal interface described in [10]. It combines natural input modes of context sensitive interaction by voice, hand gestures and direct manipulation of virtual 3D objects. We called this interaction mode the 'Virtual Operating Theatre', as a user 'plays' a role of a vascular surgeon applying a treatment of a vascular disease on a simulated patient [2].

The functionality of the VRE has been described in earlier publications [2, 3, 10]. Data conversion, segmentation, LBM-grid [5] generation, fluid flow simulation, surface and volume rendering are complicated computational tasks that need additional resources. But they are not very interesting in respect to human-machine interaction, as most of them are non-interactive or just can be run, paused or stopped by a user.

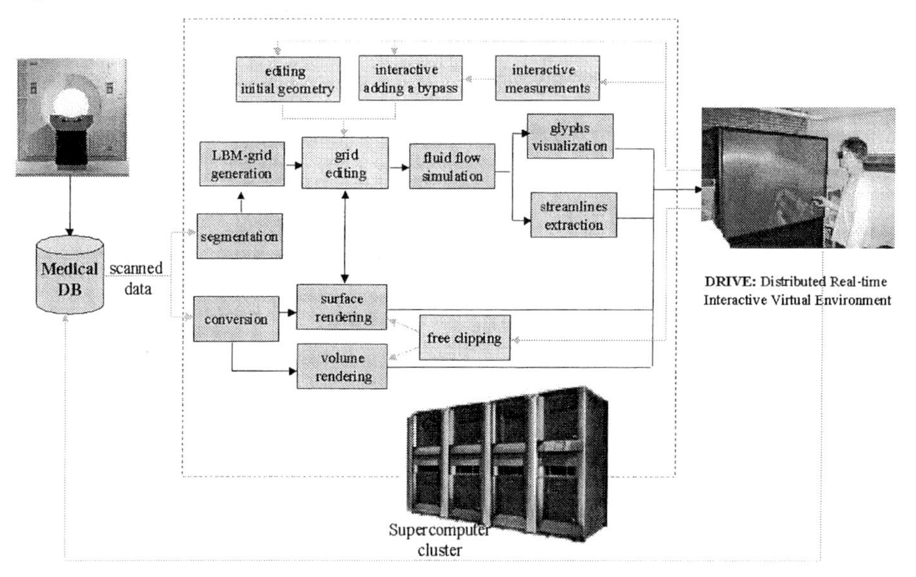

Fig. 2. Experimental set-up for the vascular reconstruction

2.3 Interaction Capabilities

Human-machine interaction within the VRE is concentrated mainly on retrieving data about a patient from a Medical DataBase (DB), conducting interactive measurements and grid editing.

Grid editing provides the possibility to edit the initial geometry of an artery: to remove insignificant elements or restore the fragments lost during the segmentation. Grid editing permits a user to create a bypass and place it on an artery. A bypass is a graft rerouting a blood flow around blockages. Usually it is a piece of vein taken from elsewhere in the body or an implant made from an organic material.

Measurements are important for diagnostics. Clinical decision-making relies on evaluation of the vessels in terms of a degree of narrowing for stenosis and dilatation (increase over normal arterial diameter) for aneurysm. The selection of a correct bypass (its shape, length, diameter) also depends on sizes and geometry of an artery. The interactive measurement component of the VRE [3] provides the possibility to measure quantitatively a distance, angle, diameter and some other parameters characterizing a 3D object in a virtual world. For conducting a measurement, a user has to position a corresponding number of active markers on an object.

A free clipping engine, which we are developing currently, is also an interactive component of the VRE. It permits to build clipping planes of different orientations in a 3D space. Having made a section of a 3D object by means of a clipping plane, a user can look inside an artery. Moving a clipping plane he will see the original data obtained from a scanner slice by slice.

3 User Attitudes to Different Projection Modalities

3.1 Users

The VRE users are divided into two main groups (see Fig. 3):

- System managers;
- Primary users.

A system manager is responsible for maintenance and support of primary users, local software and hardware availability, remote data transfer and management of distributed computational resources. A system manager deals with system and

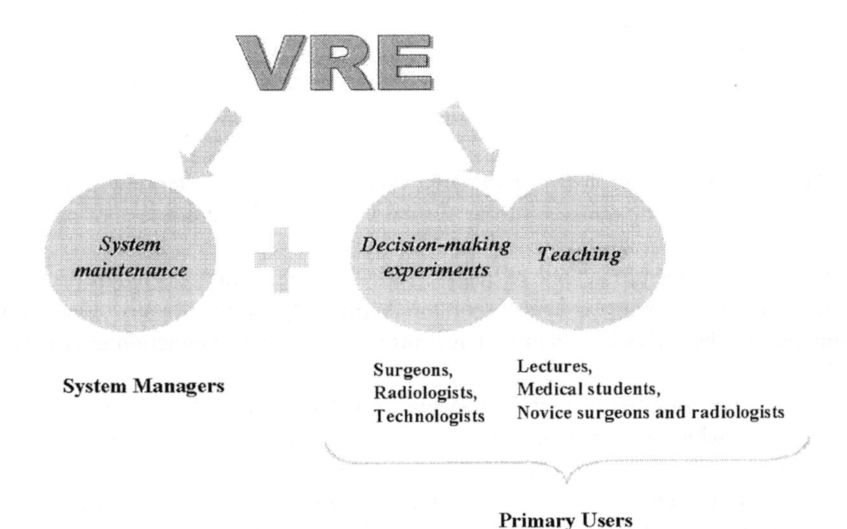

Fig. 3. User Classification

resource management: reliable network connection, secure authentication of primary users, access to computational-visualisation resources, availability of staff for specific equipment (MRA/MRI/CT scanners), etc.

Primary users are people, who use the VRE as a tool for conducting experiments. It is expected that the VRE will be used for the interactive decision support by vascular surgeons, radiologists (both diagnosis and interventional) and technologists. Vascular technologists are people from scientific or radiography background. They conduct patients' testing using special equipment, including MRA/MRI/CT scanners, for diagnosis of arterial and venous diseases. The VRE can also be used for training of medical students and novice physicians.

Potential users of the VRE differentiate by their preferences, skills, motivation, cognitive features and stress factors [11]. As a result, their expectations about the interaction within the VRE also vary.

User comfort is very important for the success of any software application. The working prototype of the VRE is almost finished. That's why we focus our today's efforts mostly on the improvement of the interaction and visualisation capabilities within the VRE. The primary users are of the most interest to us because most of them are inexperienced computer users and they expect that a PSE will provide a human-machine interaction similar to the interaction in a real world.

3.2 Users' Needs: Virtual Operating Theatre or Personal Desktop Environment?

A human-machine interaction depends to a great extent on projection equipment selected for deploying an interaction-visualisation medium. Today's existing solutions differentiate by information and visual design, provided forms of navigation, locomotion, selection and manipulation.

The research presented in this paper is focused on comparison of two projection modalities: the Virtual Operating Theatre introduced in section 2 and Personal Desktop Environment, which is an interaction-visualisation mode supported by standard PC/PDA applications.

The heuristic evaluation [6] of the VRE and the first user interviews, conducted recently, show that the concept of the Virtual Operating Theatre does not satisfy all VRE users.

It was already mentioned above that the DRIVE system was selected as projection equipment for deploying a concept of the Virtual Operating Theatre. One of the challenges of the DRIVE system is that it provides a shared interaction environment [12].

'One-to-one' semi-structured interview, conducted recently to identify the important variables for user profiling, shows that there is a possibility that some surgeons and interventional radiologists will prefer to use a personal desktop version of the exploration environment for accomplishing every-day tasks. As for the shared environment, it may happen, that it will be used only for the collaborative work, i.e. for training or medical conferences when the diagnosis and possible treatments are discussed within a group of people. Both hypotheses will be checked via series of semi-structured interviews. The preferences of technologists and diagnosis radiologists with respect to different projection modalities are not obvious currently and are also part of future analyses.

Like it was mentioned above, the VRE system can be used as a training environment for novice physicians.

Two modes of interactive training exist [11]:

- Lecture mode;
- Tutorial mode.

If it is a 'lecture mode', a lecturer, responsible for all training components, including theoretical, demonstrational and practical parts of a course, guides a class. In this case the concept of the Virtual Operating Theatre is a good solution accepted under the condition that the number of students in a group is optimal. A 'tutorial mode' is oriented to students who want to study material independently. The individual environment will satisfy them most of all. At the moment the VRE can be used only in a lecture mode because the concept of the Virtual Operating Theatre does not fit well to a tutorial mode of interactive training.

Such factors as users' preferences and motivation, discussed above, have an influence on the selection of an appropriate projection modality. But there exist one more factor, which is the most important in this respect. This factor is called 'simulator sickness'.

A simulator sickness [4, 9] is a kind of motion sickness except that it occurs in a simulated environment without actual physical motion. The simulator sickness occurs in conjunction with the virtual reality exposure. Users having simulator sickness cannot work in a virtual reality. According to [9] almost a quarter of computer users have simulator sickness. So approximately the same proportion of the VRE users will be unable to work with its virtual environment. For this type of users desktop remains the only possible solution.

4 A Personal Space Station as a Combined Interaction Medium

A Personal Space Station (PSS) is a relatively new concept for the implementation of an interaction-visualisation environment that may help to solve problems mentioned above [4].

The goal of a PSS is to allow a user to interact directly with a virtual world. A PSS consists of a mirror in which a stereoscopic image is reflected. The user reaches under the mirror to interact with the virtual objects directly with his hands or by using task-specific input devices. The main advantage of a PSS is that it combines both elements of desktop and virtual projection modalities within the same environment. And it is possible to switch in between if necessary.

A PSS is an individual environment by definition, but there is a possibility to build a shared environment where users can manipulate the same virtual objects working on different PSSs. More information about the PSS concept can be found in [7].

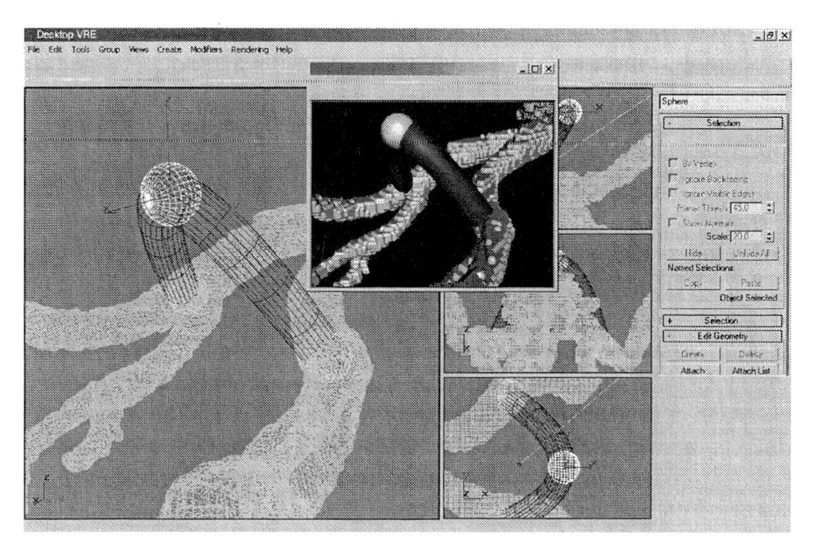

Fig. 4. Grid-editing tool of the desktop version of the VRE

The idea to combine virtual and desktop projection modalities within the same interaction-visualisation medium looks very attractive. In this case different types of users will be able to work within the same environment having a minimal feeling of discomfort. But building such combined medium is not an easy task. First of all, we have to take into account, that interaction in virtual and desktop projection modalities is different in respect to navigation, locomotion and manipulation capabilities [4, 8, 9]. To provide the possibility to switch between virtual and desktop interaction modes within the VRE, its current interface has to be changed significantly, as it is oriented currently mostly on the interaction in a virtual reality.

Let us compare a procedure of adding a bypass in virtual and desktop projection modalities. In a virtual environment a user manipulates 3D objects. He deals with 3D representations of an artery and a bypass. For building a bypass within the VRE a spline primitive is used. The procedure of adding a bypass in a virtual environment comes down to re-scaling of a spline and its correct positioning on an artery. For these manipulations a user has a wand. To manipulate successfully in a 3D virtual world a user should possess some special motor skills of navigation and manipulation, which sometimes are not trivial and depend on the level of implemented interaction capabilities.

As for the interaction in a desktop environment, user does not need additional motor skills. The main problem here is that within desktop applications we cannot manipulate 3D objects directly, we always deal with 2D projected representations of these objects [9]. Fig. 4 presents a mock-up of the graphical editor for adding a bypass within a desktop version of the VRE. 3D representation of objects does not have a big sense for a desktop environment. It is just a passive viewer, while a user adds a bypass operating with several projections of an artery, which orientations are pre-defined manually.

The same situation we have with a clipping procedure. In the case of virtual reality a user can change orientation of a clipping plane using a wand. He can navigate in a virtual world, look inside an artery and even walk through. In a desktop version additional interface capabilities have to be applied. For instance, a user can select slices of interest be means of a menu or a slider. A unique identification number will help to define a concrete slice.

As for interactive measurements, like it was already mentioned in section 2, a user first has to add active markers. Markers are building blocks of distance, angle and linestrip measurements [3]. The number of necessary active markers depends on a measurement to be done. For measuring a distance, a user has to add 2 markers, if it is angle – 3, for conducting linestrip or tracing measurements – at least 2. At the moment the interactive measurement component is available only for the virtual version of the VRE. Switching to a desktop projection modality leads to a necessity of deploying extra menus and sliders to simplify the work of a user in a projected 3D world.

5 Discussion and Future Work

In this paper a PSE for simulated vascular reconstruction has been introduced. The work is still in progress and one of the crucial issues to focus on is the improvement of interaction capabilities within it. The heuristic evaluation conducted recently and the initial stage of user profiling show that the human-machine aspects depend to a great extent on a projection modality selected for deploying an interaction-visualisation medium.

As the primary users of the VRE are physicians who are usually not very familiar with modern computer technologies, it is very important to make the process of their interaction within a PSE as much comfortable as possible.

Two concepts – the Virtual Operating Theatre and the Individual Desktop Environment – have been presented. It has been shown that both virtual reality and desktop solution are claimed by the VRE users. That's why we decided to combine virtual and desktop interaction capabilities within one medium. A PSS may help to bring this idea to life. Its main advantage is that it permits to switch from one to another projection modality if necessary. The experimental version of a PSS is currently being built in the University of Amsterdam within the Token 2000 project "Distributed Interactive Medical Exploratory for 3D Medical Images".

At the same time we are deploying a desktop version of a grid editing tool and free clipping engine. It is also planned to build a desktop version of an interactive measurement component of the VRE.

The final goal of this research is to investigate navigation, locomotion, manipulation and measurement capabilities within the Virtual Operating Theatre and the Individual Desktop Environment. The VRE running on a PSS will be an experimental environment for this research. Of course, of most interest to us are the attitudes of real users exploiting the VRE in their daily activities. For these case studies we will collaborate with the Amsterdam Medical Centre, the Leiden University Medical Centre and the Rotterdam Medical Centre.

Acknowledgements. The authors would like to thank Robert Belleman, Roman Shulakov, Hans Ragas Denis Shamonin and Daniela Gavidia (Section Computational Science, UvA) for the contribution to the development of the VRE; and Henriette Cramer and Dr. Vanessa Evers (Social Informatics Department, UvA) for the input to the usability studies.

References

1. Belleman R.G., Stolk B., de Vries R.: Immersive Virtual Reality on commodity hardware, Proceedings of the 7th annual conference of the Advanced School for Computing and Imaging, pp. 297–304. ASCI, May 2001 (2001).
2. Belleman R.G., Sloot. P.M.A.: Simulated Vascular Reconstruction in a Virtual Operating Theatre. CARS 2001 Conference (CARS2001), Berlin, Germany, June 2001 (2001).
3. Belleman R.G., Kaandorp J.A., Dijkman D., Sloot P.M.A: GEOPROVE: Geometric Probes for Virtual Environments, Proc. of HPCN Europe '99, Amsterdam, The Netherlands, pp. 817–827 (1999).
4. Bowman D.G., Hodges L.F.: User Interface Constraints for Immersive Virtual Environment Applications. Graphics, Visualization and Usability Center Technical Report GIT-GVU-95–26 (1995).
5. Kandhai B.D., Koponen A., Hoekstra A.G., Kataja M., Timonen J. and Sloot P.M.A.: Lattice Boltzmann Hydrodynamics on Parallel Systems, Computer Physics Communications, vol. 111, pp. 14–26 (1998).
6. Nielsen J.: Usability Engineering, Academic Press (2000).
7. Mulder J.D., van Liere R.: The Personal Space Station: Bringing Interaction within reach. Center for Mathematics and Computer Science, CWI, Amsterdam, The Netherlands (2001) http://www.cwi.nl/~robertl/papers/2002/laval/.
8. Pierce J., Forsberg A., Conway M.J., Hong S., Zeleznik R.: Image Plane Interaction Techniques in 3D Immersive Environments, Proceedings of 1997 Symposium on Interactive 3D Graphics, pp. 39–43 (1997).
9. Raskin J.: The Humane Interface: New Directions for Designing Interactive Systems, Addison-Wesley Pub Co (2000).
10. Zudilova E.V., Sloot P.M.A., Belleman R.G.: A Multi-modal Interface for an Interactive Simulated Vascular Reconstruction System, Proc. of the IEEE International Conference on Multimodal Interfaces, Pittsburgh, Pennsylvania, USA, October 2002, pp. 313–319 (2002).
11. Zudilova E.V., Sloot P.M.A.: A First Step to a User-Centered Approach to a Development of Adaptive Simulation- Visualization Complexes, Proc. of the International Conference of the Systemics, Cybernetics and Informatics, Orlando, Florida, USA, July 2002, V. V, pp. 104–110 (2002).
12. The University of Amsterdam Distributed Real-time Interactive Virtual Environment (UvA-DRIVE) webpage. http://www.science.uva.nl/~robbel/DRIVE/

Online Trajectory Classification

Corina Sas[1], Gregory O'Hare[1], and Ronan Reilly[2]

[1] Department of Computer Science, University College Dublin
Belfield, Dublin 4, +353 1 716 {2922, 2472}
{corina.sas, gregory.ohare}@ucd.ie
[2] Department of Computer Science, National University of Ireland
Maynooth, Co. Kildare, +353 1 708 3846
ronan.reilly@may.ie

Abstract. This study proposes a modular system for clustering on-line motion trajectories obtained while users navigate within a virtual environment. It presents a neural network simulation that gives a set of five clusters which help to differentiate users on the basis of efficient and inefficient navigational strategies. The accuracy of classification carried out with a self-organizing map algorithm was tested and improved to above 85% by using learning vector quantization. The benefits of this approach and the possibility of extending the methodology to the study of navigation in Human Computer Interaction are discussed.

1 Introduction

This study is part of ongoing research whose purpose is to identify the procedural and strategic rules governing navigational behaviour within virtual worlds. The present paper investigates the motion trajectories of a set of subjects while they accomplish spatial tasks within a Virtual Environment (VE). Prior studies in the area of spatial cognition were concerned with testing hypotheses about the impact of various factors on spatial knowledge acquisition [3]. However, none of them tried to investigate holistically the motion trajectories themselves. By providing a rich set of primary data, trajectory analysis can support the extraction of valuable information regarding the rules users employ in accomplishing spatial tasks. Moreover, when this analysis is performed in the light of some performance criterion (e.g., time required to perform a search task) it could provide valuable insights into discriminating efficient and inefficient navigational strategies and clustering the users accordingly.

Trajectory classification provides the benefits of reducing the huge amount of information stored in raw data and once a typology has been created it can be used to assess any new trajectory by associating it with an appropriate class. On-line trajectory classification would allow the identification of user's in terms of good or poor performers of spatial tasks. This identification could represent an essential initial step in designing the VE. Thus, the VE could be dynamically reconfigured in order to enable poor users to learn the efficient navigation procedures, while for good performers it can be redesigned in order to challenge users' spatial skills.

P.M.A. Sloot et al. (Eds.): ICCS 2003, LNCS 2659, pp. 1035–1044, 2003.

Attempts to cluster trajectories have been carried out primarily in the area of visual surveillance, especially novelty detection, with the purpose of identifying suspicious behaviour of pedestrians within an outdoor open area [5], [14]. This goal is directly linked to the idea of automatic surveillance, which would allow the replacement of human operator. In their study, Owens and Hunter have shown that the self-organizing feature map neural network could be successfully employed to perform trajectory analysis by both identifying the characteristics of normal trajectories and detecting novel trajectories [14].

However, trajectory analysis performed on a spatial cognition task represents a novel approach. The objective of this study involves identifying the *good* and *poor* motion trajectory and their associated characteristics. What is good and poor is determined in the light of both users' performance and findings of spatial cognition studies.

Without underestimating the role of traditional clustering methods, we propose the use of Artificial Neural Networks (ANN) as an alternative tool for trajectory classification. Neural networks provide a very powerful toolbox for modelling complex non-linear processes in high dimensionalities [11]. ANNs have many advantages over the traditional representational models, particularly distributed representations, parallel processing, robustness to noise or degradation and biological plausibility [6]. We consider that at least part of these strengths can be harnessed to model user's navigational behaviour.

2 Cluster Analysis Performed by Artificial Neural Networks

The main goal of cluster analysis is to reduce the amount of data, by subdividing a set of objects into (hierarchical arrangement of) homogeneous subgroups. A significant outcome is reduced complexity with a minimal loss of information which allows a better understanding of the analysed data [12].

An important aspect of any clustering method is the minimisation of classification errors. As Kaski pointed out, [7] one problem usually associated with clustering methods is the interpretation of clusters. Due to their ability to extract patterns and to visualise complex data in a two-dimensional form [7], Self-Organizing Feature Map (SOM) are used to perform the trajectory cluster analysis. Like many other clustering techniques, SOM reduces representations to the most relevant facts, with minimum loss of knowledge about their interrelationships [7].

The SOM is a neural network algorithm with several advantages over other clustering techniques [7], [14]. The mapping from a high dimensional data space onto a two-dimensional output map is effectively used to visualise metric ordering relations of input data. Reducing the amount of data allows comprehensible cluster identification and interpretation, which is a difficult task in the case of traditional clustering methods [7]. As any other ANN, SOM has a considerable potential to generalise, meaning that once it is trained, SOM is able to classify new data within the set of clusters previously identified.

Features like the approximation of the probability density function of input space, the identification of prototype best describing the data, the visualisation of the data and the potential to generalise, highly recommend SOM as a basis for on-line automatic extraction of trajectory clusters.

Furthermore the basic features of SOM and Learning Vector Quantization (LVQ) as unsupervised and supervised learning processes respectively are outlined. LVQ is a supervised learning algorithm related to SOM. The SOM and LVQ algorithms have been developed by Teuvo Kohonen and implemented by his team from Helsinki University of Technology, in the form of SOM_pak [9] and LVQ_pak [10]. These comprehensive software package are available online and were used in this study. SOM is based on an unsupervised learning process, allowing both the cluster identification within the input data and the mapping of an unknown – not previously seen – data vector with one of the clusters. This process is carried out without any prior knowledge regarding number and content of the clusters to be obtained [7]. When a set of already clustered input data is available, a supervised learning process can be employed to identify to which class an unknown data vector belongs.

2.1 Self Organizing Maps

A basic SOM consists of an input layer, an output map and a matrix of connections between each output unit and all the input units. The input is usually represented by a multidimensional vector with each unit coding the value from one dimension. Every node from the two-dimensional output layer is associated with a so-called *reference vector* (m_i), consisting of a set of weights from each input node to the specified output node. In a simplistic way, each input vector is compared with all the reference vectors and the location of *best match* in some metric, usually the smallest of the Euclidean distances, is defined as the winner. Around the maximally responding unit, a topological neighbourhood is defined and the weights of all units included in this neighbourhood are adjusted, according to equation (1), where m_i is the weight at time $(t+1)$ and η is the learning rate.

$$m_i(t+1) = m_i(t) + \eta\,[\,x(t) - m_i(t)\,]\,. \quad (1)$$

The topological neighbourhood should be quite large at the beginning, to enable a global order of the map, while in the subsequent stages its values decreases as a function of time. Accordingly, the learning rate varies in time from an initial value close to unity, to small values over a long time interval. Training is performed during two phases: an ordering phase during which the reference vectors of the map units are ordered (neurons in different areas of the network learn to correspond to coarse clusters in the data), and a much longer fine-tuning phase during which the reference vectors in each unit converge to their correct values (neurons adjust to reflect fine distinctions).

The learning process consists of a "winner-takes-all" strategy, where the nodes in the output map compete with each other to represent the input vectors. For this reason, the output layer is also called the competitive layer. Competitive learning is an adaptive process, through which the neurons from the output layer become slowly sensitive to the input data, learning to represent better different types of inputs.

As Kohonen pointed out, [8] a significant property of SOM is the tendency to preserve continuity in terms of mapping similar inputs to neighbouring map locations influenced by the weight vectors trying to describe the density function of the input vectors. As a result of these antagonistic tendencies, the distribution of reference vectors is rather smooth, given the search for an optimal orientation and form to match those of the input vector density. In addition, the greater the variance between the input vector features, the better their representation on the output map. It is expected that these features correspond to the most important dimensions of the inputs.

2.2 Learning Vector Quantization

LVQ consists of an input layer comprising multidimensional vectors described by their features and an output layer whose neurons correspond to the predefined classes. There is also a matrix of connections between each output unit and all the input units, consisting of weights vectors. Since each weight vector corresponds to a class, they are considered as labelled. The basic idea is that input vectors belonging to the same class will cluster in data space, in a form of a normal distribution around a prototype vector. Classifying an input vector consists of computing the Euclidean distance between the considered input vector and all the weight vectors, followed by assigning it to the class associated with a weight vector for which the Euclidean distance is minimum [10].

During training, an adaptive process occurs with respect to the closest weight vector, also called the winning neuron. When both the input vector and the weight vector belong to the same class, meaning that the input vector was correctly classified, the weight vector is modified in order to become a better approximation of the input vector. However, when the input vector is incorrectly classified, the weight vector is adjusted in a way which increases its distance of the input vector (since they belong to different classes).

2.3 SOM versus LVQ

While the SOM algorithm strives to approximate the weight vectors to the input ones, LVQ tries to lead to weights that effectively represent each class. The process of adjusting the weights, without respect to any topological neighbourhood differentiates LVQ from SOM. The performance of LVQ can be increased by initialising the codebook vectors with those values obtained by training the SOM [7]. Variants of SOM have been successfully applied to a large number of domains, ranging from monitoring and control of industrial tasks, to robot navigation, from data processing to machine vision, from image analysis to novelty detection [7]. However, their adoption within the frame of spatial cognition in VE constitutes a novel approach.

3 Procedure

Virtual Environments (VE) have become a rich and fertile arena for investigating spatial knowledge. Within the VE, the user set of *actions* is restricted, consisting mainly of navigation and locomotion, object selection, manipulation, modification and query [4]. Through their powerful tractable characteristic [1], VEs enable accurate spatio-temporal recording of users' trajectory within the virtual space. Attempts to understand spatial behaviour in both real and artificial worlds were primarily concerned with highlighting the symbolic representation of spatial knowledge.

In this study we utilised ECHOES[1] [13], [2] as an experimental test-bed. It is a virtual reality system which offers a small-scale world, dense, static and with a consistent structure. Adopting a physical world metaphor, the ECHOES environment comprises a virtual multi-story building, each one of the levels containing several rooms: conference room (Fig.1), library (Fig.2), lobby etc. Users can navigate from level to level using a virtual elevator. The rooms are furbished and associated with each room there is a cohesive set of functions provided for the user. These features enable ECHOES to offer an intuitive navigational model.

A sample of 30 postgraduates was asked to perform two tasks within the VE. The first, an exploratory task, provided the primary data for the trajectory classification, while the second, a searching task, offered a basis for assessing the quality of exploration and the efficiency of the exploratory strategy. The time needed to search for a particular room acts as a *performance indicator.*

A comprehensive set of data was recorded throughout the experiment. Each movement greater than half a virtual meter, and each turn greater than 30°, were recorded. This was achieved by the inclusion of a rich set of virtual sensors together with an odometer and rotational event listener [2].

Fig. 1. Virtual Conference Room

Fig. 2. Virtual Library

4 Data Analysis and Results

The use of SOM [9] and LVQ [10] for performing the trajectory cluster analysis requires several steps: data collection, construction and normalisation of data set, unsupervised training, visualisation of the resulting map, cluster identification, obtaining a set of trained labelled codebook vectors to be used in supervised training

---

[1] ECHOES (European Project Number MM1006) is partially founded by the Information Technologies, Telematics Application and Leonardo da Vinci programmes in the framework of Educational Multimedia Task Force.

and measuring classification accuracy. When all these steps are performed in order to classify online trajectories, they should be automatic and seamlessly intertwine. For this we used the developed several modules serially connected as presented in Fig. 3.

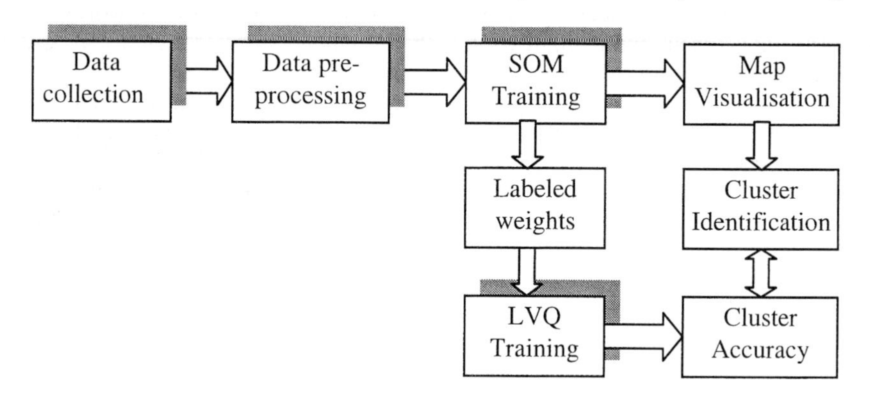

Fig. 3. The modular system for on-line trajectory classification

Collecting Data
The data collection module is based on the listener agent previously developed by Delahunty [2]. The Listener Agent gathers information about user behaviour within the virtual world. The data captured when the user interacts with the ECHOES virtual space contains details of navigation paths through the world and time spent in different rooms.

Pre-processing Data
Data pre-processing consisted of transforming the raw data into a suitable form to be fed into the SOM. In the raw data each trajectory is represented by a multivariate time series. However, we were interested in testing if a static representation of trajectory is sufficient to perform a classification. For this, the reduction of the raw data should be done by preserving their significant features. We choose to represent each trajectory by the degree of occupancy of a predefined set of spatial locations and an extra input node representing the degree of rotating in VE (29 nodes). For the SOM analysis, we overlaid the virtual space with a grid composed of 28 squares of 4×4 virtual meters. Each trajectory was converted to a succession of locations on the grid. The next step necessitated the mapping of each trajectory into a sequence of 28 neurons (one for each location), according to equation (2), where NV is the input node value and LOC is location occupancy expressed as how many times the user revisited that location.

$$NV = \log_{10}(9 \times LOC + 1)\ . \tag{2}$$

The above transformation allows a clear differentiation between non-visited (NV = 0) and visited locations. In the later case the NV is within the range 1–2, 1 for only one visit and 2 for 11 visits, 11 being the maximum number of times for revisiting a location. Apart of the previous encoding which features the space covering, the trajectories were characterised by the amount and size of users' rotations. We considered that trajectories characterise by rotation angles greater than 90° present an interesting feature. If a trajectory has more than 10% of the rotation angles equal or greater than 90°, the 29th node of the input vector was set to 3, otherwise it was set on 0.

SOM Training
Once the data were pre-processed, we randomly divided them in two equal subsets, keeping one for training and the other one for the testing. Each set consisted of 63 vectors, comprising encoded trajectories covered by the users on each level.

A SOM of 16×12 neurons was used to perform a topology-preserving mapping. The first phase of training was carried out for 1000 epochs, a radius of 16 and with a learning rate of 0.8, while the second phase lasts 120000, with a learning rate of 0.01 and a radius of 2. The random seed was 275 identified by using the vfind program. These parameters were retained, after we tried more than 50 trainings, with different architectures and learning rates, because they led to the smallest quantization error for the testing set (1.97), while for training set it was 0.35. Quantization error represents the norm of difference of an input vector from the closest reference vector [9].

Maps Visualisation
The resulting organisation of the map, shown in Fig. 4 and 5 shows five clusters of users, where clustering is on the basis of their navigational pattern within the VE. Figure 4 is associated with the training set of trajectory while Figure 5 with the testing set. Numbers which were associated with the winner neurons within each cluster are replaced by the original corresponding trajectory.

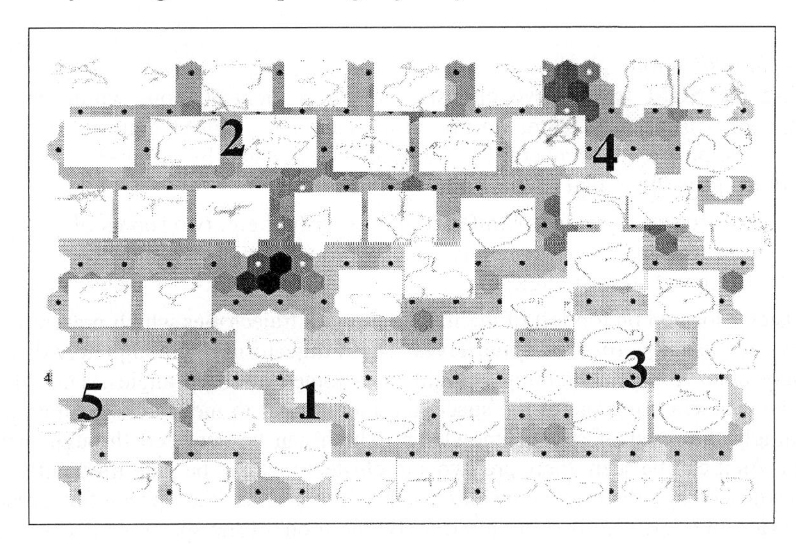

Fig. 4. SOM Map obtained from the training set

With respect to visualisation, the SOM provides an additional benefit: the clusters boundaries are represented by darker shades of grey, since they represent larger distances between adjacent neurons.

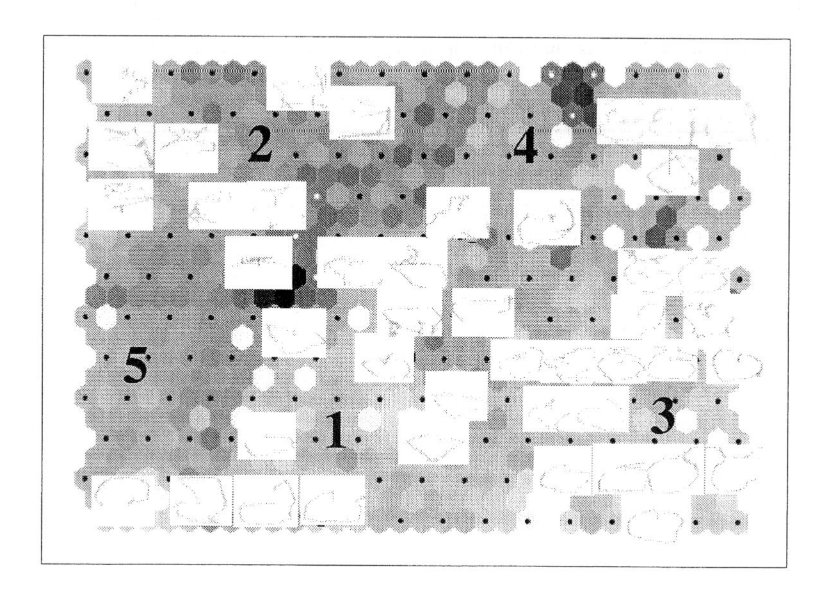

Fig. 5. SOM Map obtained from the testing set

Cluster Identification

Training the SOM led to five clusters. For their identification, within the area corresponding to each of them, we placed the associated cluster number e.g. cluster number 1 comprises the trajectories within area designated by number 1, located in the middle of the lower half of the map.

Cluster 1 groups trajectories limited to the lower half e.g. two rooms of the spatial layout. These trajectories are generally circular. Some of these trajectories are smooth while other present sharper angles.

Cluster 2 located on top left of the map comprises trajectories which present lots of turnings and usually crossover themselves. These trajectories are completely different than any other group, containing straight lines joined at sharp angles. They do not allow an efficient coverage of the space, are more likely to induce disorientation and accordingly the level of spatial knowledge which can be acquired through them is limited. As it can be seen, there are two sub-clusters that can be identified within this class, whose main distinction resides in the coverage of the space. For some of trajectories, the coverage is restricted to only one room of the space, while the rest of them cover larger space, but rarely going circular. Actually these trajectories are erratic and the user seems anxious to explore the space, e.g. he/she rather moves *in the same area* or covers larger space but in this case, it is likely that the returning to the starting point is achieved through approximate the same path.

Cluster 3 located on the right part of the second half of the map, consists of very smooth circular trajectories, which have at least one direction towards the centre of the spatial layout. Cluster 4 comprises longer trajectories, which cover most of the spatial layout. They present the "going around the edge" feature, more pronounced than other clusters, except cluster 2. Cluster 5 presents circular trajectories perform within the first half of the spatial layout.

Each set of trajectories, with the exception of cluster 2, proves beneficial along the temporal dimension. Carefully selected and ordered, they enable users to acquire particular spatial knowledge, with a minimum investment of resources.

Previous work in classifying trajectories, performed only on the basis of locations, led to a more detailed classification [15]. However, since the purpose of this analysis is to discriminate between users employing efficient strategies and those navigating through a set of inefficient strategies, we do have to take the rotation angle into account. This leads to a more detailed representation of cluster 2.

Training LVQ

Once the SOM was trained, the codebook vectors could be used for initialising the weights for LVQ algorithm. This did indeed lead to increased classification accuracy from 72% obtained using random initialisation to 87%. In other words, each trajectory from the testing set was correctly classified by the LVQ with 87% accuracy. Within each class, the classification accuracy is slightly different: cluster 1 – 86%, cluster 2 – 100%, cluster 3 – 63%, cluster 4 – 87 % and cluster 5 – 100%. As it can be seen, the trajectories belonging to cluster 2, and which require special attention are correctly classified in each case. This is an important outcome, supporting the goal of our study aiming to discriminate users in terms of good and poor performers of the spatial tasks. This finding should be also emphasised in the light of the fact that more than 50% of trajectories composing cluster 2 are covered by the subjects with worst performance in the searching tasks (bottom 10% female and bottom 10% male).

5 Conclusion

The study shows that the ANN could be successfully employed in modelling spatial behaviour in VE, in terms of classifying users' motion trajectories performed on-line. Based on this classification, each new user can be associated with one of the clusters, and accordingly identified as employing efficient or inefficient navigational strategies.

The SOM and LVQ analysis led to the identification of five user trajectory clusters within the same VE. The accuracy of classification is above 85% which is a significant outcome given the relatively limited size of our training and testing sets. Within each cluster, trajectories share common features. Some of them were already identified while the others request further analysis. A future direction will be to extract the quantitative rules governing the clusters and to express them in a symbolic manner. The study findings could provide insights in understanding what do the efficient and inefficient strategies mean, by interpreting them through theoretical aspects of spatial cognition described by environmental psychology. Moreover, the study indicates that using neural networks as a tool in studying navigation can be beneficial for user modelling in the area of spatial knowledge acquisition. Permitting a comparative analysis between efficient and inefficient navigational strategies, this methodology could suggest how VEs might be better designed. Based on these results, further work will focus on assisting new users to improve their spatial abilities in exploring a new virtual environment. After a period of navigation, SOM would be able to integrate the online trajectory within the appropriate cluster. If the user's trajectory history matches, for example, cluster no. 2, the system will assist the user in his/her further exploration. Thereafter this guidance will improve user exploration. Alternatively real-time dynamic reconstruction of the VE could assist the user in their tasks.

References

1. Amant, R.S., Riedl, M.O.: A practical perception substrate for cognitive modelling in HCI. International Journal of Human Computer Studies 55(1) (2001) 15–39
2. Delahunty, T.: ECHOES: A Cohabited Virtual Training Environment. Master Thesis, Department of Computer Science, University College Dublin (2001)
3. Freksa, C., Habel, C., Wender, K.F. (eds.): Spatial Cognition, An Interdisciplinary Approach to Representing and Processing Spatial Knowledge. Lecture Notes in Computer Science, Vol. 1404. Springer-Verlag, Berlin (1998)
4. Gabbard, J., Hix, D.: Taxonomy of Usability Characteristics in Virtual Environments, Final Report to the Office of Naval Research (1997)
5. Grimson, W., Stauffer, C., Lee, L., Romano, R.: Using Adaptive Tracking to Classify and Monitor Activities in a Site, Proceedings IEEE Conf. on Computer Vision and Pattern Recognition (1998) 22–31
6. Haykin, S.: Neural Networks: A Comprehensive Foundation. Prentice-Hall New Jersey (1994)
7. Kaski, S.: Data exploration using self-organizing maps. Acta Polytechnica Scandinavica, Mathematics, Computing and Management in Engineering Series No. 82, Finnish Academy of Technology (1997)
8. Kohonen, T.: Self-organizing maps. Springer Series in Information Sciences, Vol. 30, Springer-Verlag Berlin (2001)
9. Kohonen, T., Hynninen, J., Kangas, J., Laaksonen, J.: SOM PAK: The Self-Organizing Map program package, Report A31, Helsinki University of Technology, Laboratory of Computer and Information Science (1996)
10. Kohonen, T., Hynninen, J., Kangas, J., Laaksonen, J., Torkkola, K.: LVQ PAK: The Learning Vector Quantization Program Package. Version 3.1. Helsinki University of Technology, Laboratory of Computer and Information Science (1995)
11. Lint, H. van, S.P. Hoogendoorn, H.J. van Zuylen: Freeway Travel Time Prediction with State-Space Neural Networks, Preprint 02-2797 of the 81st Annual Meeting of the Transportation Research Board, Washington D.C. (2002)
12. Lorr, M.: Cluster analysis for social scientists. Jossey-Bass Publishers, San Francisco (1983)
13. O'Hare, G.M.P., Sewell, K., Murphy, A.J., Delahunty, T.: ECHOES: An Immersive Training Experience. Proceedings of Adaptive Hypermedia and Adaptive Web-based Systems (2000) 179–188
14. Owens, J., Hunter, A.: Application of the self-organizing map to trajectory classification. Proc. of the 3rd IEEE Workshop on Visual Surveillance (2000) 77–83
15. Sas, C., O'Hare, G.M.P., Reilly, R.G.: A Connectionist Approach to Modelling Navigation: Trajectory Self Organization and Prediction. Proceedings of 7th ERCIM Workshop, User Interfaces for All. Carbonell, N. and Stephanidis, C. (eds.) (2002) 111-116

Trajectory Mapping for Landmine Detection Training

Yang Cai

Human-Computer Interaction Institute,
School of Computer Science,
Carnegie Mellon University,
5000 Forbes Avenue, Pittsburgh, PA 15213, USA
ycai@cmu.edu

Abstract. A head-mounted camera is a useful tool for studying the usability of mobile devices in the field. In this paper, a computerized visualization method is presented. It includes the target trajectory mapped with the deformable template-based tracking algorithm and landmarks-based relative object registration. A landmine detection training video is used for the case study. The results show that this approach has advantages over optical flow and overhead camera methods.

1 Introduction

Human field performance has been studied for decades, from golfing to landmine detection. It has become a renaissance area because of: 1) emerging mobile computers for field applications, such as Ground Penetration Radar (GPR) for landmine detection and handheld training computers for Navy personnel, etc., 2) emerging remote control through telepresence, such as robotic rescue systems and capsule medical cameras, 3) traditional manned field missions with new situations, such as landmine detection for peace-keeping and vehicle driving studies, etc. In light of this, human field performance study is an "old field" that is redeemed with new technologies.

Video cameras have been widely used in human performance studies, such as surveillance camera, infrared camera, high-speed camera, microwave imaging camera, etc. It is common to keep vision systems static while tracking human subjects' movement. In these cases, human tracking is relatively easy during the video post-processing phase. For example, we can use the background subtracting method to separate the human subjects and the static background. Also, it is easy to measure the distance or track the motion speed. However, in many situations, the static camera-based approach is rather expensive or difficult to use in the field. For example, it is very hard to use a single static camera to track human activities in an obscured scene or multiple rooms. In addition, it is hard to track the human operation in a very large open field without an overhead camera or multiple cameras. If we use multiple cam-

P.M.A. Sloot et al. (Eds.): ICCS 2003, LNCS 2659, pp. 1045–1053, 2003.

eras, then we have to add image fusion, object registration and synchronization functions. In these cases, a head-mounted video camera seems a reasonable choice.

Head-mounted video cameras have been used in special research projects for example, "augmented reality" by video overlaying on head-mounted displays [1,2,3,4,5,6], lip movement tracking [7] and eye movement tracking [8]. The "augmented reality" registers computer generated graphics to a video image. In this paper, we attempt to do the opposite: to register an object from the video image to computer generated graphics. The current head-mounted camera-based lip and eye movement tracking systems only observe a single component on human face. In this paper, the author focuses on a broader problem: "how to visually evaluate human field performance with a head-mounted video camera?" The goal of this study is to develop a computerized object tracking and mapping system that can automatically register the moving target to a trajectory map. This study uses military landmine detection video as a case study and uses computer vision algorithms to map the original video data to a dynamic tracking graph. It is expected that the method can be applied to other fields, such as behavior measurement for elderly in nursing homes, user performance modeling for airplane inspection, etc.

2 Trajectory Map

A trajectory map is a 2D or a 3D space that is registered with a dynamic trace of a target. It is a visual model of human field performance dynamics, for example, search patterns, pace, and sweeping patterns, etc. Fig.1 shows a comparison of a trace of a sweeping metal detector from an expert and a novice. From the map we discovered that the trace of the expert is uniform and thorough. However, the trace of the novice is uneven that contains missing spots.

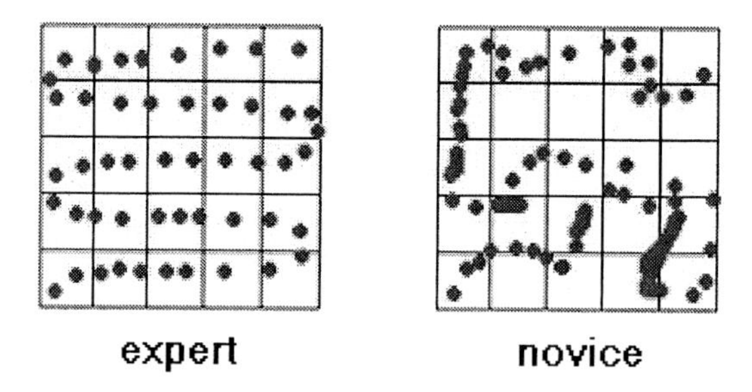

Fig. 1. *"Sweeping Patterns" of a metal detector head. Note the expert's pattern is uniform and thorough. The novice's pattern is uneven and random*

Visualization helps the analysts to understand the dynamics of the human behavior in the trajectory. The methods include: 1) time-stamping the trajectory points with "temperature colors", which maps a duration time with a color, etc., 2) plotting the effective halo envelope for the trajectory, which reveals the overlapping patterns. In many cases, adding verbal protocols which are aligned to the trajectory points would help analysts to understand the subject's motivation, cognition and decision making process.

The challenge for processing the data from a head-mounted video is how to register the trajectory map. A head-mounted camera has four degrees of freedom (DOF): pitch, yaw, tilt and zoom, which make the registration rather cumbersome. To make efficient target tracking and registration, landmarks are recommended in the head-mounted video tracking. Fig 2 shows an example of the usage of the measurement tapes as landmarks.

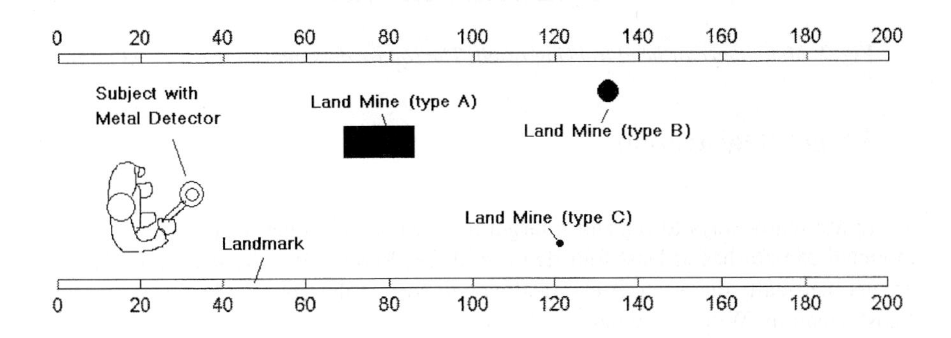

Fig. 2. *Landmine detection training field*

3 Target Tracking

The moving target normally has a defined shape, e.g. circle, etc. In this study, an active contour model, "Snake" [9] is used for tracking. It is a general algorithm for matching a deformable model to an image by means of energy minimization. The energy function is a weighted combination of internal and external forces. The snake is defined parametrically as $\mathbf{v}(s) = [x(s), y(s)]$, where x(s), y(s) are x, y co-ordinates along the contour and $s \in [0,1]$. The energy functional to be minimized as

$$E_{\text{snake}} = \int_0^1 \{[E_{int}(\mathbf{v}(s)] + [E_{image}(\mathbf{v}(s)] + [E_{con}(\mathbf{v}(s)]\}\, ds$$

where E_{int} represents the internal energy of the spline due to bending, E_{image} denotes image forces, and E_{con} external constraint forces. In this case, $\mathbf{v}(s)$ is approximated as a spline to ensure desirable properties of continuity.

Fig.3 shows a result from the algorithm for tracking a metal detector head from the head-mounted video. After nearly ten iterations for each frame, the deformable "snake" successfully located the metal detector head in the video.

Fig. 3. *Metal detector head tracking result. The square dots indicate the tracked target*

4 Target Registration

There are many ways to register a target to a two-dimensional map. Because a head-mounted camera has at least four degrees of freedom (DOF), the images in the video are geometrically distorted. A classical way to correct the image is linear or non-linear transformation. With a few pairs of "control points", a warping function can be used to transform the distorted raw images from pixels in the source video to the destination defined by a transformation function. [10] We call the transformation-based methods "*absolute registration*" methods since they generate "absolute" coordinates of the tracked object in the map.

However, in reality, there are only very few video frames containing required control points for the linear transformation. In many cases, the video only shows partial landmarks. What do we do to fill in these gaps? In addition, it is not necessary to transfer all pixels from one to another. For target tracking purposes, we only need the relative coordinators and distances referenced to the landmark and the plane of the landmarks. In light of the shortcomings of the transformation, this study focuses on the "*relative registration.*"

Relative registration is a non-metric measurement method in which a target is not only just an object but also a reference itself. It is an approximate way for a quick measurement of the object's size and the distance between things. For example, artists often use "number of heads" to measure human figure's height and use "number of eyes" to measure the width of face. This is based on observations of our daily life. Our perception systems have "internal yardsticks" for qualitative measurement. Our eyes do not make absolute measures of characteristic of the subject, but instead detect these characteristics only in a relative way. We do not see the true color of a thing, but rather an apparent color, which is our sensation of how a color is different from the colors surrounding it. As Van Gogh said, "There is no yellow if there is no blue if there is no red." The same principle applies to our perception of edges, patterns, and shapes. We may focus our attention solely on a part of the subject or notice this part peripherally as we scan our eyes around the whole visual field. More evidence has been found from artists' painting textbooks and the landmine detection tapes.

The relative registration procedure in this study is based on a few assumptions. For example, we assume that the metal-detector head is near parallel to the ground and the subject looks at the near field ground, etc. The registration heuristics is as followings:

- Determine a template (box) of the object (metal detector head) in the image.
- Locate a feature point on the landmark (e.g. the numerical mark)
- Find the reference lines that are either perpendicular or parallel to the landmark on the plane.
- Measure the distances (perpendicular and parallel directions) from the object (metal-detector head) to the landmark with the template. For example, in Fig. 4 on the right, the relative coordinate to the mark "100" for the metal-detector head object is (0.8,0) in terms of "number of templates"

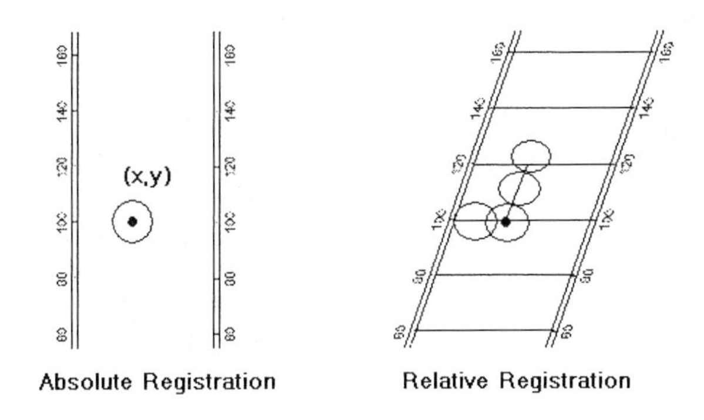

Fig. 4. *Illustration of "Relative Registration" versus "Absolute Registration." In relative registration, the template of the target is used as a measurement reference.*

5 Results

A preliminary experiment has been conducted based on a 60 minute test video from the field. It was stored on MiniDV and converted to AVI files at 30 frame per second with the frame size of 177 x 172. Before the video analysis, we removed about 20% of "irrelevant" clips, e.g. vomiting and resting. Those irrelevant clips are easy for humans to understand but difficult for computational processing. Fig.5 shows a result of the trajectory mapping from a video clip, where the dots are reference points with intervals of 10 frames and the trace of the metal detector head indicates a normal sweeping pattern, which is uniform and thorough. Fig.6 and Fig. 7 show examples of the traces of a metal detector head while making decisions to determine the location and type of the landmine. The closer dots, the more decision time that the subject spent. Also, from the trajectory maps, we found a few decision making "styles." For example, the pattern in Fig.5 shows a circular search style. The pattern in Fig.6 shows a cross-shape search style.

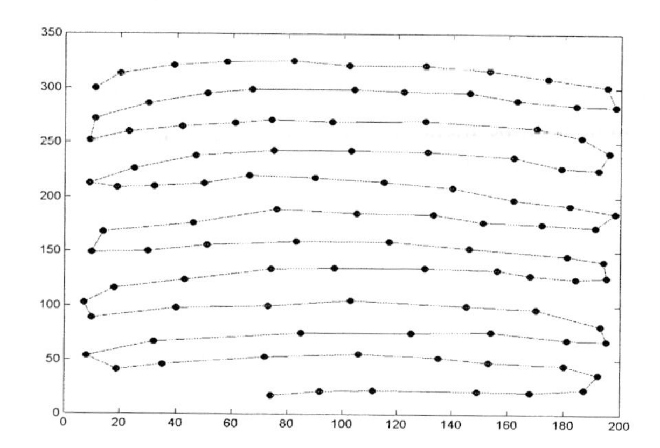

Fig. 5. *An output of the "sweeping" patterns. Dots are reference points with an interval of 10 frames and the trace of the metal detector head indicates a normal sweeping pattern, which is uniform and thorough.*

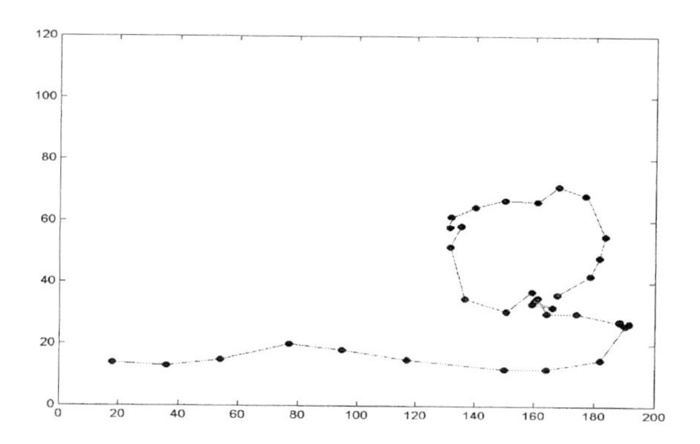

Fig. 6. *A circular shape searching style. The trace of the metal detector shows how the subject makes decisions to determine the location and type of landmines.*

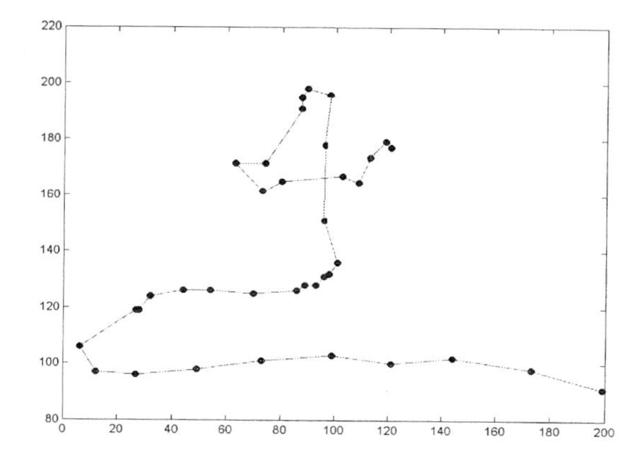

Fig. 7. *A cross-shape searching style. The trace of the metal detector shows how the subject makes decisions to determine the location and type of landmines.*

Table 1. *Performance summary*

Items	Performance
Registration accuracy	¼ to ½ of metal detector head diameter
Trajectory mapping speed	Near real-time
Object tracking speed	3 times real-time on PC (700 MHz,256MB RAM)
Landmark visibility	57% single lane, 19% both lanes, 24% none
Manual reinitiating	35% (e.g. subject looks at sky, drinks water, etc.)

Table 2. *Comparison between head-mounted camera and overhead camera*

Items	Head-Mounted Camera	Overhead Camera
Protocol sound tracks	yes	Wireless needed
Camera tower	no	yes
Object registration	Relative registration	Absolute registration
Object tracking	Software-based	Light-bulb-based
Distance distortion	less	yes

Table 3. *Comparison between the Landmark-based and Optical Flow-based methods*

Items	Landmark-based	Optical Flow-based
"drafting"	no	yes
generalization	no	yes
computation	simple	intensive

6 Discussions

A preliminary experiment has been also conducted to apply an Optical Flow algorithm to estimate the moving directions. Unfortunately, because there is a moving target in addition to the moving background, it is rather hard to separate the target from the background.

At this stage, manual initiation is used in tracking. Since the manual initiation is just to put a circular shape near the center of the target, and the manual and automation ratio is significant, it is tolerable to use the manual initiation.

The main problem for tracking based on the head-mounted camera is the "drifting problem". The trajectory would accumulate the error by time and eventually drift away from the correct course. To avoid the drifting problem, we have to use frequent landmark checking, introduce additional sensors, such as magnetic sensors, etc. to calibrate the ground truth at a certain duration. Besides, dual-camera registration, e.g. one for head-mounted and another for overhead, is desirable to minimize drifting.

Tracking a target with the current algorithm is time-consuming. The video processing time over the real-time is 3:1 on a PC with 700MHz, 256MB RAM. To improve this, development of faster tracking algorithms is underway.

At this stage, the optimal tracking accuracy is ¼ to ½ of the size of the metal detector head template. A 2D trajectory map may also lose the 3D information. For example, landmine detection experts sometimes turn the metal detector head at 45 degree to identify the halo envelop of the mine or just test the head. It is hard to observe this on a 2D map.

7 Conclusions

In this paper, a visual trajectory model for field performance evaluation is presented. The author addresses the approach for a trajectory map, target tracking and the "relative object registration" method. A landmine detection training video is used as a case study.

Although head-mounted video cameras have been used for field performance observation for long time, very little has been done in trajectory mapping. It is concluded that the "snake" algorithm can be used for deformable target tracking for the head-mounted video camera. However, it is rather computation intensive. The *relative registration* method is a novel approach to map the target from incomplete larndmarks in video scenes. It provides approximate and fast qualitative measurement. Compared to the overhead camera and optical flow methods, this approach is inexpensive and flexible.

It is also suggested that computerized head-mounted video analysis has its great potential in studying human field performance. To improve the accuracy of the image registration, physical trackers, such as magnetic sensors, are recommended in addition to the landmarks in the video. In addition, since video analysis involves both human and computational collaborative efforts, a well-designed human-computer interface would significantly increase the productivity.

Acknowledge. The author thank Dr. James Staszewski for providing the field data and insightful advice and Dr. Margaret Nasta for her review and comments on this paper.

References

1. Chai, L., W.A. Hoff, and T. Vincent, "3-D motion and structure estimation using inertial sensors and computer vision for augmented reality," *Presence: Teleoperators and Virtual Environments*, 2000.

2. Chai, L. and K. Nguyen, W. Hoff, and T. Vincent, "An adaptive estimator for registration in augmented reality," *Proc. of 2nd IEEE/ACM Int'l Workshop on Augmented Reality*, San Francisco, Oct. 20–21, 1999.
3. Hirota, G. et al, "Hybrid tracking for augmented reality using both camera motion detection and landmark tracking", US Patent 6064749, May 16, 2000
4. Hoff, W. A. and T. Vincent, "Analysis of Head Pose Accuracy in Augmented Reality," *IEEE Trans. Visualization and Computer Graphics*, Vol. 6., No. 4, 2000.
5. Hoff, W. A. "Fusion of Data from Head-Mounted and Fixed Sensors," Proc. of First International Workshop on Augmented Reality, IEEE, San Francisco, California, November 1, 1998.
6. Hoff, W. A., Lyon, T., and Nguyen, K., "Computer Vision-Based Registration Techniques for Augmented Reality," Proc. of Intelligent Robots and Computer Vision XV, Vol. 2904, in Intelligent Systems & Advanced Manufacturing, SPIE, Boston, Massachusetts, Nov. 19-21, pp. 538–548, 1996.
7. Takaaki, K, et al. "Principal Components Based Lip Contour Extraction from Head-Mounted Camera and Cross-Subject Facial Animation", IPSJ SIGNotes Computer Graphics and cad Abstract No.100 - 014, 2000
8. Sodhi, M., B. Reimer, JL. Cohen, E. Vastenburg, R. Kaars, S. Kirchenbaum. "On-Road Driver Eye Movement Tracking Using Head-Mounted Devices". Proceedings of the Eye Tracking Research and Applications Symposium, March 2002
9. Trucco, E. and Verri, A. "Introductory Techniques for 3-D Computer Vision", Prentice Hall, 1998
10. Lillesand, T. et al, Remote Sensing and Image Interpretation, fourth edition, John Wiley & Sons, Inc. 2000

A Low-Cost Model Acquisition System for Computer Graphics Applications

Minh Tran, Amitava Datta, and Nick Lowe

School of Computer Science & Software Engineering
University of Western Australia
Perth, WA 6009
Australia
{tranm03,datta,nickl}@csse.uwa.edu.au

Abstract. Most 3D objects in computer graphics are represented as polygonal mesh models. Though techniques like image-based rendering are gaining popularity, a vast majority of applications in computer graphics and animation use such polygonal meshes for representing and rendering 3D objects. High quality mesh models are usually generated through 3D laser scanning techniques. However, even the inexpensive laser scanners cost tens of thousands of dollars and it is difficult for researchers in computer graphics to buy such systems just for model acquisition. In this paper, we describe a simple model acquisition system built from web cams or digital cameras. This low-cost system gives researchers an opportunity to capture and experiment with reasonably good quality 3D models. Our system uses standard techniques from computer vision and computational geometry to build 3D models.

1 Introduction

Polygonal mesh models are widely used in computer graphics for representing and rendering complex 3D objects. The surface of a 3D object is usually represented as a triangulated mesh in such models. While most users and researchers in computer graphics routinely use such models, quite often it is difficult for them to acquire their own models according to their specific requirements. The most popular model acquisition system is a 3D laser scanner which is usually an expensive device. Even an inexpensive laser scanner may cost tens of thousands of dollars. Hence, it is difficult for researchers in computer graphics to buy such systems for model acquisition. Most researchers depend on the models available from a few research labs such as the Stanford 3D scanning repository [4] and Georgia Tech large model archive [6] where high quality models have been acquired through laser scanning. However, sometime this is too restrictive since the researchers do not have the freedom to experiment with specific models according to their requirements. In many cases, it is necessary to experiment with models with specific topological features for designing efficient data structures and techniques in computer graphics.

In this paper, we discuss a simple model acquisition system built from web cams and digital cameras that can be used as a low-cost alternative for model acquisition. This low-cost system gives researchers an opportunity to capture and experiment with reasonably

P.M.A. Sloot et al. (Eds.): ICCS 2003, LNCS 2659, pp. 1054–1063, 2003.

good quality 3D models. Our system employs standard techniques from computer vision and computational geometry to build 3D models.

A major research area within computer vision is stereo reconstruction from images taken from monocular (multiple images with a single camera) or polynocular views (single images with multiple cameras) [3]. Three-dimensional (3D) reconstruction from multiple images attempts to simulate human perception of our 3D world from disparate two-dimensional (2D) images. A realistic representation of objects from camera images can provide an inexpensive means of object or scene modeling compared to specialized hardware such as laser scanners.

Automatic object modeling has well-known applications including the construction of 3D polygonal models for computer graphics applications, which is the focus of this paper. Once a camera captures an object, by identifying and exploiting specific scene information in the image, we can retrieve the depth, i.e., the third dimension. Next, we produce a cloud of 3D points by locating the depth values for various points of interest. An appropriate visualization technique is then employed to establish connectivity between the unstructured point cloud representing the object surface. The quality of object surface representation is dependent on the accuracy of depth retrieval. It is possible to reconstruct an object from a single image, but having multiple images of a scene improves the accuracy at which depth of object points can be determined. Further, matching identical points between images relies on the exploitation of image information including camera parameters, intensity, edge pixels, lines and regions. Analyzing and identifying areas or features, whose characteristics are preserved among disparate views, can establish point correspondences between successive images. Camera parameters give information regarding the transformation or projection of the object from the 3D world onto a 2D image. Its knowledge improves the accuracy and efficiency of point matching and can be determined before or during the matching process. In our system, the task of 3D reconstruction can be divided into roughly four stages: *camera calibration, correspondence, recovery of depth,* and *visual representation.*

The rest of the paper is organized as follows. We describe our methodology in Section 2. The results obtained from our system are discussed in Section 3 along with an example. Finally we conclude with some remarks and possible future work in Section 4.

2 Our Methodology

In this section, we discuss the techniques we have used for implementing the model acquisition system. Most of these techniques are taken from the existing computer vision and stereo vision literature with modifications to suit our needs. For standard techniques in stereo vision, see the book by Hartley and Zisserman [3]. The stereo reconstruction problem has been discussed in several papers [1,5,2,7]. In particular, Beardsley *et al.* [1] and Mandal *et al.* [5] discuss the problem of extracting a 3D model from multiple stereo images. They use multiple cameras and focus mainly on reconstructing outdoor scenes. In this paper, we concentrate on extracting 3D models for indoor objects. Hence, our techniques are considerably simpler than those in [1,5] since the images are limited to local objects captured in front of a plain background.

When a camera captures an object, a projective transformation from 3D space to 2D takes place. A typical model of the camera is displayed in Figure 1(a). In Figure 1(a), an object point in 3D space, (X, Y, Z), is mapped onto the image plane according to the line joining (X, Y, Z) and the camera's center of projection, P. The image point $\mathbf{p}$, coincides with the position where this ray intersects the image plane, (x, y) in Figure 1(a).

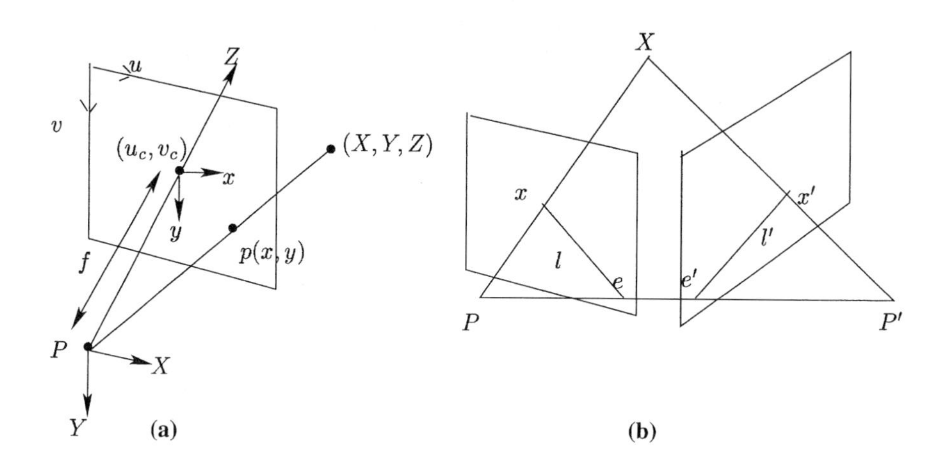

Fig. 1. (a) This is a typical camera model where a point of an object in 3D space, (X, Y, Z), is projected onto a viewing plane, $\mathbf{p}$, by a viewing ray that converges at the center of projection, P. **(b)** Two views of an object possess a unique geometric relationship expressed by the epipolar constraint. The plane formed by the two center of projections and the 3D object point of interest is called the epipolar plane. Where this plane intersects with the viewing planes are the images' respective epipolar lines.

The coordinates of the point $p(x, y)$ can be found through the simple geometric calculation. $x = \frac{Xf}{Z}$ and $y = \frac{Yf}{Z}$, where f is the focal length of the camera. The image coordinates u, v of p can be described in terms of x and y such that $u = u_c + \frac{x}{p_w}$ and $v = v_c + \frac{y}{p_h}$, where p_w is pixel width, p_h is pixel height, and (u_c, v_c) represents the principle point formed by the perpendicular projection of the camera center, P, onto the image plane. The mapping of the object from 3D space to 2D can be expressed in homogeneous coordinates as follows. $s[x, y, 1]^T = K[X, Y, Z, 1]^T$, where the intrinsic camera calibration matrix K can be represented as

$$K = \begin{bmatrix} \frac{f}{p_w} & 0 & 0\,0 \\ 0 & \frac{f}{p_h} & 0\,0 \\ u_c & v_c & 1\,0 \end{bmatrix} \tag{1}$$

and s is an arbitrary scalar value. The images acquired in our experiments are assumed to have negligible lens distortion therefore no rectification techniques were applied. The focal point P is located at a distance perpendicular to the image center i.e., principle point, (u_c, v_c). This distance is known as the focal length f. The principle point, focal length and P are collectively referred to as the intrinsic parameters of the camera along with effective pixel size and any radial distortion coefficient of the lens. Intrinsic parameters are characterized by their invariance to the camera's geographic position and orientation.

On the other hand, extrinsic parameters are dependent on the camera's orientation relative to the world reference frame and consist of a rotation matrix and a translation vector. Object points such as (X, Y, Z) in Figure 1(a) are described with respect to some world reference frame. These points must be transformed to coincide with the camera axes made possible by a rotation followed by a translation. The transformation of a 3D point to a 2D coordinate can now be represented as $\mathbf{x} = K\ [R|\mathbf{t}]\ \mathbf{X}$ where K represents the intrinsic camera parameters, R is a 4×3 rotation matrix, and $\mathbf{t}$ a 4×1 translation vector. The relationship between the projections of the object from the world coordinates into image coordinates can be found in the camera parameters, which can be solved via various camera calibration techniques. Cameras can be actively calibrated using calibration targets or passively from image correspondences.

2.1 Camera Calibration

A typical calibration method uses a calibration target. Depth is inferred by finding image points and using the known camera parameters to solve for (X, Y, Z). Camera calibration requires the known 3D positions of certain known pixels on the calibration target, and their respective image coordinates to solve linearly for C, the 3×3 camera calibration matrix composed of $K[R|\mathbf{t}]$. In this paper, the cameras are calibrated actively using a calibration target. This enables a metric reconstruction and also the recovery of epipolar geometry.

The retrieval of camera parameters unlocks the epipolar geometry between two images, which can be exploited during the matching process. The relationship between two images and the 3D object is illustrated in Figure 1(b). Image epipoles, $\mathbf{e}$ and $\mathbf{e}'$, are located at the intersection of the baseline joining the cameras' optical centers with the image plane. An epipolar plane is formed by the cameras' center of projection and the 3D point of interest, $\mathbf{X}$ in Figure 1(b). The epipolar line is the intersection of this plane with the image plane. From Figure 1(b), it is clear that a point in one image corresponding to a point in 3D space has its matching point contained in the epipolar line in the second image. Therefore, the epipolar constraint not only reduces the match search from the image area to a line, but also improves the robustness of the matching process. However, in 3D reconstruction, the value $\mathbf{X}$ is unknown and it is what we are trying to determine. We identify the epipolar line through the fundamental matrix described by Hartley and Zisserman [3] as the algebraic representation of epipolar geometry.

2.2 The Fundamental Matrix

The fundamental matrix encapsulates the relationship between image points in a stereo pair. This relationship is represented in the equation $\mathrm{x}'^T F\mathbf{x} = 0$ where x' is the correspondence of $\mathbf{x}$. In accordance with projective geometry, the dot product between a point x' located on a line l' and l' is 0. Thus, $\mathrm{x}'.\mathrm{l}' = \mathrm{x}'^T\mathrm{l}' = \mathrm{x}'^T F\mathrm{x} = 0$ resulting in $\mathrm{l}' = F\mathbf{x}$ where l' is the epipolar line. Conversely, $\mathrm{l} = \mathrm{F}^T\mathrm{x}'$. Therefore, the epipolar line in the second image, l', corresponding to a point in the first image, $\mathbf{x}$, can be identified through the fundamental matrix, F and vice versa.

The camera projects a point in 3D space into 2D image coordinates. The fundamental matrix can be described in terms of camera matrices. In projective geometry, the cross product of two points returns a line containing the two points. Therefore, $\mathbf{l}' = \mathbf{e}' \times \mathbf{x}' = [\mathbf{e}']_x \mathbf{x}$ where $\mathbf{e}'$ is the epipole and $[\mathbf{e}']_x$ is a 3×3 skew-symmetric matrix of $\mathbf{e}'$. Let $\mathbf{e}' = (e_1, e_2, e_3)$, then $[\mathbf{e}']_x$ is defined as:

$$[\mathbf{e}']_x = \begin{bmatrix} 0 & -e3 & e2 \\ e3 & 0 & -e1 \\ -e2 & e1 & 0 \end{bmatrix}. \tag{2}$$

The cross product of the two vectors $\mathbf{e}'$ and $\mathbf{x}'$, can be expressed as $\mathbf{e}' \times \mathbf{x}' = [\mathbf{e}']_x \mathbf{x}'$ $= (\mathbf{e}'^T [\mathbf{x}']_x)^T$. If H represents the homography, mapping $\mathbf{x}$ with $\mathbf{x}'$ such that $\mathbf{x}' = H\mathbf{x}$, we obtain the following relationship. $\mathbf{l}' = [\mathbf{e}']_x H\mathbf{x} = F\mathbf{x}$, and thus $F = [\mathbf{e}']_x H$. The fundamental matrix can be obtained by identifying point correspondences ($\mathbf{x}' F \mathbf{x} = 0$) or from camera parameters ($F = [\mathbf{e}'] C' C^+$). Since F will be used in the matching process, that is to find $\mathbf{x}$ and $\mathbf{x}'$, F is derived from camera matrices found from image calibration.

2.3 Point Matching

Exploitation of epipolar geometry, through the estimation of the fundamental matrix (ideally) enables robust establishment of point correspondences whilst improving the efficiency of each search. A stereo pair of pixels enables the recovery of depth through triangulation. Two popular approaches to stereo matching are the intensity-based and feature-based methods. Typically, feature-based methods are robust to significant change of viewpoints and depth disparity, but are ineffective in matching areas of smooth changing intensity values. On the other hand, intensity-based methods work well in textured areas, but large depth discontinuity along with change of viewpoint can lead to mismatches. Since depth values of objects typically do not vary drastically and in the hope that dense matches will improve structure reconstruction, we use a cross-correlation intensity-based matching method to locate point correspondences.

Cross correlating matches between image pairs improves the accuracy of matches identified. Pixels are matched according to the similarity or correlation between neighborhood intensity values. To improve accuracy of the matching process, once a match is found in the second image, a corresponding match is searched for in the first image (hence the name cross-correlation). If the match identified in the first image is different from the initial pixel, the match is rejected. This was a common occurrence when the algorithm attempted to match featureless areas in our experiments. The user determines the neighborhood size or window size, centered at the point of interest. Region W_1 in image 1 is matched with a region W_2 in image 2 according to their correlation coefficient given by the following equation:

$$c(W_1, W_2) = \frac{2cov(W_1, W_2)}{var(W_1) + var(W_2)}. \tag{3}$$

This equation is referred to by Sara [8] as the modified normalized cross-correlation algorithm, which has an advantage of tending to zero when similarly textured areas have different contrasts. The correlation coefficient $c(W_1, W_2)$ ranges between 0 and 1. The higher the coefficient the higher the correlation between the two regions W_1 and W_2.

To preserve the structure of the object surface, edge pixels are matched first and then point correspondences between image pairs are searched for at regular intervals. The search is conducted along the pixel's corresponding epipolar line to improve the accuracy and efficiency of each matching attempt. Accepting matches above a certain threshold and imposing the similarity constraint where a pixel and its match will have similar intensity values further improves the accuracy of the matching process. All image processing was conducted with grey image values. Therefore, all of the images were converted to grey scale before processing. Since the search was conducted along the epipolar line, if a match was found, it would be restricted to lie within that line. Therefore, the two rays back-projecting from the respective camera centers lie on the epipolar plane (Figure 1(b)) and will intersect at a point in 3D space.

2.4 Depth Inference

Since a correspondence pair back-project to the same point in 3D space, its depth can be approximated through geometric triangulation. Linear inference of depth is employed since metric reconstruction is assumed. The relationship between the common (X, Y, Z) point in 3D space and the point correspondences (x_1, y_1) and (x_2, y_2) is expressed homogeneously as:

$$s_i[x_i, y_i, 1]^T = C_i[X, Y, Z, 1]^T \; for \; i = 1, 2 \tag{4}$$

where C_1, C_2 and s_1, s_2 are the camera calibration matrices and scalar factor of image 1 and image 2 respectively. The camera calibration matrix C is a 3×4 matrix. The (X, Y, Z) values can be obtained by substituting C_1 and C_2 (the two calibration matrices) in Equation 4 and solving. The resulting structure is a cloud of points in the $3D$ space representing the surface of the object.

We use Delauney triangulation for generating a triangular mesh from this cloud of points. A $3D$ Delauney triangulation satisfies the following two conditions :

- three points are connected to form a triangle such that no neighboring points can be contained within or on the circle circumscribed by the vertex points,
- the outer edges of the union of all triangles form a convex hull.

3 Results

We now discuss the results obtained by our low-cost acquisition system. Image noise is inevitably introduced through the digitization of continuous scene information during the image acquisition process. Matching images taken from a 1600×1200 digital camera produced denser matches compared to several 640×480 web cameras, given the same matching parameters (i.e. window size, edge detection threshold, number of images matched and correlation co-efficient threshold).

3.1 Point Matching

Finding point correspondences between images is by far the most significant bottleneck in the reconstruction process. Attempting to match four 640×480 pictures using epipolar and similarity constraints, edge detection threshold of 0.1, window size of 10, and cross-correlation coefficient of 0.8, matching every fifth pixel can take hours to complete. For the web cameras, the object was taken with a uniform or featureless background. Since an intensity-based matching method was employed, featureless areas were not matched. However, the background was not completely featureless and non-object matches were found. Therefore, we found it necessary to find a 'mask' corresponding to the object for eliminating the background.

A window-based approach possesses some deficiencies when attempting to match depth discontinuities and textureless areas. Mismatches occur where the change in view has changed the neighborhood of the corresponding pixel in other images. A typical instance of this is where the pixel cannot be seen in other images because it is blocked by another part of the scene. Particularly, if the pixel corresponds with depth discontinuities in the image, the view disparity can change the neighborhood intensity values or occlude the pixel of interest. It is clear that area-based matching would fail if many pixels along the epipolar line have neighborhoods of the same intensities. Textureless and patterned areas cause this problem of ambiguity but are combated by the cross-correlation approach.

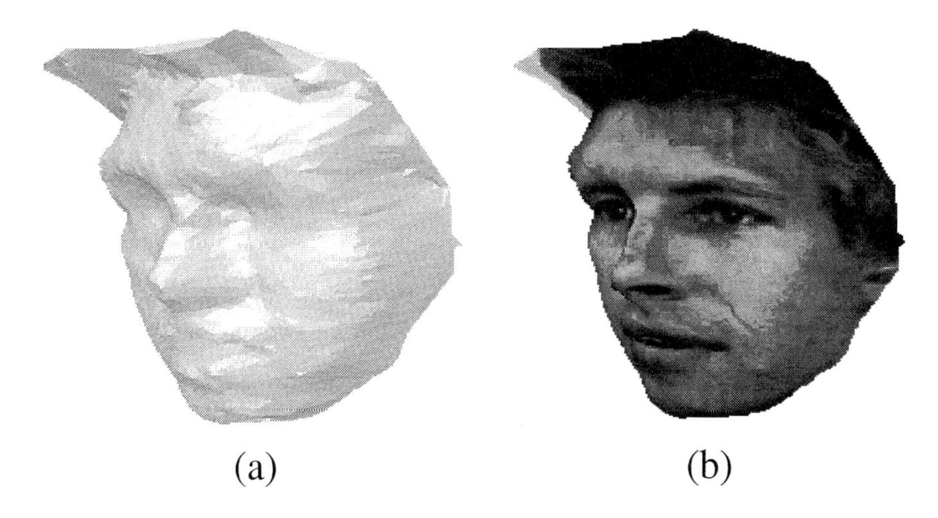

Fig. 2. (a) The recovered structure of the face from Figure 3. The face is shown with lighting and shading. **(b)** Surface details can be enhanced by texture mapping the image onto the retrieved surface. We have used the original image in Figure 3(a) as the texture.

Mismatches can be prevented during the matching phase of the reconstruction process. This can be achieved by setting the correlation coefficient to a high value in conjunction with epipolar constraints. In our experiments, there was a high correlation between matching accuracy and accurate matches. Having a high correlation coefficient threshold prevents the occurrence of false positives but may also reject accurate matches. This is deemed appropriate, since the depth retrieval step is sensitive to errors.

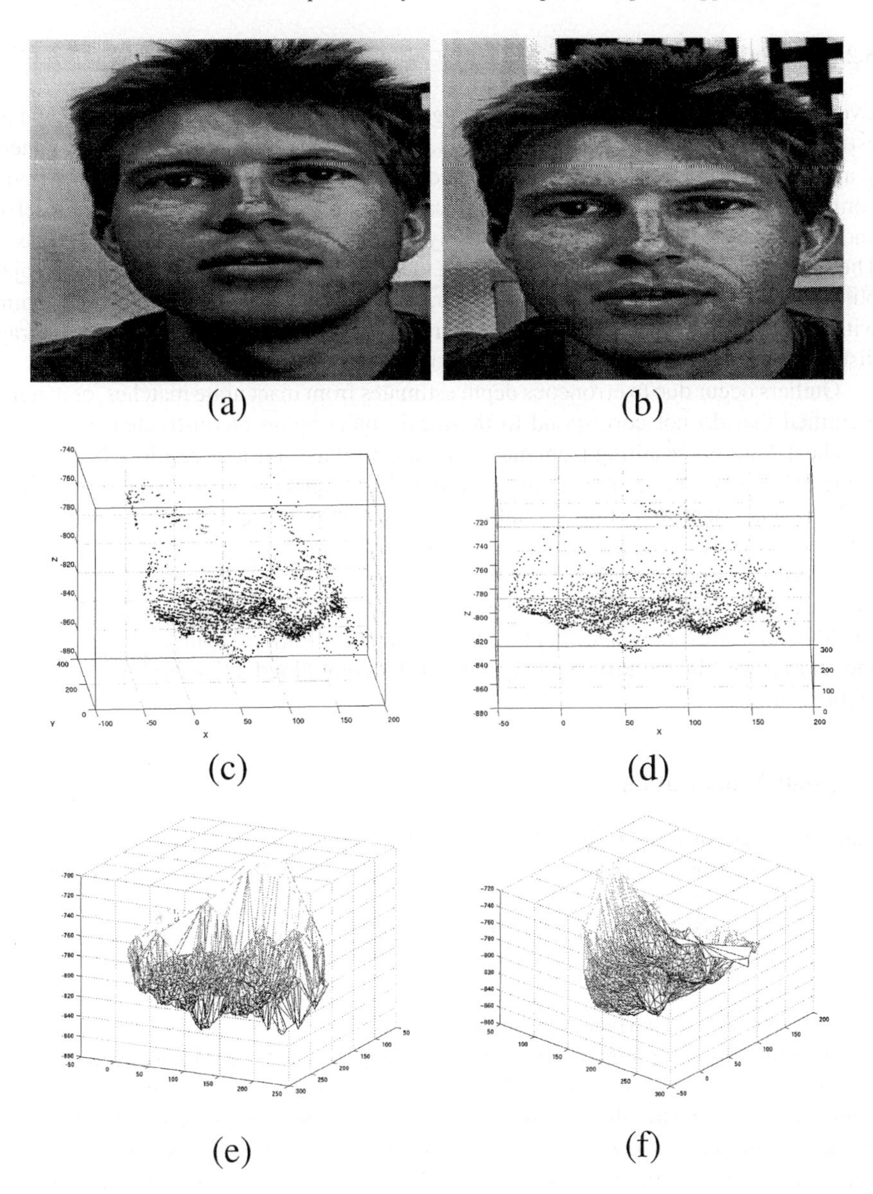

Fig. 3. Results from our model acquisition system. **(a),(b)** The stereo image pairs used in one of our reconstructions. **(c)** The initial point cloud obtained after matching. After applying a median filter over the face data points, the outliers located in front of the face are adjusted to the facial surface. **(d)** The point cloud after applying median and averaging filters. The resulting point cloud represents a smooth surface with little or no outliers. **(e)** The recovered structure after triangulating point cloud in (c). **(e)** The recovered structure after triangulating the point cloud in (d). All images are reproduced with permission from Frédéric Devernay.

3.2 Depth Inference

Even though the matches obtained are 'roughly' accurate, mismatches still occur and the estimated 3D point array contains outliers. Typically the 3D points reflect the general shape of object, but this is much maligned by a litter of outliers. Since object points congregate in a localized cluster, outliers can have large distances between themselves and their nearest neighbor and are sparsely populated relative to object point density. Therefore points that are far away from the majority of the points (assumed to be object points) can be identified and removed from the data set. In our implementation, points with the mean distance of its 11 or so nearest neighbors that deviate from the average distance by more than the standard deviation are removed.

Outliers occur due to erroneous depth estimates from inaccurate matches, or matches identified that do not correspond to the local object being reconstructed (non-object matches). Outliers resulting from matches often produce spurious results when triangulating 3D points. This can result in recovered object points being squashed into a sheet of 3D points rather than a polygonal mesh. After the removal of spurious outliers, 3D data points that lie near but not on the object surface still exist. Applying a median filter can adjust these points back to the object surface without losing the underlying object structure. The results of median-filtering the facial data points, using a neighborhood of the 20 or so nearest points, is displayed in Figure 3(c). Further, applying both a median and an average filter improves the quality of the point cloud considerably as shown in Figure 3(d).

3.3 Data Visualization

Data visualisation aims to establish connectivity between an unstructured set of 3D points. Implementing 2D Delauney triangulation is simpler than the 3D approach especially if the data points are aligned with the axis. Otherwise, the data set would need to be rotated by an appropriate angle to align the depth axis with one of the axes in the world reference frame. If the object points are not rotated so that the depth values are aligned with a world reference frame axis, 2D triangulation can result in a 'spiky' object structure or a skewed surface representation.

Triangulating the points in 2D can have a jagged effect once returned to 3D. This occurs when neighboring points in 2D are triangulated have large discrepancies in depth values. The resulting topology, looking along the depth axis, clearly represents the object structure. But once turned on its side is populated with a series of peaks and troughs. This is illustrated in Figures 3(e) and 3(f). The resulting 3D point set can be further smoothed by an averaging filter applied to all of the points with the filter the size of its connected neighbors.

4 Conclusion

We have implemented a system for low-cost model construction using inexpensive web cams and digital cameras. We have experimented with many objects extensively and one of our reconstructions is shown in Figure 3. The triangulated mesh in Figure 3 has several thousand triangles which is quite good for computer graphics applications. The resulting models can be used for computer graphics applications as shown in Figure 2. The reconstructed model with shading, lighting and texture mapping is shown in Figure 2.

Currently, our system uses only two cameras. As a result, it is not possible to acquire a complete 3D model. We plan to extend the system so that we can place the object within a circular array of web cams or digital cameras. We plan to build the model piecewise from pairs of cameras and then reconstruct a complete model from these partial models.

Acknowledgments. The research of the second and third authors is supported by Western Australian Interactive Virtual Environments Centre (IVEC) and Australian Partnership in Advanced Computing (APAC).

References

1. P.A. Beardsley, P.H.S. Torr, and A. Zisserman. 3D model acquisition from extended image sequences. In *ECCV (2)*, pages 683–695, 1996.
2. G. Farneback. The stereo problem. Technical report, Computer Vision Laboratory, Linköping University, February 2001.
3. R. Hartley and A. Zisserman. *Multiple View Geometry in Computer Vision*. Cambridge University Press, 2000.
4. Stanford Graphics Laboratory. The stanford 3d scanning repository. `http://graphics.stanford.edu/data/3Dscanrep/`.
5. C. Mandal, H. Zhao, B. C. Vemuri, and J.K. Aggarwal. 3D shape reconstruction from multiple views. Available: `http://citeseer.nj.nec.com/149335.html`.
6. Georgia Institute of Technology. Georgia tech large model archive. `http://www.cc.gatech.edu/projects/large_models/`.
7. Marc Pollefeys. 3D modeling from images. `http://www.esat.kuleuven.ac.be/~pollefey/SMILE2/tutorial.html`, June 2000.
8. R. Sara. Accurate natural surface reconstruction from polynocular stereo. In F. Solina A. Leonardis and R. Bajcsy, editors, *Proceedings NATO Advanced Research Workshop Confluence of Computer Vision an Computer Graphics*, number 84 in 3, pages 69–86. Kluwer Academic Publishers, 2000.

How Many Pixels Do We Need to See Things?

Yang Cai

Human-Computer Interaction Institute,
School of Computer Science,
Carnegie Mellon University,
5000 Forbes Avenue, Pittsburgh, PA 15213, USA
ycai@cmu.edu

Abstract. Today's computer display devices normally provide more information than we need. In this paper, the author presents an empirical model that shows minimal pixel requirements for computer users to recognize things from digital photos under different contextual conditions. It is found that face recognition alone needs far fewer pixels than people normally thought. However, more pixels are needed for users to recognize objects within outdoor scenes and paintings. Color and age have effect on object recognition but the differences are not significant. The results can be applied to adaptive display design, computer vision, adaptive human-computer interaction and telecommunication system design.

1 Introduction

One of the challenges in today's human-computer interaction design is that the electronic components become smaller and smaller but users want the display to be larger and larger. Increasing the size of images will increase the data communication traffic and [illegible]e versa, since the bandwidth is normally limited. Also, processing large images would [illegible]rease the computing time in orders of magnitude in those systems, such as machine [illegible]ion, visualization, game engine, etc.

Studie[illegible] show that photos normally provide more information than we need. The redun[illegible] can be as high as 99%! It is found out that the number of stimuli per display had [illegible]ct when the display time required to reach 75% accuracy was determined.[2] [illegible] communication, dramatic reductions in spatial and resolution of images can be [illegible] by viewers.[3]

Therefore, fro[illegible] [illegible]nomics point of view, the resolution in image transmission can be greatly re[illegible] [illegible] example, photos in newspapers normally only have two values for each dot [illegible]ith ink or without ink. With grid screen processing, the size of the smallest p[illegible] [illegible]eased so that the dots per area can be reduced greatly. However, the picture [illegible]nizable. Increasing the resolution of the grid screen can make the image mo[illegible] [illegible]but it doesn't increase the information content.

P.M.A. Sloot et al. (Eds.): ICCS 2[illegible] [illegible]pp. 1064–1073, 2003.

The resolution of an image can be represented by pixels. For example, how well subjective impressions of amount of architectural details can be predicated by objective measurement of the percentage of pixels covered by small elements.[4]

The famous face in Fig. 1 can be recognized in both resolutions a) 300 x 200 pixels and b) 150 x 100 pixels, but hardly recognized in c) 75 x 50 pixels. If we are asked to identify who is in the picture, then the 150 x 100 pixel image should be good enough. The background is redundant to the face recognition but certainly helpful in this case. Recent studies [5] show that lower resolutions of images actually are better for computer vision! For many high resolution images the process of finding the symmetry or the reflection plane of an object does did not converge to the correct solution, e.g., the process converged to local minima due to the sensitivity of the symmetry value to noise and digitization errors. To overcome this problem, a multi-resolution scheme is often introduced, where an initial estimation of the symmetry transform is obtained at low resolution and is fine-tuned using high-resolution images.

a) image with resolution 300 x 200 pixels

b) 150 x 100 pixels

c) 75 x 50 pixels

Fig. 1. Examples of the redundancy in a Black and White image

For decades, computer scientists and engineers have been focused on high-end imaging and high-resolution display technologies. It seems that very few people are interested in the low-end imaging and low-resolution display technologies. In practice, scientists and engineers use *ad hoc* methods to come up the minimal pixel requirements. For example, 32x32, as a starting point.

From the visual cognition and vision study point of view, the question like "how many pixels do we really need to see things?" actually is an important one that is related to human pattern recognition, attention, and visual information processing. As we know, human visual attention is sensitive to the purposes of seeing and the demand of visual information resources is different from task to task. [6] Many pattern recognition processes are measured by reaction time or error rate. In this study, we use *number of pixels* as a measurement. Pixel is the smallest unit of an image element. In this study, we assume that all the pixels are square. The main goal of this study is to explore the limitations of minimal resolution for people to see things under various contextual conditions.

2 User Modeling Tool Design

The purpose of the tool for the lab experiment is to show: 1) the average minimal pixels of images (face, indoors, outdoors, etc.) that subjects can recognize things, 2) the effect of questions for subjects to determine the minimal pixels for face recognition, 3) the effect of age, and 4) the differences between the recognition with color images and black and white images.

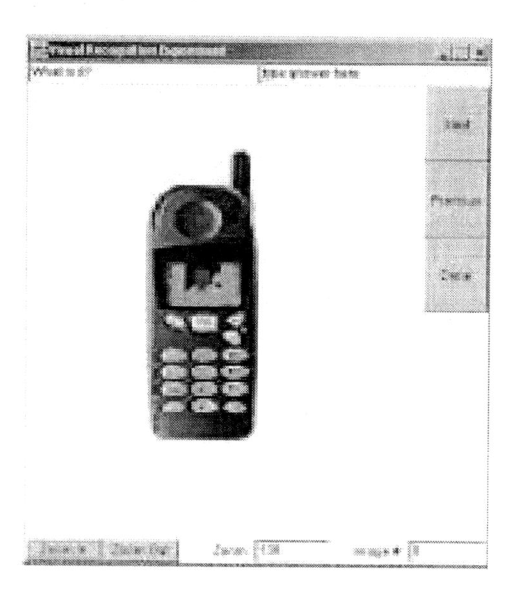

Fig. 2. A screen shot of the experimental panel. (The size of the image on the cellular phone panel can be modified by pushing the buttons on the windows.)

The subjects were 19 university students and 4 faculty members. All subjects have had vision check and had no vision problems, such as color blindness, or low vision, etc. Ten unique images in both color and black & white formats were randomly chosen to cover 4 categories: (1) faces, (2) indoor scenes, (3) outdoors scenes, and (4) complex images. These images were also randomly ordered and presented individually with an interval via a simple computer program. For a color image, there are 24-bit colors for each pixel. For a gray image, there are 8-bit gray levels.

A Java-based software has been developed for a regular PC. The resolution of a hand-held prototype screen can be modified by pushing "Zoom In" or "Zoom Out" buttons. It allows a subject to enlarge a given image in the miniature display area *until* the threshold of a *correct* recognition is reached. The subject would then be asked to answer a question accompanying the given image. We did not include the false recognition data.

The program consisted of a main image area that displayed the given image initially at 10x10 pixels. Subjects were asked to "zoom in" on the image using a button on the bar at the bottom of the window until the moment he/she recognized the contents of the image. The subjects were also asked to answer and type in their replies to a question presented in the upper left corner immediately upon image recognition. The subjects were allowed to continue zooming in on the image until the question could be answered. The subjects proceeded through the 20 images (labeled 0 through 19) by use of the "Next" button in the upper right corner of the window.

We assumed that given a set of randomly selected images, those containing human faces would be recognized at smaller resolutions, followed by simple commonly known objects, and then more-complex indoor and outdoor scenes. Regarding facial recognition, we believed that simple recognition of a face would require the simplest features, while gender identification and recognition of a well-known individual (i.e. former President Bill Clinton) would require more pixels. We also assumed that the subject's age had no effect on required image size, and that an image's being in black & white or color would make a negligible difference, with a slight advantage toward color images.

3 Results

First, we asked subjects "What is this?" about photos of face-only, indoors, outdoors, figures, and complex scenes (such as oil paintings). Subjects adjusted the size of the images until they could recognize in the image. Facial recognition required significantly fewer pixels than for human figures, indoor scenes, and outdoor scenes. As expected, complicated scenes required the largest number of pixels for identification. See the data in Table 1 and Fig. 3.

Second, we tested a set photos of faces and asked subjects with three questions: "Who's this?" "What is this?" "Male or female?" respectively. The results show that the minimal resolution in corresponding to the question "Who's this?" is the smallest. To identify male or female needs more resolution, since it's hard to distinguish in many cases in real world. To some extent, the number of pixels reflects the difficulty of the cognitive task. See Table 2 and Fig. 4 for the results.

Table 1. Minimal Pixels for Identifying Objects

Catalog	Minimal Size	Minimal Pixels
Face	17 x 17	289
Outdoor	32 x 32	1024
Figure	35 x 35	1225
Indoor	40 x 40	1600
Complex	47 x 47	2209

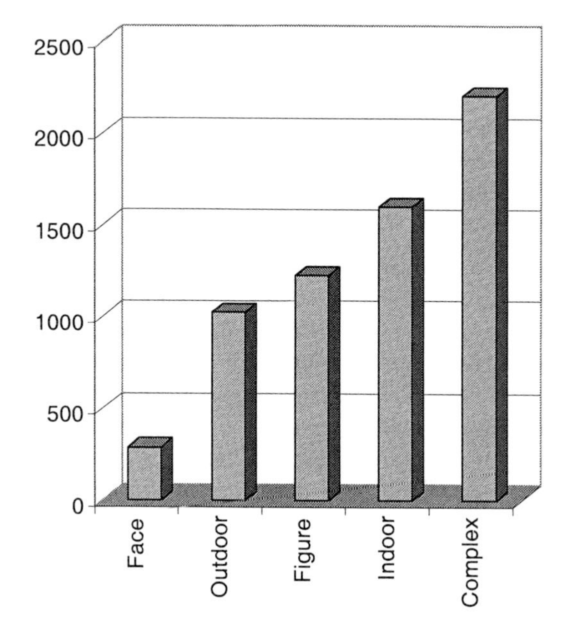

Fig. 3. Minimal pixels for object recognition

Third, we showed subjects images with color or black and white to compare the differences. Given that black & white and their color counterparts were randomized in order of presentation, the subject's short-term memory would not have altered these findings. Interestingly enough, black and white images need fewer pixels than color. See Table 3 and Fig. 5 for the test results.

Finally, we showed the images to different age groups to see whether age is an effect for determining the minimal resolution of images. We use age 21 as a cutting point, since it is a normal line to separate undergraduate students and post-graduate students as well as other "adults". It is amazing to find that younger subjects actually use more pixels to recognize objects than elder subjects. Experience and patience might play a role here. However, the differences are not significant statistically. See Table 4 and Fig. 6 for the test results.

Table 2. Minimal Pixels for Identifying Faces

Question	Minimal Size	Minimal Pixels
Who is this?	17 x 17	289
What is this?	32 x 32	1024
Male or female?	35 x 35	1225

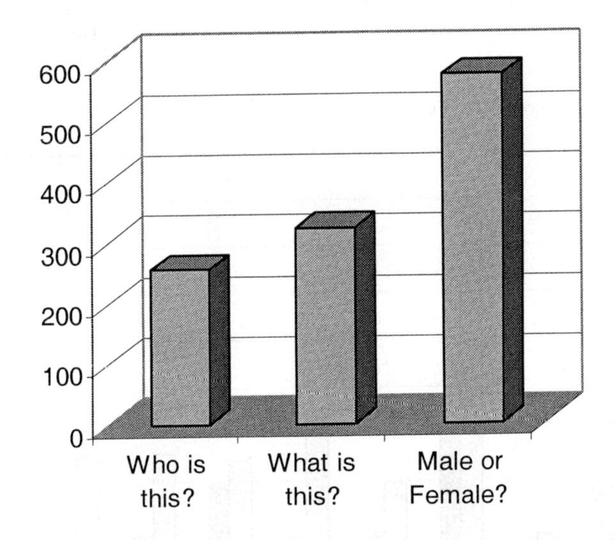

Fig. 4. Minimal pixels for facial recognition under various inquiries

4 Conclusions

In this paper, the author presents an empirical study of the minimal resolution of images in terms of pixels for computer users to recognize visual objects. Here are preliminary conclusions:

Table 3. Color images versus B&W images

Image No. Color / B &W	Color Image Minimal Size	B & W Image Minimal Pixels
[0] / [19]	40 x 40	30 x 30
[8] / [1]	20 x 20	18 x 18
[2] / [14]	63 x 63	38 x 38
[3] / [17]	31 x 31	21 x 21
[4] / [6]	42 x 42	20 x 20
[5] / [10]	35 x 35	31 x 31
[7] / [13]	36 x 36	21 x 21
[9] / [18]	36 x 36	19 x 19
[12] / [16]	15 x 15	12 x 12
[15] / [11]	28 x 28	46 x 46

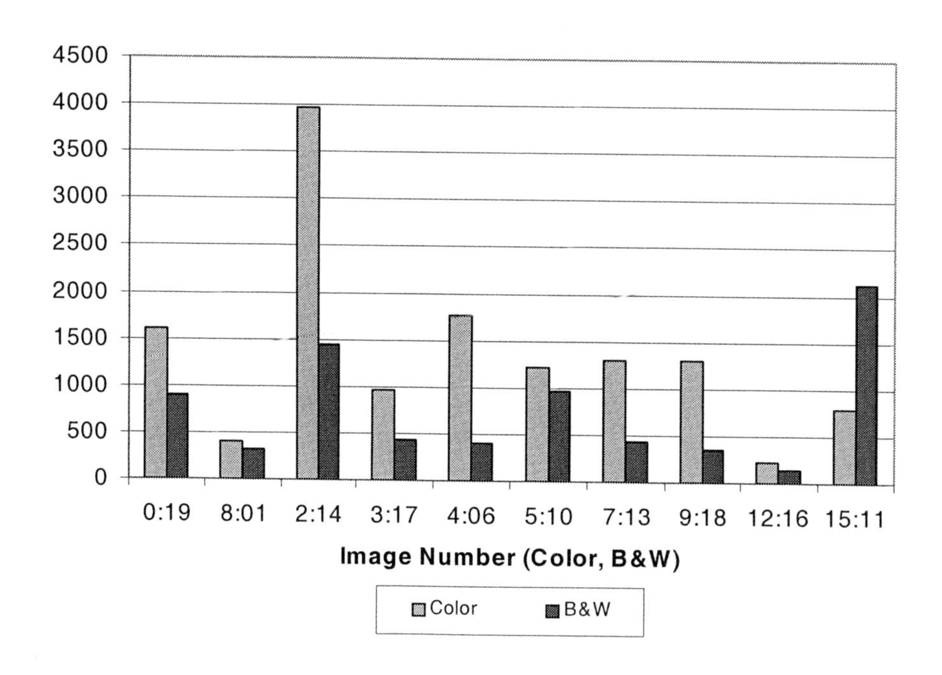

Fig. 5. Minimal pixels for color vs. B&W

Table 4. Minimal sizes versus ages

Image	Less than age 21	Age 21 or elder
[0]	42 x 42	38 x 38
[2]	75 x 75	49 x 49
[3]	32 x 32	31 x 31
[4]	15 x 15	14 x 14
[5]	40 x 40	30 x 30
[7]	25 x 25	21 x 21
[8]	21 x 21	19 x 19
[9]	40 x 40	31 x 31
[12]	17 x 17	14 x 14
[15]	29 x 29	28 x 28

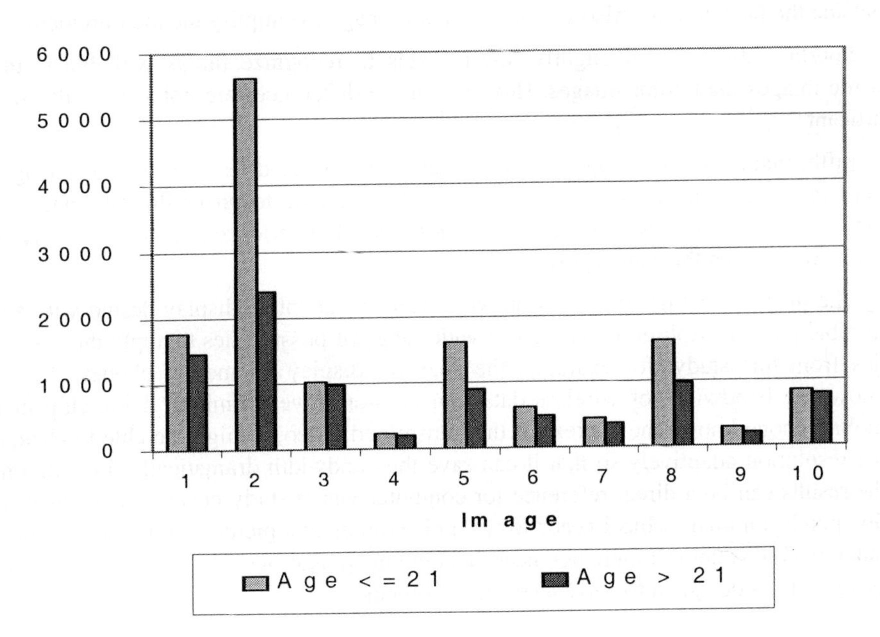

Fig. 6. Minimal pixels for age groups

First, face recognition needs far fewer pixels than people normally thought, especially, if the subject knows what he/she is looking for. The experiment agrees with the theory that human visual attention is sensitive to the purposes of seeing and the demand of visual information resources is different from task to task. [6] When we think about our faces, they are well-structured compared to other objects. They are also highly featured so that it is easy to identify a face from an image relatively. Also, it's still under investigation, whether or not a human has special "wired connections" to recognize a face.

Second, although it is context dependent, we found a general trend of the order of the complexity of human visual information processing. The order, from less pixels to more pixels, is "faces", "outdoors", "figure", "indoors", and "complex scenes." Complex scenes, such as oil paintings contain more vague objects that confuse viewers and make it hard to identify things.

Third, we found that "pixel" can be a numerical measurement of visual information processing. Traditionally, cognition scientists use reaction time, number of entities, error rate, etc. to measure the visual information processing. Pixel is a simple way to capture and compute within normal human-computer interaction environment. However, pixels of an image may not always represent the amount of visual information, because there are redundant pixels in a picture if we don't preprocess the image carefully. For example, for face recognition tasks, we cut off the background that is outside the face outline. Also we used a square image to simplify the measurement.

Fourth, subjects need slightly fewer pixels to recognize things with black and white images than color images. However, those differences are not statistically significant.

Fifth, mature subjects (age 21 and up) appeared to need less pixels than younger ones in recognition tasks. However, we need more data to prove this finding. One explanation is that mature subjects have more visual experience and more stability in perception, given the same eyesight.

The purpose of this study is to move toward the adaptive display design with adjustable picture resolutions. There is a wide range of possibilities to apply the heuristics from this study, for example, the adaptive display on mobile phones. As we know, the bandwidth of wireless data transmission is very limited. If the chip in a mobile phone knows the context of the transmitted video, it might be able to change the resolution adaptively so that it can save the bandwidth dramatically. In addition, the results can be a direct reference for computer vision study because it shows how few pixels a human subject needs to recognize things in a picture. In the future, computer vision systems might not need a high-end image acquisition and processing system, if it's designed to solve a specific problem.

Acknowledgement. The author thanks to Mr. Peter Hu for his assistance in data collection and data processing.

References

1. Strobel, L, et al, Visual Concepts for Photographers, ISBN 7-80007-236-3,1997
2. Petersik, J.T. *The Detection of Stimuli Rotating in Depth Amid Linear Motion and Rotation Distractors*, Vision Research, August 1996, vol.36, no.15, pp.2271–2281(11)
3. Bruce, V. *The Role of the face in communication: implications for video phone design*, Interaction with Computers, June 1996, vol. 8, no.2, pp. 166–176(11)
4. Stamps III, A.E. *Architectural Detail, Van der Laan Septaves and Pixel Counts*, Design Studies, January 1999, vol.20, no.1, pp. 83–97 (15)
5. Zabrodsky, H. and et al, *Symmetry as a Continuous Feature*, IEEE Trans. On Pattern Analysis and Machine Intelligence, Vol. 17, No.12, December, 1995
6. Solso, R, Cognition and Visual Art, The MIT Press, 1999
7. Brand, S. The Media Lab, Penguin Books, 1988, pp. 170–172
8. Cai, Y., Pictorial Thinking, Cognitive Science, Vol.1, 1986
9. Cai, Y., *Texture Measurement of Visual Appearance of Recognition of Oil Paintings*, IEEE IMTC, Anchorage, May, 2002
10. Buhmann, J.M., et al, *Dithered Color Quantization*, Computer Graphics Forum, August 1998, vol. 17, no.3, pp.219–231
11. Batchelor, D. Minimalism – Movements in Modern Art, University of Cambridge, 1997

Appendix A. Test Images

Bio-Feedback Based Simulator for Mission Critical Training

Igor Balk

Polhemus, 40 Hercules drive,
Colchester, VT 05446
+1 802 655 31 59 x301
balk@alum.mit.edu

Abstract. The paper address needs for training and simulation tools needed to address a multitude of threats to national security. In order to most effectively prevent terrorist threats, government agencies need to be able to train for extreme event scenarios that are very difficult to realistically replicate. A crucial aspect of the training challenge is to provide scalable, real-time hands-on training capability that includes not only visual demonstration and description of possible threats but also the capability for the individuals to interactively practice specific threat scenarios and be evaluated on their performance. This paper describes a bio-feedback system incorporated into the training simulator for the mission critical practices.

1 Identification and Significance

One of the major goals of human computer interaction (HCI) studies is to develop adaptive, smart computer systems, which are reconfigured based on user response. Such systems use biometric identification such as facial recognition, fingerprint recognition and eye tracking in order to customize access to the system. However, such systems do not adapt to the user need based on their psycho-physiological conditions. This paper describes development of non-invasive touch based system, which will allow tuning a computer based training system to the user emotional and stress levels.

The system is based upon the fact that during the operation of a computing device, the user stays in almost permanent manual contact with an input device, such as a mouse, tracking ball, joystick or for more advanced version of the simulators integrated with motion or eye tracking devices. This paper will describe the incorporation of sensors to monitor the physiological conditions, such as galvanic skin response (GSR), electric skin resistance (ESR), and temperature. Based on the sensory response, the computer system would be able to determine stress level and emotional condition of the user, visual-somatic response and adjust itself to provide better training and determine if the user is suitable to the task assigned.

P.M.A. Sloot et al. (Eds.): ICCS 2003, LNCS 2659, pp. 1074–1080, 2003.

There are many practical applications for such a system. For example, it could allow parental control of minors, allowing them to play computer games at an emotionally safe level. The described system can enable user interface developers to better monitor the human response to a prototype. Also, the system will enable game developers to dynamically adjust the difficulty level based upon the player's condition. The system can also be integrated into a car steering wheel along with an onboard computer to monitor the motorist's condition during the trip and prevent him/her from falling asleep at the wheel. But one of the most important applications of the system is to monitor the work related stress and enable failure prevention in mission critical tasks in nuclear stations, hazardous chemical, oil factories and other high risk areas.

Today's governments face the monumental challenge of addressing a multitude of threats to national security. In order to most effectively prevent terrorist threats, government agencies need to be able to train for extreme event scenarios that are very difficult to realistically replicate. A crucial aspect of the training challenge is the ability to provide a large number individuals amongst numerous agencies with a scalable, real-time hands-on training capability that includes not only visual demonstration and description of possible threats but also the capability for the individuals to interactively practice specific threat scenarios and be evaluated on their performance.

At present, such systems do not exist in the market. However, active research is being performed by several groups, such as, IBM BlueEyes team [1, 9], MIT Media Lab [2] and other organizations [1]. The major advantage of the described technique compared to those reported in literature is the monitoring of visual-motor reactions and attention distributions, both very important parameters for training simulator and game developers. For example, the "Emotional mouse", the experimental system developed at IBM is limited to measuring six emotional states – anger, disgust, fear, happiness, sadness and surprise.

This paper describes a system which is significantly more advanced than those being investigated. The system will have a significant impact on the human computer interaction development in the next decade and will lead to more expressive communication with machines.

2 Overview

Paul Ekman's facial expression work demonstrated that there is a correlation between a person's psychological state and a person's physiological measurements. Based on this work he developed Facial Action Coding System [3]. Later, Dryer [4] determined how physiological measures could be used to distinguish the various emotional states. He also demonstrated that people consider a computer system as having personality [5], and a system having a similar personality as the user leads to better collaboration [6]. In his experiments Johnson [7] used video cameras to determine users psychological state and concluded that there is a correlation between patterns in user behavior and users psychological state.

Wendy Ark [9] in her work reported that during normal computer use (creating and editing documents and surfing the web) people spend about 30% of their total computer time touching their input device. In addition, it has been demonstrated that the major physiological parameters, associated with physiological state, are Galvanic Skin Response (GSR), Electric Skin Resistance (ESR), General Somatic Activity (GSA), temperature (TEM), heart rate (HR) and blood pressure (BP) [3, 4].

Based on the prior art we decided to integrate a bio-sensing device into motion or eye tracking hardware and make it in a way, that it actively monitors the above listed variables — GSR, ESR, GSA and HR.

3 Technical Characteristics

The major objective was to develop a scanning device, which would be able to monitor the real time psycho-physiological state of the trainee. We term such a system a 'bio-sensor'. The bio-sensor will enable the study of

- Psychosomatic and psychotherapeutic issues related to the HCI
- Psychological state and personality traits of the subject
- Characteristic peculiarities
- Mental performance level
- Central nervous system state
- Functional abilities of central regulation and peripheral blood circulation of cardio-vascular system
- Psychological tension and stress level

In order to complete these tasks three major objectives need to be achieved (Figure 1). First algorithms to process analog signals from the sensors should be developed. Second, a set of psycho-physiological tests has to be developed to collect and analyze data. And third, an interface should be developed for both report generation and other software integration.

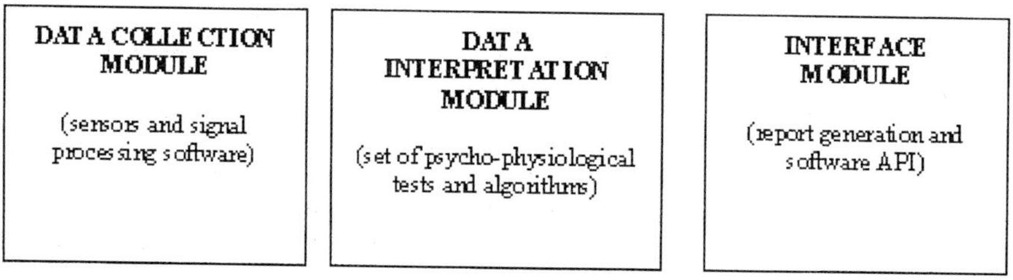

DATA COLLECTION MODULE (sensors and signal processing software)	DATA INTERPRETATION MODULE (set of psycho-physiological tests and algorithms)	INTERFACE MODULE (report generation and software API)

Fig. 1. Major modules of the system

Each of these three modules is described in more detail below.

3.1 Data Collection Module

The data collection module consists of two parts – hardware sensors and control software. A data collection module has following characteristics:

- Include sensors of photo-plethysmogram and GSR
- Tunable discretization frequency up to 1k Hz
- Bandwidth of bio-amplifiers not less then 0.1-1000 Hz
- Electric shock protection
- Power supply from computer

Control software properties:

- Real time signal filtering
- Spectral and autocorrelation analysis algorithms
- Bio-amplifiers bandwidth control
- Signal intensity control
- Automatic detection of photo-plethysmogram parameters

3.2 Data Interpretation Module

This is the main analytical module of the system. The goal of this module is to manage data collection and to extract meaningful information from the data. It should provide a set of tests and analysis algorithms, which will study different aspect of HCI. The main functions of the data interpretation module will include:

- Chrono-cardiography analysis based of variation principle
- Simple and complex visual-semantic response study
- Attention distribution test
- Associative test
- Response speed test
- Stress level determination

3.3 Interface Module

In order to be a useful tool the system has to be able to generate reports and communicate to other software. This brings us to the third objective – to generate an easy to use report generation system and application programmer interface (API). Report generation system should be able to display graphical physiological data and test results as well as being able to save results and provide access to stored user data. The API should be easy to use and allow third party software developers easy access to collected data and diagnostic results.

4 Conclusion and Future Work

Development of a smart adaptive computer system has been one of the major challenges of computer science in the last decade. Several different tools and methodologies have been developed in order to obtain some physiological feedback from user. One of the most well known applications of such systems is the polygraph or "lie detector" which is now widely used in criminology and human resources and cardio-systems in gymnasiums. Several other techniques have been used for biofeedback in psychotherapy.

The main concept of this work was to develop an easy to use system, which includes a bio-sensing device and software package integrated with modern training simulator hardware, which will determine users psychological conditions using biofeedback. Data from the IBM research team working on "emotional mouse" (Table 1-4) shows strong correlation between emotional state and physiological data. The next step that we are planning is to extract not only the emotional data but also the stress level, activity level and attention level.

We are planning to integrate different sensors in to the motion or eye tracker hardware. First, we will use infrared detectors in order to obtain photo-plethysmogram of the computer user. This signal is depending on blood flow in fingers and allows the detection RR intervals (time between two consecutive heart pulses). Performing statistical analysis and filtering of the RR intervals, we will obtain heart rate and heart rate variability (VF) [10]. A high value of VF indicates that the user is falling asleep and a low value of VF indicates that the user is overloaded or tired. Average RR value for human is 900-1000 milliseconds (equivalent of heart rate 60-70), and average value of VF is 4-5%. Based on this data, the system will be able to warn when the user is too tired or about to fall asleep or getting too stressed to adequately complete the mission.

Another analytical technique that we are planning to incorporate into our system is the analysis of video-somatic reaction. This technique is excellent for action based computer game developers. The system will collect reaction time of the user and apply mathematical models similar to the one used to determine a heart rate variability. This will allow us to correlate to the psychological condition of the user. In this case, we will measure average response time and its variability. The advantage of this method is that no hardware modification is required.

Combination of heart rate variability, video-somatic reaction based methods and emotional state determination technique similar to one described by IBM [9] will allow error minimization and generate a system which will be able to help make a computer which will be more adjustable to users psychological state

Table 1. Difference Scores. (Source: IBM)

		Anger	Disgust	Fear	Happiness	Sadness	Surprise
GSA	Mea	-0.66	-1.15	-2.02	.22	0.14	-.1.28
	Std.	1.87	1.02	0.23	1.60	2.44	1.16
GSR	Mea	-	-53206	-61160	-38999	-417990	-41242
	Std.	63934	8949	47297	46650	586309	24824
Pulse	Mea	2.56	2.07	3.28	2.40	4.83	2.84
	Std.	1.41	2.73	2.10	2.33	2.91	3.18
Temp	Mea	1.36	1.79	3.76	1.79	2.89	3.26
	Std.	3.75	2.66	3.81	3.72	4.99	0.90

Table 2. Standardized Discriminant Function Coefficients. (Source: IBM)

	Function			
	1	2	3	4
GSA	0.593	-0.926	0.674	0.033
GSR	-0.664	0.957	0.350	0.583
Pulse	1.006	0.484	0.026	0.846
Temp.	1.277	0.405	0.423	-0.293

Table 3. Functions at Group Centroids. (Source: IBM)

EMOTION	Function			
	1	2	3	4
anger	-1.166	-0.052	-0.108	0.137
fear	1.360	1.704	-0.046	-0.093
sadness	2.168	-0.546	-0.096	-0.006
disgust	-0.048	0.340	0.079	0.184
happiness	-0.428	-0.184	0.269	-0.075
surprise	-1.674	-0.111	-0.247	-0.189

Table 4. Classification Results. (Source: IBM)

		Predicted Group Membership						Total
	EMOTION	Anger	Fear	Sadness	Disgust	Happiness	Surprise	
Original	anger	2	0	0	0	2	1	5
	fear	0	2	0	0	0	0	2
	sadness	0	0	4	0	1	0	5
	disgust	0	1	0	1	1	0	3
	happiness	1	0	0	0	5	0	6
	surprise	0	0	0	0	1	2	3

Acknowledgments. We would like to thank Vladimir Bryskin and the staff of the BioLab Inc. for the help in the work and the information and ideas provided to the authors

References

1. http://www.ibm.com
2. http://media.mit.edu
3. Ekman, P. and Rosenberg, E. (Eds.) (1997). What the Face Reveals: Basic and Applied Studies of Spontaneous Expression Using the Facial Action Coding System (FACS). Oxford University Press: New York
4. Dryer, D.C. (1993). Multidimensional and Discriminant Function Analyses of Affective State Data. Stanford University, unpublished manuscript
5. Dryer, D.C. (1999). Getting personal with computers: How to design personalities for agents. Applied Artificial Intelligence, 13, 273-295.
6. Dryer, D.C., and Horowitz, L.M. (1997). When do opposites attract? Interpersonal complementarity versus similarity. Journal of Personality and Social Psychology, 72, 592-603
7. Johnson, R.C. (1999). Computer Program Recognizes Facial Expressions. EE Times www.eetimes.com, April 5, 1999.
8. Picard, R. (1997). Affective Computing. MIT Press: Cambridge
9. Ark, W., Dryer, D.C. and Lu, J. L. Emotional Mouse. IBM White Paper, www.ibm.com
10. European Society of Cardiology and North American Society of Pacing and Elerctrophysiology. Heart Rate Variability. Standards of Measurement, Physiological Interpretation and Clinical Use.

Workshop on Innovative Solutions for Grid Computing

Proposing and Evaluating Allocation Algorithms in a Grid Environment

Salvatore Cavalieri, Salvatore Monforte, and Fabio Scibilia

Department of Computer Science and Telecommunications Engineering,
Faculty of Engineering, University of Catania
Viale A.Doria, 6 95125 Catania, Italy
Tel.: +39 095 738 2362, Fax: +39 095 738 2397
{salvatore.cavalieri, salvatore.monforte, fabio.scibilia}@diit.unict.it

Abstract: Distributed computing addresses workload challenges by aggregating and allocating computing resources to provide for unlimited processing power. In the last ten years, it has grown from a concept that would enable organizations to simply balance workloads across heterogeneous computer resources, to an ubiquitous solution that has been embraced by some of the world's leading organizations across multiple industry sectors. And while distributed computing harnesses the full potential of existing computer resources by effectively matching the supply of processing cycles with the demand created by applications, even more importantly it has paved the way for grid computing – a more powerful, yet global approach to resource sharing. The aim of this paper is to propose new allocation algorithms for workload management in a computing grid environment. A simulation tool is used to validate and estimate performance of these algorithms. The paper is organised as follows. After a brief introduction about grid environment and on the DataGrid Project in Section 1, Section 2 presents the proposal for novel allocation policies. Finally, Section 3 provides for a performance evaluation of the algorithms proposed, comparing their performances it with those offered by the actual solution adopted in the DataGrid Project.

1 Introduction

In the last years the amount of data managed and shared by scientific communities has grown very fast. Thus, the needs to create a new platform, allowing the use of intensive computational power as well as data sharing independently on its location, became indispensable. This new concept of informatics infrastructure is known in literature as *Computing Grid*.

The term *grid* is used to indicate the power-network that provides for the distribution of electrical energy supplying any electrical device, wherever it is placed. Similarly, a computing grid is an Internet-based infrastructure that allows different users to access distributed computing and storage resources on a wide area network. The heterogeneity of machines, operating systems, communication protocols and data format representation imply the necessity to study new mechanisms in order to permit a non-ambiguous communication as well as semantic understanding.

P.M.A. Sloot et al. (Eds.): ICCS 2003, LNCS 2659, pp. 1083–1092, 2003.

As an evolution of distributed computing, grid computing allows enormous opportunities for organizations to use computing and storage capabilities from networks of computers spanning multiple geographical boundaries. Grid computing elevates clusters of desktop computers, servers or supercomputers to the next level by connecting multiple clusters over geographically dispersed areas for enhanced collaboration and resource sharing. It is clear that, Grid computing is especially beneficial to the life sciences and research communities, where enormous volumes of data are generated and analysed during any given day. In this context the most improved are physics particles, earth observation and biology applications. Next researcher's generations will be able to compute a great amount of data, from hundreds of terabyte to some petabyte, per year.

Taking into account that, different tasks, with different requirements, are expected to be executed in a computational grid, both traffic management and resource allocation policy are clearly needed in order to improve the efficiency of resources usage and to balance the resources allocation, as well.

Several activities aimed to realise a Grid environment are currently present in literature [1][2]. One of this is relevant with the European DataGrid Project [3].

The DataGrid project was born by the desire of some of the most important science communities of UE like INFN, PPARC etc. Aim of the project is to realise a new computing grid that provides informatics tools and services for different applications enabling organizations to aggregate resources within an entire IT infrastructure no matter where in the world they are located. Figure 1 summarises the structure of the DataGrid Project, highlighting the layers providing different services.

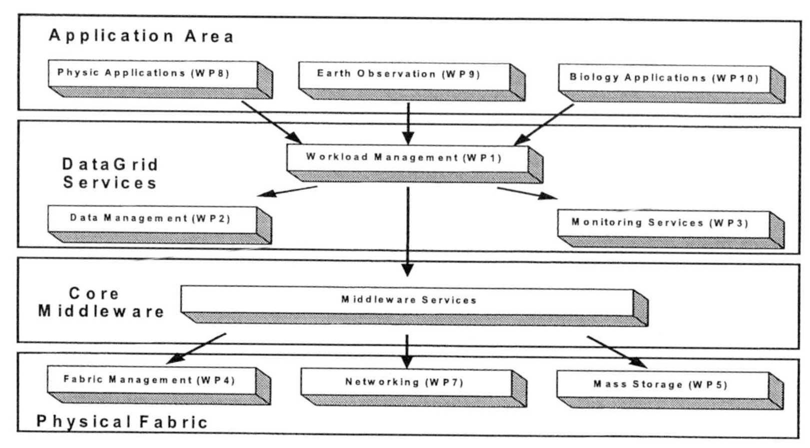

Fig. 1. Grid layers diagram

Lowest layer, *Physical Fabric Layer*, provides management of some infrastructures like computing resources (Fabric Management), mass storage management system (MSMS) and networking (management of quality of services and information collection).

Services concerning security management and information collection are mainly provided for by the *Core Middleware layer*. All these services are based on the Grid

Information Service (GIS), which is the Grid component responsible to collect and publish information about Grid status [4].

Workload and data management as well as monitoring are the most relevant services provided by the *DataGrid Services Layer*.

The *Application Area Layer* represents the upper layer of the architecture and comprises some applications that use DataGrid services for running: biology, particle physics and earth observation.

The DataGrid project is divided in some different activities called *Work Packages* (WPs) according to functional criteria, as can be seen from Figure 1. In particular, the main task of WP1 "Workload Management" concerns the definition of a suitable architecture for workload management, distributed scheduling and resource allocation for realising an optimal allocation of resources in order to improve performance of the Grid. The authors of the paper are currently involved in this WP.

The aim of this paper is to describe in detail the main feature of the Workload Management functionalities, focusing on the allocation strategy here adopted. Then new allocation algorithms will be presented and evaluated. Comparison with the current allocation strategy used in DataGrid will allow pointing out the increase in performance that can be achieved by the adoption of the strategies here proposed.

2 Brokering Policy

Brokering policy allows allocation of jobs submitted by users into resources in a grid environment, aiming to respect users' requirements and to balance the workload. Choices of computing resources where jobs has to be allocated for execution, is a complex task. Many factors can affect the allocation decision and their importance may even differ depending on the job. These factors include: location of input and output data requested by the job, authorisation to use a resource, allocation of resources to the user or to the groups the user belongs to, requirements and preferences of the user, status of the available resources, as well.

The DataGrid component which realises the brokering policy is called the Resource Broker (RB), whose development is actually the core of WP1. The RB interacts with several components. A User Interface (UI) passes all the users' requests of job submission to the RB. A Replica Catalog (RC) is used by the RB to resolve logical data set names as well as to find a preliminary set of computing resource (called Computing Elements or CEs) where the required data are stored. A Grid Information Index (GII) allows the RB to know information about Grid status. Finally, a Job Submission Service (JSS) receives from the RB the job and the name of CE where the job must be submitted.

2.1 State of the Art

The aim of this section is to provide for a description at the state-of-the-art of the matchmaking algorithm performed by the RB in order to select the most suitable Computing Element where a given job must be executed.

It is worth to identify two main different possible scenarios depending on the requirements of the user who submitted the job. It is clear that the simplest possible scenario is when the user specifies the Computing Element where the job has to be executed. In such a case the RB doesn't actually perform any matchmaking algorithm at all, but simply limits its action to the delegation of the submission request to the JSS, for the actual submission. This case will no longer be considered in the analysis undertaken in this paper.

Let's do a little step onwards and let us consider the scenario where the user specifies a job with given execution requirements, but without data files to be processed as inputs. Once the job submission request is received by the RB (through the UI, as said before), it starts the actual matchmaking algorithm to find if the current status of Grid resources matches the job requirements, i.e. to find a suitable CE where to allocate the job.

The matchmaking algorithm consists of two different phases: requirements check and the rank computation. During the requirements check phase the RB contacts the GII in order to highlight a set of CEs compliant with a subset of user requirements, passed to the RB inside the job submission. Examples of these user requirements are the authorisation key, the operating system and the kind and/or the number of processors needed to execute the job. Taking into account that all these requirements are almost constant in time (i.e. it's improbable that a CE changes its operating system or its runtime environment in the very short term, e.g. every half an hour), it is clear that GII represents a good source for testing matches between job requirements and Computing Element features. It is clearly more efficient than contacting each Computing Element to find out the same information.

Once the set of the suitable Computing Elements where the job can be executed is selected, the RB starts performing the second phase of the matchmaking algorithm, which allows the RB to acquire information about the "quality" of the just found suitable Computing Elements: the *ranking phase*. In the ranking phase the RB contacts directly each of the Computing Elements selected in the previous phase, in order to obtain the values of those attributes that appears in a particular rank expression passed by the user to the RB inside the job submission request. Examples of these attributes are the number of free CPUs, the Estimated Traversal Time of the input queue of the CE (i.e. the estimated time interval a job must wait in the input queue before the beginning of its execution), and the amount of free memory. It should be pointed out that conversely to the previous phase, it is better to contact each suitable Computing Element, rather than using the Grid Information Index as source of information, since the rank attributes represents variables varying in time very frequently.

Once the rank expression has been evaluated by the RB, it performs a deterministic choice of the Computing Element basing its decision on the value obtained. Namely, given a list of CEs matching user requirements CE_1, CE_2, ... , CE_n (found by the RB during the requirements check phase) and their correspondent rank values R_1, R_2, ... , R_n (evaluated during the second phase of the matchmaking algorithm), the selected CE where job is really submitted is the one with higher rank value.

Finally, let's consider the scenario where the user specifies a job with given execution requirements, together with the input data files needed by the job during its execution. The main two phases of the match making algorithm performed by the RB remain unchanged, but the RB has also to interact with the Replica Catalog in order to find out the most suitable Computing Element where the requested input data set is physically stored. If the RB isn't able to find any CE with all the files needed by the job, job submission must be delayed until all the files needed by the job have been transferred to the CE.

Actually, the Estimated Traversal Time is currently assumed by the RB as the default ranking attribute, when the user doesn't specify any rank expression ,and none network parameter, such as file locations or bandwidth, is considered. Thus, Resource Broker decisions are based on CE parameters without considering network performances.

The Job-Burst Problem

As previously said the RB uses services provided by GIS, which collects information about Grid status and maintains this information in the GII. Each Grid resource publishes, periodically, its status. Published values remain unchanged till the expiration of their time-to-live (TTL). All the Grid resources are periodically queried in order to collect all published information and store their values in the GII. It is clear that, due to hierarchical architecture described above, the higher is the destination the older is the information.

Moreover, it is clear that if the job submission rate was higher than the TTL this would lead to inconsistency between the real Grid and its representation by the GII. For the sake of understanding let's consider the scenario where a very huge set of job submission requests, featured by the same user requirements, is issued to the RB. Let's also assume that the rank expression is the default one, i.e. made up only by the Estimated Traversal Time. It is clear that inside a time interval lasting TTL, the matchmaking algorithm performed by the RB will yields the same result and thus the RB will submit the burst of jobs to the same Computing Element for all those requests issued within the same time interval. But submission of the very huge set of jobs to the same CE may cause a very strong increase in the Estimated Traversal Time parameter for that CE, but it will remain unchanged in the GII until the TTL interval expires, leading to a "wrong" allocation choice by the RB. Moreover, to submit the job the RB uses services provided by JSS, that puts the job in queue waiting for input file data transfer and than submits the job to the final CE. This behavior adds new delays to the RB-CE job migration.

2.2 Contribution to the State of the Art

In this section novel ranking policies (ranking expressions) as well as a new criteria used by the RB for choosing the Computing Elements will be presented. The main aim of the proposed ranking policy regards the minimization of job lifetime, which is defined as time spent between submission of the job and its successful completion, including queuing delays as well as input file transfer time.

Let us consider the following ranking expression:

$$\text{rank(CE, filelist)} = \alpha\ \text{rank\_expr(CE)} + (1-\alpha)\ \text{file\_rank(CE,filelist)} \qquad (1)$$

where *CE* is the Computing Element candidate for executing the job, *filelist* is the list of input files specified in the submission request and $\alpha \in [0,1]$.

The *rank_expression(CE)* is the rank expression already described. As said before, its deafult value is given by the Estimated Traversal Time, i.e.:

$$\text{rank\_expr(CE)=- EstimatedTraversalTime(CE)} \qquad (2)$$

The *file_rank(CE,filelist)* function indicates goodness of the Computing Element CE with respect to both file access and transfer time, and can be expressed as:

$$\text{file\_rank(CE,filelist)=-[}\ \beta\ \text{file\_access\_time} + (1-\beta)\ \text{file\_transfer\_time]} \qquad (3)$$

where $\beta \in [0,1]$.

Values of α e β parameters could be or not specified by user in the classAd during submission phase. In not specified by user default values are used. Default values of α e β parameters will be find by using offline simulation tools in which configuration files describe, with most possible adherence to the reality, current Grid architecture and workload conditions. Their values could be updated by novel simulations periodically (every month, every year ..), or when Grid architecture and/or workload conditions change in a significantly manner.

Let us consider a set of *n* suitable Computing Elements CE_1, CE_2, ... , CE_n and define R_i as the rank for the *i-th* Computing Element. Two different allocation strategies are here proposed.

The first one is very simple and features the choice of the CE with the lower value of the rank given by (1). This strategy will be called *deterministic* one.

Let's define K_i as inverse of absolute value of rank R_i , i.e. $K_i = 1/|R_i|$. According to the second allocation strategy here proposed, the probability that a CE_i is selected for job execution can be given by $P_i = K_i/K_{tot}$ where $K_{tot} = \sum_{i=1}^{n} K_i$. This second strategy will be called *stochastic* one. It associates different probabilities to all CEs that are proportional to their rank values. In this mode is possible to minimise problems derived by presence of *job bursts*. Let us image that a burst of 100 jobs is submitted to Resource Broker. With deterministic choice criteria all job submission requests issued within a time interval featured by a duration equal to TTL, would be submitted to the same ComputingElement. On the other hand, using stochastic choice criteria, jobs would be distributed to all the suitable CEs, according to their probability that the value given by the evaluated rank expression yields.

2.3 Evaluation of Brokering Policy

In order to evaluate the proposed brokering policies described above, a simulation tool was purposely developed in C++, which models in a very detailed manner Grid components and their interactions. The simulator engine is event driven. Each event is characterized by a timestamp, which is used to sort event list with growing time criteria, by a message, which identifies the kind of message and the actions to do, and by an owner, which is the generator and handler of the event.

The network infrastructure of the Grid is the GARR-B net used by Italian Science Communities [5]. Network parameters used as input for simulation were measured directly on the real networks using services provided by GARR-B routers, and thus they are fixed for all simulations.

Life cycle of a job is evaluated by the simulator. As said before, life cycle begins in a User Interface that issues job submission requests to a Resource Broker as shown in Figure 2.

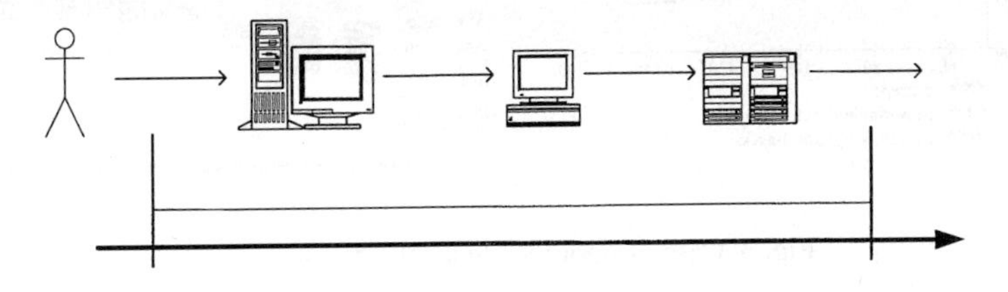

Fig. 2. Job life cycle

Interarrival time between two consecutive job submissions follows the negative exponential distribution. Moreover, the User Interface submits continuously several instances of jobs of the same kind. Therefore, a simulation is both related to a given workload, which is specified by number of average submissions per second, and characterized by a unique kind of job submissions.

In the following sub sections results obtained by several simulations will be presented.

Jobs with Light Execution Time

A first simulation set regards job submission requests featured by a light execution time, which is comparable with both file access and file transfer time. In particular, it's assumed that each job is featured by an estimated execution time given by $10^5/CE_{MIPS}$ sec (CE_{MIPS} is the number of MIPS featured by the CE) and requires, for execution, two files as input data set whose size are 1GB and 50MB, respectively. The overall computing power of the grid consists of sixteen CEs spread with different computational power: 300 MIPS (high), 200 MIPS (medium) and 100 MIPS (low).

The first scenario is characterised by a computational power uniformly distributed all over the Grid. The same apply for the sizes of files. Thus, this scenario can be considered as a balanced case.

Figure 3a shows a direct comparison between current brokering policy and the proposed brokering policy with both deterministic and stochastic choice criteria, varying workload conditions. Values of parameter α and β are, respectively, 0.75 (high importance to EstimatedTraversalTime of the candidate CE) and 0.25 (low importance to file_access_time respect with file_transfer_time). These values have been selected after several simulations, as they are able to give the best performances.

On X-axis average interarrival time (seconds) between two job submissions is represented, thus the left side of the diagram presents values relevant to the highest workload conditions. On Y-axis average job life time of submitted jobs is represented.

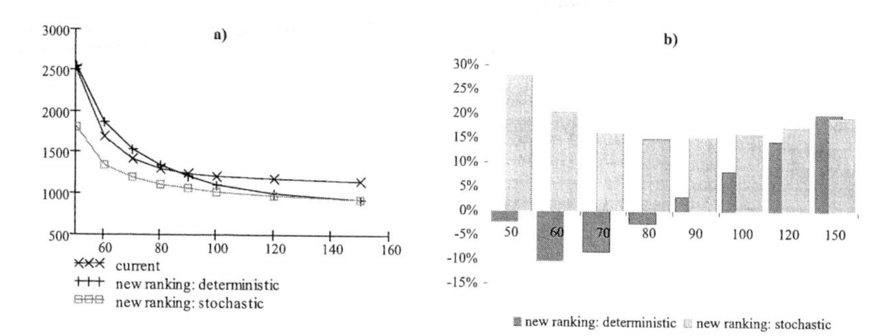

Fig. 3. Case of light jobs in a balanced scenario

Figure 3b depicts the percentage improvement of job-lifetime obtained using the two proposed policies varying workload conditions, with respect to current scheduling policy. New ranking policy with deterministic choice criteria improves performances, respect with current scheduling policy, for interarrival submission time greater than 80 seconds (lower workload conditions). New ranking policy with stochastic choice criteria improves performance for any workload conditions, especially for high workload conditions where the job burst problem occurs. For lower workload conditions new deterministic policy gives better results than current policy, cause the first considers network parameters, such as both file transfer and file access time, that are not used in the latter one. Thus, Grid performances are sensitive to network conditions.

Additionally, a second set of simulations was carried out. The characteristics of job submission requests are the same of the first set of simulations described above. The two sets of simulations only differ in the distribution of computational and storage resources. Namely, the higher the computational power of a site, the larger the size of the file neighbouring to such a site. Thus, this scenario could be considered such as a worst case just to evaluate behavior of novel scheduling policies.

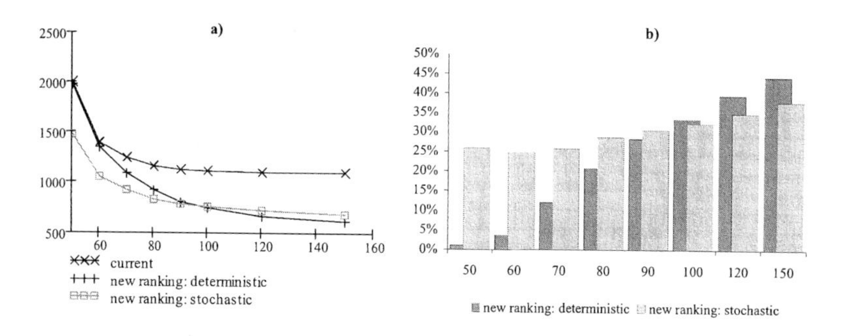

Fig. 4. Case of light jobs in an unbalanced scenario

Figure 4a shows a direct comparison between the three brokering policies in this case.

Referring to the previous figure, it is possible to notice that again both proposed brokering policies, the one with deterministic choice criteria and the one with stochastic choice criteria, give results that are better than current brokering policy for all workload conditions. Lower is grid workload higher is improvement of both new scheduling policies.

Intensive Computational Jobs

Another set of simulations regards job submission requests featured by a heavy execution time, in which both file access and transfer time are negligible when compared with job execution time. Therefore, the meaningful parameter for brokering phase is the computational power of the eligible CEs. Each job is featured by an estimated execution time given by $10^7/CE_{MIPS}$ sec and requires, for execution, two files as input data set whose size are 1MB and 300KB, respectively. The overall computing power of the simulated grid consists of five sites having a number of CPUs between 20 to 30 and featured by a computational power uniformly distributed in the interval [100, 300] MIPS.

Graphics shown in this section refer to simulation having α=0,25 and β=0,25. Again these values have been chosen after several simulations, as they allow achieving the best performance.

Figure 5a shows a direct comparison between the different policies in the scenario described above. Grid performances are improved by both proposed brokering policy. In this case improvement is less than other cases described in previous sections cause the job burst problem doesn't occur for two different considerations described below.

Average job execution time is higher than previous cases. Thus, Grid resources reach saturation with an higher average interarrival job submission time (720 seconds vs 50 seconds of previous cases). Therefore the rate of job bursts is very low.

JSS queuing time is lower than previous cases due to a lower input file transfer. Thus, time spent between the election of the destination CE and the following job submission through JSS is lower than previous cases.

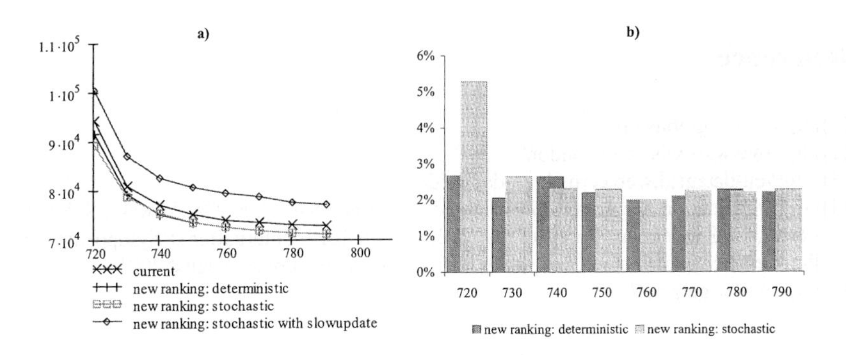

Fig. 5. Case of intensive computational jobs

Figure 5b, as said before, depicts the percentage improvement of job-lifetime obtained using the proposed policy varying workload conditions, with respect to current scheduling policy.

In this scenario, where no overhead due to file transfer is present, the job-burst problem, highlighted in a previous sub-section, plays a very heavy impact in the performance of the system. In fact, in this case, when a job submission request is received by the RB, once this last has chosen the best CE for the job, it can immediately pass the job to the JSS in order to be submitted to the CE. So, the submission of the job is immediate because no file transfer is needed (it's important to remember that file transfer requires a lot of time when files are very huge and the number of files needed is great, too). This means that a very high job submission rate to a particular CE may cause a very sudden increase in the Estimated Traversal Time. If the updating rate of the GII is not enough high, the job-burst problem described before arises. In order to evaluate the impact of the presence of this problem onto the performance of the system, a set of simulations has been carried out. Two different refresh rates of the Grid Information Index were considered. The first one was fixed equal to the job submission frequency. The other was set equal to 1/10 of this last frequency. In Figure 5a a new curve labelled "stochastic with slow update" refers to the stochastic allocation strategy used by the RB in the case the frequency of job submission is ten times greater that the updating frequency of the GIS. As expected, this figure confirms that, due to the inconsistencies arising, the slower the GII is refreshed the worst the average job life time is.

3 Conclusions

This paper has given a state of the art of one the current project aimed to define a Computational Grid, the DataGrid European Project. After a detailed description of the allocation strategy here adopted, new allocation algorithms were proposed and studied. It was pointed out that an allocation policy based on stochastic choices could improve grid performances.

Reference

[1] http://www.globus.org/
[2] http://www.cs.wisc.edu/condor/
[3] http://eu-datagrid.web.cern.ch/eu-datagrid/
[4] Grid Information Services for Distributed Resource Sharing. K. Czajkowski, S. Fitzgerald, I. Foster, C. Kesselman. Proceedings of the Tenth IEEE International Symposium on High-Performance Distributed Computing (HPDC-10), IEEE Press, August 2001.
[5] http://www.garr.net/

Hierarchical Addressing and Routing Mechanisms for Distributed Applications over Heterogeneous Networks

Damien Magoni

Université Louis Pasteur – LSIIT
magoni@dpt-info.u-strasbg.fr

Abstract. Although distributed applications such as grids and peer-to-peer technologies are quickly evolving, their deployment across heterogeneous networks such as non-globally addressable networks has not received much attention. The lack of IP addressing space for instance has created an explosion of network address translators and thus a large amount of non-globally addressable hosts. Whereas these boxes do not disturb much the behavior of a client-server communication, this is no longer true in a distributed application built on the peer-to-peer model where any entity can play both roles and thus needs a globally known address. Application level gateways are proposed to solve this problem but there are usually specific to a given distributed application. In this paper we propose an application level addressing scheme and a routing mechanism in order to overcome the non-global addressing limitation of heterogeneous networks such as those brought by network address translation technologies. Our architecture is designed to bring increased scalability and flexibility for communications between members of a distributed application. We give simulation results concerning the efficiency of the addressing and routing mechanisms.

1 Introduction

In this paper we present the basic mechanisms of an architecture designed to overcome the limitations caused by non-globally addressable members of a distributed application such as a grid or a P2P application. The idea is to create an application level overlay to manage addresses specific to our architecture. Network level connections are managed between members and thus they can belong to different addressing spaces. Our architecture must provide some routing but in a much simpler way than what can be found in routers for instance. We are currently investigating the efficiency of our architecture by simulation and thus we have not yet started its implementation. The rest of this paper is organized as follows. In section 2 we give some information on our architecture. Section 3 contains the algorithms for providing the hierarchical addressing and routing architecture to be used in the overlay. Finally, in section 4 we give simulation results obtained by simulation on the efficiency of our hierarchical addressing and routing scheme.

P.M.A. Sloot et al. (Eds.): ICCS 2003, LNCS 2659, pp. 1093–1102, 2003.

2 Architecture

Our architecture consists in creating a network overlay that has his own addressing scheme. Each host willing to use our architecture, will run our network middleware (to be implemented soon). Any distributed application on the host can use our middleware and there will be (at least) one overlay per distributed application instance. A new member, for a given distributed application, will connect to one or more members by using an end-to-end network connection such as a TCP connection and then it will be addressable by any host in the same overlay. Thus an overlay member being a NAT box [1] will be able to connect to members located in the public Internet while also connecting members in a private non-globally addressable network and it will forward messages by using our application level routing. Furthermore every member whatever its type of network address (e.g. public IP, private IP, IPX, etc.) will be globally addressable in the overlay. This can be very useful for many distributed applications. Notice that NAT boxes as-is do not provide this important property (i.e. a NAT box just lures both hosts by making them believe that itself is the end-point of both connections). Thus high-level protocols such as TCP/IP only serve for connecting neighbor members of an overlay (thus behaving like data-link protocols). We have chosen a hierarchical addressing scheme in to order to make the architecture scalable to a large number of members (i.e. overlay size). We also have chosen to allow the overlay to have a graph structure instead of a tree structure (typically found in overlays) in order to introduce some robustness through the use of alternate paths and better path lengths. This, in turn, increases the complexity of the overlay routing mechanism. Please note that the mechanisms needed to pipe connections in members acting like routers in the overlay is not described here due to space limitations.

3 Hierarchical Addressing and Routing

As hosts can use different protocols to communicate, the address attribution process to the hosts must be protocol independent and uniform. In order to make the addressing as well as the routing scalable, we define a hierarchical addressing. Figure 1 illustrates a network of hosts with their corresponding addresses. An address is composed of one or several fields separated by dots. The level of the address is equal to the number of fields in the address. The prefix of an address is equal to the address without the latest field.

3.1 Hierarchical Addressing

We have designed a centralized algorithm (for the moment) to address all the hosts of a distributed application overlay in order to build a hierarchical structure. This algorithm is given in the `Create-Hierarchy` procedure below. This algorithm uses one parameter called `zone-size`. This parameter controls the number of hosts in the top level zone (i.e. hosts having a level 1 address) as well as the

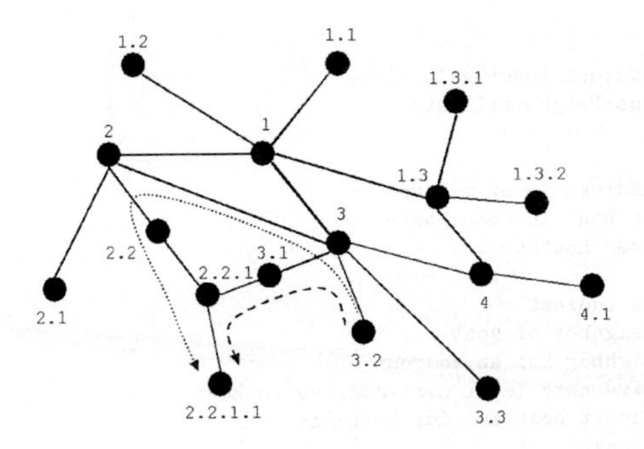

Fig. 1. Hierarchical addressing and routing example.

number of hosts in each sub level zone. First the biggest host with respect to the degree (i.e. number of communication routes) is addressed as host 1. All his neighbors are added to a `degree-pick-list`. Then, the highest degree host of the list is picked and given the same level next address (e.g. 2 for the second one). The neighbors of the next chosen host are added to the list (if they are not in it already). The process is repeated until a top level zone having `zone-size` hosts is obtained. Top level hosts form a connected graph. Then each top level host addresses its `zone-size` highest degree unaddressed neighbors with level 2 addresses where the prefix is its own address. This process is done recursively until no more host can be addressed. So when all level 1 hosts have addressed their `zone-size` highest degree unaddressed neighbors, level 2 hosts do the same, etc. Of course in an implementation, this algorithm will have to be distributed. The top level address attribution protocol may be a bit tricky, but the protocol for sub levels address attribution should be fairly easy to implement. After this phase, some hosts may still not be addressed especially if the `zone-size` parameter is small. We call them leftovers. Each unaddressed remaining host picks its first found addressed neighbor and take the next level next address (e.g. if the neighbor has address 4.5 and has addressed 7 neighbors, host will take address 4.5.8). This means that the number of hosts per zone prefix is not strictly limited to `zone-size` hosts. The `addr-host-list` is a list of the currently addressed hosts.

```
Create-Hierarchy(zone-size)
1.  Empty addr-host-list
2.  Empty degree-pick-list
3.  Give the level 1 address 1 to the host with the highest degree
4.  Add its neighbors to the degree-pick-list
    // select top level hosts
5.  while zone-no <= zone-size
6.      Remove of the degree-pick-list the host with the highest degree
7.      Give the level 1 next address to this host
8.      Insert this host in addr-host-list
9.      Add its neighbors to the degree-pick-list if not already in it
    // select sub level hosts
10. level-no = 1
11. while addr-host-list < number of hosts
12.     if previous loop addressed no host
13.         break
```

```
14.       foreach host
15.           if host address level = level-no
16.               Address-Neighbors(host)
17.       level-no++
18.       foreach host
19.           if host address level = level-no
20.               Insert host in addr-host-list
     // add non addressed hosts
21. foreach host
22.       if host has no address
23.           foreach neighbor of host
24.               if neighbor has an address
25.                   Give next level next address to host
26.                   Insert host in addr-host-list
27.                   break
```

The `Address-Neighbors(host)` function is called by our addressing algorithm. The host given in parameter will address its `zone-size` highest degree unaddressed neighbors with the next level next address. For instance, if host has address 4.5, the first highest degree unaddressed host will be given address 4.5.1, the second 4.5.2, and so on until the last 4.5.`zone-size`.

```
Address-Neighbors(host)
1.    Empty degree-pick-list
2.    foreach neighbor of host
3.        if neighbor does not have an address
4.            if degree-pick-list size < zone-size
5.                Add neighbor to the degree-pick-list
6.    foreach neighbor of host
7.        Remove the highest degree neighbor from the degree-pick-list
8.        Give to neighbor the next level next address
```

3.2 Hierarchical Routing

We propose an algorithm for routing inside our hierarchical addressing structure. The algorithm is distributed (i.e. each host will run this algorithm in its routing module). The `Hierarchical-Routing` function returns true if the next-hop host given in argument does lead to the destination host by a shortest path or false otherwise. The algorithm uses the address of the host running it (`host-addr`), the address of the destination host (`dest-addr`, found in the messages to route) and the address of the next-hop host (`next-hop-host`). Level 1 routing is different because hosts do not have a tree-like relation. It is explained later in this section. In the code, address A is included (strictly) in address B means that A is at most a prefix of B. For instance addresses 1, 1.3 and 1.3.5.2 are all included in address 1.3.5.2.1. Figure 1 illustrates the effects of hierarchical routing and flat routing between the hosts 3.2 and 2.2.1.1 on path length. The hierarchical routing forces the message to be routed via 2 thus giving a path length of 5 hops while a flat shortest path routing requires only 4 hops to reach the destination.

```
Hierarchical-Routing(next-hop-host, dest-addr) returns bool
1.    if host-addr level = dest-addr level
2.        if host-addr prefix = dest-addr prefix
3.            if host-addr label = dest-addr label
4.                // we are at dest!
5.            else
6.                Up-Routing(next-hop-addr, dest-addr)
```

```
7.          else
8.              return (next-hop-addr is included in host-addr)
9.      else
10.         if host-addr is included in dest-addr
11.             if next-hop-addr = dest-addr
12.                 return true
13.             else if next-hop-addr is included in dest-addr
                  and host-addr is included in next-hop-addr
14.                 return true
15.             else
16.                 return false
17.         else
18.             Up-Routing(next-hop-addr, dest-addr)
```

The `Level-Routing(next-hop-addr, dest-addr)` function is called by our routing algorithm when the message to be routed is in a level 1 host. The host needs to know the distance between its neighbors and all the level 1 hosts. In our simulation we run a centralized Floyd-Warshall algorithm on the top level zone to obtain this topology knowledge. In an implementation we will propose a path-vector distributed routing protocol (i.e. a very lightweight BGP-like protocol). The number of hosts involved in level 1 routing will be, by design, limited to `zone-size` thus making the job of the protocol easier. Experiments presented in the next section tend to show that this architecture can scale. In the code, `level-1-addr` of the destination address is the value of the first field of the destination address. The `Up-Routing(next-hop-addr, dest-addr)` function is called by our routing algorithm when the message to be routed must go towards lower level hosts (i.e. hosts located higher in the hierarchy).

```
Level-Routing(next-hop-addr, dest-addr) returns bool
1.      if next-hop-addr level = 1
2.          d1 = hops from next-hop-addr host to
              level-1-addr host of dest-addr host
3.          foreach neighbor of host
4.              if neighbor-addr level = 1
5.                  d2 = hops from neighbor-addr host
                      to level-1-addr host of dest-addr host
6.                  if d2 < d1
7.                      return false
8.          return true
9.      else
10.         return false
```

```
Up-Routing(next-hop-addr, dest-addr) returns bool
1.      if host-addr level = 1
2.          return Level-Routing(next-hop-addr, dest-addr)
3.      else
4.          return (next-hop-addr is included in host-addr)
```

4 Experiments

In this section we present preliminary results obtained by simulation for evaluating the feasibility and efficiency of our hierarchical addressing and routing architecture. We show that the overhead incurred on path length is quite low compared to the gain in routing scalability.

4.1 Simulation Settings

We have used power law graphs for modelling distributed application overlay inter-host connections in accordance with our reasonable assumption that hosts will connect freely and thus will most probably create this kind of topology. We have used a huge 280k-node router-level Internet map [2] collected by Govindan *et al.* and we have sampled it to produce peer-to-peer overlay maps of sizes ranging from 1000 to 5000 nodes which can be considered as large scale distributed application overlays. We have used the *nem* software [3,4] to produce these overlays. We have constructed various addressing plans on them by using zone sizes ranging from 8 to 512. Table 1 shows the parameter space of our simulations.

Table 1. Simulation Parameters

Parameter	Values
Number of members / Overlay size	1000, 2k, 3k, 4k, 5k
Zone size	8, 16, 32, 64, 128, 256, 512

Then we have randomly picked up source and destination host pairs in the distributed application overlays and have computed the shortest path length and the hierarchical path length between the source and destination. They can differ as shown for instance in figure 1. In our simulations, we have measured all the metrics shown in table 2. The number of levels and leftovers are measured after each addressing plan creation and the path length and ratio are measured after each source and destination host pair selection.

Table 2. Measured Metrics

Metric
Flat path length (in hops)
Hierarchical path length (in hops)
Ratio hierarchical/flat path length
Number of levels in hierarchy
Number of leftovers

As the process of generating distributed application overlays and selecting hosts involves random selection (and thus random rolls), we have used a sequential scenario of simulation [5] to produce the results shown in the next section. We have used the Mersenne Twister code [6] for producing the random numbers needed in our simulation. As the random rolls are the only source of randomness in our simulation, we can reasonably assume that the simulation output data obey the central limit theorem. We have performed a terminating simulation where each run is made of picking a source and a destination and determining the number of hops between them using flat (shortest path) and hierarchical routing (i.e. one run is the time horizon). Ideally we should have created a distributed application overlay, planned the addressing, selected two hosts in this overlay, measured the metrics and carried out a sequential checkpoint. However this would have been too costly in terms of computing power. Thus we have performed a sequential checkpoint each time after having created a distributed application overlay, planned the addressing and selected and measured routing distances for 100 pairs of hosts in this overlay. All the simulation results have been obtained assuming a confidence level of 0.95 with a relative statistical error threshold of 1% for the path length distances and ratio and 5% for the number of levels and leftovers.

4.2 Results

Figure 2 shows the ratio between the path length provided by the hierarchical routing and the shortest path length. We call this ratio the *routing overhead*. The routing overhead is shown as a function of the zone size for several overlay sizes. Very surprisingly, the overhead is quite low for this kind of overlay routing, especially for zone sizes above 64. A 4000-node overlay yields an overhead of 32% when the zone size is equal to 128. Furthermore the routing overhead becomes really low (around 10%) when the zone size is equal to 512. We believe that the routing power necessary to manage a 512-node top level zone in a host should not be overwhelming especially when compared to the 120000+ network prefix entries that a BGP core router has to handle! The results do however slightly depend on the distributed application overlay size which can tend to question the scalability. We discuss this issue in the next paragraph. Finally the wavy aspect of some parts of the plots are probably an artefact of our simulation. Our technique consisting in running a number (i.e. 100) of runs per overlay and then performing a sequential checkpoint may slightly affect the statistical results. If we had chosen other values for the number of runs we may have obtained smoother plots. Globally, the overhead cost of using the hierarchical routing is always below 50% and it can be kept around 10% for large zone sizes. As a comparison, in [7] Govindan *et al.* have shown that the routing overhead incurred by routing policies in the Internet is 25% for half of the IP level paths. They also have found that 20% of the paths are inflated by more than 50%! We have also shown in [8] that the routing overhead at the Autonomous System level of the Internet was 8.7% on the average in year 2000 and that 25% of the AS paths were inflated (i.e. longer than their corresponding shortest path). As any BGP router can be considered as knowing the whole AS topology, these results are not good and clearly demonstrate the impact of routing policy in the Internet. Concerning other kinds of routing, Waldvogel *et al.* have shown by simulation in [9] that their topology aware routing architecture for overlays yields a routing overhead of more than 200% for 40% of the paths when using 4 dimensions (however they use the delay as a metric instead of the hop count).

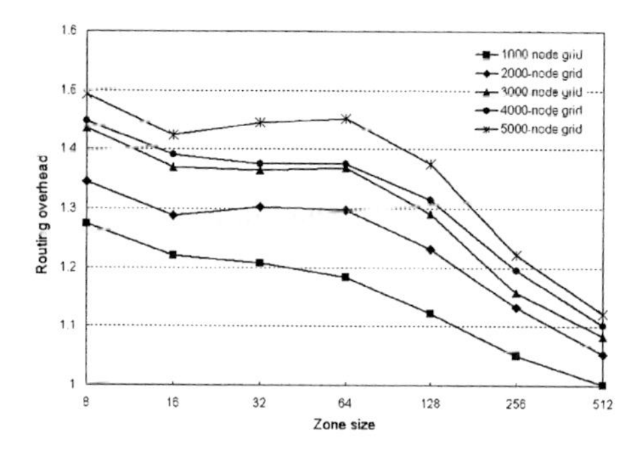

Fig. 2. Average path length inflation as a function of the zone size.

In figure 3, the overhead is shown as a function of the overlay size for several zone sizes. We can see that the routing overhead does increase with the overlay size, topping at 50% for a 5000-node overlay with a zone size of 8. However, the overhead is diminishing when it gets towards the right part of the plots which could show that the overhead increase is not linear. This is a positive indication of scalability. Furthermore for overlays with a zone size of 512, the overhead increases very slowly although in this case we do not know if it diminishes at some point. We are confident that large enough zone sizes can handle large scale overlays for a reasonable routing overhead cost. We plan to do more simulations on larger graphs to confirm this point.

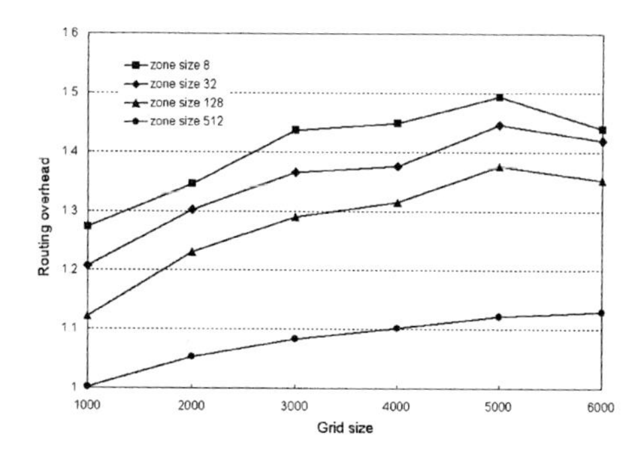

Fig. 3. Average path length inflation as a function of the overlay size.

Figure 4 shows the number of levels used to hierarchically address all the hosts of a distributed application overlay. This number seems to be exponentially inversely proportional to the zone size. This property can be related to the property that the depth of a tree is proportional to the logarithm of its number of nodes. The number of levels seems to be loosely dependent (maybe by a log function) on the overlay size. We can see that for large overlays, the number of levels needed may be quite high thus producing long addresses. A solution to reduce this is to build the sub-levels like the top level and not only by using the neighbors. This will reduce the number of levels and the routing overhead, however it will increase the complexity of the architecture. We plan to develop and analyze this scheme in a future work.

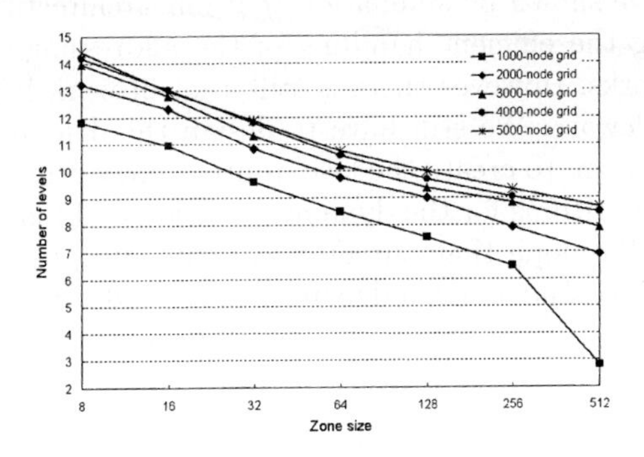

Fig. 4. Average number of levels as a function of the zone size.

The number of leftovers produced during the addressing phase is shown in figure 5 as a function of the zone size. It can be very high especially when the zone size is small. The very important thing to notice is that leftovers do require more addresses and thus can inflate the size of a given zone (i.e. defined as the number of hosts having the same prefix and address level) but they do not imply any penalty on the routing architecture as they can not be part of the top level hosts. Thus the routing information to collect and manage by the top level hosts is always bounded by a number of entries equal to zone size.

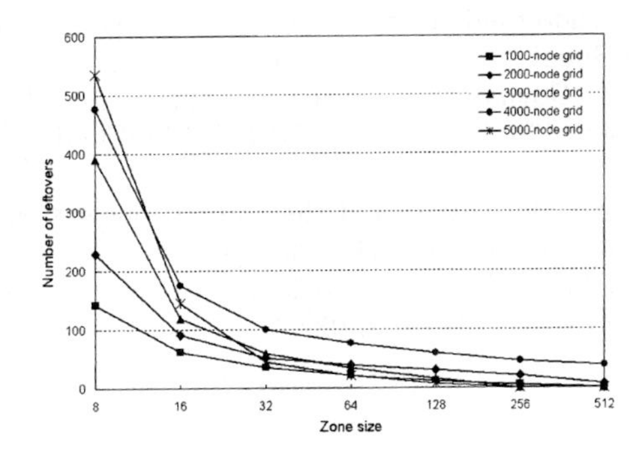

Fig. 5. Average number of leftovers as a function of the zone size.

5 Conclusions

Designing an application layer addressing and routing overlay network for distributed applications in order to overcome the addressing limitations found in heterogeneous networks is not an easy task. And making it scalable and flexible is even more challenging. In this paper we have proposed such an architecture that will be implemented as a communication middleware for distributed applications. We have shown by simulation that our architecture yields promising results concerning the efficient handling of the addressing across non-globally addressable networks. However there is still a lot of work left to build a prototype of our middleware. We still have to design the distributed version of our addressing mechanism, to create the level routing protocol needed inside the top zone and define algorithms for the dynamic addition and removal of hosts within our architecture. We hope that our proposal will bring to distributed applications more flexibility and more independence towards the underlying network addressing.

References

1. Srisuresh, P., Egevang, K.B.: Traditional ip network address translator. Request For Comments 3022, Internet Engineering Task Force (2001)
2. Govindan, R., Tangmunarunkit, H.: Heuristics for internet map discovery. In: Proceedings of IEEE INFOCOM'00, Tel Aviv, Israël (2000)
3. Magoni, D.: nem: a software for network topology analysis and modeling. In: Proceedings of the 10th IEEE/ACM International Symposium on Modeling, Analysis and Simulation of Computer and Telecommunication Systems, Fort Worth, Texas, USA (2002)
4. Magoni, D., Pansiot, J.J.: Internet topology modeler based on map sampling. In: Proceedings of the 7th IEEE Symposium on Computers and Communications, Giardini Naxos, Sicily, Italy (2002) 1021–1027
5. Law, A., Kelton, W.: Simulation Modelling and Analysis. 3rd edn. McGraw-Hill (2000)
6. Matsumoto, M., Nishimura, T.: Mersenne twister: a 623-dimensionally equidistributed uniform pseudo-random number generator. ACM Transactions on Modeling and Computer Simulation **8** (1998) 3–30
7. Tangmunarunkit, H., Govindan, R., Shenker, S., Estrin, D.: The impact of routing policy on internet paths. In: Proceedings of IEEE INFOCOM'01, Anchorage, Alaska, USA (2001)
8. Magoni, D., Pansiot, J.J.: Analysis of the autonomous system network topology. Computer Communication Review **31** (2001) 26–37
9. Waldvogel, M., Rinaldi, R.: Efficient topology-aware overlay network. In: Proceedings of ACM SIGCOMM HotNets'02. (2002)

Active Network Approach to Grid Management

Alex Galis[1], Jean-Patrick Gelas[2], Laurent Lefèvre[2], and Kun Yang[1]

[1] University College London, Department of Electronic and Electrical Engineering
Torrington Place, London WC1E 7JE, UK
agalis@ee.ucl.ac.uk, kyang@ee.ucl.ac.uk

[2] INRIA RESO / LIP, Ecole Normale Supérieure de Lyon
46, allée d'Italie, 69364 Lyon Cedex 07, France
Jean-Patrick.Gelas@ens-lyon.fr, Laurent.Lefevre@inria.fr

Abstract. Due to the large-scale Grid environments and their rapid expansion, both in Grid and network resources, it is getting imperative to provide Grid management mechanisms that can enable Grid Computing to adapt to various application requirements in a flexible and automated way. This paper proposes an Active Grid architecture and middleware for rapid and autonomic GRID service creation, deployment, activation and management. Proposed solutions are based on active networks (Tamanoir Execution Environment) which support dynamic Grid services deployment for the management of Grid architectures.

1 Introduction

The widespread Internet is the living environment of Grid computing. As described in [10], a basic premise of Open Grid Services Architecture (OGSA) is that everything is represented by a service, which is a network-enabled entity that provides capability to users. In addition there is a network paradigm shift dictated by the need of rapid and autonomic service creation, deployment, activation and management combined with context customization and customer personalization. Such motivation can be traced in different organizations, fora, and research activities as well as market forces. This paper aims to contribute to the state of the art of Grid research field by focusing on network, especially the enhancement of the network by using active networks technology to open up the functionalities of networks. And particularly, the policy-based Grid management largely maximizes the efficiency of this enhancement.

In a highly connected Internet world the needs for integrated services across distributed, heterogeneous, dynamic domains of different resources are getting more imperative. The provision of integrated services is very challenging due to the large varieties of the underlying proprietory platforms and their management systems. Research on Grid computing was originated to cope with the first challenge, which has resulted in a long list of Grid supporting platforms, among which Globus [13] attracts more attention and has been widely adopted as a Grid technology solution for scientific and technical computing [10]. But Globus Toolkit is less concerned about the underlying network issues that are

P.M.A. Sloot et al. (Eds.): ICCS 2003, LNCS 2659, pp. 1103–1112, 2003.

vital to a successful Grid environment. On the other hand, the complexity of Grid supporting environment, together with the various resources, also cast serious management issues. All these challenges must be addressed intelligently and effectively. This paper tends to initially practice the applicability of active networks technology to Grid supporting environment and the use of policy-based management method to Grid management.

While Active Network research has precisely tackled that problem domain in fixed and wireless network environments, the particular requirements of Grid network services in terms of management (OS heterogeneity, dynamic topology, efficient failure detection ,fault tolerance) and data transport (reliable collective communications, Quality of Service, streams adaptation) have not sufficiently been taken into account. Tamanoir [11] aims to provide such a case study by focusing on the significant active network support for Grid computing from the network engineering's point of view.

Most challenging problems in Grid context come from heterogeneity of Grid resources and network elements, sheer largeness and inter-domain complexities of programming environment and service deployment [9]. These problems can be addressed by using active network technology. This paper proposes solutions for rapid and autonomic GRID network services creation, deployment, activation and management based on an Active Grid architecture [12].

The paper is structured as follows. Based on the discussion in this section, Section 2 analyses the requirements for Grid Management and Services that highlights the use of policy-based management method for overall Grid management and the use of active networks technology for speeding up and opening up the network layer. Then based on the analysis of OGSA, an active Grid architecture (together with its PBM middleware and active network middleware) is described in Section 3. Section 4 details the active networks support for policy-based Grid management and active networks support for Grid middleware services. And finally, Section 5 concludes the paper.

2 Requirements for Grid Network Services and Management

2.1 Requirements for Grid Network Services

A distributed application running in a Grid environment requires various kinds of data streams: Grid control streams and Grid application streams. First of all, we can classify two kinds of basic Grid infrastructures:

Meta cluster computing. A set of parallel machines or clusters are linked together with IP networks (Internet) to provide a very large parallel computing resource. Grid environments like [13] or [3] are well designed to handle meta-cluster computing session to execute long-distance parallel applications. We can find various network needs for meta-clustering sessions : Grid environment deployment (for OS heterogeneity support, dynamic topology re-configuration, fault tolerance), Grid application management (collective communications like multicast and gather for deployment of binaries of applications, parameters and collection of results of distributed tasks), Grid support (collection of data control, node synchronization, node workload information). The information exchanged is also needed to provide high-performance communications between nodes inside and outside the clusters.

Large scale computing. These environments usually provide support on thousand of connected machines (like [14], [1] or [15]). We can find various network needs for large scale computing sessions: Grid environment deployment (dynamic enrollment of unused machines), Grid application deployment (fault tolerance support, check-pointing protocols), Grid support (workload information of subscribed machines). These two Grid infrastructure can support various usage : computational Grid, Data Grid...

2.2 Policy-Based Grid Management

End-to-end Grid services can be very complex in the Grid computing environment, and this raises the increasing requirement for the management of Grid system as a whole. Most current researches with this goal are carried out from the Grid resources themselves' point of view, with examples as Condor-G system [14] and Nimrod-G Grid resource broker [15]. The research towards flexible Grid Services from the network point of view has yet been significantly taken into consideration. But network, as the transporting media for Grid services, is critical to guarantee fully efficient Grid services. Obviously, the bad quality of service in the networks can significantly obstruct the efficient provisioning of Grid services. Due to the complexity of Grid system, and the trend of getting more complex in both hardware/software and service requirements, the management of the overall Grid system itself and the services it provides in a flexible way is getting more and more important. It is time-consuming and error-prone for Grid administrator or resource manager/broker to configure his system manually. And it is extremely hard for him to configure his local resource while considering other domains in the whole Grid system.

Policy-based management (PBM) is a good candidate for such complex management environment. Policies are seen as a way to guide the behaviour of a network or distributed system through high-level declarative directives. An example of policy is as follow: IF (sourceHost==Camden) and (destHost==skyfire) THEN provideGoldService, which specifies the QoS for specific user. In comparison with previous traditional network management approaches, such as [5] or [4], PBM offers a more flexible, customizable management solution that allows controlled elements to be configured or scheduled on the fly, for a specific requirement tailored for a customer [8],[16]. The aim of PBM is to apply integrated management system so that system management, network management, and service management can cooperate in Grid computing. PBM method has been widely used in the IP network management field, whereas the application of PBM to the Grid management field has yet attracted much attention. Yang, et al [17] presented a policy-based Grid management architecture supervising the overall Grid management, but without considering the Grid services.

2.3 Active Network for Grid Management and Grid Services

Based on the requirement analysis given above, we can see that the requirements cast by Grid services can be satisfied when dynamic injection of new functionalities into current Grid architecture is enabled; whereas the flexibility promised by PBM doesn't come without the automation of policy transit, policy enforcement and code downloading. Active Networks (AN), as an enabling technology, have been proposed as an architectural solution for the fast and flexible deployment of new network services. The basic idea of active networks is to enable third parties (end users, operators, and service providers) to inject application-specific services (in the form of code) into the networks. Applications are thus able to utilize these services to obtain required support in terms of element and network management resources, thus becoming network-aware. This code is dispatched and executed at designated (active) nodes performing operations to change the current state of the node. Active network is distinguished from any other networking environment by the fact that it can be programmed. In our approach, this programmability is provided by the management policies.

To support most of Grid applications, active nodes must deal with the two main Grid configurations:

- Meta cluster computing : in this highly coupled configuration, an active node is mapped on network head of each cluster or parallel machine. This node manages all data streams coming or leaving a cluster. All active nodes are linked with other AN mapped at backbone periphery. An Active node delivers data streams to each node of a cluster and can aggregate output streams to others clusters of the Grid.
- Large scale computing : in this loosely coupled configuration, an active node can be associated with each Grid node or can manage a set of aggregated Grid nodes. Hierarchies of active nodes can be deployed at each network heterogeneity point. Each AN manages all operations and data streams coming to Grid Nodes: subscribing operations of voluntary machines, results gathering, nodes synchronization and check-pointing. For both configurations, active nodes will manage the Grid environment by deploying dedicated services adapted to Grid requirements: management of nodes mobility, dynamic topology re-configuration and fault tolerance.

3 Active Grid Architecture and Its Management

3.1 Overall Architecture

Even though Open Grid Services Architecture (OGSA) is still in its draft and continues to be revised, it has attracted a lot of attention and is regarded as a promising means for providing pervasive services across the Internet. Building on concepts and technologies from the Grid and Web services communities, OGSA architecture defines a uniform exposed service semantics (the Grid service) and standard mechanisms for creating, naming, and discovering transient Grid service instances; it also provides location transparency and multiple protocol bindings for service instances and supports integration with underlying native platform facilities [10]. Due to its many benefits, OGSA is adopted in this paper as a guideline for Grid services. It is further integrated with both policy-based management method and active network technology thus resulting an active Grid architecture as depicted in Figure 1. This architecture aims to provide mechanisms automatically adapting Grid network elements to different Grid services and the management of the Grid system itself.

In this paper, the method to add programmability to Grid management is to extend the widely used Grid supporting tool, Globus, which is also the powerful supporting tool for OGSA [10]. Both policy-based Grid management middleware and active network middleware can be used by Grid supporting environment to facilitate the corresponding functionality so as to achieve better usage and management of different Grid resources such as massive storage resources, computing resources and special scientific instruments.

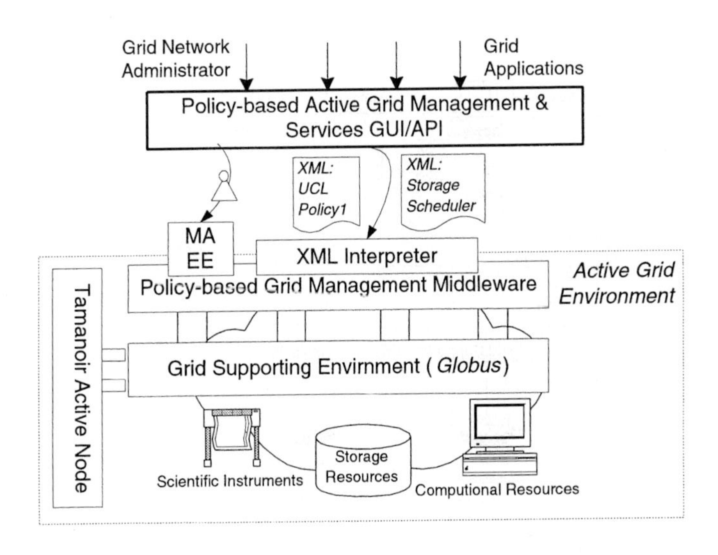

Fig. 1. Overall Active Grid Architecture

In OGSA, the user interface to an OGSA-enabled system will typically be structured as a registry, one or more factories, and a handle mapper service. Each factory is recorded in the registry, to enable clients to discover available factories. When a factory receives a client request to create a Grid service instance, the factory invokes hosting-environment-specific capabilities to create the new instance, assigns it a handle, registers the instance with the registry, and makes the handle available to the handle mapper service. The implementations of these various services map directly into local operations [10].

Active nodes are based on the Tamanoir Execution Environment. The Tamanoir [11] suite is an active network framework that primarily addresses the network management challenges. Based on standards, Tamanoir can easily be deployed in Grid networks.

3.2 Policy-Based Management Middleware

Policy-based Grid management middleware is part of the Active Grid Management Environment and is used to control and manage the Grid environment by defining new policies, e.g., to apply a new DiffServ shaper, or modifying existing policies, e.g., to add a new massive storage accessing role.

In order to deploy PBM technology, a standardization process should be followed to ensure the inter-operability between equipment from different vendors and, furthermore, PBM systems themselves from different developers. The framework and policy information model defined by Internet Engineering Task Force (IETF) Policy Framework Group [2] gains wider popularity and is adopted as the baseline for the PBM system used in this paper.

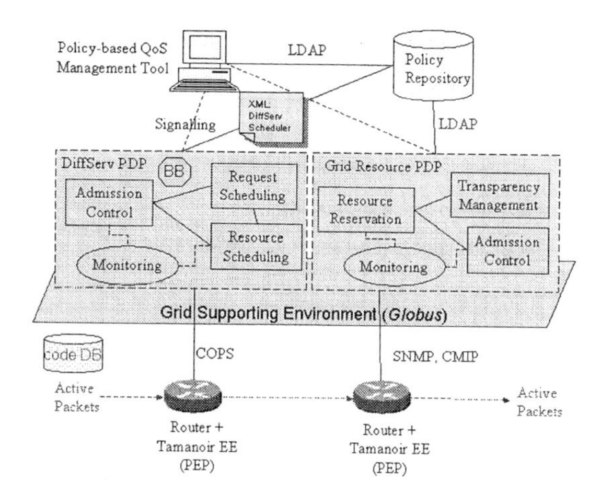

Fig. 2. PBM Middleware Architecture

As illustrated in Figure 2 from top down, the PBM system for Grid management mainly includes four components: policy management tool, policy repository, Policy Decision Point (PDP) and Policy Enforcement Point (PEP). Policy management tool serves as a policy creation environment for the administrator to define/edit/view policies in a high-level declarative language. After validation, new or updated policies are translated into a kind of object oriented representation or so-called information objects and stored in the policy repository. The policy repository is used for the storage of policies in the form of LDAP (Lightweight Directory Access Protocol) directory. Once the new or updated policy is stored, signalling information is sent to the corresponding PDP, which then retrieves the policy and enforces it on PEP. There is a need of a transport protocol for communication between PDP and PEP so that PDP can send policy rules or configuration information to PEP, or read configuration and state information from the device. A wide range of protocols can be used here, such as SNMP, CMIP or COPS (Common Open Policy Service), among which COPS is becoming the standard.

A draft version of object oriented information model has been designed to represent the Grid policies. This information model is based on the IETF PCIM (Policy Core Information Model) [6] but with addition of Grid resources management information and deduction of some rarely used classes to make the whole information model easy to implement. Due to the space limitation, this model will be introduced in this future paper. Furthermore, policies are represented by XML during its transit due to XML's built-in syntax check and its portability across the heterogeneous platforms.

All PDPs, such as DiffServ PDP and Grid Resource PDP, are integrated with Grid supporting tool, Globus [13]. There is also a PDP manager to coordinate the cooperation among different PDPs so as to support some complex Grid services which requirement the cooperation of more than one PDP. More detailed information about this PBM middleware was described in [16]

Various services for Grid application can be introduced by defining new policies, e.g., to apply a new massive storage scheduler, or modifying existing policies. Then the Java classes for fulfilling these policies, which abide by class hierarchy and naming rules of policy information model developed within this system, can be instantiated by storage PDP according to these policies. And this Java bytecode is referenced into Tamanoir active packets and is delivered to the corresponding active nodes to fulfill the management tasks.

4 Active Networks Support

4.1 Active Networks Supporting Environment: Tamanoir

The Tamanoir [11] suite is a complete software environment dedicated to deploy active routers and services inside the network. Tamanoir Active Nodes (TAN) provide persistent active routers, which are able to handle different applications and various data stream (audio, video,..) at the same time (multi-threaded approach). The both main transport protocol TCP and UDP are supported by TAN. The Execution Environment (EE) relies on a demultiplexer receiving active packets and redirecting these packets towards the adapted service in function of a hash key contained in packets header. New services are plugged in the TAN dynamically.

Grid active services can be deployed on a Tamanoir node on various levels :

- in kernel space for lightweight management services without strong memory requirements. Tamanoir allows the deployment of small active modules inside the Linux kernel by using NetFilter toolbox;
- in user space for management and data services requiring storage facilities. Tamanoir execution environment is embedded in a Java Virtual machine. Active services are executed inside distinct Java threads;
- in clustered architecture for data active services requiring high processing power and storage facilities. Tamanoir relies on the Linux Virtual Server (LVS) where a dedicated front-end node distributes streams among back-ends nodes running Tamanoir execution environment.

4.2 Active Networks Support for Dynamic Service Deployment

The injection of new functionalities, called services, is independent from the data stream: services are deployed on demand when streams reach an active node, which does not hold the required service. Two services deployment are available: by using a service repository, where TANs send all requests for downloading required services, by deploying service from TAN to TAN (TAN query the active node that sends the stream for the service). In order to avoid single point of failure service repository can be distributed among sites.

Until the required service be downloaded and ready to process the data stream efficiently we provide a cache mechanism in order to remove any overhead from the sender point of view. This data cache inside the network is done with IBP technology (which provides best effort data storage server called IBP depots[7]).

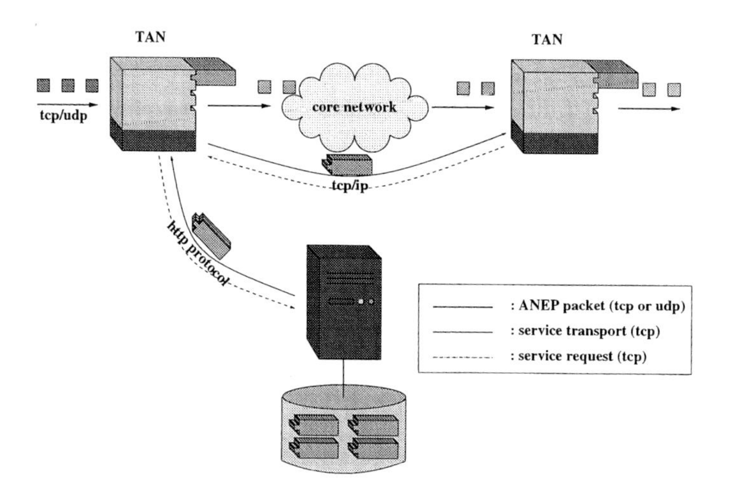

Fig. 3. Active Grid services deployment

4.3 Supporting High Performance Grid Management and Data Streams

Tamanoir execution environment has been extensively experimented on local 100Mbits and Gbit platforms.

With Grid active service running in user space, experiments demonstrate that one Tamanoir active node can support 100Mbits of active packets in several streams. Reactivity and latency of Grid management services are also improved when running in active node kernel space.

Experiments of Tamanoir on Gbit networks show the ability to support large number of active services deployed on TAN and to provide raw performances sufficient for high performance Grid around backbones. With simplified active packets encapsulation, active services running in user space support around 300 Mbits of bandwidth on a GigaEthernet platform. In order to fully support Gbit/s data streams, Tamanoir active node can be deployed on a cluster with LVS.

5 Conclusions and Future Works

This paper contributes to the definition and development of the active Grid architecture and technologies needed for rapid and autonomic GRID service creation, deployment, activation and management. It analyses the requirements for Grid Management and Services that highlights the use of policy-based management method for overall Grid management. It highlights the use of active networks technology in Grids for opening up the network layer by the use of the Tamanoir active network node systems. An OGSA compatible active Grid architecture, together with its Policy Based Management middleware and active network middleware is proposed and currently under development.

Experiments of Tamanoir show that this high performance execution environment can support Grid management streams and data active streams for Grid deployed around high performance backbones.

By providing support for Grid network services, the active network infrastructure becomes a part of Grid resources (like computational or storage resources). We are currently designing a complete framework for efficient management of active nodes and active services deployment.

Acknowledgements. This paper partly describes work in progress in the context of the EU IST project Context. This research is also partly supported by French RNTL Etoile project and ACI-GRID.

References

[1] Entropia : high performance internet computing. http://www.entropia.com.
[2] Ietf policy workgroup. http://www.ietf.org/html.charters/policy-charter.html.
[3] Netsolve project. http://icl.cs.utk.edu/netsolve/.
[4] Tina. www.tinac.com.

[5] Tmn. www.itu.int/TMN/.
[6] Ietf pcim draft, 2002. http://www.ietf.org/internet-drafts/draft-ietf-policy-pcim-ext-08.txt.
[7] Alessandro Bassi, Jean-Patrick Gelas, and Laurent Lefèvre. Tamanoir-IBP : Adding Storage to Active Networks. In *Fourth Annual International Workshop on Active Middleware Services (AMS 2002), 11th IEEE International Symposium on High Performance Distributed Computing*, pages 27–34, Edinburgh, Scotland, jul 2002. IEEE Computer Society. ISBN 0-7695-1721-8.
[8] N. Damianou, N. Dulay, E. Lupu, and M. Sloman. The ponder specification language. In *Workshop on Policies for Distributed Systems and Networks (Policy2001*, HP Labs Bristol, jan 2001.
http://www.doc.ic.ac.uk/ mss/Papers/Ponder-Policy01V5.pdf.
[9] I. Foster, C. Kesselman, and S. Tuecke. The anatomy of the grid: Enabling scalable virtual organizations. *International Journal of High Performance Computing Applications*, 15(3):200–222, 2001.
http://www.globus.org/research/papers/anatomy.pdf.
[10] Ian Foster, Carl Kesselman, Jeffrey M. Nick, and Steven Tuecke. The physiology of the grid: An open grid services architecture for distributed systems integration. In *Open Grid Service Infrastructure Working Group (OGSI-WG) of Global Grid Forum*, 2002. http://www.ggf.org/ogsi-wg/drafts/ogsa_draft2.9_2002-06-22.pdf.
[11] Jean-Patrick Gelas and Laurent Lefèvre. Tamanoir: A high performance active network framework. In C. S. Raghavendra S. Hariri, C. A. Lee, editor, *Active Middleware Services, Ninth IEEE International Symposium on High Performance Distributed Computing*, pages 105–114, Pittsburgh, Pennsylvania, USA, aug 2000. Kluwer Academic Publishers. ISBN 0-7923-7973-X.
[12] Jean-Patrick Gelas and Laurent Lefèvre. Towards the design of an active grid. In Lecture Notes in Computer Science, editor, *Computational Science - ICCS 2002*, volume 2230, pages 578–587, Amsterdam, The Netherlands, apr 2002. ISBN 3-540-43593-X.
[13] Foster I. and Kesselman C. *The Grid: Blueprint for a New Computing Infrastructure.* Morgan Kaufmann, 1999. Globus: A Toolkit-Based Grid Architecture.
[14] Frey J., Tannenbaum T., Foster I., Livny M., and Tuecke S. Condor-g: A computation management agent for multi-institutional grids. In *Proceedings of the Tenth IEEE Symposium on High Performance Distributed Computing (HPDC10)*, San Francisco, USA, aug 2001.
[15] Buyya R. Nimrod/g: An architecture for a resource management and scheduling system in a global computational grid. In *Proc. 4th Int'l Conf. on High Performance Computing in Asia-Pacific Region (HPC Asia 2000)*, Los Alamitos, USA, 2000. IEEE CS Press.
[16] Kun Yang, Alex Galis, Telma Mota, and Stylianos Gouveris. Automated management of ip networks through policy and mobile agents. In *Proceedings of Fourth International Workshop on Mobile Agents for Telecommunication Applications - MATA2002*, pages 249–258, Barcelona, Spain, oct 2002. Springer. LNCS-2521.
[17] Kun Yang, Alex Galis, and Chris Todd. Policy-based active grid management architecture. In *Proceedings of 10th IEEE International Conference on Networks (ICON02)*, pages 243–248, Singapore, aug 2002. IEEE Press. ISBN: 0-7803-7533-5.

The PAGIS Grid Application Environment

Darren Webb and Andrew L. Wendelborn

Department of Computer Science
University of Adelaide
South Australia 5005, Australia
{darren,andrew}@cs.adelaide.edu.au

Abstract. Although current programming models provide adequate performance, many prove inadequate to support the effective development of efficient Grid applications. Many of the hard issues, such as the dynamic nature of the Grid environment, are left to the programmer. We are developing a programming model that incorporates a familiar, formal computational model and a reflective interface. The programming model, called PAGIS, provides a desirable abstract computer with an interface to introduce and customize Grid functionality. Using PAGIS, an application programmer constructs applications that are implicitly parallel and distributed transparently. This paper describes the basic components of the PAGIS framework for constructing and executing applications, and the reflective techniques to customize applications for computation on the Grid.

1 Introduction

The Grid is used in numerous scientific disciplines to solve large, complex problems. The Grid provides pervasive access to a geographically-dispersed, large scale execution environment. It integrates heterogeneous resources with dynamic availability and capability connected by an unreliable network. Computational scientists and engineers use the Grid to distribute computation to hosts offering various computation and data services.

Grid-enabled programming systems enable familiar programming models to be used in Grid environments[1]. A programming model defines an abstract machine, user libraries and software tools a programmer uses to interact with a system. Grid programming models are generally adapted from established parallel programming models designed for homogeneous, tightly coupled supercomputers. However, there is growing consensus that these programming models are inadequate to support the effective development of efficient Grid applications[2]. Models, such as MPICH-G2[3], are typically low-level, restricted in their applicability, and lack many essential properties and capabilities required for Grid applications.

We believe that current Grid programming models are too complex for many potential Grid programmers. While providing high-performance, the responsibility for managing the hard issues, such as adapting to the dynamic nature of

P.M.A. Sloot et al. (Eds.): ICCS 2003, LNCS 2659, pp. 1113–1122, 2003.

the Grid environment, fall to the programmer. As a result, most programmers will lack the skills necessary to exploit the full power of the Grid.

Many projects[4–9] are developing high-level component-based programming models suitable for the Grid. In these models, the programmer's focus is on *what* program components are executed. Our approach is different, focusing on the techniques used to describe *how* these program components are executed.

PAGIS is a middleware that enables scientists to tap into the Grid with little or no Grid programming skills. PAGIS provides two interfaces that emphasise a separation of what an application does from how it does it. The *application programming interface* uses a formal yet simple, abstract computational model to describe what is executed. The *reflective interface* enables the introduction of new behaviour to the application, such as its execution in a Grid environment and its dynamic control.

In this paper, we present the PAGIS framework. We describe its basic components for building and executing applications, and discuss the reflective techniques used to customize applications for computation on the Grid.

2 A Grid Programming Model

Our work emerges from the PAGIS project[10], an architecture for Grid programming. For Grid computing to become widely accessible, its users must be able to communicate with it on their own terms. To this end, PAGIS is middleware that uses composition of processes (a process network[11]) to specify the execution and remote processing of a set of computational tasks. The architecture is generic, although our prototype implementations focus on the domain of satellite image processing for Geographical Information Systems (GIS).

The PAGIS middleware enables a user to specify the set of tasks as a process network. The middleware supports higher-level tools including a domain-based visual programming environment for the visualization of applications in design and execution. Fig. 1 shows a process network to access a data set, crop, and visualize the result. Other operations might include to geo-rectify and scale the data.

Pragmatically speaking, the process network model is compositional, and provides an intuitive notation for building Grid applications. Our work uses the semantic foundation of the process network model to explore the theme of safe adaptation via reconfiguration, which we believe to be of great importance in a flexible web services architecture. Our compositional framework lends itself also to supporting web services[12], an aspect under investigation.

2.1 PAGIS Middleware

The PAGIS middleware consists of a Java-based computational engine that schedules and manages the process network. The PAGIS middleware is an abstract machine with a language that defines the interface for communicating

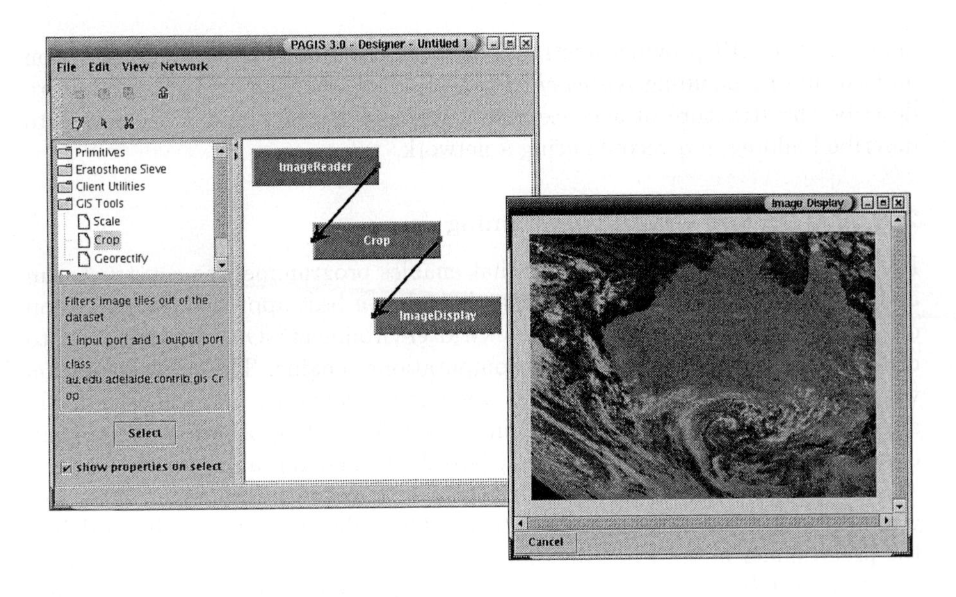

Fig. 1. The PAGIS Visual Programming Environment.

with the machine. The programmer uses the language to instruct the machine to perform certain tasks.

The computational model used in PAGIS is Kahn's Process Networks[11]. Process networks is an intuitive computational model for parallel programming. The model describes a collection of autonomous computational resources connected in a network of unidirectional communication links. A given computational resource computes on data coming along its input communication links, using memory of its own, to produce output on some or all its output links.

The structure of a process network is a directed graph where a node represents a process and an edge represents a channel able to carry data of a given type. A process is a sequential program representing successive passes each of which incrementally transforms a stream of data. Processes communicate through first-in–first-out streams of tokens such that each token is produced exactly once and consumed exactly once. Production of tokens is non-blocking while consumption from an empty stream is blocking.

The model has a number of desirable characteristics. A process network program is deterministic. Computation depends only on program state. Program structure determines the schedule, relieving the programmer of the burden of scheduling and guaranteeing consistent behaviour across implementations[13]. In addition, the process network model allows pipelined parallel computation. Future input concerns only future output, hence a process may produce output before all its input is available.

The Process Network Application Programming Interface (PNAPI) defines a graph-based syntax for describing the structure and behaviour of a process network. The API provides abstractions to describe both network construction and execution, including *Network*, *Process*, *Port* and *Channel* abstractions to describe the structure of a process network, and *Builder* and *Framework* to describe building and reconfiguring a network.

2.2 PAGIS and Grid Programming

PAGIS is a lightweight middleware that enables programmers to construct, run and reconfigure process networks. But how do we best approach the execution of process network applications in a Grid environment? One alternative is to change the implementation of the computational engine. The implementation will undoubtedly benefit some Grid application domains, but will very likely adversely affect others. A new implementation for each application domain will diversify the code base, complicating portability and reusablity of the system. A second alternative is to leave Grid concerns to the programmer. The programmer contorts application code to use Grid libraries directly. In so doing, the programmer mixes what the application does (its functional concerns) with how it does it (its non-functional concerns), thereby increasing complexity, and reducing portability and reusability.

In investigating a way to alleviate such problems, we chose to explore the application of reflective techniques. We will see that such techniques allow programmers to separate out, and focus on, the characteristics of particular grid behaviours one at a time, and define them as customizations of a well-understood base model. The reflective technique we have used (an example of metalevel programming) allows us to expose selected aspects of the programming model the application programmer is likely to need to introduce such behaviour. In the next section, we describe the technique of metalevel programming, the metalevel architecture introduced to our programming model, and how the metalevel is used to introduce Grid behaviours.

3 Customizing PAGIS Applications

In this section, we report our experience in metalevel design and reflective programming for Grid applications. The semantics of the process network model are defined formally, and execution can be performed by any computational engine conformant with those semantics. Hence, it is desirable to express any given computational engine as a customization of a generic computational engine. In this section, we discuss reflective infrastructure for customizing how a PAGIS application executes. We introduce metalevel architectures, and describe a metalevel architecture designed for the development of Grid applications.

3.1 Metalevel Programming

One view of metalevel architectures, shown in Fig. 2, is to treat the system as a black box, with separate interfaces that address the functional and non-functional concerns of an application. The *baselevel* interface, implemented by the system, describes the functional behaviour of a program, and comprises "ordinary" objects and classes called baseobjects and baseclasses. The *metalevel* interface exposes particular aspects of the underlying system to customization. The metalevel consists of objects that *reify* elements of the system, and carry out *reflective computation* on them.

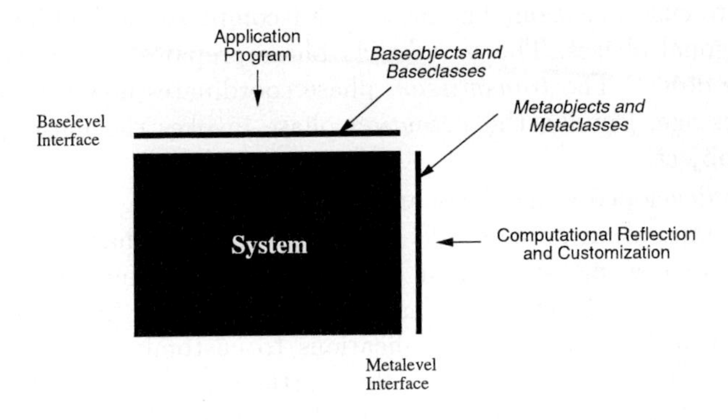

Fig. 2. A system with baselevel and metalevel interfaces.

Reflection is the ability of a system to observe and change its own execution[14]. It is described by Maes [15] as being "the activity performed by a computational system when doing computation about its own computation". The goal of reflection is to allow programs to reason about their own execution state and alter it to change its own meaning. Reification is essentially the process of converting some component of system state into a representation that may be computed upon[16], using *metaobjects*[15]. Modifications to the metaobject result in actual modifications to the behaviour of the system, through the property of causal connectivity[15].

There are two forms of reification: *structural reification* and *behavioural reification*. Structural reification is the process of converting some element of a system into a metaobject, and might include reification of an object, class or method. Behavioural reification is the process of converting computation or behaviour into a metaobject, such as a method invocation. These metaobjects constitute an interface for introspection and customization called a *Metaobject Protocol*.

In this paper, we describe our work toward the development of a framework for Grid application development that incorporates metalevel programming techniques. We present a metalevel architecture influenced by, and the result of the analysis of a number of other metaobject protocols including ProActive[17], FRIENDS[18] and Coda[19] (further details are available in [20]). We show how this metalevel architecture, called *Enigma*, is suitable for Grid programming and proceed to show how it benefits the development of component-based Grid applications.

3.2 Enigma Metalevel

Enigma is a metalevel architecture that decomposes the method invocation process into several operational phases, and allows the composition of new metabehaviour at each phase. The metalevel reifies various structures (e.g. objects and classes) and behaviours (method invocations or messages), and exposes these reifications to customization. The metalevel decomposes method invocation into three orthogonal phases. The *marshaling* phase prepares a message to be sent from a base object. The *transmission* phase coordinates how a base object receives a message. Finally, the *execution* phase invokes the message upon the target baseobject.

We have developed a metalevel architecture called Enigma that decomposes the method invocation process into several operational phases, and allows the composition of new metabehaviour at each phase. The metalevel reifies various structures (e.g. objects and classes) and behaviours (method invocations or messages), and exposes these reifications to customization. The metalevel decomposes method invocation into three orthogonal phases. The *marshaling* phase prepares a message to be sent from a base object. The *transmission* phase coordinates how a base object receives a message. Finally, the *execution* phase invokes the message upon the target baseobject.

Figure 3 shows the course of a method invocation through the Enigma metalevel. The metaobject `MetaInstance:a` intercedes messages sent from baseobject `Object:a`. The metaobject marshals the message and directs it to its target, in this case `Object:b`. The message is transmmitted to `MetaInstance:b` which executes the method invocation upon the baseobject `Object:b`.

Metalevel programmers introduce new behaviour by customizing one of these phases. New metabehaviour is introduced by adding a handler to a phase. Many handlers can be added to each phase and applied as one combined customization.

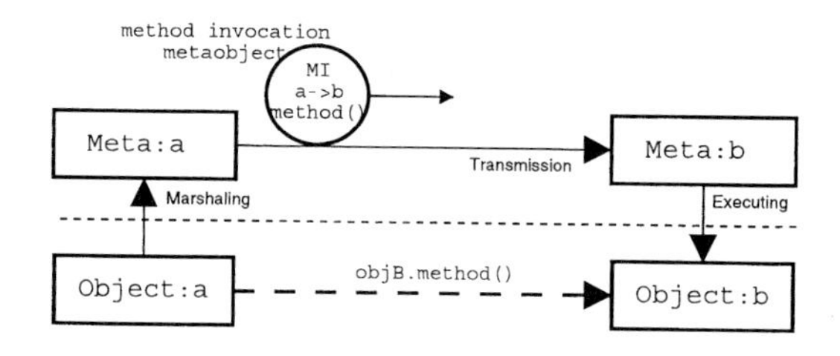

Fig. 3. A model reifying the three phases of method invocation.

To show how to design a new metabehaviour, we now briefly describe a tracing metabehaviour. A tracing behaviour, useful for debugging, is readily defined (similar to [15], for example) in terms of the execution handler: the metalevel program can simply extract and print the method name, invoke the method code, then extract the results and print those. We now discuss how we apply Enigma to process network applications.

3.3 Enigma Applied to Process Networks

One critical design decision in applying Enigma to process networks is: what is the best way to represent, from the point of view of reflective programming in Enigma, the various aspects of a process network? To make a poor decision leads to a cumbersome and complex meta-program; a separate reification the individual process network components turned out to be the wrong approach.

Foote[21] suggests a key requirement of selecting a set of metaobjects be to mirror the actual structure of the language model. Kahn[11] describes the process network model as a network of autonomous computational resources connected by unidirectional links. Rather than reifying each process network element individually, we instead mirror this structure of computational resources, introducing a new metaobject defined as the composition of these baselevel elements. We call this metaobject a `Computation`.

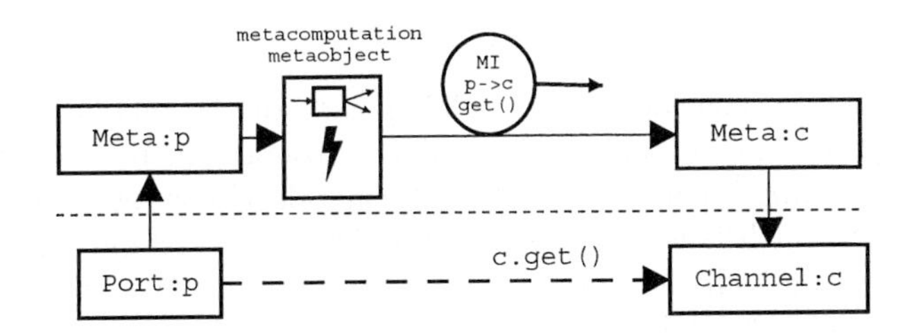

Fig. 4. The `Computation` metaobject reifies a computation's components.

The `Computation` is introduced into the metalevel to compose the process, port and channel elements into a single structure. We use the Enigma metalevel to insert the `Computation` in the transmission phase of relevant metaobjects. The `Computation` metaobject then intercedes all method invocations for the metaobjects of `Computation` elements. Consider Fig. 4. *MI* is the reification at the metalevel of the baselevel method call `c.get()`. Figure 4 illustrates the route of a method invocation *MI* from a port object *Port:p* to a channel object *Channel:c*. This forms the basis of customization.

The advantage of our approach is that we can customize all elements of the `Computation` together as a whole, rather than the more cumbersome approach of applying individual customizations one at a time. By customizing the `Computation` metaobject, which in itself represents such baselevel aspects as process code, ports, channels and threads, we influence baselevel behaviour. The mechanism that we use to customize the `Computation` metaobject is another metalevel (a meta-metalevel) which treats the `Computation` metaobject itself as a baseobject and makes it customizable by reapplying the view of Fig. 3. The `Computation` serves a functional purpose (the composition of metaobjects) at the first metalevel, and a second metalevel helps us to separate new behaviour from the `Computation`. Figure 5 illustrates the introduction of this second metalevel.

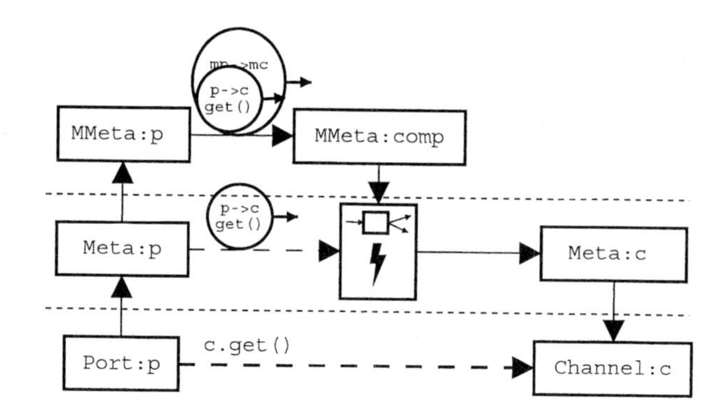

Fig. 5. A meta-metalevel for customizing all `Computation` components.

Finally, we have considered a simplified view of the metalevel, where the metalevel exposes only `Computation` and the interaction between them. The advantage of this approach is simplicity – the metalevel consists of a minimal set of metaobjects, which is less complex and easier to understand, and consistent with Kiczales' minimalistic view of metaobject protocol[22]. The cost of this simplified approach is flexiblity – the simplified view does not allow fine-grain customization of individual baselevel elements. We are still investigating which approach is best for defining Grid metabehaviours.

4 Grid-Related Metabehaviours

In this section, we discuss the implementation of several metabehaviours to deploy a PAGIS application on the Grid. We show how each metabehaviour type can be used to introduce new behaviour relevant to Grid programming. For example, consider the deployment of a satellite image processing application to the Grid. In this example, we may wish to introduce behaviours to migrate computations to Grid hosts, compress the communication of image data between these hosts, and monitor performance to induce adaptation. We now discuss the implementation of these behaviours.

- **Compression.** The compression metabehaviour intercedes method invocations at the marshaling phase of the `Computation` and compresses the reified method invocation. The metabehaviour first serializes the reification to a byte stream, then GZIP compresses the byte stream, and finally encapsulates the compressed data within a new method invocation object. The new

method invocation object decompresses the original method invocation as required, such as when the method is executed. A similar approach could be followed for data encryption.
- **Migration.** The migration metabehaviour affects the transmission phase of the `Computation` to allow the remote deployment of a `Computation`. The metabehaviour enables a third-party (a user or software "agent") to initiate the migration of the `Computation` and its elements. The metabehaviour implements a migration protocol that delegates method invocations to another (possibly remote) metaobject.
- **Performance Monitoring and Adaptation.** We now consider an execution phase metabehaviour that detects and forecasts large-scale changes in system behaviour to drive computation migration. A "performance monitoring" behaviour, based on the tracing behaviour described earlier, records information about method invocations (its destination, timing, frequency and so on) into a suitable repository. An agent may use this information as the basis for an adaptation event, and trigger a computation to migrate. At the CCGrid 2002 GridDemo Worshop, we demonstrated an Application Builder interface through which a user can generate these adaptation events. The interface enables introspection of implementation attributes such as physical location, searching Grid Information Services for potential hosts, and triggering adaptation events. At this time, the generation of events is invoked through user action, and is a very manual process. In the near future, a software agent using the performance monitoring behaviour and Grid Information Services will initiate this automatically.

5 Conclusions

In light of the low-level nature of current Grid programming models, we have developed a high-level programming model with an interface to a simple, intuitive computational model and an interface to customize aspects its implementation. Programmers use the application interface to describe the functional requirements of their application. Programmers optionally use the metalevel interface describe non-functional requirements, such as the application's execution on the Grid. The model forms a unifying framework for describing a computational engine independent of its deployment.

We have developed a metalevel with an operational decomposition and form of metabehaviour composition that we believe is useful for deploying existing systems to the Grid, for building new Grid applications, and especially for supporting adaptive software. We intend to continue to build better abstractions for Grid metabehaviour composition, and improve our understanding of how to use them.

References

1. Foster, I.: The Anatomy of the Grid: Enabling Scalable Virtual Organizations. International Journal of Supercomputer Applications (2001)

2. Lee, C., Matsuoka, S., Talia, D., Sussman, A., Mueller, M., Allen, G., Saltz, J.: A grid programming primer. Technical report, Global Grid Forum Programming Models Working Group (2001)
3. Foster, I., Geisler, J., Gropp, W., Karonis, N., Lusk, E., Thiruvathukal, G., Tuecke, S.: Wide-Area Implementation of the Message Passing Interface. Parallel Computing **24** (1998)
4. Bhatia, D., Burzevski, V., Camuseva, M., Fox, G., Furmanski, W., Premchandran, G.: WebFlow - a visual programming paradigm for Web/Java based coarse grain distributed computing. Concurrency: Practice and Experience **9** (1997) 555–577
5. Shah, A.: Symphony: A Java-based Composition and Manipulation Framework for Distributed Legacy Resources. PhD thesis, Virginia Tech (1998)
6. Neary, M., Phipps, A., Richman, S., Cappello, P.: Javelin 2.0: Java-based parallel computing on the Internet. In: Proc. of Euro-Par 2000, Munich, Germany (2000)
7. Armstrong, R., Gannon, D., Geist, A., Keahey, K., Kohn, S., McInnes, L., Parker, S., Smolinski, B.: Toward a Common Component Architecture for High-Performance Computing. In: Proc. of HPDC8. (1999)
8. Krishnan, S., Bramley, R., Gannon, D., Govindaraju, M., Alameda, J., Alkire, R., Drews, T., Webb, E.: The XCAT Science Portal. In: Proc. of SC2001. (2001)
9. Welch, P., Aldous, J., Foster, J.: CSP networking for Java (JCSP.net). In P.M.A.Sloot, C.J.K.Tan, J.J.Dongarra, A.G.Hoekstra, eds.: Computational Science - ICCS 2002. Volume 2330 of LNCS., Springer-Verlag (2002) 695–708
10. Webb, D., Wendelborn, A., Maciunas, K.: Process Networks as a High-Level Notation for Metacomputing. In: Proc. of 13th Intl. Parallel Processing Sym. Workshops: Java for Distributed Computing, Springer-Verlang (1999) 718–732
11. Kahn, G.: The Semantics of a Simple Language for Parallel Programming. In: Proc. of IFIP Congress 74, North Holland Publishing Company (1974) 471–475
12. Foster, I., Kesselman, C., Nick, J., Tuecke, S.: The physiology of the grid. Technical Report Draft 2.9, Globus (2002)
13. Stevens, R., Wan, M., Lamarie, P., Parks, T., Lee, E.: Implementation of Process Networks in Java. Technical report, University of California Berkeley (1997)
14. Smith, B.: Reflection and Semantics in a Procedural Language. Technical report, MIT (1982)
15. Maes, P.: Concepts and Experiments in Computational Reflection. In: Proc. of OOPSLA 87. (1987)
16. Sobel, J.: An introduction to Reflection-Oriented Programming. In: Proc. of Reflection'96. (1996)
17. Caromel, D., Klauser, W., Vayssiere, J.: Towards seamless computing and metacomputing in Java. Concurrency: Practice and Experience **10** (1998) 1043–1061
18. Fabre, J.C., Pèrennou, T.: FRIENDS: A Flexible Architecture for Implementing Fault Tolerant and Secure Distributed Applications. In: Proc. 2nd European Dependable Computing Conf. (EDCC-2), Taormina, Italy (1996)
19. McAffer, J.: Meta-level Programming with CodA. In: Proc. of ECOOP85. Volume 952 of LNCS. (1985)
20. Webb, D., Wendelborn, A.: The pagis grid application environment. Technical Report DHPC TR-139, Adelaide University (2003)
21. Foote, B.: Object-oriented reflective metalevel architectures: Pyrite or panacea? In: Proc. ECOOP/OOPSLA '90 Workshop on Reflection and Metalevel Architectures. (1990)
22. Kiczales, G.: Towards a New Model of Abstraction in the Engineering of Software. In: Proc. of Intl. Workshop on New Models for Software Architecture (IMSA): Reflection and Meta-Level Architecture. (1992)

Visual Modeler for Grid Modeling and Simulation (GridSim) Toolkit

Anthony Sulistio, Chee Shin Yeo, and Rajkumar Buyya

Grid Computing and Distributed Systems (GRIDS) Laboratory,
Department of Computer Science and Software Engineering,
The University of Melbourne, Australia
ICT Building, 111 Barry Street, Carlton, VIC 3053
{anthony, csyeo, raj}@cs.mu.oz.au
http://www.gridbus.org

Abstract. The Grid Modeling and Simulation (GridSim) toolkit provides a comprehensive facility for simulation of application scheduling in different Grid computing environments. However, using the GridSim toolkit to create a Grid simulation model can be a challenging task, especially when the user has no prior experience in using the toolkit before. This paper presents a Java-based Graphical User Interface (GUI) tool called Visual Modeler (VM) which is developed as an additional component on top of the GridSim toolkit. It aims to reduce the learning curve of users and enable fast creation of simulation models. The usefulness of VM is illustrated by a case study on simulating a Grid computing environment similar to that of the World-Wide Grid (WWG) testbed [1].

1 Introduction

Grid computing has emerged as the next-generation parallel and distributed computing that aggregates dispersed heterogeneous resources for solving all kinds of large-scale parallel applications in science, engineering and commerce [2]. This introduces the need to have effective and reliable resource management and scheduling systems for Grid computing. There is also the need to administer resources and application execution depending on either resource users' or owner's requirements, and to continuously keep track of changes in resource availability.

Managing various resources and applications scheduling in highly distributed heterogeneous Grid environments is a complex and challenging process [3]. A generic view of the World Wide Grid (WWG) computing environment is shown in Figure 1. The Grid resource broker hides the complexities of the Grid computing environment from a user. It discovers resources that the user can access using information services, negotiates for access costs using trading services, maps application jobs to resources, starts execution and monitors the execution progress of tasks.

Different scenarios need to be evaluated to ensure the effectiveness of the Grid resource brokers and their scheduling algorithms. Given the inherent heterogeneity of a Grid environment, it is difficult to produce performance evaluation in a

P.M.A. Sloot et al. (Eds.): ICCS 2003, LNCS 2659, pp. 1123–1132, 2003.

repeatable and *controllable* manner. Therefore, the GridSim toolkit is developed to overcome this critical limitation. The GridSim toolkit is a Java-based discrete-event Grid simulation toolkit that provides features for application composition, information services for resource discovery, and interfaces for assigning application tasks to resources and managing their execution. The GridSim toolkit has been applied successfully to simulate a Nimrod-G [4] like Grid resource broker and to evaluate the performance of deadline and budget constrained cost- and time- optimization scheduling algorithms [3].

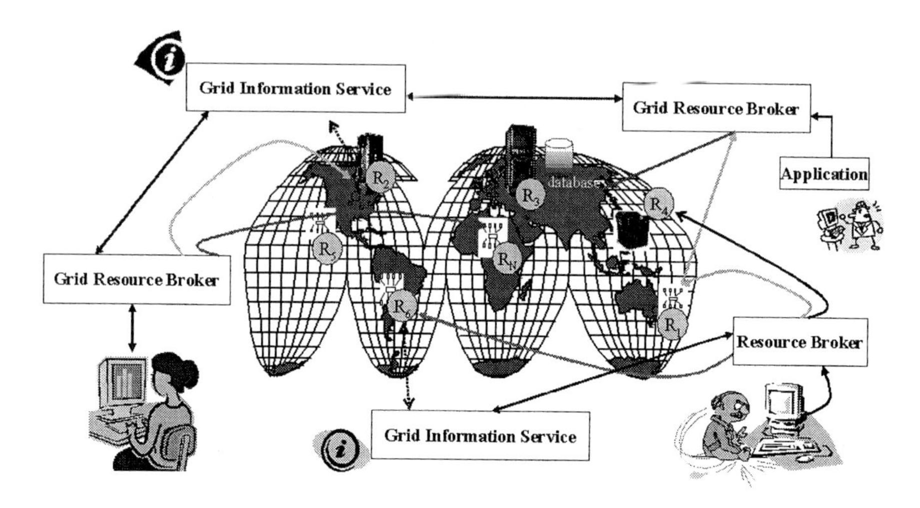

Fig. 1. A generic view of World-Wide Grid (WWG) computing environment

Since the GridSim toolkit is an advanced and powerful simulation toolkit, its users will experience a high learning curve in order to utilize the toolkit functionalities for effective simulations. In addition, the users need to write Java code that use the toolkit packages to create the desired experimental scenarios. This process is repetitive and tedious. It is specially disadvantageous to those who have little or no experience in Object-Oriented (OO) concepts in general, and Java programming in specific. Therefore, it is necessary to build a Graphical User Interface (GUI) for GridSim that enables its users to create and modify simulation models easily and quickly as many times as required.

This paper presents a Java-based GUI tool called Visual Modeler (VM) that is designed as a separate additional component residing in a layer above the GridSim toolkit. Figure 2 illustrates the role of VM in creating a simulation model. A GridSim user utilises VM as a tool to create and modify the Grid simulation model (see Figure 2 step 1A). VM will then generate Java code for the simulation (see Figure 2 step 1B). In the absence of VM, the user needs to write Java code manually using a text editor or development tools such as JBuilder (see Figure 2 step 2). However, this approach is prone to create programs with syntax errors.

VM provides a simple and user-friendly GUI to facilitate its user to be able to create and modify different simulation models easily. This relieves the users from spending a lot of time and effort trying to understand the GridSim toolkit code. VM is designed to enable the users to create simulation models without the need to know the actual Java code behind it or the GridSim toolkit code. In addition, it automatically generates Java code that uses the GridSim toolkit, so users can compile and run the simulation. Therefore, the aim of VM is to reduce the learning curve of GridSim users and to enable them to build simulation models easily and effectively. The initial prototype of VM has been implemented and bundled together with the new release version of the GridSim toolkit 2.0 in November 2002.

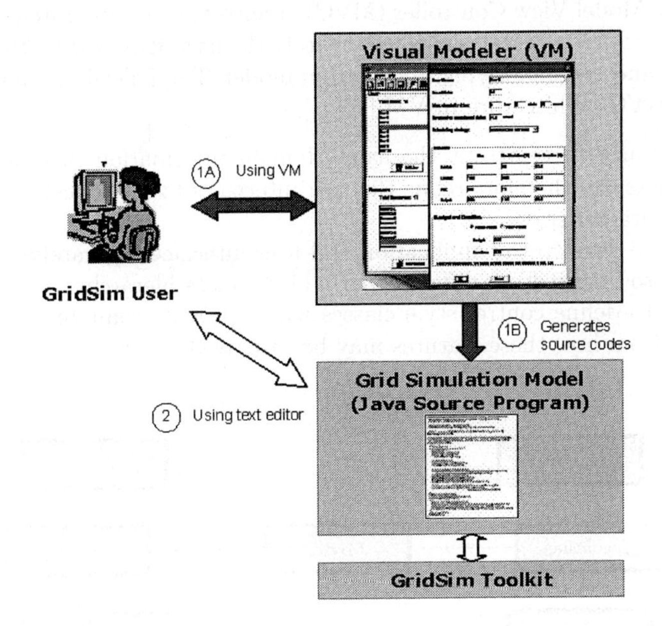

Fig. 2. Relationship between *Visual Modeler*, *Grid Simulation Model* & *GridSim toolkit*

This paper is organized as follows: Section 2 mentions related work. Section 3 presents the architecture and features of VM, while Section 4 discusses the design and implementation of VM. Section 5 illustrates the use of VM for simulating a Grid computing environment. Section 6 concludes the paper and suggests some further work to be done on VM.

2 Related Work

The GridSim 1.0 toolkit does not have any additional GUI tools that enable easier and faster modeling and simulation of Grid environments. It only comprises the Java-based packages that are to be called by users' self-written Java simulation programs. VM has now been included in the recently released GridSim 2.0 toolkit. There are many similar tools like VM that generate source code, but are not related to Grid simulation in specific, such as SansGUI™ [5] and SimCreator [6].

Other than the GridSim toolkit, there are several other tools that support application scheduling simulation in Grid computing environments. The notable ones include Bricks [7], MicroGrid [8] and SimGrid [9]. For a brief comparison of these tools with the GridSim toolkit, please refer to [3].

3 Architecture

VM adopts Model View Controller (MVC) architecture as illustrated in Figure 3. The MVC architecture is designed to separate the user display from the control of user input and the underlying information model. The following are the reasons for using MVC architecture in VM [10]:

- opportunity to represent the same domain information in different ways. Designers can therefore refine the user interfaces to satisfy certain tasks and user characteristics.
- ability to develop the application and user interface separately.
- ability to inherit from different parts of the class hierarchy.
- ability to define control style classes which provide common features separately from how these features may be displayed.

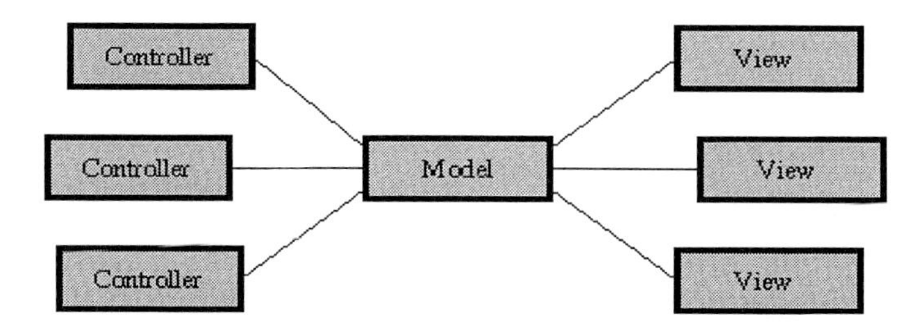

Fig. 3. The basic *Model View Controller* architecture

VM is programmed using Java as Java supports powerful Swing libraries that facilitate easy GUI programming. In addition, Java works well with the MVC architecture. VM support the following main features:

- enables the creation of many Grid testbed users and resources,
- generates the simulation scenario into Java code that the users can compile and run the simulation with the GridSim toolkit,
- saves and retrieves a VM project file that contains an experiment scenario in eXtensible Markup Language (XML) format, and
- works on different operating system platforms (as it is implemented in Java).

4 Design and Implementation

The MVC paradigm proposes three class categories [10]:

1. **Models** – provide the core functionality of the system.
2. **Views** – present models to the user. There can be more than one view of the same model.
3. **Controllers** – control how the user interacts with the view objects and manipulates models or views.

Java supports MVC architecture with two classes:

- **Observer** – any object that wishes to be notified when the state of another object changes.
- **Observable** – any object whose state may be of interest to another object.

4.1 Model

VM consists of three model classes, whose relationships are shown in Figure 4. The models are:

- **FileModel** – deals with UserModel and ResourceModel for saving and retrieving an XML file format for the VM project file.
- **UserModel** – stores, creates and deletes one or more user objects.
- **ResourceModel** – stores, creates and deletes one or more resource objects.

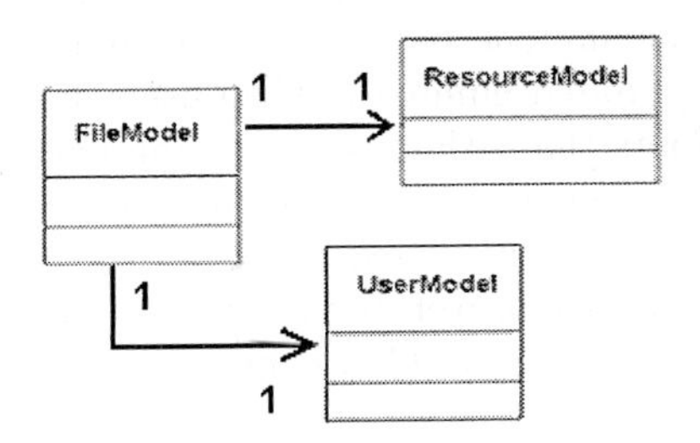

Fig. 4. Class diagram that contains the model classes of VM and their relationships

Both UserModel and ResourceModel contain objects that are hybrids of both model and view, i.e. they have their own widgets to display the stored values. This technique reduces the number of classes needed and the overall design complexity, thus saving significant development time. All models extend Java's Observable class, which provides the behaviours required to maintain a list of observers and notify them when there are changes to the model. The initialization for UserModel and ResourceModel is straightforward with both classes not requiring any parameters. However, FileModel needs to request a reference to each of the UserModel and ResourceModel object. It thus becomes an observer of UserModel and ResourceModel when dealing with the saving and retrieving project file.

4.2 View

Views are registered as dependents of a model. The model broadcasts a message to all dependents when it changes. The main views of VM are:

- **MenuView** – creates GUI components for menu bar.
- **IconView** – creates GUI components for icon toolbar.
- **DisplayView** – creates the main window to display the lists of Grid user and resource.

These three main views are created first, by constructing their widgets independently and then adding their sub-views. This design ensures that the sub-views are separate from their parents' view [10]. Each main view contains one or more references to the model since the model contains user/resource objects that also display widgets. This design reduces the number of method calls and any overheads associated with the interaction between the object and its view.

There are two possible approaches for creating user/resource objects: *with* or *without* their widgets. These approaches are tested by using VM to create 500 Grid users and 1,000 Grid resources. Table 1 shows that creating objects without their widgets requires much less time. However, Lines of Codes (LOC) inside the object class is slightly larger than that of the approach with widgets.

The approach without widgets requires less passing of reference objects and messages in comparison to the other method. In addition, the time taken and the amount of memory needed to create components and to register their event listeners are minimal. VM can thus supports fast creation of a large number user/resource objects at any one time. This is helpful since a Grid simulation model does not have a limited number of Grid users and resources.

Therefore, VM adopts the without-widgets approach by displaying its GUI upon the user's request. When the user closes its GUI window, the GUI components are retained so that they need not be created again for future access (similar to the cache concept in web browsers). This reduces the utilization of memory and enables fast display of GUI components repeatedly.

Table 1. Comparison for the creation of 500 Grid users and 1,000 resources using VM, running on a Pentium II 300 MHz with 64MB RAM

Approach	Time completion	Average LOC
Creating objects *without* their widgets	1 min	950
Creating objects *with* their widgets	20 mins	600

4.3 Controller

The controller needs to be informed of changes to the model as any modifications of the model will also affect how the controller processes its input. The controller may also alter the view when there is no changes to the model. The primary function of the controller is to manage mouse/keyboard event bindings (e.g. to create pop-up menus). If an input event requires modification of application-specific data, the controller notifies the model accordingly.

In this implementation, controllers which relate to views have several responsibilities. Firstly, they implement Java's event-handler interfaces to listen for appropriate events, e.g. the icon toolbar controller detects the clicking of the save toolbar button and notifies FileModel to save a project file. Secondly, views delegate the construction and maintenance of button panels to controllers. Thirdly, controllers can broadcast semantic events (e.g. save file), to objects who have informed the controller that they are interested in these events [10].

5 Use Case Study – Grid Computing Environment Simulation

This section describes how a simulated Grid computing environment is created using VM. First, the Grid users and resources for the simulated Grid environment have to be created. This can be done easily using the wizard dialog as shown in Figure 5. The VM user only need to specify the required number of users and resources to be created. Random properties can also be automatically generated for these users and resources. The VM user can then view and modify the properties of these Grid users and resources by activating their respective property dialog.

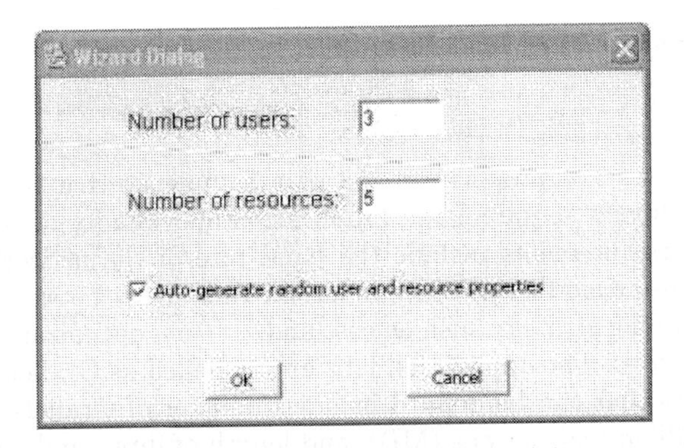

Fig. 5. *Wizard dialog* to create Grid users and resources

Figure 6 shows the property dialog of a sample Grid resource. VM creates Grid resources similar to those present in the WWG testbed [1]. Resources of different capabilities and configurations can be simulated, by setting properties such as cost of using this resource, allocation policy of resource managers (time/space-shared) and number of machines in the resource (with Processing Elements (PEs) in each machine and their Million Instructions Per Second (MIPS) rating).

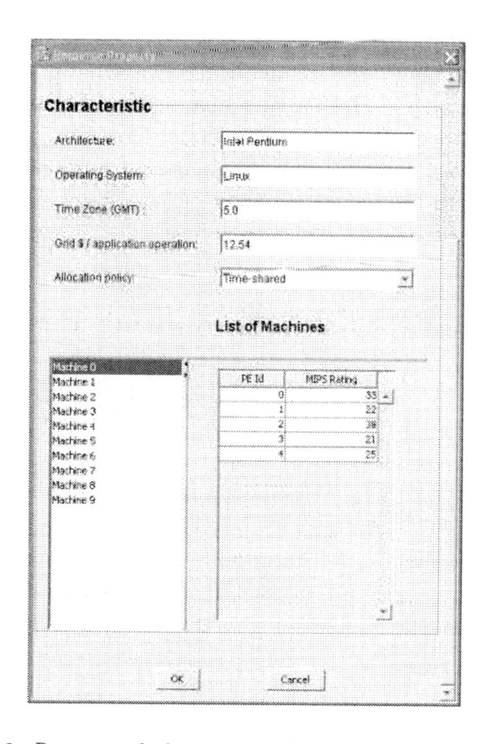

Fig. 6. *Resource dialog* to view Grid resource properties

Figure 7 shows the property dialog of a sample Grid user. Users can be created with different requirements (application and quality of service requirements). These requirements include the baud rate of the network (connection speed), maximum time to run the simulation, time delay between each simulation, and scheduling strategy such as cost and/or time optimization for running the application jobs. The application jobs are modelled as Gridlets. The parameters of Gridlets that can be defined includes number of Gridlets, job length of Gridlets (in Million Instructions (MI)), and length of input and output data (in bytes). VM provides a useful feature that supports random distribution of these parameter values within the specified derivation range. Each Grid user has its own economic requirements (deadline and budget) that constrains the running of application jobs. VM supports the flexibility of defining deadline and budget based on factors or values. If it is factor-based (between 0.0 and 1.0), a budget factor close to 1.0 signifies the Grid user's willingness to spend as much money as required. The Grid user can have the exact cost amount that it is willing to spend for the value-based option.

VM will automatically generate Java code for running the Grid simulation. This file can then be compiled and run with the GridSim toolkit packages to simulate the required Grid computing environment.

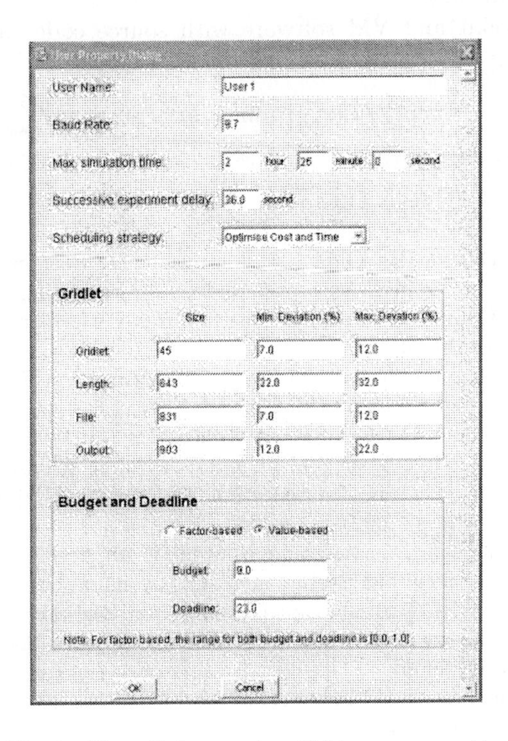

Fig. 7. *User dialog* to view Grid user properties

6 Conclusion and Further Work

This paper describes a Java-based GUI tool called Visual Modeler (VM) that facilitates GridSim users in creating and modifying Grid simulation models easily and effectively. It incorporates features such as easy-to-use wizards that enables users to create simulation models, and automatic source code generation facility that outputs ready-to-run simulation scenario code.

The implementation of VM in Java is ideal since Java supports powerful GUI components with its Swing packages. Moreover, the MVC architecture can be adapted in Java using Observable and Observer class. Hence, Java and MVC provide a perfect combination in creating a GUI for GridSim.

Possible improvements to be implemented include incorporating the GridSim run-time environment into VM, and generating visual diagrams such as graphs based on the GridSim simulation results. This will enable users to receive dynamic run-time feedback from GridSim through VM.

Acknowledgement. We thank Srikumar Venugopal for his comments on the paper. We also thank anonymous reviewers for providing excellent feedbacks.

Software Availability
The GridSim toolkit and VM software with source code can be downloaded from the following website:

`http://www.gridbus.org/gridsim/`

References

1. Buyya, R., Stockinger, H., Giddy, J., Abramson, D.: Economic Models for Management of Resources in Peer-to-peer and Grid Computing. SPIE International Conference on Commercial Applications for High-Performance Computing, Denver, USA (August 20–24, 2001)
2. Foster, I., Kesselman, C. (eds.): The Grid: Blueprint for a Future Computing Infrastructure. Morgan Kaufmann Publishers, USA (1999)
3. Buyya, R., Murshed, M.: GridSim: A Toolkit for the Modeling and Simulation of Distributed Resource Management and Scheduling for Grid Computing. The Journal of Concurrency and Computation: Practice and Experience, Vol. 14, Issue 13–15. Wiley Press (2002)
4. Buyya, R., Abramson, D., Giddy, J.: Nimrod-G: An Architecture for a Resource Management and Scheduling System in a Global Computational Grid. Proceedings of 4th International Conference and Exhibition on High Performance Computing in Asia-Pacific Region (HPC Asia 2000). IEEE Computer Society Press, Beijing, China (May 14–17, 2000)
5. ProtoDesign Inc. SansGui^{TM}. `http://protodesign-inc.com/sansgui.htm`
6. Realtime Technologies Inc. SimCreator.
`http://www.simcreator.com/simtool.html`
7. Aida, K., Takefusa, A., Nakada, H., Matsuoka, S., Sekiguchi, S., Nagashima, U.: Performance Evaluation Model for Scheduling in a Global Computing System. The International Journal of High Performance Computing Applications, Vol. 14, No. 3. Sage Publications, USA (2000)
8. Song, X., Liu, X., Jakobsen, D., Bhagwan, R., Zhang, X., Taura, K., Chien, A.: The MicroGrid: A Scientific Tool for Modelling Computational Grids. Proceedings of IEEE Supercomputing (SC 2000). Dallas, USA (Nov 4–10, 2000)
9. Casanova, H.: Simgrid: A Toolkit for the Simulation of Application Scheduling. Proceedings of the First IEEE/ACM International Symposium on Cluster Computing and the Grid (CCGrid 2001). IEEE Computer Society Press, Brisbane, Australia (May 15–18, 2001)
10. Mahemoff, M. J., Johnston, L. J.: Handling Multiple Domain Objects with Model-View-Controller. In: Mingins, C., Meyer, B. (eds.): Technology of Object-Oriented Languages and Systems 32. IEEE Computer Society Press, Los Alamitos, USA (1999) 28–29

Layered Resource Representation in Grid Environment: An Example from VEGA Grid

Fangpeng Dong, Yili Gong, Wei Li, and Zhiwei Xu

[1]Institute of Computing Technology of CAS
Beijing China, 100080
{fpdong, gongyili, liwei, zxu}@ict.ac.cn

Abstract. Resource discovery in wide-area distributed environments is a great challenge for the large quantity and volatility of resources. To solve this problem, mechanisms effectively representing resources are very necessary. In this paper, a resource representation scheme is proposed, which embodies the principles of VEGA Grid. This scheme is designed for routing-transferring model that is implemented as the resource discovery mechanism in VEGA Grid. The scheme is based on a three-layer model: the top user layer, on which specialized representation can be deployed according to the demand of users; the middle router layer, on which resource routers work for globally resource discovery; and the bottom provider layer for providers to publish their resources. We explain our choices of resource representations in each of these layers, and evaluate the usability, scalability and performance of the approach.

1 Introduction

The dramatic increase in demand for resource sharing and cooperating in wide-area environments motivates the emergence of new technologies, e.g., computational Grid [1] and Web Service are invited to enhance support for e-science and e-business. Although different in architectures, these technologies are confronted with resources having features of large quantity, wide distribution, mobility and multiple policies.

In VEGA Grid, we adopt a main idea of constructing Grid as a virtual computer in wide area. VEGA Grid makes a mapping between the Grid system and the traditional computer system. We use the virtual computer concept to build the VEGA Grid on legacy systems. Although the architecture adopted by Vega Grid is very simple, it can solve the problem of resource sharing and efficient cooperating. The benefit of the mapping method used here is that we can borrow many mature technologies from the traditional computer architecture design. Many important ideologies and theories are helpful to our design, but here we only introduce those related to resource discovery mechanism used by VEGA:

A *Grid processor* is an abstraction of a resource consumer that controls the activities of Grid virtual hardware [2]. A Grid processor is a logical concept, and it can be either software or special hardware.

A *Resource router* [2] is a component in Grid system bus [2]. The basic idea of the resource router is the routing-transferring method, that is, a resource router can

P.M.A. Sloot et al. (Eds.): ICCS 2003, LNCS 2659, pp. 1133–1142, 2003.

choose a route for a resource request and forward it one by one until the request reaches a Grid virtual device that provides the proper resource.

A *Grid virtual device* is an abstraction of a resource provider in Grid. In our design, different resource providers are viewed as different virtual devices that provide various services. For each virtual device, there is a specified *virtual device driver* installed in Grid process end to provide necessary accessing interface to the device.

In an overview, VEGA Grid is constructed as Fig. 1.

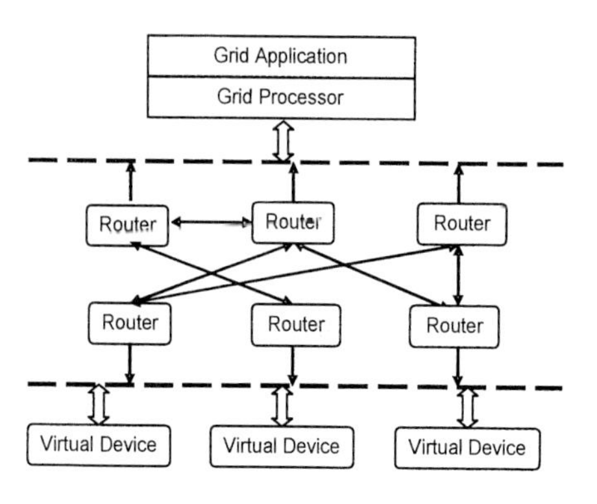

Fig. 1. The layered structure of Vega Grid virtual computer.

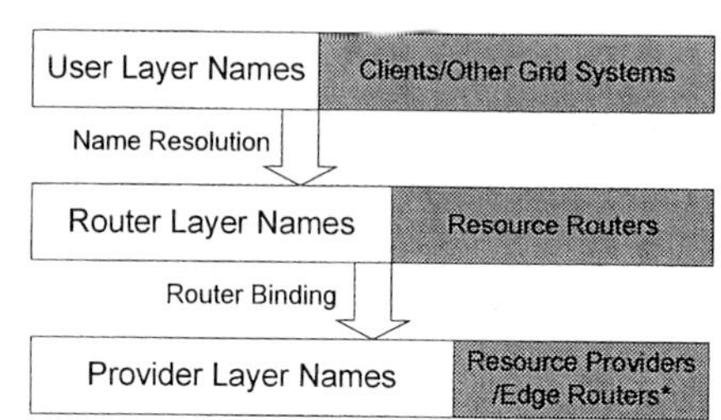

Fig. 2. The VEGA resource representation stack. *: Routers directly connected with resource providers are called edge routers.

A resource discovery process is a series of operations initiated by the top Grid Processor layer requests and completed by the bottom Virtual Device layer responses. To adapt this layered structure of VEGA Grid, we take a natural way to represent resources in three layers:

Grid applications in Grid processors are various in resource using patterns and representations, e.g., a Globus application which gets resource information from MDS [4] in LDAP is very different from a Web Service application which gets it from UDDI [12] in XML. So, if we want to integrate applications from different systems into ours, the Grid processor layer resource representations should have the capability of customizing according to application requirements. In other words, the Grid processor layer representations are not fixed but changeable. To make the definition generally understandable, we call representation in this layer user-layer name.

To locate resources required from Grid processors in a wide distributed environment cooperatively, resource routers need a global uniform mechanism to identify resources with different attributes. When the performance of routers is considered, resource representation in this layer should be stored, transferred, and identified efficiently. Here, we call resource representation in this layer router-layer name, a sequence of integral and string values.

In the bottom layer of VEGA, resource providers need a location-aware mechanism to make their resources accessible. Representation of resources at this layer is called provider-layer name. Thus, resource representation is also organized into three layers and forms a stack shown in Fig. 2. We also give the mapping between layers and typical roles working on each layer.

The rest of this paper is organized as follows: in the second section, we introduce related works. The third section explains VEGA principles, which guide our work. The fourth section gives a brief explanation for the resource router architecture, analyzes its requirements and describes our naming method in details. We report the implementation of this method in VEGA Grid in the fifth section. At last, we draw a conclusion and point out further work.

2 Related Works

In Grid environments, two strategies are mainly used to identify resources: in some cases, resources are explicitly named using either names or identifiers; some other systems provide catalogs that allow users to describe resources using attributes, and by querying these catalogs, users can select resources with particular properties [6]. The former is an appealing idea for its simpleness, but it makes the resource discovery a painful mission for its hardness: resource attributes' volatility is difficult to express [13] [14]. The latter is based on the attribute-value-matching resource discovery mechanism and it works well in an environment in which resource types are relatively stable and few, but it is difficult to define and maintain numerous types of resources and serve the matching process in a wide-area environment.

Secure Grid Naming Protocol [5] defines a scheme for location-independent logical naming of grid resources. A SGNP name is a Location-independent Object Identifier (LOID) that uniquely identifies a Grid resource. The actual position of a Grid resource is located by associating the LOID with one or more communication protocols and network endpoints. In SGNP, a LOID is hierarchically resolved, and the result of resolution is the client-to-resource binding, which means wherever a resource moves, it must connect to the same *Domain Resolver* and *Binding Resolver*.

GLOBE [9] realizes a system that assigns resources location-independent names that are resolved to contact addresses by a location-service as a means for retrieving mobile resources. In [10], the author argues the scalability problems of present object reference and symbolic names, and discusses the solutions in Globe.

In UDDI, information about business entities and services are classified in White page, Yellow page and Green page. Each data element has a globally unique primary key, for example, a business identifier can be a D&B number, tax number, or other information types via which partners will be able to uniquely identify a business.

Systems mentioned above are examples that adopt resembling ID-based resource naming and discovering in Grid environments. Name resolution is popular in these systems, and a service to locate a resource entity is required to map location-independent resource identifiers to physical resources.

Condor's Matchmaker [7] is a typical distributed resource sharing system that does not use global names for resource discovery: resource description and request are sent to a central server responsible for matching the advertisement between the resource requestor and provider. This architecture is based on the assumption that an organization will be willing to operate the central server in wide area.

Another relevant experience comes from Globus MDS, which defines for each object type an object class specification declaring the required and optional attributes of a particular object. GGF has proposed a Grid Object Specification (GOS) [8] draft to formalize object class definitions. Although MDS avoids name resolution, user-visible name space is limited: object class names are globally defined and universal, that is, all Grid users can only see one common name space. Systems adopting other name schemes can hardly merge into MDS. Further, the information service is configured into a tree-like architecture, thus restricts both the resource representation and discovery process flexibility.

3 Naming Principles in VEGA Grid

In our designing of VEGA Grid architecture, we propose the following principles called *VEGA* to evaluate our designing work. These principles are also embodied in the resource representation and discovery approach.

Versatile Services. The Grid should be constructed as an infrastructure that provides a developing and running environment supporting various applications, using patterns, platforms and compatibilities. To achieve this goal, resource representation at application level must be flexible enough to adapt various name schemes and using patterns.

Enabling Intelligence. The Grid should have the ability to delegate users to make decisions in some circumstances. Given some polices, the Grid also should have the ability to aggregate, produce or filter information about itself. Resource router is such a component, which delegates users to find resources of interest and collects resource status in the Grid. This principle also requires support to friendly user patterns.

Global Uniformity. To work corporately and find resources globally, representation at router layer must be globally uniform. Additionally, Global uniformity gives resources the ability to join the Grid anywhere, anytime and its accessibility can be promised.

Autonomous Control. The Grid should not be governed by a central administration. For the large quantity and wide distribution, resources should not be named and resolved centrally in order to gain promising performance and scalability as the growth of Grid.

4 Resource Representation Scheme

In VEGA Grid, resource routers are deployed to address the resource registry and discovery problems. Resource routers take account of the merits of resource attributes, and adopt a proper routing policy as well as an information-aggregating algorithm to overcome some performance drawbacks in the attribute matching strategy, such as the directory service. Proper representation schemes in multiple layers play an important role in promoting routers' performance, giving upper users a friendly way to access Grid resources and facilitating resource providers to make their resources accessible.

4.1 Resource Representation in User Layer

In this layer, our goal is to provide mechanisms to applications/users so that they can join VEGA and utilize resources conveniently. Guided by the principle of *Versatile Service*, it is not applicable to deploy resource name space uniform in this layer, but changeable according to the requirements of different applications. In other words, users should have the ability to customize their own resource representations in the Grid processor they connect to, so it is hard to give a generic view of user-layer representations. But some examples may be helpful to demonstrate the features of this mechanism and discuss related problems. We have defined a naming scheme for user querying computing resources that we deployed in VEGA prototype. The syntax of this scheme is as following [15]:

```
<name>::=(<class_name>|<class_id>)["?" <specification>]
<class_name>::=<string>
<class_id>::=<number>
<specification>::=(<attribute><op><value>){<op>(<attrib
ute><op><value>)}
<attribute>::=<string>
<value>::=<string>|<name>
<op>::=">"|"="|"<"|">="|"/="|"<="|AND|OR|NOT
```

Using this scheme, a user can query resources he wants by giving a "name" like:

```
ComputingResource ? (CPU > "933MHz") AND (memory =
"256M") AND ((OperatingSystem ? (type = "Linux")
```

GOS [8] provides another example of user layer resource representation. An instance of GridCompute- Resource maybe looks like:

```
dn: hostname=burns.csun.edu, o=Grid
objectclass: GridComputeResource
hostname: burns.csun.edu
......
diskDrives: /dev/dsk/dks0d4s
```

Although friendly to users, resource representation in these formations can hardly be used by resource routers because of the complexity. Moreover, it is not necessary for routers to know the meaning of each attribute-value pair. A mechanism is required to map the user-layer representation to the router-level representation. This process is similar to the domain name resolution in Internet But it is obviously too costly to deploy a specialized global "Name Resolver" for each type of representation in the Grid. Fortunately, most of representations in this layer are defined by some globally-obeyed specifications in which the meaning of each attribute is stated. This means if we understand these specifications, we can locally resolve the representations they have defined into any formation we want to. In VEGA Grid, virtual devices that want to support a particular resource representation should implement the resolver which maps the user-layer representation to the router-layer representation. These resolvers are parts of the *virtual device drivers* deployed in Grid processors. This mechanism does not exclude the adopting of the global name resolution. Resolvers like DNS can be integrated into VEGA as a special kind of services whose locations are static indicated by their virtual drivers and need not to be located via resource routers.

4.2 Resource Representation in Router Layer

In the middle layer of the architecture, resource routers work corporately to locate resources globally, receive registry from resource providers and collect resource information from neighbors. To support attribute-value based qucries from users, routers construct their common name space based on resource types. This is not difficult to understand because attribute-value pairs are defined according to different resource types. This layer is the kernel of our model, for the following reasons: 1) to work globally, router-layer names are uniform in the whole Grid, that is, in this layer, the representation formation of a type of resources is immutable and can be recognized by all routers; 2) the quantity of router-layer representation formations finally determines the actual number of resource types that can be supported by the Grid, which influences the scalability of the whole Grid.

Router-layer Resource Representation Syntax. As mentioned in Section 4.2, user-layer representations will be mapped into router-layer ones when users submit their requests to routers. Different from users, routers do not concern the meanings of attributes. In the user-layer representations, matching patterns are added to attribute-value pairs, e.g., CPU>933MHz. Routers must use these matching patterns when they decide whether or not a particular resource can satisfy a user's request. Therefore, the router-layer representations converted by resolvers in virtual device drivers at Grid processor ends should contain matching operations. The syntax in BNF is:

```
<name> ::= (<class_id>) ["/" <specification> ]
<class_id> ::= <number>
<specification> ::= [<op> <value>] {"/" <op> <value>  }
<value> ::= <string> | <name>
<op> ::= ">" |"=" | "<" | ">=" | "/=" | "<="|AND|OR|NOT
```

While representations collected from neighbors and resource providers only indicate the status of resources, which are independent from matching operations. These representations are recorded in routers' routing tables, and refreshed periodically. The syntax is:

```
<name> ::= (<class_id>) ["/" <specification> ]
(<direction>)
<class_id> ::= <number>
<specification> ::= [<value>] {"/"<value>  }
<direction> ::= <URL>
<value> ::= <string> | <name>
```

class_ids of each resource type are assigned by a global administration, and form a hierarchical global name space. So, it supports the naming service well both in management and in scalability. A naming service in wide-area distributed system should have the mechanism to generate a globally unique name as locally as possible. In hierarchical architecture, the namespace can be divided into different-leveled domains, each of which administers a portion of the whole namespace with local autonomy (DNS is an example of such systems). The naming service in each domain can be assigned a locally unique name. This name with the path from the root down to the domain node constitutes a globally unique name. In a hierarchical namespace, the increase in resource types and quantity can be distributed to different domains. The probability of a single node taking all the penalty of system growth is quite low, so a bottleneck can be avoided. This brings the namespace good scalability.

Key Features of the Routing-Transfer Model. As the backbone of VEGA Grid, the resource router plays a vital role and deserves explanation in details. Here, three key features of routers will be discussed: the structure of routing tables, the updating policy and the request routing algorithm. These features bring resource routers good performance at a relatively low cost [3].

The routing table of a resource router is hashed with resources' *class_ids* used as the variable of the hashing function. From the hash table entries, aggregated resource information can be accessed according to the formation of the resource type. An important character of a routing table is that it does not have the information about all of resource instances in the Grid. This is because a router does not record the physical location of a resource instance, but records to which direction, can it find a resource instance having those attribute values it records. That is, no matter how many resource instances of the same type and equal attribute values, a router only records one of them, and usually, it is the one "nearest" to the router. Therefore, theoretically, the size of a route table is independent of the scale of resource quantity, but related to the quantity of types and instance states. For example, if there are k types of resources, each type of resource has l attributes and an attribute has m possible values, the spacial complex of the routing table will not exceed $k*(l^m)$ items.

Another factor influencing on routing tables is the updating policy. In our implementation, resource instance entries not updated after a specific interval will be regarded unavailable and removed from the routing table so that the size of the routing table will not grow too large as time goes by. To keep the performance as good as possible, when a nearer resource instance is known, the corresponding item in the table will be updated to make the direction value point to the new router.

VEGA Grid resource routers adopt a routing algorithm called SD-RT (Shortest Distance Routing-Transferring) [3]. SD-RT algorithm is designed to locate a resource as soon as possible. When a resource router receives a resource request, SD-RT algorithm will choose a router by which the resource is accessible through the shortest path from the local resource router. Especially, when the distance is 1, the resource provider is the neighbor of the router. The whole routing process is the process of mapping from the location-independent router-layer representations to location-aware provider-layer ones. Using SD-RT algorithm, a resource request can arrive at the resource provider as soon as possible. If there is more than one provider supplying the same resource, SD-RT algorithm can guarantee the request arriving at the provider nearest to resource requestors.

4.3 Provider-Layer Names

The router, which a resource is registered to, needs the location information for client-to-resource binding. So, when a provider registers its resource to a neighbor router, the location information should also be delivered. That is, the resource providers and the routers, to which providers register their resources, need a location-aware mechanism to represent location information. When a resource is replicated to different addresses, it is also required to distinguish these replicas. Location information can be a key attribute in the resource description [8]. In our model, we put it in the resource identifier to form the bottom layer name. The reason is, for some special resources that can be accessed only via IDs, e.g., a particular file that has only one copy in the Grid, we can leave their attributes null to omit router processing. Representations in this layer are formed by binding the location information, such as URLs, to the resources' router-layer. The syntax in BNF is:

```
<name> ::= (<class_id>) ["/" <specification> ]
(<location>)
<class_id> ::= <number>
<specification> ::= [<value>] {"/" <value> }
<location> ::= <URL>
<value> ::= <string> | <name>
```

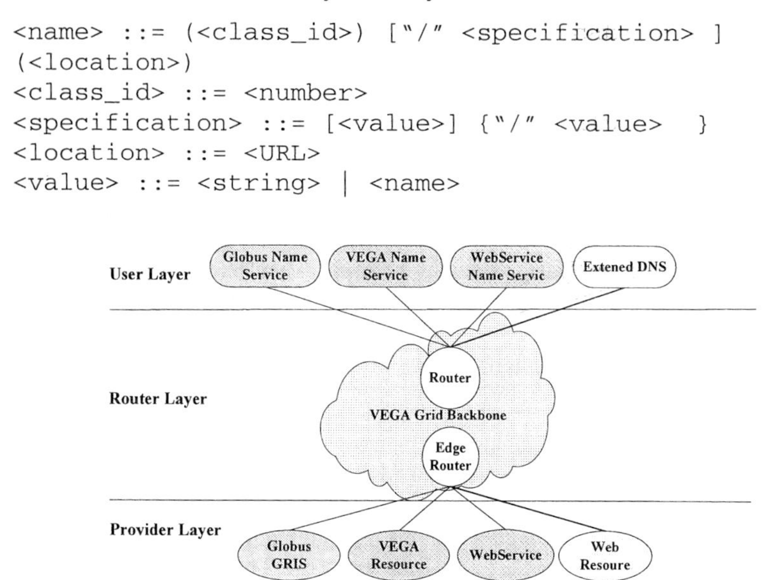

Fig. 3. Various resources integrated in VEGA Grid.

5 Evaluation

Resource representation is a very important topic in Grid environments. It has vital effect on both the systems and users. In this section, our approach is evaluated from three respects: usability mainly concerned by resource consumers, scalability mainly concerned by resource provider, and performance concerned by the whole system.

Usability: The explicit separation of the user layer from the router layer facilities the deploying of the variable user customized representations. By implementing specialized name resolvers plugged into virtual device drivers, user-layer names can carry enough information to describe users' requirements. This improves the usability, and also makes other Grid systems merge into VEGA easily. When another Grid system having its own naming scheme wants to join VEGA Grid, it only needs to deploy a new name resovler that converts its local names to VEGA's router-layer representations. Because the original name space is reserved, applications based on the former system can be easily replanted in VEGA, which accommodates VEGA's versatile service principle. Fig. 3 gives a view of other Grid systems integrated with VEGA. UDDI in Web Service and GIIS (Grid Information Index Service) in Globus can be replaced with resource routers. By installing VEGA virtual device drivers, Globus and Web Service applications can run on VEGA Grid infrastructure.

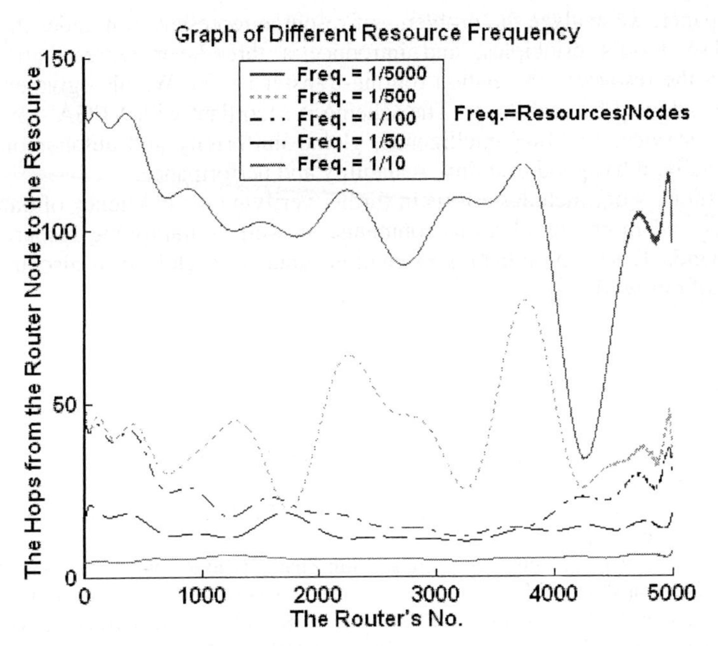

Fig. 4. The x Axis gives router IDs from 1 to 5000, and y Axis shows the hops needed to access a particular type of resources from each router under different resource frequencies .

Scalability: Three factors enhance scalability of our scheme. 1) Location independence: The resource representations in the upper two layers are not bound to resources' physical locations, and it is the routers' responsibility to locate the proper resources. Therefore, resources can transfer and be replicated in wide area without imposing additional restrictions to users and providers. 2) Resource router features: As analyzed in 4.3, the size of routing table is independent of the quantity of resource instances in the Grid. This means the scalability in resource instance quantity will not be limited by resource routers' capacity. Additionally, the resource type name space is constructed hierarchically, which also benefits resource type's scalability. 3) The ability of top-layer namespace partition relieves the burden of name resolution and allows VEGA to expand by merging other Grid systems.

Performance: We do beneficial work in both user layer and router layer to promote the performance of resource discovery. In the user layer, remote name resolution, which is usually a bottleneck of wide-area distributed systems, is avoided. Local resolvers based on resource representation specifications are implemented in virtual device drivers. In the router layer, resource information routing-transferring model provides good performance to locate resources in environments that have large scale of resources and nodes. Fig. 4 shows the experiment results [3] in an environment consisting of 5000 nodes, where resource frequency varies from 1/5000.to 1/10

6 Conclusion and Future Work

In this paper, we analyze the problems of resource representation under the guidance of VEGA Grid's principles, and introduce a three-layer representation scheme adapting the resource information routing-transfer model. We also give examples of syntaxes adopted by each layer. The scheme is accordant with VEGA's principles of versatile service, enabling intelligence, global uniformity and autonomous control. Additionally, it has good usability, scalability and performance.

Our future work includes efforts in further verifying the efficiency of our approach in larger and more complex environments, as well as improving resource routers' performance. Research on the representation syntax of each layer is also an important content of our work.

References

1. I. Foster and C. Kesselman (Eds): The Grid: Blueprint for a New Computing Infrastructure, Morgan Kaufmann Publishers, 1998.
2. Wei Li, Zhiwei Xu, Bingchen Li and Yili Gong: The Vega Personal Grid: A Lightweight Grid Architecture, PDCS 2002.
3. Wei Li, Zhiwei Xu, Fangpeng Dong and Jun Zhang: Grid Resource Discovery Based on a Routing-Transferring Model, 3rd International Workshop on Grid Computing (Grid 2002).
4. K. Czajkowski, S. Fitzgerald, I. Foster, and C. Kesselman: Grid Information Services for Distributed Resource Sharing, Proc. of HPDC-10, IEEE Press, August 2001.
5. Joshua Apgar, Andrew Grimshaw, Steven Harris, Marty Humphrey and Anh Nguyen-Tuong: Secure Grid Naming Protocol (SGNP), Avaki Corporation, University of Virginia, GGF Working Group Chairs and Steering Group, January, 2002.
6. Ann Chervenak (ISI/USC): Naming and Information Management in Grid Systems, http://www.sdsc.edu/GridForum/RemoteData/Papers/chervenak.pdf.
7. R. Raman, M. Livny and M. Solomon, Matchmaking: Distributed Resource Management for High Throughput Computing, Proc. of IEEE Intl. Symp. High Performance Distributed Computing, July 1998.
8. Steven M. Fitzgerald, Gregor von Laszewski and Martin Swany: GOSv2: A Data Definition Language for Grid Information Services, Feb. 2001.
9. Maarten Van Steen, Philip Homburg and Andrew S. Tanenbaum: Globe: A Wide-Area Distributed System, IEEE Concurrency (1999), pp. 70–78.
10. G. Ballintijn, M. van Steen and A.S. Tanenbaum: Scalable Naming in Global Middleware, Proc. 13th Int'l Conf. on Parallel and Distributed Computing Systems.
11. M. van Steen, F. Hauck, P. Homburg and A.S. Tanenbaum: Locating Objects in Wide-Area Systems, IEEE Commun. Mag., vol.36, nr.1, pp.104–108, Jan. 1998.
12. UDDI Technical White Paper, http://www.uddi.org.
13. Adriana Iamnitchi and Ian Foster: On Fully Decentralized Resource Discovery in Grid Environments, GRID 2001, 2nd International Workshop on Grid Computing.
14. Amin Vahdat, Michael Dahlin, Thomas Anderson and Amit Aggarwal: Active Names: Flexible Location and Transport of Wide-Area Resources, Proceedings of the 2nd USENIX Symposium on Internet Technologies & Systems, October 11–14, 1999.
15. Yili Gong, Fangpeng Dong, Wei Li and Zhiwei Xu: A Dynamic Resource Discovery Framework in Distributed Environments, Proc. of International Workshop on Grid and Cooperative Computing, (GCC 2002).

TCM-Grid: Weaving a Medical Grid for Traditional Chinese Medicine

Huajun Chen, Zhaohui Wu, Chang Huang, and Jiefeng Xu

Grid Computing Lab,College of Computer Science, Zhejiang University,P.R.China,
{huajunsir,wzh,changhuang,xujf}@zju.edu.cn
http://grid.zju.edu.cn

Abstract. We present a TCM-Grid for Traditional Chinese Medicine (TCM). The purpose of the TCM-Grid is to aid the development of distributed systems that help health professionals, researchers, enterprizes and personal users to retrieve, integrate and share TCM information and knowledge from geographically decentralized TCM database resources and knowledge base resources in China. Our approach involves developing a Database Grid for discovering and accessing TCM database resources coordinately and a Knowledge Base Grid supporting TCM knowledge sharing globally. With our application experience, we argue that nowadays' Grid architecture is not enough: we need Database Grid to support finely granular data sharing and integration, and we also need a Knowledge Base Grid to support knowledge-intensive task. We also recommend a Grid Ontology effort to enable Grid intelligence.

1 Introduction

Traditional Chinese Medicine (TCM), with a history that spans thousands of years, has provided us a wide variety of resources for biomedical and health science. These resources include TCM literatures, medicinal materials, Chinese herbs, TCM compounding rule, Chinese medical formula and so on, both ancient and present. In China, there have been thousands of TCM hospitals, enterprizes and research institutes who have developed various TCM information products to serve people's health care and research demands such as TCM self-health consultation, TCM literature analysis and retrieval, new medicine design,etc. For example,by cooperating with China Academy of Traditional Chinese Medicine, we have set up Traditional Chinese Medical Database System, a series of databases of TCM [10] .

In both e-business and e-science for TCM, we have an urgent need to integrate TCM systems or services across distributed, heterogeneous, dynamic "virtual organizations" formed from the disparate TCM resources within a single organization (hospitals, enterprizes and research institutes) and/or from external resource sharing and TCM information resource provider relationships. Fortunately, we have found that the OGSA (Open Grid Service Architecture)[1] has promised to addresses above challenges. Therefore, with the cooperation with China Academy of Traditional Chinese Medicine, we began to develop

P.M.A. Sloot et al. (Eds.): ICCS 2003, LNCS 2659, pp. 1143–1152, 2003.

a TCM-Grid to achieve our goals. However, with our first inspection and efforts, we have found that OGSA is not enough to satisfy our real needs and expectations.

At first, OGSA dose not provide proper way of database resource integration, registration and discovery. Second, medical informatics is a knowledge-intensive domain, we have requirements of represent TCM information at a knowledge level and construct high-level expert system such as TCM new medicine design expert system and TCM tutor expert system,etc. That urged us to provide basic infrastructure supporting TCM knowledge base creation, maintaining, utilizing and knowledge sharing across China. At last, sometimes we want autonomic discovery , binding and then integration of resources, in other word, we need some intelligence to help us discover and locate proper resources (both database resources and knowledge base resources) in such a information overwhelming environment,but OGSA care nothing on that at all(dynamic dose not mean intelligence).

Therefore, we introduce an architecture in term of Database Grid to support smaller granularity data sharing (database resource sharing). And with our research on semantic web [11] area, we have also present an architecture in term of Knowledge Base Grid to support knowledge sharing in semantic web. We also argue that to bring the Grid into its full potential, we should develop Grid Ontology corresponding to the Web Ontology effort in Semantic Web communities to address the complexity , dynamic, diversity and heterogeneity of a Global Grid and support autonomic grid service discovering,locating and executing .

2 The Backbone of TCM-Grid: DB-Grid and KB-Grid

2.1 Database Grid

Comparing to the DataGrid [2] with the aim of the sharing of huge amounts of distributed data files that we call as coarsely granular data resources over the network infrastructure, the Database Grid (DB-Grid) initiative is to develop and test the technological infrastructure that will enable the sharing of databases that contain more finely granular data resources.

A large number of existing database are important information resources, which contain rich domain specific information. Take the TCM area as an example, through years of work on collection, translation and compilation of ancient books of TCM, hundreds of TCM databases have been formed which collects numerous clinic cases, medicine prescriptions, diseases records and other precious experience in the TCM field. These databases constitute a basic information platform that needs to be accessed in many TCM research activities. Database Grid should get each of these database resources accessible on the Grid. Furthermore, Database Grid should provide users with an efficient mechanism for coordinated use of these content-related databases, for example providing the integration access service by constructing a virtual database so that users can access distributed data sources with single access point.

In a service-oriented view, we should respectively define some generic database services. These generic services define the functionalities that a qualified database resource must provide. In [3] we have proposed the following basic services:

- Database Grid Information Service: support the initial discovery and ongoing monitoring of the existence and characteristics of resources, services, computation and other entities
- Database statement service: support the typical operations on database contents, including retrieval, insertion, deletion, and modification of the data schema;
- Database Management Service : enables DBA to perform a remote control;
- Database Accounting Service : enables database users to be charged of resource usage;
- Database Directory : Database directory is an aggregate registry that collects, manages and indexes individual database information models;
- Virtual Database : A virtual database is a middleware that integrates a number of databases but does not actually store any data;
- Database Market : data market is an open place for databases to interchange database units.

2.2 Knowledge Base Grid

TCM and medical informatics is a knowledge-intensive domain. In a service-oriented perspective, we want to build large-scale knowledge-based service such as self-health consultation services.Underlying those services are there a variety of TCM knowledge base resources such as TCM Ontology, TCM compounding rule bases, TCM therapeutic principle knowledge bases, etc., which maybe be maintained by different organizations and/or individuals all around China.

Nevertheless, how could we integrate so many decentralized knowledge base resources to support constructing intelligent services? We argue that OGSA is not enough here for its aim to integrate information systems but not knowledge-based systems. In the following, we extend this argument in two respects and then introduce the core components to define more precisely how our KB-Grid functions and how it could be implemented and applied.

2.2.1 Standard Knowledge Representation for Web

To overcome the heterogeneity of web,the first thing we need is a standard method for knowledge representation and information description in web. With this effort, we can enable the web to be an intelligent information space, which is machine-understandable. That's the aim of the W3C's activity on Semantic Web [5]. They have recognized the importance of the standardization of knowledge representation and recommended the RDF [5] as the basic model for web knowledge representation .Above the RDF, DAPPA has presented the DAML+OIL [12] for Web Ontology description.

Based on above efforts, we have proposed a hybrid knowledge markup language (KML) as our standard method for TCM knowledge representation. KML consists of a D-Box (Description Logic Box) for TCM-Ontology definition, R-Box (rules box) for TCM compounding rule definition, and a C-Box (case knowledge box supporting case-based reasoning) for TCM clinical cases description. The KML schema draft has been available from our website (http://grid.zju.edu.cn).

2.2.2 Standard Knowledge Protocols

In an environment with intelligence enabled , that means there are great of knowledge services distributed on the Grid , those knowledge services must coordinate their activities with each other to further their own interests or satisfy group goals. The fundamental concerns lying here are distributed control and semantic interoperability, that means we need knowledge level protocols. The design of knowledge protocol is more difficult and complex than simple information protocols such as LDAP or SOAP. As all known, computing grid has chosen the LDAP protocol for their GRIP/GIIS [6], and maybe they will choose SOAP as the substitute. We argue that both LDAP and SOAP is not enough for knowledge level communication.

At first, at the knowledge level, the message format for communication should contain the knowledge interchanged. The knowledge interchange format is very different with the data model adopted in LDAP. It should take some form like the first order logic syntax at the semantic level. At second, the control process is more complex here for that we should model the BDI (Belief, Desire, Intention) of the intelligent services by epistemic logic. Those are topics of multi-agents system in distributed artificial intelligence. We have made some suggestion on designing the knowledge protocol for KB-Grid in [4]. Such knowledge protocols will play core role not only for knowledge system cooperation but also for autonomic application integration in large-scale environment.

2.2.3 Core Components of KB-Grid

A typical KB-Grid consist of the following core components:

Shared Ontologies: Ontology lay down the ground concepts and rules for their domain. One could view the ontology as some controlled vocabulary which could facilitate the semantic interoperability between heterogeneous knowledge base resources .[14]

Knowledge Server: Our Knowledge Server is the web container for knowledge which is represented by RDF/DAML+OIL/OWL/RuleML. It is the runtime environment for webKBs. The key characteristic of knowledge server is that it could process semantic query, then do some inference within its KBs and then return the answers in a semantic form.

KB-MDS: KB-MDS is provided for model the meta-information about the organization and discovery of the KB resources.It act as the meta-directory for webKBs

Ontology Browser: that's the user interface for KB-Grid. It supports semantic browsing against the knowledge server. We have finished the prototype of a TCM-Ontology browser. It supports TCM-Ontology (represented by RDF) and semantic query against the TCM-Ontology server.

For more detail about KB-Grid, please refer [4].

2.3 Grid Ontology

The Web communities such as W3C has initialed a Web Ontology effort [13]. Ontologies figure prominently in the emerging Semantic Web as a way of representing the semantics of documents and enabling the semantics to be used by web applications and intelligent agents. Ontologies can prove very useful for a community as a way of structuring and defining the meaning of the metadata terms that are currently being collected and standardized.Ontologies are critical for applications that want to search across or merge information from diverse communities. Although XML DTDs and XML Schemas are sufficient for exchanging data between parties who have agreed to definitions beforehand, their lack of semantics prevent machines from reliably performing this task given new XML vocabularies. [7] tells us more about the importance and the usage of Ontology in Web.

We argue that OGSA will face the same problem when it scales up globally. We really have used an ontology for the Computing Grid. That ontology includes the concepts and vocabularies for computing grid domain, for example CPU, Memory, Storage device and their combination to construct high-level virtual machines. Yes, that is the MDS data model existing in the LDAP server. For Computing Grid, the data model for LDAP is enough, because the resource model of computing grid is simple. However, when we extend the Grid to global application or service integration , the data model of LDAP or SOAP will fail because the global web is really much more complex than we have imagined : so many heterogenous web resources (web databases,web pages,web services) cover almost all domain of human knowledge and distribute all around the world. To enable Grid intelligence, we need build Grid Ontology to enable autonomous service discovery and integration.

Generally, a domain of virtual community such as TCM will build their own shared Ontology. Then the organization and/or individual who want to join such community will obey that Ontology to build their information or knowledge systems. Section3 will introduce the TCM Ontology which we have built for TCM community.

3 Implementation of TCM-Grid

To verify the ideas presented above, this section will discuss some key issue with respect to the implementation of TCM-Grid.

3.1 TCM Ontology

The Ontology we want to build include related terms/concepts, their definitions and/or meanings, their relationships with each other, and some basic rules of the domain. We use Protg 2000, which is developed by Medical Informatics of Stanford University, to build TCM ontology. Generally, Protg 2000 uses the Semantic Web language RDFS as basic data storage format, but it can be easily adapted and extended to the other Semantic Web languages such as DAML+OIL.

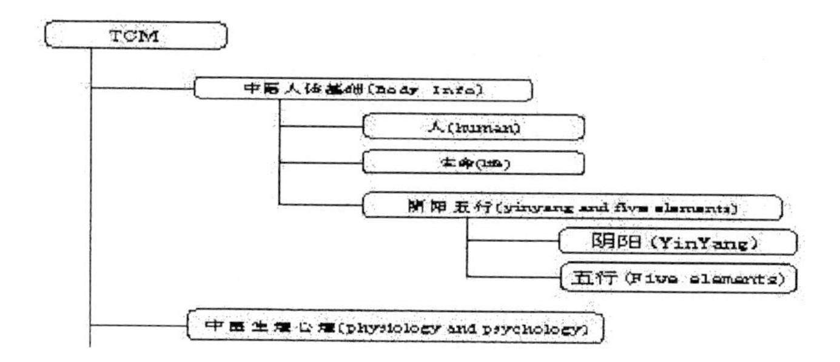

Fig. 1. The snapshot of TCM Ontology

Fig. 1 depicts the snapshot of TCM ontology class hierarchy. It shows that TCM is an abstract base class in TCM Ontology, and there are about 13 basic slots (attributes) of it. Then there exist about 20 subclasses like Human Info, psychology and physiology, therapeutics, and Chinese Medical Formula in TCM Ontology. TCM ontology has contained much more special concepts and classes such as yin-yang symptom complex, five-elements,body point, Chinese Medical Formula, Meridians, etc. Now we have finished building the whole class definition of TCM Ontology and edited about 100,000 records of TCM Ontology instances.

UMLS (Unified Medical Language System) is a web-accessible medical thesaurus developed by NLM (National Library of Medicine) of America as a basic knowledge source in medicine. It has 135 kinds of semantic classes and about 53 semantic relationships defined. Because TCM is one of the members of world medicine, we adopt this formal and regular definition in medicine as the basis of TCM ontology design.

3.2 TCM-Grid Infrastructure

Just as the figure2 has illustrated, we divide the TCM-Grid infrastructure into three levels. The backbone consists of a variety of data sources and knowledge sources. Above those sources, we could construct high-level services including some high-level database service and knowledge-based services. Above those services, we could form the virtual hospitals, virtual enterprizes and virtual research institutes.

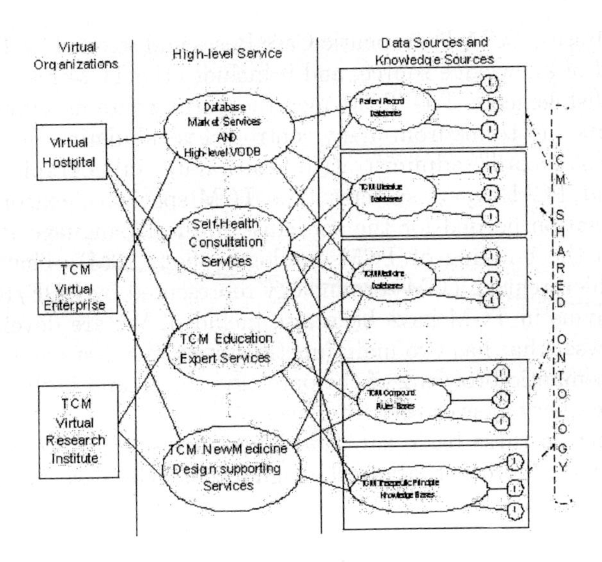

Fig. 2. TCM-Grid Infrastructure

Database Grid is used to manage database resources in a TCM virtual oraginzation. In this case, there are a number of basic databases, including Patient Record Databases, TCM Medicine Databases, TCM OTC Databases, TCM Literature Databases, Traditional Chinese Drug Database, Traditional Tibetan Drug Database and so on. Those databases serve as public information sources, which are geographically distributed and owned by different institutions and/or individuals. For example, the patient record databases maybe consist of hundreds of databases belonging to different hospitals distributed all around China, and we could call such a database group a sub-db-grid of TCM-Grid. Above that, a virtual database is constructed to integrate those basic TCM databases and provide federated access service. One can access one another's local data in two modes. One way is through the database access service, which deals with remote requests and return records. The other is downloading the application data units from a database market and loading them into its own database for local access. We have finished the design of the database resource model, and are developing a component which could run as daemon at the database client, when some modifications (adding a table for example) have happened in the member database, it will update the information about himself in the central database directory server.We are also developing the virtual database component which act as a uniform DB-Grid query interface. User sends their query to the virtual database service, and the virtual database service will dispatch the query to proper member databases by the meta-information stored by the central directory server.

KB-Grid is used to facilitate knowledge sharing among TCM VOs. In this case, there are also a number of knowledge bases including TCM symptom KBs, TCM Therapeutic Principle KBs, TCM medical formulas, TCM Compound Rule Bases, TCM Therapeutic Case Bases and so on. The TCM Ontology is also a kind of knowledge source, and it includes the TCM metathesaurus and TCM specialist lexicon. The TCM metathesaurus contains information about TCM concepts and terms from many controlled vocabularies and classifications used in patient records, administrative health data, bibliographic and full-text databases and TCM expert systems. The TCM specialist lexicon provides the lexical information needed for the specialist Natural Language Processing. We have finished the building of TCM ontology (about 10M). That ontology has contained a hierarchical TCM terminology represented by RDF/RDFS. Almost all the synonyms in TCM have been also included. We are developing a TCM ontology browser that has two main functionalities: the first one is the graphical display and semantic browsing of the ontology; the second is that it acts as the ontology-base query answer user interface.

Relying on these basic data and knowledge sources, we could construct high-level service. Moreover, above the high-level services, we could form virtual organizations.

3.3 A Working Scenario: TCM New Medicine Design

To make our discussion more clearly, we give a simple working scenario for TCM-Grid.

Relying on the basic data and knowledge sources, a clinical development research for a new medicine is carried out, which can be deployed into the following phrases: TCM compound rule analysis, new medicine and medical formula design, preclinical testing, clinical testing and new medicine production. The above process maybe involves a research institute, a few hospitals for clinical test and some enterprizes to produce and sell the new medicine. At first, the research institute initiates a new medicine design process. With the assistance of the new medicine design supporting service, they make analysis of the TCM compound rule base and TCM therapeutic principle knowledge base, and design a new TCM medical formula. Some hospitals are involved in the preclinical test and clinical test phrase. With their clinical decision supporting service, they make the clinical test and record the results. When the new medicine is appropriate, some TCM enterprise will be involved and put the new medicine into production.

We have seen a working scenario which have a urgent need for dynamic integration of geographically distributed database resources, information system and knowledge-based system across the Internet environment.

4 Related Work

For database resources sharing across Grid, DAI-WG (Database Access and Integration Working Group) in GGF [8] is currently identifying consistent and effective ways of making existing, autonomously managed databases available within a Grid setting. Most requirements identified by them are applied to our proposal. Their approach to database access and integration is constructing OGSA compatible service specification. However, we think protocol is a fundamental issue in defining any networked computing model no matter what form of implementation is finally taken. So our efforts will focus on the formal representation of the related protocols.

For the large-scale web-based knowledge-base system , The Knowledge Grid effort in Chinese Academy of Sciences [9] has also proposed a worldwide resource (knowledge, information, and services) sharing and management platform. However, we argue that their platform has no awareness of developing knowledge protocols and high-level knowledge services. Another work worthy to mention is the DARPA's DAML-S [12] effort which supplies Web service providers with a core set of markup language constructs for describing the properties and capabilities of their Web services in unambiguous, computer-intepretable form. DAML-S markup of Web services will facilitate the automation of Web service tasks including automated Web service discovery, execution, interoperation, composition and execution.

5 Conclusion and Future Work

This paper describes our experience with building a medical grid for Traditional Chinese Medicine. Our ultimate goal is to aid the development of distributed systems that help health professionals, researchers, enterprizes and personal users to retrieve, integrate and share TCM information and knowledge from geographically decentralized TCM database resources and knowledge base resource all around China.

Based on our experience, we have found that nowadays' Grid architecture such as OGSA does not satisfy our real need and expectations. With our effort on Database Grid and Knowledge Base Grid, three respects have enhanced the OGSA. First, the DB-Grid enhances OGSA by supporting the database resources sharing by its database services and protocols, because most of the e-businesses involve the database. Second, KB-Grid enhances OGSA by knowledge sharing and knowledge-based services . Third, Grid Ontology enhances OGSA for its usage as the central control vocabulary which will enables the automatic Grid service dicovering, locating and executing.

In the future, we plan to finish the design and implementation of core database services and protocols and then deploy those components to the Traditional Chinese Medical Database System [10] . In the knowledge-based system aspect, we plan to design and implement a knowledge server as the run time environment for RDF-based webKBs, and high-level knowledge query language will be designed to support ontology query . All the above efforts will be integrated into the TCM-Grid effort.

Acknowledgement. We gratefully acknowledge helpful discussion with other members in the Grid Computing Lab of Zhejiang University. This work is supported in part by the Grid-Based TCM Dynamic Information Resource Management and Knowledge Service subprogram of the Foundational Technology and Research Program, China Department of Science and Technology, and the China

863 Research Program on Intelligent Workflow Technologies supporting Creditable E-Commerce under Contract 2001AA414320, and the China 863 Research Program on Core Workflow Technologies supporting Components-library-based Coordinated Software Development under Contract 2001AA113142.

References

1. The Physiology of the Grid: An Open Grid Services Architecture for Distributed Systems Integration. I. Foster, C. Kesselman, J. Nick, S. Tuecke, Open Grid Service Infrastructure WG, Global Grid Forum, June 22, 2002.
2. The Data Grid: Towards an Architecture for the Distributed Management and Analysis of Large Scientific Datasets. A. Chervenak, I. Foster, C. Kesselman, C. Salisbury, S. Tuecke. Journal of Network and Computer Applications, 23:187–200, 2001.
3. Zhaohui Wu, Changhuang, Guozhou Zhen, Database Grid: An Internet Oriented Database Resource Management Architecture, in Proceedings of GCC 2002, the International Workshop on Grid and Cooperative Computing of Chinese Academic of Science..
4. Zhaohui Wu, Huajun Chen, Jiefeng Xu, The Anatomy of Knowledge Base Grid, in Proceedings of GCC 2002, the International Workshop on Grid and Cooperative Computing of Chinese Academic of Science.
5. W3C Semantic Web Activity:http://www.w3.org/2001/sw/Activity [6]
6. Karl Czajkowski, Steven Fitzgerald, Ian Foster, Carl Kesselman : Grid Information Services for Distributed Resource Sharing: Proc. 10th IEEE International Symposium on High-Performance Distributed Computing (HPDC-10),IEEE Press,2001;
7. Deborah L. McGuinness. "Ontologies Come of Age". In Dieter Fensel, J im Hendler, Henry Lieberman, and Wolfgang Wahlster, editors. Spinning the Semantic Web: Bringing the World Wide Web to Its Full Potential. MIT Press, 2002.
8. Database Access and Integration Services WG of GGF,website, http://www.ggf.org/6_DATA/dais.htm;
9. H.Zhuge, A Knowledge Grid Model and Platform for Global Knowledge Sharing, Expert Systems with Applications, vol.22, no.4, 2002, pp.313-320, (Elsevier Science), 2002.
10. Traditional Chinese Medical Database System ,website, http://www.cintcm.com/e_cintcm/index.htm;
11. Tim Berners-Lee, James Hendler, Ora Lassila, the Semantic Web, Scientific American, May 2001
12. The DARPA Agent Markup Language Program: http://www.daml.org/
13. W3C Web-Ontology Activity: http://www.w3c.org/2001/sw/WebOnt/
14. Stanford Knowledge Interchange Format: http://www-ksl.stanford.edu/knowledge-sharing/kif/

Author Index

Lecture Notes in Computer Science

For information about Vols. 1–2584

please contact your bookseller or Springer-Verlag

Vol. 2585: F. Giunchiglia, J. Odell, G. Weiß (Eds.), Agent-Oriented Software Engineering III. Proceedings, 2002. X, 229 pages. 2003.

Vol. 2586: M. Klusch, S. Bergamaschi, P. Edwards, P. Petta (Eds.), Intelligent Information Agents. VI, 275 pages. 2003. (Subseries LNAI).

Vol. 2587: P.J. Lee, C.H. Lim (Eds.), Information Security and Cryptology – ICISC 2002. Proceedings, 2002. XI, 536 pages. 2003.

Vol. 2588: A. Gelbukh (Ed.), Computational Linguistics and Intelligent Text Processing. Proceedings, 2003. XV, 648 pages. 2003.

Vol. 2589: E. Börger, A. Gargantini, E. Riccobene (Eds.), Abstract State Machines 2003. Proceedings, 2003. XI, 427 pages. 2003.

Vol. 2590: S. Bressan, A.B. Chaudhri, M.L. Lee, J.X. Yu, Z. Lacroix (Eds.), Efficiency and Effectiveness of XML Tools and Techniques and Data Integration over the Web. Proceedings, 2002. X, 259 pages. 2003.

Vol. 2591: M. Aksit, M. Mezini, R. Unland (Eds.), Objects, Components, Architectures, Services, and Applications for a Networked World. Proceedings, 2002. XI, 431 pages. 2003.

Vol. 2592: R. Kowalczyk, J.P. Müller, H. Tianfield, R. Unland (Eds.), Agent Technologies, Infrastructures, Tools, and Applications for E-Services. Proceedings, 2002. XVII, 371 pages. 2003. (Subseries LNAI).

Vol. 2593: A.B. Chaudhri, M. Jeckle, E. Rahm, R. Unland (Eds.), Web, Web-Services, and Database Systems. Proceedings, 2002. XI, 311 pages. 2003.

Vol. 2594: A. Asperti, B. Buchberger, J.H. Davenport (Eds.), Mathematical Knowledge Management. Proceedings, 2003. X, 225 pages. 2003.

Vol. 2595: K. Nyberg, H. Heys (Eds.), Selected Areas in Cryptography. Proceedings, 2002. XI, 405 pages. 2003.

Vol. 2596: A. Coen-Porisini, A. van der Hoek (Eds.), Software Engineering and Middleware. Proceedings, 2002. XII, 239 pages. 2003.

Vol. 2597: G. Păun, G. Rozenberg, A. Salomaa, C. Zandron (Eds.), Membrane Computing. Proceedings, 2002. VIII, 423 pages. 2003.

Vol. 2598: R. Klein, H.-W. Six, L. Wegner (Eds.), Computer Science in Perspective. X, 357 pages. 2003.

Vol. 2599: E. Sherratt (Ed.), Telecommunications and beyond: The Broader Applicability of SDL and MSC. Proceedings, 2002. X, 253 pages. 2003.

Vol. 2600: S. Mendelson, A.J. Smola, Advanced Lectures on Machine Learning. Proceedings, 2002. IX, 259 pages. 2003. (Subseries LNAI).

Vol. 2601: M. Ajmone Marsan, G. Corazza, M. Listanti, A. Roveri (Eds.) Quality of Service in Multiservice IP Networks. Proceedings, 2003. XV, 759 pages. 2003.

Vol. 2602: C. Priami (Ed.), Computational Methods in Systems Biology. Proceedings, 2003. IX, 214 pages. 2003.

Vol. 2603: A. Garcia, C. Lucena, F. Zambonelli, A. Omicini, J. Castro (Eds.), Software Engineering for Large-Scale Multi-Agent Systems. XIV, 285 pages. 2003.

Vol. 2604: N. Guelfi, E. Astesiano, G. Reggio (Eds.), Scientific Engineering for Distributed Java Applications. Proceedings, 2002. X, 205 pages. 2003.

Vol. 2606: A.M. Tyrrell, P.C. Haddow, J. Torresen (Eds.), Evolvable Systems: From Biology to Hardware. Proceedings, 2003. XIV, 468 pages. 2003.

Vol. 2607: H. Alt, M. Habib (Eds.), STACS 2003. Proceedings, 2003. XVII, 700 pages. 2003.

Vol. 2609: M. Okada, B. Pierce, A. Scedrov, H. Tokuda, A. Yonezawa (Eds.), Software Security – Theories and Systems. Proceedings, 2002. XI, 471 pages. 2003.

Vol. 2610: C. Ryan, T. Soule, M. Keijzer, E. Tsang, R. Poli, E. Costa (Eds.), Genetic Programming. Proceedings, 2003. XII, 486 pages. 2003.

Vol. 2611: S. Cagnoni, J.J. Romero Cardalda, D.W. Corne, J. Gottlieb, A. Guillot, E. Hart, C.G. Johnson, E. Marchiori, J.-A. Meyer, M. Middendorf, G.R. Raidl (Eds.), Applications of Evolutionary Computing. Proceedings, 2003. XXI, 708 pages. 2003.

Vol. 2612: M. Joye (Ed.), Topics in Cryptology – CT-RSA 2003. Proceedings, 2003. XI, 417 pages. 2003.

Vol. 2613: F.A.P. Petitcolas, H.J. Kim (Eds.), Digital Watermarking. Proceedings, 2002. XI, 265 pages. 2003.

Vol. 2614: R. Laddaga, P. Robertson, H. Shrobe (Eds.), Self-Adaptive Software: Applications. Proceedings, 2001. VIII, 291 pages. 2003.

Vol. 2615: N. Carbonell, C. Stephanidis (Eds.), Universal Access. Proceedings, 2002. XIV, 534 pages. 2003.

Vol. 2616: T. Asano, R. Klette, C. Ronse (Eds.), Geometry, Morphology, and Computational Imaging. Proceedings, 2002. X, 437 pages. 2003.

Vol. 2617: H.A. Reijers (Eds.), Design and Control of Workflow Processes. Proceedings, 2002. XV, 624 pages. 2003.

Vol. 2618: P. Degano (Ed.), Programming Languages and Systems. Proceedings, 2003. XV, 415 pages. 2003.

Vol. 2619: H. Garavel, J. Hatcliff (Eds.), Tools and Algorithms for the Construction and Analysis of Systems. Proceedings, 2003. XVI, 604 pages. 2003.

Vol. 2620: A.D. Gordon (Ed.), Foundations of Software Science and Computation Structures. Proceedings, 2003. XII, 441 pages. 2003.

Vol. 2621: M. Pezzè (Ed.), Fundamental Approaches to Software Engineering. Proceedings, 2003. XIV, 403 pages. 2003.

Vol. 2622: G. Hedin (Ed.), Compiler Construction. Proceedings, 2003. XII, 335 pages. 2003.

Vol. 2623: O. Maler, A. Pnueli (Eds.), Hybrid Systems: Computation and Control. Proceedings, 2003. XII, 558 pages. 2003.

Vol. 2625: U. Meyer, P. Sanders, J. Sibeyn (Eds.), Algorithms for Memory Hierarchies. Proceedings, 2003. XVIII, 428 pages. 2003.

Vol. 2626: J.L. Crowley, J.H. Piater, M. Vincze, L. Paletta (Eds.), Computer Vision Systems. Proceedings, 2003. XIII, 546 pages. 2003.

Vol. 2627: B. O'Sullivan (Ed.), Recent Advances in Constraints. Proceedings, 2002. X, 201 pages. 2003. (Subseries LNAI).

Vol. 2628: T. Fahringer, B. Scholz, Advanced Symbolic Analysis for Compilers. XII, 129 pages. 2003.

Vol. 2631: R. Falcone, S. Barber, L. Korba, M. Singh (Eds.), Trust, Reputation, and Security: Theories and Practice. Proceedings, 2002. X, 235 pages. 2003. (Subseries LNAI).

Vol. 2632: C.M. Fonseca, P.J. Fleming, E. Zitzler, K. Deb, L. Thiele (Eds.), Evolutionary Multi-Criterion Optimization. Proceedings, 2003. XV, 812 pages. 2003.

Vol. 2633: F. Sebastiani (Ed.), Advances in Information Retrieval. Proceedings, 2003. XIII, 546 pages. 2003.

Vol. 2634: F. Zhao, L. Guibas (Eds.), Information Processing in Sensor Networks. Proceedings, 2003. XII, 692 pages. 2003.

Vol. 2636: E. Alonso, D, Kudenko, D. Kazakov (Eds.), Adaptive Agents and Multi-Agent Systems. XIV, 323 pages. 2003. (Subseries LNAI).

Vol. 2637: K.-Y. Whang, J. Jeon, K. Shim, J. Srivastava (Eds.), Advances in Knowledge Discovery and Data Mining. Proceedings, 2003. XVIII, 610 pages. 2003. (Subseries LNAI).

Vol. 2638: J. Jeuring, S. Peyton Jones (Eds.), Advanced Functional Programming. Proceedings, 2002. VII, 213 pages. 2003.

Vol. 2639: G. Wang, Q. Liu, Y. Yao, A. Skowron (Eds.), Rough Sets, Fuzzy Sets, Data Mining, and Granular Computing. Proceedings, 2003. XVII, 741 pages. 2003. (Subseries LNAI).

Vol. 2641: P.J. Nürnberg (Ed.), Metainformatics. Proceedings, 2002. VIII, 187 pages. 2003.

Vol. 2642: X. Zhou, Y. Zhang, M.E. Orlowska (Eds.), Web Technologies and Applications. Proceedings, 2003. XIII, 608 pages. 2003.

Vol. 2643: M. Fossorier, T. Høholdt, A. Poli (Eds.), Applied Algebra, Algebraic Algorithms and Error-Correcting Codes. Proceedings, 2003. X, 256 pages. 2003.

Vol. 2644: D. Hogrefe, A. Wiles (Eds.), Testing of Communicating Systems. Proceedings, 2003. XII, 311 pages. 2003.

Vol. 2645: M.A. Wimmer (Ed.), Knowledge Management in Electronic Government. Proceedings, 2003. XI, 320 pages. 2003. (Subseries LNAI).

Vol. 2646: H. Geuvers, F, Wiedijk (Eds.), Types for Proofs and Programs. Proceedings, 2002. VIII, 331 pages. 2003.

Vol. 2647: K.Jansen, M. Margraf, M. Mastrolli, J.D.P. Rolim (Eds.), Experimental and Efficient Algorithms. Proceedings, 2003. VIII, 267 pages. 2003.

Vol. 2648: T. Ball, S.K. Rajamani (Eds.), Model Checking Software. Proceedings, 2003. VIII, 241 pages. 2003.

Vol. 2649: B. Westfechtel, A. van der Hoek (Eds.), Software Configuration Management. Proceedings, 2003. VIII, 241 pages. 2003.

Vol. 2651: D. Bert, J.P. Bowen, S. King, M, Waldén (Eds.), ZB 2003: Formal Specification and Development in Z and B. Proceedings, 2003. XIII, 547 pages. 2003.

Vol. 2653: R. Petreschi, Giuseppe Persiano, R. Silvestri (Eds.), Algorithms and Complexity. Proceedings, 2003. XI, 289 pages. 2003.

Vol. 2656: E. Biham (Ed.), Advances in Cryptology – EUROCRPYT 2003. Proceedings, 2003. XIV, 649 pages. 2003.

Vol. 2657: P.M.A. Sloot, D. Abramson, A.V. Bogdanov, J.J. Dongarra, A.Y. Zomaya, Y.E. Gorbachev (Eds.), Computational Science – ICCS 2003. Proceedings, Part I. 2003. LV, 1095 pages. 2003.

Vol. 2658: P.M.A. Sloot, D. Abramson, A.V. Bogdanov, J.J. Dongarra, A.Y. Zomaya, Y.E. Gorbachev (Eds.), Computational Science – ICCS 2003. Proceedings, Part II. 2003. LV, 1129 pages. 2003.

Vol. 2659: P.M.A. Sloot, D. Abramson, A.V. Bogdanov, J.J. Dongarra, A.Y. Zomaya, Y.E. Gorbachev (Eds.), Computational Science – ICCS 2003. Proceedings, Part III. 2003. LV, 1165 pages. 2003.

Vol. 2660: P.M.A. Sloot, D. Abramson, A.V. Bogdanov, J.J. Dongarra, A.Y. Zomaya, Y.E. Gorbachev (Eds.), Computational Science – ICCS 2003. Proceedings, Part IV. 2003. LVI, 1161 pages. 2003.

Vol. 2663: E. Menasalvas, J. Segovia, P.S. Szczepaniak (Eds.), Advances in Web Intelligence. Proceedings, 2003. XII, 350 pages. 2003. (Subseries LNAI).

Vol. 2665: H. Chen, R. Miranda, D.D. Zeng, C. Demchak, J. Schroeder, T. Madhusudan (Eds.), Intelligence and Security Informatics. Proceedings, 2003. XIV, 392 pages. 2003.

Vol. 2667: V. Kumar, M.L. Gavrilova, C.J.K. Tan, P. L'Ecuyer (Eds.), Computational Science and Its Applications – ICCSA 2003. Proceedings, Part I. 2003. XXXIV, 1060 pages. 2003.

Vol. 2668: V. Kumar, M.L. Gavrilova, C.J.K. Tan, P. L'Ecuyer (Eds.), Computational Science and Its Applications – ICCSA 2003. Proceedings, Part II. 2003. XXXIV, 942 pages. 2003.

Vol. 2669: V. Kumar, M.L. Gavrilova, C.J.K. Tan, P. L'Ecuyer (Eds.), Computational Science and Its Applications – ICCSA 2003. Proceedings, Part III. 2003. XXXIV, 948 pages. 2003.

Vol. 2670: R. Peña, T. Arts (Eds.), Implementation of Functional Languages. Proceedings, 2002. X, 249 pages. 2003.

Vol. 2674: I.E. Magnin, J. Montagnat, P. Clarysse, J. Nenonen, T. Katila (Eds.), Functional Imaging and Modeling of the Heart. Proceedings, 2003. XI, 308 pages. 2003.

Vol. 2675: M. Marchesi, G. Succi (Eds.), Extreme Programming and Agile Processes in Software Engineering. Proceedings, 2003. XV, 464 pages. 2003.

Vol. 2692: P. Nixon, S. Terzis (Eds.), Trust Management. Proceedings, 2003. X, 349 pages. 2003.

Vol. 2707: K. Jeffay, I. Stoica, K. Wehrle (Eds.), Quality of Service – IWQoS 2003. Proceedings, 2003. XI, 517 pages. 2003.